Eon	Era	Epoch or sub-era	Algal evolutionary events	Environmen...
Precambrian (4500)	Proterozoic (2500)	Late Proterozoic (900)	Stromatolites decrease in abundance; acritarchs diversify; early multicellular green algae	Oldest trace...
		Middle Proterozoic (1600)	Early multicellular red algae and photosynthetic stramenopiles; diverse filamentous and colonial cyanobacteria, fossil terrestrial cyanobacteria; acritarchs (fossils of putative unicellular eukaryotic algae) present	Relatively cool conditions
		Early Proterozoic (2500)	Stromatolites abundant; geochemical evidence that cyanobacteria had colonized land; possible early photosynthetic eukaryotes	Geochemical evidence for oxygenic photosynthesis; earliest geological evidence of eukaryotes
	Archean (4500)	Mid–Late Archean (3300)	Abundant stromatolites; geochemical evidence for cyanobacteria	
		Early Archean (3900)	Oldest cyanobacterial fossils and stromatolites	Oldest geologic evidence of early life
		Hadean (4500)	The world before algae	Period of intense bombardment of Earth by debris left over from formation of solar system; life originates toward end of period as bombardment decreases.

Note: Numbers in parentheses indicate the age, in millions of years (rounded to nearest million), at which the geologic time period begins.

ALGAE

SECOND EDITION

ALGAE

SECOND EDITION

Linda E. Graham
University of Wisconsin, Madison

James M. Graham
University of Wisconsin, Madison

Lee W. Wilcox
University of Wisconsin, Madison

Benjamin Cummings

San Francisco   Boston   New York
Cape Town   Hong Kong   London   Madrid   Mexico City
Montreal   Munich   Paris   Singapore   Sydney   Tokyo   Toronto

Editor-in-Chief: Beth Wilbur
Executive Director of Development: Deborah Gale
Acquisitions Editor: Star MacKenzie
Editorial Assistants: Erin Mann and Nina Sparer
Managing Editor: Michael Early
Production Supervisor: Camille Herrera
Production Service: GGS Book Services PMG
Illustrator and Compositor: Lee W. Wilcox
Interior and Cover Designer: Lee W. Wilcox
Cover Production: Seventeenth Street Studios
Manufacturing Buyer: Michael Penne
Director of Marketing: Christy Lawrence
Executive Marketing Manager: Lauren Harp
Text printer and Binder: Courier Kendallville
Cover printer: Phoenix Color Corp.

Cover Photo Credit: Martha E. Cook (digital editing by Lee W. Wilcox)

Graham, Linda E., 1946-
  Algae / Linda E. Graham, Lee W. Wilcox, James Graham. -- 2nd ed.
     p. cm.
  Includes bibliographical references and index.
  ISBN-13: 978-0-321-55965-4
  ISBN-10: 0-321-55965-7
  1. Algae. 2. Algology. I. Wilcox, Lee Warren. II. Graham, James M., 1945-
III. Title.
  QK566.G735 2009
  579.8--dc22

                                                                          2008030812

ISBN-13: 978-0-321-55965-4
ISBN-10: 0-321-55965-7

Benjamin Cummings
is an imprint of

Brief Contents

Contents

Chapter 1

Introduction to the Algae 1

Chapter 2

The Roles of Algae in Biogeochemistry 18

Chapter 3

Algae in Biotic Associations 38

Chapter 4
Technological Applications of Algae   61

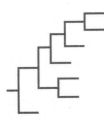

Chapter 5
Algal Diversity and Relationships   78

Chapter 6

Cyanobacteria (Chloroxybacteria)   **94**

Chapter 7

Endosymbiosis and the Diversification of Eukaryotic Algae—With a Focus on Glaucophytes and Chlorarachniophytes   **122**

Chapter 8
Euglenoids   146

Chapter 9
Cryptomonads   159

Chapter 10
Haptophytes   170

Chapter 11
Dinoflagellates 186

Chapter 12
Photosynthetic Stramenopiles I— Introduction and Diatoms 217

Chapter 13

Photosynthetic Stramenopiles II— Chrysophyceans, Synurophyceans, Eustigmatophyceans, Raphidophyceans, Pelagophyceans, and Dictyochophyceans   **247**

Chapter 14

Photosynthetic Stramenopiles III— Xanthophyceans, Phaeophyceans, and Their Close Relatives   **272**

Chapter 15
Red Algae   309

Chapter 16

Green Algae I—Introduction and Prasinophyceans   **353**

Chapter 17

Green Algae II—Ulvophyceans   **373**

Chapter 18
Green Algae III—
Trebouxiophyceans **404**

Chapter 19
Green Algae IV—Chlorophyceans **412**

Chapter 20
Green Algae V—Charophyceans
(Streptophyte Algae, Charophyte
Algae) **443**

Chapter 23

Terrrestrial Algal Ecology   587

Preface

This book is designed to be used by university undergraduate and graduate students taking courses in algal biology and to be a resource for researchers and professionals in the fields of aquatic ecology, protist phylogeny, and technological applications of the algae. We aimed for this book to be useful worldwide. Thus, we have endeavored to provide comprehensive coverage of the algae, including species of diverse habitats and their numerous ecological, evolutionary, and technological roles. This book covers the spectrum of topics in algal biology from genomics, biochemistry, and cell biology to past and present global biogeochemical impacts.

Special Features

To survey algal biodiversity, this text profiles nearly 300 representative genera in an evolutionary context. These genera were selected for their ecological, evolutionary, or technological significance. Each profile includes one or more illustrations, descriptions of major features, and summaries of recent research on these organisms. In organizing algal species, we have primarily used informal taxonomic terms, and in controversial cases we have described alternative views rather than advocating any particular formal classification scheme. Given our goal of worldwide relevance, we chose not to include identification keys, because these are widely available for specific geographical regions. However, we have included many references to recent keys, particularly those for harmful algal species.

This book is extensively referenced, with reference details gathered into the Literature Cited section at the end. Even so, references are often representative of larger bodies of literature, since it was not possible to cite every relevant book, chapter, or paper for lack of space. We have included numerous references from the expanding sphere of journals that are published only in electronic form.

Many terms of importance in the study of algae are highlighted in boldface in the text and defined in a glossary. The inside covers contain several useful and easy-to-find reference materials. A survey of the occurrence of algal groups and other important climatic and biotic events in geological time is provided on the inside opening cover. A useful summary of the major features of plastids of the eukaryotic algal groups and a listing of metric system terms and their meanings can be found on the inside end cover.

Contents

Five introductory chapters point out the importance of algae by focusing on their ecophysiological traits, biogeochemical significance, biotic interactions, research and industrial applications, and biodiversity. These early chapters are designed to stimulate student interest and to provide an overview of these fields for professionals. Hence, the five introductory chapters do not assume taxonomic knowledge of the algae beyond that provided in an overview of the major algal groups in chapter 1. Chapter 1 provides general background sufficient to allow the reader to approach the other chapters in any order.

A core of 14 chapters focuses on the unique features and important ecophysiological traits of 10 algal lineages, beginning with the cyanobacteria. A chapter on endosymbiosis reviews the recent research in this very active field and provides useful background for the study of eukaryotic algal groups. Chapters covering groups composed primarily of microalgae precede chapters on groups that include both microalgae and macroalgae. We chose this organization for its utility to users mainly interested in phytoplankton on the one hand and those primarily focused on seaweeds on the other. We have avoided using phylogeny to determine the order of the core chapters because the deepest branching patterns of the eukaryotic portion of the tree of life are still being clarified.

This book concludes with three chapters that synthesize recent research in the fields of phytoplankton ecology, periphyton and macroalgal ecology, and terrestrial algal ecology. These final three chapters extend and integrate information provided in earlier portions of the text and are written in a more synthetic mode. These three final chapters pay greater attention to the development of ecological theory and the use of quantitative methods.

New Features of the Second Edition

Every chapter has undergone extensive revision in organization and content, and a final chapter on terrestrial algal ecology has been added to reflect modern advances in this important area. There is an increased philosophical focus on the importance of algal traits. This second edition also features updated concepts of evolutionary diversification of each major lineage. There is increased coverage of harmful algal blooms, as well as greater focus on the utility of algal genomics and algae as energy resources. The content has been streamlined to conserve energy, paper, and reader time.

Acknowledgements

We thank the staff at Benjamin Cummings and Pearson Education, including Star MacKenzie, Camille Herrera, Becky Ruden, Claudia Trotch, and Nina Sparer, as well as Becky Giusti and associates at GGS Book Services and our excellent copy editor Kitty Wilson and proofreader Jeff Georgeson. We owe a debt of thanks to Andrew Gilfillan for commissioning the second edition of *Algae*, and Lisa Tarabokjian (both of Prentice Hall) for obtaining expert reviews of the first edition, which greatly facilitated production of this second edition. Beth Wilbur helped shepherd the transition of this project from Prentice Hall to Benjamin Cummings, for which we are most grateful. We also wish to express our deep appreciation to Prentice Hall editor Theresa Ryu, who encouraged us to produce the first edition of *Algae*, without which this second edition would not exist.

This book features many photographs and detailed descriptions of representative algal genera from diverse habitats. Professor Paul Silva, University of California–Berkeley, graciously researched the etymologies of the generic names for these representatives. In addition, numerous phycologists from around the world graciously contributed original photographs to this project. Contributors are acknowledged in the figure captions. As a token of our appreciation for these contributions from the community, we will continue a previous practice of donating a portion of book royalties to one or more funds that support student travel to international phycological conferences.

The original of the color image on the cover was made by Martha Cook and subjected to artistic manipulations by Lee Wilcox. The image is of the green macroalgal genus *Anadyomene*, from a collection made in Florida.

We greatly appreciate reviews of the first edition provided by the following experts at the publisher's request:

JoAnn Burkholder, North Carolina State University
Hudson DeYoe, University of Texas, Pan American
David Domozych, Skidmore College
Paul Falkowski, Rutgers University
Paul W. Gabrielson, University of North Carolina, Chapel Hill
Sean Patrick Grace, Southern Connecticut State University
Curt Pueschel, SUNY–Binghamton
John Stiller, East Carolina University

We are also grateful for reviews of first-draft revisions of particular chapters provided at the authors' request by:

JoAnn Burkholder, East Carolina University
Martha Cook, Illinois State University
Eunsoo Kim, Dalhousie University, Canada
John Raven, University of Dundee, Scotland
John Stiller, East Carolina University
Robert Waaland, University of Washington

We also thank Elsa Althen, director of the UW–Madison Biology Library, for assistance in finding reference materials, and the UW Botany Department office staff for many types of assistance, including maintaining and un-jamming the photocopy machine.

Last, but not least, we thank our friends, lab associates, and families for putting up with our obsessive behavior while we worked on this book. We are very grateful for their patience.

Linda E. Graham
James M. Graham
Lee W. Wilcox

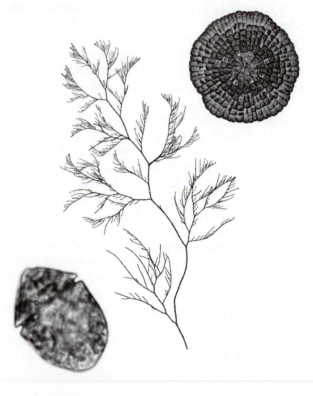

Chapter 1

Introduction to the Algae

The algae are a heterogeneous group of organisms that exert profound effects in today's world and have been doing so for billions of years. For example, as a result of photosynthetic activities, algae generate a large fraction of the oxygen present in Earth's atmosphere and produce an enormous quantity of organic carbon. Much of this organic carbon serves other organisms as food, and the expensive oil that helps power modern life largely originates from the organic components of algae that lived hundreds of millions of years ago.

Rather than waiting hundreds of millions of years for new fossil fuels, modern biotechnologists and engineers are developing new ways to generate sustainable fuels by using algae. Biofuels produced from algae and other sources may help reduce human reliance on fossil fuels. Humans use algae in many other ways. Certain algal species are widely used as convenient laboratory systems; such small and fast-reproducing green "lab rats" have revealed essential information about the biochemistry and molecular biology of photosynthesis and other cellular processes. Algae are also harvested from nature for the extraction of industrially useful products, some of which cannot be obtained in any other way. Flavorful or nutrient-rich algae are grown as aquaculture crops for direct use as human food (e.g., sushi), or as food sources for fish and shellfish farming operations. Algae can also be used to remove pollutants from sewage and agricultural effluents before they are returned to nature.

The green alga *Eremosphaera*

(Photo: L. W. Wilcox)

Unfortunately, in response to water pollution caused by humans, some algal species have become notorious for forming harmful blooms in oceans or freshwaters. Algal blooms can poison waterfowl, marine mammals, fish, domesticated animals, and humans. For these reasons, water resource managers are increasingly interested in finding ways to monitor harmful blooms and to prevent them from developing.

Algae are also major contributors to global biodiversity, with estimated numbers of species ranging from 36,000 to more than 10 million. Each species displays a unique combination of traits and thus plays one or more essential roles in ecosystems. Algal species are interconnected with other organisms in biogeochemical cycles, food webs, and symbiotic associations. For this reason, ecologists are concerned that environmental changes such as global warming are likely to perturb natural patterns of algal species in ways that have large and unexpected effects. Consequently, there are increasing efforts to catalog algal biodiversity and understand the functional traits of natural algal communities and how they might change.

This book is designed to provide information that will be useful to people interested in conserving biodiversity, comprehending the dynamics of aquatic ecosystems, improving water quality, generating new industrial products, selecting or developing laboratory model systems, and tracing major events in the history of life on Earth. This chapter provides an overview of the algae, beginning with a definition of algae and a survey of the diversity of algal bodies and reproduction. There follow brief summaries of the diverse algal groups covered in this book, algal photosynthesis, and current societal issues that involve algae.

1.1 Defining the Algae

The algae are a heterogeneous assemblage of organisms that range in size from tiny single cells to giant seaweeds and that belong to diverse evolutionary lineages. As a result, the algae are largely defined by ecological traits. The algae are mostly photosynthetic species that produce oxygen and live in aquatic habitats. In addition, the algae lack the body and reproductive features of the land plants that represent adaptations to terrestrial life. This concept of the algae includes both photosynthetic protists, which are eukaryotes, and the prokaryotic cyanobacteria, also

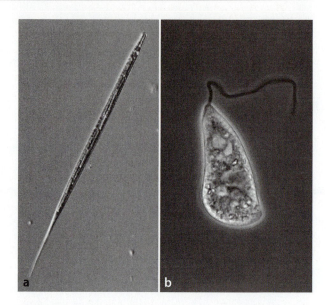

Figure 1.1 Photosynthetic and heterotrophic algae. (a) The photosynthetic euglenoid *Euglena* displays many green plastids. Storage granules of organic material produced by photosynthesis are present in the cytoplasm. (b) The heterotrophic euglenoid *Peranema* lacks plastids and is thus dependent on ingested organic food. (Photos: L. W. Wilcox)

known as the blue-green algae. Several distinctive features—including a nucleus enclosed by an envelope with pores—characterize eukaryotes, whereas prokaryotes lack such features. Though certain non-cyanobacterial prokaryotes are photosynthetic, those species do not produce oxygen—in contrast to cyanobacteria, photosynthetic protists, and land plants.

Though we can generally define **algae** as photosynthetic, oxygen-producing aquatic bacteria or protists, there are many exceptions. For example, a number of non-photosynthetic protists are included among the algae because they are closely related to photosynthetic species. As a case in point, the photosynthetic flagellate known as *Euglena*, which is commonly studied in introductory biology classes, has many close relatives that are heterotrophic, meaning that they depend entirely on ingested organic food (Figure 1.1). Together, such photosynthetic and heterotrophic protists form an algal lineage known as the euglenoids. Other algal species are exceptional because they occur in nonaquatic habitats, such as soil, rocks, and other relatively dry terrestrial habitats. Such species are able to tolerate dry or cold conditions in a metabolically

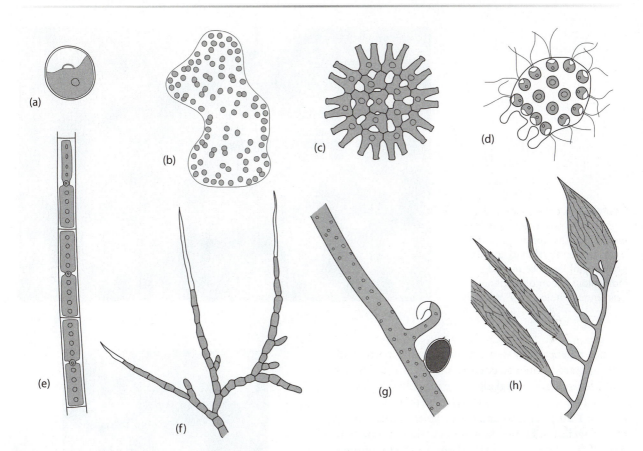

Figure 1.2 Diverse algal body types. (a) Coccoid unicell of the green algal genus *Chlorococcum*. (b) Colonies of coccoid cells held together by mucilage, demonstrated by the cyanobacterial genus *Microcystis*. (c) Coenobial colonies of the green algal genus *Pediastrum*. (d) Flagellate colonies of the green algal genus *Platydorina*. (e) A uniseriate unbranched filament of the green algal genus *Mougeotia*. (f) Part of a branched filament of the green algal genus *Stigeoclonium*. (g) Part of the coenocytic body of the photosynthetic stramenopile *Vaucheria*, showing sexual reproductive structures. (h) The apex of a frond of the parenchymatous body of the giant kelp genus *Macrocystis*, showing how new blades arise.

dormant state. Even so, sufficient moisture must be present before terrestrial algae can become metabolically active, reflecting the fundamental dependence of algae upon a watery habitat.

1.2 Algal Body Types

The algae occur in diverse body types, representatives of which are illustrated in Figure 1.2. Algal bodies can be so small that a microscope is needed to observe them, and organisms having such small bodies are known as **microalgae**. Algal bodies that are large enough to be seen with the unaided eye are referred to as **macroalgae**.

Microalgae

Many microalgal species occur as solitary cells, known as **unicells**. The previously mentioned euglenoids are examples of unicellular algae. Unicells occur in a variety of shapes, but **coccoid** algae that take the form of small round balls may have the most common body type found among the algae (Figure 1.2a). The coccoid body type has evolved independently in diverse lineages of algae. Experts sometimes colloquially refer to coccoid algae as "little round green things" or "little brown balls."

Other algal species occur as several to many cells arranged loosely or in a highly organized way to form **colonies** (Figure 1.2b). In some cases, colonies

Figure 1.3 A conspicuous algal bloom produced by a microalga. This surface bloom in a freshwater marshland consists primarily of the unbranched, filamentous green algal genus *Spirogyra.* (Photo: L. W. Wilcox)

feature a genetically defined number and pattern of cells and are known as **coenobia** (Figure 1.2c). Some unicellular and colonial algae are propelled by flagella, in which case they may be referred to as **flagellates** (Figure 1.2d). Euglenoids are examples of unicellular algal flagellates. However, it is important to note that not all flagellates are algae; there are many species of non-photosynthetic flagellates that are not closely related to photosynthesizers. Examples include choanoflagellates, the modern protists that are most closely related to the animal kingdom (metazoans).

Another common type of algal body is the **filament**, a linear array of cells joined end-to-end, often sharing a common wall (Figure 1.2e). Filaments that are composed of a single row of cells are known as uniseriate filaments, and those composed of two or more rows of cells are described as biseriate or pluriseriate filaments. While many algal species occur as unbranched filaments, others produce branches by dividing in a direction perpendicular to the main filament axis and are described as **branched filaments** (Figure 1.2f).

Branched filamentous algae often occur attached to rocks or other substrates in shallow waters, and they are common components of communities known as the **periphyton**. While microalgae may also occur in the attached periphyton, many swim or float in the open water and are collectively known as the **phytoplankton**. When present in relatively low population numbers, the phytoplankton are often so inconspicuous in their

Figure 1.4 Macroalgae in a marine habitat. Shell-shaped seaweeds of the brown algal genus *Padina* dominate a nearshore patch of coral reef habitat in the Bahamas. A large, forked gorgonian coral is in the foreground. (Photo: L. E. Graham)

habitats that plankton nets or other types of concentration methods are commonly used to collect them. On the other hand, when phytoplankton populations become very large, forming what are

Figure 1.5 A coenocytic algal body. The common macroalgal genus *Caulerpa* has a coenocytic body composed essentially of one very large, complex cell. This seaweed bears a remarkable superficial similarity to certain gorgonian corals. Angiosperm sea grasses grow in the sandy seabed nearby. (Photo: R. J. Stephenson)

Figure 1.6 Parenchymatous bodies of giant kelps. Their parenchymatous construction is instrumental in the ability of these brown macroalgae to grow to large sizes. These kelps, primarily the genus *Macrocystis*, form a forest off the Chilean coast. (Photo: R. Searles)

known as **algal blooms**, they become more conspicuous (Figure 1.3).

Macroalgae

In contrast to microalgae, algae having coenocytic, parenchymatous, or pseudoparenchymatous bodies are often **macroscopic**; that is, they can be seen with the unaided eye. Seaweeds commonly have these body types and thus may be conspicuous in their marine habitats (Figure 1.4). Algal species that have **coenocytic bodies**, also known as siphonous bodies, are essentially composed of one very large multinucleate cell. A common seaweed genus known as *Caulerpa* has a coenocytic body (Figure 1.5). Coenocytic species lack internal cell walls except when reproductive structures are produced (see Figure 1.2g). **Parenchymatous bodies** are composed of tissues—three-dimensional arrays of cells (see Figure 1.2h) that are often interconnected by intercellular connections known as plasmodesmata. The macroalgae known

as giant kelps have parenchymatous bodies that may reach 50 m in length (Figure 1.6). In some cases, branched filaments may closely adhere to form what is known as a **pseudoparenchymatous body**—i.e., one that appears to be composed of tissue but is fundamentally filamentous. Even though most coenocytic, parenchymatous, and pseudoparenchymatous algae can be seen without the use of a microscope, several types of microscopic methods are typically used to observe fine details of their structure and reproductive features.

1.3 Algal Reproductive Types

Algae reproduce by a variety of methods, both asexual and sexual. Asexual reproduction does not involve the fusion of gametes or meiosis, which are both essential features of sexual reproduction. Some algal species reproduce only asexually, but many reproduce by both sexual and asexual processes.

Sexual and asexual reproduction processes confer distinct advantages. Sexual reproduction allows populations to increase genetic variability, fostering the ability to respond to environmental change by means of evolution. In addition, many algae use tough, resistant structures generated by sexual processes to survive periods unfavorable to growth. In contrast, under conditions favorable to growth, asexual reproduction allows organisms to replicate

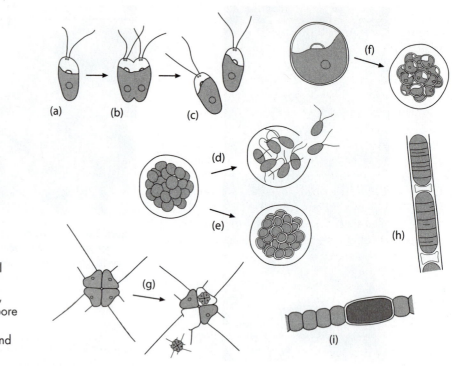

Figure 1.7 Examples of asexual reproduction in algae. (a)–(c) cellular bisection; (d) zoospore, (e) aplanospore, and (f) autospore formation; (g) autocolony formation; (h) fragmentation; and (i) akinete formation.

themselves without the need to produce gametes and find mates. Thus, asexual reproduction allows rapid population growth.

Asexual Reproduction

Examples of algal asexual reproductive processes are shown in Figure 1.7. Populations of many unicellular species increase by simple longitudinal or transverse cell division (Figure 1.7a–c).

Many algal species of diverse body types reproduce asexually by means of flagellate unicells known as **zoospores**, or non-flagellate **aplanospores** or **autospores**. During the process of zoosporogenesis, the cytoplasm of an algal cell is transformed into one to many flagellate unicells. When mature, the zoospores are released into aquatic environments, where they disperse. Each zoospore has the potential to develop into a new unicellular or multicellular individual (Figure 1.7d). Aplanospores are cells that have the genetic capacity to produce flagella, but flagella do not develop, possibly in response to environments that are low in moisture. Aplanospores disperse from the confines of the parental cell wall and are able to grow into mature bodies (Figure 1.7e). Autospores are the asexual reproductive cells of algal species that lack the capacity

to produce flagella; they disperse and develop much like zoospores and aplanospores (Figure 1.7f).

Colonies of defined cell number and shape, coenobia, reproduce by **autocolony** formation. Each cell of the colony undergoes cell divisions that give rise to a miniature version, the autocolony, of the original coenobium (Figure 1.7g). Filaments often reproduce asexually by fragmentation, a process that can be highly controlled (Figure. 1.7h). **Akinetes** are specialized asexual cells that develop from actively growing cells when environmental signals indicate impending conditions unsuitable for growth, such as winter. During akinete development, cells enlarge, accumulate storage materials, develop thick cell walls, and suspend active metabolism (Figure 1.7i). These traits allow akinetes to survive stressful conditions that are unsuitable for growth. When conditions improve, akinetes may germinate. During this process, the thick wall breaks open, and stored food materials are used to produce an actively growing individual.

Sexual Reproduction

Sexual reproduction in eukaryotes involves gamete production, gamete fusion, zygote production, and zygote development into an algal body. Although cyanobacteria display certain genetic exchange mechanisms similar to those of other eubacteria (see Chapter 6), they lack the sexual processes typical of most eukaryotes. Sexual reproduction is a feature of most algal lineages but is absent from some, the euglenoids being a prominent example.

Among sexually reproducing algae, gametes may be more or less differentiated, depending on the species. **Isogamous reproduction** describes the situation when mating gametes are structurally indistinguishable. Even so, such gametes are often biochemically differentiated, and thus described as + and − mating types (Figure 1.8a). **Anisogamous reproduction** involves the mating of two gametes that differ in size or behavior. In **oogamous reproduction**, a flagellate or non-flagellate male gamete fuses with a larger, nonmotile egg cell. Algae also differ in type of sexual life cycle.

There are three major types of sexual life cycle or life history among the algae. The principal differences among them are the point at which meiosis occurs, the types of cells produced by meiosis, and the number of multicellular life stages present in the life cycle. In this book, we refer to life cycles on the basis of the point at which meiosis occurs: zygotic, gametic, or sporic. However, other authors have named life cycles on the basis of the number of multicellular stages and their relative chromosome levels (Bold and Wynne 1985).

Zygotic meiosis characterizes the life cycles of algae in which the only diploid cells are zygotes (Figure 1.8a). Many types of algae display zygotic meiosis, sharing this feature with the fungi. Among organisms that display zygotic meiosis, non-reproductive bodies composed of **vegetative cells** are haploid, as are their gametes. Mating-type genes are segregated during meiosis, leading to the production of two populations of cells that differ in potential mating type. The expression of mating-type genes is controlled by the environment, which induces mating and zygote formation.

Gametic meiosis indicates that meiosis occurs during the production of haploid gametes from diploid vegetative cells (Figure 1.8b). These gametes fuse to form a diploid zygote that undergoes repeated mitotic divisions to form a multicellular diploid body. While gametic meiosis is also characteristic of animals and thus familiar to people, it is important to realize that relatively few groups of protists display gametic meiosis. The microalgae known as diatoms and certain macroscopic brown algae, such as the genus *Fucus*, display gametic meiosis.

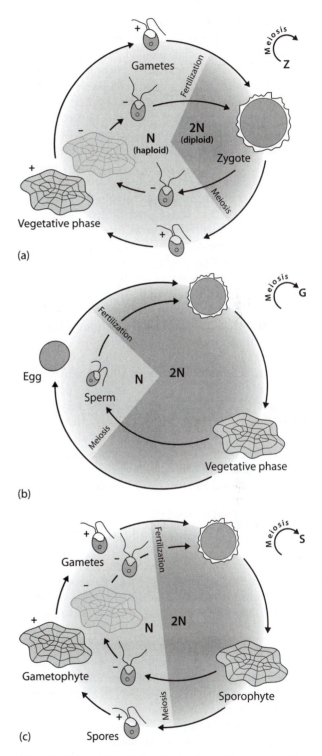

Figure 1.8 Sexual reproduction in algae. Three major life histories occur: (a) zygotic meiosis, (b) gametic meiosis, and (c) sporic meiosis, with alternation of generations. Note the icons (used throughout the text) that indicate life cycle type: Z = zygotic, G = gametic, and S = sporic.

Sporic meiosis indicates that meiosis generates haploid spores, single cells that are capable of growing into a multicellular haploid body known as a gametophyte. Under inducing environmental conditions, particular gametophyte cells produce haploid gametes that fuse to form a diploid zygote. In organisms having sporic meiosis, the diploid zygote undergoes repeated mitotic divisions to form a multicellular diploid body known as the sporophyte. Under inducing conditions, some cells of the sporophyte undergo meiotic divisions to produce spores, completing the cycle (Figure 1.8c). A life cycle involving sporic meiosis, also known as **alternation of generations**, has evolved independently in several algal lineages and in the common ancestor of the land plants.

The common marine green alga known as *Ulva* (informally, sea lettuce) displays a life cycle with two alternating generations, sporophyte and gametophyte, that look much alike. This condition is known as **isomorphic alternation of generations**. However, many algal species are like land plants in having alternating generations that do not look alike, a trait known as **heteromorphic alternation of generations**. Among the algae, most species displaying sporic meiosis have generations that live independently of each other, though exceptions are known. In the past, the independent, heteromorphic bodies of algae having a sporic life cycle have sometimes been described as different genera. Currently, culture and molecular methods are used to reveal such life cycle linkages.

1.4 A Survey of Algal Diversity

The algae, as we have defined them on the basis of ecological characteristics, cannot be classified into a single group that has descended from a common ancestor. However, the algae can be classified into a dozen or so lineages, many of which are named as phyla (divisions). Particular photosynthetic pigments, food storage materials, and type of cell covering characterize each lineage (Table 1.1). This book covers 10 algal lineages, listed here by their informal names: cyanobacteria, glaucophytes, chlorarachniophytes, euglenoids, cryptomonads, haptophytes, dinoflagellates, photosynthetic stramenopiles, red algae, and green algae. The following brief summary of each lineage focuses on distinctive features, evolutionary relationships, and major ecological impacts.

Cyanobacteria (also known as chloroxybacteria, cyanophytes, or blue-green algae) (Figure 1.9) form

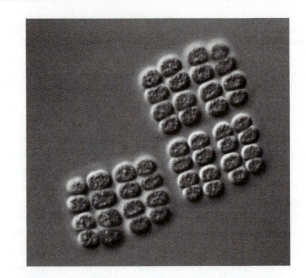

Figure 1.9 The cyanobacterial genus *Merismopedia*. (Photo: L. W. Wilcox)

a well-defined branch of the eubacteria. Cyanobacteria are unique among the algae in having prokaryotic cellular organization, and they are the most ancient algal lineage. Sexual reproduction involving gametes, zygotes, and meiosis is lacking. So far as is known, all cyanobacteria possess chlorophyll and other light-absorbing pigments and are thus photosynthetic. Unlike other prokaryotes, but like eukaryotic algae and land plants, cyanobacteria generate oxygen as a product of photosynthesis, a trait known as **oxygenic photosynthesis**. Many are able to perform **nitrogen fixation**, a particularly important ecosystem service. Cyanobacteria include unicellular, colonial, unbranched filamentous, and branched filamentous forms. Some filamentous cyanobacteria possess the defining traits of multicellular organisms: They produce cells that are specialized for particular functions, display controlled cell death during development, and possess intercellular communication systems. Cyanobacteria are common and diverse in marine and freshwater habitats and in terrestrial ecosystems, and they are well known for important biotic interactions with marine invertebrates, fungi (to form lichens), and land plants. Some cyanobacteria are notorious bloom formers and can produce harmful toxins.

Glaucophytes include several unicellular or colonial freshwater eukaryotes that have blue-green photosynthetic plastids (Figure 1.10). Glaucophytes are of particular importance in studies focused on the origin of plastids because their plastids differ

Table 1.1 Predominant photosynthetic pigments, storage products, and cell wall components of the major algal groups

Group	Photosynthetic and protective pigments	Storage products	Cell covering
Cyanobacteria (Chloroxybacteria)	Chlorophyll *a* (chlorophyll *d* instead of *a* in some; chlorophylls *a* and *b* in some), phycobilins, β-carotene, xanthophylls	Cyanophycin granules, cyanophytan starch (glycogen), plant-like starch in some	Peptidoglycan
Glaucophytes	Chlorophyll *a*, phycobilins, β-carotene, xanthophylls	Starch	Cellulose
Chlorarachniophytes	Chlorophylls *a* and *b*, β-carotene, other carotenes, xanthophylls	Carbohydrate	Naked
Euglenoids (photosynthetic forms)	Chlorophylls *a* and *b*, β-carotene, other carotenes, xanthophylls	Paramylon	Proteinaceous pellicle beneath plasma membrane
Cryptomonads	Chlorophylls *a* and *c*, phycobilins, α- and β-carotene, xanthophylls	Starch	Proteinaceous periplast beneath plasma membrane
Haptophytes	Chlorophylls *a* and *c* β-carotene, xanthophylls	Chrysolaminarin (chrysolaminaran)	$CaCO_3$ scales common
Dinoflagellates (most photosynthetic forms)	Chlorophylls *a* and *c*, β-carotene, xanthophylls	Starch	Vesicles beneath plasma membrane, often containing cellulosic plates
Photosynthetic stramenopiles	Chlorophylls *a* and *c* (chlorophyll *a* alone in some), β-carotene, xanthophylls	Chrysolaminarin, lipids	Some naked, some with silica/organic scales, cellulose and alginates in some
Red algae	Chlorophyll *a*, phycobilins, α- and β-carotene, xanthophylls	Floridean starch	Cellulose, sulfated polysaccharides, some calcified
Green algae	Chlorophylls *a* and *b*, β-carotene, lutein, other carotenes, xanthophylls	Plant-like starch	Wall of cellulose/other polymers, organic scales on some, some naked, some calcified

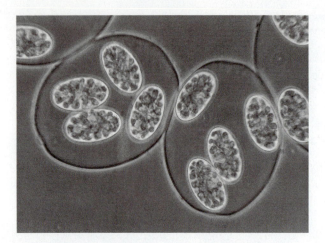

Figure 1.10 The glaucophyte genus *Glaucocystis*. (Photo: L. W. Wilcox)

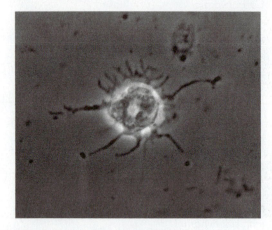

Figure 1.11 The chlorarachniophyte genus *Chlorarachnion*. (Photo: L. W. Wilcox)

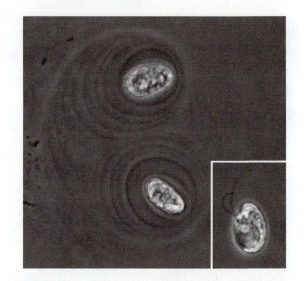

Figure 1.12 The cryptomonad genus *Cryptomonas*. Two nonmotile cells are invested with concentric layers of mucilage. The inset shows a cell with flagella. (Photo: L. W. Wilcox)

from those of other eukaryotic algae and more closely resemble cyanobacteria in some ways. It is unclear which protist group is most closely related to glaucophytes. Sexual reproduction is not known.

Chlorarachniophytes are a small group of unicellular marine species (Figure 1.11). The chlorarachniophytes—together with heterotrophic foraminifera, radiolarians, and some other protists—form a eukaryotic supergroup known as the **Rhizaria** (informally, rhizarians). Chlorarachniophytes are unusual among rhizarians in possessing photosynthetic plastids resembling those of green algae. Chlorarachniophytes are of special interest in the study of endosymbiosis and the origin of plastids. Sexual reproduction has been observed in some members.

Euglenoids are primarily unicellular flagellates (see Figure 1.1). Together, kinetoplastid protists (such as the genus *Trypanosoma*, which causes sleeping sickness) and some other heterotrophic protists and euglenoids make up a eukaryotic supergroup known as **Euglenozoa**. Some euglenoids (such as the genus *Euglena*) possess photosynthetic plastids similar to those of green algae, but many euglenoid genera are heterotrophic (see Figure 1.1b). Some euglenoids lack plastids altogether, and others have colorless plastids that no longer retain photosynthetic capacity. Most species are found in freshwaters, though marine representatives occur. Euglenoids are recognized for their diverse roles in wetland ecosystems. Sexual reproduction is unknown.

Cryptomonads (also known as cryptophytes) are freshwater or marine unicellular flagellates (Figure 1.12). The early-diverging genus *Goniomonas* lacks plastids, and some cryptomonad species have colorless plastids, but most species have photosynthetic plastids that can be colored red, blue-green, olive, or brown. Cryptomonads are closely related to a group of heterotrophic protists known as katablepharids and to the photosynthetic haptophytes. Cryptomonads are valued as particularly nutritious food for aquatic animals and are important in evolutionary studies. Sexual reproduction has been documented for some representatives.

Haptophytes (also known as prymnesiophytes) include unicellular flagellates and nonflagellate unicells

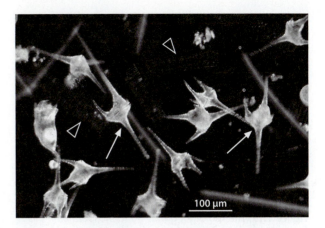

Figure 1.14 The horned dinoflagellate genus *Ceratium*. Arrows indicate the *Ceratium* cells. This phytoplankton population, which includes diatoms (arrowheads), is from a southern Wisconsin lake. (Photo: L. W. Wilcox)

Figure 1.13 The haptophyte genus *Emiliania*. (Micrograph: A. Kleijne, in Winter and Siesser 1994)

and colonies. Golden plastids are present in the cells of haptophytes. Haptophytes are closely related to cryptomonads. Haptophytes primarily occur in marine habitats, but some freshwater representatives are known. The group is named for a flagellum-like structure, known as a haptonema, which occurs on the cells of many flagellate representatives. A subgroup of haptophytes, the coccolithophorids, is famous for the production of a cellular covering of elaborately structured carbonate scales called coccoliths (Figure 1.13). Coccolithophorids are exceptionally important for their production of volatile compounds that influence climate and their past production of large carbonate deposits that are rich in petroleum. Sexual life cycles involving heteromorphic stages are common.

Dinoflagellates are primarily unicellular flagellates having two flagella of different types (Figure 1.14). Together, the dinoflagellates, heterotrophic ciliates (such as the genus *Paramecium*), parasitic apicomplexans (such as the genus *Plasmodium*, which causes malaria), and some other protists form a eukaryotic supergroup known as **Alveolata**. This supergroup is named for the common presence of membranous sacs known as **alveoli** at the periphery of these organisms' cells. In many dinoflagellates, alveoli contain material that forms a cell wall. About half of dinoflagellate genera are colorless heterotrophs, and others possess photosynthetic plastids with a distinctive type of rubisco, the enzyme

responsible for carbon fixation. Plastids are mostly golden-brown in color, owing to the presence of a distinctive xanthophyll known as peridinin. The majority of species are marine, though many freshwater representatives also occur. Dinoflagellates have exceptionally important biotic interactions as sources of marine toxins or symbionts in reef-forming corals and other marine invertebrates. Sexual reproduction has been documented for some species.

Photosynthetic stramenopiles (also known as photosynthetic heterokonts, chromophytes, or ochrophytes) include a wide range of algae displaying exceptionally diverse body types and reproductive cycles. Most algae classified as photosynthetic stramenopiles have plastids that are pigmented golden brown, with a major accessory pigment known as fucoxanthin, but some are yellow-green or have colorless plastids. The photosynthetic stramenopiles comprise more than a dozen classes of algae, including microscopic silica-walled diatoms and giant kelps (see Figure 1.6). Together with several lineages of heterotrophic stramenopiles, such as the oomycete plant parasitic genus *Phytophthora*, photosynthetic stramenopiles form a eukaryotic supergroup known as **Stramenopila** (informally, stramenopiles). The stramenopiles are named for tubular hairs that commonly occur on the flagella, and they are closely related to the supergroups Alveolata and Rhizaria.

Red algae (also known as rhodophytes) occur as unicells, filaments, aggregations of filaments, and sheets of cells (Figure 1.15). Photosynthetic pigments are present in the red (or occasionally blue-green)

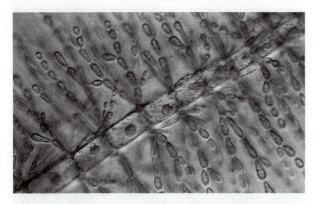

Figure 1.15 The red algal genus *Batrachospermum*. This freshwater red alga occurs as a highly branched filament whose generic name reflects its similarity to gelatinous strands of frogs' eggs. (Photo: M. E. Cook)

Figure 1.17 A green algal-fungus symbiosis, the lichen genus *Umbilicaria*. This large, flat lichen is growing on a shaded cliff with the fern *Polypodium*. (Photo: L. W. Wilcox)

Figure 1.16 Symbiotic green algae living within the body of the freshwater coelenterate *Hydra*. (Photo: L. E. Graham)

plastids of most species; certain parasitic species have non-photosynthetic plastids. Red algae are more or less closely related to the green algae and land plants but are unusual among eukaryotes in completely lacking flagella, a circumstance that has strongly influenced the evolution of sexual reproductive cycles. The majority of red algae have a sexual life cycle involving alternation of three generations: a gametophyte and two types of sporophytes. Red algae are especially diverse and abundant as macroalgae in tropical marine waters. Some species are cultivated for making industrial products or food. The coralline red algae are very important as carbonate producers, and some play an essential ecological role by consolidating coral reefs.

Green algae (also known as chlorophytes) display diverse algal body types, sexual reproductive cycles, and habitats (see Figure 1.5). Most of the green algae are grass green in color, but red protective pigments hide the chlorophylls of some, and others have colorless plastids that have lost photosynthetic capacity. Green algae are important sources of food for aquatic animals, and some representatives form significant symbiotic partnerships with freshwater protists and invertebrates (Figure 1.16) and with fungi to form lichens (Figure 1.17). Green algae share many molecular, biochemical, cellular, and reproductive traits with land plants, and the term **Viridiplantae** is widely used to name a lineage composed of green algae and land plants. The Viridiplantae is more or less closely related to red algae, and together they have been linked with one or more additional algal lineages to form a eukaryotic supergroup of varying name: Plantae (Cavalier-Smith 1993), Archaeplastida (Adl et al. 2005), or Plastidophila (Kim and Graham 2008).

1.5 An Overview of Algal Photosynthesis

As shown by the preceding diversity survey, oxygenic photosynthesis is a defining feature of the algal lineages, present in half or more of the species. Photosynthesis is so important to the algae that it is useful to take an overview of photosynthesis in general and then consider how algae in particular accomplish photosynthesis. General biology textbooks are a good source of information about the process of photosynthesis as it occurs in land plants, and a text by Falkowski and Raven (2007) focuses on algal photosynthesis.

The General Processes of Photosynthesis

Photosynthesis is the process by which light energy is harnessed to produce organic compounds that serve as cellular building blocks and energy reserves (Figure 1.18). In the first phase of oxygenic photosynthesis, known as the **light-dependent reactions**, light energy that reaches reaction center chlorophyll *a* molecules is stored in ATP and NADPH. These processes take place in thylakoid membranes, in pigment–protein assemblages known as photosystems I and II (PSI and PSII). During the light-dependent reactions, sunlight energy is used to oxidize water to molecular oxygen (O_2), a waste product that mostly diffuses out of cells. The oxidation of water provides the reducing power needed in a second phase, known as the **light-independent reactions,** or Calvin cycle. In these reactions, ATP and NADPH are used to reduce carbon dioxide (CO_2), thereby forming organic

compounds in a process known as **carbon fixation**. Cells use the organic products of photosynthesis to make many other organic materials.

In their primarily aquatic habitat, the algae encounter several problems in accomplishing photosynthesis. These challenges include coping with variation in light environments and acquiring and fixing carbon dioxide in today's oxygen-rich habitats.

The Light-Acquisition Problem

Land plants receive full-spectrum sunlight, including the blue and red light maximally absorbed by chlorophyll. In contrast, full-spectrum sunlight is generally not available to aquatic algae because water and substances that may be dissolved in it absorb some wavelengths. Red light is absorbed in the few meters of water closest to the surface so that the light environment at greater depths is depleted in this wavelength. This explains why the light environment at such depths is blue-green in quality. In addition, algal cells may sink or be transported into deep, dark waters. In both cases, the amount of light that chlorophyll *a* can absorb may not be sufficient to supply the needs of algal photosynthesis.

Algae have responded to the problem of acquiring sufficient light in two major ways. First, flagellates are able to use light receptor molecules to "see" light and move into optimal regions of illumination, a process known as **phototaxis**. Cryptomonads, green flagellates, and dinoflagellates use the photoreceptor **rhodopsin** (which is structurally similar to a light receptor used by animals), while photosynthetic stramenopiles and euglenoids use flavin-based blue-light sensors such as **aureochrome** and **phototropin**. The red/far-red sensor **phytochrome** occurs widely

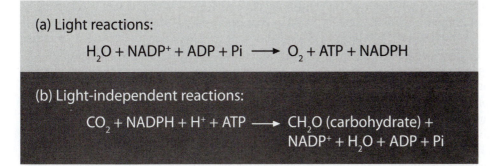

(a) Light reactions:

$$H_2O + NADP^+ + ADP + Pi \longrightarrow O_2 + ATP + NADPH$$

(b) Light-independent reactions:

$$CO_2 + NADPH + H^+ + ATP \longrightarrow CH_2O \text{ (carbohydrate) } + NADP^+ + H_2O + ADP + Pi$$

Figure 1.18 The light-dependent and light-independent reactions of photosynthesis. (a) The light-dependent reactions. (b) The light-independent reactions. Pi = inorganic phosphate.

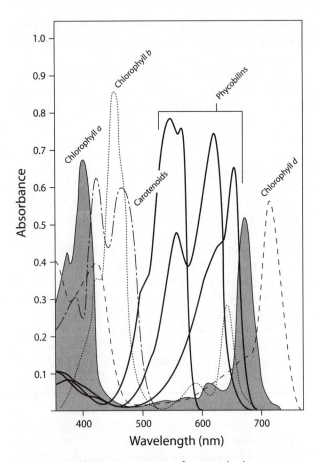

Figure 1.19 Absorption spectra of major algal accessory pigments. Accessory pigments increase the spectral range of light energy that can be used in photosynthesis beyond that absorbed by chlorophyll *a* (shaded) alone. (Phycobilin absorption data from Gantt, E. *BioScience* 25:781–788. © 1975 American Institute of Biological Sciences)

in algae, and in some green algae it is known to be involved in chloroplast movements that foster maximal use of light in photosynthesis. It has been proposed that such light sensors might also help algae detect other photosynthesizers by means of red light emitted from chlorophylls by fluorescence.

A second way that algae have coped with variable light environments is by evolving diverse accessory photosynthetic pigments having different light-absorption properties. The presence of such pigments helps explain color variations that occur among algae (see Table 1.1). **Accessory pigments** are able to absorb light outside the range of chlorophyll *a* absorption (Figure 1.19) and conduct energy to reaction center chlorophyll *a* molecules by resonance

transfer. Chlorophyll *b*, produced by green algae (and land plants) and certain cyanobacteria, is an example of such an accessory pigment. Beta-carotene, produced by many types of algae, is another example. Different algal lineages have evolved different suites of accessory chlorophylls, carotenoids (including xanthophylls), and phycobilins (see Table 1.1).

Accessory pigments occur in several types of **light-harvesting complexes** (LHCs) that are also known as antennas. LHCs are closely associated with the core pigments and proteins of PSI and PSII. LHCs gather light energy and transfer it into these cores. Variation in the structure of algal LHCs arises from the types of light-harvesting pigments present and differences in the proteins that spatially organize such pigments. Such variation has allowed the algae to exploit diverse light niches, habitats that vary in spectral composition and irradiance level.

Algae also cope with variation in the availability of light for photosynthesis by being able to take in and metabolize organic compounds from the environment, as do heterotrophs. The ability to both perform photosynthesis and acquire exogenous organic nutrients is known as **mixotrophy** (mixed nutrition). Organisms that display mixotrophy are known as **mixotrophs** (Figure 1.20). The vast majority of algal species that have been tested are able to take up and metabolize some form of dissolved or particulate organic carbon. Many algal species are thereby able to survive periods of low irradiance. In addition, particle ingestion can provide mineral nutrients, such as iron and phosphorus, that are essential to maintaining maximal photosynthetic rates in waters of low fertility.

The Photoprotection Problem

A second major problem that algae face in accomplishing photosynthesis in the water is adapting to sudden changes in light levels. This is much like the visual shock people may experience when emerging from a dark movie theater into bright sunlight. Algae may encounter similar fast irradiance changes when currents transport deep-dwelling phytoplankton cells into bright surface light or when a receding tide exposes shoreline macroalgae. Some algae live in extremely high-irradiance habitats (Figure 1.21). Light in excess of that which can be used in photosynthesis is harmful to the photosynthetic apparatus, destroying essential proteins and chlorophyll molecules, and causing the formation of destructive

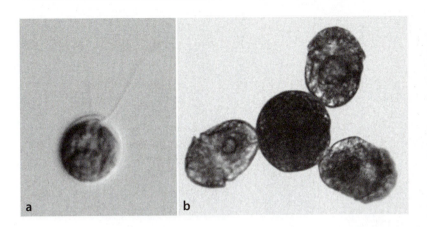

Figure 1.20 Mixotrophic algae. (a) The flagellate green alga *Chlamydomonas reinhardtii* is a mixotroph because it can absorb and metabolize acetate in addition to producing organic compounds via photosynthesis. (b) The cold-loving dinoflagellate *Amphidinium cryophilum* is a mixotroph because it is photosynthetic and also feeds on other cells. Here, three cells of *A. cryophilum* are feeding on another dinoflagellate species. (a: M. E. Cook; b: L. W. Wilcox)

oxygen radicals. Algae have responded to this threat by evolving photoprotection systems of carotenoid pigments that absorb excess light energy and dissipate it as heat (Figure 1.22); components of the LHCs may also function in photo protection. Even so, algae seem to experience a trade-off between the ability to absorb scarce light present at depth and the capacity to cope with high light levels present at the surface. A number of species have been found to include strains or **ecotypes** that are adapted to high or low light levels. For example, the coccoid marine green algal species *Ostreococcus tauri* includes one ecotype that lives in surface waters and another that occurs at greater depths. The surface-dwelling ecotype has greater photoprotection capacity but lower ability to harvest light than the deep-living strain (Cardol et al. 2008).

The Carbon-Fixation Problem

A third photosynthetic problem experienced by aquatic algae is the acquisition of carbon dioxide, which makes up less than 0.05% of Earth's atmosphere and diffuses 10,000 times more slowly in water than in air. Large amounts of carbon dioxide are needed in the light-independent phase of photosynthesis; otherwise, organic compounds cannot be made. In response, algae have evolved diverse carbon concentrating mechanisms (CCMs) that help them to obtain sufficient carbon dioxide for photosynthesis. CCMs are discussed more completely in Chapter 2.

Algae (and photosynthetic bacteria) have also evolved different forms of the enzyme commonly known as rubisco (ribulose bisphosphate carboxylase/oxygenase). **Rubisco** catalyzes the incorporation of carbon dioxide molecules into organic compounds.

Figure 1.21 Algae that live in a high-irradiance environment. The darkened areas on this Colorado snowfield are reddish patches of algae that normally grow in snow. Although they are members of the green algae and contain green plastids, the cells appear red because they contain high levels of photoprotective carotenoid pigments. (Photo: L. W. Wilcox)

Different forms of rubisco have different affinities for oxygen and carbon dioxide. Oxygen is important because it is currently much more abundant than carbon dioxide and competes with it by binding to the active site of rubisco. Oxygen binding at the active site of rubisco results in the process known as **photorespiration**, which has the net effect of reducing the amount of organic carbon that photosynthesizers can produce.

Rubisco variants are composed of different numbers and types of protein subunits and consequently display differing catalytic properties. There are four basic families of rubisco: I, II, III, and IV. Algae possess only Types I and II. Type II rubisco, composed of

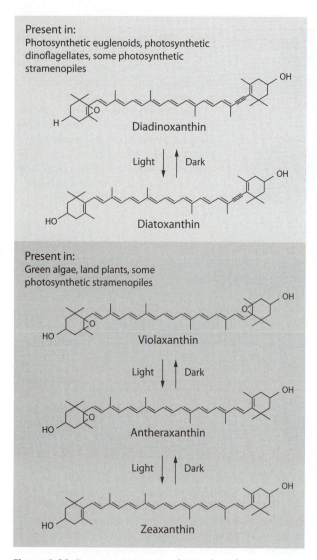

Present in:
Photosynthetic euglenoids, photosynthetic dinoflagellates, some photosynthetic stramenopiles

Diadinoxanthin

Light ↓↑ Dark

Diatoxanthin

Present in:
Green algae, land plants, some photosynthetic stramenopiles

Violaxanthin

Light ↓↑ Dark

Antheraxanthin

Light ↓↑ Dark

Zeaxanthin

Figure 1.22 Protective carotenoids. Cycles of chemical conversion among different types of the oxygen-containing carotenoids known as xanthophylls aid in dissipating excess light energy. (After After Falkowski, Paul G., *Aquatic Photosynthesis*, Second Edition. Reprinted by permission of Princeton University Press)

Figure 1.23 Cultivation of the red alga *Porphyra*. This macroalga is cultivated on submerged nets that are floated in the ocean and harvested as shown. Rows of floats supporting additional nets can be seen in the distance. The seaweed is processed into sheets of nori that are sold for use in preparing sushi and other dishes. (Photo: B. Waaland)

two large subunits, occurs in the most common type of photosynthetic dinoflagellates, those having a distinctive accessory pigment known as peridinin (Morse et al. 1995). Type I rubisco occurs in other photosynthetic algae and consists of eight large subunits and eight small subunits, encoded by different genes.

There are four forms of Type I rubisco: A, B, C, and D. Form IA occurs only in prokaryotes, including many marine cyanobacteria found in low-nutrient ocean waters. Form IB rubisco is characteristic of coastal and freshwater cyanobacteria and the plastids of green algae, euglenoids, chlorarachniophytes, and land plants. Form IC occurs only in prokaryotes that perform only anoxygenic photosynthesis (photosynthesis that does not produce oxygen). Form ID occurs in beta-proteobacteria, red algae, cryptomonads, haptophytes, and photosynthetic stramenopiles. The different forms of rubisco vary in their selectivity for carbon dioxide, a topic discussed in more detail in Chapter 2. In summary, variations in light sensing and phototaxis, accessory pigments and LHC proteins, photoprotective systems, mixotrophic capacity, and rubisco form combine to produce an amazing diversity of algal photosynthetic traits.

1.6 Societal Issues Involving Algae

As a result of their diverse traits, algae play important roles in everyday life. They provide the organic carbon that is the basis for aquatic food chains, influencing the quality of fisheries. Because Earth is a watery planet, algae play a role equal to that of land plants in generating atmospheric oxygen. Humans cultivate algae for food and other products (Figure 1.23), and algae have the potential to help meet fuel needs and generate many other useful products. Despite these

Figure 1.24 Algae in the headlines.

positive attributes, algae sometimes make headline news when they become invasive or produce harmful toxins (Figure 1.24). In consequence, knowledge of the diversity of algal species and their traits can aid in maintaining human health and sustainability.

As this chapter illustrates, the algae vary greatly in the structural and biochemical traits that influence their ability to accomplish photosynthesis in particular habitats. The other chapters of this book provide more insight into algal traits, such as diverse ways of accomplishing reproduction and interacting with other species. In the past, algal species were thought to play a fairly standard and redundant ecological role: photosynthesis in aquatic habitats. With few exceptions, algal species were regarded as functionally interchangeable, with many species substitutable for one another. Observing algae primarily by measuring chlorophyll *a* concentrations in water samples, though sometimes justified, is a reflection of such old-school views. More recently, algal species and ecotypes have been increasingly recognized as possessing unique trait combinations. As a result, algal species are increasingly less apt to be considered functionally redundant and their specific ecological roles better recognized. One challenge for aquatic biologists of the future is to catalog the diversity of algal species, ecotypes, and functional traits. Such a catalog would no doubt generate a treasure trove of new medicines, industrial processes, microfabrication strategies, and other rewards that have not yet been imagined. It will also aid our understanding of algal evolutionary patterns and processes, helping to place early branches of the tree of life in proper order and also illuminating the non-vertical pathways by which traits are transmitted among unrelated organisms.

An increased focus on traits will also be necessary for predicting responses to environmental change (Green et al. 2008). For example, Arctic phytoplankton communities are dominated by unicellular green algae that photosynthesize optimally at low temperatures (e.g., 6°C–8°C) and low irradiances. Species having these traits are particularly vulnerable to climate warming, which is ongoing in the region (Lovejoy et al. 2007). If the green cells decline, what will happen to Arctic food webs that have co-evolved with them? Will algal species that are likely to colonize warmer Arctic waters be able to fill the same ecological roles? How far will the effects propagate?

Additional challenges lie in preventing the formation of harmful algal blooms that result from overenrichment of natural waters by nutrients such as phosphorus and nitrogen. Such blooms shade aquatic plants, causing them to decline and thereby reduce habitat for fish and invertebrates. When algal blooms decay, much of the oxygen in the water is consumed, killing fish and shellfish. In fact, algal blooms are responsible for the global spread of "dead zones," coastal regions depleted of life that have substantial economic impacts on humans. As previously noted, some algal blooms produce toxins that poison animals, and experts expect that environmental change is likely to increase the incidence of such blooms. For example, the toxin-producing freshwater cyanobacterial species *Cylindrospermopsis raciborskii* has been colonizing higher latitudes in correlation with climate warming and nutrient enrichment. This species grows best in temperatures above 20°C and is adapted to light conditions present in nutrient-rich water (Wiedner et al. 2007). Does *C. raciborskii* grow in your local waters, and, if not, how soon is it likely to move in? Increased knowledge of algal traits, together with methods for reducing mineral nutrient loading to natural waters, will be necessary for managing the water resources that are essential to the survival of humans and other life.

Chapter 2

The Roles of Algae in Biogeochemistry

Biogeochemistry is the study of chemical interactions between the atmosphere, hydrosphere (aquatic systems), lithosphere (crustal minerals), and biosphere (living organisms). Algae have played significant roles in Earth's biogeochemistry for billions of years and continue to do so today. For example, ancient cyanobacteria generated Earth's first oxygen atmosphere. Ancient algae also produced important fossil fuel deposits and massive carbonate rock formations that reduced atmospheric carbon dioxide levels. Modern algae produce about half of the atmosphere's oxygen and powerfully influence the cycling of carbon, nitrogen, phosphorus, sulfur, and other elements, affecting other organisms in diverse ways. On the basis of these effects, experts have suggested that algae could be used in engineered systems to produce renewable sources of energy and remove mineral pollutants from water. Some have even proposed that algal growth could be manipulated to mitigate human effects on atmospheric chemistry, helping to moderate climate change. An understanding of algal roles in biogeochemistry is thus important in understanding Earth's past, as well as present and future global ecological issues.

This chapter begins by discussing algal production of atmospheric oxygen and then covers the roles of algae in global biogeochemical cycles. This background will be useful in understanding interactions that

Chara, Desmodesmus, and Chroococcus

(Algae photos: L. W. Wilcox; Earth: NASA)

occur between algae and other organisms (the focus of Chapter 3) and the technological applications of algae (Chapter 4).

2.1 Cyanobacteria and the Origin of an Oxygen-Rich Atmosphere

Like most eukaryotic algae and plants, modern cyanobacteria influence Earth's atmospheric chemistry by releasing gaseous oxygen as a result of photosynthesis, a process known as **oxygenic photosynthesis** (Figure 2.1). Fossil, geochemical, and molecular evidence indicates that the cyanobacteria were the first oxygenic photosynthesizers. Thus, the evolutionary origin of cyanobacteria and oxygenic photosynthesis were pivotal events in the history of life on Earth. By means of oxygenic photosynthesis, ancient cyanobacteria created an atmospheric chemistry suitable for other life forms. An oxygen-rich atmosphere fostered energy-efficient aerobic respiration and the formation of an ozone shield against ultraviolet radiation from the sun.

When oxygenic cyanobacteria first appeared more than 2.7 billion years ago (Buick 1992; Brocks et al. 1999), Earth's atmosphere was richer in carbon dioxide than it is today, but oxygen was sparse. Consequently, most organisms used relatively inefficient anaerobic (non-oxygen-using) processes to generate cellular ATP (adenosine triphosphate). During the

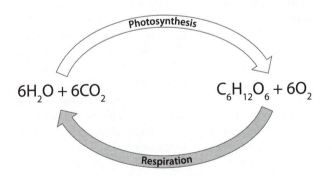

$$6H_2O + 6CO_2 \qquad C_6H_{12}O_6 + 6O_2$$

Figure 2.1 Photosynthesis and cellular respiration. The reactants in photosynthesis are the products of cellular respiration and vice versa. At the ecosystem level, if respiration were to occur at the same rate as photosynthesis, atmospheric oxygen and carbon dioxide levels would not change.

next several hundred million years, oxygen produced by early cyanobacteria accumulated in the atmosphere (Figure 2.2), with several important effects. First, by about 2.4 billion years ago, atmospheric oxygen had become abundant enough that organisms could use it as an electron acceptor in more efficient aerobic (oxygen-using) respiration (see Figure 2.1) (Eigenbrode and Freeman 2006). This change, known as the Great Oxidation Event, fostered the origin and early diversification of eukaryotes. The earliest fossils attributed to eukaryotes occur in deposits that are between 1 and 2 billion years of age (see Figure 2.2). Not only do most modern eukaryotes depend on aerobic respiration, they also use oxygen to produce unique cell membrane sterols and communication proteins (Acquisti et al. 2007). These observations suggest that the buildup of oxygen in Earth's atmosphere triggered the evolutionary processes by which multicellular animals, fungi, and plants later arose.

Oxygenic cyanobacteria were also essential to the origin of the first eukaryotic algae. These arose when one or more early heterotrophic eukaryotic cells took in cyanobacterial cells that persisted in a stable symbiotic relationship. Such cyanobacterial cells, known as endosymbionts, eventually evolved into plastids. All the plastids present in modern-day protists and plants arose as a result of evolutionary processes that started with cyanobacterial endosymbionts (see Chapter 7 for more information). Together with cyanobacteria, early eukaryotic algae continued to produce oxygen, with the result that atmospheric levels had nearly reached modern levels (21% oxygen) by 550 million years ago. This oxygen-rich atmosphere is associated with the rise of diverse communities of multicellular marine organisms, including the Ediacaran fauna (see Figure 2.2). The oxygen produced by cyanobacteria, algae, and plants remains essential to life on Earth today.

A third major impact of oxygenic photosynthesis was that by 1 billion years ago, the interaction of atmospheric oxygen with solar ultraviolet (UV) radiation had generated a stratospheric ozone shield sufficient to protect surface life from UV damage (Beardall and Raven 2004). UV radiation damages cellular DNA, and high levels of it cause organism death. Prior to the formation of the ozone shield, early life forms were likely restricted to sub-surface ocean waters, whose surface layers absorbed the harmful UV radiation. An oxygen-rich atmosphere was an essential precursor to the formation of Earth's earliest ozone shield. Once formed, the ozone shield allowed life to survive in surface waters and on land. The modern production of atmospheric oxygen by cyanobacteria, eukaryotic algae, and plants continues to build Earth's ozone shield, protecting its living things. The importance of Earth's ozone shield explains why experts are concerned about polar ozone holes that have recently appeared as a result of human activities.

2.2 Algae and the Carbon Cycle

Cyanobacteria and their descendants—the plastids of algae and plants—have been generating oxygen for billions of years. However, this process alone does not explain how oxygen accumulated in Earth's atmosphere or today's relatively high and stable levels of atmospheric oxygen. The explanation involves two important ways in which algae affect Earth's carbon cycle. The first of these is the capability of algae and plants to produce organic compounds that are resistant to microbial breakdown and thus are readily buried in anoxic ocean or lake sediments. Organic compounds that accumulate in such sediments are not oxidized back to carbon dioxide gas, with the result that atmospheric carbon dioxide levels decline over time. A second major effect of algae on the global carbon cycle is their ability to generate large deposits of sedimentary carbonate minerals such as calcium carbonate, a process that also reduces atmospheric CO_2 level. The burial of organic carbon and carbonates in sediments is an example of **carbon sequestration**, the removal of carbon from active cycling processes for long periods of time. Because the concentration of carbon dioxide in water is a function of that in air, carbon sequestration reduces the amount of CO_2 available to algae in aquatic ecosystems. Photosynthetic algae rely on an adequate supply of carbon dioxide for photosynthesis, and thus they have adapted to relatively low modern levels of atmospheric CO_2. Such adaptations include carbon concentration mechanisms and the uptake of organic carbon.

Algae and Organic Carbon Sequestration

Consider the reactions that describe oxygenic photosynthesis and aerobic respiration (see Figure 2.1).

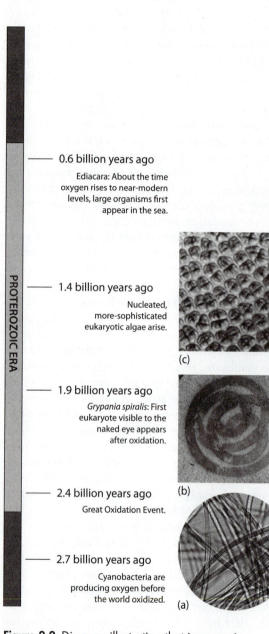

PROTEROZOIC ERA

— 0.6 billion years ago

Ediacara: About the time oxygen rises to near-modern levels, large organisms first appear in the sea.

— 1.4 billion years ago

Nucleated, more-sophisticated eukaryotic algae arise.

(c)

— 1.9 billion years ago

Grypania spiralis: First eukaryote visible to the naked eye appears after oxidation.

(b)

— 2.4 billion years ago

Great Oxidation Event.

— 2.7 billion years ago

Cyanobacteria are producing oxygen before the world oxidized.

(a)

Figure 2.2 Diagram illustrating that increase in atmospheric oxygen (first produced by ancient cyanobacteria) is correlated with the appearance and radiation of eukaryotes. (a) The modern cyanobacterium *Oscillatoria*. (b) *Grypania*, a fossil interpreted as an early eukaryotic alga. (c) The modern eukaryotic red alga *Porphyra*. (a: L. W. Wilcox; b: Reprinted with permission from Knoll, A. H., The early evolution of eukaryotes: A geological perspective. *Science* 256:622–627. © 1992 American Association for the Advancement of Science; c: L. E. Graham)

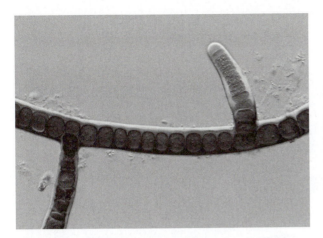

Figure 2.3 Cyanobacterium with sheath containing scytonemin. This degradation-resistant compound is often present in the extracellular mucilage or sheaths of cyanobacteria. With the exception of the tip of the branch growing upward, the sheath of this cyanobacterium has a dark yellow-brown color that is due to the presence of scytonemin. (Photo: L. W. Wilcox)

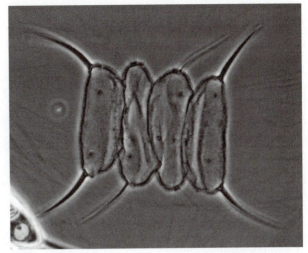

Figure 2.4 Algaenan-containing, decay-resistant cell walls of *Desmodesmus*, which often persist in the water column and sediments at the bottoms of lakes. (Photo: L. W. Wilcox)

If aerobic breakdown of organic carbon occurred at the same rate as oxygen liberation and the formation of organic compounds in photosynthesis, atmospheric oxygen would not accumulate. However, organic carbon is degraded to CO_2 more slowly than it is produced. This occurs because some proportion of dead algae, or organic carbon compounds produced by algae, can sink through the water column to the bottom without being completely decomposed. Thus algal organic carbon can be sequestered in deep anoxic sediments where it is sheltered from microbial oxidation. Because oxygen is not being used to degrade sedimentary organic carbon, global photosynthetic oxygen production can exceed or balance the rate of oxygen consumption, allowing atmospheric oxygen to accumulate or remain steady.

Algae produce several types of organic compounds that are resistant to chemical breakdown and decay and are thus described as **refractory carbon**. This refractory carbon, together with more degradable organic compounds that may be associated with it, can build up as hydrocarbon-rich sedimentary deposits known as **kerogens**. Decay-resistant compounds often occur in mucilage sheaths or cell walls, thus preserving the form of the algae after death and contributing to the formation of fossils. For example, **scytonemin**, which is present in the sheaths of many cyanobacteria, shields cells from UV radiation and is also degradation-resistant (Figure 2.3). The presence of scytonemin may explain ancient cyanobacterial fossils. Certain green algae produce tough cell wall materials known as **algaenans**. Composed of cross-linked hydrocarbons, algaenans are thought to be relatively waterproof, allowing microalgae to disperse from one body of water to another without perishing from desiccation (Versteegh and Blokker 2004). Algaenans also effectively resist microbial attack, explaining the common presence of algal cell wall remains in modern water samples (Figure 2.4). Algaenans also account for certain coal deposits; these originated from cellular organic compounds that were protected from decay after algal death and sedimentation. Green algae that are closely related to land plants produce zygotes that have degradation-resistant cell wall layers. Such wall layers are similar to the tough **sporopollenin** characteristic of plant spore walls. Sporopollenin confers mechanical stability and resistance to microbial attack, explaining the occurrence of plant spores in the fossil record. Dinoflagellates often produce resistant-walled cyst stages, which enable survival through stressful conditions, and, in some cases, these cyst walls contain a resistant substance that is distinct from sporopollenin and algaenan (Kokinos et al. 1998). The enigmatic fossil microorganism known as *Gloeocapsamorpha prisca*, which is a major component of marine oil shales formed during mid-Ordovician times,

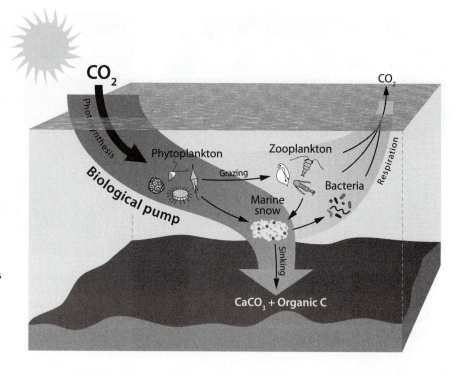

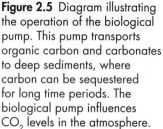

Figure 2.5 Diagram illustrating the operation of the biological pump. This pump transports organic carbon and carbonates to deep sediments, where carbon can be sequestered for long time periods. The biological pump influences CO_2 levels in the atmosphere.

contains a resistant polymer composed of resorcinol units (Versteegh and Blokker 2004).

In addition to the resistant algal materials described previously, aquatic bacteria can convert degradable organic carbon exuded from living algae or released from decomposing algae into fairly decay-resistant colloids (Fry et al. 1996). Colloidal substances are defined by their size (1–1000 µm in diameter), which is larger than soluble organic materials but smaller than many organic carbon particles. Some 30% to 50% of the organic carbon found in aquatic systems—perhaps 250 gigatons on a global basis—may occur in such colloidal form (Wells 1998). Much of this colloidal material occurs as acylpolysaccharides (APS) (branched carbohydrates having two acetate groups for every five sugars) that are resistant to bacterial decomposition. APS occur in all fresh and marine waters that have been examined and in deep-sea deposits. The global reserve of APS is thought to be 10 to 15 gigatons, with algae representing the only known source (Aluwihare et al. 1997). APS also aggregate with bacteria, zooplankton remains, and fecal pellets to form larger (>500 µm) particles—known as **marine snow**—that readily sink. Marine snow is a very important mechanism by which organic carbon reaches sediments.

Though much of the organic material produced by algal photosynthesis is recycled within surface waters by grazing zooplankton and microbes (Chapter 3), some small percentage of the organic compounds present in the open ocean is transported downward into the sediments. This transport process, known as the **biological pump**, has the effect of drawing additional atmospheric carbon dioxide into the ocean,   where it is incorporated into the organic constituents of living organisms. Organic carbon transported into the ocean's depths can be sequestered for hundreds to thousands of years, and a fraction of it remains there for millions of years (Figure 2.5). As a result of operation of the biological pump for billions of years, Earth's sedimentary rocks now hold more than 10^7 gigatons of organic carbon (Reimers 1998). Rates of organic carbon sedimentation are influenced by the "rain rate" (the rate at which organic carbon sinks through the water column, which is a function of algal productivity) and the amount of time that organic carbon is exposed to oxygen and aerobic decomposers. Over long time periods, and as a result of thermal and microbial changes, organic sediments may be transformed into fossil fuels such as coal, oil, or natural gas (methane). In the past there have been periods of unusually great organic sedimentation and fossil hydrocarbon formation. For example, during the Mesozoic Era (65–250 million years ago) warm temperatures, sluggish ocean circulation, and low seawater oxygen content contributed to a particularly high organic rain rate, which resulted in the economically significant North Sea oil deposits. It is clear that algae were involved in the formation of this oil because characteristic mineral remains in the form of calcium carbonate are abundant in the North Sea deposits (Tucker and Wright 1990). The next section surveys the role of algae in transforming CO_2 into carbonates, a second mechanism by which atmospheric CO_2 can be sequestered.

The Role of Algae in Carbonate Formation

Beginning with cyanobacteria in Precambrian times, various groups of algae have transformed very large amounts of carbon from the atmosphere into carbonate sediments and rocks such as limestone. In fact, much of the carbon dioxide that once occurred

in Earth's atmosphere has been transformed into carbonate rocks, primarily by algae (McConnaughey 1994). Carbonates are also important because they contain at least 40% of the world's known hydrocarbon reserves (Tucker and Wright 1990). This is not surprising because organic constituents of algae that are associated with relatively heavy carbonate materials often sink into the deep ocean, where they can be preserved. Recently, experts have expressed concern that increases in atmospheric CO_2 caused by human activities (such as burning fossil fuels) have led to the acidification of ocean waters as the additional CO_2 dissolves, becoming H_2CO_3 (carbonic acid) (Feely et al. 2004). Carbonic acid dissolves carbonate minerals, and experiments have shown that algae produce less carbonate when atmospheric CO_2 levels are increased (Riebesell et al. 2000). Experts are concerned that if algae produce less carbonate in the future, a smaller amount of organic carbon will be sequestered in the deep ocean, which would cause atmospheric CO_2 to increase. For these reasons, the production of carbonates by algae is an important topic.

The process by which algae (and some other organisms) produce calcium carbonate is known as calcification. Algal groups that precipitate calcium carbonate on their surfaces include various cyanobacteria, some freshwater green algae, and certain green, red, and brown seaweeds. In the remote

Figure 2.6 Modern (A) and fossil (B) rhodoliths (arrowheads). (From Foster 2001, by permission of the *Journal of Phycology*)

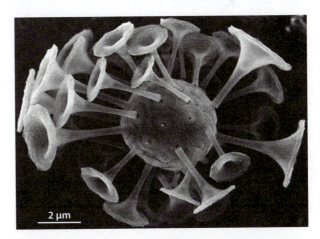

Figure 2.7 The coccolithophorid *Discosphaera tubifera* from the central North Atlantic. Members of this group of primarily unicellular marine phytoplankton have a covering of calcified scales, known as coccoliths. Certain ocean sediments are rich in the coccolith remains of these algae. (From Kleijne 1992)

Figure 2.8 A model of atmospheric carbon dioxide levels through time (black line). The model is supported by stable carbon isotope ratio measurements (white bars) extending back about 500 million years. According to this model, carbon dioxide levels were considerably higher in the remote past, when algae were the dominant autotrophs. However, considerable uncertainty (represented by the dark gray area) surrounds quantitative estimates of ancient atmospheric carbon dioxide levels. (Redrawn with permission from Berner, R. A., The rise of land plants and their effect on weathering and CO_2. *Science* 276:544–546. © 1997 American Association for the Advancement of Science)

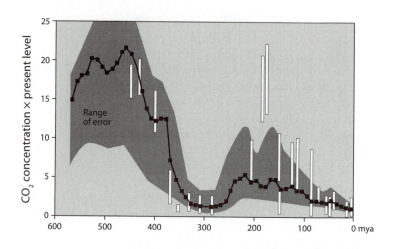

past, cyanobacteria produced extensive layered, often mound-shaped, calcium carbonate deposits known as **stromatolites** (Allwood et al. 2006), and their modern relatives still build stromatolites in some places. A number of modern tropical green seaweeds, such as *Halimeda*, are important producers of carbonate sand; their ancient relatives left calcium carbonate deposits up to 52 m thick at sites near the northern Great Barrier Reef (Australia) and other locations (Tucker and Wright 1990). Flat, hard sheets of calcified red seaweeds known as crustose corallines coat and stabilize extensive surfaces of coral reefs in warm waters around the world, as in the past. Some coralline red algae, today and in the past, have formed massive beds of **rhodoliths**, round balls of calcified algae informally known as "tumbleweeds of the sea" (Figure 2.6). Particularly large rhodolith beds, also called rhodolites or maërl, occupy the Brazilian Shelf between the latitudes of 2°N and 25°S (Foster 2001). Coccolithophorids are unicellular haptophyte algae that produce miniscule calcium carbonate scales known as coccoliths (Figure 2.7). Coccoliths account for 20% to 40% of modern biologically produced carbonates. Ancient coccolith deposition helped to build massive chalk deposits in the Cretaceous Period (see Chapter 10).

Although the advantages of calcification and the exact mechanism by which it occurs are not completely understood, as we shall see, this process helps some algal cells to acquire carbon dioxide (McConnaughey and Whelan 1996). In some algae, calcification may help prevent photosynthesis from becoming limited by the availability of carbon dioxide. This is advantageous to modern algae because carbon dioxide is less abundant today than it was in the remote past, our next topic.

Impact of Modern Carbon Dioxide Levels on Algal Photosynthesis

As we have seen, algae and plants require CO_2 in order to accomplish photosynthesis. Geochemical evidence indicates that carbon dioxide levels in the atmosphere and water were much higher, and oxygen levels much lower, 2.2 billion years ago than at present. Over time, oxygenic photosynthesis and the sedimentation of organic carbon and carbonates caused carbon dioxide

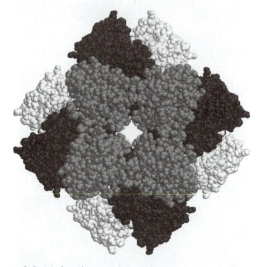

Figure 2.9 Molecular model of the enzyme complex Form I Rubisco that occurs in cyanobacteria and the plastids of most algae and plants and that catalyzes carbon fixation. Form I Rubisco includes eight small and eight large subunits. Different forms of rubisco occur in other autotrophic bacteria and some algae. (From http://www. biologie.uni-hamburg.de/lehre/bza/smap.htm. Reprinted by permission of R. Bergmann)

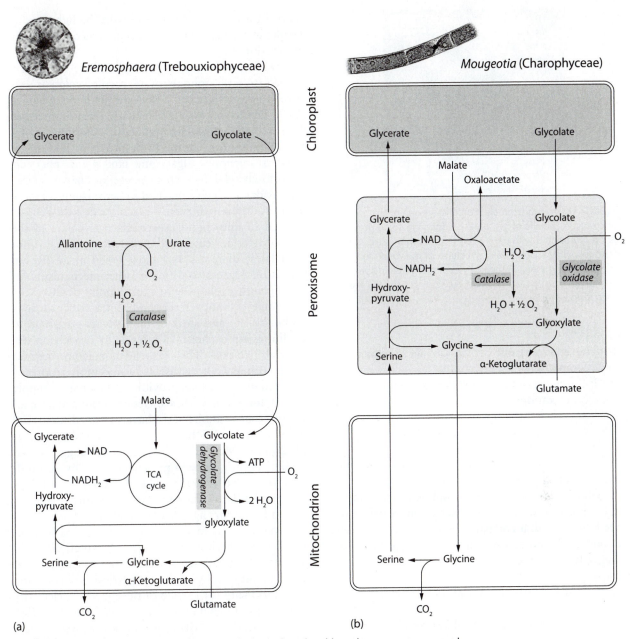

Figure 2.10 The process of photorespiration, which involves the chloroplast, peroxisome, and mitochondrion. Details vary among different groups of algae, particularly in the oxidation of glycolate, which can involve the enzyme glycolate dehydrogenase in the mitochondrion, as in the trebouxiophycean green alga *Eremosphaera* (a) or glycolate oxidase in the peroxisome of algae such as the charophycean green alga *Mougeotia* (b). The enzyme catalase, located in the peroxisome, breaks down toxic hydrogen peroxide to water and oxygen. (Re-drawn, with permission, from Stabenau and Winkler 2005. *Physiologia Plantarum* 123:235–245. © *Physiologia Plantarum*; Photos: L. W. Wilcox)

levels to decrease to today's relatively low levels (Figure 2.8). The rise of vascular land plants caused a particularly dramatic drop in carbon dioxide levels to a historic low about 300 million years ago (Berner 1997).

As a consequence of these changes, the algae acquired cellular adaptations that allowed them to cope. Many of these adaptations involve **rubisco** (ribulose bisphosphate carboxylase/oxygenase) (Figure 2.9), the enzyme

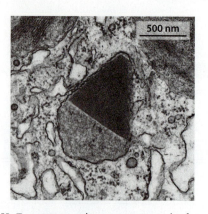

500 nm

Figure 2.11 Transmission electron micrograph of a peroxisome, the site of glycolate oxidation in some green algae. (Reprinted with permission from Graham, L. E. and Y. Kaneko 1991. Subcellular structures of relevance to the origin of land plants [Embryophytes] from green algae. *Critical Reviews in Plant Science* 10:323–42. © CRC Press, Boca Raton, Florida)

that converts inorganic carbon dioxide to reduced organic compounds.

Rubisco catalyzes the following reaction, known as carbon fixation:

$$RuBP + CO_2 + H_2O \rightarrow 2PGA$$

Rubisco is said to function as a carboxylase when it uses CO_2 as a reactant in the production of the 3-carbon compound PGA (phosphoglyceric acid), as in the reaction shown above. But when oxygen levels are relatively high compared to the levels of carbon dioxide, rubisco can operate as an oxygenase, oxidizing RuBP in the reaction:

$$RuBP + O_2 \rightarrow PGA + phosphoglycolate$$

In this case, phosphoglycolate is produced instead of PGA because oxygen competes with carbon dioxide in binding to the active site of rubisco, thereby reducing the rate of carbon fixation. Phosphoglycolate can be excreted from photosynthetic cells, which represents a loss of organic carbon and a reduction of photosynthetic efficiency. Phosphoglycolate also ties up cellular phosphate and inhibits photosynthesis. Photosynthetic organisms have adapted to such oxygen effects by means of photorespiration (Figure 2.10), a metabolic pathway that breaks down phosphoglycolate and recovers some of the ATP that the cell would otherwise lose. Photorespiration is named for the fact that it occurs in the light, and it results in the loss of organic carbon, as does cellular respiration.

During photorespiration, phosphate is first removed from phosphoglycolate, then glycolate is oxidized to glyoxylate and processed further (see Figure 2.10). Photosynthesizers display two major processes for oxidizing glycolate, reflecting alternate evolutionary responses to the same problem. Red algae, most stramenopile algae, and land plants plus certain closely related green algae use the enzyme glycolate oxidase, which is located in **peroxisomes** (Figure 2.11). A different enzyme—glycolate dehydrogenase, located in mitochondria—oxidizes glycolate in the cyanobacteria, euglenoids, cryptomonads, dinoflagellates, diatoms, and most green algae (Raven 1997a,b). Algae also differ in other mechanisms for coping with a relatively low availability of CO_2.

Some cyanobacteria and eukaryotic algae—including the ancestors of land plants—adapted to declining atmospheric CO_2 levels by colonizing terrestrial habitats. This was advantageous because gases such as CO_2 diffuse 10^4 times more readily in air than in water, thus providing an increased supply of CO_2 for photosynthesis. Geochemical evidence indicates that cyanobacteria have occupied terrestrial habitats since 2.6 billion years ago (Watanabe et al. 2000b). Molecular clock evidence suggests that plants might have been present on land by 700 million years ago (Heckman et al. 2001), and fossils suggest that land-adapted plants certainly existed by 470 million years ago (Wellman et al. 2003).

Another way in which algae have coped with altered atmospheric conditions is by evolutionary diversification in rubisco structure and function. Algal rubiscos vary in two major properties: specificity and K_{cat}. **Specificity** is the ratio of the enzyme's selectivity of CO_2 over O_2, a number that varies from 20 to 280. In contrast, K_{cat} is the rate of enzyme turnover, the rate of catalysis when the substrate is present in saturating amount. There is a trade-off in these properties: When a rubisco's selectivity for CO_2 is relatively high, its K_{cat} is relatively low, and vice versa. Red algae and groups having plastids derived from red algae (haptophytes, cryptomonads, and stramenopile algae) have comparatively high CO_2 specificity. In contrast, cyanobacteria and most photosynthetic dinoflagellates possess types of rubisco having relatively low CO_2 specificity. Variations among algal rubiscos reflect the effects of selection for higher selectivity of CO_2 or faster enzyme turnover (Tcherkez et al. 2006).

A third type of adaptation to reduced CO_2 and increased O_2 was the evolution of **carbon concentrating mechanisms** (**CCMs**). These vary among algae, but all serve to increase the supply of CO_2 to rubisco. CCMs frequently involve the use of bicarbonate ion, which is more abundant in many aquatic habitats than is dissolved carbon dioxide. At pH 8 to 9, which is characteristic of many lakes and the ocean, the concentration of bicarbonate (about 2 mM) is 200 times greater than that of dissolved carbon dioxide (only about 12–15 µM). Only at pH around 6 does dissolved CO_2 become more abundant than bicarbonate (see Figure 21.12). Algal CCMs may involve cell membrane inorganic carbon transporters, enzymes that interconvert CO_2 and bicarbonate, calcification-linked processes, and specialized cellular structures (Giordano et al. 2005). The CCMs of cyanobacteria are particularly well understood.

Carbon Concentration Mechanisms of Cyanobacteria

Many cyanobacteria rely on CCMs because their rubisco has relatively low specificity for CO_2. If CO_2 is available, the gas diffuses into cells and is likely converted into bicarbonate ion at the thylakoid membranes. Cyanobacterial cells also take up bicarbonate ion from surrounding water by the use of plasma membrane transporters (Figure 2.12). Because bicarbonate is charged, it cannot diffuse rapidly out of cells, and so it is trapped. However, before bicarbonate ion can be used in photosynthesis, it must be converted to carbon dioxide because CO_2 is the only form of dissolved inorganic carbon that can be used directly in carbon fixation. Bicarbonate ions and CO_2 are interconverted via the reaction:

$$HCO_3^- + H^+ \leftrightarrow CO_2 + H_2O$$

Because the uncatalyzed rate of this reaction is too slow to supply carbon dioxide to algal photosynthesis, cyanobacteria use an enzyme known as **carbonic anhydrase** to dramatically speed the reaction. In fact, carbonic anhydrase (CA) is noteworthy among enzymes for its almost unbelievably high turnover rate—a feature that greatly facilitates provision of CO_2 to algal photosynthesis (Stemler 1997). In cyanobacteria, CA is located within carboxysomes—polygonal protein structures that can be readily identified in electron micrographs (Figure 2.13). Cyanobacterial rubisco is also located within

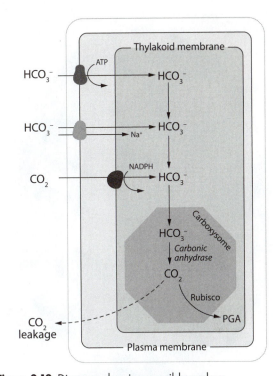

Figure 2.12 Diagram showing possible carbon concentration mechanisms in cyanobacterial cells. Because bicarbonate is a charged ion, its uptake from the surrounding water requires transporter molecules located in the plasma membrane. Cyanobacteria produce several types of bicarbonate transporters. Carbon dioxide can diffuse more freely across the plasma membrane, and its intracellular transformation into bicarbonate traps inorganic carbon within the cell, thereby concentrating it. Carbonic anhydrase at the carboxysome converts bicarbonate to carbon dioxide for fixation by rubisco. (After Giordano et al. 2005. Re-drawn, with permission, from the *Annual Review of Plant Biology* Volume 56. © 2005 by *Annual Reviews* www.annualreviews.org)

carboxysomes; co-localization of these two enzymes facilitates the capture of CO_2 by rubisco before it can diffuse out of the cell.

Carbon Concentration Mechanisms of Eukaryotic Algae

Eukaryotic algae possess a wide range of mechanisms for acquiring inorganic carbon for photosynthesis. These include membrane transporters and several types of carbonic anhydrase enzymes that usually contain zinc as a cofactor at the active site but may also use cadmium or cobalt (Figure 2.14). In addition, some algae possess plastid structures known as **pyrenoids** (Figure 2.15), which are thought to play

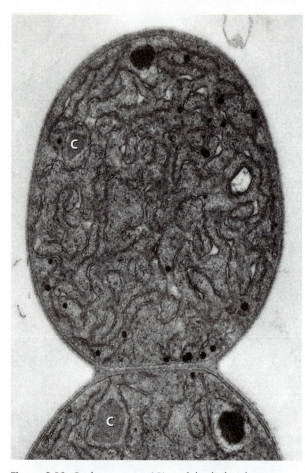

Figure 2.13 Carboxysomes (C), polyhedral inclusions in the cells of cyanobacteria, such as *Cylindrospermum*, viewed here by transmission electron microscopy. The enzymes carbonic anhydrase and rubisco have been localized to the carboxysome. (TEM: L. E. Graham)

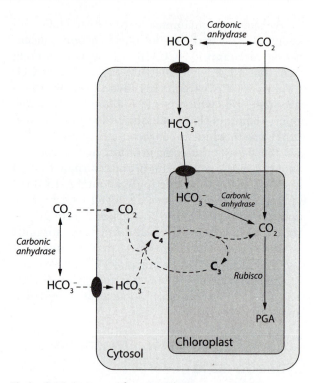

Figure 2.14 Diagram of some carbon concentration mechanisms of eukaryotic algal cells. Bicarbonate in the surrounding water can be converted into carbon dioxide by external (periplasmic) carbonic anhydrase enzyme (CA) and perhaps imported by means of hypothetical plasma membrane transporter proteins. Carbon dioxide that diffuses into cells (or possibly is imported via transporters) can be trapped within cells by conversion to bicarbonate, a process that may involve CA. Bicarbonate must be converted back to carbon dioxide for use by rubisco, which occurs in the pyrenoids present in many algal plastids. Some algae may have a C_4 mechanism (dashed lines) in which carbon dioxide is first fixed into a four-carbon organic compound and then later released for fixation by rubisco. (After Giordano et al. 2005. Re-drawn, with permission, from the *Annual Review of Plant Biology* Volume 56. © 2005 by Annual Reviews www.annualreviews.org)

a role in CCMs. Pyrenoids may be the structural equivalents of cyanobacterial carboxysomes (Badger and Price 1994), but pyrenoids do not occur in the plastids of all algae that have CCMs. Some eukaryotic algae may have a C_4-like photosynthesis, in which the enzyme PEP carboxylase traps inorganic carbon into a 4-carbon organic compound. CO_2 is later released from this compound for fixation by rubisco.

Many eukaryotic algae are known to secrete carbonic anhydrase into the periplasmic space, which is located between the plasma membrane and the cell wall. The role of this periplasmic CA is to catalyze the conversion of extracellular bicarbonate into carbon dioxide, which then moves into the cell via passive or active transport (see Figure 2.14). Immunolocalization at the transmission electron microscopic

level can be used to localize periplasmic CA (Figure 2.16). In some eukaryotic algae, periplasmic CA is constitutive (always present), whereas in other cases, CA is induced by low levels of environmental carbon dioxide. Still other species completely lack periplasmic CA, even when grown in a medium having a low concentration of inorganic carbon. In a study of 18 species of marine phytoplankton, Nimer et al. (1997) found that extracellular CA was constitutive in four of the five dinoflagellates examined and that external CA was inducible in several ecologically important

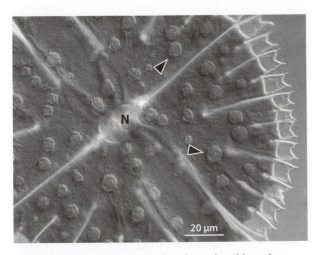

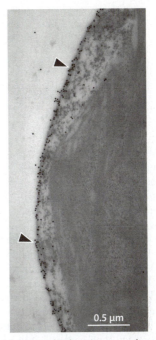

Figure 2.15 Pyrenoids (arrowheads) in the chloroplast of the green alga *Micrasterias*, as viewed by differential interference contrast light microscopy. Many kinds of eukaryotic algae contain pyrenoids in their plastids; these are often visualized at the light microscopic level by staining the starch shell that often surrounds pyrenoids with a solution of iodine and potassium iodide. The nucleus (N) of this unicellular alga is also evident. (Photo: L. W. Wilcox)

Figure 2.16 Carbonic anhydrase immunolocalized to the cell wall and periplasmic space (the region between the cell wall and the cell membrane—arrowheads) of the green alga *Chlamydomonas reinhardtii*. Here, a thin section of algal cell prepared for examination by transmission electron microscopy was treated with an antibody to the enzyme carbonic anhydrase. A second antibody labeled with small gold particles was then used as a method for visualizing the location of the first antibody and the antigen to which it is bound. The distribution of gold particles thus reflects the distribution of carbonic anhydrase. (Unpublished micrograph by P. Arancibia, L. Graham, and W. Russin using an antibody provided by J. Coleman)

algae, such as the diatom *Skeletonema costatum* and the coccolithophorid *Emiliania huxleyi*.

Yet another way in which algae can generate CO_2 from environmental bicarbonate is to excrete protons (H^+) across cell membranes. The protons react with bicarbonate to yield carbon dioxide and water, whereupon the CO_2 can be taken into cells and plastids (McConnaughey 1998). Such processes are often linked to calcification. In stonewort green algae (*Chara* species), the import of hydrogen ions from the external medium creates alkaline bands where calcification occurs. Compensatory export of hydrogen ions at other regions of the cell surface generates acid bands where carbon dioxide can be produced from bicarbonate (Figure 2.17a,b). This CO_2 may move into cells via ATP-mediated active transport (Mimura et al. 1993). In the haptophytes known as coccolithophorids, calcification occurs in intracellular vesicles across the surface of which hydrogen ions may be exchanged with calcium cations (Figure 2.17c). Interaction of protons with bicarbonate then generates carbon dioxide that can be used in photosynthesis. Some open-ocean coccolithophorids (*Coccolithus pelagicus* and *Gephyrocapsa oceanica*), which do not encounter much variation in bicarbonate concentrations, rely

primarily on this means of obtaining CO_2, so they lack extracellular carbonic anhydrase (Nimer et al. 1997). At the surfaces of coralline red algae, exchange of protons with calcium ions (Figure 2.17d) is thought to occur. The hydrogen ions then react with bicarbonate to release CO_2 for photosynthesis. The chemical reactions that link algal calcification to CO_2 acquisition are shown in Figure 2.18.

Algal Use of Organic Carbon

Non-photosynthetic algal species rely on the uptake of organic carbon from their environment. Those that consume dissolved organic carbon are described as osmotrophs, and those taking up particles of

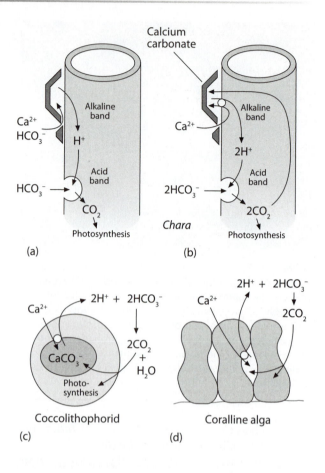

Figure 2.17 Models of the calcification process in three groups of algae: the green algal genus *Chara*, the types of haptophytes known as coccolithophorids, and the types of red algae known as corallines. Calcification of *Chara* (alternate possibilities are shown in (a) and (b)) and corallines (d) occurs externally, whereas this process occurs within intracellular vesicles of coccolithophorids (c). In each case, Ca^{2+} transport across a membrane is linked to export of H^+, which participates in the extracellular conversion of bicarbonate (HCO_3^-) to CO_2 that is subsequently taken up by cells for use in photosynthesis. Calcification can be viewed as an adaptation that allows algae living in alkaline waters to obtain sufficient inorganic carbon for photosynthesis. It should be noted that these models represent hypotheses that need further testing. (Reprinted from *Earth Science Reviews* Vol. 42, McConnaughey, T. A., and J. F. Whelan. Calcification generates protons for nutrient and bicarbonate uptake. Pages 95–118. © 1996, with permission from Elsevier Science)

organic material are phagotrophs. Examples of obligate osmotrophs include *Prototheca* and *Polytoma*, which are members of the green algae. But there is also strong evidence that many photosynthetic algae are facultative osmotrophs, particularly when light-limited (Neilson and Lewin 1974). For example, several species of the freshwater or soil diatom *Navicula* are able to grow with glucose as the sole carbon source in the light or in the dark (Lewin 1953). The unicellular red alga *Galdieria sulphuraria*

is genetically equipped to take up and use more than 50 different organic compounds, whether light is present or not (Barbier et al. 2005). Tuchman (1996) pointed out that all groups of photosynthetic algae had heterotrophic ancestors, so it should not be surprising for modern algae to have the capacity to take up organic compounds. In fact, many algae possess transport systems that allow the active transport of sugars such as glucose. In some cases these systems are constitutive, whereas in other cases uptake is

Calcification:

$$CO_2 + Ca^{2+} + H_2O \longrightarrow CaCO_3 + 2H^+$$

HCO_3^- utilization:

$$2H^+ + 2HCO_3^- \longrightarrow 2CO_2 + 2H_2O$$

Photosynthesis:

$$CO_2 + 2H_2O + >8 \text{ photons} \longrightarrow CH_2O + H_2O + O_2$$

Net:

$$2HCO_3^- + Ca^{2+} \longrightarrow CaCO_3 + CH_2O + O_2$$

Figure 2.18 Reactions linking the calcification process in algae to dissolved carbon utilization and photosynthesis.

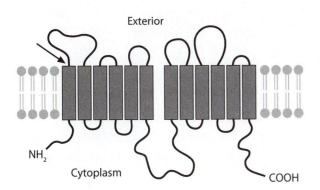

Exterior

NH₂

Cytoplasm   COOH

Figure 2.19 Model of a plasma membrane hexose transporter protein. Many such proteins are characterized by 12 transmembrane regions (dark gray boxes) consisting of α–helically coiled regions rich in hydrophobic amino acids, separated by loops of more hydrophilic amino acid sequences that extend both into the cytoplasm and outward to the exterior environment. The arrow indicates the site where sugar in the external medium is thought to bind. The common occurrence of such transporters in prokaryotes and eukaryotes suggest that they are likely to be of widespread occurrence in algae. (After *Botanica Acta* 106 [1993]:277–286. Sauer, N., and W. Tanner. *Molecular biology of sugar transporters in plants*. Georg Thieme Verlag, Stuttgart)

inducible by darkness or high levels of environmental glucose (Carthew and Hellebust 1983).

Some algae utilize dissolved organic carbon only in the light. An example is the diatom *Cocconeis*, which was isolated from sediments of Biscayne Bay, Florida that were rich in decaying sea grasses (Bunt 1969). The green alga *Chlorella*, which may occur in organic-rich habitats such as soils and sugar mill effluents, is another such example. In light, the green alga *Coleochaete* is able to utilize glucose, but only when dissolved inorganic carbon is limiting to growth (Graham et al. 1994). Lewitus and Kana (1994, 1995) demonstrated that axenic (bacteria-free) cultures of several phytoplankton species common to estuaries—including the green alga *Closterium*, two cryptomonads, two stramenopiles, and a haptophyte—utilized glucose. This was accomplished by both mitochondrial respiration in low light levels and chlororespiration—a form of respiration that occurs in chloroplasts—at high irradiance. The first of these processes allows the algae to overcome growth restrictions resulting from reduced light, and the latter provides the algae with greater ability to use saturating levels of light in photosynthesis.

Studies of model species have revealed the molecular and biochemical basis for uptake of organic compounds by algae and its interaction with CCMs. For example, the green alga *Chlorella* has at least three well-characterized cell membrane hexose transporter proteins (Stadler et al. 1995) that are homologous to members of the major facilitator superfamily of transporters found in plants, animals, fungi, other protists, and bacteria (Griffith et al. 1992). Transporters of this type possess 12 α-helical transmembrane regions (which are embedded within the cell membrane) connected by internal and external loops. Will and Tanner (1996) used molecular techniques to demonstrate in *Chlorella* that the likely site for substrate recognition and binding occurs at the first external loop (Figure 2.19). Other workers have shown that glucose uptake by *Chlorella* suppresses expression of carbonic anhydrase and hence the CCM, even when ambient CO_2 levels were low (Shiraiwa and Umino 1991; Villarejo et al. 1997).

In summary, the presence of diverse CCMs and the widespread ability to use organic carbon mean that growth of most modern algae is not generally limited by inorganic carbon availability, despite today's relatively low atmospheric CO_2 levels. If atmospheric CO_2 levels continue to increase in the future, the utility of algal CCMs might decline. However, even for dinoflagellates and cyanobacteria—which are extremely reliant on CCMs—atmospheric levels of CO_2 would have to increase several-fold before they could rely on diffusive CO_2 supplies. Thus nitrogen, phosphorus, iron, silica minerals, and light probably limit algal growth more often than does inorganic carbon (Raven 1997b). The next sections address these mineral nutrients and their relevance to algal roles in global biogeochemical cycles.

2.3 Mineral Limitation of Algal Growth

For growth to occur, algae (and plants) require a wide range of mineral nutrients (Table 2.1) in addition to light, water, and carbon. Such mineral nutrients are commonly taken up by means of transport proteins located in the plasma membrane, though some algae also obtain minerals by engulfing particles such as bacterial cells. Combined (fixed) nitrogen, iron, phosphate, and sometimes silica are essential nutrients needed by algae

Table 2.1	Elements commonly required by algae
Element	**Examples of function/location in algal cells**
N	amino acids, nucleotides, chlorophyll, phycobilins
P	ATP, DNA, phospholipids
Cl	oxygen production in photosynthesis, trichloroethylene, perchloroethylene
S	some amino acids, nitrogenase, thylakoid lipids, CoA, carrageenan, agar, DMSP, biotin
Si	diatom frustules, silicoflagellate skeletons, synurophyte scales and stomatocyst walls, walls of the ulvophyte *Cladophora*
Na	nitrate reductase
Ca	alginates, calcium carbonate, calmodulin
Mg	chlorophyll
Fe	ferredoxin, cytochromes, nitrogenase, nitrate and nitrite reductase, catalase, glutamate synthetase, superoxide dismutase cofactor
K	agar and carrageenan, osmotic regulation (ionic form), cofactor for many enzymes
Mo	nitrate reductase, nitrogenase
Mn	oxygen-evolving complex of photosystem II, wall-like lorica of some euglenoids and the chlorophyte *Dysmorphococcus*, superoxide dismutase
Zn	carbonic anhydrase, alcohol dehydrogenase, glutamic dehydrogenase
Cu	plastocyanin, cytochrome oxidase
Co	vitamin B_{12}
V	bromoperoxidase, some nitrogenases
Br, I	halogenated compounds with antimicrobial, anti-herbivore, or allelopathic functions

in particularly large amounts. When mineral nutrients are present in such small amounts that algae or plants do not display maximum growth, the minerals are described as limiting factors. If sufficient light is available, the growth of photosynthetic organisms is limited by the mineral nutrients whose environmental concentrations are closest to the minimal levels required. If the levels of such minerals increase, growth likewise increases, leading to larger populations. Algae differ widely in the levels at which nutrients are limiting to their growth, and nutrient ratios are also important.

Laboratory analysis of the effects of varying resource levels can be used to determine the concentrations at which specific nutrients limit algal growth (Figure 2.20). The results often help to determine why a particular alga is found in one type of aquatic environment but not another, explain algal community changes through time, and suggest ways to increase yields of cultivated algae. Differences in the growth responses of algae to low and high levels of mineral nutrients can be observed in nature by comparing

oligotrophic and eutrophic aquatic systems. **Oligotrophic aquatic systems** are characterized by having relatively low nutrient concentrations and low populations of algae and plants. They typically display clear water throughout the growing season and relatively high species diversity, and they are valued for their ecosystem services and appeal to humans. In contrast, **eutrophic aquatic systems** feature relatively high levels of nutrients, large populations of algae and/or aquatic plants, and comparatively low species diversity. Oligotrophic systems can be transformed into eutrophic water bodies when nutrients become so abundant that algae and plants are released from nutrient limitation. In such circumstances, algae and aquatic plants may form nuisance-level growths, of which algal blooms are examples. Hypereutrophic freshwaters, which arise by human actions that increase nutrient levels, often display repeated and enduring algal blooms. Such aquatic systems are less valued for their ecological properties and utility to humans. For this reason, the mineral nutrients most likely to limit algal growth

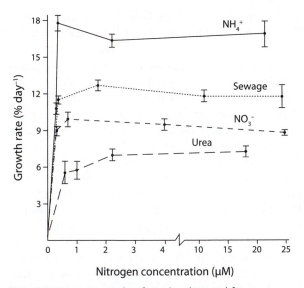

Figure 2.20 An example of results obtained from laboratory analysis of growth of an algal species (the red seaweed *Agardhiella subulata*) in response to differences in the concentrations of a variety of combined nitrogen sources in the external medium. Growth rates first exhibit rapid (logarithmic) increase and then reach a plateau stage during which increases in nitrogen concentration do not change growth rate. At the onset of the plateau stage, the algal growth rate has become limited by environmental levels of some other required resource. (After de Boer et al. 1978, by permission of the *Journal of Phycology*)

are important because additions of these minerals are most likely to cause algal blooms and other undesirable changes in aquatic systems. Fresh and marine waters differ in the nutrients most likely to limit algal growth.

In oligotrophic freshwaters, the mineral nutrient that most commonly limits algal growth is phosphorus, in the form of phosphate ion ($H_2PO_4^-$) (Vitousek et al. 1997a,b). Phosphate levels are usually low in oligotrophic freshwaters because this ion readily binds Al^{3+}, Fe^{3+}, and Ca^{2+}, forming highly insoluble complexes in soils and lake sediments. For example, relatively high levels of iron in the anoxic sediments of freshwaters of near-neutral pH allow the formation of iron oxyhydroxide phosphate precipitates (Blomqvist et al. 2004). Also, unless they are drained by volcanic soils, oligotrophic lakes usually receive sufficient nitrogen from surface water and groundwater that they are not nitrogen limited (Lampert and Sommer 1997), and freshwaters often contain large numbers and biomasses of nitrogen-fixing cyanobacteria (Falkowski

and Raven 1997). There are several exceptions to this general rule of phosphate limitation in freshwater systems. When phosphorus becomes so abundant that it no longer limits algal growth in eutrophic lakes, another nutrient (usually nitrogen, though sometimes iron) becomes limiting. In tropical lakes and lakes with high pH, carbonates bind phosphate, thereby limiting its availability to algae. Levels of bicarbonate can be so low in acidic softwater lakes of pH 5 or lower that algal growth is limited by inorganic carbon rather than phosphate (Turner et al. 1991; Fairchild and Sherman 1993; Hein 1997).

Algal growth in temperate, coastal marine waters is most commonly limited by the availability of combined nitrogen in the form of NH_4^+ or NO_3^- (Vitousek et al. 1997a,b). This occurs because marine waters are relatively high in sulfate ion (SO_4^-). Sulfate-reducing bacteria convert sulfate to sulfide ion, which readily combines with iron to form a highly insoluble precipitate, FeS. As a result, less dissolved iron is available to complex with phosphate, so dissolved phosphate levels remain relatively high. In addition, the sulfate present in marine waters inhibits nitrogen fixation and increases rates of denitrification, lowering levels of dissolved combined nitrogen (Blomqvist et al. 2004). There is considerable evidence that in the oceans, the amount of fixed nitrogen depends on the ratio of nitrogen fixation to denitrification, which in turn depends on the oxidation state of the ocean and the supply of trace elements such as iron (Falkowski 1997). Exceptions to the general concept of ocean nitrogen-limitation include some tropical lagoons where sand binds phosphate (in which case phosphate is limiting) and eutrophic estuaries and seas that receive discharges high in dissolved nitrogen. In some areas of the open ocean, there is increasing evidence that iron controls algal growth. The next two sections focus on algal interactions with nitrogen and iron.

2.4 Algae and the Nitrogen Cycle

Like plants, algae require combined nitrogen to synthesize amino acids, nucleic acids, chlorophyll, and other nitrogen-containing organic compounds. Although algae typically take up ammonium or nitrate ions directly from the surrounding water, some algae obtain ammonium by using extracellular enzymes to cleave it from dissolved compounds (Palenik and

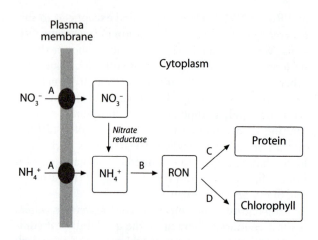

Plasma membrane

Cytoplasm

Figure 2.21 A comparison of algal cell utilization of nitrate versus ammonium. Following uptake involving membrane transport systems (A), ammonium can be enzymatically converted directly to reduced organic nitrogen (RON) (B), then utilized in protein and chlorophyll synthesis (C and D). In contrast, nitrate utilization requires an additional step involving the enzyme nitrate reductase. (After Lobban and Harrison 1994)

Henson 1997). Some algae take up organic nitrogen in the form of amino acids, small amides, and urea (see review by Tuchman 1996). For example, the diatom *Phaeodactylum tricornutum* can import 14 different amino acids from dilute solution (Lu and Stephens 1984), and the coccolithophorid *Emiliania huxleyi* absorbs small amides such as acetamide. In such cases, the needed ammonium is cleaved from these organic compounds within the cell.

Ammonium can be used more directly than nitrate to synthesize cellular N-compounds. However, aquatic nitrifying bacteria readily convert ammonium to nitrate, which often limits the amount of available ammonium. Algae are generally able to convert nitrate to ammonium by using the enzyme nitrate reductase (Figure 2.21), but this enzyme requires the cofactors iron and molybdenum, which are not always present in sufficient amounts. Not surprisingly, nitrate reductase activity is usually induced by low ammonium levels and repressed by high levels of ammonium. Ecologists find that measuring nitrate reductase activity or gene expression can be a useful way of estimating rates of nitrate incorporation into algae, which are otherwise difficult to ascertain.

Several types of bacteria, including many cyanobacteria, are the only organisms able to fix nitrogen—that is, to convert gaseous N_2 into ammonium.

This type of metabolism is known as **diazotrophy**. Because N_2 is abundant, making up some 70% of the atmosphere, diazotrophy provides cyanobacteria with a competitive edge over other algae and plants growing in nitrogen-limited waters. The ability to fix nitrogen is one of the factors contributing to development of cyanobacterial blooms in marine waters and freshwaters. For example, the cyanobacterial genus *Trichodesmium* is a common bloom-former in the tropical North Atlantic Ocean and the Caribbean, and its success is due in part to the ability to fix as much as 30 mg of nitrogen m^{-2} day^{-1} (Carpenter and Romans 1991; Capone et al. 1997). In freshwaters rich in phosphate, nuisance cyanobacterial blooms of N_2-fixing *Anabaena* and *Aphanizomenon* are common. Some 40% to 60% of the nitrogen fixed by cyanobacteria can be excreted into the open water, thus providing this essential nutrient to other algae and aquatic plants. Nitrogen fixation and excretion by cyanobacteria explain why the cells or tissues of many types of photosynthetic organisms contain symbiotic cyanobacteria. Such symbioses include certain marine and freshwater diatoms, the freshwater fern *Azolla*, and many types of lichens and terrestrial plants. More details about cyanobacterial nitrogen fixation, nitrogen-fixation symbioses, and their ecological significance can be found in Chapters 3, 6, 7, and 21.

2.5 Iron Limitation of Algal Growth in the Oceans

Dissolved iron was probably much more common (about 1 mM) in primordial anoxic oceans than in today's oxygen-rich surface waters (Falkowski and Raven 2007). Because early cells most likely had an abundant supply of Fe^{2+}, they became dependent on iron as a cofactor for many enzymes, including nitrate reductase. In nitrogen-fixing cells, each nitrogenase complex contains 32 to 36 iron atoms. Photosynthetic cells require relatively large amounts of iron for photosystem reaction centers. As Earth's atmosphere and oceans increased in oxygen content following the origin of oxygenic photosynthesis, the oceans also became saltier and richer in ions such as sulfide that precipitate iron. As a result of iron oxidation to low-solubility compounds and precipitation of iron with sulfide, dissolved iron levels decreased by six or seven orders of magnitude.

The concentration of iron in modern seawater continues to be relatively low and arises from several sources. Coastal oceans receive dissolved iron from river discharges. In addition, iron released from the sediments may rise to the surface, along with other minerals, in upwelling currents. As dead organisms are decomposed, small amounts of iron are released and can be taken up by living organisms. It is estimated that each atom of iron in seawater has been recycled through living organisms an average of 170 times before being lost to the sediments in sinking particles (Hutchins 1995). Also, winds carry iron-rich dust into the oceans from continental land masses, and this dust provides a source of iron to algae. For example, the cyanobacterium *Trichodesmium* is abundant in the Arabian and Caribbean seas and the Indian Ocean because it is able to capture iron from dust entering such waters (Hutchins 1995; Falkowski and Raven 1997). Many modern algae have acquired adaptations that allow them to harvest and store iron when it is available. When iron is relatively abundant, some algae can store it in protein aggregations known as **ferritin**. For example, the cyanobacterium *Synechocystis* stores as much as 50% of cellular iron in ferritin, and mutants in which the genes encoding ferritin were inactivated showed iron starvation symptoms even when iron was available in their environment (Keren et al. 2004). Certain small cyanobacteria, some dinoflagellates, and certain diatoms are known to harvest iron from low concentrations in seawater by producing surface iron-binding organic molecules known as **siderophores** (Murphy et al. 1976).

The arctic and equatorial Pacific Ocean and the Southern Ocean surrounding Antarctica are collectively known as "high nutrient (or nitrate)-low chlorophyll" (HNLC) regions. This is because HNLC algal populations are surprisingly low, considering the relatively high availability of most mineral nutrients there. In HNLC regions, which include about one-third of Earth's total ocean water, low levels of dissolved iron explain the low algal productivity. Phytoplankton of HNLC regions have the highest levels of iron stress known and thus are particularly responsive to iron additions. In consequence, in 1988 Martin and Fitzwater suggested that people might fertilize HNLC regions with iron in order to increase algal populations. These experts hypothesized that increased operation of the biological pump might then transport more organic carbon to deep sediments, thereby reducing atmospheric carbon dioxide levels. This was viewed as a possible engineering solution to the problem of increasing atmospheric levels of carbon dioxide caused by human activities. In order to test this idea,

several large-scale iron-enrichment experiments have been conducted. Ships have been used to distribute hundreds of kilograms of iron within HNLC regions more than 100 km². The results have demonstrated that growth of phytoplankton, especially diatoms, does increase; chlorophyll levels have been shown to rise by a factor of two or three, and phytoplankton biomass increased by a factor of four after iron fertilization (Martin et al. 1994; de Baar et al. 1995; Coale et al. 2004; Hoffmann et al. 2006). However, this increased algal growth has about the same sequestration effect as natural phytoplankton blooms and thus is relatively small by comparison with the global carbon cycle (Buesseler et al. 2004). Even so, private organizations have voiced plans to release iron to HNLC regions of the ocean as a way of selling carbon credits. Because iron fertilization has many potential ecological impacts whose magnitudes or directions are unknown, experts suggest that humans would be wise to explore these effects before attempting large-scale iron fertilization (Chisholm et al. 2001; Jickells et al. 2005; Buesseler et al. 2008).

2.6 Algae and the Sulfur Cycle

Algae require sulfur for biosynthesis of two amino acids—cysteine and methionine–and some thylakoid lipids. In marine waters, sulfate is abundant and hence does not limit algal growth, but in freshwaters, anaerobic bacteria can convert sulfate to H_2S, thereby sometimes limiting sulfate availability to algae (Falkowski and Raven 1997). It is thus not surprising that some freshwater algae have adapted by acquiring high-affinity sulfate transport systems (Yildiz-Fitnat et al. 1994).

Some types of abundant marine phytoplankton produce sulfur compounds that influence the global carbon cycle. Haptophytes such as the widespread *Emiliania huxleyi* and *Phaeocystis*, together with dinoflagellates, some diatoms, and green prasinophyceans, growing in a wide range of habitats, generate dimethylsulfonioproprionate (DMSP). This sulfur-containing compound functions in osmoregulation, in herbivore avoidance, or as a cryoprotectant in cold conditions. DMSP can be converted to volatile dimethyl sulfide (DMS), which is released from cells (Figure 2.22). In the atmosphere, DMS is transformed by free radicals into sulfur dioxide and sulfate, which cause climate cooling. These sulfur compounds scatter solar radiation, reducing the amount that reaches Earth's surface. They also

Remote Sensing of Phytoplankton

Phytoplankton pigments interact with light in ways that can be detected from space by means of satellite-borne sensors (radiometers). Sensor data can be used to construct global to local views of phytoplankton distribution patterns. Images can be used to study changes through time or to compare phytoplankton of different regions. For example, the lightest areas of the global map below (a composite image from September 1997 through July 1998) indicate regions of high chlorophyll concentrations/algal populations, most notably along the coasts of North America, northern Europe, northern Asia, Indonesia, southeastern South America, and southwestern Africa. More diffuse phytoplankton populations are indicated by bands extending through the northern and southern oceans; these contrast with mid-ocean regions where phytoplankton populations are low. A more localized example (right) reveals the extent of a phytoplankton (coccolithophorids) bloom in the Bering Strait. The bloom—showing up as the lighter gray areas off the Alaskan coast—imparted an aqua coloration to these normally dark waters.

Until mid-1986, eight years of such records were made using the Coastal Zone Color Scanner (CZCS) operated on the *Nimbus 7* environmental satellite mission. The CZCS was the first source of remote sensing data on phytoplankton but was not continuously operational, so only partial databases are available. The data obtained by the CZCS are maintained at the Goddard Distributed Active Archive Center (DAAC), which also collects all the ocean color data from NASA satellite missions. These data (and those described below) are available free of charge through the World Wide Web and by other means.

Currently operating is the Sea-viewing Wide Field-of-view Sensor (SeaWiFS). It observes more than

90% of the oceans every two days, with a resolution of 4.5 km. This technology has allowed improved measurement of phytoplankton pigment concentration in both open ocean and turbid near-shore areas. The Goddard DAAC also archives SeaWiFS images.

Since 2000, a new imaging system has been orbited: the Moderate Resolution Imaging Spectroradiometer (MODIS). This system provides higher-quality images as well as new capabilities. It is now possible to monitor chlorophyll fluorescence (an indicator of the physiological condition of phytoplankton), atmospheric levels of marine aerosols, detached coccolith concentrations, and other features. MODIS data are archived at modis.gsfc.nasa.gov.

(Satellite images: NASA)

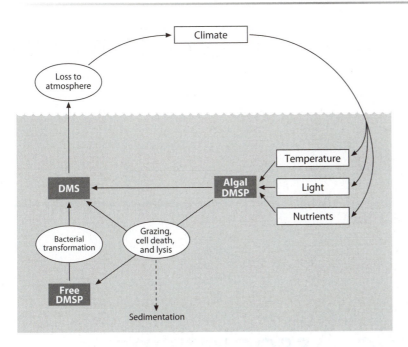

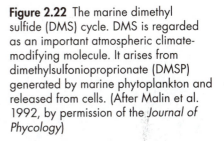

Figure 2.22 The marine dimethyl sulfide (DMS) cycle. DMS is regarded as an important atmospheric climate-modifying molecule. It arises from dimethylsulfonioproprionate (DMSP) generated by marine phytoplankton and released from cells. (After Malin et al. 1992, by permission of the *Journal of Phycology*)

act as nuclei for the formation of clouds, which reflect solar energy into space. Meskhidze and Nenes (2006) used satellite remote sensing (MODIS to sense cloud nuclei and SeaWiFS to measure chlorophyll) to determine that twice as many cloud nuclei formed in the atmosphere immediately above phytoplankton blooms than in portions of the atmosphere above ocean waters lacking such blooms (Text Box 2.1). Some algae, like certain bacteria, are able to assimilate DMSP, thereby reducing its environmental concentration (Vila-Costa et al. 2006). Some algae produce volatile halocarbon compounds that likewise influence Earth's atmosphere, our last topic in the area of biogeochemistry.

2.7 Algal Production of Halocarbon Compounds

A variety of marine phytoplankton and macroalgae produce volatile halocarbons—organic compounds that contain halogens such as chlorine or bromine. Halocarbons are generated by the enzyme bromoperoxidase, which occurs widely in algae. Halocarbons released from algae are of concern because they are oxidized in the atmosphere to hydroxyl, chlorine, and bromine radicals that can destroy stratospheric ozone (as can halocarbons originating from human activities). Methyl bromide (CH_3Br) is probably the single most abundant halocarbon in the atmosphere. Brominated halocarbons are regarded as 4 to 100 times more active than chlorine compounds in ozone destruction. The global input of methyl bromide to the atmosphere is estimated to be 97 to 300 Gg yr^{-1} (1 Gg = 10^9 grams). About 50% of this is natural production and 50% is human-made, arising from sources such as fire extinguishers, pesticides, and antifreeze. Macroalgae are estimated to generate only 0.6% of coastal production of halocarbons, and thus phytoplankton are considered to be a more significant source. A study of methyl bromide production by phytoplankton revealed that 13 of 19 species tested could generate halocarbon and that the haptophyte *Phaeocystis* was a prolific producer, generating 1.7 to 30 pg (picograms) CH_3Br per µg chlorophyll *a* per day (Sœmundsdóttir and Matrai 1998). Marine red, green, and brown macroalgae generate halocarbon compounds such as trichloroethylene and perchloroethylene.

Chapter 3

Algae in Biotic Associations

Crustose lichen

Algae interact with other organisms in an amazing variety of ways, collectively known as **biotic associations**. For example, drifts of seaweeds that accumulate on beaches serve as food for amphipods and other terrestrial animals and also support the growth of decay microorganisms. If you take a sample of seaweed drift back to the lab and look at it with a microscope, you will see many types of small algae living on or near the seaweed. In fact, samples from diverse habitats feature algal actors involved in dramatic biotic associations. Some are voracious predators that consume their close relatives, while hungry protozoa and small animals prey on other microscopic algae. Attacked algae defend themselves by ejecting spear-like projectiles, flashing bright light, or producing foul-tasting or toxic chemicals that cause predators to reject them. The shapes and sizes of other algae make them difficult to ingest. Algae can thwart killer viruses, bacteria, or fungi with tough cell walls or by secreting coats of slimy mucilage or antibiotics. More peaceful scenes are provided by the many types of algae that live in beneficial symbiotic associations with prokaryotes, protists, fungi, animals, or plants. This chapter describes food web and symbiotic associations involving algae and their ecological, medical, and economic importance. An understanding of such biotic associations is important because human actions that disturb them can reduce global biodiversity (McCann 2007).

(Photo: L. W. Wilcox)

3.1 Algae in Food Webs

Algae play multiple roles in aquatic and terrestrial food webs. A **food web** is a model that illustrates the feeding interactions occurring among diverse types of organisms in a particular habitat. During the past half-century, concepts of aquatic food webs have become more complex as more types of feeding connections have been discovered (Figure 3.1). Some eukaryotic algae function in food webs as **predators,** organisms that feed by ingesting all or parts of other organisms. For example, a number of algal species are **bacterivores** because they ingest and digest large populations of bacteria (Figure 3.2). Other eukaryotic algae prey on both bacteria and small eukaryotic cells or prefer eukaryotic prey. Certain saprophytic algae ingest organic materials in the form of nonliving particles (particulate organic material [POM]) or dissolved molecules (dissolved organic material [DOM]) (see Figure 3.1). Most algae are **photoautotrophs,** which are able to produce all or most of their own organic compounds via photosynthesis, using light as an energy source.

Living algae often exude organic compounds that feed heterotrophic microbes in the **microbial loops** of food webs (see Figure 3.1). Heterotrophic microbes also consume dissolved and particulate organic carbon released from algae when they die. Larger organisms that feed on decaying microalgae or seaweeds are classified as **detritivores.** Algae are often the prey of **herbivores** (Figure 3.3), organisms that consume algae (or plants). The cultivation of fish and shellfish in commercial aquaculture operations requires a supply of appropriate algae as a basic food source for herbivores. The quality of the algal food is of great importance in such applications, as well as in the natural world. Algae play an additional role in food webs when they defend themselves against herbivores by means of toxins. When such toxins reach high concentrations, they influence food webs even more dramatically by causing illness and death of nontarget organisms.

The particular roles that algae play in food webs can be ascertained by direct observation, such as video microscopy, or inferred by the use of stable isotopes (Text Box 3.1). We next take a closer look at the roles of algae as sources of dissolved organic carbon and detritus, major types of herbivores that consume algae as food, the quality of algae as

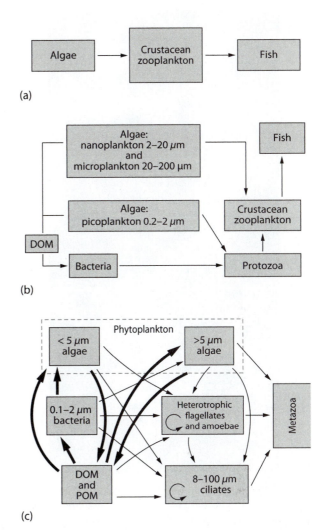

(a)

(b)

(c)

Figure 3.1 Changing concepts of aquatic food webs through time. (a) In the mid-1960s through the early 1970s prevailing views of food webs were relatively simple, involving linear transfer of food energy from algal producers to consumers (zooplankton) to secondary consumers (fish). (b) In the 1970s researchers realized that various size classes of phytoplankton were consumed by herbivores of different sizes and that bacteria were part of the food web as well. (c) Bacteria and algae were recognized as important members of a previously underappreciated microbial loop, indicated by heavier lines. Food web models arising in the 1980s emphasized diversity of connections among members of the microbial community. (After Graham 1991)

food for herbivores, and ways in which algae defend themselves against herbivores, with a particular focus on toxins that affect the health of humans and other vertebrates.

Figure 3.2 Bacterivorous algal flagellates. (a) *Dinobryon*, an example of a mixotrophic algal flagellate. Individual biflagellate cells are arrayed in a dichotomously branched, treelike colony. Cells contain golden-brown plastids but are also capable of ingesting bacterial prey. (b) *Chrysosphaerella*, another colonial flagellate whose cells contain golden plastids but which can also ingest bacteria. (a: C. Taylor; b: L. W. Wilcox)

Algae as Sources of Dissolved Organic Material and Detritus

Many algae are known to secrete dissolved organic compounds or produce detritus that supports the growth of heterotrophic protists and prokaryotic populations. Algae excrete a variety of organic compounds, including amino acids, amino sugars, peptides, carbohydrates, lipopolysaccharides, and DMSP (see Chapter 2), but the precise mixture appears to differ from one algal type to another. Different exudates support the growth of distinct types

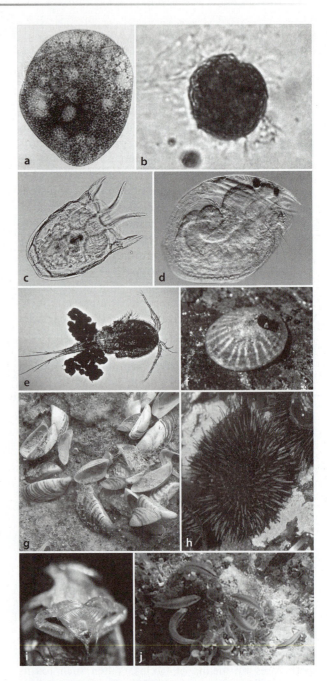

Figure 3.3 Examples of herbivores that consume algae as major portions of their diets include (a) ciliates, (b) amoebae, (c) rotifers, (d) cladocerans, (e) copepods (dark bodies to sides of animal are egg sacs), (f) limpets, (g) mussels, such as zebra mussels, (h) sea urchins, (i) water boatmen (note forelegs, which are modified to scoop up algae), and (j) fish. (a, f, j: L. E. Graham; b: J. M. Graham; c, d, e, g, i: L. W. Wilcox, h: C. Taylor; i: specimen provided by the University of Wisconsin Insect Research Collection)

Stable Isotopes in Food Web Studies

Stable—that is, nonradioactive—isotopes are chemically identical to one another but have different physical properties. Isotopes of carbon and nitrogen are widely used for analyzing food web relationships because they provide an integrated view over an extended time period. By contrast, analysis of herbivore digestive system contents gives a more defined picture of feeding during a limited time period.

Stable isotopes of carbon can be used to infer whether a particular alga is able to use bicarbonate ion (HCO_3^-) as an inorganic carbon source for photosynthesis in addition to CO_2 (see Chapter 2). Because algae differ in this capacity, their carbon isotope signatures vary. Such variations can sometimes be used to determine an herbivore's preferred food source (Pel et al. 2003). The two most abundant carbon isotopes are ^{12}C and ^{13}C. ^{12}C is by far the more abundant, accounting for approximately 99% of stable carbon. Photosynthetic organisms discriminate between these isotopes, preferentially using ^{12}C because at a given temperature the lighter atom collides more frequently with other atoms and thus is more reactive. Discrimination primarily occurs at the active site of rubisco during carbon fixation (carboxylation). Algal biomass or other samples to be tested are first dried and then combusted to CO_2, which is fed to an isotope ratio mass spectrometer. For each sample, $\Delta^{13}C$ is calculated according to the following formula:

$$\delta\,^{13}C = \frac{R_{sample} - R_{standard}}{R_{standard}} \times 1000$$

where R_{sample} is the $^{13}C{:}^{12}C$ ratio of the sample and $R_{standard}$ is the $^{13}C{:}^{12}C$ ratio of the PDB standard—a calcium carbonate deposit from the Cretaceous "Pee Dee" formation, located in South Carolina. The standard has a $^{13}C{:}^{12}C$ ratio of 0.011237. Algae that are able to use bicarbonate as a source of inorganic carbon in photosynthesis are characterized by $\Delta^{13}C$ values that are significantly less negative than those of algae that can use only CO_2. In an application of stable C isotopes, France (1995) observed that in several lakes, the attached nearshore algae contained more ^{13}C than phytoplankton, and as a result, herbivores that consumed the attached algae were likewise enriched in ^{13}C compared to herbivores that consumed phytoplankton. Pinnegar and Polunin (2000) found similar results in a study of food webs in nearshore waters of the Mediterranean Sea; seaweeds and their herbivores were enriched in ^{13}C compared to planktonic organisms.

Stable isotopes of nitrogen (^{14}N and ^{15}N) are used to trace energy flow through food webs because the concentration of the ^{15}N isotope generally increases at each trophic level of a food chain. For example, the $\Delta^{15}N$ in consumer tissue is about 3% greater than that of their prey. This allows ecologists to infer the food web positions of organisms. Samples for study are weighed and ground, and then they are oxidized to N_2 gas, which is passed to an isotope ratio mass spectrometer. Atmospheric N_2 provides a standard reference. Catenazzi and Donnelly (2007) applied C and N isotopic ratio analysis to a study of terrestrial animals of the hyperarid Atacama Desert in Peru. Such an environment offers little terrestrial plant food but faces the highly productive Peru–Chile cold current. Upwelling of nutrient-rich bottom waters supports a high biomass of seaweed, notably the green alga *Ulva*. Wave action deposits conspicuous amounts of *Ulva* on beaches, and isotopic ratios revealed that amphipods consume the seaweed and are consumed by geckos. In this study, the stable isotopic data explained how geckos and other terrestrial animals were able to survive in such an arid place.

of prokaryotes. For example, DMSP and amino acids secreted from the dinoflagellate *Pfiesteria piscicida* attract the bacterium *Silicibacter*, which takes up and metabolizes these compounds (Miller et al. 2004). As a result, seasonal and other variation in algal communities affects prokaryotic community composition (Graham et al. 2004; Pinhassi et al. 2004).

In open oceans, amorphous aggregates of detritus known as marine snow arise from algae and other organisms. Such detritus plays an important role in the operation of the biological pump that conveys carbon to the ocean depths (see Chapter 2). During transport, aggregates of algal detritus may be completely or partially consumed by decomposer bacteria and fungi, as well as animals such as ostracods, amphipods, copepods, salps, and fish. Entire microbial communities may exist within and upon fecal pellets produced by zooplankton and larger aquatic animals. Such microbes help to recycle mineral nutrients that are necessary for algal growth in subsequent seasons. In Lake Tahoe in the United States, for example, sinking detritus originating from spring phytoplankton blooms was the chief source of nitrate available for the next year's bloom (Paerl et al. 1975).

Herbivores

Herbivores feed on particles whose sizes fall within a range that can be accommodated by the predators' body size, ingestion apparatus, or other aspects of feeding. Predator and food size relationships are particularly important for small organisms that swim or float in the open water, collectively known as **plankton**. Aquatic ecologists group planktonic organisms into several size classes. The **picoplankton** includes organisms that are 0.2 to 5 μm in diameter; this encompasses most bacteria, including some cyanobacteria, and very small eukaryotes. Recent studies suggest that picoplanktonic algae play much more important roles in ocean food webs than previously thought (Richardson and Jackson 2007). **Nanoplankton** ranges from 5 to 20 μm, **microplankton** 20 to 200 μm, and **mesoplankton** 0.2 to 2 mm in diameter. (In some of the older literature, the term nanoplankton was used to cover the size ranges now known as picoplankton, nanoplankton, and part of the microplankton.)

Diverse types of organisms consume algae as a source of food. Among these are microscopic protists

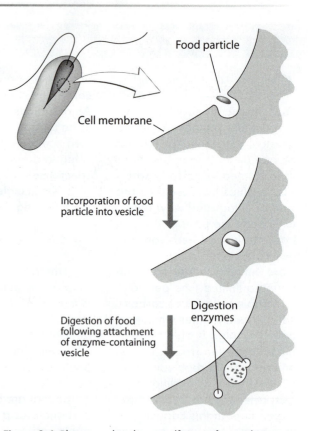

Figure 3.4 Phagotrophy, the engulfment of particles at a cell surface.

(Tillman 2004) that use the process of endocytosis to engulf prokaryotes and small algae into cellular digestion structures known as **food vacuoles**. Such feeders are known as **phagotrophs**, and this type of feeding is termed phagotrophy (Figure 3.4). Phagotrophs often have specialized cellular structures that aid in consuming prey, and they typically lack rigid cell coverings that would interfere with prey engulfment. Phagotrophic protists occur as flagellates, ciliates, or amoebae, distinguished by their type of mobility and food gathering. **Flagellates** are unicellular or colonial, photosynthetic or non-photosynthetic protists that swim or gather prey by means of one or more long eukaryotic flagella (see Figure 3.2). Unicellular **ciliates** move or collect prey by means of surface cilia, which are internally similar to flagella but shorter and more numerous (see Figure 3.3a). Together, ciliates and heterotrophic dinoflagellates are often referred to as **microzooplankton**. **Amoebae** move and obtain prey by extending pseudopodia from their cells (Figure 3.3b). Small animals that occupy open water habitats, the zooplankton, include

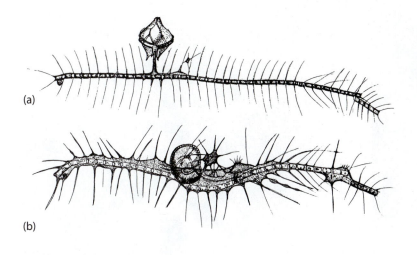

(a)

(b)

Figure 3.5 Illustration of a pallium, a feeding structure that can be extended from the cells of certain dinoflagellates. (a) *Protoperidinium spinulosum* (above) is shown extending its pallium around the filamentous diatom *Chaetoceros* (below). (b) Within the extended pallium, the diatom cell contents are liquefied and then taken into the feeding cell. (From Jacobson and Anderson 1986, by permission of the *Journal of Phycology*)

herbivorous **rotifers** and **crustaceans**, the latter including **cladocerans** and **copepods** (Figure 3.3c–e).

Mesograzers are somewhat larger animal herbivores such as oligochaete worms, freshwater dipteran larvae, and marine amphipods and pteropods. Larger, more conspicuous herbivores include limpets (Figure 3.3f), mussels (Figure 3.3g), crabs, sea urchins (Figure 3.3h), insects (Figure 3.3i) and fish (Figure 3.3j). More information about the roles of mesograzers and larger herbivores can be found in Chapter 22, which focuses on attached algae. Here, we look more closely at microscopic herbivores: flagellates, ciliates, amoebae, rotifers, and crustaceans.

Flagellates

Many types of freshwater and marine flagellates consume bacterial or algal prey by means of phagotrophy (see Figure 3.4). Flagellate algal grazers include certain euglenoids, cryptomonads, dinoflagellates, green algae, and chrysophyceans. Some of these grazers lack chloroplasts or possess non-photosynthetic plastids. However, many algal flagellates are **mixotrophic**, meaning that they are able to photosynthesize as well as ingest prey or organic materials (see Chapter 1). Mixotrophy allows algae to obtain organic food under conditions that limit photosynthesis, such as low light, and can provide minerals when dissolved levels are low. For example, in low-iron regions of the Pacific Ocean, the chrysophycean flagellate *Ochromonas* obtains iron by consuming bacterial cells (Maranger et al. 1998). In Antarctic lakes that are ice covered for all or most of the year, and in which minerals are low, the green flagellate *Pyramimonas gelidicola* can consume 16% of available bacterial biomass. In addition to minerals, this flagellate may obtain 30% of its daily carbon needs from the ingested bacteria (Bell and Laybourn-Parry 2003). Cryptomonads that occur in perennially ice-covered lakes are dependent on getting bacterial prey year-round (Marshall and Laybourn-Parry 2002). Fluorescence *in situ* hybridization techniques are useful for quantifying prokaryotic prey cells located within mixotrophic protists (Medina-Sánchez et al. 2005).

Some algal flagellates consume larger, eukaryotic prey in addition to or instead of bacterial prey. For example, the chrysophyceans *Ochromonas*, *Poterioochromonas*, and *Chrysamoeba* consume both bacteria and various types of eukaryotic cells, such as small diatoms, green algae, and other chrysophyceans. In fact, *Poterioochromonas malhamensis* grows better as a mixotroph than it does as an autotroph when prey is absent, and it can ingest prey several times its own size (Zhang and Watanabe 2001). The euglenoid *Peranema trichophorum* can also dilate—python-like—to ingest similar-sized, closely related cells of *Euglena*. Many photosynthetic and non-photosynthetic dinoflagellates are capable of feeding on other algae. For example, the widespread and common marine *Protoperidinium depressum* produces an extracellular "fishing net," an extension of the cell membrane known as a pallium. This net is used to collect bacterial and phytoplankton prey (Figure 3.5). Using transmission electron microscopy (TEM), Jacobson and Anderson (1996) found that many photosynthetic marine dinoflagellates contain food vacuoles whose contents include the remains of other dinoflagellates and ciliates. The

Figure 3.6 The fish-eating dinoflagellate *Pfiesteria,* viewed with scanning electron microscopy (SEM). (From Lewitus et al. 1999, by permission of the *Journal of Phycology*)

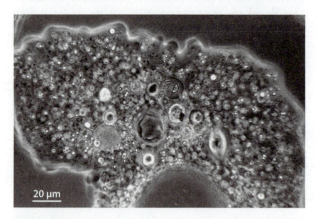

Figure 3.7 An amoeba that is actively feeding by extending pseudopodia. A variety of pigmented algae are present in its cytoplasm. (Photo: L. W. Wilcox)

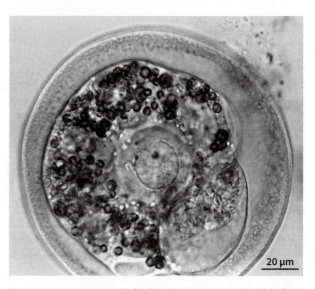

Figure 3.8 A testate (shelled or thecate) amoeba that has fed on algae. Remains of the algal food can be observed within the cytoplasm through the translucent test wall. (Photo: L. W. Wilcox)

non-photosynthetic dinoflagellate *Pfiesteria shumwayae* (Figure 3.6) practices the protistan equivalent of mammoth hunting. Populations of this dinoflagellate kill fish by both physical attack and the release of neurotoxins (Burkholder et al. 2005; Gordon and Dyer 2005). The *Pfiesteria* cells ingest tissues that slough off the dying animals. *Pfiesteria* has been associated with major fish kill events, decimating populations of up to 1 million, including commercially valuable species (Burkholder et al. 1992; Burkholder et al. 1998).

Amoebae

Diverse marine and freshwater amoebae ingest unicellular and filamentous algae (Figure 3.7). Amoebae appear to discriminate among algal food choices. Some amoebae ingest only particular algae, while others digest only some types of the algal food they ingest. Even closely related amoebae may have different food preferences. For example, *Amoeba proteus* feeds on such algal flagellates as the cryptomonad *Chilomonas* and the euglenoid *Euglena,* while *A. discoides* engulfs green flagellates such as *Chlamydomonas* and *Pandorina* but rejects living cells of the nonflagellate greens *Chlorella* and *Staurastrum* (Ho and Alexander 1974). A species of *Naegleria* was observed to consume several types of unicellular and filamentous cyanobacteria but to excrete unicellular cyanobacterial prey such as *Synechococcus* or *Aphanocapsa* undigested (Xinyao et al. 2006). Even testate amoebae—those living within enclosures called tests—can ingest algal prey by extending feeding pseudopodia into the water through a pore in the test (Figure 3.8). *Thecamoeba verrucosa* ingests filaments of *Oscillatoria* that are several hundred micrometers long by winding them up like spaghetti within its cells.

Ciliates

Algivorous ciliates, those that eat algae, include *Balanion planctonicum, B. comatum,* and *Urotricha farcta.* These and other heterotrophic protists can play important roles in controlling phytoplankton

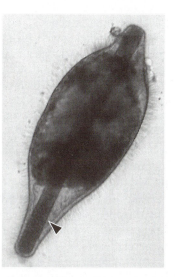

Figure 3.9 One of many ciliates known to consume algae via a cytostome (cell mouth). In this individual, the remains of an algal meal—a cyanobacterial filament (arrowhead)—are visible within a cytoplasmic food vacuole. (Photo: J. M. Graham)

Figure 3.10 *Vorticella,* a sessile (attached) ciliate that feeds on algae and other particles by using undulating membranes and cilia to generate water currents. This particular species contains endosymbiotic green algae. (Photo: L. E. Graham)

populations. For example, extensive springtime phytoplankton blooms in northern oceans likely occur in large part because co-occurring ciliates grow more slowly than their prey in cold temperatures. When temperatures rise, the ciliates grow faster and reduce phytoplankton populations (Rose and Caron 2007). Most ciliates possess a cell mouth known as a cytostome, which is specialized for intake of food to a cytoplasm-filled pharynx, but vary in other aspects of

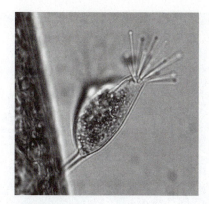

Figure 3.11 A suctorian ciliate; these protozoa attach themselves to solid substrata, then use tentacles to snare food particles from the surrounding water. (Photo: C. Taylor)

feeding. For example, the pharynxes of some ciliates are reinforced by rodlike structures to form a gullet known as a cytopharyngeal basket, which helps them ingest large diatoms, dinoflagellates, and filamentous cyanobacteria (Figure 3.9). Others use cilia to create water currents that carry food to the cell mouth. In some cases, as with the attached, stalked ciliate *Vorticella* (Figure 3.10), cilia in the mouth region are fused into undulating membranes that generate food-collection currents. Food is enclosed in cell vacuoles that form at the base of the cell mouth and is then digested as the vacuoles move through the cell. The marine and freshwater ciliates known as tintinnids have several undulating food-collecting anterior membranes. They can remove as much as 41% of the standing crop of algal chlorophyll-*a* produced in a day and can consume as much as 52% of the nanoplankton (Verity 1985). Suctorians are ciliates that use retractable tentacles to ingest food material, incorporating it into vacuoles at the base of the tentacles (Figure 3.11).

Herbivorous rotifers and crustaceans

Rotifers are important members of the zooplankton in lakes and oceans, where they consume diverse types of bacterial and protist prey ranging from <1 to 200 μm in diameter. *Keratella quadrata* and *Brachionus calyciflorus* are examples of common freshwater rotifers that are known to consume algal food. Cladocerans such as *Daphnia* and *Bosmina* (see Figure 3.3) are very important freshwater grazers. Copepods (see Figure 3.3), though present in freshwaters, are particularly significant in marine food webs. *Acartia*

Figure 3.12 Components of the filtering apparatus of the common cladoceran *Daphnia*, viewed with SEM. The spacing of the mesh determines the size of particles that can be retained and consumed. (Micrograph: L. E. Graham)

tonsa is an example of a widespread marine copepod. Cladocerans reproduce asexually most of the year in warmer waters, with generation times on the order of weeks. In contrast, copepods reproduce sexually, and generation times are longer. Two major ways in which crustaceans feed are **filter feeding** and **raptorial feeding**.

During filter feeding, cladocerans and some copepods use specialized feeding structures (Figure 3.12) to filter phytoplankton from large volumes of water and sweep food into the mouth. Cladocerans can filter about five times the volume of a eutrophic lake in one day, though filtering rates are lower in oligotrophic waters (Porter 1977). Some studies indicate that *Daphnia* may be able to detect the odor of preferred food such as the green alga *Scenedesmus* (van Gool and Ringelberg 1996). *Daphnia* uses this ability to find patches of water that are relatively rich in edible algae, thus effectively avoiding patches of less-desirable algae. Ghadouani and associates (2004) used video recordings and computerized image analysis to compare the effects of *Scenedesmus* and the toxin-producing cyanobacterium *Microcystis aeruginosa* on feeding behavior of *Daphnia pulicaria*. They found *Daphnia* able to detect even low levels of *Microcystis*, perhaps by "tasting" chemicals produced by it, and respond by reducing food ingestion. These investigators also discovered that even low levels of *Microcystis* toxin reduce feeding, and high toxin levels (5000 ng ml^{-1}) caused feeding to be compromised beyond recovery, helping to explain why cladoceran populations often decline following blooms of toxic algae.

In raptorial feeding, copepods and some cladocerans use feeding appendages to pick up pieces of algal food, taste it, and select types that are high in nutritional quality, digestible, and nontoxic (Porter 1977). For example, marine herbivorous copepods such as *Acartia* preferentially feed on nutritious, nontoxic nanoplankton such as coccolithophorids. Freshwater copepods select and consume filaments of the cyanobacterium *Oscillatoria tenuis* but reject the nearly identical but toxic *Planktothrix rubescens* (DeMott and Moxter 1991). The upper end of the algal size range that can be consumed by crustaceans is determined by the opening width of the mandibles or the carapace gape. This is about 50 μm for copepods and some large cladocerans (Lampert and Sommer 1997). Thus some algae are too large to be consumed by particular herbivores, a feature that affects their quality as food.

Algal Food Quality

Experiments have shown that neither the marine copepod *Diaptomus* nor the freshwater cladoceran *Daphnia* absolutely requires algae as food, but in both cases, more viable young are produced when algae are included in their food supply (Sanders et al. 1996). Rotifers are known to grow better on some types of algal food than others (Rothhaupt 1995). Observations such as these indicate that algae can serve as high-quality food for herbivores yet may differ in **food quality**—the extent to which algae are able to provide essential nutrients. Algal species vary widely in their ingestibility, digestibility, content of mineral nutrients and essential organic compounds, and toxin content. Algae that have features that prevent ingestion by herbivores, are toxic or indigestible, or if digested do not match the nutrient requirements of herbivores are said to display low food quality. Algae that are ingestible, nontoxic, and when digested match the nutrient requirements of herbivores are considered to be high in food quality.

The cyanobacteria *Planktothrix rubescens* and *Microcystis aeruginosa* are examples of low-quality algal food for zooplankton. Not only do their shapes make ingestion difficult, but they also often contain high levels of microcystin, a potent neurotoxin (Ghadouani et al. 2004). *Cryptomonas phaseolus* and other photosynthetic cryptomonads are generally considered to be superior food sources for zooplankton because these unicellular, wall-less flagellates are readily ingested and digested, lack toxins, and contain relatively high proportions of two essential, highly unsaturated fatty acids (HUFAs) (eicosapentaenoic and docosahexaenoic acids). Other biochemical constituents affecting algal food quality include polyunsaturated fatty acids (PUFAs), sterols, particular mineral ratios, and amino acids such as leucine. Some of these constituents may be more important for reproduction than in other phases of herbivore life (Boëchat and Adrian 2006). Algae may be high in food quality under some conditions but poor in others. The green flagellate *Chlamydomonas*, for example, is a high-quality food for zooplankton when grown with adequate N and P, but it possesses decreased amounts of essential polyunsaturated fatty acids and is thus of lesser food quality when it is P limited (Weers and Gulati 1997). Poor food quality can reduce herbivore growth rate, body size, and reproduction rate or prevent growth and development altogether. Food quality is thus important in both natural ecosystems and managed aquaculture systems.

The importance of algal food quality is illustrated by studies of the role of diatoms as food for marine copepods that are consumed by fish. Although copepods readily feed on diatoms, laboratory experiments have shown that copepod reproduction and development are not well supported by a diatom diet. Some experts have suggested that diatoms might be toxic to copepods (Ianora et al. 1996; Ban et al. 1997; Ianora et al. 2004), but others contend that diatoms are instead deficient in some essential nutrient. To address this issue, Jones and Flynn (2005) studied the effects on the copepod *Acartia tonsa* of a diet of single diatom species (*Thalassiosira pseudonana* or *T. weissflogii*) and a diet of diatoms mixed with a dinoflagellate (*Aureodinium pigmentosum*). Within 15 days, juvenile copepods that had been fed only on diatoms had died, but the copepods grew well when more than 70% of their ingested food consisted of dinoflagellates. This investigation indicated that diatoms are more likely to be incomplete nutrient sources for copepods than toxic to them. This conclusion is supported by a field study conducted by Vargas and associates (2006) of three species of copepods native to Chilean coastal waters. The Chilean copepods consumed dinoflagellates and ciliates during the autumn and winter but switched to abundant diatoms during the spring and summer. The study showed that a diet primarily composed of diatoms was correlated with increased copepod egg production but reduced levels of egg hatching and larval

Figure 3.13 *Pediastrum*, a green alga that occurs commonly in freshwaters, which is characterized by an elaborate, lacy colonial structure that may function in herbivore avoidance and help to reduce settling in the water column. (Photo: L. W. Wilcox)

survival. Together, these and other recent studies indicate that a mixed diet of diatoms and other protists seems to be important to copepod growth and reproduction.

Algal Defenses Against Herbivory

Algae display a variety of mechanisms that deter herbivores. For example, some members of nearly all major algal groups display both very small cell size and rapid cell division (on the order of hours to days), thereby generating large populations. Because herbivores are unable to find and consume all such cells, some are able to generate new populations when growth conditions allow. Ability to grow in the cold season, when populations of herbivores are relatively low, can be another way to avoid herbivores. Some algal species produce life stages or cellular features that are resistant to herbivore attack. Others startle herbivores with brilliant light flashes or produce chemical deterrents or toxins.

Structural defenses

Relatively large size helps filamentous freshwater algae and seaweeds to escape predation from smaller herbivores. Likewise, many phytoplankton species produce large cells or multicellular bodies (Figure 3.13), gelatinous coatings (Figure 3.14), horns (Figure

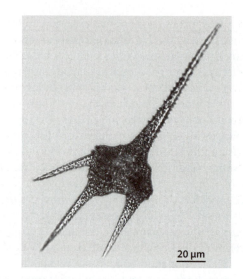

Figure 3.14 Mucilage sheaths. The often extensive mucilaginous sheaths of various algae, such as the green alga *Desmidium majus*, are thought to reduce the chance of ingestion by herbivores. Such sheaths are often transparent and difficult to visualize unless a contrast method is used. Here, dilute india ink has been added to the preparation. The ink particles cannot penetrate the sheath, and thus they reveal its limits. The presence of such a sheath can have the effect of making the alga appear larger to potential herbivores. (Photo: M. Fisher and L. W. Wilcox)

Figure 3.15 The dinoflagellate *Ceratium hirundinella*, with horns that may help it avoid being eaten. (Photo: C. Taylor)

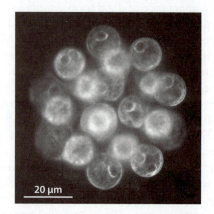

Figure 3.16 *Sphaerocystis*, a colonial green alga found in freshwaters that can survive ingestion by zooplankton by virtue of copious mucilage and a tough cell wall. (Photo: L. W. Wilcox)

3.15), or spiny projections that make them too large for particular predators to consume. Some phytoplankton species have such tough cell walls that they resist digestion even when consumed by herbivores. *Sphaerocystis*, for example, not only survives gut passage but also benefits from the opportunity to obtain mineral nutrients during the ingestion process (Figure 3.16) (Porter 1977). Such structural defenses are often constitutive, meaning that they are usually present. However, some structural defenses are induced by the presence of herbivores or low nutrient conditions. For example, P-limitation is thought to cause chemical changes in the walls of the unicellular green algae *Chlamydomonas reinhardtii* and *Selenastrum capricornutum*, making them more resistant to digestion by herbivores during periods of low growth (van Donk et al. 1997). Exposure to living *Daphnia* or a medium in which *Daphnia* have been grown induces the highly edible green alga *Desmodesmus* (formerly *Scenedesmus subspicatus*) to form larger colonies composed of cells with tougher walls and more spiny projections (Figure 3.17) (Hessen and van Donk 1993; van Donk et al. 1999). A study of 40 different strains of *Scenedesmus* and related taxa revealed that inducible colony formation is a common response to the presence of herbivorous rotifers or cladocerans

(Verschoor et al. 2004). Such induced defenses may favor zooplankton such as *Daphnia* that can consume larger algal prey, thereby changing the structure of food webs (Van der Stap et al. 2006).

Bioluminescence as an herbivory defense

There is evidence that the bioluminescence exhibited by many marine dinoflagellates has a defensive function. When such algal cells are agitated, they give off a flash of blue-green light that results from reaction of the substrate luciferin with the enzyme luciferase, as in fireflies and various luminous bacteria (Sweeney

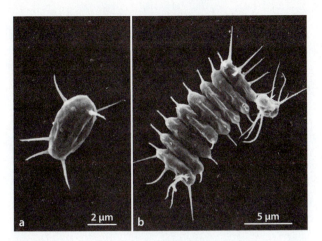

Figure 3.17 Putative inducible morphological defenses exhibited by *Desmodesmus subspicatus* in cultures that contain exudates from the cladoceran herbivore *Daphnia* or the animals themselves. The alga was primarily one or two celled when not exposed to *Daphnia* exudate (a), whereas eight-celled colonies became significantly more abundant when the alga was exposed to *Daphnia* or its exudate (b). (From Hessen and van Donk 1993)

1987) (see Chapter 11 for more details). In the case of dinoflagellates, bioluminescence is thought to deter copepod feeding by two mechanisms. The first is a direct "startle" effect on the herbivores. The second is an indirect effect; bioluminescence makes copepods that have fed on glowing dinoflagellates more visible to their own predators, with the result that the copepod populations decline. An experimental test of the latter hypothesis showed that bioluminescent dinoflagellates increased the efficiency with which stickleback fish were able to capture copepods (Abrahams and Townsend 1993).

Chemical deterrents and toxins

Many algal species produce defensive chemical compounds that impair the growth of competing algae or deter herbivores (Ianora et al. 2006; Leflaive and Ten-Hage 2007). For example, freshwater diatom films produce fatty acids that discourage grazers (Jüttner 2001), and the cyanobacterial genus *Cylindrospermum* excretes the compound DMDP (dihydroxymethyldihydroxypyrrolidine), which inhibits digestive enzymes of crustacean zooplankton and macrograzers (Jüttner and Wessel 2003). Seaweeds produce a variety of chemical deterrents, including terpenoids, acetogenins, and polyphenols such as phlorotannins (McClintock and Baker 2001) (Chapter 22).

Certain cyanobacteria, dinoflagellates, diatoms, and haptophytes are major producers of **toxins**, compounds that poison and often kill other organisms. Toxic algae most often occur as phytoplankton, suspended in the water column or forming surface scums. However, toxin-producing cyanobacteria can also occur in benthic mats, such as those found in drinking water reservoirs in southern California (Izaguirre et al. 2007) and other habitats (Cox et al. 2005). Toxins may occur at high concentrations during or after algal blooms, released from large populations of dying algal cells. In such cases, vertebrates such as birds, dogs, farm animals, or sea mammals may be sickened or killed. Humans may also be affected: Dinoflagellate toxins cause diarrhetic, paralytic, or amnesiac shellfish poisoning, and cyanobacterial toxins have been linked to human cancer and death (Codd et al. 2005a,b). Algal toxins can be classified into four general types, based on the animal tissues or cells affected: neurotoxins, hepatotoxins, cytotoxins, and toxins affecting the skin and gastrointestinal system (Codd et al. 2005a,b).

Neurotoxins. Neurotoxins, which act on nervous tissues and block neuromuscular activity, occur in several forms. Examples include anatoxin-a (Figure 3.18a) and homoanatoxin-a, secondary amine alkaloids produced by certain strains of *Anabaena flos-aquae*, some strains of *Planktothrix*, and other cyanobacteria. These toxins act by irreversibly inhibiting acetylcholinesterase, thereby interfering with normal muscle contraction and causing severe nervous system dysfunction, which may lead to death. An electrochemical method has been developed for detecting anatoxin-a in water samples (Villatte et al. 2002).

A second type of neurotoxin, the amino acid β-N-methylamino-L-alanine (BMAA) (Figure 3.18b), is produced by all known groups of cyanobacteria, including *Nostoc* living within the roots of cycads. BMAA has been linked to some cases of dementia in humans, particularly among the Chamorro people of Guam (Cox et al. 2005). The affected Chamorro may have unsuspectingly ingested BMAA in traditional foods, cycad seed flour and the meat of animals that had fed upon cycads. Cyanobacteria that produce BMAA have been found in a wide variety of habitats, and the extent to which BMAA might affect people elsewhere is under investigation.

Yet another group of neurotoxins, the saxitoxins, are produced by both cyanobacteria and dinoflagellates. Saxitoxins (Figure 3.18c) are carbamate

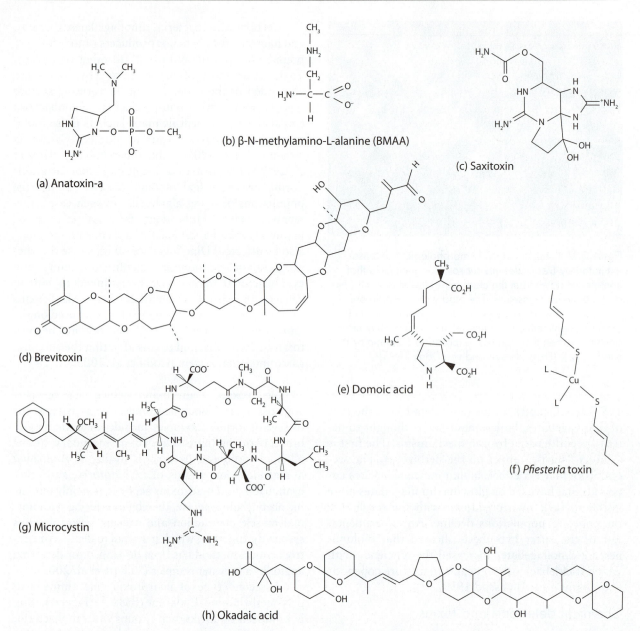

Figure 3.18 The chemical structures of major algal toxins: (a) anatoxin-a, (b) BMAA, (c) saxitoxin, (d) brevitoxin, (e) domoic acid, (f) *Pfiesteria* toxin, (g) microcystin, cyclic peptide, and (h) okadaic acid (a: After Villatte et al. 2002 by permission of Springer-Verlag; c: After Gawley et al. 2002. Reprinted with permission from the *Journal of the American Chemical Society.* © 2002 American Chemical Society; d, h: After Taylor 1990; e: After Baden et al. 1998 with kind permission of Kluwer Academic Publishers; f: After Moeller et al. 2007, by permission of the American Chemical Society; g: After Bourne et al. 1996)

alkaloids that block plasma membrane sodium-channels, thereby stopping nerve impulses to muscles, which become paralyzed. Saxitoxins are so potent that lab mice are killed in minutes when a dose of 10 μg kg^{-1} is injected. Whale deaths and the occurrence in humans of paralytic shellfish poisoning (PSP) have been linked to saxitoxins produced by marine dinoflagellates. Brevitoxins (Figure 3.18d) are larger and differently structured molecules that also affect sodium-channel function, likewise causing paralytic poisoning. Fluorescence sensor methods have been developed for saxitoxin (Gawley et al. 2002), and a rapid test strip for PSP has become commercially available.

Domoic acid (Figure 3.18e) is a type of neurotoxin produced by some marine diatoms and haptophytes. This toxin can be transferred through food webs to seabirds such as pelicans, causing death. Domoic acid also causes amnesiac shellfish poisoning (ASP)—also known as toxic encephalopathy—in humans. Diatoms that produce domoic acid may be consumed by shellfish such as mussels; the shellfish are not themselves affected, but their tissues may concentrate the toxin to levels that are poisonous to humans (Perl et al. 1990). Domoic acid binds to kainate receptors in the central nervous system, causing depolarization of neurons, which leads to their degeneration. Permanent loss of recent memory in human victims results from degeneration of the hippocampus. A rapid and easy-to-use test strip for ASP has become commercially available.

Pfiesteria neurotoxins are not only associated with dramatic fish kills but are also linked to amnesia and other symptoms (Grattan et al. 1998; Schmechel and Koltai 2001). Though the chemical nature of the toxin has been unclear, several types of spectroscopy have been used to identify a new type of toxin produced by *Pfiesteria piscicida*. The new toxin consists of copper compounds ligated to thiol-containing organic compounds (Figure 3.18f). Such a toxin kills cells by producing free radicals, which damage DNA, oxidize proteins and lipids, and stimulate apoptosis, leading to cell death. This *Pfiesteria* toxin is labile—easily broken down by light, variation in pH, and heat—helping to explain why it has previously been difficult to identify (Moeller et al. 2007).

Hepatotoxins. Produced by certain cyanobacteria and eukaryotic algae, hepatotoxins are cyclic peptide polymers that inhibit protein phosphatases, which are involved in many essential cellular processes in eukaryotes. Thus, in nature such toxins may deter the growth of competing organisms (Windust et al. 1996). Hepatotoxins are named for their effect on animals: irreversible inhibition of liver-cell protein phosphatases. Symptoms include weakness, heavy breathing, pallor, cold extremities, vomiting, diarrhea, and massive bleeding in the liver that can lead to death within 2 to 24 hours after toxin ingestion. In addition to causing liver damage, hepatotoxins are tumor promoters.

Microcystin (Figure 3.18g) and nodularin are cyanobacterial hepatotoxins. Microcystin occurs in more than 70 different types, with microcystin-LR, microcystin-RR, and microcystin-YR being the most common. Though widely produced among cyanobacteria, microcystin is most commonly associated with blooms of *Microcystis*, *Planktothrix*, and *Anabaena*. When cyanobacteria are abundant, microcystin concentrations of 5.0 µg per liter or so are not unusual, and the toxin can persist in freshwaters for two or more weeks before being degraded by bacteria. The World Health Organization has recommended a guideline of 1.0 µg of microcystin per liter of drinking water. Though municipal water treatment plants effectively remove microcystins from water through ozonation and sand filtration, monitoring microcystin in drinking water is recommended (Hoeger et al. 2005). Microcystins can be analyzed by means of high-performance liquid chromatography with photodiode array detection (HPLC-PDA), HPLC with online mass spectral analysis (HPLC-MS), commercial antibody-based enzyme linked immunosorbent assay (ELISA) kits, the protein phosphatase inhibition assay (PPIA) (McElhiney and Lawton 2005), electrochemical immunosensors (Campàs and Marty 2007), and molecular techniques. The latter involve the use of molecular methods to detect cyanobacteria, followed by molecular methods designed to detect microcystin synthetase genes (Saker et al. 2007). Microcystin is of particular concern in parts of the world where untreated natural waters are used for drinking; monitoring such situations requires inexpensive detection methods that are accurate yet easy to use.

Cytotoxins. Cytotoxins include cylindrospermopsin, an alkaloid that inhibits protein synthesis and can also cause genetic damage to cells. In mammals, cytotoxins cause death of cells in the liver, kidneys, lungs, spleen, and intestine (Codd et al. 2005a,b). **Okadaic acid** (OA) (Figure 3.18h) and dinophysistoxin are cytotoxins responsible for diarrhetic shellfish poisoning (DSP) in humans. A rapid test has been developed for use by regulatory laboratories as a screening method to eliminate negative samples, thereby identifying positive samples for more detailed analysis (Laycock et al. 2006).

Toxins affecting the skin and gastrointestinal system. Marine mat-forming cyanobacteria produce lyngbyatoxin and other toxins known to irritate skin or promote tumor formation. **Lipopolysaccharides**

(LPS) produced by freshwater planktonic cyanobacteria are implicated in cases of fever, inflammation, and gastrointestinal irritation in humans.

In summary, algal food web associations include organic exudates and detritus that fuel microbial loops, herbivory interactions, and algal defensive reactions that sometimes affect nontarget organisms such as humans. In addition, diverse types of microorganisms attack algae, causing disease and death. Such disease microbes recycle the organic carbon of their algal hosts, thereby influencing food webs. As we shall see in the next section, the relationships between algae and disease microbes are examples of symbiotic relationships, those in which two or more organisms live in close proximity.

3.2 Algae in Symbiotic Associations

Algae are involved in diverse types of symbiotic relationships, living in close association with one or more other organisms: bacteria, other protists, fungi, animals, and plants. The partners involved in symbiotic associations may have positive, negative, or neutral influences on each other. Algae that live within the cells of other organisms in a stable association reflect cases of endosymbiosis, a phenomenon that has strongly affected algal evolution and is covered in Chapter 7. Here, we focus on other types of symbiotic associations: parasites, pathogens, epibionts, and endobionts.

Parasites are organisms that obtain nutrition from living organisms, known as hosts, without ingesting them. Parasites that cause disease symptoms are **pathogens**. Although parasites of diverse types attack algae, it is also the case that certain algae themselves function as parasites or pathogens. Parasitic relationships are often relevant to food webs, as well as aquaculture, agriculture, and human health. Algal hosts often defend themselves against pathogens by producing antibiotic compounds that are potential sources of new pharmaceuticals.

Epibionts grow on the surfaces of other organisms, and **endobionts** live within the tissues of other organisms, seemingly without causing serious harm. There are many examples of algae that live as epibionts and endobionts, as well as algae that serve as hosts for algal or other types of epibionts and endobionts.

Parasites and Pathogens of Algae

Microorganisms that attack algae, sometimes causing disease symptoms, include viruses, bacteria, various protists, and fungi. Although a typical milliliter of water contains an estimated 10^3 fungal cells, 10^6 bacteria, and 10^7 viruses (Rheinheimer 1991), many algae avoid infection by using structural or chemical defenses.

Viral pathogens

Lytic viruses—viruses that result in cell lysis—are a major cause of phytoplankton mortality. Present in levels greater than 10^4 and less than 10^8 viruses ml^{-1}, they can result in the decline of phytoplankton blooms. For example, viral attack seems to limit bloom formation by the common marine haptophytes *Phaeocystis pouchetii* and *Emiliania huxleyi* (Jacobsen et al. 1996). The cosmopolitan and abundant marine green flagellate *Micromonas* can lose 2–10% of its population each day to viral lysis (Cottrell and Suttle 1995), and it is estimated that about 3% of marine phytoplankton primary productivity is lost to viral lysis (Suttle 1994). Viruses likewise attack bloom-forming freshwater algae, and some experts propose that viruses might be useful in preventing toxic cyanobacterial blooms (Yoshida et al. 2006b). Viruses also attack seaweeds, such as the South American coastal brown seaweed *Myriotrichia clavaeformis*. The virus causes the seaweed to become sterile because the structures that would normally produce flagellate spores (known as plurilocular sporangia) are co-opted by massive production of viruses. Cell lysis results in a whitish discharge of viruses to the surrounding seawater (Müller et al. 1996).

Viruses have been reported to infect at least some members of all major classes of freshwater and marine algae (Van Etten et al. 1991). Viruses specific to cyanobacteria are known as cyanophages, whereas those of eukaryotic algae are called phycoviruses. Both DNA and RNA viruses are known to infect algae (Tai et al. 2003). Most virus-like particles have polyhedral capsids (viral coats) that are five- or six-sided in cross-section (Figure 3.19). In the past, most reports on algal viruses consisted of accidental observations of virus-like particles within algal cells, viewed with TEM (see Figure 3.19), but now viruses can be observed directly and counted after being stained with fluorescent dyes such as DAPI (4,6-diamidino-2-phenylindole) or Yo-Pro, which bind viral DNA (Maier and Müller 1998; Hennes

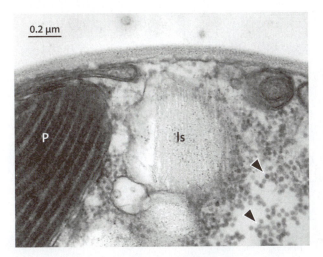

Figure 3.19 Viruslike particles with polyhedral capsids (arrowheads) within the cells of a eustigmatophycean algal cell. P = plastid, ls = lamellate storage material. (TEM: L. E. Graham)

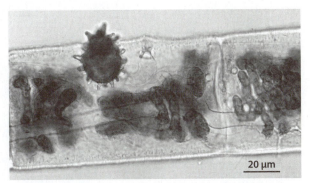

Figure 3.20 A filament of the green alga *Spirogyra* that has been attacked by a water mold, a protistan parasite. Note the distorted plastids and parasitic hyphae ramifying throughout the host cell's cytoplasm. A spiny spore is present. (Photo: L. W. Wilcox)

and Suttle 1995). A recent approach to detecting and identifying algal viruses is the combination of ultrafiltration to isolate and concentrate viral particles from natural waters, with comparative analysis of nucleic-acid sequences obtained from concentrated samples. Nucleotide sequences specific for the ends of genes known to generally occur in viruses are used as primers to amplify viral DNA (Chen et al. 1996). This approach has allowed the detection of previously unknown phycoviruses.

Bacterial pathogens

The most common genera of algicidal bacteria (that is, those that kill algae) are *Cytophaga* (also known as *Cellulophaga*), *Saprospira* (in the phylum Bacteriodetes), and the gamma-proteobacteria *Alteromonas* and *Pseudoalteromonas*. *Cytophaga* kills algae by attaching to and degrading their cell surfaces, while other bacteria secrete lethal chemical compounds such as proteases.

In reviewing marine algicidal bacteria, Mayali and Azam (2004) considered the evidence to be circumstantial that they might cause marine phytoplankton blooms to decline. However, other experts have suggested that bacteria cause declines of freshwater phytoplankton blooms. For example, Gram-negative myxobacteria attack and lyse several types of bloom-forming cyanobacteria (Daft et al. 1975), and *Bdellovibrio*-like bacteria lyse *Microcystis*, contributing to the decline of blooms of this

cyanobacterium (Grilli Caiola and Pellegrini 1984). The bacterial agent of CLOD (coralline lethal orange disease) attacks and destroys multicellular coralline red algae, such as *Porolithon onkodes*, that consolidate the surfaces of coral reefs. The numerous bright orange bacteria occur within a gelatinous matrix that is mobile because the bacteria are capable of gliding. These bacterial aggregations destroy the coralline algae as they move across their surfaces (Littler and Littler 1995).

Protistan parasites

A number of stramenopile protists are heterotrophic parasites of algae. *Ectrogella perforans*, for example, is a virulent parasite of marine diatoms such as *Licmophora* (van Donk and Bruning 1995). *Pirsonia diadema* is a related flagellate that infects several species of the diatom *Coscinodiscus*. It penetrates cells by using pseudopodia to pierce tubular passages in the diatom frustule. Part of the parasite invades and digests the algal cytoplasm, while part remains outside the diatom, eventually producing many infective progeny (Kühn et al. 2004). *Olpidiopsis*, another stramenopile, parasitizes *Bostrychia* and other marine seaweeds (West et al. 2006a). Other protists attack freshwater algae, including *Spirogyra* (Figure 3.20). *Stylodinium* (Figure 3.21) and *Cystodinedria* are freshwater dinoflagellates that have amoeboid stages that consume cells of freshwater filamentous green algae such as *Oedogonium* and *Mougeotia*. Parasitic *Amoebophrya* dinoflagellates infect other dinoflagellates, such as *Alexandrium*

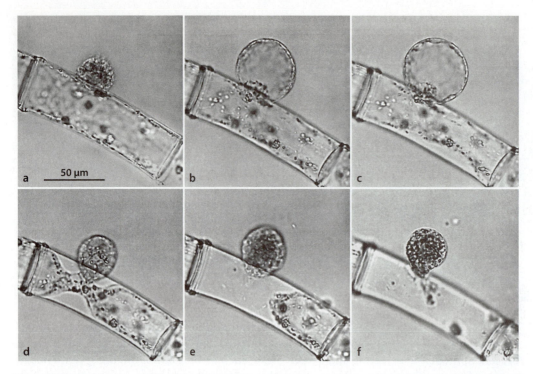

Figure 3.21 *Stylodinium*, a dinoflagellate with amoeboid stages that consume the cytoplasm of algal cells. Shown here is a sequence of an amoeba feeding on the green alga *Oedogonium*. The entire process took only a few minutes. (Photos: G. J. Wedemayer and L. W. Wilcox)

affine and *Gonyaulax spinifera*, potentially helping to control the formation of blooms (Kim et al. 2004b).

Fungal parasites

Chytrids, consisting of multiple lineages of aquatic fungi, are important parasites of algae because they have the potential to greatly reduce algal populations. Most chytrid species attack only one or a few closely related algal species. Some experts suggest that fungal epidemics might play a role in ending or preventing the formation of algal blooms or in the timing and pattern of phytoplankton succession (see Chapter 21). Chytrids reproduce by means of flagellate spores that seem to locate preferred host algae through chemical cues. The zoospores then settle on the algal cell surface, invade the cytoplasm, and produce a nonmotile stage that uses algal cell components for growth and production of more flagellate chytrid spores, which are released into the water. In the last stages of an epidemic, sexual processes generate chytrid resting stages. Chytrids have been observed to attack freshwater cyanobacteria, dinoflagellates, green algae such as *Oedogonium* (Figure

3.22) and *Staurastrum paradoxum*, chrysophyceans, and diatoms.

Algal defenses against pathogens

Algal defenses against pathogens and parasitic attack include structural barriers such as resistant cell wall polymers or antibiotic compounds. The production by the freshwater green alga *Spirogyra* of pentagolloyl glucose, which inhibits microbial α-glucosidase, is one such example (Cannell et al. 1988). Seaweeds produce many types of antibacterial and antiviral compounds that provide resistance against a wide variety of microbes. For example, the brown seaweed *Cystoseira tamariscifolia* produces an antifungal and antibacterial merditerpenoid (Bennamara et al. 1999). Another brown seaweed, *Lobophora variegata*, produces a cyclic lactone (lobophorolide) that was active against at least one pathogenic fungus and at least one saprophytic fungus, but not a chytrid, a pathogenic bacterium or herbivorous fish. These observations suggest that *Lobophora* targets chemical defenses against specific types of pathogens (Kubanek et al. 2003). Such defenses help to explain why algae rarely display obvious signs of infection.

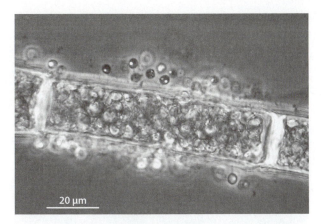

Figure 3.22 Chytrids, aquatic members of the Kingdom Fungi that, like many algae, reproduce by means of flagellate zoospores. The chytrids shown here have settled on and invaded cells of the common freshwater filamentous green alga *Oedogonium*. Algal cells that have been parasitized by chytrids can be recognized by pigment loss, evidence of cellular disorganization or senescence, and by colorless, spherical chytrid walls, which remain attached to the outer surface of the algal victim. (Photo: L. W. Wilcox)

Figure 3.23 Parasitic growths of the green alga *Cephaleuros* on leaves of the flowering plant *Magnolia*. (Photo: R. Chapman and C. Henk)

Algae as Parasites or Pathogens

Algae of diverse types may function as parasites or pathogens. For example, the cyanobacterium *Phormidium corallyticum* is a dominant component of the pathogenic microbial consortium known as black band disease, which kills coral colonies on reefs worldwide (Richardson 1997). Various corals (including brain and star corals) are affected, and the disease, once started at injury sites, spreads over the corals, moving a few millimeters per day. The cyanobacterial filaments facilitate the introduction of sulfide-oxidizing and sulfide-reducing bacteria into coral tissues. These create an anoxic environment and produce hydrogen sulfide, killing coral tissue, which is then degraded. It has been suggested that black band disease is correlated with water pollution in low-turbulence environments, as well as with other stresses to the corals (Peters 1997).

Nearly 15% of red algal species are parasites of other red algae (Zuccarello et al. 2004). Parasitic red algae establish cellular connections with host cells and transfer parasite nuclei into host cells, transforming them. Reproductive cells that are subsequently produced may carry the parasitic genome, allowing the relationship to persist into the next generation (Goff et al. 1996) (Chapter 15).

Some green algae appear to parasitize vascular plants, though usually causing only yellowing of host leaves (Chapman and Good 1983). Examples include *Rhodochytrium*, *Synchytrium*, and *Phyllosiphon*. *Cephaleuros* (Figure 3.23) grows within the leaves of hundreds of host plant species, including coffee, tea, and other crops, but can easily be controlled by spraying plants with a copper-sulfate solution (Chapman and Waters 1992).

Prototheca wickerhamii and *P. zopfii* are nonphotosynthetic members of the green algae that can cause infections of humans and cattle (Figure 3.24). They are common osmotrophs in soil, sewage, and water samples. In humans they sometimes cause skin infections and occasionally bursitis and peritonitis, but these can be cured through antibiotic treatment (Gibb et al. 1991; Sands et al. 1991). In cattle, *Prototheca* may cause a form of mastitis that is highly contagious, with the result that infected herds are usually destroyed. *Prototheca* is also associated with poor growth of anuran tadpoles (Wong and Beebee 1994). *Helicosporidium*, a close relative of *Prototheca*, is a parasite of diverse invertebrates.

Algae as Epibionts

Epibionts are organisms that spend most of their life cycle attached to the surface of another organism,

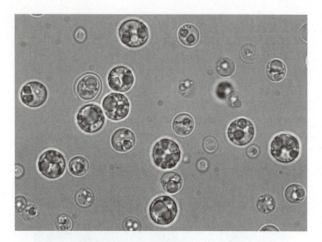

Figure 3.24 Colorless cells of the green algal relative *Prototheca*. (Photo: L. W. Wilcox)

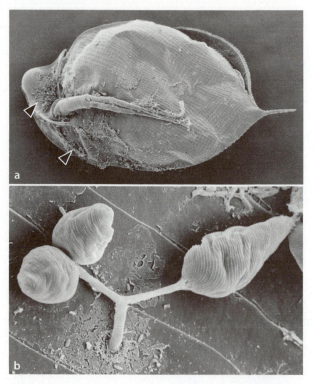

Figure 3.25 Scanning electron micrographs showing the epibiontic euglenoid *Colacium vesiculosum* growing on the cladoceran *Daphnia pulex*. (a) Low-magnification view (arrowheads point to clumps of *Colacium*). (b) In higher magnification, a branched stalk, by which the algal cells are attached, is evident, as is the spiral cell covering. (Micrographs: R. Willey)

which serves as a substrate. Algae can be epibiontic on the surfaces of other organisms or may themselves support algal or bacterial epibionts.

Many algal species grow attached to the surfaces of bacterial biofilms, other algae, protozoa, plants, zooplankton, and larger animals, such as crabs, fish, turtles, whales, tree sloths, and polar bears. In most cases, it is not clear whether or how the organisms involved in epibiontic associations might benefit; it is possible that further study might reveal that many are mutually beneficial symbioses. Examples of epibiontic algae include dinoflagellates of the genus *Gyrodinium* that live on the surfaces of up to 84% of the large planktonic protists known as foraminifera and radiolaria. Up to 20,000 dinoflagellates may occupy the surface of a single substrate cell (Spero and Angel 1991). Biofouling algae, which must be removed from the hulls of ships, dock surfaces, or other constructions, can be regarded as epibionts because these algal growths start when their reproductive cells attach to bacterial biofilms on nonliving surfaces. In the case of the green biofouling seaweed *Enteromorpha*, the alga senses the biofilm by its production of molecules of N-acylhomoserine lactone used for interbacterial communication (quorum sensing) (Joint et al. 2002).

In some cases, algal epibionts may become so numerous as to weigh down or shade the organism on which they grow. One well-studied example is the euglenoid *Colacium vesiculosum*, which commonly grows on the cladoceran *Daphnia* (Figure 3.25) (Bartlett and Willey 1998). Several hundred algal cells or colonies may be attached via carbohydrate

stalks to a single animal. Heavy infestations may deleteriously affect the *Daphnia*, unless it is able to molt—shed the exoskeleton and regenerate a new one. When molting occurs, *Colacium* produces flagellate reproductive cells that disperse and perhaps attach to another animal (Al-Dhaheri and Willey 1996). In a high mountain lake in Spain, the green alga *Korshik-oviella gracilipes* preferentially occurs on surfaces of *Daphnia pulicaria*. The algal epibiont coordinates its production of reproductive cells with molting of the crustacean (Pérez-Martinez et al. 2001). *Daphnia* grazes on these algal dispersal stages, thereby benefiting from the association (Barea-Arco et al. 2001).

Epizoic algae are those living as epibionts on animals. Examples are provided by 40 epizoic algal species (4 cyanobacteria, 20 red algae, 3 green algae, 3 brown algae, and 10 diatoms) found on surfaces of the predatory scorpion fishes *Scorpaena grandicornis* and *S. plumieri* in Puerto Rico. These epizoic algae are thought to provide camouflage that keeps prey from noticing these scorpion fishes until it is too late to avoid capture. While mucilage produced by fish skin normally prevents the attachment of epizoic algae, it is thought that these species' skin differs in some way that allows colonization of algal epibionts (Ballantine et al. 2001).

In terrestrial habitats and wetlands, diverse types of epibiontic algae occur on moist plant surfaces and thus are also described as epiphytic algae. For example, cyanobacteria commonly live on the surfaces of mosses in late successional boreal forests, where they may be important sources of nitrogen compounds (DeLuca et al. 2002; DeLuca et al. 2008; Houle et al. 2006). In arid habitats, cyanobacteria and green algae often associate with certain mosses and lichens to form microbial crusts (Chapter 23). In aquatic systems, the surfaces of plants, seaweeds, and microalgae can become coated with epibiontic algae (Figure 3.26). Such algae can account for up to 22% of system primary production. Coralline red algae common along southern California coasts may have as many as 30 species of algal epibionts. Under some conditions, epibionts may decrease the growth rate of substrate algae by shading and impeding nutrient uptake. A heavy load of epibionts may also increase the drag forces that tend to tear seaweeds from the rocks (Lobban and Harrison 1994) (Chapter 22). Certain red algae produce halogenated organic compounds (see Chapter 2) that prevent algal epibionts from colonizing their surfaces, and brown algal cells contain phenolic compounds (Chapter 14) that similarly serve as antifouling compounds. Some seaweeds eliminate epibionts by shedding their surface cell layers or through copious production of surface mucilage (Round 1992).

Algae in Mutualistic Symbioses

Algae are involved in a wide variety of **mutualisms**, associations in which all partners benefit from the partnership. While some mutualistic symbioses appear to primarily involve two partners, **multispecies consortia** involving three or more partners are increasingly being recognized. Mutualisms can help to ensure the survival of the partners by providing conditions suitable for growth and reproduction that are superior to those available when growing alone. Mutualistic associations often maintain stable environmental conditions necessary for survival

Figure 3.26 A golden-brown algal epiphyte (*Peroniella* sp.) growing on the green alga *Desmidium majus*. This epibiont does not appear to adversely affect the desmid substrate. (Photo: L. W. Wilcox)

that are not present outside the association (Paerl 1992). For example, aquatic mutualisms involving algae and heterotrophs are particularly common in well-lighted clear waters of low-productivity habitats, such as coral reef communities or oligotrophic lakes. In such environments, prey availability may be low, limiting survival or reproduction of heterotrophs unless they have algal partners that provide organic nutrients.

Proof of mutualistic association is usually obtained by tracking the movement of radioisotope-labeled molecules from one organism to another. For example, $^{14}CO_2$ provided to a photosynthetic algal partner might later be detected as carbohydrates or other organic molecules in cells of a heterotrophic partner. Biologists also test for mutualistic benefit by cultivating the partners separately in the laboratory and comparing their growth and/or reproduction to that observed in the mutualistic association. If the partner organisms grow or reproduce better in the association than they do alone, mutualism is indicated. Often, efforts are made to reassociate the separated partners in the laboratory and understand the nature of chemical signals passed between partners, which seem essential to mutualisms.

Endosymbiotic associations occur when algae live within the cells or tissues of other organisms, or when the cells of other organisms live within those of algae. Endosymbiosis and its role in algal evolution is the topic of Chapter 7. Mutualistic relationships discussed here include algal associations with heterotrophic bacteria, fungi, and land plants.

Bacterial associations

Bacterial–algal relationships fall into three broad categories: (1) close associations between microalgae and bacterial cells, (2) macroalgal–bacterial partnerships, and (3) highly structured benthic microbial mats.

Various types of exudates from microalgae and macroalgae provide organic carbon needed for the growth of attached bacteria, often known as **epibacteria**. Such bacteria may, in turn, provide algal partners with growth factors, vitamins, chelators, or inorganic nutrients such as Fe, CO_2, NH_4^+, NO_3, or PO_4^{3-} that result from bacterial degradation of organic materials. For example, many eukaryotic algae cannot synthesize vitamin B_{12} yet require it for methionine biosynthesis. Laboratory cultures of the dinoflagellate *Amphidinium operculatum* and the unicellular red alga *Porphyridium purpureum* obtain vitamin B_{12} from associated cells of the bacterium *Halomonas* (Croft et al. 2005). In the case of *Porphyridium*, *Halomonas* cells occupy mucilage that surrounds the algal cells and hence are well positioned to receive organic exudates from the photosynthetic algae. Various kinds of bacterial cells that occur in the mucilage sheaths of green algae that grow in low-nutrient bog waters have been hypothesized to provide mineral nutrients while receiving organic exudates (Fisher and Wilcox 1996) (Figure 3.27). Bacteria also coat swimming cells of the euglenoid flagellate *Trachelomonas* (Rosowski and Langenberg 1994), and certain *Cryptomonas* species possess dense coatings of the bacterium *Caulobacter* (Klaveness 1982).

Species of the seaweed *Ulva* and close relatives (*Enteromorpha* and *Monostroma*) grow abnormally in bacteria-free culture but develop normal morphology in the presence of their bacterial floras (Provasoli and Pintner 1980). Nakanishi et al. (1996) isolated hundreds of bacterial strains from the surfaces of macroalgae, and, of these, about half influence *Ulva*

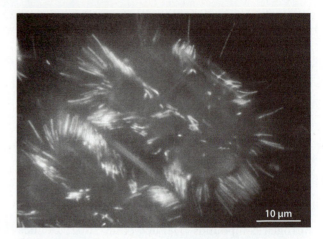

Figure 3.27 Epibacteria growing within the mucilaginous sheath of the green alga *Desmidium grevillii*. The bacteria are primarily long, thin forms; here they were rendered more visible through staining with the DNA-specific fluorescent dye, DAPI. (From Fisher and Wilcox 1996, by permission of the *Journal of Phycology*)

morphology, suggesting that bacteria produce diffusible chemical morphogens. This work suggests that the incidence of epibacterial–macroalgal developmental symbioses may be more widespread than previously recognized.

Multispecies microbial mat communities consisting of diatoms, cyanobacteria, and other bacteria are often conspicuous features of the surface sediments of streams, hot springs, deep-sea vents, polar lakes, hypersaline lagoons, the littoral zone of hardwater lakes, coral reefs, sewage treatment plants, and estuaries. Such microbial mat communities are typically layered into several chemical and light microenvironments, each providing specific conditions required for major biogeochemical transformations conducted by individual mat community members. Sunlight trapped by the autotrophic algal members provides the energy for mat production and chemical cycling activities. Photosynthesis by the cyanobacteria and diatoms produces O_2 and dissolved organic compounds that serve as carbon sources for heterotrophic components of microbial mats (Paerl and Pinckney 1996).

Fungal associations

Lichens are stable, self-supporting associations between fungi known as mycobionts and green algae and/or cyanobacteria known as **phycobionts** or photobionts (Figure 3.28). An important feature of

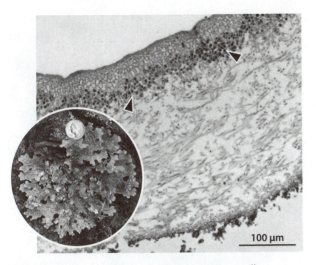

Figure 3.28 Cross-section of *Lobaria*, an internally stratified lichen, showing fungal hyphae and algal cell layer (arrowheads). Inset shows a habit shot of the lichen with quarter- dollar coin for scale. (Photos: L. W. Wilcox)

lichens is that their overall structure is typically distinct from that of the fungal partner growing alone.

Molecular data indicate that lichens have arisen multiple times and include an estimated 17,000 species. More than 20% of all fungi form lichens. The heterotrophic mycobiont obtains fixed carbon from the phycobiont in the form of glucose or sugar alcohols. Some 85% of lichens contain unicellular or filamentous green algae, 10% contain cyanobacterial partners, and 4% or more contain both green and cyanobacterial algae. In the latter case, the green algae provide a source of fixed carbon, while the cyanobacterial partner provides fixed nitrogen. Although lichens have often been regarded as partnerships between one fungus and one alga, triple and greater multiple symbioses may be more common than previously recognized. Because lichens also incorporate heterotrophic bacteria, some experts regard lichens more as microbial consortia than as individuals (Honegger 1992; Cardinale et al. 2006).

The degree of mutualism appears to vary widely among lichens. The 22% or so of lichens that are morphologically complex are considered highly mutualistic because greater success of all partners is achieved in the association than when components are grown separately. However, more simply constructed lichens may reflect controlled parasitism of the fungal component on the algae. Although some

lichens can produce asexual reproductive structures that include small amounts of algae and fungi, in many cases the associations must re-form in nature by accidental contact between fungi and appropriate algae. This offers the opportunity for changing partners, and molecular analyses suggest that partner switching has commonly occurred during evolution (Piercey-Normore and DePriest 2001; Nelsen and Gargas 2008). The unicellular green algae *Trebouxia* and *Asterochloris* are by far the most common phycobionts, and in 98% of lichen species the fungal partner is an ascomycete (Rambold et al. 1998). Lichens produce a variety of secondary metabolites, known as lichen substances, lichen acids, or lichen products, and production of these substances is influenced by carbohydrate supplied by the phycobiont (Honegger 1992).

Plant associations

A number of liverwort, hornwort, moss, fern, cycad, and angiosperm species are closely associated with nitrogen-fixing cyanobacteria—typically species of *Nostoc* (Dodds et al. 1995; Rai et al. 2000). The occurrence of cyanobacterial associates within ventral cavities of most hornworts and two liverwort genera belonging to an early-diverging lineage suggests that early land plants may also have relied on such associations for a supply of fixed nitrogen. The neurotoxin BMAA is produced by *Nostoc* isolated from a hornwort, a cycad, and *Gunnera* (Cox et al. 2005).

Nitrogen-fixing cyanobacteria occur in cavities on the undersides of leaves of the aquatic fern *Azolla*, commonly used as a biofertilizer in rice paddies (Figure 3.29). *Azolla* does not grow on nitrogen-free media without its algal partner, and nitrogen-fixation activity of the alga is as much as 20 times higher within *Azolla* than in free-living cyanobacteria. Cycads typically produce aboveground roots whose cortex contains *Nostoc* colonies (Figure 3.30). (Such roots are known as coralloid roots because their branching pattern resembles that of corals.) *Nostoc* also occurs in mucilage-filled glands in the stems (Bergman et al. 1992) or leaf petioles (Grilli Caiola 1992) of the tropical flowering plant *Gunnera*. In the symbiotic association, though *Nostoc* rubisco levels remain high in at least some cases (Bergman et al. 1992; Rai et al. 2000), cyanobacterial photosystem II (PSII) and carbon fixation are downregulated. This effect has been suggested to result from signaling by organic compounds imported from the plant host, during

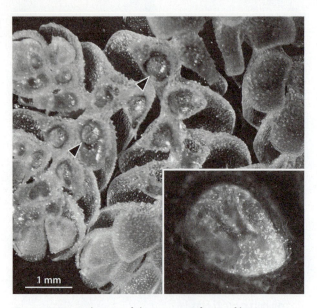

Figure 3.29 Colonies of the nitrogen-fixing, filamentous cyanobacterium *Anabaena* within cavities (arrowheads) on the undersides of leaves of the aquatic fern *Azolla*. In the inset, one cavity is viewed at higher magnification, using fluorescence microscopy. The *Anabaena* filaments have a bright appearance due to autofluorescence of photosynthetic pigments. (Photos: L. W. Wilcox)

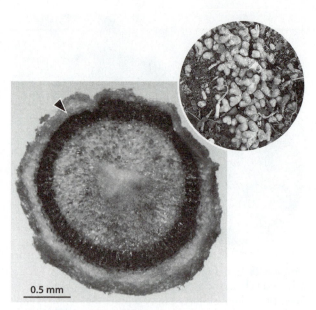

Figure 3.30 Colonies of the nitrogen-fixing, filamentous cyanobacterium *Nostoc*, which form a distinct dark ring (arrowhead) within the aboveground coralloid roots (inset) of the cycad *Cycas revoluta*. (Photos: L. W. Wilcox)

a shift from autotrophy to heterotrophic nutrition (Black et al. 2002). Although some experts have suggested that PSII does not occur in symbiotic *Nostoc*, a study of the *Gunnera tinctoria* symbiosis revealed that PSII units are likely present, but their photochemical efficiency is reduced greatly (Black and Osborne 2004). In spite of this fact,

light stimulates nitrogen fixation by cyanobacterial symbionts. It has been assumed that symbiotic cyanobacteria use cyclic photophosphorylation (cyclic electron flow) to generate ATP that helps to fuel nitrogen fixation. This light enhancement effect may explain the presence of cyanobacteria in aboveground coralloid roots of cycads and other plant tissues that have access to light.

Technological Applications of Algae

Humans use algae in a wide variety of technological applications. For example, algae provide essential laboratory tools in genomics, proteomics, and other research applications. People also use algae as environmental monitors, both to assess the health of modern aquatic systems and to infer environmental conditions of the past. Algae are the sources of numerous types of food and industrial products, including renewable fuels. Algae have also been incorporated into engineering systems designed to purify water and air. Genetic engineering techniques allow algae to be modified in ways that increase their technological utility. Technological processes involving algae may be patented and form the basis for new companies or industries.

Algal bioassay

(Photo: L. W. Wilcox)

4.1 Algae as Research Tools

The history of basic biological discoveries includes many examples in which algae played critical roles. For example, the familiar 9 doublets + 2 singlet microtubule pattern characteristic of nearly all eukaryotic cilia and flagella was first observed in algae. The existence of messenger RNA was first postulated from studies of development in the seaweed *Acetabularia*. The absorption spectrum for photosynthesis—the determination of the wavelengths of visible light that are captured by chlorophyll—was first demonstrated with the use of an alga. The first products of photosynthetic carbon fixation were also unknown prior to Melvin Calvin's Nobel Prize–winning studies on the green alga *Chlorella* (Figure 4.1). Diverse small algal species continue to be useful laboratory organisms because they grow rapidly, have short generation times, and are easily cultivated under laboratory conditions. For this reason, hundreds of microalgal species are maintained in culture collections as research resources.

Algal Culture Collections

Culture collections that focus on algae and sometimes other protists and are located in the United States include the University of Texas Algal Culture Collection (UTEX), the American Type Culture Collection (ATCC), and the CCMP-Provasoli-Guillard National Center for Culture of Marine Phytoplankton. Norton et al. (1996) provided a list of international culture collections. These organizations maintain web sites that include lists of available cultures and their original sources, recommended growth conditions, culture media composition, and other useful information. In research reports, researchers cite the algal cultures that were used by collection catalog numbers, allowing others to know exactly which algal strains were used to perform experiments. Such information can be essential to the replication of scientific results and is necessary when products or processes involving algae are submitted for patenting. When new microalgal species are formally described in the literature, it is customary and often required that the discoverers deposit cultures of the new species in a culture collection. Examples of the many types of uses for algae obtained from culture collections include genomic and proteomic studies. Because the properties of cultured algae may change over time, many cultures are kept frozen until needed (Brand and Diller 2004; Rhodes et al. 2006).

Algal Genomics and Proteomics

Genomics is the application of techniques for study of an organism's entire genome, though research may focus on nuclear, plastid, or mitochondrial DNA sequence information. Genomic information provides the basis for understanding gene expression by means of the transcriptome—the assemblage of messenger RNAs produced by cells exposed to particular conditions. **Proteomics** is concerned with the types and relative amounts of proteins produced by a cell under defined conditions. Proteins determine a cell's metabolome, its small molecules such as lipids and sugars. Such studies of algae help to explain their ecological function or industrial applications.

Genomic studies are also used in evolutionary studies to explore how genes differ among organisms and how gene expression networks operate during development and respond to environmental changes. Genomic data are collected into web-based databases, such those maintained by the Department of Energy–sponsored Joint Genome Institute (JGI). Species of eukaryotic algae are chosen for genome sequencing projects on the basis of several factors, including their ecological and economic importance and the size of the nuclear genome (Grossman 2005).

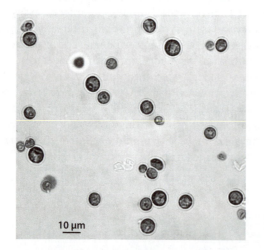

Figure 4.1 *Chlorella*, a small, unicellular green alga that has been widely used as a model system to enhance understanding of photosynthesis and other aspects of algal physiology. (Photo L. W. Wilcox)

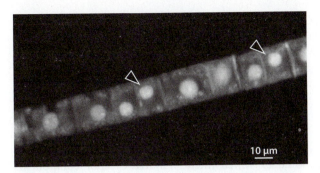

Figure 4.2 Specific binding of the fluorescent dye DAPI to genomic DNA in the cell nuclei (arrowheads), used in microfluorometry to estimate genome size. The size of the haploid nuclear genome of the green alga *Entransia fimbriata*, shown here, is estimated to be 539 Mbp. (Photo M. E. Cook)

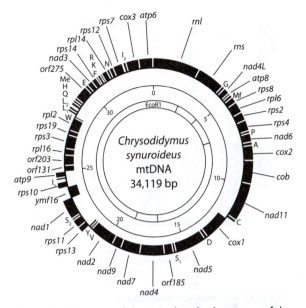

Figure 4.3 Diagram of the mitochondrial genome of the stramenopile *Chrysodidymus synuroideus*. (After Chesnick et al. 2000, by permission of Oxford University Press)

With a small genome, it is easier to obtain and describe (annotate) the full nuclear genome sequence. Genome size can be estimated in several ways, and DNA microfluorometry has proven useful for many algae (Kapraun 2007). This process involves the use of a spectrofluorometer that is attached to a light microscope. The device is used to measure the brightness of a DNA-binding fluorescent dye such as DAPI (4,6-diamidino-2-phenylindole) that has bound to the DNA within a nucleus in a cell (Figure 4.2). The level of brightness in comparison to that of nuclei of measured DNA content provides an estimate of the amount of DNA present in algal nuclei.

Full genome sequences are now available or in development for multiple species of cyanobacteria and eukaryotic algae belonging to diverse groups. The unicellular red alga *Cyanidioschyzon merolae*; the diatoms *Thalassiosira pseudonana* and *Phaeodactylum tricornutum*; and the green algae *Chlamydomonas reinhardtii*, *Ostreococcus tauri*, and *O. lucimarinus* are examples of eukaryotic algae whose genomes have been sequenced. Genome comparisons have many potential evolutionary applications. For example, Renner and Waters (2007) used these five eukaryotic algal genomes to identify and compare genes encoding the DnaK subfamily of heat shock 70 proteins, finding distinct evolutionary patterns in green versus red plastids. Comparisons of gene content also reveal how algae are adapted to their environments in distinctive ways. For example, studies of *O. lucimarinus* revealed that it absorbs iron more like prokaryotes than other eukaryotes and that this alga uses the protein ferritin to store iron

(Palenik et al. 2007). The *T. pseudonana* genome uncovered genes involved in the development of diatom silica cell walls (Armbrust et al. 2004; Mock et al. 2008). The *C. merolae* genome revealed the presence of genes encoding several types of blue light receptors, thereby helping to explain how this organism detects light in its environment (Matsuzaki et al. 2004). Plastid and/or mitochondrial genomes have been sequenced for a number of eukaryotic algae, including the green algae *Nephroselmis olivacea*, *Chlorella vulgaris*, *Chaetosphaeridium globosum*, and *Mesostigma viride* and the stramenopile *Chrysodidymus synuroideus* (Figure 4.3).

Collections of cDNAs (complementary DNAs) or ESTs (expressed sequence tags) also aid genomic research by providing fast and inexpensive ways to discover new genes and their functions and map their positions on chromosomes. Although cDNAs and ESTs lack introns or regulatory portions of genes, they can be used to study organisms for which full genomes are not yet available. EST libraries are particularly useful for algae such as dinoflagellates that have very large genomes. To produce cDNA, mRNA is isolated from cells, and the enzyme reverse transcriptase uses it as a template to make a complementary strand of DNA, and then a double-stranded cDNA is made. This is done because DNA is more stable than RNA. To produce ESTs, the cDNA is

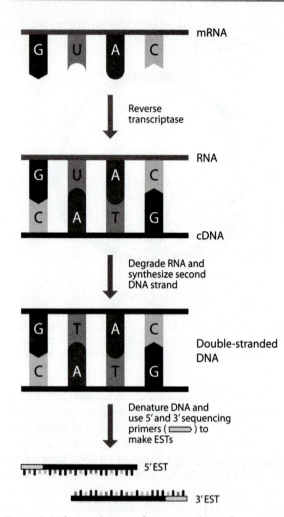

Figure 4.4 The production of cDNA and ESTs from mRNA.

sequenced for a few hundred nucleotides from each end, yielding a 5' EST and a 3' EST (Figure 4.4). These pieces of sequence are "tags" that can be used to search online databases such as GenBank for genes having similar sequence and known function. GenBank is operated by the National Center for Biotechnology Information (NCBI). ESTs are catalogued in a GenBank database known as dbEST. As an example of how the database can be used, Kajikawa and associates (2006) used an EST collection made from the green alga *Chlamydomonas reinhardtii* to identify an enzyme involved in the production of commercially valuable fatty acids.

Proteomic studies involve cleaning, freezing, and grinding algae and then extracting their proteins. The proteins are then separated using two-dimensional electrophoresis and identified using mass spectrometry

and web-based tools such as those described by Wong et al. (2006). Examples of proteomic studies of algae include analysis of the proteins present in plastid ribosomes (Yamaguchi et al. 2002), flagella (Pazour et al. 2005), and centrioles (Keller et al. 2005). Schmidt et al. (2006b) conducted a study of the proteins present in *C. reinhardtii* eyespots—red-colored structures that are associated with light detection and responses. This investigation identified 202 eyespot proteins, some of which likely stabilize eyespot components and hold them together.

Additional Uses of Algae as Research Tools

In addition to their importance as genomic and proteomic model systems, algae have been used as research tools in a wide variety of other ways. In cell biology, *C. reinhardtii* (Figure 4.5) has been a model system for understanding how protists sense mechanical stimuli such as sound, touch, or gravity (Nakayama et al. 2007). Tam and colleagues (2007) used *C. reinhardtii* to identify an enzyme that regulates the length and development of flagella. Grossman et al. (2004) reviewed ways in which *C. reinhardtii* has been useful in understanding photosynthetic pigment and photoreceptor biosynthesis. Merchant and associates (2006) reviewed the use of *Chlamydomonas* to understand trace metal metabolism, thereby explaining why cells become chlorotic (yellowish rather than green) when iron is deficient. *Chlamydomonas* cells have also been used as "microoxen," to move tiny loads as far as 20 cm at a rate of 100 to 200 µm per second, which may have potential uses in nanotechnology (Weibel et al. 2005).

The fluorescent phycobiliproteins of red algae are useful as labels in flow cytometry (a method for counting and separating algae), in immunofluorescence microscopy, and as photosensitizers in the treatment of cancer by means of light (Isailovic et al. 2006). Curacin A, a lipid derived from cyanobacteria, is sold for use in studying tubulin, the protein that makes up cellular microtubules. Under certain conditions, the cyanobacterium *Synechocystis* produces electrically conductive "nanowires" in the form of thread-like pili that extend from cellular surfaces, which may prove useful as a means of delivering electrons in nanotechnology (Gorby et al. 2006). Bao et al. (2007) used diatoms to demonstrate a new process for constructing nanoscale structures useful in making sensors for gases such as NO (nitrous oxide).

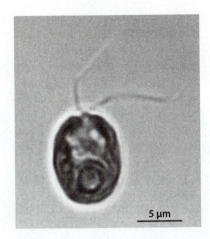

Figure 4.5 The unicellular green flagellate *Chlamydomonas reinhardtii*, which is widely used as a genetic model system. (Photo L. W. Wilcox)

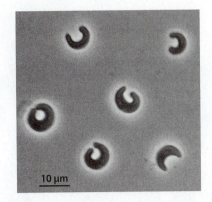

Figure 4.6 *Selenastrum capricornutum* is used in bioassays to assess the nutrient or toxin status of natural waters. (Photo L. W. Wilcox)

4.2 Algae as Environmental Monitors

Changes in algal community composition can be used as indicators of environmental change. Regional indexes of pollution impacts have been developed, and an example of their use is stream water-quality monitoring, through evaluation of diatom communities (Lavoie et al. 2008). Algae can also be used in laboratory bioassays to monitor the quality of water that will be used for drinking or other purposes. In addition, the microscopic remains of certain algae are used in paleoecological studies to infer changes in water quality of lakes over time. Such remains are also used to detect instances of climate change affecting lake or ocean algal communities.

Algae in Bioassays

A **bioassay** is a procedure that uses organisms and their responses to estimate the effects of physical and chemical agents in the environment. Algal biomonitors are widely used to monitor both nutrients that would foster algal blooms and substances that are toxic to algae and other organisms. Algae are effective in bioassays because they are more sensitive than animals to some pollutants, including detergents, textile-manufacturing effluents, dyes, and especially herbicides. Algal toxicity tests have become important components of aquatic safety assessments for chemicals and effluents and are required by Section 304(h) of the U.S. Federal Water Pollution Control Act and in the registration of pesticides (Lewis 1990). Some other countries have similar requirements.

The unicellular green alga *Selenastrum capricornutum* (culture identification numbers UTEX 1648 and ATCC 22662) is the algal species most widely used in laboratory bioassays (Figure 4.6). This and some other algal species were selected for use in tests such as the U.S. Environmental Protection Agency's Algal Assay Procedure, the similar Algal Growth Potential Test, and the Algal Growth Inhibition Toxicity Test. Such procedures are based on the premise that cell yield is proportional to the amount of a limiting nutrient or toxin that is present in water samples. Although chemical methods can also be used to determine nutrient or toxin concentrations in water samples, these may not accurately reflect biological responses. For example, a chemical measurement of total phosphorus may overestimate the amount of this element that is biologically available to organisms (Skulberg 1995).

Because algae vary widely in their responses to nutrients and toxins, standard procedures commonly recommend using a battery of test species from different algal groups. To conduct a bioassay, the algal cells are dispensed into individual flasks, tubes, or wells of microtiter plates along with water samples that have been amended with varying levels of nutrients or toxins. Initial cell density, temperature, pH, and irradiation are controlled. After a period of growth, microscopic, electronic particle counting, or spectrophotometric methods are used to estimate the growth response of the bioassay alga. For toxins, an "effective concentration" value EC_x, where x reflects

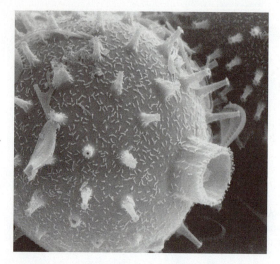

Figure 4.7 Silicified resting cyst of freshwater chrysophyceans, which are widely used to assess past ecological conditions affecting lakes and their biota. (SEM: P. Siver)

Figure 4.8 Sampling of lake sediments with tubular coring devices to obtain algal remains useful in reconstructing the history of environmental change in a region. (Photo L. E. Graham)

the percentage of reduction, usually 50, and a "no observed effect concentration" (NOEC) are calculated based on growth inhibition. Some procedures also require determination of the area under the growth curve or particular statistical procedures. Because the biological relevance of EC_x and NOEC are largely undefined, the results should be interpreted only as an indication of effects on algae in the natural system (Lewis 1990). Haglund (1997) reviewed the use of algae to assess toxicity.

Algae as Paleoecological Indicators

As algae die, their decay-resistant remains may accumulate in lake or ocean sediments. Examples of algal materials that often occur in lake sediments include distinctively ornamented silica scales and walls of resting stages (known as stomatocysts) of freshwater chrysophyceans (Figure 4.7) and diatoms. Together with diatom walls, calcified scales of coccolithophorid algae, silica skeletons of silicoflagellates, and decay-resistant cysts of dinoflagellates persist in ocean sediments for millions of years. These algal remains pile up in layers that can be used to deduce past environmental changes, such as climate shifts or other ecological disturbance by humans (reviewed by Smol and Cumming 2000). In lakes, for example, the topmost centimeter of sediment may have accumulated over the past five years. The more distant from the sediment surface, the longer ago the algal

remains were deposited. This layering of algal fossils in sediments, along with other information, allows paleolimnologists or paleooceanographers to infer the relative age of sedimentary deposits (Smol 2007). Such scientists take samples of sediments with hollow, tubular devices from which long cores of sediment are extruded (Figure 4.8). The cores can be cut in half lengthwise so that one-half can be archived for later reference and the other half can be sliced crosswise for analysis.

A database of modern algal species' responses to environmental conditions of various types is necessary in order to use the remains of ancient algae to infer ecological conditions of the past. Rühland and associates (2003) provided an example of how such databases are constructed. They examined the relationships between measured environmental variables and modern diatom communities in 77 lakes across the central Canadian Arctic and found that the diatom communities of tundra lakes are distinctly different from those of boreal forests. Databases of this type can be used to infer the ecological conditions that were present in the past. For example, Quinlan et al. (2005) examined the diatoms in sediments dating back 1000 years in three ponds on Ellesmere Island. They found shifts in the algal communities, beginning about 200 years ago, that indicated reduced ice cover resulting from climate warming. Changes in diatom communities in sediments taken from Canadian lakes studied by Harris et al. (2006) also indicated that climate warming began in about 1900 in this region.

4.3 Algae as Sources of Food and Other Products

Algae are primary producers in aquatic food chains (see Chapter 3), and some species are preferred food for aquatic animals that are farmed in aquaculture operations or maintained in display aquaria. Certain algae are valued food for humans, while others are sources of gelling compounds widely used in biological research and in industry. Algae of diverse types are known to produce compounds useful as antibiotics or in cancer treatments. In recent years, the sustainable production of biofuels from algae has been explored as a replacement for increasingly scarce oil supplies; this is logical because oil is largely derived from the organic remains of ancient algae (see Chapter 2). The major products produced from algae, and their commercial value, have been reviewed by Radmer and Parker (1994) and Radmer (1996).

Uses of Algae in Aquaculture and as Human Food

Information gained in studies of algal herbivory (see Chapter 3) has proven useful in devising aquaculture systems for the cultivation of shellfish and other aquatic species. Many marine animals cannot synthesize certain essential long-chain fatty acids in quantities high enough for growth and survival and thus depend on algal food to supply them. Nontoxic marine microalgae, including the stramenopiles *Isochrysis*, *Pavlova*, and *Nannochloropsis*, as well as various diatoms, represent the primary food source for at least some stages in the life cycle of most cultivated marine animals. For example, the diatom *Thalassiosira pseudonana* is widely cultivated to feed a variety of molluscs, including the Pacific oyster *Crassostrea gigas* and rock scallops. Algae are

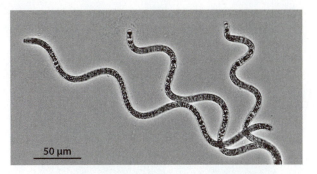

Figure 4.9 *Spirulina* (*Arthrospira*), a filamentous cyanobacterium commonly grown for use in human health-food products or collected directly from natural environments for food use. (Photo L. W. Wilcox)

also grown for production of food additives, such as fatty acids that improve the nutritional quality of baby formula.

Microalgae grown for food or food additives

Although the high nucleic acid content of many microalgae limits their use as human food, several species are cultivated for production of nutritional supplements or food additives such as β-carotene (Ben-Amotz et al. 1982). For example, the cyanobacterium *Spirulina* (also known as *Arthrospira*) (Figure 4.9) has traditionally been harvested from African lakes and consumed in sauces at the rate of some 9 to 13 grams per meal. The protein level of *Spirulina* can be as high as the levels in nuts, grains, and soybeans, ranging from 50% to 70% of algal dry weight, and thus shows promise for use as a protein supplement for malnourished populations. *Spirulina* is also naturally high in B vitamins and essential unsaturated fatty acids (Richmond 1988), as well as high levels of β-carotene, which is converted to vitamin A during digestion. Vitamin A can prevent xerothalmia, a form of blindness that arises in malnourished children. *Spirulina* can be commercially cultivated in shallow raceways, 500 to 5000 m² in area, while being stirred with paddles, or grown in closed outdoor bioreactors (Richmond 1988, 1990). When it is cultivated in open systems, care must be taken to avoid contamination of the product with toxin-producing cyanobacteria (see Chapter 3). Vonshak and associates (2000) found that *Spirulina* (A.) *platensis* grew faster and achieved higher biomass in closed systems when grown mixotrophically with glucose as a carbon source, as compared to cultures grown in light without added sugar.

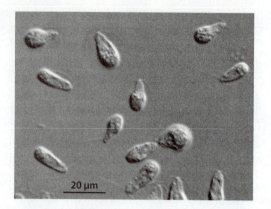

Figure 4.10 *Dunaliella,* a green algal flagellate used in the industrial production of β-carotene, an essential component of the diet of humans and other animals. Cells may appear bright orange in color because they are rich in carotene. (Photo L. W. Wilcox)

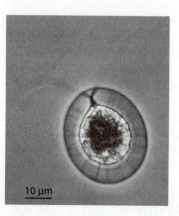

Figure 4.11 *Haematococcus,* a green algal flagellate used in industrial production of the carotenoid pigment astaxanthin. (Photo L. W. Wilcox)

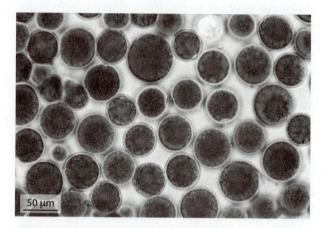

Figure 4.12 *Haematococcus* cells that have been exposed to low-nutrient, high-light conditions. These cells have lost their flagella, enlarged, and produced large amounts of bright red astaxanthin. (Photo L. W. Wilcox, courtesy AgResearch International, Inc.)

Dunaliella, a green flagellate (Figure 4.10), can manufacture 50 times the β-carotene produced by *Spirulina.* Hence, *Dunaliella* is often cultivated for industrial production of β-carotene, which is sold as a nutritional supplement. *Dunaliella* and other microalgae are often cultivated in open-air commercial ponds that may be several thousand square meters in surface area but less than a meter deep. Such algal farms can be located on land that is unsuitable for other crops. Problems may arise, however, if the ponds become contaminated with heavy metals, chemicals, insects, disease microbes, or weedy algae such as toxic cyanobacteria. *Dunaliella* can tolerate higher salinities than many contaminating organisms, and algal farmers use this as a means of preventing growth of contaminating organisms. Another strategy for reducing contamination is the use of enclosed bioreactors. For example, the diatom *Phaeodactylum tricornutum* produces high biomasses when grown outdoors in photobioreactor vessels. It is then harvested for extraction of long-chain polyunsaturated fatty acids used as a human food supplement (Alonso et al. 1996). Genetic selection programs have been successful in enhancing fatty acid content of this and other algae (Galloway 1990).

The green flagellate *Haematococcus* (Figures 4.11, 4.12) is cultivated in mass quantities for extraction of **astaxanthin,** a carotenoid valued for addition to aquaculture feeds as an antioxidant or use as a coloring agent in food and cosmetics (Guerin et al. 2003). *Haematococcus* can be grown in large-scale outdoor systems or in indoor or outdoor enclosed bioreactors. It was once thought that astaxanthin was produced in large quantities only by nonflagellate resting stages (akinetes) (see Figure 4.12), but culture experiments have shown that the flagellate stage (see Figure 4.11) can accumulate carotenoids as rapidly as resting cells (Lee and Ding 1995). Decreasing the nitrogen content of the growth medium and increasing irradiance induces astaxanthin production (Grünewald et al. 1997).

Macroalgae harvested or grown as food

In both ancient and modern times, humans have harvested some 500 species of macroalgae, also known as seaweeds, for food, fodder, and chemicals. Written records confirm that humans have harvested seaweeds in China for more than 2000 years, and today the Chinese people collect 74 species of red, green, brown, and blue-green algae in 36 genera (Xia and Abbott 1987).

A few macroalgae, mostly the reds *Porphyra*, *Kappaphycus* (synonym *Eucheuma*), and *Gracilaria*, and the brown kelps *Saccharina*, *Laminaria*, and *Undaria*, are cultivated in aquaculture operations for use in human foods or for extraction of gelling compounds (described in the next section). Dried *Porphyra* (nori) and *Laminaria* (*kombu* or *haidai*) obtained from aquaculture operations are important crops in Asia. In China alone, hundreds of thousands of people are occupied in growth and harvest of these seaweeds, and it is estimated that the Chinese consume well over 100 million pounds of fresh and dried seaweeds each year (Xia and Abbott 1987). The annual *Porphyra* harvest worldwide has been estimated to be worth US $2.5 billion (van der Meer and Patwary 1995). Seaweed cropping provides a valued source of employment in coastal areas and does not typically degrade the natural environment or the seascape. The economic success of these crops depends greatly on detailed basic knowledge of the algae.

The food value of *Porphyra* lies primarily in provision of essential vitamins, such as B and C, as well as minerals, including iodine. A single sheet of high-grade *nori*—a chopped, pressed, and dried seaweed product—contains 27% of the USDA recommended daily allowance of vitamin A, as β-carotene (Mumford and Miura 1988). *Porphyra* also contains high proportions of digestible protein (20% to 25% wet weight) (Lobban and Harrison 1994) but contributes a relatively low percentage of protein to human diets. The distinctive taste of *Porphyra* is due to the presence of free amino acids. The cultivation of *Porphyra* originated in Tokyo Bay about 300 years ago. Nets were strung between poles in areas where spore attachment occurred and then transported to the cultivation site. The farmers did not know where the spores came from or that the habitat and structure of the spore-producing form of the seaweed was very different from the harvested blades, thus limiting success. The situation changed dramatically after British phycologist Kathleen Drew Baker used culturing techniques to discover that the separately named filamentous, shell-inhabiting red algal genus *Conchocelis* was in fact the organism producing the spores that attached to nets. Baker thereby revealed

that *Conchocelis* was actually a survival phase in the life cycle of *Porphyra* (Figure 4.13) rather than a separate organism. Her report, published in *Nature* in 1949, transformed the industry (Drew 1949).

Since Baker's discovery, the filamentous conchocelis phase has been mass cultivated in sterilized oyster shells and then used to seed nets. The conchocelis is grown indoors, in greenhouse tanks, from spores produced by blades (Nelson et al. 1999). If the details of development or DNA level of sexually produced spores are not known, the spores are called phyllospores. Conchocelis can also arise from asexual agamospores. Spores are released from *Porphyra* blades after brief drying and reimmersion in seawater for a few hours. The spores attach to shells or artificial substrates and are stimulated to grow into conchocelis filaments by exposure to bright sunlight, sufficient nutrients, and aeration. Short day length and reduced temperature stimulate conchocelis to release spores known as conchospores. The conchospores are collected on nets—either by running nets through the indoor tanks or by placing conchocelis-bearing shells underneath nets mounted on poles or rafts in the sea. The typical farmer buys conchospore-seeded nets or conchocelis-occupied shells and installs them in the sea, where conchospores grow into blades. Young blades produce asexual spores known as archeospores or neutral spores that also attach to the nets and grow into additional blades, thereby increasing the crop. When mature, blades are removed from the nets, washed to remove epiphytes, chopped into a slurry, spread over screens, and dried.

Although there are about 70 species of *Porphyra* worldwide, occurring in both tropical and temperate waters, *P. yezoensis* is the most important commercial species. Because frond size is heritable, individual nori farmers practice selective breeding. Particularly large blades exhibiting delayed sexual maturity and rapid growth rates—on the order of 3 to 4 cm day^{-1}—have become the major cultivated varieties. Nearly all *Porphyra* blades are genetic mosaics derived from up to four different genotypes. This is reflected in the patterning of male and female gamete production commonly observed in sexually mature blades (Figure 4.14), which results from meiotic segregation of sex-determining alleles in the germinating conchospores (Burzycki and Waaland 1987). The first four cells of the young gametophytes represent a linear genetic tetrad (Mitman and van der Meer 1994). Separate regions of the resulting blade are thus clones composed of mitotic descendants from the first four cells

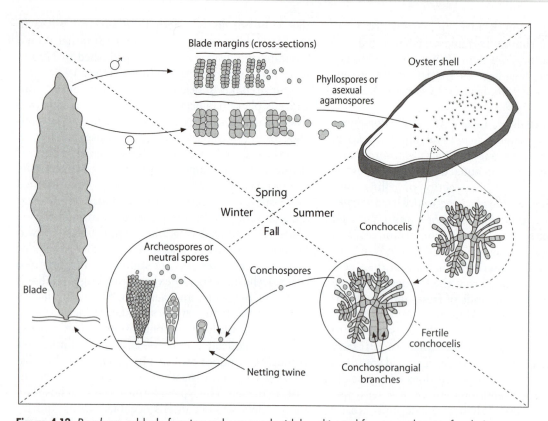

Figure 4.13 *Porphyra,* a blade-forming red seaweed widely cultivated for use as human food. An understanding of its life history—involving both an edible blade phase and a microscopic shell-dwelling "conchocelis" phase—is essential to effective cultivation. As a result of sexual reproduction, blades produce single-celled phyllospores or asexual agamospores (top) that colonize shells and grow into small filaments that generate conchospores. The latter can be induced to colonize twine that is suspended in the sea, and produce blades. Blades may also arise from an asexual process involving single-celled archeospores or neutral spores. (After Mumford and Miura 1988)

and hence may be genetically distinct. Pest problems in *Porphyra* cultivation can include herbivorous fish, which are usually deterred by protective nets, weedy epiphytes such as the diatom *Licmophora* or green algae, and bacterial, fungal (chytrid blight), or oomycete infections (Lobban and Harrison 1994).

Cultivation of edible *Saccharina* (syn. *Laminaria*) *japonica* began in China and now occurs in Japan, Korea, and other countries. The high iodine content of kelps is a useful addition to the human diet. The kelp aquaculture process begins with selection of wild or cultivated sporophytes having darkly pigmented, mature **sori** on their blades. Sori are tissues bearing large numbers of meiosporangia (Figure 4.15) (see Chapter 14). The sori are first cleaned by briefly wiping or immersing them in bleach, and then they are left in a cool, dark location for up to 24 hours prior to reimmersion in seawater. This treatment causes the release of numerous flagellate zoospores, which are then allowed to attach to twine in greenhouses. The strings are suspended in the ocean, where spores grow into microscopic filamentous gametophytes. Gamete production is followed by fertilization of eggs by motile sperm and the subsequent development of zygotes into young sporophytes. The time from gametogenesis to production of young blades 4 to 6 mm long is 45 to 60 days. Strings bearing young sporophytes are cut into pieces and attached to ropes that are suspended in the sea. In winter, the ropes are kept about 5 m below the water surface to avoid wave damage. They are raised to a 2 m depth in summer for greater illumination. Two years are usually required to achieve sufficiently large blades. Harvest occurs

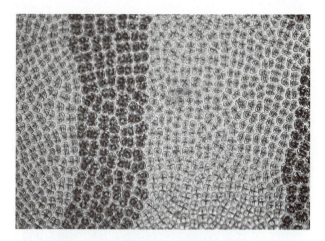

Figure 4.14 Sexually reproducing blade of *Porphyra*, which exhibits mosaics of male and female gamete-producing cells. The darker areas produce female gametes, and the lighter areas produce male gametes. This patterning reflects the genetic heterogeneity of the cells composing the *Porphyra* body. (Photo L. E. Graham)

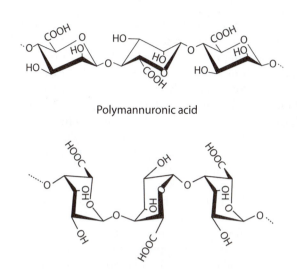

Polymannuronic acid

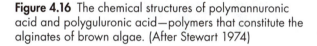

Polyguluronic acid

Figure 4.16 The chemical structures of polymannuronic acid and polyguluronic acid—polymers that constitute the alginates of brown algae. (After Stewart 1974)

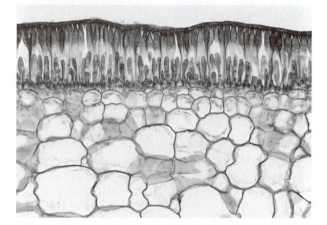

Figure 4.15 The brown alga *Laminaria* may bear patches (sori) of epidermal cells that have differentiated into spore-producing cells. The patches often appear darker than surrounding, nonreproductive tissues and thus can be easily identified. A section through a sorus is shown here. (Photo L. W. Wilcox)

in summer, and most *Saccharina* is dried for use in foods or in the alginate industry (see below).

Undaria, also known as *wakame*, is grown and used in Asia very much like *Saccharina*. Aquaculture operations have also been established off the coast of western Europe, but efforts to expand the industry are controversial because this kelp is not native to

Atlantic waters. It apparently arrived in the Mediterranean sometime during the 1980s on oysters imported from Japan for cultivation purposes. Some phycologists are concerned that *Undaria* might escape from cultivation and compete with native kelp populations, possibly causing negative effects on lobster fishing (Lobban and Harrison 1994).

Gelling Agents from Seaweeds

The gelling agents alginic acid (or its mineral salt, alginate), carrageenan, agar, and agarose are produced from certain brown and red seaweeds. In general, these products are useful because they stiffen aqueous solutions. In some cases the seaweeds are harvested from wild stocks; in other cases, aquaculture operations have been established.

Alginates

Alginates are structural sulfated polysaccharides that confer strength, flexibility, and toughness by forming gels and sols in the matrix between cells of brown algae. These molecules, which occur as mineral salts, help large seaweeds to cope with mechanical stresses generated by waves and currents (see Chapter 22). Alginates make up some 20% to 40% of the dry weight of brown seaweeds. They are polymers of D-mannuronic and L-guluronic acids (Figure 4.16), with the ratios of these monomers varying among algal species. Regions of the polymer that are high in polymannuronic acid are known as M-blocks, regions rich in polyguluronic acids are G-blocks, and regions consisting of alternating monomers are M-G-blocks. Each type has different properties: M-blocks have low affinities for metals; G-blocks bind alkaline earth elements; and M-G-blocks are soluble at low pH, unlike the other two types. An M:G ratio of 1:1 gives a soft gel, whereas a stiffer gel is formed when the M:G ratio is less than 1:1. Because alginates are too complex to be synthesized chemically, the natural source is valuable, and markets are expected to increase. Alginates are used in the textile industry; in manufacturing wrapping papers, dental creams and impression materials, shoe polish, welding electrodes, and oil drilling muds; and in food processing to thicken ice cream, jams, puddings, sauces, mayonnaise, custards, fillings, and decorations for bakery products (Jensen 1995).

Commercially important sources of alginates are the brown macroalgae *Saccharina japonica*, *Laminaria digitata*, *Laminaria hyperborea*, *Ascophyllum nodosum*, *Ecklonia maxima*, *Lessonia nigresens*, some *Fucus* species, and the giant brown kelp *Macrocystis pyrifera*, which reaches lengths of 60 m and may weigh more than 300 kg. Extensive natural *Macrocystis* beds off the coast of California are "mowed" using special barges (Figure 4.17). This process removes only the top meter or so, leaving most of the seaweed intact. Regeneration occurs within a matter of months, whereupon the macroalgae can be recropped. Harvested seaweeds are delivered to the extraction plant on shore, where they are treated with preservatives that prevent destabilization and discoloration of the product by polyphenolics. Dilute acid is used to extract inorganic salts and water-soluble carbohydrates and proteins, yielding insoluble alginic acid. Solubilization is accomplished by alkali treatment, after which the viscous product is dried and milled before storage (Jensen 1995).

Carrageenan, agar, and agarose

As is the case with alginates, there are a large number of industrial and scientific applications for gelling compounds derived from red algae, and a market will continue for these natural products because the polymers are too complex for chemical synthesis. Agars and carrageenans are linear polymers of alternating molecules of $(1{\rightarrow}3)$-β-galactose and $(1{\rightarrow}4)$-α-galactose. In agar, the $(1{\rightarrow}4)$-bonded

Figure 4.17 A barge harvesting the giant kelp *Macrocystis* for alginate production. The ship is equipped with a cutting device and a conveyor belt, which transports harvested seaweeds to the storage area in the hull. (Photo L. E. Graham)

subunits are L-galactose, whereas in carrageenans, both types of subunits are D-galactose. The polymers are extracted from dried seaweeds with warm water and then subjected to filtration, alternating with acetone, alcohol, and water washes. After grinding and extraction in hot water, alcohol precipitation yields carrageenans. Agars are obtained through filtration under pressure (Cosson et al. 1995).

Carrageenans are processed from a relatively few genera of red algae. *Kappaphycus* and *Hypnea* account for much of the commercial production in tropical regions, with about half of the world's carrageenans being produced in the Philippines and Indonesia. *Kappaphycus* aquaculture requires clear and saline waters, substantial water movement but no large waves, and temperatures between 25°C and 30°C (i.e., tropical waters). The presence of a coarse sand bottom reduces suspension of sediments and is suitable for anchoring stakes from which monofilament lines are suspended. Algal bodies are cut into small pieces, tied to the lines, and harvested in about

(a)

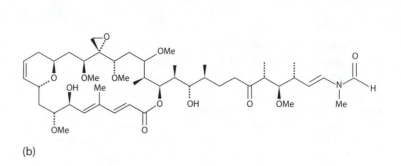

(b)

Figure 4.18 The cyanobacterium *Tolypothrix* (a) produces tolytoxin (b), a compound having low toxicity to humans that is a potent antifungal agent. (a: Photo: L. W. Wilcox; b: Re-drawn from Patterson and Bolis 1997, by permission of the *Journal of Phycology*)

two or three months, when thalli have grown to a mass of 1 kg or greater. About five harvests per year can be achieved, with low capital costs. Cultivation is labor intensive but is viewed as a valuable way to provide employment for coastal populations. The major cultivation problems are diseases that bring about pigment loss and undesirable hybridization with wild seaweeds (Lobban and Harrison 1994).

Wild populations of *Gelidium* are harvested on a worldwide basis for extraction of high-quality agar. About 35% of the world's agar production comes from this seaweed. In addition, it is the source of low-sulfate agarose for use in biotechnology, such as in gels used to visualize DNA (van der Meer and Patwary 1995).

Pharmaceuticals from Algae

Various types of algae are the sources of compounds having antibiotic and anticancer activity (Gerwick et al. 1994; Patterson et al. 1994), as well as other pharmaceuticals. This is not surprising because algal lineages are very old and have thus been subjected to microbial attack for hundreds of millions of years. During this time, algae have adapted by producing diverse protective chemical compounds. Screening programs are used to survey cultivable organisms whose medicinal properties are unknown. Water-soluble and lipid-soluble extracts of algae are initially tested for the ability to reduce pathogen effects on animal-cell cultures grown in multiple-well plates. For extracts showing activity, dilution studies

are done to determine relative potency. Finally, efforts are made to identify the chemical structures of pharmacologically active compounds. For example, a study by Patterson et al. (1993) found that a wide variety of cyanobacteria produce compounds, including sulfolipids, that were active against a herpes virus, a pneumonia virus, and HIV. A potent antifungal antibiotic, tolytoxin, which has low toxicity to humans, is produced by the cyanobacteria *Tolypothrix* (Figure 4.18) and *Scytonema* (Patterson and Carmeli 1992). *Lyngbya majuscula* collected in Papua New Guinea produces aurilides B and C, which are active against human lung tumor and mouse neuroblastoma cell lines (Han et al. 2006). Gross and colleagues (2006) described the anticancer effects of lophocladine alkaloids extracted from the marine red alga *Lophocladia* on lung tumor and breast cancer cell lines. These alkaloids appear to act by inhibiting microtubule function and are examples of a wider diversity of potentially useful defensive compounds produced by algae. Once algal sources of useful medicinal compounds are identified, the algae can be cultivated on a large scale in bioreactors, they can be harvested, and the active materials can be extracted, purified, and marketed.

Algae as Sources of Biofuels

As a result of concerns about the ecological and economic impacts of reliance on fossil fuels, people are increasingly interested in sustainable sources of hydrocarbons for energy and industrial synthesis.

Hydrogen Production by Algae

Hydrogen gas is an attractive alternative to fossil fuels, whose oxidation produces CO_2 and other atmospheric pollutants. Burning hydrogen in automobile engines, for example, yields only water as a by-product. However, supplies of hydrogen are typically generated by the electrolysis of water, a process that itself requires energy. Certain green algae and nitrogen-fixing cyanobacteria may help to solve this problem because they are able to produce hydrogen gas via their photosynthetic electron transport systems by using the energy of sunlight. Such algae can thus directly produce the energy carrier H_2.

What is so special about these algae that they can produce hydrogen gas? The answer is that at least some green algae and cyanobacteria possess distinctive hydrogenase enzymes that interact with photosynthetic electron transport systems. During hydrogen production, electrons derived from water molecules are transferred from the electron carrier ferredoxin to iron atoms located within hydrogenases, with the result that H_2 is released. Algal hydrogenases are thus pivotal to hydrogen production and have been studied intensively, with the goal of using them to produce renewable fuel or in fuel cell applications (Ghirardi et al. 2007; Melis 2007).

A number of green algae, including *Scenedesmus obliquus* and *Chlamydomonas reinhardtii*, possess hydrogenases having two iron atoms in the active site and are thus known as [FeFe]-hydrogenases. These hydrogenases are encoded in the nuclear genome, but the proteins function in the chloroplast. Unfortunately, [FeFe]-hydrogenases are irreversibly inhibited by oxygen gas, so hydrogen gas cannot be produced in the presence of oxygen. However, growing cultures of *C. reinhardtii* in a closed container with supplies of CO_2 and acetate, but without sulfur (thereby turning off photosynthetic oxygen production), allows H_2 production (Melis and Happe 2001; Kosourov et al. 2007). In addition, *Chlamydomonas* has been genetically engineered so that an addition of copper can be used to repress oxygen production and induce hydrogenase activity (Surzycki et al. 2007).

Nitrogen-fixing cyanobacteria produce two types of hydrogenases, both known as [NiFe]-hydrogenases because the metals nickel and iron are both present in the active site. Such [NiFe]-hydrogenases are not related to the green algal [FeFe]-hydrogenases; they are encoded by different genes. One type of cyanobacterial [NiFe]-hydrogenase oxidizes the H_2 produced as a side reaction of nitrogenase, the enzyme that catalyzes conversion of atmospheric N_2 into ammonium ion (see Chapter 2). By converting H_2 to water, this enzyme reduces the amount of H_2 produced by cells. However, a second [NiFe]-hydrogenase can both oxidize H_2 to water and produce H_2, and it is thus known as a bidirectional enzyme. When cyanobacterial cells are grown under conditions of low light and continuous oxygen removal, the bidirectional enzyme produces H_2. This bidirectional hydrogenase is inhibited by oxygen but not irreversibly so; when oxygen is removed, the enzyme can recover its N_2-producing activity.

Sensitivity of the green algal and cyanobacterial hydrogenases to oxygen is the major impediment to their application in sustainable fuel production. Thus, genetic engineering is being used to modify the structure of these hydrogenases, with the goal of reducing their sensitivity to oxygen. Both the [FeFe]- and [NiFe]-hydrogenases display a structural pathway by which oxygen can reach the active site, disabling it. If these enzymes' structures can be changed so that oxygen is less able to reach the metal-containing active site, they will be much more useful as systems for hydrogen fuel production. In the case of cyanobacteria, genetic engineering efforts to disable the type of hydrogenase that oxidizes H_2 may also be needed.

While corn or other plants have been used to generate ethanol for use as fuel, Hill and associates (2006) have pointed out that such sources would not meet the demand and would compete with food production. On the other hand, the cultivation of algae for biofuel production does not require the use of agricultural land resources. Algae can be used to produce biofuels in three major ways: fermentation of biomass, industrial growth of algae for lipid extraction, and hydrogen generation systems. Hydrogen

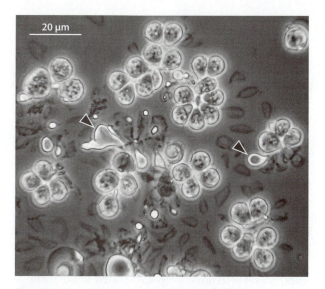

Figure 4.19 *Botryococcus*, a common planktonic freshwater green alga that is unusual in producing large amounts of extracellular hydrocarbons (arrowheads). (Photo L. W. Wilcox)

generation systems have not matured to the point of application as yet, but they offer potential (Text Box 4.1). Biomass fermentation and lipid extraction systems have been developed into large-scale commercial applications.

Much of the biomass of brown and green algae consists of cellulose-rich cell walls. These and other cellular carbohydrates can be fermented to alcohol using processes similar to those applied to corn. The use of algal biomass in fermentation systems has the advantage that algae lack lignin and other large biopolymers that can interfere with the fermentation of higher plant biomass.

Diatoms and some other algae produce relatively large amounts of lipid as a photosynthetic storage. Such algae often produce larger amounts of lipid in response to nitrogen limitation of their growth (Hu et al. 2008). For example, McGinnis et al. (1997) found that under conditions of nitrogen limitation, the diatom *Chaetoceros muelleri* produces more than 400 mg of lipid per liter of culture, which is five to seven times the amount produced when nitrogen is abundant. Diatoms can be grown in large populations in industrial-size bioreactors and lipids harvested from them. More than 300 lipid-producing algal strains have been identified as having economic potential (Roessler 1990). The common freshwater colonial green alga *Botryococcus braunii* (Figure 4.19) is an example of another alga that produces large amounts of lipid. *Botryococcus* cells generate unique C_{17}–C_{34} unsaturated polyhydrocarbons, known as botryococcenes, which constitute more than 30% of the dry weight of this alga. These lipids are produced internally and excreted to the cell surface, where they permeate the matrix between cells of the colony, often making it appear yellowish-brown. Oil shales, as well as low percentages of some crude oils, are derived from *Botryococcus* hydrocarbons. Thus, *Botryococcus* is a candidate for cultivation systems designed to produce lipids, as well as genomic and proteomic studies focused on lipid biosynthesis. Modern entrepreneurs are developing bioreactor systems for growing high-lipid algal strains in systems that are linked to point sources of CO_2 or wastewater, thereby also reducing the release of harmful materials to the atmosphere or natural waters.

4.4 The Use of Algae in Wastewater Treatment

Decades ago, wastewater effluents such as human sewage and industrial and agricultural wastes were routinely discharged to natural waters for purification. As population densities increased, effluent treatment plants were built, with a primary goal of removing particulates and pathogenic microorganisms so that water could be reused. However, even today, many sewage treatment processes still discharge effluents that are relatively high in phosphate and combined nitrogen, primarily because the incorporation of procedures that more completely remove these compounds is expensive. Agricultural runoff and dairy wastes represent non-point sources of contaminants to natural waters. Such effluents may contribute excess algal nutrients to natural waters, causing undesirable—and sometimes toxic—algal blooms. Algae have been engineered into systems designed to remove nutrients from effluents before they are discharged, thereby helping to prevent eutrophication. Hoffmann (1988) reviewed experimental studies focused on the development of such systems, of two major types: (1) ponds that make use of suspended microalgae and (2) artificial streams containing turfs composed primarily of larger, attached algae.

Sewage treatment facilities often include ponds that utilize algae to remove nutrients (Oswald 1988), and algal ponds have also been designed to treat manure wastes on dairy farms (Craggs et al. 2004).

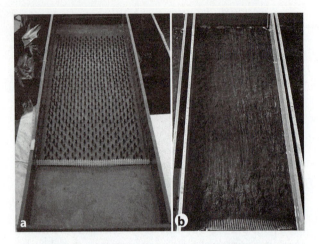

Figure 4.20 An algal turf-scrubber system developed for removal of nutrients from aquatic systems. The support screen is shown without algae (a), and with algae (b). (Photos: J. Hoffmann)

"High-rate ponds" are level and shallow (0.1–0.3 m) to encourage the growth of suspended microalgae, such as the green algae *Scenedesmus* and *Micractinium*. Such ponds can be mechanically mixed by a paddlewheel that keeps the algae from settling. The algae not only oxygenate the water but also take up mineral nutrients such as phosphate and nitrate by incorporating them into cellular biomass, a process known as assimilation. Some materials adsorb to the surfaces of algal cells. In addition, when bicarbonate ions are removed, algal photosynthesis raises the pH of water, a process that helps to precipitate phosphate with cations such as Ca^{2+}, Mg^{2+}, or Al^{3+}. The suspended algae then flow to a sloped pond in which they settle and can be removed and applied to land as a fertilizer. The water that results from such algal treatment is much improved in quality before it is discharged to the natural environment.

In algal turf systems, effluent flows in gently sloped, artificial streams across mats of natural assemblages of filamentous algae and associated biota that grow attached to plastic screening (Figure 4.20). Such mats usually contain the green algae *Cladophora*, *Spirogyra*, *Stigeoclonium*, or *Oedogonium* and/or cyanobacteria such as *Oscillatoria*, together with epiphytic diatoms, bacteria, and fungi (Davis et al. 1990a,b). The algae increase oxygen in the water by 100% to 300%; assimilate N, P, and other minerals; and help to precipitate phosphate. A patented algal turf scrubber system (ATS) was first developed for treating water for aquarium displays at the Smithsonian Institution (Adey et al. 1993). The system was later expanded to a larger-scale outdoor system in Florida to remove nutrients from effluent draining from fertilized sugarcane fields and added to a sewage treatment facility in Patterson, California. The Patterson ATS was 152 m long and more than 6 m wide, constructed on compacted soil covered with an impervious liner. Flow rates varied from more than 100 m^3 to more than 1000 m^3 per day. Harvesting algae from the screening within a week during summer or a month in winter removed nearly 1 g of N m^{-2} day^{-1} and 0.5 g of P m^{-2} day^{-1}. The harvested algae could be dried for use as fertilizer or fermented to ethanol or methanol for use as fuel (Craggs 2001). Kebede-Westhead and associates (2003) used algal turfs to treat dairy manure effluents, thereby removing up to 52% of wastewater N and up to 59% of P.

Nutrient removal/oxygenation systems employing attached algae have a number of advantages: They are odorless, no energy is required for suspending cells, accumulated algal biomass is easily harvested, and harvesting leaves the lower portions of the algal filaments attached to substrates, which allows natural regrowth to occur. Harvesting both stimulates production by removing the shade canopy and keeps grazer populations low so that any pollutants that might be present are not transmitted to higher trophic levels (Craggs et al. 1996). Turf systems that incorporate algae-grazing fish (the cichlid *Tilapia mossambica*) have also been developed (Drenner et al. 1997).

4.5 Algae in Space Research

Algae have played a significant role in space research for several decades and will have future important applications in spacecraft life-support systems and human colonization of other planets. Experiments designed to study the growth of algae in the zero-gravity conditions of space have flown aboard rockets, satellites, the space shuttle, and the *Salyut* and *Mir* space stations. In general, algae grown in space seem to suffer few adverse affects, a fact that has engendered plans for further use of algae in life-support systems. Most life-support research involving algae has focused on their use in bioreactors to remove CO_2 and add O_2 (Wharton et al. 1988). Several ground-based studies have included humans in life-support systems involving algal air-regeneration systems. For example, BIOS-3 consisted of a 315 m³ stainless steel hull containing a hydroponic garden, an algal compartment, and room for three crew members. In this system, the algae effectively processed some wastes, and plants and algae together provided sufficient oxygen (Fogg 1995).

Algae figure prominently in plans for one of the most ambitious space projects conceivable, modifying the climate of Mars to facilitate human colonization. After engineering the atmosphere of Mars to increase warming, experts propose seeding the planet with pioneering microorganisms that have adapted to the stresses of Antarctica or desert habitats. The cyanobacterium *Chroococcidiopsis*, for example, does not grow in mild environments but thrives under extremely arid, high- or low-temperature terrestrial conditions (Friedmann and Ocampo-Friedmann 1995). Because they can engender and enrich soil, natural or genetically engineered terrestrial algae of diverse types (Chapter 23) could be used in terraforming to pave the way for higher plants and animals (Graham 2004).

4.6 Genetic Engineering of Algae

Algae are currently being genetically engineered with the goal of improving their technological applications. These include hydrogen and biodiesel production for energy, pharmaceuticals and food additives, and nanotechnology. DNA sequences can be introduced into algal cells to modify biochemical pathways, either by changing the expression of existing genes or by adding genes that yield new or different products. Patents and algal biotechnology companies have resulted from such technologies. The ability to transform algal cells with foreign genes has been made possible by the development of procedures for incorporating foreign DNA into algal cells, promoter systems allowing gene expression, and reporter genes, which allow a genetic engineer to detect transformed cells.

DNA transformation techniques that have been used with algae include bombardment of cells with tiny DNA-coated particles fired by a biolistic device (gene gun), microinjection of DNA into cells via fine glass needles, electroporation (the use of electrical change to temporarily open pores in the cell membrane), and agitation of cells with DNA and glass beads or silicon carbide whiskers (Stevens and Purton 1997). Recombinant viruses and plasmid vectors represent two additional means of transferring DNA. Promotors such as CaMV (cauliflower mosaic virus) 35S and SV (simian virus) 40 and algal promoters such as those of the *aphVIII* gene (Hallmann and Wodniok 2006) have been used in algal genetic engineering. A number of reporter/marker genes have been used in algae; these include genes conferring antibiotic resistance, herbicide resistance, and other detectable traits (Walker et al. 2005).

Examples of algae for which genetic transformation has been demonstrated include the cyanobacteria *Synechococcus* and *Synechocystis*; the green algae *Chlamydomonas*, *Volvox*, *Dunaliella*, and *Chlorella*; the diatoms *Cyclotella cryptica*, *Navicula saprophila* (Dunahay et al. 1995), and *Phaeodactylum tricornutum* (Zaslavskaia et al. 2000, 2001); and the brown alga *Saccharina japonica* (Qin et al. 2005). The green alga *Haematococcus pluvialis* is a species that has been genetically modified for greater economic utility. From this alga Steinbrenner and Sandmann (2006) isolated the gene coding for phytoene desaturase, an enzyme in the carotenoid biosynthesis pathway. They mutated the gene in such a way that the enzyme retained activity but was 43 times more resistant than normal to the herbicide norflurazon. The modified gene could thus serve as a selective marker. Then they coated tungsten pellets with the modified DNA and used them to bombard *H. pluvialis* cells. Transformed cells could be identified by their ability to grow on nutrient media that contained norflurazon, and some of the engineered cells produced higher-than-normal amounts of astaxanthin in response to light stress.

Chapter 5

Algal Diversity and Relationships

Ecologists worldwide are greatly concerned about changes in species diversity arising from global climate change, nutrient pollution of aquatic ecosystems, and other large-scale human disturbances (Chapin et al. 2000). Change in algal species diversity is a concern because many algal species play essential roles in global biogeochemistry, biotic associations, and human technology (see Chapters 2–4). Learning more about these roles and understanding how algal communities respond to environmental change are thus important challenges for phycologists. Another major goal is cataloging species diversity, an activity that has led to the discovery of many new algal species. Many phycologists study the pattern and process of algal evolution by evaluating degrees of relationship among species and their genomes. Measuring levels of genetic variation within algal populations and detecting toxic algal species in natural waters are also active fields of investigation. This chapter provides an overview of algal taxonomy and systematics— the ways in which algal species are identified, named, and grouped by relationship—with an emphasis on the use of molecular methods.

5.1 The Diversity and Importance of Algal Species

As noted in Chapter 1, the cyanobacteria are a lineage of eubacteria, and the eukaryotic algal phyla are distributed among several eukaryotic supergroups (Figure 5.1). Several algal phyla are particularly rich in species. The cyanobacteria include about 2000 recognized species, there are at least 900 species of euglenoids, red algal species number 6000 or more, green algae comprise an estimated 17,000 species, and some 4000 species of dinoflagellates occur. Together, the photosynthetic phyla of the stramenopiles supergroup number more than 15,000 species. There are also about 300 species of haptophytes, 200 cryptomonads, and 13 or so glaucophyte species. Estimates of the total number of algal species range from 36,000 (John and Maggs 1997) to more than 10 million (Norton et al. 1996). New species of algae are continually being found, and they are typically named and described according to rules established in the International Code of Botanical Nomenclature.

The correct determination of algal species is important in obtaining patents that describe new biotechnological applications and in ecological studies. For example, some algal species function as "keystone" taxa whose presence in a community is unusually influential, affecting a wide variety of other species (Andersen 1992). Examples include species of the dinoflagellate *Symbiodinium* that are symbionts of many corals, and the red alga *Porolithon onkodes*, which stabilizes coral reef ridges against erosion by waves. When used as bioindicators (see Chapter 4), diatom species determinations are necessary to reveal differences in water quality (Stoermer 1978). There are many other situations in which it is crucial

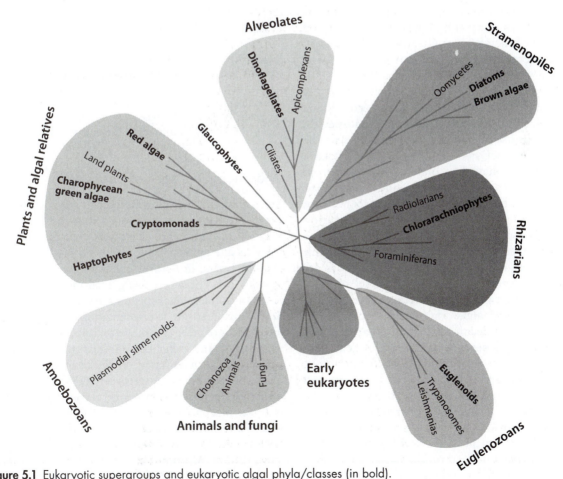

Figure 5.1 Eukaryotic supergroups and eukaryotic algal phyla/classes (in bold).

to recognize distinct algal species, but methods for doing so vary because phycologists apply different species concepts (Manhart and McCourt 1992).

Species Concepts

Three main species concepts are commonly applied to the algae: biological, morphological, and phylogenetic. Each offers particular benefits and limitations.

Biological species concept

The **biological species concept**, as defined by the noted evolutionary biologist Ernst Mayr, uses interbreeding potential as a criterion. This concept is widely used to define animal species. If two organisms are actually or potentially able to interbreed and produce viable offspring, they are classified within the same species (Figure 5.2). If interbreeding does not occur because hybrids are infertile or populations are separated by other boundaries that prevent interbreeding, they are regarded as separate species. John and Maggs (1997) described some cases of successful application of the biological species concept to algae. Interbreeding experiments with algae can be difficult to perform, for several reasons. First, such experiments normally require **unialgal cultures**—laboratory containers that contain a population of only one algal species. Cultures for use in mating experiments may be available in collections (see Chapter 4). If they are not, researchers have to isolate populations from natural mixtures and grow them under controlled conditions in the laboratory, in order to perform mating trials. Although isolation and cultivation methods are available for many types of algae (Stein 1973; Andersen 2005), it can be difficult to cultivate algae whose growth requirements are incompletely known. In addition, the set of environmental conditions that induces gamete production varies among algae and often is not known. Consequently, molecular techniques are widely used to infer the presence or absence of gene flow within and among algal populations (Section 5.2). Finally, sexual reproduction has not been detected in some algae, and sex seems to occur relatively rarely in others; such species primarily reproduce by asexual means. As a result of such challenges in applying the biological species concept, many algal species have been defined on the basis of morphological (structural) differences.

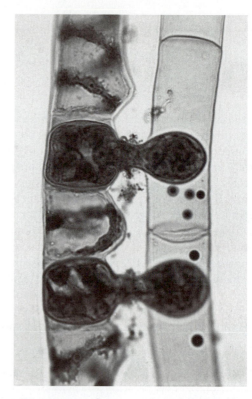

Figure 5.2 Mating in the green alga *Spirogyra*. The entire contents of the cells behave as gametes. Those of the right-hand filament are moving to and fusing with the contents of cells of the filament on the left, forming zygotes. (Photo: L. W. Wilcox; specimen courtesy Illinois State University Department of Biological Sciences)

Morphological species concept

According to the **morphological species concept**, species are the smallest groups that can be reliably defined by structural characters that are relatively easy to distinguish. Ecologists who study algal species occurring in natural habitats often rely on this species concept when using identification keys. In such keys, microalgal species are often distinguished using such features as pigmentation, number and arrangement of cells, cell shape and size, flagellar characteristics, and other cellular features. Macroalgal species are differentiated on the basis of branching patterns, reproductive features, and size, as well as other characters. Examples of taxonomic keys that rely on morphological characteristics are *Algae of the Western Great Lakes Region* (Prescott 1951), *The Freshwater Algal Flora of the British Isles* (John et al. 2002), *Marine Algae of the Eastern Tropical and Subtropical Coasts of the Americas* (Taylor

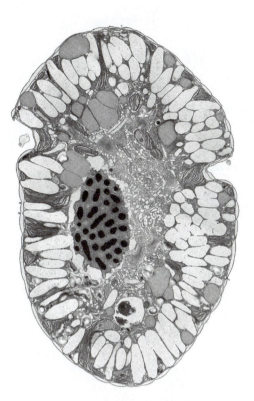

Figure 5.3 TEM of the freshwater dinoflagellate *Amphidinium cryophilum.* Details of cell ultrastructure were used in the description of this species. (Micrograph: G. J. Wedemayer and L. W. Wilcox)

1972), and *Keys to the Seaweeds and Seagrasses of Southeast Alaska, British Columbia, Washington, and Oregon* (Gabrielson et al. 2006).

Until recently, most new species of algae were detected and described using the morphological species concept. However, some microalgae may have few discernible structural differences when viewed with a light microscope (LM). In addition, different lineages of algae have often adapted in similar ways to analogous habitats. As a result, there are many examples of algal species that look very similar at the LM level but are actually not closely related. Such species are examples of parallel or convergent evolution. Visualizing structural characters that distinguish species may require specialized instruments and preparation methods. For example, scanning electron microscopy (SEM) was essential in describing a new species of sand-dwelling dinoflagellate (Murray et al. 2006). Transmission electron microscopy (TEM) was used to determine that cultures of marine microalgae represented a new chlorarachniophyte genus and

species—*Bigelowiella natans* (Moestrup and Sengco 2001). TEM was also needed to describe a new species of the dinoflagellate genus *Amphidinium* (*A. cryophilum*) (Wilcox et al. 1982) (Figure 5.3) and the flagellate stramenopile *Haramonas* (*H. viridis*) (Horiguchi and Hoppenrath 2003). Yet another challenge to the morphological species concept is that structurally distinguishable species and genera can produce fertile hybrids, a process illustrated by giant kelps (Lewis and Neushul 1995). In consequence, many phycologists use molecular characters as well as structural features and rely on the phylogenetic species concept for taxonomic descriptions of algal species.

Phylogenetic species concept

The **phylogenetic species concept** is that a species is the smallest group of organisms that exhibits at least one distinctive and unifying structural, biochemical, or molecular character. Such features are known as shared, derived characters, or **synapomorphies**. Computer software is widely used to compare the characters of organisms and model the most likely patterns of relationship, depicted as branching diagrams known as **phylogenetic trees** (Figure 5.4). In the phylogenetic species concept, individual species occupy the tips of **branches** of such trees, with hypothetical common ancestors occurring at branch points known as **nodes**. Phylogenetic tree diagrams reflecting recent concepts of algal relationships occur throughout this book, reflecting the importance and many applications of the phylogenetic species concept. However, it is important to realize that relationship concepts commonly change when new information or new analytical methods become available. This explains why phylogenetic diagrams shown in this text may differ in various ways from those of older references, as well as newer ones.

Today, new algal species are commonly defined by means of the phylogenetic species concept, often using a combination of morphological and molecular characters. For example, the unicellular green flagellate *Hemiflagellochloris kazakhstanica* was defined on the basis of LM, TEM, and molecular characteristics (Watanabe et al. 2006b). A macroalgal example is the recently described brown seaweed *Newhousia imbricata*, defined by closely overlapping blades that are cemented together with calcium carbonate (Kraft et al. 2004). SEM was useful in understanding *N. imbricata*'s distinctive structure, and molecular sequence data revealed its relationships. *Newhousia* was discovered on corals and wood pilings outside Hanauma Bay on

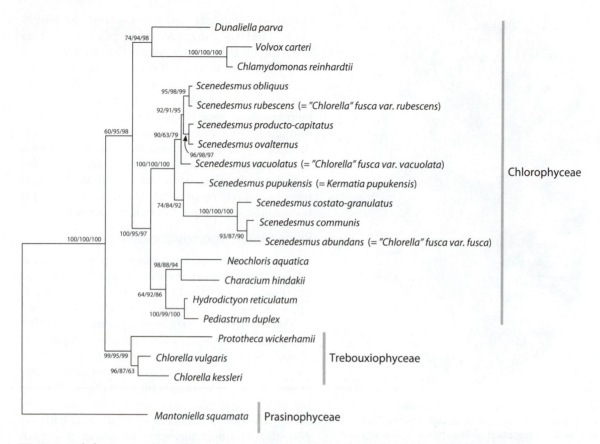

Figure 5.4 Phylogenetic relationships of some green algae, based on SSU rRNA sequence data. Separate bootstrap values are given for distance, parsimony, and maximum likelihood analyses (in that order) at each branch. (After *Botanica Acta* 110 [1997]:244-250. Kessler, E., M. Schäfer, C. Hümmer, A. Kloboucek, and V. A. R. Huss. Physiological, biochemical, and molecular characters for the taxonomy of the subgenera of *Scenedesmus* [Chlorococcales, Chlorophyta]. Georg Thieme Verlag, Stuttgart)

the Hawaiian Island of O'ahu, and *Hemiflagellochloris* was isolated from irrigated, saline soil in Kazakhstan. These examples indicate that algal species remain to be discovered in both common and unusual habitats.

Diverse phylogenetic and molecular approaches are widely used to identify and classify algae, explore their diversity in nature, and detect toxic species. The next section focuses on these methods.

5.2 Algal Phylogeny

The goal of phylogeny reconstruction is to understand patterns and processes of evolution—that is, to explain the occurrence of particular characteristics in specific groups of organisms. In the absence of an extensive fossil record, which is the case for

most algal groups, evolutionary history must be inferred from the characters of modern organisms. The characteristics used in such studies are typically variations in molecular sequences; biochemical pathways; or the structure of organisms, cells, or genomes. Because phylogeny has many useful applications, it is important for phycologists to have a general understanding of the major methods used in phylogeny reconstruction and their underlying assumptions. For example, most phylogenetic methods are applied under the assumption that the characters used in the analysis provide independent information. If an analysis violates this assumption, the results are questionable. Character independence is a major issue in phylogenetic analyses conducted with all types of characters, including molecular ones.

Another important goal of phylogenetic work is the application of appropriate methods for assessing the reliability of phylogenetic trees and their branches. This is because each branch of a phylogenetic tree represents a **clade**, a group of organisms descended from a single common ancestor (see Figure 5.4). Nested clades often form the basis of hierarchical taxonomic systems. For example, a phylum is ideally a clade that contains one or more smaller clades known as classes, with each encompassing one or more even smaller clades known as families, and so on. Thus, it is important that clades be well supported by statistical or other analytical measures. If the branches of a phylogenetic tree are well supported, the phylogeny of the group under study is considered to be well resolved. If two or more branches of a phylogenetic tree are not well supported, those particular clades are not well resolved and require additional study.

Many phylogenetic studies of algae aim to understand the branching structure of an evolutionary tree of life that includes all of Earth's organisms. However, microbiologists have noticed that the evolutionary patterns of prokaryotes and protists often seem more like networks than trees. Here, we first consider the application of a tree of life concept to algae and then survey major methods used to discern evolutionary patterns at different taxonomic levels.

Algae and the Tree of Life Concept

Charles Darwin (1859) first suggested that the process of organismal evolution, by descent with modification, yields a pattern much like the forked branching of a tree. His tree of life concept is valuable because a correctly resolved tree should depict the evolutionary history of Earth's life forms. Also, as earlier noted, the branching pattern of a tree of life provides a logical basis for organism classification; the larger branches represent domains and supergroups, while phyla, classes, orders, families, genera, and species form finer branches and twigs.

Today, many evolutionary biologists think of the tree of life as a biological fact and therefore aim to define the tree's architecture as precisely as possible (e.g., Keeling et al. 2005). In contrast, other evolutionary biologists suggest that the tree of life is a conceptual model that explains many but not all aspects of life's evolutionary history. This alternative view comes from emerging evidence for substantial effects of horizontal gene transfer (HGT)—also known as lateral gene transfer (LGT)—and hybridization on evolutionary pattern (Doolittle and Bapteste 2007). Horizontal gene transfer is the movement of genes between different species. Defined broadly, HGT includes endosymbiosis, a process that has been and still is a major force in algal evolution (Chapter 7). During endosymbiosis, organisms of distinct type, known as endosymbionts, become incorporated into the cells of a different species, known as the host. After a period of time, the endosymbiont cells, organelles, or genomic components become integrated into the host, forming a genetic chimera. HGT by endosymbiosis and other means is thus a way in which organisms can acquire genes in addition to vertical inheritance from parent to progeny. Horizontal transmission of genes is increasingly observed among eukaryotes, including the algae (Raymond and Blankenship 2003). Hence, it is useful to take a closer look at the mechanisms and roles of HGT in cyanobacterial and eukaryotic algal evolution.

HGT in cyanobacteria

Cyanobacteria are vulnerable to attack by viruses known as cyanophages (Figure 5.5). Such viruses typically inject their genetic material into host cells, where the viruses replicate, eventually causing host cells to break open, releasing many new viruses to the environment. During their reproduction, viruses can incorporate cyanobacterial genes and later transfer them to other cyanobacteria, in the process known as

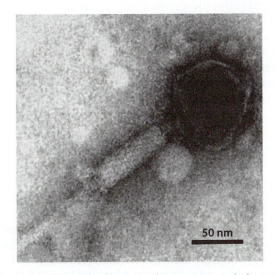

Figure 5.5 TEM view of a cyanophage. (From Yoshida et al. 2006b. Reprinted by permission of the American Society for Microbiology)

transduction. For example, *psbA*, a gene that encodes the photosystem II core reaction center protein D1, is known to move among cyanobacterial cells by means of viruses. During their stay in viral genomes, such genes may evolve and then return to host cells in modified form, a process that increases host genetic diversity (Lindell et al. 2004). Viruses are also known to transfer clusters of genes, known as genomic islands, into cyanobacteria. The acquisition of such gene clusters allows the marine cyanobacterium *Prochlorococcus* to adapt to high or low light or low nutrient conditions (Coleman et al. 2006; Martiny et al. 2006). HGT helps explain why *Prochlorococcus* occurs so widely in oceans. As a result of HGT, *Prochlorococcus* occurs in genetically and physiologically distinct strains, known as **ecotypes**, each of which is able to grow well in particular conditions.

HGT in eukaryotic algae

Eukaryotic algae receive genes horizontally by two major mechanisms: viruses and phagotrophic ingestion. Viruses have been reported to occur in more than 40 eukaryotic algal species of diverse types (see Figure 3.19 for an example). Some of the viruses that attack eukaryotic algae function as vectors in horizontal gene transfer. One example is *Coccolithovirus*, a large viral pathogen of the widespread marine haptophyte *Emiliania huxleyi* (Wilson et al. 2005). Similar large viruses are known to infect brown and green algae (Van Etten and Meints 1999).

Eukaryotic cells that feed on other cells, known as **phagotrophs**, can acquire new genes from their prey. **Phagotrophy,** the ingestion of particles by cellular endocytosis (see Figure 3.4), is the mechanism by which early eukaryotes obtained prokaryotic and eukaryotic endosymbionts that evolved into modern mitochondria and plastids. Endosymbiosis has been a major mechanism of horizontal gene transmission in the algae because it is so common and transfers many genes at the same time. In addition, many modern eukaryotic algae retain phagotrophic capacity, even though they possess plastids, and are hence known as **mixotrophs.** Mixotrophic cells can retain some of the DNA released by prey cells during digestion. For example, the mixotrophic chlorarachniophyte *Bigelowiella natans* has acquired several genes that encode plastid proteins from other organisms, presumably its prey (Archibald et al. 2003). In this case, the new genes affect chlorophyll biosynthesis, carbon fixation, and ribosome structure, thereby having substantial effects on the host's genome and metabolism. Additional examples of HGT that may have occurred via phagotrophy include the acquisition of a gene encoding ∂-aminolaevulinic acid dehydratase by dinoflagellates from green algae (Hackett et al. 2004). Waller and associates (2006) inferred that the dinoflagellate *Heterocapsa triquetra* obtained a gene encoding the oxoglutarate/malate translocator by HGT from green algae and one for acetolactate synthase from a proteobacterium.

In a number of cases, genes acquired by HGT have replaced vertically inherited genes having the same general function. For example, gene sequence comparisons show that the rubisco genes of red plastids likely originated from a proteobacterium rather than a cyanobacterium, as expected. This observation suggests that proteobacterial genes entered the common ancestor of red algae and replaced ancestral cyanobacterial rubisco genes (Delwiche and Palmer 1996). Similarly, a bacterial gene (*rpl36*) obtained by HGT has replaced the eukaryotic form of this gene in cryptomonad and haptophyte plastids (Rice and Palmer 2006). Because all algal lineages have been affected by endosymbiosis, and many modern algal species are phagotrophic and experience viral transduction, it could be argued that algae and related protists have been and still are affected by HGT more than any other group of eukaryotic organisms. As noted earlier, the algae also display cases of hybridization. Therefore, branching trees, though extremely useful, are not completely adequate as models of algal evolution.

Molecular phylogenetic approaches

Over time it has become clear that phylogenies usually provide better resolution of species relationships when larger numbers of characters and taxa are used in analyses. Molecular approaches are commonly used in modern phylogenetic studies because they provide many characters, and molecular techniques used in phytoplankton research have been reviewed (de Bruin et al. 2003). Here, we survey major molecular phylogenetic approaches used to study both microalgae and macroalgae, providing examples. Procedures used to evaluate diversity within and among species are considered first. These are followed by methods used to infer diversification at higher taxonomic levels such as phyla.

RFLPs

Restriction fragment length polymorphisms (RFLPs) estimate DNA sequence divergence by detecting

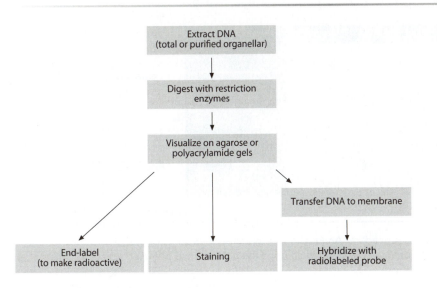

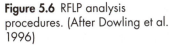

Figure 5.6 RFLP analysis procedures. (After Dowling et al. 1996)

variation resulting from nucleotide base substitutions, deletions, or insertions across whole genomes or defined parts of them. RFLP analyses have been used to link structurally different life phases of the red alga *Gymnogongrus* (Parsons et al. 1990) and to evaluate species and population variation in algae. DNA is extracted from different populations of a species or closely related species, and then the DNA is digested with a battery of restriction enzymes. Each enzyme cuts the DNA at a particular sequence (the restriction site) of four or more nucleotides. The variously sized DNA fragments that result are separated by electrophoresis on agarose gels and stained with a dye that allows DNA to be visualized as patterns of bands. When a restriction site has been altered by nucleotide substitutions or when insertions or deletions occur between restriction sites, different DNA fragments are produced, and a different band pattern results (Figure 5.6). It is important to establish that similarly sized fragments from different organisms represent homologous DNA sequences. To establish homology, bands can be identified with the use of DNA probes in the process of Southern hybridization. Alternatively, RFLP fragments can be cloned and sequenced. PCR (polymerase chain reaction), a technique for making many copies of a particular DNA sequence, can be incorporated to reduce the amount of labor and DNA required. PCR reactions are performed in a thermocycler device, which changes reaction temperature in a cyclical manner (Figure 5.7). The gene amplification products are separated and visualized as bands by performing electrophoresis in agarose gels. In PCR-RFLP, PCR is first used to amplify a particular DNA sequence, which is then cut with restriction enzymes and visualized on gels.

RAPDs

In addition to RFLPs, randomly amplified polymorphic DNAs (RAPDs) provide another method for analyzing variation within and between species. The use of randomly chosen primers in PCR produces many copies of different DNA sequences that vary in size. When visualized on gels, these DNA products form a banding pattern that is characteristic for each organism. A high degree of similarity in banding patterns indicates close relationship, whereas dissimilarity suggests more distant relationships. Advantages of RAPDs include the need for only small amounts of DNA and their relative ease, speed, and low cost. Disadvantages include repeatability issues and concerns about band homology unless sequencing is also done. RAPDs were used to detect geographically distinct populations of the diatom *Fragilaria capucina* sampled on a latitudinal gradient across North America (Lewis et al. 1997b). Barreiro et al. (2006) used RAPDs to distinguish two morphologically similar red algal species occurring on the Iberian peninsula, non-invasive *Grateloupia lanceola* and invasive *G. turuturu* (Figure 5.8). RAPDs have also been effective in estimating genetic variation within populations of the red alga *Gelidium canariense* (Bouza et al. 2006). Such measures can provide information about the extent of asexual versus sexual reproduction, and degree of population inbreeding (Faugeron et al. 2001, 2004). AFLPs and microsatellites/SSRs,

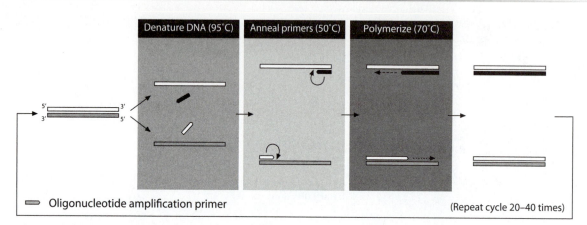

Figure 5.7 Polymerase chain reaction (PCR) procedure. Annealing temperatures, number of cycles, and times the sample spends at each temperature vary depending upon the nature of the primers and target DNA and instrumentation.

described next, provide similar, but more consistent data.

AFLPs

Like RFLPs, amplified fragment length polymorphisms (AFLPs) represent genomic characters in the form of DNA fragments of differing sizes. To generate AFLPs, genomic DNA is first treated with restriction enzymes; then PCR is used to amplify some of the resulting fragments. Products are viewed on gels or separated by electrophoresis using automated sequencing instruments. Though reproducibility of this technique requires high-quality DNA, and band homology can be an issue, AFLPs are useful in evaluating relationships among species and within populations. For example, Müller and colleagues (2005) used AFLPs to examine 29 strains of *Chlorella vulgaris* obtained from 5 culture collections. They found 11 distinct genotypes, which clustered into 5 species-level groups. Such hidden diversity is known as **cryptic speciation**. Molecular tools have been instrumental in revealing the occurrence of such genetically distinct yet morphologically indistinguishable **cryptic species**. In the case of *C. vulgaris*, the identification of cryptic species is important because this organism has been so widely used in experimental studies. It is possible that past investigators, aiming to use the same species of *Chlorella* for different studies, inadvertently used different cryptic species. This is an important reason to include information about culture collections and strain numbers in research publications.

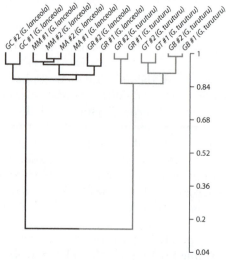

Figure 5.8 Cluster analysis of RAPD data that distinguishes representatives of *Grateloupia lanceola* from *G. turuturu*. (Data from Barriero et al. 2006)

Microsatellite DNA/simple sequence repeats (SSRs)

Microsatellites are DNA segments that consist of repeated sequences of one to six nucleotides. The dinucleotide CA repeated 10 times (CACACACACA-CACACACACA) and GT repeated 18 times are examples. Such repeats result from polymerase slipping during replication. Microsatellites are also known as simple sequence repeats (SSRs or ISSRs). SSRs are isolated by DNA digestion, cloned, and sequenced. The resulting sequences are used to develop microsatellite-specific PCR primers, and the fragments are then amplified and separated on gels. Expressed sequence tags (ESTs) (see Chapter 4) were used to generate SSRs for the green algal species *Chlamydomonas reinhardtii* and *Mesostigma viride* (von Stackelberg et al. 2006). This process has an important advantage: SSR genomic locations are known.

Generating microsatellites is laborious and expensive when first applied to a species, and cross-species applications are limited. However, once developed, SSRs can be very useful in ecological studies. For example, Nagai and colleagues (2006) identified 13 variable microsatellites from the toxin-producing stramenopile *Heterosigma akashiwo*. These markers can be used to understand the genetic structure and extent of gene flow among *H. akashiwo* populations, enabling better understanding of harmful algal blooms. SSRs and their flanking sequences were used to define five phylogenetic types (phylotypes), possible species-level taxa within the B1/B184 clade of *Symbiodinium*, the dinoflagellates most commonly present as endosymbionts in Caribbean corals (Santos et al. 2004). This study also revealed that different coral species select specific endosymbiont phylotypes. Rynearson and Armbrust (2005) isolated more than 600 cells and established clonal cultures of the diatom *Ditylum brightwellii* from an ocean bloom. Microsatellite studies revealed a high level of genetic variation among the clones. Provan et al. (2005) used SSRs from the plastid genome to trace the evolutionary history of the invasive green seaweed *Codium fragile* subspecies *tomentosoides*. Native to Japan, one genetic variant invaded the Caribbean, whereas a different genotype invaded coastal regions of the northwest Atlantic, northern Europe, and the South Pacific. Microsatellite sequence data also distinguished the brown macroalga *Fucus spiralis* from its close relative *F. vesiculosus* (Engel et al. 2005) and revealed the first-discovered occurrence of asexual reproduction in the brown algae (Tatarenkov et al. 2005). *F. vesiculosus* normally reproduces by sexual means, but dwarf populations found in the Baltic Sea reproduce only asexually, likely because low salinity retards sexual reproduction. The dwarf *F. vesiculosus* populations produce branches that break off, attach to surfaces, and grow into clonal populations (Figure 5.9).

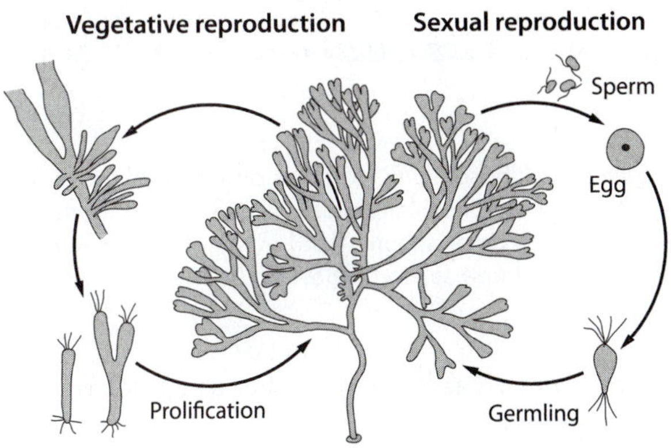

Molecular markers for phylogeny reconstruction

DNA coding sequences and non-coding spacer regions are widely used in reconstructing phylogenies. DNA can be extracted from living cells or from old herbarium specimens (Hughey et al. 2001). Whole

Figure 5.9 Asexual and sexual reproduction by *Fucus vesiculosus*. (After Tatarenkov et al. 2005, by permission of the *Journal of Phycology*)

genome sequences for many cyanobacteria and increasing numbers of eukaryotic algal species are available in databases (see Chapter 4). In addition, PCR can be used to amplify particular DNA sequences from other species inexpensively and rapidly. Some genes or regions within them display variation sufficient to resolve relationships within and among species and genera. More conservative sequences of other genes or regions are useful in understanding relationships between families, orders, and phyla. However, some genes or parts of them can be inappropriate for use in phylogenetic analyses. This is the case for several reasons.

If selection has operated similarly on different regions of the encoded protein or RNA, the corresponding DNA sequences may not have changed independently. This may occur when intramolecular associations are involved in molecular folding to form a functional structure, a property of nearly all molecular structures. In the functional large subunit of rubisco, for example, more than 30% of the 476 amino acids are physically associated (Kellogg and Juliano 1997), so the DNA bases that encode these associated amino acids do not represent independent characters. Some genes of unrelated organisms may display similar sequences because of a high degree of parallel functional constraint on the proteins they encode. Cryptic covariation can cause organisms that are not actually closely related to each other to branch together in a phylogenetic tree. Many existing phylogenies have been constructed with the use of gene sequences likely to include covariant regions (Stiller and Harrell 2005). In consequence, phylogenetic trees that are based on gene sequence data

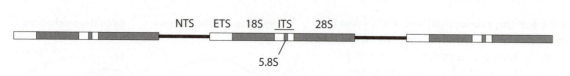

Figure 5.10 Ribosomal RNA genes and internal transcribed spacer regions whose sequences are widely used in molecular systematics. NTS = nontranscribed spacer, ETS = external transcribed spacer, 18S = small subunit (18S) rRNA gene, ITS = internal transcribed spacer, 5.8S = 5.8S rRNA gene, 28S = large subunit (28S) rRNA gene.

are best viewed as hypotheses that can guide further investigation.

Genes and sequences to be used in studies of algal relationships should also occur as single copies rather than as members of gene families. This avoids the unintended comparison of different members of the same gene family, which are known as paralogous genes. Because the paralogous gene sequences of a single species often differ to some degree, their use in phylogenetic analyses could yield misleading results. In addition, genes having extremely divergent sequences are usually avoided. Such genes, which occur at the tips of very long branches in phylogenies, are described as long branched. Two or more long-branching gene sequences can accidentally accumulate similarities that cause them to cluster together in a phylogenetic analysis when they don't actually have common ancestry. This phenomenon, known as long-branch attraction, can also generate misleading phylogenies.

DNA sequences that encode ribsosomal RNA and intergenic spacer regions (Figure 5.10) offer opportunities to examine species-level questions as well as deeper divergences, depending on the particular sequence region investigated. For example, the portions of the nuclear-encoded ribosomal RNA genes known as the internal transcribed spacers (ITS) evolve quickly and thus are particularly useful in species-level studies. Marks and Cummings (1996) used ITS data to demonstrate low genetic diversification among freshwater isolates of the common and widespread green alga *Cladophora* obtained from a wide range of habitats and geographical locations. Peters et al. (1997) used ITS sequences to compare species within the widespread brown algal family Desmarestiaceae, and the phylogeny of the brown genus *Sargassum* was explored with the use of ITS (Stiger et al. 2003). The ITS region has also been used to detect cryptic speciation in the terrestrial cyanobacterium *Microcoleus steenstrupii* (Boyer et al. 2002), the freshwater dinoflagellate *Peridinium*

limbatum (Kim et al. 2004a), the terrestrial green alga *Scenedesmus obliquus* (Lewis and Flechtner 2004), and aquatic *Scenedesmus* and *Desmodesmus* (Johnson et al. 2007a).

In contrast, DNA sequences encoding RNA of the small subunit of cytoplasmic ribosomes (SSU rDNA) (16S rDNA in prokaryotes; 18S rDNA in eukaryotes) is particularly useful in evaluating relationships among higher taxa, including algal classes and phyla, and also in biodiversity studies. For example, in a study of 18S rDNA sequences of green algae cultured from the northern Midwestern United States, Fawley et al. (2004) found 89 organisms new to science that had not previously been recognized on the basis of morphological features.

For evaluation of deep divergences among algal phyla and associated supergroups (see Figure 5.1), recent studies emphasize the value of multigene analyses. Protein-coding genes having the appropriate amount of variation are used in addition to rRNA genes. Because the algae encompass a wide taxonomic range, including prokaryotes and diverse protist groups, broad taxon sampling is also needed. However, as the number of genes and species used in an analysis increases, computation also becomes more intensive. Genomic structural characters can also be useful and include intron insertions or deletions, gene fusions, whole genome duplications, and movement of genes from one cellular compartment to another. Such variations in genomic architecture can be detected through comparative analysis of whole genome sequences and other methods.

Generating and Evaluating Phylogenetic Trees

A basic understanding of how phylogenetic trees are generated from gene sequences and other data and how such trees can be evaluated allows one to determine whether the phylogenetic conclusions made in a publication are well supported. The most commonly employed tree-generating methods are distance methods, maximum parsimony (MP), maximum likelihood (ML), and Bayesian analysis. These operations are performed with the use of a database of character information and computer software. Many studies utilize more than one of these methods, as each has advantages and limitations (Semple and Steel 2003; Felsenstein 2004).

In distance methods, the number of differences between all pairs of taxa are determined, and then these numbers are used to group taxa into a tree that attempts to accommodate all the pairwise distances. Parsimony methods operate under the assumption (which is not always correct) that evolution operates in the most efficient (parsimonious) manner. In other words, the phylogeny requiring the fewest number of changes is viewed as the most accurate one. A drawback of parsimony methods is that they do not consider sites that are invariable or those with unique changes. As a result, they can be misled by multiple superimposed changes that have occurred in ancient lineages, yet archaic protists belong to such long lineages. Another criticism of parsimony methods is that a variety of weighting schemes can be applied to the data, and the one chosen may affect the structure of the resulting phylogeny.

The application of ML methods begins with construction of a model of the evolutionary process (a tree and nucleotide substitution probabilities) that hypothetically resulted in the conversion of one sequence into another. The tree/model with the highest likelihood of yielding the sequences that are actually observed is considered the ML estimate of the group's phylogeny. An advantage of the approach is that estimation of branch lengths is an important part of the process. In contrast to parsimony methods, ML methods attempt to account for unobserved base substitutions. Although demanding in terms of computational effort, ML provides a probability estimate, i.e., a measure of the statistical significance of the phylogenetic hypothesis (Huelsenbeck and Rannala 1997). Bayesian analysis, like ML, is based on likelihood, and it uses a well-established and computationally efficient statistical process, the Markov chain Monte Carlo method (MCMC), to estimate the probability of phylogenetic trees (Huelsenbeck et al. 2002).

The number of possible evolutionary trees generated by typical analyses can be very great, and it increases with the number of taxa examined. For example, there are 2 million possible bifurcating trees linking a group of 10 taxa; for 50 taxa, the number of possible trees is 3×10^{74}. There are two basic approaches to identifying optimal trees. Every possible tree could be evaluated in an **exhaustive search**, but this is computationally demanding for trees involving 12 or more taxa. For studies involving larger numbers of taxa, **heuristic searches** (trial-and-error approaches) are more efficient; these sacrifice certainty regarding the optimal tree for a reduced computing effort.

The **bootstrap** value is a way to express the relative uncertainty of branch points on a phylogenetic tree (Felsenstein 1985, 1988). This term is derived from the colloquial phrase "to pull oneself up by the bootstrap." It is calculated by simulating the collection of replicate data sets through repeated resampling of the data and determining the number of times a particular branch appears. Branches having 90% to 100% bootstrap support are considered well supported, while values lower than 50 reflect low support. It should be noted that although bootstrap values are widely used in both phylogenetics and ecology, the statistical meaning of this procedure remains unclear. The phylogeny shown in Figure 5.4 illustrates separate bootstrap values for distance, MP, and ML analyses. Bayesian methods generate a phylogenetic uncertainty estimate known as the posterior probability (PP). The PP is the probability that a clade is accurate, given a model of the evolutionary process, prior assumptions, and the data. However, in analyses of the same data, the PP is commonly higher than MP and ML bootstrap values. Thus some biologists contend that the PP can overestimate branch reliability.

Commonly, a **consistency index** (CI) is calculated for phylogenetic trees; this is an estimate of the amount of **homoplasy** (parallel or convergent evolution) exhibited by the characters used to construct the tree recovered by parsimony. This quantity will be equal to 1 if there is no homoplasy; the lower the CI, the higher the level of homoplasy. Phylogenetic trees that have been inferred by parsimony methods may sometimes include **decay values** as indices of the degree of support for particular branches. These values indicate the number of additional evolutionary changes that would have to occur before a monophyletic group becomes subsumed into a larger set of taxa (i.e., the branch "collapses") (Bremer 1988). The larger the decay value, the more confidence is placed upon the branch.

5.3 The Application of Phylogeny

Phylogenetics, and molecular phylogenetics in particular, have many applications. These include studies of evolutionary process, taxonomy, and ecology.

Evolutionary Process

Structural, ecological, physiological, or biochemical traits that were not used in constructing a phylogeny can be mapped onto well-resolved phylogenies, thereby revealing clade-specific attributes. Such information can suggest how traits have changed through time. For example, mapping biochemical and reproductive traits onto a molecular phylogeny of the brown seaweed group Desmarestiaceae revealed that sulfuric acid–containing vacuoles appeared at the origin of a single clade (Peters et al. 1997). In contrast, dioecy—the production of male and female gametes on separate thalli—evolved at least twice, with losses of this trait also occurring more than once (Figure 5.11). Molecular phylogenetics can also aid in estimating the time of origin and diversification of groups (Britton et al. 2007). For example, SSU rDNA sequences have been used to infer the dates by which several lineages of dinoflagellates had originated (John et al. 2003).

Application in Classification

Classifications can be constructed to better reflect evolutionary process. For example, **monophyletic** groups are defined by phylogeny to include an ancestor and all of its descendants. In contrast, many taxonomic groups are inconsistent with phylogeny because they were established on the basis of morphological traits. **Paraphyletic** groups do not include all of the descendants of a common ancestor, and **polyphyletic** groups include some members that are more closely related to taxa outside the group (Figure 5.12). Phylogenetic analysis can identify monophyletic, polyphyletic, or paraphyletic groups. For example, molecular phylogenetic analysis indicated that the widespread green algal genus *Chlorella* is polyphyletic (Huss and Sogin 1990) and that the green flagellate genus *Carteria* is paraphyletic (Buchheim and Chapman 1992). The occurrence of polyphyletic or paraphyletic taxa means that taxonomic schemes

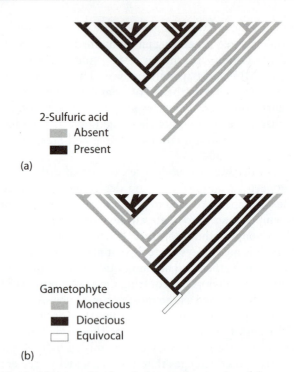

Figure 5.11 A molecular phylogeny for species of the brown seaweed family Desmarestiaceae, onto which has been mapped the occurrence of sulfuric acid (a), revealing that this character evolved only once within this group. In contrast, when dioecy (production of separate male and female thalli) is mapped onto the phylogeny (b), it appears that this trait has been gained at least twice and lost at least twice. (After Peters et al. 1997, by permission of the *Journal of Phycology*)

need to be changed if a classification system reflecting phylogeny is desired.

Ecological Applications

Molecular phylogenetic methods have been widely applied in ecological studies and promise to be even more useful in the future. For example, particular DNA sequences can be collected from as many organisms as possible, accumulated in online databases, and used as **molecular barcodes**. Preliminary studies suggest that the gene encoding mitochondrial cytochrome oxidase subunit 1 (*cox1*), widely used to barcode animals, also displays sufficient variability to work in red algae (Saunders 2005; Robba et al. 2006). ITS barcoding has been recommended for dinoflagellates (Litaker et al. 2007). Such barcodes would help in cataloging Earth's biodiversity and identifying new species. The development of portable

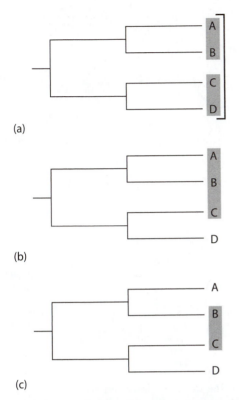

(a)

(b)

(c)

Figure 5.12 (a) monophyletic, (b) paraphyletic, and (c) polyphyletic groups. (Redrawn from *Trends in Ecology and Evolution.* Vol. 10. McCourt, R. M. Green algal phylogeny. Pages 159–163. © 1995, with permission from Elsevier Science)

devices for accomplishing PCR, sequencing, and barcode database comparisons could allow ecologists to make accurate and rapid identifications from remote locations. Current ecological applications of molecular phylogenetics include environmental genomics and the rapid detection of harmful algal species.

Environmental genomics

Environmental genomics is a set of procedures that employ phylogenetic information and molecular methods to explore the species diversity of natural microbial communities without first growing the organisms in culture. Such processes are widely used to compare microbial communities of different habitats, evaluate the effects of diverse environmental factors on communities, and estimate the effects of disturbance and environmental change.

Environmental genomic studies of microalgae typically begin by filtering water samples onto micropore filters to concentrate cells. DNA is often extracted from cells and purified. PCR is used to amplify some of this DNA by using the enzyme DNA polymerase and algae-specific primer sequences that specify a gene of interest (see Figure 5.7) (Betournay et al. 2007; Sherwood and Presting 2007). Ribosomal RNA genes are often chosen for environmental genomics because they are present in all organisms. PCR reactions are performed in a thermocycler device, which changes reaction temperature in a cyclical manner (see Figure 5.7). The gene amplification products are separated and visualized as bands by performing electrophoresis in agarose gels.

Cloning and sequencing the DNA fragments provides information that can be used to identify and classify the organismal sources of the DNA, and detect the presence of new sequence types (Figure 5.13). For cloning, DNA vectors, which are usually bacterial plasmids, are used to transfer foreign DNA into bacterial cells. Vectors are typically engineered to include a gene that confers antibiotic resistance and a gene that promotes the development of blue color at the site where foreign DNA is inserted. These features aid in later stages of the cloning process. DNA bands produced by PCR are cut out of the gel and inserted (ligated) into the vector, which is then used to transform *Escherichia coli* bacterial cells. Such bacteria are then placed onto nutrient agar containing an antibiotic and a compound that turns bacteria blue if they have incorporated plasmids not containing DNA inserts. Bacterial cells that have not taken in the vector DNA will not be able to grow because they lack resistance to the antibiotic. In contrast, bacterial cells that have received the vector multiply, each producing a colony that is blue or white. The white colonies are populations of bacterial cells that contain the foreign DNA inserts because such inserts disrupt the gene needed for production of blue color. In aggregate, the white bacterial colonies comprise a **clone library** because together they contain clones of the various types of DNA present in the original sample. DNA inserts are harvested from the bacterial cells and prepared for automated sequencing, which is usually performed at a specialized facility. The resulting sequence data can be compared with database sequences for species identification. Examples of environmental genomic work performed in this general way include studies of marine (Fuller et al. 2006) and freshwater (Richards et al. 2005) picoeukaryotes. Such organisms are otherwise difficult to detect and identify because they are so small.

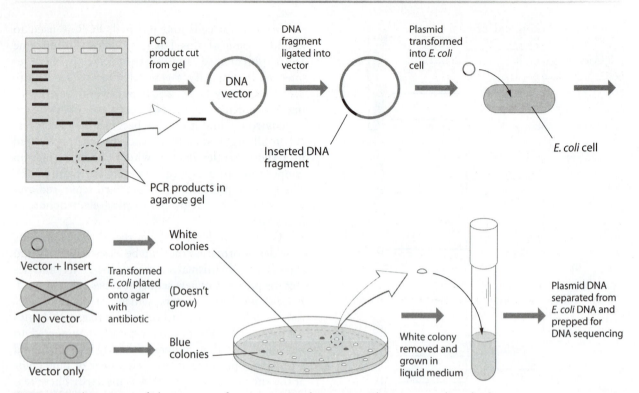

PCR product cut from gel

DNA vector

DNA fragment ligated into vector

Inserted DNA fragment

Plasmid transformed into *E. coli* cell

E. coli cell

PCR products in agarose gel

Vector + Insert

Transformed *E. coli* plated onto agar with antibiotic

No vector

(Doesn't grow)

Vector only

White colonies

Blue colonies

White colony removed and grown in liquid medium

Plasmid DNA separated from *E. coli* DNA and prepped for DNA sequencing

Figure 5.13 The process of cloning genes for environmental genomics. The steps are described in the text.

Because cloning and sequencing are labor intensive, most environmental genomics projects evaluate 100 or fewer clones per library. Thus, several more rapid methods have been developed for gaining information about the diversity of the dominant organisms in a community. In these methods, fragments of gene sequences that have been amplified from natural collections are observed as bands on gels in the case of DGGE (denaturing gradient gel electrophoresis) or peaks produced by an automated sequencing instrument in the cases of ARISA (automated ribosomal intergenic spacer analysis) and T-RFLP (terminal restriction fragment length polymorphism). These techniques provide an overview of community diversity without the need to clone and sequence DNA. However, they do not allow phylogenetic placement of the DNA fragments or detect rare sequence types.

Because different methods for evaluating diversity have both advantages and disadvantages, combinations of methods are often used. Countway and colleagues (2005), for example, used both cloning/ sequencing and T-RFLP to assess protist diversity in the western North Atlantic ocean, finding evidence for many new taxa. Savin and associates (2004)

conducted a comparative study of morphological and molecular methods for determining the community diversity of phytoplankton in the Bay of Fundy (Nova Scotia, Canada). Morphological analysis was conducted with the use of LM and SEM, and DGGE and sequencing were used to evaluate 18S rDNA (eukaryotic SSU rDNA) sequence types. Both morphological and molecular methods revealed high community diversity, but the two methods detected different communities. Though diverse diatoms were observed using traditional microscopy, few were detected with molecular methods. Few of the 18S rDNA sequences could be related to those of known organisms in genomic databases, indicating the presence of many unknown taxa. These investigators concluded that traditional morphological methods are not obsolete, and may better capture the diversity of photosynthetic plankton, but that molecular methods are essential for accurate assessment of community diversity and finding new types of protists.

An instrument known as a **flow cytometer** is widely combined with molecular methods for detecting new types of algae. As a stream of sample flows through it, this instrument uses a laser and detectors to measure the optical properties of cells, such as pigment fluorescence. When such instruments have cell-sorting capacity, cells having defined optical properties can be diverted from the main flow and collected for analysis. The new dinoflagellate *Lessardia elongata* was first collected by flow cytometry and was then described with the use of LM, TEM, SEM, and SSU rDNA sequence data (Saldarriaga et al. 2003). Environmental genomics coupled with flow cytometry has also revealed a likely new phylum of small eukaryotic algae (Not et al. 2007). A population of the new algae was collected by flow cytometry for microscopic observation. Because the plastid of the new algae displayed orange fluorescence, suggesting the presence of phycobilin pigments, the new group has been informally termed phycobiliphytes. Clone libraries suggest that the new phycobiliphytes are widely present in polar and cold temperate coastal ocean waters. After the new algae have been isolated into culture, they can be formally described and named.

Monitoring harmful algae

Phylogenetic information, molecular tools, and instrumentation are widely used to monitor environmental levels of harmful algal species (Rublee et al. 2005). For example, both flow cytometry and the newer solid-phase cytometry can be used for this and other ecological purposes. A flow cytometer equipped with continuous imaging has been used with image recognition software to detect harmful algae more precisely and in less time than traditional microscopic methods (Buskey and Hyatt 2006). The solid-phase cytometer is an instrument that allows automatic counting of algal cells by means of a microscope equipped with a motorized stage. Samples containing algal cells are concentrated onto a micropore filter. Cells on the filter are then labeled with a fluorescing antibody raised to a specific algal species, or a short piece of DNA known as an **oligonucleotide probe** that specifically binds the ribosomal RNA of particular algal species (West et al. 2006b; Töbe et al. 2006).

PCR-based methods have been developed to detect harmful dinoflagellate species such as *Pfiesteria piscicida*, even when cells are present in very low numbers (Saito et al. 2002). Real-time PCR, in which the fluorescence of stained, amplified DNA specific to a particular algal species is measured at the end of each PCR cycle, has also been used to detect low levels of *P. piscicida* (Lin et al. 2006). NASBA (nucleic acid sequence-based amplification) is a portable method for detecting toxic algal species in the field. This process uses an enzyme cocktail that includes reverse transcriptase to produce and amplify RNA and then monitors the fluorescence that occurs when a labeled oligonucleotide probe binds to its RNA target. Casper and associates (2004) used NASBA to detect mRNA for *rbcL*, the gene that encodes the large subunit of rubisco (ribulose-1,5-bisphosphate carboxylase-oxygenase), produced by the harmful marine dinoflagellate *Karenia brevis*. The rubisco mRNA was chosen because high levels of it are present in living algal cells. Another new approach monitors multiple algal species at once. Multiple types of oligonucleotide probes are attached to the ends of individual fibers in fiber-optic cables to form detector arrays. When ribosomal RNAs released from lysed algal cells bind these probes, fluorescence signals can be detected by means of a sensitive CCD (charge-coupled device) camera (Ahn et al. 2006). Such methods are able to distinguish toxic species from similar nontoxic ones and offer the prospect of early warning of harmful bloom development.

6 Chapter

Cyanobacteria
(Chloroxybacteria)

Anabaena

Though microscopic in size and simple in cellular structure, the cyanobacteria played dramatic roles in Earth's remote past. As noted in Chapter 2, photosynthetic cyanobacteria produced Earth's first oxygen atmosphere, which fostered the rise of eukaryotes. In addition, ancient cyanobacteria produced massive undersea carbonate deposits, thereby reducing atmospheric carbon dioxide levels and influencing global climate. The first plastids arose from cyanobacteria that forged endosymbiotic partnerships with early protists.

Cyanobacteria remain important today, having both positive and negative impacts from the human perspective. Recognized for their ability to occupy diverse aquatic and terrestrial habitats, cyanobacteria produce organic compounds used by other organisms, and they stabilize sediments and soils. Nitrogen-fixing cyanobacteria increase soil and water fertility and foster the growth of certain plants and fungi in symbiotic associations. Some cyanobacteria have potential biotechnological applications in the production of medicinally useful compounds or hydrogen-based energy (see Chapter 4). Yet the tendency of many cyanobacteria to produce toxins and form harmful blooms is an increasingly serious concern worldwide. This chapter provides a survey of cyanobacterial structure, physiology, ecology, and evolution, designed to be useful in understanding both beneficial and harmful aspects of cyanobacteria.

(Photo: L. W. Wilcox)

6.1 Structure, Motility, and Photosynthesis

Cyanobacteria are named for the blue-green (cyan) photosynthetic pigments commonly found within their cells and their classification as a phylum within Domain Bacteria (Figure 6.1). Like other bacteria and members of Domain Archaea, cyanobacteria have prokaryotic cells (Figure 6.2). **Prokaryotic cells** are defined by the absence of organelles—such as nuclei with porous envelopes and mitochondria—and other distinctive features of eukaryotic cells. Uniquely among bacteria, cyanobacteria produce the green photosynthetic pigment **chlorophyll *a*** and release oxygen as a result of photosynthesis, a process known as **oxygenic photosynthesis**. Consequently, cyanobacteria are also widely known as the **chloroxybacteria**. Because most algae and plants also feature chlorophyll *a* and oxygenic photosynthesis, cyanobacteria are often referred to as **blue-green algae** or **cyanophytes**. In this section we will consider the structure of cyanobacterial bodies and photosynthetic cells, which explains their motility, and then we will explore photosynthetic pigments and adaptation to light environments.

Body Types

Cyanobacteria occur in a diversity of body forms, from single cells to more complex forms that display the hallmarks of multicellularity (Figure 6.3). Some species occur as solitary **unicells**, or pairs of cells representing the products of a recent cell division (Figure 6.3a). Other species produce glue-like mucilage that holds together groups of cells to form **colonies** (Figure 6.3b). Cells joined end-to-end in a row take the form of a **filament**, also known as a **trichome**. Tiny channels known as microplasmodesmata link the cells of some filaments and allow cell-to-cell communication and transport. Filaments may be coated with a layer of polysaccharide mucilage known as a **sheath**, or EPS (extracellular polysaccharide) (Figure 6.3c). A common mucilage sheath encloses several filaments of some species (Figure 6.3d).

Many filamentous cyanobacteria produce specialized cells known as **heterocytes** (formerly heterocysts) that function in nitrogen fixation (Figure 6.3e,f) (see Section 6.3). Different species often produce

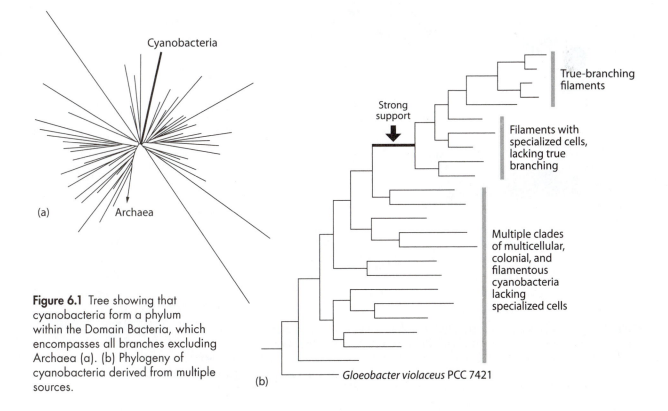

Figure 6.1 Tree showing that cyanobacteria form a phylum within the Domain Bacteria, which encompasses all branches excluding Archaea (a). (b) Phylogeny of cyanobacteria derived from multiple sources.

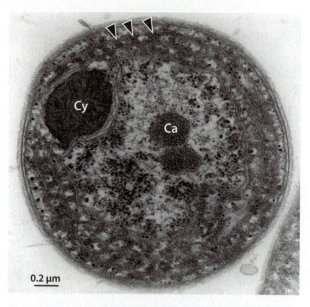

Figure 6.2 *Pseudanabaena galeata,* as seen with TEM. Note thylakoids bearing phycobilisomes (arrowheads) (see Section 6.1), cyanophycin granule (Cy), carboxysomes (Ca), and polyphosphate (arrows) . (From Romo and Pérez-Martínez 1997, by permission of the *Journal of Phycology*)

heterocytes in distinctive patterns or locations, a process that involves chemical communication among cells. For example, in the PCC 7120 strain of *Anabaena,* developing heterocytes secrete a peptide that prevents nearby photosynthetic cells of the same filament from differentiating. In contrast, cells lying outside the sphere of peptide influence are able to develop into heterocytes. This process produces a regular spacing of specialized cells every 10 cells or so along the filament (Yoon and Golden 1998). Cyanobacteria that are able to produce heterocytes usually also produce cells specialized for dormancy, known as **akinetes** (see Section 6.2). Some filamentous species display distinctive types of branches, whose development is sometimes associated with controlled cell death (see Figure 6.3e). Some cyanobacteria thus meet major criteria used to define **multicellularity**: cells attached to each other, specialized cells, intercellular communication, and the involvement of controlled cell death in development. Next, we consider the structure of unspecialized photosynthetic cyanobacterial cells, a topic related to the ability of cyanobacteria to move in their environments.

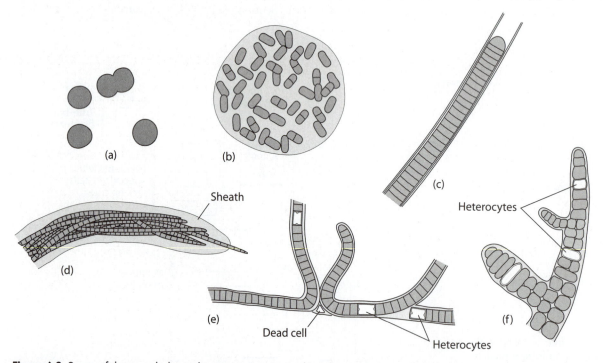

Figure 6.3 Some of the morphological types present in modern-day cyanobacteria include (a) unicells such as *Synechocystis,* (b) colonies of individual cells such as *Aphanothece,* (c) unbranched filaments including *Lyngbya,* (d) aggregations of multiple trichomes in a common sheath, as in *Microcoleus,* (e) false-branched forms including *Scytonema,* (f) true-branched forms such as *Stigonema.* (f: After Smith, G. M. *Fresh-Water Algae of the United States.* © 1950 McGraw-Hill Companies)

Structure of Photosynthetic Cells and Motility Mechanisms

Cyanobacterial cells are usually spherical or cylindrical in shape, with a rigid cell wall (see Figure 6.2). Here, we first consider the outermost mucilage layer, then the cell wall, and finally the cytoplasmic features of unspecialized cyanobacterial cells.

Mucilage

Mucilage that envelops the cell walls of many cyanobacteria may play roles in addition to the colony formation described above. For example, the sheaths of filamentous cyanobacteria often contain compounds that absorb ultraviolet light (UV radiation), thereby protecting cells from potential damage. Such protective compounds include oligosaccharide mycosporine amine acids (OS-MAAs), or **scytonemin**. These sunscreen materials allow sheathed cyanobacteria to withstand UV exposure and thus live in exposed habitats. Named after the cyanobacterial genus *Scytonema*, which often produces such protective compounds, scytonemin is an indole-alkaloid (Figure 6.4) whose synthesis starts from aromatic amino acids and is induced by UV-A radiation. Scytonemin, which gives cyanobacterial sheaths a yellow-brown color, absorbs maximally at 370 nm but also more broadly in UV. *Nostoc punctiforme* (ATCC 29133, PCC 73102) has been used as a model system to study the molecular genetics of scytonemin biosynthesis (Soule et al. 2007). The colorless, pectin-like mucilage of colonial planktonic cyanobacteria such as *Microcystis* has the capacity to capture scarce iron and other micronutrients from the surrounding water (Plude et al. 1991). Mucilage may also aid buoyancy and deter herbivores (Reynolds 2007) or harbor large populations of smaller bacterial associates (Figure 6.5).

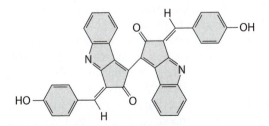

Figure 6.4 Biochemical structure of the cyanobacterial sunscreen scytonemin. (After Soule et al. 2007, by permission of the American Society of Microbiology)

Cell wall

The cyanobacterial cell wall, like that of other Gram-negative bacteria, is largely composed of **peptidoglycan**, a carbohydrate substance cross-linked by peptides. The peptidoglycan layer of cyanobacterial walls is often thicker than that of other Gram-negative bacteria. As is generally the case for Gram-negative bacteria, a lipopolysaccharide envelope encloses the peptidoglycan wall layer of cyanobacteria (Figure 6.6). However, the cyanobacterial outer envelope may contain carotenoids and

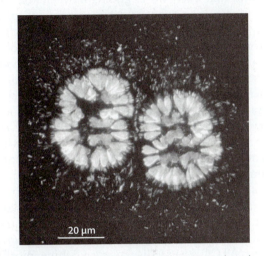

Figure 6.5 Two *Gomphosphaeria* colonies, each with a halo of bacteria in the surrounding mucilage. Note the heart-shaped cells. (Photo: L. W. Wilcox)

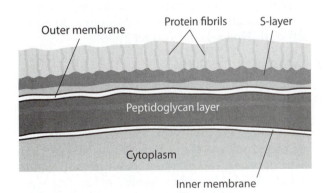

Figure 6.6 Diagram of cyanobacterial cell wall features. A relatively thick peptidoglycan layer is sandwiched between outer and inner membranes. In some cyanobacteria, such as *Phormidium uncinatum*, diagrammed here, an S-layer is found external to the outer membrane. In this species, a coating of protein fibrils is borne on the S-layer.

other chemical constituents not usually found in those of other Gram-negative bacteria (Hoiczyk and Hansel 2000).

Cell walls may display additional features related to motility or other functions. For example, the external surface of some cyanobacterial walls contains an S-layer composed of glycoprotein. S-layers may foster the development of surface mineral deposits, such as calcium carbonate on *Synechococcus* (Figure 6.7). The S-layer of the filamentous cyanobacterium *Phormidium uncinatum* bears a surface coating of protein fibrils (see Figure 6.6) that likely foster gliding movements. Various filamentous cyanobacteria are able to glide, and such motions may help them to escape from harmful levels of solar radiation. Other cyanobacteria display swimming motions made possible by surface proteins. For example, *Synechococcus* unicells isolated from a number of marine habitats were able to move at a rate of 25 µm s^{-1} (Waterbury et al. 1985), and further study showed that this motility depends on the *swmA* gene, which encodes a surface motility protein. This protein is not present on surfaces of nonmotile strains of *Synechococcus*, and when *swmA* of motile strains is inactivated, the cells become unable to change position (Brahamsha 1996). *Synechococcus* strain PCC 6803 and short filaments of *Nostoc punctiforme* ATCC 29133 display a twitching motility conferred by **pili**—thread-like structures located on the surfaces of many Gram-negative bacteria (Duggan et al. 2007). In addition, some cyanobacteria are thought to move by squirting mucilage through small pores that extend through the entire cell wall (Hoiczyk and Hansel 2000).

Cytoplasmic features

Structures that typically occur within photosynthetic cyanobacterial cells are most easily visualized with the use of transmission electron microscopy (TEM) (see Figures 6.2 and 6.7). Prokaryotic ribosomes are abundant, and the delicate threads of DNA can sometimes be observed. More conspicuous structures include thylakoids, carboxysomes, and several types of storage particles. **Thylakoids** are photosynthetic membranes that bear chlorophyll *a* and other photosynthetic components described in the next section. Hemispherical **phycobilisomes**, which contain particular accessory photosynthetic pigments linked by proteins, occur on thylakoid surfaces (see Figure 6.2). The thylakoids of cyanobacterial cells are typically

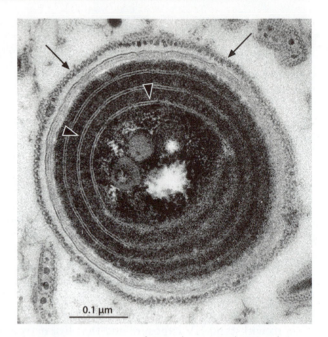

Figure 6.7 A TEM view of *Synechococcus* showing the thickened surface layer (S-layer) (arrows), upon which carbonates are deposited. Thylakoids are indicated by arrowheads. (From Thompson et al. 1997)

not stacked and are arranged concentrically at the cell periphery (see Figure 6.7). Exceptions to this generality include *Gloeobacter violaceus* PCC 7421, which contains chlorophyll *a* but lacks thylakoids, and the chlorophyll *b*-containing cyanobacteria, whose thylakoids lack phycobilisomes and occur in stacks of two or more. Thylakoid stacking has also been observed in a genetically altered *Synechocystis* species lacking phycobilisomes (Swift and Palenik 1993), suggesting that thylakoids will tend to stack when there is no spatial interference from phycobilisomes. **Carboxysomes** are polygonal aggregations of the enzyme rubisco and thus are the location of carbon fixation in cyanobacterial cells (see Figure 6.2). Storage particles include **cyanophycin**, polymers of the amino acids asparagine and arginine, which function as nitrogen storages. In TEM images of cyanobacterial cells, cyanophycin appears as moderately dense lumps (see Figure 6.2). Particles of **cyanophycean starch** are carbohydrate storages, α-1,4-linked polyglucans similar to the linear amylose portion of higher plant starch. Small granules of β-hydroxybutyrate polymer may be abundant in some cyanobacteria. Lipid may be stored as cytoplasmic droplets. Phosphate is stored as polyphosphate, which in TEM view appear as small globules that stain more intensely than ribosomes.

The cells of aquatic cyanobacteria often include a large number of the structures known as **gas vesicles** or **gas vacuoles**, which confer vertical mobility in the water column. Such up-and-down movement allows cyanobacterial cells to obtain light for photosynthesis at the surface, avoid harmful amounts of solar radiation, and obtain mineral nutrients. It is important to note that the terms *gas vesicles* and *gas vacuoles* are misleading in two ways. First, these structures do not actually contain gas at higher concentration than is present elsewhere in cells. Second, these structures are not delimited by membranes, as is suggested by the terms *vesicle* and *vacuole*. Rather, they are assemblies of hollow, pointed cylinders (Figure 6.8) whose 2 nm-thick walls are constructed of gas-permeable aggregations of protein. Gas vesicles vary in length from 62–110 μm and in width from 0.3–1 μm. While length doesn't affect strength, narrower gas vesicles are stronger than wider ones (Oliver 1994).

Gas vesicle production is induced by low light levels, which inhibit photosynthesis. Gas vesicles help cells move upward, toward the light, where photosynthesis can more readily occur. In eutrophic lakes, gas vesicles can allow cyanobacteria to outcompete eukaryotic algae (Klemer et al. 1996). Such structures also explain the ability of the important marine cyanobacterium *Trichodesmium* to accomplish vertical migration to the surface from as deep as 70 m (Villareal and Carpenter 2003). Gas vesicles work by excluding water and heavier cellular constituents. Since the density of gas vesicles (120 kg m^{-3}) is much less than that of other cell constituents, when many vesicles are present, cyanobacterial cells tend to float. If photosynthesis increases as a result, cells contain higher concentrations of sugars. This increase, together with uptake of mineral ions, generates higher cellular turgor pressure. Such high osmotic pressure may collapse the gas vesicles. Collapse of gas vesicles and/or production of ballast in the form of storage particles cause cyanobacteria to sink in the water column. In deeper, darker waters, cellular carbohydrates may be used up in respiration, allowing gas vesicles to re-form. As a result, the buoyancy cycle may continue (Figure 6.9).

However, the buoyancy cycle can be interrupted, trapping cyanobacterial blooms at the surface. Several factors may contribute to this effect. Cyanobacterial cells within thick blooms may be unable to obtain sufficient resources to manufacture sugars, which allows gas vesicles to remain intact.

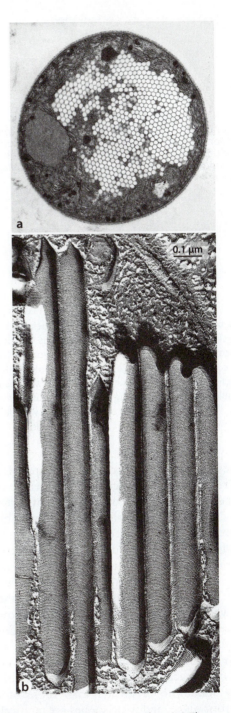

Figure 6.8 TEM views of gas vesicles. (a) Those of *Nostoc coeruleum* are shown in cross-section. (Courtesy T. Jenson) (b) Those of *Nostoc muscorum* are shown in longitudinal view. (Reprinted with permission from Waaland, J. R. and D. Branton, Gas vacuole development in a blue-green alga. *Science* 163:1339–1341. © 1969 American Association for the Advancement of Science)

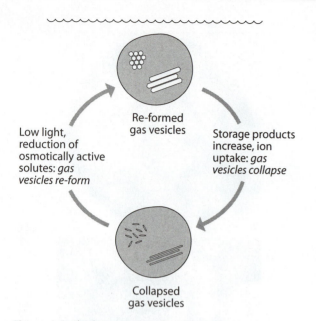

Figure 6.9 The buoyancy cycle of cyanobacteria having gas vesicles. Under low-light conditions, gas vesicles form, causing the cells/colonies to rise upward in the water column. At the surface, enhanced photosynthetic production may give rise to higher turgor pressures and accumulation of dense storage particles, causing the cells to sink. As nutrients are utilized by cells at lower depths, gas vesicles may re-form, reinitiating the cycle.

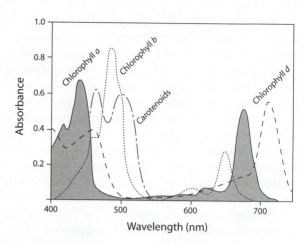

Figure 6.10 Absorption spectra for chlorophyll *a* (shaded curve), carotenoids, chlorophyll *b*, and chlorophyll *d*. Carotenoids and chlorophyll *b* act as "accessory" pigments, absorbing additional wavelengths of light and passing the energy along to chlorophyll *a*.

Gas vesicles may be too strong to collapse. Lack of wind reduces vertical circulation of surface waters and the accompanying downward transport of cyanobacteria. Consequently, cyanobacterial blooms may persist at the surface. Exposed to high light levels for extended periods, the cyanobacteria die, which releases toxins from cells, and decay, which reduces oxygen concentration in surface waters. Other aquatic organisms and humans who use such waters may be seriously affected (Dow and Swoboda 2000; Oliver and Ganf 2000; Codd et al. 2005a,b).

To summarize our survey of structure and mobility, cyanobacterial bodies display a level of complexity similar to or greater than those of many eukaryotic algae, and some species are multicellular. Conspicuous components of photosynthetic cells are thylakoid membranes and various types of storage granules. Cyanobacterial motility mechanisms aid in obtaining resources or avoiding harmful light levels, and they include gliding, swimming, twitching, and vertical movement in the water column by means of buoyancy changes.

Photosynthetic Pigments in Light-Harvesting Systems

Cyanobacteria are able to absorb light energy because they contain chlorophyll *a* or *d* and accessory pigments. Cyanobacterial accessory pigments include phycobilins, carotenoids, and, in some cases, chlorophyll *b*. The chemical structures of these pigments include rings and double bonds that interact with light and allow energy transfer by resonance, much as a struck tuning fork causes neighboring forks to also sound. Chlorophyll *a* is the pivotal photosynthetic pigment in photosynthetic light reactions, absorbing red and blue components of the visible spectrum. Accessory pigments extend the ability to harvest light for photosynthesis by absorbing additional parts of the visible spectrum (Figure 6.10). Carotenoids also protect against light-induced cell damage.

Chlorophylls

The chlorophylls of cyanobacteria likely evolved from precursor bacteriochlorophylls present in other types of bacteria (Xiong et al. 2000). As is the case for plant and algal plastids in general, cyanobacterial antenna chlorophyll *a* molecules associate with proteins within reaction centers of photosystems I and II (PSI and PSII). Antenna chlorophyll *a* molecules gather light energy and resonance-transfer it to special chlorophyll *a* molecules, which donate excited

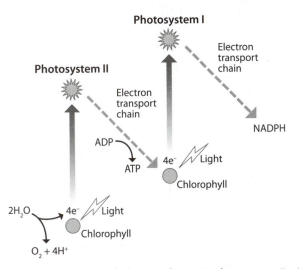

Figure 6.11 Diagram of the light reactions of photosynthesis. In photosystem II, electrons derived from the splitting of water replace those transferred from a special chlorophyll *a* molecule to an acceptor molecule when light energy is absorbed. Oxygen is produced as a by-product, along with protons. Electrons then pass along an electron transport chain, producing ATP in the process. In photosystem I, electrons don't come from water but, rather, from the electron transport chain of photosystem II. They are similarly passed from a chlorophyll *a* molecule to an electron acceptor and along another electron transport chain, resulting in the production of NADPH, which, together with ATP, is used in the dark reactions of photosynthesis to produce chemical energy in the form of sugar.

electrons to acceptor molecules. Electron transport chains then produce ATP or NADPH used in carbon fixation (Figure 6.11).

Comparative analysis of genome sequences from photosynthetic bacteria suggests that PSI is the ancestral reaction center (Mulkidjanian et al. 2006). PSII is thought to have originated from PSI by gene cluster duplication and divergence. The ancestors of cyanobacteria may have used these two photosystems alternately, in response to different environmental conditions. Origin of the water-splitting manganese-calcium (Mn_4Ca) cluster in PSII allowed early cyanobacteria to use the abundant substance water as an electron donor (see Figure 6.11), having the added advantage that the gaseous waste product O_2 was easily removed from cells. This evolutionary event also allowed cyanobacteria to use PSI and PSII in sequence, thereby improving photosynthetic efficiency (Allen and Martin 2007).

Chlorophyll *d* is the major photosynthetic pigment in *Acaryochloris marina*, a cyanobacterium that has very little chlorophyll *a* and lacks phycobilisomes and phycobilin pigments (Miyashita et al. 1996; Miller et al. 2005). In this cyanobacterium, chlorophyll *d* replaces chlorophyll *a* as the special chlorophyll in PSI (Hu et al. 1998) and PSII (Toma et al. 2007). The

presence of chlorophyll *d* allows *A. marina* to utilize far-red light, an advantage in shaded habitats. In the past, it was thought that some red algae also produced chlorophyll *d*, but more recent studies suggest that such previous studies had been made of red algae on which *Acaryochloris* occurred as an epiphyte. Presently, chlorophyll *d* appears to be a distinctive feature of *Acaryochloris* (Murakami et al. 2004) and seems to be widespread globally (Kashiyama et al. 2008).

Chlorophyll *b* is structurally very similar to chlorophyll *a* and likely evolved from it. Chlorophyll *b* is widely known to function as an accessory pigment in the green plastids of many eukaryotic algae and plants, and it also occurs in the cyanobacterial genera *Prochloron*, *Prochlorococcus*, and *Prochlorothrix*. In the past, this similarity was thought to represent close relationship, but more recent phylogenetic studies suggest that chlorophyll *b* evolved independently in these three cyanobacteria (Urbach et al. 1992; Palenik and Swift 1996; Litvaitis 2002). The chlorophyll *a/b* binding proteins of *Prochloron*, *Prochlorococcus*, and *Prochlorothrix* are closely related to cyanobacterial proteins, and not to eukaryotic chlorophyll *a/b* or *a/c* light-harvesting proteins, suggesting independent evolutionary origin of these proteins (LaRoche et al. 1996). Chlorophyll *b*

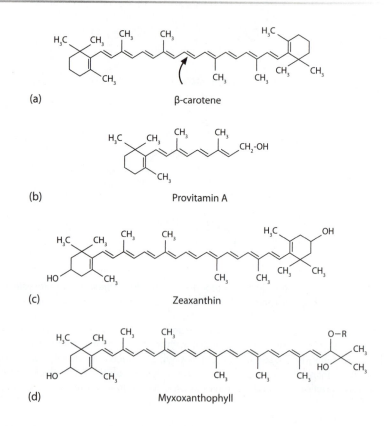

Figure 6.12 Structures of (a) β-carotene, (b) provitamin A, (c) zeaxanthin, and (d) myxoxanthophyll. Provitamin A is formed by splitting β-carotene into two equal parts (at arrow). Zeaxanthin is a common xanthophyll and is also synthesized from β-carotene. (a–c: Redrawn from *Biology of Plants* by Raven, Evert, and Eichhorn © 1971, 1976, 1981, 1986, 1992, 1999 by W. H. Freeman and Company/Worth Publishers. Used with permission; d: After Mohamed and Vermaas 2004, by permission of the American Society of Microbiology)

broadens the range of light that can be used in photosynthesis (see Figure 6.10) and transfers absorbed light energy to reaction-center chlorophyll *a*. Chlorophylls fluoresce red light, a property that allows the monitoring of algal populations in water samples or natural habitats with the use of commercial fluorescence detectors.

Carotenoid pigments

Carotenoids found in cyanobacteria include a variety of **xanthophylls**, which have oxygen in their molecular structure, and β-**carotene**, which does not. These pigments may be located in thylakoids alongside chlorophyll *a*, where they increase the ability of cyanobacteria to harvest wavelengths of light that are not directly absorbed by chlorophyll *a* (see Figure 6.10). In addition, carotenoids associated with soluble proteins provide protection from harmful photoinhibition, the loss of or damage to proteins essential to the function of photosynthetic reaction centers (Wilson et al. 2006a).

β-carotene (Figure 6.12a) occurs in nearly all photosynthetic algae and is particularly important as the main source of provitamin A (Figure 6.12b), which animals require for synthesis of the visual pigment rhodopsin and in the regulation of genes involved in limb and skin development. β-carotene is synthesized from lycopene, which does not possess terminal rings, by the enzyme lycopene cyclase (LYC). The cyanobacterial gene for LYC, *crtL*, has been cloned and sequenced, and this information has been essential to the molecular analysis of carotenoid synthesis in higher plants. The cyanobacterial xanthophyll zeaxanthin (Figure 6.12c) is synthesized from β-carotene by hydroxylation and epioxidation reactions (Bartley and Scolnick 1995). Myxoxanthophyll (Figure 6.12d) is another common cyanobacterial xanthophyll whose biosynthesis has been studied (Mohamed and Vermaas 2004).

Phycobilins

Cyanobacteria typically possess large amounts of water-soluble accessory pigments collectively known as phycobilins, which occur in phycobilisomes. These pigments allow cyanobacteria to absorb blue-green, green, yellow, or orange light and pass the energy to chlorophyll *a* (Figure 6.13). Cyanobacterial phycobilins include the open-chain tetrapyrroles **phycoerythrobilin** and **phycocyanobilin** (Figure 6.14), which are

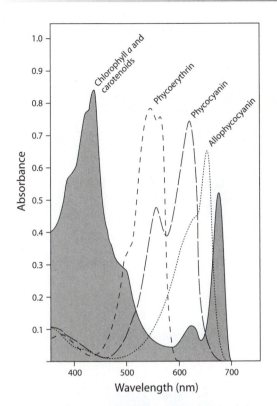

Figure 6.13 Absorption spectra for the phycobilin pigments found in non-chlorophyll *b*-containing cyanobacteria. These accessory pigments extend the range of wavelengths that can be used for photosynthesis, covering wavelengths where neither chlorophyll nor carotenoids (shaded curve) show strong absorbances. (Redrawn from Gantt, E. *BioScience* 25:781–788. © 1975 American Institute of Biological Sciences)

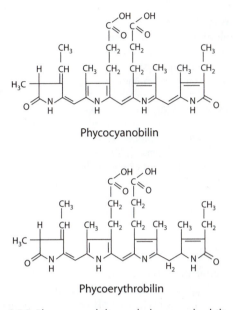

Figure 6.14 Phycocyanobilin and phycoerythrobilin. Note the open-chain tetrapyrrole structure.

bound to proteins, and as such are known as **phycobiliproteins**. Three types of phycobiliproteins are produced by various cyanobacteria: **phycocyanin** and **allophycocyanin** (which contain phycocyanobilin), and **phycoerythrin** (containing phycoerythrobilin). However, not all of these phycobilins are produced by all cyanobacterial species. Freshwater species tend to produce more phycocyanin, while phycoerythrin is often abundant in marine species. Each phycobiliprotein has a distinctive absorbance and fluorescence profile. For example, phycoerythrin in open-ocean *Synechococcus* fluoresces bright orange after absorbing green light, and this type of fluorescence increases with depth. *Synechococcus* cells collected from deep waters may fluoresce 100 times brighter than surface cells, reflecting increased levels of phycoerythrin (Olson et al. 1990). Such fluorescence properties form

the basis for commercial sensor devices that can be deployed to evaluate cyanobacterial populations in natural waters. The characteristic arrangement of phycobiliproteins in phycobilisomes (Figure 6.15) maximizes the efficiency of energy transfer to chlorophyll *a*. Energy transfer occurs from phycoerythrin (when present) to phycocyanin to allophycocyanin to chlorophyll *a* in photosystem II (Figure 6.16).

Chromatic Adaptation

Some cyanobacteria can adjust their pigment composition in response to changes in light quality. For example, exposure to orange light increases synthesis of the blue-colored phycocyanin, whereas exposure to blue-green light increases synthesis of phycoerythrin. Alteration in pigment composition, known as **chromatic adaptation**, provides an adaptive advantage to cyanobacteria whose light environment may change over their lifetime. *Anabaena* senses the color of its light environment by means of rhodopsin, a pigment-protein complex (Vogeley et al. 2004).

Anoxygenic Photosynthesis

Photosynthetic bacteria other than cyanobacteria display **anoxygenic photosynthesis**, meaning that photosynthesis does not produce oxygen as a by-product.

Figure 6.15 A model of the phycobilisome, as inferred from several lines of evidence. Note the positions of the three types of phycobilin pigment molecules relative to the energy transfer indicated in Figure 6.16. The pigments are held together by linker proteins (striped boxes). (Redrawn with permission from Glazer, A. N., C. Chan, R. C. Williams, S. W. Yeh, and J. H. Clark. Kinetics of energy flow in the phycobilisome core. *Science* 230:1051–1053. © 1985 American Association for the Advancement of Science)

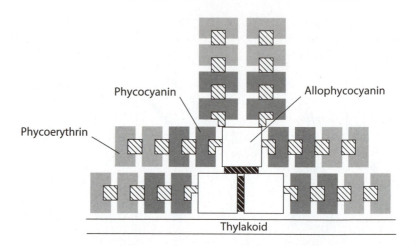

Anoxygenic bacteria cannot release O_2 from water because they lack the water-breaking manganese-calcium cluster present in PSII of cyanobacteria and plastids. Instead, anoxygenic bacteria use reduced compounds such as hydrogen sulfide (H_2S), hydrogen gas (H_2), or organic compounds as electron donors in photosynthesis. Interestingly, some cyanobacteria also perform anoxygenic photosynthesis in anaerobic environments rich in hydrogen sulfide, if light is available. An example is *Oscillatoria limnetica* isolated from anaerobic H_2S-rich bottom waters of the hypersaline Solar Lake, in Israel. As a result of the reaction:

$$2H_2S + CO_2 \rightarrow CH_2O + 2S° + H_2O$$

elemental sulfur (S°) is excreted from cells and forms conspicuous granules on the cyanobacterial filament surfaces (Cohen et al. 1975).

Carbon Fixation and Heterotrophic Growth

In cyanobacteria, as in photosynthetic plastids, the ATP and NADPH produced during the light reactions of photosynthesis (see Figure 6.11) is used to transform CO_2 into organic compounds, a process known as **carbon fixation**. The first step in carbon fixation is catalyzed by the enzyme ribulose-1,5-bisphosphate carboxylase/oxygenase, commonly known as **rubisco**. Four basic types of rubisco occur in photosynthetic organisms. Type I rubiscos display eight large subunits and eight small subunits; this type of rubisco is found in most oxygenic organisms, including cyanobacteria. Type II rubisco is composed of just two large subunits and is found in most

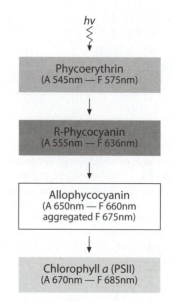

Figure 6.16 Energy transfer in phycobilisomes. In phycobilisomes, light energy is transferred from phycoerythrin to phycocyanin to allophycocyanin and finally to chlorophyll *a* in photosystem II. (Redrawn from Gantt, E. *BioScience* 25:781–788. © 1975 American Institute of Biological Sciences)

dinoflagellates, Type III has 10 large subunits and is found in Archaea, and Type IV is an enzyme used in methionine metabolism that might represent the ancestral form (Ashida et. al 2005). Type I rubisco also occurs in four forms, of which two occur in cyanobacteria. Type IA has been found only in prokaryotes, including many of the marine cyanobacteria of oligotrophic oceans. Type IB occurs in coastal and freshwater cyanobacteria and in green plastids.

While all known cyanobacterial species are capable of photosynthesis, several also display heterotrophic growth in the dark by taking up and metabolizing organic compounds. For example, several strains of *Nostoc* are known to utilize the sugars glucose, fructose, ribose, or sucrose (Dodds et al. 1995), and *Cyanothece* ATCC 51142 metabolizes glycerol (Schneegurt et al. 1997). *Microcystis*, like a number of other cyanobacterial genera, is known to take up several types of amino acids (Comte et al. 2007).

6.2 Reproduction

Cyanobacteria lack sexual reproduction involving specialized gametes and meiosis, typical of many eukaryotes. Neither does bacterial conjugation appear to occur among cyanobacteria. However, gene exchange can occur among cyanobacteria by means of viral **transduction**. For example, genomic studies of ocean-dwelling *Prochlorococcus* indicate that cyanophage gene transfer allows cyanobacterial populations to adapt to habitat variations in light and nutrients (Coleman et al. 2006).

Cyanobacterial cells divide by **binary fission**, the production of two equal-sized cells from a progenitor, without the involvement of microtubular spindles or other common attributes of eukaryotic mitosis. Binary fission and separation of division products allow populations of unicellular and colonial cyanobacteria to increase. Filamentous cyanobacteria often reproduce by means of short, motile filaments known as **hormogonia** (Figure 6.17a) that result from breakup of longer filaments. Hormogonium formation is promoted by the controlled death and collapse of certain cells, which are known as **separation disks,** or **necridia** (Figure 6.17b).

Filamentous cyanobacteria that are able to produce heterocytes (involved in nitrogen fixation) are usually also capable of producing specialized resting cells known as **akinetes** (Figure 6.18). Such cells allow cyanobacteria to survive adverse conditions and resume growth when conditions improve. Heterocytes and akinetes share a number of features that distinguish them both from photosynthetic cells. For example, particular glycolipids and polysaccharides occur in the cell walls of both heterocytes and akinetes, but not in typical vegetative cells. Also, PSII is inactivated in the akinetes of at least some cyanobacteria, as is typically the case for heterocytes. Finally, levels of the enzyme superoxide dismutase appear to be lower in akinetes and heterocytes than in nonspecialized cells of the same species. These observations suggest that early stages in the differentiation of akinetes and heterocytes may be under the control of similar genetic elements.

Akinetes become more abundant near the end of a growing season. The environmental stimulus for akinete formation in cyanobacteria varies. Low levels of cellular phosphorus may trigger akinete production in *Anabaena circinalis* (van Dok and Hart 1997), while temperature may be more important in other cases (Li et al. 1997). Akinete development includes decrease in number of gas vesicles, increase

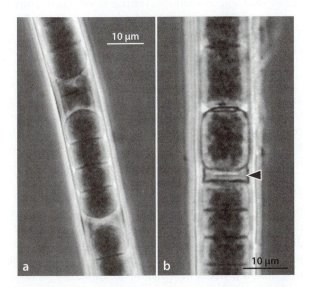

Figure 6.17 *Tolypothrix.* (a) Hormogonia. (b) Separation disk (arrowhead) adjacent to a heterocyte; this is where false branches usually occur in this organism. (Photos: L. W. Wilcox)

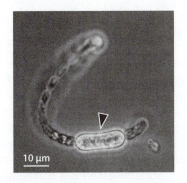

Figure 6.18 An akinete (arrowhead) in *Anabaena.* (Photo: L. W. Wilcox)

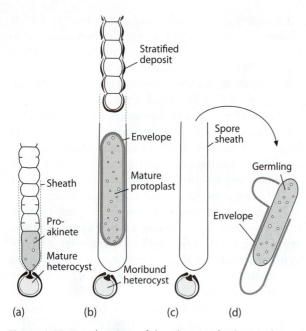

Figure 6.19 Development of the akinete of *Gloeotrichia echinulata*. (a) Differentiation begins near a heterocyte. (b) Akinetes enlarge, develop thick walls, and (c) separate from subtending vegetative filaments and heterocytes, leaving an empty sheath. (d) Germination of akinetes. (After Cmiech et al. 1984)

in size, production of a thick wall, and deposition of abundant storage granules (Figure 6.19a–c). The resulting increase in cell density contributes to rapid settling to the sediments, where akinetes remain during winter. During akinete germination in spring, cyanophycin and glycogen are metabolized, cell division occurs to form short filaments, and gas vesicles reappear, permitting filaments to ascend into the water column (Figure 6.19d) (Wildman et al. 1975; Moore et al. 2004).

6.3 Nitrogen Fixation

Many cyanobacteria and certain other prokaryotes are the only organisms known to be capable of transforming molecular nitrogen gas (N_2) into ammonia. This is important because ammonia can then be assimilated into amino acids, proteins, and other nitrogen-containing cellular constituents. Nitrogen-fixing cyanobacteria can thus avoid nitrogen-limitation of growth, a competitive advantage in many habitats. The global biogeochemical significance of cyanobacterial nitrogen fixation was

introduced in Chapter 2, and symbiotic associations of nitrogen-fixing cyanobacteria were surveyed in Chapter 3. Additional ecological aspects of nitrogen fixation are included in Chapters 21–23. Here the focus is on cellular and biochemical aspects of nitrogen fixation.

Nitrogen fixation is an inducible process, meaning that environmental levels of ammonium (NH_4^+) or nitrate (NO_3^-) regulate it. This exceptionally energy-demanding process therefore does not occur if environmental levels of fixed nitrogen are sufficiently high, but it is triggered by low levels. Nitrogen starvation causes the Krebs cycle intermediate 2-oxoglutarate to increase, which is involved in triggering heterocyte development (Laurent et al. 2005). Calcium is also important in regulating heterocyte development (Zhao et al. 2005; Shi et al. 2006). Nitrogen fixation is constrained by the level of environmental oxygen, which is often high enough to inhibit activity of nitrogenase, an enzyme essential to the nitrogen fixation process. Nitrogenase consists of an iron-containing protein and a molybdenum–iron (Mo-Fe) protein. The iron protein donates electrons to the Mo-Fe protein, the site of nitrogen fixation. Cellular production of nitrogenase requires sufficient environmental levels of sulfur, molybdenum, magnesium, cobalt, and iron (and sometimes vanadium), which function as cofactors. Many of the genes that encode the biosynthesis of nitrogenase have been identified (e.g., Dominic et al. 2000). The acetylene reduction test is used to detect the occurrence of nitrogen fixation and measure its rate (Stewart et al. 1967, 1969).

Cyanobacteria that produce heterocytes are assumed to be capable of nitrogen fixation even in aerobic habitats because heterocyte adaptations protect nitrogenase from oxygen (Haselkorn 1978). PSII expression does not occur in heterocytes, so oxygen is not generated. However, PSI is present, allowing photophosphorylation to produce ATP. Heterocytes also import carbohydrates from neighboring photosynthetic cells. Such carbohydrates are used in respiration to produce both ATP and reducing equivalents (Figure 6.20). Intracellular respiration also consumes oxygen. The oxyhydrogen reaction, a process in which hydrogen gas produced within heterocytes reacts with oxygen to generate water, also helps to reduce internal oxygen concentration. The thick walls that normally characterize heterocytes help to reduce diffusion of oxygen into

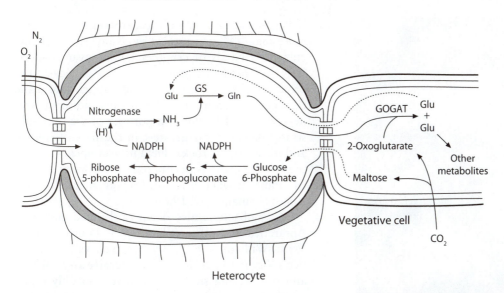

Figure 6.20 Diagram of a heterocyte and adjacent vegetative cells, illustrating the relationships between photosynthetic and nitrogen-fixation metabolism. Thick walls of heterocytes reduce diffusion of O_2 (and N_2) directly into heterocytes. Hence, entry of these gases occurs via vegetative cells and microplasmodesmata (channels) in the cell walls connecting heterocytes and vegetative cells. Such pores also allow movement of fixed nitrogen (Gln = glutamine) from heterocytes into vegetative cells. The enzyme GOGAT regenerates glutamate (Glu), some of which diffuses into heterocytes. Addition of NH_3 to glutamate is accomplished by glutamine synthetase (GS). The NH_3 originates from the activity of nitrogenase, using reducing equivalents (NADPH) generated from carbohydrates such as maltose or glucose-6-phosphate. These originated from photosynthetic carbon fixation in vegetative cells. (After Haselkorn 1978, with permission from the *Annual Review of Plant Physiology and Plant Molecular Biology* Volume 29. © 1978 by Annual Reviews [www.annualreviews.org])

cells. Bacterial cells associated with the surfaces of heterocytes—commonly found at the junctions between heterocytes and adjacent cells—may consume oxygen, thus preventing its entry into heterocytes. Mucilage-coated heterocytes of *Nostoc cordubensis* cultured from rapidly flowing streams in Argentina were able to fix nitrogen at significantly higher oxygen concentrations than were those lacking mucilage (Prosperi 1994).

Some cyanobacteria that lack heterocytes can fix nitrogen under anaerobic conditions or at night. For example, a study of a tropical marine community dominated by filamentous cyanobacteria lacking heterocytes revealed that nitrogen fixation occurred at night (Steppe et al. 2001). The unicellular cyanobacterium *Cyanothece* copes by down-regulating PSII during N_2 fixation, with the result that little O_2 is produced for a few hours of every 24-hour period, even in continuous light (Meunier et al. 1998). The filamentous cyanobacterium *Trichodesmium*, which is abundant in tropical oceans, lacks heterocytes but is able to fix nitrogen even in the daytime. It

accomplishes this by performing nitrogen fixation in cells located in the middle parts of filaments, while cells at the ends engage in photosynthesis (Berman-Frank et al. 2001).

6.4 Cyanobacteria of Extreme Habitats

Cyanobacteria occupy a diverse array of habitats, including the plankton of ponds, lakes, rivers, and oceans; temperate soils; geothermal waters; desert soils and rocks; polar regions; and hypersaline waters (Whitton and Potts 2000). Environmental genomics (Chapter 5) is increasingly used to evaluate environmental biodiversity, with the result that many cyanobacteria are known only as a distinctive DNA sequence. Many aspects of cyanobacterial ecology are described in Chapters 2, 3, 21, and 23. Here, we focus on extremely hot, cold, and saline habitats.

Figure 6.21 Yellowstone thermal pools. A community of filamentous algae is evident in the lower-left portion of the photograph. (Photo: D. Derouen)

A number of cyanobacteria inhabit hot springs and thermal pools (Figure 6.21), tolerating temperatures as high as 74°C (Ward and Castenholz 2000). In alkaline and neutral hot springs and streams flowing from them, cyanobacteria can form thick, colorful mats that exhibit banding patterns representing the distribution of species with different temperature tolerances (Darley 1982). Such forms, known as thermophiles, are defined by a temperature optimum for growth that is 45°C or greater. The unicellular *Synechococcus lividus* can not only tolerate temperatures of 70°C, but appears to be adapted to such conditions, since optimal temperatures for several field populations have been determined to be close to that of the environment, and the alga will not grow at temperatures below 54°C. Cyanobacteria are not characteristic of acidic hot springs, those having pH lower than 5.

In hot, arid desert habitats, cyanobacteria typically occupy the subsurface spaces within porous crystalline sandstone and limestone rocks. Surveys conducted within arid regions of the United States indicate very widespread occurrence; in one study, more than 90% of sampled sandstone sites were occupied by bands of cryptoendolithic cyanobacteria (*crypto* = hidden, *endo* = inside, *lith* = rocks). The proximity of the bands to the surface and band thickness varied from site to site, depending in large degree upon the color and porosity of the rock substrate. Biomass was greatest in the lightest-colored and most porous rocks. Field measurements suggest that photosynthesis occurs only when water is present and that when it is not, cellular metabolism becomes dormant (Bell 1993).

Cyanobacteria also occur in the cold deserts of Antarctica and other polar regions (Vincent 2000; Wynn-Williams 2000). The dry valleys of Antarctica's Victoria Land consist of extensive areas of rock and soil lacking snow or ice cover, possibly representing the most extreme environment on Earth. Here, cyanobacteria occur in the mostly frozen lakes and melt streams that form in the austral summer. Filamentous cyanobacteria form mats on the bottom of the shoreline zone and sometimes also occur in the water column (Spaulding et al. 1994). Cyanobacteria also are found at higher polar altitudes, living as cryptoendoliths at sites where summer high temperatures reach only about 0°C, and winter temperatures drop to −60°C.

Another type of extreme habitat that cyanobacteria tolerate is hypersaline water, where salinities are considerably higher than seawater. These include saline lakes, hypersaline marine lagoons, and solar evaporation ponds (salterns). *Aphanothece halophytica* is a unicellular, rod-shaped halophile (salt-lover) that often occurs in such habitats (Yopp et al. 1978). *Synechococcus*-like forms are reported to survive and remain metabolically active within crystalline salt deposits for as long as 10 months (Rothschild et al. 1994).

6.5 Evolution and Diversity

The evolution and diversity of cyanobacteria has been studied with two main approaches: fossils and phylogenetic analyses of modern representatives.

The Fossil History of Cyanobacteria

Fossil evidence indicates that major lineages of cyanobacteria had originated by 2.3–2.45 billion years ago (Eigenbrode and Freeman 2006; Tomitani et

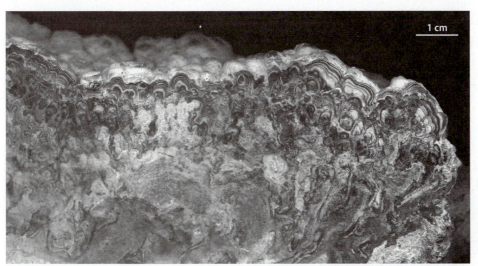

Figure 6.22 A stromatolite (seen actual size) that has been split open, revealing the layering typical of these formations. (Photo: L. W. Wilcox)

al. 2006), well before the establishment of eukaryotic cells some 1.5 billion years ago (Javaux et al. 2001). But it is not clear when oxygenic photosynthesis originated, even though this is important as a feature that distinguishes cyanobacteria from other prokaryotes. Diverse fossils resembling cyanobacteria have been found in ancient stromatolites (Figure 6.22), layered calcareous mounds as much as 3.5 billion years old. These ancient fossils occur in six structural types: small solitary, paired, or clustered rods; small single, paired, or clustered spheres; large solitary or clustered spheres; narrow filaments; broad filaments (Figure 6.23); or broad filaments having an enclosure resembling a sheath (Schopf 1993, 2006). Similar structural variation can be observed among modern cyanobacteria, and layered stromatolites are produced by cyanobacteria in some shallow ocean waters today. However, skeptics argue that such very ancient fossils and stromatolites could be the remains of non-photosynthetic or anoxygenic photosynthetic bacteria or could have resulted from processes that did not involve living organisms. The recent application of 2D Raman spectroscopy to the ancient fossil-like structures reveals the presence of carbon, indicating that they are likely to be the remains of life (Schopf et al. 2007). In addition, Allwood and associates (2006) studied several types of 3.4-billion-year-old stromatolites and concluded that living organisms produced them. Even so, the nature of the organisms that produced the most ancient stromatolites remains open to question.

Chemical fossils known as 2-methylhopanes, dating from 2.8 billion years ago, have been regarded

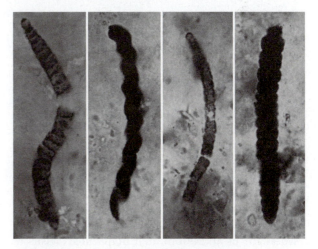

Figure 6.23 Filaments discovered in the 3.5-billion-year-old Apex Basalt in Western Australia that are thought to represent cyanobacteria. (Reprinted with permission from Schopf, J. W., Microfossils of the Early Archean Apex chert: New evidence of the antiquity of life. *Science* 260:640–646. © 1993 American Association for the Advancement of Science)

as evidence for the presence of cyanobacteria because these compounds derive from 2-methylhopanoids (Figure 6.24), which are largely produced by cyanobacteria today (Buick 1992; Brocks et al. 1999, 2003). However, it is possible that such chemical fossils were produced by anoxygenic cyanobacteria. Some have proposed that oxygenic photosynthesis first arose some 2.3 billion years ago, in the process causing a global glaciation event known as the Makganyene snowball Earth (Kopp et al. 2005).

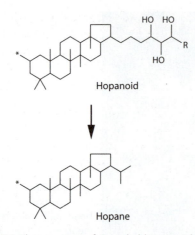

Figure 6.24 The structure of 2-methyl hopanoid and the 2-methyl hopane molecular fossil derived from it. (After Dietrich et al. 2006, by permission of Elsevier Science & Technology Journals)

Collectively, the available data suggest that oxygenic photosynthesis arose sometime between 3.2 and 2.4 billion years ago (Falkowski 2006) or between 3.4 and 2.3 billion years ago (Allen and Martin 2007). Tomitani et al. (2006) estimated that by 2.4 to 2.3 billion years ago, oxygen levels had become high enough to select for heterocytes as a way of protecting nitrogenase from oxygen inactivation. This inference is supported by the appearance of fossil akinetes in deposits about 1.6–1.4 billion years old—and perhaps 2.1 billion years ago.

Geochemical evidence indicates that cyanobacteria had colonized land by 2.6 billion years ago (Watanabe et al. 2000b), and undisputed fossils of terrestrial cyanobacteria are known from 1.2 billion-year-old rocks found in Arizona in the United States (Horodyski and Knauth 1994). Meanwhile, in the oceans, the number of stromatolite types rose to a maximum some 700–800 million years ago, in the Cambrian Epoch, and declined precipitously after that time. Some experts suggest that this decrease was caused by the evolution of numerous types of herbivorous gastropods, which graze on cyanobacteria, thereby preventing the buildup of stromatolites. The fact that modern stromatolites develop best in hypersaline areas where gastropods cannot survive supports this hypothesis. Other hypotheses that might explain the post-Cambrian decline of stromatolites include the rise of burrowing organisms that may have disrupted growth or the appearance of red and green calcareous seaweeds and sessile animals that competed with stromatolite-forming cyanobacteria

for available space. Even so, cyanobacterial stromatolites have persisted throughout the intervening time to the present, where they are known from about 20 locations (Tucker and Wright 1990).

The fossil record also demonstrates that the structural and reproductive complexity of cyanobacteria has increased over time. Fossils whose morphology is diagnostic for cyanobacteria (i.e., could not be confused with nonphotosynthetic or anoxygenic photosynthetic bacteria) include 2-billion-year-old remains of *Eoentophysalis belcherensis* from the Gunflint Formation in Canada. This fossil is very similar to the modern cyanobacterial genus *Entophysalis*, a colonial form that lacks specialized cells (Knoll 1996). The earliest known fossil remains of cyanobacterial specialized cells are 1.4–2.1-billion-year-old *Archeoellipsoides*, which resemble the akinetes of modern *Nostoc* (Knoll 1996, Tomitani et al. 2006). Distinctive cyanobacteria left 700-million-year-old fossils (Figure 6.25a) very similar to certain modern forms (Figure 6.25b) that grow in shallow marine waters (Green et al. 1987). These and earlier fossils suggest that cyanobacterial structure has not changed much for hundreds of millions of years (Schopf 1996; Knoll 2003).

Phylogeny and Diversity of Modern Cyanobacteria

In the past, taxonomists primarily used morphological features such as body form to define cyanobacterial taxa. Although such criteria remain in modern use (e.g., Anagnostidis and Komárek 1985, 1988; Komárek and Anagnostidis 1986, 1989), species delineation and classification systems increasingly depend upon molecular phylogenetic methods (see Chapter 5). Such methods indicate that a clade of cyanobacteria that produces specialized heterocytes and akinetes, consistent with the classical order Nostocales, is likely monophyletic (Tomitani et al. 2006 and previous work cited therein). However, other orders, as well as a number of genera defined by morphological characteristics, do not form monophyletic groups. For example the filamentous cyanobacteria that lack heterocytes, classically called Oscillatoriales, do not form a lineage distinct from unicellular and colonial cyanobacteria (classical Chroococcales) (Wilmotte 1994). This phylogenetic pattern indicates that filamentous bodies have evolved multiple times. The filamentous, heterocyteous, branched cyanobacteria classically known as Stigonematales (Anagnostidis and Komárek 1990) likewise appear to be polyphyletic (Gugger and Hoffman 2004).

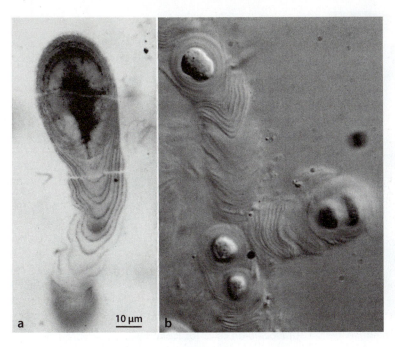

Figure 6.25 Fossil cyanobacterium and modern analogue (a) *Polybessurus bipartitus*, a 700–800-million-year-old fossil cyanobacterium. (b) An modern equivalent whose existence was predicted based on discovery of the fossil form. (a: From Knoll 1996; b: Courtesy A. Knoll)

Consequently, the most recent edition of *Bergey's Manual of Systematic Bacteriology* classifies cyanobacteria into 5 subsections, rather than classical orders (Wilmotte and Herdman 2001). A comparative analysis of 13 cyanobacterial genomes revealed the occurrence of a core set of 323 genes that seem to evolve in tandem, thus providing conserved features useful in phylogenetic studies. Together, 16S rDNA gene sequences and these core genes indicate that the earliest cyanobacteria were not nitrogen fixers and may have been thermophilic in habitat (Shi and Falkowski 2008). As the result of continued flux in cyanobacterial classification, the following examples of cyanobacterial diversity are organized informally.

Unicellular and colonial forms lacking specialized cells or reproductive processes

SYNECHOCOCCUS (Gr. *synechos*, in succession + Gr. *kokkos*, berry) (Figure 6.26) is a tiny (1 μm in diameter) cylindrical unicell that is an important primary producer in the plankton of fresh and marine waters, and it has also been collected from surfaces of algae and plants, films of algae on sandy beaches, outflow channels of hot springs and thermal pools, and endolithic habitats. Cell lengths are typically two to three times their width. There is no mucilaginous sheath. Cells may exhibit swimming

motility. Molecular phylogenetic analysis indicates that this genus is not monophyletic and should be broken up into multiple genera (Honda et al. 1999). Both marine and freshwater forms of *Synechococcus* have been associated with "whiting" events, the production of suspended fine-grained carbonates that eventually contribute to sedimentary carbonate deposition (Hodell et al. 1998). Such late-summer calcite precipitation and cell sedimentation also increase the flux of organic carbon to the sediments of lakes (Hodell and Schelske 1998). A specialized cell

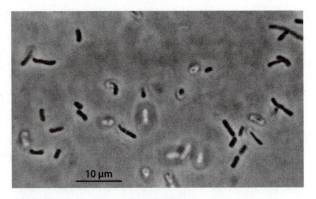

Figure 6.26 The small cylindrical cells of *Synechococcus*. (A TEM view of *Synechococcus* is shown in Figure 6.7.) (Photo: L. W. Wilcox)

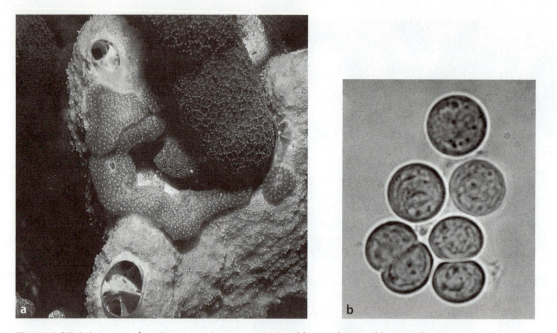

Figure 6.27 (a) An ascidian (sea squirt) containing *Prochloron*. (b) *Prochloron* cells. (a: From Lewin and Cheng 1989; b: R. Hoshaw, from Phycological Society of America slide collection)

surface layer functions as a template for carbonate deposition and is shed after it has become coated with calcium carbonate (see Figure 6.7). A new surface layer then forms, and the process is repeated (Thompson et al. 1997). *Synechococcus elongatus* PCC 7942 has been used as a model system for study of the molecular basis of circadian rhythms (Smith and Williams 2006) and a bicarbonate transporter (Omata et al. 1999). Group I introns, which are common in fungal, protist, and organellar genomes yet rare in bacteria and other organisms, have been found in two strains of *Synechococcus* (Haugen et al. 2007).

PROCHLORON (Gr. *pro*, before + Gr. *chloros*, green) is a unicell that lives primarily in association with marine didemnid ascidians, colonial tunicates also known as sea squirts (Figure 6.27). Ascidians are small, flattened animals that attach themselves to coral rubble, mangrove roots, sea grass blades, and other shallow marine substrates in tropical regions throughout the world. *Prochloron* occurs only in coastal waters within the temperature range of 21°C–31°C and ceases photosynthesis below 20°C. The alga inhabits the animals' cloacal cavities and upper surfaces and sometimes is also found intracellularly. Although

the algae release glycolate, a photosynthetic product that contributes to the animals' nutrition, the didemnids are probably not dependent upon the algae, and they require additional foods. In contrast, *Prochloron* probably does require its host in ways that are not yet understood; by itself the alga has been difficult to maintain in the laboratory. Some forms of *Prochloron* can fix nitrogen in the light, but only in association with their didemnid hosts (Paerl 1984).

Prochloron, like the unicellular *Prochlorococcus* and filamentous *Prochlorothrix*, is unusual in possessing chlorophyll *b* in addition to chlorophyll *a* and lacking phycobilisomes. These cyanobacteria thus appear grass green. Green pigmentation is believed to be adaptive in low-N habitats because production of the chlorophyll *a/b* protein complex requires only about one-third as much fixed nitrogen as do phycobiliproteins. The green pigmentation may also allow enhanced absorption of orange-red wavelengths of light present in shallow waters (Alberte 1989). Pigments occur in stacked thylakoids, and the central portion of *Prochloron* cells is occupied by a large central space, both unusual for cyanobacteria. In other ways, including cell wall and storage biochemistry and absence of organelles, *Prochloron* resembles other cyanobacteria.

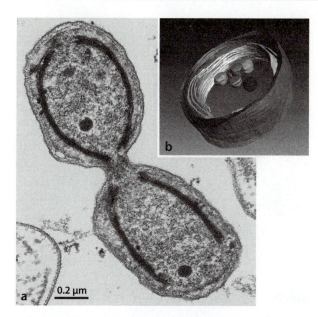

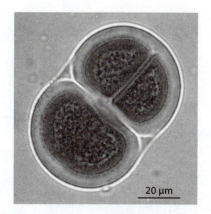

Figure 6.28 *Prochlorococcus.* (a) TEM view of a dividing cell. (b) A surface-rendered model, which was visualized by cryo-electron microscope tomography in a near-native state. (Images: C. Ting, Department of Biology, Williams College)

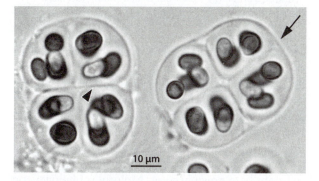

Figure 6.29 *Chroococcus turgidis,* a common bog-dwelling cyanobacterium. Following cell division, a few cells tend to remain attached to one another within the firm sheath. (Photo: L. E. Graham)

Figure 6.30 *Gloeocapsa.* Note that cells occur inside of concentric sheaths (arrow and arrowhead), reflecting the pattern of cell division. (Photo: L. W. Wilcox)

PROCHLOROCOCCUS (Gr. *pro*, before + Gr. *chloros*, green + Gr. *kokkos*, berry) (Figure 6.28) is one of the most numerous components of open-ocean plankton and was discovered using flow-cytometric methods (Chisholm et al. 1992). Superficially it is similar to *Synechococcus*, but lacks phycobilisomes and contains chlorophyll *b*. In the case of *Prochlorococcus*, green pigmentation may be adaptive in harvesting the blue light that penetrates relatively deeply into ocean waters. At least six known ecotypes occur in environments differing in temperature, irradiance, nutrients, and presence of competitors (Moore et al. 1998; Johnson et al. 2006). Genetic adaptation for phosphate acquisition (Martiny et al. 2006) and sulfolipid production as a phosphorus sparing adaptation (Van Mooy et al. 2006) have been studied in this genus.

CHROOCOCCUS (Gr. *chroa*, color of the skin + Gr. *kokkos*, berry) occupies freshwaters (particularly soft waters, including bogs), marine waters, and moist terrestrial locations. *Chroococcus* occurs as single cells or colonies of 2, 4, 16, or, less frequently, 32 hemispherical cells resulting from the adherence of daughter cells after division (Figure 6.29). The algae may be free floating, mixed with tangles of filamentous algae, or adherent to substrates. Many

species are deep blue-green in color, but a central, paler region is often distinguishable in cells. A gelatinous sheath may or may not be noticeable. The cells of one species, *C. giganteus*, are unusually large for cyanobacteria, about 50–60 μm in diameter.

GLOEOCAPSA (Gr. *gloia*, glue + L. *capsa*, box) lives in freshwater lakes or on moist soil and other terrestrial surfaces. It can form black bands on high intertidal seacoast rocks and live beneath rock surfaces in arid regions. It is a phycobiont in certain lichens (Rambold et al. 1998). *Gloeocapsa* forms colonies having more cells than *Chroococcus*. The cells are oval or ellipsoidal, with rounded ends (Figure 6.30). Each cell has a distinct mucilaginous sheath that is surrounded by older sheath material. Sometimes the sheath is colored yellow or brown, but it is most often colorless.

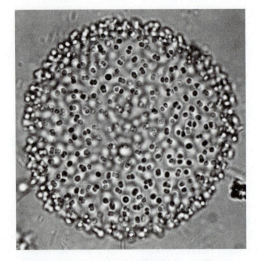

Figure 6.31 *Aphanocapsa.* Cells of this colonial cyanobacterium are spherical and without individual sheaths. (Photo: L. E. Graham)

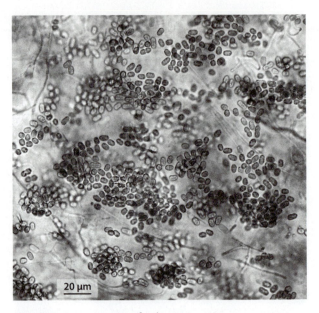

20 µm

Figure 6.32 A portion of a large (ca. 2 cm in diameter), mucilaginous colony of *Aphanothece.* Note the cylindrical shape of the cells. (Photo: L. W. Wilcox)

APHANOCAPSA (Gr. *aphanes*, invisible + L. *capsa*, box) is a free-floating member of the freshwater plankton, particularly in bog waters. Very large, usually irregularly shaped colonies consist of dozens of spherical cells evenly distributed within a colorless, or sometimes yellowish, gelatinous sheath (Figure 6.31). The individual cells are quite small, one to several micrometers in diameter, and can be pale to bright blue-green in color. Molecular phylogenetic analysis suggests that this genus is not monophyletic (Willame et al. 2006).

APHANOTHECE (Gr. *aphanes*, invisible + Gr. *theke*, sheath or box) occurs in the plankton of hard or soft freshwaters, on muddy bottoms of fresh and marine waters, and sometimes in moist terrestrial

sites. Very large, even macroscopic colonies contain many cylindrical cells embedded in transparent mucilage (Figure 6.32). The mucilage of individual cells is usually indistinct. One species grows within the mucilage of other colonial cyanobacteria. *Aphanothece* is probably not a monophyletic genus (Willame et al. 2006).

MERISMOPEDIA (Gr. *merismos*, division + Gr. *pedion*, plain) is found floating or sedentary in freshwaters and marine waters, including sand flats. It is a typical constituent of bog-water communities. The ovoid or spherical cells are compactly arranged in orderly rows

Figure 6.33 *Merismopedia,* whose cells occur in a flat sheet, often rectangular in outline (a and b). In (b), a colony is shown adjacent to a two-celled *Chroococcus,* illustrating the considerable size variation that occurs among cyanobacterial cells. (Photos: L. W. Wilcox)

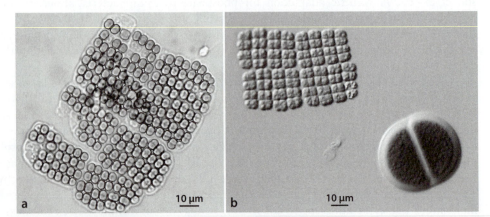

a   10 µm   b   10 µm

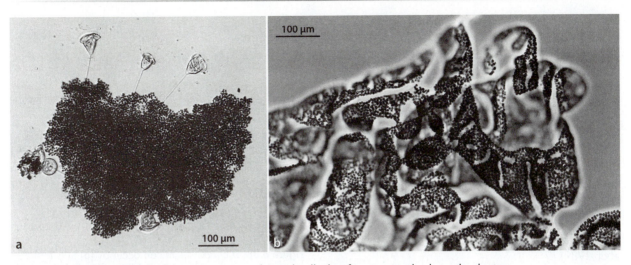

Figure 6.34 *Microcystis*, which has numerous spherical cells that form variously shaped colonies. Several vorticellid ciliates are attached to the relatively small colony shown in (a). In (b), a tortuous, perforated colony having a firm sheath is shown in an india ink preparation. Note the dark appearance of the cells, which contain numerous gas vesicles. (Photos: L. W. Wilcox)

in flat, rectangular colonies one cell thick (Figure 6.33). A mucilaginous matrix holds the colony together but does not extend far beyond the cell limits, and the mucilage enclosures of individual cells are indistinct.

MICROCYSTIS (Gr. *mikros*, small + Gr. *kystis*, bladder) exists as colonies of small cells that are evenly distributed throughout a gelatinous matrix. Molecular systematic studies suggest that *Microcystis* is a monophyletic genus (e.g., Willame et al. 2006). When young, the colonies are spherical, but older colonies are typically irregular in shape and frequently perforated (Figure 6.34). Numerous gas vesicles in the cells of *Microcystis* commonly give a black appearance and influence buoyancy (Visser et al. 2005). Low numbers of *Microcystis* may occur in soft or hard oligotrophic waters. Eutrophic freshwaters often contain high populations, which can develop from benthic reservoirs (Brunberg and Blomqvist 2003). *Microcystis aeruginosa* and *M. flos-aquae* can be very common in nuisance blooms in eutrophic lakes; Doers and Parker (1988) elucidated differences between these two species in culture. The highest photosynthetic rates for *Microcystis* occur at surface irradiance levels, and it is resistant to photoinhibition (Paerl et al. 1985). Thus *Microcystis* may be particularly well adapted to grow at the air–water interface, where it may be able to escape carbon limitation. The gas vacuoles of *Microcystis* are often too strong to be collapsed by turgor pressure. These factors help to explain this organism's frequent tendency to form thick

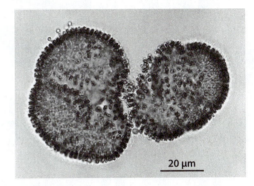

Figure 6.35 The hollow, colonial *Coelosphaerium*. (Photo: C. Taylor)

surface blooms. *Microcystis* produces the hepatotoxin microcystin and thus is of considerable health concern. PCR-based methods have been developed for monitoring the presence of *Microcystis* and genes required in the biosynthesis of microcystin in natural waters (Ouellette et al. 2006).

COELOSPHAERIUM (Gr. *koilos*, hollow + Gr. *sphairion*, small ball) (Figure 6.35) occurs in the plankton of freshwaters, often co-occurring with *Microcystis*. *Coelosphaerium* colonies are spherical or irregular in shape, and, as with *Microcystis*, the cells may be so full of gas vesicles as to give them a black appearance. Unlike with *Microcystis*, the cells of *Coelosphaerium* are distributed in a single layer,

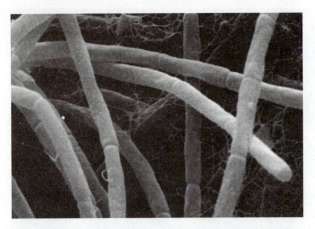

Figure 6.36 *Prochlorothrix*. (SEM: S. Seufer, from Matthijs et al. 1989)

Figure 6.38 *Trichodesmium* can occur as (a) tufts formed by parallel arrays of filaments or (b) puffs—radial arrays of filaments. (Photos: S. Janson)

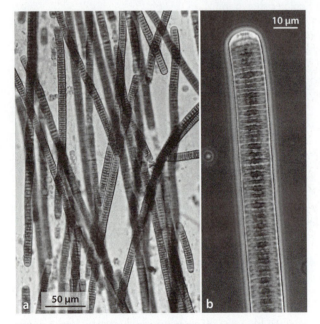

Figure 6.37 (a) A number of *Oscillatoria* filaments. (b) The tip of a filament at higher magnification. The cells are much wider than they are long. (Photos: L. W. Wilcox)

forming a hollow balloon-like structure. *Coelosphaerium naegelianum* is a common constituent of nuisance surface blooms in eutrophic waters.

Filamentous cyanobacteria lacking spores, heterocytes, or akinetes

PROCHLOROTHRIX (Gr. *pro*, before + Gr. *chloros*, green + Gr. *thrix*, hair) (Figure 6.36) has been collected from shallow lakes arising from peat extraction in the Netherlands. Filaments are composed of elongated cells. *Prochlorothrix* is notable for having chlorophyll *b* in addition to chlorophyll *a* and for the absence of phycobilisomes, as is also the case for *Prochloron* and *Prochlorococcus* (Turner et al. 1989). However, *Prochlorothrix* does not appear to be closely related to either. It can readily be cultured in the laboratory (Burger-Wiersma et al. 1986; Matthijs et al. 1989).

OSCILLATORIA (L. *oscillator*, something that swings) has been found in a wide variety of environments, including hot springs, marine habitats, and lakes of temperate, tropical, and polar regions, as well as moist terrestrial substrates. The name reflects this organism's ability to rotate, or oscillate. Gliding motility also occurs. Disk-shaped cells that are typically wider than they are long are attached end-to-end to form filaments (Figure 6.37). The end cell may be rounded or distinctive in ways that are used to define some of the species. Breakage of filaments between separation disks releases shorter hormogonia during vegetative reproduction. Filaments may sometimes adjoin in parallel to form thin films. The planktonic species *O. rubescens* and *O. agardhii* var. *isothrix* have been changed to *Planktothrix rubescens* and *P. mougeotii*, respectively. Some other former species of *Oscillatoria* have been transferred to *Limnothrix*, *Tychonema*, and *Trichodesmium*. Some *Oscillatoria*, *Planktothrix*, and other filamentous, freshwater cyanobacteria are notorious for producing the tertiary alcohols geosmin and 2-MIB, which

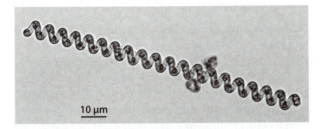

Figure 6.39 *Spirulina.* (Photo: L. W. Wilcox)

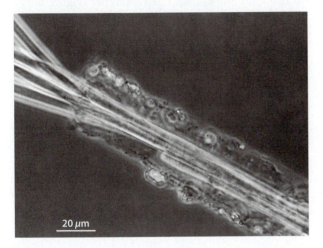

Figure 6.40 *Microcoleus.* Note the multiple filaments within the wide sheath. (Photo: L. W. Wilcox)

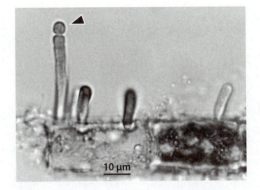

Figure 6.41 Cyanobacterial exospore. A single exospore (arrowhead) has formed on the largest of these *Chamaesiphon* individuals. (Photo: L. W. Wilcox)

cause drinking water to smell or taste muddy. Jüttner and Watson (2007) provide a list of cyanobacterial (and other bacterial) species known to produce these compounds.

TRICHODESMIUM (Gr. *thrix*, hair + Gr. *desmos*, bond) is common in tropical open-ocean waters, sometimes forming visible blooms. Colonies consisting of aggregated filaments (Figure 6.38) may be visible with the naked eye, at 0.5–3 mm in diameter. It is an important component of the phytoplankton in the tropical North Atlantic Ocean in terms of standing crop and productivity. Because it fixes significant amounts of nitrogen (30 mg m^{-2} day^{-1}), it is a major source of combined nitrogen in oligotrophic waters. The presence of abundant gas vesicles reduces algal density, and most of the biomass of this alga occurs in the upper 50 m of water. Some copepods and mesograzers feed on *Trichodesmium* (Carpenter and Romans 1991). Capone et al. (1997) reviewed the biology of *Trichodesmium*.

SPIRULINA (Gr. *speira*, coil) is found in marine and fresh waters, including lakes high in sodium carbonate (natron lakes) and bog lakes, and on moist mud. *Spirulina* (Figure 6.39) looks like a highly coiled version of *Oscillatoria* and exhibits similar oscillatory and gliding motility. *Spirulina* has been used as a human food supplement.

MICROCOLEUS (Gr. *mikros*, small + Gr. *koleos*, sheath) occurs on the surface of silty sand in marine salt marshes and lagoons, in freshwater lakes and on their muddy banks, and in terrestrial cryptogamic crust communities. Several to many trichomes occupy a single, relatively stiff communal sheath (Figure 6.40). *Microcoleus chthonoplastes*, identified by SSU rDNA sequence signature, occurs in marine microbial mats around the world (Stal 1995).

Exospore-producing cyanobacteria

CHAMAESIPHON (Gr. *chamai*, dwarf + Gr. *siphon*, tube) (Figure 6.41) is a common inhabitant of the surfaces of aquatic plants, algae, and nonliving substrates such as stones or shells in freshwater lakes and streams. The cells are cylindrical, with an open, widened sheath at the top, from which a sequence of rounded spores bud off. The pigmentation is often grayish green.

Baeocyte- (endospore-) producing cyanobacteria

CHROOCOCCIDIOPSIS (*Chroococcidium*, a genus of cyanobacteria [Gr. *chroa*, color of the skin + Gr. *kokkos*, berry] + Gr. *opsis*, likeness) occurs in extremely arid, cold, or hot terrestrial environments

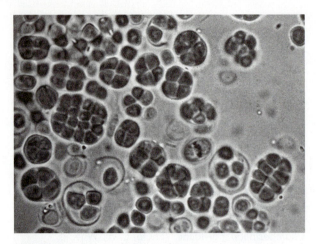

Figure 6.42 *Chroococcidiopsis.* Note the presence of baeocytes. (Photo: E. I. Friedmann)

as unicells (Figure 6.42). Reproduction is by subdivision of the cytoplasm into numerous small **baeocytes**. These escape by breakage of the parental cell wall and each can develop into a mature vegetative cell.

Heterocyte and akinete-producing cyanobacteria

NOSTOC (name given by Paracelsus, Swiss physician) primarily occurs in terrestrial habitats, frequently in association with fungi in lichens and with some bryophytes and vascular plants. It occurs on moist rocks and cliffs, on alkaline soils, and in wet meadows and at the edges of shallow lakes. *Nostoc* is frequently abundant in flooded rice paddies, where it contributes to the fertilization of some 2 million hectares. The ecology of *Nostoc* has been reviewed (Dodds et al. 1995). Characteristically bent or kinked filaments of round cells are held in a firm mucilaginous matrix to form colonies that can reach 50 cm in diameter but are more commonly marble size or smaller (Figure 6.43). The sheaths can be colored yellow, brown, or black. Transiently motile hormogonia are produced; these typically have heterocytes at the ends. The genome of *N. punctiforme* PCC 73102 (ATCC 29133) (isolated from the cycad *Macrozamia*) has been sequenced and is available at the Joint Genome Institute (JGI) web site. A molecular phylogenetic analysis of cultures obtained from symbioses and classified as *Nostoc* found that one clade was monophyletic and distinguishable from *Anabaena* but that others were not (Svenning et al. 2005). Although the strains examined in another study suggested that the genus *Nostoc* was monophyletic, they could be clustered into more than one genus (Rajaniemi et al. 2005).

ANABAENA (Gr. *anabaino*, to rise), sometimes spelled *Anabaina*, is primarily planktonic in freshwaters and marine waters such as the Baltic Sea, where some forms can be nuisance bloom-formers and toxin-producers. Like *Nostoc*, *Anabaena* consists of

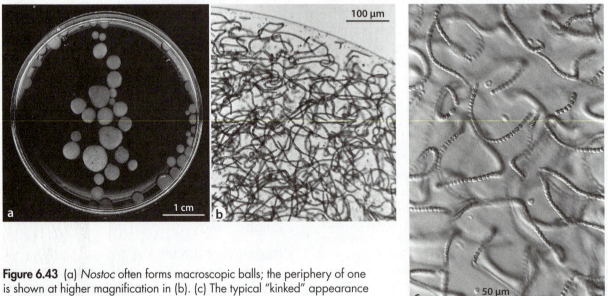

Figure 6.43 (a) *Nostoc* often forms macroscopic balls; the periphery of one is shown at higher magnification in (b). (c) The typical "kinked" appearance of the filaments is evident at greater magnification. (Photos: L. W. Wilcox)

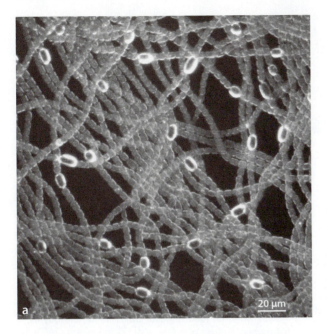

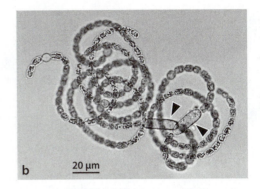

Figure 6.44 *Anabaena.* In (a), filaments are seen under dark-field illumination, where they often fluoresce a bright red color; the clear heterocytes stand out against this background. In (b), two elongate akinetes (arrowheads) are visible in a specimen taken from a freshwater bloom of the three notorious bloom-forming taxa, *Anabaena, Aphanizomenon,* and *Microcystis.* (Photos: L. W. Wilcox)

filaments of spherical cells resembling closely strung beads embedded in a mucilaginous matrix (Figure 6.44). When classical morphological features are used to differentiate *Anabaena* from *Nostoc*, the former is said to have a looser, more indistinct sheath and less contorted filaments. Molecular phylogenetic analyses indicate that *Anabaena* is not a monophyletic group and could be separated into multiple clades (Rajaniemi et al. 2005; Willame et al. 2006).

APHANIZOMENON (Gr. *aphanizomenon*, that which makes itself invisible) (Figure 6.45) consists of distinctive aggregations of filaments that can be so large as to be visible with the naked eye. Large blooms may form in the Baltic Sea and in eutrophic inland waters. Molecular phylogenetic analysis indicates that *Aphanizomenon* is nested within *Anabaena*, as the latter is classically defined (Rajaniemi et al. 2005) and may not be monophyletic (Willame et al. 2006).

CYLINDROSPERMUM (Gr. *kylindros*, cylinder + Gr. *sperma*, seed) occurs in freshwaters including soft, acid lakes, often forming dark green patches on submerged vegetation and on moist soil.

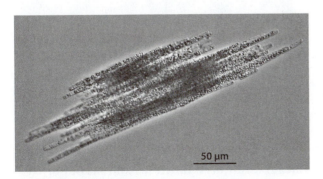

Figure 6.45 A small colony of *Aphanizomenon*, whose filaments are straight and tend to lie in parallel. (Photo: L. W. Wilcox)

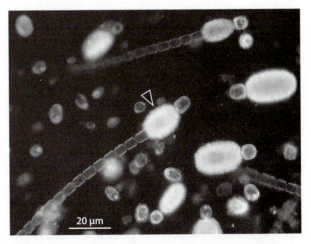

Figure 6.46 A large, ornamented akinete (arrowhead) lies adjacent a basal heterocyte in *Cylindrospermum*. (Photo: L. W. Wilcox)

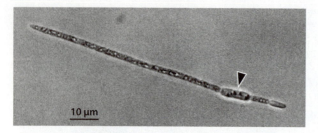

Figure 6.47 *Cylindrospermopsis.* Note the somewhat more rectangular cells than *Cylindrospermum* and the akinete (arrowhead) one cell removed from the basal heterocyte. (From Chapman and Schelske 1997, by permission of the *Journal of Phycology*)

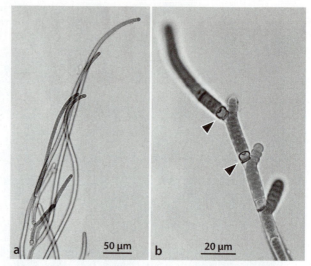

Figure 6.48 *Tolypothrix,* which exhibits single false branching, shown in (a) low and (b) higher magnification views. Branches tend to occur adjacent to heterocytes (arrowheads) (see also Figure 6.17). (Photos: L. W. Wilcox)

Filaments of vegetative cells, whose profiles are square or rectangular, with a heterocyte at one end, are enclosed in loose mucilage. A large, highly ornamented akinete may develop from the lowermost vegetative cell (Figure 6.46).

CYLINDROSPERMOPSIS (*Cylindrospermum* + Gr. *opsis,* likeness) is an important nitrogen-fixing, planktonic bloom-former in eutrophic temperate and tropical freshwaters around the world. Rectangular cells occur in linear or coiled filaments having a basal heterocyte. Akinetes may form basally, but not immediately adjacent to the heterocyte (Figure 6.47). If akinetes are not present, correct species identification may be difficult. *Cylindrospermopsis raciborskii* is of particular concern because it has in recent years formed prodigious growths with concentrations reaching nearly 200,000 filaments ml^{-1} (Chapman and Schelske 1997). This species produces the hepatotoxin cylindrospermopsin that has caused poisonings. A PCR-based procedure that targets a region of the *rpoC1* gene unique to *C. raciborskii* provides an accurate identification method (Wilson et al. 2000).

TOLYPOTHRIX (Gr. *tolype,* ball of yarn + Gr. *thrix,* hair) is planktonic or found entangled among submergent vegetation in freshwater lakes, including soft or acidic waters. Filaments are enclosed in a sheath of variable consistency and are highly **false-branched**, usually at a heterocyte (Figure 6.48). Single false branches are formed by the continued growth of the filament on one side of the heterocyte but not the other. Sometimes double false branches occur when both ends of an interrupted filament continue to grow. The frequency of false branching

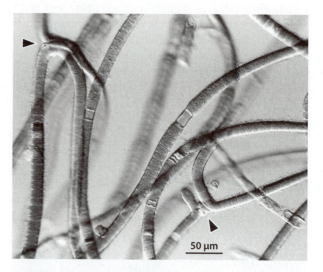

Figure 6.49 *Scytonema.* Note two occurrences of double false branching (arrowheads). (Photo: L. W. Wilcox)

gives these algae a woolly appearance, hence the name.

SCYTONEMA (Gr. *skytos,* leather + Gr. *nema,* thread) forms dark tufted mats in masses of other algae or vegetation in lakes of various types or on terrestrial surfaces, including stones, wood, and soil. It is a phycobiont in several orders of lichens

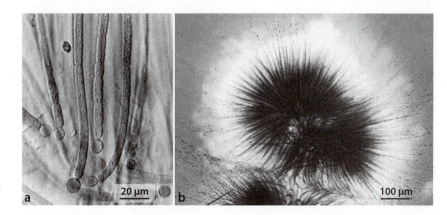

Figure 6.50 *Gloeotrichia.* In (a), the basal heterocytes are evident. The specimen in (b) was stained with india ink to show the extensive mucilage surrounding the colony. (Photo: L. W. Wilcox)

(Rambold et al. 1998). Often, but not always, double false branches occur. This results from the outgrowth from both ends of a filament that has been interrupted by death of a cell or, less commonly, by heterocyte differentiation (Figure 6.49). Heterocyte walls may be darkly pigmented. A tough and sometimes clearly layered, brown or orange-colored sheath is common.

GLOEOTRICHIA (Gr. *gloia*, glue + Gr. *thrix*, hair) occurs in freshwater habitats, attached to submerged substrates. Akinetes occur adjacent to basal heterocytes (sometimes appearing in chains) (Figure 6.50). *Gloeotrichia* colonies often detach from substrates, becoming planktonic. Growths can reach bloom proportions.

True-branching cyanobacteria

STIGONEMA (Gr. *stizo*, to tattoo + Gr. *nema*, thread) most commonly inhabits moist rocks and soil, but some species are aquatic, attached to submerged wood or entangled among other algae. It may form brown tufted mats or cushions on submerged portions of lake macrophytes. *Stigonema* is the primary phycobiont in certain lichens (Rambold et al.

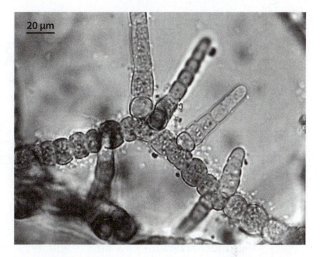

Figure 6.51 *Stigonema.* Note the true branches. (Photo: L. E. Graham)

1998) and is associated with boreal mosses, where it contributes to nitrogen availability. The filaments are more than one cell in width, a condition described as multiseriate, and are **true-branched**, i.e., branches arise by division of cells in a direction perpendicular to that of the main filament axis (Figure 6.51). The branches may also be pluriseriate.

7 Chapter

Endosymbiosis and the Diversification of Eukaryotic Algae

With a Focus on Glaucophytes and Chlorarachniophytes

If you could travel back in time 3 billion years or so, prokaryotes would likely be the only life forms you would find on Earth. You would have to point the time machine ahead 1.5 billion years to find the earliest eukaryotes. But if you could fast-forward about 500 million years—to 1 billion years or so prior to the present day—you could find diverse phyla of eukaryotic algae, which are known from fossils. This evolutionary blossoming of eukaryotes was fostered by the development of a stratospheric layer of ozone that protected life in shallow waters from UV damage. The early ozone layer, generated from oxygen released from cyanobacterial photosynthesis, helped the first eukaryotes to thrive and diversify, just as today's ozone layer protects modern life on Earth. An equally important role was played by endosymbiosis, a state in which organisms live within the cells or bodies of others. For example, cyanobacteria living symbiotically within the cells of early, non-photosynthetic eukaryotes endowed them with the power of photosynthesis and became the first plastids. Since then, endosymbiosis has continued to be an important ecological and evolutionary force.

This chapter begins by briefly considering the origin of eukaryotic algae and defining several types of endosymbiosis that have influenced algal plastid evolution. We then focus on the widespread occurrence of endosymbiosis in the modern world and the roles of endosymbiosis in the diversification of algal phyla, highlighting the Glaucophyta and Chlorarachniophyta.

Glaucocystis

(Photo: L. W. Wilcox)

7.1 Origin of Eukaryotic Algae

Eukaryotic algal cells are typically larger and more highly compartmented than cyanobacterial cells. Mitochondria (or structures related to them) and nuclei, whose envelopes are both perforated by complex pores and continuous with the endoplasmic membrane, are hallmarks of eukaryotic cells (Figure 7.1). The secretory Golgi apparatus, centrioles, and the motility structures known as 9+2 flagella, cilia, or undulipodia are also unique to eukaryotes. In addition, eukaryotic cells have more elaborate cytoskeletons and cell division processes than do prokaryotes. Many evolutionary biologists are interested in illuminating the process by which eukaryotic cells and their components arose. Fossil and molecular data are used to infer stages in this process and the times at which the earliest eukaryotic cells and the earliest eukaryotic algae appeared.

Fossil Evidence for Early Events in the Diversification of Eukaryotic Algae

Differences in the cellular size and complexity of modern prokaryotes and eukaryotes have been used to infer the identities of ancient fossils, including those interpreted as eukaryotic algae. For example, distinctive coiled fossils as old as 2.1 billion years, known as *Grypania,* were suggested to be the remains of eukaryotic algae because of their dimensions, up to 0.5 m long and 2 mm in diameter (Han and Runnegar 1992) (Figure 7.2). However, whether *Grypania* was photosynthetic is unknown. Other fossils large enough to possibly represent eukaryotic cells occur in deposits as old as 1.8 billion years (Tappan 1980), but they don't have features distinct enough to allow paleontologists to link them with known species. By 1.45 billion years ago, there were diverse types of unicellular organisms known as acritarchs that produced decay-resistant, highly ornamented exterior surfaces. These fossils outwardly resemble the resting stages of

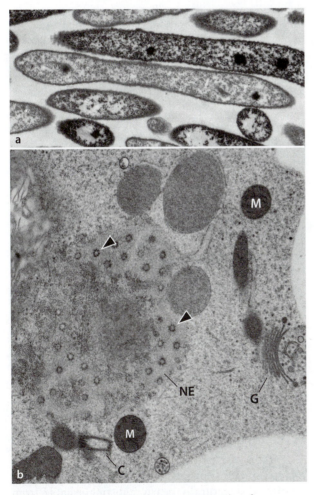

Figure 7.1 Transmission electron micrographs of prokaryotic cells (a) and a eukaryotic cell (b). Note the presence of mitochondria (M), a nuclear envelope (NE) with complex pores (arrowheads), Golgi apparatus (G), and a centriole (C) in the eukaryotic cell. (a: L. W. Wilcox; b: M. E. Cook)

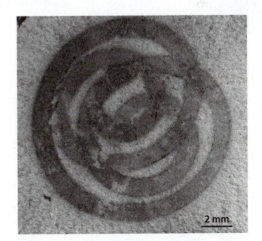

Figure 7.2 The fossil known as *Grypania* from the 1.4- billion-year-old Geoyuzhuan Formation in China, a putative early eukaryote. (Reprinted with permission from Knoll, A. H., The early evolution of eukaryotes: A geological perspective. *Science* 256:622–627 © 1992 American Association for the Advancement of Science)

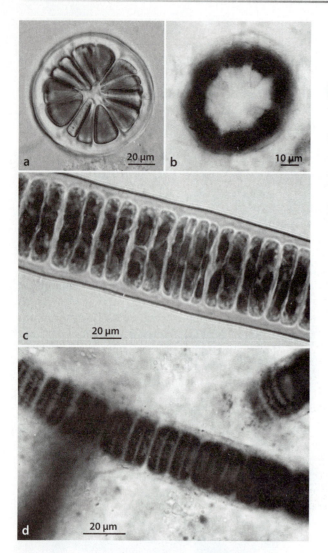

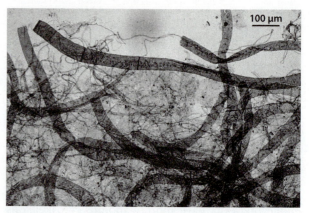

Figure 7.4 Fossils closely resembling the modern stramenopile *Vaucheria* from nearly 1-billion-year-old deposits in eastern Siberia. (Reprinted with permission from Knoll, A. H., The early evolution of eukaryotes: A geological perspective. *Science* 256:622–627. © 1992 American Association for the Advancement of Science)

Figure 7.3 Fossil *Bangiomorpha* and extant *Bangia*.
(a) Cross-section of extant *Bangia,* showing triangular outlines of cells. (b) 750–1250-million-year-old fossil that is interpreted as a cross-sectional view of an ancient *Bangia*-like organism because the cells have a similar shape.
(c) Lengthwise view of a young filament of extant *Bangia*.
(d) Precambrian fossil interpreted as an ancient *Bangia*-like red algal filament because of its similarity to modern forms.
(a: Courtesy A. Knoll; b and d: Reprinted with permission from Butterfield, N. J., A. H. Knoll, and K. Swett. A bangio-phyte red alga from the Proterozoic of arctic Canada *Science* 250:104–107 © 1990 American Association for the Advancement of Science; c: L. E. Graham)

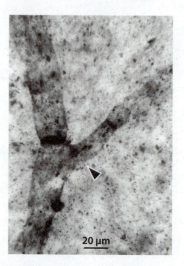

Figure 7.5 Fossils resembling the modern branched green alga *Cladophora* from 750-million-year-old deposits from Spitsbergen. Arrowhead points to branch. (Reprinted with permission from *Nature.* Butterfield, N. J., A. H. Knoll, and K. Swett. Exceptional preservation of fossils in an Upper Proterozoic shale. 334:424–427. © 1988 Macmillan Magazines Ltd.)

modern green algae known as prasinophytes (Chapter 16) (Javaux et al. 2001). The oldest fossils that can confidently be assigned to a modern algal group, based on distinctive structural features, are approximately 1.2-billion-year-old, unbranched filaments known as *Bangiomorpha* (Butterfield 2000). These multicellular fossils are structurally nearly indistinguishable from the modern red alga *Bangia* (Figure 7.3). Additional fossils that share distinctive features with modern algae include 1-billion-year-or-so-old remains (Figure 7.4) linked to the modern filamentous stramenopile *Vaucheria* (Knoll 1992; Xiao et al. 1998; Butterfield

2004) and 700–750-million-year-old branched filaments much like the modern green alga *Cladophora* (Knoll 1992) (Figure 7.5). Though earlier forms likely existed, dinoflagellate fossils became abundant about 200 million years ago, and those of haptophyte coccolithophorids and stramenopile diatoms were diverse and abundant in deposits around 100 million years ago (Chapters 10–12). Microfossil records of dinoflagellates, diatoms, and coccolithophorids were used to calibrate a molecular clock analysis that shows eukaryotes originating about 1100 million years ago. This analysis has challenged the identification of older fossils as members of modern red, brown, or green lineages (Berney and Pawlowski 2006).

Molecular Evidence Bearing on the Origin of Eukaryotic Cells and Mitochondria

Comparative analysis of modern genome sequences indicates that eukaryotic genomes contain many genes inherited from both bacterial and archaeal genomes. Most eukaryotic operational genes—those involved in everyday cellular maintenance—were obtained from bacterial ancestors. By contrast, most eukaryotic informational genes—involved in converting DNA information into proteins—were obtained from archaeal ancestors (Lake 2007). Such data show that the earliest eukaryotes arose from prokaryotic ancestors, but exactly which ones and how the transition occurred remain unknown. Several hypothetical models have been proposed to explain the origin of the first eukaryotes (summarized by Embley and Martin 2006). Most of these models invoke **endosymbiosis**, the process by which one organism becomes stably resident within the cell or body of another to form a chimera. For example, Margulis and associates (2000) proposed that early eukaryotic cells resulted from an endosymbiotic merger between an archaean and a motile eubacterium. The formation of such endosymbiotic partnerships could explain the archaeal/bacterial gene combinations found in modern eukaryotic genomes.

Mitochondria are organelles involved in cellular metabolism that occur widely in eukaryotes. Mitochondria likely arose from an endosymbiotic oxygen-consuming alpha-proteobacterium, in a common ancestor of all extant eukaryotes (Gray et al. 1999). The protein Isd11, which characterizes modern eukaryotes and functions with

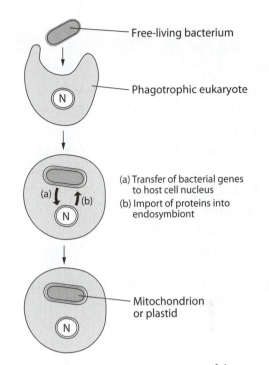

Figure 7.6 A diagrammatic representation of the process of primary endosymbiosis, in which a free-living bacterium is incorporated into a phagotrophic eukaryotic cell and eventually transformed into an organelle.

mitochondrial proteins, provides evidence that the last common eukaryotic ancestor possessed mitochondria (Richards and van der Giezen 2006). The origin of mitochondria likely stimulated the early radiation of eukaryotes (Embley and Martin 2006) and possibly origin of the nuclear envelope (Martin and Koonin 2006). The mitochondrial envelope is composed of two membranes, the inner one folded inward to form cristae (see Figure 7.1). Phyla of eukaryotic algae vary in the form of these cristae, which may be shaped like tubes, or flattened ledges or disks (Table 7.1). Mitochondria contain genetic material and protein synthesis machinery, including ribosomes, features inherited from free-living prokaryotic ancestors. Genetic data indicate that over time much of the genetic material originally present in the bacterial symbiont was transferred to the host genome and lost from the symbiont, rendering it dependent on the host for gene products. Such a dependent entity is an organelle (Figure 7.6). Proteins encoded by the host genome but required for metabolism of modern mitochondria are thus produced in the host cytoplasm and imported into

Table 7.1 Characteristics of algal mitochondria and endosymbiotic origin of algal plastids

Group	Mitochondrial cristae	Plastid origin(s)
Glaucophytes	Flattened	Primary
Cryptomonads	Flattened	Secondary (red)
Red algae	Flattened	Primary
Green algae	Flattened	Primary
Euglenoids	Disk-shaped	Secondary (green)
Chlorarachniophytes	Tubular	Secondary (green)
Haptophytes	Tubular	Secondary (red)
Dinoflagellates	Tubular	Secondary or tertiary (various sources)
Apicomplexans	Tubular	Secondary (red)
Photosynthetic stramenopiles	Tubular	Secondary (red)

the mitochondria. Import occurs by means of protein complexes, known as TIM and TOM, which are located in the mitochondrial envelope (Dolezal et al. 2006) (Figure 7.7). The study of modern organisms reveals that such proteins are targeted to the mitochondrion by means of short amino acid sequences, known as targeting sequences, which occur at the polypeptide's N-terminal end. Mitochondrial envelope proteins recognize these targeting sequences, or **transit peptides**, allowing protein import to occur. Hence, the evolutionary transition from endosymbiont to mitochondrion involved the addition of targeting sequence coding information to symbiont genes present in host nuclear DNA and protein import systems.

Together, past gene transfer and present need to import proteins explain why modern mitochondria are unable to live outside host cells. Gene transfers and organelle protein import were likewise involved in the endosymbiotic processes that gave rise to algal plastids. Consequently, plastids cannot live outside host cells either. Because mitochondria and plastids encode and synthesize some of their own proteins but obtain most essential proteins from the host, they are known as semi-autonomous organelles.

The Origin of Plastids

The acquisition of plastids by several types of endosymbiosis has played an extremely important role in the diversification of algal phyla and was a critical early step in the origin of land plants. The eukaryotic algal phyla are distinguished from other protists by having a preponderance of photosynthetic species, and the land plants inherited their plastids from green algae. Over time, some eukaryotic algae and land plants have lost photosynthetic pigments but have retained colorless plastids because these play essential roles in synthesizing fatty acids, isoprene, starch, and other organic constituents. Because of the enormous ecological importance of plastids, many biologists are interested in their origin. The diversity and evolution of plastids and their genomes have been reviewed by Kim and Archibald (2008). Genetic data indicate that the evolutionary history of all plastids reaches back to a cyanobacterial endosymbiont.

Phagotrophy, the ingestion of particulate food (see Chapter 3), is likely the main way in which eukaryotic host cells take in cells that become endosymbionts. Three important cellular features allow eukaryotic cells to ingest prey and potential endosymbionts: (1) a means for capturing food particles, which may involve flagella and other components of the cytoskeletal apparatus, (2) a flexible cell membrane, and (3) endocytosis, the process by which vesicles formed by invagination of host plasmalemma engulf external particles. As a result of phagotrophy, ingested cells occur within food vacuoles whose delimiting membrane has originated from the host's plasma membrane (see Figure 3.4). Prey cells are typically digested when lysosomes fuse with the food vacuole, releasing hydrolytic enzymes into it.

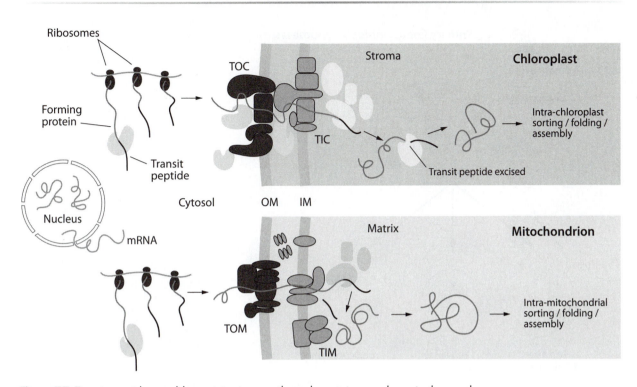

Figure 7.7 Transit peptides enable proteins to pass through protein complexes in the envelopes of mitochondria and primary plastids. IM = inner envelope membrane, OM = outer envelope membrane. (After Stiller et al. 2003, by permission of the *Journal of Phycology*)

However, endosymbiotic cells, typically maintained within an enclosing membrane, are protected from digestion. The adaptive advantage to host cells of maintaining photosynthetic endosymbionts is likely that the latter can produce organic carbon and energy over the long term, whereas their digestion would release nutrients only on a one-time basis. Endosymbionts may eventually lose the enclosing vacuole and the ability to live independently, and they may divide under the control of the host's cell cycle program. As a result, host cells become able to bequeath the useful endosymbionts to their progeny. Some experts consider that such endosymbionts qualify as plastids. However, other experts contend that endosymbionts do not qualify as organelles unless there is evidence that at least some symbiont genes have been transferred to the host nucleus (see Figure 7.6) and host proteins are imported via TIC and TOC protein systems (Soll and Schlieff 2004) located in plastid envelopes (see Figure 7.7). As more endosymbiont genes are transferred to the host nucleus and more host proteins are imported into endosymbionts, plastids become increasingly well integrated into host cells. Three major forms of endosymbiosis are important

in the evolutionary origin of algal plastids: primary endosymbiosis, secondary endosymbiosis, and tertiary endosymbiosis.

Primary endosymbiosis is the process whereby ingested cyanobacterial cells become endosymbionts within a eukaryotic host. Such symbionts may evolve into primary plastids having two envelope membranes (see Figure 7.6). Transit peptides target proteins to the envelopes of primary plastids. Protein complexes in the plastid envelope recognize the transit peptides, allowing proteins to move across (transit) the plastid envelope to the thylakoids or stroma (Figure 7.7). Primary plastids characterize three algal lineages: the red algae, the green algae, and the glaucophytes (see Table 7.1).

The number of times that primary plastids originated has been controversial. Many experts think that primary plastids arose just once in a common ancestor of the red, green, and glaucophyte algae (Keeling 2004): These phyla have consequently been aggregated into the supergroup Archaeplastida (Adl et al. 2005). However, some evolutionary biologists are skeptical that Archaeplastida is a monophyletic group, based on points discussed in

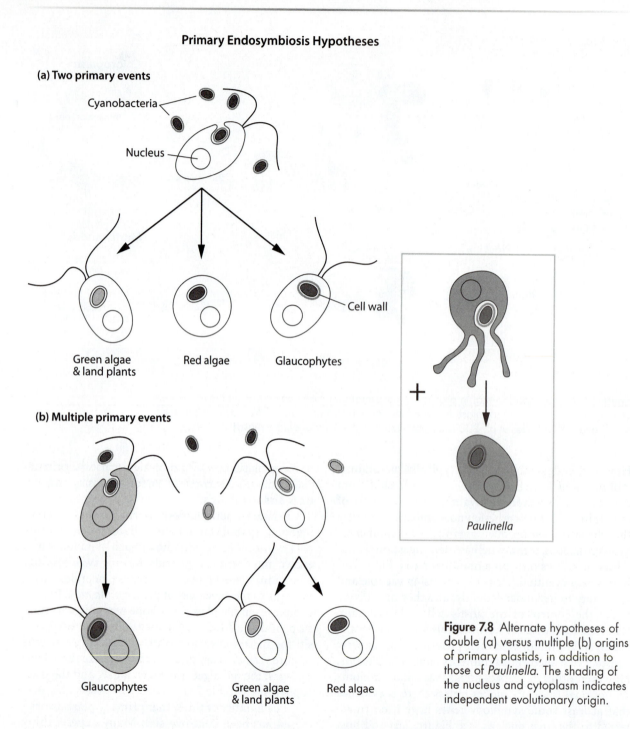

Primary Endosymbiosis Hypotheses

(a) Two primary events

Cyanobacteria

Nucleus

Green algae & land plants

Red algae

Glaucophytes

Cell wall

(b) Multiple primary events

Glaucophytes

Green algae & land plants

Red algae

+

Paulinella

Figure 7.8 Alternate hypotheses of double (a) versus multiple (b) origins of primary plastids, in addition to those of *Paulinella*. The shading of the nucleus and cytoplasm indicates independent evolutionary origin.

Section 7.3. Primary endosymbiosis has occurred independently in the rhizarian amoeba *Paulinella chromatophora*, an inhabitant of pond and lake sediments (Marin et al. 2005) (Figure 7.8). *Paulinella*'s photosynthetic organelles possess a cyanobacterial wall, are derived from cyanobacteria related to modern *Synechococcus* and *Prochlorococcus*, and likewise have a proteobacterial form of rubisco acquired by horizontal gene transfer. Absence of an enclosing vacuole, division under the control of the host cell cycle, inability to live outside the host, elevated genome AT content, and isolated position in molecular phylogenies collectively suggest that *Paulinella*'s photosynthetic endosymbionts have undergone evolutionary changes typical of primary plastids (Marin et al. 2007). Further, *Paulinella*'s photosynthetic structure—the chromatophore—has undergone genome reduction, as have plastids. For example, the *Paulinella* chromatophore genome lacks genes for essential metabolic pathways (Nowack et al. 2008).

Secondary endosymbiosis is the process by which eukaryotic cells become endosymbionts within a eukaryotic host cell. Such symbionts may evolve into secondary plastids. Sarah Gibbs first discovered secondary endosymbiosis when she noticed that *Euglena* chloroplasts had three enclosing membranes, rather than the two envelope membranes of plants. Now we know that because secondary plastids lie within the host's endomembrane system, such plastids characteristically possess three or four envelope membranes. Thus, as genes were transferred from the eukaryotic endosymbiont's nucleus to that of the host, protein targeting became more complex. Secondary plastid-targeted proteins possess both a **signal peptide** directing them to the endomembrane system and a transit peptide (Rogers et al. 2004). Haptophytes and chlorarachniophytes; most apicomplexans and cryptomonads; and many euglenoids, stramenopiles, and dinoflagellates possess secondary plastids. Plastids of haptophytes, cryptomonads, and stramenopiles arose from a red algal endosymbiont. In addition, most photosynthetic dinoflagellates have a particular type of plastid widely thought to have originated from a red alga by secondary endosymbiosis (see Table 7.1). The number of times that red algal endosymbionts have been transformed into plastids is controversial (Palmer 2003), as discussed in Section 7.4. One group of dinoflagellates obtained their plastids from a green alga, likely by secondary endosymbiosis (Schnepf and Elbrächter 1999). Molecular evidence indicates that euglenoids and chlorarachniophytes obtained secondary green plastids independently (Rogers et al. 2007). Thus, secondary plastids have originated from green eukaryotic endosymbionts on at least three occasions.

Tertiary endosymbiosis has occurred when eukaryotic cells contain a plastid that has been derived from a eukaryotic endosymbiont that possessed a secondary plastid. More than two membranes envelop tertiary plastids. In such cases, plastid-targeting systems can become complex (Patron et al. 2006b). Several dinoflagellate species possess tertiary plastids derived independently from haptophyte, cryptomonad, or diatom endosymbionts that had secondary red plastids. Thus, tertiary plastids have originated several times (Delwiche 1999).

7.2 Endosymbiosis in the Modern World

There are many modern examples of endosymbiotic cyanobacteria and eukaryotic algae in diverse hosts. These endosymbioses are often ecologically important, and they also help to illuminate the selective forces and adaptive processes that influenced more ancient endosymbiotic events. Certain modern endosymbiotic associations have been used as model laboratory systems for understanding the molecular and cellular processes that were involved in ancient plastid origins.

Prokaryotic Endosymbionts

Prokaryotic endosymbionts occur widely within cells of eukaryotic algae and include heterotrophic bacteria and cyanobacteria.

Heterotrophic bacterial endosymbionts

The application of molecular phylogenetics has revealed that a variety of heterotrophic bacteria occur within host cells of diatoms and other marine protists (Foster et al. 2006b). Electron and fluorescence microscopy have also been used to detect heterotrophic endobacteria in eukaryotic algal cells; examples include the green alga *Pleodorina japonica* (Nozaki et al. 1989), *Durinskia baltica* (= *Peridinium balticum*) (Chesnick and Cox 1986) and other dinoflagellates, and the diatom *Pinnularia*. The case of *Pinnularia* shows that endobacteria may be closely integrated into host cell processes. The Gram-negative, rod-shaped endosymbionts of *Pinnularia* occur in the lumen of the endoplasmic reticulum associated with the secondary plastid. During interphase of the host cell cycle, the bacteria bore cavities into the chloroplast as far as the thylakoids, without breaking intervening plastid membranes. In early stages of host nuclear mitosis, the bacteria unscrew themselves from the plastid and, remaining within the ER, cluster near the nucleus, where they divide. The bacteria eventually return to their interphase positions within plastids, and this cycle of behavior repeats during the next diatom cell cycle. Although the function of the endobacteria is not known, close cell cycle coordination indicates that they have become integrated into host cellular processes (Schmid 2003).

Cyanobacterial endosymbionts

Cyanobacteria occur within the cells of a variety of eukaryotic organisms in diverse habitats (Kneip et al. 2007). Some endosymbiotic cyanobacteria function much like plastids to provide fixed carbon, while others primarily supply fixed nitrogen to host cells. Yet others may provide both organic carbon and fixed nitrogen. For example, in the terrestrial habitat, nitrogen-fixing *Nostoc punctiforme* filaments occur within the cytoplasm of the soil fungus *Geosiphon*, representing the only known occurrence of an intracellular photosynthetic symbiont in fungi (Mollenhauer et al. 1996). Non-photosynthetic marine protists (including radiolarians, tintinnids, silicoflagellates, and many dinoflagellate species) often harbor unicellular cyanobacterial cells. Such endosymbionts can be detected by epifluorescence microscopy, transmission electron microscopy, and molecular methods (Foster et al. 2006a,b). These protist hosts occur in well-lit upper ocean waters where light is available for endosymbiotic cyanobacteria related to *Synechococcus* and *Prochlorococcus*.

Cyanobacterial cells also commonly occur in the intercellular matrix or cytoplasm of sponges in warm marine waters worldwide (Wilkinson 1992). Some sponge cyanobacteria may not be able to grow outside the sponge host, and some sponges may require cyanobacterial symbionts. The sponge *Chondrilla australiensis* transmits cyanobacterial endosymbionts to its progeny via eggs. Phylogenetic analysis of sponge–cyanobacterial associations shows that multiple independent symbiotic events have occurred (Steindler et al. 2005). Certain cyanobacterial species are widespread and occur in different sponge species, while some sponges contain several species of cyanobacterial endosymbionts. *Aphanocapsa feldmannii* (a complex of cryptic species) and *Synechococcus spongiarum* are common sponge endosymbionts that can be distinguished morphologically (Usher et al. 2004a, 2006) (Figure 7.9).

In freshwaters, cells of the non-photosynthetic euglenoid *Petalomonas* typically contain one or more cyanobacterial unicells. The cyanobacteria occur within individual vacuoles and persist within *Petalomonas* cells for at least several weeks. While the cyanobacteria are able to divide within host cells, they may also be digested or die, suggesting that the

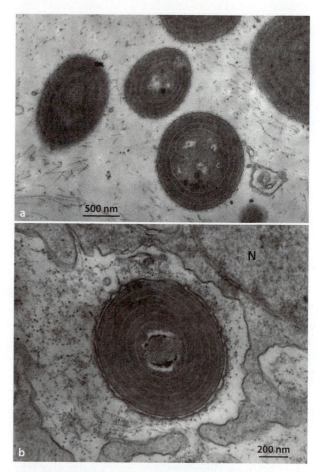

Figure 7.9 Cyanobacterial endosymbionts from marine sponges. (a) *Aphanocapsa feldmannii* in the sponge *Ircinia variabilis*. (b) *Synechococcus spongiarum* in *Chondrilla nucula*. N = host cell nucleus. (a: From Usher et al. 2006. Reprinted by permission of Taylor & Francis; b: From Usher et al. 2004b. Reprinted with the permission of Cambridge University Press)

Petalomonas endosymbiosis is in an early stage of integration (Schnepf et al. 2002).

Other cyanobacterial associations seem to be primarily based upon the nitrogen-fixation capacity of the endosymbiont. The land plant–cyanobacterial symbioses described in Chapter 3 are examples, and there are many additional aquatic cases. For example, unicellular photosynthetic and nitrogen-fixing cyanobacteria occur within host cells of the Caribbean coral *Montastrea cavernosa*, where they compensate for low nitrogen availability in surrounding waters (Lesser et al. 2004). (Such corals obtain organic compounds from symbiotic photosynthetic dinoflagellates.)

Plastid-containing diatoms belonging to the genera *Epithemia*, *Rhopalodia*, and *Denticula* contain

1–10 endosymbiotic cyanobacteria (Kies 1992) that are capable of nitrogen fixation. An experimental study showed that the number of endosymbionts per cell increased as the combined nitrogen level of the surrounding water decreased (DeYoe et al. 1992). Cells of the freshwater *Rhopalodia gibba* typically contain four or so spherical structures bound by two membranes, regarded as vertically inherited symbionts. These endosymbionts do not seem to be photosynthetic and have not been cultivated independently, but they fix nitrogen during the day. The *R. gibba* symbionts are closely related to the free-living cyanobacterium *Cyanothece* ATCC 51.142, which is able to fix nitrogen at night (Prechtl et al. 2004).

An endosymbiotic filamentous cyanobacterium known as *Richelia intracellularis* lives just inside the silica wall of the common marine, planktonic, chain-forming diatom genera *Rhizosolenia*, *Hemiaulus*, and *Bacteriastrum*. Different *Richelia* strains are associated with different diatom hosts (Foster and Zehr 2006). *Richelia* produces heterocytes in which nitrogen fixation occurs, an advantage in oligotrophic ocean waters. Up to 98% of the *Rhizosolenia* cells sampled from the central north Pacific gyre and the Caribbean area contain at least two *Richelia* filaments of about 4–15 cells each. Although *Rhizosolenia* grows well without *Richelia* in nitrogen-rich coastal waters, *Richelia* is apparently dependent upon its host (Paerl 1992).

Eukaryotic Algal Endosymbionts

A variety of eukaryotic algae commonly occur in marine or freshwater protists, sponges, flatworms, coelenterates, molluscs, and other animals. In such associations, the algal symbionts primarily provide organic carbon derived from photosynthesis. Such endosymbioses are ecologically important in marine and freshwater environments.

Marine hosts having eukaryotic endosymbionts

Among the marine protists that commonly possess endosymbiotic algae are the foraminifera. These marine rhizarians have calcareous shells, have existed for at least 500 million years, and include at least 4000 species. Many of the modern foraminifera are large enough to see with the naked eye, and these larger forms generally host endosymbiotic algae. Most foraminifera are selective in symbiont type, but some may have more than one kind of

algal endosymbiont. Symbionts may include unicellular red algae, unicellular green algae, more than 20 species of diatoms, dinoflagellates (Lee 1992a,b), or haptophytes (Gast et al. 2000). Radiolarians are another group of marine planktonic protists that often possess endosymbiotic algae. Radiolarians may be single celled or occur in colonies, build shells of crystalline strontium sulfate, and occur in all of the major oceans. Up to 100,000 algal cells may be associated with each radiolarian, often occurring within a halo of surface cytoplasm. Photosynthetic eukaryotic symbionts of radiolarians include dinoflagellates, prasinophycean green algae, haptophytes, or stramenopiles (Anderson 1992; Gast and Caron 1996; Gast et al. 2000).

The cells of marine sponges, especially those occurring in coral reef communities, may (in addition to the cyanobacterial endosymbionts described earlier) contain diatoms, cryptomonads, or dinoflagellates. Sponges that contain endosymbiotic eukaryotic algae seem better able to compete with corals and even overgrow them. After a hurricane decimated a sponge population, sponges bearing algal endosymbionts grew back more rapidly than those lacking endosymbionts (Wilkinson 1992). Marine flatworms (turbellarians) may also harbor eukaryotic algal endosymbionts. For example, *Convoluta roscoffensis* cells contain the unicellular green alga *Tetraselmis*, *C. convoluta* contains the diatom *Licmophora*, and *Amphiscolops* retains cells of the dinoflagellate *Amphidinium*. The eggs of these animals are free of algae; thus, juvenile animals must acquire symbionts from free-living communities.

Marine animals and protists may possess endosymbiotic dinoflagellates known as **zooxanthellae** (Figure 7.10) (Chapter 11). For example, zooxanthellae occur within the common green sea anemone *Anthopleura xanthogrammica* (Figure 7.11) and marine ciliates such as *Maristentor* (Lobban et al. 2002). Reef-forming corals typically possess zooxanthellae, and the carbon and nitrogen metabolisms of the associates are tightly linked (Allemand et al. 1998). One square millimeter of coral tissue can contain 1–2 million dinoflagellate cells (Muller-Parker and D'Elia 1997), and the global net photosynthetic productivity of zooxanthellae is estimated to be more than 4.6×10^8 metric tons of carbon per year. Much of the photosynthate produced by zooxanthellae is released to host corals, which helps to explain high rates of reef accumulation; corals

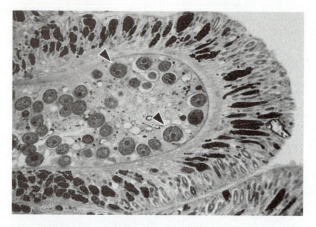

Figure 7.10 Zooxanthellae (arrowheads) in a sectioned and stained coral polyp. Zooxanthellae are readily recognized by their darkly stained spherical nuclei. (Photo: E. Newcomb)

Figure 7.11 A green sea anemone in which zooxanthellae are endosymbionts. (Photo: L. W. Wilcox)

lacking zooxanthellae rarely make substantial contributions to formation of the reef. Coral zooxanthellae release glycerol, glucose, and alanine, which are used by the host cells in respiration and growth. In the case of the Hawaiian coral *Pocillopora damicornis*, host cells release amino acids that stimulate both algal photosynthesis and the release of photosynthates to host tissues (Gates et al. 1995). Cells of the sea anemone *Aiptasia pulchella* release the amino acid taurine, which stimulates release of organic acids from symbiotic dinoflagellates (Wang and Douglas 1997). Such signalling processes demonstrate a high degree of integration between host and endosymbiont.

Freshwater hosts having eukaryotic endosymbionts

In freshwaters, diverse protists, sponges, or coelenterates such as *Hydra* ingest and maintain endosymbiotic cells of the green alga *Chlorella* or eustigmatophyceans (Frost et al. 1997) (Figure 7.12). The algal cells typically occur in **perialgal vacuoles** (Figure 7.13), which protect the enclosed endosymbionts from digestion (Karakashian and Rudzinska 1981). The low pH of host perialgal vacuoles induces release of the sugar maltose from *Chlorella* cells. Maltose inhibits the fusion of lysosomes with the phagocytic vacuole, thereby preventing digestion of the algal cell. Some 40%–60% of the carbon fixed by *Chlorella* may be released as maltose. The excretion of large amounts of fixed carbon limits the division rate of endosymbiotic *Chlorella* cells. This helps to hold algal population growth in check (Reisser 1992) but does not affect the photosynthetic capacity of cells. Similar genetically controlled chemical feedback processes probably also developed during the evolution of primary and secondary plastids and for the same reasons. *Paramecium bursaria*, a species that characteristically contains endosymbiotic *Chlorella* cells (Figure 7.14), is used as a model system to understand how endosymbioses become established. Grown in darkness, *P. bursaria* will digest its algal endosymbionts, and such hosts can be experimentally re-infected by *Chlorella*. Using such a system, Kodama and associates (2007) found that *Chlorella* cells taken up singly or in low numbers into small food vacuoles were more vulnerable to digestion than those occurring in larger numbers in larger food vacuoles. Some of the cells in these larger aggregates may be able to escape digestion and gain the protection of individual perialgal vacuoles.

Kleptoplastids

One way for a heterotrophic organism to obtain autotrophic and other capacities of plastids is to maintain plastids from digested prey within the host cytoplasm for a time. Such plastids are known as kleptoplastids, since they are "stolen" by their hosts. Chloroplasts are harvested from a variety of kinds of algae by a range of heterotrophs, including dinoflagellates (e.g., Takishita et al. 2002), ciliates, and ascoglossans (Sacoglossan molluscs). However, because plastids are unable to synthesize all of the

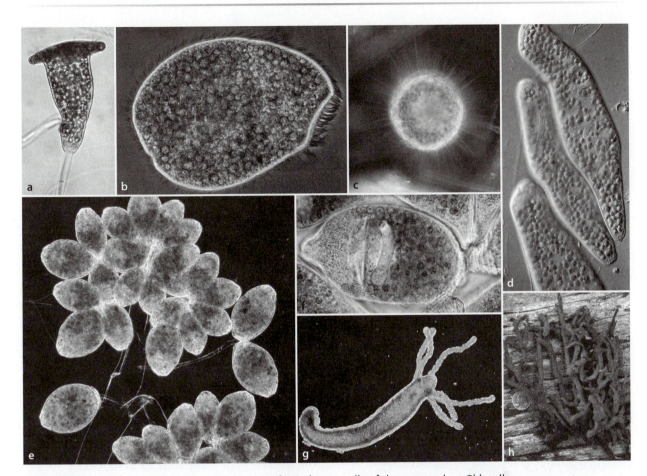

Figure 7.12 Freshwater organisms containing endosymbiotic cells of the green alga *Chlorella*.
(a) *Vorticella*, an attached ciliate with an unbranched contractile stalk. (b) *Climacostomum*,
a free-swimming ciliate. (c) *Acanthocystis*, a heliozoan ("sun animal") that has many radiating
axopodia—thin cytoplasmic extensions made rigid by internal columns of microtubules.
(d) *Ophridium*, a macroscopic colonial ciliate whose numerous cells are embedded in
massive amounts of gelatinous material. (e) *Carchesium*, an attached stalked, colonial ciliate;
individual cells can contract within the colony without causing the whole colony to contract.
(f) *Opercularia*, a noncontractile colonial ciliate that attaches to substrates via a branched
stalk. (g) *Hydra*, a coelenterate. (h) A freshwater green sponge. (a, b, e, f, g: L. E. Graham;
c, d, h: L. W. Wilcox)

proteins needed for their maintenance, and foreign
hosts lack the suite of genes necessary to compensate, such associations are more or less temporary.
Kleptoplastids eventually die and must be replaced,
so such relationships have not been considered to
be cases of endosymbiosis. However, cases of kleptoplastidy are attracting attention because they are
ecologically important and shed light on cellular
processes related to endosymbiosis.

For example, a heterotrophic marine flagellate
belonging to Katablepharidophyta captures and
maintains the plastid, mitochondria, and nucleus
of a green alga, representing a possible case of an
early stage in an evolving secondary endosymbiosis
(Okamoto and Inouye 2005a, 2006). The marine
ciliate *Myrionecta rubra* maintains not only plastids but also some nuclei of its cryptomonad (*Geminigera cryophila*) prey for up to 30 days (Johnson
et al. 2007b). The captured cryptomonad nuclei remain able to transcribe genes needed to maintain
the kleptoplastids longer than otherwise possible.
Maintenance of prey nuclei in an active condition helps to explain why this ciliate is widespread
and abundant. *Strombidium* is another ciliate that

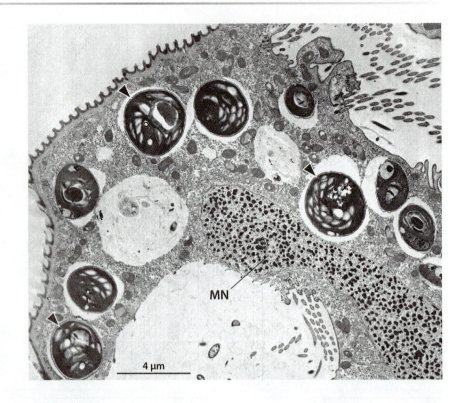

Figure 7.13 TEM of a green *Vorticella* showing numerous *Chlorella* cells within perialgal vacuoles (arrowheads). The host macronucleus (MN) is visible, as are many of its numerous cilia. (TEM: L. E. Graham)

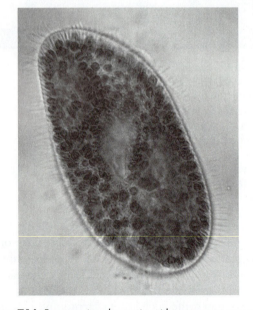

Figure 7.14 *Paramecium bursaria* with numerous green *Chlorella* endosymbionts. (Photo: L. W. Wilcox)

harvests and maintains plastids from a variety of eukaryotic algae. Sometimes these ciliates contain just one type of plastid, but mixtures of plastid types can also occur. Stoecker and associates (1987)

demonstrated that the kleptoplastids of *Strombidium* were capable of photosynthesis and that during the height of population growth, nearly 50% of *Strombidium* cells contained kleptoplastids.

The sea slug *Elysia* and its close relatives harvest plastids from siphonous green algae or the stramenopile *Vaucheria*. These ascoglossans pierce the algal cell wall with a specialized needlelike tooth and then suck out large volumes of cytoplasm containing many plastids. During *Vaucheria* plastid captivity in *Elysia*, transcription and translation of plastid genes continue (Mujer et al. 1996). In addition, at least one of the genes that encode plastid proteins is encoded by the nucleus of *Elysia crispata* cells. Such genes were likely transferred horizontally from an algal nucleus, possibly by the activity of a retrovirus (Pierce et al. 2003). The animals produce a host factor that elicits leakage of photosynthates such as glycolate, glucose, and amino acids from plastids.

A freshwater example of kleptoplastidy is provided by the dinoflagellate *Gymnodinium acidotum*, which ingests cryptomonads, keeps their plastids for at least 10 days, and digests them (Fields and Rhoades 1991). In other cases, the occupation of *Gymnodinium* cells by plastids derived from ingested cryptomonads appears to be more stable and possibly represents early stages in tertiary endosymbiosis (see Figure 7.22) (Wilcox and Wedemayer 1984).

7.3 Primary Endosymbiosis, with a Focus on the Glaucophyta

Numerous genetic commonalities indicate that all modern plastids have cyanobacterial ancestors (Gray 1992; Morden et al. 1992). Examples of genetic similarities between cyanobacteria and plastids include organization of DNA into various types of nucleoids (Coleman 1985), transcription and translation systems, ribosomal sizes, and the spectrum of antibiotic sensitivities. Even so, a number of major issues surrounding primary endosymbiosis remain to be resolved.

Unresolved Issues Concerning Primary Endosymbiosis

One as yet unanswered question is the nature of the cyanobacterial ancestor of primary plastids, because no single modern species exhibits the expected suite of ancestral traits. Experts suggest that isolation, culture, and genome sequencing of new cyanobacterial species are needed. A second major issue is the number of times that primary endosymbiosis has occurred. Because the primary plastids of red, green, and glaucophyte algae have in common envelopes composed of only two membranes, and genetic features such as similar gene order, a widely accepted hypothesis holds that these three phyla arose from a common ancestor in which a unique event of primary endosymbiosis had occurred (Keeling 2004). However, one study suggested that convergent evolution was a better explanation for common gene content than was a hypothesis of common evolutionary history (Stiller et al. 2003) (see Figure 7.8), and there are other difficulties in reconciling traits present in glaucophytes, red algae, and green algae (Stiller 2003). While a phylogenetic analysis that included 50 genes from 16 plastid genomes and 15 cyanobacterial genomes, and 143 nuclear genes from 34 eukaryotes strongly supported a hypothesis of primary plastid monophyly, the authors noted that the analysis lacked data from some major groups of

eukaryotes (Rodríguez-Ezpleta et al. 2005). Several recent multigene phylogenies did not show strong support for the monophyly of the algal groups that contain primary plastids (Harper et al. 2005; Nozaki et al. 2007; Yoon et al. 2008). In addition, an analysis of *EEF2* gene sequences indicated that algal groups with primary plastids do not have a single common ancestor (Kim and Graham 2008). Hence, there remains some controversy about the relationships of the primary plastid algal phyla.

Efforts to use molecular sequence comparisons to understand the history of primary plastids may be confounded by the extreme degree to which cyanobacterial sequences have invaded the genomes of plastid-bearing hosts. In the much-studied flowering plant *Arabidopsis*, 18% or so of genes, most encoding proteins targeted to sites other than the plastid, are thought to have originated from the cyanobacterial ancestor of a primary plastid (Martin et al. 2002). The inclusion in phylogenetic analyses of nuclear gene sequences derived from cyanobacteria could make red, green, and glaucophyte algae appear to be more closely related than they actually are (Stiller 2007). In addition, these host genomes may have been similarly influenced by the evolution of many proteins needed to regulate gene transcription in primary plastids. By evaluating the *Arabidopsis* genome, Wagner and Pfannschmidt (2006) estimated that at least 48, and perhaps as many as 100, eukaryotic transcription factors are imported into plastids, reflecting the evolution of massive host control of plastid function.

The glaucophytes have been of particular interest because their phylogenetic relationships are not clear and because their plastids have retained various cyanobacterial features, suggesting a relatively early stage in primary plastid evolution. The next section focuses on the Glaucophyta.

A Focus on the Glaucophyta

The glaucophytes, also known as glaucocystophytes, are a small group of structurally diverse freshwater microalgae. Glaucophytes characteristically contain blue-green plastids often referred to as **cyanelles** because they differ in some ways from other plastids. For example, glaucophyte plastids resemble cyanobacteria and differ from plastids of other algae in having a peptidoglycan wall (Figure 7.15). The walls of both cyanobacteria and glaucophyte plastids can be destroyed with lysozyme and their assembly prevented

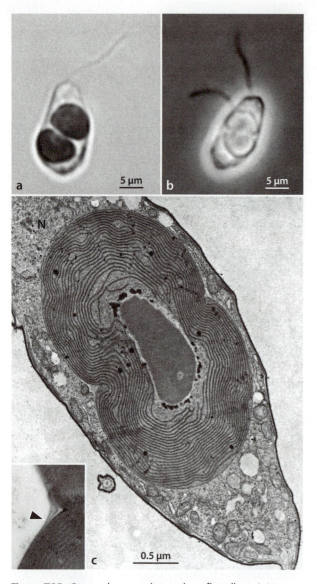

Figure 7.16 Portion of a plastid of the glaucophyte *Cyanophora paradoxa* viewed by transmission electron microscopy, showing phycobilisomes on thylakoids. (TEM: L. E. Graham)

Figure 7.15 *Cyanophora,* a glaucophyte flagellate. (a) In bright-field microscopy, the cyanelle is obvious. (b) The two flagella are evident in phase-contrast microscopy. (c) With TEM, the plastid and its thin surrounding wall (arrowhead) are visible, as are concentric thylakoids bearing phycobilisomes and a central carboxysome. (a, b: L. W. Wilcox; c: Micrograph by R. Brown, in Bold and Wynne 1985, Inset: L. E. Graham)

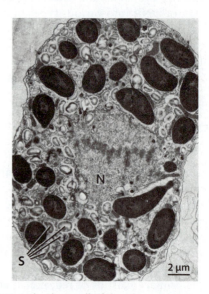

Figure 7.17 A dividing cell of the glaucophyte *Gloeochaete wittrockiana* showing a central nucleus (N) and chromosomes surrounded by darkly stained plastids. Numerous starch granules (S) occur in the cytoplasm. (From Kies 1976)

by penicillin treatment. In addition, ultrastructural study of freeze-fractured plastids of the glaucophyte *Cyanophora* revealed that they are surrounded by a layer very similar to the lipopolysaccharide envelope characteristic of Gram-negative eubacteria. Also, the inner envelope membrane of *Cyanophora* plastids is more similar to the plasma membrane of cyanobacteria than to other plastid envelope membranes (Giddings et al. 1983). These features, together with pigment composition and other attributes, demonstrates conclusively that glaucophyte plastids originated by primary endosymbiosis involving a cyanobacterium (Sitte 1993).

Glaucophyte plastid DNA exhibits a number of gene clusters typical of cyanobacteria and encodes proteins not encoded by most algal plastids. Like the

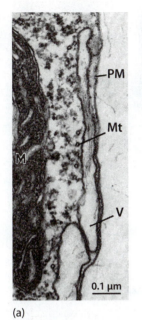

(b)

(a)

Figure 7.18 Features of glaucophyte motile cells. (a) Flattened vesicles (V) occur at the periphery of motile glaucophyte cells beneath the plasma membrane (PM), as in *Gloeochaete wittrockiana*. Fibrillar contents of the vesicles and underlying microtubules (Mt) are distinctive. M = mitochondrion. (b) TEM of the apical region of a *Gloeochaete wittrockiana* zoospore, showing two of the four multilayered flagellar roots (GW1 and GW2—from the German Geisselwurzel, meaning flagellar root). Such multilayered structures are characteristic of glaucophytes. A flagellar base occurs at the center. (From Kies 1976)

plastids of other algae, those of glaucophytes cannot be grown separately from host cells. About 90% of the soluble proteins of *Cyanophora* plastids are encoded by the host nuclear genome, a ratio similar to that of other plastids. As in the cases of other plastids, the nuclear genome of *Cyanophora paradoxa* contains the prokaryote and plastid division gene *FtsZ*, which possesses a typical chloroplast transit peptide. In *C. paradoxa* plastids, the FtsZ protein forms a division ring similar to that of other plastids (Sato et al. 2005). Glaucophyte plastid DNA encodes the small subunit of rubisco (*rbcS*), as do plastids of non-green algae. Plastid-encoded SSU rDNA sequence evidence indicates that glaucophyte plastids are monophyletic and form a clade distinct from cyanobacteria and plastids of other eukaryotic algae (Medlin and Simon 1998).

Glaucophyte plastids contain chlorophyll *a* and the phycobiliproteins phycocyanin and allophycocyanin, as well as β-carotene. Phycoerythrin appears to be absent from glaucophytes, as are typical cyanobacterial xanthophylls. Phycobiliproteins occur in typical phycobilisomes on the surfaces of thylakoids, which are not stacked (Figure 7.16). The plastids contain polyphosphate granules and conspicuous central structures resembling cyanobacterial carboxysomes (see Figure 7.15). Burey and associates (2005) found that these structures contain rubisco and rubisco activase and suggested that they might represent

an evolutionary transition between cyanobacterial carboxysomes and pyrenoids, which occur in the plastids of many algae. Glucose is the main photosynthetic product of glaucophyte plastids. As in the case of red algae, glaucophyte starch is not produced inside plastids, but, rather, by the host cytoplasm, where numerous starch granules can be observed ultrastructurally (Figure 7.17). *Cyanophora* is not able to grow in the dark when supplied with glucose. The plastids of glaucophytes are unable to fix nitrogen.

Phylogenetic analyses of nuclear genes have produced different inferences of glaucophyte relationships. Nuclear-encoded SSU rDNA (Medlin and Simon 1998) and heat shock protein 70 data (Rensing et al. 1997) suggest that the glaucophyte nucleocytoplasm is related to that of cryptomonads. An analysis of *RPBI*, which encodes the largest subunit of RNA polymerase II, suggested that glaucophytes might be closely related to certain heterotrophic protists (Stiller and Harrell 2005).

Sexual reproduction is not known for glaucophytes, but nonflagellate forms may produce flagellate asexual reproductive cells (zoospores). Flagellate glaucophyte cells exhibit a series of flattened vesicles lying beneath the cell membrane that have contents varying with the genus examined; some contain a scalelike structure, others possess a loose fibrillar material (Figure 7.18a), and yet others appear to be empty. Glaucophyte flagellate cells are also characterized by distinctive multilayered structures near flagellar bases (Kies 1976) (Figure 7.18b). Similar structures are present in flagellate cells of certain green algae and a few other protists. There are no known fossil representatives of the glaucophytes.

Diversity of glaucophytes

Glaucophytes include unicellular flagellates, planktonic colonies, and attached colonies that inhabit freshwaters, particularly soft waters such as bogs or acid swamps. The three most common genera are *Cyanophora*, *Glaucocystis*, and *Gloeochaete*, which are described below. Several additional genera have been described; these include *Archeopsis*, *Glaucystopsis*, *Peliania*, *Strobilomonas*, *Cyanoptyche*, and *Chalarodora* (Seckbach 1994).

CYANOPHORA (Gr. *kyanos*, blue + Gr. *phoras*, bearing) possesses two unequal flagella, one extending forward from a subapical depression at the cell apex and the other emerging from the same point but extending toward the cell posterior. Typically two rounded plastids are present per cell (see Figure 7.15). Reproduction is by longitudinal division into two daughter cells. *Cyanophora* is able to swim away from very bright light; its plastids have been implicated as the photoreceptors. Ultrastructural information was used to define the species *C. biloba* and compare it to *C. paradoxa* (Kugrens et al. 1999).

GLAUCOCYSTIS (Gr. *glaukos*, bluish-green or gray + Gr. *kystis*, bladder) is a colony formed by retention of ovoid daughter cells within the persistent parental cell wall, which is made of cellulose. Within each pigmented cell are two starlike aggregations of several long, thin cyanelles (Figure 7.19). Cell division results in subdivision of the cytoplasm into four autospores, which are nonflagellate, smaller versions of the parental cell.

GLOEOCHAETE (Gr. *gloia*, glue + Gr. *chaite*, long hair) occurs as a unicell or colony of two to four spherical cells, each having two distinctive long (20 times the cell diameter), thin gelatinous hairs sometimes called pseudoflagella (Figure 7.20). These hairs have internal microtubules similar to those of flagella but lack the two central microtubules. *Gloeochaete* cells are enclosed in mucilage and attached to walls of filamentous algae, leaves of aquatic mosses, or submerged macrophytes. The delicate cell wall is

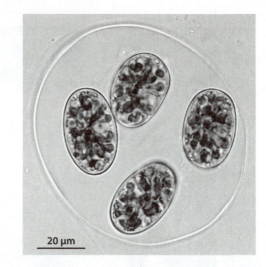

Figure 7.19 The glaucophyte *Glaucocystis* has ovoid cells containing several deeply lobed blue-green plastids. (Photo: L. W. Wilcox)

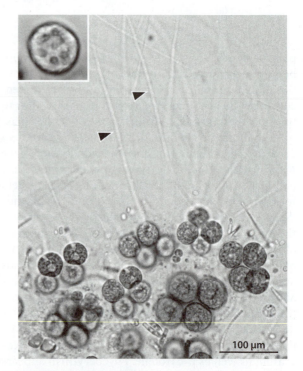

Figure 7.20 The glaucophyte *Gloeochaete*, with long pseudoflagella (arrowheads) extending from each of the cells. The inset is a higher-magnification view of a single *Gloeochaete* cell showing several distinct plastids. (Photos: L. W. Wilcox)

not composed of cellulose. The numerous plastids tend to form a cup-shaped assemblage in the basal portion of the cell. Asexual reproduction is by flagellate zoospores. Immediately upon zoospore settling, the germlings begin secretion of the hairs, during which time the protoplast rotates.

7.4 Secondary Endosymbiosis with a Focus on Chlorarachniophytes

Many algae possess plastids that arose via secondary endosymbiosis—incorporation of a eukaryotic endosymbiont having a plastid (Figure 7.21). The euglenoids, chlorarachniophytes, and dinoflagellates include representatives having secondary plastids derived from green algae. Algal groups having secondary plastids derived from red algae include haptophytes, cryptomonads, and photosynthetic stramenopiles. Many experts consider that most photosynthetic dinoflagellates have secondary plastids derived from red algae. In addition, most of the parasites known as apicomplexans, which are related to dinoflagellates, possess structures derived from red or green plastids (see Table 7.1). These organisms are not only important from an ecological point of view but also provide insight into the evolution of secondary plastids. In addition, apicomplexan plastids are targets for the development of drugs to control serious diseases of humans and animals.

Nucleomorphs as Evidence of Secondary Endosymbiosis

The cryptomonads and chlorarachniophytes provide particularly compelling evidence of the secondary origin of plastids in the form of cellular structures known as nucleomorphs. First described in cryptomonads by Greenwood (1974) and later characterized (Gillott and Gibbs 1980; McKerracher and Gibbs 1982; Ludwig and Gibbs 1985), **nucleomorphs** are bound by two membranes and occur between the inner and outer two pairs of plastid membranes (Figure 7.22). Nucleomorphs resemble nuclei in possessing nuclear pores in their envelopes, some chromosomes, and a nucleolus-like structure and thus are interpreted as the relics of the nuclei of red or green algal endosymbionts. Nucleomorphs divide simply by pinching into two equal halves. Cryptomonad plastid nucleomorphs are the remnants of a red algal nucleus, while

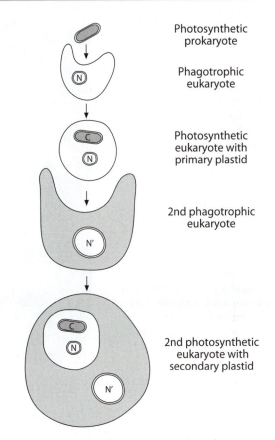

Photosynthetic prokaryote

Phagotrophic eukaryote

Photosynthetic eukaryote with primary plastid

2nd phagotrophic eukaryote

2nd photosynthetic eukaryote with secondary plastid

Figure 7.21 Diagram of secondary endosymbiosis, whereby a eukaryote that had earlier acquired a plastid via primary endosymbiosis is itself taken up by a second eukaryote. (After McFadden 1993)

chlorarachniophyte nucleomorphs are the remnants of a green algal nucleus (Figure 7.23).

Nucleomorphs illustrate an evolutionary stage between the acquisition of photosynthetic eukaryotic endosymbionts and the evolution of more typical secondary plastids, which lack nucleomorphs. During the evolution of most secondary plastids, genes originally present in the eukaryotic endosymbiont's nucleus that were essential for plastid function were transferred to the new host's nucleus, and the endosymbiont nucleus has disappeared. In cryptomonads and chlorarachniophytes, the nucleomorph encodes genes that are essential for plastid function. Lane and associates (2006) demonstrated that several cryptomonads possess nucleomorphs having three chromosomes but that nucleomorph genome size varies among cryptomonads. The nucleomorph of the cryptomonad *Guillardia* encodes 464 genes (Douglas et al. 2001). The genome sequence of the nucleomorph

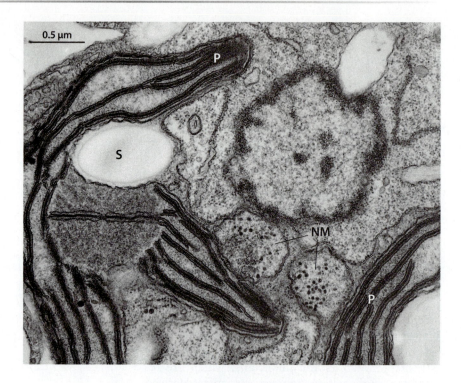

Figure 7.22 Cryptomonad nucleomorph. TEM of cryptomonad found within cells of the dinoflagellate *Gymnodinium acidotum*, where it may represent a kleptoplastid or a more stable endosymbiont in the process of becoming a plastid. There are two nucleomorph (NM) profiles in the periplastidal compartment—the space between the outer and inner pairs of membranes bounding the plastid (P). Also occurring in this space are eukaryotic-sized ribosomes and starch (S). (From Wilcox and Wedemayer 1984, by permission of the *Journal of Phycology*)

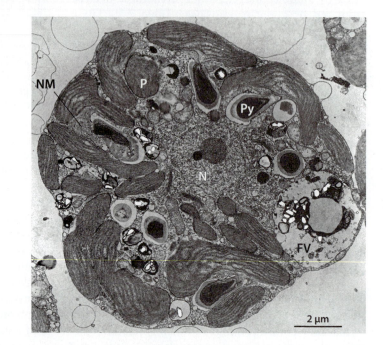

Figure 7.23 Chlorarachniophyte nucleomorph. TEM of *Gymnochlora stellata* showing numerous plastids with pyrenoids (Py). A nucleomorph is labeled NM. N = nucleus, FV = food vacuole. (From Ishida et al. 1996)

of the chlorarachniophyte *Bigelowiella natans* is even smaller, having only 331 genes on three chromosomes. Seventeen of these genes encode proteins needed by the plastid. The need for these genes has prevented nucleomorph elimination and forced the nucleomorph to retain gene expression machinery (Gilson et al. 2006). The existence of cryptomonad and chlorarachniophyte nucleomorphs suggests that similar structures once occurred in the secondary plastids of other algae but that, in these cases, all

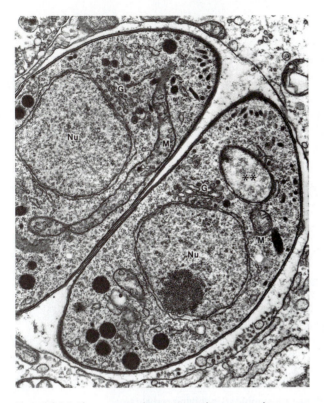

Figure 7.24 The apicomplexan *Toxoplasma gondii* apparently once possessed photosynthetic ability, based upon the presence in each cell of a small plastidlike body, the apicoplast (asterisks), which is bounded by four membranes. G = Golgi body, M = mitochondrion, Nu = nucleus. (Micrograph courtesy L. G. Tilney and D. S. Roos; reprinted with permission from Vogel, G., Parasites shed light on cellular evolution. *Science* 275:1422. © 1997 American Association for the Advancement of Science)

vestiges of the eukaryotic endosymbiont's nucleocytoplasm have been lost. A similar fate may await cryptomonad and chlorarachniophyte nucleomorphs.

Apicomplexan Plastids

Several non-photosynthetic, intracellular parasites form the phylum Apicomplexa. Apicomplexans are classified together with dinoflagellates and ciliate protozoa into a supergroup known as Alveolates because they all display membranous structures known as alveoli at the cell periphery. Many, though not all, apicomplexans possess a colorless vestigial plastid known as an **apicoplast**. ATPase gene cluster analysis suggests that apicomplexan plastids were derived from red algae (Leitsch et al. 1999). Apicomplexan

plastids are thought to have been derived from plastids similar to those of a newly discovered photosynthetic alveolate named *Chromera velia* (Moore et al. 2008). Examples of apicomplexan species for which apicoplasts are known include *Toxoplasma gondii* (Figure 7.24) (which causes toxoplasmosis), the malarial parasite *Plasmodium falciparum* (McFadden et al. 1996; Köhler et al. 1997), *Eimeria tenella*, and the coccidian pathogen *Cyclospora* (Chiodini 1994). Because of its medical importance, the genome of *Plasmodium falciparum* has been completely sequenced (Gardner et al. 2002). This sequence reveals that more than 500 nuclear-encoded genes (about 10% of the nuclear genome) encode proteins that are targeted to the apicoplast. The apicoplast of *P. falciparum* is essential for production of fatty acids, isoprene, heme, and iron-sulfur clusters. Fatty acid synthesis by the apicoplast of *Toxoplasma gondii* is likewise essential for organismal survival (Mazumdar et al. 2006). Because the apicoplast fatty acid pathway differs from that present in humans, drugs that inactivate the apicoplast pathway could be used to cure diseases such as malaria without harming the patient.

Secondary Green Plastids Arose More Than Once

An analysis of the complete chloroplast genome of the chlorarachniophyte *Bigelowiella natans* provided evidence that secondary green plastids arose independently in chlorarachniophytes and in plastid-bearing euglenoids. The *B. natans* plastid is more closely related to plastids of particular green algae than to plastids of euglenoids (Rogers et al. 2007). These results are not consistent with a previous hypothesis that euglenoids and chlorarachniophytes arose from a single common ancestor having a secondary green plastid that arose only once (Cavalier-Smith 1999; Cavalier-Smith and Chao 2003). Instead, secondary green plastids have arisen more than once, as was first suggested by Ishida et al. (1997). As noted earlier, an evolving endosymbiosis between a heterotrophic katablepharid flagellate and the unicellular green alga *Nephroselmis* (Okamoto and Inouye 2005a, 2006) and green dinoflagellates provides additional cases of secondary green endosymbiosis.

Hannaert and associates (2003) suggested that trypanosomes might represent an additional example. Trypanosomes are heterotrophic protists that

Secondary Endosymbiosis Hypotheses

(a) Chromalveolate hypothesis—only one ancient secondary endosymbiosis event in the red lineage

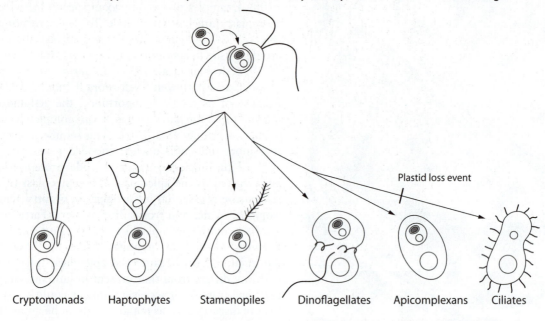

(b) Portable plastid hypothesis—multiple events of secondary endosymbiosis involving red algae

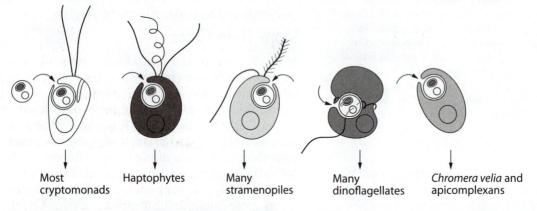

Figure 7.25 Two alternative views of the origin of secondary red plastids.

include human parasites such as *Trypanosoma* species that cause sleeping sickness and Chagas disease, and *Leishmania*, the agent of leishmaniasis. Certain photosynthesis genes, including those encoding synthesis of the sedoheptlose-1,7-bisphosphatase (SB-Pase) enzyme, which is specific to the Calvin cycle of plastids, occur in trypanosomes even though they lack plastids. However, it is likely that such genes were horizontally transmitted to trypanosomes. Horizontal gene transfer seems to be an ongoing

evolutionary process among protists (Oudot-LeSecq et al. 2007). (See Chapter 5 for a more complete discussion of horizontal gene transmission and its impacts on algal evolution.)

Did Secondary Red Plastids Arise More Than Once?

While it is now clear that secondary green plastids arose more than once, whether secondary red plastids arose once or more than once has remained an open question. Cavalier-Smith (1999, 2003, 2006) proposed that secondary red plastids, which contain chlorophyll *c* in additoin to chlorophyll *a*, originated only once in a common ancestor of cryptomonads, haptophytes, stramenopiles, and alveolates, and that these groups together form a monophyletic super-group known as the Chromalveolates. The hypothesis of single red secondary plastid origin is known as the **chromalveolate hypothesis** (Figure 7.25a). Similar replacements of plastid-targeted genes in these algae have been cited as support for the chromalveolate hypothesis (Fast et al. 2001; Patron et al. 2004; Petersen et al. 2006), and similar endoplasmic reticulum proteins are reported to convey plastid-targeted proteins in cryptomonads, apicomplexans, and diatoms (Sommer et al. 2007). However, while a multi-gene phylogeny designed to test the chromalveolate hypothesis found strong support for monophyly of alveolates (ciliates + apicomplexans + dinoflagellates), monophyly of these groups plus stramenopiles, cryptomonads, and haptophytes was not well supported (Harper et al. 2005). The Chromalveolata was also not recovered as a monophyletic group in multigene studies by Burki et al. (2007), Iida et al. (2007), or Yoon et al. (2008) or by an analysis of *EEF2* gene sequences (Kim and Graham 2008). In addition, a study of the evolution of plastid-targeting sequences in the malaria parasite has revealed that transit peptides/sequences are easily derived from existing sequences (Tonkin et al. 2008). The surprising ease with which such targeting sequences can be acquired indicates that such would not be rare events. This would facilitate multiple occasions of endosymbiotic organelle evolution.

An alternative hypothesis contends that secondary red plastids have originated more than once, a concept known as the **portable plastid hypothesis** (Figure 7.25b). Red plastids are considered portable because they contain a larger set of genes than do green plastids. Red plastids thus have greater genetic autonomy, meaning that they depend less on host proteins, and consequently are more easily incorporated into new host cells. The small subunit of rubisco provides an example of this greater red plastid autonomy. The red plastid genome encodes both the rubisco small subunit protein and the large rubisco subunit. By contrast, in green algae, the small subunit of rubisco is encoded by the nuclear genome, and corresponding genes have been lost from green plastid genomes. Hence, red algal plastids may be better

able to adjust to life within different types of plastidless host cells (Grzebyk et al. 2003, Falkowski et al. 2004). According to the portable plastid hypothesis, the ancestors of cryptomonads, haptophytes, photosynthetic stramenopiles, and photosynthetic dinoflagellates were distinct lineages of plastidless cells. In support of this concept, Müller and collaborators (2001) found molecular phylogenetic evidence that stramenopile plastids arose from different unicellular red algal ancestors than did the plastids of cryptomonads and haptophytes.

Evidence often cited against a hypothesis of single early red plastid acquisition includes the fact that plastids are absent from the ancestral lineages of the "chromalveolate" groups (Bodyl 2005). The phylogeny of stramenopiles includes several early-diverging lineages without plastids. The earliest-diverging cryptomonads and their sister group, the katablepharids (Okamoto and Inouye 2005b), are plastidless. Among alveolates, the earliest-diverging apicomplexan (*Cryptosporidium*) lacks an apicoplast, many early-diverging dinoflagellates lack plastids (Saldarriaga et al. 2003), and the complete nuclear genomes of two ciliates were reported to show no evidence of photosynthetic genes (Aury et al. 2006; Eisen et al. 2006). Extensive microscopic studies of many types of ciliates have revealed no trace of remnant plastids. Either ciliates never possessed plastids, or they have completely lost plastids. As earlier noted in the case of apicoplasts, plastids may be maintained even when photosynthetic pigments have been lost because plastids perform several essential biosynthetic functions distinct from photosynthesis. Additional examples of colorless plastids occur among cryptomonads, stramenopiles, red algae, euglenoids, green algae, and land plants. The conditions that foster complete loss of plastids are not understood.

Diversity of Chlorarachniophytes

Chlorarachniophytes are an algal phylum (Chlorarachniophyta) (Hibberd and Norris 1984) characterized by secondary green plastids within unicellular hosts related to filose amoebae (the supergroup Rhizaria). Molecular phylogenetic analyses confirm that they form a monophyletic group (Ishida et al. 1999). Chlorarachniophytes are named for the green color of their plastids and the spider shapes of the amoeboid cells. *Chlorarachnion reptans* has spider-shaped cells arranged in a network. Other species primarily occur as flagellates (e.g., *Bigelowiella natans*) or round, walled

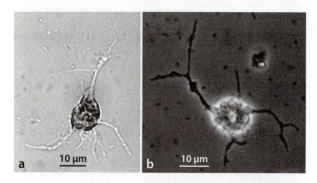

Figure 7.26 *Chlorarachnion* as seen with (a) bright-field and (b) phase-contrast light microscopy. (Photos: L. W. Wilcox)

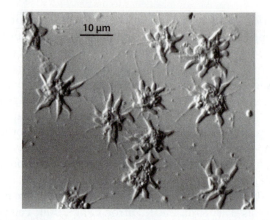

Figure 7.27 The chlorarachniophyte *Gymnochlora stellata* viewed with light microscopy. (From Ishida et al. 1996)

cells that can transform into amoebae or flagellate cells (e.g., *Lotharella*). Transmission electron microscopic studies show that the ultrastructure of flagellate chlorarachniophyte cells is distinctive (Moestrup and Sengco 2001). Chlorarachniophytes occur in temperate and tropical marine waters, growing among sand grains, on mud, in tidepools, on seaweeds, or in plankton. *Bigelowiella natans* and *Gymnochlora stellata* are mixotrophic, consuming bacterial cells that are digested in food vacuoles (Moestrup and Sengco 2001; Ishida et al. 1996).

One or many bilobed plastids occur in each chlorarachniophyte cell. Each plastid has a large pyrenoid—an accumulation of rubisco and other proteins. The form of the pyrenoid is an important classification character. The plastids of the chlorarachniophytes are similar to those of green algae in SSU rDNA (van de Peer et al. 1996) and elongation factor-Tu amino acid (Ishida et al. 1997) sequence data, and in containing chlorophylls *a* and *b*. However, chlorarachniophyte plastids have four envelope membranes, while plastids of green algae have only two. As noted earlier, a nucleomorph occurs within a **periplastidal compartment**—the region between the inner and outer pairs of chloroplast envelope membranes (see Figure 7.23).

CHLORARACHNION (Gr. *chloros*, green + Gr. *arachni*, spider) has spider-shaped cells that contain many green plastids (Figure 7.26).

GYMNOCHLORA (Gr. *gymnos*, naked + Gr. *chloros*, green) is a green, star-shaped amoeba with several filopodia—filamentous extensions of the cytoplasm (Figure 7.27). The cells do not form networks as do those of *Chlorarachnion*. Cells are 10–20 µm

in diameter and often attach themselves to surfaces. The pyrenoids are capped with a photosynthetic reserve material that is not starch. The periplastidal envelope differs from that of some secondary plastids in not being connected to the nuclear envelope and lacking ribosomes on the surface. The nucleomorph is located near the pyrenoid. There is no cell surface covering. Spherical resting stages may occur (Ishida et al. 1996).

7.5 Tertiary Endosymbiosis

Tertiary endosymbiosis involves the incorporation by a host of an endosymbiont that possesses a secondary plastid (Figure 7.28). Several examples of tertiary endosymbiosis are known from the dinoflagellates.

As noted earlier, some dinoflagellates have green plastids derived by secondary endosymbiosis involving a green alga; *Lepidodinium chlorophorum* (= *Gymnodinium chlorophorum*) and *Lepidodinium viride* (Watanabe et al. 1990) are examples. However, most photosynthetic dinoflagellates possess plastids that contain large amounts of a golden secondary pigment known as peridinin and are thus known as the peridinin dinoflagellates. Most experts consider that peridinin plastids arose by secondary endosymbiosis from a red alga (e.g., Keeling 2004), but others have proposed that such plastids descended by tertiary endosymbiosis involving a haptophyte (Yoon et al. 2002a) or a photosynthetic stramenopile (Bodyl and Moszczynski 2006). Other dinoflagellate plastids have clearly arisen by

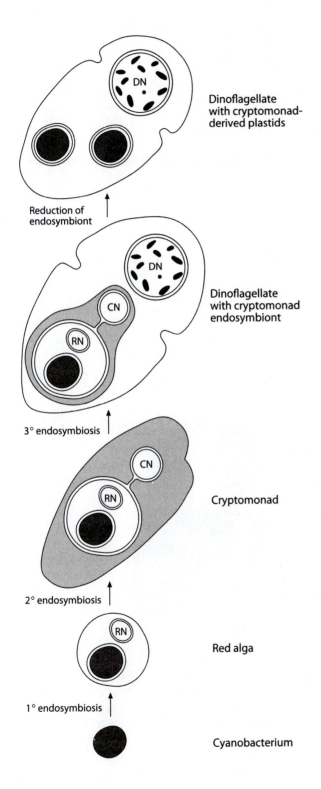

Dinoflagellate
with cryptomonad-
derived plastids

Reduction of
endosymbiont

Dinoflagellate
with cryptomonad
endosymbiont

3° endosymbiosis

Cryptomonad

2° endosymbiosis

Red alga

1° endosymbiosis

Cyanobacterium

tertiary endosymbiosis involving a cryptomonad, haptophyte, or diatom (Palmer and Delwiche 1996). These are often considered to have replaced an ancestral peridinin plastid.

Immunolocalization of phycoerythrin in chloroplasts of two *Dinophysis* species coupled with other ultrastructural data suggest that the plastids of these marine dinoflagellates originated from a cryptomonad (Vesk et al. 1996). This conclusion is supported by detection in *Dinophysis norvegica* of alloxanthin, an accessory pigment that marks cryptomonads (Meyer-Harms and Pollehne 1998). Molecular analysis of *psbA* sequences provided evidence that the plastids of three *Dinophysis* species were derived from a cryptomonad (Janson and Graneli 2003).

A clade of several related marine dinoflagellates contains plastids derived from diatoms (Hansen et al. 2007; Pienaar et al. 2007). Chesnick and colleagues (1996, 1997) found that in *Kryptoperidinium foliaceum* and *Durinskia baltica* (= *Peridinium balticum*), the diatom endosymbiont contributed a membrane-bound cytoplasm that contains not only plastids but also mitochondria and an extra nucleus, which is not reduced in size as are nucleomorphs. Their study demonstrated that the plastids and extra nucleus were derived from a pennate (bilaterally symmetrical) diatom. Although the dinoflagellate *Peridinium quinquecorne* is closely related to other dinoflagellates whose plastids are derived from diatoms, molecular data suggest that a different diatom endosymbiont has replaced the original one (Horiguchi and Takano 2006). In such amazing cases, many cellular genomes have participated in the evolution of a single species.

Figure 7.28 Diagram of the acquisition by a dinoflagellate cell of plastids (or kleptoplastids) derived from a cryptomonad by tertiary endosymbiosis. (Redrawn with permission from Wilcox, L. W., and G. J. Wedemayer. Dinoflagellate with blue-green chloroplasts derived from an endosymbiotic cryptomonad. *Science* 227:192–194. © 1985 American Association for the Advancement of Science)

Chapter 8

Euglenoids

Phacus

The euglenoids, also known as euglenids, euglenophytes, or euglenoid flagellates, are among the most ancient lineages of eukaryotic algae. Many euglenoids lack plastids, while others possess secondary green or colorless plastids (Whatley 1993) and thus vary in nutritional type. Non-photosynthetic euglenoids feed upon organic particles such as bacteria or unicellular eukaryotes or absorb dissolved organic compounds, and all require vitamins B_1 and B_{12}. Such nutritional variability explains why diverse euglenoids often occur in habitats rich in decaying organic matter, where dissolved organic compounds and bacteria are abundant. Euglenoids can be abundant in farm ponds, marshes, swamps, fens, bogs, oceanside beaches, and mud flats, even populating the hindguts of dipteran larvae and tadpole rectums. Their nutritional diversity also explains how euglenoids can ecologically link components of aquatic microbial food webs. In this chapter we look at how comparative studies of euglenoid nutrition, cellular structure, behavior, and molecular gene sequences have revealed the relationships and evolutionary history of euglenoids.

(Photo: L. W. Wilcox)

8.1 Euglenoid Relationships and Evolutionary History

Fossil euglenoids are relatively rare, likely because few ancient euglenoid cells produced structures that were resistant to decay. However, unicellular fossils known as *Moyeria* about 410–460 million years old are regarded as euglenoids (Gray and Boucot 1989), and the modern genera *Phacus* and *Trachelomonas* are similar to fossils produced in the Tertiary, beginning about 60 million years ago (Taylor and Taylor 1993). Most of what we know about euglenoid evolution is based on comparative ultrastructural and molecular studies of modern representatives.

Euglenoid Cellular Features Reveal Relationship to Trypanosomes and Diplonemids

A series of transmission electron microscopic studies (e.g., Leedale 1967; Vickerman and Preston 1976; Willey et al. 1988; Walne and Kivic 1990; Triemer and Farmer 1991; Walne and Dawson 1993; Dawson and Walne 1994; Farmer and Triemer 1994; Linton and Triemer 1999) revealed that euglenoids share several structural features with two other groups of unicellular flagellates: diplonemids and kinetoplastids. All three groups have flagella that emerge from a cellular depression known as a **pocket**, and they reproduce primarily by longitudinal mitotic division. In addition, a column of protein, known as a **paraxonemal rod** or **paraflagellar rod**, stiffens the hair-covered flagella of euglenids and kinetoplastids. These two groups also display disk- or paddle-shaped mitochondrial cristae and nuclear chromosomes that are condensed throughout the cell cycle or much of it (Figure 8.1). Kinetoplastids, diplonemids, and euglenoids were classified together as the Euglenozoa (Cavalier-Smith 1981) (Figure 8.2).

The kinetoplastids are of great medical importance because they include parasitic trypanosomes such as *Trypanosoma*, which is transmitted by insects and causes sleeping sickness in Africa and Chagas disease in Latin America. The trypanosome *Leishmania*, also transmitted by insects, causes 2 million cases of disease in many countries each year. Kinetoplastids also include the mostly free-living bodonids, such as the common genus *Bodo*, which can contaminate outdoor algal pond cultivation systems. Kinetoplastids feature a large mass of DNA, known as a kinetoplast, in their single mitochondrion, and their glycolytic enzymes occur within a modified peroxisome known as a glycosome (Vickerman 1990). In contrast, euglenoid

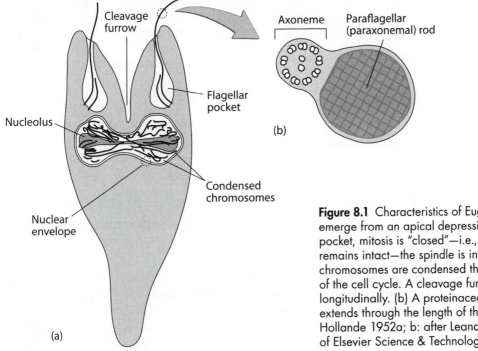

Figure 8.1 Characteristics of Euglenozoa. (a) Flagella emerge from an apical depression known as a pocket, mitosis is "closed"—i.e., the nuclear envelope remains intact—the spindle is intranuclear, and nuclear chromosomes are condensed throughout all or most of the cell cycle. A cleavage furrow splits the cell longitudinally. (b) A proteinaceous paraflagellar rod extends through the length of the flagella. (a: after Hollande 1952a; b: after Leander 2004, by permission of Elsevier Science & Technology Journals).

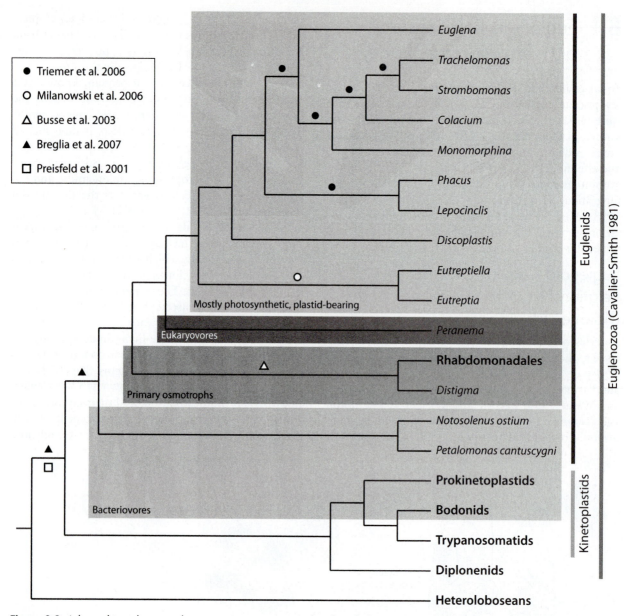

Figure 8.2 A branching diagram that summarizes recent molecular phylogenetic studies of euglenozoan evolution.

glycolysis occurs within the cytoplasm, as it does in most eukaryotes, and most euglenoids lack kinetoplasts. Diplonemids are heterotrophic flagellates that primarily consume particles, though some can be parasitic. Diplonemids and kinetoplastids are sister groups and together are sister to the euglenoids (Preisfeld et al. 2001; Breglia et al. 2007) (see Figure 8.2). Heterotrophic eukaryotic jakobids (Simpson et al. 2006) and heteroloboseans (Baldauf et al. 2000) have been proposed as sisters to the Euglenozoa.

Cellular Features That Distinguish Euglenoids from Kinetoplastids and Diplonemids

Two major cellular features distinguish euglenoids from kinetoplastids and diplonemids. First, euglenoids produce carbohydrate storage granules known as **paramylon**, a β-1,3-linked glucan that is found in the cytoplasm of even colorless euglenoid species (Figure 8.3). Though

Figure 8.3 Paramylon granules (arrowheads) in the anterior end of a *Euglena* cell. The arrow points out several mucocysts that have been visualized by neutral-red staining. (Photo: L. W. Wilcox)

paramylon is composed of the same monomers as plant starch, the different way in which glucose molecules are linked does not allow iodine solution to stain paramylon. The enzyme that synthesizes paramylon has been isolated and characterized (Bäumer et al. 2001).

Second, euglenoids uniquely display a surface structure, known as a **pellicle**, that is composed of parallel ribbonlike strips. Such strips are primarily composed of proteins known as articulins (Huttenlauch and Stick 2003). Pellicular strips run lengthwise down euglenoid cells, giving them a faintly striated appearance when viewed with a light microscope (Figure 8.4). Pellicular strips are S-shaped in cross-sectional view by transmission electron microscopy, and the point at which they join with their neighbors is known as the **articulation zone** (Figure 8.5). In preparation for cell division, new pellicular strips appear between older ones, so the two

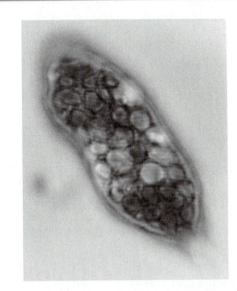

Figure 8.4 A surface view of a *Euglena* cell in which the spirally arranged pellicular strips are evident. (Photo: L. W. Wilcox)

cellular products of division possess an equal number of old and new strips (Leadbeater and Green 1993). During the evolutionary history of euglenoids, the number and structure of pellicular strips has changed in parallel with nutritional diversification.

Euglenoid Diversification

Comparative study of modern euglenoids reveals that cell structure and behavior have evolved together with changes in mode of nutrition (see Figure 8.2).

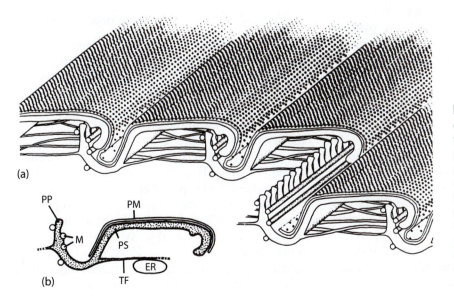

Figure 8.5 (a) Three-dimensional diagram of the interlocking strips that make up the euglenoid pellicle. (b) An individual pellicular strip (in cross-section) and associated structures. ER = endoplasmic reticulum tubule, M = microtubules, PM = plasma membrane, PP = periodic projections of upturned part of strip, PS = pellicular strip, TF = traversing filaments. (From Suzaki and Williamson 1986)

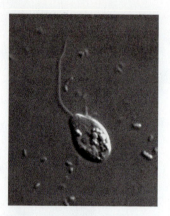

Figure 8.6 The early diverging, plastidless euglenoid *Petalomonas*. (Photo: R. Triemer)

The earliest euglenoids likely consumed bacteria and gave rise both to a lineage of osmotrophs and a lineage that was able to consume eukaryotic prey cells. The latter group was ancestral to the plastid-bearing euglenoids, some of which subsequently lost photosynthetic pigments (but retained their plastids).

Early bacteria-consuming euglenoids

Early-diverging euglenoids such as *Petalomonas* (Figure 8.6) and *Notosolenus* illustrate ancestral euglenoids. These plastidless heterotrophs feed by ingesting bacterial cells and thus display **bacterivory**. Such euglenoids have a relatively simple feeding pocket located below the flagellar pocket and only a few (4–12) thin, longitudinal, pellicular strips that are fused at their edges, making the cells rigid. Bacterivorous euglenoids glide along surfaces, a behavior likely to be adaptive in obtaining bacterial food. They generally have two flagella, which display different motions that facilitate gliding; one flagellum is directed forward, while the other moves against the substrate.

Plastidless, osmotrophic euglenoids

A clade of plastidless euglenoids has lost the capacity for bacterivory. These organisms rely upon osmotrophic nutrition, the absorption of dissolved organic compounds. Such plastidless euglenoids are thus said to be primary osmotrophs, in contrast to plastid-bearing, colorless euglenoids, which are known as secondary osmotrophs. The primary osmotrophs include *Distigma* and the order Rhabdomonadales (including genera such as *Rhabdospira*, *Rhabdomonas*, *Monoidium*, *Gyropaigne*, and *Parmidium*) (Preisfeld et al. 2001).

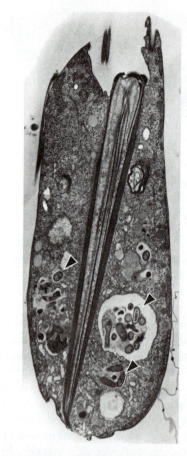

Figure 8.7 Transmission electron micrograph of the plastidless bacteriovore *Entosiphon*, showing dark ingestion rods and vanes, and bacteria that are being digested (arrowheads). (Micrograph: P. Walne)

Euglenoids that consume eukaryotic prey

Some bacterivorous euglenoids, illustrated by modern *Entosiphon*, developed more elaborate feeding structures that include pinwheel-like vanes and rods supported by microtubules (Figure 8.7). Such complex feeding structures are viewed as preadaptations that allowed the evolution of **eukaryovory**, the ability to consume eukaryotic cells, which are typically much larger than bacteria (Leander 2004). Euglenoids such as *Peranema* are able to ingest large cellular prey because they possess not only complex feeding structures but also pellicles composed of 40 or more helically arranged pellicular strips that are able to slide against each other (Murata and Suzaki 1998) (see Figure 8.4). This large number of strips means that the number of articulation zones has also increased, making

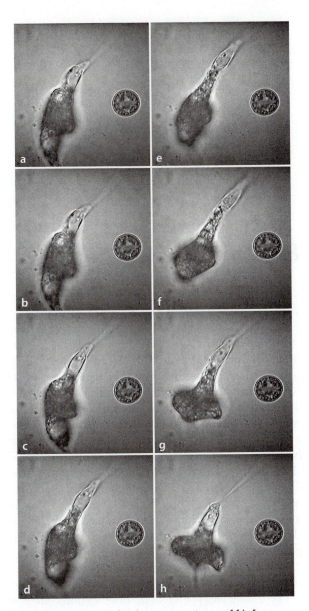

Figure 8.8 A series of video images (every fifth frame shown) of a non-photosynthetic euglenoid moving metabolically. Note the considerable shape changes that occurred while the cell moved slowly upward. (Photos: L. W. Wilcox)

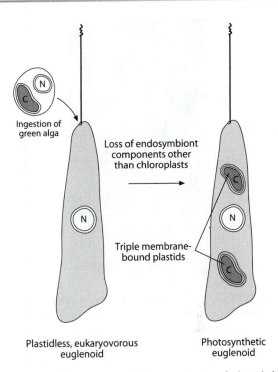

Figure 8.9 A diagram modeling acquisition of plastids by euglenoids through secondary endosymbiosis.

such cells very flexible. Consequently, *Peranema* and some other euglenoids are able to move by a form of cellular plasticity known as **euglenoid movement** or **metaboly** (Figure 8.8). Such movement enhances the capacity for eukaryovory and hence the acquisition of endosymbionts and secondary plastids (Figure 8.9). Many, if not all, photosynthetic euglenoids retain a vestigial digestion system, providing strong evidence that they arose from a phagotrophic ancestor (Willey and Wibel 1985; Shin et al. 2002). Molecular phylogenetic studies indicate that a eukaryovorous cell that had acquired an endosymbiotic green alga was ancestral to a diverse clade of green plastid-bearing euglenoids (Leander 2004). Phylogenetic analysis indicates that euglenoid plastids did not arise from the same type of green alga that donated its plastid to the chlorarachniophytes (Rogers et al. 2007).

Plastid-bearing euglenoids

Most euglenoids that contain plastids are green because their plastids contain photosynthetic pigments similar to those of green algae, namely chlorophylls *a* and *b* and β-carotene. However, some xanthophylls not typical of green algae, such as diadinoxanthin, may be present. Reflecting their secondary origin from an endosymbiotic green algal cell, euglenoid plastids are enclosed by three envelope membranes. However, in the case of euglenoid plastids, the outer chloroplast envelope membrane is not lined with ribosomes, nor is it usually connected to the nuclear envelope, as is the case for some other algae that acquired plastids by secondary endosymbiosis.

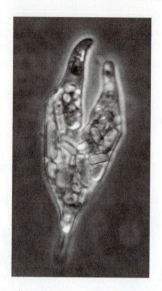

Figure 8.10 Dividing *Euglena* cell with characteristic two-headed appearance. Note also the large number of paramylon granules. (Photo: L. E. Graham)

At least one cytoplasmically synthesized protein is transported to the Golgi apparatus, where it is encased in a membrane prior to import by plastids (Sulli and Schwartzbach 1995).

The phylogeny of plastid-bearing euglenoids has now been well resolved (Milanowski et al. 2006; Triemer et al. 2006) (see Figure 8.2). The acquisition of photosynthetic plastids was accompanied by changes in the pellicle, with implications for motility. For example, relatively early-diverging photosynthetic euglenoids retain euglenoid movement, while later-diverging species have lost this capacity and become rigid. Such non-plastic species display fewer, thicker pellicular strips having projections that limit flexibility (Leander 2004). While many euglenoids may secrete mucilage, this trait is particularly evident in some plastid-bearing species. For example, the cells of some green euglenoids occur within a hard casing of mucilage impregnated with minerals, which is known as a **lorica**. Other green species produce mucilaginous stalks, which allow them to take up a largely sedentary habit. Certain plastid-bearing species have lost photosynthetic capacity and thus rely upon osmotrophy as a nutritional mode (Montegut-Felkner and Triemer 1997). Phylogenetic analyses provide evidence that osmotrophy evolved independently in several plastid-bearing clades, indicating that loss of photosynthetic pigments occurred multiple times (Nudelman et al. 2003).

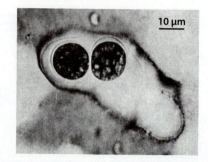

Figure 8.11 Euglenoid cysts. (From Bold and MacEntee 1973, by permission of the *Journal of Phycology*)

8.2 Euglenoid Reproduction

Sexual reproduction does not occur in euglenoids with regularity, if at all. Asexual reproduction occurs by longitudinal division, proceeding from apex to base, such that euglenoids in the process of cytokinesis appear to be "two-headed" (Figure 8.10). Prior to mitosis, the nucleus migrates to the region just below the flagellar pocket. The nuclear envelope does not break down during mitosis, as it does in some other protists, animals, and land plants. Often, a pair of basal bodies that had undergone replication prior to nuclear division can be observed at each of the spindle poles. The spindle develops within the confines of the nuclear envelope, lying at right angles to the long axis of the cell. Following chromosomal separation, daughter nuclei form by constriction of the parental nucleus.

In response to changing environmental conditions, euglenoids may form resting cysts (Figure 8.11). Laboratory and field observations suggest that cyst production is triggered by low nutrient levels or by low nitrogen to phosphorus ratios (Triemer 1980; Olli 1996). The formation of cysts involves loss of flagella, increase in the number of paramylon granules, swelling and rounding of the cells, and deposition of a layered mucilaginous enclosure that consists primarily of polysaccharides. Although not present in all euglenoid cysts, the unusually large red-orange eyespots and distinctive plastids of green forms can be helpful in deducing the euglenoid identity of nonmotile spherical structures in natural samples. Euglenoid movement sometimes occurs within the confines of the cyst wall.

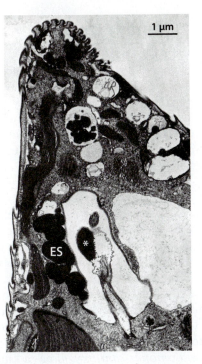

Figure 8.12 Photosensing system at the anterior end of *Euglena granulata* cell, as seen with TEM. Note eyespot granules (ES) lying opposite the paraflagellar body (asterisk). (From Walne and Arnott 1967)

8.3 Euglenoid Plastids and Light-Sensing Systems

In cells of plastid-bearing species, many plastids typically occur and plastid division is coordinated with nuclear and cytoplasmic division. Euglenoid plastids can take diverse shapes, such as plates, shallow cups, or ribbons, and are sometimes arranged in star-shaped aggregations. The edges of plastids may be dissected into lobes. Plastid thylakoids typically occur in stacks of three. Although some larger stacks can be observed, the grana typical of land plants and certain green algae are not present. Most euglenoid plastids contain pyrenoids, aggregations of rubisco and other proteins. In some species pyrenoids are embedded within the plastids, but in others the pyrenoid occurs at the end or edge of plastids or extends from the main body of the plastid on a stalk. A shell of paramylon granules sometimes covers the portions of pyrenoids that are in contact with the cytoplasm (Whatley 1993). Treatment of euglenoids with heat, extended periods of darkness, or certain antibiotics results in plastid bleaching, i.e.,

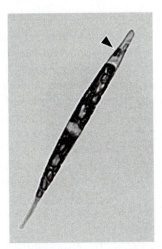

Figure 8.13 Eyespot (arrowhead) of a photosynthetic euglenoid viewed by light microscopy. (Photo: L. W. Wilcox)

loss of pigmentation. The plastids dedifferentiate into a proplastid state in which thylakoids are few or absent but proplastids continue to divide. When exposed to light, bleached cells may regain thylakoids and chlorophyll pigmentation. Ortiz and Wilson (1988) studied changes in plastid proteins that are associated with heat-induced bleaching.

Plastid-bearing euglenoids typically possess a light-sensing system. This allows cells to move away from light that is so bright that it could cause damage to plastids or move from shade toward irradiance levels optimal for photosynthesis. For example, *Euglena proxima* living in intertidal sand flats moves upward toward the light during daytime low tides but can avoid excessive light by downward migration (Kingston 1999). The light-sensing system of euglenoids consists of two structures. The first of these is the paraflagellar body (PFB), a swelling at the base of one of the flagella (Figure 8.12). The PFB contains blue, light-sensitive flavoproteins that appear green when viewed by fluorescence microscopy. A major PFB flavoprotein of *Euglena gracilis* has been identified as an adenylyl cyclase, an enzyme that synthesizes cyclic AMP when it is light activated. The cyclic AMP likely directly affects flagellar motion, allowing cells to move toward light (Iseki et al. 2002). The second component of the light-sensing system is an eyespot, also known as a stigma. The eyespot is located in the cytoplasm adjacent to the flagellar pocket, just opposite the basal PFB (see Figure 8.12). The eyespot shades the PFB when the cell is in certain positions but not when the cell lies in other orientations. This allows the cell to determine the direction of a light source. While eyespots occur in other algae, they are particularly conspicuous in euglenoids, being as large as 8 μm in diameter and therefore quite noticeable at the light microscopic level (Figure 8.13). Ultrastructural examination reveals that euglenoid eyespots consist of 50–60 globules arranged in a single layer, often bound by a membrane (see Figure 8.12). These globules contain the carotenoids astaxanthin and/ or echinenone, which give the eyespot its orange-red coloration.

Some non-photosynthetic euglenoids possess light-sensing systems that do not involve eyespots. For example, *Peranema trichophorum* uses curling behavior to change direction during swimming or while eating. In this species, light stimulation of rhodopsin, located in one of the flagella, appears to cause the opening of a calcium channel. The resulting calcium influx causes a contraction of the cell on one side, curling it (Saranak and Foster 2005).

8.4 Euglenoid Ecology

At least one expert has suggested that there are probably no truly planktonic euglenoid species (Lackey 1968), and others suggest that euglenoids are fundamentally occupants of interfaces, such as the air–water and sediment–water boundaries (Walne and Kivic 1990). In such habitats euglenoids can be infected by chytrids and consumed by herbivores, including the predator *Peranema*. Marine populations of *Eutreptiella gymnastica* are readily grazed by mesozooplankton in the coastal Baltic Sea (Olli et al. 1996).

Certain euglenoids are known for tolerating extreme conditions of desiccation, low pH, and heat. Some seem able to migrate into soils and persist there for long periods in a dormant state. Though comparatively rare, *Euglena gracilis* has been recovered from cryptogamic crusts of semiarid and arid lands of North America (Johansen 1993). *Euglena mutabilis* is able to grow in extremely low-pH waters, such as streams draining coal mines and the acidic, metal-contaminated ponds of the Smoking Hills region of the Canadian Arctic. The optimal pH for growth of this species is 3.0, but *E. mutabilis* tolerates pH values lower than 1.0. *Euglena pailasensis* occurs at the surface of boiling mud at temperatures up to 98°C and pH as low as 2, in a volcanic region of Costa Rica (Sanchez et al. 2004).

8.5 Euglenoid Diversity

There are more than 40 genera of euglenoids and 800–1000 species. Phylogenetic studies have revealed that some classically defined genera such as *Euglena* and *Phacus* were actually polyphyletic, with the result that new genera such as *Discoplastis* (Triemer et al. 2006) and *Monomorphina* (Marin et al. 2003) have been erected.

PETALOMONAS (Gr. *petalon*, leaf + Gr. *monas*, unit) is a colorless, relatively rigid cell that occurs in freshwaters, moving along surfaces where it feeds on bacteria (see Figure 8.6). The single emergent flagellum is held forward and moves only at the tip. Cells possess no plastids, eyespot, or paraflagellar body but commonly contain one or more endosymbiotic cyanobacterial cells (Schnepf et al. 2002). This early-diverging genus is thought to reflect features of ancestral euglenoids (Leander 2004). *Petalomonas cantuscygni* has a kinetoplast-like structure in mitochondria, similar to that of the kinetoplastids but unlike other euglenoids (Leander et al. 2001).

PERANEMA (Gr. *pera*, pouch + Gr. *nema*, thread) is a colorless eukaryovore (Figure 8.14). Cells can swim in fluids and also move along surfaces, by the surface motility of a rigid, forward-directed flagellum (Saito et al. 2003). The forward-facing flagellum waves at the tip, thereby possibly sensing the presence of food particles. A second flagellum is directed backward, appressed to the cell body, and has rhodopsin light-sensing capacity (Saranak and Foster 2005). Hilenski and Walne (1985a,b) studied the ultrastructure of the flagellar root system and flagellar hairs of *Peranema*. *Peranema* exhibits euglenoid motion. A specialized set of ingestion rods at the cell anterior aids in prey capture and ingestion. *Peranema* is a voracious consumer of other euglenoids, cryptomonads, yeasts, green algae, and other particles. The use of video microscopy and scanning electron microscopy (SEM) reveals that *Peranema* has two feeding modes. Particles can be ingested entirely or rasped open by grating action of the feeding apparatus, after which the contents are sucked into the cell and deposited in a food vacuole. The latter process can be accomplished in less than 10 minutes (Triemer 1997).

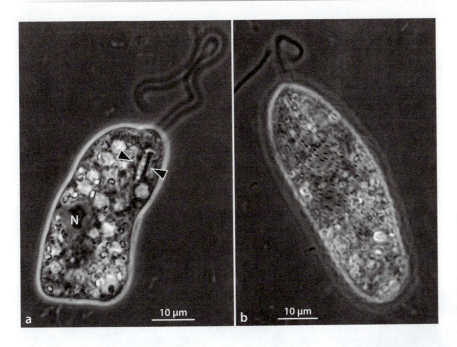

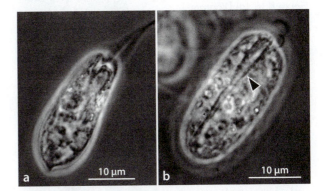

Figure 8.14 The eukaryovore *Peranema*, which has very plastic cells. In (a), the thick, swimming flagellum is particularly obvious, as are the nucleus (N) and ingestion rods (arrowheads). In (b), the ornamented pellicle is evident. (Photos: L. W. Wilcox)

ENTOSIPHON (Gr. *entos*, inside + Gr. *siphon*, tube) is a colorless phagotrophic euglenoid with a rigid cell that is longitudinally furrowed such that the anterior end has a scalloped appearance (Figure 8.15). It has a large, funnel-shaped ingestion apparatus that extends the length of the cell (see Figure 8.7). Two flagella are present; the shorter extends forward and confers motility, while the longer flagellum trails.

EUGLENA (Gr. *eu*, good, true, or primitive + Gr. *glene*, eye) is probably the best-known euglenoid genus. Marin et al. (2003) transferred *E. spirogyra*, *E. acus*, and *E. tripteris* into the genus *Lepocinclis*, and Triemer et al. (2006) transferred *E. spathirhyncha* and *E.* cf. *adunca* into the new genus *Discoplastis*. With these changes, the genus *Euglena* is regarded as monophyletic (Triemer et al. 2006).

Euglena possesses only a single emergent flagellum, and euglenoid movement occurs. Plastid shape is variable but often discoidal, and cells are cylindrical (not flattened) in cross-section. Most *Euglena* species are elongated, with a rounded anterior and the posterior tapered to a point (Figure 8.16). Some species produce red granules in numbers sufficient to give cells a bright- or brick-red appearance (Figure 8.17), resulting from large amounts of the carotenoid astaxanthin. When large populations of cells are present, they may form dramatic, blood-

Figure 8.15 The plastidless bacteriovore *Entosiphon*. In (a), the scalloped appearance of the anterior end of the cell is evident. The funnel-shaped ingestion apparatus (arrowhead) is shown with light microscopy in (b). (Photos: L. W. Wilcox)

red surface scums on ponds or other water bodies. The formation of such scums is favored by the presence of high levels of dissolved organic compounds and high temperatures. Although rain can break up the scum, it can re-form when the disturbance ends. Cells of some species appear red most of the time, but those of several species can change from green to red within 5–10 minutes in response to increased light intensity, such as at sunrise, then return to green coloration at the appearance of a cloud or at

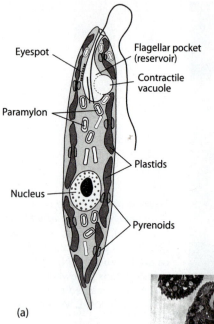

Figure 8.16 (a) Diagram illustrating typical features of *Euglena*, which include the flask-shaped pocket (reservoir), from which one of the two flagella emerges. Adjacent the reservoir are the eyespot and a contractile vacuole. Also typically visible with the light microscope are paramylon granules, plastids, and the nucleus with nucleolus and relatively large chromosomes. (b) TEM view of longitudinally sectioned *Euglena* cell. Note eyespot (ES) and plastids (P) with pyrenoids (Py) and nucleus (N). (a: After Gojdics 1953; b: From Walne and Arnott 1967)

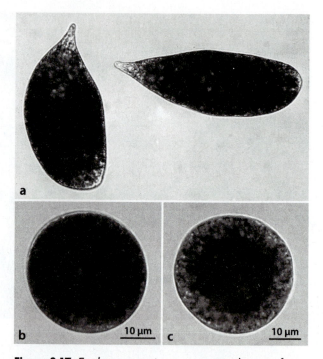

Figure 8.17 *Euglena sanguinea*, a species that can form bright-red blooms. Motile cells are shown in (a). In (b) and (c), two rounded-up cells are pictured. The dark areas represent the red pigment, while green areas appear as a lighter gray. The cells are able to control the distribution of plastids and pigment globules, such that they may appear more red or green, depending upon which bodies are closest to the cell surface. (Photos: C. Taylor)

sunset. This color change involves differential positioning of the red globules and green plastids at the center and periphery of cells. When the red globules are at the cell periphery, cells appear red; when the red globules occupy a central position, surrounded by green plastids, cells appear green. *Euglena sanguinea* is said to be the most common of the red species (see review by Walne and Kivic 1990). Toxins associated with fish kills have been isolated from unialgal isolates of *E. sanguinea* and *E. granulata* (Triemer et al. 2003).

Euglena species can secrete polysaccharide and/or glycoproteins (Cogburn and Schiff 1984) from mucocysts lying just beneath pellicular strips (Hilenski and Walne 1983). Mucocysts appear to arise from the Golgi apparatus. Each mucocyst is connected to the exterior via a canal that opens into the groove between adjacent pellicular strips. Mucocysts are transparent and generally not visible at the light microscopic level until stained with a 0.1% solution of neutral red dye (see Figure 8.3). Mucilage can cover the entire cell in layers of varying thickness. Motile cells usually have only a thin mucilage layer, whereas immobile cells lacking flagella may be embedded in relatively thick layers of jelly to form a monolayer scum on water or other surfaces.

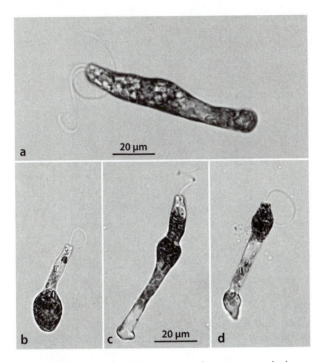

Figure 8.18 Live cells of *Eutreptia* undergoing metaboly. In (a), the two flagella are visible. Some of the different shapes assumed by cells are shown in (b)–(d). (Photos: L. W. Wilcox)

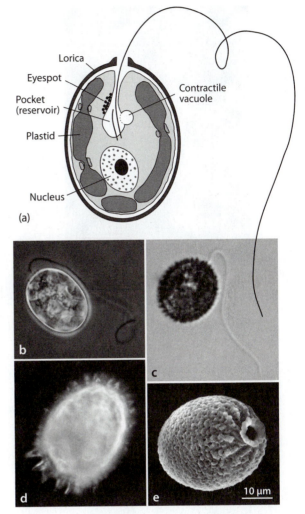

Figure 8.19 (a) *Trachelomonas*, a euglenoid with a lorica—a mineral-impregnated, mucilaginous shell—shown here in diagrammatic, cutaway view. The species shown in (b) has a thin and translucent lorica, while the species shown in (c) has a dark and somewhat warty lorica. The species shown in (d) possesses characteristic spines on its lorica (the cell's anterior is toward the upper right). A lorica is seen with SEM in (e). (b–d: L. E. Graham, e: From Dunlap et al. 1983)

EUTREPTIA (Gr. *eu*, good, true, or primitive + Gr. *treptos*, turned) is found in fresh and marine waters, where it sometimes forms blooms. The ribbonlike plastids are distinctively arranged in stellate groups of 25–30 radiating from a central pyrenoid. There are two emergent flagella (Figure 8.18), and an eyespot and adjacent paraflagellar body are present. Euglenoid movement occurs. The ultrastructure of *E. pertyi* was studied by Dawson and Walne (1991a,b). A similar form, *Tetraeutreptia*, described from eastern Canadian marine waters, is also green but has four flagella, two of which are long and two short (McLachlan et al. 1994).

TRACHELOMONAS (Gr. *trachelos*, neck + Gr. *monas*, unit) is a widespread genus whose green-pigmented cells are encased within a rigid, mineralized, sometimes highly ornamented lorica, which may vary in color from colorless to black-brown (Figure 8.19). The minerals are ferric hydroxide or manganese compounds and the intensity of color depends upon mineral concentration. When the lorica is faintly tinted, it is easy to see the green protoplast within, but when the lorica is darkly colored, it may be impossible to detect any green coloration. The cells may exhibit euglenoid movement within the lorica, but this is detectable only when the lorica is not highly colored. A single emergent flagellum protrudes through the open neck region of the lorica. Reproduction is accomplished by emergence of one or both daughter cells from the lorica, followed by the development of a new lorica. The species are differentiated by lorica

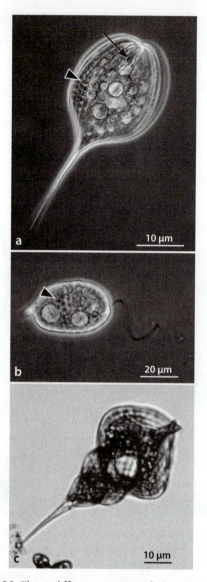

Figure 8.21 *Colacium*, a stalked euglenoid, growing epibiotically on *Daphnia*. (Micrograph courtesy R. Willey)

Figure 8.20 Three different species of *Phacus*. The small spherical plastids (arrowheads) are evident in (a) and (b). Note the eyespot (arrow) in (a), flagellum in (b), and twisted cell body of *P. helikoides* (c). (a: L. W. Wilcox; b, c: L. E. Graham)

ornamentation. One species (*T. grandis*) does not have a spiny lorica, but appears spiny because of the attachment of numerous bacteria to the lorica surface (Rosowski and Langenberg 1994). *Strombomonas* is a related loricate genus.

PHACUS (Gr. *phakos*, lentil) cells are oval to nearly circular and are highly flattened (Figure 8.20a,b). They have a pellicle sufficiently rigid that euglenoid

motion does not occur. Plastids are numerous. The fairly common *Phacus helikoides* (Figure 8.20c) is distinctive in being twisted throughout; *P. tortus* is noticeably twisted just at the posterior. There is just a single emergent flagellum, as in *Euglena*. Leander and Farmer (2001) considered taxonomic issues in the genus. *Hyalophacus* is a non-photosynthetic genus that arose from a photosynthetic *Phacus* ancestor (Nudelman et al. 2003).

COLACIUM (Gr. *kolax*, flatterer) is an attached (sessile) euglenoid of widespread occurrence on aquatic substrates, most noticeably on planktonic zooplankton (Figure 8.21). It forms colonies whose cells occur at the ends of branched, gelatinous strands. In the sessile stage, flagella are non-emergent, but individual cells may produce an emergent flagellum and swim away to generate a new colony elsewhere. The cells attach by the anterior end and secrete a stalk composed of carbohydrate; subsequent cell division yields new branched colonies. The characteristic stalk is formed of carbohydrate extruded from the cell anterior in the form of Golgi-generated mucocysts. More than 100 mucocysts may accumulate within the anterior of each cell prior to their excretion (Willey 1984). A discussion of *Colacium*'s role as an epibiont on surfaces of *Daphnia* and other aquatic animals can be found in Chapter 3.

Chapter 9

Cryptomonads

Cryptomonads, their name literally meaning "hidden single cells," are among the most inconspicuous of the algae. There are several reasons for this. First, cryptomonads are relatively small flagellates, ranging from 3 to 50 μm in length. In addition, cryptomonads are readily eaten by a wide variety of herbivores and so they tend to be found in low numbers. Further, they are often most numerous in cold or deep habitats that people sample less often than warmer surface waters. Finally, cryptomonad cells tend to burst when subjected to environmental shock or preservatives, and thus may not persist for long in water samples. Despite their low profile, cryptomonads are important in both natural systems and aquaculture operations as high-quality food for zooplankton (Chapter 3). In addition, evolutionary biologists recognize the relevance of cryptomonad cell structure to our understanding of secondary endosymbiosis (see Chapter 7). This chapter surveys the relationships, cell biology, reproduction, ecology, and diversity of cryptomonads.

Rhodomonas

(Photo: L. W. Wilcox)

9.1 Cryptomonad Relationships

Cryptomonads are unicellular flagellates (Figure 9.1) that are classified into the phylum Cryptophyta and class Cryptophyceae. Their cells do not produce structures that can be preserved and recognized as fossils, so the time of origin of this group is unclear. Though molecular phylogenetic analyses indicate that the cryptomonads form a monophyletic group, the precise location of their branch on a tree of life has been uncertain.

Cryptomonads are closely related to two groups of plastidless flagellates, the katablepharids (Okamoto and Inouye 2005b) and telonemids (Shalchian-Tabrizi et al. 2006a), as well as a recently discovered group of photosynthetic marine planktonic algae known as picobiliphytes (Not et al. 2007). Molecular analyses based on ribosomal RNA genes also suggest some degree of relationship between cryptomonads and glaucophytes (Bhattacharya et al. 1995; Rensing et al. 1997; Van de Peer and De Wachter 1997; Medlin and Simon 1998; Okamoto and Inouye 2005b). An analysis of multiple nuclear gene sequences indicated that cryptomonads and haptophytes together form a monophyletic group (Hackett et al. 2007). Cryptomonads and haptophytes also share an unusual plastid gene (*rp136*) that a common ancestor may have acquired from a bacterium by horizontal gene transfer (Rice and Palmer 2006).

Gene sequence and pigment data indicate that cryptomonad plastids arose from an endosymbiotic, unicellular red alga (Gillott and Gibbs 1980; McFadden 1993; Douglas et al. 2001). The earliest-divergent modern cryptomonad is the plastidless *Goniomonas* (*Cyathomonas*) (Medlin and Simon 1998; von der Heyden et al. 2004). This fact suggests that a later-diverging cryptomonad acquired a plastid ancestral to those of other modern cryptomonads. Evidence for this scenario would support the portable plastid hypothesis, which proposes that different types of host cells independently acquired red algal plastids (see Chapter 7). Alternatively, red plastids might have been present in the common ancestor of all modern cryptomonads and the plastid was lost from *Goniomonas* (Hackett et al. 2007). Evidence in support of this concept would help to strengthen the chromalveolate hypothesis that a single red plastid-bearing ancestor gave rise to cryptomonads, haptophytes, stramenopiles, and alveolates (see Chapter 7). Thus, cryptomonads are

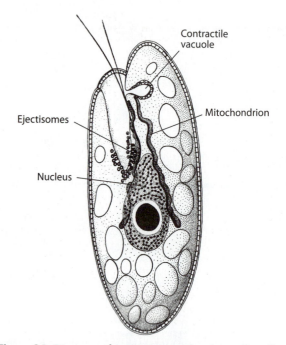

Figure 9.1 Diagram of a cryptomonad with two flagella emerging from a depression. (From Hollande 1952b)

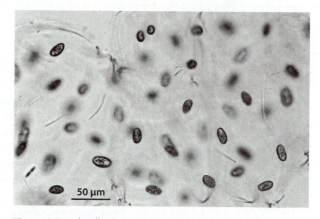

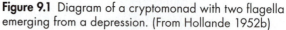

Figure 9.2 Palmelloid cryptomonad with more-or-less equally spaced cells in a gelatinous mass. (Photo: L. W. Wilcox)

central to answering important questions about the diversification of the eukaryotic algal phyla.

9.2 Cryptomonad Mobility and Cellular Structure

Although cryptomonads may produce colonies of non-flagellate cells embedded in mucilage (Figure 9.2), they more typically occur as motile flagellates (see Figure

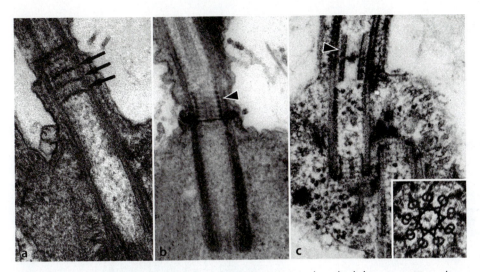

Figure 9.3 Flagellar transition region of (a) cryptomonads, which have partitions (arrows); (b) stramenopiles, which have a helical structure (arrowhead); and (c) green algae, which have a stellate structure (arrowhead). Inset in (c) shows stellate structure in cross-section. (a: From Gillott 1990, b: From Graham et al. 1993, by permission of the *Journal of Phycology*, c: From Floyd et al. 1985)

9.1). As they swim, cryptomonad cells rotate such that their flattened, asymmetrical cell shape can be readily observed with the light microscope.

Flagella, Cytoskeleton, and Light-Sensing System

The two flagella are relatively short and slightly unequal in length. The motion of one flagellum actively propels the cell, while the other flagellum is held more stiffly. Transmission electron microscopy reveals that flagellar surfaces are covered with hairs and sometimes small organic scales shaped like rosettes. Cryptomonad flagella also have a distinctive **transition region**, which occurs between the emergent portion of the flagellum and the intracellular basal body. When viewed with a transmission electron microscope, the cryptomonad transition region can be seen to contain two or more platelike partitions just below the point where the two central axonemal microtubules appear (Figure 9.3a). By contrast, in many stramenopiles this region contains a helix (Figure 9.3b), in green algae has a star-shaped structure (Figure 9.3c), and in euglenoids is comparatively simple.

As in the case of euglenoids and other unicellular flagellates, cryptomonads are able to change their position in response to light. This behavior arises from alternate shading and illumination of photoreceptor molecules as the cell swims along a helical path. Two distinct **rhodopsins** are known to serve as photoreceptors in the cases of cryptomonads that have been studied (Sineshchekov et al. 2005). An eyespot composed of lipid droplets containing carotenoids occurs at the cell periphery and acts as a shade for the rhodopsin photoreceptors.

Extending from the base of the flagella into the cytoplasm are elements of the flagellar root system, composed of microtubules and other cytoskeletal proteins (Figure 9.4). Components of the flagellar root system are believed to contribute to cell development and maintenance of cell shape, and portions of it may also be involved in cell division. The most prominent feature of the cryptomonad flagellar root system is the **rhizostyle**, a cluster of posteriorly directed microtubules that extends from the flagellar base. Other flagellar roots are known as **compound roots** because they consist of both microtubules and banded (striated) structures. A striated band links the flagellar bases of cryptomonads. A contractile protein known as centrin is associated with the basal bodies, rhizostyle, and striated band of cryptomonads (Melkonian et al. 1992). Variations in cellular features such as flagellar transition regions and root systems are very useful in deducing relationships among protist lineages.

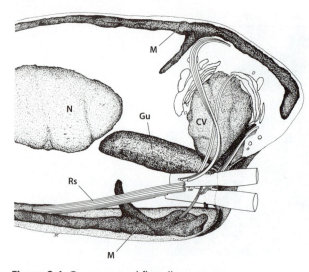

Figure 9.4 Cryptomonad flagellar apparatus components shown in diagrammatic form. CV = contractile vacuole, Gu = gullet, M = branched mitochondrion, N = nucleus, Rs = rhizostyle. (From Roberts 1984, by permission of the *Journal of Phycology*)

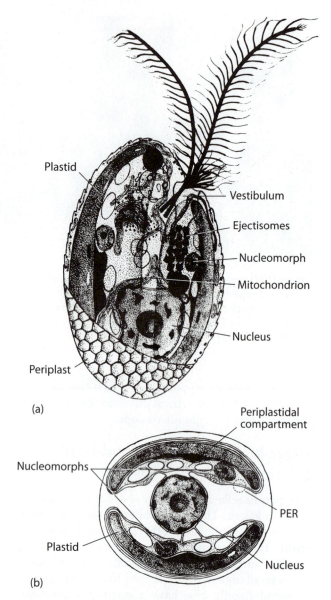

Figure 9.5 Diagram of a *Cryptomonas ovata* cell in (a) longitudinal and (b) cross-sectional views. The inner portion of the periplast is composed of polygonal plates. The two flagella emerge from the vestibulum. The two plastids are each associated with a nucleomorph. Also shown are ejectisomes, mitochondria, and the nucleus. PER = periplastidal endoplasmic reticulum. (From Santore 1985)

One of the main characteristics of cryptomonads is that their flagella emerge from an anterior depression on the ventral (front or belly) cell surface (Figure 9.5). This depression is known as a **vestibulum**, and a prominent contractile vacuole empties into it. The vestibulum also forms the anterior end of longer structures known as gullets or furrows. Furrows occur where the depression is open along its length. In some species, an enclosed tube-shaped gullet is open only at its anterior end (Gillott 1990).

Periplast

The cryptomonad plasma membrane is sandwiched between protein layers that together are known as the **periplast** (Brett et al. 1994). The innermost proteinaceous components of the periplast may take the form of hexagonal, rectangular, oval, or round organic plates (Kugrens et al. 1986) (see Figure 9.5). The pattern of these plates can be observed on surfaces of cells viewed with SEM and freeze-fracture TEM (Figure 9.6). Sometimes layers of scales or other extracellular materials coat the outer periplast surface.

Ejectisomes and Mitochondria

Rows of ejectisomes lining the vestibulum and elongate mitochondria are prominent features of cryptomonad cells. **Ejectisomes** are cytoplasmic structures that can be violently discharged from cells, probably as a defensive response to herbivore disturbance. Rows of ejectisomes can usually be seen lining the vestibulum. Ejectisomes appear vaguely square when viewed with a light microscope and

Figure 9.6 Freeze-fracture TEM view of a cryptomonad cell (*Storeatula*) showing the periplast plates. (Micrograph: P. Kugrens)

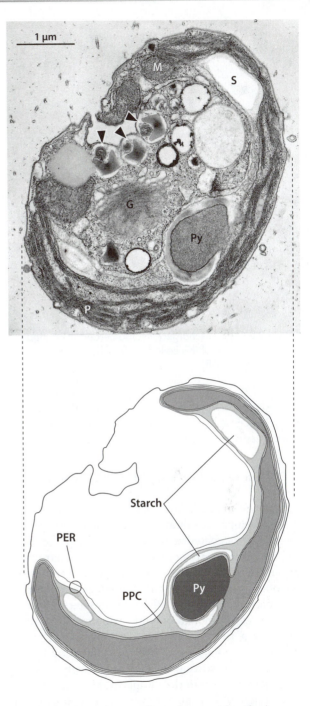

look butterfly-shaped in longitudinal-section when viewed with transmission electron microscopy (Figure 9.7). Each cytoplasmic ejectisome consists of two coiled ribbons, one smaller than the other, with the smaller ribbon in a depression formed by narrow portions of the larger ribbon. When discharged, both ribbons unfurl, forming a narrow, kinked barb (Figure 9.8). The smaller ribbon forms the shorter arm of the released ejectisome; the longer ribbon forms the longer arm. When cryptomonads are more seriously irritated or experience a sudden environmental shock, such as a change in pH, osmotic conditions, or temperature, massive discharge of ejectisomes may lead to rapid cell disintegration. This is one reason why natural collections of cryptomonads can be difficult to preserve in chemical solutions for transport back to the laboratory. The use of a low concentration (1%–3%) of fast-acting buffered glutaraldehyde (Wetzel and Likens 1991) favors the preservation of cryptomonads and other delicate algal cells. Cryptomonads are also unusually sensitive to the critical-point drying process widely used to prepare specimens for scanning electron microscopy, and freeze-drying methods are reported to work better for cryptomonads (Kugrens et al. 1986).

Cryptomonad mitochondria have flattened mitochondrial cristae (see Figure 9.7), a feature shared

Figure 9.7 Cross-section through an unidentified cryptomonad, viewed with TEM. Note the peripheral plastid (P), pyrenoid (Py), and starch (S); Golgi body (G); mitochondria (M) with flattened cristae; and three ejectisomes (arrowheads) lying next to the furrow. Accompanying diagram indicates the location of the periplastidal endoplasmic reticulum (PER), periplastidal compartment (PPC), and starch within the PPC. (TEM: L. W. Wilcox)

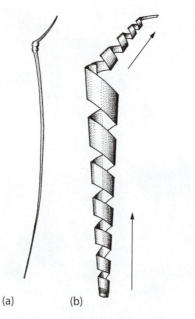

(a)   (b)

Figure 9.8 (a) A discharged ejectisome and (b) a model for the discharge of this coiled, ribbonlike structure. (From Mignot et al. 1968)

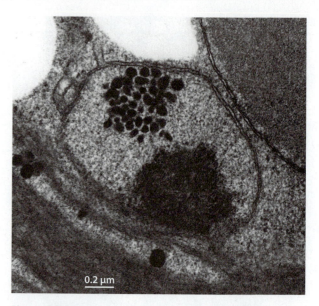

0.2 µm

Figure 9.9 A cryptomonad nucleomorph as seen with TEM. (From Gillott and Gibbs 1980, by permission of the *Journal of Phycology*)

with glaucophytes, red algae, and green algae (see Table 7.1). The mitochondrial genome of *Rhodomonas salina* has been sequenced (Hauth et al. 2005).

Plastids and Photosynthetic Pigments

Most cryptomonads possess plastids, and these have four envelope membranes. The outermost plastid membrane is coated with ribosomes, and the outer two membranes are connected to the cellular endoplasmic reticulum (Gibbs 1979, 1981). Consequently, the outer two membranes are known as **periplastidal endoplasmic reticulum** (PER). The outer and inner pairs of cryptomonad plastid membranes are separated by a space known as the **periplastidal compartment** (see Figures 9.5 and 9.7). Signaling factors involved in the sorting of proteins through the outer two membranes into the periplastidal compartment, or through all four membranes into the plastid stroma, have been studied (Gould et al. 2006a,b; Sommer et al. 2007). The periplastidal compartment is characterized by several features typical of eukaryotic cytoplasm, particularly that of red algae. Such features include 80S ribosomes with eukaryotic-type ribosomal RNAs, starch grains (see Figure 9.7), and a **nucleomorph** (Figure 9.9). The nucleomorph exhibits several features in common with typical eukaryotic nuclei: a double membrane envelope with pores, chromosomal DNA, self-replication (i.e., nuclear division) and a nucleolus where ribosomal RNA genes are transcribed. The cryptomonad nucleomorph genome occurs as three small chromosomes encoding genes required for nucleomorph maintenance (Douglas et al. 2001). Comparative molecular analyses suggest that cryptomonad nucleomorph genomes are not evolving rapidly at present (Patron et al. 2006a). Nucleomorph 16S-like rRNA genes group with those of red algal nuclei, reflecting origin of the nucleomorph from the nucleus of an endosymbiotic red alga. Cryptomonad plastid pigments also reflect red algal ancestry.

Photosynthetic plastids of cryptomonads include chlorophyll *a* and several types of accessory pigments: the phycobiliproteins phycoerythrin or phycocyanin (Hill and Rowan 1989), α- and β-carotenes, xanthophylls, and a pigment known as chlorophyll *c*. An ancestral phycoerythrin molecule gave rise to three types of red phycoerythrins and four kinds of blue phycocyanins (Glazer and Wedemayer 1995; Apt et al. 1996). Cryptomonad phycobiliprotein uniquely occurs within the lumen of plastid thylakoids, giving this region a distinctive electron-opaque appearance (Figure 9.10) (Spear-Bernstein and Miller 1989). The absence of phycobilisomes on thylakoid surfaces allows cryptomonad thylakoids to stack in pairs, in

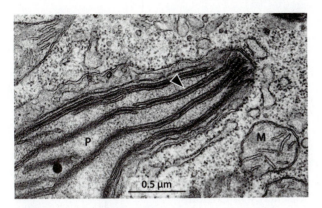

Figure 9.10 Section of a cryptomonad plastid (P) in the blue-green dinoflagellate *Gymnodinium acidotum*. Note the paired thylakoids (arrowhead) and their electron-opaque contents. Also note the mitochondria (M) with flattened cristae. (TEM: L. W. Wilcox and G. J. Wedemayer)

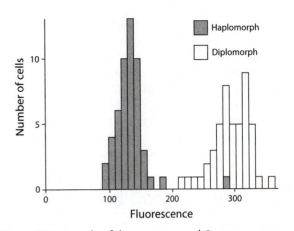

Figure 9.11 A study of the cryptomonad *Proteomonas* demonstrated the presence of two distinctive life cycle stages, one of which (the diplomorph) contains twice the amount of DNA as the other (the haplomorph). This provides evidence for some form of sexual reproduction in this organism. (After Hill and Wetherbee 1986)

contrast to the unstacked, phycobilisome-bearing thylakoids of most cyanobacteria and red algae. Unlike chlorophylls *a*, *b*, and *d* (derived from chlorins), cryptomonad chlorophyll *c* is derived from porphyrin and lacks a phytol tail (Falkowski and Raven 2007). The characteristic cryptomonad xanthophyll **alloxanthin** is chemically stable and can thus be used to infer the presence of cryptomonads within water samples or sediments. In consequence of differences in the proportions of their diverse pigments, photosynthetic cryptomonads may exhibit a wide range of colors, including blue, red, olive, and brown.

Photosynthetic storage granules of starch are located in the periplastidal space. Cryptomonad starch resembles that of green algae and plants (but not red algal starch) in staining blue-black when a solution of iodine is applied. Expression of the gene encoding cryptomonad starch synthase has been elucidated in the cryptomonad *Guillardia theta* (Haferkamp et al. 2006).

9.3 Cryptomonad Reproduction

Cryptomonad reproduction primarily occurs by asexual mitosis and cytokinesis, during which the cells continue to swim. Cells that are about to divide are distinctively rounded in shape and display four flagella in two pairs (Oakley and Dodge 1976). Following basal body replication, division of the plastid nucleomorph occurs in preprophase by a simple

pinching in two. No microtubules are involved in nucleomorph division. Daughter nucleomorphs migrate to opposite ends of the plastid so that during subsequent plastid division, each daughter plastid receives one. Mitotic division of the cell nucleus follows plastid division. Early in the mitotic process partial breakdown of the nuclear envelope occurs, and a barrel-shaped spindle forms without having centrioles at the poles (Gillott 1990; McFadden 1993; Sluiman 1993). Cytokinesis occurs by means of a cleavage furrow (i.e., an invagination of the cell membrane) that typically begins development at the posterior end of the cell.

Cryptomonads may produce resting cysts that are rounded and have a thick extracellular matrix. They can often be recognized by a pinkish coloration and the presence of large numbers of starch grains. Production of cryptomonad resting stages seems to be induced by high light levels coupled with nitrogen deficiency.

At least some cryptomonads engage in a sexual reproductive process. Fusion of similarly sized gametes was observed in cultures of one cryptomonad (Kugrens and Lee 1988). *Proteomonas sulcata* has a life history involving two morphologically distinct forms, one of which has twice the DNA level of the other, as determined by measurements of nuclear fluorescence in the presence of a DNA-specific fluorescent dye (Figure 9.11). The two forms also differ in size and periplast structure, as well as in the configuration of the flagellar apparatus (Hill and Wetherbee 1986). Some *Cryptomonas* species may also have a sexual cycle that involves cells of a different morphology that were formerly classified as the genus *Campylomonas* (Hoef-Emden and Melkonian 2003). These two cases suggest the possibility that additional cryptomonad species may be involved in life cycles involving morphologically distinct haploid and diploid stages.

9.4 Cryptomonad Ecology

Cryptomonads occur in the plankton of a variety of freshwater and marine environments, where they are exceptionally important in biotic associations. For example, cryptomonads are generally regarded as particularly high-quality food for aquatic herbivores, in part because most contain high proportions of polyunsaturated fatty acids needed for animal growth and development (Dunstan et al. 2005). Some cryptomonads prey on bacteria. Certain dinoflagellates and ciliates steal plastids and other organelles from cryptomonads as a way of accessing photosynthetic capacity. Some evidence suggests that cryptomonads can produce toxins that harm or kill fish, but more study is needed on this issue.

Nutrition, Including Bacterivory

Cryptomonads can utilize ammonium and organic sources of nitrogen, but marine forms in particular seem less able than other algae to utilize nitrate or nitrite. Most cryptomonads require vitamin B_{12} and thiamine, and some also require biotin.

Bacterivory has been documented for the plastidless *Goniomonas*, as well as plastid-bearing genera such as *Cryptomonas* (Gillott 1990; Tranvik et al. 1989; Urabe et al. 2000). A laboratory study revealed that *Goniomonas* is able to select bacterial prey, preferring *Aeromonas hydrophila* over *Pseudomonas fluorescens* (Jezbera et al. 2005). *Chroomonas pochmanni* possesses a specialized vacuole used for capturing and retaining bacterial cells. Bacteria are drawn into the vacuole through a small pore formed in the vestibulum, where periplast plates are absent. When it is full of bacteria, this vacuole can be seen by light microscopy, appearing as a transparent bulge at the anterior end of the cell. The bacteria are morphologically similar, suggesting selectivity on the part of the cryptomonad.

Bacterial cells appear to be digested within smaller food vacuoles in the cryptomonad cells (Kugrens and Lee 1990). Urabe et al. (2000) used bacteria-sized fluorescent spheres to estimate the feeding rate of a *Cryptomonas* species in Lake Biwa, in Japan, under conditions in which mineral nutrient concentrations were limiting to cryptomonad growth. The results indicated that the *Cryptomonas* consumes bacteria at a greater rate during the daytime than at night, indicating that the alga primarily ingests bacteria as a means of obtaining minerals rather than for energy or carbon.

Temperature and Light as Factors in Cryptomonad Distribution

Cryptomonads seem to be particularly important in colder, deeper waters of both freshwater and marine systems, typically becoming abundant in winter and early spring, when they can begin growth under the ice. For example, cryptomonads may dominate the spring phytoplankton bloom in the North Sea, where they are believed to make significant contributions to net primary productivity. Localized blooms of cryptomonads also occur in Antarctic waters, correlated with the influx of water from melting glaciers (Lizotte et al. 1998). In perennially ice-covered Antarctic lakes, *Chroomonas lacustris* or a *Cryptomonas* species may dominate the algal flora during the austral summer, contributing more than 70% of the total phytoplankton biomass (Spaulding et al. 1994). The maximum growth rate for many cryptomonads is one division per day and occurs at a temperature of about 20°C, with growth declining rapidly at higher temperatures. Cryptomonads seem to occur only rarely at temperatures of 22°C or higher, and they are absent from hot springs and hypersaline waters.

In oligotrophic freshwater lakes, cryptomonads may form large populations in deep waters (15–23 m) at the junction of surface oxygen-rich and bottom anoxic (oxygen-poor) zones, where light levels are much lower than in surface waters. Deep-water accumulations of photosynthetic organisms form what are generally known as **deep-chlorophyll maxima**. (In eutrophic lakes deep-chlorophyll maxima are often produced by filamentous cyanobacteria, and in other aquatic systems purple bacteria may constitute such growths.) Ecologists have wondered what characteristics enable cryptomonads to thrive in deep waters, suspecting both highly efficient light harvesting abilities and heterotrophic capacities. To test these hypotheses, Gervais (1997) isolated and grew cultures of several species of *Cryptomonas* from a deep-chlorophyll maximum layer for comparison to cryptomonad isolates from surface waters of the same lake. Two of the deep isolates grew best under light-limited conditions and were able to survive long periods of complete darkness. Neither species was able to ingest fluorescently labeled beads or bacteria, suggesting that they were probably incapable of phagotrophy. Uptake of radioactive tracer-labeled glucose was small relative to total cell carbon, suggesting that osmotrophy was probably not being used by these cryptomonads as a survival mechanism in low-light environments. Photosystem adaptation to low light levels, together with absence of predation, access to mineral nutrients regenerated by benthic decomposition processes, and tolerance of sulfide probably explain the occurrence of cryptomonad species in deep waters. Cryptomonads may also form deep-chlorophyll maxima in marine waters (see references in Klaveness 1988).

In view of cryptomonads' high food quality as prey for crustaceans and other small aquatic herbivores, it is possible that cold or deep waters may serve as refugia from predators. This may explain why cryptomonads seem to be adapted to such conditions.

Kleptoplastidy

The plastids and other organelles of both marine and freshwater cryptomonads can be stolen by protists and used to accomplish photosynthesis on a more-or-less temporary basis, a process known as kleptoplastidy (see Chapter 7). For example, the ciliate *Myrionecta rubra* (= *Mesodinium rubrum*) ingests free-living cryptophytes such as *Geminigera* and harvests their plastids, nuclei, and mitochondria (McFadden 1993). The stolen cryptomonad plastids remain photosynthetically active for relatively long periods. Such photosynthetic capacity allows the ciliate to form dramatic, red-colored blooms in waters off the coast of Peru, Baja (California), and other locations, typically in upwelling conditions that bring mineral nutrients to surface waters. Ciliates with kleptoplastids are able to synthesize chlorophyll (Gustafson et al. 2000; Johnson and Stoecker 2005), the plastids can divide under regulation by the ciliate (Johnson et al. 2006), and gene transcription by enslaved cryptomonad nuclei helps to keep the plastids

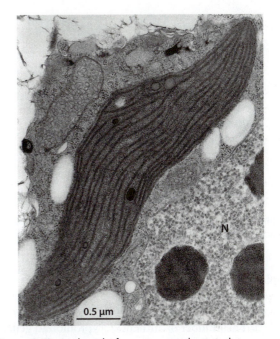

Figure 9.12 A plastid of cryptomonad origin lying adjacent the nucleus (N) in the dinoflagellate *Amphidinium wigrense*. (TEM: L. W. Wilcox and G. J. Wedemayer)

functional longer than would otherwise be possible (Johnson et al. 2007b).

The marine dinoflagellate *Dinophysis* also steals plastids from cryptomonad prey that are closely related to *Geminigera* (Takishita et al. 2002). Certain freshwater dinoflagellates, including *Gymnodinium acidotum* and *Amphidinium wigrense*, also regularly contain plastids and other portions of cryptomonad cells (Wilcox and Wedemayer 1984, 1985) (Figure 9.12; see also Figure 9.10). Ciliates and dinoflagellates having kleptoplastids (and other captured organelles) may be in the process of acquiring stable cryptomonad-derived endosymbionts.

9.5 Representative Diversity of Cryptomonads

In the past, cryptomonad genera were differentiated on the basis of phycobilin pigments and cellular features, such as periplast structure, that were observable with the use of light microscopy, TEM, or SEM. More recently, monophyletic groups indicated by

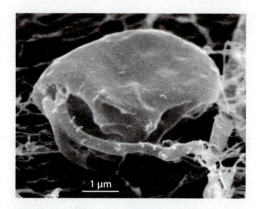

Figure 9.13 SEM view of *Goniomonas* (*Cyathomonas*). (Micrograph: P. Kugrens)

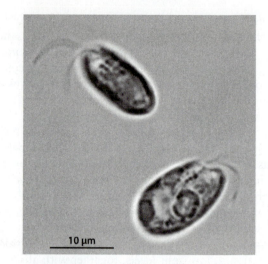

Figure 9.14 *Rhodomonas*. (Photo: L. W. Wilcox)

molecular systematic analyses were found to be more consistent with pigment variation than ultrastructural features (Deane et al. 2002; Hoef-Emden et al. 2002). The number of authentic cryptomonad genera has been influenced by recent discoveries that some genera, such as *Cryptomonas* and *Campylomonas*, are linked in the same life cycle (Hoef-Emden and Melkonian 2003) and that the genus formerly known as *Chilomonas* actually consists of non-photosynthetic species of *Cryptomonas* (Hoef-Emden 2005). The number of described species is increasingly influenced by the discovery that cryptic speciation (see Chapter 5) is common and the use of molecular signatures to define species.

GONIOMONAS (*Cyathomonas*) (Gr. *gonia*, angle + Gr. *monas*, unit) (Figure 9.13) lacks plastids and has been suggested as a model of the kind of heterotrophic host that gave rise to plastid-containing cryptomonads by secondary endosymbiosis (McFadden et al. 1994). A molecular phylogenetic analysis of multiple strains of the freshwater species *G. truncata* and two marine species, *G. pacifica* and *G. amphinema*, revealed the presence of many cryptic species (von der Heyden et al. 2004). The same analysis supported previous findings (Medlin and Simon 1998) that *Goniomonas* is sister to the plastid-bearing cryptomonads and suggested that the evolutionary divergence of freshwater and marine *Goniomonas* lineages occurred several hundred million years ago.

RHODOMONAS (Gr. *rhodon*, rose + Gr. *monas*, unit) (Figure 9.14) has a single boat-shaped, red-colored plastid with a pyrenoid. Several to

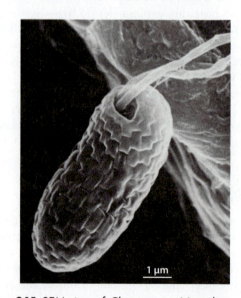

Figure 9.15 SEM view of *Chroomonas*. Note the periplast plates. (From Kugrens et al. 1986, by permission of the *Journal of Phycology*)

many large ejectisomes line the vertically oriented furrow. *Rhodomonas* sometimes forms noticeable red blooms in freshwater lakes in early spring. Although these growths may appear alarming, they are not known to be harmful. Klaveness (1981) described the ultrastructure of *R. lacustris*. Though there has been some concern about the monophyly of the genus *Rhodomonas*, a reappraisal by Hill and Wetherbee (1989) suggested that the genus should be retained.

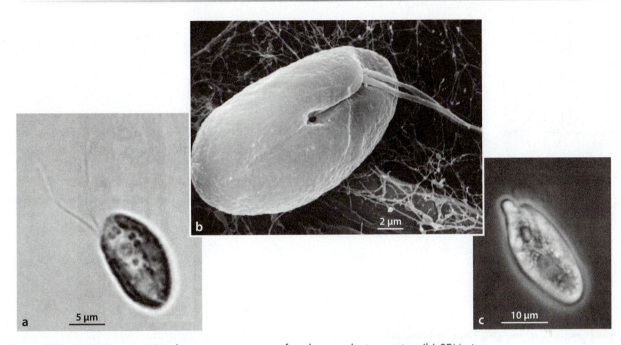

Figure 9.16 *Cryptomonas.* (a) Light microscopic view of a photosynthetic species. (b) SEM view. (c) A colorless species. (a, c: L. W. Wilcox; b: From Kugrens et al. 1986, by permission of the *Journal of Phycology*)

CHROOMONAS (Gr. *chroa*, color of the skin + Gr. *monas*, unit) (Figure 9.15) has a single blue-green, H-shaped plastid with a pyrenoid on the bridge. Periplast plates are rectangular, and there are two large ejectisomes in the vicinity of the shallow vestibular depression. A typical gullet or furrow is absent (Kugrens et al. 1986). Typically, the anterior end is somewhat broader than the posterior. Hill (1991) studied the ultrastructure of *Chroomonas* and some other blue-green cryptomonads.

CRYPTOMONAS (Gr. *kryptos*, hidden + Gr. *monas*, unit) (Figure 9.16) is a very common and widespread freshwater genus that has recently been revised on the basis of molecular data (Hoef-Emden and Melkonian 2003; Hoef-Emden 2007). There

is an open, vertical furrow lined with ejectisomes. The brown photosynthetic species (Figure 9.16a) are characterized by a phycoerythrin having an absorption maximum of 566 nm. The genus includes several plastid-bearing but colorless species (Figure 9.16c) that arose on at least three separate occasions from photosynthetic species (Hoef-Emden 2005) and were previously classified as *Chilomonas*. Colorless plastids are known as **leucoplasts** and include nucleomorphs. The colorless *Cryptomonas* species are osmotrophic and typically occur in waters rich in organic compounds released from decay processes. The genus *Campylomonas* appears to be an alternate stage in the life cycle of *Cryptomonas* and thus has been reduced to synonymy (Hoef-Emden and Melkonian 2003).

10 | Chapter

Haptophytes

Haptophyte algae—also known as prymnesiophytes—are most numerous and diverse in marine waters. Most haptophytes are unicellular, and many species are flagellate, though many others lack flagella and some produce filamentous stages. Haptophytes play important roles in global biogeochemistry and in the food webs of both natural and aquaculture systems. In biotechnology, emerging genomic information for haptophytes offers the promise of new ways to achieve crystal microfabrication. This chapter begins with a brief overview of haptophyte relationships and continues with a consideration of haptophyte cell structure, photosynthetic pigments, and reproduction. These topics serve as the basis for a closer focus on ecological issues and diversity.

Calcidiscus

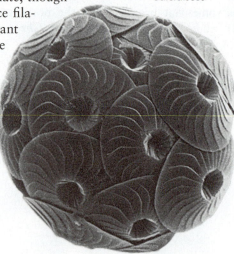

(SEM: A. Winter and P. Friedinger)

10.1 An Overview of Haptophyte Relationships

The haptophytes are named for the presence of a distinctive flagellum-like feature known as a **haptonema** (from the Greek word *hapsis*, touch) (Figure 10.1). Though not present on all haptophyte cells, a haptonema is characteristic of many species, including the earliest-diverging forms. Thus, the haptonema is thought to have evolved very early in the history of these algae, shortly after their divergence from an ancestor that also gave rise to the cryptomonads (see Chapter 9). Cavalier-Smith (1994) suggested that the haptonema arose by duplication and modification of a flagellar root. Recent molecular phylogenetic analyses, which generally support previous classifications based on morphological features such as the haptonema, confirm that the haptophytes are a monophyletic group. Many experts formally refer to the haptophytes by the phylum (division) name Haptophyta. However, others prefer the phylum name Prymnesiophyta and the colloquial term prymnesiophytes, both derived from *Prymnesium*, a common haptophyte genus.

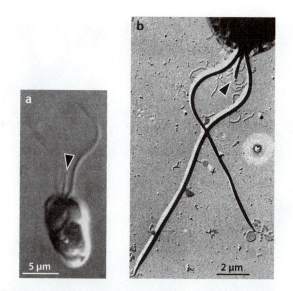

Figure 10.1 Haptonema. (a) Flagellate cell of *Phaeocystis* showing haptonema (arrowhead) between the two flagella. (b) A shadow-cast whole mount (TEM view) of a motile *Phaeocystis* cell illustrating the two flagella and short haptonema (arrowhead) between them. (From Marchant and Thomsen 1994, by permission of Oxford University Press)

Based on structural and molecular characters, haptophytes can be grouped into two classes (Edvardsen et al. 2000; Fujiwara et al. 2001; de Vargas et al. 2007) (Figure 10.2). The class Pavlovophyceae (= Pavlovaphyceae) (Cavalier-Smith 1993), which takes its name from the genus *Pavlova*, includes unicellular species having distinctive sterols and pigments (Van Lenning et al. 2003), a relatively simple cytoskeleton, and two unequal flagella. The longer flagellum often bears distinctive small, knobby scales. By contrast, the cells of a much larger group of haptophyte species typically have a more complex cytoskeleton and a scaly cell coat, and they possess two equal flagella or lack flagella altogether. Although such species have been grouped into the class known as Prymnesiophyceae, after reviewing the history of naming haptophytes, Silva et al. (2007) recommended the class name Coccolithophyceae. The latter name reflects the presence of beautifully ornamented mineralized scales, known as **coccoliths**, on the outer cell surfaces of many (though not all) of the included species (Figure 10.3). A surface covering of coccoliths is known as a **coccosphere**, and haptophytes whose cells are enclosed by coccospheres are known as **coccolithophorids**. Composed largely of calcium carbonate crystals in the form of calcite, coccoliths do not readily degrade at normal ocean pH and thus accumulate on the ocean floor. Beginning about 200 million years ago and continuing to the present time, coccolithophorids have produced huge amounts of sedimentary carbonates, thereby influencing Earth's atmospheric chemistry and climate (see Chapter 2). It is important to note that while Coccolithophyceae contains several orders of coccolith-producing genera, it also includes relatively early-diverging members that generally lack calcified scales but often have lacy organic scales made of cellulose on their cell surfaces. In the next section, we consider features of haptophyte cellular structure, with an emphasis on the haptonema and body scales.

10.2 Haptophyte Cellular Structure

Haptophyte cells display mitochondria with tubular cristae and usually have one or two yellow or golden-brown plastids enclosed by endoplasmic reticulum. Cytoplasmic carbohydrate storage vacuoles contain dissolved short β-1,3 glucan molecules known

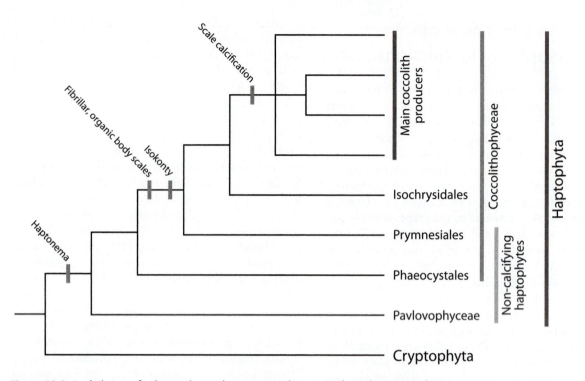

Figure 10.2 A phylogeny for haptophytes showing two classes: Pavlovophyceae and Coccolithophyceae. (Based on Edvardsen et al. 2000, Saez et al. 2004, and Nakayama et al. 2005)

as chrysolaminarin (chrysolaminaran) or leucosin. Cells typically have two flagella or none, though some species of *Chrysochromulina* possess four flagella. In addition, we have previously noted that flagellate cells may also bear a distinctive haptonema and that a covering of organic and/or calcified body scales is common.

Haptonema

The haptonema emerges from the cell apex, between the flagella. Because it is about the same thickness as a flagellum, the haptonema was mistaken for a flagellum until Parke and associates (1955) discovered its distinctive structure with the use of transmission electron microscopy. The structure of haptophyte flagellar and haptonemal basal bodies and their associated roots is shown in Figure 10.4. The haptonemal shaft includes six or seven singlet microtubules arranged in a ring or crescent, surrounded by a ring of endoplasmic reticulum (Figure 10.5). The haptonema may be quite short or considerably longer than the cell body. *Chrysochromulina camella*'s haptonema, for example, is an amazing 160 µm long. Unlike a

Figure 10.3 SEM view of *Emiliania huxleyi*, showing the coccosphere made up of interlocking coccoliths. (Micrograph: A. Kleijne, in Winter and Siesser 1994)

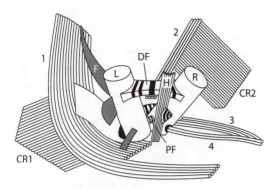

Figure 10.4 A diagram of the flagellar apparatus components of *Pleurochrysis*. There are four main microtubular roots (labeled 1 through 4). A fibrous root (F) is associated with root 1. Roots 1 and 2 have "compound components" (CR1 and CR2), which represent bundles of additional microtubules that branch off perpendicularly from the root microtubules. Basal bodies (L and R) are connected by a distal fiber (DF) and proximal fiber (PF). An intermediate fiber (shown, but not labeled) is also present. The microtubules indicated by the H represent the haptonemal base. (After Inouye and Pienaar 1985)

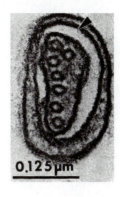

Figure 10.5 A haptonema (of *Syracosphaera pulchra*) seen in cross-section with TEM. Note the microtubules surrounded by endoplasmic reticulum (arrowhead), which is in turn bound by the cell membrane. (From Pienaar 1994)

flagellum the haptonema cannot beat, but it can bend and usually also coil.

Haptonemal coiling seems to function in obstacle avoidance. Upon contact with an obstacle, the haptonema coils very quickly (within 1/60 to 1/100 s), and the flagella change their direction of beat, propelling the cell quickly backward. Experiments have shown that coiling is induced by the rapid influx of Ca^{2+} into the cell from the environment. The endoplasmic reticulum present in the haptonema is thought to be involved in the calcium-induced response. In both the feeding process and collision avoidance, flagellar and haptonema behavior seem to be highly coordinated. In some cases the haptonema helps to attach haptophyte cells to substrates (Inouye and Kawachi 1994).

High-speed video has revealed that the haptonema plays a role in phagotrophy in *Chrysochromulina*. As a cell swims forward, prey particles—such as bacterial cells—attach to the forward-projecting haptonema. Particles are moved downward to a point about 2 μm distal to the haptonemal base, known as the particle aggregating center (PAC). Once formed, the particle aggregation moves up to the haptonemal tip, at which point the flagella stop beating while the haptonema bends toward the posterior surface, where the particle aggregation is ingested into a food vacuole (Inouye and Kawachi 1994) (Figure 10.6). In contrast, *Prymnesium*, which has a short haptonema, ingests prey at the cell posterior without the involvement of the haptonema (Tillman 1998).

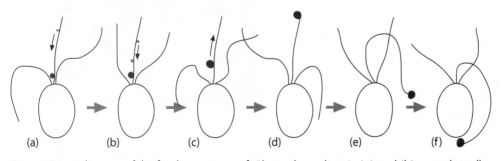

(a)   (b)   (c)   (d)   (e)   (f)

Figure 10.6 A diagram of the feeding process of *Chrysochromulina*. In (a) and (b), particles adhere to the haptonema and are translocated downward to a particle aggregating center (PAC). In (c) and (d), the aggregated, captured particles are moved to the tip of the haptonema. In (e) and (f), the aggregate is delivered to the cell surface (at the posterior end of the cell) through bending of the haptonema. Here it is taken into the cell. (Based on photomicrographs in Inouye and Kawachi 1994)

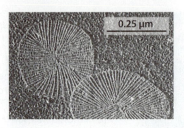

Figure 10.7 Organic fibrillar scales found on the haptophyte *Pleurochrysis*, as seen in a heavy-metal–shadowed TEM preparation. This organism also produces coccoliths. (From Pienaar 1994)

Scale Formation and Deposition

Silica scales are present on cells of at least one haptophyte, *Hyalolithus neolepis* (Yoshida et al. 2006a) and certain resting cyst stages. The most common types of scales produced by haptophytes are thin oval or circular organic scales and coccoliths.

Organic scales

Phylogenetic data indicate that organic scales are the most ancient (plesiomorphic) scale type for haptophytes. Such scales, made largely of cellulose, have a fibrillar structure when viewed with electron microscopy (Figure 10.7). As previously noted, organic scales occur on the cell surfaces of many coccolithophycean flagellates that lack coccoliths, but they are also present on the surfaces of some species that also produce coccoliths. The organic scales originate within the cisternae of a single, very large Golgi body that lies near the flagellar basal bodies, with its forming face toward the cell anterior (Figure 10.8). This unusual Golgi apparatus is characteristic of the haptophytes. Mature organic scales are secreted to the external surface.

Coccoliths

As noted earlier, coccoliths are calcite scales that form the coccospheres of coccolithophorids. Coccoliths (Gr. *kokkos*, berry + *lithos*, rock) were named in 1857 by T. H. Huxley, who observed them in samples of deep-ocean sediments. Coccoliths develop on the surfaces of organic base plates that resemble organic scales, indicating that cellulosic scales were essential to the evolution of coccoliths. Coccolithophorids produce two main types of coccoliths. **Heterococcoliths** form internally and then are secreted to the surface, and they consist of relatively few, interlocked calcite crystals. **Holococcoliths**

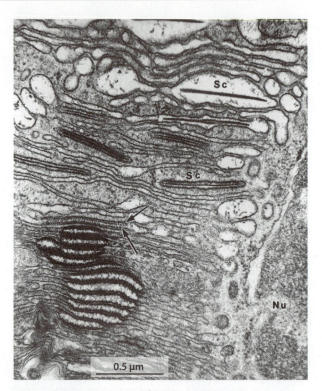

Figure 10.8 A TEM view of the large haptophyte Golgi body, shown here in *Hymenomonas lacuna*. Scales are apparent in different stages of formation. Arrows point to earliest stage of scale formation. Nu = nucleus, Sc = forming scale. (From Pienaar 1994)

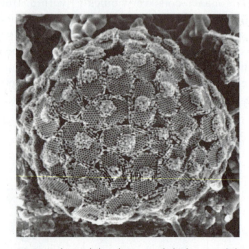

Figure 10.9 *Calyptrolithophora*, with holococcoliths. Holococcoliths are made up of smaller, similarly sized crystals (versus heterococcoliths). (SEM: S. Nishida)

form extracellularly and consist of a large number of tiny calcite crystals that are not interlocked but rather held together by organic material (Henriksen

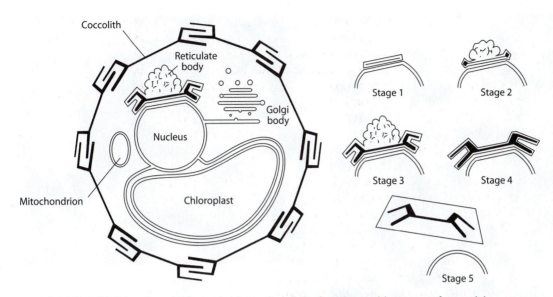

Figure 10.10 A diagram of an *Emiliania huxleyi* cell and the five discernible stages of coccolith formation that occur within the coccolith production compartment adjacent the nucleus. Stage 1: The Golgi-derived vesicle within which a coccolith will form adheres to the nuclear envelope. Stage 2: The reticulate body attaches to the nonnuclear vesicle face. Stage 3: Calcification becomes more extensive. Stage 4: At coccolith maturity, the reticulate body is no longer attached to the vesicle. Stage 5: The vesicle becomes less adherent to the coccolith, detaches from the nuclear surface, and moves toward the cell surface, where the coccolith is extruded. The reticulate body is thought to supply precursors to the developing coccolith. The coccoliths are shown edge-on in this illustration. (After deVrind-deJong et al. 1994, by permission of Oxford University Press)

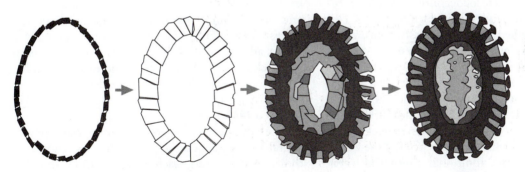

Figure 10.11 A diagram of coccolith formation in *Emiliania huxleyi*. A rim of calcite crystals is formed first. These crystals grow and interlock, and the complex shape of the mature coccolith gradually becomes apparent, with the knobby, spokelike outer (distal) shield elements being illustrated with the dark shades of gray (see Figure 10.3). (Based on electron micrographs in deVrind-deJong et al. 1994)

et al. 2004) (Figure 10.9). As we shall later see in a consideration of haptophyte sexual reproduction, an increasing number of species have been observed to produce both diploid heterococcolith-producing stages and haploid holococcolith-producing cells. *Coccolithus pelagicus,* for example, produces a nonmotile, heterococcolith-bearing stage, as well as a motile holococcolith-covered form (Parke and Adams 1960).

In the case of holococcoliths, an organic scale that has been produced by the Golgi body is first secreted to the cell surface, and then rhomboidal or hexagonal crystals of calcium carbonate are deposited on the scale within a space covered by an organic

Figure 10.12 The relatively simple coccoliths found on *Syracosphaera* (*Coronosphaera*), with a single disk having a wall at its periphery. (From Siesser and Winter 1994)

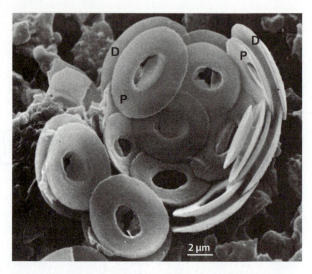

Figure 10.13 Interlocking coccoliths of *Gephyrocapsa*. D = distal shield of coccolith, P = proximal shield. (From Siesser and Winter 1994)

envelope (deVrind-deJong et al. 1994; Young et al. 2004). In contrast, heterococcoliths develop on organic scales before being secreted from the cell. This process begins when Golgi vesicles fuse to form a large, flat compartment that adheres to the nuclear envelope (Figure 10.10). A thin organic plate is then deposited within the vesicle. A reticulate body composed of a network of tubules—which is thought to provide precursors for heterococcolith development—then attaches to the exposed face of the vesicle. Crystal deposition begins around the rim of the organic base, and additional alternating vertical and radial crystals are added (Figure 10.11). By the time heterococcolith development is complete, the reticulate body is no longer present. The vesicle, which previously fit closely around the developing coccolith, loses its close fit and detaches from the nuclear surface. The newly formed heterococcolith is then extruded onto the cell surface near the flagella, where it joins other coccoliths to form the coccosphere. An example of coccolith development is provided by a TEM study of the haptophyte *Algirosphaera* that had been isolated from 37-m-deep waters of the Alboran Sea (Probert et al. 2007).

Variations in patterns of crystal deposition generate a wide variety of heterococcolith types that are characteristic of different coccolithophorid species. For example, the relatively simple coccoliths of *Syracosphaera* (*Coronosphaera*) take the form of cylinders of calcite built up at the edges of a base scale (Figure 10.12). More complex coccoliths, produced by the common bloom-formers *Emiliania huxleyi* and *Gephyrocapsa oceanica* and some other species, consist of two shields joined by a short column. This allows adjacent coccoliths within the coccosphere to interlock (Figure 10.13), thereby increasing the cohesion of the coccosphere. Other species produce coccoliths that overlap but do not interlock (Siesser and Winter 1994).

Calcification in coccolith-forming haptophytes is highly dependent upon photosynthesis. Photosynthesis provides energy-rich molecules (ATP and NADPH) needed for transport processes at the cell and Golgi membranes. Photosynthesis also acts as a sink for carbon dioxide, thus driving the net reaction for calcification:

$$2HCO_3^- + Ca^{2+} \rightarrow CaCO_3 + CO_2 + H_2O$$

In as little as one hour, a cell can produce a coccolith containing $3.5–5.5 \times 10^{-13}$ g of calcium (Brownlee et al. 1994). In experiments with *Emiliania huxleyi*, which is typically covered by about 15 coccoliths, the coccosphere can be dissolved from the surface of cells by growing them in a low-pH medium, and within 12 hours after return to high-pH conditions,

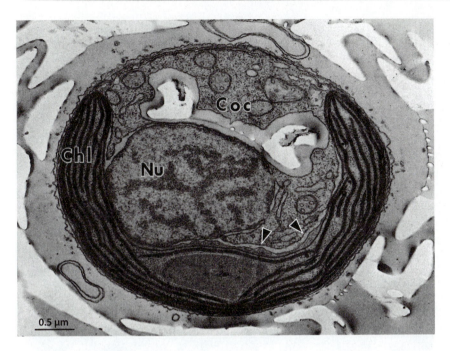

Figure 10.14 TEM of an *Emiliania huxleyi* cell. Note the periplastidal endoplasmic reticulum (PER) (arrowheads), which is continuous with the nuclear envelope. Plastid (Chl) and a developing coccolith (Coc) are also shown. (From Pienaar 1994)

the entire coccosphere will have been replaced (Pienaar 1994).

Coccoliths have been proposed to serve several functions: focusing light into cells (Brand 1994), limiting access to cells by pathogenic microorganisms and viruses, and affording some degree of protection from predation by small herbivores such as protozoa. Under conditions in which low levels of inorganic carbon limit photosynthesis, the CO_2 produced as the result of calcification has been proposed to facilitate photosynthesis (see Chapter 2), though this function has been questioned in the case of the common bloom-forming coccolithophorid *Emiliania huxleyi* (Herfort et al. 2004; Rost and Riebesell 2004). Regulated production and loss of heavy coccoliths might allow changes in cellular buoyancy that foster acquisition of nutrients such as phosphorus. Seawater phosphorous concentration is known to affect coccolith mineralization in *E. huxleyi*, a model system for genomic studies of calcification; more than 150 genes are differently expressed in cells that are grown in phosphorus-limited versus nonlimiting conditions (Nguyen et al. 2005; Quinn et al. 2006). When it is better understood, the genetic basis of cellular control over coccolith production and carbonate crystal structure may have useful applications in industrial fabrication at the nanometer scale.

Plastids and Photosynthetic Pigments

Most haptophyte cells possess one or two photosynthetic plastids, though a few reportedly lack plastids (Marchant and Thomsen 1994). Haptophyte plastids are similar to plastids of photosynthetic stramenopiles, cryptomonads, and chlorarachniophytes in having four envelope membranes, the outer two forming a periplastididal endoplasmic reticulum (PER), also known as chloroplast endoplasmic reticulum (CER) (Figure 10.14). This feature, together with pigment and genomic data (Li et al. 2006; Bachvaroff et al. 2005; Sanchez-Puerta et al. 2005), indicates that haptophyte plastids arose from those of a red alga via secondary endosymbiosis (Chapter 7). Haptophyte plastids often take the distinctive form of a butterfly when viewed with epifluorescence microscopy and contain the pigments fucoxanthin, diadinoxanthin, β-carotene, chlorophyll c_2, chlorophyll *a*, and a form of chlorophyll *a* known as MgDVP. Chlorophyll *c* is thought to function as an intermediate in energy transfer between carotenoids and chlorophyll *a* within the carotenoid/chlorophyll antenna complex. At least eight types of pigment associations incorporating different forms of chlorophyll *c* and fucoxanthin occur among haptophytes, and such variation correlates well with molecular-based phylogenies. Such pigment information is useful to oceanographers in detecting haptophytes and distinguishing them from

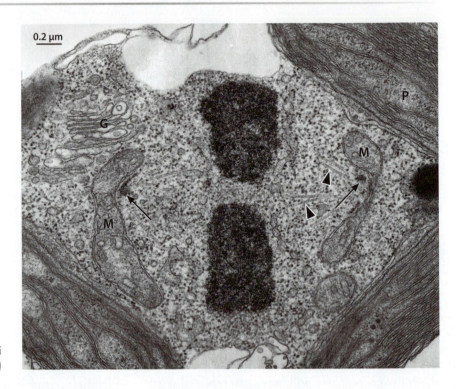

0.2 µm

Figure 10.15 TEM view of mitosis (metaphase) in *Phaeocystis*. The microtubule organizing centers at the spindle poles (arrows) are located on the surface of mitochondria (M). Arrowheads point to microtubules. G = Golgi body, P = plastid. (TEM: T. Hori)

diatoms with the use of remote sensing techniques (Zapata et al. 2004). The carotenoid known as 19HF (19'-hexanoyloxyfucoxanthin) is considered to be a diagnostic marker for coccolithophytes (Hooks et al. 1988).

10.3 Cell Division

Haptophyte populations increase by means of mitotic cell division, reviewed by Hori and Green (1994). Mitosis commonly occurs at night, presumably because such timing results in the least disruption to photosynthesis.

Events Preceding Nuclear Division

If the nuclear envelope was confluent with that of the PER, it separates from the PER prior to onset of nuclear division. Replication of flagellar and haptonemal basal bodies also precedes mitosis. Cells that are still in interphase or prophase may contain four flagellar and two haptonemal bases. The first indicator of incipient plastid division is elongation of the pyrenoid along the long axis of the plastid and the appearance of a slight depression on the plastid surface in the area of the pyrenoid midpoint. Next,

both the PER and the chloroplast envelope invaginate at their midpoints. Both the pyrenoid and the plastid then complete their division. Division of the single large Golgi body also begins prior to nuclear division and has been completed by metaphase.

Nuclear Division and Cytokinesis

In the Pavlovophyceae, the nuclear envelope remains intact throughout mitosis. The flagellar basal bodies remain aggregated until relatively late in the division process, and then they move to the spindle poles. A fibrous flagellar root acts as a spindle microtubular organizing center. The spindle has an unusual V-shaped orientation. By contrast, in most haptophytes (Coccolithophyceae), the nuclear envelope partially or more completely breaks down by late prophase. Pairs of flagellar basal bodies, together with an associated haptonemal base, then move to the vicinity of the developing spindle poles. In *Phaeocystis globosa*, the spindle microtubular organizer is located on the mitochondrial surface (Hori and Green 1994) (Figure 10.15). At the conclusion of mitosis, nuclear envelopes and continuity with the PER are reestablished; then cytokinesis occurs by invagination of scale-free cell membrane or by fusion of vacuoles in the interzonal region. Each daughter cell receives one-half of any body scales originally present on the parental cell and also generates new scales.

10.4 Sexual Reproduction and Life Cycles

Techniques for isolation and culture of haptophytes (Probert and Houdan 2004), together with flow cytometric analysis of cellular DNA content, have illuminated haptophyte sexual life cycles, revealing that distinctive cell types may be stages in the life history of the same species. Both haploid and diploid life cycle stages can reproduce asexually by means of mitotic cell division and in many cases were originally described as separate species. For example, the species formerly known as *Prymnesium patelliferum* is now known to be a sexual life history stage of *Prymnesium parvum* (Larsen and Edvardsen 1998). Several cases are known of alternation between diploid heterococcolith-bearing cells and haploid stages that bear holococcoliths or tiny structures known as nannoliths, or are naked. Examples include *Syracosphaera pulchra–Daktylethra pirus*, *Calcidiscus leptoporus–Syracolithus quadriperforatus*, *Coccolithus pelagicus–Crystallolithus braarudii*, and *Coronosphaera mediterranea–Zygosphaera hellenica*. It is now assumed that cells bearing only heterococcoliths are diploid life cycle stages, while those bearing only holococcoliths are haploid stages. Linkages between such life stages are revealed by the presence on diploid zygotes of a combination coccosphere composed of a new layer of heterococcoliths produced beneath a layer of holococcoliths contributed by the parental gametes (Geisen et al. 2002).

10.5 Haptophyte Ecology

While at least one primarily freshwater haptophyte has been described (the coccolithophorid *Hymenomonas roseola*) (Manton and Peterfi 1969), most haptophytes are marine, occurring in significant numbers in polar, subpolar, temperate, and tropical waters. In the latter, a number of strange and beautiful forms mysteriously occur in nearly dark ocean waters more than 200 m deep. Haptophytes are ecologically significant in terms of both biotic interactions and biogeochemistry. Biotic interactions of haptophytes include their roles as sources of food in natural and aquaculture food webs and toxin production by certain species. Haptophytes influence biogeochemistry and climate by sequestering carbon in deep ocean deposits and by producing volatile organic compounds.

Haptophytes in Food Webs

Haptophyte algae, because of their small size, fast growth rates, digestibility, and nutritional content, are considered to be high-quality foods for marine zooplankton such as copepods. Pavlovophyceae such as the genus *Pavlova* and certain non-calcified coccolithophytes such as *Isochrysis* are widely recognized for their high food quality. *Pavlova lutheri*, for example, contains high levels of polyunsaturated fatty acids (PUFAs) such as eicosapentaenoic acid (EPA) and docosahexanenoic acid (DHA), compounds that animals obtain from their diets. For example, the value of fish and other seafoods in the human diet is partly based on such omega-3 fatty acids, which fish obtain from algae, directly or indirectly. *Pavlova* and other microalgae are used directly as food for fish, crustaceans, and bivalves or as food for rotifers, copepods, and brine shrimp that are consumed by larger animals. Random mutagenesis has been employed to increase the EPA and DHA content of *P. lutheri* (Meireles et al. 2003) used in aquaculture systems or in bioreactor systems for the industrial production of omega-3 fatty acids.

Toxin-Producing Haptophytes

Some species of several genera of haptophytes influence biotic systems by producing toxins or other harmful substances. For example, the widespread flagellate species *Chrysochromulina polylepis* can produce toxic offshore marine blooms that cause the death of fish and invertebrates and inhibit the growth of diatoms and dinoflagellates. *Prymnesium parvum* causes similarly toxic blooms in brackish waters. In 1989, a *P. parvum* bloom along the Norwegian coast caused a US$5 million loss of salmon (Nicholls 1995). This haptophyte has also caused massive fish kills in inland ponds in Texas, causing millions of dollars of economic losses (Baker et al. 2007). Cultured cells of *P. parvum* produce two compounds—known as prymnesin-1 and prymnesin-2—whose stereochemistry and synthesis are understood (Sasaki et al. 2001). These toxins destroy membranes of both prokaryotic and eukaryotic cells and can thus also affect the haptophytes themselves (Olli and Trunov 2007). Nutrient limitation increases *P. parvum* toxin production (Legrand et al. 2001), which is related to observations that this mixotrophic haptophyte uses its toxins to immobilize prey cells (Skovgaard and Hansen 2003). These observations suggest that such toxins may have evolved in association with phagotrophy, but when bloom-level populations of these haptophytes are present, toxin levels become high enough to impact non-target animal species. Monoclonal antibodies have been developed for use in solid-phase cytometry as a rapid and specific way to detect *P. parvum* in water samples (West et al. 2006b; Töbe et al. 2006).

Phaeocystis, which often occurs as large colonies, produces copious amounts of organic slime and foam that clog fish gills and fishing nets and is reported to produce compounds that inhibit mitosis in other flagellate algae. Some species of the coccolithophore genera *Pleurochrysis* and *Jomonlithus* are toxic to the brine shrimp *Artemia salina* (Houdan et al. 2004). The physiology and bloom dynamics of *Prymnesium* and *Chrysochromulina* were reviewed by Edvardsen and Paasche (1998), and the taxonomy of toxic haptophytes was reviewed by Moestrup and Thomsen (2003).

Biogeochemical Impacts of Haptophytes

Coccolithophorids influence global carbon cycling by means of coccoliths, which contribute importantly to deep-sea carbonate accumulation (Rost and Riebesell 2004). Coccoliths contribute at least 25% of the total annual vertical transport of inorganic carbon to the deep ocean. Heavy coccoliths also serve as ballast for organic-containing aggregates such as marine snow, helping to sediment organic carbon as well (see Chapter 2). Thus, coccolith synthesis and burial help to sequester atmospheric CO_2 for long time periods. This process is quantitatively important because deep-sea carbonate deposits cover about one-half of the world's seafloor, an area that represents one-third of Earth's surface.

Haptophytes, including the bloom-forming species *Emiliania huxleyi* and *Phaeocystis pouchetii*, are also known for production of large amounts of dimethyl sulfide (DMS), a volatile sulfur-containing molecule that increases acid rain (Figure 10.16) (Malin and Steinke 2004). Together, coccoliths, which readily reflect light, and DMS, which enhances cloud formation, contribute to increased albedo (reflectance of the earth's surface) and thus have a cooling influence

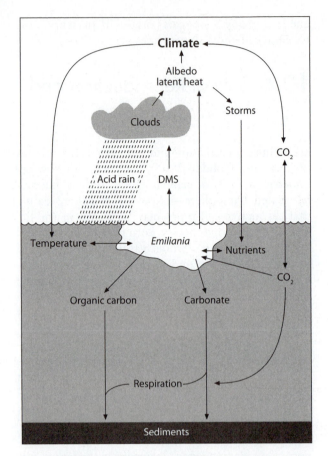

Figure 10.16 Blooms of *Emiliania huxleyi* can have important effects on Earth's climate in a variety of ways, summarized here in diagram form. (After Westbroek et al. 1994, by permission of Oxford University Press)

on the climate. Some experts warn that increasing ocean acidification as a result of rising atmospheric CO_2 threatens the survival of coccolithophorids and other marine organisms that have external calcium carbonate coverings or skeletons (Riebesell et al. 2000; Orr et al. 2005; Stanley et al. 2005). On the other hand, laboratory studies showed that both net primary production and calcification increased in *Emiliania huxleyi* grown under high carbon dioxide levels (Iglesias-Rodriquez et al. 2008).

10.6 Haptophyte Fossil Record

Molecular clock data suggest that photosynthetic haptophytes may have existed in the Neoproterozoic period, perhaps 600 million years ago (Berney

Figure 10.17 Chalk cliffs on the southern coast of England, which consist largely of coccoliths deposited during the Late Cretaceous. (Photo: L. E. Graham)

and Pawlowski 2006). If so, such ancient haptophytes might have resembled modern Pavlovophyceae and organic scale-producing members of the Coccolithophyceae because they left no recognizable fossils. The fossil record for haptophytes begins with fossil coccoliths that are well over 200 million years old (Bown et al. 2004). From that time, the coccolithophorids have an excellent fossil record (reviewed by Young et al. 2005) despite the fact that some types of coccoliths probably did not survive to the present time. The abundance of coccolith fossils peaked during the Late Cretaceous (63–95 million years ago), when very extensive chalk deposits were laid down across much of northern Europe and other sites around the world (Figure 10.17). In fact, the geological term *Cretaceous* refers to this chalk, and some blackboard chalks that are derived from such deposits contain coccolith remains. Although such Cretaceous chalk deposits formed in relatively shallow waters over continental shelves, similar deposits have been formed since the Cretaceous and are being formed today beneath large areas of the deep ocean. The fossil record of haptophytes is also notable for demonstrating the destructive power of the bolide impact/volcanic events that caused the famed K/T mass extinction event 65 million years ago. Coccolith fossils indicate a K/T coccolithophorid species extinction rate of 93% and a generic extinction rate of 85%. Coccolithophorid diversity rebounded thereafter but declined again 26–29 million years ago. Following a period of recovery

5–11 million years ago, diversity decline has been the more recent pattern (Bown et al. 2004).

Because coccoliths are common, small, and exhibit low endemism (restriction of certain species to particular locales), they are widely used as stratigraphic indicators to match rocks of equivalent ages from different locales. Fossil coccoliths are widely used as bioindicators in the oil industry and as indexes of past climate and ocean chemistry conditions (Young et al. 1994). Coring devices are used to obtain fossils. Light microscopy and a counting chamber such as a hemacytometer are commonly employed to count fossil coccoliths in filtered samples. Scanning electron microscopy is a major tool for comparing the structures of fossil and modern coccoliths.

10.7 Diversity of Living Haptophytes

As many as 280 species of haptophytes have been described, many on the basis of distinctive features of their organic or mineralized scales. For example, the new genus *Solisphaera* and three new species were erected on the basis of coccolith structural features visualized using SEM (Bollmann et al. 2006). However, establishing the number of haptophyte species is complicated by the involvement of more than one morphologically defined species in the same life cycle and by the occurrence of cryptic speciation (see Chapter 5) or pseudocryptic speciation (Sáez et al. 2003). Further, substantial haptophyte diversity may occur in the form of extremely tiny, picoplanktonic forms known only from SSU rDNA sequences amplified from ocean water samples (Moon-van der Staay et al. 2000). In addition, molecular evaluation of genera may reveal a degree of polyphyly that requires the establishment of new genera. For example, the genus *Chrysochromulina*, which includes about 60 species, is polyphyletic, consisting of multiple lineages that did not share a common ancestor. Hence, when a new species resembling *Chrysochromulina* was discovered, it was placed into the new genus *Chrysoculter* (*rhomboideus*) and a new family on the basis of molecular and other features that separate it from the type species of *Chrysochromulina*, *C. parva* (Nakayama et al. 2005). Likewise, distinctive molecular features caused Edvardsen et al. (2000) to transfer some former species of *Pavlova* into the new genus *Rebecca*. These examples indicate that concepts of haptophyte diversity are rapidly changing.

Figure 10.18 *Pavlova.* (From Throndsen 1997)

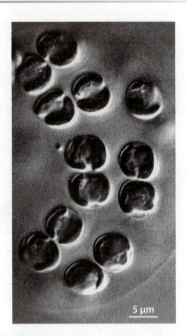

Figure 10.19 The nonmotile stage of *Phaeocystis*. The spherical cells are embedded in mucilage to form an often very large colony, the edge of which is evident at the bottom of this micrograph. (From Marchant and Thomsen 1994, by permission of Oxford University Press)

For taxonomic or ecological analyses, water samples of a liter or more are collected from known depths by use of sampling devices that can be remotely opened and closed. For relatively shallow sampling, water can be pumped through plastic tubing that has been lowered to the desired depth. Samples are obtained from deeper water with the use of ship-deployed sediment traps designed to collect particles that sink through the water column. Water samples are then concentrated by centrifugation or filtration through micropore filters. Following a rinse with alkaline water (because low-pH water may dissolve coccoliths) to remove sea salt, concentrated algal samples can be preserved in 1% glutaraldehyde and enumerated with the use of a fluorescence microscope, where the characteristic butterfly shape of the plastid allows discrimination of haptophytes from other algae. Single drops of preserved haptophytes can also be placed onto grids that have been coated with a thin film of plastic (Formvar) and then dried for examination by transmission electron microscopy (TEM). If only coccolithophorids are to be examined, samples can be air-dried onto filters and stored for later microscopic examination. For scanning electron microscopy (SEM), filters need only to be cut to size, attached to SEM stubs, and coated with metal. An SEM atlas of modern coccolithophorids can be found in Winter and Siesser (1994). Molecular sequences for use as fluorescence *in situ* hybridization (FISH) probes have been defined for the two classes of haptophytes and some additional clades (Eller et al. 2007). A technique known as COD-FISH combines the use of cross-polarizers to observe crystal orientations of coccoliths (carbonate optical detection) with FISH to simultaneously determine life cycle status (haploid or diploid) and taxonomic identity of individual cells (Frada et al. 2006).

Pavlovophyceae

PAVLOVA (named for Anna Pavlova, a famous Russian ballerina) primarily occurs as a unicellular flagellate (Figure 10.18) but also as a non-swimming stage. *Pavlova* can be found in freshwater lakes but is primarily marine, mostly occurring in brackish environments. This genus and other members of the class Pavlovophyceae possess a short haptonema and distinctively have two unequal flagella, the long flagellum's surface decorated with particles or fine hairs. Similar to cryptomonad relatives, a gullet-like region is present near the flagella. Organic knob- or mushroom-shaped scales may occur on the body surface. An eyespot may be present, which is unusual for haptophytes. The two plastids are lemon yellow rather than brown, as are many haptophytes.

Coccolithophyceae

Phaeocystales

PHAEOCYSTIS (Gr. *phaios*, dusky + Gr. *kystis*, bladder) typically occurs as large gelatinous colonies (up to 8 mm in diameter) (Figure 10.19) or as unicells of

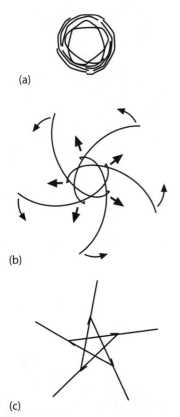

Figure 10.20 Mesoflagellate chitin fibrils. The ring of fibrils (a) uncoils (b) upon release from the vesicle and cell to form a five-rayed star (c). The biological function of these stars is unknown. (From Chrétiennot-Dinet et al. 1997, by permission of the *Journal of Phycology*)

several types (see Figure 10.1). Growth experiments indicate that cells occurring in colonies have much higher division rates than single cells grown under identical conditions (Veldhuis et al. 2005). The colonial form is regarded as an adaptation that reduces herbivory. Cells of the colonial stage lack scales and a haptonema. The hollow colonies consist of a single surficial layer of cells embedded in transparent polysaccharide mucilage that does not fill the colony center (van Rijssel et al. 1997). The mucilage may function as an energy reserve that can be metabolized by cells when they are light limited or as a reservoir for phosphate and trace minerals such as iron and manganese. The mucilage also reduces density, contributing to buoyancy. Unicellular forms of *Phaeocystis* may be non-flagellate or occur as three types of flagellates that differ in size: microflagellates, mesoflagellates, and macroflagellates. Micro- and mesoflagellates are haploid stages of the life cycle, whereas macroflagellates are diploid, are the same size as non-flagellate unicells, and can form new colonies. Micro- and mesoflagellates are thought to arise by meiosis from the macroflagellate (Peperzak et al. 2000). Mesoflagellates uniquely possess large vesicles within which are five chitin filaments that can be discharged from the cells (Figure 10.20). As they are released, the filaments uncoil and extend to form a five-rayed star (Chrétiennot-Dinet et al. 1997). Experimental grazing studies of nauplii of the calanoid copepod *Temora longicornis* on these unicell types revealed that the herbivore avoided only the mesoflagellates, suggesting a

Figure 10.21 Thick accumulation of *Phaeocystis* slime on a beach near Scheveningen, the Netherlands. Note the two dogs on the beach, one of which is nearly covered by the foam. (Photo: P. Morehead)

defensive function for the stars (Dutz and Koski 2006). *Phaeocystis* has been estimated to generate 10% of the total global flux of dimethyl sulfide (DMS) to the atmosphere. Fish seem to be repulsed by *Phaeocystis* blooms, possibly by the smell of DMS. Massive slimy growths of *Phaeocystis* clog fishing nets and wash up onto beaches, sometimes producing meter-thick foam that deters recreational use (Figure 10.21) (Marchant and Thomsen 1994). The ecology of *Phaeocystis* has been reviewed by Lancelot et al. (1998).

Prymnesiales

PRYMNESIUM (Gr. *prymnesion*, stern-cable) is a single-celled flagellate (Figure 10.22) noted for tolerance of an extremely broad range of salinities and production of toxins that cause fish kills. There is a short haptonema that does not coil. Species are distinguished by small ultrastructural differences in their organic body scales or by molecular methods.

CHRYSOCHROMULINA (Gr. *chrysos*, gold + *Chromulina*, a genus of chrysophyceans) is a polyphyletic group of 60 or so species of primarily marine flagellates (Figure 10.23), though several freshwater forms are known, and some produce amoeboid stages. The genus occurs worldwide, in polar as well as warmer waters. There is typically a very long haptonema that functions in prey capture. Although golden-brown plastids are present, phagotrophy is common (Hansen 1998). Experimental studies have shown that low cell-phosphate content is correlated with bacterivory, suggesting that phagotrophy may be a method of acquiring phospholipids from bacteria as a source of needed phosphate (Jones et al. 1994). Many species are also photoheterotrophic, that is, able to take up dissolved organic carbon in the presence of light. This is thought to allow survival in low-light polar environments. Species are differentiated by molecular methods or on the basis of the structure of elaborate fibrillar organic scales, several types of which may occur on the same cell, in more than one layer. As noted earlier, some marine species produce toxins that are known to kill fish; the freshwater forms are associated with odor production and tadpole deaths.

Isochrysidales

EMILIANIA (named for Cesare Emiliani, an Italian-American micropaleontologist and physicist) is

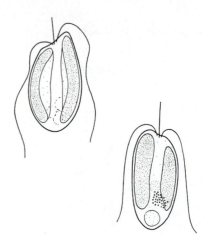

Figure 10.22 *Prymnesium.* (From Throndsen 1997)

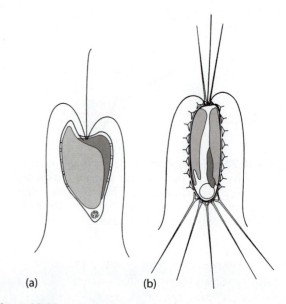

(a)   (b)

Figure 10.23 Diagrammatic view of two species of *Chrysochromulina*: (a) *C. polylepis.* (b) *C. pringsheimii.* (From Throndsen 1997)

ubiquitous but most abundant in nutrient-rich temperate and subpolar waters (see Figure 10.1). There are three cell types, all about 5–7 μm in diameter: coccolith-bearing cells that are non-motile; naked, nonmotile cells; and motile, flagellate cells that have organic body scales but no coccoliths. All of these life stages reproduce asexually by cell constriction. Motile cells lack a haptonema. *Emiliania huxleyi* is found in nearly every sample of ocean water and sediments from the Late Quaternary to the present. The organism has a very wide temperature range, 1°C–30°C. Although no single isolate can tolerate

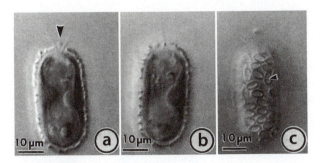

Figure 10.24 Through-focus series of a motile cell of *Pleurochrysis*, whose scale covering is evident. Arrowhead in (a) points to haptonema, small arrowhead in (c) to coccolith. (From Pienaar 1994)

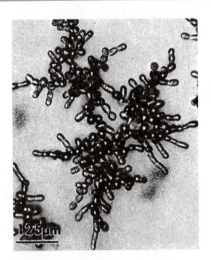

Figure 10.25 The benthic, filamentous (apistonema) stage of *Pleurochrysis*. (From Pienaar 1994)

this range, there appear to be temperature ecotypes (genetically discrete populations that are physiologically adapted to local conditions). *Emiliania* also grows in a wide range of nutrient levels, ranging from eutrophic waters to very low-nutrient, subtropical open-ocean habitats. *Emiliania* will tolerate high salinities, up to 41 ppt, as in the Red Sea. It grows throughout the top 200 m of the water column and can survive at irradiance levels less than 1% that of the surface. Fossil evidence suggests that this coccolithophorid originated approximately 278,000 years ago, acquired the ability to form blooms 85,000 years ago, and has been a dominant member of the phytoplankton for the past 73,000 years (Thierstein et al. 1977). *Emiliania huxleyi* is widely known for bloom formation today, attaining 60%–80% abundance in phytoplankton communities. Such blooms may have a density of 1×10^8 cells l^{-1} and extend over areas greater than 50,000 km². *Emiliania* blooms are readily detected by satellite remote sensing. Although the cells themselves are pigmented yellow or golden brown, when viewed by satellite, blooms have a milky-green appearance due to reflectance of light from coccoliths. Agglutination of cells, viral and bacterial infections, and grazing are the major

factors that can terminate blooms. In healthy cells, DMSP (dimethylsulfoniopropionate) is likely segregated from enzymes that cleave DSMP to the gas DMS (dimethyl sulfide) and acrylic acid (AA). When cells are eaten, the enzymes become mixed with substrate, producing DMS and AA at a level 10 times that produced by viral infection. Thus, even though DMS and AA are both produced as a result of herbivory and viral attack, herbivory is quantitatively more important (Evans et al. 2007).

Other coccolithophorids

PLEUROCHRYSIS (Gr. *pleuron*, rib + Gr. *chrysos*, gold) occurs as a marine flagellate (Figure 10.24). One species is known to also have a branched, filamentous, benthic stage known as an apistonema stage) that does not possess coccoliths (Figure 10.25). The motile cells are spherical to ovoid, 5–10 µm in length, with two equal flagella. Several layers of organic scales occur between the cell membrane and an outer layer of coccoliths. *Pleurochrysis* is readily eaten by copepods.

11 Chapter

Dinoflagellates

Peridiniopsis

Most dinoflagellates are photosynthetic or heterotrophic uni-cells having two distinctive flagella whose motion causes the cells to rotate as they swim. In fact, the term *dinoflagellate* originates from the Greek word *dineo*, meaning "to whirl." However, dinoflagellates also occur as several types of nonflagellate single-cell or filamentous forms whose relationships to other dinoflagellates are revealed by the characteristic structure of their flagellate reproductive cells. One fascinating example is *Blastodinium*—a multicellular, worm-shaped parasite occurring in the guts of marine copepods. This parasite's reproductive cells are recognizable as dinoflagellates by their close resemblance to the common free-living flagellate *Peridinium*, an important primary producer in fresh or brackish waters.

Many other examples of dinoflagellates' ecological importance occur in diverse aquatic habitats. These include photosynthetic dino-flagellates that live as endosymbionts within the reef-building corals so prominent in coastal tropical and subtropical oceans. Stimulated by chemical signals from their animal hosts, dinoflagellate partners provide corals with essential organic food. This relationship explains why corals are so vulnerable to the loss of their photosynthetic symbiont—a process known as coral bleaching that occurs as a result of climate warming. Because coral reefs harbor a very high diversity of

(SEM: G. J. Wedemayer and L. W. Wilcox)

additional marine species, many of which have not been adequately catalogued and their ecological roles defined, conservation biologists are very concerned about the impact of coral bleaching on global biodiversity. Certain planktonic dinoflagellates are also globally important as toxin producers in coastal marine waters that have been affected by human influences. Dinoflagellate toxins have been responsible for the deaths of sea birds and mammals and can affect the health of humans.

In order to better understand the ecological functions of dinoflagellates, in this chapter we first survey dinoflagellate relationships to other protists and the unusually complex structure of dinoflagellate cells. Such information helps to explain the importance of dinoflagellates in today's world.

11.1 Evolutionary History and Relationships

The evolutionary history and relationships of dinoflagellates, like those of other organisms, are deduced from chemical and structural fossil remains and comparative analyses of gene sequences from modern organisms.

Fossil Record

Chemical compounds thought to be specific to dinoflagellates, known as dinosteranes and dinosterols, have been found in Proterozoic (Precambrian) rocks (Moldowan and Talyzina 1998; Brocks and Summons 2003). Such chemical fossils suggest that dinoflagellate-like protists existed more than 600 million years ago. The fossil record indicates that many types of single-celled fossils (known as **acritarchs** because their relationships are uncertain) also occurred at that time. However, the Proterozoic acritarchs lack distinctive structural features that would unequivocally link them with dinoflagellates. Undisputed structural fossils of dinoflagellates occur much later, beginning some 200 million years or so ago. These more recent fossils, known as **hystrichospheres** (Figure 11.1), are very similar to resting cyst stages of some modern dinoflagellates. For example, the resting stage of the modern dinoflagellate *Pyrodinium bahamense* is identical to the hystrichosphere *Hemicystodinium zoharyi*, whose history can be traced back to the Eocene (40–50

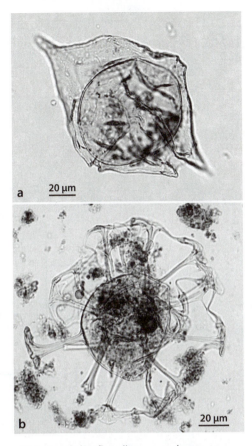

Figure 11.1 Fossil dinoflagellate cysts, known as hystrichospheres, from Late Cretaceous Arctic sediments. (a) *Alterbidinium*, (b) *Adnatosphaeridinium*. (Photos: L. W. Wilcox; specimens provided by D. Clark)

million years ago) (Wall and Dale 1969). Some modern dinoflagellate species produce a layer of organic material known as **dinosporin** just beneath the outer cell wall of resting cysts or zygotes. Dinosporin is resistant to decay and chemical treatments and thus aids the survival of dinoflagellate fossils, of which a considerable diversity has been described (Fensome et al. 2003). On the whole, the fossil record suggests that dinoflagellates are likely a very ancient group but that features such as dinosporin-containing cyst walls likely evolved more recently, revealing a major diversification event in the Jurassic. Modern dinoflagellates produce more than 35 types of sterols, some of which have specific chemical features linking them to certain oil deposits (Robinson et al. 1984). Thus it is thought that past dinoflagellate blooms may have contributed to the formation of fossil carbon deposits.

Figure 11.2 Evolutionary relationships of dinoflagellates (Based on Kuvardina et al. 2002, Taylor 2004, and Moore et al. 2008)

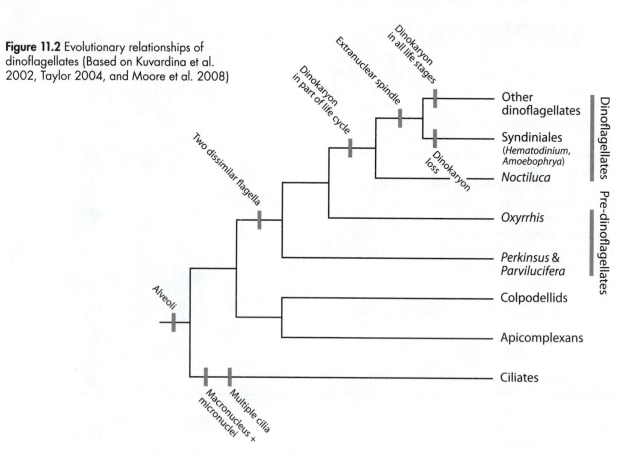

Relationships of Living Dinoflagellates

Dinoflagellates have been described as a fundamentally heterotrophic, phagotrophic (particle-feeding) lineage (Taylor 2004). This concept is supported by the fact that about 50% of dinoflagellate species are heterotrophs (Kofoid and Swezy 1921; Schnepf and Elbrächter 1992) that generally seem to lack plastids. On the other hand, a related protist recently discovered on Australian coral reefs, named *Chromera velia*, has photosynthetic plastids, suggesting the possibility that plastidless dinoflagellates arose from a plastid-bearing ancestor (Moore et al. 2008). Although the divergence patterns of dinoflagellates are not well understood (Delwiche 2007), molecular phylogenetic analyses indicate that plastidless genera such as *Noctiluca*, *Hematodinium*, and *Amoebophyra* are the earliest-diverging modern dinoflagellates (Figure 11.2). Dinoflagellates' closest living relatives are heterotrophic protists (*Perkinsus*, *Parvilucifera*, and *Oxyrrhis*) that have been described as "pre-dinoflagellates" because they lack a distinctive type of nucleus—known as a **dinokaryon**—which is shared by true dinoflagellates (Taylor 2004). The pre-dinoflagellates possess two dissimilar flagella, as do true dinoflagellates, but not their next-closest relatives: a clade consisting of colpodellids + apicomplexans + gregarines + *Chromera velia* + ciliates. With the exception of *Chromera velia*, which has plastids, and some apicomplexans and *Perkinsus* (whose cells contain a nonphotosynthetic plastid), these dinoflagellate relatives are mostly plastidless (Teles-Grilo et al. 2007; Stelter et al. 2007). Together, dinoflagellates, colpodellids, gregarines, *Chromera velia*, apicomplexans, and ciliates form a supergroup (or superphylum) known as the **alveolates** (formally Alveolata) because peripheral vesicles known as **alveoli** commonly occur in their cells. The alveolates appear to be closely related to the stramenopiles, another lineage characterized by a diversity of early-divergent, heterotrophic, apparently plastidless members (see Chapter 12).

Some experts have proposed that alveolates and stramenopiles, together with haptophytes and

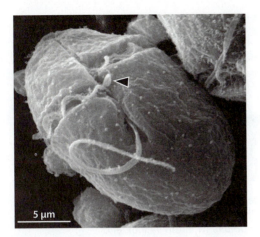

Figure 11.3 SEM of ventral view of *Amphidinium cryophilum*. Note the anterior epicone and larger posterior hypocone. The transverse flagellum emerges from a pore at the junction of the cingulum and sulcus and lies coiled within the cingulum. The longitudinal flagellum extends from the same pore and along the sulcus. This species also features a structure known as a peduncle. (arrowhead) that likewise emerges from the cell at the junction of the cingulum and sulcus (This structure is involved in feeding in some dinoflagellates. A different structure is utilized by *A. cryophilum*, however.) (From Wedemayer et al. 1982, by permission of the *Journal of Phycology*)

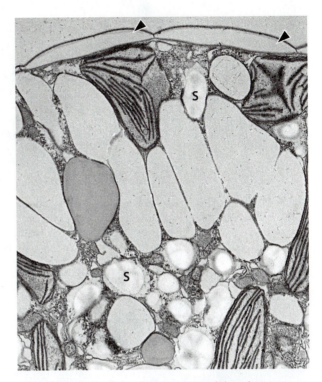

Figure 11.4 TEM of *Amphidinium cryophilum* showing cytoplasmic starch (S) and surface alveoli (arrowheads). The alveoli of this dinoflagellate lack thecal plates. (Micrograph: L. W. Wilcox and G. J. Wedemayer)

cryptomonads, share a common ancestor that possessed a photosynthetic plastid obtained from a red alga by a single secondary endosymbiotic event (see Chapter 7). This concept, known as the chromalveolate hypothesis, implies that plastids were lost from diverse, early-diverging lineages of alveolates and stramenopiles. Whether such extensive plastid loss actually occurred is the subject of ongoing research. In the next section, we survey features of dinoflagellate cell structure, including plastids. Such a survey not only reveals a surprising level of cellular complexity but also helps to explain dinoflagellate lifestyles.

11.2 Dinoflagellate Cell Biology

Viewed from the outside, most dinoflagellate cells consist of two parts, an anterior (top) **epicone** (or **epitheca**) and a posterior **hypocone** (or **hypotheca**) (Figure 11.3). The two parts are separated by a groove that encircles the cell, known as the **cingulum** (meaning "girdle"). A smaller groove, known as the **sulcus**,

extends posteriorly (and anteriorly in some, e.g., Figure 11.3) into the hypocone from the cingulum. At the intersection of the cingulum and sulcus is a pore from which the two flagella emerge. The region of the cell from which the flagella emerge is defined as the ventral side. Thus, dinoflagellate cells display conspicuous anterior–posterior and dorsal–ventral differentiation.

Internally, dinoflagellate cells are also complex because they contain both typical eukaryotic cell structures and a variety of features unique to dinoflagellates (Figure 11.4). This section continues with a focus on common cellular features of dinoflagellates and describes cellular features occurring in many, but not all, dinoflagellate species.

Features Commonly Found in Dinoflagellate Cells

Consistent features of dinoflagellate cells include mitochondria with tubular cristae, cytoplasmic starch granules and lipid droplets, which serve as

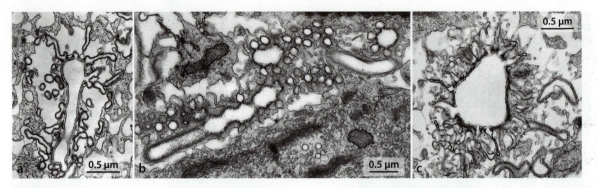

Figure 11.5 Examples of pusules from three freshwater dinoflagellates. Note the variety of sacs and tubules. The membrane lining the pusule tubules is continuous with the cell membrane. (a) *Amphidinium cryophilum*. (b) *Katodinium campylops*. (c) *Gymnodinium acidotum*. (a: From Wilcox et al. 1982; b: L. W. Wilcox; c: From Wilcox and Wedemayer 1984, by permission of the *Journal of Phycology*)

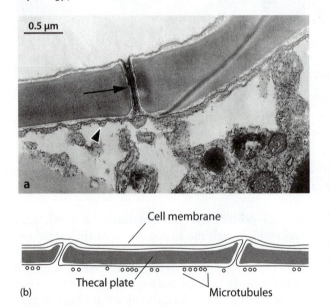

Figure 11.6 Amphiesma of armored (thecate) dinoflagellates. (a) TEM view of the cell covering of *Diplopsalis acuta*, a freshwater colorless dinoflagellate. Note the relatively thick thecal plates within alveoli (thecal vesicles). Also present is a trichocyst pore (arrow) and a few microtubules (arrowhead) lying beneath the thecal vesicles. (b) Diagram of part of the cell covering of a dinoflagellate whose alveoli contain a thick cellulose plate (a: G. J. Wedemayer and L. W. Wilcox; b: After Dodge and Crawford 1970)

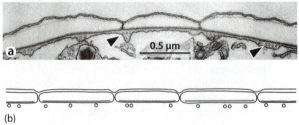

Figure 11.7 Amphiesma of unarmored (naked, nonthecate) dinoflagellates. (a) TEM view of the cell covering of *Gymnodinium acidotum* showing the nearly empty alveoli, which are overlain by the cell membrane. Cytoskeletal microtubules (arrowheads) are evident beneath the vesicles. (b) Diagram of part of the cell covering of a dinoflagellate whose alveoli contain little or no cellulose (a: From Wilcox and Wedemayer 1984, by permission of the *Journal of Phycology*; b: After Dodge and Crawford 1970)

storage materials, and accumulation bodies thought to store wastes (see Figure 11.4). Many dinoflagellate cells also display an unusual internal array of tubular membranes known as a **pusule** (Figure 11.5), whose function is unclear. Other general features of dinoflagellate cells include surface coverings and associated defensive projectiles, two structurally distinct flagella and their effects on motility, the dinokaryon, and a unique cell division.

Cell covering

The dinoflagellate cell surface features an external cell membrane, below which lies a single layer of the membrane sacs known as alveoli, or amphiesmal or thecal vesicles. Together, this collection of cell surface features is known as the **amphiesma**. In many species each alveolus contains a flat **thecal plate** composed of cellulose; such species are described as armored or thecate dinoflagellates (Figure 11.6). In other species the alveoli are devoid or nearly devoid of contents

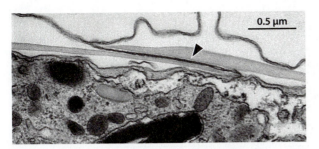

0.5 μm

Figure 11.8 A section through the junction (suture) (arrowhead) between two overlapping thecal plates of *Peridiniopsis berolinensis*. (TEM: G. Wedemayer)

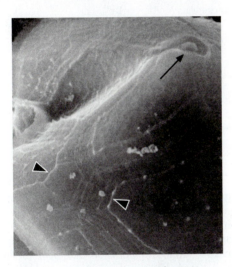

Figure 11.9 SEM view of broad intercalary (growth) bands (arrowheads) between thecal plates and the apical pore (arrow). (Micrograph: G. J. Wedemayer and L. W. Wilcox)

(Figure 11.7); such species are referred to as naked, unarmored, or non-thecate dinoflagellates.

The thecal plates of armored cells usually fit very closely together, overlapping slightly to form a continuous surface (Figure 11.8). The regions between adjacent plates are known as **sutures**. Cell growth occurs by addition of material along the margins of the plates, forming regions known as **intercalary bands**, or growth bands (Figure 11.9). Dinoflagellate genera and species vary predictably in the numbers, sizes, and shapes of thecal plates, so these features have been widely used in taxonomy. Some major types of plate organization found among living dinoflagellates are shown in Figure 11.10. Each species has a characteristic plate formula that begins with the anterior and moves toward the posterior of the cell. The terminology used in compiling tabulations is shown in Figure 11.11, and a similar tabulation of alveoli can also be accomplished for unarmored dinoflagellates. Dinoflagellate identification often requires breaking cells open and spreading the plates flat in order to tabulate them.

Defensive projectiles

Dinoflagellate cells often contain peripheral structures that can be discharged from the cell body into the environment as a defensive response, and these occur in several types. **Trichocysts** (also known

Figure 11.10 The major types of dinoflagellate thecal organization.
(a) Prorocentroid (e.g., *Prorocentrum*);
(b) Dinophysoid (e.g., *Dinophysis*);
(c) Gonyaulacoid (e.g., *Gonyaulax*);
(d) Peridinioid (e.g., *Peridinium*); and
(e) Gymnodinoid (e.g., *Gymnodinium*). (Redrawn from *BioSystems* 13, Taylor, F. J. R. *On dinoflagellate evolution*, © 1980, pages 65–108, with permission from Elsevier Science)

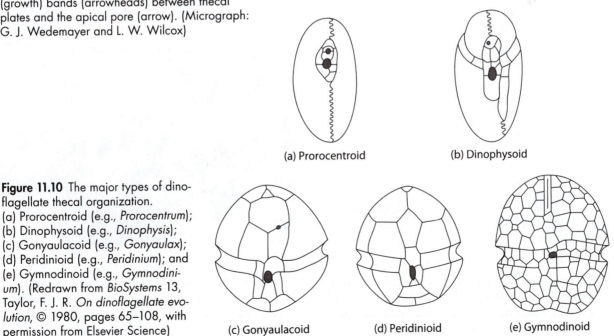

(a) Prorocentroid

(b) Dinophysoid

(c) Gonyaulacoid

(d) Peridinioid

(e) Gymnodinoid

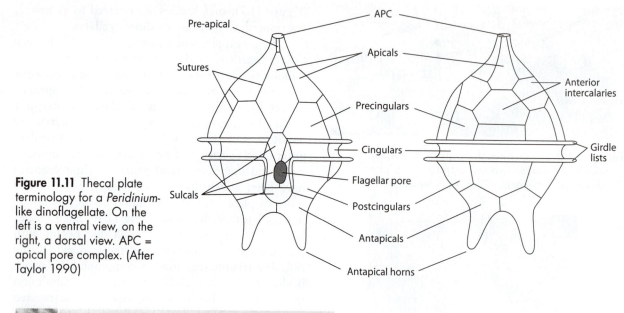

Figure 11.11 Thecal plate terminology for a *Peridinium*-like dinoflagellate. On the left is a ventral view, on the right, a dorsal view. APC = apical pore complex. (After Taylor 1990)

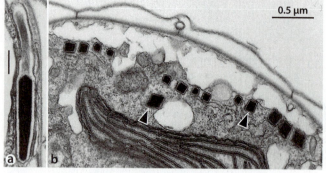

Figure 11.12 (a) Dinoflagellate trichocyst seen in longitudinal view in *Amphidinium cryophilum*. Note that there is a dense basal region and a fibrillar portion (scale bar = 0.25 µm). (b) Trichocysts in cross-section in *Gymnodinium acidotum*. (a: From Wilcox et al. 1982; b: From Wilcox and Wedemayer 1984, by permission of the *Journal of Phycology*)

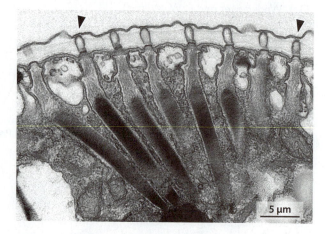

Figure 11.13 A row of trichocysts found along the cingulum of *Peridiniopsis berolinensis*. Note that each lies beneath a pore (arrowheads) in the theca. (From Wedemayer and Wilcox 1984)

as extrusomes) are ejectile rods that occur almost universally at the periphery of dinoflagellate cells (Figure 11.12). They are very similar to the trichocysts of ciliate protozoa, except that the latter are capped with a spine, which is absent from dinoflagellate trichocysts. Dinoflagellate trichocysts lie within the amphiesma, generally oriented perpendicularly to the cell surface. In armored species,

Figure 11.14 A "negatively stained" TEM preparation of a discharged trichocyst showing the fine crossbanding that is present. (TEM: L. W. Wilcox)

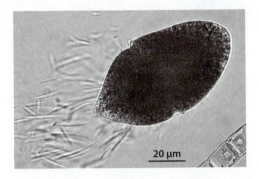

Figure 11.15 A disturbed *Gymnodinium fuscum* cell that has ejected a large number of mucocysts. Treatment of such cells with NiSO$_4$ to stop swimming can cause this to occur. (Photo: L. W. Wilcox)

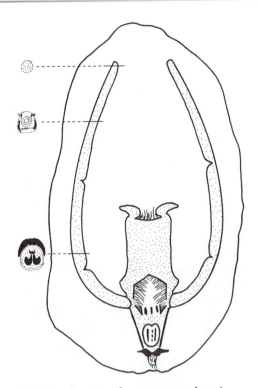

Figure 11.16 A drawing of a nematocyst found among certain dinoflagellates. (From Mornin and Francis 1967)

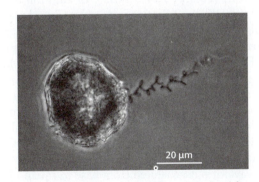

Figure 11.17 The transverse flagellum dislodged from the cingulum of a *Peridinium* cell. Its coiled nature is evident. (Photo: L. E. Graham)

trichocysts lie directly beneath pores in the thecal plates (Figure 11.13; see also Figure 11.6a), and almost all of such pores are preformed sites for trichocyst discharge. Trichocysts develop within the Golgi apparatus and are produced within a sac. They consist of a protein rod that is a few micrometers long and rectangular in cross-section. The distal end consists of twisted fibers. When dinoflagellates are irritated by temperature changes, turbulence, or other disturbance, the sac ruptures. This allows the entrance of water, which causes a change in the conformation of the protein, resulting in elongation of the trichocyst by eight times, and explosive release of the trichocyst from the cell. Some experts hypothesize that trichocyst release causes a jet-propulsive response that is useful in escaping from predators. Upon discharge, trichocysts become much longer and thinner and exhibit a crossbanded structure when viewed with TEM (Figure 11.14). **Mucocysts** are relatively simple sacs that release mucilage to the cell exterior, often in the form of rather thick rod-shaped bodies such as those extruded by *Gymnodinium fuscum* when it is perturbed (Figure 11.15). **Nematocysts** (Figure 11.16), produced by a few dinoflagellate

genera, including *Polykrikos* and *Nematodinium*, are larger (up to 20 µm long) and even more structurally elaborate than trichocysts.

Flagella and motility

Dinoflagellates are considered to be the champion swimmers among flagellate algae, achieving rates of 200–500 µm s^{-1}. In most dinoflagellates, two flagella—one transverse and one longitudinal—emerge close

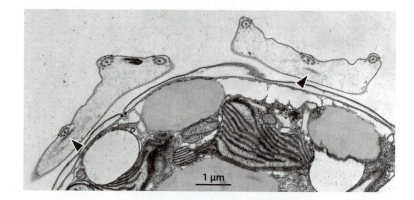

Figure 11.18 A cross-section through an *Amphidinium cryophilum* cell, in which two profiles of the transverse flagellum are seen lying in the cingulum. Note the outer, axonemal (9 + 2) strand and the finer striated strand to its inside (arrowheads). (From Wilcox et al. 1982, by permission of the *Journal of Phycology*)

together from a pore on the ventral side of the cell. The ribbon-shaped transverse flagellum lies in the cingulum, curving around the cell in a counterclockwise direction (as seen by a viewer looking down on the anterior end of the cell) and is usually difficult to visualize with light microscopy. If the flagellum becomes dislodged from the cingulum when the cell is disturbed, its coiled appearance can be seen (Figure 11.17).

Following treatment with appropriate chemical fixatives, dinoflagellate flagella can be best viewed with SEM (Figures 11.18, 11.19). In addition to the usual 9 doublet + 2 singlet array of axonemal microtubules, the transverse flagellum contains a striated, contractile strand composed of the protein centrin (see Figure 11.18). The transverse flagellum is helically coiled in such a way that each turn appears to be stretched where it lies closest to the cingulum and compressed where it is most exposed. The outermost edge of the flagellum projects out of the cingulum, where the ribbon edge appears to undulate. A single row of fine hairs occurs on the transverse flagellum. The waveform is generated from base to tip and propels the cell forward, accounting for about half the forward speed, as well as rotating the cell at a rate of about 0.7 s per turn. Cells that lose the longitudinal flagellum may continue to rotate and move forward by the action of this transverse flagellum. The early-divergent genus *Noctiluca* is unusual in lacking a transverse flagellum and does not rotate while swimming (Goldstein 1992).

The longitudinal (trailing) flagellum (see Figure 11.19) may extend more than 100 μm beyond the cell body and bears two rows of fine hairs. This flagellum accounts for about half of the forward swimming speed and apparently also has a steering function. Dinoflagellates can stop or reverse their swimming

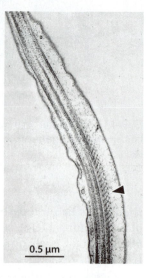

Figure 11.19 A section through a longitudinal flagellum. Note the crosshatched appearance of the fine (centrin-containing?) strand (arrowhead) that lies adjacent the axonemal microtubules. (From Wilcox et al. 1982, by permission of the *Journal of Phycology*)

direction by changing the position of the longitudinal flagellum. The flagellum stops beating, points in a different direction by bending at the site where it exits from the sulcus, and then resumes beating at the new position. This steering ability may be related to the presence within it of a fine filament that may be composed of the contractile protein centrin (Melkonian et al. 1992). The longitudinal flagellum is also responsible for rotation of cells perpendicular to the longitudinal axis. This motion, combined with the rotation conferred by the transverse flagellum, causes cells to swim in a helical path (Fenchel 2001).

Flagellar roots are intracellular cytoskeletal features that are associated with the flagellar apparatus, and their structure is fairly consistent within the dinoflagellates (Figure 11.20). A regular feature is a multimembered longitudinal microtubular root that extends posteriorly along the left side and beneath the sulcus. A second microtubular root is associated with the

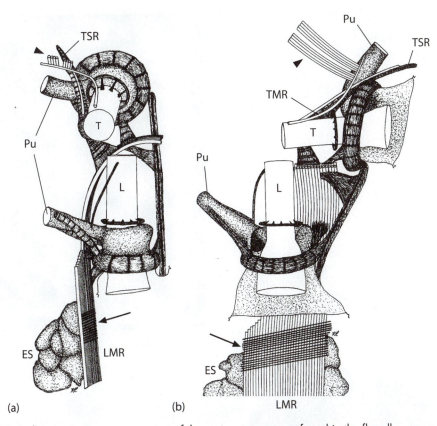

Figure 11.20 A diagrammatic representation of the major components found in the flagellar apparatus of dinoflagellates. (a) View from the cell's right (the ventral surface of the cell is toward the right). (b) Ventral view. Each of the flagella is surrounded by a striated collar. The substantial longitudinal microtubular root (LMR) originates near the longitudinal basal body (L) and extends posteriorly beneath the sulcus, where it is overlain by subthecal fibers (arrows). An eyespot (ES) often underlies the LMR. The prominent transverse striated root (TSR) and transverse microtubular root (TMR) extend from the transverse basal body (T). Microtubules (arrowheads) extend from the TMR like a flag from a pole. A variety of striated and fibrous connectives are typically present, interconnecting the various components. Pusules (Pu) typically connect to the outside of the cell near the point of emergence of the flagella. (From Roberts et al. 1995, by permission of the *Journal of Phycology*)

transverse flagellar basal body; it consists of a "flag" of microtubules that are attached to a "pole" of one or two microtubules. In addition to striated roots that are associated with the basal bodies, a fibrous connective links the two flagellar bases; this structure contains centrin, as do fibrous connectives of other algae.

Dinokaryon

The dinoflagellate nucleus is known as a dinokaryon because it has distinctive chromosomes and mitotic features. Throughout the cell cycle, the chromosomes of most dinoflagellates lack the histone proteins that characterize other eukaryotes. As a result, dinoflagellate chromosomes are condensed most of the time

(Figure 11.21) and do not undergo a cycle of coiling prior to mitosis and uncoiling thereafter, as do most eukaryotic chromosomes. Dinoflagellate chromosomes uncoil for only a brief time to allow DNA replication to occur. The dinokaryon is thought to have evolved by loss of histone proteins, a process that may have occurred by stages. For example, *Oxyrrhis* and other pre-dinoflagellate protists lack a dinokaryon, but *Noctiluca* has a dinokaryon in some parts of its life cycle.

Mitosis and cytokinesis

The dinoflagellate nuclear envelope remains intact throughout mitosis (as in a variety of other algae), but, uniquely in most dinoflagellates, the microtubular

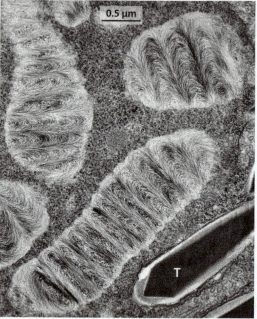

Figure 11.21 The permanently condensed chromosomes of the dinoflagellate *Prorocentrum micans*, showing the swirled appearance of the "naked" (histone-lacking) DNA. Such an arrangement of DNA is exhibited by dinoflagellates as well as the nucleoids of certain bacteria. Also note portions of trichocysts (T). (TEM: L. W. Wilcox)

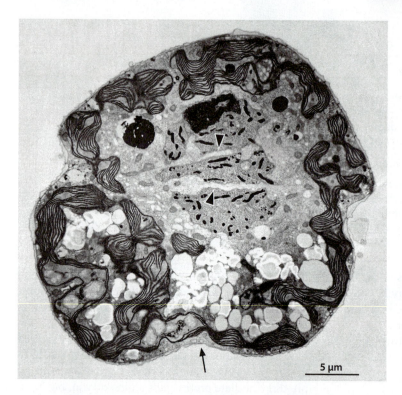

Figure 11.22 A mitotic cell of the thin-walled, photosynthetic freshwater dinoflagellate *Woloszynskia*. This species remains motile throughout mitosis and cytokinesis (which occur concurrently in this species). Note the cytoplasmic channels (arrowheads) passing from left to right through the nucleus, each of which contains spindle microtubules. Note also that a number of the chromosomes appear to be under tension—particularly near the lower cytoplasmic channel—and are seemingly being pulled toward the right. The cleavage furrow (arrow) proceeds from the posterior toward the anterior end of the cell. (TEM: L. W. Wilcox)

spindle is entirely extranuclear. Bundles of spindle microtubules pass through tunnels in the mitotic nucleus (Figure 11.22), and each chromosome attaches to microtubules via contact with the nuclear envelope at the same site (Oakley and Dodge 1974)

(Figure 11.23). Extension of the nuclear envelope and extension of interzonal spindle microtubules are involved in chromosomal separation and movement to the spindle poles, after which the nucleus is pinched into two at the midpoint. The extranuclear spindle

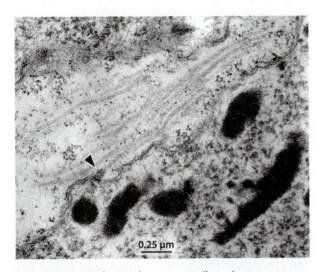

Figure 11.23 *Woloszynskia* mitotic cell. A close-up view of a cytoplasmic channel through which the extranuclear spindle microtubules pass. Note the attachment of a microtubule to a kinetochorelike structure (arrowhead) on the nuclear envelope, at a point where a chromosome is presumably also attached. (TEM: L. W. Wilcox)

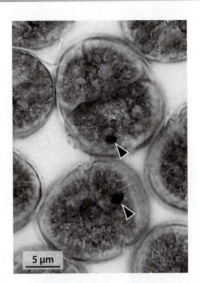

Figure 11.25 A dinoflagellate (*Woloszynskia* sp.) that divides while immobile (the top cell has recently divided). Each daughter cell makes a new theca following cell division. A large bright-red eyespot (arrowheads) is found in this species. (Photo: L. W. Wilcox)

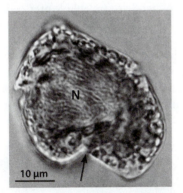

Figure 11.24 *Amphidinium cryophilum* is an unarmored dinoflagellate that undergoes mitosis and cell division while remaining motile. The cleavage furrow (arrow) proceeds from the antapex toward the apex of the dividing cell. The numerous chromosomes are visible in the large central nucleus (N). (From Wedemayer et al. 1982, by permission of the *Journal of Phycology*)

first appears in an early-diverging group of dinoflagellates known as the Syndiniales (see Figure 11.2).

Cytokinesis in unarmored cells may involve a simple pinching in two along an oblique division line, with continuous synthesis of new amphiesma parts. Unarmored cells commonly remain motile while they divide, as in *Amphidinium cryophilum* (Figure 11.24). In contrast, some armored species

shed the cell covering prior to mitosis, and then each daughter cell resynthesizes a new theca; mitosis in such species generally occurs while cells are nonmotile (Figure 11.25). In other armored forms, such as *Ceratium*, oblique fission of the theca occurs along a predetermined sequence of suture lines. Each daughter cell receives half of the parental wall components, and regenerates the missing portion (Figure 11.26). In nature, *Peridinium limbatum* populations double approximately every 11 days (Graham et al. 2004).

Cellular Features of Some, but Not All, Dinoflagellates

Some, but not all, dinoflagellates possess specialized structures that aid in prey-capture, light-emitting scintillons that aid in predator avoidance, plastids of several types, or variously structured eyespots.

Prey-capture adaptations

Phagotrophy, the process of feeding on particles such as the cells of other organisms, is very common among dinoflagellates. It occurs among marine and freshwater dinoflagellates, photosynthetic and plastidless dinoflagellates, and armored and unarmored types. Dinoflagellates primarily prey on eukaryotes, which can include other dinoflagellates, members of other algal groups, large ciliates, nematodes, polychaete larvae, and even

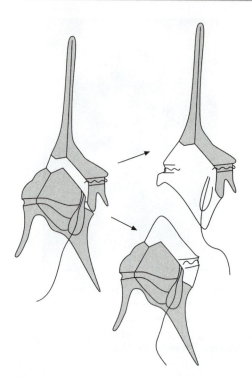

Figure 11.26 A diagrammatic view of cell division in *Ceratium hirundinella*, a common armored dinoflagellate. In this dinoflagellate, the original theca (shaded gray) is split among the two daughter cells, each of which must synthesize the remaining portions following division. (After Fensome et al. 1993)

fish. Reflecting this diversity of prey organisms is a diversity of feeding structures and methods. There are three main feeding mechanisms—engulfment by dinoflagellate cells of whole prey cells, use of a feeding tube known as a **peduncle** or **phagopod**, or extension of a feeding veil of cytoplasm known as a **pallium**.

Engulfment. Unarmored dinoflagellates may be able to stretch as they engulf intact prey cells and then digest them inside an internal food vacuole. Amazingly, at least some armored dinoflagellates can also ingest whole cells. *Fragilidium subglobosum*, a predator of *Ceratium*, is one such example. About 15 minutes is required for *Fragilidium* to engulf cells of a large *Ceratium* species having volumes of over 1×10^5 μm³, though smaller *Ceratium* species can be ingested within 5 minutes. *Fragilidium* breaks down its prey's theca very quickly, allowing the feeding cell to pack the large prey cell into a manageable food vacuole; a portion of the cellulose remains is discharged as a fecal pellet. It is thought that armored dinoflagellates that have many small plates, such as *Fragilidium*, can more easily expand around whole prey cells than those having few but large plates (Skovgaard 1996).

Feeding tubes. Many dinoflagellates feed through an extensible peduncle, which is bound by the cell membrane and contains tens to hundreds of microtubules (Figures 11.27, 11.28). This structure emanates

Figure 11.27 The internal portion of the peduncle of *Peridiniopsis berolinensis*, a common freshwater heterotrophic dinoflagellate. Numerous overlapping rows (arrowheads) of microtubules splay off from a dense cluster of microtubules at the top of the micrograph. When first described in dinoflagellates, this structure was termed a "microtubular basket," a term that may still be encountered in the literature. These microtubules originate near the cingulum on the dorsal side of the cell, curve along the periphery of the cell, passing near the cell's apex, and form the "backbone" of a membrane-ensheathed external peduncle that exits the cell on the ventral surface, near the cingular–sulcal intersection (see chapter- opening photo). (From Wedemayer and Wilcox 1984)

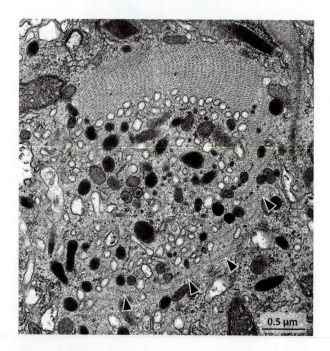

0.5 μm

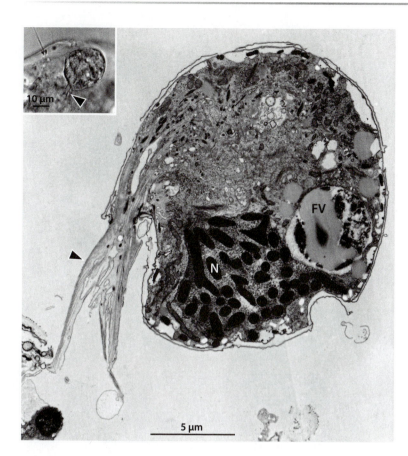

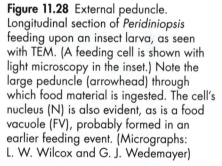

Figure 11.28 External peduncle. Longitudinal section of *Peridiniopsis* feeding upon an insect larva, as seen with TEM. (A feeding cell is shown with light microscopy in the inset.) Note the large peduncle (arrowhead) through which food material is ingested. The cell's nucleus (N) is also evident, as is a food vacuole (FV), probably formed in an earlier feeding event. (Micrographs: L. W. Wilcox and G. J. Wedemayer)

from the ventral side of the cell in the cingular–sulcal region. In addition to its role in feeding, the peduncle may also be used to attach to surfaces. Dinoflagellates feeding through peduncles may engulf their prey whole or take in food in the form of smaller particles or liquefied, digested prey material. *Pfiesteria*, which consumes fish flesh, is an example of a dinoflagellate having thin thecal plates that feeds through a peduncle (see Figure 3.6). Peduncles are also produced by armored species such as the marine *Dinophysis* and the freshwater *Peridiniopsis berolinensis* (Wedemayer and Wilcox 1984; Calado and Moestrup 1997). The latter species first uses a fine cytoplasmic filament (a "capture" or "tow" line) to establish contact with potential food items, and then it ingests the prey whole or ingests its contents. The cold-water mixotrophic *Amphidinium cryophilum*, an unarmored freshwater dinoflagellate, employs a phagopod that extends from the posterior pole of the cell and contains no microtubules (Wilcox and Wedemayer 1991) (Figures 11.29, 11.30). This feeding structure is apparently formed anew each time the organism feeds and is left behind after feeding.

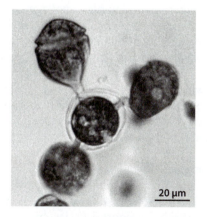

Figure 11.29 Three *Amphidinium cryophilum* cells feeding upon a prey cell whose protoplast is shrinking as its cytoplasm is being drawn into the *Amphidinium* cells. Food material passes through a feeding tube (phagopod), which extends from the posterior end of the *A. cryophilum* cell and attaches to the amphiesmal membranes of the prey cell. (From Wilcox and Wedemayer 1991, by permission of the *Journal of Phycology*)

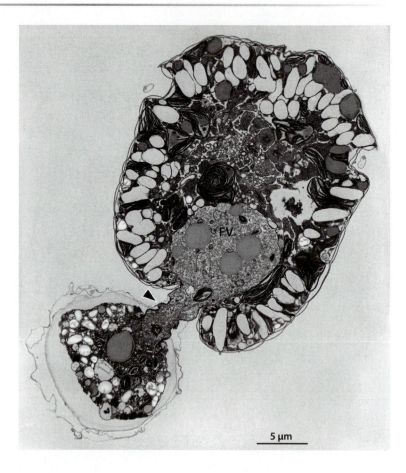

Figure 11.30 Feeding *Amphidinium* cell shown with TEM. The prey cell cytoplasm is in continuity with a forming food vacuole (FV) in the feeding cell. The maintenance of intact prey cell amphiesmal membranes appears necessary for feeding to occur in this system. Arrowhead points to phagopod. (From Wilcox and Wedemayer 1991, by permission of the *Journal of Phycology*)

A number of nonmotile dinoflagellates in addition to *Blastodinium* (Figure 11.31), such as *Protoodinium*, *Crepidoodinium*, and *Piscinodinium*, contain plastids but are parasites of zooplankton or fish. They produce motile spores, known as **dinospores**, which have typical dinoflagellate structure. The dinospores use peduncle-like structures to attach to food sources, where they develop into the mature, nonmotile form.

Feeding veil. Two dozen species of the armored dinoflagellate *Protoperidinium*, along with several *Diplopsalis* species, feed by the extension of a pallium, or "feeding veil" (see Figure 3.5) (Jacobson and Anderson 1986; Naustvoll 1998). This is a sheetlike extension of the cytoplasm that develops from a peduncle-like tube that emerges from the cingular–sulcal region of the cell. The pallium encloses the prey, which are first captured with a filament. Enzymes produced by the pallium then digest prey cytoplasm, and the products of digestion are transported back into the feeding dinoflagellate. *Protoperidinium* species feed mainly on

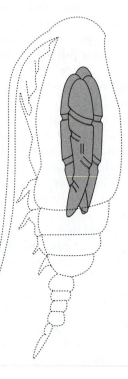

Figure 11.31 Diagram of a four-celled stage of the parasitic *Blastodinium* (gray area) living in the gut of a copepod. (After Fensome et al. 1993)

diatoms and dinoflagellates, and their populations are typically most abundant following diatom blooms occurring in coastal temperate and even polar waters. *Diplopsalis* feeds on a wider variety of prey, including haptophytes, green algae, and cryptomonads, as well as diatoms and dinoflagellates. At 31 µm in diameter, *D. lenticula* consumes cells between 3 and 78 µm in diameter, including motile cells. Individually, bacteria are too small for capture with a pallium, but if aggregated in detritus (marine snow), they can be ingested by pallium-feeding dinoflagellates (Naustvoll 1998).

Some additional feeding structures include the tentacle with which *Noctiluca* snares diatoms and other planktonic particles such as copepods and fish eggs, which are then conveyed to a cell mouth similar in function to the cytostome of ciliates. *Erythropsidinium* and related forms have a thick feeding "piston" that can be extended from the cell and retracted back into the cell within a fraction of a second (Greuet 1976).

Bioluminescence and scintillons

Bioluminescence occurs in approximately 30 photosynthetic dinoflagellates, including *Lingulodinium*, *Protogonyaulax*, *Pyrodinium*, *Pyrocystis*, and *Ceratium*, as well as some non-photosynthetic marine forms, such as *Noctiluca* and *Protoperidinium* (Taylor 1990). There are bioluminescent and non-luminescent strains of the same dinoflagellate species. As discussed in Chapter 3, bioluminescence is regarded as an adaptation that reduces attack on dinoflagellates by their predators (see review by Lewis and Hallett 1997).

Bioluminescent dinoflagellates possess spherical intracellular structures known as **scintillons** or **microsources**. These are about 0.5 µm in diameter and are arrayed at the cell periphery. In transmission electron micrographs, scintillons appear as dark rodlike structures surrounded by a membrane. They are derived from the Golgi apparatus and contain luciferin, luciferase, and, in some cases, a luciferin-binding protein (LBP). Immunogold labeling was used to establish the presence in scintillons of the enzyme luciferase. In *Lingulodinium polyedrum* (= *Gonyaulax polyedra*), luciferin is bound to LBP at pH 8, but it is released at pH 6, becoming available for reaction with luciferase. Luciferase oxidizes the luciferin with molecular O_2, causing a 0.1 s flash of blue light. This reaction is triggered by environmental stimuli that cause an influx of protons across the scintillon membrane, thus lowering pH to the critical level. The number of scintillons in *L. polyedrum* decreases from 540 per cell in the night phase to just 46 in day-phase cells, and the amount of bioluminescence is two orders of magnitude greater in night-phase cells. There is a daily (circadian) rhythm

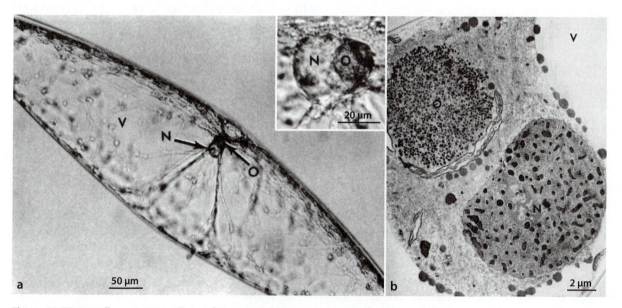

Figure 11.32 Scintillons. (a) Scintillons of *Pyrocystis* forming an orange body (O) near the nucleus (N) during the daytime. They are seen at higher magnification in the inset. (b) The orange body seen adjacent the nucleus with TEM. During the night scintillons disperse throughout the highly vacuolate (V) cell. (From Sweeney 1982, by permission of the *Journal of Phycology*)

in synthesis and destruction of scintillons, luciferin, and luciferase. This is viewed as an adaptation that conserves energy, as bioluminescence would not be visible in the daytime (Fritz et al. 1990). In contrast, *Pyrocystis*, which can emit about 1000 times more light than *Lingulodinium* (Swift et al. 1973), does not undergo a daily cycle of destruction and resynthesis of scintillons, and it lacks LBP (Knaust et al. 1998). However, there is a diurnal change in the positions of *Pyrocystis* scintillons: during the day they occur as a spherical mass of vesicles near the nucleus, and at night they disperse to the cell periphery, returning to the nucleus at daybreak (Figure 11.32); plastids exhibit the opposite pattern of movement (Sweeney 1982).

Lingulodinium polyedrum genes for LBP and luciferase have been cloned and sequenced (Lee et al. 1993; Bae and Hastings 1994), and homologous luciferase genes of several other photosynthetic dinoflagellates are similar (Liu et al. 2004). However, the structural organization of the luciferase gene from the early-divergent, heterotrophic species *Noctiluca scintillans* is distinctive (Liu and Hastings 2007); homologous gene sequences encode two distinct proteins in *L. polyedrum* but encode a single, two-domain polypeptide in *Noctiluca*. The biochemical pathway leading to luciferin is not known for dinoflagellates. Bioluminescence and many other dinoflagellate processes are known to involve endogenous circadian rhythms, studied intensively by Sweeney (1987) and also reviewed by Lewis and Hallett (1997).

Plastids and photosynthesis

The 50% or so of dinoflagellates that possess plastids acquired them from a variety of photosynthetic eukaryotes, including haptophytes, cryptomonads, diatoms, and green algae. This reflects dinoflagellates' fundamentally phagotrophic lifestyle and kleptoplastidy (tendency to retain plastids stolen from ingested prey) (see Chapter 7). Even so, most plastid-containing dinoflagellates have golden-brown plastids with a unique accessory pigment, the xanthophyll **peridinin** (Figure 11.33). Such plastids are referred to as **peridinin plastids**. Together with chlorophyll *a*, peridinin forms the light-harvesting complex of most photosynthetic dinoflagellates. Peridinin plastids also contain chlorophyll *c*, have thylakoids stacked in threes, and are bound by an envelope of three (occasionally two) membranes that are not connected to the nuclear envelope or the endoplasmic reticulum (Figure 11.34). In addition, the peridinin plastid of at least some species

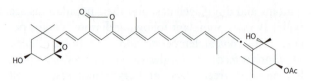

Figure 11.33 The structure of peridinin, a common accessory pigment in dinoflagellates. (Redrawn with permission from Hofmann, E., P. M. Wrench, F. P. Sharples, R. G. Hiller, W. Welte, and K. Diederichs. Structural basis of light harvesting by carotenoids: Peridinin–chlorophyll–protein from *Amphidinium carterae*. *Science* 272:1788–1791. © 1996 American Association for the Advancement of Science)

is notable for near total loss of the plastid genome. The genes present within peridinin plastids occur on 2–3 kilobase minicircles (Zhang et al. 1999; Nelson et al. 2007), and nuclear genes may actually encode some of the minicircles (Laatsch et al. 2004). Uniquely among oxygenic photosynthesizers, peridinin plastids utilize nuclear-encoded, Form II rubisco. This type of rubisco, which consists of two subunits of the large rubisco subunit (and no small subunits), otherwise occurs only in certain prokaryotes and was thus likely acquired by horizontal gene transfer. Form II rubisco has relatively low selectivity for CO_2 and is thus vulnerable to O_2 binding, which promotes photorespiration. In at least some dinoflagellates, a carbon concentration mechanism (see Chapter 2) helps to prevent photorespiration (e.g., Leggat et al. 1999). The ancestry of peridinin-containing plastids is as yet unclear. These plastids are generally considered to have descended from red algal plastids via secondary endosymbiosis (Delwiche 2007), but it has also been proposed that the ancestral peridinin plastid originated via tertiary endosymbiosis involving a plastid-bearing stramenopile. The latter hypothesis would explain the presence of chlorophyll *c*, which is absent from red algal plastids but is typical of photosynthetic stramenopiles (Bodyl and Moszczynski 2006).

Studies of energy transmission through the light-harvesting system (accomplished by measuring changes in fluorescence) have revealed that the peridinin-containing dinoflagellates are amazingly efficient in light capture and energy transmission. The reason for this high efficiency is the occurrence within thylakoid lumens of a water-soluble peridinin–chlorophyll–protein light-harvesting complex (PCP LHC), in addition to a more typical membrane-bound LHC similar to that of plants.

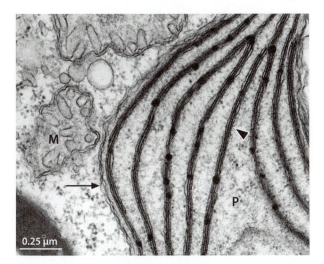

Figure 11.34 A section through the plastid (P) of *Woloszynskia pascheri*, showing the three membranes (two of which are often closely appressed to one another) bounding the plastid (arrow); the thylakoids (arrowhead) occur in stacks of three. The dark globules found on the thylakoids result from the particular preservation technique used in preparing this specimen for TEM. Note also the tubular cristae in the adjacent mitochondria (M), a feature typical of the dinoflagellates and other protist groups (see Chapter 7). (TEM: L. W. Wilcox)

Peridinin captures light energy in the blue-green range of 470–550 nm, which is present in aquatic habitats and inaccessible to chlorophyll alone. X-ray crystallographic studies have shown that the soluble LHC protein forms a boat-shaped structure with hydrophobic cavities that are filled by two lipid, eight peridinin, and two chlorophyll *a* molecules. The chlorophylls are completely buried within the hydrophobic environment, with half their surface area covered by peridinins and the remainder covered by protein and fatty acid chains of the lipids. The main function of the protein is to provide a hydrophobic environment for the pigments (which would otherwise be insoluble in the watery lumen environment), but another function is to hold the peridinins at the most appropriate distance for effective energy transfer to chlorophyll. Peridinin must be located particularly close (0.33–0.38 nm) to chlorophyll *a* because the half-life of the excited carotenoid is very short. The PCP complex allows an energy transfer efficiency of nearly 100% (Hofmann et al. 1996).

As noted earlier, a number of dinoflagellates have stable or temporary plastids obtained from haptophytes (e.g., Gast et al. 2007), diatoms (e.g., Imanian et al. 2007; Imanian and Keeling 2007), green algae (e.g., Hansen et al. 2007), or cryptomonads (e.g., Hackett et al. 2003). These represent examples of secondary or tertiary endosymbiosis (Chapter 7). In some cases, phylogenetic patterns have been used to infer that such plastids replaced preexisting peridinin plastids (Shalchian-Tabrizi et al. 2006b). Some researchers hypothesize that some or all heterotrophic dinoflagellates evolved from photosynthetic dinoflagellates by loss of pigments, or the plastid itself. For example, in searching a genetic (expressed sequence tag or EST) database obtained from the relatively early-diverging heterotrophic dinoflagellate *Crypthecodinium cohnii*, Sanchez-Puerta and associates (2007) found evidence for the presence of genes encoding proteins with plastid-targeting regions. These results suggest that *C. cohnii* descended from photosynthetic ancestors, though it is unclear if colorless plastids were retained. Other examples of heterotrophic species that evolved from photosynthetic ancestors can be found among the euglenoids, cryptomonads, stramenopiles, red algae, and green algae; in these cases, the cells retain colorless plastids (see Chapters 8, 9, 12, 15, and 18).

Phototaxis and eyespots

Dinoflagellates can swim toward light (phototaxis), and their light-sensing system is thought to be a protein-bonded carotenoid. Dinoflagellates lack an autofluorescent, flavin-based light detector such as is associated with the flagella of euglenoids, brown algae, and some other algal groups (Kawai and Inouye 1989). Though most dinoflagellates lack eyespots, others display one of several types of eyespots. Some dinoflagellates possess cytoplasmic eyespots composed of aggregated globules (Figure 11.35). *Glenodinium*, *Gymnodinium*, and *Woloszynskia*, which have peridinin plastids, have eyespots consisting of a single or double layer of carotenoid-containing lipid droplets located between the three-membrane plastid envelope and the outermost layer of thylakoids (Figure 11.36). In *Peridinium* (*Kryptoperidinium*) *foliaceum* and *Peridinium balticum*, whose photosynthetically active plastids were acquired from a diatom endosymbiont (Chesnick et al. 1996), the eyespots occur as carotene droplets surrounded by three membranes. Such eyespots are thought to represent the remains of highly reduced peridinin plastids whose photosynthetic function was replaced by diatom plastids.

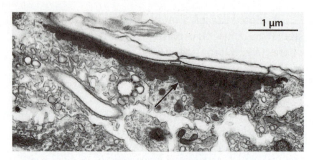

Figure 11.35 The eyespot (arrow) of the colorless dinoflagellate *Katodinium campylops*. The granules making up this eyespot are found free in the cytoplasm rather than within a plastid or surrounded by any membranes. (TEM: L. W. Wilcox)

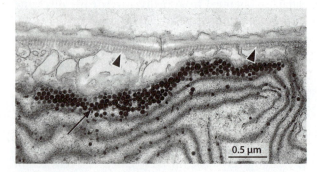

Figure 11.36 The eyespot (arrow) within the plastid of *Woloszynskia pascheri*. The eyespot is located beneath the sulcus. Numerous microtubules are seen beneath the thecal vesicles. These are part of the longitudinal microtubular root, which extends posteriorly from the flagellar apparatus. (TEM: L. W. Wilcox)

A group of unicellular, predatory dinoflagellates known as the Warnowiaceae possess unusually complex, conspicuous eyespots known as **ocelli**. One example is the genus *Erythropsidinium*, whose name means "red-eyed dinoflagellate" (Figure 11.37). The subcellular components of ocelli bear an extraordinary resemblance to the parts of metazoan eyes. Ocelli include a lens-like refractile **hyalosome** (meaning "clear body") that is constructed within the endoplasmic reticulum. The hyalosome becomes surrounded by mitochondria and constricting fibers that can move, changing the shape of the lens. Ocelli also include an "ocular chamber," backed by a cup-shaped, darkly pigmented, retina-like structure (Figure 11.38) (Greuet 1968). This assembly presumably allows the formation of images upon the retinoid (Francis 1967), but the mechanism of sensory transfer, with resulting changes in dinoflagellate behavior,

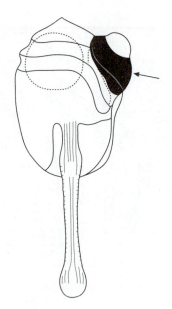

Figure 11.37 A drawing of the ocellus-containing *Erythropsidinium*, showing the large size of the ocellus and dark pigment cup (arrow) in relation to the rest of the cell. The long structure extending posteriorly is the "piston," or "tentacle," thought to be involved in phagotrophy. (After Kofoid and Swezy 1921)

if any, is not understood. It has been proposed that this specialized visual system has evolved in response to selective pressures of the phagotrophic habit, allowing predatory dinoflagellates to "see" their prey.

11.3 Sexual Reproduction and Cyst Formation

Dinoflagellates that spend most of their life cycle as unicellular flagellates undergo population increases by mitosis, and several have been documented to undergo sexual reproduction as well. Dinoflagellates also produce various types of cysts, which are unicellular nonflagellate stages that are able to survive conditions unfavorable for population growth. Cysts often play important roles in dinoflagellate ecology, and as noted earlier, resistant dinoflagellate cysts occur in the fossil record.

Sexual Reproduction

Although sexual reproduction has been studied in relatively few dinoflagellates (reviewed by Pfiester and

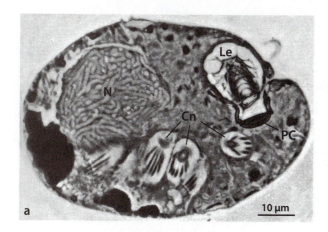

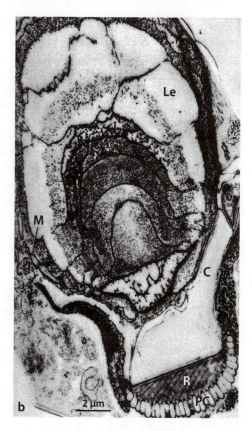

Figure 11.38 Longitudinal views of the ocellus of *Nematodinium*. (a) Low-magnification view of the entire cell. Several cnidocysts (Cn) are evident, as is the ocellus, with its clear lens (Le) and pigment cup (PC). N = nucleus. (b) The outer, hyaline lens (Le) is at the top and the pigment cup (PC) and retinoid (R) toward the bottom of this TEM image. M = mitochondrion, C = canal open to outside of cell. (From Mornin and Francis 1967)

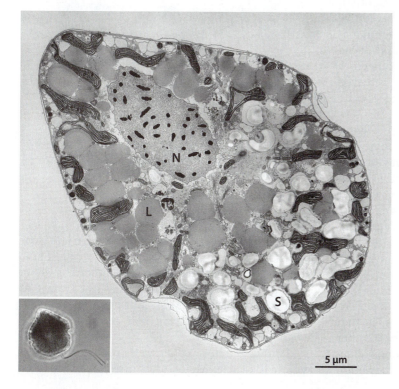

Figure 11.39 Planozygotes of the freshwater dinoflagellate *Woloszynskia pascheri*, as seen with light (inset) and transmission electron microscopy. Planozygotes are often identifiable due to the presence of two "ski-track" longitudinal flagella and a more angular outline than is seen in vegetative cells. There are also large stores of lipids (L) and starch (S). In this case, the starch is found primarily in the hypocone, and lipids, in the epicone, along with the nucleus (N). (TEM: L. W. Wilcox)

Anderson 1987; Pfiester 1988), it is thought to occur much more widely than actually observed. Explanations for the cryptic nature of dinoflagellate sexual reproduction include the facts that gametes resemble vegetative cells, gamete fusion is so slow that it is difficult to distinguish from cell division, and gamete fusion seems to occur at night in the photosynthetic species. Further, while early stages of zygote development can be recognized by the presence of more than two flagella, later stages can difficult to recognize. Flagellate zygotes, known as **planozygotes** (Figure 11.39), of some species undergo meiosis rather quickly, whereas those of other dinoflagellates transform into nonflagellate **hypnozygotes** ("sleeping" zygotes) (Figure 11.40) and thus resemble resting cysts produced by asexual means.

Sexual reproduction can be induced in the laboratory by reducing the nitrogen concentration of the growth medium or by changing the temperature. It is assumed that these factors are also related to induction of sexual reproduction in nature. In some dinoflagellates sexual reproduction involves isogametes—gametes that closely resemble each other in size and shape. In *Scrippsiella* (Xiaoping et al. 1989), isogametes may be about the same size as vegetative cells or noticeably smaller. In other cases (e.g., *Ceratium*), gametes are anisogamous, that is, one gamete of each fusing pair is much smaller than the other, the larger being about the same size as a vegetative cell (von Stosch 1964). Sometimes sexual reproduction is **homothallic**, meaning that gametes produced within a clonal population will fuse, and sometimes it is **heterothallic**, meaning that syngamy is achieved only when mating individuals arise from genetically different clones.

The mating process often begins when gametes dance around each other before they come into contact, but a diversity of subsequent behaviors has been observed. In *Scrippsiella* and some other dinoflagellates, the transverse flagellum of one gamete moves out of the cingulum, grasps the longitudinal flagellum of the other gamete, and then returns to the cingulum. Fusion then begins at the sulcal region at the point where flagella emerge, and fusion of the hypocones is completed before fusion of the epicones (Xiaoping et al. 1989) (Figure 11.41). In other dinoflagellates the epicone of one gamete attaches to the sulcal region of the other, such that the cells are oriented perpendicular to one another (Chesnick and Cox 1987, 1989). Sexual reproduction in the nonmotile *Gloeodinium montanum*

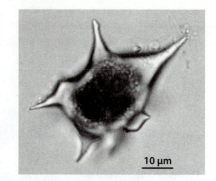

Figure 11.40 A hypnozygote of *Woloszynskia*, with an extremely thick wall bearing a number of projections. (Photo: L. W. Wilcox)

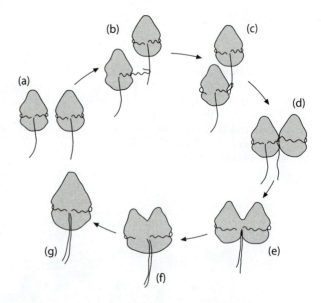

Figure 11.41 Diagram of mating in *Scrippsiella*. Using its transverse flagellum, one gamete lassos the longitudinal flagellum of the other (b–d), pulling it close, after which fusion occurs (e, f). The resulting planozygote has two longitudinal flagella. (After Xiaoping et al. 1989)

involves production of two to four biflagellate isogametes or anisogametes per vegetative cell. These fuse in pairs, forming a large, nonmotile zygote. After a dormancy period that may last from two months to more than a year, four nonmotile vegetative cells are produced by zygote germination (Kelley and Pfiester 1990).

Dinoflagellates are generally thought to have a zygotic life cycle. Meiosis has rarely been reported

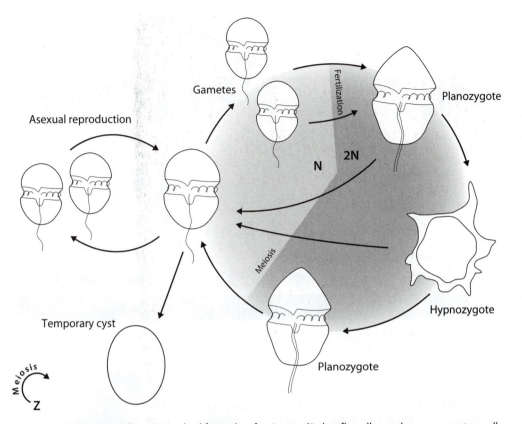

Figure 11.42 Diagram illustrating the life cycle of a "typical" dinoflagellate whose vegetative cells are haploid. Diploid, motile zygotes (planozygotes) may be present following gamete fusion and again following germination of nonmotile hypnozygotes. Temporary asexual cysts are formed in some species. The vegetative stage of some dinoflagellates may be nonmotile and/or strikingly different in appearance from the typical dinoflagellate motile cell morphology. However, such organisms are recognizable as dinoflagellates due to the presence of dinospores (asexual or sexual motile cells with a dinoflagellate morphology) at some point in their life cycle.

and is in need of further study but is inferred to occur at zygote germination. This would imply that cells other than zygotes are haploid (Figure 11.42). The onset of the first meiotic division can be inferred by swirling of chromosomes within the diploid zygote nucleus, a process known as nuclear cyclosis. This process has been documented to occur in dividing zygotes of *Pfiesteria shumwayae* and related estuarine cryptoperidiniopsoid dinoflagellates (Parrow and Burkholder 2003a,b, 2004). Most often nuclear division is followed by two cytokineses, but sometimes cytokinesis is delayed until after the second meiotic division (Beam and Himes 1974). The early-diverging *Noctiluca* has been reported to be exceptional in that meiosis appears to be gametic; vegetative cells are regarded as diploid in this case. A meiotic division giving

rise to four daughter nuclei is followed by repeated mitoses, ultimately yielding more than 1000 uniflagellate, flattened isogametes. These may fuse, with the flat side of one gamete joining the narrow edge of the other, to form zygotes that apparently differentiate into vegetative cells (Zingmark 1970). On the other hand, other authors observed that *Noctiluca* has a typical zygotic life cycle (Schnepf and Drebes 1993).

The life histories of some dinoflagellates have been reported to involve multiple stages involving amoebae and other cellular forms unusual for dinoflagellates (Pfiester and Popovsky 1979; Pfiester and Lynch 1980; Popovsky and Pfiester 1982). Molecular studies are needed to confirm the genetic identity of diverse cell types thought to represent different aspects of the same dinoflagellate life cycle.

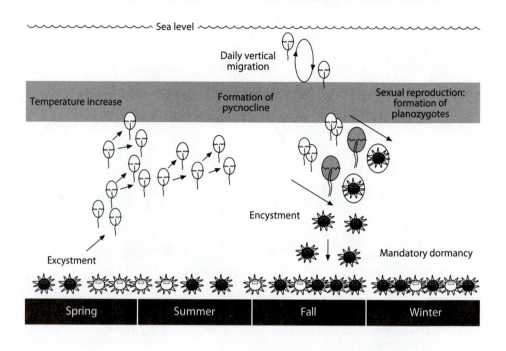

Figure 11.43 Diagram illustrating the cycle of cyst production and excystment in the bloom-forming *Lingulodinium polyedrum* (= *Gonyaulax polyedra*). Dormant cysts are shown in black, and excysted hypnozygotes are shown in white. A pycnocline is a zone separating water of different temperatures. (After Lewis and Hallett 1997)

Cysts

Some dinoflagellates are known to produce nonmotile resting cysts by sexual or asexual processes. In *Lingulodinium polyedrum*, cyst formation typically follows sexual reproduction in the autumn (Figure 11.43). Cyst development in this species begins with formation of a colorless peripheral region of the cytoplasm. The planozygote stops swimming and sheds its flagella, after which the thecal plates separate and pull away from the cell surface. The cyst wall, including spiny outgrowths, then forms. There is a necessary dormancy period that lasts several months, during which this species cannot excyst. Cyst germination is also inhibited by darkness and low O_2. Excystment in the spring occurs through an aperture in the cyst wall known as an **archeopyle** (which is often a distinctive feature of fossil cysts). Within a few hours the resulting cells have developed a normal theca. Cysts can survive in storage for as long as 12 years (see Lewis and Hallett 1997). Like the vegetative cells of this species, they are toxic and may be ingested by shellfish (Dale et al. 1978).

Formation of resting cysts has been attributed to changes in nutrients, irradiance, photoperiod, or temperature. In some cases, particular bacterial associates seem to promote the formation of cysts (Adachi et al. 2004). Walls of resting cysts typically contain cellulose and dinosporin as well as mucilage, arranged in several layers. Many dinoflagellate cysts

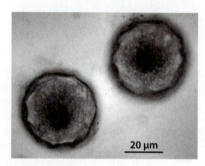

Figure 11.44 Thick-walled cysts with bright-red bodies (seen here as dark areas), a common feature of dinoflagellate cysts. (Photo: L. W. Wilcox)

are ornamented with spiny processes (see Figure 11.1b). Photosynthetic pigments may be reduced and storage products increased in resting cysts. A conspicuous red-pigmented accumulation body that often occurs within dinoflagellate cysts (Figure 11.44) can sometimes be used to distinguish the cyst forms of dinoflagellates from similar-appearing cysts or vegetative cells belonging to other algal groups. Matsuoka and Fukuyo (2003) provided a more detailed treatment of cyst structure and taxonomy.

Large numbers of cysts have been found in sediments below coastal waters that previously supported dinoflagellate blooms. Cysts may be transported via water currents or ship ballast water (Hallegraeff and Boalch 1991, 1992; Carlton and Geller 1993) to new geographical locations, where they may serve as inocula for the formation of harmful blooms. The piscivorous (fish-eating) dinoflagellate *Pfiesteria* is reported to spend most of the time as benthic resting cysts, until the presence of live finfish or their fresh excreta induces cysts to germinate into flagellate stages within minutes to days, depending upon the age of cysts. The dynamics of cyst formation by *Scrippsiella trochoidea* in Daya Bay in the South China Sea were described by Wang and associates (2007).

11.4 Ecology

Common in both marine waters and freshwaters, dinoflagellates are found in both pelagic (open-water) and benthic habitats such as sand. Ecological constraints faced by dinoflagellates in such habitats include sensitivity to turbulence and their relatively large size and consequent low surface area-to-volume (SA/V) ratio, which influence their ability to obtain nutrients. A number of adaptations foster the ability of mostly photosynthetic dinoflagellates to form harmful blooms when nutrients are abundant. Photosynthetic dinoflagellates may also live within the cells or bodies of various kinds of protists and invertebrate animals, notably corals, where they play essential roles.

Nutrition

There is a nearly universal requirement among dinoflagellates for vitamin B_{12}; in addition, the many completely heterotrophic species must obtain sufficient organic carbon to support swimming, division, and reproduction. The high frequency of mixotrophy indicates that photosynthesis does not completely provide the energetic needs of many plastid-bearing species. Photosynthetic dinoflagellates are generally regarded as at least potentially capable of osmotrophy or phagotrophy or both. In addition to organic carbon, dinoflagellates must obtain inorganic nutrients such as phosphorus and combined nitrogen. Phagotrophic forms can obtain these in particulate form, whereas species that do not feed on organic particles must take up inorganic nutrients at the cell surface. Here, because of their relatively large sizes and compact shapes, dinoflagellates are at a competitive disadvantage in comparison to nanophytoplankton, which have higher SA/V ratios and thus relatively high nutrient uptake capacities. In terms of organic carbon produced per unit of organic carbon

biomass, dinoflagellates are 10 or more times less efficient than are nanoplankton, primarily because of this difference in SA/V ratio. As a result, dinoflagellate population growth rates are generally lower than those typical for nanoplanktonic algae (Pollingher 1988). Low SA/V ratio is also regarded as a reason why osmotrophy alone is unlikely to provide sufficient organic carbon to support growth of heterotrophic dinoflagellates (Gaines and Elbrächter 1987). Prey choices and prey ingestion and growth rates of the heterotrophic dinoflagellate *Luciella masanensis* were elucidated by Jeong and associates (2007).

Impact of Turbulence

Dinoflagellates in general are much more vulnerable than many other algae, particularly nanophytoplankton, to high levels of turbulence (Juhl et al. 2001). Though turbulence is beneficial in the dispersal and resuspension of benthic cysts, flagellate vegetative cells do not require turbulence to remain suspended and are rather easily damaged by it. Storm events that cause massive turbulence can be associated with destruction of large numbers of dinoflagellate cells. Because of this sensitivity, researchers who must grow dinoflagellate cultures in the lab are careful not to shake or stir them vigorously. Experimental studies have shown that moderate levels of turbulence impede cell division, but when turbulence is reduced, dinoflagellate cells can resume cell division. In both marine and freshwater habitats, conspicuous populations of dinoflagellates are more likely to occur in calm, windless conditions (Pollingher 1988).

Bloom Formation

Dinoflagellates may form large populations known as blooms in freshwaters (Graham et al. 2004) but are more famous for this behavior in marine systems, where such blooms are called red tides. Dinoflagellates achieve their greatest levels of population growth, 10^7–10^8 cells per liter, in calm waters or along boundary fronts formed at the junction of stratified open-ocean waters and the coastal mixed zone. Most bloom-forming dinoflagellates are photosynthetic species, but seasonally high populations of the heterotroph *Protoperidinium depressum* can occur in estuaries and other nearshore waters.

A number of features contribute to the ability of dinoflagellates to form blooms. Efficient flagellar motion coupled with phototactic capacity provides dinoflagellates with much greater ability to migrate vertically through great depths in the water column than most other planktonic algae. Dinoflagellates can thus harvest organic particles and/or inorganic nutrients from much of the water column. Photosynthetic dinoflagellates can often be observed to ascend to surface waters during the day, where they harvest photons, and then descend or disperse at night, a behavior that contributes to both nutrient acquisition and predator avoidance. Swimming also confers the ability to avoid excessively high light levels. In addition, dinoflagellates can scavenge phosphorus from dissolved organic sources such as ATP and store relatively large amounts of phosphorus for later use when ambient levels are lower. Dinoflagellate blooms can therefore occur at times when dissolved nutrient levels would seem too low to support large algal populations. Aided by their relatively large size, motility, and cell-wall features, dinoflagellates can avoid predation by most predators, the mesoplankton and planktivorous fish being exceptions (Pollingher 1988). A study of grazing by the marine copepod *Acartia tonsa* on the bloom-forming species *Karenia brevis* suggested that the dinoflagellate was not toxic to the herbivore but did not stimulate a grazing response either. In contrast, *Peridinium* (*Kryptoperidinium*) *foliaceum* did stimulate grazing and fostered higher herbivore reproduction (Breier and Buskey 2007). Freshwater dinoflagellates can become infested with fungi, and the parasitic dinoflagellate *Amoebophrya* and pre-dinoflagellate *Parvilucifera* can attack marine species. Even so, many dinoflagellates and their dinosporin-lined cysts seem to be relatively immune to pathogenic attack by bacteria and viruses (Park et al. 2004). Nutrient-acquisition adaptations also allow dinoflagellates to occupy notoriously nutrient-poor, clear tropical and subtropical open-ocean waters, where they may form deep chlorophyll maxima some 75–150 m below the surface.

Both non-photosynthetic (e.g., *Amphidiniopsis* and *Roscoffia*) and pigmented (e.g., *Spiniferodinium*, *Amphidinium*, and *Prorocentrum*) dinoflagellates occur within sand, and the pigmented forms may form blooms sufficient to color sand flats. *Spiniferodinium* has a nonmotile vegetative phase with a transparent, rigid, spiny, helmet-shaped shell. It reproduces by motile dinospores that regenerate the shell after settling (Horiguchi and Chihara 1987). One can find sand-dwelling dinoflagellates by shoveling into a beach to the point where seawater seeps into the hole, and then collecting the seep water (Horiguchi and Kubo 1997).

Toxin Production

About 60 species of marine dinoflagellates produce toxins that affect many types of animals, including humans (see Chapter 3 for more information about the toxins and their effects). The majority of toxin-producing dinoflagellates are photosynthetic, estuarine, or coastal shallow-water forms that are capable of producing benthic resting cysts, and they tend to form monospecific populations. Some species of *Dinophysis* and *Prorocentrum* are known for producing okadaic acid, and saxitoxins are produced by various species of *Alexandrium*, *Pyrodinium bahamense*, *Lingulodinium polyedrum* (= *Gonyaulax polyedra*), and *Gymnodinium catenatum* (reviewed by Steidinger 1993). *Karenia brevis* (= *Gymnodinium breve*) is the source of both illness-inducing brevetoxins and a potentially useful antidote (Text Box 11.1). Ciguatoxins cause a type of poisoning known as ciguatera. This condition affects people who eat fish that have consumed dinoflagellates such as *Ostreopsis siamensis*, *Prorocentrum lima*, and *Gambierdiscus toxicus*. *Gambierdiscus* lives epiphytically on a variety of tropical red, green, and brown macroalgae (seaweeds) that grow on coral reefs or on sea grasses (Bomber et al. 1989). There is some evidence that the seaweeds may promote *Gambierdiscus* growth. The dinoflagellate cells occur within a mucilaginous matrix on the seaweed/sea grass surface or may be attached by a short thread. When there is little water motion, the *Gambierdiscus* cells may detach and swim around their host, but when water motions increase, they reattach themselves (Nakahara et al. 1996). Molecular methods are increasingly being used to detect toxin-producing dinoflagellates (e.g., Litaker et al. 2003; Rublee et al. 2005) (see Chapter 5).

Symbiotic Associations

A number of photosynthetic dinoflagellates form symbiotic associations with many types of marine protists and invertebrate hosts. Such hosts include giant clams, anemones, and several types of corals. Among these, the stony/hard corals (scleractinians) are particularly important in forming large reefs that support many other life forms. The genus *Symbiodinium* is the major endosymbiont in corals, where its spherical cells live within protective vacuoles inside endodermal cells. Ocean waters also contain free-living *Symbiodinium*.

A Dinoflagellate Toxin and Its Antidote

The dinoflagellate *Karenia brevis* is notorious for producing dramatic, toxic blooms along coastlines in the Gulf of Mexico. Commonly known as "Florida red tides," the blooms are associated with massive fish kills and also poison birds, sea turtles, manatees, and dolphins. Florida red tide algae are also responsible for cases of neurotoxic shellfish poisoning when people consume contaminated shellfish. In addition, the dinoflagellates and their toxins can occur in aerosols, causing respiratory problems in humans by constricting bronchioles in the lungs. These effects result from several types of polyether toxins known as brevitoxins (see Figure 3.18d) that are produced by *K. brevis*. Brevetoxins work by activating voltage-sensitive sodium channels, which are examples of cell membrane ion transporters that open and close in response to change in membrane potential. By binding to a specific site on sodium channel proteins, brevetoxins cause the channels to stay open longer than normal, thereby disrupting the normal ionic composition of cells in neural, respiratory, and other tissues (Baden et al. 2005).

Karenia brevis (Photo: S. Pate May)

During studies of brevetoxin effects on fish, Andrea Bourdelais and associates discovered that *K. brevis* also produces a compound known as brevenal. Although the role of brevenal in dinoflagellate biology remains unclear, in animals brevenal works as a kind of antidote to brevetoxins. Brevenal works in this way by competing with brevetoxin for its binding site on sodium channel proteins, thereby mitigating its toxic effects. Preparations of brevenal made from laboratory cultures of *K. brevis* or water collected from a red tide are able to protect fish from the neurotoxic effects of brevitoxin (Bourdelais et al. 2004). Natural brevenal and a synthetic form of it were able to prevent respiratory symptoms that would otherwise result from brevetoxin exposure in sheep (Abraham et al. 2005). These results suggest that brevenal may be useful in treating wild animals such as manatees that have been poisoned by

brevenal

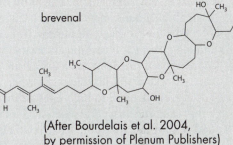

(After Bourdelais et al. 2004, by permission of Plenum Publishers)

brevetoxins. In addition, brevenal might be useful in treating human patients with cystic fibrosis by relieving respiratory symptoms.

From hosts, the endosymbionts gain inorganic nutrients such as CO_2 and phosphorus. The dinoflagellates provide host cells with as much as 90% of their organic food and oxygen, which is used in respiration and also helps to protect hosts from infection. The golden-brown dinoflagellate endosymbionts may also help to camouflage hosts from their predators. The nudibranch mollusk *Phyllodesmium*, for example, is partially colored by its numerous dinoflagellate symbionts (Burghardt and Wägele 2004). Another example is *Maristentor dinoferus*, a giant ciliate that has only recently been discovered on the coral reefs of Guam (Mariana Islands). Each of this ciliate's cells is intensely colored by 500–800 *Symbiodinium* endosymbionts similar to those present in nearby corals, making the protist difficult to see unless it perches on a light-colored background. Endosymbiotic dinoflagellates are also very numerous in coral tissues; a healthy reef is estimated to have more than 10^{10} symbionts m^{-2}.

Molecular techniques are widely used to evaluate *Symbiodinium* diversity in ocean waters and in hosts because few structural characters are available for use in distinguishing species (LaJeunesse 2001; Santos et al. 2004; Coffroth and Santos 2005). Hundreds of unique genotypes of *Symbiodinium* have been found, and they cluster into several major clades (Baker 2003). While some corals transmit endosymbionts of a particular genotype to their progeny, the larvae of other corals must take up *Symbiodinium* from ocean water via phagocytosis. In the case of the coral *Fungia scutaria*, coral larvae seem to use lectins to bind sugars on the surfaces of preferred dinoflagellate species, thereby selecting them for incorporation (Wood-Charlson et al. 2006). In at least some cases, larvae may not be particularly selective, taking up a variety of *Symbiodinium* genotypes, one of which may eventually become dominant (Little et al. 2004). Although most corals seem to select one major dinoflagellate strain and retain it over many generations (Goulet 2006), some corals are able to maintain more than one *Symbiodinium* genotype. Because some forms of *Symbiodinium* appear to be more tolerant of thermal stress than others, corals that are able to harbor more than one symbiont genotype may be better able to cope with environmental changes by shifting populations (Berkelmans and van Oppen 2006).

The major environmental impact upon reef-building corals is bleaching associated with global climate change. Bleaching occurs when corals jettison their symbiotic dinoflagellates or when algal pigments are lost in response to high light or thermal stress. In either case, the host coral loses much of its food source and often dies. In 1998, bleaching caused by rise in sea surface temperature killed 90% of shallow corals of most Indian Ocean reefs (Sheppard 2003). The mechanism by which even small increases in sea surface temperature cause coral bleaching has been elucidated. Heat-damaged dinoflagellate plastids remain able to use photosystem II (PSII) to split water, thereby producing O_2, but energy coupling to photosystem I (PSI) is lost. In this case, PSI electrons reduce O_2 to highly reactive oxygen species (ROS). The ROS may accumulate to levels that result in death or expulsion of endosymbionts. Studies of thylakoid membrane lipids have revealed that more highly saturated lipids confer greater stability to heat and increased resistance to damage by reactive oxygen. Hence, corals bearing *Symbiodinium* rich in such lipids may be able to tolerate higher temperatures without experiencing bleaching (Tchernov et al. 2004). After a bleaching event that does not kill the host, some corals are able to recover endosymbionts by taking them up from the water (Lewis and Coffroth 2004). An EST library was used to evaluate a transcriptome of *Symbiodinium* in

an intact symbiosis versus a clone isolated from the blue morph of the coral *Acropora aspera* under various types of stress conditions, including elevated temperature (Leggat et al. 2007). This study revealed the expression of heat-shock and other stress-related genes that may be important in understanding the process of coral bleaching and recovery.

Corals that do not directly die as a result of bleaching, but instead lose 50%–80% of their symbionts, experience reduced growth rates and increased vulnerability to pathogens (Venn et al. 2006). Several major diseases of corals are caused by bacterial or fungal infections; in addition to climate change, water pollution and overfishing may also foster coral disease (Rosenberg et al. 2007). Coral reefs can also be asphyxiated by giant red tides (Abram et al. 2003). This apparently happened in the Indian Ocean in 1997, as a result of Indonesian wildfires that had the effect of fertilizing nearby ocean waters with iron, causing free-living dinoflagellates to bloom. Bacterial decay of the huge bloom presumably depleted oxygen in the water, causing coral death.

11.5 Dinoflagellate Diversity

Dinoflagellates exhibit high levels of living and fossil biodiversity; more than 550 genera and 4000 species have been described. The dinoflagellate genera featured here were chosen because they are common or exhibit interesting or unusual traits. Because the overall phylogeny of dinoflagellates is not yet well understood (Murray et al. 2005), these genera have not been clustered into classes, orders, or families.

NOCTILUCA (Latin *nox*, night + Latin *lux*, light) has very large (up to 2 mm diameter) cells that are spherical (Figure 11.45). Although the cells are unarmored, empty thecal vesicles are present (TEM is required to visualize this feature). A single, relatively short longitudinal-type flagellum and a thicker feeding **tentacle** emerge from a ventral groove. The cytoplasm includes a large vacuole that likely serves to increase buoyancy. Typical plastids are absent, but photosynthetic endosymbionts or vestigial plastids have been reported. The cells are phagotrophic, often consuming other dinoflagellates. Although vegetative cells have a nucleus much like that of other eukaryotes in that chromosomes are not permanently condensed, nuclei of the gametes are similar to those of other dinoflagellates in containing condensed

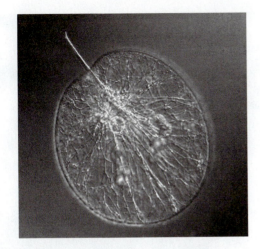

Figure 11.45 The early-diverging dinoflagellate *Noctiluca*, seen with differential-interference contrast optics. The cells possess a tentacle used to transfer food materials to a cytostome. (Photo: C. Taylor)

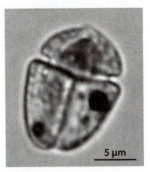

Figure 11.47 *Amphidinium* is a genus with a markedly smaller epicone than hypocone. Shown here is the freshwater species *A. wigrense*, which contains crypto-monad-derived plastids (or kleptoplastids). (Photo: L. W. Wilcox)

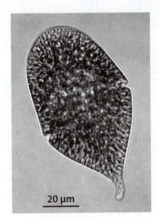

Figure 11.46 *Gymnodinium caudatum* is an unarmored dinoflagellate with a distinct "tail" that is common to acidic freshwater habitats. (Photo: G. J. Wedemayer)

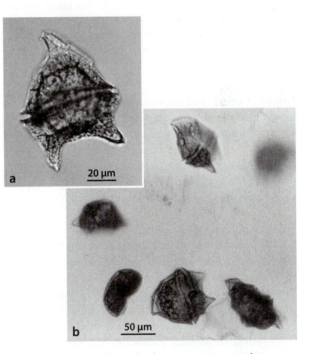

Figure 11.48 *Peridinium limbatum*, a common bog-dwelling dinoflagellate with prominent antapical horns. A single cell is seen in dorsal view in (a). In (b), a number of cells are shown, illustrating the overall shape of the cells, which have a convex dorsal and concave ventral surface. (Photos: L. W. Wilcox)

chromosomes. *Noctiluca scintillans* is widespread in coastal marine waters; it is bioluminescent and sometimes forms red tides. Elbrächter and Qi (1998) reviewed the population dynamics of *Noctiluca*.

GYMNODINIUM (Gr. *gymnos*, naked + Gr. *dineo*, to whirl) has only very thin thecal plates and thus appears to be unarmored. The cingulum occurs at the midpoint of the cell so that the epicone and hypocone are approximately the same size. The cells may be dorsiventrally compressed. Not all species contain chloroplasts. Most species are marine; many of these have variously colored pusules. There

are several important freshwater forms, including *G. fuscum* and the closely related *G. caudatum* (Figure 11.46), which occur in ponds and acid bogs.

AMPHIDINIUM (Gr. *amphi*, on both sides + Gr. *dineo*, to whirl) is unarmored, and the cingulum is positioned close to the cell apex, so that the epicone is small. Most species are photosynthetic, though some contain kleptoplastids. The genus can be found in freshwaters,

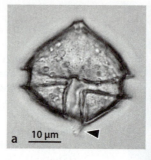

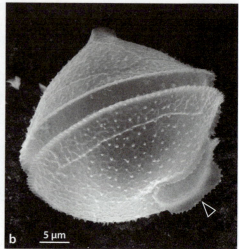

Figure 11.49 *Diplopsalis,* a dinoflagellate genus common in marine habitats but represented in freshwater by *D. acuta,* shown here with LM (a) and SEM (b). A prominent list occurs along the left side of the sulcus (arrowheads). The flagella were lost during the preparation of these specimens for microscopic observation. (SEM: G. J. Wedemayer and L. W. Wilcox)

brackish waters, and oceans. *Amphidinium carterae* is known to form extensive blooms in marine waters, and several species occur abundantly on sand flats. *Amphidinium cryophilum* (see Figure 11.24) occurs beneath the ice of temperate lakes in winter. Freshwater *Amphidinium wigrense* (Figure 11.47) contains cryptomonad-derived plastids (Wilcox and Wedemayer 1985).

PERIDINIUM (Gr. *peridineo,* to whirl around) is heavily armored with thick thecal plates and conspicuous sutures. The cingulum is more-or-less median. Cells tend to be flattened, and some species, such as *P. limbatum* (Figure 11.48), have conspicuous horns. Most species occur in freshwater or brackish waters. *Peridinium cinctum* occurs on a worldwide basis in freshwaters. *Peridinium gatunense* is a large form that blooms February through June each year in Lake Kinneret (Israel), forming 90% of the total phytoplankton biomass during the bloom (Berman-Frank et al. 1995).

DIPLOPSALIS (Gr. *diplos,* double + Gr. *psalis,* scissors) is an armored dinoflagellate lacking plastids that occurs in both marine waters and freshwaters (Figure 11.49). It has a prominent sulcal **list** (a flattened extension of a thecal plate) and apical pore.

GONYAULAX (Gr. *gony,* knee + Gr. *aulax,* furrow) (Figure 11.50) is armored and characterized by a longitudinal furrow that extends from the anterior to the posterior of the cell. The plastid-containing cells may be solitary or remain together following cell division, forming short chains. The genus is widespread in warm and temperate marine waters.

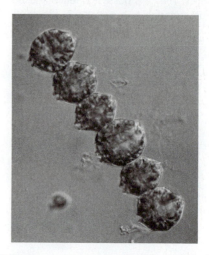

Figure 11.50 *Gonyaulax,* a dinoflagellate of freshwater and marine habitats. Shown here is the chain-forming *G. catenella.* (Photo: C. Taylor)

Spiny benthic cysts are produced. *Gonyaulax hyalina* is implicated in the production of large amounts of carbohydrate mucilage that have sometimes had harmful effects in coastal waters of New Zealand (MacKenzie et al. 2002). *Gonyaulax polyedra* is now referred to as *Lingulodinium polyedrum.*

CERATIUM (Gr. *keration,* small horn) has armored cells with three or four elongate horns, one anterior and two or three posterior. The horns of some species can be very elaborate. The sutures are comparatively narrow, and the surface is ornamented with fine polygonal markings. Most species are

Figure 11.51 *Ceratium*, a dinoflagellate common to both marine waters and freshwaters. The bog-dwelling *C. carolinianum* is shown here. (Photo: L. E. Graham)

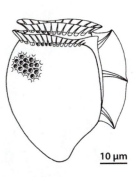

Figure 11.53 The marine *Dinophysis norvegica*. Note the large cingular lists at the top of the cell and the left sulcal list extending longitudinally. (From Steidinger and Tangen 1997)

10 µm

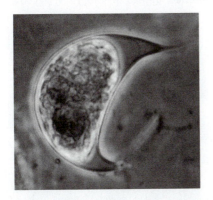

Figure 11.52 The vegetatively nonmotile freshwater dinoflagellate *Cystodinium*. (Photo: L. E. Graham)

10 µm

Figure 11.54 *Ornithocercus*, an impressive marine dinophysoid dinoflagellate with large cingular lists and a pronounced left sulcal list. (SEM: C. Taylor)

photosynthetic, but the presence of food vacuoles suggests that phagotrophy also occurs. Most of the species are marine or occur in brackish waters, but there are several important freshwater forms. *Ceratium hirundinella* can occur abundantly in hard water or eutrophic lakes and may be 100 µm long and wide. *Ceratium carolinianum* (Figure 11.51) is characteristic of soft bog waters.

CYSTODINIUM (Gr. *kystis*, bladder + Gr. *dineo*, to whirl) (Figure 11.52) is a member of a group of dinoflagellates that are nonmotile except for a dinospore phase. This genus is encountered quite often in bog habitats.

DINOPHYSIS (Gr. *dineo*, to whirl + Gr. *physis*, inborn nature). The cingulum of these cells is located very close to the apex, and thecal plates in the vicinity of the cingulum and the sulcus extend outward from the cell, forming distinctive funnel-shaped, sail-

like lists (Figure 11.53). Some species contain plastids (sometimes of cryptomonad origin), and all are marine. Phagotrophy is known for at least some representatives of both colorless and plastid-containing species (Jacobson and Andersen 1994).

ORNITHOCERCUS (Gr. *ornis*, bird + Gr. *kerkos*, tail) (Figures 11.54, 11.55) is similar to and related to *Dinophysis* but is distinguished by an extremely well-developed anterior cingular list. Species of this genus do not have plastids, but some harbor photosynthetic symbionts in a chamber or pouch located in the cingular area. The genus is widespread in warm marine waters.

PROROCENTRUM (Gr. *prora*, prow + Gr. *kentron*, center). This dinoflagellate has a theca composed of two halves (valves), both more-or-less flattened, so that cells appear narrow in edge view at

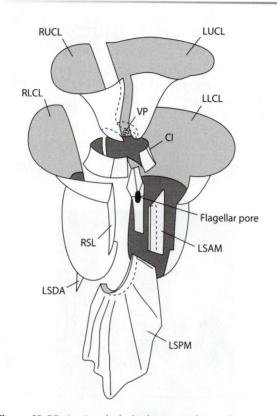

Figure 11.55 An "exploded" diagram of *Ornithocercus*, showing the various components of the elaborate theca. CI = cingulum, LLCL = left lower cingular list, LUCL = left upper cingular list, LSAM = left sulcal list, anterior moiety, LSDA = left sulcal list, posterior moiety, LSPM = left sulcal list, posterior moiety, RLCL = right lower cingular list, RUCL = right upper cingular list, RSL = right sulcal list, and VP = ventral pore. (After Taylor 1971, by permission of the *Journal of Phycology*)

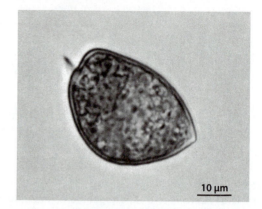

Figure 11.56 Light micrograph of *Prorocentrum micans*, a dinoflagellate having two large thecal plates and apically inserted flagella. (Photo: L. W. Wilcox)

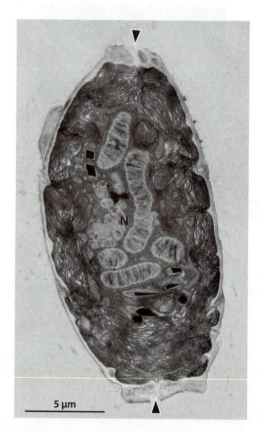

Figure 11.57 A longitudinal section through *Prorocentrum micans*. The suture (arrowheads) between the two large thecal plates (valves) is evident at the top and bottom of the cell. Note also the large chromosomes in the nucleus (N) and dark trichocyst profiles. (TEM: L. W. Wilcox)

the sutures and spherical or oval in valve view (Figure 11.56). Additional small plates and sometimes a spinelike process occur at the point where flagella emerge. One flagellum encircles the other, which extends outward from the cell. Most if not all species have one or more chloroplasts per cell. The genus is of widespread occurrence in fresh, brackish, or marine waters. Some species are so abundant on tidal sand flats that they color the sand brown. *Prorocentrum micans*, a TEM view of which is shown in Figure 11.57, and *P. minimum* (Heil et al. 2005) are known to form harmful blooms.

Photosynthetic Stramenopiles I

Introduction and Diatoms

Centric diatom

In contrast to the leafy green environment experienced by humans, many other organisms occupy a predominantly brown or golden-brown world. A brown canopy of giant kelps shelters sea otters, urchins, many kinds of fish, and other inhabitants of kelp forests. Bushy thickets of brown *Turbinaria* seaweed growing on tropical reefs are favored by the dinoflagellate *Gambierdiscus*. Other microscopic organisms live within films of golden diatoms growing on shoreline rocks of diverse aquatic habitats. Such brown seaweeds, golden-brown diatoms, and similarly pigmented relatives are classified together with several clades of heterotrophic protists into the supergroup **Stramenopila.** This chapter begins with an overview of the photosynthetic stramenopiles and then focuses on the diatoms. Additional groups of photosynthetic stramenopiles are the subjects of Chapters 13 and 14.

(Photo: L. E. Graham)

12.1 Introduction to Photosynthetic Stramenopiles

The exceptionally diverse supergroup Stramenopila is named for distinctive straw-like hairs on one of the two flagella found on swimming cells of most representatives (Figure 12.1). The term *stramenopiles* is derived from the Greek words *stramen*, meaning "straw," and *pila*, referring to "hairs." These hairs, composed of glycoproteins, are described as tripartite because they display three main parts: a basal attachment region, a long tubular shaft, and terminal fibrils (Figure 12.2). Functioning something like

oars, tripartite flagellar hairs help the long flagellum to pull cells through the water (Goldstein 1992). The shorter flagellum is known as the smooth flagellum because it lacks tripartite hairs, though hairs of a different type may be present. A light-sensing flagellar swelling often occurs at the base of the smooth flagellum (see Figure 12.1). Because the flagella of stramenopiles are so distinctive, these protists are also known as heterokonts, which means "different flagella." However, the term *stramenopiles* is somewhat more precise because two dissimilar flagella also characterize dinoflagellates (see Chapter 11). Stramenopiles are amazingly diverse, ranging from tiny flagellates to giant kelps (Figure 12.3). More than a dozen clades of primarily photosynthetic stramenopiles are known (Table 12.1), though relationships among them are not yet well resolved. As a group, photosynthetic stramenopiles are known by a variety of other names: heterokont algae, heterokontophytes, stramenopile algae, stramenochromes, and ochrophytes (Cavalier-Smith and Chao 1996).

Relationships of Photosynthetic Stramenopiles to Other Protists

Although photosynthetic stramenopiles are structurally diverse, they are thought to have originated from a common ancestor having a plastid derived from a secondary red algal endosymbiont (Potter et al. 1997; Vanderauwera and Dewachter 1997; Daugbjerg and Andersen 1997a; Guillou et al. 1999; Karpov et al.

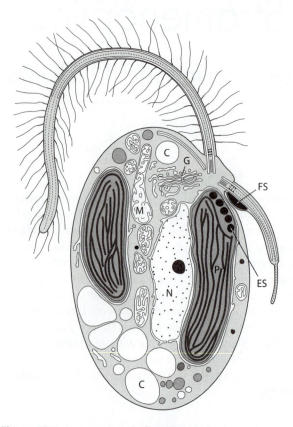

Figure 12.1 A stramenopile flagellate cell. The swimming cells of stramenopiles possess flagella like those of the tribophycean zoospore illustrated here. The longer anterior flagellum bears two rows of stiff hairs, but the shorter posterior flagellum is smooth and often bears a swelling (FS) that is part of the light-sensing system. ES = eyespot, G = Golgi body, M = mitochondrion, N = nucleus, P = plastid, C = chrysolaminarin vacuole.

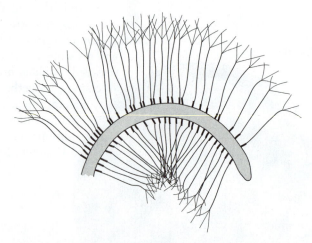

Figure 12.2 The flagella of stramenopiles bear glycoproteinaceous tripartite hairs, having a basal attachment region, a long tubular shaft, and terminal fibrils.

Figure 12.3 Diversity of stramenopiles. Plastidless stramenopiles include small, colorless flagellates such as (a) the bicosoecid *Cafeteria*, the somewhat more complex *Labyrinthula* (b), and the oomycete *Phytophthora* (c). The photosynthetic stramenopiles include (d) radially symmetrical and (e) bilaterally symmetrical diatoms, (f) yellow-green algae (tribophyceans) such as *Vaucheria*, and (g) enormous kelps, including *Macrocystis*, as well as numerous other forms. (a: After Throndsen 1997; b: Drawn from micrograph in Grell 1973; c: After Alexopoulos, C. J., C. W. Mims, and M. Blackwell. 1996. *Introductory Mycology.* John Wiley and Sons, Inc. Reproduced with permission of The McGraw-Hill Companies)

Table 12.1 Major classes of photosynthetic stramenopiles and major pigments

Common name	Formal name	Chlorophylls	Fucoxanthin	Chapter
Diatoms	Bacillariophyceae	a, c_1, c_2	+	12
Bolidophytes	Bolidophyceae	a, c_1, c_2, c_3	+	12
Raphidophytes	Raphidophyceae	a, c_1, c_2	+/−	13
Chrysophytes	Chrysophyceae	a, c_1, c_2	+	13
Synurophytes	Synurophyceae	a, c_1	+	13
Eustigmatophytes	Eustigmatophyceae	a	−	13
Dictyochophytes	Dictyochophyceae	a, c_1, c_2	+	13
Pelagophytes	Pelagophyceae	a, c_1, c_2	+	13
Pinguiophytes	Pinguiophyceae	a, c_1, c_2	+	14
Phaeothamniophytes	Phaeothamniophyceae	a, c_1, c_2	+	14
Chrysomerids	Chrysomerophyceae	a, c_1, c_2	+	14
Xanthophytes/Tribophytes	Xanthophyceae/Tribophyceae	a, c_1, c_2	−	14
Phaeophytes	Phaeophyceae	a, c_1, c_2	+	14

2001; Moriya et al. 2002; Kühn et al. 2004). The monophyly of photosynthetic stramenopiles is also supported by the presence in diverse clades of a distinctive blue-light photoreceptor, **aureochrome**. This type of photoreceptor seems to be absent from heterotrophic stramenopiles, as well as cryptomonads and haptophytes (Takahashi et al. 2007).

Photosynthetic stramenopiles are closely related to several earlier-diverging clades of plastidless, heterotrophic stramenopiles, including opalozoans, bicosoecids, labryrinthulids, thraustochytrids, hyphochrytrids, oomycetes, and others (Figure 12.4). This fact has suggested to some experts that stramenopiles are a fundamentally heterotrophic lineage (Leipe et al. 1996). An alternative "chromalveolate" hypothesis posits that stramenopiles arose from a photosynthetic ancestor, and that the many early-diverging heterotrophic clades subsequently lost plastids (see Chapter 7). Although the presence of many genes associated with photosynthesis in the genomes of two species of the heterotrophic oomycete *Phytophthora* has been interpreted as evidence that these heterotrophic plant parasites once possessed plastids but lost them (Tyler et al. 2006),

massive horizontal gene transfer from infected plant cells is an alternate explanation. Stramenopiles' close relationship to the protist supergroup Rhizaria argues against the chromalveolate hypothesis (Burki et al. 2007). Rhizarians are mostly plastidless heterotrophs, the exceptions being chlorarachniophytes, which have secondary green plastids (see Chapter 7). Stramenopiles are also closely related to alveolates, another clade containing large numbers of both plastidless and plastid-bearing species (see Chapter 11).

Cell Biology of Photosynthetic Stramenopiles

Cells of photosynthetic stramenopiles uniformly have mitochondria with tubular cristae. Another common feature is the carbohydrate storage product known as **chrysolaminarin (chrysolaminaran)**, **laminarin (laminaran)**, or leucosin. Chrysolaminarin consists of 20–60 glucose units with β-1,3 linkages and some β-1,6-linked branches. It is dissolved within vacuoles often located in the posterior portion of the cell (see Figure 12.1). Cytoplasmic lipid droplets are also common. Distinctive features of plastids and the flagellar

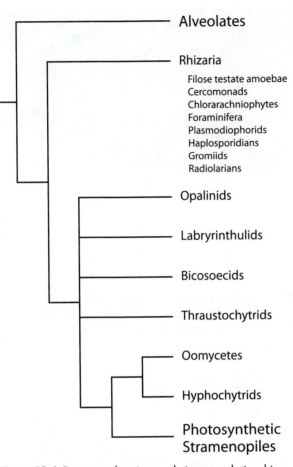

Figure 12.4 Diagram showing evolutionary relationships among stramenopiles.

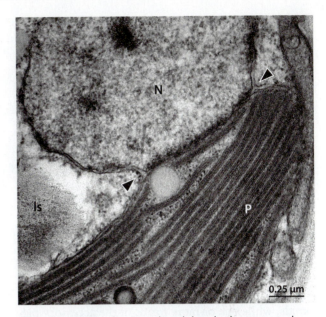

Figure 12.5 PER. The periplastidal endoplasmic reticulum (PER), characteristic of photosynthetic stramenopiles, links the plastid (P) to the nucleus (N) in some taxa (arrowhead), such as the eustigmatophycean shown here. ls = lamellate storage. (From Frost et al. 1997 © Blackwell Science, Inc.)

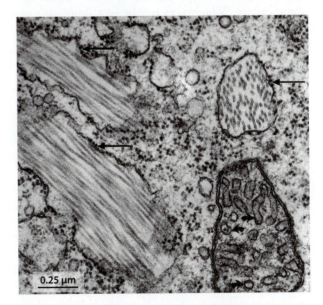

Figure 12.6 Flagellar hairs of photosynthetic stramenopiles are produced within the periplasmic endoplasmic reticulum (arrows) or the nuclear envelope. In this transmission electron micrograph of the raphidophycean *Vacuolaria*, flagellar hairs are seen in both cross-section and longitudinal section. Curved arrows point to the tubular cristae in the mitochondria. (From Heywood 1983)

apparatus are also common to many photosynthetic stramenopiles.

Plastids

Cells of photosynthetic stramenopiles possess one or more plastids, characteristically located within the membranes of the **periplastidal endoplasmic reticulum (PER)** (also known as the chloroplast endoplasmic reticulum, CER). As a result, plastids of photosynthetic stramenopiles have four enclosing membranes, across which cytoplasmically synthesized proteins destined for plastids must be transported. The PER is often, but not always, connected to the nuclear envelope (Figure 12.5). The tripartite flagellar hairs so characteristic of photosynthetic stramenopiles are produced within the PER or adjacent nuclear envelope (Figure 12.6). Thylakoids typically occur in stacks of three. In most photosynthetic

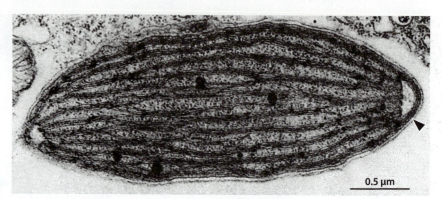

Figure 12.7 Stramenopile plastid. The plastids of many groups of photosynthetic stramenopiles are characterized by a girdle lamella (arrowhead) that lines the plastid periphery, just beneath the envelope, as in the raphidophycean *Vacuolaria*. (From Heywood 1983)

stramenopiles, plastids display a **girdle lamella**, a thylakoid that runs around the periphery of the plastid just beneath the innermost plastid membrane (Figure 12.7). Chlorophyll *a* occurs bound to thylakoids, while one or more forms of the accessory pigment chlorophyll *c* occur in the stroma.

The major yellow accessory pigment in photosynthetic stramenopiles is β-carotene. Large amounts of the xanthophyll **fucoxanthin** (Figure 12.8) confer golden-brown or brown pigmentation to diatoms, chrysophyceans, brown algae, and some others. **Vaucheriaxanthin** is a dominant accessory pigment in raphidophyceans, eustigmatophyceans, and tribophyceans (= xanthophyceans), which tend to have yellow-green or yellow-brown pigmentation. The pigment combinations present in the light-harvesting complexes of photosynthetic stramenopiles extend the ability of these algae to collect light beyond what is possible with chlorophyll *a* alone. Xanthophylls also protect the photosystem from the deleterious effects of high-intensity light. Some lineages of photosynthetic stramenopiles include species whose cells contain **leucoplasts**, plastids that have lost photosynthetic lamellae and pigments and are thus incapable of photosynthesis. Such colorless plastids presumably continue to function in amino acid, fatty acid, or heme synthesis, as is the case for other non-photosynthetic plastids (e.g., Mazumdar et al. 2006).

Flagellar basal bodies and cytoskeleton

The structure of the flagellar apparatus—consisting of basal bodies and associated cytoskeletal features—varies among photosynthetic stramenopiles. Even so, there are common features, illustrated by the flagellar apparatus of the mixotrophic genus *Ochromonas* (Figure 12.9). In this genus, the long anterior

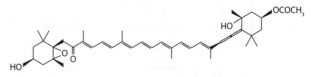

Figure 12.8 Chemical structure of fucoxanthin. (After Falkowski, Paul G., *Aquatic Photosynthesis*, Second Edition. Reprinted by permission of Princeton University Press)

flagellum and the posteriorly directed short flagellum diverge perpendicularly from each other. Within the cell, two fibrous bands connect the two flagellar basal bodies, and a contractible rhizoplast connects basal bodies to the nuclear envelope. Each basal body is associated with two microtubular roots. The four microtubular roots are designated R1, R2, R3, and R4. Of these, R1 and R2 are associated with the anterior flagellar basal body, and R1 may organize the microtubular cytoskeleton, since numerous microtubules emerge from it. R3 and R4 are associated with the posterior flagellar basal body. R3 is a looplike root that is associated with the ability to capture and ingest particulate prey. Particles are captured with a "feeding cup" that is formed when the particles contact the *Ochromonas* cell membrane. Prey particles are engulfed into a food vacuole that forms by the sliding of a particular R3 microtubule. Though all or most of these four roots occur in flagellate cells of other photosynthetic stramenopiles, differences in their structure and organization affect aspects of cellular behavior. For example, changes in the structure of R3 are correlated with loss of the ability to ingest particulate prey.

This overview of features common to photosynthetic stramenopiles provides a basis for a focus on individual clades. We begin with diatoms because they are familiar photosynthetic stramenopiles and are of great ecological importance.

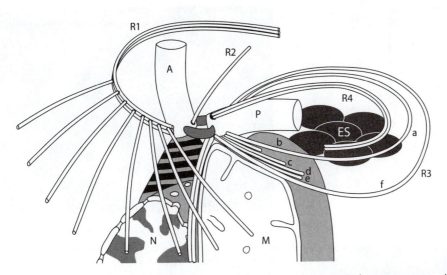

Figure 12.9 Diagram of the flagellar root system of the chrysophycean *Ochromonas*, regarded as the primitive type for photosynthetic stramenopiles. Flagellar apparatuses that have a sliding, looplike root (R3) confer the ability to ingest particulate prey; such roots are typical for mixotrophic photosynthetic stramenopiles. a–f, microtubules of R3 root, A = anterior flagellar basal body, ES = eyespot, M = mitochondrion, N = nucleus, P = posterior flagellar basal body, R1–R4 = flagellar roots. (Redrawn with permission from Inouye 1993, in *Ultrastructure of Microalgae*, edited by T. Berner. © CRC Press, Boca Raton, Florida)

12.2 Diatoms

In terms of evolutionary diversification, the diatoms have been wildly successful. Though occurring only as nonflagellate single cells or cell aggregations, some 285 genera (according to Round et al. 1990) and 10,000–12,000 species are recognized (Norton et al. 1996). New diatom species are continuously being described from fossil deposits as well as modern habitats. Diatoms are also very abundant and are thus important primary producers, responsible for an estimated 20% of global carbon fixation. They dominate the phytoplankton of cold, nutrient-rich waters, such as upwelling areas of the oceans and recently circulated lake waters. Their abundance has resulted in massive sedimentary accumulation of diatom silica walls. Such deposits are now mined for diatomite, a silicate mineral having many industrial applications. Diatom remains are also useful in assessing past environmental conditions and change (Harris et al. 2006). In addition, diatoms have diverse industrial applications, including the emerging use in producing renewable biodiesel (Lebeau and Robert 2003).

Diatoms share several features with other photosynthetic stramenopiles: the accessory pigments fucoxanthin and chlorophyll *c*, plastids occurring within periplastidal endoplasmic reticulum, plastid girdle lamellae, chrysolaminarin, and lipid reserves. In addition, tripartite tubular hairs occur on the single flagellum of sperm cells produced by some diatoms. However, diatoms display unique features that readily distinguish them from other photosynthetic stramenopiles. For example, diatoms have a gametic life cycle, uncommon among protists. Another distinctive feature is that diatom cells are enclosed within a rigid lidded box, the **frustule** or **theca**, which is composed of amorphous opaline silica with organic coatings. Silica is often plentiful in natural waters and is thus an energetically inexpensive source of wall material (Falkowski and Raven 1997). A relatively heavy material, silica may ballast diatom cells, allowing them to take advantage of mineral nutrients available in deeper waters (Raven and Waite 2004). In addition, silica is inert to enzymatic attack, so the frustule may protect diatoms from attack by microbes or digestion within the guts of herbivores (Pickett-Heaps et al. 1990). There is also evidence that the silica frustule acts as a pH buffer that increases the activity of

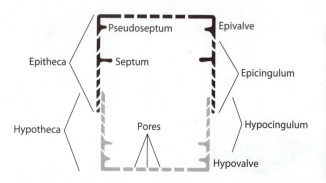

Figure 12.10 Diagrammatic representation of the diatom frustule, showing major features viewed in cross-section. (After Hasle and Syvertsen 1997)

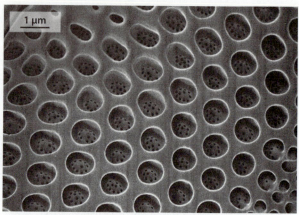

Figure 12.12 High-resolution, low-voltage scanning electron microscopic view of pores in *Mastogloia angulata*. (From Navarro 1993)

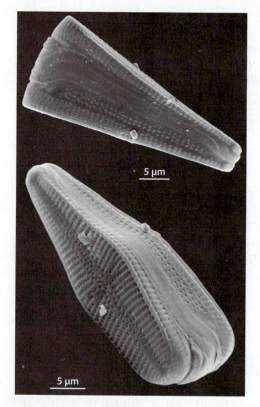

Figure 12.11 Valve and girdle views of the same diatom (a *Gomphonema* species is shown here). Diatoms often have quite different shapes when seen in valve or girdle view. (Micrograph by O. Cholewa)

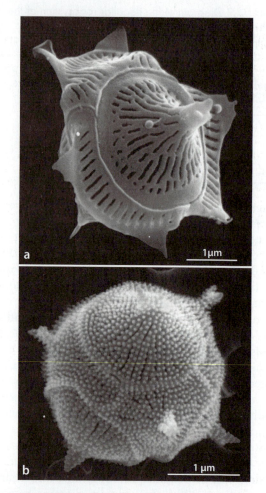

Figure 12.13 Scanning electron microscopic images of parmalean algae, showing their distinctive polygonal silica plates. (From Marchant and McEldowney 1986, © Springer-Verlag)

diatom carbonic anhydrase, thereby aiding CO_2 uptake (Milligan and Morel 2002) (see Chapter 2).

The bottom and the top of the diatom frustule are known as the **hypotheca** and **epitheca**. Because the epitheca overlaps the hypotheca, the latter is slightly smaller in size. The hypotheca and epitheca are each

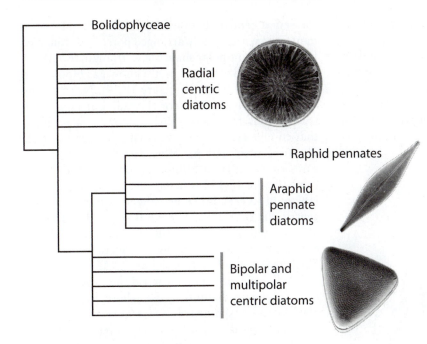

Figure 12.14 Diagram showing relationships of diatoms, with many as-yet-unresolved branches. (SEM of pennate and multipolar diatoms: O. Cholewa; LM of centric diatom: L. W. Wilcox)

composed of a more-or-less flattened valve (hypovalve, epivalve) rimmed by silica bands that make up the hypocingulum and the epicingulum. These overlapping components make up the frustule **cingulum**, also known as a **girdle** (Figure 12.10). Diatoms seen from the top or bottom are in **valve view**, whereas diatoms seen from their sides are in **girdle view**. The same diatom frustule often has quite a different appearance when viewed from the valve view as compared to girdle view (Figure 12.11). The diatom frustule is characteristically ornamented with narrow pores known as aerolae (Figure 12.12), slits, or tubes. Such openings aid buoyancy by lightening the heavy frustule, provide increased contact between the cytoplasm and environment, and often contribute to cell motility.

In the following sections, we first explore what is known about the origin and phylogeny of diatoms. We then consider diatom cell division, frustule development, reproduction, motility, ecology and nutrition, and diversity.

Diatom Evolutionary History

Molecular phylogenetic studies and an excellent fossil record provide insights into diatom evolutionary history (Medlin and Kaczmarska 2004; Alverson et al. 2006; Sims et al. 2006). The silica wall of diatoms is soluble in strong alkaline solutions or hydrofluoric acid but resistant to dissolution and to

biological decay in natural settings. Thus, as diatoms die and settle to the bottoms of water bodies, their silica walls accumulate in the sediments. Such remains can be used to understand relatively recent environmental changes as well as more ancient ecosystems (see Chapter 4). Fossil diatoms are known from deposits extending from recent times back to about 180 million years ago (reviewed by Kooistra et al. 2007). Such fossil data are consistent with molecular clock evidence suggesting that diatoms appeared no earlier than 240 million years ago (Kooistra and Medlin 1996). Chemical fossils, particular isoprenoid alkanes characteristic of diatoms, began to appear about 92 million years ago (Sinninghe-Damsté et al. 2004). The fossil evidence suggests that the first diatoms lived in marine waters but sheds no light on diatoms' immediate ancestors or how they acquired their distinctive silica wall. Molecular phylogenies to date indicate that diatoms are closely related to Bolidophyceae, a group of tiny photosynthetic marine flagellates that lack silica coverings (Guillou et al. 1999; Daugbjerg and Guillou 2001). Photosynthetic marine nonflagellates known as the Parmales have walls composed of polygonal plates of silica (Booth and Marchant 1987) (Figure 12.13) and thus have been suggested as possible diatom relatives (Mann and Marchant 1989). However, this idea has not yet been tested with molecular phylogenetic methods.

Fossil and molecular evidence indicates that the earliest diatoms had frustules whose valves were radially symmetrical, as are those of some modern diatoms (Figure 12.14). The term *radially symmetrical* refers to the arrangement of pores in lines (**striae**) that run from the center of the valve to the rim. Fossil and modern diatoms having radially symmetrical valves of more-or-less circular outline are known as the **radial centric diatoms**. Diatoms with valves ornamented with radially arranged pores and elongate, bipolar, or multipolar shapes (e.g., triangular outline) evolved later. Such diatoms are known as **bipolar**, or **multipolar, centrics** (see Figure 12.14). Neither centric diatoms as a whole nor the radial or polar centric diatoms form monophyletic groups, a problem that complicates diatom systematics (Alverson et al. 2006). Centric diatoms often display tubular structures known as **labiate (lipped) processes** or **rimoportulae** (Figure 12.15). Mucilage may be secreted from labiate processes, allowing slow movement. Certain centric diatoms produce **central strutted processes**, also known as a **fultoportulae** (see Figure 12.15). Strutted processes are involved in secretion of chitin fibrils that aid in flotation (Figure 12.16).

A lineage of bipolar centric diatoms gave rise to the **pennate diatoms**, those whose frustules display bilateral symmetry (see Figure 12.14). Such diatoms typically have elongate valves with a longitudinal line or sternum, from which parallel ribs extend perpendicularly toward the edge. These ribs cause pores to occur in parallel lines, giving valves a feathery ("pinnate") ornamentation (Figure 12.17). The pennate diatoms are thought to form a monophyletic group formally known as class Bacillariophyceae. Some 50–60 million years ago, one clade of pennate diatoms acquired a critical innovation, a new feature that fostered exceptional diversification. This new feature was an elongate slit in one or both valves, known as the **raphe** (see Figures 12.17, 12.18). The raphe is thought to have evolved from

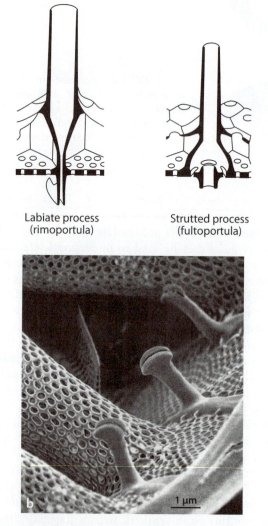

Labiate process
(rimoportula)   Strutted process
                       (fultoportula)

(a)

Figure 12.15 (a) Diagrams of longitudinal cutaway views of two major types of tubular processes that pass through the valves of centric diatoms. Labiate processes (rimoportulae) are so called because the internal portion of the tube is compressed to form liplike structures. Strutted processes (fultoportulae) are not compressed internally, and there are other structural differences. (b) Internal openings of labiate processes of *Isthmia enervis*. (a: After Hasle and Syvertsen 1997; b: From Navarro 1993)

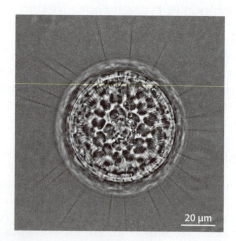

Figure 12.16 Chitin fibrils extending from a living cell of *Stephanodiscus*. (Photo: L. W. Wilcox)

a labiate process. Diatoms having raphes, known as **raphid pennates**, are able to secrete mucilage from this slit, allowing them to move relatively rapidly. This new feature allowed raphid pennate diatoms to occupy new types of habitats. By contrast, pennate diatoms without raphes, known as **araphid pennates**, and centric diatoms do not display such rapid movement. The evolution of the raphe in diatoms has been compared to the evolution of flight in birds, so great was its ecological impact. Despite being the youngest group of diatoms, raphid pennate diatoms are the most diverse (Kooistra et al. 2007).

The evolution of rapid mobility via the raphe is thought to have led to other evolutionary changes in pennate diatoms. For example, they have relatively few, larger plastids (one, two, or four), compared to many smaller plastids in araphid pennate and centric diatoms (see Figure 12.16). Deployment of many plastids throughout the cell may help araphid diatoms acquire light, since they cannot move as quickly as raphid diatoms in response to changes in their light environment. Next, we will examine how diatoms divide and the frustule develops.

Cell Division and Frustule Development

Several features of diatom cell division and frustule development are particularly important to diatom ecology and evolution. First, diatom cell division depends on a sufficient supply of dissolved silica, $Si_nO_{2n-x}(OH)_{2x}$, where n and x are whole numbers

(Darley and Volcani 1969). With the exception of some endosymbiotic forms, diatoms are not able to live without their silica wall and thus cannot reproduce when the silicate concentration falls below critical levels (Martin-Jézéquel et al. 2000). Although silicon is one of the most abundant elements in the earth's crust, spring diatom blooms can deplete levels of dissolved silica in illuminated waters, thereby limiting continued diatom growth. A second important feature of diatom cell division is that it always occurs in the plane of the valve, with the result that diatoms occur only as single cells or chains of cells joined at their valves. Thus, diatom bodies are simple, and there are no truly multicellular species. Finally, because the two cells resulting from a mitotic division each retain one-half of the parental frustule as an epitheca and both must synthesize a new, smaller hypotheca, one of the progeny cells is often smaller than the other. In consequence, over time many clonal diatom populations experience a reduction in mean cell size. A closer view of diatom mitosis, cytokinesis, and cell division is useful in understanding these constraints on diatom ecology and evolution.

Mitosis and cytokinesis

Diatoms that are preparing to divide lack centrioles, which organize the microtubular spindles of many other organisms. Rather, in diatoms an electron-opaque body thought to function as a microtubule organizing center (MTOC) replicates just prior to mitosis, and one copy occupies each spindle pole. The spindle begins its development outside the nucleus and then enters the nucleus via a breach in the nuclear envelope, which eventually disappears. The

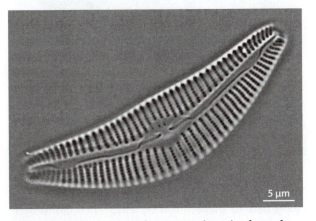

Figure 12.17 Striae (rows of pores) on the valve face of an acid-cleaned specimen of *Cymbella*, viewed with light microscopy and digitally processed. (Photo: L. W. Wilcox)

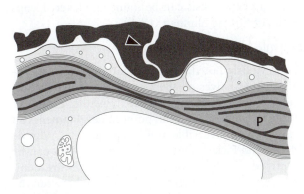

Figure 12.18 Diagrammatic and generalized representation of a cross-sectional view through the raphe region of a raphid diatom. Note the zigzag pattern of the slit (arrowhead).

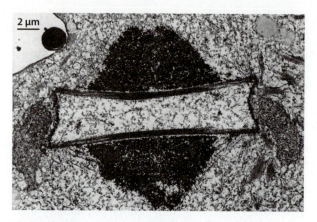

Figure 12.19 The mitotic spindle of *Surirella robusta*. Chromosomes are grouped around the hollow spindle, composed of two interdigitating half spindles. (From Pickett-Heaps et al. 1984. Reproduced from the *Journal of Cell Biology* 1984, Vol. 99:137–143, by copyright permission of The Rockefeller University Press)

early spindle consists mostly of microtubules that run continuously from pole to pole. By prometaphase, however, the spindle consists of two half spindles whose microtubules interdigitate in the region of overlap (Figure 12.19). Early in anaphase, chromosomes move along these microtubules toward the poles. Later in anaphase, the spindle elongates, reducing the amount of overlap between the half spindles. The two half spindles continue to slide past each other, eventually separating completely. By this time, cytoplasmic cleavage has begun by ingrowth of a furrow extending from the cell membrane. Nuclear envelopes then re-form around nuclei, and cytokinesis is completed. Additional details about diatom mitosis and cytokinesis can be found in Pickett-Heaps et al. (1990) and Sluiman (1993).

Frustule development

As previously noted, dividing diatoms retain parental frustule halves as their epithecae (frustule lids) and produce new hypothecae (frustule bottoms). While various aspects of hypotheca development can be intriguingly variable among species (e.g., Van de Meene and Pickett-Heaps 2004), generalities can be identified. When mitosis and cytokinesis are complete, the MTOCs migrate to a position between the two new nuclei (Figure 12.20). A flattened **silica deposition vesicle (SDV)** develops by fusion of Golgi vesicles and lies close to the cell membrane along the cleavage furrow (Drum and Pankratz 1964). The

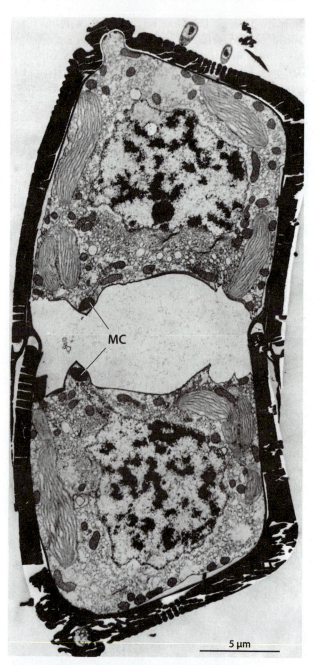

Figure 12.20 A recently divided *Hantzschia amphioxys*, showing expansion of the silica deposition vesicle across the cleavage plane. Microtubule centers (MC) occur in each daughter cell. (From Pickett-Heaps and Kowalski 1981)

MTOC and associated microtubules and microfilaments help to shape the SDV, which then molds new frustule parts and influences the formation of pores, raphe slits, and other features (Pickett-Heaps et al. 1990). Peptides known as silaffins and associated

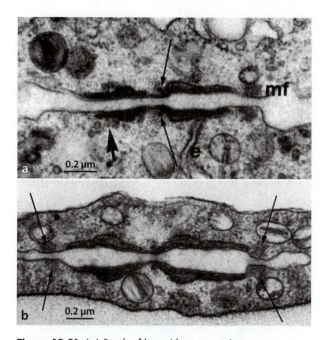

Figure 12.21 (a) Raphe fibers (thin arrows) cut transversely, associated with the developing raphe of *Navicula* (*Craticula*) *cuspidata*. Thick arrow = vesicles associated with one of the two ribs forming the raphe, e = endoplasmic reticulum, mf = microfilaments. (b) Later stage, where ribs have expanded. Arrows=microfilaments. (From Edgar and Pickett-Heaps 1984b, by permission of the *Journal of Phycology*)

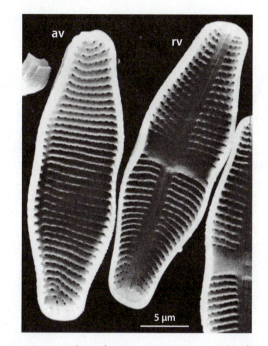

Figure 12.22 *Achnanthes coarctata*, an example of a diatom having one araphid valve (av) and one valve with a raphe (rv). (From Boyle et al. 1984, by permission of the *Journal of Phycology*)

polyamines that are embedded within frustules are thought to be largely responsible for initiating silica formation and influencing its nanostructure (Kröger et al. 2002, Sumper 2002, Sumper et al. 2005). Frustule density appears to be influenced by external salinity, with higher-density silica forming at lower ionic strengths (Vrieling et al. 2007). The recent use of microarray technology to explore diatom frustule development has revealed specific genes responsible for silicification, information that may prove useful in industrial applications (Text Box 12.1).

Valves are generally produced before girdle bands. Silicification in centric diatoms often starts very close to the point of emergence of a labiate process, whose development at an earlier point in time is associated with a specialized organelle. In many pennate diatoms the valve begins development from an axial area that runs the length of the valve, from the center outward. In raphid pennate diatoms, silicification starts along the site of the raphe, whose development is associated with a structure called a

raphe fiber (Figure 12.21). In some diatoms silicification occurs from the margins inward, but it occurs uniformly across the valve in others. When the valve is mature, fusion of Golgi vesicles generates a separate SDV in which girdle bands are produced. Once the frustule is mature, diatoms grow by sliding the epitheca and hypotheca apart at the same time as new cingular bands are produced. Once produced, valves of diatoms do not increase in circumference; cell expansion can occur only at the region of their overlap, the cingulum.

Usually the two new valves of sibling cells are identical. However, in a number of diatoms one new valve develops a raphe while the other produces a **pseudoraphe** (false raphe) in which the fissure is filled in with silica (Figure 12.22). As a result of an ultrastructural analysis of valve development in *Achnanthes coarctata*, Boyle et al. (1984) were able to explain how such differences occur. Up to a point, development in the sibling cells is alike. After cytokinesis, both daughter nuclei and their associated MTOCs move closer to one side of the cell. Nearby, the two new tubular SDVs each form a central rib running the length of the cells. Then both SDVs generate

How Is Diatom Frustule Development Genetically Controlled?

A distinctive pattern of pores, tubes, and/or slits characterizes the frustules of each of the many thousands of modern diatom species. Such variation raises the issue of how frustule development is genetically controlled. Although silaffin proteins and polyamines are known to be involved in fine-scale silica polymerization, the formation of pores, slits, and tubes by alterations in silica deposition has remained mysterious. Yet such information would not only be valuable in understanding diatom evolution and diversification but could also prove useful to the semiconductor industry by suggesting new ways to produce fine detail on silica computer chips.

As an approach to identifying genes involved in silicification, a team of biologists used microarray technology to analyze the *Thalassiosira pseudonana* transcriptome, all of the genes that are expressed differently in controls versus particular environmental perturbations, such as silica limitation (Mock et al. 2008). The use of this diatom species was facilitated by a previous genome-sequencing project that had identified new genes involved in silicic acid uptake and many other cellular processes (Armbrust et al. 2004). Genomic sequences related to signaling and regulation have also been identified for this species (Montsant et al. 2007). In addition, complete sequences were available for the plastid genomes of *T. pseudonana* (and the pennate diatom *Phaeodactylum tricornutum*, for which the nuclear genome sequence is available at the Joint Genome Institute web site, www.jgi.doe.gov) (Oudot-Le Secq et al. 2007).

The analysis revealed that expression of at least 75 genes increased only under low silicon levels but not other conditions tested (low N, low Fe, low temperature, or alkaline pH). For example, only silicon limitation induced the transcription of two genes encoding silicon transporters. This response would help natural diatoms to harvest more silicon from low-silicon water, thereby allowing them to continue to divide. Importantly, expression of a common set of 84 genes increased only when both silicon and iron were limiting. For example, expression of one of the genes encoding a silaffin was increased only by iron limitation. These results indicate that iron plays a greater role in diatom silification than has previously been recognized and may explain why diatoms growing in natural Fe-limited waters have more heavily silicified walls. The next step is to evaluate the roles of each of these genes, as well as their function in diatoms that differ in frustule traits.

another longitudinal rib next to the first. The gap between the two ribs becomes the future raphe fissure. From this point, development of the two new valves differs. In the cell that will form a raphid valve, the SDV, flanked by microtubules, expands laterally in both directions. In the sister cell, the MTOC and its microtubules withdraw from the SDV, and the SDV moves laterally until the double rib is at the corner of the cell. Silica is then secreted into the expanding SDV, and the raphe fissure fills in (Figure 12.23). This explains the occurrence of a pseudoraphe at the edge of one valve of this diatom.

The parental epitheca can be recycled several times but seems to have a determinate life span. In one form of *Stephanodiscus*, the parental valve was observed to last only six to eight generations, ultimately disappearing from the population (Jewson 1992). As noted earlier, each new valve that is formed is always the smaller hypotheca. Because the parental hypotheca serves as the epitheca of one of the two new cells resulting from cell division, such a cell will often be slightly smaller than its sibling cell. Consequently, the mean cell size in a clonal population will decline over time. This size reduction resulting from continued cell division in diatoms is known as the McDonald/Pfitzer rule after its two discoverers, who independently reported this phenomenon in 1869 (Figure 12.24). In exception to the general rule, some diatoms do not exhibit size reduction; they may have more elastic cingula that are able to expand more greatly after cell division than most diatoms. In the more general case, when diatoms reach a critical

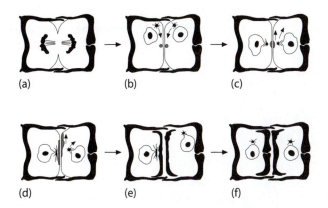

Figure 12.23 Diagram of the process by which a raphe and pseudoraphe develop in only one of the two new valves produced after a cell division in *Achnanthes coarctata*. (After Boyle et al. 1984, by permission of the *Journal of Phycology*)

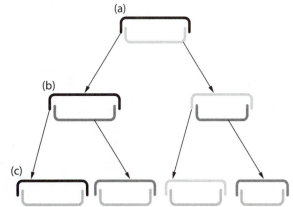

Figure 12.24 Diagrammatic view of three generations of a diatom, illustrating reduction in the mean cell size as the population increases through time.

size, typically about one-third their maximal size, they may undergo sexual reproduction. This process, which is also influenced by environmental conditions and described in more detail below, restores maximal size for the species. In nature, diatoms that are smaller than one-third the maximal size for the species and are unable to restore cell size by means of sexual reproduction eventually die (Mizuno and Okuda 1985). However, such small-sized diatoms can persist in laboratory cultures, and it has been pointed out that many physiological and ecological experiments have probably been done with atypically small diatoms (Edlund and Stoermer 1997).

Sexual Reproduction

Vegetative cells of diatoms are diploid, and gamete production involves meiosis. Thus, like animals, diatoms have what is known as a gametic life cycle (Figure 12.25). The diatom life cycle was first described by Klebahn in 1896, working with *Rhopalodia*, and later by Karsten (1912), in *Surirella*. As noted earlier, sexual reproduction is important in regenerating maximal size in diatom populations, and, as in other eukaryotes, sexual reproduction generates genetic variability.

Environmental cues that may induce diatom sexual reproduction vary among species, but changes in temperature, day length, and nutrient availability, as well as presence of a mate are common triggers. Such factors control the timing of diatom sexual reproduction, of which four types have been defined

(Edlund and Stoermer 1997): (1) Synchronous sexuality occurs under favorable conditions for growth. In this case, large numbers of cells engage in sexual reproduction within a relatively short period of time. Synchronous sexuality may occur among freshwater epiphytic or epilithic diatoms such as *Cymbella* and *Gomphonema* in the springtime, when increasing irradiance and temperature serve as cues. Fifty percent of the competent cells in populations of the freshwater planktonic diatom *Stephanodiscus* may undergo sexual reproduction when nitrate levels in the water rise above 10 µM. A marine example is the mass occurrence of sexuality in Antarctic oceanic blooms of *Corethron criophilum* (Crawford 1995). (2) Synchronous sexuality may also occur under conditions that do not favor vegetative growth, such as nitrogen or light limitation. (3) Asynchronous sexuality, in which relatively few members of a population undergo sexual reproduction over an extended time period, may occur in conditions favorable for growth. (4) In other cases, asynchronous sexual reproduction occurs in poor growth conditions, a strategy that may prove advantageous in unpredictable or fluctuating environments.

Centric diatoms are oogamous, producing one or two egg cells per parental cell. During this process, two or three of the four meiotic products die. Other diatoms of the same species produce from 4 to 128 flagellate sperm per parent cell as a result of mitotic divisions following meiosis. Production of a large number of sperm is regarded as a strategy for increasing the efficiency of fertilization in open water

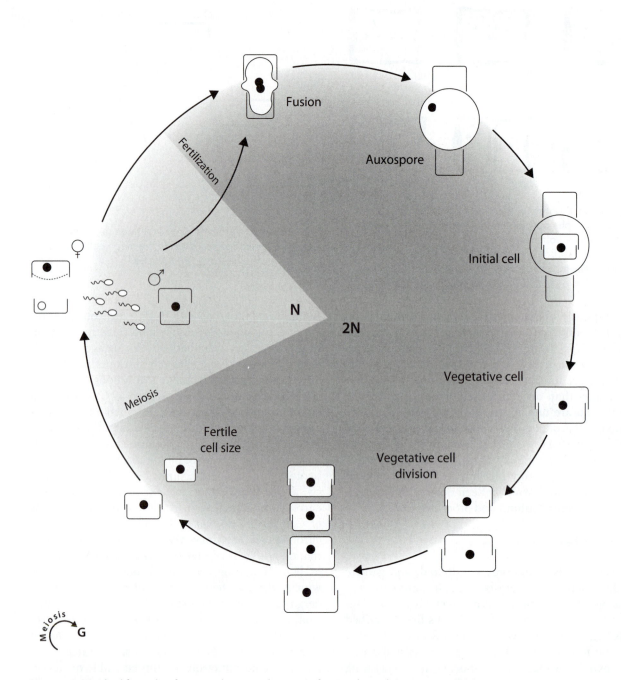

Figure 12.25 The life cycle of a typical centric diatom. (After Hasle and Syvertsen 1997)

environments, where many centric species live. Centric diatom sperm have only a single flagellum that is unusual in having no central singlet microtubules, such as occur in the vast majority of other flagella/cilia. Diatom flagellar basal bodies are also unusual in consisting of microtubule doublets rather than triplets, as in other basal bodies and centrioles. When gametes are mature, the valves surrounding sperm separate, allowing the flagellate cells to escape. The silica wall surrounding

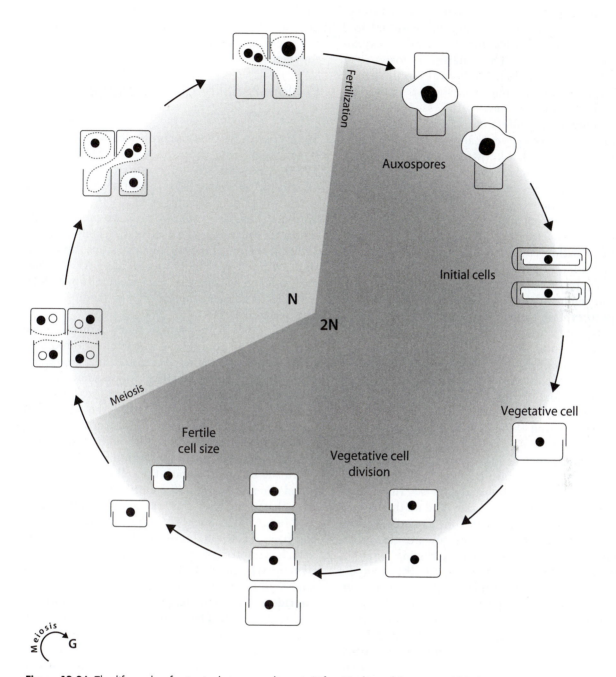

Figure 12.26 The life cycle of a typical pennate diatom. (After Hasle and Syvertsen 1997)

eggs also opens, exposing the protoplast. Gamete fusion results in formation of a zygote known as an **auxospore** (Figure 12.25). It has been pointed out that auxospores aren't really typical spores; rather, their function is to shape the new, relatively large frustules that develop from them (Medlin and Kaczmarska 2004; Sims et al. 2006). Young auxospores generally possess a first (primary) organic wall ornamented with silica scales (thought to reflect features of an ancient diatom ancestor).

Auxospores of radial centric diatoms go on to develop a secondary wall of organic material and scales, which is flexible enough to allow expansion in all directions. Thus, mature auxospores of radial centric diatoms are spherical or ellipsoidal, and new radially symmetrical silica frustules develop inside the confines of this secondary wall (see Figure 12.25). Bipolar and multipolar centric diatoms have auxospores that develop delicate siliceous bands, together known as a **properizonium**, on part of the primary wall. Such bands restrict expansion in some directions, thereby controlling the shapes of new bipolar or multipolar frustules.

Sexual reproduction in pennate diatoms (see Figure 12.26) often begins with pairing of parental cells within enclosing mucilage (Mann 1993). Such pairing of gamete-producing cells (gametangia) is known as gametangiogamy. The ability of raphid pennate diatoms to move toward each other helps to explain gametangiogamy, but this process has also been observed to occur in some araphid pennate diatoms (Sims et al. 2006). An unequal cytokinesis follows the first meiotic division, and the smaller of the two cells dies. The frustules then gape open, and the protoplasts begin to emerge prior to completion of the second meiotic division. Within each of the two large protoplasts, one haploid nucleus resulting from the second meiotic division dies, and the other survives. Post-meiotic protoplasts thus contain a single haploid nucleus. Pennate diatoms are almost always isogamous: The two gametes are similar in size, and neither is flagellate. Even so, in some cases, one gamete—known as the active gamete—moves toward the other, passive, gamete. The toxin-producing *Pseudonitzschia multiseries* is one example (Kaczmarska et al. 2000), and the attached *Licmophora communis* is another (Chepurnov and Mann 2003). In pennate diatoms, gamete fusion results in a young auxospore similar to that of centric diatoms. However, in pennate diatoms, the primary wall often ruptures during the development of a bilateral, tubular **perizonium** made of silica bands. These bands constrain the development of the new silica frustule into an elongate shape. Caps of primary auxospore wall are thought to control the width of developing pennate frustules, which are generally narrow (Sims et al. 2006). The development of two frustule components, new hypotheca and new epitheca, occurs as a result of two successive mitotic divisions without an intervening cytokinesis. The

two extra nuclei produced during this process degenerate. In many cases, mature diatom cells that are released from auxospores are the maximal size for their species.

Diatom Motility and Mucilage Secretion

Some centric diatoms can move slowly by secretion of mucilage from the rimoportula (Medlin et al. 1986). An example is *Odontella*, which undergoes "shuffling" movements as a result of such mucilage extrusion (Pickett-Heaps et al. 1986). However, diatoms that have raphes can move more rapidly along the surfaces of substrates, usually with a rather jerky motion and frequent stops, at speeds of $0.2–25$ µm s^{-1} (Häder and Hoiczyk 1992). Diatom motility is regarded as a means by which to avoid local nutrient limitation and shading (Cohn and Weitzell 1996).

A number of hypothetical mechanisms have been proposed to explain motility of raphid diatoms. Edgar and Pickett-Heaps (1984a) suggested that vesicles having fibrillar polysaccharide contents are liberated from the cell by exocytosis at the raphe fissure, at which point the fibrils become hydrated, transforming into rods that project toward the substrate. Although it was first thought that intracellular bundles of actin microfibrils located beneath the cell membrane provided the motive force for motility, a more recent hypothesis suggests that the actin bundles actually function in anchoring the cell membrane to the frustule. Raphe-associated microtubules and attached motor proteins such as kinesin and/or dynein are currently regarded as a more likely source of the motive force that drags the polysaccharide rods along, resulting in diatom motility (Schmid 1997). The latter hypothesis more adequately explains the ability of diatoms to stop and reverse the direction of their motion, as well as the fact that motility ceases during mitosis, when microtubule proteins are recruited for spindle formation.

In the case of some raphid pennate diatoms, such as *Achnanthes*, mucilage secreted from the raphe can also form a stalk that is used to form a more stable attachment of cells to substrates (Figure 12.27) (Wustman et al. 1997, 1998). Application of certain polysaccharide inhibitors to *Achnanthes longipes* results in both loss of motility and loss of ability to

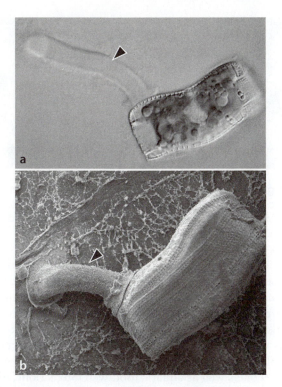

Figure 12.27 Stalk (arrowheads) production by *Achnanthes longipes.* (a) LM view, (b) SEM view. (Images: Y. Wang and M. Gretz)

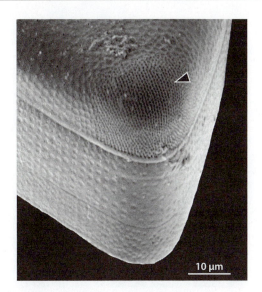

Figure 12.28 Ocellus of *Triceratium*, SEM view. (Micrograph: O. Cholewa)

adhere to surfaces (Wang et al. 1997). Other stalked diatoms generate attachment mucilage from aggregations of polar pores known as ocelli (Figure 12.28) or from tubular processes. Mucilage stalks are thought to confer the ability to overcome shading and possibly also allow greater access to water-column nutrients (Cohn and Weitzell 1996). Diatom mucilages are also recognized for their importance in stabilizing sediments (de Brouwer et al. 2005). Mucilage glues sediment particles together, giving rise to a gel-mud. Such sediment stabilization is thought to provide adaptive advantage to benthic diatoms and is ecologically important in reducing erosion. A review of diatom mucilage production and its relationship to motility is provided by Wetherbee et al. (1998).

Diatom Spores and Resting Cells

Diatom spores and resting cells are modified cells that serve a perennation function—i.e., they allow diatoms to survive periods that are not suitable for growth and then germinate when conditions improve (McQuoid and Hobson 1996). Conditions that

reduce growth may include ice cover, nutrient depletion, or stratification (layering) of the water column. The development of diatom resting cells and spores does not usually involve sexual reproduction. Both resting cells and spores are characteristically rich in storage materials that supply the metabolic needs of germination. For example, the freshwater diatoms *Stephanodiscus*, *Fragilaria*, *Asterionella*, *Tabellaria*, *Diatoma*, and *Aulacoseira* produce resting cells by condensing cytoplasm into a dark brown mass that contains many large lipid droplets and polyphosphate granules (Sicko-Goad et al. 1986). Furnished with such provisions, resting cells of these diatoms can survive in sediments for years and perhaps decades. Some resting cells are known to require a period of dormancy before they will germinate.

Resting cells (Figure 12.29) are morphologically similar to vegetative cells, whereas the frustules of spores (Figure 12.30) become very thick and may assume a rounder shape and exhibit less elaborate ornamentation than vegetative cells of the same species. Hence, resting cells can form under conditions of low dissolved silicate, unlike spores, which require silicate for frustule thickening. Freshwater diatoms and pennate taxa tend to produce resting cells, whereas spore production is more common by coastal marine centric diatoms. Spores are believed to be capable of surviving for decades in benthic sediments. If resuspended to surface waters from the sediments, resting cells and spores can germinate given sufficient

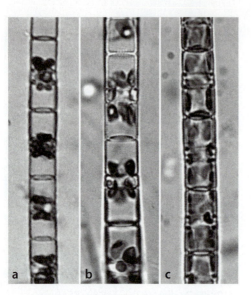

Figure 12.29 Light micrographs of resting and rejuvenating cells of *Aulacoseira granulata*. (a) Resting cells with cytoplasm aggregated in midregion. (b) Rejuvenating resting cells showing elongating plastids. (c) Fully rejuvenated cells having expanded plastids. (Photos: L. Goad and M. Julius)

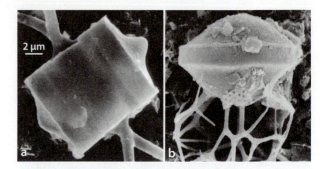

Figure 12.30 Diatom spores from Saanich Inlet, British Columbia, showing thickened frustules. (a) *Chaetoceros didymus*, (b) *C. diadema*. (Micrographs: M. McQuoid)

light and nutrients. Two successive mitotic divisions are necessary to generate the normal form of the epitheca and hypotheca. The spores of marine diatoms often occur in marine snow (see Chapter 2) and are important in transporting organic carbon and silica to the sediments. If not resuspended, diatom spores become part of the fossil record.

Ecology and Nutrition

Diatoms occur in diverse habitats. They occur suspended in the plankton and attached to living and non-living surfaces in shallow waters of oceans, lakes, rivers, and streams—in some cases forming harmful growths. For example, some species of the planktonic diatom *Pseudo-nitzschia* and some other diatoms (Fryxell and Hasle 2003) are notorious for producing the toxin domoic acid, which causes amnesiac shellfish poisoning in humans and affects marine animals (see Chapter 3). The invasive diatom *Didymosphenia geminata* forms extensive nuisance growths on streambeds in many countries, displacing natural communities. Diatoms are well known for growing beneath the ice during the cold season, and rich communities of diatoms are adapted in ways that foster their growth in polar ice formations

(Lizotte et al. 1998, Janech et al. 2006). Some diatoms are epizoic, living on the surfaces of aquatic animals such as crustaceans and whales. Other diatoms are endosymbiotic within other protist cells. Diatoms also occur in moist terrestrial habitats and have been isolated from air samples. Diatoms play important roles in past and present global cycling of silica and carbon as well as aquatic food-web interactions, discussed in more detail in Chapter 21. Here we focus on the ecological impacts of diatom inorganic and organic nutrition.

Inorganic nutrients

The availability of several inorganic nutrients strongly influences relative levels of diatom growth in nature. In addition to silicon, iron and nitrate also commonly limit diatom growth in open ocean waters. Low levels of iron particularly characterize the equatorial Pacific Ocean, the subarctic Pacific Ocean and the Southern Ocean, and upwelling areas off the coast of California. Some diatom species have adapted to low iron conditions by decreasing their cellular need for it (Strzepek and Harrison 2004). For example, *Thalassiosira oceanica* has replaced the iron-containing electron transport substance cytochrome c_6, which is typical of other photosynthetic stramenopiles, with copper-containing plastocyanin, otherwise known to occur only in cyanobacteria, green algae, and plants (Peers and Price 2006). Experimental additions of iron to ocean water can increase the abundance of large chain-forming diatoms such as *Chaetoceros* and the pennate *Pseudo-nitzschia* by six times over controls, as well as increase silicification (Hutchins and Bruland 1998).

Low nitrate concentrations can also limit diatom growth in ocean waters, and diatoms display adaptations that aid in coping with this problem. In the warmer waters of all oceans, multiple species of the diatom *Rhizosolenia* occur in mats that sink below the lighted zone to exploit subsurface nitrate pools, and then nitrate-replete cells rise to the surface, where light levels facilitate photosynthesis. Such buoyancy regulation is postulated to occur via changes in ionic balance. A comparison of chemical composition of sinking and rising mats showed that sinking mats were more nitrogen stressed than those rising to the surface. The maximum ascent rate was 6.4 m hr^{-1}, and a complete migration cycle required 3.6–5.4 days. These diatoms provide a previously unrecognized mechanism for the transport of nitrate from deep to surface waters. It has been estimated that *Rhizosolenia* mats could convey to the surface as much as 40 µmol N m^{-2}day^{-1} (Villareal et al. 1996). Another way in which oceanic diatoms have coped with low N is by forming symbiotic associations with the heterocyte- (heterocyst-) producing, nitrogen-fixing cyanobacteria *Richelia* and *Calothrix*. The marine diatoms *Rhizosolenia*, *Hemiaulus*, *Bacteriastrum*, and *Chaetoceros* are known to possess such symbionts (Foster and Zehr 2006).

Organic nutrition

Many diatoms occur in environments that are low in irradiance and relatively high in dissolved organic compounds. Such habitats include polar sea-ice communities, periphyton mats in shaded streams, and sediments or sand that can shift, covering the algal inhabitants. Reduced irradiance is known to induce osmotrophy (uptake of organic carbon) in a number of diatoms found in these habitats, and laboratory experiments have demonstrated their ability to grow in the dark when organic compounds are supplied. Many diatoms can grow as rapidly heterotrophically as autotrophically, and some species grow faster. A survey of the heterotrophic capabilities of eight benthic diatoms revealed that all were able to metabolize 94 different organic compounds, suggesting the presence of multiple types of cell membrane transport systems (Tuchman 1996). Many marine benthic pennate diatoms migrate back and forth between the sediment–water interface where light is available and the dark, nutrient-rich sediments; such diatoms rely on heterotrophy (Lewin and Lewin 1960). A few marine pennate diatoms, such as *Nitzschia alba*, are colorless obligate heterotrophs. Although they possess chloroplasts, photosynthetic pigments are not expressed. Such

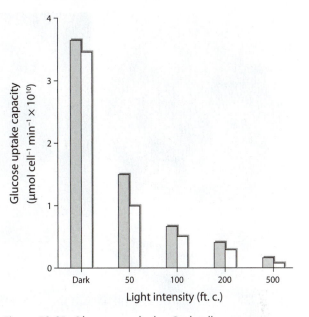

Figure 12.31 Glucose uptake by *Cyclotella cryptica*. Low illumination stimulates glucose uptake, whereas this process is reduced in the light. Pairs of bars represent replicate experiments. (From Hellebust 1971, by permission of the *Journal of Phycology*)

diatoms are analogous to various colorless, but plastid-containing, members of other protist groups. Several centric diatoms are also able to grow in the dark, using dissolved organic compounds (DOC) such as glucose. One example is *Cyclotella meneghiniana*, which grows heterotrophically in the dark if glucose concentrations between 5 mg l^{-1} and 10 g l^{-1} are provided. Light above a certain low irradiance level probably retards glucose uptake by this diatom, and 12–14 hours of darkness are required to induce glucose uptake (Lylis and Trainor 1973). Similarly, the glucose transport system is downregulated in high-light conditions and upregulated in low light and darkness in the centric diatom *Cyclotella cryptica* (Hellebust 1971) (Figure 12.31).

Even when light is not limiting to photosynthesis, the light-independent reactions of photosynthesis (the "dark reactions") are shut off, and carbon fixation ceases when sufficiently high levels of DOC such as glucose are available. This allows photosynthetic cells to use the products of the light reactions (ATP and NADPH) for energy-dependent uptake of organic carbon and for other cellular processes. The uptake of DOC in the light is known as photoorganotrophy or photoheterotrophy, and

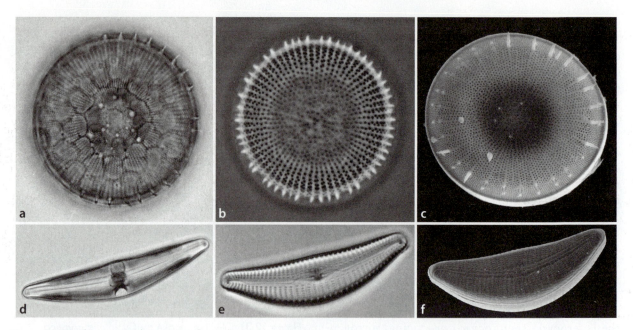

Figure 12.32 The centric diatom *Stephanodiscus* (top panel) and the pennate diatom *Cymbella* (bottom panel), viewed in three ways: (a, d) the living condition, viewed with bright-field microscopy (note chloroplasts and lipid droplets); (b, e) after having been subjected to a cleaning process, viewed with phase or bright-field optics (note absence of plastids and other cytoplasmic constituents); (c, f) after having been chemically cleaned and gold-coated for SEM. Each visualization process yields useful information and has advantages and disadvantages. A disadvantage of extensive processing is loss of easily damaged parts such as peripheral spines of *Stephanodiscus* (c), unless great care is taken in handling frustules. (a, d, e: L. W. Wilcox; b: C. Taylor; c: L. E. Graham; f: O. Cholewa)

it occurs in diatoms (and some other algae) occupying illuminated, DOC-rich environments such as the surfaces of dead or living animals, aquatic plants, and other algae. Organic compounds are readily available as feces, decaying organic matter, or organic exudates, the latter known to be produced by various living algae and aquatic plants. Diatoms are very common inhabitants of such surfaces, but the extent to which they use algal and plant organic exudates has not been determined (Tuchman 1996).

Diatom Collection, Identification, and Diversity

Diatoms can be collected using fine-meshed plankton nets, water-collection devices that can be remotely opened and closed, or pumps. Collections made by the latter two methods must be concentrated by allowing the diatoms to settle or by membrane filtration. They can be preserved in formaldehyde or Lugol's solution

prepared as described by Hasle and Syvertsen (1997). Preservation at pH < 7 helps prevent dissolution during long-term storage. Species of some diatoms can be identified by gross morphology, shape of valves, and processes, but when details of the frustule are needed, organic constituents must be removed because they interfere with visualization of fine features by various microscopic methods (Figure 12.32). The cleaning process separates the valves, which is also helpful in making identifications. Cleaning methods can include treatments with strong acids, solutions of potassium permanganate and acid, UV radiation, hydrogen peroxide, or heating in a muffle furnace. The refractive index of the silica frustule is similar to that of water and glass slides and cover slips, making it difficult to see details. Use of phase-contrast, Nomarski differential interference contrast, or dark-field microscopy can be helpful. Cleaned diatoms can be permanently mounted on glass slides by use of a mounting medium whose refractive index is higher than that of silica. Transmission or scanning electron microscopy is often required to sufficiently visualize frustule details in order to make species determinations. A drop of cleaned diatoms can be placed on a formvar-coated grid and then air-dried for TEM examination. For SEM, cleaned diatoms are dried onto a cover slip and then gold coated. Round et al. (1990) provided an atlas of SEM images of freshwater and marine diatoms. An excellent source of information regarding identification of marine planktonic diatoms can be found in Hasle and Syvertsen (1997). Diatoms of the United States are described by Patrick and Reimer (1966, 1975). Molecular techniques are used to identify toxic diatoms, detect cryptic species, and clarify phylogenetic relationships (Eller et al. 2007; Rynearson and Armbrust 2005; Rynearson et al. 2006; Kaczmarska et al. 2005).

Radial centric diatoms

CYCLOTELLA (Gr. *kyklos*, circle) occurs as single cells, 3–5 µm in diameter (Figure 12.33). Chitin fibrils over 150 µm long may be produced from strutted processes that occur in a ring at the valve margin and sometimes also scattered on the valve surface. Rows of pores are grouped. There are approximately 100 species. *C. meneghiniana* is a commonly encountered form. *Cyclotella* is unusual in its tendency to perennate in the deep chlorophyll layer of freshwater lakes, rather than in benthic sediments. *Cyclotella cryptica* has been used as a model system to elucidate the composition of diatom photosynthetic pigment protein complexes (Brakemann et al. 2006).

STEPHANODISCUS (Gr. *stephanos*, crown + Gr. *diskos*, disk) is a single-celled diatom characterized by a ring of marginal spines (Figures 12.32, 12.34). Chitin fibrils are produced from strutted processes that occur at the margin, beneath some of the spines.

THALASSIOSIRA (Gr. *thalassa*, sea + Gr. *seira*, chain) occurs as cells in chains or embedded in mucilage (Figure 12.35). Cells are linked into chains by chitin threads extending from marginal strutted processes. Cells are about 3–5 µm or larger in diameter and circular in valve view. Valves are ornamented with numerous pores arranged in arcs and marginal spines. Each cell has one labiate process at the rim. There are more than 100 species. The genome of *T. pseudonana* has been sequenced (Armbrust et al. 2004).

Figure 12.33 *Cyclotella*. (SEM: C. Cook)

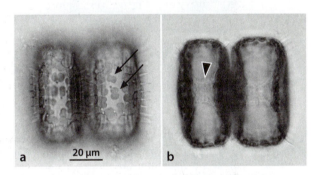

Figure 12.34 Recently divided cells of *Stephanodiscus* sp. in girdle views at two focal planes. Note the marginal spines, multiple plastids (arrows), and nucleus suspended in cytoplasm (arrowhead). (Photos: L. W. Wilcox)

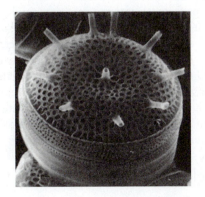

Figure 12.35 *Thalassiosira*. (SEM: G. Fryxell, in Bold and Wynne 1985)

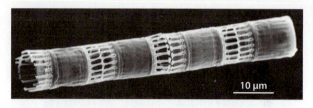

Figure 12.36 *Skeletonema* sp. Note association of cells into a linear array by means of marginal rings of tubular processes. (SEM: O. Cholewa)

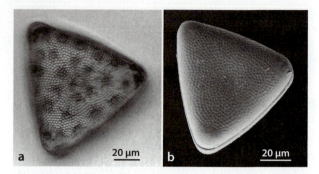

Figure 12.38 *Triceratium* sp. viewed with (a) light and (b) scanning electron microscopy. Note that ocelli at the corners are somewhat more conspicuous in the SEM view. (a: L. W. Wilcox, b: O. Cholewa)

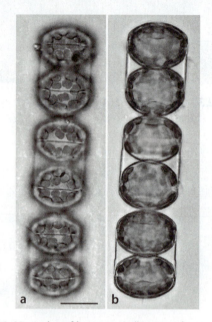

Figure 12.37 A short filament (girdle views focused at two planes) of *Melosira* sp. whose cells are linked by short spines. Scale bar = 20 μm. (Photos: L. W. Wilcox)

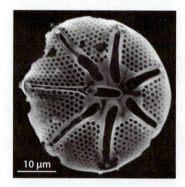

Figure 12.39 *Asteromphalus*. Note the conspicuous and distinctive raised rays; rimoportulae are located at the marginal end of each ray. (SEM: O. Cholewa)

SKELETONEMA (Gr. *skeleton*, skeleton + Gr. *nema*, thread) cells occur in chains formed by strutted tubular processes arranged in a marginal ring (Figure 12.36). One labiate process occurs inside a ring of strutted processes or close to the valve center.

MELOSIRA (Gr. *melon*, apple + Gr. *seira*, chain) occurs as chains of cylindrical cells joined valve to valve (Figure 12.37). Most of the freshwater species formerly bearing this name have been transferred to *Aulacoseira* (Gibson et al. 2003).

Multipolar centric diatoms

TRICERATIUM (Gr. *trias*, in threes + Gr. *keras*, horn) (Figure 12.38) is usually triangular (or sometimes quadrangular) in valve view and is thus a multipolar centric diatom. There are ocelli and sometimes also labiate processes at each edge. More than 400 species have been described, mostly from coastal waters. Not all triangular diatoms are *Triceratium*, however.

ASTEROMPHALUS (Gr. *aster*, star + Gr. *omphalos*, navel) (Figure 12.39) cells have hollow rays that are open to the valve interior as slits and to the exterior through holes at the ends of rays. Labiate processes occur at the marginal ends of rays. The surface of the discoid cells undulates because of the raised rays. There are more than 10 species.

CHAETOCEROS (Gr. *chaite*, long hair + Gr. *keras*, horn) is found as chains of multipolar centric cells attached by intercalary setae extending from

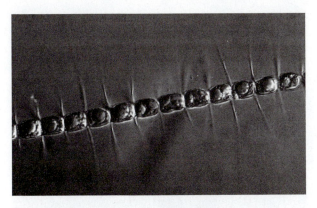

Figure 12.40 *Chaetoceros* sp. Note the distinctive elongate spinelike setae that help link adjacent cells into linear colonies. (Photo: C. Taylor)

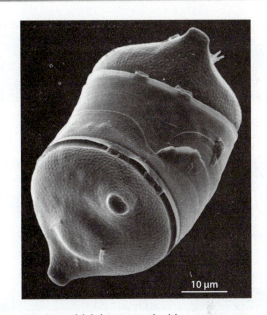

Figure 12.41 *Biddulphia.* Note the blunt processes on the valve faces. (SEM: O. Cholewa)

adjacent cells and touching one another near their point of origin (Figure 12.40). **Setae** are hollow outgrowths of the valve that project beyond the valve margin. Internally, they have a structure different from that of the valve. There are two setae per valve, formed after cell division (Rogerson et al. 1986). Terminal cells of the chain typically have distinctive setae. The cells are rectangular in girdle view and elliptical or nearly circular in valve view. Some species of *Chaetoceros* have cells connected by silica bridges that arise following incomplete cytokinesis (Pickett-Heaps et al. 1990). There are hundreds of described species; Rines and Hargraves (1988) is a good source of additional information.

BIDDULPHIA (named for Susanna Biddulph, a British botanist) (Figure 12.41), shown here as an individual cell, often occurs in zigzag chains attached to substrates such as seaweeds but also occurs in the nearshore phytoplankton. The multipolar, elliptical valves bear prominent blunt processes with clusters of pores that decrease in diameter from the periphery toward the center (pseudocelli). One to several labiate processes occur near the valve center. Cells appear rectangular in girdle view, particularly when observed with the light microscope. Some authorities cite the occasional occurrence of *Biddulphia* in freshwaters (Smith 1950).

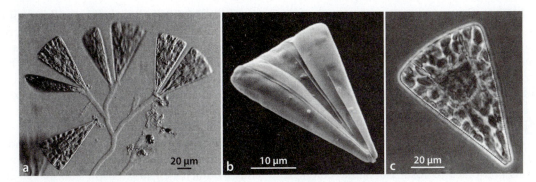

Figure 12.42 (a) A colony of *Licmophora* showing the branched mucilage stalks by which this diatom is attached to substrates. (b) SEM of *Licmophora* sp. in girdle view. (c) Live cell seen with phase-contrast light microscopy. (a, c: L. W. Wilcox; b: O. Cholewa)

Figure 12.43 *Synedra.* (SEM: O. Cholewa)

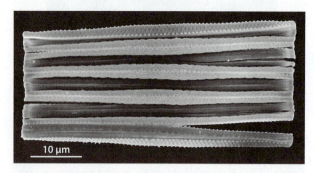

Figure 12.44 *Fragilaria* sp. SEM view of adherent cells in girdle view. (Micrograph: O. Cholewa)

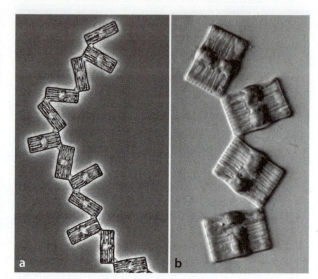

Figure 12.45 (a) Low and (b) higher magnification views of *Tabellaria* sp. colonies, with cells in girdle view. (a: C. Taylor, b: L. W. Wilcox)

Figure 12.46 *Tabellaria* sp. showing attachment of cells by means of mucilage pads (arrowheads). (Photo: L. W. Wilcox)

Araphid pennate diatoms

LICMOPHORA (Gr. *likmos*, winnowing fan + Gr. *phoras*, bearing) is wedge-shaped in girdle view and oar-shaped in valve view (Figure 12.42). The cells are typically attached to marine substrates such as seaweeds, marine plants, shells, stones, or animals by branched mucilage stalks or pads (Figure 12.42a). Mucilage is extruded through a row of slits at the frustule base. A single labiate process occurs at one cell pole or the other. *Licmophora* is cosmopolitan in coastal areas of the world's oceans.

SYNEDRA (Gr. *synedria*, sitting together) is found as single, long, needle-shaped cells (Figure 12.43) or arrays of cells clustered at one pole. Freshwater and marine species occur. A marine form is known to glide, without the use of a raphe or labiate processes, by secretion of mucilage through marginal grooves, in twin trails from either end of the cell (Pickett-Heaps et al. 1990).

FRAGILARIA (L. *fragilis*, easily broken) (Figure 12.44) cells are joined to form ribbonlike colonies, presenting girdle views. There is a single labiate process at one end of each valve. There are two platelike plastids. Common in freshwaters.

TABELLARIA (L. *tabella*, small board) (Figure 12.45) cells are joined valve to valve in short stacks that are further connected by frustule edges to form zigzag patterns. Cell–cell attachment is by mucilage pads (Figure 12.46). There is one labiate process per valve, located near a centrally expanded region. Common in freshwaters.

ASTERIONELLA (Gr. *asterion*, a kind of spider) (Figure 12.47) cells are elongate and joined at the ends by mucilage pads to form star-shaped colonies. A labiate process occurs at both ends of both valves. Short spines occur along the valve edges. Plastids occur as many small plates. Common in freshwaters.

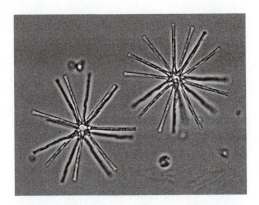

Figure 12.47 *Asterionella* sp. (Photo: L. W. Wilcox)

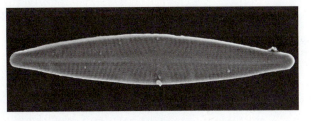

Figure 12.48 *Navicula* sp., as viewed by SEM. (Micrograph: L. Goad and M. Julius)

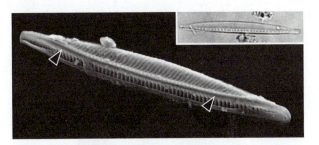

Figure 12.49 A freshwater *Nitzschia* species viewed by SEM, showing eccentric keel with raphe (arrowheads). Inset: LM view. (Micrographs: L. Goad and M. Julius)

Raphid pennate diatoms

NAVICULA (L. *navicula*, small ship) exists as single cells or ribbons of cells (Figure 12.48). The valves are boat shaped, and each bears a raphe. *Navicula* is probably the most species rich of all diatom genera, with nearly 2000 widely accepted species, most of which are bottom-dwelling forms. Some species are common on sea ice. *Navicula thallodes* is a blade-forming diatom that occurs in the Bering Sea. The blades can reach 50 cm in length, the greatest length known for colonial diatoms (Kociolek and Wynne 1988).

NITZSCHIA (named for Christian Ludwig Nitzsch, a German naturalist) cells are linear in both valve and girdle views (Figure 12.49). The valve face is marked by parallel linear striae composed of small pores. Marginal markings reveal the location of the **canal raphe**. Raphes occur on both valves, at the margin, and on the same side of the frustule. The related diatom *Hantzschia* differs in that canal raphes occur on the opposite side of the frustule.

PSEUDO-NITZSCHIA (Gr. *pseudes*, false + *Nitzschia* [see above]) was recognized as separate from *Nitzschia* by Hasle (1994). It is a marine planktonic form that occurs in colonies whose cells are overlapped at their ends. It often is weakly silicified, and the raphe is located at the extreme edge of the valve (Figure 12.50). Some *Pseudo-nitzschia* species produce the toxin domoic acid.

COCCONEIS (Gr. *kokkos*, berry + Gr. *neos*, ship) cells are often attached to the surfaces of submerged plants and other substrata (Figure 12.51) in marine waters or freshwaters. Cells are oval in valve view (Figure 12.52). A raphe occurs only on one valve; this valve is the one appressed to substrates. Auxospores may be observed near substrates on which vegetative cells grow (Figure 12.53).

10 μm

Figure 12.50 *Pseudo-nitzschia*, a toxin-producing marine diatom. (From Hasle 1994, by permission of the *Journal of Phycology*)

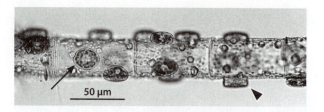

Figure 12.51 Numerous cells of *Cocconeis* attached to the green alga *Oedogonium*, in both valve (arrow) and girdle (arrowhead) views. (Photo: L. W. Wilcox)

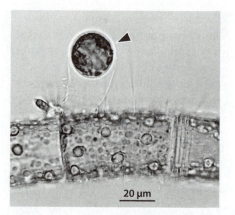

Figure 12.53 An auxospore of *Cocconeis* (arrowhead) that is adherent to a filament of the green alga *Oedogonium* by means of mucilage (the "copulation sheath") extending from the frustules of the fused cells. (Photo: L. W. Wilcox)

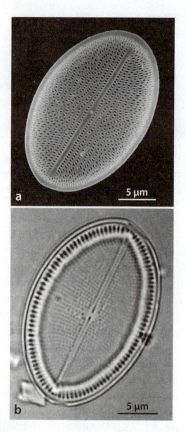

Figure 12.52 *Cocconeis* spp. (a) SEM view of the rapheless valve; (b) LM view of a cleaned raphe-bearing valve. (a: O. Cholewa; b: L. W. Wilcox)

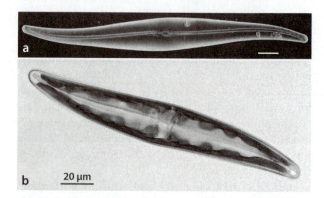

Figure 12.54 *Gyrosigma* spp. (a) Viewed by SEM (scale bar = 10 μm) and (b) in the living condition. Note that the raphe system is S-shaped, as is the frustule. (a: O. Cholewa; b: L. W. Wilcox)

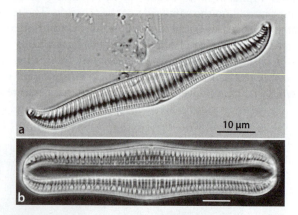

Figure 12.55 Cleaned frustules of *Rhopalodia* sp. in (a) valve and (b) girdle views (scale bar = 10 μm). (Photos: L. W. Wilcox)

GYROSIGMA (Gr. *gyros*, circle + Gr. *sigma*, the letter *S*) (Figure 12.54) cells occur singly or within mucilage tubes. The valves are curved into an S-shape (sigmoid). There is a raphe system, also S-shaped. Two lobed plastids are present at the cell periphery. Most species are brackish in habitat, with some occurring in marine environs and some in freshwaters.

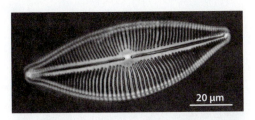

Figure 12.56 A cleaned frustule of *Cymbella* sp., in valve view. Note the bilaterally asymmetric valve and curved raphe of this form. (Photo: L. Wilcox)

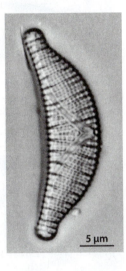

Figure 12.57 *Epithemia* sp. Valve view of a cleaned frustule showing the distinctive V-shaped raphe. (Photo: L. W. Wilcox)

Figure 12.58 SEM images of *Epithemia* sp. showing primarily valve view (top) and girdle view (bottom). (Micrographs: O. Cholewa)

RHOPALODIA (Gr. *rhopalo*, club) (Figure 12.55) valves are linear or curved into a strongly asymmetrical shape. A raphe system is present in an eccentric position, often raised in a keel and bordered by flanges. Cells contain a single plate-like plastid and endosymbiotic cyanobacteria.

CYMBELLA (Gr. *kymbe*, cup) (Figures 12.32, 12.56) often occurs at the ends of branched gelatinous stalks that are attached to submerged substrata of freshwaters. In stalked species, the stalk material is secreted through apical pores. A raphe is present; this is curved in species having bilaterally asymmetric valves. There is a single, H-shaped plastid, consisting of two plates joined by a bridge in which the pyrenoid is located. A form that occurs within mucilaginous tubes is classified as *Encyonema*.

EPITHEMIA (Gr. *epithema*, cover) is another common epiphyte. A distinctive V-shaped raphe occurs on both valves (Figures 12.57, 12.58). There is

a single, large platelike plastid with lobed margins. A few small endosymbiotic cyanobacteria per cell is typical (Round et al. 1990).

PINNULARIA (L. *pinnula*, small feather) cells are often quite large (Figure 12.59). Frustules are elongate-oval in valve view (a) and rectangular in girdle view (b). Valves are ornamented with **costae** that take the form of chambers within walls, with internal openings. There is a linear raphe. There are two platelike plastids or there is a single H-shaped plastid composed of two plates linked by a bridge. Most species are freshwater, but the genus also occurs rarely in marine habitats.

GOMPHONEMA (Gr. *gomphos*, nail + Gr. *nema*, thread) frustules are bottle shaped in valve

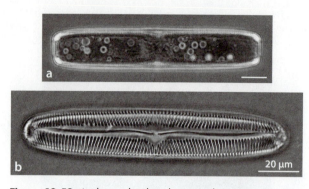

Figure 12.59 A cleaned valve showing distinctive costae (a). A living *Pinnularia* cell in girdle view (b). Scale bar in (a) = 20 μm. (Photos: L. W. Wilcox)

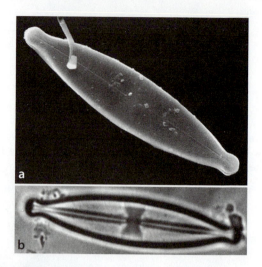

Figure 12.61 *Stauroneis* sp. (a) Valve view in SEM. (b) Valve view in LM, showing conspicuous cross for which the genus is named. (Micrographs: L. Goad and M. Julius)

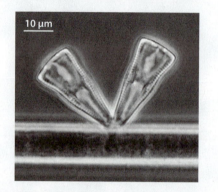

Figure 12.60 *Gomphonema* in the living condition (girdle view), attached to a substrate, as is typical. See also Figure 12.11. (Photo: L. W. Wilcox)

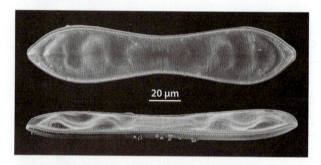

Figure 12.62 SEM images of *Cymatopleura* sp. in valve (top) and girdle (bottom) views. Note the distinctive valve surface undulations. (Micrographs: O. Cholewa)

view and bluntly triangular in girdle view (see Figure 12.11). Cells may be attached directly to substrates, including algae (Figure 12.60), or via gelatinous stalks. There is a single plastid that is indented longitudinally beneath the raphe of both valves and the girdle midline. A single pyrenoid is present.

STAURONEIS (Gr. *stauros*, cross + Gr. *neos*, ship) (Figure 12.61) valves are much like those of *Navicula*. This diatom occurs singly or in colonies. A crosspiece of non-porose silica lies perpendicular to the longitudinal raphe. Together they form a cross-

shaped "stauros." *Stauroneis* occurs in freshwaters, on soil, and in mosses.

CYMATOPLEURA (Gr. *kyma*, wave + Gr. *pleuron*, rib) (Figure 12.62) has peanut-shaped valves with a raphe system that runs around the valve circumference, within a keel. Mature valve faces are undulate (wavy), and the wave pattern is bilaterally symmetrical. There is a single plastid consisting of two plates joined by a narrow isthmus.

Photosynthetic Stramenopiles II

Chrysophyceans, Synurophyceans, Eustigmatophyceans, Raphidophyceans, Pelagophyceans, and Dictyochophyceans

The diversity of photosynthetic stramenopiles is truly amazing. Such diversity includes the diatoms and bolidophyceans (Chapter 12), brown algae and their close relatives (Chapter 14), and the six classes described in this chapter: chrysophyceans, synurophyceans, eustigmatophyceans, raphidophyceans, pelagophyceans, and dictyochophyceans (Figure 13.1). Although a few of the species classified in these six groups live in the periphyton (attached to submerged surfaces), most occur in the phytoplankton as swimming or floating single cells or colonies. Certain species are notorious for forming harmful marine blooms, while others bloom in freshwaters, sometimes affecting the taste of drinking water.

Synura

(Photo: L. W. Wilcox)

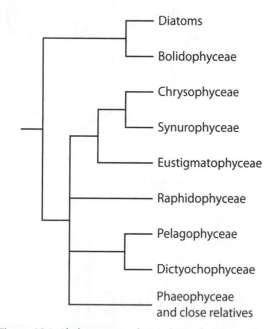

Figure 13.1 Phylogenetic relationships of photosynthetic stramenopiles. (Based on Andersen 2004, by permission of the Botanical Society of America, Inc.)

13.1 Chrysophyceans

Chrysophyceans, also known as chrysophytes, are named for their golden-brown pigmentation; the Greek word for gold is *chrysos*. This golden color arises from large amounts of the accessory pigment fucoxanthin in their plastids, which also contain chlorophylls *a*, c_1, and c_2 (see Table 12.1). Chrysophyte cells may be naked, enclosed within cell walls, or covered by muci- lage or a casing of organic or silica scales.

Other features of chrysophyceans include one or two emergent flagella of the typical stramenopile types, open mitosis in which the nuclear envelope breaks down, isogamous sexual reproduction with zygotic meiosis, and chrysolaminarin (chrysolami- naran) stored in vacuoles (Bouck and Brown 1973) (Figure 13.2). Chrysophyceans, like many other pho- tosynthetic stramenopiles, have a light-sensing pho- toreceptor swelling on the smooth flagellum coupled with a plastid eyespot. Further, the blue-light recep- tor chromoprotein **aureochrome**, which seems to be distinctive to photosynthetic stramenopiles, has been found in at least one chrysophyte (the genus *Ochromonas*) (Takahashi et al. 2007).

Chrysophyceans primarily occur in freshwater habitats, though marine representatives are known.

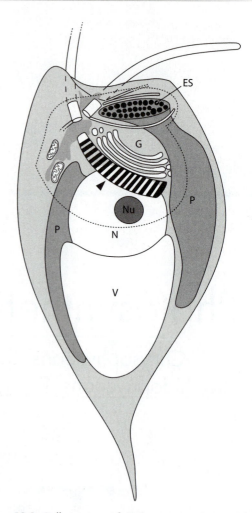

Figure 13.2 Cell structure of *Ochromonas*, showing the relative positions of the posterior vacuole (V), central nucleus (N) with nucleolus (Nu), lateral plastids (P) with apically positioned eyespot (ES), and anterior Golgi (G). The flagellar apparatus includes a robust, banded rhizoplast (arrowhead). (After Bouck and Brown 1973. Reproduced from *The Journal of Cell Biology*, 1973, 56:340–359, by copyright permission of the Rockefeller University Press)

A major adaptation to such environments is a silica- walled resting stage known as a **stomatocyst**, which is characteristic for chrysophyceans and their close relatives the synurophyceans.

Chrysophycean Stomatocysts

The term *stomatocyst* comes from the Greek word *stoma*, meaning mouth, which reflects the occur- rence of an opening that is unique to this type of

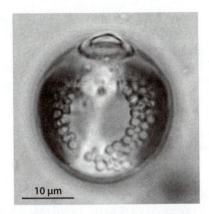

Figure 13.3 Stomatocyst from a sample collected in northern Wisconsin (United States). (Photo: L. W. Wilcox)

Figure 13.4 Stomatocysts of *Dinobryon* collected in February from beneath the ice of a pond in southern Wisconsin (United States). (Photo: L. W. Wilcox)

resting cell (Smol 1995) (Figure 13.3). Stomatocysts are also known as statospores or siliceous cysts.

The walls of stomatocysts are so heavily silicified that they resist silica dissolution processes and thus accumulate in the sediments of freshwater lakes. Fossils thought to represent stomatocysts extend back to the lower Cretaceous (which started about 150 million years ago), suggesting that chrysophyceans have existed for at least this length of time (Harwood and Gersonde 1990). Different species produce structurally distinctive stomatocysts, allowing their use by paleolimnologists to deduce environmental conditions at the time of sedimentary deposition. Paleolimnologists start this process by making calibration data sets, which are statistical correlations between patterns of species present in recent sediments and modern physical, chemical, and biological factors. Such calibration sets are then used to infer past environmental conditions from patterns of stomatocysts in more ancient samples obtained by using sediment-coring devices. For example, an analysis of arctic peat samples more than 7000 years of age revealed 161 morphologically distinct stomatocyst types, including 52 that had previously been unknown (Gilbert et al. 1997). Such data have been used to trace the onset of biotic effects resulting from acid rain and eutrophication (Smol 1995; Zeeb et al. 1996).

Most stomatocysts are spherical, usually 2–30 μm in diameter, with a single pore whose margin is often elevated into a collar. Scanning electron microscopy (SEM) of stomatocysts cleaned with hydrogen peroxide is usually required to visualize species-specific differences in adequate detail (Cronberg 1995). An atlas of chrysophycean cysts that describes more

than 240 types of stomatocysts has been published (Duff et al. 1995).

Stomatocysts can result from asexual or sexual reproduction, and the sexual and asexual statospores of individual species are indistinguishable (Sandgren 1988). Sexual reproduction involves fusion of isogametes that are structurally similar to vegetative cells. In *Dinobryon*, the female strain produces a pheromone that attracts a male flagellate cell to swim to the vicinity of a female cell. The zygote resulting from syngamy loses its flagella and forms a stomatocyst that remains attached to the colony in the place of the cell that served as the female gamete (Sandgren 1981) (Figure 13.4). Whether cysts arise sexually or asexually, the developmental process is the same. The cyst wall is produced inside the cell membrane, within an extensive peripheral vesicle formed by coalescence of Golgi vesicles into a silica deposition vesicle (SDV) similar to that of diatoms. A region at the apex remains unsilicified, forming a pore that becomes plugged with polysaccharide when stomatocyst maturation is complete. Stomatocyst formation does not appear to be correlated with any particular environmental factor but rather with cell density, suggesting production of chemical inducers. Cyst germination occurs continually, with small numbers of cysts germinating at any one time (see review by Sandgren 1988). Germination involves dissolution of the pore plug, which is often followed by mitotic division of the cytoplasmic contents to form two or four cells that develop flagella.

Chrysophycean Ecology

Chrysophyceans typically favor slightly acid, soft (low alkalinity and conductivity) waters of moderate to low productivity. Preference for low-pH waters has been related to their production of acid, but not alkaline, phosphatases. These enzymes are released into the water, where they liberate phosphate from organic compounds. Chrysophycean abundance and species richness increase with lake trophic status up to the point of slightly eutrophic conditions and then decrease dramatically in more-eutrophic waters (Elloranta 1995). Such a decrease may be related to differences in herbivore presence and numbers, competition with other algae, or water chemistry. Water chemistry and herbivory are probably the most important factors explaining the distribution of planktonic chrysophyceans (Sandgren 1988). Chrysophyceans may have strong ecological impacts because many are mixotrophs, able to take up and metabolize dissolved organic compounds and/or particulate food as well as photosynthesize.

Mixotrophy

Mixotrophy allows some chrysophyceans to be nutritional opportunists, able to switch between photoautotrophy, mixotrophy, and heterotrophy, depending upon cellular conditions and environmental circumstances (Sandgren 1988). There is also wide variation in the degree to which chrysophycean species rely upon photoautotrophy, osmotrophy, or phagotrophy. Certain *Ochromonas* species and unpigmented *Spumella* are strict phagotrophs. Other *Ochromonas* species and *Uroglena* are both obligately phagotrophic and photoautotrophic (i.e., they cannot ingest particulate food unless light is also present). *Dinobryon* has plastids but is facultatively phagotrophic, and *Chrysamoeba* and *Ochromonas danica* seem to be facultatively photoautotrophic and facultatively phagotrophic. *Poterioochromonas*, though capable of photoautotrophy, depends on photosynthesis only when the supply of particulate food is low. This species so strongly tends toward phagotrophy that the organism will engage in cannibalism rather than rely

upon photosynthesis (Caron et al. 1990). Light seems to be necessary for phagotrophy in *P. malhamensis* (Zhang and Watanabe 2001). The genera *Catenochrysis*, *Chromulina*, *Chrysococcus*, *Chrysosphaerella*, *Chrysostephanosphaera*, and *Phaeaster* are also plastid-bearing chrysophyceans that are known phagotrophs, consuming bacteria, yeasts, small eukaryotic algae, and nonorganismal particulate foods such as starch grains (reviewed by Holen and Borass 1995). *Paraphysomonas imperforata* is omnivorous, capable of consuming algal prey such as the diatom *Phaeodactylum tricornutum* and the green flagellate *Dunaliella tertiolecta* (Caron 1990). Members of the chrysophycean clade usually should be regarded as fundamentally capable of mixotrophy under at least some conditions.

Bacterivorous chrysophyceans are particularly important in oligotrophic lakes, where they are the primary consumers of prokaryotes. These include four species of *Dinobryon* and two species of *Uroglena*, both common colonial phytoplankters (Bird and Kalff 1986). Each *Dinobryon* cell consumes an average of three bacterial cells every five minutes, removing more bacteria than crustaceans, rotifers, and ciliates combined. The maximal rate of bacterivory for *Ochromonas* has been measured at 182–190 bacteria consumed $cell^{-1}$ hr^{-1} (Holen and Borass 1995). Populations of chrysophyceans such as *Dinobryon* can occupy lake waters as deep as 7 m, forming extensive metalimnetic growths. In the dimly lit metalimnion—the middle layer of a water column that is overlain by less dense, warmer water and underlain by more dense, cooler water—they obtain about 80% of their total carbon from phagotrophic activity (Bird and Kalff 1986). Phagotrophy provides not only fixed carbon but also phosphorus and vitamins. Ingesting bacteria satisfied 77%–100% of the phosphorus need of a *Dinobryon* species from Lake Constance (Germany) (Kamjunke et al. 2007). Most chrysophyceans are auxotrophic, requiring several B vitamins (Holen and Borass 1995). Chrysophyceans also have an unusually high requirement for iron, needed for the synthesis of an essential cytochrome in photosynthesis, and phagotrophy is thought to be useful in harvesting this nutrient as well (Raven 1995). Even so, nutritional flexibility has a price; phototrophic growth rates of mixotrophs can be lower than those of obligate photoautotrophic algae of comparable size, and the metabolic costs of mixotrophy can be higher than those of obligately phagotrophic flagellates.

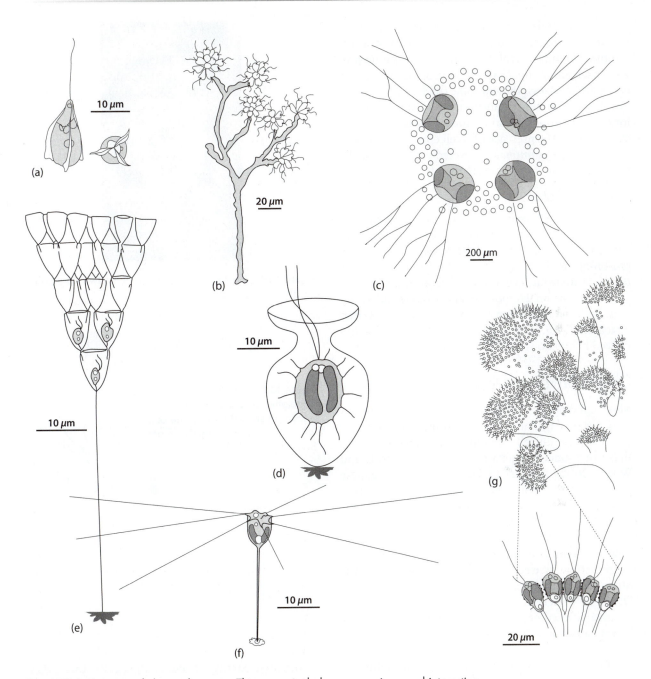

Figure 13.5 Diversity of chrysophyceans. This group includes many unique and interesting morphologies. Some examples include (a) *Pyramidochrysis splendens*, a flagellate with a distinctive winged cell covering; (b) the sessile, colonial form *Anthophysa vegetans*; (c) *Chrysostephanosphaera globulifera*, a colony of nonflagellate cells that have distinctive projections and occur in a granular matrix; (d) *Derepyxis crater*, whose cells possess what appear to be two equal flagella and are enclosed in a sessile lorica; (e) *Stylobryon*, with a colonial structure somewhat resembling that of the more familiar *Dinobryon* but attached to substrates by a long stalk; (f) *Rhizaster crinoides*, a nonflagellate unicell that is attached by means of a stalk and possesses long cellular extensions; and (g) *Mycochrysis oligothiophila*, a large colonial aggregate of *Ochromonas*-like cells with a branched gelatinous stalk, said to have the appearance of a miniscule mushroom. (After Bourrelly 1968)

Phagotrophic cells may attach to particulate food via anterior pseudopodia, and then they use a distinctive looplike microtubular root to consume prey within a feeding cup. Many phagotrophic chrysophyceans are wall-less, though some (such as *Dinobryon* and *Lagynion*) may occur in organic, open vase-like loricas. The food ingestion apparatus is thus unencumbered by the presence of a cell covering. The surface of *Paraphysomonas* is covered with silica scales that apparently do not interfere with phagotrophy.

Bloom formation

Several chrysophyceans are associated with the formation of undesirable blooms. These may have indirect or direct effects on lake biota, such as toxin poisoning or clogging of fish gills. *Uroglena volvox* produces toxic fatty acids that affect fish. Living cells of *Uroglena* and *Dinobryon* can excrete aldehydes and ketones (mainly n-haptanal) into the water, which can give it an unpleasant taste and odor (reviewed by Nicholls 1995).

Chrysophycean Diversity

Most chrysophyceans are unicellular flagellates or colonies of flagellate cells, but some species occur as nonmotile unicells, colonies, or amoeboid forms that can produce flagellate zoospores (Nicholls and Wujek 2002) (Figure 13.5). Iron and manganese precipitates may color the stalks, loricas, and gelatinous extracellular matrices of some chrysophyceans (Leadbeater and Barker 1995). The bodies of many chrysophyceans are covered with species-specific types of scales whose detailed structure requires examination by electron microscopy. Such detail is often needed to distinguish species of chrysophytes. Drops of preserved chrysophyceans are dried onto Formvar-coated grids or cover slips for examination by transmission or scanning electron microscopy.

OCHROMONAS (Gr. *ochros*, pale yellow + Gr. *monas*, unit) is a wall-less flagellate with one or two golden-brown, platelike plastids (Figure 13.6). There are two unequal flagella of the typical stramenopile type, and at least one protein associated with the tubular shafts of the tripartite hairs has been identified in *Ochromonas* (Yamagishi et al. 2007). An eyespot contained within a plastid and contractile vacuoles may be visible at the anterior. The posterior portion of the cell typically comes to a point. *Ochromonas* cells may contain the remains

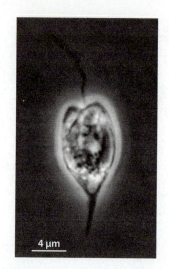

Figure 13.6 *Ochromonas.* Only one of the two flagella is visible. (Photo: L. W. Wilcox)

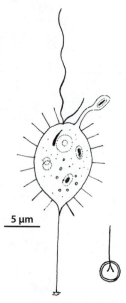

Figure 13.7 *Paraphysomonas.* Vegetative cells are covered with a layer of spined scales, one of which is shown separately. (From Bourrelly 1968)

of ingested particles such as algal cells. In *O. danica*, a banded rhizoplast runs from the basal body of the anterior flagellum to the space between the nucleus and Golgi body. This rhizoplast duplicates prior to cell division, and each of the two products serves as a pole of the mitotic spindle (Uemori et al. 2006). *Ochromonas* occurs most abundantly in oligotrophic freshwaters. Stomatocysts are often produced. There are about 80 species of *Ochromonas*, but the genus is probably not monophyletic (Andersen et al. 1999).

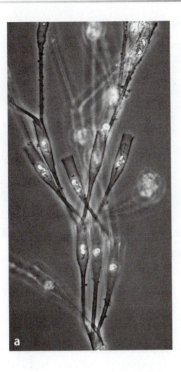

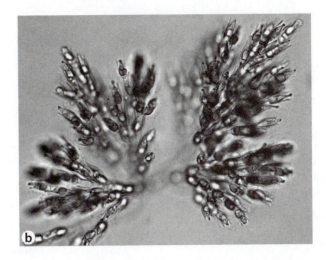

Figure 13.8 Two species of *Dinobryon*, illustrating the treelike arrangement of the flagellate cells in the colony, as viewed by light microscopy. (Photos: L. W. Wilcox)

PARAPHYSOMONAS (Gr. *para*, near + *Physomonas*, a genus of flagellates [Gr. *physa*, bellows + Gr. *monas*, unit]) is a single-celled flagellate with two flagella whose surface is covered with silica scales (Figure 13.7). Although it is not pigmented, there is a leucoplast (a colorless plastid), indicating that *Paraphysomonas* is derived from an ancestral form that was pigmented. It is osmotrophic or phagotrophic. There are about 50 species, whose identification requires electron microscopic examination of scale structure. *Paraphysomonas* is of common occurrence in tropical freshwaters (Wujek and Saha 1995).

DINOBRYON (Gr. *dineo*, to whirl + Gr. *bryon*, moss) is a colonial organism; each cell has two flagella that contribute to the motility of the whole colony, which swims slowly. Each cell is contained in a vase-shaped cellulosic lorica. Cells are attached to the base of their lorica by a thin cytoplasmic thread. Loricas are arranged to form a dendroid (tree-shaped) colony. This colony structure arises as a result of longitudinal mitosis and subsequent migration of one of the daughter cells to the lorica aperture, where it attaches and generates a new lorica. This process can be repeated so that two daughter cells protrude from the parental lorica. Thus one or two loricas can be

Figure 13.9 Colonial habit of *Dinobryon* viewed by SEM. The cells (whose flagella have been lost) are visible through the electron-translucent loricas. (Micrograph: L. W. Wilcox)

observed to extend from the mouth of the lorica just posterior in the colony (Figures 13.8, 13.9). Loricas can be colorless or brownish in color, and the shapes of loricas vary among species. Cells contain one or two brown plastids. An eyespot and two contractile vacuoles can usually be detected at the apical end of the cell. Colonies exhibit phototaxis (Canter-Lund

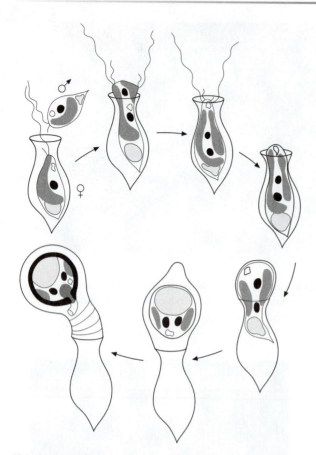

Figure 13.10 The sexual reproductive process in *Dinobryon*. Fertilization occurs within the lorica of one gamete (termed the female) by a gamete that has swum out of its lorica (termed the male). The resulting zygote emerges from the lorica and forms a stomatocyst at the lip. (After Sandgren 1981, by permission of the *Journal of Phycology*)

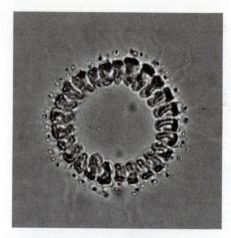

Figure 13.11 *Cyclonexis*, a relatively rare, ring-shaped swimming colony. (Photo: L. W. Wilcox)

and Lund 1995). A chrysolaminarin vacuole occurs in the cell posterior. *Dinobryon* is widely recognized as a bacteria-consuming mixotroph. Asexual reproduction occurs when cells swim out of their lorica and start new colonies. Sexual reproduction involves isogametes and formation of statospores that are attached to colonies (Sandgren 1981) (Figure 13.10). *Dinobryon* may occur in the plankton of either soft or hard (high-alkalinity) freshwaters. Some species grow in the mucilage of cyanobacterial colonies or on the frustules of the diatom *Tabellaria* (Prescott 1951). *Dinobryon* can be very common in the plankton of temperate and boreal lakes that are poor in nutrients and not very alkaline. It is also common in mountainous regions and in pools in areas where

the soil is acidic (Canter-Lund and Lund 1995). Environmental concentrations of potassium can be toxic to a number of freshwater *Dinobryon* species, limiting their seasonal and geographical occurrence. *Dinobryon sertularia* is limited to waters where temperatures are no higher than 20° C (Lehman 1976). *Dinobryon balticum* is a dominant species in cold North Atlantic coastal marine waters, for which phagotrophy has been demonstrated. It is associated with bacteria-rich fecal pellets and other detrital aggregations and is probably bacterivorous (McKenzie et al. 1995).

CYCLONEXIS (Gr. *kyklos*, circle + L. *nexis*, act of swimming) is a motile colony of golden-pigmented cells arranged in a ring held together by mucilage (Figure 13.11). The colony rotates as it swims. It occurs in freshwater ponds and pools, but is rarely observed.

UROGLENOPSIS (*Uroglena* + Gr. *opsis*, likeness) is a motile, spherical, mucilaginous colony of hundreds of ovoid cells evenly distributed along the periphery (Figure 13.12). Colonies can reach 500 μm in diameter, somewhat resembling the green colonial flagellate *Volvox*. Individual cells are 3–7 μm in diameter and have one or two golden-brown plastids shaped like plates or disks. Each cell has two flagella that extend through the mucilage and contribute to colony motility. *Uroglenopsis* can be a common component of the phytoplankton in late summer, especially in lakes of fairly high alkalinity (Prescott 1951). A key to the species and

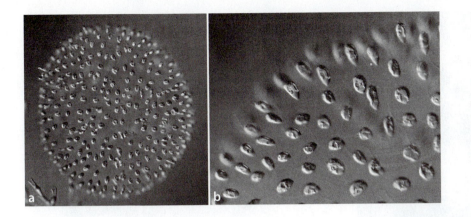

Figure 13.12 *Uroglenopsis*. (a) *Uroglenopsis* colonies contain hundreds of cells, (b) each of which has flagella, and hence superficially resembles nonreproductive colonies of the green alga *Volvox*. (Photos: C. Taylor)

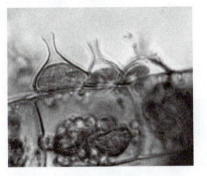

Figure 13.13 *Lagynion*. (Photo: L. E. Graham)

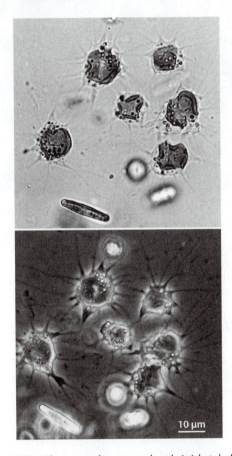

Figure 13.14 *Chrysamoeba*, viewed with (a) bright-field optics, which reveal plastids effectively, and (b) phase optics, where the colorless rhizopodia are better seen. (Photos: L. W. Wilcox)

illustrations of stomatocysts is provided by Wujek and Thompson (2002).

LAGYNION (L. *lagenion*, small flask) (Figure 13.13) includes epiphytes whose cells are enclosed by a flask or bottle-shaped transparent or, more often, brown lorica of organic composition. A cytoplasmic extension may protrude through the narrowed lorica neck. Flagellate zoospores may be produced; when they settle onto a substrate, they first produce several robust cytoplasmic extensions, and then they generate the lorica (O'Kelly and Wujek 1995). The lorica base is typically flattened against the substrate (often larger, filamentous algae) in soft-water lakes or bogs, especially in areas rich in organic material (Prescott 1951).

CHRYSAMOEBA (Gr. *chrysos*, gold + Gr. *amoibe*, change) is a globose or discoid amoeboid cell with an irregular cell outline and radiating cytoplasmic protrusions that can vary in length by active extension and retraction (Figure 13.14). There may be a thin mucilage investment. Cells can become motile by production of a single emergent flagellum. There

are two platelike golden plastids per cell. Cells may occur singly or as aggregates in the plankton of freshwater lakes and in muddy areas but are of rare occurrence (Prescott 1951). *Chrysamoeba pyrenoidifera*,

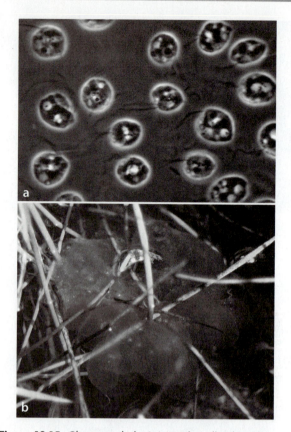

Figure 13.15 *Chrysonephele*. (a) Motile cells. (b) A large gelatinous colony. (Photo: P. Tyler)

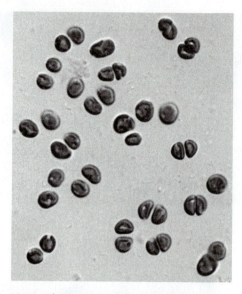

Figure 13.16 *Chrysocapsa*. (Photo: R. Andersen)

having noticeable pyrenoids, occurs within the hyaline cells of *Sphagnum* moss gametophores (O'Kelly and Wujek 1995).

CHRYSONEPHELE (Gr. *chrysos*, gold + Gr. *nephele*, cloud) (Figure 13.15) is a hollow, mucilaginous colony including hundreds to thousands of cells, each with functional flagella. Although the flagella beat, the colony is not motile. The cells have typical chrysophycean eyespots, plastids, contractile vacuoles, and chrysolaminarin vacuoles (Pipes et al. 1989). *Chrysonephele palustris* is known from only one freshwater swamp in central Tasmania and is thus highly endemic.

CHRYSOCAPSA (Gr. *chrysos*, gold + L. *capsa*, box) is a planktonic colony as large as 250 µm in diameter and composed of up to 64 spherical or elliptical cells enclosed by mucilage (Figure 13.16). The cells lack flagella but contain one or two goldenbrown peripheral platelike plastids. *Chrysocapsa* is common in both hard- and soft-water lakes.

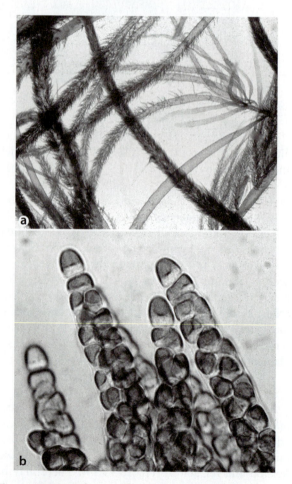

Figure 13.17 *Hydrurus* seen in low magnification (a) and at higher magnification (b). (Photos: P. Kugrens)

HYDRURUS (L. *hydor*, water + Gr. *oura*, tail) bodies (Figure 13.17) can be relatively large and conspicuous to the naked eye, occurring in fast-flowing, cold (less than 10°C) streams in mountainous regions, attached to submerged rocks or other rigid substrates. *Hydrurus foetidus* emits a strong, unpleasant odor that can be detected from a distance. Spherical cells about 7–15 µm in diameter—each containing one cup-shaped golden-brown plastid—occur within a gelatinous investment that can take a pseudobranched, feathery, or leafy form.

13.2 Synurophyceans

Synurophyceans are unicellular or colonial silica-scaled flagellates defined by flagellar root system features that correlate with loss of phagotrophic capability and lack of chlorophyll c_2 (Andersen 1987). Synurophyceans also have distinctive oval-shaped organic scales on the flagella (in addition to the tripartite hairs on the anterior flagellum so common to stramenopiles) (Hibberd 1973). Molecular sequence data and comparative study of scales suggest that synurophyceans are a monophyletic group and that *Tessellaria* is an early-divergent genus (Lavau et al. 1997). Synurophyceans are named for the widespread colonial genus *Synura*, but the spiny unicellular *Mallomonas* and bi-celled *Chrysodidymus* are additional examples of widespread and well-studied representatives. Fossil scales similar to those of modern *Synura* and *Mallomonas* have been found in deposits about 47 million years of age (middle Eocene), indicating that the synurophyceans are at least this old and have changed little during the intervening time (Siver and Wolfe 2005).

Synurophycean Cell Biology

Synurophycean cells contain one large bilobed plastid or two plastids, each with a girdle lamella. Pyrenoids are rare but present in some species. There is a periplastidal endoplasmic reticulum (PER), as in other plastid-bearing stramenopiles, but it is not connected with the nuclear envelope.

Flagella and flagellar roots

Synurophyceans are distinctive in having parallel orientation of flagella and flagellar basal bodies. As a result of this orientation, synurophycean cells or colonies swim with both flagella held out in front.

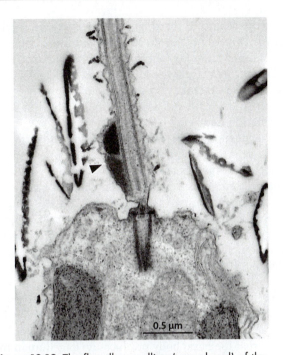

Figure 13.18 The flagellar swelling (arrowhead) of the scale-bearing synurophycean *Chrysodidymus synuroideus* consists of two parts, a more electron-opaque upper portion, and a somewhat less opaque lower region. (From Graham et al. 1993, by permission of the *Journal of Phycology*)

By contrast, as noted earlier, most stramenopiles have an anterior-directed hairy flagellum and a posterior-directed smooth flagellum. However, the synurophycean smooth flagellum exerts little or no motive force; the hairy flagellum does all the propulsive work, as in other stramenopiles (Goldstein 1992). Like many other photosynthetic flagellates, synurophyceans rotate during swimming, a behavior thought to maximize photosynthetic gain by taking advantage of variation in the light field (Raven 1995). Plastid-based eyespots are absent from synurophycean cells, but *Synura* and *Chrysodidymus* (Graham et al. 1993) have a flagellar swelling (Figure 13.18) that autofluoresces green in blue or blue-violet irradiance, indicating presence of a flavin, as is found in several other groups of photosynthetic stramenopiles (Kawai 1988). *Mallomonas splendens*, which has no emergent smooth flagellum, lacks a flagellar swelling (Beech and Wetherbee 1990b).

The flagellar root system includes a crossbanded rhizoplast that extends from the flagellar bases to the nucleus, whose anterior surface is covered with a cone of rhizoplast fibrils. The contractile protein centrin occurs in the rhizoplast area of synurophyceans (Melkonian et al. 1992). In addition, there is an R1 microtubular root that extends from the rhizoplast, forms a clockwise loop around the flagella, and generates numerous cytoskeletal microtubules. Some of these microtubules appear to be involved in scale formation (see below).

Synurophyceans exhibit a phenomenon that is widespread among flagellate eukaryotes, known as **flagellar transformation**. In this process, flagellar maturation, judged by changes in length or addition of ornamentation to the flagellar surface, requires more than one cell generation. Thus, the two or more flagella of a given cell are of different ages. Prior to cell division, synurophycean cells produce two new long flagella. In forms such as *Synura* that typically have both long and short emergent flagella, the parental long flagellum is then transformed into a new short flagellum. In other cases, including certain species of *Mallomonas*, the parental long flagellum retracts such that the basal body produces no flagellum in the next and subsequent generations (Beech and Wetherbee 1990a). As a result of the discovery of flagellar transformation, a system of flagellar numbering has been developed that can be applied to all flagellates. The most mature flagellum is numbered 1, and successively younger flagella are numbered 2 and higher. This system makes it possible to compare flagella in equivalent stages of development among different taxa. In the case of *Synura*, the smooth, swelling-bearing flagellum is the oldest (number 1), whereas the younger, longer, tripartite hair-bearing flagellum is number 2 (see references in Beech and Wetherbee 1990b).

Silica scale structure, development, and deployment

Overlapping silica scales, sometimes with spiny bristles, cover the surfaces of synurophycean cells. The scales are perforated (Figure 13.19), which reduces their weight and allows access for exchange of material at the cell surface. Differences in perforation patterns of scales are used as taxonomic features in the delineation of species. The scales may have a rim that facilitates overlapping of scales to form a casing. A number of adhesive polysaccharides are involved in scale attachment: separate scale–scale, bristle–scale, and scale–cell membrane adhesives have been identified (Ludwig et al. 1996).

Silica body scales and bristles are formed in silica deposition vesicles (SDVs) located on the surface

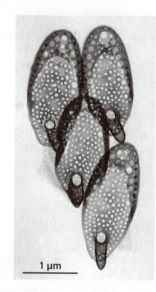

Figure 13.19 TEM view of some of the overlapping silica scales from the surface of *Chrysodidymus synuroideus*. Note the occurrence of perforations, rims of perforated silica, and short spines. (From Graham et al. 1993, by permission of the *Journal of Phycology*)

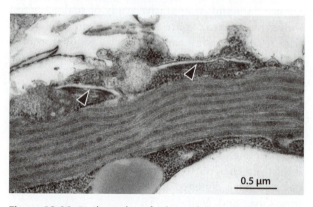

Figure 13.20 Body scales of *Chrysodidymus synuroideus* forming within silica deposition vesicles (arrowheads) located on the surface of a plastid. (TEM: L. E. Graham)

of one of the plastids, closely adjacent to the PER (Mignot and Brugerolle 1982) (Figure 13.20). The functional and biochemical basis of this unique association is unclear, but it has been hypothesized that the PER may serve as a template for scale shaping (Leadbeater and Barker 1995). Proteins that are associated with scales are also suspected to play roles in scale development (Nachname 2001). Large cells of *Synura petersenii* possess about 90 scales (Leadbeater and Barker 1995). At cell division each daughter

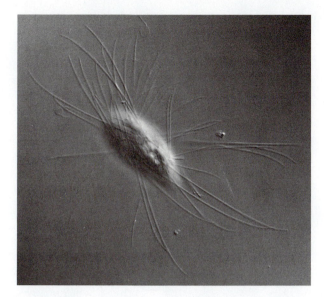

Figure 13.21 *Mallomonas* is a unicellular flagellate that bears silica scales with very long silica bristles. (Photo: C. Taylor)

cell receives about half the scales, then generates additional scales until the surface is completely covered. However, if dissolved silicate levels are insufficient, incompletely covered or naked cells will result. Unlike diatoms, synurophyceans can continue to divide and function in the absence of an external silica covering (Sandgren et al. 1996). Cultures of naked cells will grow indefinitely in media having undetectable levels of dissolved silicate if light and other nutrients are provided. When silicate is resupplied to silica-depleted cells of *Synura*, most cells will regenerate a complete cell covering within 24 hours (Leadbeater and Barker 1995). Mature scales migrate to the cell membrane and are extruded to the surface.

Mallomonas scales typically bear long silica bristles, usually in localized positions (Figure 13.21). Differences in the distribution of bristles are used as taxonomic characters in species differentiation. Bristles are produced in SDVs separate from those where scales are generated, but bristle SDVs are also associated with PER. In *Mallomonas splendens*, bristles destined for attachment to the cell posterior are extruded with their basal ends leading but attached to the cell membrane by a fibrillar complex. Once free of the cell, the basal portion of the bristles is drawn back to the cell surface; this process involves a 180° reorientation of the bristle while it is outside the cell, mediated by a thin protuberance of cytoplasm (Figure 13.22). New base-plate scales are then deposited at

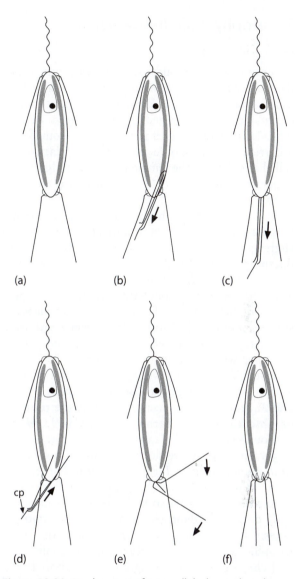

Figure 13.22 Deployment of intracellularly produced bristles to the posterior surface of *Mallomonas splendens*. Following extrusion to the outside of the cell (b, c), bristles are rotated 180°, with the involvement of a cytoplasmic protrusion (cp) (d), then attached to newly deposited base-plate scales (e, f). (After Beech et al. 1990, by permission of the *Journal of Phycology*)

the cell posterior, and the fibrillar complex is involved in attaching the bristle bases to the base-plate scales. Anterior bristles are produced similarly but are extruded tip first, such that no reorientation is involved (Beech et al. 1990). A review of synurophycean scale structure and development and their relevance to synurophycean phylogeny is provided by Wee (1997).

Synurophycean Reproduction and Ecology

Synurophyceans produce stomatocysts much like those of chrysophytes. Sexual reproduction in *Synura* is similar to that in *Dinobryon*; zygotic stomatocysts can be recognized by the presence of four plastids and two nuclei, and the mature stomatocyst is carried along by motile colonies in the place of the cell that served as the female gamete (Sandgren and Flanagin 1986) (Figure 13.23). Meiosis is assumed to be zygotic.

Like their chrysophycean relatives, synurophyceans are most abundant and diverse in neutral to slightly acidic waters of low alkalinity and conductivity. As a general rule, scaly synurophyceans are indicators of low levels of pollution. However, a number of species are characteristic of eutrophic lakes or lakes of different pH. Such environmentally specific species are useful in long-term monitoring of phytoplankton aimed at detecting and understanding water-quality decline (Siver 1995). The inverse relationship between lake phosphorus availability and the abundance of taxa such as *Synura* and *Mallomonas* has been ascribed to the presence or absence of large herbivores such as *Daphnia*. Although small cladocerans that may be present in oligotrophic lakes are deterred from eating bristly *Mallomonas* species, *Daphnia* are not. Fortunately for synurophyceans and chrysophyceans, *Daphnia*'s food requirements are so high that it does not typically occur abundantly in cold and/or low-nutrient lakes. Thus seasonally cold periods and oligotrophic lakes provide a refuge for synurophyceans and chrysophyceans (Sandgren and Walton 1995). *Mallomonas* and *Synura* can also escape potential predators and acquire nutrients by vertically migrating to deep waters at night and well-lighted surface waters during the day. It has also been suggested that the greater availability of dissolved iron in low-pH, highly colored bog and pond waters may help explain the common occurrence of *Synura* and its relatives, which have a higher demand for iron used in photosynthetic electron transport than some other algae (Raven 1995).

Synurophycean Diversity

Synurophycean genera are easily distinguished on the basis of light microscopic characters, but species determination requires electron microscopic

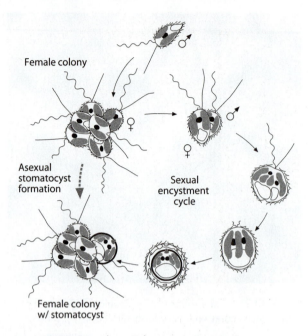

Figure 13.23 Sexual reproduction in *Synura*. One gamete, defined as the female because the cell remains associated with a colony, is fertilized by a flagellate cell that is released from its colony (and therefore designated male). The fertilized cell then develops into a stomatocyst that remains associated with the motile colony, as in *Dinobryon*. Stomatocysts may also develop asexually. (After Sandgren and Flanagin 1986, by permission of the *Journal of Phycology*)

examination or molecular characterization. There are approximately 120 described species of the widespread genera *Mallomonas* and *Synura*. *Tessellaria*, which appears to be endemic to Australia, is unique in that scales cover the entire colony surface rather than individual cells (Pipes et al. 1991; Pipes and Leedale 1992).

MALLOMONAS (Gr. *mallos*, wool + Gr. *monas*, unit) (Figures 13.24) is a single-celled flagellate with (usually) only one emergent flagellum. The cell membrane is covered with overlapping silica scales, at least some of which bear long silica bristles. Cells contain a single deeply divided plastid that may give the appearance of two plastids. *Mallomonas* is associated with taste and odor problems with drinking water. Local and regional diversity of *Mallomonas* in the Czech Republic was studied as a test of this alga's ability to disperse widely (Rezacova and Neustupa (2007).

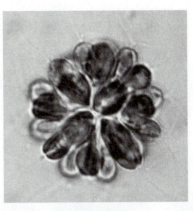

Figure 13.25 *Synura*, which occurs as a swimming colony of cells adjoined at their posterior ends. Note flagella at the periphery of the colony. (Photo: L. E. Graham)

Figure 13.24 *Mallomonas*, showing characteristic elongate silica bristles attached to flattened silica scales. (Photo: C. Taylor)

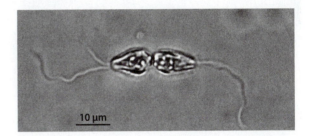

Figure 13.26 *Chrysodidymus*, a two-celled colony of *Synura*-like cells. Note the heterokont flagella. (Photo: L. E. Graham)

SYNURA (Gr. *syn*, together + Gr. *oura*, tail) is a colonial organism formed by variable numbers of cells held together at the posterior ends to form a tightly coherent spherical colony (Figure 13.25). Colonies grow by addition of new cells through longitudinal cell division. New colonies may form by fragmentation. Cells bear two flagella each and are covered with overlapping silica scales. Spines are present on anterior scales. There are two parietal golden-brown plastids per cell. *Synura ulvella* is a very common species found in the plankton of many kinds of lakes (Prescott 1951). *Synura petersenii* is often associated with taste or odor problems that affect drinking water (Nicholls 1995).

CHRYSODIDYMUS (Gr. *chrysos*, golden + Gr. *didymos*, twins) is a two-celled colony of *Synura*-like cells, each bearing two heterokont flagella (Figure 13.26). The two cells are adherent at their bases. Cells are covered with silica scales very much like those of *Synura*. *Chrysodidymus* is found in low-pH *Sphagnum* bogs, as is *Synura sphagnicola*. Ultrastructural similarities suggest that *Chrysodidymus* may be related to *Synura sphagnicola* (Graham et al. 1993). *Chrysodidymus* is geographically widespread but is not often locally abundant. The mitochondrial genome of *C. synuroideus* has been sequenced, showing (together with other data) that stramenopile

mitochondrial genomes display low variability in size and gene content (Chesnick et al. 2000).

13.3 Eustigmatophyceans

Eustigmatophyceans are named for a large, orange-red eyespot located in the cytoplasm of flagellate cells (Figure 13.27) (Hibberd and Leedale 1970; Hibberd 1990a). This characteristic is unique among the photosynthetic stramenopiles, whose eyespots, if present, are typically located within plastids. SSU rDNA sequence data suggest that eustigmatophyceans are a monophyletic group that is sister to a clade containing chrysophyceans and synurophyceans (see Figure 13.1) (Andersen et al. 1998b; Hegewald et al. 2007). There are no known fossils.

The eustigmatophyceans are small (2–32 μm), unicellular, and primarily coccoid in shape, with a

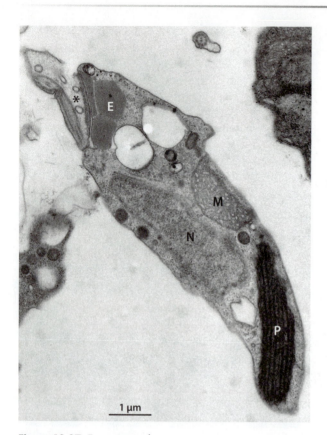

Figure 13.27 Eustigmatophycean zoospore. Eustigmatophyceans that produce zoospores exhibit the diagnostic feature known as a eustigma (E), an eyespot that occurs outside the plastid. This eyespot occurs at the anterior of flagellate cells, near an expanded region of the flagellum (*). M = mitochondrion, N = nucleus, P = plastid. (TEM: L. Santos)

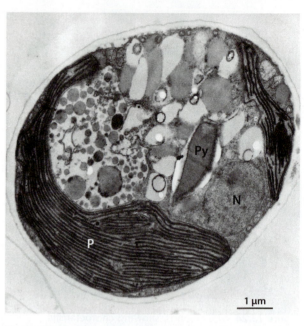

Figure 13.28 TEM view of *Eustigmatos vischeri*. Note the stalked pyrenoid (Py) extending into the cytoplasm. N = nucleus, P = plastid. (Micrograph: L. Santos)

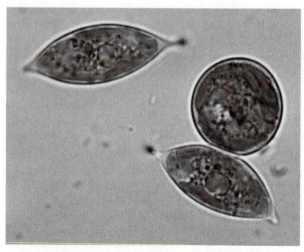

Figure 13.29 *Pseudocharaciopsis* cells, which are often spindle shaped. (Photo: L. Santos)

cell wall of uncertain composition. Reproduction is either by production of two to four autospores (small, walled versions of nonmotile parental cells) or flagellate zoospores. Zoospores may have two flagella of the typically stramenopile type or only one emergent anterior flagellum.

Eustigmatophyceans are thought to be obligate photoautotrophs because there is no evidence for their use of exogenous organic compounds or ability to ingest particles (Santos 1996). Cells have one or more yellow-green plastids that contain chlorophyll *a* but not chlorophyll *c* or fucoxanthin. Violaxanthin is the major accessory pigment (Whittle and Casselton 1975) and appears to play a prominent role in light harvesting. Although thylakoids occur in stacks of three, as is common in photosynthetic stramenopiles, there

is no girdle lamella. PER is present but is not always connected with the nuclear envelope. Unusual stalked pyrenoids occur on the plastids of some, but not all, eustigmatophyceans. Pyrenoids don't seem to occur in zoospores—they disappear during zoospore development and then regenerate after zoospores settle onto a surface (Santos

Figure 13.30 *Nannochloropsis.* (a) Diagram of three different species. (b) Light microscopic view of another species. Inset: Two plastid lobes are evident, as is the so-called red body, which appears as a dark spot in the cell center. (a: From Andersen et al. 1998a; b: L. W. Wilcox)

1996). Other typical cellular features include a red-pigmented body whose function is unknown and storage vesicles having lamellate contents of uncertain chemical constitution.

The systematics of the Eustigmatophyceae have been described by Hibberd (1981) and Santos (1996). A single order (Eustigmatales) comprises five families (Hegewald et al. 2007) that are distinguished on the basis of cell shape and size, presence or absence of zoospores, and number of flagella on zoospores. Most occur in freshwater or soil, but there are also some marine forms. A eustigmatophycean whose features do not match those of known genera occurs as a symbiont within the freshwater sponge *Corvomyenia everetti* (Frost et al. 1997).

Eustigmatophycean Diversity

EUSTIGMATOS (Gr. *eu*, well equipped + Gr. *stigma*, eyespot) (Figure 13.28) is a unicellular, coccoid soil alga having a single parietal plastid. Reproduction is by two to four autospores and uniflagellate zoospores.

PSEUDOCHARACIOPSIS (Gr. *pseudos*, false + *Characiopsis*, a genus of tribophyceans) consists of ovoid to pointed cells commonly attached to a substrate by means of a short stipe or attachment pad (Figure 13.29). Biflagellate, naked zoospores are produced. Older cells can contain several nuclei and chloroplasts (Lee and Bold 1973). *Pseudocharaciopsis* is very similar to, and easily confused with, the tribophycean (xanthophycean) alga *Characiopsis* and the green alga *Characium*.

Pseudocharaciopsis can be distinguished from *Characiopsis* and *Characium* by its possession of a conspicuous stalked pyrenoid that is visible with light microscopy.

NANNOCHLOROPSIS (*Nannochloris*, a genus of green algae [Gr. *nannos*, dwarf + Gr. *chloros*, green] + Gr. *opsis*, likeness) (Figure 13.30) is a marine and freshwater coccoid form that resembles *Chlorella* and lacks a pyrenoid (Santos and Leedale 1995). It produces a relatively high content of the polyunsaturated fatty acid eicosapentaenoic acid and thus is used as a food for aquacultured (maricultured) marine animals. A surprisingly high level of SSU rDNA sequence diversity was found in a study of 21 strains (Andersen et al. 1998a). *Nannochloropsis* diversity in freshwater systems has been evaluated by culture and molecular approaches (Fawley and Fawley 2007).

13.4 Raphidophyceans

Raphidophyceans have also been called chloromonads, a term that is no longer in use but appears frequently in the literature. They are relatively large (30–80 µm in their largest dimension), naked, unicellular flagellates (Figure 13.31) but may sometimes occur as masses of nonmotile cells embedded in mucilage. There is a long, forwardly directed, tripartite hair-bearing flagellum and a backward-directed smooth flagellum, typical of stramenopiles. However, in some cases the posterior flagellum is very short and thus may appear to be missing. Raphidophycean cells "spurt" forward

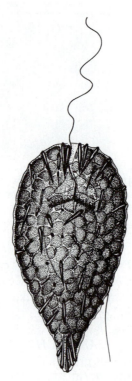

Figure 13.31 A raphidophycean, *Gonyostomum semen*, illustrating two flagella (one directed forward and one trailing) and trichocysts. (From Mignot 1967)

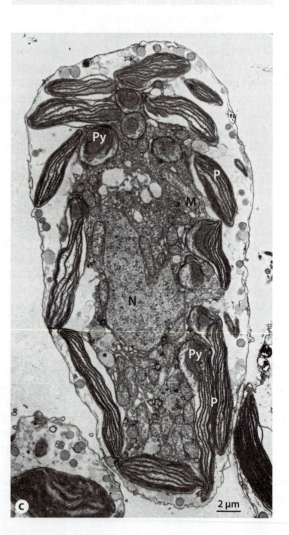

in a straight path or rotate, propelled primarily by the anterior flagellum. The posterior flagellum is either motionless or moves only slightly, possibly serving a steering function (Goldstein 1992). Both flagella emerge from a shallow anterior groove (gullet) on the ventral side of the often-flattened cells. In addition to flagellar-driven motility, metaboly has been observed in some forms (Horiguchi 1996).

Raphidophyceans have no eyespots or other noticeable evidence of a photoreceptor system, such as a flagellar swelling. Several large Golgi bodies typically occur per cell. Rod-like trichocysts (also known as extrusomes) can be observed with the light microscope. These may be discharged from

Figure 13.32 *Haramonas.* (a, b) Through-focus pair. Note two flagella (arrow) in (a) and overlapping plastids (arrowheads) in (b). (c) Transmission electron micrograph. Note that cells lack a cell wall and that there are multiple plastids (P), each with a pyrenoid (Py). The centrally located nucleus (N) is conspicuous, as are mitochondria (M). (From Horiguchi 1996, © Blackwell Science Asia)

cells, forming long mucilaginous strands. In addition, cells contain mucocysts that can produce extracellular mucilage (Heywood 1983, 1990).

The nuclei are unusually large and conspicuous (Figure 13.32). A fibrous flagellar root (rhizostyle or rhizoplast) contacts the nuclear envelope, as in some other stramenopiles. The PER does not connect with the nuclear envelope. Mitosis in *Vacuolaria* begins with production of an array of microtubules from the flagellar basal body region. These microtubules extend over the nuclear envelope and at prophase pass through the envelope (which remains intact throughout mitosis) via polar gaps (Figure 13.33). Cytokinesis occurs by formation of a cleavage furrow from the cell periphery (see review of algal mitosis by Sluiman 1993). Sexual reproduction has not been reported.

Cells are photoautotrophic, typically containing multiple discoid plastids. Thylakoids occur in stacks of three, and a girdle lamella may or may not be present. Plastids are bright glistening green, yellow-green, or yellow-brown and contain chlorophyll a and usually both chlorophyll c_1 and c_2 (Chapman and Haxo 1966). However, *Fibrocapsa* lacks chlorophyll c_2, and *Haramonas* does not possess chlorophyll c_1 (Mostaert et al. 1998). The marine and freshwater forms differ in their accessory pigments. Marine raphidophyceans resemble chrysophyceans, eustigmatophyceans, synurophyceans, and phaeophyceans in having fucoxanthin and violaxanthin. *Fibrocapsa* and *Haramonas* contain fucoxanthin derivatives not found in the rest, and fucoxanthinol is a good pigment marker for *Fibrocapsa* (Mostaert et al. 1998). In contrast, the freshwater forms resemble tribophyceans (xanthophyceans) (Chapter 14) in having diadinoxanthin, heteroxanthin, or vaucheriaxanthin (or a derivative). Further, pyrenoids occur in the plastids of marine species but not in those of freshwater forms. These differences in plastid structure and pigmentation have led some experts to recommend separate classification. However, SSU rDNA and *rbcL* sequence analyses robustly unite the marine and freshwater raphidophyceans as a monophyletic group (Potter et al. 1997, Daugbjerg et al. 1997b, Figueroa and Rengefors 2006). It is intriguing that such dramatic pigment differences occur in plastids that appear, on the basis of sequence evidence, to be closely related. Oil bodies represent the major photosynthetic storage product in both marine and freshwater raphidophyceans.

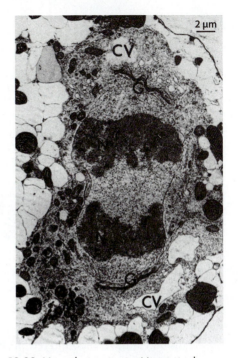

Figure 13.33 *Vacuolaria* mitosis. Mitosis in the raphidophycean *Vacuolaria* involves an intranuclear spindle (developed within the nuclear envelope). (From Heywood 1983)

The marine raphidophyceans can be found in the open sea or in brackish estuaries. *Chattonella* may bloom in organic-rich brackish waters, where it can cause extensive losses to fisheries. The killing mechanism is thought to be production of large amounts of reactive oxygen species such as superoxide anion and hydrogen peroxide (Kim et al. 2007), together with toxic fatty acids (Marshall et al. 2003). Harmful blooms of this raphidophycean are favored by water-column stratification, when winds are weak, and by water-temperature increases to 20°C–22°C. Such conditions favor germination of benthic cysts (Amano et al. 1998). *Heterosigma*, *Fibrocapsa*, *Olisthodiscus*, and *Haramonas* have also been associated with blooms harmful to finfish (Hallegraeff and Hara 2003).

Freshwater taxa (*Gonyostomum*, *Merotricha* [*Merotrichia*], and *Vacuolaria*) typically occur in waters of neutral to acidic pH where vegetation is abundant, either in the plankton or growing on mud (Heywood 1983, 1990). *Gonyostomum* may produce local blooms. Waters affected by acid precipitation have been reported to show an increase in numbers of *Gonyostomum* cells, and bathers at

Figure 13.34 *Heterosigma akashiwo.* (After Throndsen 1997)

beaches in Scandinavia have encountered problems with slime and skin irritation that may be associated with such blooms (Canter-Lund and Lund 1995). Sexual reproduction by *Gonyostomum semen* and the roles of resting cysts have been described (Cronberg 2005, Figueroa and Rengefors 2006).

Raphidophycean Diversity

Commonly recognized genera of raphidophycean algae include the marine genera *Chattonella*, *Fibrocapsa*, *Heterosigma*, *Haramonas*, and *Olisthodiscus* and the freshwater forms *Gonyostomum*, *Merotricha* (*Merotricha*), and *Vacuolaria* (Heywood 1990, Mostaert et al. 1998). *Haramonas* has only recently been described from mangrove swamps in Australia (Horiguchi 1996).

HETEROSIGMA (Gr. *heteros*, different + Gr. *sigma*, the letter *S*) (Figure 13.34) is a marine flagellate having ovoid cells with flagella inserted laterally and plastids arranged along the cell periphery. Trichocysts are not present. This marine toxic bloom former has been the subject of physiological and molecular studies. It is an obligate photoautotroph, and light changes regulate many aspects of its physiological ecology (Doran and Cattolico 1997).

GONYOSTOMUM (Gr. *gony*, knee + Gr. *stoma*, mouth) cells are motile flagellates that are dorsiventrally compressed (Figure 13.35). The dorsal

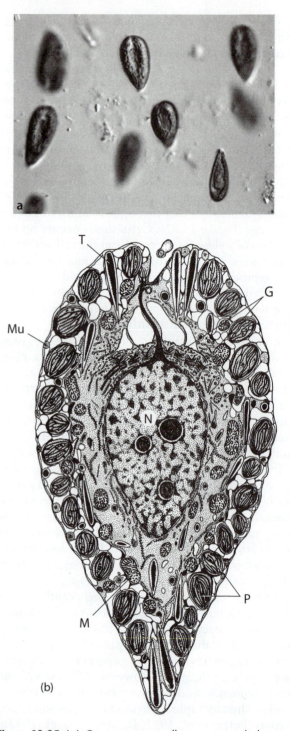

(b)

Figure 13.35 (a) *Gonyostomum* cells, as seen with the light microscope. (b) Diagram of a *Gonyostomum* cell, based on TEM data. A large central nucleus (N) is present, with numerous Golgi bodies (G) situated anterior to it. Peripheral cytoplasm is vacuolate, with numerous plastids (P), trichocysts (T), mitochondria (M), and mucocysts (Mu). (a: L. E. Graham; b: From Mignot 1967)

surface is convex and the ventral surface flattened. A longitudinal ventral furrow extends from an anterior opening to a colorless, three-cornered cavity. The long flagellum projects forward, and the shorter flagellum trails posteriorly. Both flagella are about as long as the cell body (36–92 μm). The numerous discoid plastids occur at the cell periphery (Prescott 1951). Clusters of trichocysts can occur at either end of the cells. The genus occurs in freshwater swamps, acid bogs, and other relatively low-pH waters. Thiamine, biotin, and vitamin B_{12} are required for growth in culture.

13.5 Pelagophyceans

Pelagophyceans usually occupy pelagic (open-water) marine habitats, though some are attached to shoreline rocks. They display a diversity of body types: flagellates, nonmotile single cells, colonies, or filaments. Cells can be naked or have an organic covering. Plastids contain chlorophylls a, c_1, and c_2, as well as fucoxanthin and other accessory pigments. A rod-shaped structure (paraxonemal rod) sometimes occurs within the hair-bearing anterior flagellum. This feature is important because it links pelagophyceans with their sister group the dictyochophyceans (see Figure 13.1).

Cell division in the genus *Pelagococcus* has been reviewed by Sluiman (1993). The outer membrane of the nuclear envelope is connected to the PER, as in a number of other photosynthetic stramenopiles, but pelagophyceans are unusual in that this connection is retained throughout the process of mitosis. There are no flagella or basal bodies (flagellar bases or centrioles) in the interphase cells, but two centrioles make their appearance in prophase and subsequently serve as the organizing centers for the development of an unusually small extranuclear spindle. The spindle forms next to the nuclear envelope, and then the nearby portion of the envelope breaks down, allowing the spindle access to the chromosomes. The nuclear envelope remains intact except at the spindle poles. Following chromosomal separation by spindle elongation, the nuclear envelope breaks and re-forms around each daughter nucleus. Cytokinesis occurs by cell constriction.

The pelagophyceans *Aureococcus anophagefferens* and *Aureoumbra lagunensis* are serious marine bloom formers. These small golden-brown cells can occur in populations as dense as 3×10^9 cells l^{-1}, forming summer "brown tides" along the New England and Texas coasts of the United States. Although these pelagophyceans are not known to produce toxins and do not pose a direct threat to humans, the blooms have had disastrous effects on thousands of hectares of coastal shorelines. The blooms exclude light from sea grasses, causing loss of sea grass beds. They are also associated with major declines in scallop harvest. By physically interfering with filter feeding, the algal blooms cause scallops to starve to death. Fish-eating birds cannot see their prey through the dense blooms and thus leave the area. *Aureococcus* has also been responsible for an estimated several millions of dollars worth of damage to blue mussel and cultured oyster production (Nicholls 1995). In addition, negative effects on egg production by copepods and reduced populations of ciliates are associated with dense pelagophycean blooms. The factors that stimulate pelagophycean blooms are as yet poorly defined. Although increased levels of phosphorus or nitrogen do not seem to be implicated, watershed input of iron is a possible correlate (Bricelj and Lonsdale 1997). *Aureococcus* can utilize exogenous glutamic acid (glutamate) and the sugar glucose (Nicholls 1995), as well as organic phosphates (Raven 1995); these abilities may enhance its success in waters having higher-than-normal levels of organic compounds.

The pelagophyceans were formally described as a new class (Pelagophyceae) (Andersen et al. 1993) on the basis of ultrastructural characters and SSU rDNA sequence data. The order Pelagomonadales includes the nonflagellate, coccoid unicells *Pelagococcus*, *Aureococcus*, and *Aureoumbra* (DeYoe et al. 1997) and the unicellular flagellate *Pelagomonas*.

AUREOUMBRA (L. *aureus*, golden + L. *umbra*, shadow), specifically *A. lagunensis*, is the Texas brown tide organism (Figure 13.36) (DeYoe et al. 1997).

PELAGOMONAS (Gr. *pelagos*, sea + Gr. *monas*, unit) (Figure 13.37) cells are very small, 1–3 μm in diameter. The flagellar apparatus is highly reduced. There is only one flagellum, and the second basal body and flagellar roots found in most other stramenopiles are absent. Further, the flagellar hairs of *Pelagomonas* are bipartite. Although the flagellar hairs occur in two opposite rows, they are flexible

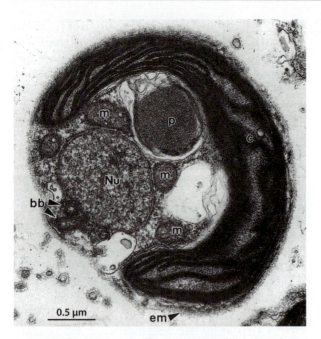

Figure 13.36 TEM view of *Aureoumbra*. bb = basal bodies, c = plastid, em = extracellular matrix, Nu = nucleus, m = mitochondrion, p = pyrenoid. (From DeYoe et al. 1997, by permission of the *Journal of Phycology*)

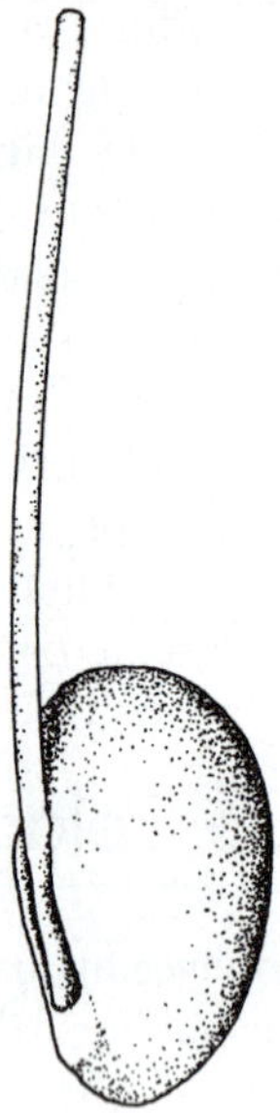

Figure 13.37 *Pelagomonas*. (From Andersen et al. 1993, by permission of the *Journal of Phycology*)

rather than stiff, raising the question of whether their hydrodynamic function is similar to that of other stramenopiles. The flagellum contains a rod-like structure (the paraxonemal rod) (Andersen et al. 1993) that increases flagellar thickness. The flagellum of *Pelagomonas* does not bear a swelling that autofluoresces green as do those of some other photosynthetic stramenopiles, and the cells of *Pelagomonas* are not known to be phototactic (Andersen et al. 1993).

13.6 Dictyochophyceans

Named for the genus *Dictyocha*, the class Dictyochophyceae encompasses single cells, flagellates, and amoebae that primarily occupy marine waters. The cells are naked or covered with organic scales. There are three orders: Dictyochales, whose members are also known as silicoflagellates; the flagellate Pedinellales (pedinellids); and the amoeboid Rhizochromulinales (rhizochromulinids). Flagellate cells of silicoflagellates and pedinellids (but not rhizochromulinids) display a

paraflagellar rod in the flagellum, a feature also found in pelagophyceans.

Silicoflagellates

Silicoflagellates are defined by the occurrence in at least one life cycle stage of a distinctive, one-piece external silica skeleton that is highly perforate and has been described as a basket (Figure 13.38). Although there are only a few living taxa of silicoflagellates (species of the genera *Dictyocha* and *Verrucophora*), they have a substantial fossil record extending back to the Cretaceous (about 120 million years ago). Silicoflagellates reached their peak diversity during the Miocene; 100 fossil species have been described. The fossil remains of the silica exoskeletons are used as micropaleontological indicators of ancient temperature change. Modern silicoflagellates are widespread in modern oceans but tend to be most abundant in colder waters, sometimes growing to bloom proportions. Blooms of *V. farcimen* cause fish kills (Edvardsen et al. 2007).

DICTYOCHA (Gr. *diktyon*, net + Gr. *ochos*, holding) cells have a distinctive, highly perforate cytoplasm that has been described as foamy or frothy (Figure 13.39). Typical stramenopile flagella are present: a long, anterior, hair-bearing

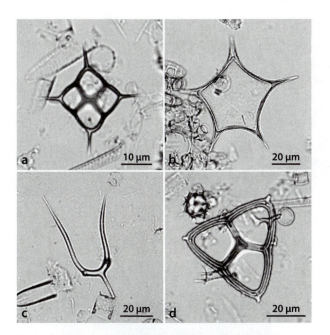

Figure 13.38 The highly perforate silica skeletons of silicoflagellates. (a) *Dictyocha*, (b) *Vallacerta*, (c) *Lyramula*, (d) *Corbisema*. (Photos: L. W. Wilcox, of Late Cretaceous specimens provided by D. Clark)

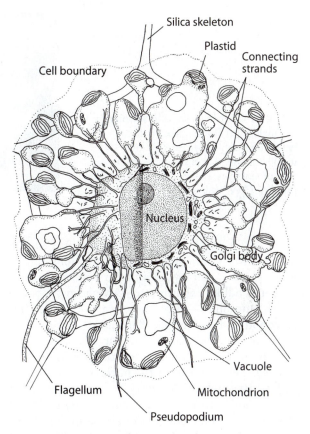

Figure 13.39 Diagram of *Dictyocha*. The cytoplasm is highly reticulate and vacuolate. (From Van Valkenberg 1971, by permission of the *Journal of Phycology*)

flagellum and a very short, smooth flagellum. The long flagellum is extended laterally and contains a paraxonemal rod. *Dictyocha* has no eyespot, no contractile vacuoles, and no nuclear envelope connection to the PER. *Dictyocha* is presumed to be photoautotrophic because mixotrophy has not been reported. *Dictyocha* has three life-history stages, each being a unicellular flagellate with numerous (30–50) yellow plastids: (1) a uninucleate, silica skeleton–producing form; (2) a uninucleate, naked stage; and (3) a multinucleate, amoeboid stage. Cell structure was reviewed by Moestrup (1995).

Pedinellids

The pedinellids are unicellular flagellates having a single anterior flagellum that emerges from an anterior pit and has a paraxonemal rod located in a winglike extension (Figure 13.40). The flagellum bears two rows of tripartite hairs, as is typical for heterokonts. Pedinellids are covered with unmineralized (organic) body scales that are produced in the Golgi apparatus (Moestrup, 1995). Cells

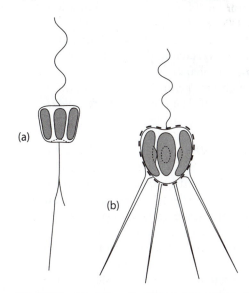

Figure 13.40 Diversity of pedinellids. (a) *Pseudopedinella* and (b) *Apedinella* represent some of the variation known for pedinellids. Plastids (shaded) typically occur in a ring. (After Throndsen 1997)

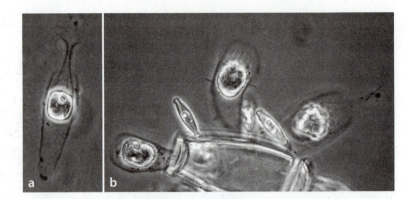

Figure 13.41 *Palatinella*, a freshwater pedinellid. (a) The single cells occur within an open-ended lorica that is usually attached to a substrate (b). A ring of tentacles emerges from the lorica. (Photos: L. W. Wilcox)

are radially symmetrical along their longitudinal axis. There is usually an anterior ring of **rhizopodia**—threadlike extensions of cytoplasm supported by microtubules—also known as tentacles. The rhizopodia surround the flagellum. There are no flagellar microtubular roots, but cytoskeletal microtubules occur in characteristic groups of three (triads). These run from the nuclear surface to the cell surface, where they may extend into the tentacles. Some forms—*Pedinella, Pseudopedinella, Apedinella,* and *Mesopedinella*—contain several (three to six) golden-brown plastids, arranged in a circle at the cell periphery. These plastids have a girdle lamella; chlorophylls a, c_1, and c_2; and fucoxanthin. *Pteridomonas danica* possesses a colorless plastid (leucoplast) bound by four membranes, the outermost continuous with the outer nuclear membrane (Sekiguchi et al. 2002). Cells of plastidless genera are phagotrophic, capturing bacterial cells. Although the pedinellids are primarily marine, a few—such as *Palatinella* (Figure 13.41)—occur in freshwaters.

APEDINELLA (Gr. *a*, not + *Pedinella*, a genus of pedinellids [Gr. *pedon*, oar]) (Figure 13.42) cells are covered by two types of body scales that occur in overlapping layers. In addition, there are six elongate spine scales (longer than the cells that produced them). Spine scales are generated within Golgi vesicles in the posterior cytoplasm. As the spine elongates, it protrudes from the end of the cell, surrounded by both the deposition vesicle and cell membrane. The spines are deposited just below the cell apex, and they move when the cell is swimming. *Apedinella* has one of the most complex cytoskeletal apparatuses ever described for a single-

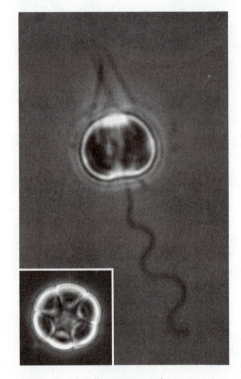

Figure 13.42 *Apedinella*. Spiny scales are at top and flagellum at bottom. Inset: Apical view of cell in which the six radially arranged plastids are evident. (Photos: A. Koutoulis)

celled protist; it is thought to participate in spine movement (Wetherbee et al. 1995).

MESOPEDINELLA (Gr. *mesos*, middle + Gr. *Pedinella*, a genus of pedinellids [Gr. *pedon*, oar]) (Figures 13.43, 13.44) was discovered in Canadian arctic ocean waters. Apple-shaped cells possess

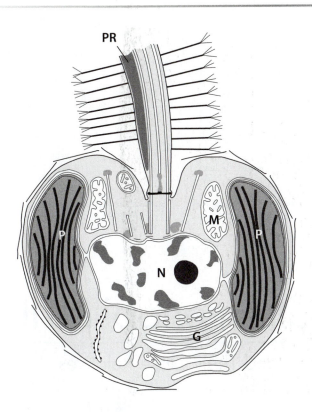

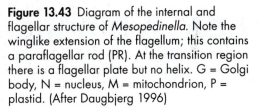

Figure 13.43 Diagram of the internal and flagellar structure of *Mesopedinella*. Note the winglike extension of the flagellum; this contains a paraflagellar rod (PR). At the transition region there is a flagellar plate but no helix. G = Golgi body, N = nucleus, M = mitochondrion, P = plastid. (After Daugbjerg 1996)

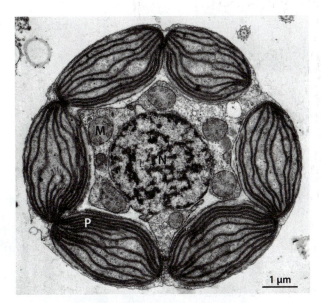

Figure 13.44 TEM view of cross-section of *Mesopedinella* cell. Note the ring of plastids (P) and mitochondria (M) surrounding the central nucleus (N). (From Daugbjerg 1996)

golden plastids that lack pyrenoids; these are arranged in a ring, as is typical of pedinellids. *Mesopedinella* cells rotate about the longitudinal axis while swimming, similarly to *Pseudopedinella*. Tentacles are absent. A cyst stage possesses a layered wall that does not include silica. *Mesopedinella* does not survive in temperatures above 8°C–10°C and is therefore restricted to cold waters (Daugbjerg 1996).

Chapter 14

Photosynthetic Stramenopiles III

Xanthophyceans, Phaeophyceans, and Their Close Relatives

Conspicuous growths of large brown seaweeds often dominate the rocky intertidal and subtidal regions of tropical, temperate, boreal, and polar oceans worldwide. Among the best known of brown algae are the giant kelps, some of which are harvested for extraction of industrially useful alginates (see Chapter 4) or for food uses. In Chile, for example, street peddlers sell the kelp *Durvillaea antarctica*, locally known as *cochayuyo*, for use in soups and other dishes. A few brown algae, transplanted from their native waters, have become notorious invasives that cause ecological damage by displacing native species. These and other brown algae, classified as Phaeophyceae, are closely related to the yellow-green algae, formally known as the Xanthophyceae or Tribophyceae. The best-known xanthophycean is the genus *Vaucheria*, which occurs in both freshwater and marine habitats and has a distinctive cylindrical body containing numerous disk-shaped plastids and many nuclei. As noted in Chapter 7, certain marine sea slugs puncture the cell wall of *Vaucheria*, suck out its plastids, and retain the plastids as photosynthetic organelles. This chapter focuses on the brown and yellow-green algae, as well as several related groups that are much less noticeable in nature.

Durvillaea antarctica

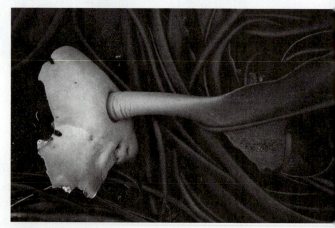

(Photo: L. W. Wilcox)

14.1 Relationships of the Brown and Yellow-Green Algae

Putative xanthophycean fossils resembling modern *Vaucheria* are known from deposits extending back to the Mesoproterozic (Butterfield 2004). These are the earliest-known fossil stramenopiles. Though certain Neoproterozoic fossils (about 555–590 million years old) are much like modern brown algae in size and shape, these fossils lack characteristics that distinguish them from red algae (Xiao et al. 2002). Fossils having distinctive features that are more confidently linked to modern brown algae include the Miocene (5–25 million year old) genera *Zonarites*, *Limnophycus*, *Cystoseirites*, *Cystoseira*, *Paleocystophora*, and *Paleohalidrys* (Parker and Dawson 1965) (Figure 14.1). Molecular phylogenetic analyses indicate that the brown Phaeophyceae and yellow-green Xanthophyceae are closely related to several smaller classes of photosynthetic stramenopiles: Chrysomerophyceae, Pinguiophyceae, Phaeothamniophyceae, and Schizocladiophyceae (Figure 14.2).

Chrysomerophyceae

The genus *Giraudyopsis* is the best-known member of the class Chrysomerophyceae. It resembles certain simple brown algae in having branched erect filaments produced by a disk-shaped filamentous structure that attaches to substrates. Because the *Giraudyopsis* body is composed of both erect and prostrate (growing along the substrate surface) filaments, it is said to be **heterotrichous**, meaning "different hairs." *Giraudyopsis* also resembles certain brown algae in having filaments composed of more than one row of cells, a condition known as **pluriseriate** filaments. However, *Giraudyopsis* lacks alginates in the cell walls, in this respect differing from brown algae.

Pinguiophyceae

Microscopic marine flagellates or non-flagellate unicells known for unusually high cellular content of omega-3 fatty acids have been grouped into the class Pinguiophyceae (*pingue* is Latin for "fat" or "grease"), primarily on the basis of molecular analysis (Kawachi et al. 2002a). The cells have one or two brown plastids with chlorophyll *a* and *c*, together with fucoxanthin and other carotenoid pigments. Plastids contain pyrenoids. The group includes *Phaeomonas*, which occurs as a flagellate or nonmotile

Figure 14.1 *Paleocystophora* (left) and *Paleohalidrys* (right) are examples of fossil brown algae that are thought to be related to modern *Fucus* and *Ascophyllum*. (a: From Parker and Dawson 1965; b: B. Parker)

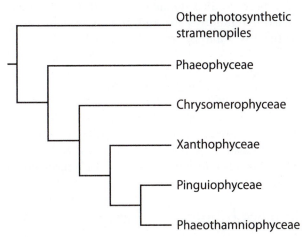

Figure 14.2 Relationships of xanthophyceans, phaeophyceans, and allied classes.

unicell (Honda and Inouye 2002); the nonmotile unicells *Pinguiochrysis* (Kawachi et al. 2002b) and *Pinguiococcus* (Andersen et al. 2002); unicellular or colonial *Glossomastix* (O'Kelly 2002); and the loricate unicell *Polypodochrysis* (Kawachi et al. 2002c).

Phaeothamniophyceae

Named for the freshwater genus *Phaeothamnion* (Figure 14.3), the class Phaeothamniophyceae was established on the basis of distinctive molecular sequence data, pigment composition, and cellular characteristics such as lack of obvious chrysolaminarin (chrysolaminaran) vacuoles (Bailey et al. 1998). Cellulose occurs in cell walls. *Phaeothamnion* (Gr. *phaios*, dusky + Gr. *thamnion*, small shrub) is a 1 cm-long branched filament whose cells contain one to several platelike, olive-brown plastids. It grows attached to freshwater substrates such as submerged mosses. Zoospores, released from cells through lateral pores, have two laterally inserted flagella; the longer has tripartite tubular hairs and is directed toward the cell anterior, while the shorter, smooth flagellum faces toward the posterior. The flagellar root system resembles those of zoospores produced by brown and yellow-green algae. Resting cysts may be produced, but they lack silica. Other phaeothamniophyceans are *Chrysapion*, *Chrysoclonium*, *Chrysodictyon*, *Phaeobotrys*, *Phaeogloea*, *Phaeoschizochlamys*, *Selenophaea*, *Sphaeridiothrix*, *Stichogloea*, *Tetrachrysis*, *Tetrapion*, and *Tetrasporopsis*. The cellulose-synthesizing complexes of *Phaeothamnion confervicola*, *Phaeoschizochlamys mucosa*, and *Stichogloea doederleinii* have been studied (Okuda et al. 2004). Together, phaeothamniophyceans and pinguiophyceans may be sister to Xanthophyceae (Andersen 2004).

14.2 Xanthophyceae

Xanthophyceae (xanthophytes or xanthophyceans) are also known as Tribophyceae (tribophytes or tribophyceans), after the common genus *Tribonema*. They are also informally called the yellow-green algae.

Xanthophycean Cell Biology

Cell walls of xanthophyceans are composed primarily of cellulose, with silica sometimes also present,

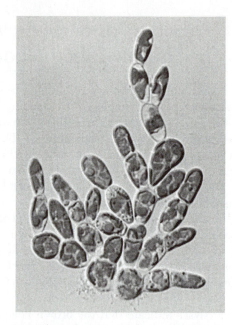

Figure 14.3 *Phaeothamnion*, a filamentous freshwater alga, previously thought to be a chrysophycean but now regarded as a close relative of tribophyceans. (Photo: R. Andersen)

and sometimes consist of two overlapping halves. The main photosynthetic storage product is lipid, in the form of cytoplasmic droplets. The soluble polysaccharide chrysolaminarin is probably also present within a cytoplasmic vacuole. The cells usually possess several to numerous discoid plastids arranged at the periphery of the cytoplasm, and many species display pyrenoids within the plastids. In addition to chlorophyll *a*, the porphyrin derivative chlorophyll *c* is also present, though in low amounts. Accessory pigments include β-carotene, the xanthophylls diatoxanthin and diadinoxanthin, and other minor pigments. Fucoxanthin, an accessory pigment otherwise widely present among photosynthetic stramenopiles, is absent from xanthophyceans. Consequently, their chloroplasts appear green or yellow-green, with the result that xanthophyceans are sometimes difficult to distinguish from morphologically similar green algae. Plastids of green, but not yellow-green, algae produce the carbohydrate reserve starch, which stains blue-black when a solution of iodine and potassium iodide is added. This test can often be used to distinguish starch-negative yellow-green algae from morphologically similar but starch-positive green algae. Some xanthophyceans may also resemble certain eustigmatophycean taxa (Chapter 13) so closely that they generally cannot be distinguished without the use of chromatography to assess pigment composition, ultrastructural comparison, or molecular phylogenetic analysis.

Xanthophycean Reproduction

Asexual reproduction in xanthophyceans can occur by means of nonflagellate autospores, aplanospores, or thick-walled cysts or by flagellate zoospores, depending upon the genus or environmental conditions. Autospores are developmentally distinct from zoospores, aplanospores develop similarly to zoospores up to the point of flagellar production, and zoospores are flagellate asexual reproductive cells (see Chapter 1). Such reproductive cells are typically produced by division of cellular cytoplasm and ultimately released from the confines of parental cell walls. Their function is to increase and disperse the population. Cysts form by modification of entire vegetative cells and help species survive conditions unsuitable for growth of the vegetative phase or populations. Silica may occur in cyst walls and cysts may consist of two overlapping parts. Sexual reproduction involving isogamy or oogamy has been observed in a few cases. Flagellate species, zoospores, and flagellate gametes usually have two typical stramenopile flagella—a longer anterior one with tubular tripartite hairs occurring in two opposite rows, and a shorter posterior one lacking stiff hairs. *Vaucheria* is an exception in that it has multiflagellate zoospores.

   Most flagellate cells have a photoreceptor swelling on the posterior flagellum, and it is aligned such that the swelling lies near an eyespot located within the plastid (Hibberd and Leedale 1971; Hibberd 1990b). However, *Vaucheria* zoospores and sperm lack such eyespots. The blue-light receptor aureochrome occurs in *Vaucheria frigida* and some other photosynthetic stramenopiles (Takahashi et al. 2007).

Xanthophycean Diversity

There are about 90 genera and 600 species of xanthophyceans, occurring primarily in freshwaters or soil. Many forms are considered to be rare (Hibberd 1990b). Xanthophyceans include a variety of morphological types, including flagellates such as *Chloromeson*, non-flagellate unicells such as *Pseudopleurochloris antarctica* (Andreoli et al. 1999), colonies of cells embedded in mucilage, amoeba-like

Figure 14.4 *Chloridella.* (Photo: L. W. Wilcox)

rhizopodial forms, filaments such as *Tribonema*, and multinucleate bodies known as **siphonaceous coenocytes**, illustrated by *Botrydium* and *Vaucheria*. However, phylogenetic analysis indicates that morphological convergence has commonly occurred during xanthophycean evolution. Siphonous forms, for example, have evolved at least twice (Maistro et al. 2007). Such studies also suggest that Xanthophyceae probably originated from a unicellular ancestor (Nigrisolo et al. 2004). A number of genera not covered here have been demonstrated to belong to the Xanthophyceae by SSU rDNA sequence analysis (e.g., Potter et al. 1997), ultrastructural examination (e.g., Hibberd and Leedale 1971), or both. These include *Aeronema*, *Capitulariella*, *Chaetopedia*, *Botrydiopsis*, *Brachynema*, *Bumilleria*, *Heterococcus*, *Heterotrichella*, *Mischococcus*, *Xanthonema*, *Bumilleriopsis*, *Pleurochloris*, *Sphaerosorus*, and *Chlorellidium*. Several of these genera are known to be polyphyletic (Maistro et al. 2007).

CHLORIDELLA (Gr. *chloros*, green) (Figure 14.4) is a unicellular, spherical form that reproduces by autospore formation, much like the green alga *Chlorella*.

OPHIOCYTIUM (Gr. *ophis*, serpent + Gr. *kytos*, cell) (Figure 14.5) has distinctive elongate cells that may be straight, curved, or spirally twisted, sometimes with spines at the ends. Cells are attached to surfaces by small stalks. Sometimes cells are aggregated into a colony-like arrangement. Such cell clusters result from the attachment and germination of multiple zoospores at the distal end of a

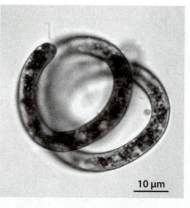

Figure 14.5 A highly elongate and coiled cell of *Ophiocytium* with many plastids. (Photo: L. W. Wilcox)

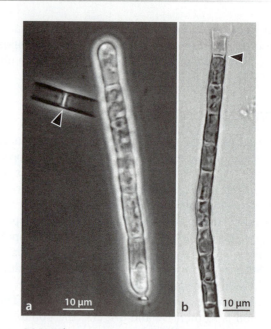

Figure 14.6 *Tribonema* is an unbranched filament characterized by cell walls that occur as two overlapping halves that look like the letter H when focused midway through the cell. The distinctive wall (arrowheads) can be detected by observing individual cell wall pieces that have separated from filaments (a) or broken ends of filaments (b). (Photos: L. W. Wilcox)

parental cell. The cell wall is composed of two interlocking parts of unequal size, the apical portion being smaller. Cells can elongate by the addition of more cell wall units at the distal end. Zoospores or aplanospores are released by detachment of the lid-like apical wall piece. There are multiple discoid plastids per cell. *Ophiocytium* is widely distributed in freshwaters but is not a monophyletic group (Maistro et al. 2007).

TRIBONEMA (Gr. *tribo*, to rub or wear down + Gr. *nema*, thread) is an unbranched filament having a single row of cells (Figure 14.6). *Tribonema* may occur as bright green floating mats in still waters, including bog pools, and is more abundant in cooler seasons (spring and fall) (Hibberd 1990b). There is a worldwide distribution in freshwaters. A distinctive feature is the cell wall, which consists of units composed of two rigid, basally attached, open-ended cylinders. Each cell is enclosed by two such cylinders that overlap at their open ends (Figure 14.7). The closed ends of two back-to-back cylinders form the cross-wall between adjacent cells. If a filament is broken, the damaged end cells will typically show what are known as H-shaped pieces (see Figure 14.6). These are views of the double cylinder wall unit described above, in optical section. A new double cylinder is constructed at each cell division, beginning with development of a ring of wall material deposited at the cleavage plane. The ring ultimately closes to generate the intermediate piece (the crossbar of the H). The cell then elongates until the lower half of the upper cell and the upper half of the lower cell occupy the new cell wall cylinders

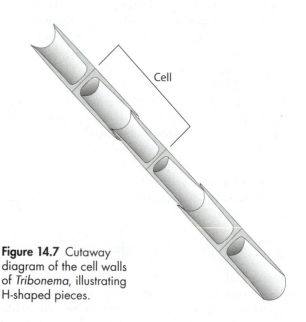

Figure 14.7 Cutaway diagram of the cell walls of *Tribonema*, illustrating H-shaped pieces.

(Sluiman 1993). (Cell wall development is similar in the green alga *Microspora*, which also exhibits H-shaped pieces at the light-microscopic level.)

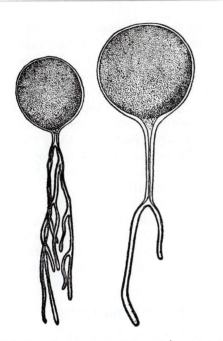

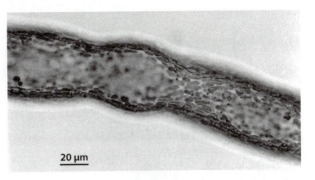

Figure 14.9 A portion of the coenocytic, tubular body of *Vaucheria*. The numerous plastids are located in the thin layer of cytoplasm surrounding a large central vacuole. (Photo: L. W. Wilcox)

Figure 14.8 *Botrydium,* a coenocytic vesicle containing many plastids and nuclei. Two species are shown. (From Smith, G. M. *Fresh-Water Algae of the United States.* © 1950 McGraw-Hill Companies. Reproduced with permission of the McGraw-Hill Companies)

One to several plastids without pyrenoids occur in each cell. Asexual reproduction is by one or two zoospores produced per cell; these are released by separation of parental cell walls at the region of overlap. The ultrastructure of *Tribonema* zoospores and development of their flagellar hairs has been studied (Massalski and Leedale 1969; Leedale et al. 1970). Aplanospores and resting cysts are also known, and isogamous sexual reproduction has been reported. There are at least two clades of *Tribonema* that differ in filament width, and many of the morphologically defined species are not monophyletic (Zuccarello and Lokhorst 2005).

BOTRYDIUM (Gr. *botrydion,* small cluster of grapes) have small saclike bodies that grow in clusters on damp soil, attached by means of subterranean colorless rhizoidal filaments (Figure 14.8). Individual vesicles may reach several millimeters in size and thus be visible to the unaided eye. *Botrydium* and *Vaucheria* (described next) are exceptions to the general case that tribophyceans are microscopic. The cytoplasm occupies a thin peripheral layer and contains many plastids and nuclei. Hence

the body morphology is described as coenocytic and siphonous (multinucleate and tubelike, without cross-walls). When a water film is present, many uninucleate, biflagellate cells (zoospores or gametes) are released from the sac. When water is less abundant, aplanospores or resistant cysts may be produced. Sexual reproduction is by isogametes or anisogametes; gametes are not formed in specially differentiated structures (compare to *Vaucheria,* below).

VAUCHERIA (named for Jean Pierre Etienne Vaucher, a Swiss clergyman and botanist) has a large, cylindrical, coenocytic (multinucleate) body (Figure 14.9), with cross-walls formed only where zoospores or gametes are produced. The coenocytes are branched and have indeterminate apical tip growth. Cell wall extensibility during the process of branching has been analyzed (Mine et al. 2007). There is a peripheral layer of cytoplasm with numerous nuclei and discoid plastids surrounding a large central vacuole. Organellar streaming is a conspicuous attribute of *Vaucheria,* and presumably functions to circulate materials within *Vaucheria*'s very large cells. Chloroplasts reorient according to changes in the light environment. Ott and Brown (1972) observed that the nuclear envelope remains intact during mitosis.

Vaucheria zoospores are distinctive in being very large, multinucleate, and multiflagellate. They result from incomplete cleavage and are described as compound zoospores or synzoospores (Hibberd 1990b). Zoospore production occurs in the apical regions of vegetative filaments, where organelles accumulate, and then the tip cytoplasm is segregated from the

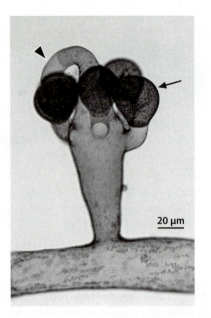

Figure 14.10 A species of *Vaucheria* having gametangia borne on a specialized portion of the coenocyte. A whorl of oogonia (arrow) surrounds an antheridium (arrowhead). (Photo: L. W. Wilcox)

rest of the coenocyte by cross-wall production. The segregated region functions as a zoosporangium; the process of zoospore development is known as zoosporogenesis. A pair of centrioles is associated with each of the numerous nuclei within the zoosporangium. These centrioles generate a pair of flagella that extend into internal vesicles. These vesicles fuse, resulting in the formation of a large pool of flagella that then migrate to the surface of the developing zoospore. At maturity, naked zoospores are released by rupture or disintegration of the parental wall (Ott and Brown 1975). The presence of the cross-wall prevents loss of additional body cytoplasm upon zoospore release. Upon settling to a substrate, flagella retract into the cell body, and a new cell wall develops. Asexual reproduction can also involve large multinucleate aplanospores. Whereas asexual reproduction of *Vaucheria sessilis* is stimulated by cool temperatures (12°C–15°C), sexual reproduction appears to be controlled by photoperiod, occurring more readily under long-day (18-hour days) than short-day (8-hour days) conditions, probably as a result of greater production of photosynthate under long-day conditions (League and Greulach 1955).

Sexual reproduction is oogamous. Specialized antheridia generate large numbers of colorless,

uninucleate sperm. Oogonia each contain a single uninucleate egg, borne either on special branches (Figure 14.10) or directly upon the main axis. Antheridial structure may be quite variable and useful in defining species. Sperm flagella are atypical for stramenopiles but somewhat similar to sperm of certain brown algae in that the anterior flagellum that bears tripartite hairs is relatively short, and the smooth posterior flagellum is longer. Sperm enter an egg via a pore in the oogonial wall. A thick-walled resting zygote typically remains within the oogonial wall, attached to the parental body.

Vaucheria may form extensive dark-green mats on soil or sand, on brackish banks of salt marshes (Simons 1974), or on submerged substrates in freshwaters or shallow marine waters. There are 70 or more species.

14.3 Phaeophyceans (Brown Algae)

The brown algae include more than 250 genera and more than 1500 species (Norton et al. 1996), ranging in structure from microscopic filaments to giant kelps many meters in length. Some phaeophycean species display the greatest organ, tissue, and cellular specialization known among protists. Many have photosynthetic blades, as well as specialized blade-bearing stipes and attachment structures known as holdfasts. Some brown algae possess specialized transport cells bearing a striking resemblance to sieve elements of vascular land plants.

Brown algae also have important ecological impacts. *Petroderma maculiforme* occurs both free-living and as a symbiont in the marine lichen *Verrucaria tavaresiae* (Sanders et al. 2005). Large kelps can form underwater forests that harbor many ecologically and economically important species. Such forests can have productivity rates as high as 1 kg C m^{-2} yr^{-1}, with growth rates highest in the cool season of the year. Whereas some phaeophycean species are annual, others are perennial—some living up to 15 years. As a consequence of their size, productivity, and longevity, brown algae form large biomasses in intertidal and subtidal coastal regions throughout the world, including surprisingly abundant growths in deep tropical waters (Graham et al. 2007). The common genus *Fucus* (Figure 14.11) is abundant in the intertidal region of temperate rocky shores of the Northern Hemisphere and has been used as a model system to

Figure 14.11 A monospecific population of *Fucus* on a rocky temperate shoreline. (Photo: C. Taylor)

study polarity formation in zygotes (e.g., Hadley et al. 2006). Although most brown algae are marine, some occur in freshwaters. Some freshwater brown algae are closely related to marine species, but *Bodanella*, *Heribaudiella*, and *Pleurocladia* seem to be restricted to freshwaters (McCauley and Wehr 2007).

Phaeophycean Cell Biology

Phaeophyceans have distinctive cell walls, characteristic plastids, and cytoplasmic physodes, features that help to explain their ecological and economic importance. Cellular differences also distinguish phaeophyceans from close relatives known as the Schizocladiophyceae.

Cell walls

The cell walls of brown algae generally contain three components: alginic acid, cellulose, and sulfated polysaccharides. **Alginic acid** (alginate) is a polymer of mannuronic and guluronic acids and their Na^+, K^+, Mg^{2+}, and Ca^{2+} salts. Alginates are located primarily within the intercellular matrix, where they confer flexibility upon the body, help prevent desiccation, and function in ion exchange. Certain brown algae

are harvested for extraction of alginates because these materials have many useful applications (see Chapter 4). Alginates are the primary component of the brown algal wall matrix, in some cases constituting up to 35% of body dry weight. The bodies of most brown seaweeds contain a 1:1 ratio of mannuronic and guluronic acids, but the proportions of these acids can vary with season, age and species, tissue type, and geographic location (Kraemer and Chapman 1991). Several investigators had proposed that alginate compositional differences might confer distinct mechanical properties, providing adaptive advantage in different environments, and Kraemer and Chapman (1991) tested this hypothesis. They grew the phaeophycean *Egregia menziesii* for 6–10 weeks in flowing seawater at three different current levels and then tested body strength and performed correlative analysis of alginic acid composition. Although *Egregia* grown in the highest-energy environments were about twice as strong and stiff as those growing in the lowest current level, there was no correlation with change in percentages of mannuronic and guluronic acids. Thus, alternate explanations for such differences in mechanical strength must be sought.

Brown algal cell walls are also notable for containing cellulose. The brown algae are closely related to branched filaments such as *Schizocladia ischiensis* (Schizocladiophyceae), whose cell walls possess alginates but lack cellulose (Kawai et al. 2003). Cellulose provides structural support but is usually a minor constituent of the brown algal cell wall, making up 1%–10% of body dry weight. Cellulose occurs as ribbonlike microfibrils that are generated at the cell surface by protein complexes embedded in the cell membrane (Tamura et al. 1996). Such protein complexes are deposited in the cell membrane by fusion of Golgi vesicles with it. Brown algal cell membranes also contain arrays of protein particles hypothesized to produce cell wall matrix polysaccharides (Reiss et al. 1996). Actin microfibrils probably control the orientation of cellulose microfibrils in brown algae (Katsaros et al. 2002; Katsaros et al. 2006).

Fucans (also known as fucoidins or ascophyllans) are polymers of L-fucose and additional sugars that are sulfated. Antibodies have been used to demonstrate the presence of fucose-rich fucans in the cell walls of *Fucus*. In these molecules, there may be a 1:1 occurrence of fucose and sulfate, suggesting that each sugar monomer is sulfated. Sulfate linkage appears to take place in the Golgi apparatus (Callow et al. 1978) and may occur only in specialized cells

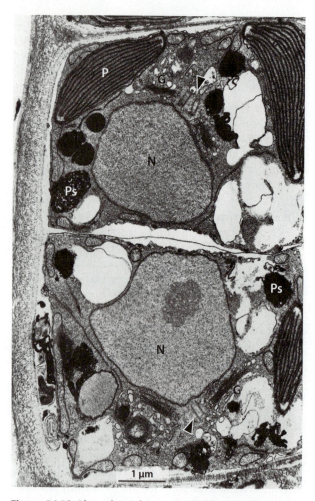

Figure 14.12 Physodes. Ultrastructural features of recently divided cells of the brown alga *Pylaiella littoralis* include irregular electron-opaque structures that contain tannins and that are known as physodes (Ps). Note centrioles (arrowheads) located near nuclei (N), and Golgi bodies (G), on each side of the centrioles. P = plastid. (From Markey and Wilce 1975, © Springer-Verlag)

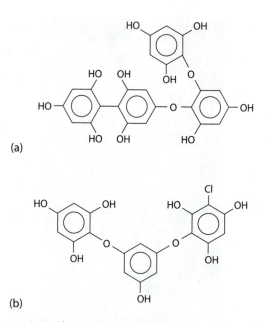

Figure 14.13 Chemical structures of some polyphenolic tannins found in brown algae. (a) A polyphloroglucinol having both biphenyl and ether linkages and (b) a chlorinated polyphloroglucinol. (From Ragan 1981, in Wynne, M. J. and C. S. Lobban (eds.), *Biology of Seaweeds.* © University of California Press and Blackwell Scientific, Inc.)

in kelps. The functions of fucans are not well understood, but fucans are thought to play an important role in firmly anchoring *Fucus* zygotes and germlings to intertidal substrates so that they are not washed off in heavy waves. It is possible that fucans serve a similar function in other brown algae.

Plastids and photosynthetic reserve products

Depending upon the genus, there may be from one to many plastids per cell. Plastids contain chlorophyll *a*, two forms of chlorophyll *c*, β-carotene, violaxanthin, and relatively large amounts of fucoxanthin, which confers a brown color. Plastids typically possess a girdle lamella, a thylakoid stack that runs continuously beneath the plastid surface. As is more generally true for photosynthetic stramenopiles, there is a periplastidal endoplasmic reticulum (PER), and the phaeophycean PER is continuous with the nuclear envelope. Pyrenoids occur in the plastids of some brown algae, but not others.

The main photosynthetic reserve material is **laminaran**, a β-1,3-glucan similar to the chrysolaminarin of other photosynthetic stramenopiles. Laminaran is water soluble and is stored in cytoplasmic vacuoles. Phaeophycean cells may also contain the low-molecular-weight compounds mannitol, sucrose, and/or glycerol, which are thought to balance osmotic pressure and lower the cytoplasmic freezing point, an advantage in cold conditions. Mannitol, a 6-carbon sugar alcohol, may account for 20%–30% of the dry weight of brown algal bodies and may constitute up to 65% of the sap translocated in the specialized translocatory cells of kelps (Schmitz 1981).

Physodes

Brown algal cells often contain cytoplasmic spheres known as **physodes** (from a Greek word meaning

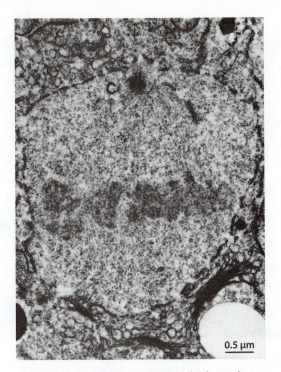

Figure 14.14 Metaphase in *Cutleria cylindrica*, showing one of the polar centrioles (at top) and the nearly intact nuclear envelope. (From La Claire 1982b)

darkly in transmission electron micrographs (Figure 14.12) and are refractile when viewed with light microscopy. Physodes contain polyphenolic polymers of phloroglucinol known as phlorotannins (Figure 14.13). Such compounds would precipitate cellular proteins were they not separated from the cytoplasm by an enclosing membrane. Physodes can secrete their contents by exocytosis, and they are commonly thought to function in herbivore deterrence because the phlorotannins can interfere with normal nerve and muscle function. These compounds are also constituents of the brown algal cell wall, may serve as UVB (280–320 nm) screens that protect brown algae from radiation damage, and have additional functions (see review by Schoenwaelder 2002).

The cytoskeleton and cell division

Cells of brown algae have not been observed to contain a cortical (peripheral) microtubular cytoskeleton (Reiss et al. 1996; Katsaros et al. 2006), and microtubules are generally few in nondividing cells. However, a pair of centrioles is thought to generally occur in interphase cells of brown algae (Figure 14.14), often occurring in a cleft in the nuclear envelope (Brawley and Wetherbee 1981). The centrioles can generate flagella during development of flagellate reproductive cells, a process that appears to occur readily in all brown algae. Just prior to mitosis, centrioles are duplicated, and then pairs of centrioles migrate to opposite poles of the cell early in prophase. Centriolar

"bubble" or "bladder"). Often abundant in epidermal and dividing cells, physodes may also occur in reproductive cells. The Golgi apparatus and endoplasmic reticulum produce physodes, which stain

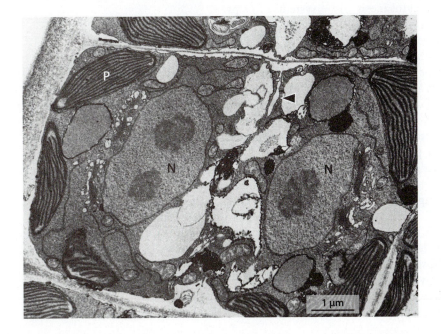

Figure 14.15 Cytokinesis in *Pylaiella littoralis*, illustrating the process of furrowing (arrowhead). N = nucleus, P = plastid. (From Markey and Wilce 1975 © Springer-Verlag)

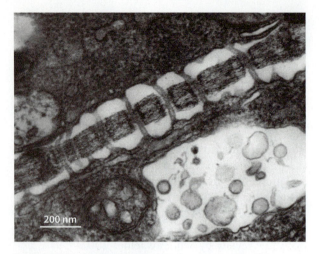

Figure 14.16 Plasmodesmata in the cross-walls of a brown alga. (TEM: J. La Claire)

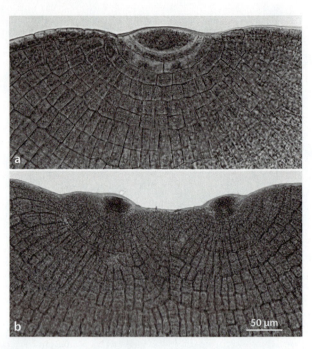

Figure 14.17 Apical meristem. (a) A parenchyma (tissue)-forming apical cell of *Dictyota*. (b) Daughter apical cells resulting from the division of the apical cell. Further production of separate tissues by each daughter cell will result in dichotomous branching of the body. (Photos: L. W. Wilcox)

migration is an early event in the establishment of division polarity, and it is critical in determining the direction of cytokinesis. In *Dictyota dichotoma*, two sets of microtubules diverge from the daughter centrioles and interdigitate, forming a plane that will be the eventual site of cytokinesis (Katsaros and Galatis 1992).

The nuclear envelope of dividing brown algal cells usually remains intact (see Figure 14.14) (except for small open regions at the pores known as polar fenestrations) until anaphase, when the envelope disintegrates completely (Brawley and Wetherbee 1981). Following the end of mitosis, cytokinesis in most brown algae occurs by infurrowing of the cell membrane (Figure 14.15), and the developing furrow is associated with microtubules. In *Macrocystis pyrifera*, actin filaments develop just beneath the cell membrane in the plane of cytokinesis and seem to be involved in the process of cytokinesis in highly vacuolate cells (Varvarigos et al. 2005). In *Fucus* and the related genus *Ascophyllum*, a cell plate may develop from the center of the cell toward the periphery. Mitosis and cytokinesis in some brown algae (*Petalonia fascia* and *P. zosterifolia*) are known to peak two to four hours after sunset and to be completed in about two hours (Kapraun and Boone 1987).

Plasmodesmata are generated during cytokinesis and probably occur in the cross-walls of all brown algae (Figure 14.16), but they are absent from their relatives the Schizocladiophyceae (Kawai et al. 2003). Plasmodesmata confer a high degree of intercellular continuity and allow symplastic (cytoplasm to cytoplasm) communication. For example, modified plasmodesmata allow for the cell-to-cell transport of photosynthates through the internal translocatory systems of kelps.

Growth Modes and Meristems

The bodies of brown algae may be filamentous, aggregates of filaments, or composed of tissues analogous to those of land plants. Bodies composed of aggregated filaments (lacking true tissues) are described as **pseudoparenchymatous**. In contrast, brown algae composed of tissues are described as **parenchymatous** and can form solid axes, blades, or complex bodies with specialized blade, stipe, and holdfast regions. **Blades** are photosynthetic organs that may also bear reproductive structures. **Stipes** are cylindrical structures to which blades are attached. **Holdfasts** are structures specialized for attachment to substrates such as rocks.

Some brown algae display **diffuse growth**, meaning that there is no localized meristematic region. Rather, cell division occurs throughout the body.

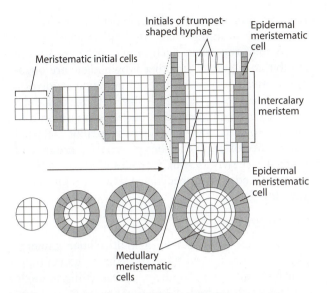

Figure 14.18 Development of the intercalary meristem of *Chorda filum*, as viewed in longitudinal section (top row) and in cross-section (lower row). Meristematic cells (shaded) differentiate from unspecialized initials. (After Kogame and Kawai 1996, © Blackwell Science Asia)

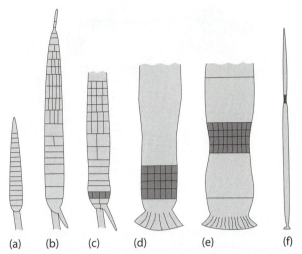

Figure 14.19 Growth stages in *Chorda filum*. Growth begins with diffuse (nonlocalized) cell divisions (a, b), is followed by a stage of basal meristematic activity (shaded region) (c, d), and culminates in development of a well-defined intercalary meristem (shaded region) (e, f). (After Kogame and Kawai 1996, © Blackwell Science Asia)

Other brown algae have **meristems**, which are localized regions of cell division. The simplest type of brown algal meristem is a single apical cell that repeatedly divides transversely, yielding a filament that is one cell in width, known as a **uniseriate filament**. Other brown algal apical meristems are composed of one or more cells that can divide in additional directions, generating three-dimensional cylindrical, ribbonlike, or thickened and branched bodies such as that of *Fucus*. A row of apical cells occurs in *Padina*, generating a flat, fanlike body, and similar apical cell arrays occur in related algae, such as *Dictyota* (Katsaros 1995) (Figure 14.17). In the kelps there is typically an **intercalary meristem**, located between stipe and blade tissues. This type of meristem generates tissues in two directions, increasing the length of the body. Kelps also possess a surface meristematic region whose activity increases girth; this is known as the **meristoderm**. Meristoderm activity could be considered somewhat analogous to that of the cambia of higher land plants. Development of the intercalary meristem in *Chorda* has been intensively studied (Figure 14.18) (Kogame and Kawai 1996). *Chorda* growth occurs in three distinct stages: early diffuse growth, growth from a basal meristem, and intercalary meristematic growth (Figure 14.19). Finally, some brown algae produce an unusual type of

meristem that occurs at the base of a hair. Such algae display what is known as **trichothallic growth** (derived from the Greek words *thrix*, meaning "hair," and *thallus*, meaning "body").

Reproduction

Most brown algae have a type of sexual life cycle known as **alternation of generations** that involves flagellate reproductive cells. A multicellular diploid **sporophyte** generation produces flagellate haploid meiospores by means of meiosis. Such meiospores grow by mitosis into multicellular haploid **gametophyte** generations. At maturity, and under appropriate environmental conditions, gametophytes produce flagellate gametes that mate. A successful mating results in a diploid zygote, which may grow by mitosis into a new sporophyte. The sporophyte generation thus alternates with the gametophyte generation. Asexual reproduction may also occur, and this involves the production of asexual zoospores.

Flagellate reproductive cells

As we have noted, brown algae display three types of flagellate reproductive cells: meiospores, asexual zoospores, and gametes. Often these possess typical stramenopile flagella, but in some cases the anterior

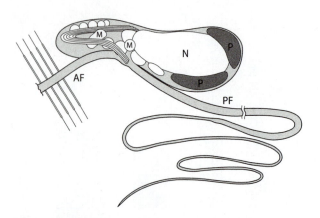

Figure 14.20 Diagrammatic representation of a laminarialean sperm cell, showing lateral emergence of heterokont flagella. AF = anterior flagellum, M = mitochondrion, N = nucleus, P = plastid, PF = posterior flagellum. (From Henry and Cole 1982b, by permission of the *Journal of Phycology*)

flagellum is shorter than the posterior one, and in other cases there is only a single anterior flagellum. Flagella are lacking altogether from the female gametes of some species. Phaeophycean motile cells are distinctive in that their flagella typically emerge laterally, rather than apically or subapically, as in other photosynthetic stramenopiles (Figure 14.20). Swimming involves beating of the anterior but not the posterior flagellum. The anterior flagella of zoospores, meiospores, and settling gametes possess a coiled region (or knob) that is involved in attachment to the substrate (Goldstein 1992). Flagellate reproductive cells of brown algae usually have eyespots, located within the chloroplast, but the sperm of one group lack eyespots, and both the sperm and the meiospores of certain forms lack eyespots. The ultrastructure of *Ectocarpus* male gametes was studied in detail by Maier (1997a,b).

Some of the brown algae are **isogamous**, which means that the flagellate gametes are morphologically indistinguishable while motile. Such brown algae are said to exhibit **isogamy**. Even though they look alike, one gamete typically settles soon after it is released, whereas the other gamete actively swims for a longer period of time (see, e.g., the study by Clayton [1987] of *Ascoseira mirabilis*). Thus, gametes can be functionally differentiated. The zygotes of isogamous species can often be recognized by the presence of two eyespots, one contributed by each gamete. Isogamous brown algae exhibit biparental inheritance of mitochondria and plastids. Sexual

reproduction in some other phaeophyceans is **anisogamous**, meaning that one flagellate gamete is larger than the other. Such brown algae are said to exhibit **anisogamy**. Yet other brown algae are **oogamous**, meaning that sexual reproduction involves a nonmotile or only transiently flagellate egg cell and a smaller flagellate sperm cell. Such species are said to display **oogamy**, and they typically exhibit maternal inheritance of organelles but paternal inheritance of centrioles (Lewis 1996b). Gametes of all these types are produced within specialized structures known as **gametangia**.

Gametangia

Gametes are produced by a multicellular gametophyte generation, within specialized gametangia that have multiple chambers, each holding a single gamete. Such multichambered gametangia are variously known as plurilocular (meaning "many chambered") gametangia, plurisporangia, mitosporangia, or **plurilocs** (Figure 14.21). In the case of oogamous phaeophyceans, plurilocular gametangia have been evolutionarily reduced to single cells that lack internal compartmentation.

It may seem quite confusing that gametes are generated within structures often described as "sporangia," but there is an explanation. In many cases the cells produced by plurilocs are gametes that fuse in pairs during mating. However, plurilocs sometimes produce asexual zoospores that do not mate but rather develop into bodies like those of the parent, a situation for which use of the term *sporangium* is appropriate. In *Ectocarpus* and some other genera, production of zoospores in plurilocs can result in a series of identical generations, unpunctuated by occurrence of sexual reproduction (Papenfuss 1935; Pedersen 1981). Furthermore, unmated gametes of some brown algae—*Ascoseira mirabilis* (Clayton 1987) and *Colpomenia peregrina* (Yamagishi and Kogame 1998), for example—readily develop into multicellular bodies. In the latter case, more frequent development of larger, brighter-pigmented unfertilized female gametes (versus smaller, paler male gametes) results in populations of macroscopic gametophytic seaweeds that exhibit a female:male ratio of 19:1. The development of gametes into multicellular bodies without a mating process is known as **parthenogenesis**. Brown algae that reproduce exclusively by means of plurilocular sporangia-produced zoospores are said to exhibit a direct, or monophasic, life history. This is in contrast to the more common

4 μm

Figure 14.21 A developing plurilocular gametangium of *Cutleria hancockii.* (From La Claire and West 1979, © Springer-Verlag)

occurrence of alternation of generations (see below). Parthenogenesis is relatively common among brown algae exhibiting isogamy or anisogamy, where both gametes possess flagella, and less common in oogamous species.

Mating

Reproductive cells are usually released from plurilocular gametangia through a single apical pore, though in some forms there is a series of peripheral openings through which zoospores or gametes escape. In many cases it is known that low-molecular-weight, highly volatile hydrocarbons produced by nonmotile female gametes function as pheromones, acting to attract sperm, and in some cases also inducing antheridial dehiscence. Such pheromones are known from divergent brown algal groups, including the relatively simple *Ectocarpus*, the highly specialized

Fucus, and the kelps. Thus, pheromones are probably produced by most, if not all, sexually reproducing brown algae. Video observations of *Ectocarpus siliculosus* gametes show that in the absence of pheromone, flagellate gametes swim in a straight line or wide loops, but in the presence of pheromone, they swim in smaller and smaller loops (mediated by beating of the posterior flagellum), until the source of the pheromone (the female gamete) has been contacted (Müller 1981).

Pheromones also form part of the basis for species isolation. The ability of two brown algal taxa to cross or hybridize depends upon successful completion of five stages in reproduction: (1) pheromone recognition, (2) gamete recognition, (3) plasmogamy (gamete fusion), (4) zygote growth into sporophytes, and (5) sporophyte meiosis. Flagellate gametes that are attracted to settled gametes make contact with the tip of the anterior hairy flagellum. Antibodies and lectins—carbohydrate-binding proteins with high specificity and affinity—have been used to establish that eggs or settled gametes bear specific glycoproteins, while the motile gamete carries receptors for these glycoproteins (see review by Lewis 1996b). Lectin-binding sites on gametes of *Ectocarpus siliculosus* can be visualized by the use of fluorescence microscopy (Maier and Schmid 1995). Plasmogamy occurs at its maximal level about 20 minutes after gamete contact, and karyogamy occurs about 90–120 minutes later (Brawley 1992).

Sporophyte function

Zygotes of brown algae develop by repeated mitosis and cytokinesis into a multicellular, diploid sporophyte generation. When brown algae are mature, plurilocs may develop on the sporophytes of some (but not all) brown algae. Such plurilocs produce flagellate zoospores that grow into either sporophytes resembling the parent or smaller bodies (microbodies) with plurilocs. Sporophytes are unique in producing another type of reproductive structure, **unilocular sporangia**, also known as unisporangia, meiosporangia, or **unilocs** (Figure 14.22). The term *unilocular* means "one chamber." Unilocs do not occur on the gametophyte generation.

The first division within uniloc initials is often assumed to be meiotic. However, in many cases, unilocs produce reproductive cells without completing meiosis. In such cases the reproductive cells have the same chromosome number (DNA level) as the parent sporophyte. When meiosis occurs, the resulting daughter

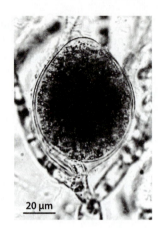

Figure 14.22 A unilocular sporangium of *Tinocladia*. (Photo: J. La Claire)

20 μm

nuclei in meiospores are haploid. Sex-determining chromosomes are known to occur in at least some brown algae, and these segregate at meiosis, determining the sex of gametophytes that develop from meiospores. In *Saccorhiza polyschides*, for example, a large X chromosome and a smaller Y chromosome pair are segregated at meiosis (Evans 1965). In some phaeophyceans only four meiospores—known as tetraspores—are produced per uniloc. In other phaeophyceans each meiotic product may continue to divide mitotically, yielding 16, 32, 64, 128, or more haploid meiospores. However, no intervening wall deposition occurs. Consequently, unilocular sporangia are not subdivided into walled chambers as are plurilocular sporangia. Environmental factors may determine whether unilocs or plurilocs are produced by sporophytes. In *Ectocarpus siliculosus*, for example, unilocular sporangia are produced when the temperature is below 13°C, whereas above 13°C only plurilocs are produced (Clayton 1990).

By producing meiospores of diverse types as a result of Mendelian processes, the sporophyte generation of brown algae increases the genetic diversity of subsequent gametophyte populations, thereby enhancing the evolutionary potential of the species. The larger the sporophyte, the more diploid cells are available to function as unilocs, with the result that more spores of greater genetic diversity can be produced. This feature helps to explain the success of the larger brown algae, such as the giant kelps.

Alternation of generations

In some exceptional situations, such as occurs in the widespread *Elachista fucicola*, flagellate cells produced in unilocs develop directly into bodies similar to that of the parent, without undergoing life history phase change. This is another example of the so-called direct or monophasic type of life history. More generally, however, haploid meiospores develop by repeated mitosis and cytokinesis into multicellular, haploid gametophytes that produce gametes in plurilocs. Such brown algae display biphasic alternation of generations. Although the sporophyte generation is commonly thought of as the diploid phase, and the gametophyte as the haploid phase, polyploidy and other chromosome-level variation may occur.

In some cases the two alternating life-history phases are nearly indistinguishable in appearance; this is known as **isomorphic alternation of generations** (Figure 14.23). In other cases the two alternating phases are structurally distinct; this is **heteromorphic alternation of generations** (Figure 14.24). Most phaeophyceans exhibiting heteromorphic alternation of generations produce sporophytes that are larger than the gametophyte. The giant kelps are notable examples of extreme size divergence, with sporophytes growing to lengths of 60 m or so and tiny gametophytes consisting of at most a few cells. In some species, the sporophyte may be smaller and less conspicuous than the gametophyte. Heteromorphic alternation of generations is regarded as an adaptive response to seasonal changes in selective pressures, such as herbivory, and thus is more common in temperate, boreal, and polar seaweeds than in tropical forms (John 1994).

Alternation of body form (also known as phase change) is not always associated with sexual reproduction or change in chromosomal level. For example, two morphologically distinct body types, known by the generic names *Feldmannia* and *Acinetospora*, were discovered to actually represent different ecological manifestations of the same taxon. In the same habitat, the *Feldmannia* form occurred only in the warm season, whereas the *Acinetospora*-form was observed only in the cold season (Pedersen 1981). Other examples include *Petalonia* (a blade) and *Scytosiphon* (a tubular form) that alternate with microscopic crust or filamentous stages (Wynne 1981); in some cases sexual reproduction is involved in alternation of these generations, and in other cases it is not. Additional examples of the production of gametophytes, sporophytes, or both that are not correlated with change in chromosome level (ploidy) are described by Lewis (1996a). Such examples emphasize that chromosome level is not the only, or perhaps even the major, determinant of morphological and reproductive expression in phaeophyceans and that

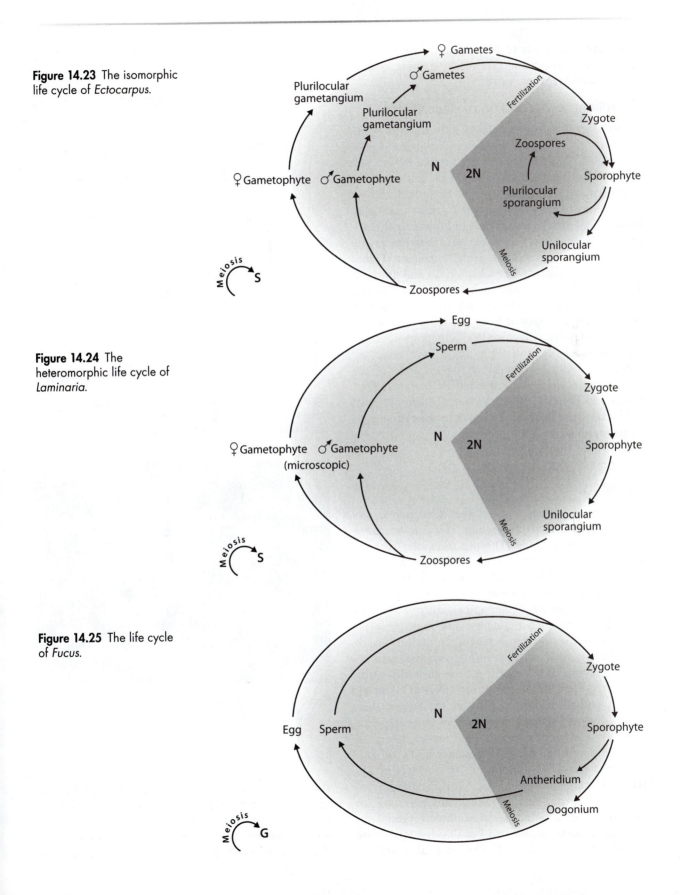

Figure 14.23 The isomorphic life cycle of *Ectocarpus*.

Figure 14.24 The heteromorphic life cycle of *Laminaria*.

Figure 14.25 The life cycle of *Fucus*.

more attention needs to be paid to interactions between environment and genome in eliciting specific structure and reproduction.

Exceptions to alternation of generations

Fucus and related genera classified in the orders Fucales (or Durvillaeales) do not exhibit alternation of generations. Rather, gametes are formed by meiosis, followed by mitotic divisions, and the multicellular, macroscopic life stage is regarded as diploid, though many cases of polyploidy occur in these and other brown algal groups (see review of brown algal chromosome numbers by Lewis 1996a). Thus *Fucus* and relatives appear to exhibit gametic meiosis (Figure 14.25), as do diatoms (see Chapter 12) and animals. The presence of tiny gametophytes that develop and generate gametes while confined within the tissues of the larger sporophytic body in Syringodermatales and Ascoseirales (Clayton 1987) suggests that the life history of fucalean algae might have arisen by reduction and retention by the sporophyte of the gametophyte phase, as suggested by Jensen (1974). Schiel and Foster (2006) reviewed the life cycles of the Fucales and kelps with respect to their ecological consequences.

Phaeophycean Diversity and Systematics

Phylogenetic studies suggest that *Choristocarpus tenellus* and *Discosporangium mesarthrocarpum* together form a clade (Discosporangiales) that is sister to all other brown algae (Kawai et al. 2007) (Figure 14.26). These species exhibit features that are considered to be basic attributes (plesiomorphies) of the brown algae: apical, diffuse growth; uniseriate, branched filaments; and multiple plastids per cell without pyrenoids. Delimitations of some brown algal orders is still in progress, new orders are being discovered (e.g., the early-diverging Ishigeales, with the single genus *Ishige* [Cho et al. 2004]), and the sequence in which the brown algal orders diverged remains uncertain in some respects. However, among the brown algal orders described in more detail below, *rbcL* sequence analyses suggest that the Dictyotales and Sphacelariales are relatively early diverging; the Sporochnales, Ectocarpales s.l. (from the Latin *sensu lato*, meaning "in the modern sense") and Laminariales diverged later; and the Desmarestiales and Fucales diverged more recently (Kawai et al. 2007; Phillips et al. 2008).

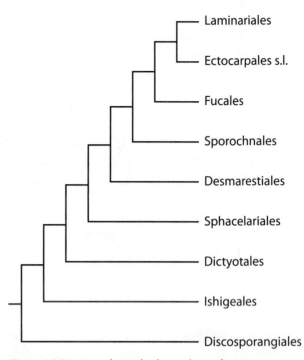

Figure 14.26 Hypothetical relationships of major brown algal classes. (Based on Kawai et al. 2007, by permission of the *Journal of Phycology*)

Figure 14.27 *Dictyota* growing on a coral head, San Salvador, The Bahamas. (Photo: R. J. Stephenson)

Dictyotales

Dictyotales have isomorphic alternation of generations; apical growth; production of tissues; cells with many discoid, pyrenoidless plastids; nonflagellate spores; and uniflagellate sperm. The genera are most diverse in tropical and subtropical waters, but they

Figure 14.28 A pressed specimen of *Zonaria*. Note fan-shaped blades (arrowheads). (Photo: L. W. Wilcox)

1 cm

extend into temperate zones. Some taxa exhibit blue-green iridescence when submerged. Some species of the genera *Dictyopteris* and *Spatoglossum* have cellular vacuoles with very low pH (0.5–0.9), explaining why these seaweeds turn green when damaged (Sasaki et al. 1999). *Lobophora variegata* is known to produce a cyclic lactone (lobophorolide) that is active in low concentrations against pathogenic and saprophytic marine fungi (Kubanek et al. 2003).

DICTYOTA (Gr. *diktyotos*, netlike) bodies are flattened and highly branched in a repeatedly dichotomous pattern (Figure 14.27) or are pinnately branched. There is a single prominent apical cell (see Figure 14.17); this cuts off derivatives that contribute to development of the surface as well as the internal tissues of the medulla. Bodies are three cells thick. Division of the apical cell leads to dichotomous branching. Changes in microtubule organization during the cell cycle in apical cells of *Dictyota dichotoma* were studied by Katsaros and Galatis (1992).

Gametophytes are usually dioecious, with sperm-producing plurilocs borne on separate bodies from those bearing female gametangia, each of which contains one egg. *Dictyota dichotoma* is famous for its regular release of gametes at two-week intervals about an hour after dawn, following the maximum

of spring tides. A sperm attractant is produced by eggs of *Dictyota* (Müller et al. 1981). Sporophytes produce unilocs containing four nonflagellate meiospores, two of which grow into male gametophytes and two into female gametophytes. *Dictyota* primarily occurs in shallow waters but has been found as deep as 55 m (Taylor 1972).

ZONARIA (Gr. *zone*, belt) (Figure 14.28) has flattened, fan-shaped blades of about eight cells in thickness, which are generated by a relatively short row of marginal apical cells (Neushul and Dahl 1972). The banded (zoned) appearance is due to the presence of parallel rows of hairs occurring at intervals. Groups of unilocs, known as sori, occur in irregular patches scattered over the body surface. Unilocs each produce eight nonflagellate spores. *Homoeostrichus* and *Exallosorus* are related genera (Phillips 1997; Phillips and Clayton 1997).

PADINA (Gr. *pedinos*, flat) bodies, like those of *Zonaria*, are flattened and fan shaped (Figure 14.29), and they develop from an extensive marginal row of meristematic cells. Although alternation of isomorphic generations is the general rule, direct development of new sporophytes from the spores of *P. pavonica* has been reported (Pedersen 1981); in this case, the gametophyte

Figure 14.29 Pressed specimen of *Padina*, seen life size. (Photo: L. W. Wilcox)

Figure 14.30 Habit of *Padina* (upper left) growing with *Dictyota* (center) and red algae on coral rubble near San Salvador, The Bahamas. (Photo: R. J. Stephenson)

generation has been bypassed. The genus is very common in tropical and subtropical waters around the world, forming large populations on rocks and coral rubble in shallow waters to depths of 14 m or so (Figure 14.30) (Taylor 1972). *Padina* may be encrusted with calcium carbonate, and the more extensively calcified specimens may not at first be readily recognizable as brown algae because they are so pale. Another brown alga known to deposit calcium carbonate is the Hawaiian *Newhousia imbricata* (Kraft et al. 2004).

Sphacelariales

Brown algae belonging to the order Sphacelariales are branched pluriseriate bodies that grow from a conspicuous apical cell. Cells generated from the apical cells by transverse division enlarge and then may undergo additional transverse and longitudinal divisions, thus generating tissues. Cells contain many discoid plastids without pyrenoids. The genera occur as small tufts growing on rocks, on other algae, or as endophytes. Alternation of generations is isomorphic, and isogamy, anisogamy, and oogamy all occur within the order. Blackening in response to bleach treatment is one of the notable characteristics of this order (Draisma et al. 2002).

SPHACELARIA (Gr. *sphakelos*, gangrene) (Figure 14.31) is widespread, occurring in polar to tropical marine waters, but bodies are rather small, occurring as tufts or spreading mats. It is characterized by production of distinctive, often triradiate, asexual propagules (Figure 14.32). *Sphacelaria fluviatile* and *S. lacustris* are notable freshwater species (Wehr 2002). The exclusively freshwater genera *Heribaudiella* and *Bodanella* are related to *Sphacelaria*, according to *rbcL* analyses and other traits (McCauley and Wehr 2007).

Desmarestiales

Most members of Desmarestiales are composed of pseudotissue (pseudoparenchyma), and growth is associated with meristematic cells located at the bases of hairs (trichothallic growth). When young, these hairs fringe the entire body, but hairs may be lost from more-mature specimens. There is a main axis from which diverge lateral "blades," or branches (Figure 14.33). Diversity is greatest in the Antarctic, where desmarestialeans dominate the seaweed flora. Molecular phylogenetic analysis supports the hypotheses that the Desmarestiales are monophyletic and that the family Desmarestiaceae most likely originated in the Southern Hemisphere (Peters et al. 1997). The sporophytes are the conspicuous generation. Gametophytes are microscopic filaments, and those of at least one species grow as endophytes within red algae. The pheromone desmarestene is produced (Müller et al. 1982). The large (10 m long and 1 m wide) bodies of the common Antarctic desmarestialean *Himantothallus* develop from a filamentous sporeling by the activity of trichothallic meristems whose location determines the number and position of blades in the mature sporophyte. The mature stipe and blade possess meristoderms that increase the thickness of these organs. **Trumpet hyphae** with perforate end walls very similar to

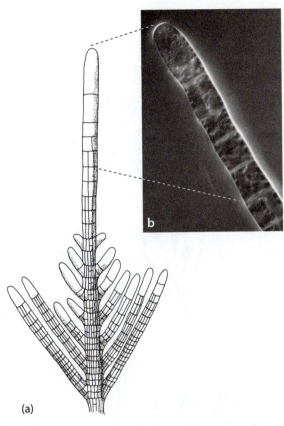

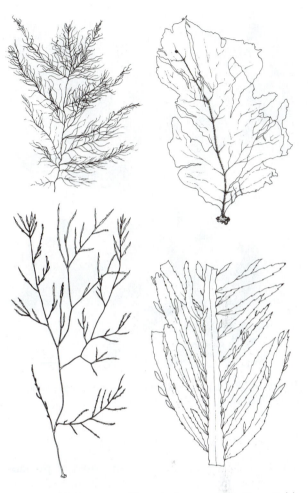

Figure 14.31 (a) Diagram of the apical portion of a *Sphacelaria* body. The tip of a filament is shown in (b). Note the nucleus and nucleolus in the apical cell. (a: W. Woelkerling; b: J. La Claire)

Figure 14.33 Diversity of form in *Desmarestia*. (From Bold and Wynne 1985)

Figure 14.32 Diagram of a typical sphacelarialean asexual propagule. (After Bold and Wynne 1985)

those of some laminarialeans occur in the medulla and are postulated to function in conduction (Moe and Silva 1981). Reproduction involves unilocular sporangia that occur interspersed with hairs (paraphyses) in aggregations (sori) as in various kelps. The meiospores produced from unilocs develop into

microscopic dioecious gametophytes, with gametes being formed under short-day conditions (Wiencke and Clayton 1990).

DESMARESTIA (named for Anselme Gaetan Desmarest, a French naturalist) (Figure 14.34) occurs in both high-north and high-south latitudes in cold waters, but also occasionally in warmer waters. One clade of species is characterized by the presence of free sulfuric acid within vacuoles having a pH of 1 or lower. If collections of these forms are mixed with other seaweeds, the unavoidable breakage of cells will release acid, causing extensive disintegration of the specimens. Collectors typically keep *Desmarestia* separate from other collections and store them under cool conditions to avoid autohydrolysis.

Figure 14.34 Pressed specimen of *Desmarestia*. (Photo: L. W. Wilcox)

Sporochnales

The bodies of this group are composed of pseudo-tissue generated by meristematic cells located at the base of a dense tuft of intensely pigmented apical hairs. Sexual reproduction is oogamous, and the life history is heteromorphic alternation of generations. Sporophytes are the larger and more conspicuous phase; gametophytes are monoecious, microscopic filaments. Most of the species of the six genera occur in the Southern Hemisphere. *Sporochnus* can occur along the southeastern and southwestern United States coastlines.

Sporochnales share with Desmarestiales and Laminariales the following features: pyrenoidless plastids, eyespotless sperm, posterior flagellum of sperm longer than anterior flagellum, oogamy, heteromorphic alternation of generations, similar female gametophytes and pheromones, and growth of zygotes in association with female gametophytes.

SPOROCHNUS (Gr. *spora*, seed + Gr. *chnous*, foam) (Figure 14.35) bodies have a wiry appearance and are highly branched. The apical tufts of

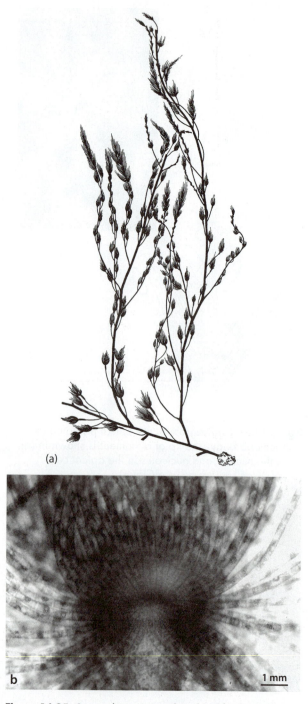

Figure 14.35 *Sporochnus.* (a) Habit. (b) Tuft of apical hairs. (a: From Taylor 1972 © The University of Michigan Press 1972; b: J. La Claire)

hairs are quite noticeable and distinctive. There is a conspicuous fibrous holdfast that attaches seaweeds to rocks in shallow water to depths of

Figure 14.36 The ectocarpalean genus *Elachista,* which often grows as tufts of filaments on the surfaces of larger seaweeds such as *Fucus* (shown here). (Photo: L. W. Wilcox)

approximately 100 m (Taylor 1972). Members of the Sporochnales produce the pheromone caudoxirene (Müller et al. 1988).

Ectocarpales

In recent years, species formerly placed in the orders Chordariales, Dictyosiphonales, and Scytosiphonales have been aggregated with classic Ectocarpales into the larger group Ectocarpales s.l. (Rousseau and De Reviers 1999; Draisma et al. 2001). Ectocarpales s.l. can be characterized as branched filaments whose cells have one to several band-shaped plastids, with plastids having a pyrenoid that is extended on a stalk (Rousseau and de Reviers 1999). The filaments often occur in tufts attached to substrates, such as larger seaweeds (Figure 14.36), and some are endophytic. Growth is typically diffuse, meaning that a well-defined meristem is generally absent. Sexual reproduction is by isogamy or anisogamy, and it involves isomorphic alternation of generations. A direct monophasic life history also occurs. Ectocarpaleans are common and widespread. The genus *Pleurocladia*, which occurs on rocks or plants in lakes and rivers in Europe and North America, is linked to Ectocarpales in *rbcL*-based phylogenetic analyses, but the freshwater brown algae *Heribaudiella fluviatilis* and *Bodanella lauterborni* are not (McCauley and Wehr 2007).

ECTOCARPUS (Gr. *ektos*, external + Gr. *karpos*, fruit) (Figure 14.37) is widespread throughout the world, growing on rocks or larger seaweeds, and

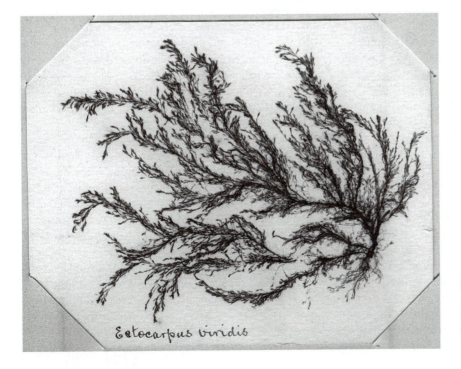

Figure 14.37 A herbarium specimen (seen life size) of *Ectocarpus* prepared in 1890 by a private collector. (Photo: L. W. Wilcox)

Figure 14.38 A light microscopic view of an *Ectocarpus* filament revealing the irregularly shaped plastids. (Photo: L. W. Wilcox)

Figure 14.39 *Streblonema* (dark filaments) creeping along the surface of a *Cutleria* gametophyte. (Photo: J. La Claire)

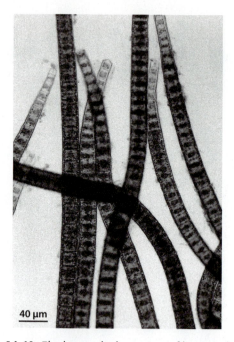

Figure 14.40 *Elachista*, which occurs as filaments that are commonly branched only at the bases. (Photo: L. W. Wilcox)

can be a significant ship-fouling organism and cause problems in mariculture. The branched filaments are uniseriate, and cells contain several elongate, irregularly shaped chloroplasts, each having several pyrenoids (Figure 14.38). Life history in *E. siliculosus* is biphasic and isomorphic, with both unilocs and plurilocs on the sporophyte generation and gametophytes producing only plurilocs (see Figure 14.23). Numerous clones of the sporophyte generation result from zoospore production. After release, zoospores readily attach to substrates via adhesive material produced in the Golgi apparatus and extruded from cells (Baker and Evans 1973).

The gamete attractant exuded by settled gametes of *Ectocarpus siliculosus* was identified by gas chromatography, mass spectrometry, infrared spectrometry, and nuclear magnetic resonance (NMR) spectroscopy as allo-cis-1-(cycloheptadien-2',5'-yl)-butene-1 (Müller et al. 1971). This was the first algal pheromone to have been chemically defined, and it is known as sirenin, ectocarpin, or ectocarpene. Viruses are known to attack *Ectocarpus*, resulting

in reduction of reproductive capacity (see review by Correa 1997). Hybridization has been extensively studied (see review by Lewis 1996b). Molecular systematic investigations conducted on nuclear ITS sequences (see Chapter 5) and the spacer region of the plastid-encoded rubisco cistron from geographically diverse strains of *E. fasciculatus*, *E. siliculosus*, and the related genus *Kuckia* confirmed the validity of these entities (Stache-Crain et al. 1997). In a molecular phylogenetic analysis, a freshwater collection of *E. siliculosus* was very similar to a marine isolate (McCauley and Wehr 2007).

STREBLONEMA (Gr. *streblos*, twisted + Gr. *nema*, thread) (Figure 14.39) usually occurs as an endophyte. *Streblonema* occurs particularly in kelps such as *Macrocystis*, or red algae such as *Grateloupia*, and is a suspected cause of diseases in some economically important seaweeds, such as *Undaria*. *Streblonema* lives within the intercellular spaces of its hosts but retains typical brown coloration. Macroscopically, it appears as discolored spots on the host, with just the reproductive structures extending from the host's surface (Correa 1997). *Laminarionema* is another ectocarpalean that occurs in sporophytes of the kelp *Laminaria*, and molecular analyses suggest that endophytism evolved more than once in the Ectocarpales (Peters and Burkhardt 1998). Most older kelp bodies appear to be infected by ectocarpalean endophytes (Kawai and Tokuyama 1995).

ELACHISTA (*Elachistea*) (Gr. *elachistos*, smallest) grows as tufts on *Fucus* (Figure 14.40) or within depressions in bodies of *Sargassum*. The macroscopic sporophyte consists of a colorless basal portion that may be embedded within the host's body, from which emerge uniseriate filaments, branched only at the base. There are numerous discoid plastids per cell. Culture studies suggest that it can reproduce asexually via zoospores, or produce meiospores within unilocs. Meiospores grow into a microscopic gametophytic stage.

SCYTOSIPHON (Gr. *skytos*, skin or hide + Gr. *siphon*, tube) is widespread but primarily occurs in temperate waters. The conspicuous gametophytic bodies are usually tube shaped (though sometimes flattened) and, when mature, are noticeably constricted at regular intervals (Figure 14.41). They are attached to substrates via a small basal disk. Uniseriate plurilocs may extend over the entire surface, interspersed with inflated hairs (paraphyses) that might be mistaken for unilocs. The sporophytic phase is an inconspicuous filamentous crust. There is evidence for day-length influence on phase change in *Scytosiphon lomentaria*, mediated by blue light (Dring and Lüning 1975). The adaptive significance of a heteromorphic life history in *S. lomentaria* was explored by Littler and Littler (1983). Although alternation of generations is not obligate, and both sporophytes and gametophytes can generate more bodies of the same type, there are adaptive advantages to each form. The crustose phase is more resistant to herbivory, whereas the upright phase is more productive. Such heteromorphic alternation of generations represents a kind of "bet hedging"—a way of spreading risks arising from different sources of mortality.

Laminariales

Laminarialean brown algae are commonly known as kelps. They are ecologically important because large populations form kelp forests that shelter many other species (Chapter 22). Kelps are also important as iodine accumulators, and in the past were used as sources of iodine for food supplements that prevent the medical condition known as goiter. Recently, the reason for kelp iodine accumulation has been elucidated. X-ray absorption spectroscopy has

Figure 14.41 *Scytosiphon*, whose bodies are tube shaped and constricted at intervals. (Photo: L. W. Wilcox)

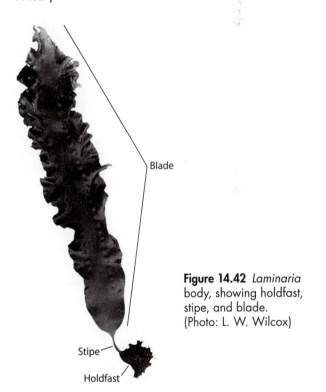

Figure 14.42 *Laminaria* body, showing holdfast, stipe, and blade. (Photo: L. W. Wilcox)

Blade

Stipe

Holdfast

revealed that iodide ion is present on the surfaces and in intercellular spaces, where it detoxifies ozone and other oxidants. Thus, iodide provides extracellular protection against oxidative stress, and is the first inorganic antioxidant discovered in a living organism (Küpper et al. 2008). Under stress conditions, kelps release iodocarbons and iodine oxide that lead

Figure 14.43 *Macrocystis* holdfast, composed of many branched haptera. (Photo: L. E. Graham)

Figure 14.44 Branched stipe with blades and floats of *Macrocystis*. (Photo: L. E. Graham)

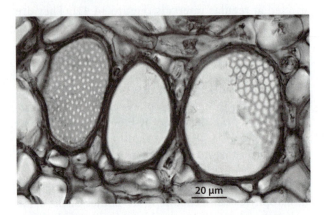

Figure 14.45 Perforated end walls of conducting cells in *Macrocystis*. (Photo: L. W. Wilcox)

to formation of cloud condensation nuclei, thereby affecting climate.

Kelps are characterized by large sporophytic bodies differentiated into holdfast, stipe, and blade regions (Figure 14.42), with an intercalary meristem located at the junction of the stipe and blade. Holdfasts are branched structures formed of thickened tissues known as **haptera** (Figure 14.43). These are essential for effective anchorage of the often large and heavy kelps. Some kelps have branched stipes, others have a single unbranched stipe, and in a few the stipe is greatly reduced or missing from mature stages. In some forms the blade is entire, but in others the blade is divided by longitudinal splits. Inflated **pneumatocysts** (floats) may occur in a lateral position on the stipe or at the base of the blade(s) (Figure 14.44). The gas bladders of *Nereocystis luetkeana* and *Pelagophycus porra* are noteworthy in that they contain up to 10% CO (carbon monoxide gas) (Carefoot 1977), but the significance of this is unknown. The structure, gas content, and function of pneumatocysts were reviewed by Dromgoole (1990).

Within stipes and blades, tissue specialization can occur. There is typically an outer layer of pigmented cells wherein cell divisions leading to increase in body circumference may occur; as previously noted, such a meristem is known as the **meristoderm**. Internal to the meristoderm are colorless cells of the cortex, within which may occur a central area known as the **medulla**. The inner cortex and/or medulla contain(s) cells specialized for conduction of solutes. Conducting cells have been reported from most of the genera in the Laminariales. They are elongate, with perforated terminal end walls (Figure 14.45), arranged in long continuous files. The ends of individual conducing cells may be expanded so that they resemble trumpets, and they are hence sometimes called trumpet hyphae (Figure 14.46). The diameter of the perforated end walls in *Macrocystis pyrifera* may be as large as 62 μm (Parker and Huber 1965). Series of conducting cells run throughout the body except for the holdfast (Figure 14.47). Solutes translocated include mannitol (about 65% of sap content), free amino acids, and inorganic ions (Parker 1965, 1966). These solutes originate in

Figure 14.46 Conducting cells of *Laminaria*. One linear array of conducting cells has been outlined in white. (Photo: L. W. Wilcox)

20 μm

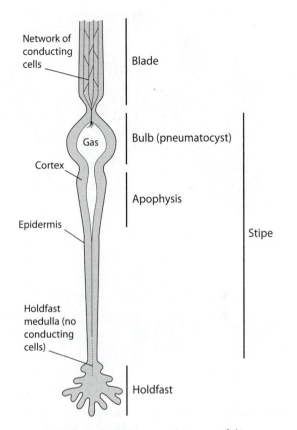

Network of conducting cells

Blade

Gas

Bulb (pneumatocyst)

Cortex

Apophysis

Epidermis

Stipe

Holdfast medulla (no conducting cells)

Holdfast

Figure 14.47 Diagrammatic representation of the conducting system of *Nereocystis luetkeana*. (After Nicholson 1970, by permission of the *Journal of Phycology*)

the photosynthetic cells of the meristoderm and outer cortex. Loading into conducting cells is thought to occur by either symplastic (involving plasmodesmata) or apoplastic (involving cell membrane transporter molecules) processes, or both. In order to understand the loading process, Buggeln et al. (1985) fed radioactive bicarbonate to blades of *Macrocystis* and followed its movement within the body using external Geiger-Müller counters. Microscopic studies revealed a possible symplastic pathway from photosynthetic cells to conducting cells of the medulla.

Translocation occurs in both the light and the dark. Experimental efforts toward understanding this process are reviewed by Schmitz (1981). Physiological studies of translocation in laminarialean conducting cells began with the pioneering radioisotope tracer work of Parker (1965, 1966), who found translocation rates of 65–78 cm hr^{-1} in *Laminaria*, with transport primarily toward the intercalary meristem. Translocation rates in *Macrocystis* are about 50 cm day^{-1}. Such solute

movement is of great advantage in dense *Macrocystis* canopies where light levels may be insufficient to allow photosynthesis to balance respiration in the lower portions of the body. Isotope tracer studies have revealed that translocation occurs from well-illuminated surface fronds to juvenile fronds of *Macrocystis*. Immature blades may primarily import photosynthates, but mature ones may only export (Lobban 1978a,b). Translocation to basal regions may also be necessary for development of fertile (uniloc-bearing) fronds at the basal region of the *Macrocystis* body. Translocation also greatly benefits growth of the annual kelp *Nereocystis*, which can grow 13 cm day^{-1} or more and can extend 25–50 m during a single growth season. The tremendous ecological significance of the giant kelps is largely based upon their capacity for rapid growth, ability to achieve large size, and translocatory competence.

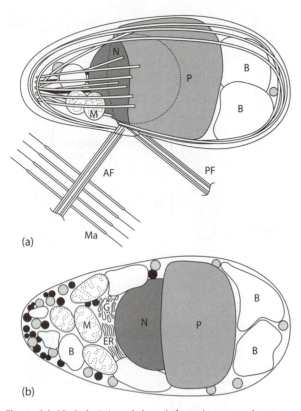

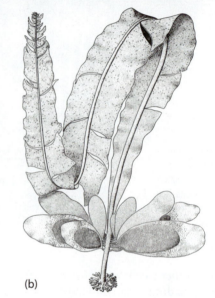

Figure 14.48 Side (a) and dorsal (from the top surface) (b) views of the zoospores of laminarialean algae, as deduced from ultrastructural studies. AF = anterior flagellum, B = vesicles, ER = endoplasmic reticulum, G = Golgi body, M = mitochondrion, Ma = mastigonemes (tripartite hairs), N = nucleus, P = plastid, PF = posterior flagellum. (After Henry and Cole 1982a, by permission of the *Journal of Phycology*)

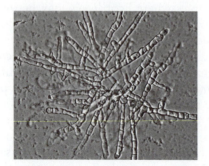

Figure 14.49 *Laminaria* gametophyte, grown in culture. (Photo: G. McBride)

Reproduction in the Laminariales occurs by means of unilocular sporangia produced on both surfaces of blades; plurilocs do not occur on the sporophytes of these taxa. The ultrastructure of the

Figure 14.50 *Alaria.* (a) Terminal portion of a blade. (b) Diagram showing the entire body, with the basal sporophylls. (a: Photo: L. W. Wilcox; b: From *Marine Algae of the Monterrey Peninsula*, 2nd ed., by Gilbert M. Smith, incorporating the 1966 supplement by George J. Hollenberg and Isabella A. Abbott. Used with the permission of the publishers, Stanford University Press. © 1966, 1969 by the Board of Trustees of the Leland Stanford, Junior, University)

zoospores produced in these unilocs was studied by Henry and Cole (1982a) (Figure 14.48). During uniloc development, a 1:1 association between nuclei and plastids occurs, which ensures appropriate packaging of organelles in meiospores (Motomura et al. 1997). In some cases the unilocs are localized in darker-brown patches known as **sori,** and sometimes unilocs occur on specialized reproductive blades known as **sporophylls.** In the case of

Figure 14.51 Portion of the ribbed blade and unbranched stipe of *Costaria costata*. (Photo: L. W. Wilcox)

Figure 14.52 A *Laminaria* species with a highly dissected blade. (Photo: L. W. Wilcox)

Macrocystis and *Alaria*, the sporophylls are located close to the body base, facilitating the ability of meiospores to reach suitable substrates for attachment. In the case of *Nereocystis*, large soral patches tear out of blades and readily sink to the bottom, thus depositing unilocs close to deep substrates where released meiospores can attach. Meiosis segregates sex-determining genes, such that meiospores germinate into unisexual gametophytes. Kelp gametophytes are microscopic branched filaments (Figure 14.49) or, in some cases, single oogonial cells. Male gametophytes produce dense clusters of antheridial cells, each giving rise to a biflagellate sperm. Cells of female gametophytes may release their contents from cell walls, and these wall-less cells function as eggs. The egg cells often remain attached to their former cell wall, however, with the consequence that fertilization and subsequent zygote development occur in association with the female gametophyte. The eggs of *Laminaria angustata* have two flagella that lack tubular hairs, and there are no flagellar roots. The flagella are abscised during egg liberation (Motomura and Sakai 1988). The pheromone produced by kelps is lamoxirene, which is structurally distinct from the pheromones produced by other brown algal groups (Müller et al. 1985). Culture studies have revealed that some percentage of unfertilized eggs is capable of developing into sporophytes. Intergeneric hybrids can form between members of the Laminariales (Lewis and Neushul 1995).

The center of diversity of the Laminariales is the Pacific coast of North America. In the past, families were distinguished by structural features, but more recent molecular phylogenetic analyses have resulted in taxonomic reorganization of the families (Lane et al. 2006), reflected below.

Alariaceae. *ALARIA* (L. *ala*, wing) has an undivided main blade with a conspicuous flat midrib (Figure 14.50) and numerous basal sporophylls. Blades can be as long as 25 m. It is perennial and occurs on exposed rocks from Alaska to California. Molecular phylogenetic analysis indicates that *Alaria* is closely related to *Lessoniopsis*, *Pleurophycus*, *Pteryogophora*, and *Undaria* (Lane et al. 2006).

Costariaceae. *COSTARIA* (L. *costa*, rib) *triplicata* has an undivided blade that has three longitudinal folds on the blade and a relatively thin, flattened, unbranched stipe (Figure 14.51). Sori occur in the regions between folds. It is perennial and common on rocks in the lower intertidal and subtidal zones of northern Japan as well as the eastern Pacific from Alaska to southern California (Abbott and Hollenberg 1976). Molecular data suggest that *Costaria* is closely related to *Agarum*, *Dictyoneurum*, and *Thalassiophyllum* (Lane et al. 2006).

Laminariaceae. *LAMINARIA* (L. *lamina*, blade) is very common worldwide on temperate and boreal rocky shorelines. There are many described species. Molecular data suggest that *Laminaria* species occur in two clades and that the one not including the type species, *L. digitata* (Figure 14.52), should be renamed *Saccharina* (Lane et al. 2006). *Cymathere japonica* and the genera *Hedophyllum* and *Kjellmaniella* have been subsumed into *Saccharina*.

Figure 14.53 *Postelsia palmaeformis.* Numerous palm tree–like bodies are typically clustered together on wave-impacted shores of the eastern Pacific. (Photo: L. E. Graham)

Figure 14.54 *Nereocystis.* This kelp is characterized by a very long stipe that is crowned with a single large float and many blades. *Nereocystis* is considerably larger than *Postelsia,* a clump of which is shown in the center of the photograph. (Photo: L. W. Wilcox)

The blade of some *Laminaria* and *Saccharina* species is entire (see Figure 14.42), but that of others can be split (see Figure 14.52); blades of neither genus possess a midrib. Most species are perennial. At the end of the growing season, the blade may either detach or persist until it is displaced by growth of a new blade at the start of the next growing season. Unilocs may be produced at any place on the blade. An ultrastructural analysis of fertilization and zygote development in *Saccharina angustata* revealed that zygotes develop cell walls immediately after gamete fusion, possibly to block penetration of eggs by more than one sperm. Further, egg centrioles disappear from zygotes, such that centrioles are inherited from the sperm-contributing parent. In contrast, mitochondria are maternally inherited since sperm mitochondria are degraded in zygotes (Motomura 1990). *Saccharina japonica* is widely cultivated in the western Pacific region for use as food.

POSTELSIA (named for Alexander Philipou Postels, an Estonian geologist and naturalist) resembles a small (to 60 cm tall) palm tree in that there is a group of terminal blades at the top of an extremely flexible, tapering stipe (Figure 14.53). The blades have toothed margins and are ridged, with linear arrays of unilocs (sori) being produced between the ridges in late spring. *Postelsia* occupies high to mid-intertidal zones of areas exposed to very heavy wave action. Physiological and transplant studies indicate that *Postelsia* populations living in the high intertidal zone are subject to heavy mortality by desiccation and high temperature and light stress but that those living in the lower intertidal zone experience physiological stress as a result of lower light availability when submerged. *Postelsia* growing in the mid-intertidal zone experience less impact from desiccation than those growing higher, and they experience less shading than those growing lower on the shore (Nielsen et al. 2006). *Postelsia* is able to cope with waves by means of its tightly adherent holdfasts and flexible stipes that resist drag forces. Its range is along the eastern Pacific coast from British Columbia to California. Analysis of population structure using molecular fingerprinting and RAPD analysis (see Chapter 5) showed that dispersal distances are short (1–5 m), so that individuals within clusters are likely to be siblings. In addition, evidence was obtained for the occurrence of distinguishable biogeographic populations (Coyer et al. 1997). Gametogenesis, development of

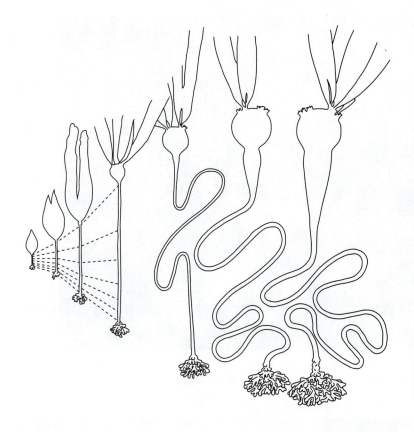

Figure 14.55 Development of *Nereocystis luetkeana*, determined by marking young bodies during a period of six weeks. (After Nicholson 1970, by permission of the *Journal of Phycology*)

the young sporophyte, and change in chromosome number during the life history of *Postelsia palmae-formis* were studied by Lewis (1995).

NEREOCYSTIS (Gr. *Nereus*, a sea god + Gr. *kystis*, bladder) (Figure 14.54) has a stipe that may be more than 30 m in length and is attached to rocky substrates at the bottom of subtidal habitats by a relatively massive holdfast. The apical end of the stipe is terminated by a large (about 15 cm in diameter) pneumatocyst or float, from which emerge four meristematic branches that generate as many as 100 blades per body. Each blade may be several meters in length. Development of the body was studied by Nicholson (1970) (Figure 14.55). At reproductive maturity, blades exhibit dark-brown patches (sori), which are groups of unilocs. Sori detach from blades and may readily sink, which aids the attachment of meiospores to deep-water substrates. *Nereocystis* is annual, so all of its growth is amazingly attained within a single growing season. Large populations of *Nereocystis* may dominate areas within northern regions of the Pacific North American coast (Figure 14.56).

MACROCYSTIS (Gr. *macros*, long + Gr. *kystis*, bladder) is a massive kelp that can form dense forests. Fronds may reach lengths of 60 m and are harvested for alginate production (see Chapter 4). Results of several studies of the biotic impact of kelp harvesting suggest that no problems arise as long as the integrity of the *Macrocystis* population itself is not affected (North 1994). *Macrocystis* beds serve as important refuges and nurseries for fish and many other ecologically and economically valuable species. They influence water motions and hence the dispersal patterns of planktonic larvae, as well as the distribution of phytoplankton. Sea otters are well known to moor themselves while sleeping by wrapping themselves with *Macrocystis* fronds. The dissolved organic carbon and particulates generated by kelp decay provide a substantial portion of the nutrient requirements of detritivores and other organisms in kelp communities (Duggins et al. 1989). This genus is thus regarded as a keystone taxon whose presence influences a cascade of ecological processes and numerous other organisms. The keystone properties of *Macrocystis* derive from its growth processes. These result in a proliferation of meristems that can

Figure 14.56 Habit of *Nereocystis luetkeana* in the Strait of San Juan de Fuca, Washington State, United States. Blades and floats of numerous individuals can be seen. (Photo: C. Taylor)

Figure 14.57 Branched *Macrocystis* bodies just below the water surface. Each blade is associated with a float. (Photo: L. E. Graham)

generate the huge body and that confer the ability to continue growth even if some meristematic tissue is removed by herbivory.

The body, with branched stipe, multiple blades (each as long as 40 cm), and floats (Figure 14.57), arises by division of a single original blade, beginning from a small hole that develops near the base. The juvenile blade undergoes a longitudinal split,

generating two primary fronds that continue to undergo longitudinal splitting until many blades have been formed (Figure 14.58). The two outermost fronds serve as frond initials, and inner fronds function as the first basal meristems. Basal meristems can produce new fronds by continuing to divide by basal clefts. The innermost blade of the resulting pair continues to function as a basal meristem, and the outermost may become either a basal meristem or a frond initial. Frond initials undergo basal splitting to produce blades, and the most distal blade then becomes capable of generating new blades (each with an individual float) and stipe tissue, thus serving as an apical meristem. Meiospores are produced in unilocs on specialized blades (sporophylls) located just above the holdfast apex; these may not possess floats. The female gametophytes are very small, consisting of just a single cell, but male gametophytes are multicellular. The consequences of inbreeding were investigated in laboratory cultures of *M. pyrifera*; these consequences may explain periodic senescence of natural populations (Raimondi et al. 2004).

There are about four species, which exhibit varying degrees of hybridization. *Macrocystis* occurs off the coasts of every major Southern Hemisphere landmass, as well as many islands, but it is much less

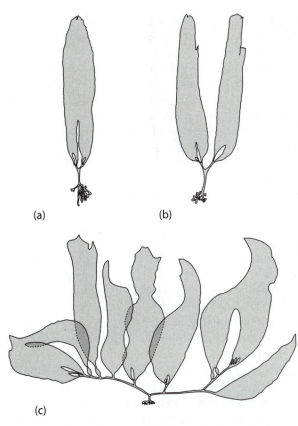

Figure 14.59 *Egregia* species having narrow blades emerging from each side of a flattened stipe. (Photo: L. W. Wilcox)

Figure 14.58 Development of the branched body of *Macrocystis*. A hole develops at the base of young blades (a), and this leads to longitudinal splitting of the single original blade into two (b). This process continues until multiple blades and branched stipes have been produced (c). (After Bold and Wynne 1985)

Figure 14.60 *Ecklonia* in Australia. (Photo: W. Woelkerling)

widely distributed in the Northern Hemisphere, occurring only along the northeast Pacific coast. Absence of this and other kelps from low latitudes is attributed to their sensitivity to water temperatures above 20°C–22°C, either as a direct metabolic effect or an indirect effect of differences in occurrence or activity of grazers and microbial pathogens. DNA fingerprinting techniques are beginning to be used to assess genetic diversity within and between populations and life-history phases of *Macrocystis* (Coyer et al. 1994). Additional information regarding the biology of *Macrocystis* can be found in an extensive review by North (1994).

Lessoniaceae. *EGREGIA* (L. *egregius*, remarkable) is known as the "feather boa" kelp because of the numerous lateral blades and elongate-to-spherical floats that emerge from a long, flattened stipe (Figure 14.59). It is perennial and endemic to the northeast Pacific. Molecular data indicate that *Egregia* is related to *Ecklonia*, *Eisenia*, and *Lessonia* (Lane et al. 2006).

ECKLONIA (named for Christian Friedrich Ecklon, a German botanical collector) (Figure 14.60) typically has blades with numerous pinnately arranged outgrowths from a flattened mid-region, arising from the top of a stipe of varying length, depending on the species. Although most *Ecklonia* species produce sori of unilocs on specialized sporophylls, restriction of sori to sporophylls does not occur in all species, and one species lacks sporophylls entirely. *Ecklonia* occurs in warm temperate waters of both hemispheres and at least one tropical area (islands off the coast of western Australia); it is considered to be the most warmth tolerant of the kelps. It forms extensive kelp forests in southern Africa, Australia, New Zealand, China, and the Galapagos Islands (Graham et al. 2007), often in association with upwelling of nutrient-rich, cool subsurface waters. Drift *Ecklonia* is collected for alginate production. A review of *Ecklonia* biology can be found in Bolton and Anderson (1994).

Fucales

The Fucales include phaeophyceans having apical growth, bodies composed of tissues, and, distinctively, a life cycle involving meiosis during gamete production. Apical growth in Fucales is thought to occur by division of an apical cell or a group of several apical initials, each having several cutting faces. Their derivatives generate a thick parenchymatous body, which, though not approaching the maximal sizes of the giant kelps, can be a meter or more in length. Division of the apical meristem produces dichotomous branching. Radioisotope tracer experiments demonstrated long-distance transport of photosynthetic products within midribs of *Fucus serratus*. The branch tips are a strong sink, and the main translocated substance is mannitol, but amino acids and low-molecular-weight organic acids also are transported (Diouris 1989).

Gametes are generated within gametangia produced in branch termini or systems of branches known as **receptacles**. Chambers known as **conceptacles** are sunken into the surfaces of receptacles, with only an open pore (ostiole) visible from the receptacle surface (Figure 14.61). Gamete discharge from conceptacles into seawater occurs through this pore in taxa such as *Fucus*. Depending upon the fucalean species, an individual conceptacle may contain both eggs and sperm or just one type of gamete. Some species produce both sperm- and egg-containing conceptacles on the same body, and others produce them on separate bodies (i.e., they are dioecious). Conceptacles are derived from single cells produced at the body surface close to the apical meristem. Some of the

cells lining the developing conceptacle differentiate into oogonia (egg-producing gametangia) or antheridia (sperm-producing gametangia) (Figure 14.62). Development of these gametangia is thought to be analogous to that of unilocular sporangia in other brown algae, and the gametophyte stage is thought to consist of only a single cell (Jensen 1974). The first divisions of the gametangia are meiotic; genera vary in the number of egg cells produced, from one to eight. In antheridia, the meiotic products continue to divide mitotically, generating 64 sperm. Fucalean sperm are distinctive in having an expanded apical region (known as a proboscis) and a posterior flagellum that is longer than the apical flagellum. Sperm possess eyespots. The wall-less gametes of *Fucus* are released when desiccated bodies are inundated with the incoming tide, and sperm are attracted to eggs as the latter release the pheromone fucoserraten (Müller 1981). Binding between sperm and egg of *Fucus serratus* is mediated by interaction between proteins derived from the sperm that recognize sulfated glycoconjugates on the egg cell membrane; such interactions are important in activating the egg (Wright et al. 1995a,b). Ultrastructure of the gametes and embryo of *Fucus* was studied by Brawley et al. (1977). Fertilization of *Fucus* occurs in the water, so zygotes must rapidly attach to substrates before they are washed out to sea. After fertilization and prior to further development, zygotes of *Fucus* secrete an adhesive mucilage (Vreeland et al. 1998).

About 2 hours after fertilization, *Fucus* zygotes begin to deposit a wall, and about 12 hours later, the rhizoid is produced. This event reflects previous establishment of zygote polarity. Light appears to be the strongest external stimulus affecting polarity establishment. The cell wall provides information to the underlying cell membrane and cytosol that influences establishment of a polar axis and orientation of the first plane of zygote division (Quatrano and Shaw 1997). The rhizoid forms on the shaded side of zygotes that are exposed to unidirectional illumination (Brawley and Wetherbee 1981). Calcium signalling is involved in induction of rhizoid development, as ascertained by patch-clamp electrophysiological techniques and laser microsurgery (Taylor et al. 1996) and by the use of inhibitors (Love et al. 1997). Rhizoid development is dependent upon the expression of fucoidans in the cell wall (Quatrano 1973).

Eggs of *Sargassum horneri* are extruded within a mass of mucilage and remain attached to the

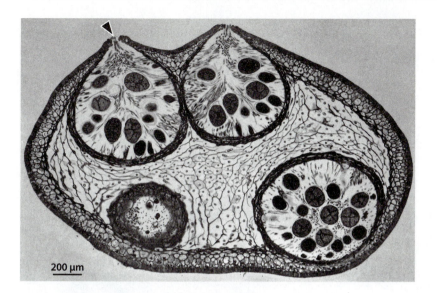

Figure 14.61 Section through a receptacle of *Fucus* showing portions of four conceptacles. This species is monoecious—that is, each conceptacle contains both oogonia and antheridia. Conceptacles open to the outside via a pore (ostiole) (arrowhead). (Photo: L. W. Wilcox)

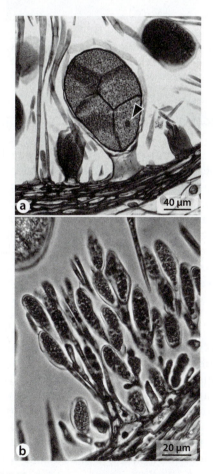

Figure 14.62 *Fucus* gametangia. (a) Oogonium with egg cells. A nucleus (arrowhead) is visible in one of the eggs. (b) Antheridia, containing sperm, photographed with phase-contrast optics. (Photos: L. W. Wilcox)

conceptacle by a stalk formed from oogonial wall material. After fertilization and a few cell divisions, germlings are released from the stalk and become attached to the substrate by means of adhesive material released from rhizoids (Nanba 1995). In genera such as *Turbinaria* and *Sargassum*, zygotes are also retained on the parental body. Such retention is regarded as an adaptation that allows reproduction and colonization to occur within very high-energy environments such as coral reefs (Stiger and Payri 1997).

FUCUS (L. *fucus*, derived from Gr. *phykos*, seaweed) has flattened, dichotomously branched or pinnate fronds with a midrib (see Figure 14.11). Air bladders occur within the fronds. Receptacles occur at the frond tips and are noticeably swollen. The holdfast is discoid. A surface layer of pigmented columnar cells encloses a medulla composed of a network of colorless filaments surrounded by a mucilaginous matrix of cellulose, alginates, and fucoidan. These water-absorbing polysaccharides are thought to help mitigate desiccation effects. The surface cell layer is involved in secretion of alginates, fucoidan, and polyphenolic compounds to the outside of the body. Comparative studies of photosynthesis in air and seawater by *Fucus spiralis* demonstrated that this species photosynthesizes effectively in air, revealing its tolerance to desiccation (Madsen and Maberly 1990). The blue-light photoreceptor aureochrome occurs in *Fucus distichus* and other photosynthetic stramenopiles (Takahashi et al. 2007).

Figure 14.64 A live specimen of *Sargassum*, showing the branching pattern and small round floats (air bladders). (Photo: L. W. Wilcox)

Figure 14.63 A pressed specimen of *Ascophyllum*, showing the branching pattern and numerous elongate floats. (Photo: L. W. Wilcox)

Species relationships have been studied with the use of mitochondrial DNA markers (Coyer et al. 2006). Genetic structure of natural populations of *F. vesiculosus* has been assessed with the use of molecular markers (microsatellite loci) (Tatarenkov et al. 2005, 2007). Molecular markers have also been used to define mating patterns in natural populations (Engel et al. 2005).

ASCOPHYLLUM (Gr. *askos*, leather bag + Gr. *phyllon*, leaf) often occurs together with or near *Fucus*, and both taxa are colloquially known as rockweeds. The bodies of *Ascophyllum* have thinner, longer axes than those of *Fucus*, and a midrib is absent (Figure 14.63). Dichotomous and lateral branching occur. Air bladders are present at intervals on the fronds.

SARGASSUM (Port. *sargaco*, name given by fishermen) bodies are highly differentiated into a holdfast, a cylindrical main axis, leaflike blades, and air bladders in the axils of blades (Figure 14.64). This genus is widespread in temperate, subtropical, and tropical waters in both intertidal and subtidal zones. Some forms are free floating, sometimes occurring in extensive rafts that harbor distinctive communities of organisms adapted to the buoyant *Sargassum* habitat. These occur in the Sargasso Sea off the western coast of Africa. During the 1940s, *Sargassum muticum* spread from Japan to the northern Pacific coast of the United States, and by the 1970s, it had made its way south to California. This species of *Sargassum* has also spread to Europe, probably on oysters destined for aquaculture operations. *Sargassum* forms nuisance growths in harbors and on beaches, and it can quickly spread to new areas due to the flotation capabilities conferred by its many air bladders. Other features contributing to rapid spread include fast growth rate, fertility in the first year, and monoecious reproduction (Lüning 1990).

HORMOSIRA (Gr. *hormos*, necklace + Gr. *seira*, chain) bodies are dichotomously branched

Figure 14.65 A dense carpet of *Hormosira* in Australia. Note the distinctive beaded necklacelike appearance. (Photo: W. Woelkerling)

arrays of hollow, round segments interconnected by short nodes (Figure 14.65). It is of common occurrence on the shores of Australia and New Zealand. Parthenogenesis (development of unfertilized eggs) has been observed in *Hormosira*. Clayton and associates (1998) observed that by 24 hours after release from the oogonium, 74% of unfertilized eggs had secreted a cell wall, and after two days following release, up to 22% had undergone polarization. However, only a few were able to divide. These, unlike eggs that did not divide, were observed to possess a pair of centrioles.

DURVILLAEA (named for Jules Sebastien Cesar Dumont D'Urville, a French explorer) is commonly known as "bull kelp." The four or five species are restricted to temperate to sub-Antarctic regions of the Southern Hemisphere, where they may form large biomasses and become dominant components of intertidal and subtidal communities. A cladistic, biogeographic analysis suggested that *Durvillaea* evolved in the Southern Hemisphere and was well established prior to the dissociation of New Zealand from Gondwana some 80 million years ago (Cheshire et al. 1995).

Figure 14.66 *Durvillaea* along the Australian coast. The blades are lacerated into straps by the action of the surf. Note the massive disklike holdfasts (arrowheads and inset). (Photo: W. Woelkerling; inset: L. W. Wilcox)

Durvillaea has a massive solid holdfast, a branched or unbranched stipe, and a large leathery blade, often ripped by wave action into several straplike segments (Figure 14.66). One species produces honeycomb-like gas-filled floats within the blade, while other species produce additional lateral blades along the stipe. The bodies are perennial and dioecious. There is a surface layer of pigmented cells that have two plastids each and a cortex and internal medulla formed of interwoven slender cells. Growth seems to occur by means of a peripheral meristoderm that adds new layers of cells at the beginning of new growth periods; an apical meristem is absent (Hay 1994). Individuals may live for seven or eight years and grow to 10 m in length, becoming fertile in the second year (Clayton et al. 1987).

Conceptacles develop on the main blade—usually in a subterminal position—beginning in the spring, and gametangial production is complete in the autumn. Conceptacle development in *Durvillaea* was studied by Clayton et al. (1987). There are separate male and female bodies; conceptacles of male bodies produce only sperm, and those of female bodies generate only eggs. There are typically equal numbers of males and females in populations. The first division leading to gamete production is meiotic. Four eggs are produced per oogonium, and 64 sperm are generated in antheridia. Egg production can be prodigious; it is estimated that a single body could produce 120 million eggs in just one night (Hay 1994). Two years' growth is required before bodies become fertile. *Durvillaea* is harvested for food and/or alginate production in Chile, Tasmania, and New Zealand (Hay 1994).

Chapter 15

Red Algae

Red algae are probably best known for their economic and ecological importance. For example, *Porphyra yezoensis* and some other species are grown in mariculture operations for use as human food, and *Kappaphycus* is among several genera that are cultivated or harvested for extraction of gel-forming agar, agarose, and carrageenan (see Chapter 4). Gelling polysaccharides, which play essential ecological roles in red algae, are widely used for laboratory cell-culture media, nucleic acids research, and food processing. Red algae are also increasingly appreciated as sources of compounds that evolved as defensive responses to microbial or herbivore attack that also have potential uses in human medicine.

Though several genera of red algae occur in freshwaters, thousands of red algae occur on tropical and temperate marine shores, where they play important ecological roles. For example, certain calcified red algae known as corallines help to build and maintain coral reefs, which harbor diverse organisms. Giant clams, banded coral shrimp, and clownfish are just a few of the fascinating animals that rely on reef habitats. By forming hard, flat sheets that consolidate and stabilize reef crests, coralline algae protect reefs from wave damage. For this reason, coralline red algae are regarded as keystone organisms, species whose decline (like the removal of the keystone from an arch) could cause the collapse of a larger-scale structure, in this case the loss of entire biotic communities. Fossil evidence indicates that coralline red algae have been playing this important role

Odonthalia

(Photo: L. W. Wilcox)

for hundreds of millions of years. In this chapter, we survey the evolutionary history, cell biology, body forms, intriguingly complex reproduction, ecology, and diversity of the red algae.

15.1 Evolutionary History of Red Algae

In the past, the evolutionary history of red algae, also commonly known as rhodophytes, has been inferred on the basis of differences in structure and reproduction. Fossils are an additional source of useful information. Molecular phylogenetic methods are increasingly being used to test hypotheses of relationship and to estimate the divergence times of modern lineages.

Fossil Red Algae

The oldest fossils that provide strong evidence of red algae, found in 1200-million-year-old (Mesoproterozoic) deposits on Somerset Island in arctic Canada, were named *Bangiomorpha pubescens* (Butterfield et al. 1990; Butterfield 2000). These fossils are one-cell-wide filaments or wider filaments composed of radially arranged wedge-shaped cells, in these distinctive ways resembling the modern red alga *Bangia* (see Figure 7.3). Other fossils interpreted as red algae occur in the approximately 600-million-year-old (Late Neoproterozoic) Doushantuo Formation in southern China, whose phosphorites preserved organic structure particularly well (Xiao et al. 2002; Xiao et al. 2004). Some of the Doushantuo fossils are thought to represent uncalcified relatives of the modern corallines. Calcified corallines left fossils beginning in the lower Ordovician nearly 500 million years ago (Riding et al. 1998) and have a continuous fossil record to modern times. The common modern genus *Lithothamnion*, for example, which is important in reef edge consolidation, is millions of years old. Although a group of fossils known as the Solenoporaceae were earlier thought to represent red algae something like modern corallines, they are now considered to be the remains of particular sponges (Riding 2004). Some fossils of non-coralline red algae are known from Miocene (5–25 million years ago) deposits in California (Parker and Dawson 1965). There are many other reports of fossil red algae, but the identity of such remains is often controversial when they do not

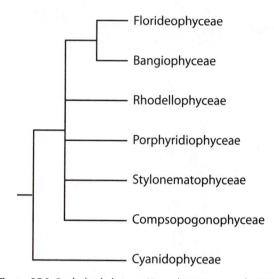

Figure 15.1 Red algal classes. (Based on Yoon et al. 2006, by permission of the *Journal of Phycology*)

possess features that clearly distinguish them from similarly shaped brown or green seaweeds (Taylor and Taylor 1993).

Molecular Evidence for Red Algal Relationships

Molecular studies indicate that red algae form a monophyletic group (Ragan et al. 1994; Freshwater et al. 1994). Red algae have commonly been linked with the ancestries of green algae (e.g., Rodríguez-Ezpeleta et al. 2005) and glaucophytes. These three algal groups possess plastids enclosed by two membranes, which are thought to have arisen by the process of primary endosymbiosis (see Chapter 7). On the assumption that a common ancestral host cell acquired a primary plastid in a single pivotal event, red, green, and glaucophyte algae have been aggregated into a supergroup called Archaeplastida (Adl et al. 2005). However, difficulty in reconciling the traits of red, green, and glaucophyte algae (Stiller 2003), some molecular evidence (e.g., Nozaki et al. 2005; Patron et al. 2007), and the observation that primary plastids appear to have evolved independently in a rhizarian (Marin et al. 2005) challenge the concept that Archaeplastida is necessarily a monophyletic group. An alternate hypothesis links red algae with green algae (+ land plants), cryptomonads (+ katablepharids), and haptophytes (but not glaucophytes) to form a "plastid-loving" supergroup Plastidophila (Kim and Graham 2008).

Figure 15.2 Commercial packaging of dried nori sheets derived from cultivated *Porphyra*, used in making sushi.

Figure 15.4 Several types of red algae, including both erect bodies and expanses of encrusting corallines, in a subtidal community on the eastern shore of Newfoundland, Canada. (Photo: L. W. Wilcox)

Figure 15.3 A bushy growth of *Chondrus crispus*. (Photo: M. Cole)

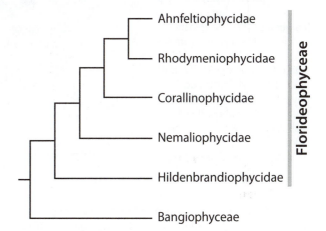

Figure 15.5 Subclasses of Florideophyceae. (Based on Le Gall and Saunders 2007, by permission of Elsevier Science & Technology Journals)

The red algae have traditionally been classified into the phylum (division) Rhodophyta (Wettstein 1901), though Saunders and Hommersand (2004) suggested the subkingdom Rhodoplantae, and Adl et al. (2005) grouped red algae into the class Rhodophyceae. Molecular evidence has been used to partition red algae into seven classes, six of which have relatively simple body structure and reproduction (Yoon et al. 2006) (Figure 15.1). The earliest-diverging red algae, informally termed **cyanidophytes,** are unicells that typically live in hot, acidic freshwaters. Cyanidophytes include *Cyanidioschyzon merolae*, whose genome has been completely sequenced (see Chapter 4). Molecular data suggest that cyanidophytes may be closely related to the plastids of photosynthetic stramenopiles (Müller et al. 2001). *Bangia* (see Figures 7.3c and 15.30) and the economically important genus *Porphyra* (Figure 15.2) belong to a clade commonly termed **bangiophytes.** Closely related to the bangiophytes are **florideophytes,** a monophyletic

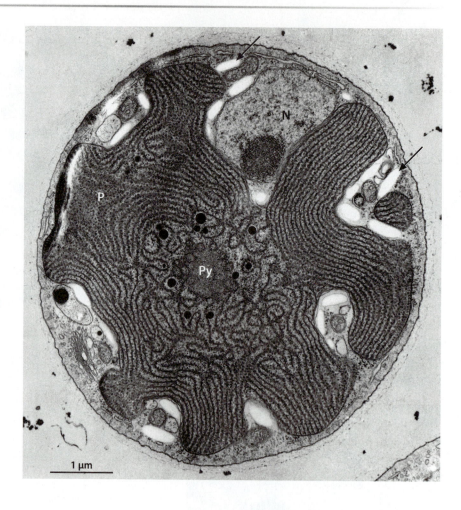

Figure 15.6 Transmission electron micrograph of *Porphyridium*, a unicellular red alga, illustrating the nucleus (N), cytoplasmic floridean starch grains (arrows), and single, lobed plastid (P) with central pyrenoid (Py) and unstacked thylakoids bearing phycobilisomes. Note the absence of a coherent cell wall and presence of a loose extracellular matrix. (From Schornstein and Scott 1982)

group of red algae having more complex bodies and reproduction (Saunders and Hommersand 2004). The economically important *Chondrus* (Figure 15.3), ecologically significant corallines (Figure 15.4), and the vast majority of other modern red algal species are florideophytes. Structural and molecular characters have been used to partition florideophytes into four (Saunders and Hommersand 2004) or five (Le Gall and Saunders 2007) monophyletic groups defined as subclasses (Figure 15.5), which include more than 20 orders and about 6,000 species.

15.2 Red Algal Cell Biology

Red algal cells display several unusual features. For example, unlike the green algal plastids to which they are supposedly closely related, red algal plastids do not contain starch. Rather, granules of a differently branched glucan, known as **floridean starch**, occur in the cytoplasm (Figure 15.6). Floridean starch stains only slightly upon iodine treatment, in contrast to the deep purple-blue staining starch in plastids of green algae and land plants. Red algae also display distinctive cell coverings, plastids, and cell division features that have strongly influenced their physiology, reproduction, ecology, and evolutionary diversification.

Extracellular Matrix

The red algal cell covering is often referred to as an **extracellular matrix (ECM)** because it is less rigid than the walls of other algae or land plants (see Figure 15.6). The red algal ECM is typically composed of a loose network of cellulose microfibrils that is filled in with an amorphous gel-like mixture of sulfated galactan polymers and mucilages. The ECM of red algae has a softer consistency than the cell walls of many algae because the gelatinous component is

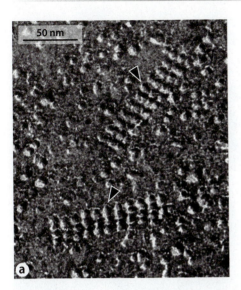

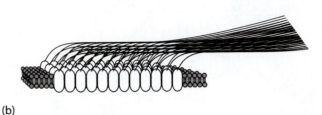

(b)

Figure 15.7 (a) Transmission electron micrograph of a freeze-fractured cell membrane of *Erythrocladia* (*Sahlingia*) *subintegra*, showing rectangular terminal complexes characteristic of red algae that spin out ribbon-shaped or cylindrical cellulose microfibrils. (b) Diagrammatic representation of ribbon-shaped cellulose microfibril production by the rectangular terminal complexes of red algae. (a: From Tsekos et al. 1996; b: After Tsekos 1996, © Springer-Verlag)

more abundant than is cellulose. The softer red algal cell covering helps to explain several unusual features of red algae, including the high degree to which non-gamete cells are able to fuse with other cells. Such fusion allows red algal cells of different types to mix nuclear and cytoplasmic information, fosters wound repair, and has been critical in the evolution of parasitic red algae. The flexible ECM fosters the amoeboid movements displayed by some asexual reproductive cells, as well as gamete fusion. The red algal ECM also allows body cells to increase greatly in length and volume, an important feature of development in many species. The flexibility of the ECM of most red algae increases body pliability, which can be an advantage in coping with water motions. However, some red algae, including corallines, possess an ECM that includes stony calcium carbonate, which stiffens the body. The surfaces of some other red algae are coated with a proteinaceous cuticle that toughens their bodies and may confer iridescence.

Microfibril components of the extracellular matrix

Although cellulose microfibrils are commonly found in the red algal ECM, they often make up a relatively small proportion of the total content. For example, cellulose occupies only 9% of the ECM of the economically important seaweed *Kappaphycus alvarezii* (Lechat et al. 2000). Freeze-fracture ultrastructural studies have revealed that red algal cell membranes are studded with protein particles and aggregations of protein particles, known as linear terminal complexes (Figure 15.7), that produce cellulose microfibrils. These complexes are thought to be clusters of enzymes that convert UDP-glucose into cellulose. Microfibrils produced by these types of complexes are ribbonlike or cylindrical (Tsekos 1996). Cellulose-synthesizing proteins are transported to the cell membrane by Golgi vesicles. A review of red algal cellulose microfibril synthesis can be found in Tsekos (1996), and a freeze-fracture study of the development of cellulose-synthesizing complexes in the cell membrane of the red alga *Erythrocladia* was done by Tsekos et al. (1996).

ECM of the gamete-producing stage of *Porphyra* contains microfibrils composed of xylans rather than glucans. Three xylan chains are aggregated to form a macromolecular triple helix that is stabilized by hydrogen bonding interactions within and between chains. Such polymers differ from cellulose in that they are not birefringent—i.e., they do not yield a characteristic bright cross-shape when examined using polarized light. A relatively insoluble structural mannan is also present. In contrast, X-ray diffraction studies showed that the spore-producing stage of *Porphyra* and *Bangia*, known as the conchocelis stage, does have cellulose in its extracellular matrix (Gretz et al. 1980).

Sulfated polygalactans and mucilages

The red algal ECM is dominated by various types of highly hydrophilic, sulfated polygalactans, which are polymers of β-(1→4) galactose and α-(1→3) linked 3,6 anhydrogalactose. These are the major constituents of

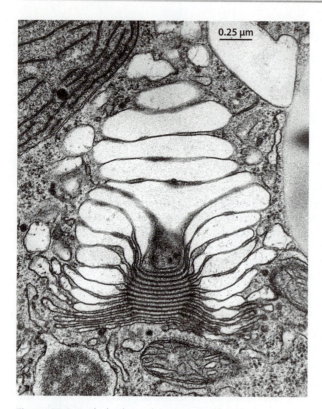

Figure 15.8 Red algal mucilages and sulfated polygalactans are produced within the large, complex Golgi apparatus, such as that of *Polysiphonia denudata*, shown here. (TEM: J. Scott, in Pueschel 1990)

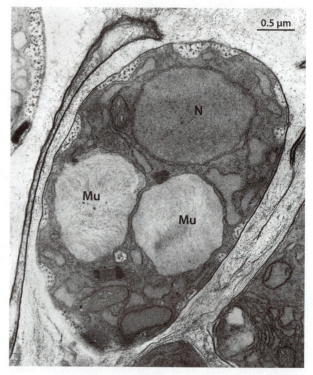

Figure 15.9 Mucilage (slime) sacs (Mu) within a reproductive cell (spermatium) of *Polysiphonia hendryi*. N = nucleus. (From Kugrens 1980)

economically valuable **agars** and **carrageenans**. The presence of D-galactose and anhydro-D-galactose distinguishes the more highly sulfated carrageenans from the less highly sulfated agars (anhydro-L-galactose). Agars are also more heavily methoxylated than are carrageenans. In the presence of cations (such as K$^+$ or Ca^{2+}), sulfated galactans form double helices that aggregate into a three-dimensional network having colloidal properties.

Carrageenans exhibit a repeating disaccharide backbone: β-(1→4)-D-galactopyranosyl-α-(1→3)-D-galactopyranosyl. This backbone is assembled from nucleotide precursors, and then it is sulfated within the conspicuous and highly developed Golgi apparatus (Figure 15.8) by sulfotransferases. There are at least 17 different types of carrageenans, which occur in different combinations in different species and can also vary among the life-history stages of a single species. Agars consist of alternating 3-O-linked β-D-galactopyranose and 4-O-linked 3,6 anhydro-α-L-galactopyranose, but the repetition can be interrupted

by blocks of repeated units of either one of the two constituents. Hybrid agarose/carrageenan combinations have sometimes been observed. Craigie (1990) and Cosson et al. (1995) reviewed the sulfated polygalactans of red algae.

Mucilages are polymers of D-xylose, D-glucose, D-glucuronic acid, and galactose, the proportions differing among taxa. Sulfated mannans occur in the mucilages of certain forms. Mucilages are produced in the Golgi apparatus and are often abundant in reproductive cells, where mucilage expansion upon hydration plays a role in reproductive cell dispersal and attachment. Conspicuous mucilage sacs are often observed within red algal reproductive cells (Figure 15.9), which may use mucilage secretion as a motility mechanism.

Calcium carbonate and protein

The majority of the ecologically important corallines and some other red algae have heavily calcified extracellular matrices. In the coralline algae, calcium carbonate is deposited in a particular crystalline form known as calcite. Some non-coralline, calcified red algae, such as *Galaxaura* and *Liagora*, deposit calcium carbonate in the form of aragonite crystals. These two forms of calcium carbonate differ in cations that substitute for calcium within the crystals. Aragonite can become enriched in strontium, which is a larger ion than the calcium that it replaces. In contrast, the Ca^{2+} of calcite can be replaced by smaller ions such as magnesium, iron, or zinc. Although the advantages of producing calcite versus aragonite are not well understood, Mg, Fe, and Zn are micronutrients that sometimes limit algal growth. Thus, it has been suggested that calcite production confers a useful nutrient harvesting/storage capacity. Calcification occurs most rapidly in growing regions of red algal bodies, with the calcium carbonate being deposited onto an organic matrix. Red algae are able to control deposition of calcium carbonate such that some cells become calcified, whereas others do not. The mechanism by which such control is exerted is unknown, but it is possible that regulation of the production of essential organic matrix materials is involved. Carbonate deposition is 10 times faster in the light than in the dark. Photosynthesis favors calcification indirectly by increasing the pH at the cell surface (see review by Craigie 1990).

A continuous layer of a densely staining, insoluble proteinaceous "cuticle" covers the surfaces of some red algae It should be noted that such layers are not biochemically similar to the cuticles of land plants. The surfaces of the red alga *Mazzaella* include as many as 17 layers of electron-opaque surface material alternating with electron-translucent layers (Figure 15.10). This layering produces light interference patterns that give the seaweed an iridescent appearance when submerged. The adaptive advantage, if any, of iridescence is not understood.

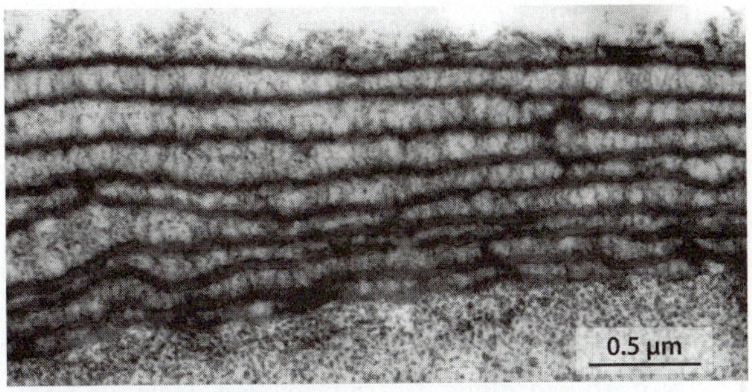

Plastids, Pigments, and Photosynthetic Storage

Plastids occur in nearly all red algal cells, even colorless cells that occur in the internal regions of thick seaweed bodies or in the cells of red algae that function as parasites (Goff 1982). Red algal plastids may be star shaped (stellate) or discoid, and they may have lobed edges (see Figure 15.6). Cells in relatively early-divergent forms may contain only a single plastid, but multiple plastids per cell is the general rule. In some red algae, plastids undergo diurnal change in position. In *Griffithsia*, for example, plastids move away from cell poles early in the light period to form

Figure 15.10 Electron micrograph of the multilayered surface "cuticle" characteristic of *Mazzaella*. This cuticle is thought to be responsible for the iridescent appearance of submerged bodies of this genus formerly known as *Iridaea*. (TEM: N. Lang)

distinctive bands in the center of the cell. Later in the light period, plastids return to the cell ends, remaining there throughout the night. The fact that the actin inhibitor cytochalasin B interferes with chloroplast banding indicates that actin mediates plastid movement in *Griffithsia* (Russell et al. 1996). Plastids of some of the early-diverging red algae contain a conspicuous pyrenoid (see Figure 15.6), but this structure is lacking in most species.

Red algal plastids are unusual in having Type ID rubisco, which is otherwise known to occur only in β proteobacteria. In contrast, Type IB rubisco, produced by coastal marine and freshwater cyanobacteria, is thought to have been present in the ancestors of primary plastids and also occurs in green algae and land plants (see Chapter 1). This difference has been explained by horizontal transmission of ID rubisco genes from proteobacteria into plastid genomes of early red algae, where the ID genes replaced the original IB genes (Delwiche and Palmer 1996; Rice and Palmer 2006). Both the large and small subunits of rubisco are encoded in the red plastid genome. Even so, some genes required for red algal photosynthesis are encoded in the nucleus, a likely consequence of post-endosymbiotic transfer (see Chapter 7), with the result that the cytoplasmic translation products must be targeted to the plastid. An example is the γ subunit of phycoerythrin (Apt et al. 1993).

Most red algae are pink to deep red in color because their plastids contain large amounts of the red accessory pigment **phycoerythrin**, which obscures chlorophyll *a*. Phycoerythrin is extremely efficient in harvesting blue and green light in the subtidal habitats occupied by most red algae, explaining its abundance within red algal plastids. There are at least five types of phycoerythrin in red algae, differing somewhat in spectral absorption but all having an absorption peak in the green wavelengths. Within phycobilisomes—located on unstacked thylakoids (see Figure 15.6)—light energy captured by phycoerythrin, **phycocyanin**, and **allophycocyanin** is efficiently transferred to chlorophyll *a*. This process allows red plastids to harvest light energy that is otherwise inaccessible to chlorophyll.

Not all red algae are red colored. Certain parasitic forms whose plastids lack photosynthetic pigments may be white, cream, or yellowish in color. Freshwater red algae are often colored blue-green because their plastids contain a preponderance of red light–absorbing phycocyanin. Rhodophytes that grow in highly irradiated habitats of upper marine shorelines may be colored yellow, deep violet, brown, or black due to the presence in cells of large amounts of photoprotective carotenoids such as α- and β-carotene, lutein, zeaxanthin, antherixanthin, and violaxanthin. Schubert and associates (2006) characterized the carotenoid composition of 65 red algal species from diverse families. They did not find a uniquely red algal carotenoid profile and observed that unicellular red algae had the simplest carotenoid composition.

Absence of Centrioles, Flagella, and Swimming Motility

Red algae are unusual among eukaryotes in lacking centrioles and flagella from their vegetative cells, spores, and gametes. In contrast, the vast majority of other eukaryotic phyla exhibit centrioles and flagella in some life stages. The red algal condition could be explained by loss of function of genes necessary for synthesis of centrioles and flagella. The absence of flagella is thought to have had profound effects on reproductive evolution in red algae, leading to the widespread occurrence of sexual life histories having two or three multicellular phases (Figure 15.11). In contrast, a maximum of two phases occur in the sexual life histories of other multicellular algae and land plants. The origin of the unique third life-history stage has been explained as an evolutionary compensation for loss of flagella, serving to enhance reproductive fecundity under this constraint (Searles 1980).

Even though red algal cells are unable to move by means of flagella, some unicellular species and reproductive spores have been observed to have a type of motility. The unicells of *Porphyridium* move at about 0.66 μm per second, and *Batrachospermum* spores

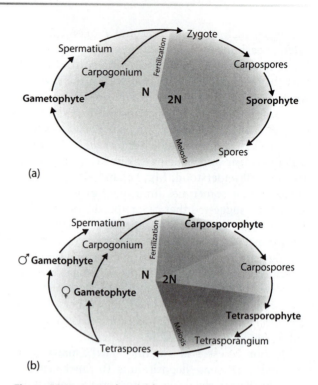

Figure 15.11 (a) Biphasic life history of early-divergent red algae contrasted with (b) the triphasic life history of later-divergent florideophytes.

glide at a rate of about 2.2 μm per second, detectable by observation with the light microscope. The mechanism of this movement is uncertain; polysaccharide mucilage secretion and amoeboid movement via pseudopodia have been observed (Pickett-Heaps et al. 2001). Amoeboid movement is possible because the ECM is so flexible. Inhibitors have been used to determine that actin and myosin regulate amoeboid and gliding movement of *Porphyra pulchella* spores (Ackland et al. 2007).

15.3 Cell Division, Pit-Plug Formation, and Development

Cell Division

Among red algae, mitosis and cytokinesis vary primarily in the structure of spindle pole regions and the extent to which endoplasmic reticulum is associated with the nucleus, but they are otherwise quite similar (Figure 15.12). The nuclear envelope remains intact throughout

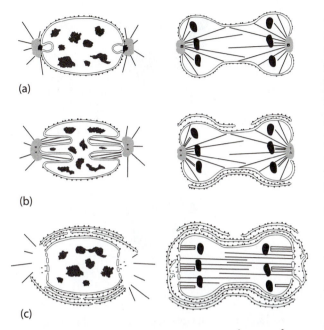

Figure 15.12 Diagrammatic representation of some of the variation in mitosis that occurs during prometaphase (at left) and anaphase (at right) in red algae. Note that the nuclear envelope remains intact except for small interruptions (fenestrations) at the poles and that centrioles are not present, but spindle microtubule organizing structures are present at the poles. (a) A putatively primitive type of mitosis illustrated by the bangiophytes unicell *Porphyridium*. (b) A somewhat more derived type (note the addition of a layer of endoplasmic reticulum) illustrated by the relatively early-divergent, multicellular florideophyte *Batrachospermum*. (c) A derived type of mitosis (note the addition of more ER layers) illustrated by *Polysiphonia*. (After Scott and Broadwater 1990)

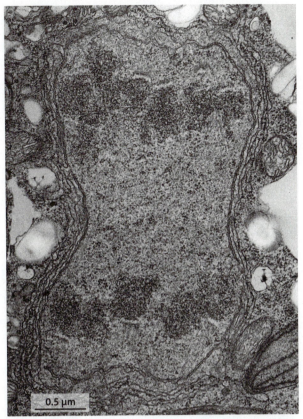

Figure 15.13 TEM view of a late-anaphase nucleus of *Dasya baillouviana*. Note the intact nuclear envelope and surrounding ER. (From Phillips and Scott 1981, © Springer-Verlag)

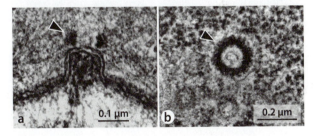

Figure 15.14 Transmission electron micrographs of the nuclear associated organelle (NAO) (arrowheads) of *Polysiphonia harveyi*. (a) Median longitudinal section through the NAO and the underlying nuclear protrusion at late prophase. (b) Transverse section of the *Polysiphonia harveyi* NAO in interphase or early prophase. (From Scott et al. 1980, by permission of the *Journal of Phycology*)

mitosis except for gaps (fenestrations) at the poles (Figure 15.13). As previously noted, centrioles are not present, but a distinctive organelle known as a **polar ring**, or **nuclear associated organelle (NAO)**, typically occurs at the mitotic spindle poles (Figure 15.14). The term NAO is preferred because not all polar structures are ring shaped. The NAO often appears as a pair of short hollow cylinders (Figure 15.15). The NAOs are about the same diameter as centrioles, but they are shorter and lack the distinctive internal microtubular structure typical of centrioles. An extension of the nuclear envelope may be associated with the NAO, sometimes appearing to extend through it (Figure 15.16). Although the chemical composition and function(s) of the NAO are not known, these structures, like centrioles, are associated with microtubules. It is possible that they, or some associated substances, function as microtubule organizing centers, as does pericentriolar material in other eukaryotes (McDonald 1972).

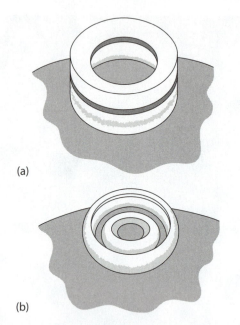

(a)

(b)

Figure 15.15 Diagrammatic representation of two of the four basic NAO types found among red algae.
(a) *Polysiphonia*. (b) *Batrachospermum*. (After Scott and Broadwater 1990)

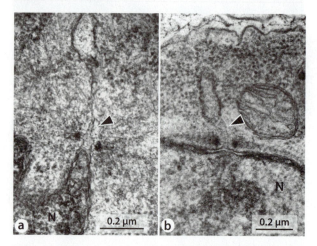

Figure 15.16 Transmission electron micrographs showing tubular extensions of the outer membrane of the nuclear envelope (arrowheads) extending through the NAO and expanding into an extranuclear cisternum.
(a) *Agardhiella subulata* in late prophase. (b) *Lomentaria baileyi* in prophase. N = nucleus.
(From Scott and Broadwater 1990)

Daughter nuclei are kept well separated at te-lophase by the large plastid in monoplastidic uni-cells such as *Porphyridium* (Schornstein and Scott 1982) or by formation of a large central vacuole in

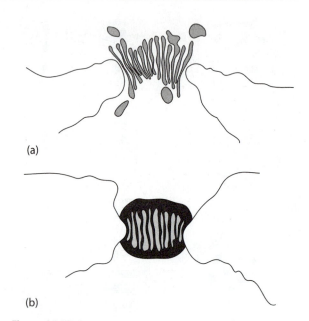

(a)

(b)

Figure 15.17 Diagrammatic representation of pit-plug formation at the ultrastructural level. At the end of cytokinesis, the septal pore becomes filled with tubular membranes (a), and then granular protein material is deposited around the tubules. Pore maturation (b) involves disappearance of the tubules and addition of additional pit-plug components, such as domes or cap membranes. (Based upon micrographs in Scott et al. 1980)

florideophytes. Cytokinesis occurs exclusively by centripetal furrowing (Scott and Broadwater 1990). In florideophytes, completion of furrowing occurs only during tetraspore formation; division of all other cells is incomplete and results in formation of structures known as primary **pit plugs**, which are described next.

Pit-Plug Formation

In the florideophytes and some stages of bangio-phytes, septum (cross-wall) formation is incomplete, leaving a membrane-lined pore in the central region. Tubular membranes appear in this region, then a homogeneously granular protein mass—the plug pore—is deposited around the tubules, followed by disappearance of the tubules (Figure 15.17). In some groups, notably bangiophytes and corallines, the plug core constitutes the pit plug. Pit plugs of other red algae display additional features such as carbohydrate domes and cap membranes, which are continuous with the cell membrane (Figure 15.18) (reviewed by Scott and Broadwater 1990; Pueschel

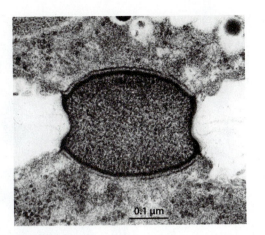

Figure 15.18 TEM view of a pit plug in the central region of a cross-wall separating adjacent sister cells of *Dumontia*. (From Pueschel and Cole 1982)

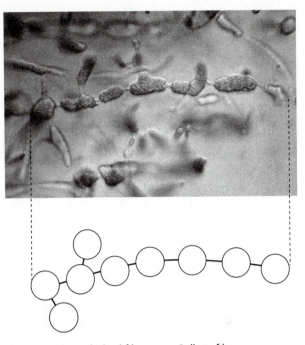

Figure 15.19 Red algal filaments. Cells in filaments of most red algae such as *Mazzaella* are separated by pit plugs that, when mature, may appear as linear connectives between sister cells. Consequently, red algal filaments are usually diagrammed as circles (representing cells) connected by lines (extended pit plugs). (Photo: L. E. Graham)

1990). Variation in pit-plug ultrastructure among red algae has been used to infer phylogenetic relationships among the florideophytes (Pueschel and Cole 1982; Pueschel 1987; Gabrielson and Garbary 1987; Saunders and Bailey 1997). The main function of pit plugs seems to be a structural link between cells. Given the comparatively loose nature of the red algal cell wall/ECM, the evolution of pit plugs that securely hold cells together may have been essential to building the multicellular bodies that most red algae display.

Although a translocatory function has been proposed for pit plugs (Wetherbee 1979), Pueschel (1980, 1990) has pointed out that a pit plug does not connect the cytoplasms of adjacent cells in the same way as do the plasmodesmata of brown and green algae and land plants. A membrane that is continuous with the cell membrane generally covers the surfaces of pit plugs (see Figure 15.18). The pit plug is clearly extracellular in most cases (Pueschel 1977, 1990). The term *pit plug* rather than its synonym *pit connection* is therefore used here in order to avoid implication of plasmodesmata-like function. In mature filaments, pit plugs may be so stretched out that they appear as threadlike linkages between adjacent beadlike cells. This explains why red algal filaments are often diagrammed as beads separated by threads (Figure 15.19). Although primary pit plugs are defined by their formation at cytokinesis, many red algae also produce secondary pit plugs whose formation is not directly linked to cytokinesis.

Secondary pit plugs can form between non-sibling cells of the same body and also between the cells of parasitic red algal species and their hosts. Secondary pit plug formation between such different cells helps to maintain structural integrity. Secondary pit plugs are typically formed at the end of a process that starts with an unequal cell division that produces a primary pit plug. The smaller sibling cell then fuses with a nearby non-kindred cell, and the secondary pit plug forms after the cell fusion (Figure 15.20).

Cell Elongation and Repair

As noted earlier, red algal cells may undergo a tremendous increase in volume through elongation. For example, the length of certain cells of *Ceramium echionotum* can increase by a factor of 100—from 4 μm to 420 μm—accompanied by a 14,000-fold increase in volume. Cells of the freshwater *Lemanea fluviatilis* may elongate 1000 times (from 8 μm to 8000 μm), with an attendant 44,000-fold increase in volume. Cell elongation may occur in apical cells by growth at the tip, or in non-apical cells by incorporation of ECM material all along the length of

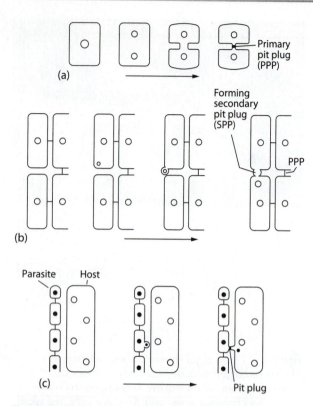

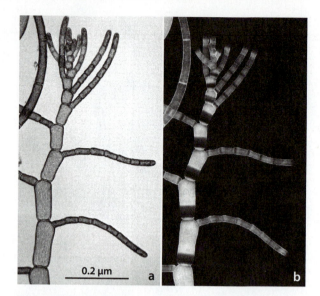

Figure 15.21 Red algal cell elongation. The process of cell elongation in red algae is illustrated by *Callithamnion* filaments that have been stained with Calcofluor White—a fluorescent, cellulose-specific reagent—and then removed from stain and allowed to continue growth for 24 hours. (a) Body viewed in bright-field microscopy. (b) Body viewed by epifluorescence microscopy with UV excitation; new wall material added to apical cells of branches and at the bases of intercalary cells during the period following removal from stain is not fluorescent (i.e., is dark). (From Waaland and Waaland 1975, © Springer-Verlag)

Figure 15.20 Diagrammatic representation of pit-plug formation. (a) Primary pit-plug (PPP) formation between sister cells (cells derived from a common cell division). (b) Secondary pit-plug (SPP) formation between non-sister cells involves an unequal division leading to formation of a small cell termed a conjunctor cell; fusion of the conjunctor cell with a non-sister cell results in transfer of a nucleus to the recipient cell and development of the secondary pit plug. (c) Secondary pit-plug formation also occurs between genetically unrelated cells during red algal parasitism; the parasite initiates the formation of a small conjunctor cell by asymmetric division, and then this cell fuses with a host cell. (After Goff and Coleman 1990)

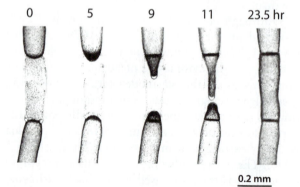

Figure 15.22 Cell repair by cell fusion in *Griffithsia pacifica*, where an intercalary cell has died. A rhizoid from the cell immediately above the dead cell and a repair shoot from the cell below it grow toward each other and fuse. (From Waaland and Cleland 1974, © Springer-Verlag)

the extending cell wall or in one or more localized bands, known as zones of extension (Figure 15.21) (Waaland 1990).

When a cell in the middle of a filament dies and the remains of its ECM are intact, the cells immediately above and below it, which normally would not divide, each divide to produce narrow-tipped, rhizoid- and apex-like cells that grow toward each other and eventually fuse, effecting a repair of the wounded filament (Figure 15.22). In *Griffithsia pacifica*, the cell immediately above the wounded cell produces a diffusible, species-specific glycoprotein wound-repair hormone.

Known as rhodomorphin, this hormone is active at very low concentrations (10^{-13}–10^{-14} M). A signal glycoprotein analogous to rhodomorphin appears to mediate wound healing in another red

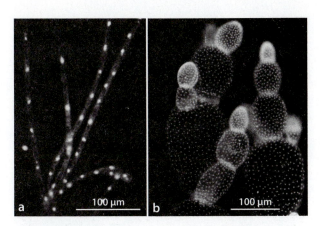

Figure 15.23 Fluorescence microscopy of DAPI-stained red algae showing uninucleate cells of *Acrochaetium pectinatum* (a) and multinucleate cells of *Griffithsia globulifera* (b). (From Goff and Coleman 1990)

alga, *Antithamnion nipponicum*, suggesting that such hormones may be common in red algae (Kim and Fritz 1993). Tagged lectins—molecules that exhibit strong specific binding to particular carbohydrate structures—were used to demonstrate that *Antithamnion* repair hormone binds a wall glycoprotein having alpha-D-mannosyl residues (Kim et

al. 1995). Glycoproteins may mediate other occurrences of cell fusion in red algae, including fertilization, secondary pit-plug formation, and fusions between cells of different life-history stages, or genotypes, during parasitism (Waaland 1990). Further study of diffusible red algal glycoproteins may yield explanations for many of the unusual features of red algae.

Development of Multinucleate Cells and Polyploid Nuclei

In many of the more highly derived red algae, nuclei continue to undergo mitosis without intervention of cytokinesis, giving rise to multinucleate cells. The uninucleate or multinucleate condition of cells can be detected by Feulgen staining, which makes nuclei visible in bright-field light microscopy or by staining with a DNA-specific fluorochrome that forms a complex detectable with fluorescence microscopy (Figure 15.23). The presence of numerous nuclei allows the maintenance of very large cells whose cytoplasm would unlikely be efficiently served by a single nucleus, particularly since red algal cells lack cytoplasmic streaming. There is often a tight correlation between the number of

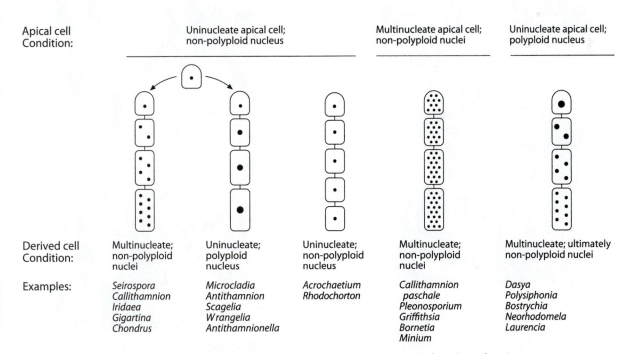

Apical cell Condition:	Uninucleate apical cell; non-polyploid nucleus		Uninucleate apical cell; non-polyploid nucleus	Multinucleate apical cell; non-polyploid nuclei	Uninucleate apical cell; polyploid nucleus
Derived cell Condition:	Multinucleate; non-polyploid nuclei	Uninucleate; polyploid nucleus	Uninucleate; non-polyploid nucleus	Multinucleate; non-polyploid nuclei	Multinucleate; ultimately non-polyploid nuclei
Examples:	*Seirospora* *Callithamnion* *Iridaea* *Gigartina* *Chondrus*	*Microcladia* *Antithamnion* *Scagelia* *Wrangelia* *Antithamnionella*	*Acrochaetium* *Rhodochorton*	*Callithamnion paschale* *Pleonosporium* *Griffithsia* *Bornetia* *Minium*	*Dasya* *Polysiphonia* *Bostrychia* *Neorhodomela* *Laurencia*

Figure 15.24 Diagrammatic representation of variation among red algae in terms of number of nuclei in apical and other cells, and in whether nuclei are polyploid. (After Goff and Coleman 1990).

nuclei in a red algal cell and the number of plastids (Goff and Coleman 1990).

Red algal nuclei can also undergo a process known as **endoreduplication**—repeated replication of the entire nuclear genome without intervening mitosis—resulting in polyploidy. Such genome amplification is also known as nuclear polygenomy, endopolyploidy, or polyteny, and it is common in red algae. Genome amplification is thought to serve as a buffer against mutation of essential genes. Endoreduplication and other changes in nuclear DNA level, such as those resulting from gamete fusion or meiosis, can be detected by use of microspectrophotometry or microfluorometry that measure the extent of DNA staining. In some red algae the apical meristematic cell of filaments is also multinucleate, although apical cells are not typically polyploid (Goff and Coleman 1990). Examples of variation among red algae in numbers of nuclei and polyploid nuclei in apical and other cells of filaments are shown in Figure 15.24.

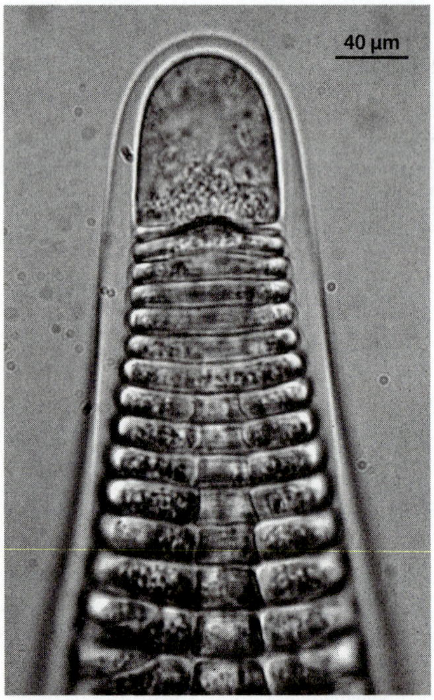

15.4 Body Organization of Red Algae

Early-diverging rhodophytes occur as unicells, colonies, filaments, or sheets of cells that usually lack pit plugs. Even in the multicellular bangiophytes that display pit plugs, there is no specialized meristematic region; cell division may occur almost anywhere in the body, and growth is thus said to be diffuse. In contrast, the bodies of florideophytes, which are fundamentally filamentous, grow via division of an apical cell located at the terminus of a filament. Florideophytes thus have **apical growth**. Apical cells cut off derivatives at their bases in a single linear series (Figure 15.25). Divisions of apical cells are responsible for increases in the number of cells in filaments and, together with cell enlargement, increase florideophyte body size.

Delicate florideophyte bodies may be **uniaxial**, composed of a single branched filament (Figure 15.26). Commonly, subapical cells divide several times radially to form whorls of branch initial cells

Figure 15.25 Apical meristem of the uniaxial red alga *Platysiphonia*, showing the way in which derivatives are cut off from the base of the apical cell. The subapical cells have undergone further divisions to form whorls of periaxial cells that may serve as branch initials. (Photo: R. Norris, in Bold and Wynne 1985)

Figure 15.26 An example of a uniaxial, branched filamentous red alga, *Aglaothamnion*, in microscopic view. (Photo: C. Taylor)

known as **periaxial cells**, which divide to form branch filaments, each with a terminal apical cell. Branches are typically determinate, meaning that only a genetically determined number of cell divisions occur, thereby limiting the length of the branch filaments. Subapical cells also generate rhizoids—single cells or filaments that are involved in attachment (Coomans and Hommersand 1990).

More-robust and fleshy florideophyte bodies may be **multiaxial**, composed of multiple filamentous axes, each derived from a terminal apical cell (Figure 15.27). These multiple axes arise during early development through the transformation of determinate branches into indeterminate axes. This means that the branch apical cells become less constrained in the number of cell divisions they can undergo. Fleshy macroscopic red algae, some reaching 2 m in width (Figure 15.28) and exhibiting considerable body and cellular differentiation, are typically formed by the coalescence of the branched filaments making up a multiaxial body. Such bodies are said to be composed of **pseudoparenchyma**. Differences in division-plane orientation are thought to explain variations in the branching patterns of different rhodophyte genera (Coomans and Hommersand 1990).

Thick, fleshy red algal bodies are commonly composed of highly pigmented filaments that form an outer **cortex** and colorless filaments that make up a central region known as the **medulla** (Figure 15.29). Specialized basal holdfasts often anchor red algae to substrates. In addition to apical meristematic cells, some other types of specialized cells occur in florideophytes. Thin, elongate **hair cells** that may extend from the body surface are thought to function in nutrient uptake. **Gland cells**, also known as vesicle cells, have specialized secretory functions. In other red algae, such as *Asparagopsis armata*, gland cells possess crystals or halogenated (bromine or iodine-containing) compounds that are thought to have antimicrobial, anti-herbivore, or allelopathic functions (Paul et al. 2006).

Figure 15.28 *Schizymenia borealis*, which is one of the largest of the red algae. Blades can be as long as 2 m. (Photo: L. E. Graham)

Figure 15.27 An example of a multiaxial red alga, *Agardhiella*, in microscopic view. (From Gabrielson and Hommersand 1982, by permission of the *Journal of Phycology*)

Figure 15.29 *Mazzaella*, a common example of a multiaxial, thalloid red alga. As shown in this section, the body is composed of aggregations of filaments (pseudoparenchyma), and there are specialized regions of pigmented surficial cortical cells that are densely packed as well as more loosely arranged, colorless filaments of the central medulla. (Photo: L. E. Graham)

Some authorities maintain that blade-forming bangiophytes such as *Porphyra* and certain lineages of florideophytes display true tissue, known as **parenchyma** (Bold and Wynne 1985). Parenchyma is a feature of some brown algae, notably kelps and fucoids (see Chapter 14), as well as land plants of the green lineage. However, unlike in red algae, the cells of land plants and parenchymatous brown algae are linked by plasmodesmata, which allow intercellular movement of materials, including information molecules. Further, some experts suggest that early in their evolutionary diversification, red algae underwent a developmental commitment to the production of male and female gametangia from apical cells. This key evolutionary event is thought to have limited growth possibilities, preventing red algae from evolving parenchyma (Hommersand and Fredericq 1990). The absence of true tissue from most or all red algae probably explains why even the largest and most intricate species do not approach the sizes and internal complexity of brown kelps and land plants. Even so, red algae have achieved an amazing diversity of body types by varying the way in which their filamentous bodies develop.

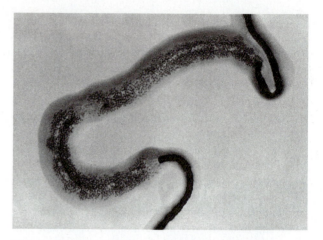

Figure 15.30 Asexual spore production by *Bangia* growing on an agar surface. Transformation of vegetative cells into asexual spores and their release have occurred within an internal portion of the filament. (Reprinted from *Aquatic Botany*, Volume 28. Graham, J. M., and L. E. Graham. Growth and reproduction of *Bangia atropurpurea* [Roth] C. Ag. [Rhodophyta] from the Laurentian Great Lakes. pp. 317–331. © 1987, with permission from Elsevier Science)

15.5 Reproduction and Life Histories of Red Algae

Red algae may reproduce by asexual means that do not involve gamete fusion and meiosis, or by sexual processes characterized by gamete fusion and meiosis. Zygotic and gametic life cycles (see Chapter 1) are not known for red algae. The simplest type of sexual reproduction known to occur in red algae is the alternation of two heteromorphic, multicellular generations. Florideophytes display a uniquely complex type of life cycle that involves alternation of three multicellular generations.

Asexual Reproduction

Many red algae reproduce asexually by discharging unicellular spores, such as archeospores of bangiophytes (Figure 15.30) or monospores of florideophytes. In general, asexual spore production is more common in early-diverging red algal groups than in later-diverging forms, where it may be rare or absent (Hawkes 1990). If conditions are suitable, asexual spores attach to a substrate and grow by repeated

mitosis into new seaweeds similar to the parent. Red algal spores sink relatively slowly, and a surrounding coat of mucilage aids in the initial attachment to substrates. Completion of the attachment process requires hours to days. Once attached, spores are quite resistant to shear stresses of water movement (Kain and Norton 1990). Asexual spores are produced singly within sporangia, which may be produced in clusters. The ultrastructure of *Porphyra* archeospores was studied by Hawkes (1980), and genes that are preferentially expressed during archeospore production have been identified in *P. yezoensis* (Kitade et al. 2008). Production and swelling of mucilage is the mechanism for release of asexual spores. Vegetative fragmentation, "stolon" production (analogous to asexual propagation in plants such as the strawberry), and formation of specialized propagules may also occur. A list of genera exhibiting such variations in asexual reproduction is provided by Hawkes (1990).

Sexual Reproduction

Sexual reproduction is characteristic of the vast majority of red algae. Though sexual reproduction has not been observed in the early-diverging cyanidophytes or the unicellular *Porphyridium*, it has been well documented in some other early-diverging red

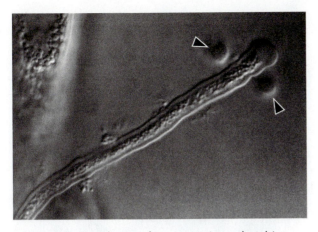

Figure 15.31 Attachment of spermatia (arrowheads) to the tip of a trichogyne (the elongate terminal portion of a carpogonium) of *Aglaothamnion*. (Photo: C. Taylor)

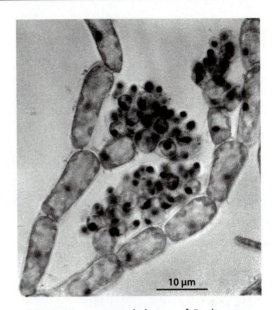

10 μm

Figure 15.32 Spermatangial clusters of *Dudresnaya crassa*. Nuclei were stained with iron hematoxylin. (From Hommersand and Fredericq 1990)

algal lineages, including *Rhodochaete*, *Erythrotrichia*, *Porphyrostromium*, *Smithora*, *Bangia*, and *Porphyra* (see references in Garbary and Gabrielson 1990). In all known cases, sexual reproduction of red algae is oogamous, involving fusion of a relatively small, nonflagellate **spermatium** (the male gamete) with a larger, nonflagellate female gamete. In florideophytes, the female gamete, known as a **carpogonium**, is typically flask shaped, with an inflated base and an elongated neck called the **trichogyne** (Figure 15.31). All sexually reproducing red algae display an alternation of distinct gamete-producing **gametophyte** bodies and one or two spore-producing **sporophyte** generations or phases. The gametophyte body and gametes produced by it are assumed to be haploid, while sporophyte bodies are assumed to be diploid. Meiosis occurs during the production of spores by a sporophytic stage. The life cycle of red algae is thus known as **biphasic** or **triphasic alternation of generations** (see Figure 15.11).

Sexually reproducing bangiophytes display biphasic alternation of generations. The most famous example is provided by the economically important seaweed *Porphyra*. By performing careful culture work in the laboratory, British phycologist Kathleen Drew Baker linked the life history of inconspicuous spore-producing red filaments previously known by the generic name *Conchocelis* with the edible blade-forming, gamete-producing seaweed (see Chapter 4). Drew Baker's work formed the basis for the modern billion-dollar-per-year *nori* production industry in Japan, Korea, China, and the United States. In recognition of the importance of her work, a memorial

park was established in Drew Baker's honor in Kumamoto Prefecture, Japan, where she is annually revered with a ceremony dedicated to "the mother of the sea."

In contrast to bangiophytes, florideophytes typically produce and release spermatia from a male gametophyte, while a separate female gametophyte produces a carpogonium that remains attached throughout the processes of fertilization and postfertilization development of the zygote. Triphasic alternation of generations is a distinctive feature of the florideophytes.

Spermatangia and spermatia

In florideophytes, spermatia are produced singly within spermatangia, which often occur in clusters (Figure 15.32). Spermatangia originate as subapical protrusions that are cut off by oblique cell walls during division of apical initial cells. In fleshy forms, spermatangia may occur in clusters at the surface of the body. In other cases, spermatangia may line cavities, such as the conceptacles of the coralline red algae (Figure 15.33). *Polysiphonia*, which is widely used in classrooms to demonstrate sexual reproduction of red algae, displays groups of spermatangia embedded within a common matrix (Hommersand and Fredericq 1990). As mature spermatia are released, younger ones are produced in the same area

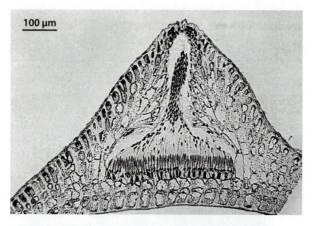

Figure 15.33 Light micrograph of a section through a spermatial conceptacle of the coralline red alga *Mastophora*. (From Woelkerling 1988, by permission of Oxford University Press)

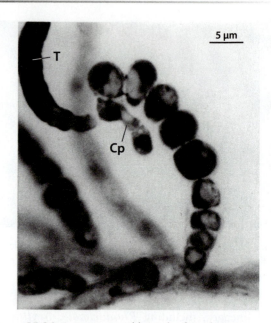

Figure 15.34 A carpogonial branch of *Dudresnaya crassa* showing division of the fertilized carpogonium (Cp) and the trichogyne (T). (From Hommersand and Fredericq 1990)

(Kugrens 1980). Spermatium release in *Ptilota densa* was studied at the ultrastructural level by Scott and Dixon (1973). Red algal spermatia had been thought to move toward female gametes passively, by means of water currents. However, in *Tiffaniella*, spermatia are released in mucilaginous strands that coalesce, extend, contract, and rotate in the water until they make contact with a female body (Hommersand and Fredericq 1990). The absence of flagella from the male gametes of red algae is thought to reduce fecundity because spermatia are less able than the flagellate gametes of other eukaryotes to transport themselves to females. This feature is thought to have strongly influenced the evolutionary diversification of red algae.

Carpogonia and fertilization

The female gametes known as carpogonia are also produced singly from precursor cells. In most red algae, carpogonia are borne at the tips of specialized branches known as **carpogonial branches** (Figures 15.34 and 15.35), which can arise either laterally from the main axis or terminally, by transformation of a cell at the tip of a vegetative filament (Hommersand and Fredericq 1990).

Fertilization is initiated by adhesion of a spermatium to the carpogonial trichogyne (see Figure 15.31). Upon contact of spermatia of the appropriate type, enzymes digest a channel through the carpogonial extracellular matrix, allowing entry of a spermatial nucleus. Although more than one male nucleus may enter the carpogonium, only one fuses

with the female nucleus. The carpogonial vacuole then contracts, which helps to pull the male nucleus into contact with the female nucleus, located at the carpogonial base. The cytoplasm then pinches closed at the top of the inflated base of the carpogonium, preventing other spermatia from entering (Broadwater and Scott 1982). Time-lapse video has been used to document fertilization in the red alga *Bostrychia* (Pickett-Heaps and West 1998). In red algae, the diploid zygote resulting from fertilization develops by mitotic divisions into a spore-producing diploid multicellular body.

Postfertilization Development and Life History in Bangiophytes

In *Porphyra* and *Bangia*, the fertilized zygote undergoes mitotic divisions to produce several diploid **spores**. These diploid spores are released into the water, settle, and grow into an independent, diploid sporophyte filament often referred to as the **conchocelis phase**, based on Kathleen Drew Baker's path-making work. In nature, the conchocelis phase often grows within mollusc shells or the tubes of polychaetes and may live in water as deep as 78 m (Lüning 1990), far deeper than gametophytes can survive. The conchocelis phase may thus extend the

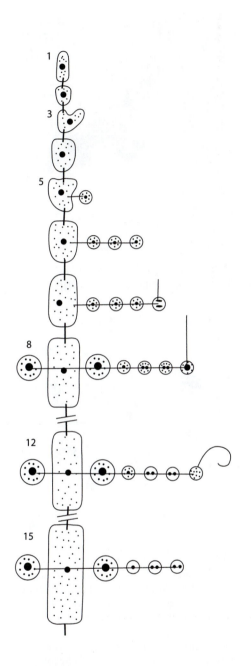

range of the species beyond that occupied by the more conspicuous blade-forming life-history phase. The conchocelis phase may be perennial and proliferate by means of asexual spores produced in terminal cells. Under appropriate environmental conditions, it produces spores that grow into the more conspicuous gametophytic blade. Some other bangiophytes likewise exhibit heteromorphic alternation of generations; *Porphyrostromium* gametophytes are erect filaments, whereas its sporophytes are flattened (Kornmann 1984, 1987).

During early development of the sheetlike body of *Porphyra*, sporic meiosis first yields a four-celled linear germling, each cell of which has become haploid. The meiotic process segregates alleles for sexual determination. Thus, two of the resulting cells contain genes encoding male gamete development and the other two, female development (Ma and Miura 1984). Subsequent mitotic proliferation of each of these genotypically distinct cells may result in a genetically sectored body. In some species, a patchwork nature of the body is revealed where some sectors produce spermatia, whereas others generate carpogonia (see Chapter 4).

Postfertilization Development and Triphasic Life History of Florideophytes

The fertilized carpogonium of florideophytes, rather than producing spores directly, generates one or more multicellular, diploid life phases known as **carposporophytes**. A unique and derived feature of florideophytes, the carposporophyte consists of a mass of filaments that can eventually produce many diploid spores known as **carpospores**. When mature, carpospores are released into the water, settle, and germinate into a second multicellular, diploid generation known as the **tetrasporophyte** because it produces **tetraspores**. Alternation of generations in florideophytes is triphasic because there is sequential alternation among three multicellular generations—gametophyte, carposporophyte, and tetrasporophyte (see Figure 15.11).

Carposporophyte development

Carposporophytes develop from fertilized carpogonia or other female gametophyte cells that have received a copy of the zygote, by processes described later. Nutritional resources provided by female gametophyte cells fuel mitotic division of such zygote nuclei (Turner and Evans 1986), generating

Figure 15.35 Carpogonial branch development in the female gametophyte of *Callithamnion cordatum*. For ease in illustration, stages of carpogonial branch development are shown on adjacent cells in the axis. Numbers are ages of axial cells, in days. Number 1 is the apical cell, which divides once per day. Vegetative laterals are initiated in 3-day-old cells; 5-day-old cells may initiate carpogonial branches, and by 8 days, mature carpogonial branches are present. At 12 and 15 days, carpogonial branches are senescent and degrading, respectively. (After O'Kelly and Baca 1984)

filaments of diploid cells that each contain a copy of the zygote nucleus. A mass of such filaments that has developed from a single cell containing a diploid nucleus is known as **gonimoblast**. A single gonimoblast is equivalent to a carposporophyte. The gonimoblast uses additional nutritional resources obtained from the female gametophyte to produce carposporangia, which release diploid carpospores into the water. This nutritional dependency explains why carposporophytes do not exist independently of the female gametophyte.

Evidence for a nutritional relationship between the female gametophyte and carposporophyte includes various types of cellular specializations that suggest nutritional function, such as wall and membrane proliferations in regions known as placentae. In addition, radioactive tracer work suggests that photosynthates can be translocated from parental gametophyte cells to those of developing carposporophytes (Hommersand and Fredericq 1990). A genetic analysis of reproductive success, in which hybrids of parents from geographically distinct populations of *Gracilaria verrucosa* were studied, revealed that the female parent has a strong effect on crossing success (Richerd et al. 1993). These results suggest that gametophyte–carposporophyte nutritional/developmental interaction has a strong effect upon fecundity. Kamiya and Kawai (2002) creatively demonstrated the nutritional dependence of carposporophytes on maternal gametophytes in several florideophytes by comparing the numbers and sizes of carpospores liberated from detached gametophytic branches of varying size. Longer branches having more nutritive gametophytic cells produced more and larger carpospores.

In some florideophytes, the carposporophyte develops directly from the fertilized carpogonium or from a cell within the carpogonial branch that has fused with the carpogonium and is thus known as a **fusion cell**. Such fusion presumably supplies nutrients that are used to produce more carpospores than would otherwise be possible. Many florideophytes take this process even further by transferring mitotically produced copies of the diploid nucleus into other gametophyte cells by cell–cell fusion. Cells outside the carpogonial branch that will or have received a copy of the zygote nucleus are termed **auxiliary cells** (Figure 15.36). Fusion cells and auxiliary cells serve as hosts and nutritional sources for repeated divisions of the adopted copy of the zygote nucleus. Fusion cells and auxiliary cells are also thought to be sources of morphogenetic compounds that influence postfertilization

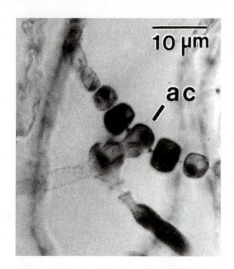

Figure 15.36 Auxiliary cell (ac) in *Dudresnaya crassa*. (From Hommersand and Fredericq 1990)

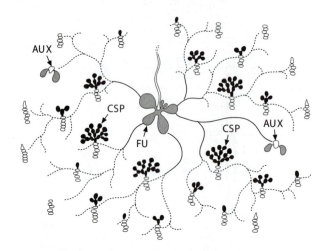

Figure 15.37 Diagrammatic representation of the transfer of diploid nuclei (which are mitotic copies of a single zygote) from fusion cells (FU) to auxiliary cells (AUX) scattered throughout the body of *Hommersandia*. Successful transformation of auxiliary cells results in the formation of carposporophytes (CSP) and production of numerous carpospores. (After Hansen and Lindstrom 1984, by permission of the *Journal of Phycology*)

development and to function as part of an isolating mechanism by which incompatible fertilizations can be rejected. It is emphasized that the incorporated diploid nuclei do not fuse with gametophytic cell nuclei; these retain their separate identities. Gametophytic cells that contain both their original nucleus and a nucleus derived from the zygote are heterokaryons and are regarded as having been genetically transformed.

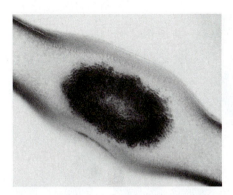

Figure 15.38 Cystocarp. A single cystocarp, which includes masses of bright-red developing carpospores, is revealed within a bladelike body of *Mazzaella* that has been thinly sliced for viewing by light microscopy. (Photo: L. E. Graham)

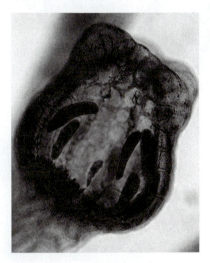

Figure 15.39 An urn-shaped pericarp, formed of gametophytic cells, surrounds developing carposporophytes of *Polysiphonia*. Photograph from a stained and mounted specimen. (Photo: L. W. Wilcox)

2 cm

Figure 15.40 Macroscopic view of cystocarps on a fertile body of *Mazzaella* (a dried specimen scanned as transparency), showing multitudinous dark spots (cystocarps) in peripheral portions of the blade. (Photo: L. W. Wilcox)

Copies of the diploid zygote nucleus can often be transmitted to auxiliary cells for long distances within the body by means of long tubular cells, variously known as **ooblast/connecting cells** or **ooblast/connecting filaments** (Figure 15.37). This process allows red algae to amplify the zygote nucleus produced by a single fertilization into many carposporophytes, each of which at maturity generates numerous carpospores.

Many red algae protect carposporophytes by producing an enclosure of gametophyte cells. Together, a single carposporophyte plus surrounding gametophyte cells is known as a **cystocarp** (Figure 15.38). The protective components of cystocarps can develop

from the connecting filament, cells adjacent to an auxiliary cell, or the auxiliary cell itself. Cystocarps typically have three layers: (1) an outermost region of photosynthetic gametophytic cells that may be modified into a distinctive **pericarp**, (2) an interior region of non-photosynthetic gametophyte cells regarded as a nutrient provision and processing center, and inside that, (3) a developing carposporophyte whose cells sometimes have secondary pit plugs in common with gametophytic cells (Figure 15.39). Close examination of fertile red algal gametophytes, such as those of *Mazzaella*, may reveal the presence of many small, densely pigmented cystocarps (Figure 15.40).

As a result of the nuclear transfer activities of connecting filaments, each diploid nucleus produced by fertilization may be capable of generating many cystocarps, each of which can release large numbers of carpospores into the water. Each zygote of *Schmitzia sanctae-crucis* generates some 4500 carpospores (Wilce and Searles 1991). A single body of *Gelidium robustum* releases 34,000–300,000 carpospores per month, and the mean number of carpospores released by *Chondrus crispus* is estimated to be nearly 10 billion month^{-1} m^{-2} (see references in Kain and Norton 1990). Evolution of the carposporophyte has allowed florideophytes to overcome fecundity constraints imposed by the inability of sperm to actively swim to female gametes by means of flagella (Searles 1980). If they settle in favorable habitats, diploid carpospores develop into diploid tetrasporophytes, which further amplify the products of sexual reproduction. One group of the florideophyte red algae (Palmariales) is unusual in that the fertilized carpogonium develops directly into tetrasporophytes, so that a carposporophyte is said to be lacking (van der Meer and Todd 1980). However, phylogenetic evidence indicates that this group is derived from ancestral forms that possessed carposporophytes.

Tetrasporophytes and tetraspore production

Multicellular tetrasporophyte bodies generally develop from carpospores by mitosis and cytokinesis. Tetrasporophytes are usually free living, occurring independently from gametophytes and carposporophytes. However, in *Gymnogongrus* and *Phyllophora*, the tetrasporophyte lives upon the gametophyte generation, the latter being the only free-living generation in the triphasic life cycle. When mature, tetrasporophyte bodies produce tetraspores in many **tetrasporangia**. Each tetrasporangium develops by modification of a branch apical cell. Usually four tetraspores are produced in each tetrasporangium (Figure 15.41). Tetrasporangia are regarded as a shared, derived feature of the florideophytes.

It has been assumed that meiosis occurs within the tetrasporangium during the formation of tetraspores, and recent comparative measurements of nuclear DNA levels by DAPI microfluorometry have confirmed that this is indeed the case for some red algae, such as *Wrangelia plumosa* (Figure 15.42). When meiosis occurs, sex-determining alleles are segregated such that two tetraspores of each quartet are able to develop into male gametophytes and the other two, female gametophytes. However, similar DNA-level measurements in *Scagelia pylaisaei* revealed that meiosis occurs during tetraspore germination rather than tetraspore formation and that tetraspores are formed mitotically (Goff and Coleman 1990).

A single red algal tetrasporophyte can release prodigious numbers of tetraspores. For example, individuals of *Gelidium robustum* produce 11,000–27,000 tetraspores per month, bodies of

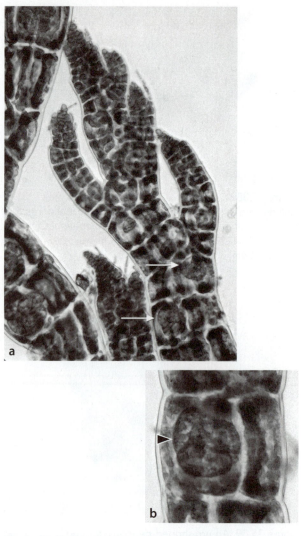

Figure 15.41 Tetraspore production in *Polysiphonia*. (a) Four tetraspores are generated terminally on one-celled lateral filaments (arrows). (b) Higher-magnification view of a single tetrasporangium (arrowhead) with four tetraspores. (Photos: L. W. Wilcox)

Botryocladia pseudodichotoma, nearly 4 million tetraspores per day, and *Chondrus crispus*, more than 200 million tetraspores month^{-1} m^{-2} (see references in Kain and Norton 1990). After their release into the water, tetraspores attach to substrates and grow into gametophytes. These numbers illustrate that the occurrence of the diploid carposporophyte and tetrasporophyte generations in the life cycle of florideophytes provides the potential to dramatically amplify the ecological impact of a single fertilization event. One zygote nucleus can potentially give rise to billions of gametophytes.

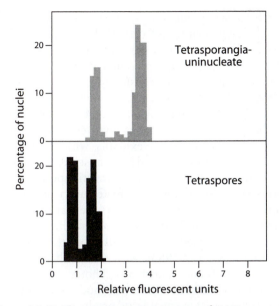

Figure 15.42 Fluorometric measurements of DAPI-stained nuclei of *Wrangelia plumosa* tetrasporangia and tetraspores provide evidence that meiosis occurs prior to tetraspore formation in at least some red algae. (From Goff and Coleman 1990)

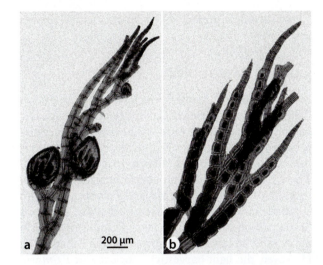

Figure 15.43 Isomorphic alternation of generations, illustrated by *Polysiphonia*, whose (a) gametophytes and (b) tetrasporophytes are indistinguishable until they become fertile—and then only by microscopic observation (as shown here). (Photos: L. W. Wilcox)

The reproductive amplification process present in florideophytes also contributes to greatly increased genetic diversity. Many more genetic recombinants will result from the combined meiotic activities of the millions (or even billions) of tetrasporophytes that could potentially arise from the carpospore progeny of a single carposporophyte than could be produced from a fertilized bangiophyte zygote. Hommersand and Fredericq (1990) hypothesized that red algal diversification has involved an evolutionary trend toward increased carposporophytic reliance upon gametophytic resources and defined several major types of postfertilization development of increasing specialization. In some cases (such as *Liagora tetrasporifera*), the tetrasporophyte generation seems to have been lost, and meiosis apparently occurs during sporogenesis in the carposporophyte (see Hawkes 1990).

Isomorphic and Heteromorphic Alternation of Florideophyte Generations

In most florideophyte red algae, the gametophyte and tetrasporophyte generations resemble one another closely and are therefore said to be isomorphic, as illustrated by *Polysiphonia* (Figure 15.43). Corallines also have isomorphic life cycles. However, more heteromorphic life cycles are being discovered in which the tetrasporophytes are microscopic and filamentous or consist of a thin crust that is tightly attached to the rock substrate (Hawkes 1990). In some cases, the alternating life phases had previously been given distinct generic names. Cultivation in the laboratory is usually necessary to establish the occurrence of such heteromorphic life histories. Some experts speculate that the selective advantages of differences in life-cycle phase appearance may include adaptation to seasonal variation in habitat and herbivory pressure, topics discussed in more detail in Chapter 22. The main basis for this conjecture is that heteromorphic alternation of generations in red algae seems to occur more commonly in seasonal temperate and polar habitats than in the tropics (John 1994).

15.6 Red Algal Physiology and Ecology

Red algal physiology and ecology are similar to those of other seaweeds in some ways, such as carbon metabolism, but also display some distinctive features.

These include a greater diversity of calcified species and parasitic forms.

Carbon Metabolism

Evaluation of the literature suggested to Raven et al. (1990) that with the exception of rhodophyte-specific secondary compounds and some unusual osmoregulatory carbohydrates described below, carbon metabolism within red algae is similar to that of other algae. There is evidence that a number of marine red algae, including *Ceramium rubrum* and *Palmaria palmata* (Maberly et al. 1992), can use bicarbonate, and thus have a carbon concentrating mechanism (CCM), as do many other marine and freshwater algae (Chapter 2). However, the light-harvesting efficiency of *Palmaria palmata* was determined to be inversely proportional to the extent of bicarbonate use, indicating that there is a considerable physiological cost of bicarbonate acquisition (Kübler and Raven 1995).

Some marine subtidal red algae, such as *Delesseria sanguinea*, intertidal forms such as *Lomentaria articulata* (Maberly et al. 1992), and red algae of freshwater streams, appear to lack CCMs, and are therefore dependent upon dissolved carbon dioxide. Carbon dioxide dependence of the freshwater *Lemanea mamillosa* has been extensively explored (Raven 1997b). In the well-aerated waters of fast-flowing streams, CO_2 is not limiting to algal growth. Marine red algae that occupy deeply shaded habitats experience limitation of photosynthesis by light availability, with the result that ambient dissolved CO_2 levels saturate photosynthesis (Raven 1997b).

The calcification process, such as occurs in corallines, may be a consequence of inorganic carbon uptake (see Chapter 2). The hypothesized mechanism for this process is that photosynthetic CO_2 fixation results in an increase in pH near the site of calcification, facilitating $CaCO_3$ precipitation. Coupled acidification processes are thought to be necessary (McConnaughey 1994), and a calcium-dependent ATPase that might be involved in such calcification-associated proton pumping has been isolated from the coralline *Serraticardia maxima* (Mori et al. 1996).

Red Algal Parasites

Approximately 10% of red algal genera are completely or partially parasitic on other red algae (Goff 1982). Such parasites can often be identified by their lack of pigmentation, penetration into the host's body, and body-size reduction. In some cases radiolabeled tracers have been used to establish net flow of photosynthate from host to parasite (Goff 1979, 1982). Microfluorometry and molecular markers have also been used to detect the transfer of parasite nuclei and mitochondria into host cells, a hallmark of red algal parasitism (Goff and Coleman 1984). Establishment of the parasitic relationship involves host–parasite cell–cell fusions, which explains why rhodophyte parasites occur only on red algal hosts. As a result of cell fusions with the host, the parasitic species transforms the host in ways that foster nutrient transfer from the photosynthetic host to the parasite (Goff and Coleman 1987). In at least one studied case, the parasites harvested only a small proportion of host resources (Kremer 1983), with the result that hosts were able to function normally. Though often colorless when associated with a host body, several parasitic red algae are capable of producing pigmented cells when excised and grown separately in culture, suggesting that the genetic information for pigment production is present but that expression is regulated by the host (Goff and Zuccarello 1994). Most red algal parasites infect hosts that are members of the same family (Zuccarello et al. 2004).

An interesting example is the tiny parasite *Neotenophycus ichthyosteus* that occurs on the host *Neosiphonia poko* in shallow lagoons of the U.S. territory Johnston Atoll, in the central Pacific Ocean. The parasite and host are both classified into the family Rhodomelaceae, but they differ in important features. Infective cells of the parasite first connect to a cell of the central axis of the host and then grow into few-celled male or female gametophytes. Carpospores produced by carposporangia on the female gametophytes germinate into small tetrasporophytes that parasitize the same host species (Kraft and Abbott 2002).

Responses to Drought and Osmotic Stress

Many red algae are highly sensitive to dehydration, and a number of growth habits have been interpreted as adaptations that favor maintenance of a well-hydrated state. Many reds are obligate understory species, living beneath a cover of larger brown seaweeds, which provides a high-humidity habitat upon emersion. Corallines are particularly sensitive to drought stress and thus tend to inhabit tide pools (see Figure 15.4) rather than nearby surfaces subject to drying during low tide.

A number of red algae occupy shoreline environments, including mangrove forests, where they are alternately submerged and exposed, or estuaries, in which they experience mixing of fresh- and seawater, creating severe osmotic stress. Following changes in salinity, ions such as sodium, potassium, and chloride; soluble carbohydrates such as floridoside and digeneaside (Figure 15.44); and amino acids are adjusted to restore normal turgor pressure to cells (Reed 1990). Mannitol is the major osmolyte in *Dixoniella grisea* (Eggert et al. 2007). *Bangia*, an inhabitant of rocky shorelines of both marine and freshwaters, can be gradually acclimated to increased or decreased salinity (Geesink 1973; Sheath and Cole 1980), by use of floridoside as an osmolyte (Lüning 1990). *Bangia atropurpurea* can tolerate exposure to air for as long as 15 consecutive days, then revive rapidly upon resubmergence (Feldmann 1951). *Bostrychia* and *Stictosiphonia*, common epiphytes of mangroves around the world, primarily use D-sorbitol (as well as dulcitol in warm waters) rather than digeneaside as osmolytes (Karsten et al. 1995; Karsten et al. 1996). In contrast, another common mangrove epiphyte, *Caloglossa leprieurii*, accumulates potassium and chloride ions and actively extrudes sodium to adjust osmotic conditions within cells. Mannitol is also involved but contributes much less to salinity adjustment (Mostaert et al. 1995). *Polysiphonia paniculata* is an intertidal form that can produce relatively large amounts of the osmolyte DMSP, which is then enzymatically degraded into the volatile, climatically active gas DMS (see Chapter 2). The enzyme DSMP lyase has been isolated, purified, and characterized from *P. paniculata* (Nishiguchi and Goff 1995).

Ecology of Red Algae

Most red algae occupy intertidal and subtidal regions of marine shorelines of tropical, subtropical, temperate, and boreal latitudes. Even so, a variety of red algae occupy unusual and even extreme habitats. These include the deepest-growing photosynthetic eukaryote, a crustlike coralline found by the use of a submersible. This record-setting rhodophyte lives on 210 m-deep seamounts near San Salvador, The Bahamas, where the irradiance is only 0.0005% of surface levels (Littler et al. 1985, 1986). Some red algae survive in Arctic and Antarctic waters, where they are covered by 2 m of sea ice for 10 months of the year. As noted earlier, cyanidophytes grow in acidic hot springs, and molecular methods have detected red

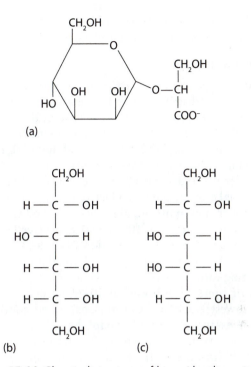

Figure 15.44 Chemical structures of low-molecular-weight carbohydrates that are produced and used by various red algae to regulate osmotic status of cells: (a) digeneaside, (b) D-sorbital, (c) dulcitol. (After Karsten et al. 1995, © Blackwell Science Asia)

algae in extremely acidic (pH 0.8–1.38), hot (30°C–50°C), and metal-rich mine drainage (Baker et al. 2004). Although most red algae are rather intolerant of salinity changes, growing best in normal seawater, a few grow best at lower salinities, and there are about 150 species of freshwater red algae.

A number of red algae are obligate epiphytes on other seaweeds or marine animals. Some red algae produce halogenated terpenoids, alkaloids, and other secondary compounds that may function to inhibit herbivory or have antimicrobial properties, and they have potential uses as pharmaceuticals (Gross et al. 2006). In contrast to most other algal clades, among the Rhodophyta there are few examples of nuisance growth-formers. However, some red algae are regarded as invasive species (see Chapter 22). For example, *Acanthophora spicifera* was introduced to the Hawaiian Islands in 1952 from a barge that had originated in Guam (Smith et al. 2002). This species has brittle, spiny bodies that readily break apart and then snag on objects. As a result, *A. spicifera* has become the most common alien species in Hawaii, and its population-level genetic variation has been studied (O'Doherty and Sherwood 2007).

Marine red algae

Water temperature appears to be a major factor controlling the distribution and diversity of marine red algae. Rhodophytes are most diverse in the tropics. Red algae occupy a wide range of irradiance environments, including high-latitude and high-intertidal habitats subjected to long periods of full sunlight. Day length and temperature are the main cues for seasonal growth and reproductive behavior of red algae. Although most red algae live attached to rocky substrates, more than 100 species grow free floating or entangled in other vegetation (Kain and Norton 1990). An example of a particularly abundant free-floating form is *Phyllophora*, which may occupy areas of some 15,000 km² at 10–60 m depth in the northwest Black Sea. Certain unattached corallines form extensive beds of **rhodoliths**—slow-growing, rounded nodules known as "tumbleweeds of the sea" (Foster 2001). In the case of rhodolith red algae, calcification helps to prevent cell damage as the nodules roll around on the seabed. Rhodoliths serve as habitat for several types of invertebrate animals. While rhodoliths occur in a variety of geographical locations, they are most prominent off the coast of Brazil, where a number of species occur (Amado-Filho et al. 2007; Foster et al. 2007).

Some red algae grow epizootically, on the surfaces of animals. *Rhodochorton concrescens* is notable for its tendency to colonize hydroids, ectoprocts, and crustaceans having chitinous surfaces (West 1970). Red algae often occur as epiphytes, attached to other seaweeds or sea grasses that provide a means of support. Several cases of specific epiphytism are known, including *Porphyra nereocystis* on the kelp *Nereocystis luetkeana*, *P. subtumens* occurring obligately on species of the bull kelp *Durvillaea* (Nelson and Knight 1996), and *Smithora naiadum* on the sea grasses *Zostera* and *Phyllospadix* (Figure 15.45). *Microcladia californica* attaches only to the kelp *Egregia menziesii*, but a related species, *M. coulteri*, colonizes a variety of seaweeds, including the red macroalga *Prionitis* and the green *Ulva* (Gonzalez and Goff 1989a,b). In some cases there is evidence for more than a simple epiphytic relationship. *Polysiphonia lanosa*, whose rhizoids penetrate surfaces of the fucoid brown alga *Ascophyllum* (and, less frequently, *Fucus*), probably receives amino acids and mineral nutrients from *Ascophyllum*. Radiolabeled

Figure 15.45 Numerous blades of *Smithora naiadum* growing on a slender leaf of the marine angiosperm *Phyllospadix*. (Photo: L. W. Wilcox)

tracer experiments have revealed reciprocal exchange of about the same amount of fixed carbon between *Polysiphonia* and *Ascophyllum*, suggesting a hemiparasitic relationship (Ciciotte and Thomas 1997).

Molluscs, crustaceans, sea urchins, and fish consume various red algae. Toughness of the outer body is considered to be a major factor affecting edibility, and difference in toughness may occur between life stages and reproductive versus non-reproductive tissues. In laboratory studies, the snail *Tegula funebralis* preferred gametophyte reproductive material over sporophytes and non-reproductive biomass of the red alga *Mazzaella flaccida*, and field surveys showed the same pattern of selective damage (Thornber et al. 2006). Calcified red algae and low-growing crusts appear to be more resistant to herbivory than non-calcified and erect forms (Littler et al. 1983; Watson and Norton 1985), but some authors have pointed out that calcified rhodophytes are not invulnerable (Padilla 1985). It has been suggested that calcification prevents damage caused by herbivores or mechanical stress from being propagated throughout an algal body (Padilla 1989). When damaged, many rhodophytes are able to persist and later regenerate from lower, prostrate portions of the body. Herbivory resistance and environmental persistence may contribute to the abilities of some calcified or crustose red algae to live to great ages. For example, some *Petrocelis middendorfii* crusts are estimated to be 25–87 years old (Paine et al. 1979). An individual of the crust-forming coralline *Clathromorphum nereostratum* of the Aleutian Islands, Alaska, was estimated to be approximately 700 years old, the oldest living alga known (Frantz et al. 2005).

Freshwater red algae

About 150 species and 20 genera of filamentous red algae occur in freshwaters. Most occur in the winter and spring, in mildly acidic streams located in forested watersheds that have a narrow range of flow rates (29–57 cm sec⁻¹). Light is more readily available in the winter and spring, and current flow provides resupply of minerals and carbon dioxide; inability to use bicarbonate is a frequent attribute of freshwater red algae. Many species are intolerant of pollution, and few occur in large rivers. Freshwater red algae, particularly delicate forms, are consumed by a wide variety of animals, among which are amphipods and caddis fly larvae. Some species of the latter also use red algae in the construction of their cases (Hambrook and Sheath 1987).

The vast majority of freshwater red algae have a heteromorphic life history involving a perennial chantransia or audouinella stage—a small, benthic diploid filamentous phase. No tetraspores are produced. Rather, certain cells of the diploid stage undergo meiosis, giving rise to the first cell of the larger gametophyte generations. Such a life-history type is regarded as a possible adaptation to the stream environment (Sheath and Hambrook 1988, 1990). The need for tetraspore attachment, which might prove difficult in flowing waters, is thus avoided.

15.7 Diversity of Red Algae

It is estimated that there are 500–600 genera of red algae, and estimates of the number of species range from 5000 to as many as 20,000 (Norton et al. 1996). Cryptic species are increasingly being discovered by the application of molecular methods, the genus *Bostrychia* providing examples (Zuccarello and West 2003). On the basis of molecular, structural, and reproductive evidence, Saunders and Hommersand (2004) recommended the classes Cyanidophyceae,

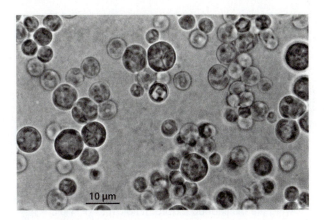

Figure 15.46 *Cyanidium*, which occurs as single cells, each having a single cup-shaped, blue-green plastid. (Photo: L. W. Wilcox)

Porphyridiophyceae, Rhodellophyceae, Compsopogonophyceae, Bangiophyceae, and Florideophyceae, and Yoon et al. (2006) added the class Stylonematophyceae. Molecular phylogenetic data indicate that Bangiophyceae is sister to Florideophyceae (Müller et al. 2001).

Cyanidiophyceae

CYANIDIUM (Gr. *kyanos*, dark blue) occurs as spherical unicells, 2–6 μm in diameter, with a single blue-green, cup-shaped plastid with phycobilisome-bearing thylakoids, but no pyrenoid (Figure 15.46). There is a distinct cell wall. Reproduction is by production of four autospores (nonflagellate unicells) per parental cell (Broadwater and Scott 1994). Cytokinesis occurs by means of a contractile ring of actin (Takahashi et al. 1998). The protein apparatus involved in plastid division was discovered in *C. caldarium*, and the protein Mda1 is similarly associated with constriction and division of mitochondria in the related *Cyanidioschyzon merolae* (Nishida et al. 2007). *Cyanidium* has been isolated from acidic hot springs around the world. Usually *Cyanidium* occupies waters of pH 2–4 with a maximum temperature of 57°C, and most forms will not grow at pH greater than 5. *Cyanidium* can be grown osmotrophically in the dark on 1% glucose, autotrophically or mixotrophically, and is used to study chloroplast and photosynthetic pigment biogenesis and regulation of chloroplast gene expression (Troxler 1994). The cyanidiophytes *Cyanidioschyzon merolae* and *Galdieria sulphuraria* also occur in acidic hot springs, and all

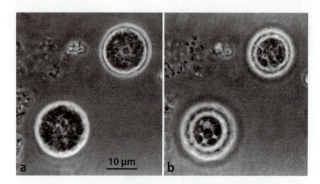

Figure 15.47 Through-focus pair of *Glaucosphaera*, showing the multi-lobed plastid. (Photos: L. W. Wilcox)

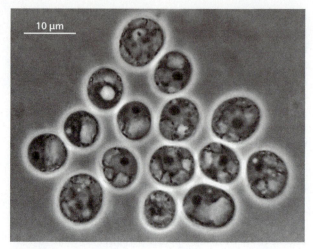

Figure 15.48 Cultured *Porphyridium purpureum* cells. (Photo: L. W. Wilcox)

three are also capable of growing in a 100% CO_2 atmosphere. *Cyanidioschyzon*, like *Cyanidium*, produces a storage material that is more like glycogen than floridean starch, but *Cyanidioschyzon* cells differ in having an oval shape and reproducing by binary fission. In contrast, *Galdieria* does possess floridean starch and produces 4–16 daughter cells (endospores) during reproduction (Seckbach et al. 1992). Molecular evidence strongly suggests that acidophilic cyanidophytes are monophyletic and represent a very early-divergent group of rhodophytes (Tsunakawa-Yokoyama et al. 1997; Yoon et al. 2002b, 2004).

Rhodellophyceae

GLAUCOSPHAERA (Gr. *glaukos*, bluish-green or gray + Gr. *sphaira*, ball) occurs as single cells, 10–20 µm in diameter, surrounded by wide mucilage sheaths (Figure 15.47). The mucilage increases the apparent volume of cells by about seven times. TEM examination reveals that a layer of fibrous, densely staining extracellular matrix material occurs just outside the cell membrane. There is a single central nucleus. Several Golgi bodies surround the nucleus; these are thought to manufacture the mucilage sheath. Numerous small vacuoles and floridean starch granules occur in the cytoplasm (Richardson and Brown 1970). Vesicle transport and the cytoskeleton were studied by Wilson et al. (2006b). There is a single, highly lobed plastid that has two envelope membranes, parallel thylakoids bearing phycobilisomes, and granular regions, but a typical pyrenoid is not present. Reproduction is by cell division to form two daughter cells (Seckbach et al. 1992). It occurs in freshwaters.

Porphryridiophyceae

PORPHYRIDIUM (Gr. *porphyra*, purple) (Figure 15.48) occurs as single cells, 5–13 µm in diameter, or as masses of such cells aggregated within a mucilaginous matrix composed of water-soluble sulfated polysaccharides. Cells are reportedly able to move slowly by mucilage excretion. The single plastid is star shaped (stellate), is located in the cell center, and possesses a central pyrenoid. *Porphyridium aerugineum* is blue-green in color and occurs in freshwater, whereas *P. purpureum* is red colored and inhabits areas of high salt content (Broadwater and Scott 1994), including soils that are exposed to high-salinity waters or the sides of flowerpots. Marine *Porphyridium* cells are preyed upon by a dinoflagellate that can consume up to 20 algal cells at a time (Ucko et al. 1997).

Compsopogonaceae

BOLDIA (named for Harold C. Bold, an American phycologist and morphologist) (Figure 15.49) has a one-cell-thick sac or tubular form that is usually about 20 cm long, though it may occasionally reach 75 cm in length. Pit plugs are not known. *Boldia* is brownish-red or olive in color. It is ephemeral in freshwater streams, usually growing on snails, attached by means of a flat, discoid system of adherent filaments. Reproduction is by means of asexual spores generated by unusual narrow filaments growing among other, more isodiametric cells. Following

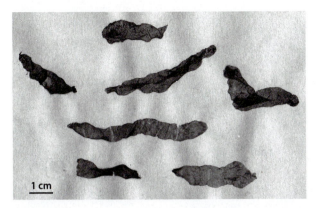

Figure 15.49 Dried and pressed tubular bodies of *Boldia* obtained from a freshwater stream. (Photo: L. W. Wilcox)

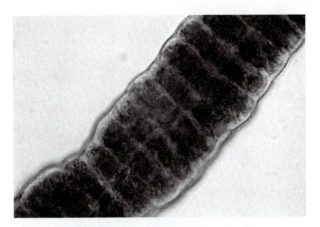

Figure 15.50 A portion of a pluriseriate filament of freshwater *Bangia atropurpurea*. (Photo: L. E. Graham)

their liberation, asexual spores attach to substrates and then generate the prostrate systems from which, first, unicellular filaments and then, subsequently, the erect tubular bodies develop (Herndon 1964). *Boldia* is considered to be heterotrichous, since it has both prostrate and upright components.

Bangiophyceae

BANGIA (named for Niels Bang, a Danish botanist) is an unbranched filament that in early developmental stages is uniseriate (see Figure 7.3c) and later becomes pluriseriate (Figure 15.50). Cells are embedded in a firm mucilaginous sheath, and there are no pit plugs. There is a single stellate plastid with a central pyrenoid per cell. *Bangia* is visible as dark-red or purple strands occurring at or above the waterline on rocks or other substrates along marine shores or freshwaters of Europe and the U.S. Great Lakes (Sheath and Cole 1980). In both marine and freshwaters, asexual reproduction is by archeospores. Spermatia are produced by male gametophytes, and phyllospores are produced on separate female gametophytes, following fertilization. Phyllospores develop into a tiny, filamentous conchocelis phase, which is distinctive in possessing pit plugs (lacking in the gametophyte). Spores liberated from the conchocelis phase settle onto substrates and undergo divisions that begin development of the gametophyte. The factors influencing growth and asexual reproduction by Great Lakes *Bangia atropurpurea* have been elucidated (Graham and Graham 1987). SSU rDNA, *rbcL* and *rbcS*, and spacer region sequence analyses of the Bangiales

ascertained that *Bangia* nests within *Porphyra*, implying that either *Porphyra* is a paraphyletic genus or *Bangia* is merely a filamentous form of *Porphyra* (Oliveira et al. 1995; Brodie et al. 1998; Müller et al. 2001). However, molecular studies suggest that freshwater *Bangia* constitutes a monophyletic group and that U.S. populations arose from a European transplant (Müller et al. 1998).

PORPHYRA (Gr. *porphyra*, purple) consists of relatively large (up to 75 cm long) cell sheets (Figure 15.51). Bodies are either one (monostromatic) or two (distromatic) cells thick. Molecular systematic studies suggest that the monostromatic condition may be ancestral and that the distromatic condition may have arisen at least twice (Oliveira et al. 1995). Regeneration studies and comparisons of mRNA expression patterns performed on different regions of the blade of *P. perforata* suggest that the blade is not as morphologically and physiologically simple as might appear. Morphologically distinct regions are characterized by specific gene expression patterns (Polne-Fuller and Gibor 1984; Hong et al. 1995).

Porphyra blades are attached to substrates by means of numerous thin, colorless rhizoidal cells. Blade cells of some species possess only one plastid, but those of other species contain two. The blade is the gametophyte generation, and this alternates with a small, branched filamentous conchocelis phase (the sporophyte generation)—similar to that of *Bangia*. Spores produced by the conchocelis phase regenerate the blade-forming phase. Additional details regarding sexual reproduction, life history, and cultivation of *Porphyra* can be found in Chapter 4. *Porphyra* is

Figure 15.51 A portion of a blade of *Porphyra miniata.* (Photo: L. W. Wilcox)

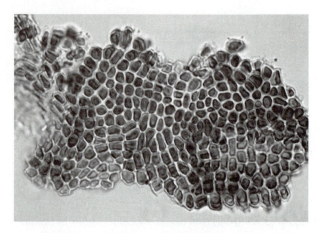

Figure 15.52 Light micrograph of *Hildenbrandia* collected from a tropical freshwater stream. (Photo: L. E. Graham, from a prepared mount made by J. Blum)

common worldwide on rocky shores in the intertidal and subtidal zones. More than 70 species have been described (Kurogi 1972).

Florideophyceae

The Florideophyceae is supported by molecular data and at least two reproductive autapomorphies (unique, derived characters): **tetrasporangia** and a **carposporophyte** composed of filamentous **gonimoblast**. Saunders and Hommersand (2004) recommended the florideophycean subclasses Hildenbrandiophycidae, Nemaliophycidae, Ahnfeltiophycidae, and Rhodymeniophycidae, and Le Gall and Saunders (2006) added the subclass Corallinophycidae. There are more than 20 orders (Le Gall and Saunders 2006), defined on the basis of molecular, reproductive, and structural characters, including pit-plug features (Pueschel and Cole 1982). According to molecular data, Hildenbrandiales is probably the earliest-diverging group of florideophytes (Freshwater et al. 1994; Le Gall and Saunders 2006). Some representative genera from a diversity of orders are described below. Molecular systematic research into florideophyte relationships is an ongoing process that is likely to result in future changes in classification schemes.

Hildenbrandiophycidae

Hildenbrandiales. The order Hildenbrandiales, consisting of crustose red algae, was established on the basis of distinctive pit-plug ultrastructure (Pueschel and Cole 1982). It shares with Corallinales the unusual feature of secondary pit plugs produced without the involvement of small conjunctor cells, in this trait contrasting with most florideophytes.

HILDENBRANDIA (named for Franz Edler von Hildenbrand, a Viennese physician) (Figure 15.52) is a pseudoparenchymatous array of filaments closely packed into a thin, flat, deep rose-colored crust growing on stones (or other substrates) in both marine and freshwater habitats. In freshwaters, *Hildenbrandia* occurs in flowing streams, particularly in shady places, or in deep lake waters (Canter-Lund and Lund 1995). The crust is several cells thick and grows by means of apical cells. The crust may be perennial, with dieback of surface cell layers during winter, a process studied ultrastructurally by Pueschel (1988). Secondary pit plugs are abundant. Reproduction appears to be exclusively asexual, by either fragmentation, specialized propagules (gemmae), or stolons (Nichols 1965). Species relationships have been evaluated by morphological and molecular characteristics (Sherwood and Sheath 2003).

Nemaliophycidae

Palmariales. The order Palmariales includes seaweeds having an unusual life cycle that lacks a carposporophyte generation, probably as a result of evolutionary

Figure 15.53 Hollow saclike bodies of *Halosaccion* are typically filled with seawater. This specimen was collected in Newfoundland, Canada. (Photo: L. W. Wilcox)

Figure 15.54 *Halosaccion* characteristically occurs in large populations in the littoral zone. This population was located in the intertidal region on the western coast of Vancouver Island, British Columbia, Canada. (Photo L. E. Graham)

loss. The life history of *Palmaria* (*Rhodymenia*) *palmata* was elucidated in culture by van der Meer and Todd (1980). Male and tetrasporangial bodies are readily observed in nature, whereas the cryptic female gametophytes are not. Laboratory germination of tetraspores obtained from field-collected tetrasporophytes established that female gametophytes remain microscopic and become sexually mature when only a few days old. In contrast, male gametophytes become macroscopic and do not attain sexual maturity until they are several months old. The tiny female gametophytes are fertilized by spermatia from males of older generations. The carpogonium has a long trichogyne, and there is no carpogonial branch or auxiliary cell. Diploid tetrasporangial bodies grow directly on the fertilized female, eventually overgrowing them. Palmariales are characterized by tetrasporangial stalk cells that can produce successive crops of tetraspores (Guiry 1978). The male gametophyte and tetrasporophyte are isomorphic. This order includes two parasitic genera, *Neohalosacciocolax* and *Halosacciocolax*. There are two families, Rhodophysemataceae (named for *Rhodophysema*) and Palmariaceae, which includes *Palmaria*, *Devaleraea*, and *Halosaccion* (described below). The Palmariales appears to be a monophyletic group (Ragan et al. 1994). Relationships within the Palmariales and related Acrochaetiales were studied by means of nuclear-encoded SSU rDNA sequences (Saunders et al. 1995).

HALOSACCION (Gr. *hals*, sea + Gr. *sakkos*, sack) has yellow-green, cylindrical, hollow, saclike bodies that are normally filled with seawater (Figure 15.53). Bodies can reach 25 cm in length and often occur in large populations in the littoral zone (Figure 15.54). If *Halosaccion* is squeezed, seawater jets out through small pores, and otherwise dignified phycologists have been known to use *Halosaccion* like waterpistols in impromptu shoreline water fights.

Batrachospermales. The red algae grouped in the Batrachospermales (named for the genus *Batrachospermum*) were previously classified in the Nemaliales until ultrastructural studies revealed unique features of their pit plugs that justified placing them into a separate order (Pueschel and Cole 1982). The pit plugs of both groups are characterized by two cap layers. In contrast to the thin outer pit-plug caps of the Nemaliales, Batrachospermales have dome-shaped outer caps. There are about 8 genera, with the paraphyletic genus *Batrachospermum* including more than 100 species (Stewart and Vis 2007). Batrachospermales diversified in freshwaters, primarily streams (Sheath 1984). Red algae in this order are freshwater, are uniaxial or multiaxial, and have determinate lateral branches. The life history involves heteromorphic alternation of a macroscopic gametophyte generation with a smaller filamentous sporophyte, known as the chantransia phase, which lacks tetrasporangia (Figure 15.55). Molecular data show that *Ptilothamnion richardsii* is a chantransia stage of *Batrachospermum* (Vis et al. 2006) and reveal that the same chantrasia morphotype (*Chantransia pygmaea* = *Audouinella pygmaea*)

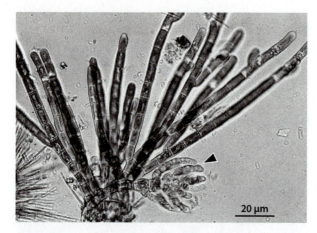

Figure 15.55 Early stage in the growth of the larger gametophyte generation (arrowhead) of *Batrachospermum* from a cell of the smaller sporophytic chantransia phase that has undergone meiosis. (Photo: R. Sheath)

is associated in the Hawaiian Islands with at least three *Batrachospermum* species, as well as members of a related order (Thoreales): *Nemalionopsis tortuosa* and *Thorea violacea* (Chasson et al. 2007). Meiosis occurs in some apical cells of the sporophyte, and the resulting haploid cells differentiate into the gametophyte. The development of the gametophyte has been studied with light and electron microscopy (Aghajanian and Hommersand 1980). Families include Batrachospermaceae (*Batrachospermum*) and Lemaneaceae (*Lemanea*). A description of *Lemanea*, *Paralemanea*, and *Psilosiphon* is found in Sheath et al. (1996a), and a description of *Nothocladus lindaueri* in Sheath et al. (1996b). SSU rDNA and *rbcL* sequences were used to evaluate the phylogeny of Batrachospermales (Vis et al. 1998). Nuclear DNA content for representatives of Batrachospermales and Thoreales has been investigated (Kapraun et al. 2007).

BATRACHOSPERMUM (Gr. *batrachos*, frog + Gr. *sperma*, seed) has a uniaxial, branched, mucilaginous body that macroscopically resembles a mass of frog eggs. Whorls of determinate lateral branches arise at intervals from four to six periaxial cells that extend in a circular manner from the central axis (Figures 15.56 and 15.57). Periaxial cells are initiated as apical protrusions that extend laterally and then curve apically before being cut off to form branch initials. The periaxial cells also generate descending rhizoidal filaments that may envelop the axis, making it appear more than one cell thick.

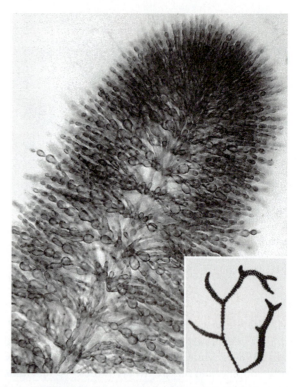

Figure 15.56 Microscopic view of the branching pattern typical of *Batrachospermum*. The alternation of branch whorls with lowermost regions of axial cells produces a beaded effect (inset) resembling strings of frog eggs. (Photos: L. W. Wilcox)

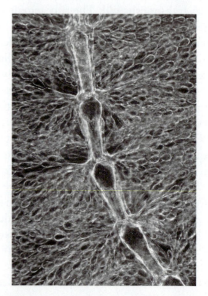

Figure 15.57 Whorls of branches in *Batrachospermum*. Branches arise from the apical portions of periaxial cells. This species was blue-green in color and found in a peat bog. (Photo: L. W. Wilcox)

Figure 15.58 Carposporophytes (arrowheads) in *Batrachospermum*. (From Bold and Wynne 1985)

Spermatia bud from terminal cells of filaments, while carpogonia (and carposporophytes) develop from a lateral branch that may be determinate or indeterminate, depending upon the species (Figure 15.58). *Batrachospermum* occurs in cold, running streams, spring-fed ponds, bogs, and lakes around the world. Bodies may be colored blue-green, olive, gray, or deep red. Molecular sequence analyses indicate that *Batrachospermum* is probably not a monophyletic genus (Vis et al. 1998).

Corallinophycidae

Corallinales. Corallines can occur on or within macroalgae or on sea grasses, animals, coral ridges,

or rocks. There are some parasitic corallines, e.g., *Choreonema thuretii*, that grow endophytically within other corallines (Broadwater and LaPointe 1997). Distinctive features of the Corallinales include reproductive organs produced within conceptacles (see Figure 15.33) and deposition of calcite in the cell walls of most forms. As noted earlier, these features are considered to be adaptations that improve herbivory resistance. Identification usually requires information on reproductive features, and gathering such data requires decalcification by pretreatment with acid, followed by the sectioning of fertile bodies. Because of their great ecological significance and extensive fossil record, there is a considerable literature on fossil and modern corallines. Major compendia of information on coralline algae include Johansen (1981) and Woelkerling (1988).

Classically, modern corallines have been divided into two morphological groups: articulated (jointed) forms and non-articulated (crustose) forms. More recently, these two structural types have come to be known as geniculate (from a Latin word meaning knee or joint) and non-geniculate, respectively. Geniculate corallines have uncalcified body regions—these are the **genicula**, or joints (Figure 15.59). Genicula confer considerable flexibility to bodies, allowing them to reach lengths up to 30 cm without suffering physical damage from water movement. Calcified regions known as intergenicula separate the genicula (Figure 15.60). Geniculate corallines are typically highly branched. Body growth occurs at branch apices and at intergenicular surfaces (Johansen 1981). The presence of genicula implies that jointed corallines are able to control the occurrence and location of calcification. An analysis of breaking force revealed that genicular

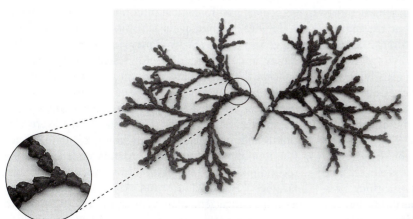

Figure 15.59 *Bossiella*, an example of a jointed or geniculate coralline. (Photo: L. W. Wilcox)

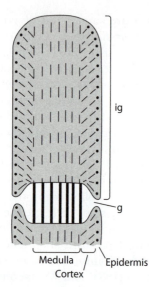

Figure 15.60 A diagrammatic representation of the flexible, non-calcified geniculate region (g) located between adjacent rigid, calcified intergeniculate regions (ig), and differentiation of intergeniculate regions of the body into epidermis, subsurface cortex, and central medulla. (Redrawn with permission from *Coralline Algae, A First Synthesis*, H. W. Johansen 1981, © CRC Press, Boca Raton, FL)

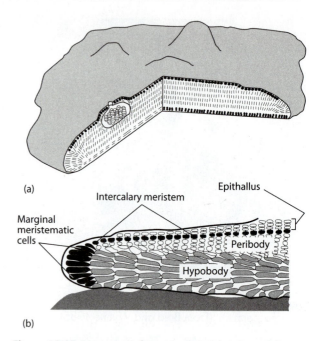

Figure 15.61 Non-geniculate corallines. (a) *Sporolithon* is an example of a crustose, or non-geniculate, coralline. (b) A diagrammatic representation of body differentiation; marginal meristematic cells generate both the hypobody and an intercalary meristem. The intercalary meristem cuts off cells in two directions; upper cells function as a protective epibody, and cells cut off from the lowermost surfaces of intercalary meristematic cells form vertically oriented filaments of the peribody. (a: Redrawn with permission from *Coralline Algae, A First Synthesis*, H. W. Johansen, 1981, © CRC Press, Boca Raton, FL; b: After Lebednik, in Bold and Wynne 1985)

material of *Calliarthron* is much stronger than tissue from fleshy algae and that the size and strength of genicula increase as bodies enlarge (Martone 2006).

Non-geniculate corallines lack uncalcified regions. Although they may be many cells thick, they are often low-growing crust-formers whose entire lower surface is adherent to a substrate (Figure 15.61). The bodies consist of tightly packed filaments and have a dorsiventral organization. The lowermost filaments are oriented parallel to the substrate, forming the **hypobody**. Branches arising from the upper layer of the hypobody are oriented perpendicularly to the substrate. An apical meristem at the margin of the crust generates cells of the hypobody. Growth of the surface cell layer occurs by division within an intercalary meristem—a region of cells lying one or more cells inward from the surface. Cells of the intercalary meristem cut off derivatives toward the surface, forming a protective **epibody**, and toward the substrate, forming the **peribody**. The thickness of the peribody generally determines the thickness of the crust. Some very thin crusts lack a peribody, and species that occur as thick crusts have a thick peribody. Pigments are present in the peribody but are typically absent from the hypobody (Johansen 1981).

The primary productivity of four species of crustose corallines associated with the Great Barrier Reef, Australia, has been estimated (Chisholm 2003).

At present, approximately 24 genera of non-geniculate and 15 geniculate corallines are recognized. A number of different classification schemes have been suggested (reviewed by Bailey and Chapman 1996). Bailey and Chapman (1996) tested the widespread concept that geniculate and non-geniculate corallines represent evolutionarily distinct groups by the use of nuclear-encoded SSU rDNA sequence analyses of 23 species of corallines. Two sister clades were distinguished, one including only non-geniculate species (the monophyletic subfamily Melobesiodeae) and the second including geniculate forms (monophyletic subfamilies Corallinoideae and Amphiroideae, as well as some other geniculate types) and a non-geniculate species, *Spongites yendoi*. The results of this analysis

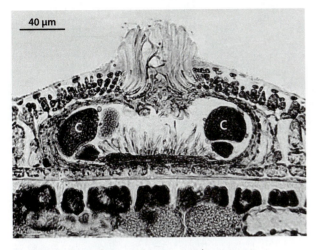

Figure 15.62 Carpogonial conceptacle containing carpospores (C) of the crustose coralline *Pneophyllum*. Carpospores will be released via a pore in the upper body surface, the ostiole. (From Woelkerling 1988, by permission of Oxford University Press)

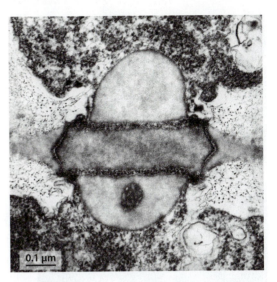

Figure 15.64 Electron micrograph of the characteristic pit plug of coralline red algae, showing dome-shaped caps. (From Pueschel and Cole 1982)

Figure 15.63 A tetrasporangial conceptacle containing tetrasporangia with four tetraspores (T) and bisporangia containing two bispores (B), in the crustose coralline *Pneophyllum*. (From Woelkerling 1988, by permission of Oxford University Press)

Morphogenesis and evolution in the Amphiroideae was studied by Garbary and Johansen (1987).

The life history of corallines is triphasic, with isomorphic alternation of gametophytic and tetrasporophytic generations. Male gametophytes produce spermatangia within conceptacles, and female gametophytes generate carpogonial conceptacles (Figure 15.62). Carpogonial branches are only two cells long, and the cells bearing them function as auxiliary cells. Similar conceptacles occur on tetrasporophytes (Figure 15.63), where tetrasporangia and/or bisporangia are produced. Conceptacles of all types generally open to the environment via a pore known as the ostiole. Distinctive domelike outer cap layers characterize the pit plugs of the Corallinales (Pueschel 1989) (Figure 15.64). Methods for collection, preservation, preparation, and identification of non-geniculate corallines, including rhodolith forms, are described by Woelkerling (1988) and Harvey and Woelkerling (2007).

LITHOTHAMNION (Gr. *lithos*, stone + Gr. *thamnion*, small shrub) (Figure 15.65), a nongeniculate, whitish-pink or pink-purple coralline, commonly exhibits distinctive surface protuberances (Woelkerling 1988). Spermatangia are produced in dendroid branching filaments (Johansen 1981). *Lithothamnion* grows throughout the world on rocks, shells, or other seaweeds, and some species produce rhodoliths.

were not completely consistent with any previous classification scheme and suggested that the assumption that all natural groups will include only geniculate or non-geniculate forms is incorrect. The study also raised the possibility that various groups of geniculate corallines originated separately from different non-geniculate ancestors, but additional data are needed to test this hypothesis (Bailey and Chapman 1996).

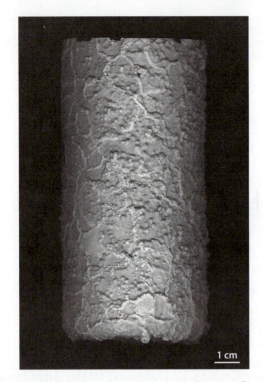

Figure 15.65 *Lithothamnion*, a non-geniculate coralline, growing on a glass bottle. (Photo: L. E. Graham)

Figure 15.66 *Corallina*. Note the characteristic planar, pinnate-branching pattern, which gives the body a flattened appearance. (Photo: L. W. Wilcox)

CORALLINA (Gr. *korallion*, coral) is a purple geniculate coralline with pinnate branching, attached to substrates by a crustose base (Figure 15.66). Branches tend to lie in the same plane, giving a flattened appearance.

LITHOTHRIX (Gr. *lithos*, stone + Gr. *thrix*, hair) is a highly branched geniculate coralline in

which genicula are made up of only a single tier of very long cells (Figure 15.67). The ratio of genicular to intergenicular cell length can be as great as 40:1. Primary branching is dichotomous, and secondary branches are usually alternate. The branches do not tend to lie in one plane, which results in a bushy appearance. It is attached to substrates, usually rocks but sometimes shells, by means of a crustose base (Johansen 1981). The reproductive and population ecology of *Lithothrix aspergillum* in southern California was analyzed by Pearson and Murray (1997).

Figure 15.67 *Lithothrix*, seen life size. Because branches do not lie in the same plane, the body has a bushy appearance. (Photo: L. W. Wilcox)

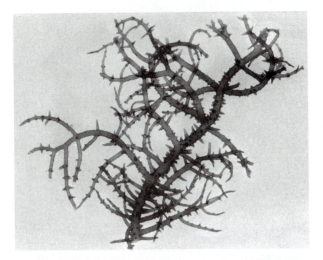

Figure 15.68 *Eucheuma,* a warm-water seaweed. (Photo: M. Hommersand)

Rhodymeniophycidae

Gigartinales. Gigartinales is a large order with many families that has been revealed by both SSU rDNA and *rbcL* sequencing studies to be polyphyletic (Ragan et al. 1994; Freshwater et al. 1994; Le Gall and Saunders 2006) and, therefore, in need of revision. Red algae that used to be classified in the Cryptonemiales have been merged into the Gigartinales (Kraft and Robins 1985). A number of economically valuable carrageenan-producing genera (carrageenophytes) are members of this group, many occurring together in a major clade (Freshwater et al. 1994). Some are uniaxial in construction, while others are multiaxial.

EUCHEUMA (Gr. *eu,* good, true, or primitive + Gr. *cheuma,* molten substance) may have blade-like or highly branched fleshy or tough bodies (Figure 15.68). Pericarp-enclosed carposporophytes are produced on stalks projecting from the body surface. Spermatia occur in groups on male gametophytes, and female bodies generate carpogonia with reflexed trichogynes. Isomorphic tetrasporophytes produce tetrasporangia in the outer cortex.

MASTOCARPUS (Gr. *masto,* breast + Gr. *karpos,* fruit) (formerly included within *Gigartina*) is a large robust blade (Figure 15.69). The female gametophytes are typically covered with distinctive rough papillae, giving it the texture of a terry-cloth towel. Carposporangial branches are produced in these papillae. Blades can be entire or divided into branches. The tetrasporophytes are dark, slippery, crustose bodies that were known as *Petrocelis* (Figure 15.70) prior to their identification as part of the life history of *Mastocarpus,* through culture studies (West 1972; Polanshek and West 1975, 1977). *Petrocelis* has a one- to several-cells-thick hypobody of prostrate branched filaments and a peribody of erect, sparingly branched filaments in a gelatinous matrix. *Mastocarpus* and *Petrocelis* both occur in the lower intertidal zone of the U.S.

Figure 15.69 *Mastocarpus.* (Photo: J. West)

Figure 15.71 *Chondrus.* A portion of the body, illustrating characteristic dichotomous branching of the flattened blade. (Photo: L. W. Wilcox)

Figure 15.70 *Petrocelis*, the tetrasporophytic phase of *Mastocarpus.* (a) Habit view showing *Mastocarpus* bodies growing from the crustose *Petrocelis* bodies. (b) An outer later of tetrasporangia. (Photos: J. West)

West Coast. Heteromorphic species of *Gigartina* were transferred to *Mastocarpus* by Guiry et al. (1984).

CHONDRUS (Gr. *chondros*, cartilage) bodies are multiaxial, bushy, dichotomously branched, flattened fronds that diverge from a tough stalk (Figure 15.71). There are separate male and female gametophytic bodies and isomorphic tetrasporophytes. Spermatangia are colorless and occur on younger branches. Carpogonial branches develop in the inner cortex, and trichogynes extend to the surface. Cystocarps do not include a pericarp. Various species occur in the lower intertidal and subtidal zones throughout the Northern Hemisphere. A key to the

species, as well as an extensive review of the biology of this genus, is provided by Taylor and Chen (1994). *Chondrus crispus*, occurring on both sides of the North Atlantic, is commonly known as Irish moss. The genus has long been used for food and extraction of carrageenans in Western countries. In fact, the term *carrageenan* derives from the Celtic name for *C. crispus*, "carragheen" (Taylor and Chen 1994). The carrageenans of the gametophytic and tetrasporophytic phases are different. ITS (internal transcribed spacer) sequences (see Chapter 5) have been used to examine species relationships (Chopin et al. 1996). EST and microarray expression studies have been performed with *C. crispus* (Collén et al. 2006a,b).

MAZZAELLA (named for Angelo Mazza, an Italian phycologist) bodies are smooth, with entire or lobed blades that exhibit brilliant iridescence when viewed underwater (Figure 15.72). The iridescence results from interaction of light with a proteinaceous surface layer (cuticle) composed of many layers (see Figure 15.11) (Gerwick and Lang 1977). Blades develop at the beginning of each growing season from a perennial holdfast, the principle mode of reproduction. Both gametophytes and isomorphic tetrasporophytes occur throughout the growing season, but the population of tetrasporophytes is usually larger. The effects of frond crowding in *Mazzaella cornucopiae* from British Columbia were studied by Scrosati and DeWreede (1998).

Figure 15.72 *Mazzaella* submerged in shallow water, where blue iridescence is normally exhibited. (Photo: L. E. Graham)

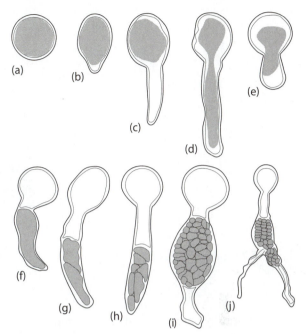

Figure 15.74 Diagrammatic representation of the characteristic tetraspore germination pattern exhibited by members of the Gelidiales. (After Ganzon-Fortes 1994)

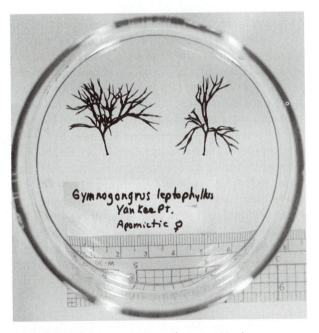

Figure 15.73 *Gymnogongrus.* (Photo: J. West)

GYMNOGONGRUS (Gr. *gymnos*, naked + *gongros*, excresence) consists of tough, branched, multiaxial bodies having thin, rounded or somewhat compressed axes (Figure 15.73). Dense clumps of branched axes arise from a single discoid base. Some species are monoecious, with spermatangia and carpogonia both produced in the outer cortex, while others are dioecious. Carposporophytes occur as wartlike **nemathecia**, which were at first thought

to be parasitic growths. Some species are known to have a separate crustose tetrasporophyte generation resembling *Petrocelis*, but in other species, morphologically reduced tetrasporangia occur within the carposporangial filaments that form the nemathecia of monoecious gametophytes. *Gymnogongrus* occurs worldwide, from Antarctica through the tropics to Alaska, in intertidal and subtidal waters. About 30 species have been described, all containing carrageenans. A list of species, information on carrageenan content, and a review of the biology of this genus are provided by Anderson (1994). Perennial *Gymnogongrus furcellatus* is commercially harvested in Chile and Peru, and ecological analyses have been used to determine the most appropriate time for harvesting (Santelices et al. 1989).

Gelidiales. The order Gelidiales includes macroscopic forms that have a uniaxial construction and are commercial sources of agar and agarose. Members of this group are characterized by a dome-shaped apical cell at the branch tips and a distinctive mode of tetraspore germination. As soon as the tetraspore attaches to the substrate, a germ tube emerges, and the cytoplasmic contents

Figure 15.75 *Gelidium.* (Photo: L. W. Wilcox)

of the spore flow into the tube. The empty spore is cut off by a wall, and the germ tube serves as the first cell of the next generation. A multicellular sporeling with rhizoids develops over several days and eventually produces an apical cell (Figure 15.74). There is typically a prostrate, branched stolon from which several erect axes develop. Stolon attachment to substrates is by unicellular or peglike rhizoids. Genera include *Acanthopeltis, Acropeltis, Beckerella, Gelidiella, Gelidium, Pterocladia, Porphryroglossum, Ptilophora, Suhria,* and *Yatabella.* Nuclear-encoded SSU rDNA and *rbcL* sequences have been used to evaluate systematics of the Gelidiales (Bailey and Freshwater 1997).

GELIDIUM (Gr. *gelidus,* congealed) bodies are pinnately branched, with cylindrical or flattened axes that are stiff and cartilaginous (Figure 15.75). Spermatangia occur in colorless aggregations at the apices of male gametophytes. Alternation of isomorphic tetrasporophytes is common. Vargas and Collado-Vides (1996) analyzed apical growth patterns in an effort to understand morphological variation in this genus. Molecular phylogenetic analysis suggests that this genus may not be monophyletic (Bailey and Freshwater 1997).

PTEROCLADIA (Gr. *pteron,* feather + Gr. *klados,* branch) (Figure 15.76) is structurally similar to *Gelidium,* the two genera differing in cystocarp structure and occurrence of carposporangia in short chains in *Pterocladia* and singly in *Gelidium.*

However, some authorities have questioned the utility of these characters. The bushy bodies may be quite small, to 60 cm or so in length. Both monoecious and dioecious species occur. Colorless spermatangial sori are found on branches at the tips of the main axes. Cystocarps occur on lateral pinnules and tetrasporangia occur in sori. *Pterocladia* is found in all the warm, temperate marine waters in the world. It grows in the lower intertidal and subtidal zones. Wounding induces a regeneration reaction, and bacterial galls are reported. *Pterocladia* is harvested from natural populations for agar production in some regions, and it is of interest in establishment of mariculture operations. Fertile female bodies are required for species determinations, but these are not present in some locations. A review of the biology, species, and utility of *Pterocladia* is provided by Felicini and Perrone (1994).

Gracilariales. The red algal order Gracilariales is well known for species that are of commercial importance as sources of agar and agarose.

GRACILARIA (L. *gracilis,* slender) bodies are cylindrical to somewhat flattened and branched, with a cartilaginous texture (Figure 15.77). *Gracilaria* has a uniaxial construction, growing from a single apical cell. Spermatia are produced in conceptacles. Carposporophytes occur within well-developed pericarps, the whole forming cystocarps that project from the body surface. The life history includes separate male and female gametophytes and isomorphic

Figure 15.76 *Pterocladia capillacea,* showing the characteristic bushy body. (Photo: L. W. Wilcox)

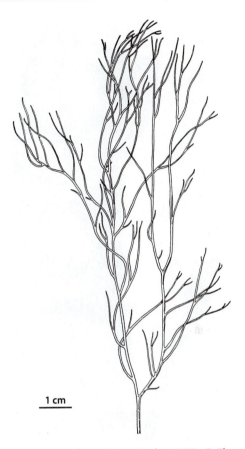

Figure 15.77 *Gracilaria.* (From Taylor 1972, © The University of Michigan Press 1960)

tetrasporophytes. It often grows in large clumps in shallow, turbid bays or lagoons. The closely related genus *Gracelariophila* is parasitic on *Gracilaria.* *Gracilaria* and *Gracilariopsis* include more than 170 species, and species-specific plasmids are present. ITS sequences have proven useful in probing species-level, but not intergeneric, relationships (see Chapter 5) (Goff et al. 1994). Proteomic studies have been done with *G. changii* (Wong et al. 2006), which is used as a source of food and for other purposes.

Rhodymeniales. Rhodymeniales is a large group of multiaxial forms having solid or hollow bodies, including some beautiful deep rose-colored blades. The auxiliary cell is characteristically located on a two-celled filament arising from the same supporting cell as the three- or four-celled carpogonial branch.

RHODYMENIA (Gr. *rhodon,* rose + Gr. *hymen,* membrane) bodies are blades that may be entire or divided dichotomously or irregularly and are attached

to substrates by a holdfast that may be disk shaped or stolonlike (Figure 15.78). Spermatangia occur on male gametophytes in small, irregular patches (sori), while carpogonial filaments occur on the innermost cortical cells of female gametophytes. The carposporophytes grow toward the body surface, forming globular masses of carposporangia. Cystocarps may be distributed fairly evenly over the blades of the female gametophyte or have a more restricted location. They possess a thick pericarp with an ostiole. On the isomorphic tetrasporophyte bodies, tetrasporangia occur over the whole blade surface or only at the tips of blade segments, either singly or aggregated into groups (sori).

BOTRYOCLADIA (Gr. *botrys,* cluster of grapes + *klados,* branch) looks much like a bunch of deep-red grapes (Figure 15.79). The lowermost part of the 15 cm tall body is a solid, branching cylinder attached to substrates by a disc-shaped holdfast. Branch tips end

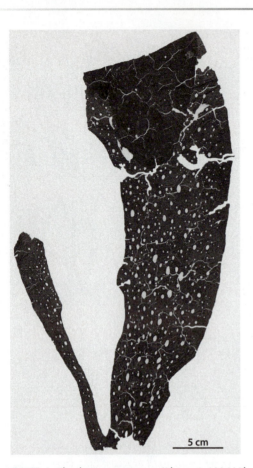

Figure 15.79 *Botryocladia pseudodichotoma*, which has a distinctive body that resembles a bunch of elongate grapes. (Photo: L. W. Wilcox)

Figure 15.78 A *Rhodymenia* species. (Photo: L. W. Wilcox)

in hollow, inflated spherical or pear-shaped vesicles that are filled with mucilage produced by gland cells within the inner cortex. Male gametophytes produce spermatangia in small groups (sori) on the vesicles. Carpogonial filaments are borne on vesicles of the female gametophytes. The carposporophytes grow toward the body surface and are enclosed in an ostiolate pericarp. Tetrasporangia occur singly on vesicles of the isomorphic tetrasporophyte bodies (Smith 1969).

Ceramiales. Ceramiales is a large order that appears to be monophyletic, according to an *rbcL* sequence analysis (Freshwater et al. 1994). All members are uniaxial and form auxiliary cells after fertilization, rather than before (as occurs in other Florideophyceae). An auxiliary cell is produced from the supporting cell of the four-celled carpogonial branch. This distinctive type of female reproductive apparatus is called a **procarp**.

The life history typically involves isomorphic gametophytic and tetrasporophytic generations and

is known as the *Polysiphonia*-type life history. There are many parasitic genera, e.g., *Sorellocolax stellaris*, a species from Japan (Yoshida and Mikami 1996). There are four families: Ceramiaceae (illustrated by *Ceramium*); Rhodomelaceae, the largest group, with about 125 genera (illustrated by *Polysiphonia*); Delesseriaceae (illustrated by *Caloglossa*); and Dasyaceae (exemplified by *Dasya*). A key to 89 genera of Delesseriaceae is provided by Wynne (1996). Preliminary molecular systematic evidence suggests that the Ceramiaceae and Rhodomelaceae, as currently delimited, are not monophyletic groups (Freshwater et al. 1994).

CERAMIUM (Gr. *keramion*, ceramic vessel) is a delicate uniaxial filament in which junctions of the large axial cells are covered by bands of corticating cells (Figure 15.80). Filaments consist of axial, periaxial, and cortical cells. A ring of periaxial cells develops from the upper portions of axial cells, and cortical cells are produced by division of periaxial cells. Four to 10 periaxial cells are produced, the number depending on the species. In some species, spines are generated by periaxial and cortical cells, and gland cells may develop

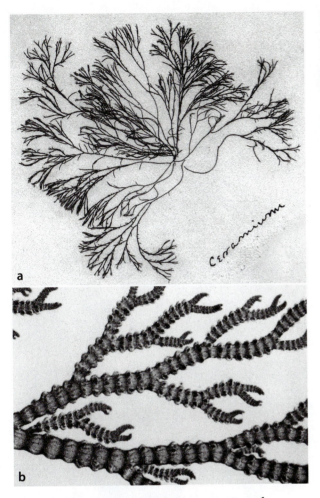

Figure 15.80 *Ceramium.* (a) Macroscopic view of a pressed specimen (life size). (b) Microscopic view showing the typical banding pattern resulting from production of corticating cells that often cover some regions of periaxial filaments more than others. (a: L. W. Wilcox; b: M. Hommersand)

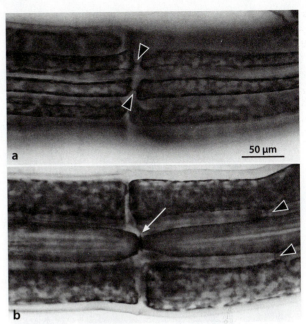

Figure 15.81 Through-focus pair of *Polysiphonia* main axis. The portions of two central cells are seen in (b), joined by a primary pit plug (arrow). Arrowhead in (a) points to a secondary pit plug between periaxial cells. Arrowheads in (b) point to secondary pit plugs between periaxial cells and a central cell. (See Figures 15.41 and 15.43 for habit views.) (Photos: L. W. Wilcox)

from cortical cells. Branches originate pseudodichotomously by division of the apical cell. The body is attached via rhizoids that develop from periaxial and corticating cells. Spermatangia occur on the cortical bands, while carpogonial branches develop from pericentral cells. Tetrasporangia develop from periaxial or cortical cells. *Ceramium* occurs on most marine coasts around the world. Some species serve as sources of agar. The biology of *Ceramium* was reviewed by Boo and Lee (1994).

POLYSIPHONIA (Gr. *poly*, many + Gr. *siphon*, tube) gives the appearance of being composed of a

cylindrical array of adherent tubes. In reality, each of the central cells of the single axis generates a whorl of periaxial cells that is aligned with periaxial whorls produced by the cells above and below in the central axis (Figure 15.81). There may be as few as 4 or as many as 24 periaxial cells per axial cell, depending upon the species. In *Polysiphonia*, the periaxial cells are the same length as the axial cells that produced them, but in some related genera, periaxial cells undergo transverse division. Spermatia and carpogonial branches are produced on special lateral branches (known as trichoblasts) that are produced by the apical cell prior to development of periaxial cells. Carposporophytes are generated within a well-developed, urn-shaped pericarp (see Figure 15.39), the whole forming the cystocarp. Tetrasporangia are produced by periaxial cells (see Figure 15.41). More than 150 species have been described.

CALOGLOSSA (Gr. *kalos*, beautiful + Gr. *glossa*, tongue) is a member of the Delesseriaceae, a family

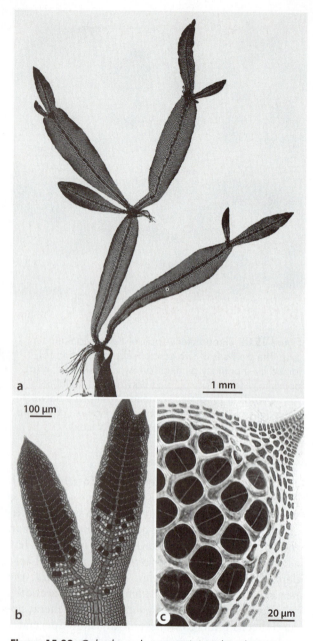

Figure 15.82 *Caloglossa leprieurii.* (a) Habit, showing rhizoids at the nodes. This specimen was cultured from material collected in Natal, South Africa. (b) Tetrasporic blade. (c) Tetrasporangia with cover cells. (a: From Kamiya et al. 1998, by permission of the *Journal of Phycology*; b and c: C. Yarish, in Bold and Wynne 1985)

Figure 15.83 *Grinnellia americana,* a member of the Delesseriaceae, having a simple, translucent blade with prominent midrib. (Photo: L. W. Wilcox)

that includes members having flat, leaflike blades with beautiful cellular patterns (Figure 15.82). Bodies are attached to substrates via numerous rhizoids. *Caloglossa* occurs in tropical to temperate waters, and one species, *C. leprieurii,* is a common inhabitant of the mangrove seaweed flora. Related genera include *Hypoglossum* (see Wynne 1988), *Polyneura, Phycodrys, Delesseria, Membranoptera,* and *Grinnellia* (Figure 15.83). The rubisco spacer region was used to evaluate relationships in *C. leprieurii* and *C. apomeiotica* (Kamiya et al. 1998).

Chapter 16

Green Algae I

Introduction and Prasinophyceans

Part I—Introduction to the Green Algae

The green algae are named for their typical color, the bright grass-green characteristic of land plants. Green algae and land plants usually appear green because abundant chlorophylls *a* and *b* are not concealed by large amounts of differently colored accessory pigments. However, certain green algae that occupy very sunny habitats don't appear green because they accumulate photoprotective orange or red pigments in amounts sufficient to obscure chlorophyll. Common examples include blankets of orange-red *Trentepohlia* on sea cliffs, purple-red films of *Haematococcus* in birdbaths, and red growths of *Chlamydomonas nivalis* in mountain snows. In addition to terrestrial habitats, green algae also commonly occupy marine shorelines and freshwater streams, ponds, and lakes. Some green algae are biogeochemically significant carbonate producers, and certain lipid-rich green algae are sources of petroleum deposits and modern renewable fuels. Green algae also participate in a wide variety of important biotic associations. Although some produce nuisance blooms under conditions of nutrient pollution, many green algae are sources of food for aquatic microbes and animals, and some are grown industrially to produce commercial food supplements.

Pyramimonas

(Photos: M. E. Cook)

Figure 16.1 Relationships of the green algae.

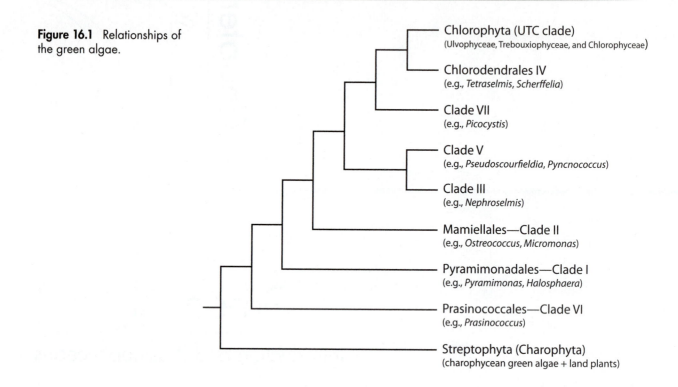

Chlorophyta (UTC clade)
(Ulvophyceae, Trebouxiophyceae, and Chlorophyceae)

Chlorodendrales IV
(e.g., *Tetraselmis, Scherffelia*)

Clade VII
(e.g., *Picocystis*)

Clade V
(e.g., *Pseudoscourfieldia, Pyncnococcus*)

Clade III
(e.g., *Nephroselmis*)

Mamiellales—Clade II
(e.g., *Ostreococcus, Micromonas*)

Pyramimonadales—Clade I
(e.g., *Pyramimonas, Halosphaera*)

Prasinococcales—Clade VI
(e.g., *Prasinococcus*)

Streptophyta (Charophyta)
(charophycean green algae + land plants)

Several green algae provide key laboratory research materials. In the past, Nobelist Melvin Calvin used laboratory cultures of the green alga *Chlorella* to elucidate the light-independent reactions of photosynthesis, now known as the Calvin cycle. Transplant experiments with the seaweed *Acetabularia* conducted in the 1930s allowed Hämmerling to postulate the existence of messenger RNA before it was chemically known. More recent examples include genetic studies conducted with collections of *Chlamydomonas* mutants, the complete genome sequencing of *C. reinhardtii* and other green species, and electrophysiological studies performed with the giant cells of *Eremosphaera, Chara,* and *Nitella.* Because green algae are more closely related than other protists to the ancestry of land plants, many green species are sources of information about the evolutionary origin of plant traits.

Green algae seem to be a monophyletic group that has diversified into a wide variety of body types: unicellular flagellates or colonies, nonflagellate unicells or colonies, unbranched and branched filaments, and multinucleate coenocytes. For many years green algae were classified according to these structural types. However, today we recognize that green algae have undergone extensive parallel evolution of

body form. As a result, species of the same body type are not necessarily closely related, while close relatives may have diverse body structures. This chapter begins with a description of green algal relationships inferred from ultrastructural analyses, comparative biochemistry, life-history studies, and molecular features. Such relationships reveal the extent to which parallel evolution of body types has occurred in the green algae and serve as the basis for a more natural classification system. The second part of this chapter focuses on the earliest-divergent green algae, those commonly known as the prasinophytes or prasinophyceans. A survey of prasinophyte features and diversity helps to explain the evolutionary origin of two major green lineages that are described in subsequent chapters.

16.1 Green Algal Relationships

The green algae are widely regarded as close relatives of red algae (Figure 16.1), in part because primary plastids featuring two enveloping membranes (Figure 16.2) are common to both groups. Primary plastids arise via host cell incorporation of a cyanobacterial endosymbiont and also occur in glaucophytes and

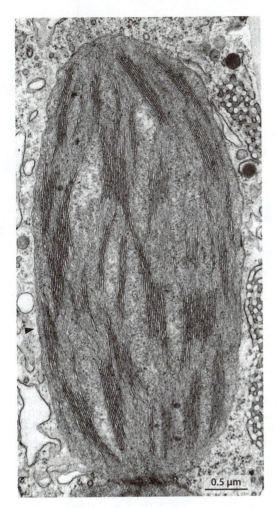

Figure 16.2 TEM view of a primary green plastid. This discoid plastid of *Chara zeylanica* illustrates the envelope of two membranes that characterizes primary plastids (arrowhead). Note the absence of a periplastidal endoplasmic reticulum such as that surrounding the plastids of ochrophytes and some other algae. This plastid also illustrates thylakoid stacks known as grana, which are common features of land plant chloroplasts. (Reprinted with permission from Graham, L. E. and Y. Kaneko. 1991. Subcellular structures of land plants [Embryophytes] from green algae. *CRC Critical Reviews in Plant Science* 10:323–340 © CRC Press, Boca Raton, FL)

the distantly related rhizarian *Paulinella chromatophora* (Marin et al. 2005). The common presence of primary plastids was a major basis for grouping green, red, and glaucophyte algae into a supergroup called Archaeplastida (Adl et al. 2005), and certain molecular features suggest that green and red algae

are sister groups derived from a common plastid-bearing ancestor. However, both of these concepts are controversial. A recent multigene analysis did not find support for monophyly of the hypothetical supergroup Archaeplastida (also known as Plantae) (Yoon et al. 2008), and some experts suggest that green and red algae might have independently acquired plastids whose genetic similarities arose via parallel adaptive processes (Stiller et al. 2003). An analysis of the gene encoding eukaryotic elongation factor 2 suggests that green algae (+ land plants), together with the red algae, cryptomonads (+ katablepharids), and haptophytes (but not glaucophytes), form a strongly supported clade. This clade was named Plastidophila to reflect the widespread presence of primary or secondary plastids among the members. Early-diverging plastidless members of Plastidophila (katablepharids and the cryptophyte genus *Goniomonas*) model the plastidless protist that was ancestral to the green algae and related eukaryotes (Kim and Graham 2008).

Together with the land plants, the green algae form a monophyletic group (Yoon et al. 2008) that has been known as Viridiplantae (Cavalier-Smith 1981), Chlorobionta (Jeffrey 1982), Chlorobiota (Kenrick and Crane 1997), and Chloroplastida (Adl et al. 2005). The green algae display both unifying features reflecting common origin and traits whose variation reflects evolutionary diversification.

Unifying Features of the Green Algae

The green algae possess several unifying traits, many of which are also shared with land plants. Flagellate cells typically possess two (or a multiple of two) flagella that are usually of equal length. Distinctive star-patterned (stellate) structures occur at the transition region of flagella. In three dimensions, stellate structures are composed of several vertically arranged plates that connect the A-tubules of alternate peripheral doublet microtubules; such plates form a star-shaped pattern when viewed in cross-section (see Figure 9.3c). Mitochondria display flattened cristae. In addition, green algae and land plants share plastid structural features, photosynthetic pigments, and light-harvesting complex proteins.

Plastids

So far as is known, all cells of green algae and land plants contain at least one plastid, though these do not always display photosynthetic pigments. Certain

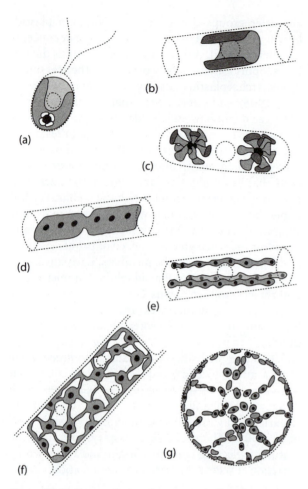

Figure 16.3 Types of plastids commonly found in green algae: (a) cup-shaped parietal (peripheral), (b) napkin ring–shaped parietal, (c) asteroidal (star-shaped), (d) axial platelike, (e) ribbonlike, (f) reticulate (netlike), and (g) multiple discoid plastids.

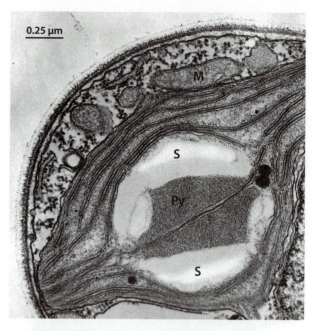

Figure 16.4 Starch production and storage in the plastid stroma. In this TEM of the green alga *Golenkinia*, starch (S) occurs as a shell around the outside of a proteinaceous pyrenoid (Py) located within the plastid stroma. In this organism, a thylakoid runs through the middle of the pyrenoid, bisecting it. Although pyrenoids do not occur within the plastids of many green algae and most land plants, starch formation and storage in the plastid stroma is a basic feature. (TEM: L. W. Wilcox)

heterotrophic green algae or plants and various tissues of green plants have non-photosynthetic plastids that function in key biochemical pathways. No cases of plastid loss have been postulated for green algae or land plants. Although plastids of some green algae resemble the multiple, discoid plastids typical for land plant cells (see Figure 16.2), by comparison to other photosynthetic eukaryotes, green algae display an unusual variability in plastid shape and number (Figure 16.3). For this reason, green algal plastid number and shape are widely used as taxonomic characters.

Green algal and plant plastids uniquely synthesize and store starch in the stroma (Figure 16.4). By contrast, other algal lineages produce and store starch or other types of carbohydrate elsewhere in the cell. Another unique feature of the green algae and land plants is nuclear coding of the small rubisco subunit: This molecule is encoded in the plastid genomes of the red lineage. The photosynthetic pigments of land plants are also similar to those of green algae, as are proteins that occur in light-harvesting complexes (LHC), also known as antennas (see Chapter 1 for an overview of photosynthetic structures and processes).

Photosynthetic pigments and light-harvesting complexes

Chlorophyll *b*, lutein, and β-carotene are important general accessory pigments of green algae and plants (Larkum and Howe 1997). Recall that accessory pigments are able to absorb light energy and transfer it by resonance to chlorophyll *a* in the reaction centers of photosystems I and II (PSI and PSII). Among eukaryotes, chlorophyll *b* otherwise occurs only in

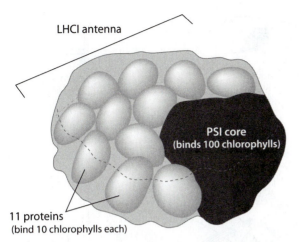

LHCI antenna

PSI core
(binds 100 chlorophylls)

11 proteins
(bind 10 chlorophylls each)

Figure 16.5 Diagram of the association between light-harvesting complex I (LHCI) and photosystem I (PSI) in *Chlamydomonas reinhardtii*. Electron microscopy and X-ray and electron crystallography were used to infer that the LHCI antenna forms a crescent around the PSI core, which is connected to the thylakoid membrane. (After Kargul et al. 2003, *J. Biol. Chem.*, with permission of The American Society for Biochemistry and Molecular Biology, Inc.)

chlorarachniophyte and euglenoid plastids, which independently originated by secondary endosymbiotic incorporation of green algal cells (Ishida et al. 1997; Rogers et al. 2007) (see Chapter 7). The plastids of green algae and land plants lack the phycobilin accessory pigments and thylakoid-bound phycobilisomes so characteristic of cyanobacteria and the plastids of red algae and glaucophytes (see Chapters 6 and 7). No genetic traces of phycobilisomes have been found in green algal or land plant genomes.

Genomic studies show that green algae share with plants a number of LHC proteins that do not occur in other organisms. Such proteins include PsbS, which appears to carry zeaxanthin to PSII and removes it during operation of the protective xanthophyll cycle under stressful high light conditions (see Chapter 1). The early appearance of PsbS in the history of green algae and plants allowed them to reduce the amount of photoprotective carotenoids needed in LHCs, thereby allowing the concentration of chlorophyll molecules to increase. PsbS also allows green algae and plants to more quickly mobilize the xanthophyll cycle (Horton and Ruban 2005). Additional LHC proteins shared by green algae and plants include a minor protein known as CP26, which binds carotenoids in PSII, and LHC proteins associated with PSI (Elrad and Grossman 2004; Koziol et al. 2007). A three-dimensional reconstruction of the *Chlamydomonas reinhardtii* PSI LHC reveals how proteins and associated antenna pigments form a flattened crescent around the core of PSI (Kargul et al. 2003) (Figure 16.5). The PSI core of *C. reinhardtii* is similar to that present in cyanobacterial thylakoids, and the crescent represents a pigment/protein antenna that was added onto the core early in the evolutionary history of green algae and inherited by plants. The major PSII LHC protein-pigment assemblages of land plants (Figure 16.6) are also

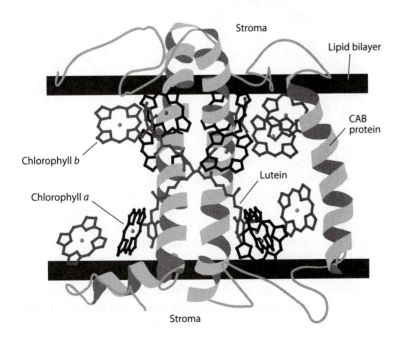

Stroma

Lipid bilayer

CAB protein

Chlorophyll *b*

Lutein

Chlorophyll *a*

Stroma

Figure 16.6 Diagrammatic view of protein 1 of light-harvesting complex II (LHCII). (After, with permission, *Nature*. Kühlbrandt, W., D. N. Wang, and Y. Fujiyoshi. Vol. 367:614–621. © 1994 Macmillan Magazines Limited)

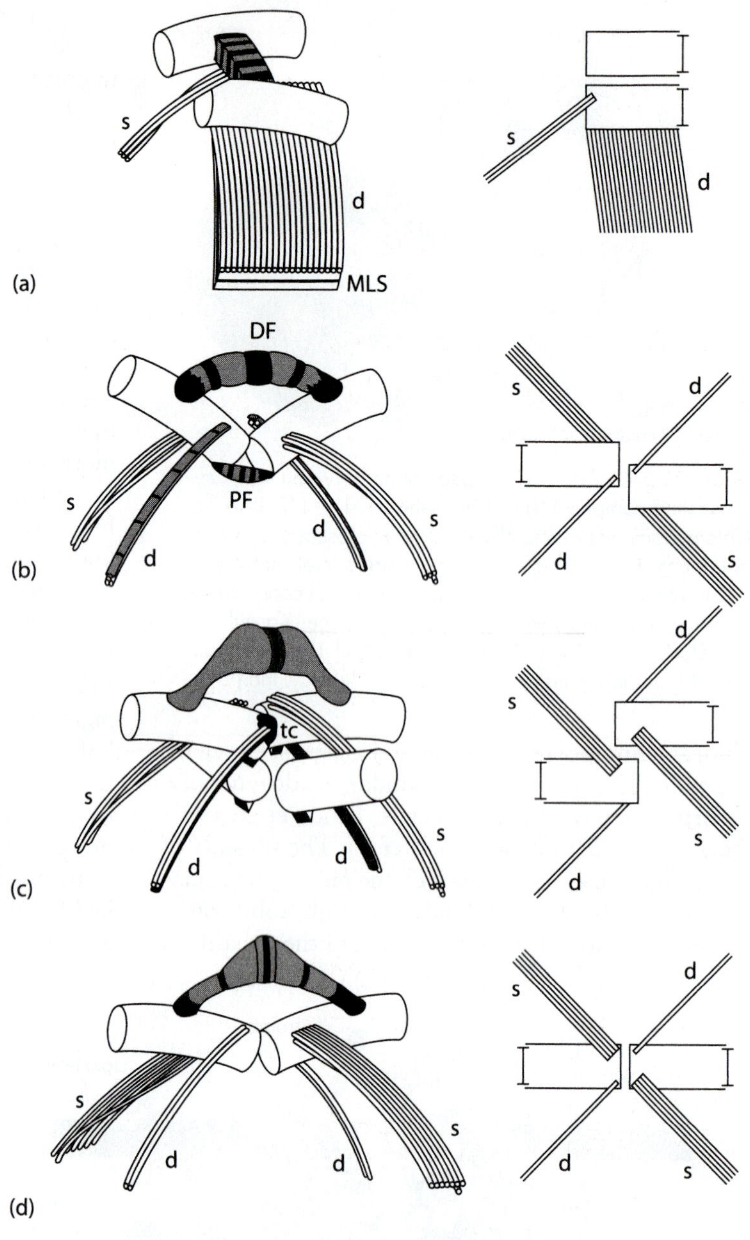

Figure 16.7 The four general types of flagellar apparatuses found among the green algae. These flagellar apparatuses are diagrammed as they would be viewed from the side (left) and from the top (right). They characterize major evolutionary lineages. The apparatuses generally include two or four basal bodies (shown here as cylinders or rectangles), microtubular roots (s or d), and distal (DF) and/or proximal (PF) connecting fibers. (a) Flagellar apparatus with asymmetrical distribution of the flagellar roots, showing the characteristic multilayered structure (MLS). (b) Flagellar apparatus with cross-shaped (cruciate) roots showing clockwise displacement of the flagellar basal bodies from an imaginary line drawn parallel to and between them (when viewed from the cell's anterior). (c) Flagellar apparatus with cruciate roots and basal bodies displaced in a counterclockwise direction. (d) Flagellar apparatus with directly opposed flagellar basal bodies.   (Reprinted with permission from Inouye, I. 1993. Flagella and flagellar apparatuses of algae. Pages 99–134 in *Ultrastructure of Microalgae*, edited by T. Berner. © CRC Press, Boca Raton, FL)

shared by most green algae (Nield et al. 2000), having first appeared during green algal diversification. The three-dimensional structure of abundant LHCII protein 1 and its associations with chlorophylls and carotenoids have been determined at a high resolution with the use of electron crystallography (Kühlbrandt et al. 1994). The 232-amino acid protein has 3 thylakoid membrane-spanning alpha-helices. Some of the protein's polar amino acids help to maintain protein shape and also attach to the magnesium atoms of at least 12 chlorophyll molecules (7 chlorophyll *a*

and 5 chlorophyll *b*). By holding chlorophyll *a* in close contact with chlorophyll *b*, LHCII protein 1 helps to facilitate rapid energy transfer (see Figure 16.7). In addition, the protein binds two carotenoids, probably lutein; these prevent the formation of toxic forms of oxygen, such as singlet oxygen, a free radical that can damage DNA and other cell components. Red algae and plastids descended from them display different major accessory pigments and light-harvesting complex proteins. However, some red-lineage proteins are similar to the minor green LHC protein LI818, suggesting that this protein might have appeared prior to a split between the red and green lineages (Koziol et al. 2007).

16.2 The Major Green Algal Lineages

Despite their many unifying features, green algae display substantial variation in features of cell biology, biochemistry, and ecology that reflect evolutionary diversification (Pickett-Heaps and Marchant 1972; Pickett-Heaps 1975; Mattox and Stewart 1984). More recent molecular systematic investigations have largely corroborated the results of these earlier studies (reviewed by Lewis and McCourt 2004). Ultrastructural, biochemical, and molecular sequence evidence suggests that the earliest-diverging modern green algae are unicellular forms known as prasinophytes. The prasinophytes do not form a monophyletic group (see Figure 16.1) but, rather, consist of about seven clades that are described more fully in Part II of this chapter. Importantly, prasinophytes are related to two larger clades that include unicellular and multicellular representatives. These two major clades are informally known as the streptophytes (or charophytes) and the chlorophytes (also known as the UTC clade).

Streptophytes

The clade informally termed streptophytes and formally Streptophyta (Bremer et al. 1987) includes all of the land plants and their closest green algal relatives. The green algae most closely related to land

plants were at one time grouped into a class formally named Charophyceae and informally charophyceans, after the genus *Chara* (Mattox and Stewart 1984). However, it is now known that such green algae do not by themselves form a monophyletic group or natural class (see Figure 16.1). Consequently, an alternate classification has been suggested in which streptophytes have been renamed as a division/phylum Charophyta (informally, charophytes), within which land plants constitute a class, Embryophyceae (Lewis and McCourt 2004). For this reason, we have not advocated any particular formal taxonomy for the green algae most closely related to land plants but, rather, continue to use the term *charophyceans* as an informal aggregate. This parallels the common term *bryophytes*, which usefully aggregates liverworts, mosses, and hornworts, even though these groups together do not form a clade.

Modern charophyceans primarily occupy freshwater habitats, though some are terrestrial. In addition to *Chara* and close relatives, collectively known as charaleans, charophyceans include coleochaetaleans, zygnematalean algae such as the familiar *Spirogyra* and related desmids, and several other taxa (Mattox and Stewart 1984). Charophyceans display a number of traits that are shared with land plants but not most other green algae. One example is an asymmetrical cytoskeletal apparatus that contains a distinctive multilayered structure known as a **multilayered structure (MLS)** (Figure 16.7a). Very similar MLSs occur in flagellate sperm produced by land plants. MLSs of charophycean green algae and plants lie adjacent to the flagellar basal bodies near the cell anterior and consist of several layers. A distal layer of microtubules lies over an array of parallel plates and

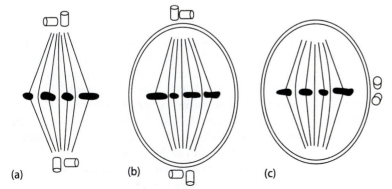

(a)   (b)   (c)

Figure 16.8 A diagrammatic comparison of mitosis in different lineages of green algae. (a) Open mitosis, characterized by dissolution of the nuclear envelope. (b) Closed mitosis, involving an intact or nearly intact nuclear envelope. (c) Metacentric mitosis, wherein centrioles are located in the plane of metaphase chromosomes rather than poles.

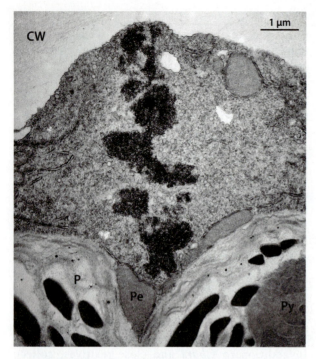

Figure 16.9 TEM of open mitosis (metaphase) in *Coleochaete*. CW = cell wall, P = plastid, Pe = peroxisome, Py = pyrenoid. (Micrograph: L. E. Graham)

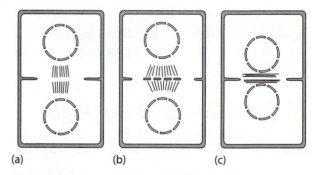

(a)　　　　　　　(b)　　　　　　　(c)

Figure 16.10 Diagrammatic comparison of specialized cytokinesis systems of green algae. (a) Early-diverging charophyceans divide by furrowing, though microtubules may be involved. (b) Later-diverging charophyceans display phragmoplasts very similar, if not identical, to those of land plants. In this cytokinesis system, little furrowing occurs, and cell plates develop from the center toward the cell periphery. Microtubules arranged perpendicular to the developing cell plate help guide vesicles to the developing cell plate and aid in the formation of plasmodesmata. (c) Chlorophyceans often produce phycoplasts, arrays of microtubules lying parallel to the developing cleavage furrow. Such microtubules are believed to keep daughter nuclei—which may lie close together—apart during cytokinesis.

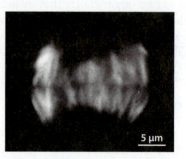

Figure 16.11 Land plantlike phragmoplast in dividing cells of *Coleochaete*. This image was made by fluorescently labeling the tubulin protein that makes up phragmoplast microtubules. (From Brown et al. 1994)

a lower layer of smaller tubules. The upper layer of microtubules extends down into the cell body, serving as a cytoskeleton. While such distinctive MLSs have been regarded as a cellular marker for the green algal groups that share ancestry with plants, it should be noted that similar structures have been observed in flagellate cells of other protists. For example, MLSs occur in glaucophytes (Kies and Kremer 1990), the euglenoid *Eutreptiella* (Moestrup 1978), and certain dinoflagellates (Wilcox 1989). Whether or not these are homologous structures is not known. MLSs are thought to function primarily as microtubule organizing centers, but this hypothesis has not been rigorously tested.

Additional distinctive features that are generally shared by green algae and land plants include open mitosis (Figures 16.8a and 16.9), Cu/Zn superoxide dismutase (DeJesus et al. 1989), class I aldolases (Jacobshagen and Schnarrenberger 1990), and peroxisomes containing glycolate oxidase (Table 16.1). Charophycean cytokinesis features a persistent spindle that helps keep daughter nuclei separate until cytokinesis has been accomplished. Early-divergent charophyceans undergo cytokinesis

by furrowing, the process by which a new wall grows inward (Figure 16.10a). Later-diverging charophyceans produce a **phragmoplast** (Figures 16.10b and 16.11), **cell plates** (Figure 16.12), and primary **plasmodesmata** formed together with the cell plate (Figure 16.13). Phragmoplasts are composed of a double set of microtubules oriented perpendicularly to the plane of cytokinesis. These microtubules organize the aggregation and coalescence of vesicles containing cell wall material to

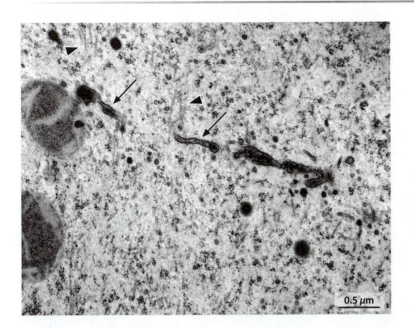

Figure 16.12 TEM view of a phragmoplast and forming cell plate (arrows) in *Chara*. All of the structures associated with phragmoplast formation in higher land plants, such as transverse microtubules (arrowheads) and coated vesicles, also occur in *Chara*. (From Cook et al. 1998, © Springer-Verlag)

form a cell plate in a process nearly indistinguishable from that occurring during plant cytokinesis (Pickett-Heaps 1975; Cook et al. 1998). Additional traits shared by later-diverging charophyceans and land plants include apical growth and oogamous sexual reproduction. Such shared features and molecular evidence strongly indicate that land plants arose from branched, filamentous charophycean ancestors (Karol et al. 2001; Petersen et al. 2003; Qiu et al. 2007). More information about charophyceans and their close relationship to land plants can be found in Chapter 20.

Chlorophytes

A second green algal lineage that includes both unicellular and multicellular representatives is informally known as the chlorophyte clade, formally Chlorophyta (Lewis and McCourt 2004). However, it should be noted that the term *Chlorophyta* is also widely used to include all green algae, a situation that can cause confusion in communication. Chlorophytes include the classes Ulvophyceae, Trebouxiophyceae, and Chlorophyceae, for which reason they are together also informally known as the UTC clade. Among these classes, the Ulvophyceae diverged earliest, and the Trebouxiophyceae and Chlorophyceae form sister groups (see Figure 16.1). These classes may or may not be monophyletic groups, but they feature some distinctive traits (see

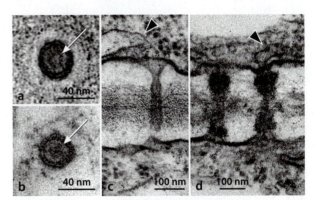

Figure 16.13 Comparison of charophycean plasmodesmata with those of an early-diverging land plant. TEM views of the plasmodesmata of *Chara* in (a) cross-section and (c) longitudinal section, with (b) cross-sectional and (d) longitudinal views of those found in germlings of the moss *Sphagnum*. Internal structure is visible in cross-sections of both *Chara* and *Sphagnum* plasmodesmata (arrows). Spokelike structures (arrowheads) connect the central structure to the cell membrane in both *Chara* and *Sphagnum*. In (c) and (d), endoplasmic reticulum (arrowheads) can be seen to be associated with *Chara* and *Sphagnum* plasmodesmata. (From Cook et al. 1997)

Table 16.1). Ulvophyceans primarily occupy marine waters and include common green seaweed genera such as *Ulva* and *Codium*. By contrast, trebouxiophyceans are mostly freshwater or terrestrial in

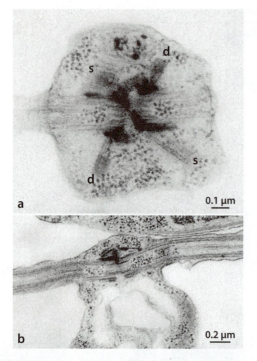

Figure 16.14 Flagellar basal bodies and associated microtubular roots of *Pediastrum duplex* gamete viewed (a) in cross-section near the cell apex and (b) in longitudinal section. (Labeling of roots follows the convention shown in Figure 16.7.) (a: From Wilcox and Floyd 1988, by permission of the *Journal of Phycology*; b: L. W. Wilcox)

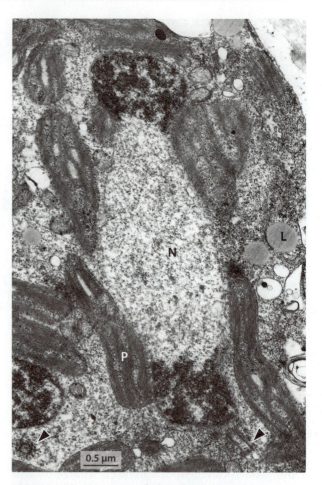

Figure 16.15 TEM view of closed mitosis (telophase in the reproductive structures of *Trentepohlia*). Arrowheads point to centrioles near bottom of image. L = lipid droplet, N = nucleus, P = plastid. (From Graham and McBride 1978, by permission of the *Journal of Phycology*)

habit and include such familiar genera as *Chlorella* and the common lichen component *Trebouxia*, for which the group is named (Friedl 1995). Chlorophyceans include *Chlamydomonas* and *Volvox*, as well as many other primarily freshwater green algae.

Flagellate cells of chlorophyte green algae have more or less symmetrical cruciate (cross-like) root systems wherein rootlets of variable (X) numbers of microtubules alternate with rootlets composed of two microtubules to form what is known as an "X-2-X-2" arrangement (Moestrup 1978) (Figures 16.7b–d and 16.14). Among the algae having cruciate root systems, there are three main variations in the orientation of the flagellar bases when cells are viewed "top-down," i.e., in an anterior-posterior direction: clockwise (CW) (see Figure 16.7b), counterclockwise (CCW) (see Figure 16.7c), and directly opposed (DO) (see Figure 16.7d) (O'Kelly and Floyd 1984). In the first, the basal bodies are shifted in a clockwise

direction from a line drawn parallel to and between their axes. Counterclockwise flagellar apparatuses have the opposite orientation, plus in these forms, the basal bodies are often overlapping. A number of unique flagellar apparatus features are correlated with each of these three basal body arrangements.

In some chlorophytes, the nuclear envelope persists throughout mitosis, a condition known as **closed mitosis** (Figures 16.8b and 16.15) The mitotic nuclei of such algae often appear dumbbell shaped just prior to completion of mitosis, and after separation, the two daughter nuclei tend to drift toward each other, sometimes even flattening against

Table 16.1 Characteristics of the five major green algal groups

	Flagellar/cytoskeletal apparatus	Photorespiratory enzymes	Mitosis	Cytokinesis	Habitat (primary)	Life history
Prasinophyceans	Cruciate roots, rhizoplasts, some with MLS, flagellar and body scales common	Variable	Variable	Furrowing	Marine	Zygotic meiosis
Ulvophyceae	Cruciate X-2-X-2 roots, CCW orientation, +/– body and flagellar scales, rhizoplast present	Glycolate dehydrogenase	Closed, persistent spindle	Furrowing, some with phragmoplast, cell plate, and plasmodesmata	Marine or terrestrial	Zygotic meiosis or alternation of generations or gametic meiosis
Trebouxiophyceae	Cruciate X-2-X-2 roots, CCW orientation, no scales, rhizoplast present	Glycolate dehydrogenase	Semi-closed, non-persistent spindle	Furrowing	Freshwater or terrestrial	Zygotic meiosis
Chlorophyceae	Cruciate X-2-X-2 roots, CW or DO orientation* scales occur rarely, rhizoplasts	Glycolate dehydrogenase	Closed, non-persistent spindle	Furrowing, phycoplast, some with cell plate and plasmodesmata	Freshwater or terrestrial	Zygotic meiosis
Charophyceae	Asymmetric roots, MLS, body and flagellar scales usually present, rhizoplast rare	Glycolate oxidase and catalase in peroxisome	Open, persistent spindle	Furrowing, some with cell plate, phragmoplast, and plasmodesmata	Freshwater or terrestrial	Zygotic meiosis

* Except *Hafniomonas reticulata*, whose basal bodies are CCW but whose SSU rDNA sequences are allied with the chlorophyceans (Nakayama et al. 1996b).

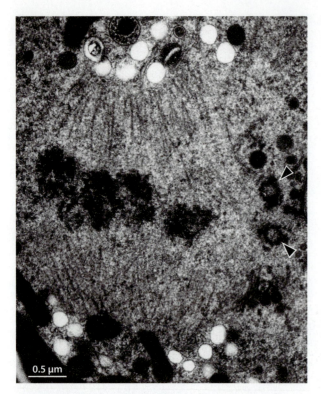

Figure 16.16 TEM of the metacentric spindle in *Friedmannia* (= *Myrmecia*; Friedl 1995). Arrowheads point to centrioles. (Reprinted with permission from Lokhorst, G. M., P. J. Segaar, and W. Star. 1989. An ultrastructural reinvestigation of mitosis and cytokinesis in cryofixed sporangia of the coccoid green alga *Friedmannia israelensis* with special reference to septum formation and the replication cycle of basal bodies. *Cryptogamic Botany* 1:275–294)

each other (Pickett-Heaps 1975). Trebouxiophyceans have an unusual and distinctive **metacentric spindle**, where the centrioles are located near the metaphase plate of chromosomes (Figures 16.8c and 16.16) rather than at the spindle poles, as is more usual. Many chlorophytes accomplish cytokinesis by simple furrowing, which is common among protists. Chlorophytes that possess a well-developed cell wall also produce a distinctive set of microtubules—the **phycoplast**—that lies parallel to the plane of cytokinesis (Figures 16.10c and 16.17). It has been proposed that the phycoplast microtubules help to separate daughter nuclei that would otherwise drift close together in the region of cross-wall development. The phycoplast is thought to ensure that after cytokinesis each daughter cell contains a nucleus. In some chlorophytes, the phycoplast helps to organize vesicles containing cell wall material in such a way that a cell plate develops outward from the center; plasmodesmata may occur in the cross-walls of such algae.

Green Algal Body Diversification

As noted earlier, the green algae have previously been classified according to body type, such as flagellate versus nonflagellate unicells, colonies, and unbranched versus branched filaments. More recently, it has become clear that charophyceans, ulvophyceans, trebouxiophyceans, and chlorophyceans each display a wide variety of such body types. Consequently, very similar-appearing forms are classified within each of these separate lineages.

Figure 16.17 TEM view of the phycoplast in *Microspora*. Microtubules (arrowheads) lie in the same plane as the developing furrow (arrow). M = mitochondrion, N = nucleus, P = plastid. (From Pickett-Heaps 1975)

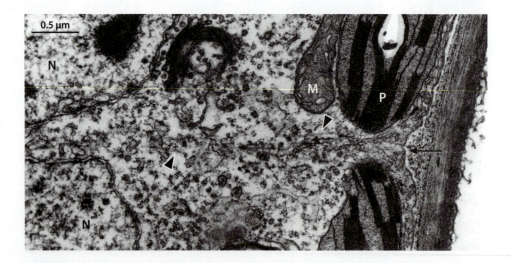

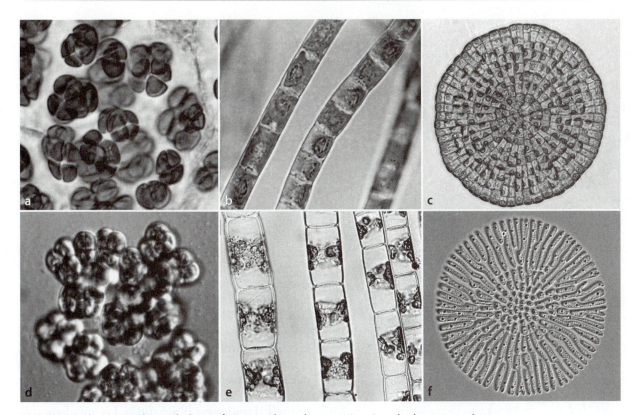

Figure 16.18 The external morphology of various charophyceans (a–c) and other green algae (d–f). (a) The sarcinoid charophycean *Chlorokybus*, compared to (d) the sarcinoid chlorophycean *Chlorosarcinopsis*. (b) The unbranched, filamentous charophycean *Klebsormidium*, compared to (e) *Ulothrix*. (c) The prostrate, radially symmetrical charophycean *Coleochaete*, compared to (f) the similar ulvophycean *Ulvella*. These similarities are examples of parallel evolution of thallus form within the green algae; similar examples can be cited between green algae and other algal groups. (a, c, d, e: L. E. Graham; b: L. W. Wilcox; f: After Floyd and O'Kelly 1990)

Examples are provided by *Chlorokybus* (charophycean) and *Chlorosarcinopsis* (chlorophycean), which are look-alike packets of cells; *Klebsormidium* (charophycean), *Ulothrix zonata* (ulvophycean), and *Uronema flaccidum* (chlorophycean) that are similar-looking unbranched filaments; and certain species of *Coleochaete* (charophycean) and *Ulvella* (ulvophycean) that occur as flat, radially symmetrical disks of branched filaments growing at their margins (Figure 16.18). Similar unicells also occur in these green algal classes. As a result, it can be difficult to identify or classify green algae based upon body form alone; molecular and ultrastructural data are often necessary. The similar bodies of distantly related green algae probably resulted from parallel adaptation to comparable selection pressures, but such selection-adaptation responses are poorly understood. Determination and study of greater numbers of green algal genomes may help to reveal the genetic basis for parallel/convergent evolution of green algal body form.

Part II—Prasinophyceans

Prasinophytes, the earliest-diverging modern green algae, take their name from the Greek word *prasinos*, meaning "green." In some literature, they have been termed micromonadophytes. This non-monophyletic group includes at least seven clades (Guillou et al. 2004). Prasinophytes display few, if any, unique and defining derived characteristics. Rather, they are characterized by a collection of traits that are considered plesiomorphic for green algae (Sym and

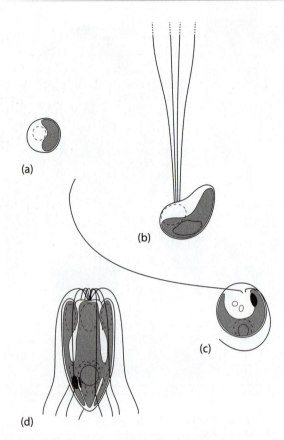

Figure 16.19 Some of the morphological variation observed within the prasinophyceans. Shown are (a) *Ostreococcus*, (b) *Pachysphaera*, (c) *Nephroselmis*, and (d) *Pyramimonas*. (b–d: After Throndsen 1997)

Chapter 5) to determine that *Micromonas pusilla* dominates the eukaryotic picoplankton (cell size fraction 0.2–3 μm) of the western English Channel year round. *Pyramimonas australis* forms blooms in sea ice holes (Moro et al. 2002). Complete genome sequences exist for some minute planktonic prasinophyceans, such as *Ostreococcus tauri* (Derelle et al. 2006) (see Chapter 4).

16.3 Cellular Features of Prasinophyceans

Because prasinophytes are unicellular, much of their variation is based on ultrastructural features. Most studies of prasinophyte cell structure have been performed using conventional techniques, but ultrastructure of the prasinophyte *Ostreococcus tauri* has been described using electron cryotomography. This process, which involves preservation of cells by rapid freezing followed by collection of multiple images made at different tilt angles, can be used to digitally reconstruct the three-dimensional contents of an entire cell (Henderson et al. 2007). Prasinophyte cells generally possess a large Golgi apparatus that lies between flagellar basal bodies and the nucleus and that plays an important role in generating the cell covering. A single, highly dissected mitochondrion is typical. Defensive structures known as **extrusomes** may be ejected. Some species produce coiled ribbons that are ejected as elongate tapering "trichocysts" following rapid hydration, much like the ejectisomes of katablepharids and cryptomonads (Lee et al. 1991). Such ultrastructural characters are consistent with recent molecular data linking the ancestry of green algae to that of katablepharids and cryptomonads (Kim and Graham 2008).

Flagella, Flagellar Apparatus, and Cytoskeleton

The flagella of flagellate species typically emerge from an apical depression or pit, and the flagellar basal bodies are often very long and parallel. Flagellar number varies from 1 in *Pedinomonas* to 16 in *Pyramimonas cyrtoptera*. The flagellar transition region—typically fairly uniform within other algal groups—is unusually variable within prasinophyceans. Three major elements can occur in prasinophycean transition regions: a stellate structure

Pienaar 1993). Fossils that are structurally similar to cyst stages of modern prasinophytes are known from Lower Cambrian sediments, and biochemical evidence suggests that early green algae were present in the Neoproterozoic (Knoll et al. 2007).

Most prasinophyceans occur primarily as flagellate or nonflagellate unicells (Courties et al. 1994) (Figure 16.19). However, certain species occur as sessile (attached) dendroid (treelike) colonies or mucilaginous aggregations of nonflagellate cells, known as palmella stages. Although all known members of the prasinophyceans possess at least one green plastid, *Pyramimonas gelidicola* (Bell and Laybourn-Parry 2003) is known to also ingest particulate food and is thus mixotrophic. Most occur in marine habitats, where they can be important components of plankton communities (Worden et al. 2004). For example, Not and associates (2007) used molecular probes (see

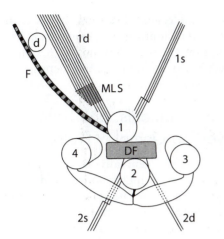

Figure 16.20 Flagellar apparatus structure of the prasinophyte genus *Halosphaera*. Four basal bodies (1–4) are shown. DF is the distal fiber that links basal bodies 1 and 2. Four microtubular roots are present (1d, 1s, 2d, 2s). The 1d root includes an MLS. A fibrous root (F) and scale-releasing duct (d) also characterize cells of this prasinophyte. (After Sym and Pienaar 1993)

that is generally present (as in most green algae) and sometimes also a plate or helix. The contractile protein centrin has been found in the transition region of some prasinophyceans (Melkonian et al. 1992).

Although in most cases the flagella of an individual prasinophycean cell are all of the same length and morphology, heteromorphic flagella are known to occur. Prasinophyte flagella display flagellar maturation, a term that describes the occurrence of flagella of different ages—or maturation states—on the same cell. As in other algae, flagellar maturation may require more than one cell generation. Though *Micromonas pusilla* has flagella that are devoid of scales, at least one layer of flagellar scales is common. *Tetraselmis* flagella, for example, have four layers of structurally distinct scales. An outermost layer of long, thin hair scales is very common. A compendium of prasinophyte flagellar scale types can be found in Sym and Pienaar (1993).

The parallel basal bodies of prasinophytes are linked by a distal connecting fiber that contains the contractile protein centrin. In fact, this protein, which occurs in all groups of eukaryotes, was first discovered in studies of prasinophyceans (see Coling and Salisbury 1992). Microtubular roots vary. In some cases, a non-contractile striated rootlet composed of a phosphoprotein and known as a system

I fiber is thought to function in absorbing stress. **Rhizoplasts**, also known as system II fibers, are striated, contractile, centrin-containing structures that often link the flagellar apparatus to the anteriormost nuclear surface (e.g., Salisbury et al. 1981). These musclelike structures sometimes also form links with the chloroplast or the cell membrane. Rhizoplast contraction requires Ca^{2+} and ATP and is a mechanical adaptation to problems arising from flagellar position at the bottom of a pit (Salisbury and Floyd 1978). It is typical for the rhizoplast to be associated in some way with the prasinophycean microbody, a structure that contains catalase. Several prasinophyceans, including *Pterosperma* and *Halosphaera*, possess a flagellar root that includes an MLS (Figure 16.20) very similar to those observed in flagellate cells of charophycean green algae and land plants. A summary of prasinophycean flagellar root variation can be found in Sym and Pienaar (1993).

Cell Covering

The cells of most prasinophyceans are enclosed by one to five layers of scales attached to the cell membrane, with the scales of each layer characteristic for the species (Figure 16.21). A compendium of these scale types and their occurrence can be found in Sym and Pienaar (1993). A continuous underlayer of small, square scales is very common but not ubiquitous. The outer scales can be amazingly complex and three dimensional, with lacy, basketlike shapes being common. Scales are produced within the Golgi apparatus, and all types of body and flagellar scales can be produced within the same cisterna. An ultrastructural study of *Cymbomonas tetramitiformis* (Moestrup et al. 2003) illustrates the scale development process. Scales are extruded to the cell exterior in the region of the flagellar pit and eventually pushed to their final position on the cell surface. In some cases, there appears to be a permanent, specialized scale-releasing duct (Figure 16.22), which has been hypothesized to be the vestigial remains of the cytopharyx (feeding apparatus) of ancestral particle-feeding cells.

Some scales are known to be composed primarily of a pectin-like carbohydrate, together with a small amount of protein. Unusual 2-keto-sugar acids are also present; their occurrence in both prasinophyceans and the walls of higher plants has

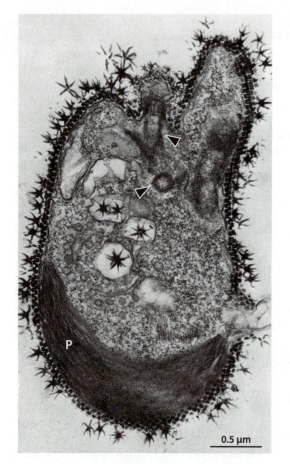

Figure 16.21 Prasinophyte scales. Body scales of the prasinophycean *Nephroselmis* are arranged in five layers. Note the scales within vesicles. Arrowheads point to basal bodies. P = plastid. (TEM: S. Sym and R. Pienaar)

been suggested as evidence that the ancestry of land plants included scaly forms (Becker et al. 1991). These acids confer a net negative charge to the scale layer that, together with calcium cations, is involved in interlocking the scales in their characteristic layers. The Golgi apparatus is also thought to produce mucilage vesicles that discharge their contents to the cell surface during cyst and palmella-stage formation.

The cells of some prasinophytes are naked, while a cell wall encloses the cells of others. Cells of *Tetraselmis* and *Scherffelia* are enclosed by a single, coherent wall composed of two or three layers, interrupted only by a slit at the point of flagellar emergence. Close examination reveals that the wall is composed of subunits having a size and

arrangement resembling those of the scale layers of other prasinophyceans (Domozych 1984), suggesting that the wall develops by fusion of Golgi-derived scales following their deposition on the cell surface.

Plastids, Pigments, and Light-Harvesting Complex Proteins

Prasinophyceans generally have a single plastid, though it may be highly lobed, and plastids usually contain at least one starch-sheathed pyrenoid. Sometimes an eyespot is present within the plastid. Chlorophylls *a* and *b* are always present, but other pigments vary and include some carotenoids that are unusual for green algae (Fawley et al. 2000). For example, a distinctive carotenoid known as prasinoxanthin is present in species of three clades, though this pigment is absent from four other prasinophyte clades (Latasa et al. 2004). Some species possess the chlorophyll *c*–like pigment MgDVP (Mg-2,4-divinyl-phaeoporphryn a_5 monomethyl ester). Isolates obtained from deep-water samples may possess an additional chlorophyll *c*–like pigment known as CS-170 that aids in harvesting blue-green light, which dominates at such depths (Latasa et al. 2005; Rodriguez et al. 2005). Prasinophytes differ from other green algae in having multiple genes that encode a prasinophyte-specific type of light-harvesting complex protein known as Lhcp. Such proteins bind chlorophylls and carotenoids. Prasinophytes display LHCI proteins similar to those of other green algae, but not the major LHCII proteins typical of most green algae and land plants. These observations suggest that green algal LHCI proteins (see Figure 16.5) originated before the major LCHII proteins (Six et al. 2005).

Cell Division

Mitosis and cytokinesis of the scale-covered or naked prasinophytes occur while the cells are swimming. The first indications that mitosis is imminent are cell enlargement and widening of the pyrenoid. Basal bodies replicate and generate flagella, and then interconnections between basal bodies break down, allowing reorganization of the flagellar apparatus that accompanies flagellar maturation cycles. The rhizoplast then divides, and the nuclear envelope develops openings at the poles. In some prasinophyceans, such as *Mantoniella*,

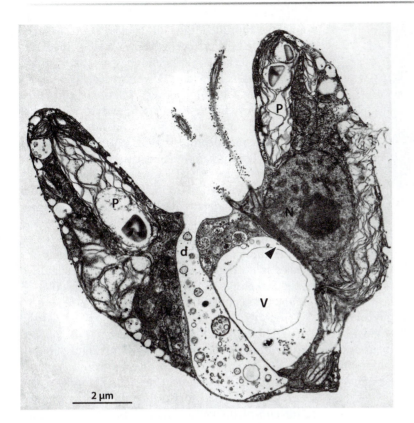

2 μm

Figure 16.22 TEM of *Halosphaera* in longitudinal view, showing the specialized scale-releasing duct (d), which is regarded as a vestigial cytopharynx (cell mouth or particulate feeding apparatus). Arrowhead points to a rhizoplast. N = nucleus, P = plastid, V = vacuole. (TEM: T. Hori)

the nuclear envelope is present throughout mitosis (Barlow and Cattolico 1981), whereas in *Pyramimonas amylifera* (Woods and Triemer 1981) and others, the nuclear envelope has almost completely broken down by mid-mitosis. Cytokinesis occurs by development of a constricting furrow from the cell periphery. The interzonal spindle may persist until it is broken by the furrowing cell membrane, suggesting that the persistent spindle characteristic of charophycean green algae is a plesiomorphy (ancestral trait).

Cell division of walled prasinophyceans differs from that in scale-covered forms in several ways. The flagella are shed prior to mitosis, and cytokinesis occurs within the parental wall. The nuclear envelope never disintegrates, and the spindle is metacentric, meaning that the basal bodies (centrioles) lie at one side of the metaphase plate, as in trebouxiophycean green algae. The spindle collapses quickly, allowing daughter nuclei to move close together, and a phycoplast system of microtubules develops parallel to the plane of cytokinesis, as in the Chlorophyta/UTC clade. Daughter cells then develop new walls and flagella. These cell division

features and molecular evidence together indicate that walled prasinophyceans are sister to the Chlorophyta/UTC clade (see Figure 16.1).

Asexual and Sexual Reproduction

A number of prasinophyceans produce resting or other stages that are walled and thus known as cysts, or phycoma stages. For example, *Pterosperma* and *Halosphaera* produce large spherical cysts up to 230 μm in diameter (Figure 16.23). Phycoma stages develop from flagellate cells by deposition of an inner wall formed from the contents of numerous mucilage vesicles that discharge to the outside, increases in cytoplasmic lipids that confer buoyancy, and the development of an outer wall. The fossil genus *Tasmanites*, found in coaly tasmanite or oil shales of Cambrian to Miocene age, is hypothesized to be the phycoma stage of early prasinophytes.

In modern species, following a period of time from two weeks to nearly four months, depending on environmental conditions, the phycoma/cyst nucleus and cytoplasm undergo division to form numerous flagellate cells that are released by rupture

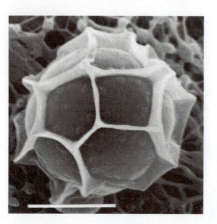

Figure 16.23 Phycoma (asexual cyst) stage of *Pterosperma*. Scale bar = 10 μm. (From Inouye et al. 1990, by permission of the *Journal of Phycology*)

of the wall. The motile cells produced by phycomata swim backward (Inouye et al. 1990). Evidence for the involvement of sexual reproduction in cyst production has been observed for *Cymbomonas tetramitiformis* (Moestrup et al. 2003) and *Nephroselmis olivacea* (Suda et al. 1989), but whether cyst formation generally results from sexual reproduction in prasinophytes is unknown. The mating process has been studied ultrastructurally in *N. olivacea* (Suda et al. 2004). The gametes of *N. olivacea* retain their scale covering while mating, and the zygotes secrete a fibrous, adhesive wall that attaches them to surfaces.

16.4 Prasinophycean Diversity

Molecular phylogenetic analyses indicate that prasinophyceans diverged relatively early, do not form a monophyletic group, and consist of at least seven clades (Steinkötter et al. 1994; Nakayama et al. 1998; Fawley et al. 2000; Latasa et al. 2004; Guillou et al. 2004) (see Figure 16.1). Moestrup (1991) recommended that the uniflagellate, scaleless genera *Pedinomonas*, *Marsupiomonas*, and *Resultor* be placed into a class Pedinophyceae. Pyramimonidales, including the genera *Halosphaera*, *Pyramimonas*, *Cymbomonas*, and *Pterosperma*, is regarded as the earliest-diverging order and sister to the rest of the prasinophytes. Additional clades include the orders Pseudoscourfieldiales (*Nephroselmis*, *Pseudoscourfieldia*, and *Pycnococcus*) and Mammiellales (*Ostreococcus*, *Micromonas*, *Mamiella*, and *Mantoniella*),

and the Chlorodendrales (*Tetraselmis* and *Scherffelia*), whose cells are enclosed by a wall or theca. As noted earlier, the Chlorodendrales are sister to the Chlorophyta/UTC clade (Ulvophyceae, Trebouxiophyceae, and Chlorophyceae) (Melkonian and Surek 1995; Fawley et al. 2000) (see Figure 16.1). Further evidence for this relationship is provided by a molecular synapomorphy, the UGUAA-motif in mRNA polyadenylation signal sequences, which is unique to *Scherffelia* and the UTC clade (Wodniok et al. 2007). Prasinophytes also include some lineages that are less well understood. Though scaly freshwater *Mesostigma viride* is sometimes included among the prasinophytes (e.g., Lewis and McCourt 2004), several lines of molecular evidence indicate that this species is an early-diverging modern representative of the Streptophyta (Kim et al. 2006; Nedelcu et al. 2006; Simon et al. 2006; Lemieux et al. 2007).

HALOSPHAERA (Gr. *hals*, sea + Gr. *sphaira*, ball) is a unicellular marine flagellate (Figure 16.24). Four flagella (rarely fewer) emerge from a pit and may be twice the length of the cell. The cells rotate as they swim, either forward or backward. The anterior end of the cell is highly lobed, as is the cup-shaped plastid, the lobes of which extend into the cytoplasmic lobes. There are two or four pyrenoids. A single eyespot occurs in the posterior portion of the cell. There are numerous mucilage vesicles arranged in longitudinal rows at the cell periphery. Their contents can be discharged as threads, rods, or spheres. A large posterior reservoir is connected to the flagellar pit by a short canal; this makes up the presumed phagocytotic apparatus of the cell (Sym and Pienaar 1993; Hori et al. 1985). There is a large (250–800 μm in diameter) planktonic phycoma/cyst stage, the outer wall of which is composed of resistant material; this wall may have surface ornamentations. The cyst stage contains a large amount of lipid, which contributes to buoyancy. The cytoplasm eventually divides to form many small uninucleate units, which then continue dividing to form flagellate cells. These remain within the cyst wall for a time, continuing to divide, before eventual release through a slit in the wall (Sym and Pienaar 1993).

PYRAMIMONAS (Gr. *pyramis*, pyramid + Gr. *monas*, unit) is a flagellate unicell found in marine, brackish, or freshwaters (Figure 16.25). The 4–16 flagella can be up to five times as long as the cells and emerge from a deep, narrow pit. Some species

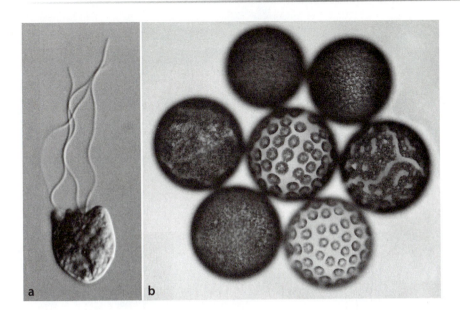

Figure 16.24 *Halosphaera.*
(a) Motile cell. The four flagella emerge from an apical pit surrounded by cytoplasmic lobes. (b) Phycomata in various stages of development.
(a: G. Floyd and C. O'Kelly; b: M. Dube)

of *Pyramimonas* are benthic, associated with sand or ice. The larger number of flagella (8 or 16) of these forms is regarded as an adaptation to benthic life, facilitating attachment (Daugbjerg et al. 1994). Another such adaptation is extensive production of mucilage from muciferous vesicles, as in the benthic phase of *P. mucifera* (Sym and Pienaar 1991). The anterior is four lobed, as is the usually single, cup-shaped chloroplast. These cell lobes and chloroplast lobes surround the pit. There is usually at least one pyrenoid and one or more eyespots in the plastid. There are several layers of body and flagellar scales. More than 70 species have been described, but many of them have not yet been subjected to critical analysis. The description of new species typically requires

both molecular information and ultrastructural characterization, as illustrated by Suda (2004) for *P. aurea* sp. nov. Ultrastructural studies of the microtubular root systems suggest that *Pterosperma* and *Halosphaera* are related to *Pyramimonas* and that all possess an X-2-X-2 root system typical of noncharophycean green algae (Inouye et al. 1990).

TETRASELMIS (Gr. *tetras*, from *tessares*, four + Gr. *selmis*, angler's noose) (Figure 16.26) may occur as a flagellate or a nonmotile cell attached by a gelatinous stalk. Flagellate cells have four flagella emerging from the pit in two pairs. A distinctive

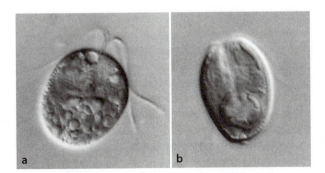

Figure 16.25 Two *Pyramimonas mucifera* cells.
(a) Note four flagella. (b) Note pit and pyrenoid with surrounding starch. (Photos: R. Pienaar and S. Sym)

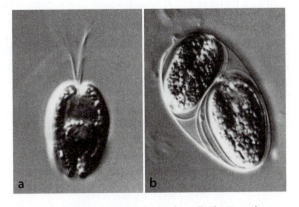

Figure 16.26 *Tetraselmis.* (a) Motile cell. (b) Daughter cells within parental wall/theca. (Photos: R. Pienaar and S. Sym)

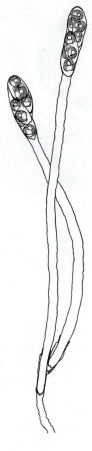

Figure 16.27 *Prasinocladus.* A diagrammatic view showing cells at the ends of elongate, branched gelatinous stalks. (From Proskauer 1950)

wall (theca) composed of small scalelike particles in a crystalline array covers cells. Motile cells often stop swimming for extended periods, and flagella are sometimes lost. Though usually green, some can become red by accumulation of carotenoids. About 26 species are reported from marine and freshwaters (Sym and Pienaar 1993). *RbcL* sequences (Daugbjerg et al. 1995) and ultrastructural data showing presence of a periplast of fused scales suggest that the sessile, stalked genus *Prasinocladus* (Figure 16.27) (Proskauer 1950), as well as the genus *Platymonas*, and some other forms may be allied with *Tetraselmis* (Norris et al. 1980).

Chapter 17

Green Algae II

Ulvophyceans

Cladophora

(Photo: L. W. Wilcox)

Ulvophyceans include some of the largest and most conspicuous of the green algae. A number of genera are notable for forming nuisance growths that interfere with human activities or harm other organisms. Examples include tubes of *Enteromorpha* that foul the hulls of oceangoing ships and form "green tides" along rocky marine shorelines affected by nutrient pollution. *Codium*, the seaweed aptly known as "dead man's fingers," can wreak havoc in shellfish beds. By growing on their surfaces, this heavy seaweed weighs shellfish down, thereby preventing their escape from predators. Vast lawns of a giant form of *Caulerpa taxifolia* have become notorious for smothering corals and other sea life in the Mediterranean Sea. Overgrowths of gray-green *Dictyosphaeria* occur on Hawaiian coral reefs as a result of human disturbances. In freshwaters, surface blankets of *Pithophora* may clog boat motors, and shoreline thickets of *Cladophora* generate terrible odors after washing up onto beaches and rotting there. On land, the ulvophycean genus *Cephaleuros* becomes a pest in tropical plantations, producing rusty spots on the leaves of tea and other plants.

Other ulvophyceans play more positive roles, providing food and habitat for aquatic animals. Beautiful tropical green seaweeds such as the mermaid's wine cup *Acetabularia* attract snorkelers and divers, while the lacy blades of *Anadyomene* decorate both coral reefs and the cover of this book. This chapter surveys the relationships, diversity of body forms, and reproductive cycles of ulvophycean green algae.

17.1 General Characteristics, Relationships, and Fossil History of Ulvophyceans

The green algal class Ulvophyceae is informally known as ulvophyceans or ulvophytes. Most representatives are marine, though some occupy freshwater or terrestrial habitats. Body types include flagellate and nonflagellate unicells and colonies, branched and unbranched filaments, membranous sheets, and siphonous coenocytes. *Pseudoneochloris marina* is an example of a unicellular member (Watanabe et al. 2000a). *Pseudendocloniopsis botryoides* is an example of a colonial (sarcinoid) genus having nonmotile vegetative cells (Friedl and O'Kelly 2002), and *Oltmannsiellopsis viridis* is a unicellular or colonial flagellate. Many filamentous and coenocytic

ulvophytes occur in nature, and diverse representatives are described in this chapter. Coenocytic, siphonous bodies are essentially very large multinucleate cells with cross-walls formed only at reproduction. *Caulerpa, Codium,* and *Acetabularia* are examples of ecologically important ulvophyceans having coenocytic bodies (Figure 17.1).

General Characteristics of Ulvophyceans

Ulvophyceans are generally characterized by a suite of plesiomorphic (ancestral) characters: closed mitosis, persistent telophase spindles and furrowing at cytokinesis, and flagellar apparatuses having cross-shaped (cruciate) X-2-X-2 roots that are overlapping and offset in the counterclockwise (CCW) direction (see Chapter 16). Some ulvophyceans produce flagellate reproductive cells whose surfaces are covered by a layer of small diamond-shaped scales like those on some prasinophytes (Sluiman 1989; Floyd and O'Kelly 1990). In addition, many ulvophyceans display the xanthophyll pigments siphonein or siphonoxanthin (Fawley and Lee 1990), which also occur in some prasinophytes (Sym and Pienaar 1993). Interestingly, the sexual life cycle of many (though not all)

Figure 17.1 The widespread green seaweeds (a) *Caulerpa*, (b) *Acetabularia*, and (c) *Codium*, which represent siphonous, coenocytic ulvophyceans. (a: R. J. Stephenson; b: C. Lipke; c: L. E. Graham)

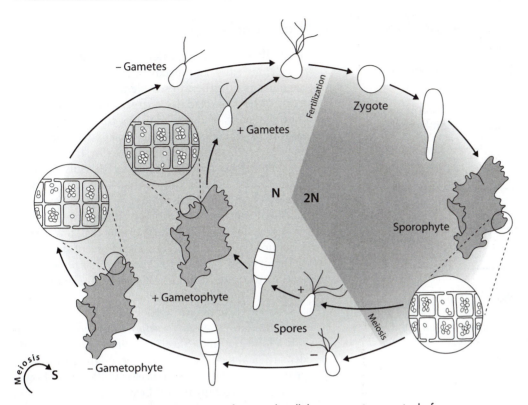

Figure 17.2 Diagram showing alternation of two multicellular generations typical of some ulvophytes, here represented by *Ulva*.

ulvophyceans involves two alternating stages, diploid sporophyte and haploid gametophyte (Figure 17.2). This type of life cycle, known as a sporic life cycle, is thought to derive from a simpler zygotic life cycle. Non-ulvophycean green algae uniformly display zygotic life cycles.

Relationships of Ulvophyceans

The Ulvophyceae has been regarded as an early-diverging lineage of the Chlorophyta/UTC clade. However, an absence of shared, derived structural or reproductive characters has led to uncertainty about the monophyly of this class (Lewis and McCourt 2004). In light of such concerns, some authorities have distinguished the classes Cladophorophyceae and Bryopsidophyceae from Ulvophyceae (van den Hoek et al. 1995; Gabrielson et al. 2006). However, on the basis of molecular evidence, other experts retain at least some genera of the former in an order Cladophorales within Ulvophyceae (Rindi et al. 2006), and others retain the latter as the order Bryopsidales within the Ulvophyceae (Lam and Zechman 2006).

In a recent analysis, monophyly of Ulvophyceae varied in level of support from strong to none, depending upon the method used to analyze SSU rDNA sequence data (Watanabe and Nakayama 2007). In the latter study, some small early-diverging clades of unicellular or colonial ulvophyceans appear to be related to a larger clade of unbranched filaments and sheets, known as Ulotrichales/Ulvales. Branched filamentous Trentepohliales and Cladophorales/Siphonocladales, and siphonous Caulerpales and Dasycladales, form additional clades whose higher-level relationships require additional study. For the time being, on the basis of recent molecular work we have diagrammed these clades within the single class Ulvophyceae (Figure 17.3).

Ulvophycean Fossils

The earliest-known fossils that have been attributed to ulvophycean green algae are branched tubes—50 to 800 µm in diameter and up to 1 mm long—that resemble modern *Cladophora* (see Chapter 7). Such fossils were recovered from 700- to 800-million-year-

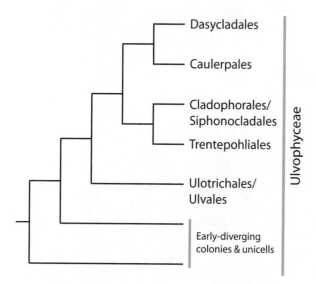

Figure 17.3 Diagram showing relationships of ulvophyceans. (Based on Friedl and O'Kelly 2002, Leliaert et al. 2003, Lewis and McCourt 2004, and Watanabe and Nakayama 2007)

old (Precambrian/Proterozoic) rocks (Butterfield et al. 1988). Some of the macrofossils recovered from nearly 600-million-year-old Proterozoic shales of the Chinese Doushantuo Formation are thought to be remains of siphonous ulvophyceans (Xiao et al. 2002).

Dasycladalean fossils

Radially symmetrical, calcified forms such as *Yakutina* from early Cambrian deposits and about 175 younger fossil genera resemble the modern ulvophycean order Dasycladales (Berger and Kaever 1992). The tendency of dasycladaleans to acquire a stony coating of calcium carbonate that is well preserved in sediments is likely responsible for the existence of a nearly continuous record of this ulvophycean class to the present time. Amazingly, many types of dasycladalean fossils do not appear to have been crushed or otherwise seriously damaged by millions of years of compressive burial, upheaval, or other alternations. For example, cross-sections of structurally intact remains of Upper Permian *Mizzia* are quite similar to those of modern *Cymopolia* (Figure 17.4) (Kirkland and Chapman 1990).

Some prominent Ordovician dasycladaleans were *Rhabdoporella*, *Vermiporella*, and *Cyclocrinites* (Johnson and Sheehan 1985). Diversity and density peaks occurred in the Permian (250 million years ago [mya]), Mid-Triassic (230 mya), Lower Cretaceous (130 mya), and Lower Tertiary (55 mya). These peaks correlate with instances of high sea levels and extensive shallow continental seas. The Cretaceous–Tertiary (KT) extinction event that occurred some 65 mya, attributed to the impact of a

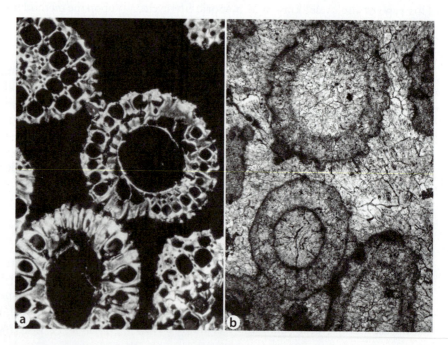

Figure 17.4 (a) Cross-section of modern *Cymopolia* compared to (b) cross-section of the fossil known as *Mizzia*. The fossils have persisted because the original seaweed was highly calcified, as is *Cymopolia*. (Photos: B. Kirkland)

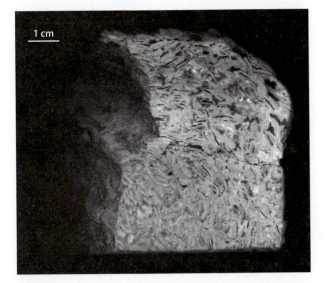

Figure 17.5 Fossil *Halimeda*. A portion of a calcareous mound formed millions of years ago by growth of the green seaweed *Halimeda*. Individual calcified segments of the algal body can be distinguished. (Photo: L. W. Wilcox)

Figure 17.6 *Halimeda* (right) and *Rhipocephalus* (left) growing in a sea grass bed near San Salvador, The Bahamas. When they die, such seaweeds contribute to the formation of carbonate sands. (Photo: R. J. Stephenson)

large meteor or comet, did not much affect dasycladaleans. However, a number of major extinction events have occurred over time. For example, during a Mid-Oligocene (33 mya) period lasting millions of years, dasycladaleans were apparently quite rare. The modern dasycladalean genus *Acetabularia* first appeared about 38 million years ago, survived the Oligocene crisis, and was joined by additional modern dasycladalean taxa about 10,000 years ago.

Halimeda fossils

In addition to dasycladalean algae, other types of calcified ulvophyceans also form fossil deposits. Giant mounds of fossil *Halimeda*, belonging to the Caulerpales, reach a thickness of over 50 m in the continental shelf waters associated with the Great Barrier Reef in Australia and other regions, at depths from 12 to 100 m. Covering hundreds of square meters, such mounds are estimated to grow by sedimentation at rates of nearly 6 m per 1000 years. Fossil remains of ancient *Halimeda* (Figure 17.5) are recognized by their similarity to the calcified parts of living *Halimeda*. In the modern Bahamas, Florida Bay, and reef flats or lagoons associated with Pacific coral atolls, remains of *Halimeda* and related algae generate much of the carbonate sediments upon their death and disintegration (Tucker and Wright 1990)

(Figure 17.6). In a Caribbean reef lagoon, *Halimeda incrassata* annually produces 815 g of calcium carbonate per square meter (van Tussenbroek and van Dijk 2007).

17.2 Ulvophycean Diversity and Ecology

More than 100 genera and 1100 species are included in Ulvophyceae (Floyd and O'Kelly 1990). These have been classified according to cellular and reproductive features and molecular sequence analyses (e.g., Hayden and Waaland 2002; Leliaert et al. 2003; López-Bautista and Chapman 2003; Pombert et al. 2004; Rindi et al. 2006; Watanabe and Nakayama 2007).

Ulotrichales/Ulvales

While some morphological traits and molecular studies (e.g., Hayden and Waaland 2002) support

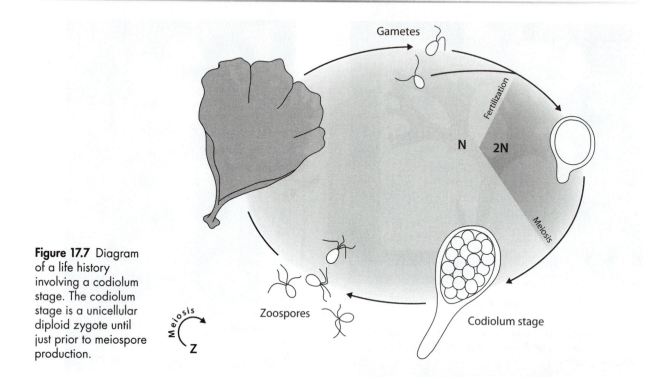

Figure 17.7 Diagram of a life history involving a codiolum stage. The codiolum stage is a unicellular diploid zygote until just prior to meiospore production.

separation of the orders Ulvales and Ulotrichales, other molecular analyses do not provide strong support for their partition. Even so, Ulotrichales and Ulvales are closely related and together appear to form a well-supported clade (Watanabe and Nakayama 2007). Gamete-producing (gametophyte) body types include unbranched filaments of uninucleate cells such as *Ulothrix*, branched filaments of uninucleate cells (e.g., *Spongomorpha*), unbranched filaments whose cells are multinucleate (e.g., *Urospora*), branched filaments of multinucleate cells (e.g., *Acrosiphonia*), and blades or tubes (e.g., *Monostroma*, *Ulva/Enteromorpha*). Most occur along rocky ocean coasts, attached to stable substrates, though a few occur in nearshore freshwaters. Flagellate zoospores and gametes are widely produced during asexual and sexual reproduction, respectively. Such flagellate cells can be consumed by planktonic species of the ciliate *Strombidium*, which retain the algal plastids for some time as a source of photosynthate (McManus et al. 2004). This behavior is a form of kleptoplastidy, described in Chapter 7.

Members of this clade that have been grouped into the order Ulvales have a life cycle involving isomorphic alternation of multicellular generations (see Figure 17.2). In contrast, the diploid, spore-producing stage of genera that have been classified

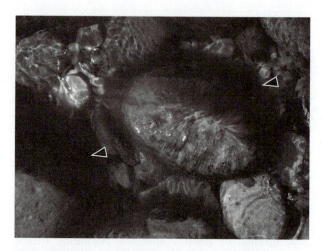

Figure 17.8 The dark-green freshwater ulotrichalean *Ulothrix zonata* (arrowheads) growing on emergent rocks at the edge of Lake Superior. *Ulothrix* can be found much of the year in this cool-water lake. In smaller northern temperate lakes, growths are common in the spring and sometimes fall but are not present during the summer. (Photo: L. W. Wilcox)

as Ulotrichales is a small, thick-walled, sac- or club-shaped unicell that is attached to substrates by means of a stalk (Figure 17.7). Such small spore-producing structures were originally thought to be a distinct unicellular species named *Codiolum* (or a

Figure 17.9 *Ulothrix zonata*, which is an unbranched filament. Light micrograph showing the parietal band-shaped chloroplast with numerous pyrenoids. (Photo: L. E. Graham)

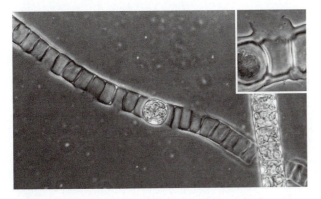

Figure 17.10 *Ulothrix zonata* filament with a single zoosporangium. The remaining cells in the filament have already discharged their flagellate reproductive cells. Inset: a pore through which flagellate reproductive cells were discharged. (Photos: L. E. Graham)

Figure 17.11 *Spongomorpha*. (a) Macroscopic view, showing tuft of filaments. (b) Microscopic view of an apical cell of a filament. (Photos: C. O'Kelly)

becomes endophytic within the bodies of seaweeds, thereby occupying highly protected and stable microhabitats. The codiolum stage can survive a period of dormancy, typically the warm season of the year. Secure attachment is a necessary adaptive response to the turbulence often encountered in nearshore waters. When conditions are favorable, the codiolum phase produces flagellate spores that are assumed to be products of meiosis and thus are known as meiospores. Following a period of motility, such meiospores settle, attach to a substrate, and grow into gametophytic filaments or blades.

ULOTHRIX (Gr. *oulos*, woolly + Gr. *thrix*, hair) can occur in marine or freshwaters, attached to rocks or submerged pieces of wood, and is particularly evident during cold seasons (Figure 17.8). Cells can divide at any point along the length of the unbranched filaments. There is a single, parietal, band-shaped chloroplast per cell, and several pyrenoids are present (Figure 17.9). Each cell can generate numerous biflagellate isogametes or quadriflagellate zoospores that are released via a pore in the cell wall (Figure 17.10), presumably generated by localized action of wall-dissolving enzymes. Zoospores are covered with a layer of small scales. The zoospores can attach to substrates and produce new filaments, rapidly generating clonal populations. Gametes fuse in pairs to form a negatively phototactic quadriflagellate zygote that settles onto substrates and develops into a thick-

species of unicellular *Chlorochytrium*), but now they are known to often be part of the life cycles of other genera. However, sometimes it is not clear that codiolum is associated with a sexual cycle. Hence, it is referred to as a codiolum phase or stage (O'Kelly et al. 2004). Codiolum stages develop from diploid zygotes produced by the fusion of gametes released from gametophytes of the genera *Acrosiphonia*, *Chlorothrix*, *Collinsiella*, *Eugomontia*, *Gomontia*, *Protomonostroma*, *Ulothrix*, and *Urospora* (Gabrielson et al. 2006). Sometimes the codiolum stage is free-living and can be so abundant that it appears as green patches in the intertidal zone in summer. In other cases, the codiolum phase bores into shells or

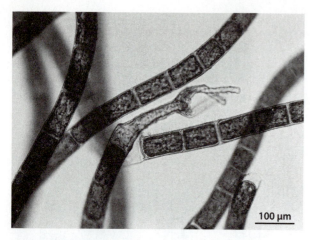

Figure 17.12 *Urospora,* an unbranched filament with multinucleate cells. A basal rhizoidal system is evident. (Photo: L. W. Wilcox)

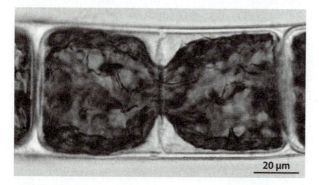

Figure 17.13 Cytokinesis in *Urospora.* The reticulate plastid is being pinched by the developing cleavage furrow. (Photo: L. W. Wilcox)

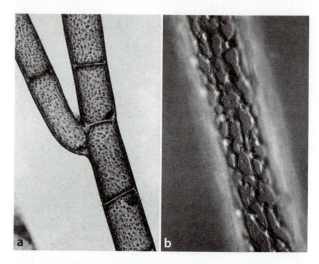

Figure 17.14 *Acrosiphonia,* which is a branched filament with multinucleate cells. A branch is shown in (a) and the reticulate plastid in (b). (Photos: C. O'Kelly)

along temperate coastlines. The uninucleate cells are much longer than they are wide, and branches arise from the apical portions of the cells via emergence of a protrusion that is later cut off by a cross-wall. Branches often appear curled or hooked. Growth is thought to be restricted to apical cells. Sexual reproduction occurs by means of biflagellate isogametes that are liberated from cells through a lidded pore similar to those generated by the gametangia (gamete-producing cells) of many other ulvophyceans.

UROSPORA (Gr. *oura,* tail + Gr. *spora,* spore) is an unbranched filament that can attain lengths of several centimeters (Figure 17.12). It attaches to substrates on rocky shores of cold marine waters with a holdfast consisting of basal, rhizoidlike cells. The cells are large and multinucleate. Nuclei in this and other algae can be visualized by staining the DNA pink via the Feulgen reaction or rendering nuclei fluorescent by treatment with DNA-specific fluorochromes. Mitosis in *Urospora* is uncoupled from cytokinesis. Prior to mitosis, nuclei congregate at the plane of cell division, then divide synchronously. Mitosis is closed, and cleavage is by ingrowth of a furrow, as in other ulvophyceans (Figure 17.13). The plastid is cylindrical, parietal, and highly dissected in mature cells; there are many pyrenoids. Sexual reproduction is by anisogametes but otherwise resembles that of *Ulothrix* and *Spongomorpha.* The flagella and flagellar basal bodies of the quadriflagellate zoospores of *U. penicilliformis* are very unusual in a number of

walled, stalked resting cell that persists throughout the warm season. Zygote germination is presumed to involve meiosis in meiospore production. Short days and low temperatures appear to induce meiosporogenesis. In nutrient-rich lakes or running waters, *Ulothrix zonata* can develop large populations in spring, but it disappears as temperatures increase above 10°C, when the parental filaments disintegrate as a result of massive zoospore production (Graham et al. 1985). *U. zonata* is believed to have invaded freshwaters from the marine habitat since most of its close relatives are marine.

SPONGOMORPHA (Gr. *spongos,* sponge + Gr. *morphe,* form) occurs as dense tufts of branched filaments, commonly attached to larger marine seaweeds (Figure 17.11). It primarily occurs in the spring

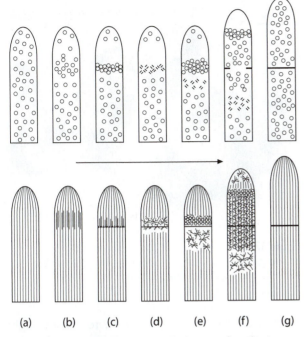

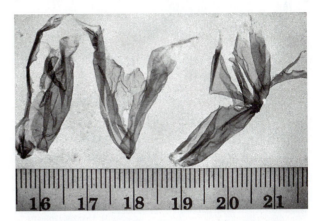

Figure 17.16 *Monostroma*, which occurs as a blade that is one layer thick. (Photo: C. O'Kelly)

Figure 17.17 *Monostroma* development series. Small filamentous germlings proceed to form hollow tubes that eventually open out into the typical blade form. (After Tatewaki 1972)

Figure 17.15 Diagrams showing behavior of nuclei (top row) and microtubules (bottom row) during cell division of *Acrosiphonia*. (a) Vegetative cells at filament tips prior to cytokinesis. (b, c) Microtubules form a band at the future site of cytokinesis and nuclei migrate to this region. (d, e, f) Stages in mitosis and onset of the formation of a septum (cross-wall). (g) Vegetative cell at the tip of a filament after septum formation has been completed. (After Aruga et al. 1996, © Blackwell Science Asia)

respects, summarized by Sluiman (1989). Lindstrom and Hanic (2005) and Hanic (2005) evaluated the phylogeny and taxonomy of *Urospora*.

ACROSIPHONIA (Gr. *akron*, apex + Gr. *siphon*, tube) possesses a branched thallus composed of multinucleate cells (Figure 17.14). Mitosis and cytokinesis have been studied using immunofluorescence methods (Aruga et al. 1996). Numerous nuclei and cortical microtubules occur in tip cells. Prior to cell division, 30%–40% of these nuclei migrate downward to the region where cytokinesis will take place (Figure 17.15), forming a nuclear ring. In this region, cortical microtubules reorient, forming a transverse band, as many mitotic spindles form. Following mitosis, nuclei migrate back into the apical region, reestablishing the parallel, longitudinal arrangement of microtubules. Cytokinesis occurs by furrowing, and the transverse microtubules occupy the leading edge of the furrow. Sussman and DeWreede (2002) determined that

diploid codiolum or chlorochytrium life cycle phases of *Acrosiphonia* occur within the bodies of fleshy or crustose red algae as well as the crustose brown alga *Ralfsia* and that host cell wall composition influences colonization. *Acrosiphonia* grows in the lower intertidal zone along rocky marine coasts of temperate and polar regions. A new bacterial genus and species, *Mesonia algae*, belonging to the family Flavobacteriaceae, was isolated from *Acrosiphonia sonderi* (Nedashkovskaya et al. 2003). Such epibacteria are increasingly recognized as symbionts that may secrete compounds that influence the growth and development of their algal hosts.

MONOSTROMA (Gr. *monos*, one + Gr. *stroma*, layer) is appropriately named because it consists of a single layer of cells arranged in a sac or blade (Figure 17.16). *Monostroma* thalli are attached to substrates via a disklike holdfast. They arise by upward protrusion of cells from an initially flat, discoid germling. Cell proliferation results in a hollow tube of cells, which in some forms persists into the adult phase. In other forms, tubes open into flat blades superficially resembling the genus *Ulva* (Figure 17.17).

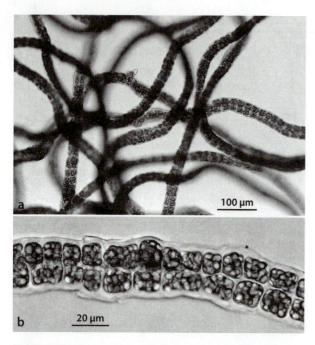

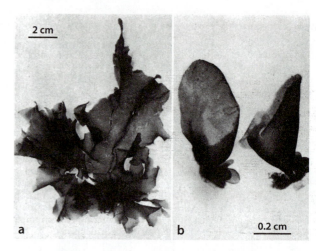

Figure 17.18 *Percursaria*, which occurs in the form of biseriate filaments, shown at two magnifications in (a) and (b). (Photos: L. W. Wilcox)

Figure 17.19 *Ulvaria*, which can occur as a blade (a) or sac (b) growing from an adherent disk. (From Dube 1967, by permission of the *Journal of Phycology*)

Normal morphological development in culture requires the presence of appropriate bacteria or medium in which bacteria-free brown or red seaweeds have been grown. This suggests that morphogenetic substances are released from bacteria and algae that could influence *Monostroma* development in nature (Tatewaki et al. 1983).

Cells contain a single parietal, cup-shaped plastid with a pyrenoid. Sexual reproduction resembles that of other ulvophyceans. Anisogametes are covered with tiny prasinophycean-like scales and zygotes are unicellular and stalked (codiolum-like). Zygote germination is induced by short-day and low-temperature conditions. Species possessing a life history involving alternation of multicellular generations have been transferred to the genus *Ulvaria. Monostroma* can be easily confused with *Ulvaria* and *Ulva*, and young stages are difficult to distinguish from the trebouxiophycean *Prasiola* (see Chapter 18). *Monostroma* can be distinguished from *Prasiola* on the basis of chloroplast morphology (*Prasiola*'s plastids are axial); *Ulva* has a two-cell-layered blade; and *Ulvaria* has basal rhizoids rather than a disk holdfast. *Monostroma* is surprisingly tolerant of variation in salinity, occurring in brackish and freshwaters (Smith 1950), as well as marine waters of normal salinity.

PERCURSARIA (L. *percurso*, to run through) may be found attached to stones or other substrates in intertidal marine habitats but is typically much less conspicuous than some other ulvophyceans (Figure 17.18). Zoospores may grow into a flat (prostrate) adherent disk from which emerge several biseriate (two-cell-thick) ribbons, or zoospores may grow directly into biseriate ribbons that attach to substrates via basal rhizoids. *Ulvella* forms prostrate, radially organized disks similar to those of *Percursaria* germlings, and *Blidingia* also has a prostrate disk, from which develop erect tubular thalli. *Blidingia dawsonii* has been recombined into *Percursaria dawsonii* (Lindstrom et al. 2006).

ULVARIA (from *Ulva*) (see below) occurs as small sacs or leaflike blades that can grow to 30 cm in length and thus resemble *Ulva* (Figure 17.19). Monostromatic blades are attached by rhizoids in rocky marine intertidal environments. Development from zoospores proceeds through successive uniseriate filamentous stages, pluriseriate stages, and tubular stages, then tubes open to form blades. Though some species appear to be entirely asexual, others are reported to exhibit isomorphic alternation of generations (Dube 1967). Hoops et al. (1982) studied the ultrastructure of biflagellate reproductive cells.

ENTEROMORPHA (Gr. *enteron*, intestine + Gr. *morphe*, form) (Figure 17.20) is the name given to elongated tubes, sometimes with constrictions,

Figure 17.20 The tubular bodies of *Enteromorpha*. *Enteromorpha* often occurs in crowded populations, attached to rocky substrates in tide pools and nearshore marine habitats. (Photo: L. W. Wilcox)

Figure 17.21 *Ulva*, which is a blade composed of two cell layers. (Photo: L. W. Wilcox)

that attach to substrates via rhizoidal branches that may form an attachment disk. *Enteromorpha* is informally known as "gut weed." Tubular thalli arise from phototactic zoospores, zygotes, or unmated gametes that attach to substrates (Callow et al. 1997) and grow into uniseriate filaments. These then form a two-cell-layered ribbon that separates in the center to generate the mature tubular body, similar to early stages in the development of *Monostroma*. The tubes are only one cell thick and may bifurcate, giving a branched appearance. The tubular construction may aid in buoyancy if gas bubbles are trapped in the central area. There is a single, parietal, platelike chloroplast per cell, but interestingly, plastids are reported to rotate in response to changes in the direction of light. Although zoospores are generated primarily from the distal ends of sporophytes—leaving much of the thallus intact—gametogenesis occurs throughout the gametophyte body, resulting in thallus disintegration. Maximum liberation of gametes occurs a few days prior to the highest tide in a lunar cycle. There is a worldwide distribution, with some species tolerating waters as warm as 30°C, and others occurring in Antarctic waters where the maximum temperature is 1.8°C. In addition to growing well at normal salinity, *Enteromorpha* may also occur in hypersaline lakes (at 52% salinity) or tide pools that have experienced evaporation, as well as brackish and freshwaters (Poole and Raven 1997). Although some species appear to reproduce only asexually, others exhibit a typical alternation of isomorphic

generations (Kapraun 1970). Molecular analyses (Hayden and Waaland 2002; Hayden et al. 2003) revealed that *Enteromorpha* is not a monophyletic group and could not be separated from all species of the genus *Ulva*.

ULVA (ancient Latin name for sedge) (Figure 17.21) forms conspicuous flat blades that are composed of two cell layers. Ulva is commonly known as "sea lettuce." Blades can be as long as 1 m. *Ulva* is attached to substrates in marine coastal waters by means of rhizoidal branches, and it also occurs in free-floating masses. The distromatic blade arises from zoospores via uniseriate, pluriseriate, and tubular stages of development, followed by collapse of the tube. If grown axenically (in the absence of bacteria and other organisms) in the laboratory, development is disrupted, and cushions of branched, uniseriate filaments are produced (Provasoli 1958). However, addition of bacteria isolated from natural *Ulva* collections will restore normal development and morphology. Similar bacterial effects upon morphogenesis of *Enteromorpha* and *Monostroma* have been observed (Nakanishi et al. 1996). The effect is caused by a compound named thallusin that epibacteria secrete (Matsuo et al. 2005).

Reproduction is exclusively asexual in some forms, but others express alternation of isomorphic generations. Both zoospores and gametes are mainly produced in cells at the edge of the thallus. This allows reproduction to occur without causing complete

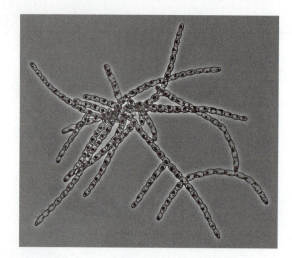

Figure 17.22 *Trentepohlia*. Branched filaments are composed of uninucleate cells that often have conspicuous red droplets composed of lipids with dissolved carotenoids. (Photo: L. W. Wilcox)

Figure 17.23 *Trentepohlia*, which forms orange-red felty growths on wood, stones, and tree bark in humid areas. (Photo: G. McBride)

disintegration of the parental thallus, thereby contributing to *Ulva*'s persistence in the environment. Sixteen gametes or 16 zoospores are produced per parental cell of *Ulva mutabilis*, and extracellular inhibitors control the release of flagellate cells (Stratmann et al. 1996). The carbon concentrating mechanisms of three species of *Ulva* were studied by Björk et al. (1993). *Acrochaete*, a genus of ulvophycean algae that grows as an endophyte within the bodies of fleshy red algae, is closely related to the *Enteromorpha/Ulva* complex (Bown et al. 2003).

Growths of *Enteromorpha*, *Ulva*, and *Ulvaria* can be so extensive that they are called "green tide" seaweeds. Green tide seaweeds may occur attached to rocky substrates or as free-floating pieces torn from attached algae. An example of a free-floating form is the newly described species *Ulva ohnoi* (Hiraoka et al. 2003). Green tides can displace native species and may have other biotic impacts. Extracts from *Ulva fenestrata* and *Ulvaria obscura* inhibit the development of *Fucus* zygotes, oyster larvae, algal epiphytes, and these seaweeds' own growth (Nelson et al. 2003).

Trentepohliales

Trentepohliales are named for the genus *Trentepohlia* but are also known as the Chroolepidaceae. There are a few genera and about 40 species, all of which appear to be terrestrial. The algae occur as branched

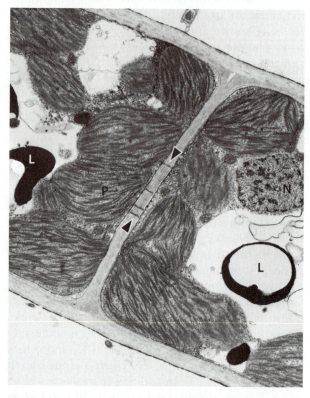

Figure 17.24 *Trentepohlia*, whose cells are interconnected by plasmodesmata (arrowheads), as seen in this TEM view. Also shown are the many discoid plastids (P), the single nucleus (N) per cell, and carotene-containing lipid droplets (L). (From *Origin of Land Plants*. L. E. Graham. ©1993. Reprinted by permission of John Wiley & Sons, Inc.)

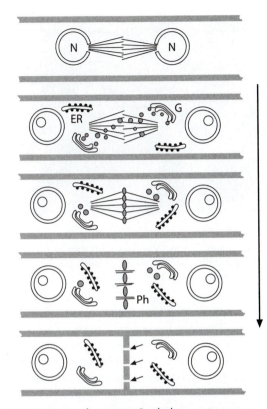

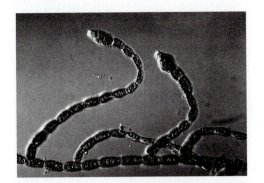

Figure 17.26 *Trentepohlia*, which produces specialized urnlike gametangia at the ends of branches. (Photo: L. E. Graham)

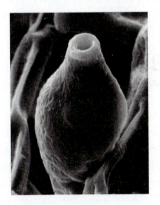

Figure 17.27 *Trentepohlia* gametangium viewed using SEM, showing specialized pore through which reproductive cells are discharged. (SEM: L. E. Graham)

Figure 17.25 Cytokinesis in *Cephaleuros virescens*. Development (top to bottom) involves a system of transverse microtubules that is analogous to the phragmoplasts of some charophycean algae and land plants. Plasmodesmata are indicated by arrows. ER = endoplasmic reticulum, G = Golgi body, N = nucleus, Ph = phragmoplast. (After Chapman and Henk 1986, by permission of the *Journal of Phycology*)

filaments (Figure 17.22) on inorganic substrates such as rocks or within the leaves or upon surfaces of vascular plants (Figure 17.23), sometimes acting as economically significant parasites (Chapter 3). Some occur as phycobionts in lichens. Trentepohlialeans are often orange or red pigmented because of the presence in cells of lipid droplets containing β-carotenes (Figure 17.24). Such pigments probably play a photoprotective role, and experiments suggest that nitrogen limitation is the major stimulus for their production (Czygan and Kalb 1966).

Trentepohlialean cells are uninucleate, but there are numerous discoid, pyrenoidless plastids per cell. Vegetative cytokinesis is quite unusual for ulvophyceans in that a phragmoplast-like array of microtubules (see Chapter 16), a cell plate (Figure 17.25), and simple plasmodesmata (see Figure 17.24) are

present. These features have been documented for *Cephaleuros parasiticus* (Chapman and Henk 1986) and *Trentepohlia odorata* (Chapman et al. 2001). Genes associated with the production of phragmoplastin, a protein linked with cell plate formation in land plants, have been identified by amplification from *Trentepohlia* and *Cephaleuros* (López-Bautista et al. 2003). Phragmoplasts are otherwise known only from certain derived charophycean algae and their relatives the embryophytes (land plants) (Chapters 16 and 20). It thus appears that a phragmoplast-like cytokinetic system has arisen independently at least twice within green algae.

Production of flagellate reproductive cells— biflagellate gametes or quadriflagellate zoospores— occurs within specialized, urn-shaped gametangia or zoosporangia (Figure 17.26). Flagellate cells are released by dissolution of a plug of material in a preformed pore (Figure 17.27). Unmated gametes are able

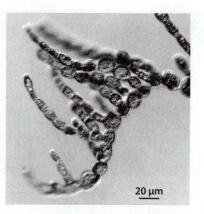

20 µm

Figure 17.28 Microscopic view of *Trentepohlia*. It is a heterotrichous filament having a prostrate basal system of filaments from which emerge numerous erect filaments. (Photo: L. W. Wilcox)

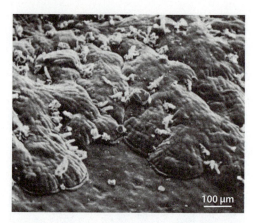

100 µm

Figure 17.29 *Cephaleuros* seen with SEM. It consists of radial, pseudoparenchymatous disks that produce reproductive cells similar to those of *Trentepohlia*. (From Chapman 1976)

to grow into new thalli, as is the case with some other ulvophyceans. Flagellate cells are unusual among ulvophyceans in lacking eyespots and in possessing distinctive, layered structures at flagellar bases; the latter superficially resemble MLSs of charophyceans and land plants (Chapter 16), but detailed examination reveals some structural differences (Graham 1984). Trentepohlialean flagella are unusual in possessing flattened, lateral "wings," or "keels," whose function is not understood. Zoosporangia may break off and be dispersed by wind, releasing zoospores when moist substrates are available. Thompson (1958) reported that at least one genus exhibits heteromorphic alternation of generations, the sporophyte being smaller than the gametophyte generation, but the life histories of the organisms in this group require additional investigation.

Monographed by Printz (1939), the order includes *Trentepohlia*, *Physolinum*, *Cephaleuros*, *Phycopeltis*, and *Stomatochroon*. Thompson and Wujek (1997) summarized the morphology, taxonomy, and ecology of the latter three taxa. A few other algae have been suggested as members of the group but have not been well studied. Molecular analyses indicate that Trentepohliales form a monophyletic group that is closely related to a clade containing the Siphonocladales/Cladophorales complex (López-Bautista and Chapman 2003) (see Figure 17.3).

TRENTEPOHLIA is named for Johann Friedrich Trentepohl, a German clergyman and botanist, but this genus and the order named for it are often misspelled (as "Trentepholia," "Trentopohlia,"

or "Trentopholia"). This genus is a branched, often red-orange filament having both a prostrate portion that is adherent to the substrate and erect branches (Figure 17.28), conferring a fuzzy appearance when viewed on terrestrial substrates in nature (see Figure 17.23). Urn-shaped reproductive structures typically occur at the ends of erect branches (see Figures 17.26 and 17.27).

CEPHALEUROS (Gr. *kephale*, head + Gr. *euros*, breadth) (Figure 17.29) has flattened thalli with branched filaments that are radial in organization, display tip growth, and produce specialized reproductive structures. *Cephaleuros* is regarded as a parasite that grows beneath the cuticles of tropical and subtropical plants such as *Magnolia* (see Figure 3.23), *Camellia*, *Citrus*, and *Rhododendron*. Erect branches bearing sporangia emerge to the surface, where zoospores can be released if moisture is available, or the sporangia abscise and are wind-dispersed. Gametes are produced on the prostrate filament system (Chapman 1980). The lichen *Strigula* contains *Cephaleuros* as a phycobiont (Chapman 1976).

PHYCOPELTIS (Gr. *phykos*, seaweed + L. *pelte*, small shield) (Figure 17.30) is an epiphyte on numerous vascular plants and some bryophytes throughout tropical and subtropical regions, and it also occurs in temperate areas (Printz 1939). Good and Chapman (1978) suggested that meiosis was probably zygotic. Bodies are radially symmetrical, with growth occurring at the margins, and adjacent filaments are closely adherent. The cell walls are impregnated with

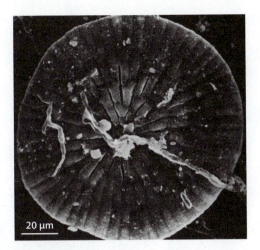

Figure 17.30 *Phycopeltis,* as seen with SEM. (From Good and Chapman 1978)

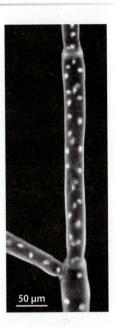

Figure 17.31 *Cladophora* nuclei stained by using the DNA-specific fluorochrome DAPI. (Photo: L. W. Wilcox)

highly resistant materials—identified on the basis of acid hydrolysis tests and infrared absorption spectra as sporopollenin. This polymer is believed to play a role in filament adherence and perhaps also in pathogen and desiccation resistance (Good and Chapman 1978).

Cladophorales/Siphonocladales

A group of some 30 genera and 425 species (Floyd and O'Kelly 1990) that occur as branched or unbranched filaments, pseudoparenchymatous blades, nets, or spherical vesicles has been classified in a variety of ways. These species have been placed into the separate orders Cladophorales and Siphonocladales (Bold and Wynne 1985), combined into a single order named either Cladophorales or Siphonocladales, or raised into a class Cladophorophyceae having either the single order Siphonocladales or the two orders Siphonocladales and Cladophorales. On the basis of a species-rich analysis of partial LSU rRNA sequences, Laliaert et al. (2003) identified a monophyletic lineage composed largely of genera such as *Anadyomene stellata* that had earlier been placed into Siphonocladales but found that the former Cladophorales were paraphyletic. On the other hand, a later SSU rDNA analysis that included far fewer representatives recovered a monophyletic Siphonocladales that included genera placed by others into Cladophorales (Watanabe and Nakayama 2007). In light of the clear need for more study of these algae, we have chosen to refer to them for the time being as Cladophorales/Siphonocladales, as in Rindi et al. (2006).

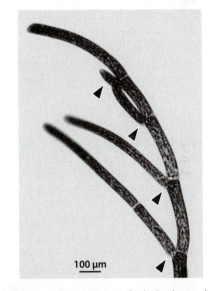

Figure 17.32 Branching pattern of *Cladophora glomerata.* Note that branches (arrowheads) originate by elongation of cellular protrusions from the distal regions of the large elongate cells of the main axis. (Photo: L. W. Wilcox)

Species included in the Cladophorales/Siphonocladales have two features in common: Vegetative cells are multinucleate (Figure 17.31), and bodies are compartmentalized into cells at maturity. In the case of species commonly classified as Cladophorales, bodies are consistently subdivided by cross-walls (septa) to form filaments (Figure 17.32). In the case of species commonly classified as Siphonocladales, bodies are

Figure 17.33 A green scum of *Pithophora* covers the surface of a lake. (Photo: C. Lembi)

coenocytic (lacking septa) early in development but later undergo a form of compartmentation known as **segregative cell division**. In this process, units of cytoplasm simultaneously become surrounded by cell wall material, with the result that mature bodies are composed of many cells. Such bodies are said to be "not persistently siphonous/coenocytic." The immature coenocytic stage is vulnerable to loss of all cytoplasm as a result of wounding. Rapid wound healing is an adaptive response.

Wound healing in *Siphonocladus*, *Valonia*, and several other related genera that undergo segregative cell division has been studied intensively (La Claire 1982a; O'Neil and La Claire 1984; La Claire 1991; Goddard and La Claire 1991). The cell contents retract from the wound site and may fragment, depending on the species. Retraction and fragmentation involve the action of actin, the associated protein calmodulin, and myosin. Each unit of the subdivided cytoplasm is viable and can grow into a new thallus. It has been proposed that segregative cell division evolved from an ancestral wound-healing process.

These green algae generally possess either highly dissected, reticulate chloroplasts or many small discoid plastids arrayed in a reticulate network. Chloroplasts usually contain pyrenoids. The antenna carotenoid siphonoxanthin, which absorbs blue-green light, occurs in certain deepwater Cladophorales/Siphonocladales (and some other green algae). In contrast, species adapted to high-light environments have lost siphonoxanthin and instead possess loroxanthin or lutein pigments (Yoshii et al. 2004). Some representatives form nuisance growths in freshwaters (Figure 17.33), and others generate conspicuous growths in tide pools or on coral reefs. The freshwater representative *Aegagropila linnaei* (= *Cladophora aegagrophila*) forms remarkable round balls, known in Japan as marimo, which may reach 30 cm in diameter. Turbulent water motions that occur at the sediment surface form these balls. An analysis of *A. linnaei* in Japan revealed close similarity to lake balls that occur in Europe (Niiyama 1989).

A newly discovered genus of unicellular algae that forms golden crusts on tree bark in the Hawaiian islands, *Spongiochrysis hawaiiensis*, is closely related to the Cladophorales/Siphonocladales (Rindi et al. 2006). This discovery reveals that ulvophycean algae have colonized terrestrial habitats at least twice (the Trentepohliales represent the other case).

CLADOPHORA (Gr. *klados*, branch + Gr. *phoras*, bearing) is a generic name given to highly branched filaments of multinucleate cells (see Figures 17.31 and 17.32). Molecular phylogenetic analyses indicate that the genus *Cladophora* is not monophyletic (Bakker et al. 1995; Hanyuda et al. 2002; Leliaert et al. 2003; Yoshii et al. 2004). Some *Cladophora* species form an early-diverging lineage; certain other *Cladophora* species form a second clade together with *Aegagropila*, *Pithophora*, *Arnoldiella*, *Wittrockiella*, and *Basicladia*; and most *Cladophora* species group with species of *Rhizoclonium* and *Chaetomorpha* and genera displaying segregative cell division (the traditional Siphonocladales of Bold and Wynne [1985]). Clearly, this genus will require

Figure 17.34 Dense masses of *Cladophora glomerata* growing on rocks at the edge of Lake Mendota, Wisconsin, U.S.A. (Photo: L. W. Wilcox)

splitting into multiple new generic entities. For the time being, the structural and ecological descriptions provided below may be generally applicable.

The chloroplast is highly reticulate and may occur as more than one piece. Many pyrenoids are present. Cells are longer than they are wide, and branches originate as cytoplasmic protrusion from the cell apex. Branches eventually are segregated from the main axis by formation of a cross-wall (septum). Attached to rocks or other substrates by rhizoidal cells, *Cladophora* may reach several meters in length (Figure 17.34). Several marine species (Bold and Wynne 1985) and at least one freshwater species (*C. surera*) (Parodi and Cáceres 1995) have been demonstrated to display isomorphic alternation of generations. The sporangia of sporophytes and gametangia of gametophytes look much like vegetative cells. Multiple biflagellate gametes are produced per gametangium, and numerous quadriflagellate zoospores are produced in each sporangium. Reproductive cells are primarily produced in cells of the branches rather than in cells of the main axis; this preserves the integrity of the main axis after reproductive cell discharge has occurred (Figure 17.35). Perennation may occur via thick-walled basal cells that persist through conditions incompatible with active growth (Dodds and Gudder 1992).

In marine communities that receive terrestrial nutrient inputs such as sewage effluents or agricultural runoff, large mats of *Cladophora* may form. During decomposition of such mats, anoxia may result, smothering nearshore invertebrates, including economically valuable species. Experiments showed that growth of *C. prolifera* is controlled primarily

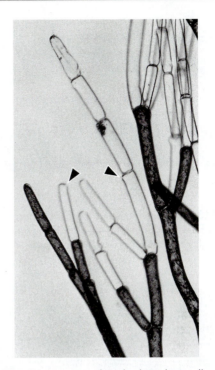

Figure 17.35 Empty terminal and subtending cells of *Cladophora glomerata*, from which reproductive cells have been released via lateral pores (arrowheads). (Photo: L. E. Graham)

by irradiance and nutrient availability, strongly implicating anthropomorphic sources as the agents of increased nuisance algal growth (Bach and Josselyn 1979). Species of *Cladophora* occur widely in temperate and tropical seas but are absent from polar waters.

Freshwater *Cladophora* occupies diverse habitats, from pristine streams to eutrophic lakes and estuaries, and is globally widespread. *Cladophora* has attracted the attention of freshwater biologists because it can attain high biomass that contributes to the structure of benthic stream communities and because it can generate nuisance-level populations in freshwater lakes having both rocky shorelines and high phosphate levels. Freshwater *Cladophora* occurs in a variety of morphological forms that vary in cell dimensions and branching patterns. Transplantation experiments have revealed that branching patterns can be influenced by environmental conditions, specifically water velocity (Bergey et al. 1995). A study of ITS regions from the nuclear ribosomal cistron suggested that there are few and possibly only one species of freshwater *Cladophora* (Marks and Cummings 1996). On the other hand,

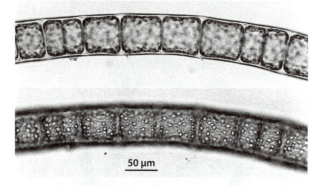

Figure 17.36 Through-focus pair of *Chaetomorpha*, which consists of unbranched filaments of multinucleate cells with a highly reticulate plastid. (Photos: L. W. Wilcox)

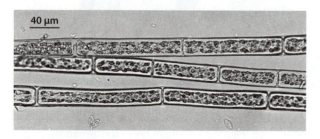

Figure 17.37 *Rhizoclonium* filaments, which have a coarse, wiry appearance and are sparsely branched, if at all. (Photo: L. W. Wilcox)

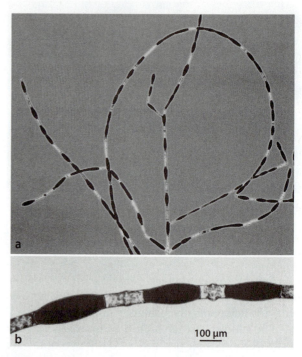

Figure 17.38 The branched filaments of *Pithophora*, seen here at two magnifications in (a) and (b). They characteristically exhibit alternation between green vegetative cells and dark green or black akinetes, which serve as resting cells and as a means of asexual reproduction. (Photos: L. W. Wilcox)

other molecular analyses suggested that *Cladophora* colonization of freshwaters from marine habitats has occurred at least twice (Hanyuda et al. 2002). Freshwater *Cladophora* commonly reproduces asexually by means of biflagellate zoospores that may represent unmated gametes capable of growth. Photoperiod is the primary factor involved in induction of zoosporogenesis in freshwater *C. glomerata*; asexual reproduction is maximal in short-day conditions. Growth of freshwater *C. glomerata* is primarily influenced by temperature, although day length, irradiance, and B-vitamin availability are also significant factors (Hoffmann and Graham 1984; Hoffmann 1990). Silicon occurs in the cell walls of *Cladophora glomerata* and is a required nutrient (Moore and Traquair 1976).

CHAETOMORPHA (Gr. *chaite*, long hair + Gr. *morphe*, form) is a marine genus that has an unbranched filament, is distributed worldwide, and grows either attached to rocks or shells by holdfast cells or is free floating (Figure 17.36). The reticulate chloroplast has many pyrenoids, and there are

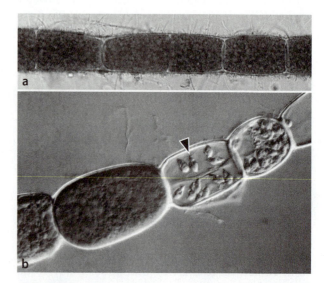

Figure 17.39 *Basicladia*, which may have a few branches at its base but is primarily unbranched (a). Some of the cells in (b) have formed zoospores (arrowhead). (a: L. W. Wilcox; b: D. Chappell)

multiple nuclei per cell, as is typical for the group. Isomorphic alternation of generations has been demonstrated for some species; typical quadriflagellate meiospores and biflagellate gametes are produced (Köhler 1956). Unmated gametes are able to regenerate the gametophytic phase of growth, producing new haploid filaments.

RHIZOCLONIUM (Gr. *rhiza*, root + Gr. *klonion*, small twig) is an unbranched or very sparsely branched filament (Figure 17.37) that is sufficiently coarse and wiry that individual filaments are readily visible with the unaided eye. Molecular phylogenetic analysis indicates that this is not a monophyletic genus (Hanyuda et al. 2002). *Rhizoclonium* often serves as a substrate for attachment of dense coatings of periphytic cyanobacteria and diatoms. *Rhizoclonium* grows in freshwaters, brackish habitats, and marine waters, either attached via lobed holdfast cells or free floating. Cells are considerably longer than they are broad and contain a reticulate plastid with many pyrenoids and many nuclei. Isomorphic alternation of generations has been demonstrated in some marine species (Bliding 1957). Freshwater forms occur in hard-water lakes or streams, entangled with other vegetation, and reproduce mainly by fragmentation or the rare production of biflagellate cells.

PITHOPHORA (Gr. *pithos*, earthen wine jar + Gr. *phoras*, bearing) is a robust, branched-filamentous form that occurs only in freshwaters. Cells are multinucleate and possess a reticulate plastid with pyrenoids. The genus is common in ponds and is found year-round in warm locales and during summer in temperate regions. There are characteristic large, dark akinetes interspersed with lighter-green vegetative cells. Akinetes form by the swelling of cell apices and subsequent formation of a septum (crosswall) that separates the akinete from the subtending cell. Because of this developmental pattern, akinetes alternate with vegetative cells in both the main axis and branches (Figure 17.38). The akinete cell walls are resistant to degradation and protect the cellular contents during periods of environmental stress. Akinetes germinate when growth conditions improve; germination was studied in culture by O'Neal and Lembi (1983). Because benthic sediments may contain very large numbers of akinetes, large nuisance-level populations may suddenly appear. These are undesirable because they form floating mats that clog boat motors and otherwise interfere with recreational use of inland waters (Lembi et al. 1980).

BASICLADIA (Gr. *basis*, base + Gr. *klados*, branch) (Figure 17.39) superficially appears to be an unbranched filament, but a few branches are produced

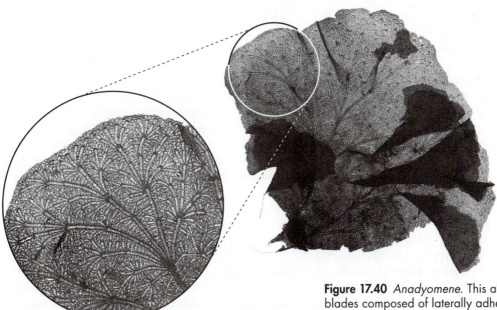

Figure 17.40 *Anadyomene.* This alga has fan-shaped blades composed of laterally adherent filaments, which form intricate patterns. (Photo: L. W. Wilcox)

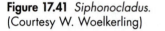

Figure 17.41 *Siphonocladus.*
(Courtesy W. Woelkerling)

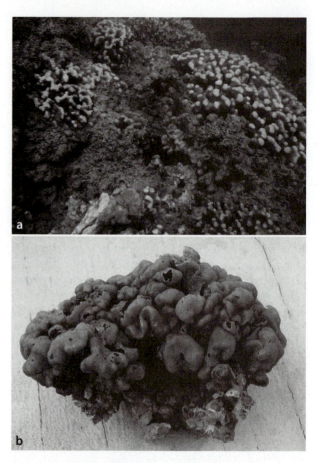

Figure 17.42 *Dictyosphaeria*, which occurs as hollow, gray-green spheres that often break open. (a) Underwater view of numerous thalli growing on coral reef. (b) Close-up view. (Photos: S. Larned)

at the base, just above the rhizoidal holdfast cells. This genus is often found attached to the backs of snapping turtles. Because the alga will attach to inorganic substrates when grown in the laboratory, the basis for its preferred substrate in nature is unknown. The multinucleate cells are long (approximately 50 times their width) and cylindrical, with thick lamellate walls, and plastids are parietal reticulate networks. Sexual reproduction is reported to occur by production of isogametes in unspecialized cells at the apex, and asexual reproduction by zoospores is also known (Prescott 1951). Like *Pithophora*, *Basicladia* is apparently restricted to freshwaters.

ANADYOMENE (named for the Greek goddess Aphrodite, when represented as arising from the sea) is one of the most beautifully formed seaweeds. The one-cell-thick, fan-shaped blades are composed of adherent branched filaments wherein the cells become progressively smaller toward the thallus periphery (Figure 17.40). Small cells also fill spaces between the large-celled filaments (known as veins). In some species the blades are perforate, but in others blades are entire. Colorless rhizoidal filaments at the base attach the blades to substrates. *Anadyomene* can be found attached to rocks on tropical coral reefs and has been dredged from deeper tropical waters. A phylogenetic analysis based on morphological characters suggested that *Anadyomene* is a monophyletic genus (Littler and Littler 1991).

SIPHONOCLADUS (Gr. *siphon*, tube + Gr. *klados*, branch), which grows attached to substrates in tropical marine waters, begins as a saccate multinucleate cell, a coenocyte. At some point, the cytoplasm cleaves internally and simultaneously to form individual walled units, the process of segregative cell division. At first the subunits are spherical, but later each protrudes beyond the confines of the original sac, forming an elongated branch. This form of branching can be repeated to form several orders of branches. The resulting body appears as a central axis that is densely tufted (Figure 17.41).

DICTYOSPHAERIA (Gr. *dictyon*, net + Gr. *sphaira*, ball) grows attached to substrates such as coral heads in shallow tropical waters; rhizoids serve an attachment function. It begins development as small balloon-shaped cells having a large central vacuole.

The peripheral, multinucleate cytoplasm cleaves into numerous small, walled cells. As these enlarge, they contact their neighbors, forming a tough layer at the periphery of the spherical bodies. Older individuals may crack open, forming pebbly, pale-green sheets (Figure 17.42). *Dictyosphaeria cavernosa* competes with corals on patch and fringing reefs and can form large growths when grazing is reduced due to overfishing and human-generated nutrient inputs (Stimson et al. 1996).

VALONIA (Venetian vernacular name) is a balloon-shaped vesicular form that is attached to substrates in tropical waters (Figure 17.43). Similarly to *Dictyosphaeria*, *Valonia* undergoes segregative cell division at the periphery such that many small cell units are formed just beneath the parental wall. However, these small cells are not as noticeable as those of *Dictyosphaeria*, giving the false impression that *Valonia* is composed of a single marble-sized cell. *Valonia* is typically brighter green in color than is *Dictyosphaeria*. Depending upon the species, *Valonia* can occur as single vesicles or aggregations of vesicles.

Figure 17.43 *Valonia*, which forms bright-green, balloon-shaped vesicles on coral reefs. (Photo: R. J. Stephenson)

Caulerpales

Named for the genus *Caulerpa*, the order Caulerpales includes about 26 genera and 350 species of multinucleate, siphonous ulvophyceans (Floyd and O'Kelly 1990). The group has also been referred to as Bryopsidaceae (van den Hoek et al. 1995), Bryopsidales (Vroom et al. 1998), Codiales, or Siphonales. The vast majority of caulerpaleans are marine, primarily occurring in warm tropical or subtropical waters of normal salinity. They may be easily observed in shallow waters—in sea grass beds or attached to coral heads—although some occur in much deeper waters. The widespread, and sometimes nuisance, genus *Codium* commonly extends into temperate coastal waters, and *Dichotomosiphon* occurs in sandy areas at the edges of freshwater lakes. Caulerpalean taxa have in common a body that is coenocytic, lacking cross-walls except at sites of reproductive cell formation.

Uniaxial versus multiaxial body structure of caulerpaleans

Caulerpalean bodies may be uniaxial or multiaxial. Uniaxial forms are composed of a single branched siphon and are consequently rather delicate. In contrast, multiaxial forms are more robust. Multiaxial bodies arise from extensive early dichotomization of a single siphon into multiple, branchlike, siphonous proliferations (Figure 17.44). Each siphonous proliferation then enlarges and undergoes further dichotomization. The mature thallus is thus composed of multiple, aggregated siphons having a common developmental origin. Unless they become calcified, multiaxial forms have a robust and spongy texture. The peripheral portions of the siphons may be inflated and aggregated to form a coherent outer surface. Such inflated areas are known as **utricles**, and reproductive structures originate from them. Multiaxial forms can be differentiated into an internal colorless region known as the medulla and a green photosynthetic surface region. In some cases, chloroplasts are withdrawn from the surfaces into the medulla during the night and then redeployed to the surface at daybreak. This behavior is regarded as a means of protecting chloroplasts from herbivores.

Some multiaxial forms precipitate calcium carbonate on the siphon surfaces, including those of the medulla. The function of such carbonate precipitation is not known, but some researchers have suggested that it may be a method of acquiring carbon dioxide (see Chapter 2) or that calcification might retard herbivory. When fish consume calcified algae, the carbonate reacts with stomach acid to produce unpleasant amounts of carbon dioxide gas (Hay et al. 1994). Among Caulerpales, *Halimeda* is distinctive

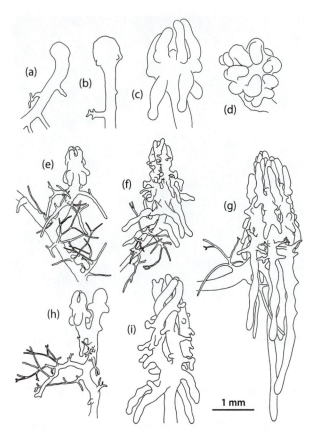

Figure 17.44 Development of a multiaxial thallus from a uniaxial germling. (From Friedmann and Roth 1977)

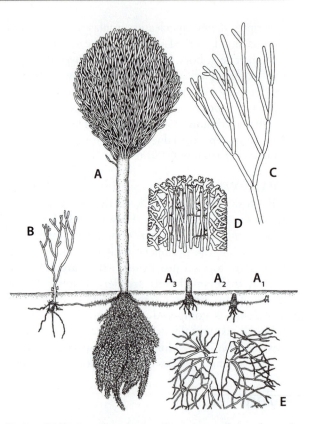

Figure 17.45 Development and anatomy of a multiaxial caulerpalean, *Penicillus*. Subsurface siphonous runners interconnect thalli of differing developmental ages (A_1–A). Young thalli are uniaxial (B, C) but rapidly undergo dichotomous branching to form numerous aerial filaments that interweave to form a thick thallus (D). The subsurface siphonous system ramifies into an extensive holdfast (E). (From Friedmann and Roth 1977)

in having calcified segments that alternate with uncalcified joints, thereby conferring flexibility helpful in avoiding damage from wave or current action.

Both uniaxial and multiaxial forms attach themselves to the substrate, typically sand, by colorless siphonous branches called rhizoids. The rhizoid aggregates of sand-dwelling forms bind sand particles to form a massive holdfast region (Figure 17.45).

Wound healing in caulerpaleans

Although each composed of but a single cell, even uniaxial caulerpaleans can be quite large, even several meters in length. Such bodies are vulnerable to extensive loss of cytoplasm in the event of cell wall damage. Consequently, wound healing is extremely efficient, occurring within a matter of seconds by actin-mediated contraction of the protoplast and construction of a plug of cell wall material to seal the wound. In *Caulerpa taxifolia*, immediately after wounding, an esterase transforms a terpenoid known

as caulerpenyne into oxytoxin 2, an extremely reactive 1,4-dialdehyde (Figure 17.46). This compound rapidly cross-links cellular proteins to form the wound plug (Adolph et al. 2005). The caulerpenyne also functions in chemical defense against herbivores, and by cross-linking cellular proteins, the wound response itself may contribute to herbivore defense by reducing herbivore access to nutrients (Weissflog et al. 2008).

Cell wall chemistry and plastids of caulerpaleans

In some taxa (or particular developmental or life-history stages), cell walls may be composed primarily of noncellulosic polysaccharides, namely mannans, xylans, or xyloglucans. The adaptive

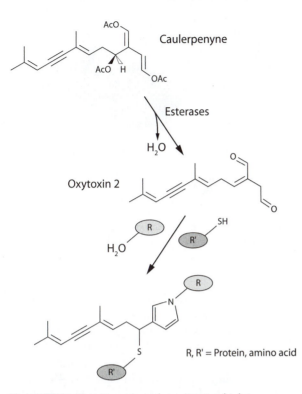

Caulerpenyne

Esterases

H_2O

Oxytoxin 2

H_2O

SH

R, R' = Protein, amino acid

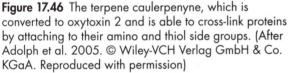

Figure 17.46 The terpene caulerpenyne, which is converted to oxytoxin 2 and is able to cross-link proteins by attaching to their amino and thiol side groups. (After Adolph et al. 2005. © Wiley-VCH Verlag GmbH & Co. KGaA. Reproduced with permission)

advantage of these variations in cell wall chemistry is not understood.

Chloroplasts are discoid and numerous. Pyrenoids may be either present or lacking. In forms such as *Udotea* and *Caulerpa*, there may be two different populations of plastids, some pigmented and others lacking pigments but containing starch. These unpigmented, starch-rich plastids are referred to as amyloplasts by analogy to those of land plants. When both pigmented and unpigmented plastids are present, the algae are said to exhibit **heteroplastidy**. The accessory xanthophyll pigments siphonein and siphonoxanthin may be present.

Caulerpalean reproduction

The freshwater genus *Dichotomosiphon* is unusual in being oogamous, producing larger eggs and smaller sperm. Reproduction typically involves sexual fusion of biflagellate anisogametes; the larger gametes are usually brighter green than the smaller ones. In some genera, gamete production and release uses up all of the coenocyte's cytoplasm, a phenomenon known as **holocarpy**. In other genera, gametes are produced in regions separated from the rest of the cytoplasm by a wall or septa so that not all of the cytoplasm is used up during reproduction. Such genera are non-holocarpic.

Zoospores are rarely produced by caulerpaleans, but the uniaxial genus *Ostreobium* is exceptional in releasing quadriflagellate zoospores similar to those generated by other ulvophyceans. Other cases of zoospores are the unusual multinucleate, stephanokont (bearing a ring of flagella) zoospores generated by *Derbesia*, *Bryopsis*, and *Bryopsidella* (Calderon-Saenz and Schnetter 1989).

In most cases, the life history of caulerpaleans is poorly understood, and little information on the timing and location of meiosis is available. In the past, experts thought that most caulerpaleans had diploid vegetative thalli that underwent meiosis in gametangia, but this concept is now being challenged. Zygotes, though sometimes long-lived, do not serve as resting stages, and in some cases are thought (though not proven) to undergo meiosis prior to the formation of the macroscopic vegetative thallus (van den Hoek et al. 1995). Caulerpalean zygotes typically have very large nuclei.

Relationships and diversity of caulerpaleans

A cladistic analysis of the Caulerpales based on morphological, reproductive, and ultrastructural features suggested that it is monophyletic, that heteroplastidy and multiaxial thalli are derived, and that *Codium* is an early-divergent form (Vroom et al. 1998). Based on absence or presence of heteroplastidy, absence or presence of holocarpy, and geography, Hillis-Colinvaux (1984) recommended establishment of the two suborders Bryopsidineae and Halimedineae. More recent phylogenetic analyses of rubisco large subunit gene sequences support the concept that these suborders are monophyletic lineages (Lam and Zechman 2006).

Bryopsidineae. BRYOPSIS (Gr. *bryon*, moss + Gr. *opsis*, likeness) (Figure 17.47) is uniaxial. A prostrate portion of the coenocyte extends horizontally over substrates and gives rise to erect systems of main axes having many long, branched, featherlike divergences known as laterals or pinnae. Gametangia develop from the laterals, and anisogametes are discharged into the water. Planozygotes resulting from fertilization attach to

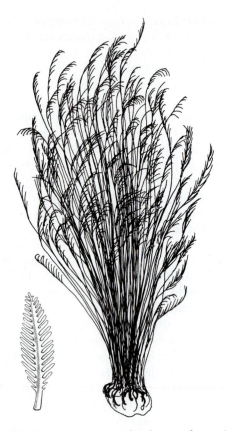

Figure 17.47 *Bryopsis*, a uniaxial siphonous form whose erect parts appear feathery. (Courtesy W. Woelkerling)

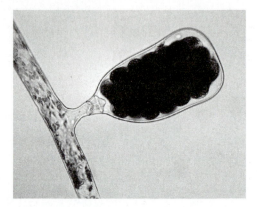

Figure 17.48 Portion of a *Derbesia* siphon bearing a sporangium with numerous round, dark spores. (Photo: L. E. Graham)

substrates and grow into tiny filaments that each contain a single large nucleus. Nuclear division then occurs, forming many stephanokont zoospores that regenerate the vegetative thalli (van den Hoek et al. 1995). *Bryopsis*-like phases may be part of the life cycles of other genera (Bold and Wynne 1985). Microspectrophotometric measurements of nuclear DNA levels in *Bryopsis hypnoides* suggest that gametogenesis does not involve meiosis (Kapraun and Shipley 1990). *Bryopsis* occurs in both temperate and tropical seas. At least one species has been renamed *Bryopsidella* (Calderon-Saenz and Schnetter 1989).

DERBESIA (named for August Alphonse Derbes, a French phycologist) is a uniaxial, branched 1–10 cm coenocyte with a single plastid type. Only during development of saclike sporangia are septa formed (Figure 17.48). A thin layer of cytoplasm surrounds a large internal vacuole. In 1938, Kornmann observed in cultures that *Derbesia* is the sporophytic stage

of another marine algal genus, the small balloon-shaped genus *Halicystis* (Gr. *hals*, sea + Gr. *kystis*, bladder). *Halicystis* has a thin cytoplasmic layer and a large central vacuole and grows attached to substrates such as crustose red algae. The life cycle of *Derbesia-Halicystis* is shown in Figure 17.49.

Halicystis bodies are of two types: One produces small, pale gametes, and the other generates larger, dark green gametes. *Derbesia* is the sporophytic stage in an alternation of heteromorphic generations. The stephanokont zoospores produced by *Derbesia* are multinucleate but contain only one nuclear type. They are produced in and discharged from specialized sporangia separated from the rest of the thallus by a wall (septum). Thus, the body is non-holocarpic and able to persist after release of reproductive cells. Zoospores grow into the vesiculate *Halicystis* stage (the gametophyte). Unmated gametes of the larger type can germinate directly into *Derbesia*-like bodies, but these are haploid.

CODIUM (Gr. *kodion*, small sheepskin) includes numerous types of dark-green, spongy, multiaxial thalli up to 1 m in length. It grows attached to substrates such as rocks or shells in temperate or tropical regions. There are about 120 species, and their external morphology can be quite variable, from flat, crustlike forms to spheres and the more common dichotomously branched forms (Figure 17.50). Highly interwoven siphons are expanded at the surface to form utricles whose form is important in distinguishing the species. Numerous discoid chloroplasts lacking pyrenoids are present in utricles; there are no amyloplasts. Internal

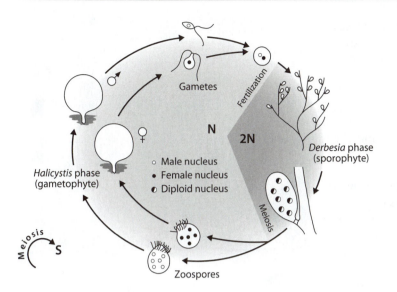

Figure 17.49 Diagram of the life history of *Derbesia-Halicystis*. The *Derbesia* phase produces stephanokont zoospores that grow into the *Halicystis* phase. Anisogametes are explosively discharged from the *Halicystis* form; their fusion leads to development of the *Derbesia* phase. (After Eckhardt et al. 1986)

medullary portions of the thallus are colorless. Cell walls do not contain cellulose but rather mannans or sulfated arabinogalactans.

Gametangia develop as branches from the utricles; they may be dark in color or pale because they contain two different types of biflagellate anisogametes (as in *Derbesia-Halicystis*). Gametangia are separated from the rest of the thallus by cross-walls (septa); thus the body is non-holocarpic and can persist after gamete discharge. Gametes are thought to arise by meiosis (Vroom et al. 1998).

Microspectrophotometric studies of DNA content in the life cycle of three *Codium* species were performed by Kapraun, et al. (1988). No stephanokont zoospores are produced. However, as in *Derbesia-Halicystis*, unmated gametes of the larger, darker type can germinate into vegetative thalli; this may be viewed as a form of asexual reproduction. The *Codium fragile* subspecies *tomentosoides* is an invasive alga that has spread from its native Japan through Europe, both the east and west coasts of the United States, and into South America, South

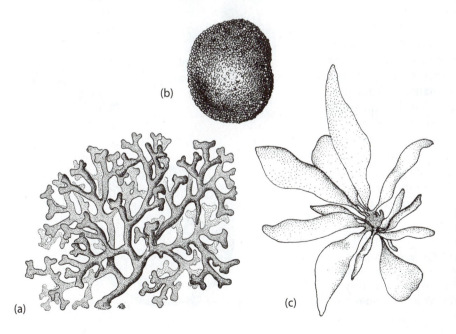

Figure 17.50 Diagrammatic views of several species of *Codium*, illustrating some of the known diversity of morphological types. (a: From Taylor 1972, © The University of Michigan Press 1972; b and c: From Bold and Wynne 1985)

Africa, Australia, and New Zealand. It displaces native species and can overgrow and smother oysters in aquaculture operations. Provan et al. (2005) used plastid microsatellite markers and DNA sequence data to trace the history of invasions, finding evidence for at least two independent invasions, one localized to the Mediterranean and the other to the Northwest Atlantic, northern Europe, and the South Pacific.

Halimedineae. *CAULERPA* (Gr. *kaulos*, stem + Gr. *herpo*, to creep) is a uniaxial siphonous form that occurs as a series of erect, highly differentiated photosynthetic shoots as well as anchoring rhizoids, both of which emerge from a horizontal siphonous axis. Biologists have been impressed by *Caulerpa*'s plantlike organization since it was first described about 150 years ago; sometimes they refer to *Caulerpa*'s organs as "roots, stem, and leaves" (Jacobs 1994), though of course the positioning of these structures is merely analogous, rather than homologous, to that of terrestrial plants. The horizontal system may extend for meters along the surface of sea grass beds or other substrates in tropical or subtropical waters. New shoots and rhizoids are produced from the growing end of the horizontal system. If the bodies are experimentally inverted, the next-produced rhizoids and green shoots will occur on the opposite sides of the thallus from their older peers, suggesting the occurrence of gravity detection during morphogenesis (Jacobs 1993).

There are about 70 species, differentiated by shoot-structure differences. Shoots can be leaflike or have radiately or bilaterally symmetrical laterals; laterals can resemble the pinnae of feathers or be spherical, resembling bunches of grapes (Figure 17.51). Ingrowths of wall material, known as **trabeculae**, extend across the siphon lumen, probably providing structural support. The cell wall is composed of a polymer of xylose, as are those of *Halimeda* and *Udotea*. Amyloplasts as well as green discoid plastids are present. Reproduction occurs by liberation of the entire protoplast in the form of numerous biflagellate anisogametes. *Caulerpa* is thus

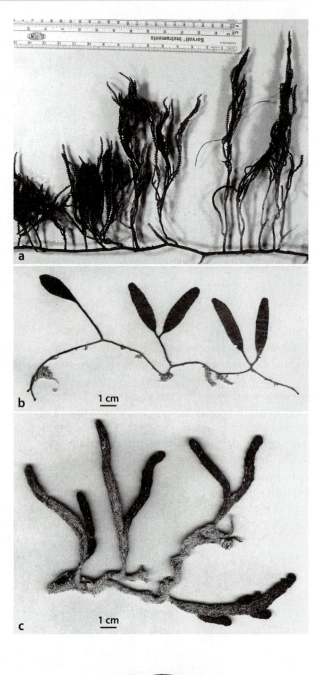

Figure 17.51 (a–c) Three species of *Caulerpa*, illustrating some of the diversity of photosynthetic shoots within this genus. (d) Drawing of thallus in cross section, showing trabeculae. (a: C. Taylor; b,c: L. W. Wilcox; d: from W. Woelkerling)

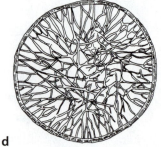

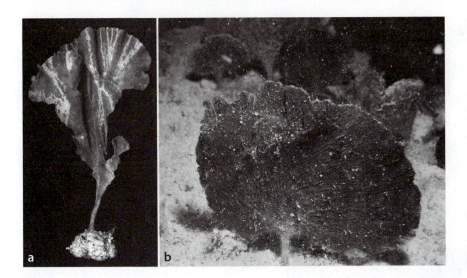

Figure 17.52 *Udotea*, a fan-shaped inhabitant of tropical sea grass beds. (a: L. W. Wilcox; b: R. J. Stephenson)

holocarpic. Gametes of thousands of individuals are released simultaneously, a case of "mass spawning" (Clifton 1997). Zygotes generate single-celled uninucleate "protospheres" from which new vegetative thalli arise. The center of diversity is the southern Australian coast, where the genus is thought to have originated; species relationships have been studied by nuclear rDNA ITS sequences (Pillmann et al. 1997).

An invasive strain of *C. taxifolia* that escaped from the Monaco Oceanographic Museum in 1984 now dominates large nearshore underwater habitats of six Mediterranean countries. Subsequently, this strain was identified in waters off southern California, presumably having been dumped there from aquaria. An intensive control effort was instituted, with the goal of eliminating the pest before it could spread. Patches of algal growth were covered with tarps weighted by sandbags, and then toxic chlorine gas was pumped beneath the tarps to kill the seaweed. Accidental introductions remain a threat to natural ecosystems, with the result that experts recommend a complete ban on the sale of *Caulerpa* for use in aquaria. Reasons for such a ban include difficulty in accurately identifying *C. taxifolia* unless DNA methods are used and the unknown potential for other species to exhibit invasive behavior (Stam et al. 2006).

UDOTEA (named for a sea nymph in Greek mythology) occurs as fan- or funnel-shaped multiaxial aggregates of siphonous branches (Figure 17.52). There is typically a holdfast consisting of aggregated colorless siphonous branches. The cell wall is composed of xylans. Calcium carbonate is deposited on the outside of the thallus and in the spaces between siphons. Cells at the periphery of the fan or funnel may be partitioned off as gametangia in some species, but in other forms the entire protoplast is converted into biflagellate anisogametes, and upon their discharge the thallus becomes white and dies (holocarpy) (Meinesz 1980). Zygotes develop into a nonmotile, unicellular "protosphere" stage in which there is but a single large nucleus. This is regarded as the only diploid stage in the life cycle, and meiosis is surmised to occur prior to development of the macroscopic vegetative thallus. However, this hypothesis requires testing by microspectrophotometric measurements of relative DNA levels. *Udotea* is common in sandy sea grass beds in tropical nearshore waters and lagoons. Molecular analysis of SSU rDNA sequences suggests that *Udotea* is not a monophyletic genus (Kooistra 2002).

PENICILLUS (L. *penicillus*, painter's brush) is a multiaxial form whose dichotomously branched siphons are aggregated at the base into a stipe but occur in a free, brushlike arrangement at the apex (Figure 17.53). Thus the thallus looks like a handled brush and is commonly known as "Neptune's shaving brush." There is a bulbous holdfast much like that of *Udotea*. The thallus is lightly encrusted with calcium carbonate. A series of *Penicillus* thalli of different ages can be generated from a horizontal "runner," as in *Caulerpa*. Sexual reproduction resembles that of *Udotea*. In at least one well-studied

Figure 17.53 *Penicillus*, which has a brushlike array of photosynthetic branches emerging from a stipe having a massive holdfast. (Photo: L. W. Wilcox)

Figure 17.54 *Halimeda*, which consists of flattened, calcified segments interspersed with flexible joints. (Photo: J. Hackney)

form, gamete production involves the entire cytoplasm, such that reproduction results in the death of the vegetative body (Friedmann and Roth 1977). The results of a molecular analysis suggest that *Penicillus* and the similar genus *Rhipocephalus* (see Figure 17.6) are actually forms of *Udotea* that have evolved by loss of siphon adherence (Kooistra et al. 2002).

HALIMEDA (named for a sea nymph in Greek mythology) has a flattened, fan-shaped thallus composed of dichotomously branching, segmented erect portions and a large holdfast composed of aggregated rhizoidal siphons (Figure 17.54) with extensive calcification of the shoot segments but not joints. Sexual reproduction involves conversion of the entire biomass into anisogamous gametes whose discharge from cells at the periphery leaves the thallus empty, whereupon it dies. *Halimeda* is widespread in warm tropical or subtropical waters, growing as deep as 50 m. In part because it can produce several crops per year, as noted earlier, *Halimeda* is a major contributor to carbonate sediments. The genus has been monographed by Hillis-Colinvaux (1980).

Dasycladales

A unique molecular deletion present in representatives suggests that the order Dasycladales, which consists of 11 living genera with 38 species, is monophyletic (Olsen et al. 1994), and pa *rbcL*-based phylogeny supports this hypothesis (Zechman 2003). Dasycladaleans are characterized by siphonous vegetative thalli (Floyd and O'Kelly 1990, Berger and Kaever 1992). Some authorities have elevated this group to class status (Dasycladophyceae) (van den Hoek et al. 1995). Two families have been defined; an SSU rDNA–based phylogeny and extensive structural analysis has been conducted of one of these families (Polyphysaceae) (Berger et al. 2003).

Some forms are uninucleate until reproductive development begins, but others appear to have multinucleate vegetative cells. There is typically a basal holdfast region consisting of rhizoidal branches in which the diploid or polyploid nucleus resides, an elongate erect axis, and lateral branches occurring in whorls (Figure 17.55). In *Acetabularia*, adherent reproductive laterals form distinctive caps. Based on comparisons of the ultrastructure of flagellate cells, Floyd et al. (1985) suggested that Dasycladales were closely related to Cladophorales/Siphonocladales. In contrast, some molecular data indicate that Caulerpales are sister to Dasycladales (see Lewis and McCourt 2004).

The single-celled thalli possess a thin layer of cytoplasm with numerous plastids, surrounding a large central vacuole. As in all other green algae, starch

Figure 17.55 The dasycladalean genus *Dasycladus* (for which the order is named). *Dasycladus* exhibits whorled branching typical of the group. These thalli are attached to carbonate sand grains by a system of rhizoids. (Photo: R. J. Stephenson)

occurs in dasycladalean plastids, but surprisingly, cytoplasmic polysaccharide reserves are also reported. The cell walls are composed of a mannose polymer, but cellulose occurs in the walls of reproductive structures. Many forms are encrusted with calcium carbonate; calcification is reduced under conditions of reduced irradiance or temperature (Berger and Kaever 1992).

Reproduction is by biflagellate gametes produced in small spherical, walled cysts formed in the reproductive laterals. Zygotes possess a single large primary nucleus and develop into vegetative thalli. Spectrophotometric measurements of nuclei in *Acetabularia* and *Batophora* indicate that the zygote undergoes meiosis sometime prior to the onset of reproductive development (Koop 1979; Liddle et al. 1976). Subsequent repeated mitotic divisions produce hundreds of small secondary nuclei that are incorporated into cyst cytoplasm and eventually partitioned into gametes. Cyst formation requires the arrival into reproductive laterals of secondary nuclei. A cage of microtubules surrounds each nucleus and rings of actin filaments are involved in the cleavage of cytoplasm to form gametes (Menzel et al. 1996). Microtubules also specify the location of a circular region of the cyst wall that will become the lid or operculum. This lid opens during gamete discharge (Berger and Kaever 1992). Gamete discharge thus does not result in the demise of the entire thallus, as occurs in many caulerpaleans.

Dasycladaleans, especially the genus *Acetabularia*, are extensively used for laboratory study of cell and developmental biology (Mandoli 1998).

Acetabularia's utility is based upon ease of cultivation, large cells (up to 200 mm in length), and high tolerance to surgical manipulation, thanks to rapid wound-healing processes. Cells can be enucleated by removing the rhizoids with scissors. Nuclei can be isolated and then implanted into enucleated cells. Basal, nucleated rhizoidal regions from which the rest of the thallus has been removed are capable of regenerating the entire thallus. Based on the results of nuclear transplantation experiments conducted in the 1930s, Hämmerling postulated the occurrence of messenger RNA prior to its chemical characterization. More recent experiments have demonstrated the role of localization of cell membrane calcium-ion binding sites in morphogenesis; this determines the pattern of whorl development (Berger and Kaever 1992). An easily prepared, highly effective growth medium has been developed for *Acetabularia* (Hunt and Mandoli 1996).

Dasycladaleans occur in shallow (to 10 m) waters of tropical or subtropical shores in protected areas. Dasycladaleans are distributed within a broad belt from 25° N and 35° S but can occur as far north as 32° in Bermudan waters and to 40° S on the west coast of South Africa. They can frequently be found on mangrove roots. Tolerance of most dasycladaleans to salinity variation is generally low. Thus dasycladaleans commonly occur in open lagoons but not generally closed ones where salinities can vary because of evaporation or rainfall. However, the genus *Batophora* is more tolerant of salinity changes than are other dasycladaleans (Berger and Kaever 1992).

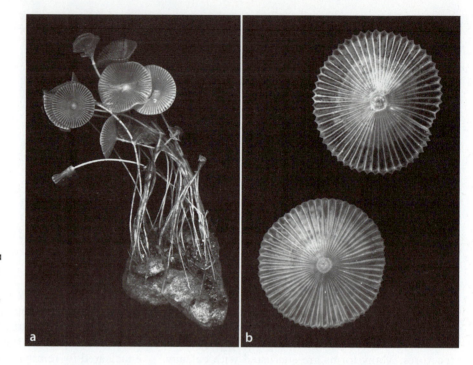

Figure 17.56 *Acetabularia.* Thalli often grow as clumps of individuals, attached to hard substrates such as coral rubble. (a) The unicells have a delicate holdfast, an elongate axis, and a terminal cap, wherein reproductive cells are generated. (b) Close-up view of two caps. (Photos: L. W. Wilcox)

ACETABULARIA (L. *acetabulum*, vinegar cup) (Figure 17.56a), commonly known as the "mermaid's wine glass," typically grows in large clumps, sometimes forming lawns. The unbranched, cylindrical stalk is anchored to substrates by a holdfast of rhizoidal branches. When thalli reach about 1 mm in length, the first whorls of lateral branches develop at the growing tip; these then degenerate, leaving a ring of scars on the stalk. The cap (Figure 17.56b) is formed of adherent branches known as cap rays, which are **gametophores** (gametangia-/cyst-containing structures). After development of the cap rays, protoplasm containing numerous secondary nuclei then streams into the gametophores, imparting a dark-green color. A septum divides each gametophore from the rest of the thallus, and cyst walls develop around each cytoplasmic unit. When the cysts are mature, they are released from gametophores. Later, gametes are released from cysts as they press on the operculum, forcing it open. The biflagellate gametes possess eyespots and are phototactic. Gametes can swim for up to 24 hours and mate by agglutination of flagellar tips. Zygotes are negatively phototactic and thus settle onto a substrate, lose their flagella, and generate a cylindrical structure. Rhizoids develop from one end and penetrate the substrate, whereas the other end becomes tapered and elongates from the

Figure 17.57 *Batophora.* Main axes of this alga bear whorls of lateral branches. (Photo: L. W. Wilcox)

tip, beginning development of the next generation of vegetative thalli. The mechanics of the cytoskeleton during morphogenesis in *Acetabularia* are discussed by Goodwin and Briére (1994), and aspects

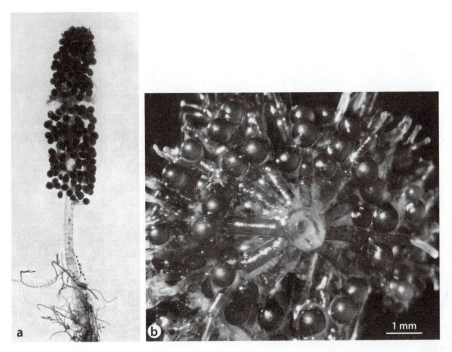

Figure 17.58 Gametophores of *Batophora*. (a) Longitudinal view of thallus. (b) Cross-sectional view showing the gametophores, which are borne on lateral branches. (Photos: L. W. Wilcox)

of morphogenesis are also reviewed by Mandoli (1998) and Kratz et al. (1998). There are about eight species (Berger and Kaever 1992), but the genus *Acetabularia* is not monophyletic (Zechman 2003).

BATOPHORA (Gr. *batos*, bramble + Gr. *phoras*, bearing) has many whorls of repeatedly branched laterals on vertical axes that do not themselves branch (Figure 17.57). Gametophores are produced at nodes of the laterals (Figure 17.58). Thalli are not calcified. *Batophora* occurs in shallow waters of lagoons and has been found in waters of low salinity (Berger and Kaever 1992).

CYMOPOLIA (named for a daughter of Neptune in Roman mythology) has a repeatedly and dichotomously branched construction (Figure 17.59). Branching occurs in a single plane. Calcified segments alternate with uncalcified joints, a pattern that confers flexibility to the thallus. Tufts of uncalcified laterals occur at the growing tips. Gametangia are produced individually at the tips of primary lateral branches surrounded by inflated secondary laterals in uncalcified apical regions of the thallus. Gametangia are not operculate and are produced continuously throughout the life of the thallus, in contrast to those

Figure 17.59 Underwater photograph of the ulvophycean *Cymopolia* growing on a coral reef off San Salvador, Bahamas. Note the white calcified segments separated by a more flexible uncalcified joint. The hazy clouds at branch tips are many thin green filaments that function in photosynthesis. (Photo: R. J. Stephenson)

of *Acetabularia* and *Batophora* (Berger and Kaever 1992). According to an *rbcL*-based phylogenetic analysis, *Cymopolia* is the earliest-diverging modern member of the Dasycladales (Zechman 2003).

18 Chapter

Green Algae III

Trebouxiophyceans

Trebouxiophyceans occur most commonly in freshwater or terrestrial habitats. They include some familiar and widespread genera that are biogeochemically, technologically, or biotically significant. For example, *Botryococcus* is a common member of the freshwater phytoplankton whose ability to produce large amounts of lipid is unusual among the green algae. The lipid functions as a flotation device, which allows these relatively large colonies to remain suspended in well-illuminated surface waters. Fossil remains similar to modern *Botryococcus* indicate that this alga has been the source of significant fossil fuel deposits for hundreds of millions of years. Today, *Botryococcus* is being utilized in biotechnological efforts to produce renewable oil supplies.

Several genera of trebouxiophyceans function as photosynthetic symbionts in lichens. *Trebouxia*, for which the class is named, is the single-most-common algal component of lichens. Even so, *Trebouxia* occurs in many genotypes that lichen fungi readily trade, possibly as a way of adapting to changing conditions. In contrast to most trebouxiophytes, the genus *Prototheca* is colorless and obligately heterotrophic. It lives in soil, where it survives by absorbing organic compounds, but it can also infect dairy cattle, causing an incurable disease. Humans

Botryococcus

(Photo: M. E. Cook)

sometimes suffer *Prototheca* infections of the skin, which may require surgical treatment. *Prototheca* is related to what may be the strangest of all the algae, the parasitic genus *Helicosporidium*. This alga disperses as tough, drum-shaped spores that open only after entering the guts of insects and other invertebrate animals. In this location, *Helicosporidium* unfurls a barbed needle that stabs into gut cells, infecting the animal with parasitic algal cells that ultimately kill the animals. These examples illustrate that trebouxiophyceans can be more exciting than you might at first expect. This chapter surveys the general features of Trebouxiophyceae and provides examples of diversity.

18.1 General Features of Trebouxiophyceae

Trebouxiophyceans occur as non-flagellate unicells or colonies, unbranched or branched filaments, or small blades similar to those found among ulvophyceans. Trebouxiophyceans commonly produce asexual spores in the form of non-flagellate autospores or flagellate zoospores. Sexual reproduction involving flagellate sperm and nonmotile eggs is known for some representatives.

A unique combination of plesiomorphic ultrastructural features defines this class: counterclockwise-oriented flagellar basal bodies (see Figure 16.7c), nonpersistent metacentric spindles (see Figure 16.8c), and presence of a phycoplast at cytokinesis (Figure 16.10c). This set of traits has been used to classify species within Trebouxiophyceae, an example being the filamentous soil alga *Jaagiella alpicola* (Porta and Hernandez-Marine 1998). Trebouxiophyceans share basal body orientation with ulvophyceans, metacentric spindles with walled prasinophytes, and non-persistent spindles and phycoplasts with chlorophyceans.

In most cases, membership in this class is determined by molecular phylogenetic studies (e.g., Friedl 1995, 1996; Tartar et al. 2002; Ueno et al. 2003; Krienitz et al. 2004; Senousy et al. 2004; Henley et al. 2004; Karsten et al. 2005; Rindi et al. 2007). Molecular data are necessary to evaluate the relationships of green algae that do not produce flagellate cells. Trebouxiophyceans are commonly regarded as the sister group to chlorophyceans (e.g., Lewis and McCourt 2004), but not all molecular

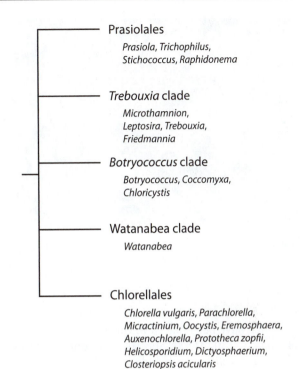

Figure 18.1 Major clades of Trebouxiophyceae. Relationships among these clades are not yet clearly resolved. (Based on Tartar et al. 2002, Ueno et al. 2003, Krienitz et al. 2004, Senousy et al. 2004, Henley et al. 2004, Karsten et al. 2005, and Rindi et al. 2007)

phylogenies support this concept. Unique structural or reproductive features are not known, and molecular phylogenetic studies vary in level of support for monophyly of trebouxiophyceans. Such analyses suggest that at least five trebouxiophycean lineages exist (Figure 18.1), and the following survey of diversity has been organized accordingly.

Diversity of Trebouxiophyceans

Chlorellales

CHLORELLA (Gr. *chloros*, green) (Figure 18.2) occurs as small spherical (coccoid) unicells that reproduce by autospores. Sexual reproduction is not known. Molecular phylogenetic studies and biochemistry of cell walls (glucose plus mannose vs. glucosamine) (Takeda 1991) reveal that this genus is polyphyletic, and some species have been renamed as genera such as *Auxenochlorella* (Friedl 1995) and *Parachlorella* (Krienitz et al. 2004). Ellipsoidal species having walls of a single smooth layer and lacking

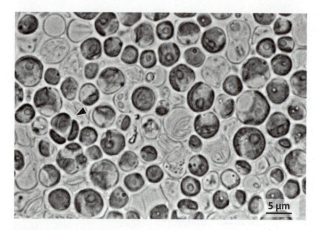

Figure 18.2 *Chlorella.* There is a single cup-shaped parietal plastid that possesses a pyrenoid. The arrowhead points to a cell that has cleaved into autospores. (Photo: L. W. Wilcox)

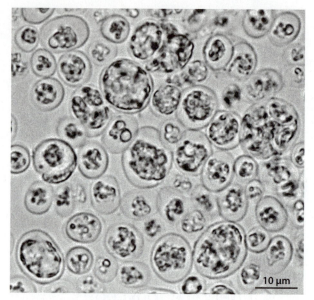

Figure 18.3 *Prototheca.* Although a plastid is present, it is not pigmented and therefore is difficult to identify unless the starch contained within is stained with an I₂KI solution. (Photo: L. W. Wilcox)

pyrenoids have been placed in the genus *Watanabea* (Hanagata et al. 1998). Species that continue to be classified as *Chlorella* include *C. vulgaris, C. lobophora,* the polyphyletic *C. sorokiniana* (Krienitz et al. 2004), and *C. kessleri* (Ustinova et al. 2001). Analysis of many *C. vulgaris* isolates revealed a surprising level of variation, indicating a level of cryptic speciation that should be taken into account when reporting experiments performed with this "species" (Müller et al. 2005).

PROTOTHECA (Gr. *protos,* first + Gr. *theke,* sheath or box) (Figure 18.3) is a spherical unicell that reproduces via autospores. It is an obligate osmotroph that descended from photosynthetic, chlorella-like ancestors. The genus *Prototheca* may not form a monophyletic group; *P. wickerhamii* is closely related to the extant autotrophic *Auxenochlorella protothecoides,* and that clade is sister to a clade containing several other *Prototheca* species and *Helicosporidium* (Ueno et al. 2003). Starch-containing plastids are present and can be used to differentiate *Prototheca* from yeasts, with which it can be confused in medical contexts. Borza et al. (2005) studied the nuclear-encoded plastid-targeted proteins of *Prototheca wickerhamii,* finding evidence that plastids conduct carbohydrate, amino acid, lipid, tetrapyrrole and isoprenoid metabolism and synthesize purine. This alga is capable of growing with adenine as the sole nitrogen source (Sarcina and Casselton 1995). The outer walls, which have a three-layered structure, contain a hydrolysis-resistant, UV-fluorescent

compound whose infrared spectrum differs from that of sporopollenin. Presence of this wall material is suspected to contribute to drug resistance (Puel et al. 1987).

Prototheca occurs in soils and freshwater environments, particularly sewage-contaminated waters and decaying manure. It has also been isolated from diverse locations in cattle housing and dairies. Cattle acquire infections through mammary glands, and milk from infected cattle can be unacceptable for consumption. Cattle from which *Prototheca* is cultured are typically culled because the algal infection is not susceptible to commonly used antibiotics. Humans may acquire skin infections with the formation of nodules that may have to be surgically removed. *Prototheca* has also been associated with tree pathology (Pore et al. 2003; Pore 1986).

HELICOSPORIDIUM (name referring to the helical coiling of filaments released from spores) occurs primarily as a drum-shaped cyst (Figure 18.4) enclosing three ovoid cells and a coiled elongate cell (Figure 18.5). When the cysts open in the guts of invertebrate animals, a barbed elongate cell uncoils (Figure 18.6) and penetrates host intestinal cells, infecting them (Boucias et al. 2001). It has been found in cladocerans, mites, trematodes, collembolans,

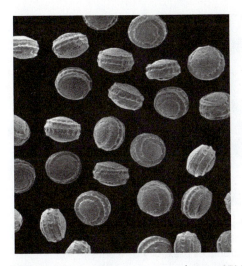

Figure 18.4 *Helicosporidium* cysts viewed using SEM. (From Boucias et al. 2001, by permission of the Society of Protozoologists)

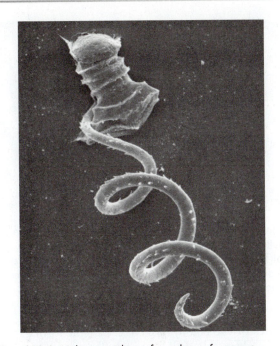

Figure 18.6 *Helicosporidium* after release from cyst wall. The long, sharp-ended, barbed filamentous cell has uncoiled, and the three ovoid cells have expanded in preparation for attack. (From Boucias et al. 2001, by permission of the Society of Protozoologists)

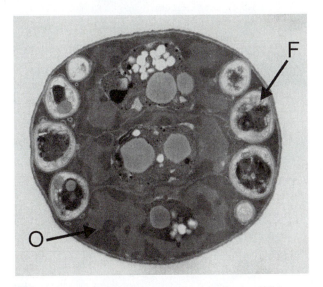

Figure 18.5 *Helicosporidium* cysts viewed using TEM. Cross-sections of the coiled filamentous cell (F) surround profiles of the three ovoid cells (O). (From Boucias et al. 2001, by permission of the Society of Protozoologists)

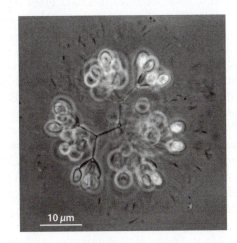

10 μm

Figure 18.7 *Dictyosphaerium* colony. (Photo: L. W. Wilcox)

scarab beetles, mosquitoes, simuliid flies such as black flies, and pond water. Molecular phylogenetic analyses link *Helicosporidium* to *Prototheca* (Tartar et al. 2002). Although a plastid has not been visualized in the cells, molecular evidence indicates that *Helicosporidium* maintains a functional plastid genome (Tartar and Boucias 2004). A study of the complete plastid genome revealed that it is highly reduced, encoding 26 proteins, 3 rRNAs, and 25 tRNAs (de Koning and Keeling 2006).

DICTYOSPHAERIUM (Gr. *diktyon*, net + Gr. *sphaira*, ball) cells are embedded in mucilage but are distinctive in that each occurs at the end of a transparent branching structure, radiating from the

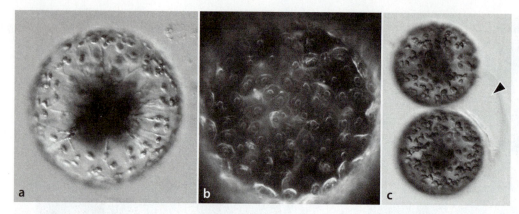

Figure 18.8 *Eremosphaera.* (a) *Eremosphaera* cells are unusually large and are characterized by multiple plastids and long cytoplasmic strands that are visible even at the light microscopic level. (b) The plastids contain a single pyrenoid each. (c) *Eremosphaera* reproduces asexually by means of autospores, two of which are shown along with remains of the parental cell wall (arrowhead). (Photos: L. W. Wilcox)

center of the colony, which is thought to originate from older cell wall material (Figure 18.7). Cells and colony shape may be either spherical or ovoid. Cells are rather small, ranging from 3–10 μm in diameter, and have one or two parietal chloroplasts. *Dictyosphaerium* occurs in acid bogs, soft-water lakes, and hard-water lakes. Bacteria often occur in the mucilage. This alga can dominate the plankton during late spring or early summer.

EREMOSPHAERA (Gr. *eremos*, solitary + Gr. *sphaira*, ball) consists of spherical unicells that can reach diameters of 200 μm (Figure 18.8). Cells have a highly layered wall and contain many small discoid chloroplasts, each with a single small pyrenoid. Chloroplasts are arranged in strands of cytoplasm radiating from the cell center to the periphery. Asexual reproduction is by production of two or four autospores (Figure 18.8c), and the remains of parental cell walls can often be observed in the vicinity of the daughter cells. Oogamous sexual reproduction is known. Up to 64 small sperm can be produced per parental cell; eggs resemble autospores in size. Zygotes produce thickened walls, but germination has not been well studied. *Eremosphaera* is a common inhabitant of soft freshwaters and acid bogs.

Botryococcus clade

BOTRYOCOCCUS (Gr. *botrys*, cluster of grapes + Gr. *kokkos*, berry) consists of colonies of oval or spherical cells embedded in tough, irregular, amber-

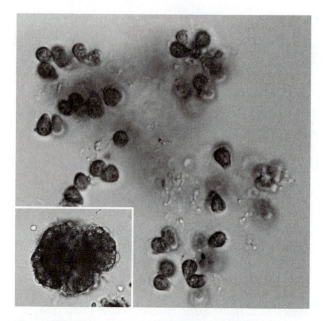

Figure 18.9 *Botryococcus* colony viewed with light microscopy. Inset shows colony with cells obscured by amber-colored mucilage. (Photo: M. E. Cook; Inset photo: L. E. Graham)

colored mucilage (Figure 18.9). This mucilage contains numerous lipid globules excreted from cells (see Figure 4.19). This lipid, and its production, have been the subject of investigations focused upon development of renewable sources of energy (Tenaud et al. 1989). Metabolic commitment to extensive lipid production is cited as an explanation for the relatively slow growth rate of this alga. Senousy et al (2004) demonstrated that four isolates of *B. braunii* formed a clade within the Trebouxiophyceae but that *B. sudeticus* is actually a member of Chlorophyceae. The types of lipids produced by *B. sudeticus* (triacylglycerols) differ from those of *B. braunii* (alkadienes, triterpenoids, or tetraterpenoids, depending on the strain). Lee et al. (1998) compared gravimetric and spectrophotometric methods for assessing total lipid content of *B. braunii*.

The chloroplast is netlike and parietal and contains one pyrenoid and starch. Reproduction is by fragmentation and production of autospores. *Botryococcus braunii* can be common and abundant in the plankton of moderately alkaline lakes, including eutrophic and brackish to marine waters, sometimes forming surface blooms. *Botryococcus* outer cell walls are electron-opaque and lamellate, suggesting the occurrence of resistant biopolymers (Wolf and Cox 1981) similar to algaenans (Gelin et al. 1997). Such resistant polymers may help to explain the occurrence of fossil deposits displaying recognizable remains of *Botryococcus* from the Carboniferous Era to the present. *Botryococcus* is believed to have contributed substantially to deposition of high-grade oil shales and coals around the world. A study of Cenozoic petroleum deposits in Thailand by Petersen et al. (2006) illustrates the geochemical methods used to analyze telalginites, formations that originate from *Botryococcus*.

Trebouxia clade

TREBOUXIA (named for Octave Treboux, an Estonian botanist) is a spherical unicell with an axial (suspended in the center of the cell) chloroplast containing a pyrenoid (Figure 18.10). Wall-less zoospores are produced during asexual reproduction. *Trebouxia* occurs either free-living on terrestrial substrates such as tree bark or in lichens. Lichens with *Trebouxia* phycobionts possess an active carbon concentrating mechanism (CCM; see Chapter 2), as determined by carbon isotope discrimination methods (Smith and Griffiths 1996). This is in contrast to lichens whose phycobiont is the pyrenoidless trebouxiophycean *Coccomyxa*. Thus the presence of a CCM has been correlated with occurrence of pyrenoids. SSU rDNA sequences have been used to assess species relationships in *Trebouxia* (Friedl and Rokitta 1997). Some forms previously classified as *Trebouxia* have been transferred to the genus *Asterochloris*

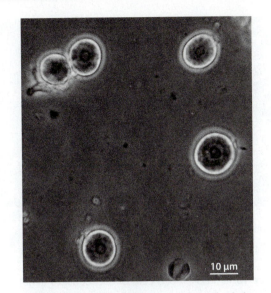

Figure 18.10 *Trebouxia* cells extracted from the lichen *Physcia*. (Photo: L. W. Wilcox)

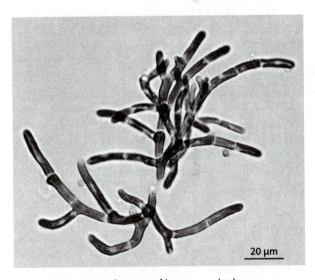

Figure 18.11 *Microthamnion* filaments, which are composed of thin, elongate cells. (Photo: L. W. Wilcox)

(Tschermak-Woess 1980). Another coccoid unicellular trebouxiophycean having a pyrenoid and wall-less zoospores is *Parietochloris*. It is distinguished from *Trebouxia* by the location of the plastid—parietal (though deeply incised) in *Parietochloris*, axial in *Trebouxia* (Watanabe et al. 1996). Comparative analysis of *Trebouxia* phycobiont phylogenies with those of fungal partners in 33 lichens (Cladoniaceae) revealed that these lichens often switch algal partners (Piercey-Normore and DePriest 2001).

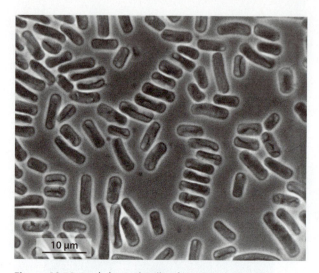

Figure 18.12 Rod-shaped cells of *Stichococcus*. (Photo: L. W. Wilcox)

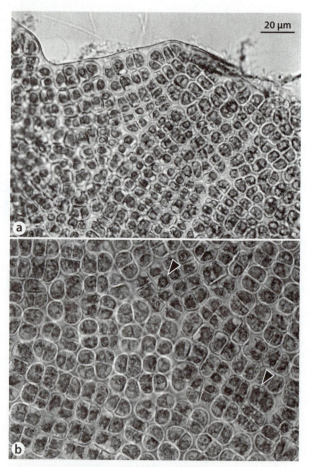

Figure 18.13 *Prasiola.* (a) Edge of bladelike thallus of an individual collected from a Colorado, U.S.A., mountain stream. Note that cells occur in packets. (b) At higher magnification, each cell can be seen to contain a single stellate plastid (arrowheads). (Photos: L. W. Wilcox)

MICROTHAMNION (Gr. *mikros*, small + Gr. *thamnion*, small shrub) is a highly branched filament of cells that are substantially longer than they are wide (Figure 18.11). It grows on soil or in slightly acidic freshwaters and attaches to substrates via a specialized holdfast cell. Chloroplasts are parietal and lack pyrenoids. Biflagellate zoospores arise in pairs from the vegetative cells; their ultrastructure has been studied (Watson and Arnott 1973; Watson 1975).

Prasiolales

STICHOCOCCUS (Gr. *stichos*, row or line + Gr. *kokkos*, berry) consists of rod-shaped cells that can occur singly or be arranged end to end to form short filaments (Figure 18.12). The cells are quite small, only 2–3 µm in diameter. The single chloroplast is parietal, with or without pyrenoids. *Stichococcus* has been isolated from soils and other terrestrial habitats (Graham et al. 1981), as well as freshwaters and estuaries. A close relative of *Stichococcus chodatii* was detected in desert crust communities by the use of molecular methods, which revealed that both taxa were closely related to *Prasiola* (Lewis and Flechtner 2002). Sorbitol and proline maintain osmotic balance in such desiccating habitats (Brown and Hellebust 1980). At least 11 amino acids can be taken up by *Stichococcus bacillaris*, by means of several specific carrier systems involving active transport (Carthew and Hellebust 1982). Some lichens possess

Stichococcus as phycobionts. Those lichens with pyrenoid-containing *Stichococcus* cells exhibited a more effective means of carbon concentration than lichens possessing pyrenoidless *Stichococcus* (Smith and Griffiths 1996). Reproduction is by fragmentation of filaments and division of single cells into two progeny. No zoospores or sexual reproduction are known. A new species, *S. ampulliformis*, was distinguished on the basis of asexual reproduction by a type of budding (Handa et al. 2003).

PRASIOLA (Gr. *prason*, leek) occurs as filaments or small (about 1.5 cm tall) blades that attach to soil, wood, salt marsh plants, or rocks by hairlike rhizoids or thicker holdfasts. The blades are often one cell thick but are sometimes thicker.

Prasiola commonly occurs on marine shoreline rocks at the high-water mark, but some species occur only in cold, fast-flowing mountain streams of southeast Asia and western North America (Figure 18.13). *Prasiola* is also widespread in Antarctica on shorelines, where it is exposed to repeated freeze/thaw cycles in spring and fall and is frozen during the winter. Amino acids such as proline are thought to serve as cryoprotectants. In addition, UV-absorbing compounds appear to help *Prasiola* adapt to high levels of solar irradiation during the austral summer (Jackson and Seppelt 1997). A UV-absorbing mycosporine-like amino acid having a distinctive absorption maximum of 324 nm occurs in *Prasiola* and some other trebouxiophyceans (Karsten et al. 2005). An evaluation of UV effects on photosynthesis of an Arctic species of *Prasiola* revealed that it seemed well adapted to cope with stressful irradiance conditions (Holzinger et al. 2006). *Prasiola* is the phycobiont in some maritime lichens (Kappen et al. 1987). Cells are conspicuously aggregated into packets, and the single chloroplast in each cell is stellate and axial. The blades are reported to be diploid and to undergo production of nonflagellate spores or biflagellate sperm and nonmotile eggs at the blade apex. Gamete production is thought to involve meiosis (Friedmann 1959) but needs to be re-evaluated. Prasiolales include the filamentous genera *Rosenvingiella*, which occurs on marine docks and pilings, and *Trichophilus welckeri*, which occurs in sloth fur (Rindi et al. 2007).

Green Algae IV

Chlorophyceans

Chlorophyceans are a diverse assemblage of mostly freshwater or terrestrial green algae. Some genera that produce decay-resistant hydrocarbon polymers known as algaenans are biogeochemically important as the sources of certain fossil carbon deposits. Many chlorophyceans are ecologically significant as primary producers in aquatic or terrestrial crust ecosystems. Some offer unique insights into evolutionary processes, one example being the origin and function of altruism genes. The genus *Chlamydomonas* is valued as a genetic system for the molecular analysis of flagella and photosynthesis and also offers industrial potential as a source of renewable hydrogen fuel. This chapter surveys the relationships, general structural and reproductive characteristics, and diversity of the chlorophyceans.

Macroscopic
Chaetophora body

(Photo: L. W. Wilcox)

19.1 Chlorophycean Relationships

The Chlorophyceae are one of the four major lineages of green algae (Chapter 16). As part of the Chlorophyta or UTC clade, chlorophyceans are more closely related to ulvophyceans (Chapter 17) and trebouxiophyceans (Chapter 18) than to streptophytes (charophyceans [Chapter 20] and land plants). The chlorophyceans are thought to form a monophyletic group that may be sister to Trebouxiophyceae (Lewis and McCourt 2004) (see Figure 16.1). Genera are assigned to the class Chlorophyceae on the basis of ultrastructural and molecular traits. In recent years, some genera have been transferred to or from other classes. In addition, a number of chlorophycean genera have been recognized as polyphyletic, with species linked to different classes. For example, the lipid-producing *Botryococcus braunii* was formerly classified in Chlorophyceae but is now placed in Trebouxiophyceae, yet *B. sudeticus* is a chlorophycean (Senousy et al. 2004). In some cases, new genera have been established to accommodate phylogenetically distinct species. For example, *Neochloris* species belonging to Sphaeropleales retain this generic name, while others were renamed *Ettlia* and *Parietochloris* and transferred to other groups (Watanabe and Floyd 1989; Komárek 1989; Deason et al. 1991). Future studies of chlorophyceans are likely to result in additional taxonomic changes.

19.2 General Features of Chlorophyceans

Chlorophyceans exhibit a range of body diversity similar to that of other green algal classes. In addition, the zygotic life cycle found among sexually reproducing chlorophyceans occurs in other green algae. However, chlorophyceans display a distinctive combination of mitotic and cytokinetic features and flagellate reproductive cell structure.

Chlorophycean Body Types

Chlorophyceans provide examples of all the major body forms found among green algae. Some are unicellular or colonial flagellates. Others occur

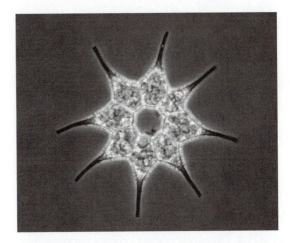

Figure 19.1 A coenobial colony, with genetically determined number and arrangement of cells (in this case, eight cells in one plane). (Photo: L. W. Wilcox)

as unicells that are nonflagellate in the vegetative (nonreproductive) state but produce flagellate reproductive cells, while other unicells never produce flagella. Certain chlorophyceans are **sarcinoid** packets of nonmotile cells formed when daughter cells fail to separate after division. Others produce a type of colony known as a **coenobium**, in which the number and arrangement of cells is under genetic control (Figure 19.1). Some chlorophyceans have unbranched or branched filamentous bodies, while others are multinucleate coenocytic siphons.

Comparatively little is known about the environmental forces that have shaped the evolution of body form in chlorophyceans and other green algae. However, a considerable body of work has shown that the chemical environment can influence body development in some chlorophycean genera. *Scenedesmus*, which has non-spiny cells, and the related, spiny-celled *Desmodesmus* may occur as single cells or multicelled colonies. In both genera, chemical compounds emitted from aquatic animals or other algae may induce colony formation. Chemical cues from herbivorous zooplankton or fish (but not carnivores) stimulate the common species *Scenedesmus obliquus* and *Desmodesmus subspicatus* to form colonies that are more difficult for the herbivores to consume (Lürling 2003a,b; Verschoor et al. 2004; Van der Stap et al. 2007) (Figure 19.2). Leflaive et al. (2008) found that epiphytic filaments of the ulvophycean *Uronema confervicolum* produce compounds that trigger colony formation in *D. quadrispina*.

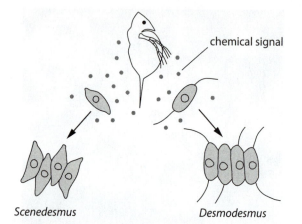

Figure 19.2 Chemical control of colony morphology. Zooplankton exudates induce morphological transformation of unicells to colonies in *Scenedesmus* and *Desmodesmus*.

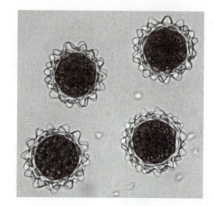

Figure 19.3 Hypnozygotes. Zygotes that have thickened, often ornamented walls allow many chlorophyceans to survive stressful environmental conditions. Hypnozygotes of the chlorophycean *Volvox* are shown here. (Photo: L. W. Wilcox)

Chlorophycean Asexual Reproduction, Sexual Reproduction, and Life Cycle

Asexual reproduction often occurs by means of zoospores, flagellate unicells that require liquid water for dispersal. Zoospores typically have an elongate, hydrodynamically adaptive shape and often possess eyespots, suggesting the presence of a light-sensing system. The zoospores of some genera have rigid cell walls, but those of others are naked; the presence or absence of zoospore cell walls can be useful in characterizing genera. Zoospores settle after a period of swimming, lose their flagella, and, in the case of naked cells, become rounded before developing into the mature vegetative form. Asexual reproduction can also occur by means of aplanospores or autospores. **Aplanospores** are nonmotile cells having some features typical of flagellate cells, such as contractile vacuoles. Aplanospores are regarded as cells whose development into zoospores has been arrested. **Autospores** are asexual reproductive cells that are apparently genetically incapable of producing flagella; they may display contractile vacuoles, but they otherwise appear to be small versions of vegetative cells. Chlorophycean genera differ in the types of asexual reproductive cells they produce; some generate only zoospores or aplanospores, and others produce only autospores.

Sexual reproduction may be isogamous, by means of gametes that are morphologically indistinguishable; anisogamous, in which case flagellate gametes are structurally distinguishable; or oogamous, with a larger nonmotile egg and smaller flagellate sperm. Chlorophyceans that are known to undergo sexual reproduction display a zygotic life cycle, in which zygotes are the only diploid cells. Zygotes often have thick and sometimes spiny walls (Figure 19.3); they germinate by meiosis when conditions allow survival of the products. Such zygotes are also known as **hypnozygotes** ("sleeping zygotes") because they function both as resting stages and in the generation of population-level genetic variation. Alternation of multicellular generations is not known to occur.

The flagellate reproductive cells of chlorophyceans usually possess two or four flagella, but in a few cases numerous flagella are present. The flagellate cells of chlorophyceans share features of the flagellar apparatus. Flagella emerge from the cell apex; in contrast, in charophyceans, flagella emerge laterally. In addition, rhizoplasts typically link flagellar basal bodies with the nucleus, and there is a cruciate (cross-shaped) X-2-X-2 arrangement of microtubular roots (see Chapter 16). The number of microtubules in the X root varies among taxa. The flagellar basal bodies are displaced in a clockwise (CW) direction or are directly opposed (DO) (see Chapter 16). This feature distinguishes chlorophyceans from ulvophyceans, whose flagellar basal bodies are arranged in a counterclockwise (CCW) manner. Molecular phylogenetic analyses indicate that the arrangement of flagellar basal bodies (CW or DO) and number of flagella reflect phylogenetic relationships in this group more accurately than do body structure, presence or absence of a cell wall, or occurrence of multinucleate cells (Booton et al. 1998a; Nakayama et al. 1996a,b).

Cell Division Features of Chlorophyceans

Chlorophycean cell division is fairly uniform in that (1) mitosis is closed, (2) the telophase spindle collapses prior to cytokinesis, and (3) a system of microtubules known as the **phycoplast** runs parallel to the cytokinetic cleavage plane. In some chlorophycean genera, cleavage takes place by furrowing—defined as the development of a centripetal ingrowth of the cell membrane from the periphery toward the center of the cell. In other chlorophyceans, cytokinesis involves production of a cell plate that develops centrifugally by fusion of Golgi-derived vesicles. The presence of plasmodesmata—channels through the cell wall that allow for intercellular communication—is correlated with the cell-plate type of cytokinesis. Plasmodesmata are absent from chlorophyceans that divide by furrowing. The presence of cell plates and plasmodesmata in some chlorophyceans, as well as certain ulvophyceans (Trentepohliales) and late-diverging charophyceans (Coleochaetales and Charales), indicates that these specialized features have evolved independently several times.

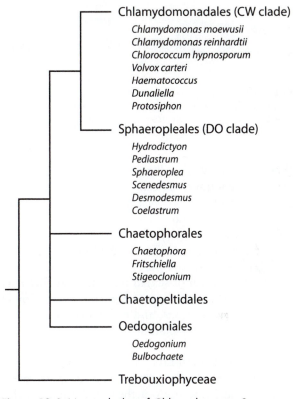

Figure 19.4 Major clades of Chlorophyceae. Some representative genera and species are listed.

19.3 Chlorophycean Diversity

In the past, chlorophycean green algae were segregated into orders primarily on the basis of organismal morphology. Flagellates, for example, were classified into Volvocales; coccoid unicells and nonmotile coenobial colonies were aggregated into Chlorococcales; most unbranched filaments into Ulotrichales; and most branched filaments into Chaetophorales (Bold and Wynne 1985). More recently, the application of molecular phylogenetic methods has revealed that extensive parallel evolution of body form has occurred. It is thus no longer possible to group the chlorophycean algae into orders based only on their morphology, if classification systems are to reflect phylogeny as closely as possible. To a very large extent, the seemingly logical and convenient system by which chlorophyceans were formerly classified has collapsed and is being replaced with a more evolutionarily accurate, but complicated, taxonomic scaffolding. In the emerging new classification, orders may contain a mixture of structural types.

Several orders have been defined (Figure 19.4). Oedogoniales, Chaetopeltidales (O'Kelly et al. 1994), and Chaetophorales are monophyletic groups that seem to diverge early in phylogenetic trees, though often with low support (Nakada et al. 2007; Alberghina et al. 2006). A major clade of unicellular, colonial, and filamentous genera having the DO pattern of flagellar bases is known as Sphaeropleales (Deason et al. 1991). Possibly the most derived chlorophycean lineage is the CW clade known as Chlamydomonadales. Alberghina and associates (2006) proposed that the DO pattern of basal bodies evolved from the CCW type found among ulvophyceans and that the CW type is an innovation from the DO type.

Oedogoniales

Three genera (and some 600 described species) of unbranched or branched filaments having some very distinctive common features are placed in this order. Oedogoniales is inferred to be monophyletic and early diverging on the basis of SSU rDNA data

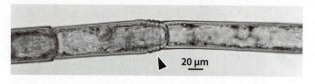

Figure 19.5 *Oedogonium*, an unbranched filament. Note the characteristic annular "rings" at the apex of the middle cell (arrowhead). (Photo: L. W. Wilcox)

(Booton et al. 1998b; Alberghina et al. 2006). All the genera have uninucleate cells with a highly dissected, parietal, netlike chloroplast that contains numerous pyrenoids. Autapomorphies of this order include an unusual form of cytokinesis involving the precocious development of a ring of wall material; a ring of flagella on both zoospores and male gametes; and a specialized form of sexual reproduction that may involve dwarf male filaments. Sexual reproduction in all genera is oogamous, and meiosis is presumed to occur at zygote germination.

OEDOGONIUM (Gr. *oidos*, swelling + Gr. *gonos*, offspring) is unbranched. Cell division generates distinctive delicate rings at the apical ends of cells.

Cells that have undergone many divisions will exhibit many such rings (Figure 19.5). The ultrastructural events surrounding ring formation, mitosis, and cytokinesis were summarized and extensively illustrated by Pickett-Heaps (1975) (Figure 19.6).

Asexual reproduction occurs by fragmentation and by zoospore production. The large, wall-less zoospores develop singly in parental cells. Centrioles replicate to form a ring of flagellar basal bodies, and these generate a ring of flagella that surrounds a clear apical dome that contains numerous vesicles. Zoospores also possess an eyespot and contractile vacuoles. They emerge from the parental cell and, after a period of motility, settle onto a submerged substrate, flagellar side down. Flagella are shed, and dome vesicles release their contents, effectively gluing the germling to the attachment surface. A cell wall then develops, and cell division gives rise to a new filament (Figure 19.7).

Sexual reproduction involves development of a single large nonflagellate egg cell within an oogonium and small multiflagellate sperm within antheridia (Figure 19.8). Division of a vegetative cell that serves as an oogonial initial generates an oogonium and a subtending supporting cell. Antheridial development

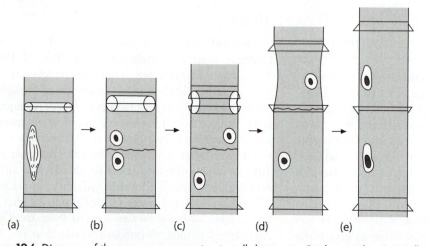

(a)   (b)   (c)   (d)   (e)

Figure 19.6 Diagram of the processes occurring in cell division in Oedogoniales. (a) Cells that are undergoing mitosis deposit a torus of wall material at the apex, on the inside of the parental cell wall. Following cytokinesis (b), the parental wall breaks at a predetermined site at the same level as the torus. Expansion of torus material, now exposed to the outside, occurs as the daughter cells elongate (c). Eventually the new wall material contributed by the torus constitutes much of the cell wall of the uppermost of the two daughter cells (d). The upper point at which breakage of the parental cell wall occurred is visible as a ring at the apex of the upper cell, whereas the lower breakage site occurs just below the new cross-wall, at the upper end of the lower daughter cell (e). If the upper cell divides again, it will acquire a second apical ring. Thus the number of times a cell has divided can be deduced by the number of apical rings it possesses. (After Pickett-Heaps 1975)

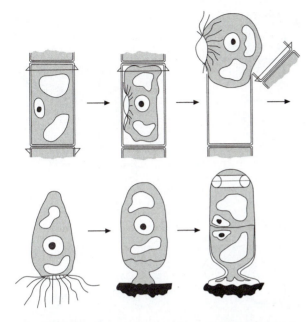

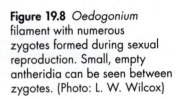

Figure 19.7 Zoosporogenesis and germling development in Oedogoniales. The entire contents of individual cells may be used in the formation of a single large zoospore having multiple flagella. Zoospores escape when the parental cell wall breaks at the anterior. Upon settling flagella-end first, attachment mucilage is exported, and mitotic divisions begin the process of filament development. (After Pickett-Heaps 1975)

Figure 19.8 *Oedogonium* filament with numerous zygotes formed during sexual reproduction. Small, empty antheridia can be seen between zygotes. (Photo: L. W. Wilcox)

occurs by asymmetrical divisions that produce very short cells and, in at least some species, may be influenced by the nearby presence of bacteria (Machlis 1973). Antheridial cells then divide further to generate two or more sperm.

Some species are monoecious; presumably self-fertilization may occur in these forms. Other species are dioecious, which may promote outcrossing. Species in which the sperm develop within cells on normal-sized filaments are called macrandrous forms. In contrast, some species are nannandrous; these species produce dwarf male filaments that are one or only a few cells in length and that grow on the supporting cells just below oogonia of normal-sized female filaments. The dwarf male filaments develop from **androspores**, cells that are intermediate in size between zoospores and sperm but likewise have numerous flagella arranged in an apical circle. The cells that produce androspores may occur on the same filament as the eggs; thus, selfing may occur in these so-called **gynandrosporous** species. Other

species produce androspores and eggs on different filaments; this facilitates outcrossing and is known as **idioandrospory**. The ultrastructure of dwarf males of *Oedogonium pluviale* was studied by Leonardi et al. (1998).

Sperm, whether produced by dwarf or normal-sized filaments, are attracted to eggs by chemical pheromones and gain access to them via a pore or fissure in the oogonial wall (Hoffman 1973; Machlis et al. 1974). A complex series of chemical signaling events is involved in the production of androspores and development of dwarf males, which is tantamount to a kind of mating ritual wherein each of a series of interchanges must succeed in order for mating to occur. Androspores are chemically attracted to oogonial initial cells of female filaments; this might be regarded as a form of mate selection. But if no female filaments are nearby, androspores can germinate into vegetative filaments, just like zoospores—a resource-saving strategy. If androspores attach and begin development into dwarf males, oogonial initials divide to produce eggs, suggesting the production of a second chemical signal. Such a signaling system may save resources; if sperm are not available, energy will not be diverted into oogonial development. Mature oogonia produce a blanket of mucilage around themselves and the nearby male filaments, presumably limiting fertilization to sperm produced by these neighboring males but also preventing them from wandering elsewhere. Sperm are then attracted to oogonia, suggesting occurrence of a third phase of chemical influence (Rawitscher-Kunkel and Machlis 1962). After fertilization, the zygotes of all species develop a thick wall and large amounts of carotenoid pigments, and they enter a dormant stage, eventually germinating by meiosis to form four zoospores that regenerate the filamentous thallus. An extensive ultrastructural survey, providing many more interesting details of cell division, asexual reproduction, and sexual reproduction in *Oedogonium*, can be found in Pickett-Heaps (1975).

There are about 400 described species of *Oedogonium*, varying in the size of vegetative cells, shape and size of oogonia, ornamentation of the zygote wall, and location of antheridia (Prescott 1951). However, recent molecular analyses indicate that the genus is not monophyletic (Alberghina et al. 2006). The alga commonly occurs attached to submerged plants and other substrates in shallow freshwaters. On occasion, *Oedogonium* can form conspicuous growths or blooms.

BULBOCHAETE (Gr. *bolbos*, bulb + Gr. *chaite*, long hair) resembles *Oedogonium* in habitat, cell structure, and details of cell division and reproduction (Figure 19.9), but differs from it in being branched and in having distinctive, colorless bulbous-based hair cells (setae) (Figure 19.10). Many details of the cell biology of *Bulbochaete* can be found in Pickett-Heaps (1975).

Chaetophorales

This order, as defined by Mattox and Stewart (1984), includes chlorophyceans with branched or unbranched filaments and quadriflagellate reproductive cells. Plasmodesmata may occur in cross-walls. *Uronema acuminata* (an unbranched form) and the branched *Fritschiella tuberosa*, *Chaetophora incrassata*, and *Stigeoclonium helveticum* are undoubted members of this monophyletic group, based on the presence of the features listed above and molecular sequence evidence (Booton et al. 1998b).

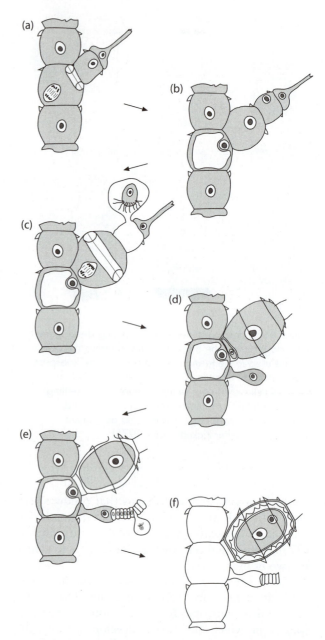

Figure 19.9 Sexual reproduction in *Bulbochaete*, which is similar to that in other members of Oedogoniales. Nannandrous sexual reproduction is illustrated here. An androspore produced by an anterior cell in a branch is released as the subtending cell undergoes division preparatory to oogonial development (c).(d) The androspore settles upon a cell near the oogonium, undergoes a few mitotic divisions to form a dwarf male filament (e), and then produces sperm. (f) Upon fertilization of the egg via a pore in the oogonial wall, the zygote develops a thick ornamented wall preparatory to a period of dormancy, while the dwarf male filament and the rest of the oogonial filament die. (After Pickett-Heaps 1975)

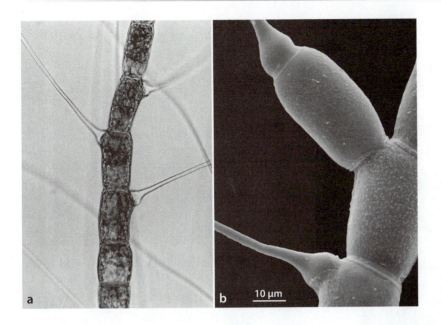

Figure 19.10 *Bulbochaete.* (a) Bulbous-based, colorless hair cells. (b) Hairs and branching viewed by SEM. (a: C. Taylor; b: S. Cook)

URONEMA (G. *oura*, tail + Gr. *nema*, thread) (Figure 19.11) filaments are typically attached to freshwater substrates by specialized disklike holdfast cells. The cells are uninucleate, and each has a single, parietal, band-shaped plastid with one to several pyrenoids. Cytokinesis involves production of a cell plate, and plasmodesmata are present (Floyd et al. 1971). One or more quadriflagellate zoospores are produced per cell. These settle apical side down onto substrates, then divide to begin holdfast and filament formation (Bold 1958). The zoospores of *Uronema belkae* (formerly known as *Ulothrix belkae*) are nearly identical to those of the branched filamentous forms *Stigeoclonium* and *Fritschiella* (Floyd et al. 1980).

CHAETOPHORA (Gr. *chaite*, long hair + Gr. *phoras*, bearing) (Figure 19.12) is often macroscopic, attached to submerged surfaces, including plants in freshwaters. The ends of its branches form multicellular pointed "hairs." Reproduction is by quadriflagellate zoospores and biflagellate isogametes. *Chaetophora* thalli are embedded in copious, firm mucilage.

STIGEOCLONIUM (L. *stigeus*, tattooer + Gr. *klonion*, small twig) (Figure 19.13) occupies similar habitats as *Chaetophora*, also produces multicellular hairlike branches, and has similar reproduction, but it has only a thin mucilage layer.

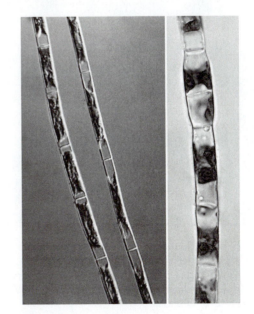

Figure 19.11 *Uronema.* The cells of unbranched filaments each have a single napkin ring–shaped chloroplast with one or more pyrenoids. There is no mucilaginous sheath. (Photos: L. E. Graham)

The chloroplast is a parietal plate much like that of *Chaetophora*, and cytokinesis involves the formation of centrifugal cell plates and plasmodesmata (Floyd et al. 1971). There is typically a prostrate system that is attached to substrates, as well as a system of erect branches. If the erect portions

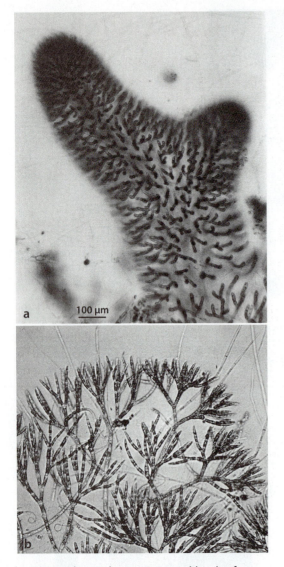

Figure 19.12 *Chaetophora*. (a) Several levels of branching occur within a firm gelatinous matrix. (b) At higher magnification, branches can be observed to terminate in multicellular, colorless hairs. (a: L. W. Wilcox; b: L. E. Graham)

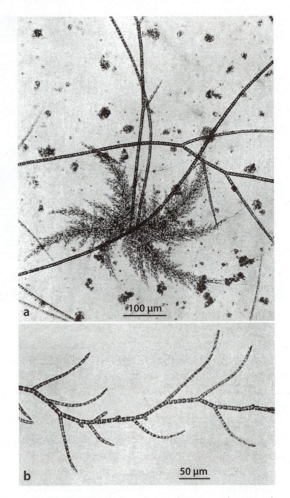

Figure 19.13 *Stigeoclonium*. (a) Habit view, showing the prostrate and erect portions. (b) Multicellular, hairlike branches. (From Cox and Bold 1966)

are eaten by herbivores, they can be regenerated from the less vulnerable, more persistent prostrate system.

FRITSCHIELLA (named for Felix Eugen Fritsch, a British phycologist) was once regarded as a model for the algal ancestors of land plants because it grows in terrestrial habitats, such as soil surfaces, and because the thallus includes tissuelike prostrate regions, as well as erect branches and colorless

rhizoids (Figure 19.14). Further, cytokinesis occurs by formation of centrifugal cell plates (McBride 1970). However, analysis of *Fritschiella*'s flagellate reproductive cells has revealed its relationship to other chlorophyceans rather than to charophyceans (Melkonian 1975). *Fritschiella* is now regarded as an example of parallel structural adaptation to terrestrial life.

DRAPARNALDIA (named for Jacques Philippe Raymond Draparnaud, a French naturalist) (Figure 19.15) is commonly found attached to rocks in cold running waters and has much larger cells in the main axes than in branches; the branch tips form multicellular "hairs." The chloroplasts are parietal bands, and thalli may be invested with soft mucilage.

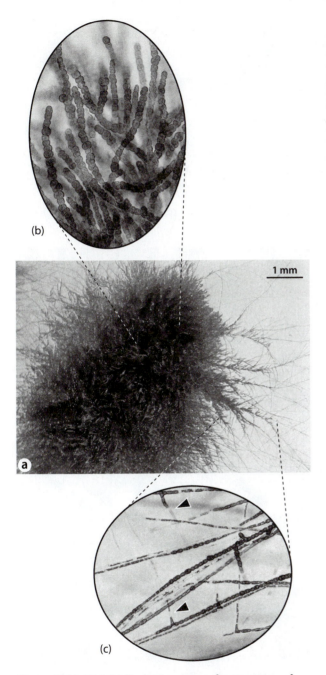

Figure 19.15 *Draparnaldia.* The main filamentous axis has relatively large cells, first-order branches have smaller cells, and higher-order branches have even smaller cells. (Photo: L. E. Graham)

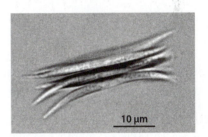

Figure 19.16 *Ankistrodesmus.* (Photo: L. W. Wilcox)

Figure 19.14 *Fritschiella.* (a) Low-magnification view of thallus growing on agar. Upright (b) and prostrate (c) portions are shown at higher magnification. Arrowheads in (c) point to colorless rhizoids. (Photos: L. W. Wilcox)

Sphaeropleales

ANKISTRODESMUS (Gr. *ankistron*, fishhook + Gr. *desmos*, bond) occurs as long, needle- or spindle-shaped cells that may occur individually or be aggregated into groups (Figure 19.16). Sometimes the cells are twisted around each other, but there is no mucilaginous covering. The chloroplast is parietal, and a pyrenoid may be present or not. Asexual reproduction occurs by cleavage of parental cells into 1–16 autospores that are released by parental wall rupture. *Ankistrodesmus* occurs in the plankton of freshwaters and in artificial ponds. *Ankistrodesmus* and the related genus *Monoraphidium* are not monophyletic (Krienitz et al. 2001).

BRACTEACOCCUS (L. *bractea*, thin metal plate + Gr. *kokkos*, berry) occurs as spherical multinucleate unicells containing many discoid plastids without pyrenoids (Figure 19.17). The biflagellate zoospores are naked. It occurs as a member of microbiotic (cryptogamic) crust communities on arid soils. A molecular phylogenetic analysis of nine strains, representing at least five species from four geographic locations, indicated that *Bracteacoccus* is

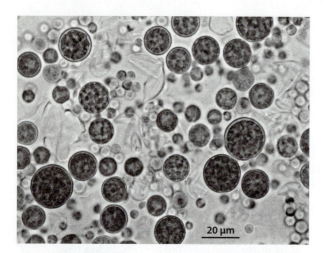

Figure 19.17 *Bracteacoccus.* Cells of various sizes and ages are evident, as are remains of parental cell walls. (Photo: L. W. Wilcox)

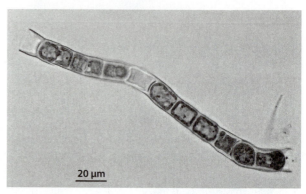

Figure 19.18 *Microspora.* (Photo: L. W. Wilcox)

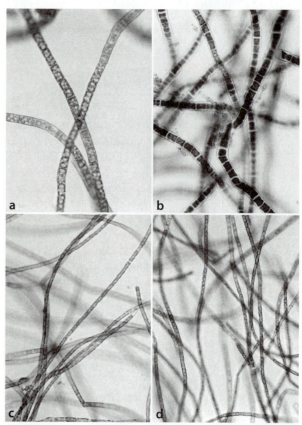

Figure 19.19 Starch test (treatment with I₂KI) employed on the morphologically similar taxa *Microspora* (a), which contains starch and stains a dark purple-blue (b), and *Tribonema* (c), a genus of Tribophyceae, whose members lack starch and do not stain (d). (Photos: L. W. Wilcox)

monophyletic (Lewis 1997). *Bracteacoccus* has the unusual trait of parallel basal bodies, which evolved independently from those of *Heterochlamydomonas* and *Dictyochloris* (Shoup and Lewis 2003).

MICROSPORA (Gr. *mikros*, small + Gr. *spora*, spore) (Figure 19.18) is an unbranched filament that may be terminated by a specialized holdfast cell that attaches the thallus to freshwater substrates. The cell walls occur as two open-ended cylinders firmly attached by an intermediate septum and appear H-shaped in optical section. Each protoplast is enclosed by portions of two of these structures, with the open cylindrical portions overlapping. Similar cell walls occur in the unbranched, filamentous stramenopile *Tribonema* (Chapter 14) but clearly evolved independently. *Tribonema* and *Microspora* can be distinguished by a starch test; *Tribonema* lacks starch, and it is present in plastids of *Microspora* (Figure 19.19). *Microspora* has a reticulate plastid but no pyrenoids. Asexual reproduction is by fragmentation or by production of flagellate zoospores. Isogamous sexual reproduction has been observed. Species of *Microspora* have been described by Lokhorst (1999). Details of the flagellar apparatus of *Microspora* zoospores link this genus with *Bracteacoccus* (Lokhorst and Star 1999).

PEDIASTRUM (Gr. *pedion*, plain + Gr. *astron*, star) is a very distinctive coenobial colony having a flattened, often starlike shape. Molecular analyses show

that *Pediastrum* is not a monophyletic genus and has thus been split into *Pediastrum* (including *P. duplex*) (Figure 19.20a), *Pseudopediastrum* (including the former *P. boryanum*) (Figure 19.20b), *Monactinus*

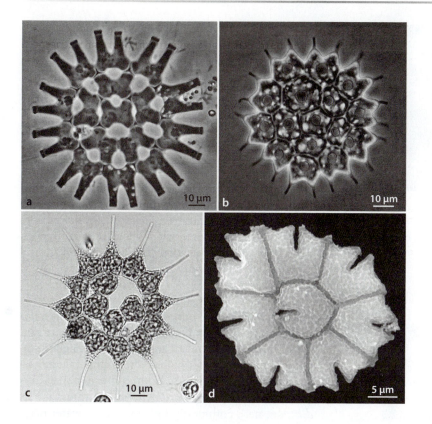

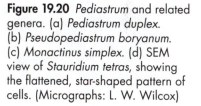

Figure 19.20 *Pediastrum* and related genera. (a) *Pediastrum duplex.* (b) *Pseudopediastrum boryanum.* (c) *Monactinus simplex.* (d) SEM view of *Stauridium tetras*, showing the flattened, star-shaped pattern of cells. (Micrographs: L. W. Wilcox)

(including the former *P. simplex*) (Figure 19.20c), *Parapediastrum* (including the former *P. biradiatum*), and *Stauridium* (including the former *P. tetras*) (Figure 19.20d) (Buchheim et al. 2005). Each species consists of a determinate number of cells arranged in a specific pattern. The peripheral cells usually possess one or two hornlike projections, whereas internal cells may have a similar or different shape. Peripheral cells may also bear clusters of very long chitinous bristles; these are regarded as buoyancy (Gawlik and Millington 1988) or herbivore deterrence devices. Cell walls contain silica and algaenans, hydrocarbon polymers that are believed to confer resistance to microbial decay and chemical hydrolysis (Gelin et al. 1997). Remains can be found in lake sediments, and ancient fossil remains are known. *Pediastrum* remains are also associated with certain fossil fuel deposits.

Asexual reproduction of *Pediastrum* occurs by autocolony formation. Cell protoplasts generate the same number of biflagellate zoospores as the typical colonial cell number for the species. These zoospores are retained within a vesicle that is derived from the inner layer of the parental cell wall and is liberated from the parental cell. After a short mobility period,

the zoospores aggregate in the same planar pattern that was present in the parental coenobium (Figure 19.21). Daughter autocolonies are then liberated from the enclosing vesicle. Each colony of *Pediastrum* is thus potentially capable of generating as many autocolonies as it possesses cells. Each daughter colony will have the same number of cells and cell pattern as the parent. Inhibitor experiments have demonstrated that the arrangement of cytoskeletal microtubules is involved in the control of cell and colony shape (Millington 1981; Marchant and Pickett-Heaps 1974).

Sexual reproduction occurs by fusion of biflagellate gametes (Figure 19.22) that are smaller than zoospores and are liberated from their parental cell. Gametes produce eyespots, unlike vegetative cells, and have mating structures at the anterior end of the cell that are similar to those of *Chlamydomonas*. Zygote germination results in production of zoospores that quickly produce thick-walled, polyhedral unicells known as **polyeders**; these then generate new coenobia. Polyeders are also known to arise from asexual reproductive processes. An outstanding ultrastructural analysis of sexual and asexual reproduction in *Pediastrum* and related algae can be

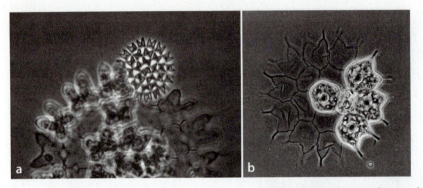

Figure 19.21 Autocolony production. (a) The small-celled colony has just been released from one of the vegetative cells of the larger parental colony of *Pediastrum duplex*. The cells of the daughter colony will expand until they reach the size characteristic for the species. (b) Autocolonies of *Pseudopediastrum boryanum* have been released from many cells of this colony through irregular openings. (Photos: L. W. Wilcox)

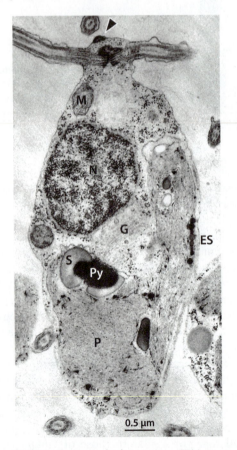

Figure 19.22 *Pediastrum duplex* gamete. TEM view showing directly opposed flagellar bases and connecting structures, nucleus (N), and plastid (P) with eyespot (ES) (which is lacking in zoospores). An electron-opaque dome-shaped mating structure (arrowhead) is evident at the cell's apex. G = Golgi body, M = mitochondrion, Py = pyrenoid, S = starch. (From Wilcox and Floyd 1988, by permission of the *Journal of Phycology*)

found in Pickett-Heaps (1975). *Pediastrum* and related genera are common in the plankton and entangled among the periphyton of many lakes, swamps, and bogs (Prescott 1951).

HYDRODICTYON (Gr. *hydro*, water + Gr. *diktyon*, net), commonly known as the "water net," occurs widely in hard-water lakes and streams, sometimes forming nuisance growths that can blanket surfaces of freshwater ponds and small lakes, accumulate rotting masses at the edges, and clog boat engine intakes. Conspicuous growths can also be observed in waters associated with rice paddies, fish farms, and irrigation ditches. Extensive growths are usually associated with some degree of eutrophication. Its significant ecological impact can partially be explained by unusual cell and colony construction and reproductive mode.

The cells of adult colonies are very large (up to 1 cm long) and are each linked to three to eight (commonly four) other cells to form a reticulate pattern (Figure 19.23). Cells are coenocytic (i.e., multinucleate), and each contains a single, highly reticulate (netlike) chloroplast with many pyrenoids (Figures 19.24 and 19.25). The internal portion of these large cells is filled by a large vacuole; the cytoplasm is arranged around the periphery (Pickett-Heaps 1975). The multiple nuclei arise from repeated closed mitoses without intervening cytokinesis. The number of cells in the colony is also very large, and colonies can be more than a meter long and 4–6 cm wide (Canter-Lund and Lund 1995).

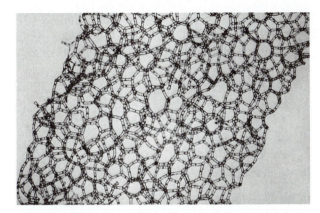

Figure 19.23 *Hydrodictyon*, which forms large netlike coenobial colonies. (Photo: L. E. Graham)

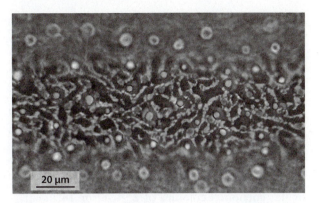

Figure 19.25 Portion of a *Hydrodictyon* cell viewed with phase-contrast optics, showing the reticulate nature of the plastid. (Photo: L. W. Wilcox)

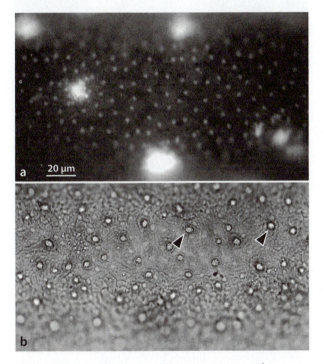

Figure 19.24 *Hydrodictyon*. The same portion of a *Hydrodictyon* cell is viewed with (a) epifluorescence microscopy, showing many bright DAPI-stained nuclei (lobes of the faintly autofluorescent chloroplast are visible in the background), and (b) bright-field microscopy, showing the reticulate plastid with numerous pyrenoids (arrowheads). (Photos: L. W. Wilcox)

Asexual reproduction occurs by means of autocolony formation. Each cell's protoplast subdivides into as many as 20,000 biflagellate zoospores, and typical microtubular phycoplasts are involved in the multiple, synchronous cytokineses (Pickett-Heaps 1975). During autocolony formation, these zoospores are not released, and within the crowded confines of the parental cell wall, following a short period of motility, they adhere to other cells to form the typical netlike colony pattern. As in *Pediastrum*, microtubules are known to mediate the pattern formation process (Marchant and Pickett-Heaps 1972). The young nets are released by disintegration of parental cell walls. Each parental net is capable of producing as many daughter nets as there are cells in the colony. As a result, if resources are plentiful, growth of *Hydrodictyon* can be explosive, explaining the occurrence of nuisance blooms.

An alternate form of asexual reproduction can occur, where zoospores are released from the parental cell and then form polygonal, thick-walled polyeders similar to those of *Pediastrum*. Polyeders can germinate to produce small spherical nets by adherence of zoospores within an extruded vesicle (reminiscent of the asexual reproductive process of *Pediastrum*). Each cell of this net can then produce a new net by the process described above (Hatano and Maruyama 1995). Sexual reproduction occurs by isogametes that are smaller than zoospores. They are released from parental cells through a pore in the wall. Gametes possess an anterior structure that is involved in the formation of a fertilization tube similar to that observed in *Chlamydomonas*. Gamete fusion produces a resting-stage hypnozygote that, after a period of dormancy, germinates (presumably by meiosis) to form four zoospores, each of which develops into a polyeder. These produce nets in a manner similar to that described above. An excellent ultrastructural survey of the events occurring in sexual and asexual reproduction of *Hydrodictyon* can be found in Pickett-Heaps (1975). The most common species is *Hydrodictyon reticulatum*, but there are several other described species. Molecular sequence analyses indicate that *Hydrodictyon reticulatum* and *Pediastrum duplex* are closely related.

SCENEDESMUS (Gr. *skene*, tent or awning + Gr. *desmos*, bond) and *DESMODESMUS* (Gr. *desmos*, bond) are related genera consisting of unicells or flat coenobial colonies of 2, 4, 8, or 16 linearly arranged cells. Trainor et al. (1976) suspected that *Scenedesmus* consisted of two genera, a group of spiny species and a group of non-spiny species, and later molecular analysis supported this observation (Kessler et al. 1997; Van Hannen et al. 2002). Species that remain in *Scenedesmus* are non-spiny (Figure 19.26) and include *S. obliquus*, *S. acutus*, *S. acuminatus*, *S. bijugatus*, and *S. dimorphus*. The genus *Desmodesmus* was formed for the spiny species (An et al. 1999) (Figure 19.27). Species transferred to *Desmodesmus* include *D. quadricauda*, *D. subspicatus*, *D. armatus*, *D. communis*, and *D. abundans*.

Cell walls may contain algaenans, which are likely responsible for the occurrence of remains in lake sediments, ancient fossil deposits, and reports of *Scenedesmus*, along with *Pediastrum* and *Tetraedron*, with certain fossil fuel deposits (Gelin et al. 1997). These compounds may play an important role in adhesion of cells into colonies (Pickett-Heaps 1975). High-molecular-weight lipids may be produced by *D. communis* (Allard and Templier 2001).

Cells contain a single plastid with pyrenoid and are uninucleate. Asexual reproduction is by autocolony formation, as in *Pediastrum* and *Hydrodictyon*, but with the difference that flagellate zoospores are not involved. Rather, parental cells divide to form nonflagellate cells that align themselves laterally, though often curled tightly within the confines of the parental wall (Figure 19.28). Cleavage is typically delayed until after at least four nuclei have been produced by successive mitoses (Pickett-Heaps 1975). Autocolonies are released by breakdown of the parental wall. A single colony is capable of producing as many autocolonies as there are coenobial cells. Biflagellate isogametes are reported to occur in nutrient-limiting conditions (Trainor and Burg 1965).

Scenedesmus and *Desmodesmus* are common inhabitants of the plankton of freshwaters and

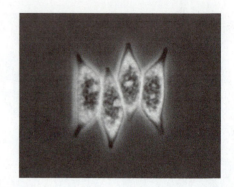

Figure 19.26 *Scenedesmus obliquus.* (Photo: L. W. Wilcox)

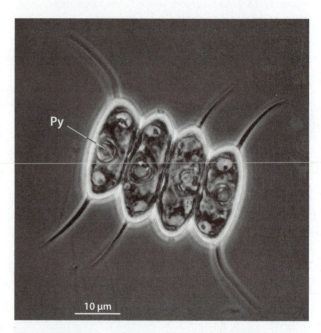

Figure 19.27 *Desmodesmus opoliensis.* Note the four adherent cells, each with a plastid and conspicuous pyrenoid (Py) and surrounding starch sheaths. Cells are ornamented with spines, which are useful in making species distinctions. (Photo: L. W. Wilcox)

brackish waters, occasionally forming dense populations, but are not typically regarded as nuisance growths. ITS-2 sequences were found to be useful in cataloging the diversity of *Scenedesmus* and *Desmodesmus* species cultured from a site in Minnesota (United States) (Johnson et al. 2007a). Cryptic speciation was observed in *S. obliquus* cultured from desert soil communities of the western United States (Lewis and Flechtner 2004). The chloroplast genome of *S. obliquus* has been completely sequenced (de Cambiaire et al. 2006).

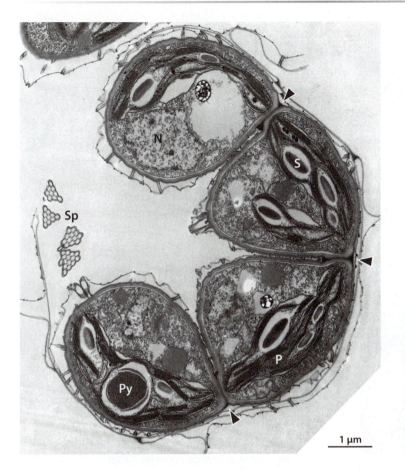

Figure 19.28 A cell of *Desmodesmus* undergoing autocolony formation. Two successive divisions of the mother cell have given rise to a linear coenobium. The cells are attached to each other by electron-opaque material (arrowheads). Note the plastid (P) with pyrenoids (Py) and intraplastidal starch (S). Cross-sections of the spines (Sp) are also apparent. N = nucleus. (From Pickett-Heaps 1975)

1 µm

COELASTRUM (Gr. *koilos*, hollow, + Gr. *astron*, star) cells are connected by blunt processes to form a hollow sphere (Figure 19.29). The number of cells (as many as 128) and their arrangement in the colony are genetically specified, as is typical for coenobial colonies. Asexual reproduction resembles that of *Scenedesmus* (i.e., autocolony formation without the involvement of flagellate zoospores). *Coelastrum* and *Scenedesmus* are also similar in being phytoplankters in freshwater lakes.

TETRAEDRON (Gr. *tetra*—from *tessares*, four + Gr. *hedra*, seat or facet) (Figure 19.30) is a single cell that may be polyhedral, pyramidal, triangular, or flat, and that often possesses distinctive spines. Dozens of species have been described from freshwater lakes and swamps. As in *Pediastrum* and *Scenedesmus*, the cell walls of *Tetraedron* contain algaenans. Chemical components of the high-molecular-weight lipids produced by *T. minimum* differ from those occurring in *Scenedesmus communis* (Allard and Templier 2001).

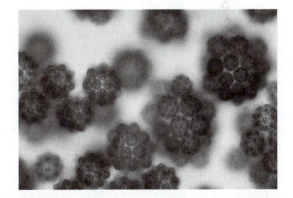

Figure 19.29 *Coelastrum*, whose cells are connected to one another by blunt processes to form hollow coenobia. Some cells have given rise to autocolonies, which can persist on the parental colony for some time. (Photo: L. W. Wilcox)

SPHAEROPLEA (Gr. *sphaira*, ball + Gr. *pleon*, many) is a free-floating, unbranched filament composed of very elongate, multinucleate cells that are 15–60 times as long as they are broad (Figure

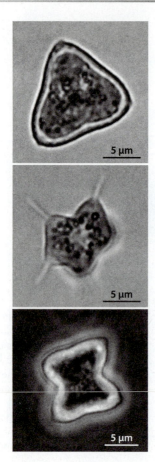

Figure 19.30 Three representatives of the genus *Tetraedron*. (Photos: L. W. Wilcox)

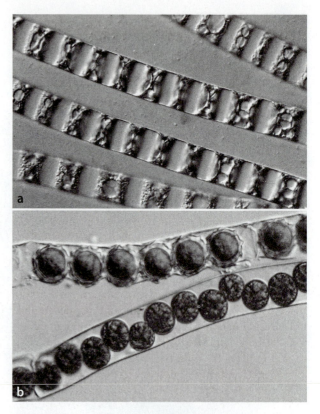

Figure 19.31 *Sphaeroplea*. (a) Vegetative cells. (b) Zygotes (top filament) and oogonia (bottom filament) are seen in this oogamous species. (Photos: L. R. Hoffman)

19.31a). Cell walls are very thin, and no mucilage sheath is present. The protoplasm is segregated into ringlike units by several large, conspicuous vacuoles. Each protoplasmic ring includes several nuclei and a ring-shaped chloroplast with several pyrenoids. Mitosis and cytokinesis are uncoupled. Cross-wall development is very unusual but thought to involve a phycoplast. Sexual reproduction can be anisogamous or oogamous, and zygote (Figure 19.31b) germination is presumed to involve meiosis; zoospores that regenerate the filament are produced. Buchheim et al. (1990) evaluated species of *Sphaeroplea*. The spindle-shaped, unicellular *Atractomorpha* is a close relative (Hoffman 1983). Buchheim et al. (2001) and Wolf et al. (2002) determined the phylogenetic positions of *Sphaeroplea* and *Atractomorpha*.

Chlamydomonadales

Chlamydomonadales, also known as the CW clade, include at least four major subclades (Watanabe et al.

2006a). Members of Clades I and IV are described here. A number of widespread genera, including *Chlamydomonas*, *Chlorococcum*, and *Chlorosarcinopsis*, are known to be polyphyletic and physiologically differentiated. For example, a group of *Chlamydomonas* species associated with Clade I is known for high tolerance of heat, desiccation, and low pH (Pollio et al. 2005; Tittel et al. 2005).

Clade I

TETRACYSTIS (Gr. *tetra*—from *tessares*, four + Gr. *kystis*, bladder) (Figure 19.32) is isolated from soils. The cells resemble those of *Chlorococcum hypnosporum* in having a single parietal chloroplast with pyrenoids. Asexual reproduction occurs by zoospores closely resembling those of *Chlorococcum hypnosporum*. Unlike in *Chlorococcum*, the cells of *Tetracystis* are aggregated into groups of four (hence the name) and even larger cell aggregations. During

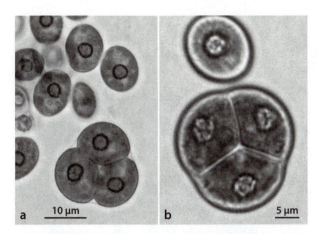

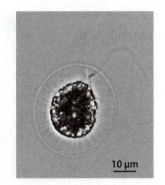

Figure 19.32 *Tetracystis* viewed at (a) low and (b) higher magnification. Cells are arranged tetrahedrally. One of the four cells is out of the plane of focus. Also note prominent pyrenoids. (Photos: L. W. Wilcox)

Figure 19.33 *Haematococcus*. The cells bear many protoplasmic extensions that traverse the space between the bulk of the cytoplasm and the cell wall. There are two flagella. (Photo: L. W. Wilcox)

the successive cell divisions that generate *Tetracystis* cell aggregates, development of daughter cell walls begins in close association with parental walls; however, each daughter cell generates a complete surficial wall, and no intercellular connections (plasmodesmata) are present.

HAEMATOCOCCUS (Gr. *haima*, blood + Gr. *kokkos*, berry) occurs as biflagellate unicells whose protoplast is connected to the wall by multiple thin strands of cytoplasm (Figure 19.33). Cell structure and reproduction appear to be similar to those of *Chlamydomonas*. *Haematococcus lacustris* commonly occurs in granitic pools or concrete birdbaths, often forming conspicuous orange-red to deep-purple growths. The coloration is based upon accumulation of astaxanthin (3,3'-diketo-4,4'-dihdroxy-beta

carotene) during formation of nonflagellate resting cells or akinetes. Astaxanthin is commercially valuable as a food colorant. Some experts suggest that carotenoids help protect the cells from the deleterious effects of high light intensity (Hagen et al. 1994) and from UV radiation in particular, but others think that the carotenoids primarily function in protection from damaging effects of highly reactive, free oxygen radicals arising from photosynthesis (Lee and Ding 1995). The akinetes are capable of surviving complete desiccation and can be transported in wind, ostensibly germinating when favorable conditions are present. Canter-Lund and Lund (1995) noted that *Haematococcus* can occur in brackish seashore rock pools but is not tolerant of high salt concentrations.

STEPHANOSPHAERA (Gr. *stephanos*, crown + Gr. *sphaira*, ball) is a motile colonial form that is phylogenetically linked with *Haematococcus*, having very similar cells with protoplastic extensions (Figure

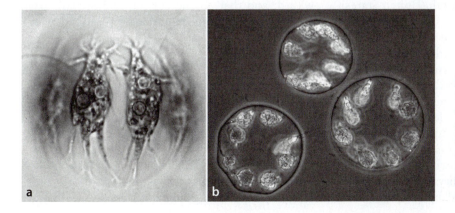

Figure 19.34 *Stephanosphaera*, whose cells have numerous cytoplasmic extensions (a). It is distinguished by arrangement of cells in a ring within a spherical mucilaginous investment (b). Each cell has two flagella that beat coordinately. (Photos: L. E. Graham)

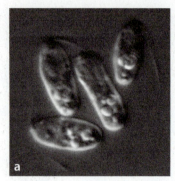

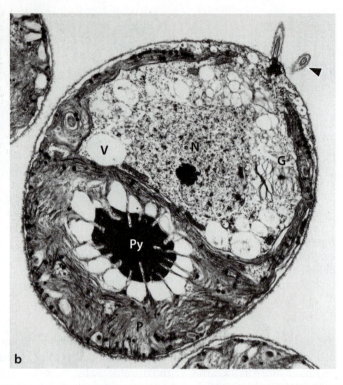

Figure 19.35 *Dunaliella*. (a) There is not a rigid cell wall, with the result that a variety of cell shapes can be seen. (b) *Dunaliella* cell viewed at the TEM level. Note the cup-shaped plastid (P) with pyrenoid (Py) and starch shell. The nucleus (N), Golgi body (G), and a vacuole (V) are also visible. Note also the delicate, extracytoplasmic matrix found on the cell body, as well as flagella (arrowhead). (a: L. W. Wilcox; b: L. W. Wilcox, of Chilean material cultured by O. Parra)

19.34). The cells are arranged in a ring having the appearance of a crown; the colony is enclosed in a globular mucilaginous matrix. Asexual reproduction occurs by autocolony formation (i.e., the division of each cell to form a colony having the same number and arrangement of cells as in the parent). Such colonies can also be described as coenobia. *Stephanosphaera* occurs in granitic pools and is reported to form reddish blooms.

DUNALIELLA (named for Michel Felix Dunal, a French botanist) has biflagellate unicells (Figure 19.35a). It occurs in extremely saline waters, including the Great Salt Lake in Utah (United States), commercial salt ponds (salterns), and the soil of salt flat communities (Kirkwood and Henley 2006). It uses photosynthetically produced glycerol to balance high external osmotic pressures and can tolerate salt concentrations above 5 M. *Dunaliella* also produces a halotolerant carbonic anhydrase (Premkumar et al. 2005). High levels of the carotenoid β-carotene are also produced by this alga under high salinity and irradiance conditions. *Dunaliella* is grown for industrial production of glycerol and β-carotene; its ability to grow in higher salinities than can be tolerated by most potential

contaminants is highly advantageous in commercial production (Chapter 4).

There is a single plastid with an eyespot, but contractile vacuoles are absent unless *Dunaliella* is grown in media of low salt concentration. Asexual reproduction is by longitudinal division, and sexual reproduction occurs by isogametes. Smooth-walled cysts can also be produced. Though *Dunaliella* is often described as naked or wall-less, electron micrographs reveal that there is a fibrous extracellular matrix on the cell surface (Figure 19.35b). Molecular phylogenetic studies suggest that *Dunaliella* is probably monophyletic, that ancestral forms were walled, and that absence of substantial walls from *Dunaliella* represents evolutionary loss (Nakayama et al. 1996a).

CHLOROCOCCUM (Gr. *chloros*, green + Gr. *kokkos*, berry) *hypnosporum*, a species commonly isolated from freshwaters and soils, occurs as single spherical or somewhat oblong cells that are sometimes aggregated into irregular masses (Figure 19.36) (Archibald and Bold 1970). There is a single cup-shaped, parietal chloroplast with at least one pyrenoid. Asexual reproduction occurs by distribution of parental cytoplasm among numerous zoospores that are

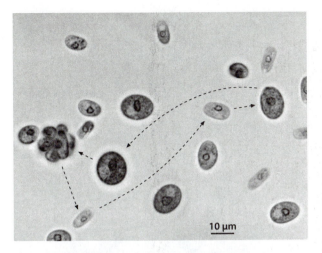

Figure 19.36 *Chlorococcum*. Arrows indicate development of elongate, recently settled zoospores into progressively larger cells that eventually cleave into zoospores or aplanospores, depending upon environmental conditions. (Photo: L. W. Wilcox)

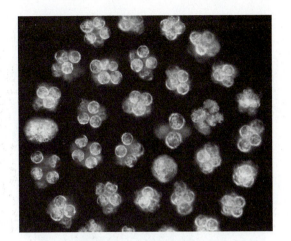

Figure 19.37 *Sphaerocystis*, viewed with dark-field microscopy. There are 4–32 vegetative cells within a mucilaginous matrix, and each is capable of producing 4–16 autospores, though often not simultaneously. In this case, most cells of the colony have done so. (Photo: L. W. Wilcox)

released from the parental cell wall. Zoospores are biflagellate and walled, with eyespots and contractile vacuoles, and thus resemble *Chlamydomonas*. If zoospore development or release is inhibited by absence of liquid water, aplanospores are produced. Both zoospores and aplanospores differentiate into mature nonmotile, spherical vegetative cells. Isogamous sexual reproduction may occur, giving rise to thick-walled zygotes or hypnospores.

SPHAEROCYSTIS (Gr. *sphaira*, ball + Gr. *kystis*, bladder) is a spherical mucilaginous aggregation of 4–32 evenly spaced, spherical cells having a single peripheral plastid with pyrenoid (Figure 19.37). Each cell is capable of generating 4–16 autospores, but not all cells reproduce at the same time. Colonies are quite large, up to 500 μm in diameter. *Sphaerocystis* is widespread in lakes of various types and is frequently dominant in early summer phytoplankton communities. It is dependent upon turbulence for suspension in the water column; the mucilage envelope is thought to reduce the sinking rate (Happey-Wood 1988).

SELENASTRUM (Gr. *selene*, moon + Gr. *astron*, star) is a colony of 4–16 sickle-shaped or curved cells, each having a single parietal chloroplast with pyrenoid (Figure 19.38). A gelatinous matrix is not present. *Selenastrum capricornutum* is commonly

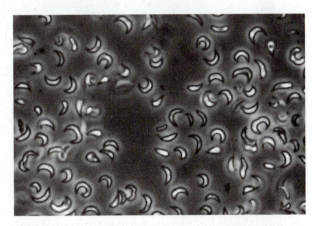

Figure 19.38 The crescent-shaped cells of *Selenastrum*. (Photo: L. W. Wilcox)

used as a laboratory bioassay test organism (Chapter 4). It occurs in the plankton of lakes and swamps.

PROTOSIPHON (Gr. *protos*, first + Gr. *siphon*, tube) is a sac-shaped, multinucleate (coenocytic) cell that may reach lengths of 1 mm and thus is visible to the naked eye (Figure 19.39a). It is common on bare soils, often forming conspicuous green or orange patches, the color depending upon whether the soil is wet or dry and whether the cells are metabolically active or in a resting stage that has accumulated carotenoids. The elongate, colorless, rhizoidlike basal portion of the cell extends into

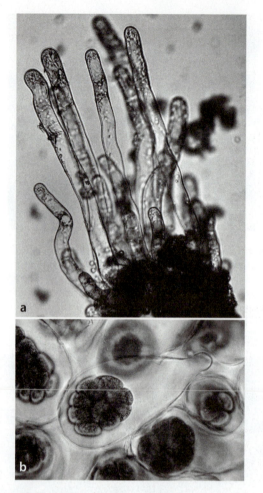

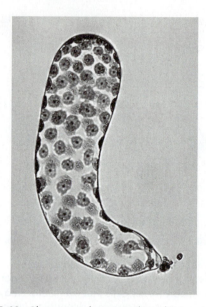

Figure 19.40 *Characiosiphon*, a multinucleate coenocyte having numerous discoid plastids. (Photo: L. E. Graham)

Figure 19.39 *Protosiphon*. (a) Young tubes growing from a clump of soil. (b) Coenocysts. (Photos: L. R. Hoffman)

the soil. The chloroplast is netlike (reticulate) and contains several pyrenoids. Asexual reproduction occurs by production of many biflagellate, naked zoospores that undergo repeated mitoses without intervening cytokinesis to generate mature coenocytes. Sexual reproduction occurs by isogametes that develop by direct cleavage of parental cells. Alternatively, cytoplasm of parental sacs may become cleaved into multinucleate units that develop thick walls and are known as coenocysts (Figure 19.39b). These serve as a desiccation-resistant resting stage that can regenerate zoospores or gametes when liquid water is available.

CHARACIOSIPHON (*Characium*, a genus of green algae [Gr. *charax*, pointed stake] + Gr. *siphon*, tube) (Figure 19.40), a coenocytic form that is even larger than *Protosiphon*, has a similar life history but differs from *Protosiphon* in having numerous discoid plastids and an aquatic (freshwater) habitat. *Characiosiphon* and the related genera *Lobocharacium* and *Characiochloris* form a monophyletic group that is distinct from *Protosiphon* (Buchheim et al. 2002).

Clade IV

GOLENKINIA (named for Mikhail Iljitsch Golenkin, a Russian phycologist) is a spherical unicell, with a parietal plastid and pyrenoid with surrounding starch shell (Figure 19.41), distinguished by the presence of several long spines per cell. Asexual reproduction is by production of 2, 4, or 8 autospores, and oogamous sexual reproduction has been described. Some cells produce 8–16 biflagellate sperm, and other cells function as eggs. Zygotes have spiny walls (Starr 1963), but the details of zygote germination and the timing of meiosis are unclear. Molecular analysis indicates that *Golenkinia* diverges at the base of the Chlamydomonadales (Wolf et al. 2003).

CHLAMYDOMONAS (Gr. *chlamys*, mantle + *monas*, unit) (Figure 19.42) is a generic name given to biflagellate unicells having a single cuplike chloroplast. The classic genus *Chlamydomonas* is known to be polyphyletic and so has undergone revision. *C. reinhardtii* is designated the type, so this species and

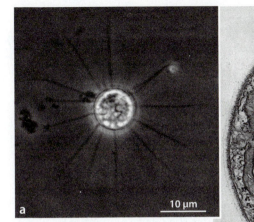

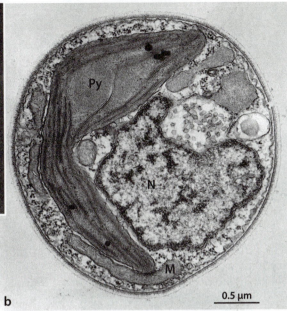

Figure 19.41 *Golenkinia.* (a) Phase-contrast LM showing several long spines per cell. (b) TEM showing part of the plastid with pyrenoid (Py) bisected by a thylakoid and surrounded by a starch sheath (S). M = mitochondrion. (Micrographs: L. W. Wilcox)

close relatives remain in the genus *Chlamydomonas,* while other species have been transferred to new genera such as the oogamous *Oogamychlamys* and *Lobochlamys,* having incised plastids (Pröschold et al. 2001). The description of *Chlamydomonas* biology that follows is based almost entirely upon studies of *C. reinhardtii,* which is particularly well understood (Harris 1989) and is a major model system for genetic (i.e., mutant-based) studies of eukaryotic cell structure and function.

Chlamydomonas reinhardtii is favored for use in laboratory work, particularly in molecular genetic studies, because it is haploid, it is easy to cultivate, it grows rapidly, and sexual reproduction can readily be induced. In addition, *C. reinhardtii* and some other species can utilize acetate as a source of exogenous dissolved organic carbon for growth in the dark. Hence, photosynthetic mutants can be rescued by growth on acetate, whereas photosynthetic mutants of obligate autotrophs would be lost. Such mutants have been used to study the structure and development of the photosynthetic apparatus (Hippler et al. 2002). The *C. reinhardtii* genome has also been sequenced, allowing comparisons with other genomes. One such comparison revealed that microRNAs (miRNAs) are involved in regulating *C. reinhardtii* gene expression by means of post-transcriptional processes (RNA silencing) (Molnar et al. 2007), which were previously thought to occur

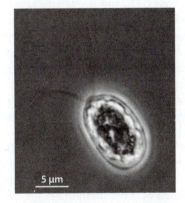

Figure 19.42 *Chlamydomonas,* which has a rigid cell wall, two equal flagella, and a parietal, cup-shaped chloroplast. (Photo: L. W. Wilcox)

only in multicellular organisms. Under some conditions, *C. reinhardtii* is capable of producing hydrogen gas, a process that is of interest in developing renewable energy sources (Ghirardi et al. 2007) (see Chapter 4).

The layered cell wall is not composed of cellulose but, rather, of polymers rich in the amino acid hydroxyproline that are linked to the sugars galactose, arabinose, mannose, and glucose, and hence known as glycoproteins. When growing under conditions in which liquid water is limiting, cells may occur as **palmelloid** stages—i.e., groups of varying

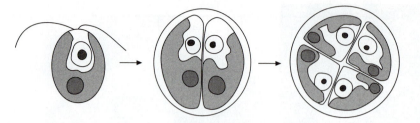

Figure 19.43 Asexual reproduction in *Chlamydomonas*. This process usually results in the production (through two successive mitotic divisions) of four daughter cells, each of which may develop two flagella.

numbers of nonflagellate cells held together by common mucilage. This mucilage, like the cell walls, is rich in hydroxyproline and sugars. When palmelloid aggregates are exposed to water, the cells typically transform to flagellates.

Chlamydomonas cells, like those of other flagellate chlorophycean cells, swim by breaststroking (Goldstein 1992). During the power stroke, the flagella are held in a nearly straight position, bending only at the bases. During the return stroke, a bending wave beginning at the base is propagated to the tip. Observations of changes in position of the eyespot reveal that the cells rotate while swimming. One of the flagella beats by itself about once in every 20 strokes, resulting in an overall helical path of motion. Internally, the flagella of *Chlamydomonas* are complex and resemble those of other eukaryotes. Much of what is known about eukaryotic flagella in general has been learned through study of *Chlamydomonas* (Qin et al. 2001). A central pair of microtubules is surrounded by nine doublets, bearing arms composed of the protein dynein, a multiprotein assembly that is responsible for generation of force. The dynein arms connect adjacent doublets, and flagellar beating results from the sliding interaction of dynein with neighboring microtubules, which is coupled to ATP hydrolysis. Radial spokes link doublets with the central microtubule pair. The flagella of *Chlamydomonas* include more than 200 polypeptides; 22 of these are associated with the central-pair complex, more than 20 with the dynein arms, and 17 with radial spokes (Kamiya 1992; Nicastro et al. 2006). Analysis of mutants has demonstrated that cells can continue to swim even if flagella lack the central microtubule pair or the radial spokes. Dutcher (1988) provided an overview of molecular analysis of flagellar basal-body function and assembly in *Chlamydomonas*. A myosinlike protein has also been identified in *Chlamydomonas*, suggesting that actin–myosin interactions occur, as in other eukaryotes (La Claire et al. 1995).

Chlamydomonas vegetative cells are typically phototactic, swimming toward moderate light but away from intense light. Phototactic behavior is based upon a rhodopsin photoreceptor located in the cell membrane just over the eyespot (Foster et al. 1984; Sineshchekov and Govorunova 1999; Nagel et al. 2002). Though the eyespot is probably not the primary light-sensing structure, it may function to shade the photoreceptor during rotational swimming, thus providing directional information. An instructional laboratory exercise designed to demonstrate phototactic behavior in *Chlamydomonas* is described by Harris (1989).

Asexual reproduction of *Chlamydomonas*. Parental cells most often produce 2, 4, 8, or 16 progeny cells by successive mitotic divisions (Figure 19.43). Division occurs longitudinally in cells whose flagellar apparatus has been duplicated. However, the sequence of divisions exhibited by *Chlamydomonas* is more sophisticated than the straightforward longitudinal division characteristic of many flagellate protists, involving changes in the polarity of successive divisions that have been compared to sequences of cell cleavages that occur in early animal embryos (Schmitt et al. 1992). As in other chlorophyceans, mitosis is closed, and cytokinesis involves phycoplast microtubules. Cytokinesis occurs by furrowing. Progeny cells are released from the confines of the parental wall by production of specialized sporangial wall autolysins that digest the cell wall. These autolysins, also known as vegetative lytic enzymes (VLE), are produced only by newly formed cells; they do not occur at any other life cycle stage, and they are not capable of digesting walls other than that of the parental cell. Such specificity is necessary to avoid digestion of the walls of the newly formed cells.

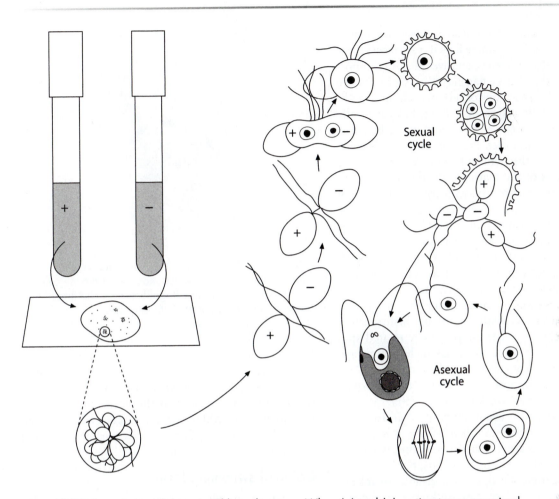

Figure 19.44 Sexual reproduction in *Chlamydomonas*. When (+) and (−) mating types are mixed on a glass microscope slide (left), the cells first clump and then mate in pairs (right, clockwise). Contact is made first by the flagella, then a mating structure links the cells, and plasmogamy begins. The quadriflagellate zygote loses its flagella and cell wall, and a new, spiny wall appears. The zygote divides by meiosis, yielding equal numbers of biflagellate daughter cells of the two mating types; these can undergo asexual reproduction repeatedly to generate large populations. (After Bold and Wynne 1985)

Sexual reproduction in *Chlamydomonas*. Gamete production, syngamy (gamete fusion), and zygote germination are known best in just a few species, including *C. reinhardtii*. Low levels of combined nitrogen (ammonium or nitrate) and blue light induce gametogenesis. Blue-light sensing, mediated by phototropin, causes gametes to not exhibit chemotaxis toward an N source, a behavior typical of vegetative cells; this change prevents gametes from being attracted to N sources that are located out of the light (Ermilova et al. 2004). The phototropin sensor has been localized to flagella (Huang et al. 2004). Some species are homothallic (monoecious)—i.e.,

mating will occur in clonal cultures. However, in species such as *C. reinhardtii*, cells of a single clone will not mate; genetically different mating types are required (i.e., they are heterothallic) (Figure 19.44). An early event in gamete development is the production of mating structures that begin as electron-opaque rings in the cell membrane at the cell apex. Filaments associated with the ring appear to be composed of actin. This region of cells of one mating type then forms a bulge at the cell anterior and extends into a mating tube that links up with the mating ring on cells of the other mating type (Goodenough et al. 1982).

In addition, linear glycoprotein molecules (agglutinins) may appear on gamete flagellar surfaces; these molecules are absent from vegetative cells. Agglutinins promote the adhesion of flagella of cells of opposite mating types. Large groups of cells may clump together during the early stages of mating, but eventually pairs separate from the agglutinated mass. Flagellar tips adhere first, and then the flagellar pairs become attached through their entire length. As a result, the anterior regions of the mating cells are brought into close proximity, and protoplasts merge via the mating tube. The flagella then disengage, and those of one gamete continue to function, propelling the zygote, known as a planozygote because it is motile.

Autolysins of a different kind from those involved in lysing parental cell walls after asexual reproduction are expressed by mating cells; these break down the gamete cell walls. Gamete autolysins are capable of breaking down cell walls of any life-history stage of the same species, or those of closely related species. These 62 kDa proteins, which include zinc, appear to act on a few cell wall polypeptides and are useful in preparing protoplasts of *Chlamydomonas* cells for experimental work (Harris 1989).

Zygotes eventually become nonmotile and develop a thick, spiny wall; they are then known as hypnozygotes. Zygote walls are rich in glycoproteins, including hydroxyproline, as are vegetative cells, but the sugars are present in different proportions. Zygotes of at least some *Chlamydomonas* species are known to include a layer of resistant material (Van Winkel-Swift et al. 1997) that may confer some degree of resistance to microbial attack during dormancy. After a period of dormancy that can be relatively short, zygotes germinate by meiosis. Typically four, but occasionally eight, cells are produced. These can develop flagella if liquid water is available, but if zygotes are germinated on an agar surface, the nonflagellate meiotic products can be separated, grown into clonal populations, and then isolated. This procedure is known as tetrad analysis and is a commonly used technique in the field of haploid genetic analysis. Unmated gametes can dedifferentiate into vegetative cells. Sometimes fused gametes do not generate typical zygotes but rather continue to exist as diploid flagellate cells. A description of the major methods used in genetic analysis of *Chlamydomonas* and a listing of mutants can be found in Harris (1989).

DYSMORPHOCOCCUS (Gr. *dys*, bad + Gr. *morphe*, form + Gr. *kokkos*, berry) consists of a

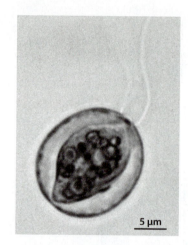

Figure 19.45 *Dysmorphococcus*, which has a mineralized lorica that is well separated from the cell membrane. There are two flagella that extend through pores in the lorica. (Photo: L. E. Graham)

Chlamydomonas-like protoplast within a lorica that is often colored by iron or manganese minerals (Figure 19.45). The structure and development of the lorica was studied at the ultrastructural level by Porcella and Walne (1980). Progeny cells, formed similarly to those of *Chlamydomonas*, are liberated by rupture of the parental lorica.

Colonial swimming forms

A dozen or so chlorophycean genera occur as colonies of *Chlamydomonas*-like cells that are found in multiples of two and are either enclosed by a common mucilaginous matrix or loosely held together by mucilage. They form a monophyletic group, with the four-celled genera *Tetrabaena* and *Basichlamys* diverging early, and are closely related to the unicellular *Chlamydomonas debaryana* and *Vitreochlamys ordinata* (Nozaki et al. 2000). These colonial forms are referred to as volvocaleans, the name referencing the well-known genus *Volvox*. The flagellar apparatus is similar to that of *Chlamydomonas* but has undergone change in relative positioning of the basal bodies. In *Chlamydomonas* and close relatives, the flagella typically lie in a V-shape arrangement and beat in opposing directions, while in volvocalean forms, the flagella are parallel and beat in the same direction. This change in orientation and beat is thought to represent a selective advantage associated with the colonial habit: When cells lie close together, beating flagella do not interfere with each other as they would if positioned as in *Chlamydomonas* (Hoops and Floyd 1982).

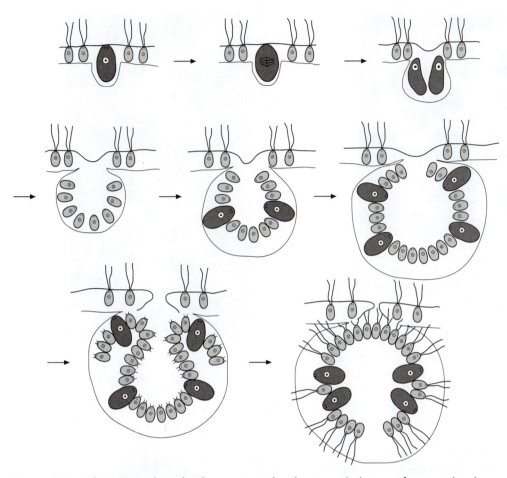

Figure 19.46 *Volvox.* Autocolony development in *Volvox* begins with division of a specialized gonidium (darkly shaded cells) to form a spherical daughter colony, some cells of which are also gonidia. When development is complete, the colony inverts so that flagella (produced by vegetative cells, but not the gonidia) are oriented toward the outside. If colony inversion did not occur, the flagella would be produced on the inside of the hollow spheres, and the colonies would be unable to swim. Mature daughter colonies (autocolonies) eventually escape from the confines of the parental sphere. (After Pickett-Heaps 1975)

Colonial swimming volvocaleans occur as slightly curved plates or hollow spheres. The number of cells in the colony and colonial shape are genetically determined and specific for each taxon. Asexual reproduction involves production of daughter colonies by successive bipartition of parental cells; these are called autocolonies (or coenobia). In some taxa, all cells of the colony are capable of autocolony formation, but in others only certain cells, known as **gonidia,** are capable of generating daughter colonies. When the organism is a hollow sphere, as in *Volvox,* daughter colonies develop with their cell apices facing inward, and the colony must go through a programmed inversion so that flagella are in contact with the external environment (Hallmann 2006) (Figure 19.46). Usually spiny, thick-walled zygotes (see Figure 19.3) are formed that germinate by meiosis.

Colonial volvocaleans commonly occur in the summer plankton of freshwater lakes, following a spring diatom bloom. Large size is regarded as adaptive in avoiding grazing pressure. Though the very large colonies of *Volvox* appear highly resistant to zooplankton grazing, young autocolonies of the smaller genus *Eudorina* are more vulnerable to predation (Happey-Wood 1988). Colonial volvocaleans often require at least one B vitamin.

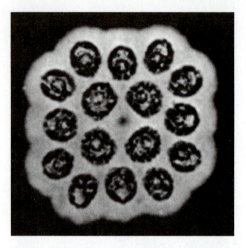

Figure 19.47 *Gonium pectorale.* Each cell of the planar colony has two flagella. This preparation was stained with india ink, which highlights the mucilage surrounding the cells. (Photo: A. Coleman)

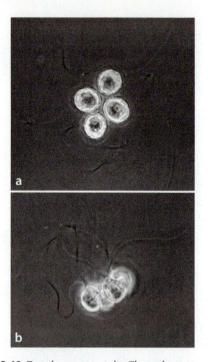

Figure 19.48 *Tetrabaena socialis.* The colony is seen in face view in (a) and side view in (b). (Photos: L. E. Graham)

GONIUM (Gr. *gonia*, angle) is a colony of 8, 16, or 32 (the number depending upon the species) chlamydomonad cells arranged in a flat plate (Figure 19.47). As they swim, the colonies rotate (Canter-Lund and Lund 1995). All cells are capable of forming autocolonies. Asexual reproduction occurs by loss of motility and subsequent division of each cell to form a colony of the size typical for the species. Sexual reproduction involves dissociation of colonies into single cells, which then function as isogametes. It is likely that different mating types are generally required for syngamy to occur. Zygotes germinate by meiosis to produce a four-celled colony. *Gonium* occurs in lakes with hard waters and high nitrogen content; it may also grow in barnyard ponds and watering troughs (Prescott 1951). The common 4-celled species *G. sociale* has been renamed *Tetrabaena socialis* (Nozaki and Itoh 1994) (Figure 19.48).

PANDORINA (named after Pandora, mythological woman) is a globular colony of 8 or 16 biflagellate cells that are closely adherent at their bases (Figure 19.49). The posterior ends of cells are somewhat narrowed. Colonies swim with a rolling motion. The eyespots of cells in the anterior of the colony are larger than those of the posterior, marking the occurrence of some degree of colony polarity. All cells are capable of forming autocolonies. Sexual reproduction is isogamous. *Pandorina* occurs in both hard and soft waters. Molecular systematic analysis suggests that with the exclusion of *Pandorina unicocca*, which has

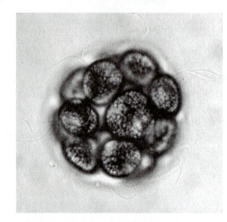

Figure 19.49 *Pandorina,* a ball of cells that are adherent at their bases and embedded in mucilage. Flagella are faintly visible in this image. (Photo: L. E. Graham)

been renamed *Yamagishiella, Pandorina* is probably a monophyletic genus (Nozaki et al. 1995). Coleman (2001) reported the occurrence of many sexually isolated groups (syngens), one of the first observations of cryptic speciation among protists.

PLATYDORINA (Gr. *platys*, flat + [Pan]*dorina*) has flattened, slightly twisted colonies of 8, 16, or 32

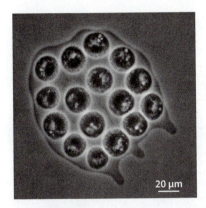

Figure 19.50 *Platydorina*, a flattened colony of cells embedded in mucilage. Flagella, borne in pairs on each cell, were not on this specimen when photographed. The posterior of the colony has distinctive lobed projections. (Photo: L. E. Graham)

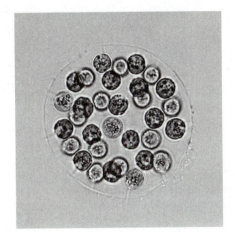

Figure 19.52 *Eudorina*. This genus consists of a hollow colony of cells that occupy the periphery of the mucilaginous sphere. Each cell bears two flagella. (Photo: L. W. Wilcox)

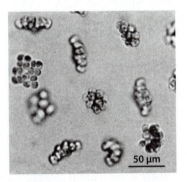

Figure 19.51 *Platydorina* colonies.These colonies undergo inversion during their development from hollow spheres to flat plates of cells. (Photo: R. Starr, in Bold and Wynne 1985)

cells. There are characteristic lobed projections of the colonial mucilage at the posterior of the colony (Figure 19.50). All cells are capable of autocolony formation, which begins with the formation of spherical colonies that must turn inside out and then flatten (Figure 19.51). Sexual reproduction is dioecious, and gametes are anisogamous. Zygote germination is by meiosis.

EUDORINA (Gr. *eu*, good, true, or primitive + [*Pan*]*dorina*) is a globular colony of spherical biflagellate cells that are not as closely adherent as those of *Pandorina* (Figure 19.52). The 16 or 32 cells in a colony first enlarge and then divide four or five times to form daughter colonies of 16 or 32 cells.

Sometimes, however, the 4 anterior-most cells do not enlarge or undergo divisions, and they eventually die. This is viewed as an early stage in the evolutionary development of the more extensive restriction of the reproductive role to posterior cells in *Volvox*. Sexual reproduction is anisogamous or oogamous. The individual cells of female colonies function as the larger female gametes, whereas cells of male colonies undergo division to form packets of adherent small sperm. These sperm packets swim as a unit to female colonies, and only upon reaching them do they dissociate into individual sperm cells. Zygotes germinate by meiosis, and variable numbers of meiotic products survive (see references in Bold and Wynne 1985). *Eudorina* is regarded as common in the plankton of hard-water lakes (Prescott 1951). Molecular phylogenetic analysis suggests that *Eudorina* is a polyphyletic species (Nozaki et al. 2000).

PLEODORINA (Gr. *pleon*, more + [*Pan*]*dorina*) differs from *Eudorina* in having both small and large cells in the same colony (Figure 19.53). All 128 or 256 cells are initially the same size and morphology, but about two-thirds of the cells in the colony posterior are able to enlarge and divide, forming autocolonies. The anterior-most cells are unable to form daughter colonies, and they ultimately die. Sexual reproduction may be dioecious (heterothallic) or monoecious (homothallic). Molecular systematic analysis suggests that this genus diverged from one clade of *Eudorina* (Nozaki et al. 2000).

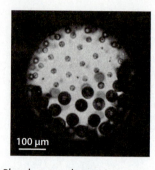

Figure 19.53 *Pleodorina* colony. At maturity, *Pleodorina* colonies are differentiated into small anterior cells unable to produce autocolonies and larger posterior cells that can produce daughter colonies. (Photo: R. Starr, in Bold and Wynne 1985)

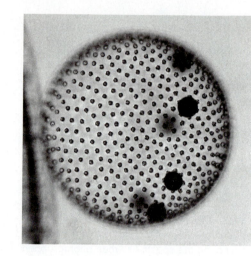

Figure 19.54 *Volvox.* (Photo: L. W. Wilcox)

VOLVOX (L. *volvo*, to roll) (Figure 19.54) is a very large motile colony that is common in ponds and lakes, in warm weather sometimes forming blooms that move up and down in the water column (Smith 1917). *Volvox* colonies contain 500 to several thousand biflagellate cells arranged at the periphery of a mucilage shell. External to the mucilage there is a boundary layer that is similar to the hydroxyprotein-rich glycoprotein wall of *Chlamydomonas reinhardtii*, but innermost regions of the extracellular matrix of *Volvox* are chemically different (see references in Schmitt et al. 1992). DNA sequence analyses suggest that *Volvox* is not monophyletic, i.e., that different species have arisen independently from ancestors resembling distinct modern groups of smaller colonial forms (Larson et al. 1992; Nozaki et al. 2000).

Volvox colonies are so large that most forms can be seen with the naked eye. Conventional preparation of microscope slides results in crushed cells, but "hanging drop" preparations allow microscopic visualization of swimming cells without crushing them. *Volvox* colonies cannot completely reverse direction while swimming, but they can change direction by stopping or slowing flagella on one side of the colony (Goldstein 1992). They are phototactic—able to swim toward light of moderate intensity. The colony has a polar organization, and the anterior-most cells have larger eyespots, suggesting enhanced phototactic ability.

In most species, protoplasmic strands interconnect adjacent flagellate vegetative cells, which are also known as **somatic cells** (Figure 19.55). Somatic cells are terminally differentiated and cannot divide, a state determined by expression of a master regulatory gene known as *regA* (Kirk et al. 1999). This gene inhibits other genes whose expression is required for cell division. Mutations in *regA* may restore the ability of somatic cells to divide and also cause them to lose flagella, reducing colony motility. Although normal somatic cells do not produce progeny, their flagellar action is essential to keep the large colony motile so that all cells are able to acquire light and mineral resources. For this reason, the somatic cells are said to display altruism, a behavior that benefits other individuals (other cells of the colony) at a cost to the altruistic individual (Nedelcu and Michod 2006; Michod 2007). Larger, nonflagellate gonidia are able to divide. Gonidia occur at the colony posterior, where they arise by asymmetrical divisions that occur early in colony development, and then enlarge further. The relatively large size of gonidia causes *regA* expression to be inhibited. Thus, gonidia are able to generate new colonies by asexual reproduction, and they also produce gametes during sexual reproduction.

Division of gonidia eventually produces hollow balls of cells—young colonies that develop inside the parental sphere. As the young colonies form, flagella are produced from the inside-facing apices of the component cells. In order for such young colonies to be able to swim, a process known as inversion must occur. (Developmental biologists are intrigued by the fact that a similar inversion process occurs during the early development of animal embryos, though the two processes evolved in parallel.) A pore, known as a phialopore, appears at the eight-celled stage, beginning as a region lacking in

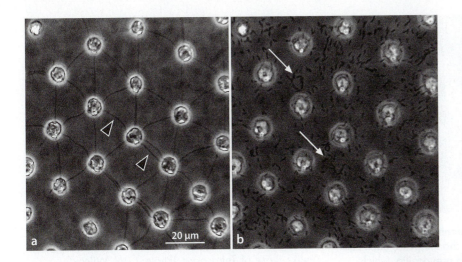

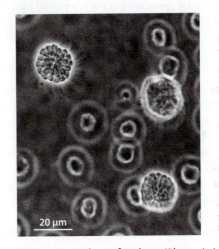

Figure 19.55 Intercellular cytoplasmic strands (arrowheads), which link each *Volvox* cell with its neighbors. Images (a) and (b) represent a through-focus series of a portion of a *Volvox* colony. (a) In one plane of focus, the intercellular strands are visible. (b) In the other, the strands are not evident, but bacterial cells (arrows) can be observed on the surface of the mucilaginous matrix. (Photos: L. W. Wilcox)

Figure 19.56 Sperm packets of *Volvox*. (Photo: L. W. Wilcox)

intercellular protoplasmic connections. When the number of cells in the young colony reaches that characteristic for the species, the colony inverts through the pore (Schmitt et al. 1992). Young colonies are eventually released by rupture of the parental colony surface.

Sexual reproduction begins when one or a few colonies are induced to develop male gametes. Sexual reproduction may be induced when water temperature reaches a critical high level, particularly in shallow pools or temporary ponds or puddles (Kirk 1997). Induced male colonies produce a potent chemical sexual attractant, or pheromone. This substance is a glycoprotein that diffuses through the water, inducing many other *Volvox* colonies to become sexually competent. It has been estimated that such pheromones can be effective at concentrations below 10^{-16} M. Mass sexual reproduction can occur if numerous colonies are exposed to the pheromone. Gonidia exposed to the pheromone undergo an alteration of their developmental fate from production of daughter autocolonies to production of a single egg cell or a flat packet of 16–64 elongate, pale, biflagellate sperm (Figure 19.56). Although some species are monoecious, most are dioecious. Upon liberation of sperm packets from parental colonies, they swim to female colonies, enzymatically lyse a hole in the colony mucilage, and then dissociate into individual sperm that fertilize the eggs. Eggs that have not been fertilized can develop into new colonies (Kirk 1997). Zygotes develop thick, spiny walls (see Figure 19.3) that serve in perennation; they may also acquire red coloration from production of carotenoid pigments. In the *Volvox* species

that have been studied, only a single meiotic product—a small biflagellate cell—survives; the other three apparently die. Successive divisions of this cell regenerate the colonial form.

Colonial nonmotile relatives of *Volvox*

TETRASPORA (Gr. *tetra*—from *tessares*, four + Gr. *spora*, spore) (Figure 19.57) is a macroscopic tubular or saclike aggregation of chlamydomonad cells within a mucilaginous matrix. *Tetraspora* is usually found growing attached to substrates, often in cold, running freshwaters. Larger stringy masses or sheets may become free floating. Cells are often arranged in groups of four, which are the products of the division of a parental cell, providing the basis for the generic

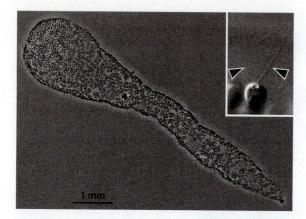

Figure 19.57 *Tetraspora*, a large nonmotile colony of *Chlamydomonas*-like cells. Rather than flagella, each cell of the vegetative colony bears two pseudocilia, which appear to have been evolutionarily derived from flagella by reduction. Inset: Higher-magnification view of a single cell with pseudocilia (arrowheads). (Photos: L. W. Wilcox)

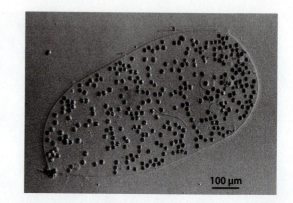

Figure 19.58 *Apiocystis*. Though much like *Tetraspora*, *Apiocystis* is smaller and has pseudocilia (not visible here) that extend beyond the limits of the mucilage, unlike those of *Tetraspora*. (Photo: L. W. Wilcox)

name. Each cell has a single, parietal, cup-shaped plastid with pyrenoid, contractile vacuoles, and two pseudocilia. Pseudocilia are long, threadlike extensions from the cell to the edge of the mucilage investment. Ultrastructural examination revealed that they are internally similar to flagella except they lack the two central microtubules (Lembi and Herndon 1966; Wujek and Chambers 1966). Pseudocilia cannot undergo swimming motions, and their function in *Tetraspora* and its relatives is unknown. Cells can transform into zoospores, but when they do, two flagella are produced *de novo*, not from modification of the pseudocilia. Isogamy and zygote formation have been observed.

APIOCYSTIS (Gr. *apion*, pear + Gr. *kystis*, bladder) (Figure 19.58) is much like a miniature *Tetraspora*, except that pairs of pseudocilia extend from each cell beyond the mucilage. Young stages of *Apiocystis* and *Tetraspora* are difficult to distinguish. Biflagellate zoospores and isogametes are known. *Apiocystis* grows attached to aquatic plants and other algae in freshwaters.

Chapter 20

Green Algae V

Charophyceans (Streptophyte Algae, Charophyte Algae)

Similarly to ulvophyceans, trebouxiophyceans, and chlorophyceans, charophyceans are green algae that display diverse body types and are defined by molecular, biochemical, and cellular traits (see Chapter 16). Although several species play ecologically significant roles, the single most important characteristic of charophyceans is that they are the modern protists most closely related to the land plants. Also known as embryophytes, the land plants consist of bryophytes (liverworts, mosses, and hornworts) and tracheophytes (vascular plants). The land plants are thought to have first appeared more than 470 million years ago—the age of the earliest fossils that are widely accepted as land plant remains (Gray et al. 1982; Wellman et al. 2003). No fossils are yet known to clearly link particular green algae to land plants. However, ultrastructural, biochemical, and molecular evidence derived from the study of modern species reveals that land plants evolved from a charophycean ancestor (Manhart and Palmer 1990; Graham 1993, 1996; Karol et al. 2001; Petersen et al. 2003; Lewis and McCourt 2004; Qiu et al. 2007). From this evolutionary perspective, the land plants are revealed to be a particular branch of the algae.

Because of their close relationship to plants, the panoply of modern charophycean species provides essential information about the evolution of fundamental land plant traits. For example, charophyceans reveal how plants acquired a key

Micrasterias

(Photo: L. W. Wilcox)

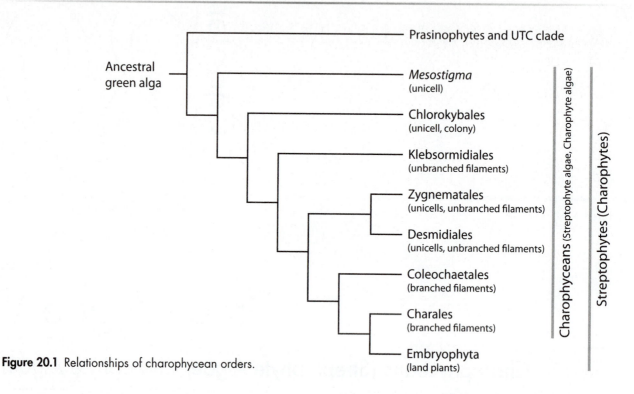

Figure 20.1 Relationships of charophycean orders.

innovation, the matrotrophic embryo. Critical to the successful colonization of land, a matrotrophic embryo is a young sporophyte whose early development is dependent on protection and nutrition provided by its maternal gametophyte (Graham and Wilcox 2000). Thus, the matrotrophic embryo also marks the origin of another common feature of the land plants, a life cycle involving alternation of diploid sporophyte and haploid gametophyte generations. Comparative studies of diverse charophyceans have illuminated the origin of many other plant traits, including MADS box genes, cytokinetic apparatus, sporopollenin-enclosed spores, and oogamous sexual reproduction. This chapter surveys charophycean diversity, emphasizing the evolutionary and ecological importance of charophyceans' traits.

20.1 General Features and Classification of the Charophyceans

Charophyceans include several early-diverging lineages of flagellate unicells, colonies, and unbranched filaments and two later-diverging lineages occurring

as branched filaments (Figure 20.1). In this book the term **charophycean** is used in the sense of (*sensu*) Mattox and Stewart (1984). These authors defined the class Charophyceae (informally, charophyceans), named for the widespread genus *Chara*, to include freshwater and terrestrial green algae sharing several distinctive traits. These traits, which also occur among land plants, include asymmetric flagellar roots with a multilayered structure (MLS), glycolate metabolism enzymes within a peroxisome, and an open and persistent spindle (see Table 16.1). In recent years, charophyceans have been found to share many other traits with land plants, and the number of shared traits has influenced classification. One example of a common feature of charophyceans and land plants is cell walls that are rich in a distinctive type of cellulose.

Charophycean and Plant Cellulose

Charophyceans and land plants both produce cellulose microfibrils at the cell membrane by membrane particles arrayed as rosettes (Figure 20.2) (Kiermayer and Sleytr 1979; Hotchkiss and Brown 1987; Giddings and Staehelin 1991; Okuda and Brown 1992). Such rosettes contain cellulose synthase, an enzyme complex that generates cellulose from precursor

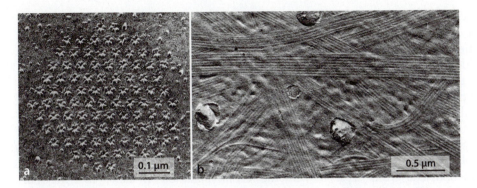

Figure 20.2 Cellulose-synthesizing complexes. (a) Hexagonal arrays of rosettes occur in the cell membranes of the charophycean alga *Micrasterias*. These are believed to be enzyme complexes involved in cellulose synthesis, spinning out ribbons of microfibrils to form the secondary cell wall (b). (From Giddings et al. 1980. Reproduced from *The Journal of Cell Biology*, 1980, 84:327–337, by copyright permission of the Rockefeller University Press)

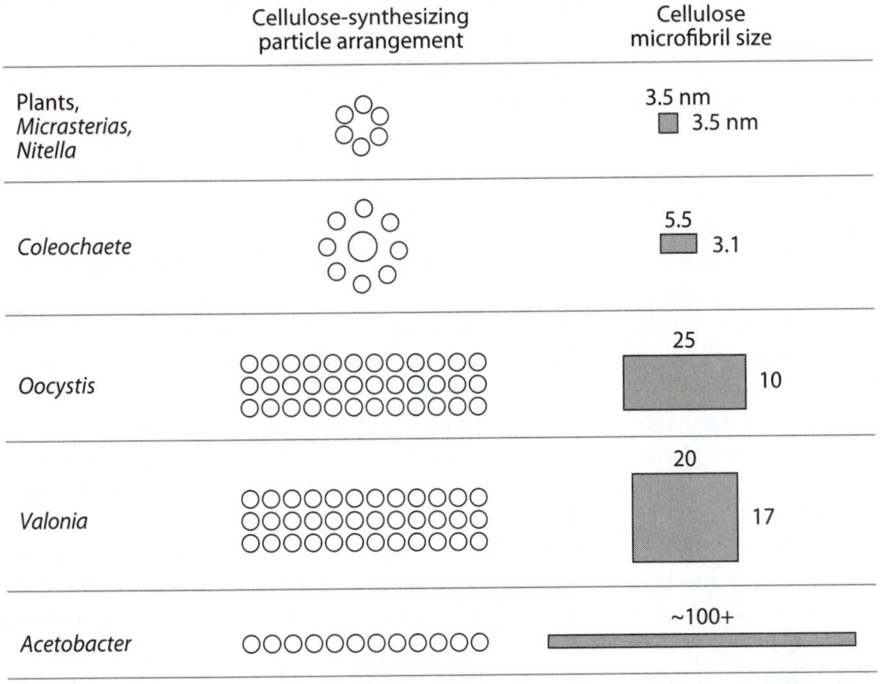

	Cellulose-synthesizing particle arrangement	Cellulose microfibril size
Plants, *Micrasterias*, *Nitella*	rosette	3.5 nm / 3.5 nm
Coleochaete	rosette	5.5 / 3.1
Oocystis	linear (3 rows)	25 / 10
Valonia	linear (3 rows)	20 / 17
Acetobacter	linear (1 row)	~100+

Figure 20.3 Diversity of cellulose-synthesizing complexes within the green algae. Rosettes similar to those of land plants occur in charophycean algae. These rosettes generate relatively thin microfibrils (shaded boxes). In contrast, cellulose-synthesizing complexes of chlorophyceans (*Oocystis*) and ulvophyceans (*Valonia*) are linear and are composed of three rows of particles that generate thicker microfibrils. The bacterium *Acetobacter* is regarded as having the ancestral type of linear cellulose-synthesizing complex. (After Okuda and Mizuta 1993)

UDP-glucose molecules. In contrast, the cellulose-synthesizing complexes of other green algae and bacteria occur in linear arrays (Figure 20.3), as do those of brown or red algae (see Figure 15.7). As a consequence of this difference, the cellulose produced by plants and charophyceans primarily occurs as smaller-diameter microfibrils and a form of cellulose known as the I_β allomorph. By contrast, other algal celluloses occur as larger-diameter microfibrils and are richer in a different cellulose allomorph, I_α.

These differences in biochemical structure cause the celluloses of plants and charophyceans to have distinctive physical properties. For example, plant cellulose is less easily degraded by fungal cellulases than is ulvophycean cellulose, a trait that has both ecological and industrial implications (e.g., Igarishi et al. 2006, 2007). Plants inherited the characteristic properties of their cellulose and many other traits from charophycean ancestors. The shared ancestry of charophyceans and land plants means that charophyceans cannot form a monophyletic group separate from the land plants. In fact, Charophyceae is now recognized as a paraphyletic group, one that does not contain all the descendants of a common ancestor (see Figure 20.1). This understanding has influenced the classification of charophyceans and collective terms used for them.

Charophycean Classification

Because charophyceans are a paraphyletic group, Bremer (1985) recommended that they be clustered with land plants to form a monophyletic phylum, Streptophyta (informally streptophytes). This classification distinguishes the streptophytes from green algae classified as prasinophytes and Chlorophyta (UTC clade). Thus, charophyceans can also be informally termed the **streptophyte algae**. More recently, Lewis and McCourt (2004) proposed an alternate formal classification scheme that groups charophyceans with embryophytes to form a phylum Charophyta with several classes, one being the land plants. Others have proposed similar ideas (e.g., Karol et al. 2001; Cimino and Delwiche 2002; Delwiche et al. 2002). Lewis and McCourt (2004) informally refer to charophyceans as **charophyte algae**; in their scheme, the informal term **charophytes** would refer to land plants plus the charophyte algae. However, because some modern authors have limited the term *Charophyta* to the genus *Chara* and a few close relatives, the precise meanings of the terms *Charophyta*, *charophytes*, and *charophyte algae* may not always be clear. Another issue to consider is that emerging concepts of eukaryotic supergroups (e.g., Burki et al. 2007; Kim and Graham 2008) may well influence future classification.

For these reasons, we have chosen for the time being to continue using the informal term *charophyceans, sensu* Mattox and Stewart (1984). Although this choice is not consistent with cladistic concepts, it does have some advantages. First, because the term *charophyceans* is widely recognized and parallels informal names given to the other green algal classes, it can be helpful to students. Further, when used as an informal collective term, *charophyceans* is consistent with the term *bryophytes*, which is widely and usefully employed to encompass the nonvascular land plants, even though liverworts, mosses, and hornworts together do not form a monophyletic group. Charophyceans are subdivided into several orders, and the ordinal names used here are consistent with the recommendations of Lewis and McCourt (2004).

Charophycean Orders and Their Evolutionary Significance

Several well-supported lines of recent molecular evidence indicate that the unicellular flagellate *Mesostigma viride* is the earliest-diverging modern charophycean known to date (Kim et al. 2006; Nedelcu et al. 2006; Simon et al. 2006; Rodríguez-Ezpeleta et al. 2007) (see Figure 20.1). The unicellular or colonial Chlorokybales diverges next in most phylogenies, though *Mesostigma* together with *Chlorokybus* formed an early-divergent clade in one study based on chloroplast genomes (Lemieux et al. 2007). Unbranched filaments that form the order Klebsormidiales diverge next, and then the Zygnematales + Desmidiales clade, which includes unicells and unbranched filaments that display isogamous sexual reproduction. The more complex charophyceans are Coleochaetales and Charales, both composed of branched filaments that display oogamous sexual reproduction, and both found primarily in aquatic habitats. An analysis based on chloroplast genome sequences indicated that Charales diverged fairly early and is sister to a clade composed of Coleochaetales, Zygnematales + Desmidiales, and land plants (Turmel et al. 2007). Other analyses, based on gene sequence data, indicate that Coleochaetales (Petersen et al. 2003) or Charales (informally charaleans) (Karol et al. 2001; Turmel et al. 2003; Luo and Hall 2007; Qiu et al. 2007) is the sister group of land plants. Work remains to be done in resolving charophycean divergence patterns, knowledge of which is important because it suggests the order in which traits first appeared.

The identity of the particular modern charophyceans that are most closely related to embryophytes has been considered important because it is assumed that such protists would best mirror the traits of the pivotal common ancestor. An important goal is to obtain genome sequences for one or more charophyceans so that comparisons can be made to early-diverging land plants. Such studies will help to reveal the genomic features of the common ancestor. However, it is important to note that more than 470 million years have passed since modern charophyceans diverged from embryophytes, and so modern charophyceans likely display many traits that do not reflect the ancestral type. For example, fossil evidence covering much of that time period indicates that charaleans have continued to undergo structural and reproductive evolution in response to aquatic selection regimes (Kelman et al. 2003; Feist et al. 2005). Further, during their evolution, the genomes of charaleans and some zygnemataleans have expanded to the point of "genomic obesity," likely in conjunction with the evolution of large cells. By contrast, *Mesostigma*, Klebsormidiales, and Coleochaetales have genome sizes within the range of those found in bryophytes; such smaller genomes are closer to that expected for the ancestral plant type and would be easier to sequence and annotate (Kapraun 2007).

Yet another issue to consider in the quest for a genomic profile of the common ancestor is raised by recent evidence for a surprising level of horizontal gene transfer (HGT) into plant mitochondrial genomes. Since genes can move from plant mitochondria to nuclei, HGT has the potential to also affect nuclear genomes (Richardson and Palmer 2007). Charophycean mitochondrial genomes are known to resemble those of land plants (Turmel et al. 2003). If charophycean mitochondria are similar to those of plants in having active DNA uptake systems and a propensity to fuse, HGT might have influenced charophycean genomes, as well as those of early-diverging land plants (Turmel et al. 2006). The possibility of HGT means that genomic components that are used for comparative analyses of charophyceans and bryophytes should be routinely tested for vertical descent.

For the reasons detailed above, the traits of a single representative charophycean or order are likely insufficient to accurately infer traits of an ancestor shared with land plants. Investigating the traits of diverse representatives will be necessary to achieve a full understanding of the evolutionary processes that gave rise to the first land plants. For example, comparative studies of chloroplast genomes show that most of the features typical of land plant

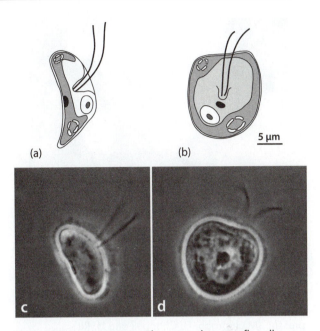

Figure 20.4 *Mesostigma*. This genus has two flagella and a single flat chloroplast that contains pyrenoids and an eyespot in the region near the flagellar basal bodies. Cells are shown in side view in (a) and (c) and face view in (b) and (d). (a, b: After Sym and Pienaar 1993; c, d: L. W. Wilcox)

chloroplast DNAs originated prior to the divergence of Zygnematales (Turmel et al. 2005). Likewise, TCP-type transcription factors characteristic of land plants are inferred to have appeared after divergence of Klebsormidiales, but prior to the divergence of Desmidiales (Navaud et al. 2007). Other important examples can be found in the following discussion of charophycean diversity.

20.2 Charophycean Diversity

MESOSTIGMA (Gr. *mesos*, middle + Gr. *stigma*, mark). The freshwater biflagellate *Mesostigma viride* (Figure 20.4) was originally recognized as a possible charophycean based on the presence in zoospores of an MLS flagellar root (Rogers et al. 1981; Melkonian 1989) (see Chapter 16). As noted earlier, several recent molecular studies have concluded that *Mesostigma* branches at the base of the streptophytes (charophytes *sensu* Lewis and McCourt 2004). The observation that *Mesostigma* and many other charophyceans live in freshwaters supports a current concept that the charophycean radiation primarily

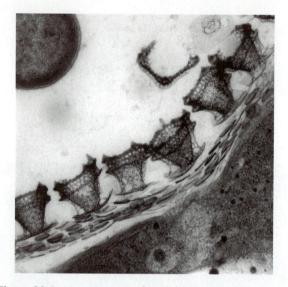

Figure 20.5 *Mesostigma* scales at the transmission electron microscope (TEM) level. Several layers are present. (Micrograph: D. Chappell and H. Ris)

occurred in freshwaters rather than marine waters (Graham 1993; Melkonian et al. 1995).

Mesostigma is disk shaped and has a central, or sometimes lateral, flagellar pit opening that penetrates quite deeply into the cell. A layer of small, flat, polygonal scales covers the flagella, and similar scales serve as the lowermost layer of body scales. *Mesostigma* has two additional layers of distinctive scales: The middle layer is composed of larger, flattened, oval scales that are ornamented with small pits; and the outer layer consists of large, distinctive basket scales (Manton and Ettl 1965) (Figure 20.5). Importantly, cellulose has not been identified in the extracellular covering of *Mesostigma*, and a study of some 3000 expressed genes did not find evidence for cellulose synthase genes in this organism (Simon et al. 2006). If future genomic analyses confirm this finding, it may be possible to infer that streptophytes acquired the ability to produce a cellulose-rich wall after the divergence of *Mesostigma*. The above-mentioned gene-expression study also found evidence that *Mesostigma* has a plastidic GAPDH (NADP+-specific glyceraldehyde-3-phosphate dehydrogenase) subunit B that seems to be unique to streptophytes (Simon et al. 2006). Glycolate oxidase is a part of the photorespiratory pathway of *Mesostigma*, a trait shared with other streptophytes (Iwamoto and Ikawa 2000). As in other streptophytes, *Mesostigma* uses both the plastidic DOXP/MEP and cytosolic

mevalonate (MVP) pathways for isoprene biosynthesis, whereas all tested chlorophytes (UTC clade) use only the plastidic pathway (Schwender et al. 2001). These examples indicate that many of the metabolic traits typical of streptophytes originated prior to the divergence of *Mesostigma*.

There is a single platelike chloroplast, which is thickened at the edges and contains several pyrenoids. A study of plastid pigments in six strains of *Mesostigma viride* identified lycopene, lutein, siphonaxanthin, siphonaxanthin C12:0 ester, siphonaxanthin C14:0 ester, γ-carotene, β-carotene, antheraxanthin, violaxanthin, all-*trans* neoxanthin, and chlorophylls *a* and *b* (Yoshii et al. 2003). In lacking 9'-cis neoxanthin, *Mesostigma* differs from other green algae and land plants. Siphonaxanthin may function as an antenna pigment in blue-green light absorption, suggesting adaptation to relatively deep water. Extensions of the chloroplast appear to connect to the flagellar basal bodies, which is unusual. An eyespot composed of two or three layers of pigmented globules lies within the plastid near the basal bodies. The occurrence of a mutant having a colorless eyespot allowed Matsunaga and associates (2003) to detect positive phototaxis of *Mesostigma* in blue light and to observe diaphototactic behavior, swimming perpendicularly to the direction of incident light in response to green light. Eyespots have not been reported in charophycean flagellate cells other than *Mesostigma*, suggesting that the eyespot must have been lost early in charophycean diversification. A large, lobed peroxisome lies between the chloroplast and the basal bodies and is attached to the latter. This association is believed to facilitate division and distribution of the peroxisome to daughter cells at cell division.

Fibrous connecting bands similar to those of other green algae link basal bodies, and each basal body is associated with a single flagellar root containing five to seven microtubules. The proximal part of each root is an MLS; thus there are two MLSs per cell. Aggregations of microtubules occupy the region of the *Mesostigma* cell beneath the flagellar pit, perhaps serving as a support system.

Chlorokybales

The order Chlorokybales includes *Chlorokybus atmophyticus* and *Spirotaenia*, a unicellular genus previously classified with Zygnematales but more recently linked with *Chlorokybus* by phylogenetic analysis. In addition to these genera being placed together by sequence analysis of SSU rDNA and *rbcL* genes, *Spirotaenia* lacks an intron typically found in zygnematalean SSU rDNA (Gontcharov and Melkonian 2004). On the basis of this evidence, we have included *Spirotaenia* in Chlorokybales, pending further study.

CHLOROKYBUS (Gr. *chloros*, green + Gr. *kybos*, cube) is a rarely encountered terrestrial or freshwater green alga (Geitler 1955). The vegetative body is a packet of rounded cells surrounded by a thick extracellular matrix (Figure 20.6) that includes cellulose (Figure 20.7). This observation suggests that cellulose biosynthesis originated in streptophytes prior to the divergence of *Chlorokybus*. Cytokinesis is followed by deposition of cell wall material only at the new cross-wall, as in filamentous and more complex charophyceans and plants. However, *Chlorokybus* is not considered to be either filamentous or parenchymatous (tissue-like) and is regarded as the simplest charophycean alga having a nonmotile vegetative stage. *Mesostigma* and *Chlorokybus* plastid genomes encode seven genes that are absent from all other green algae whose plastid genomes have been completely sequenced (Lemieux et al. 2007).

Sexual reproduction has not been observed. *Chlorokybus* cells can be induced to produce a single biflagellate zoospore during asexual reproduction. The zoospore flagella emerge laterally and are associated with a groove, in these respects resembling *Mesostigma*. Zoospore release in *Chlorokybus* is by disintegration of the parental cell wall, a process that is regarded as unspecialized compared to zoospore release mechanisms of other charophyceans (Rogers et al. 1980). The body and flagella of the zoospores are covered with small flat scales resembling those of *Mesostigma*. The flagella also possess hairs, as do those of some other charophyceans. (Such hairs are lacking from *Mesostigma*.) No eyespot is present in the zoospores, suggesting that eyespots were lost from the streptophyte lineages prior to divergence of *Chlorokybus*. There is a single peroxisome, which, like that of *Mesostigma* but not other charophyceans, is attached to the flagellar apparatus. The MLS and associated microtubular extension form the only known flagellar root. This MLS root contains fewer microtubules (10 or 11) than do those of other charophyceans (or land plants), and its microtubules do not extend as far down into the cell (Rogers et al. 1980). After the

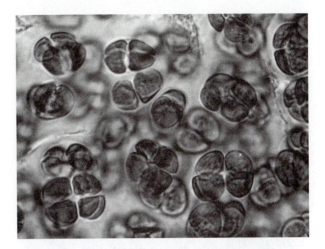

Figure 20.6 *Chlorokybus.* This genus is sarcinoid, meaning that the cells are arranged in packets of variable number, held together by mucilage. (Photo: L. E. Graham)

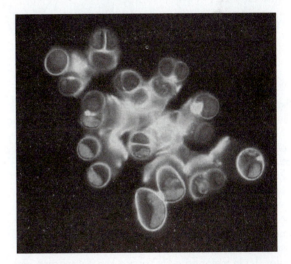

Figure 20.7 *Chlorokybus* extracellular matrix. These cells were treated with a compound that specifically binds to cellulose and fluoresces bright white. (Photo: L. E. Graham)

zoospores swim for about an hour, they become round, retract their flagella, and begin to deposit a cell wall beneath the body scales (Rogers et al. 1980). As the cell wall becomes well developed, the scale layer is lost. Each cell of *Chlorokybus* possesses a single cup-shaped chloroplast resembling those of some other charophyceans but, atypically, there are two distinct types of pyrenoids—one embedded within the plastid and containing numerous traversing thylakoids, and another located at the periphery of the plastid and lacking thylakoids. The pyrenoids

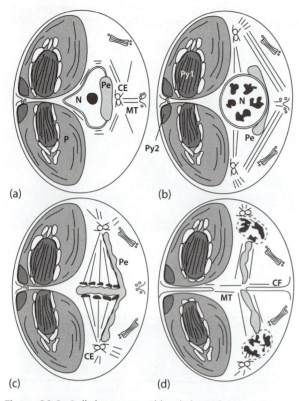

(a)   (b)

(c)   (d)

Figure 20.8 Cell division in *Chlorokybus*. This genus illustrates a number of cell division features characteristic of all charophyceans: Mitosis is open, and persistent spindle microtubules keep the daughter nuclei apart at cytokinesis. Other mitotic attributes of *Chlorokybus* occur in some other charophyceans, but not all. Mitosis in *Chlorokybus* is centric (there are centrioles at the spindle poles), and plastid and pyrenoid division occur prior to nuclear division. (a) Chloroplast (P) and pyrenoid (Py1, Py2) division precedes that of the nucleus (N). (b) Centriole pairs (CE) have moved to poles. (c) In metaphase/anaphase, the peroxisome (Pe) is closely associated with the spindle. (d) At telophase, a cleavage furrow (CF) develops alongside parallel-oriented microtubules (MT). (After Lokhorst et al. 1988, by permission of the *Journal of Phycology*)

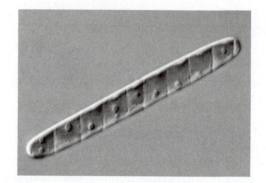

Figure 20.9 *Spirotaenia*. This genus is single celled and has a spirally twisted, ribbonlike chloroplast. (Photo: L. W. Wilcox)

move apart, forming the poles of the mitotic spindle; they are associated with material that serves as a microtubule organizing center. The centrioles are also associated with an array of astral microtubules. By metaphase, the nuclear envelope has completely broken down; mitosis is thus described as "open." The spindle persists until it is disrupted by the formation of a cross-wall (septum) as the cleavage furrow is completed. Microtubules are arrayed in the plane of the developing furrow (Lokhorst et al. 1988) (Figure 20.8).

SPIROTAENIA (Gr. *speira*, coil + Gr. *taenia*, band) (Figure 20.9) has unicells that are straight or somewhat curved, with a single spirally twisted, ribbonlike chloroplast. No flagellate cells are known. Sexual reproduction was studied in *S. condensata* (Hoshaw and Hilton 1966). This occurs by the pairing of cells of differing sizes within a blanket of mucilage. The paired cells then function as gametangia, producing non-flagellate gametes. Zygotes develop an unusual honeycomblike wall, lose pigmentation, and form large cytoplasmic oil droplets. Spontaneous germination of *Spirotaenia* zygotes begins 22 days after their formation; meiosis occurs at the onset, and four germling products are released. These may be observed in the vicinity of the empty zygote wall, surrounded by mucilage. The plastid genome of *Spirotaenia* should be examined for evidence of gene content similar to that of *Chlorokybus*. If *Spirotaenia*'s placement within Chlorokybales is upheld by future investigations, *Spirotaenia* may be of evolutionary interest as an early case of sexual reproduction in charophyceans.

of other charophyceans (and hornworts among land plants) are embedded and traversed by thylakoids; the simpler peripheral pyrenoid type is lacking (Lokhorst et al. 1988). The physiological and evolutionary relevance of these differing pyrenoids in *Chlorokybus* is not known.

At the initiation of mitosis in *Chlorokybus*, centrioles are positioned at the plane of cell division, which is indicated by a precocious (early developing) cleavage furrow. Centrioles then duplicate and

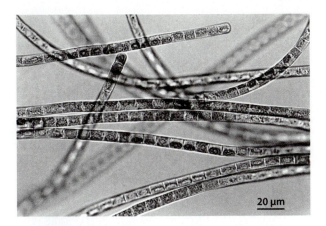

Figure 20.10 *Klebsormidium*, an unbranched filament. Cells have a single, platelike, parietal plastid. (Photo: L. W. Wilcox)

Klebsormidiales

This order includes the genera *Klebsormidium* and *Entransia* and the species currently known as *Hormidiella attenuta*. These are unbranched filaments whose cells have a single parietal plastid. *Klebsormidium* groups with *Entransia* with a high level of support (Karol et al. 2001; Turmel et al. 2002), and *H. attenuta* is sister to *Entransia* (Sluiman et al. 2008). When authentic strains of *Klebsormidium* are included in phylogenetic analyses, the order forms a well-supported clade, though it should be noted that some culture collection isolates labeled *Klebsormidium* group with Trebouxiophyceae (Sluiman et al. 2008). Sexual reproduction is not known to occur in Klebsormidiales, but asexual reproduction occurs via flagellate zoospores.

KLEBSORMIDIUM (named for Georg Albrecht Klebs, a German phycologist, + *Hormidium*, a genus of green algae [Gr. *hormidion*, small chain]) (Figure 20.10) occurs widely in terrestrial habitats such as soil and on tree trunks, as well as in splash zones of water bodies (Sluiman et al. 2008). Noting that the cells produced only a single zoospore, and that this zoospore lacked an eyespot and had two laterally emergent flagella, Marchant et al. (1973) predicted that *Klebsormidium* zoospores would possess an MLS-type flagellar root. Their subsequent ultrastructural examination confirmed this prediction. *Klebsormidium* is regarded as more specialized than *Chlorokybus* because zoospores are discharged through differentiated pores in the

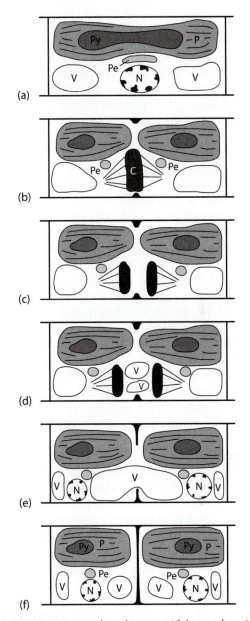

Figure 20.11 Mitosis and cytokinesis in *Klebsormidium*. (a, b) Division of the chloroplast (P), pyrenoid (Py), and associated peroxisome (Pe) is accomplished prior to completion of nuclear division. (c, d) Mitosis is open, and a precocious cleavage furrow begins to develop prior to cytokinesis. (e, f) A large vacuole (V) is instrumental in retaining separation of daughter nuclei (N) during cytokinesis, and is cleaved during cross-wall development by furrowing. (After Floyd et al. 1972, by permission of the *Journal of Phycology*)

cell wall. Production of the pore involves protrusion of the cytoplasm from a specific wall site and then deposition of presumed cell wall lytic vesicles rather than the generalized dissociation of the entire zoosporangial wall, as occurs in *Chlorokybus*. The ovoid zoospores are devoid of body scales, and the flagella also lack scales and hairs. It is presumed that scales and hairs were lost sometime after the divergence of *Klebsormidium* from the main line of charophycean evolution. After swimming for about an hour, the zoospores become round, retract their flagella, and form a cell wall without attaching to a substrate or forming a holdfast, as do some other green filamentous algae.

Klebsormidium vegetative cells and zoospores each contain a single parietal chloroplast with a pyrenoid surrounded by starch grains, but no eyespot. Each cell contains a single, land-plant-like peroxisome, but this organelle is not attached to the basal bodies as it is in *Chlorokybus* and *Mesostigma*. Rather, the peroxisome lies appressed to the midpoint of the chloroplast and is segregated at mitosis along with the plastid (Figure 20.11). Mitosis is open. Centrioles are present at the spindle poles; they are believed to be associated with spindle microtubule organizing material. An unusual large vacuole that forms in the interzonal region about halfway through anaphase appears to help complete chromosomal separation. Cytokinesis occurs by the development of a constricting furrow from the periphery (Floyd et al. 1972).

Klebsormidium may be difficult to distinguish at the light microscopic level from several other unbranched filamentous green algae that are not allied to the charophyceans. However, if it can be observed that only a single eyespotless zoospore is produced per cell, the filament is probably *Klebsormidium*. Comparative taxonomic studies of European species of *Klebsormidium* were done by Lokhorst (1996).

ENTRANSIA (named for E. N. Transeau, an American phycologist) has distinctive plastids that have multiple pyrenoids and are lobed at the edges in a way that resembles dripping paint (Figure 20.12). A tapering spine occurs at some filament tips. The unbranched filaments may spiral around each other and attach to substrates by means of basal adhesive. Cross-walls take on an H-shape when viewed in optical section. Vegetative reproduction occurs by means of zoospores that are released through a wall pore and by fragmentation achieved by programmed cell death (Cook 2004b). Programmed cell death is an important component of development in multicellular organisms; its occurrence in *Entransia* may be the earliest recorded for streptophytes.

Figure 20.12 *Entransia.* This filamentous genus has lobed plastids that resemble dripping paint. (Photo: M. Cook)

Zygnematales

Zygnematales are currently defined as unicells or unbranched filaments such as the genus *Zygnema* (Figure 20.13), whose outer walls lack pores and whose cells are not constricted (Gerrath 2002). The unicellular forms are informally known as the **saccoderm desmids**. The Zygnematales form a monophyletic sister group to the Desmidiales (Turmel et al. 2002; Gontcharov et al. 2003), and the two orders are sometimes grouped into the class Zygnemophyceae (Kenrick and Crane 1997) or Zygnematophyceae (van den Hoek et al. 1995).

No flagellate stages are known in Zygnematales or Desmidiales, and, in both orders, sexual reproduction occurs by means of a conjugation process involving non-flagellate gametes. Zygotes having distinctive shapes and wall structure (Figure 20.14) result from sexual reproduction; these typically confer the ability to survive long periods when conditions are not favorable for growth, germinating when the environment improves. Zygote persistence has been attributed to the presence of acid hydrolysis-resistant, sporopolleninlike polymers in the cell wall (DeVries et al. 1983). Such polymers

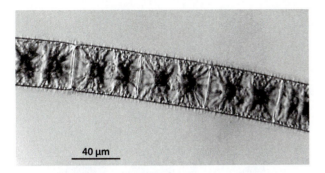

Figure 20.13 *Zygnema*. This unbranched filament lends its name to the order Zygnematales. (Photo: L. W. Wilcox)

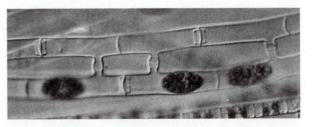

Figure 20.14 Conjugated *Spirogyra* with zygotes from a field collection. (Photo: M. Cook)

may explain why zygotes similar to those of modern zygnematalean algae have been recovered from sediments of Carboniferous age (some 250 million years old) and younger.

Zygnematalean green algae can be environmentally conspicuous. Ponds, ditches, sheltered nearshore regions of lakes, and slow-flowing streams may exhibit blooms of *Spirogyra* or related forms each spring, the growths sometimes assuming nuisance proportions (Graham et al. 1995). *Mougeotia* and some relatives can produce metaphytic (subsurface) clouds in lake waters that have been affected by acidic precipitation (acid rain) or experimental acidification (Schindler et al. 1985; Watras and Frost 1989; Howell et al. 1990; Turner et al. 1991).

Cell biology of zygnemataleans

Cell walls and mucilage. The extracellular matrix of zygnematalean cells typically includes several layers. An outer mucoid layer of calcium pectate and hemicelluloses, a thin fibrillar primary wall, and a thicker, fibrillar secondary wall are commonly produced by *Spirogyra*, *Zygnema*, *Mougeotia*, and other filamentous forms, as well as by saccoderm desmids. The fibrillar portion of the wall of one *Mougeotia* species was found to consist of both noncellulosic carbohydrates (64%) and cellulose (13%), comparable to the cell walls of *Klebsormidium* (Hotchkiss et al. 1989). The extensive mucilage envelope that characterizes many saccoderm desmids and their filamentous relatives appears to be extruded through the cell wall (Gerrath 1993). This mucilage sheath confers a characteristic slimy feel to masses of *Spirogyra* and its relatives. Possible functions for zygnematalean sheaths include water retention and resistance to desiccation, nutrient trapping, and absorption of harmful

ultraviolet radiation. Iron is sometimes deposited in the outer wall layers of zygnematalean algae; it is rather evenly distributed in some but more localized in others, giving cell walls a yellow or brown color (Gerrath 1993).

Plastid movements. The filamentous *Mougeotia* and the closely related unicell *Mesotaenium* possess the ability to orient their single, axial, platelike plastids to achieve optimal exposure to light, much as a motor-driven solar panel can be repositioned to follow changes in direction of solar radiation through the day. This process requires three components: one or more sensory pigments to perceive the wavelength and direction of the light signal, a transducer to convert the light signal into chemical information, and a mechanical effector to receive the message and move the chloroplast accordingly. In *Mougeotia scalaris*, a blue-light sensor that is likely phototropin and a red-sensing chimeric photoreceptor of phototropin and phytochrome called neochrome are involved in signal perception (Suetsugu and Wada 2007; Kagawa and Suetsugu 2007).

Red-light sensing is thought to effect changes in the binding of actin microfilaments at the cell periphery with myosin molecules that are linked to the chloroplast surface. The actomyosin motor then causes changes in positioning of the chloroplast. The result is that the platelike chloroplast is pulled into a position such that its broad face is directed toward the light when irradiance is low or optimal but so the edge faces the light when irradiance is too high (Wagner and Grolig 1992). Calcium influx, interaction of calcium ions with the calcium-binding protein calmodulin, and the binding of calcium ions with protein kinases are also thought to be involved in the light signal transduction process in these algae. Actomyosin interactions are also involved in organelle movements that occur in interphase cells of other zygnemataleans, such as *Spirogyra*, whose plastids do not rotate (Grolig 1990).

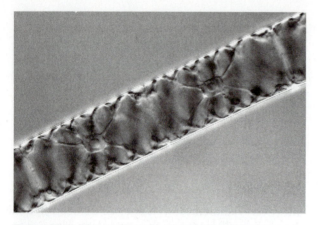

Figure 20.15 The nucleus of zygnemataleans is suspended in cytoplasm in the cell center, as illustrated here by *Spirogyra*. (Photo: M. Cook)

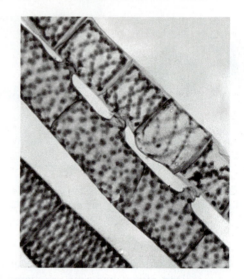

Figure 20.16 Conjugation tubes link cells of two mating filaments in *Spirogyra*. Such conjugation tubes occur commonly in related forms; flagellate gametes are not produced in any of the modern zygnemataleans. Conjugation tubes allow transfer of gamete cytoplasm from one filament to the other or to the midregion of the tube, where zygote formation may occur. (From *Origin of Land Plants*. Graham, L. E. © 1993. Reprinted by permission of John Wiley & Sons, Inc.)

Mitosis and cytokinesis of zygnemataleans. The nuclei of zygnemataleans are often large and conspicuous (Figure 20.15), consistent with the fact that large, polyploid nuclear genomes are known to occur (Kapraun 2007). Cell division in zygnemataleans typically occurs during the dark portion of the light–dark cycle. Mitosis and cytokinesis in zygnemataleans resemble those of *Chlorokybus* and *Klebsormidium*, except that centrioles are absent. In *Spirogyra* and *Mougeotia*, the nuclear envelope remains intact until metaphase, when it disintegrates, as in other charophyceans. These algae, and *Zygnema*, are also unusual among zygnemataleans in that a small phragmoplast—an array of perpendicular microtubules as well as membranous tubules and vesicles—occurs at the central region of the cell at the plane and, at the same time, as furrow extension from the periphery (Fowke and Pickett-Heaps 1969a,b; Pickett-Heaps and Wetherbee 1987). This small phragmoplast has been suggested to represent an intermediate stage in the evolutionary origin of more highly developed, plantlike phragmoplasts of *Coleochaete* and Charales (Pickett-Heaps 1975; Graham and Kaneko 1991). A study of the *Spirogyra* phragmoplast via fluorescent tagging of cytoskeletal components and video microscopy revealed several differences between phragmoplast structure and behavior as compared to those of other charophyceans and higher plants (Sawitzky and Grolig 1995). In contrast to *Coleochaete* and charalean algae, plasmodesmata are not known to occur in the cross-walls of zygnematalean algae. Furrowing in *Spirogyra* involves actin microfilaments (Goto

and Ueda 1988; Nishino et al. 1996); these presumably help to constrict the cytoplasm in a fashion analogous to the tightening of purse strings. Immunolocalization of microtubules has been followed throughout the cell cycle in *Mougeotia* (Galway and Hardham 1991).

Zygnematalean reproduction

Asexual reproduction. As noted previously, zygnemataleans do not produce flagellate zoospores, in contrast with *Chlorokybus*, *Klebsormidium*, and Coleochaetales. Such lack of flagella correlates with absence of centrioles, which are necessary for the generation of flagella (Pickett-Heaps 1975). However, filamentous zygnemataleans may reproduce asexually by fragmentation, and populations of single-celled zygnemataleans grow by means of mitosis and cytokinesis. Vegetative cells are dispersible by wind, insects, and water birds (as are desiccation-resistant zygotes) (Hoshaw et al. 1990). Zygnemataleans may sometimes produce thick-walled resting cells known as akinetes, aplanospores, asexual spores, or parthenospores, the latter originating from unpaired cells of sexual populations.

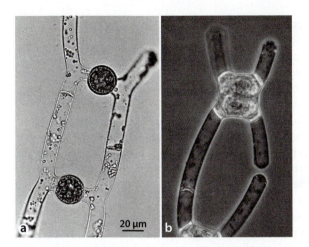

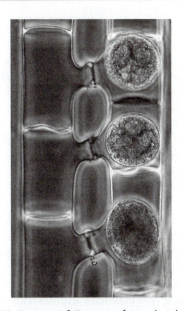

Figure 20.17 Zygotes of two species of *Mougeotia*, developed within the conjugation tube. These species both exhibit scalariform (ladderlike) conjugation. (Photos: L. W. Wilcox)

Figure 20.18 Zygotes of *Zygnema* formed within the parental walls of one of the gametes. (Photo: L. W. Wilcox)

Sexual reproduction. Zygnematalean mating is accomplished by the physical pairing of filaments or single cells, their enclosure within common mucilage, and subsequent fusion of nonflagellate gametes—a process known as **conjugation** (Figure 20.16). In the laboratory, cultures of zygnematalean algae have been induced to undergo conjugation by reducing the combined nitrogen concentration in their growth medium (Biebel 1973), by increasing carbon dioxide levels (Starr and Rayburn 1964), or by increasing temperature, light levels, and/or concentrations of Ca^{2+} and Mg^{2+} (Gerrath 1993), but little is known about the factors responsible for inducing sexual reproduction in nature.

Spirogyra, Mougeotia, Zygnema, and other filamentous forms may undergo **scalariform conjugation**, where filament pairs align themselves laterally and develop modified branches known as **conjugation tubes**, that link opposing cells (see Figure 20.16). The conjugation tube is composed of an outgrowth (papilla) from each opposing cell; when they meet, the wall at their interface is degraded to form an open tube through which gametes can move. Ultrastructural studies have suggested that papilla growth occurs by addition of new cell wall material, under the influence of enlarging vacuoles in the papilla, but the mechanism by which the end walls disintegrate is unclear (Pickett-Heaps 1975). Although gamete motion is often described as "amoeboid," there is no direct evidence that zygnematalean cells can actually move in the same way as true amoeboid cells.

Gamete protoplasts can be observed to shrink as they lose water; increased external hydrostatic pressure or mucilage accumulation in the surrounding area may be responsible for their propulsive movement (Pickett-Heaps 1975), but the mechanism is not well understood. Sometimes, both gametes move into the conjugation tube, whereupon a zygote is formed in the center of the tube (Figure 20.17). In other cases, the cytoplasm of only one of the connected cells moves across the tube, forming a zygote within the confines of the second cell (Figure 20.18). The timing of gamete nuclear fusion and meiosis in filamentous zygnemataleans appears to vary from organism to organism.

An alternate form of conjugation, known as **lateral conjugation**, occurs when gametes develop from adjacent cells within the same filament. In this case, filament pairing does not occur. Rather, a short curved tube extends from one cell to the next in the filament. Because lateral conjugation is a form of selfing, this process would be expected to result in reduced levels of genetic variability compared to scalariform conjugation, which may involve filaments that are genetically distinct. Both result in the production of resistant-walled zygotes that serve a perennation function, and during zygote germination (involving meiosis) there exists the potential for recombination events. However, little population genetic work has been done to examine the relative roles of lateral and scalariform conjugation. Among the single-celled saccoderm desmids, sexual reproduction involves cell aggregation, formation of gametes through mitotic division, papilla formation, release of gamete protoplasts from enclosing walls, and zygote formation.

Zygote development involves formation of a thick wall consisting of as many as six distinct layers. Callose and sporopollenin (DeVries et al. 1983) have been demonstrated to occur among the layers of zygote walls of at least some zygnematalean species. Mature zygotes are often highly ornamented and colored orange-brown as a result of wall formation and chlorophyll degradation. In nature, zygotes germinate in spring, or the end of a period of dry dormancy, as when a temporary pool is re-formed in the wet season. Zygnematalean zygotes can withstand burial in mud for long periods (Brook and Williamson 1988), and zygotes have been germinated after dry storage for more than 20 years (Coleman 1983). Zygote germination can be induced in the laboratory by allowing the culture medium to slowly evaporate; storing the zygotes in the dark, with or without refrigeration, for 1 to 12 months; and then rewetting zygotes with fresh culture medium. A few hours following rehydration, zygotes become green due to synthesis of chlorophyll, and one to three days later, the wall ruptures, and a germination vesicle containing the meiotic products emerges.

Zygnematalean ecology

Zygnemataleans are almost exclusively found in freshwater habitats, although a few have been collected from brackish waters. They are ubiquitous in freshwaters, occurring in pools, lakes, streams, rivers, marshes, and especially bogs and mildly acidic, nutrient-poor streams. In addition, zygnemataleans may be abundant in reservoirs, cattle tanks, roadside ditches, irrigation canals, and other water bodies of human construction. *Spirogyra*, for example, was found at nearly one-third of the more than 1000 locations sampled by McCourt et al. (1986), and in a North American continent-wide survey, Sheath and Cole (1992) located *Spirogyra* in streams from a wide variety of biomes, including tundra, temperate and rain forests, and desert chaparral. In streams and shallow lakes, *Spirogyra* is typically attached to stable substrates but can also occur as free-floating mats that originate from benthic zygotes or filaments (Lembi et al. 1988). As growth and photosynthesis

occur, oxygen bubbles become entrapped in the mats and provide flotation; at the water surface the algal mats are exposed to high temperatures and light levels. Optimal temperature and irradiance conditions for photosynthesis for one species of *Spirogyra* were determined to be 25°C and 1500 µmol photons m^{-2}, respectively. Net photosynthesis was observed to be positive at 5°C under high irradiance conditions, explaining the widespread occurrence of surface growths in the cool waters of early spring. However, the alga could not maintain positive photosynthesis at the low light levels that can result from self-shading when temperatures were high (30°C–35°C), explaining late spring and summer declines in zygnematalean mats (Graham et al. 1995).

Mougeotia, a filamentous zygnematalean, often forms large nuisance growths in subsurface freshwaters affected by acid precipitation or experimental acidification. Such waters are characterized by increases in the concentration of metals such as aluminum and zinc, reduced levels of dissolved inorganic carbon, and food web changes, including reduction in numbers of herbivores (Stokes 1983; Webster et al. 1992; Fairchild and Sherman 1993). The appearance of *Mougeotia* mats is widely regarded as an early indicator of environmental change (Turner et al. 1991). The optimal light, temperature, and pH conditions for photosynthesis, as well as effects of the metals zinc and aluminum on photosynthesis, were determined for this alga so that its common and specific association with acidification could be better understood. Net photosynthesis was high (on average over 40 mg O_2 was produced per gram dry weight per hour) over a wide range of irradiances (300–2300 µmol quanta m^{-2} s^{-1}). The optimal temperature was 25°C, and the organism exhibited tolerance of a wide range of pH (3–9) and metal concentrations. These results, together with release from herbivores, help explain the rise to dominance of large subsurface *Mougeotia* growths in acidified lakes (Graham et al. 1996a,b).

Zygnematalean diversity

SPIROGYRA (Gr. *speira*, coil + Gr. *gyros*, twisted) is a filament composed of cells having 1–16 spiral, ribbon-shaped chloroplasts per cell (Figure 20.19). The plastid edges are often beautifully sculpted, and numerous pyrenoids are present (Figure 20.20). Sometimes rhizoidal processes occur at the basal end of the filament; these are involved in attachment to substrates. Cytoplasmic streaming—based on the

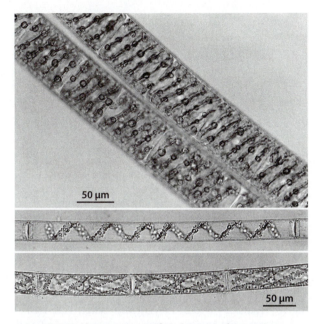

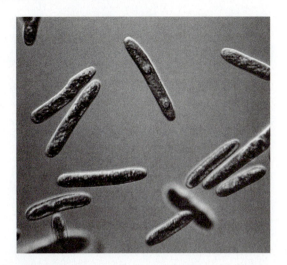

Figure 20.21 *Mesotaenium* is a unicell having just one plate-shaped axial plastid. (From *Origin of Land Plants*. Graham, L. E. © 1993. Reprinted by permission of John Wiley & Sons, Inc.)

Figure 20.19 Three species of *Spirogyra*. This genus is an unbranched filament of cells attached end to end. Note the helically twisted ribbon-shaped chloroplasts with multiple round pyrenoids. (Photos: L. W. Wilcox)

Figure 20.20 Pyrenoids. Higher-magnification view of the spiral, lobed, ribbonlike plastid of *Spirogyra* with numerous globular pyrenoids. (Photo: L. W. Wilcox)

action of actin microfibrils—can often be observed in the peripheral cytoplasm. The nucleus is suspended in the center of the cells (see Figure 20.15). Both scalariform and lateral conjugation have been observed. During conjugation, previously bright-green filamentous masses turn noticeably brownish

in color, reflecting the loss of chlorophyll pigments from zygotes and development of brown zygote walls. Zygotes germinate to form a single filament; from this it is deduced that only a single meiotic product survives. Hundreds of species have been described. *Spirogyra* is not monophyletic, nor is its close relative *Sirogonium*, but these two zygnemataleans consistently diverge closest to the base of the Zygnematales + Desmidiales clade (Drummond et al. 2005; Hall et al. 2008).

MESOTAENIUM (Gr. *mesos*, middle + Gr. *tainia*, ribbon) unicells are shaped like cylinders, each with a single platelike plastid having several pyrenoids (Figure 20.21). Culture studies have shown that sexual reproduction in *Mesotaenium kramstei* involves the formation of a broad conjugation tube that can grow from any portion of the cell wall. Mucilage is secreted inside the wall as gametes shrink during development. Mature zygotes are mahogany brown (Biebel 1973). *Mesotaenium* is not monophyletic, with some species (*M. kramstei*) more closely related to another saccoderm desmid known as *Cylindrocystis*, and others (*M. caldariorum*) linked with the filamentous *Mougeotia* (Hall et al. 2008).

MOUGEOTIA (named for Jean Baptiste Mougeot, an Alsatian physician and botanist) consists of long, unbranched, free-floating filaments, each cell of which is characterized by a single platelike

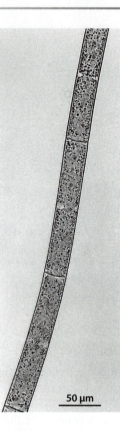

Figure 20.22 *Mougeotia.* This genus is an unbranched filament having a single platelike plastid per cell. (Photo: L. W. Wilcox)

50 μm

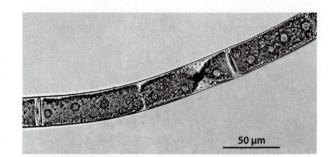

50 μm

Figure 20.23 Plastid rotation in *Mougeotia*. The *Mougeotia* plastid can rotate around a lengthwise axis, and portions of the plastid can turn independently, generating a twisted plastid. (Photo: L. W. Wilcox)

chloroplast (Figure 20.22). The plastids are suspended in the central area of the cell (i.e., are axial in location). Orientation of the plastid to achieve optimal light conditions can be commonly observed; in the same filament, some plastids may be in face view, others in edge view, and yet others twisted so that part or parts of the plastid are in face view and others in edge view (Figure 20.23). Pyrenoids are either arranged in a single row or scattered throughout the chloroplast. Conjugation in *Mougeotia* is usually scalariform, and zygotes typically form in the conjugation tube. Only a single filament is produced upon zygote germination. *Mougeotia* cells contain numerous small vacuoles filled with phenolic compounds, believed to serve in protecting the cells from herbivores (Wagner and Grolig 1992).

CYLINDROCYSTIS (Gr. *kylindros*, cylinder + Gr. *kystis*, bladder) unicells are cylindrical and contain two axial, stellate chloroplasts similar to those of *Zygnema*. Sexual reproduction of a number of species has been studied in culture. Easily grown forms that might prove useful in molecular genetics studies are strains of *C. brebissonii* (Figure 20.24)

(UTEX 1922 and 1923) isolated by Biebel (1973). They grow well in defined culture medium (Bold's Basal Medium—Stein 1973), and when one-week-old cultures of the two mating types are mixed and placed into culture medium lacking nitrogen salts, conjugating pairs appear within four days. A high yield of zygotes results within two weeks. Conjugation begins with lengthwise pairing, and wall papillae appear at the midregions of the lateral walls, later fusing to form a tube (Figure 20.24b). The plastids remain visible during zygote dormancy. Zygotes (Figure 20.24c) can be concentrated and placed on agar medium, dried, kept for two months in a dormant condition, and then induced to germinate by transfer to fresh agar medium and exposure to light. Germination (Figure 20.24d) occurs about one month later. At the onset of germination, the plastids migrate to new positions, one pair lying above the other, and meiosis is presumed to commence. All four meiotic products (gones) survive and could presumably be isolated for tetrad analysis. *Cylindrocystis* is not monophyletic (Hall et al. 2008).

ZYGNEMA (Gr. *zygon*, yoke + Gr. *nema*, thread) occurs as relatively short, unbranched filaments of cylindrical cells, enclosed by a mucilage sheath. As is the case for *Spirogyra*, there may be basal rhizoidal outgrowths that serve in attachment. There are two stellate chloroplasts per cell, each with a central pyrenoid (Figure 20.25). The chloroplast genome of *Z. circumcarinatum* is significantly larger than that of some other charophyceans and land plants, suggesting extensive expansion by addition of intergenic spacers and introns (Turmel et al. 2005). Conjugation is similar to that of

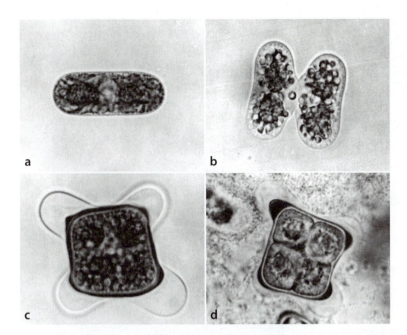

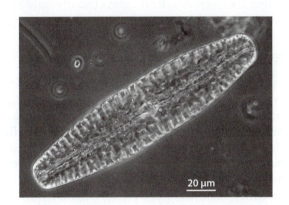

Figure 20.24 Sexual reproduction in *Cylindrocystis*. (a) Vegetative cell. (b) Mating gametes with short conjugation tube. (c) Pillow-shaped zygote with remains of parental gamete cell walls at the corners. (d) Zygote that has undergone meiosis to form four daughter products. (From Biebel 1973)

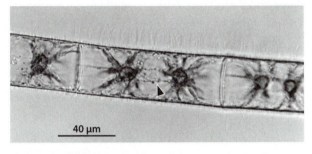

40 μm

Figure 20.25 *Zygnema*. This genus is an unbranched filament having two stellate plastids per cell, each with a large conspicuous pyrenoid. Arrowhead points to cytoplasmic bridge between plastids, which contains nucleus. (Photo: L. W. Wilcox)

20 μm

Figure 20.26 *Netrium*. This genus is a unicellular saccoderm desmid having two elaborately lobed chloroplasts, one in each cell half. (Photo: L. W. Wilcox)

Spirogyra; both scalariform and lateral conjugation are observed. Akinetes—asexually produced, thick-walled cells containing abundant oil and starch storage products—can be formed; these are able to survive for over one year before germinating into vegetative filaments. Prescott (1951) observed that in the western Great Lakes region, species growing in high-pH waters more often possessed more zygotes than forms occurring in low-pH habitats. Zygotes germinate to form a single filament, suggesting that only a single meiotic nucleus survives. *Zygnema* is frequently found in nature together with *Spirogyra* and/or *Mougeotia*. At least 80 species have been described.

NETRIUM (Gr. *netrion*, small spindle) (Figure 20.26) unicells are elongated and cylindrical, usually with rounded ends. Two large, elaborately lobed and ridged chloroplasts, each with a pyrenoid, occur per cell. Copious mucilage is excreted into the surrounding environment. A broad conjugation tube is formed, gametes shrink when they contact each other, the protoplasts then fuse, and the zygote develops in the conjugation tube. The plastid pigmentation disappears, and a golden-brown wall develops on the zygote. After two months in culture medium, the zygotes germinate spontaneously. Typical plastid

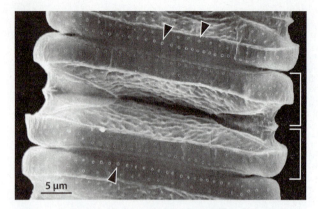

Figure 20.27 Scanning electron microscope (SEM) view of mucilage pores (arrowheads) in the semi-cell walls of *Desmidium*. The brackets indicate the two semi-cells of the cell that is shown in its entirety. (Micrograph: S. Cook)

Figure 20.28 Mucilage. Prisms of mucilage are produced from individual pores (arrows), forming a confluent sheath on *Desmidium grevillii* (shown here in a captured video image of a filament viewed with light microscopy). The small black dots (arrowheads) are end-on views of bacteria that live in the mucilaginous sheath. (Photo: L. W. Wilcox)

pigmentation reappears, and the protoplast swells, causing the zygote wall to burst. Usually only two meiotic products survive to form germlings.

Desmidiales (Placoderm Desmids)

Desmidiales, also known as placoderm desmids, have distinctive cell walls that explain characteristic features of their ecology. The cells of Desmidiales are often constricted into two parts, known as semi-cells, by a narrow region known as the isthmus. The walls surrounding such semi-cells are formed at different times, for which reason the desmidialean cell wall is said to be composed of two pieces. Desmidialean cell walls are also perforated with pores (Figure 20.27), are often highly ornamented, and sometimes contain extremely resistant compounds. The presence of wall pores helps explain the common ability of placoderm desmids to move by mucilage secretion, and mucilage production from pores may serve other functions. Some desmids produce acid and microbe-resistant polyphenolic polymers in vegetative cell walls (Gunnison and Alexander 1975a,b; Kroken et al. 1996; Graham et al. 2004). Such wall compounds are likely responsible for preservation of the old-est- known fossil vegetative cells of desmids, *Paleo-closterium leptum*, from the Middle Devonian (about 380 million years ago) (Baschnagel 1966). Cells similar to modern *Cosmarium* have been reported from amber of Triassic age (220 million years old) (Schmidt et al. 2006a). Resistant cell wall polymers may also explain the survival of placoderm desmids

such as *Closterium*, *Micrasterias*, *Pleurotaenium*, and *Euastrum* in drying mud at the edges of lakes. Such desmids have been found to be alive after three months of drying and at depths of 6 cm in sediments (Brook and Williamson 1988).

Cell wall structure, mucilage extrusion, and cell motility

Placoderm desmids have a primary wall consisting of pectins and cellulosic microfibrils that is often discarded after the development of an often highly ornamented secondary wall. However, the cells of filamentous placoderm desmids are held together by retained primary wall material (Krupp and Lang 1985). Pectinaceous mucilage extruded from wall pores can be visualized at the light microscopic (Figure 20.28) and ultrastructural levels (Figure 20.29) as "prisms" of extruded material. Ultrastructural and immunocytochemical analysis of *Closterium* mucilage revealed that mucilage vesicles are produced by Golgi bodies and then released through flask-shaped cell wall pores. Under normal conditions, each cell

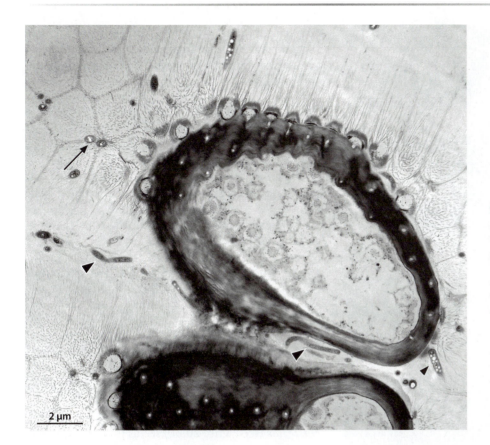

2 μm

Figure 20.29 Mucilage, pores, and epibacterial communities. Prisms of mucilage and their association with wall pores of *Desmidium grevillii*, as viewed with TEM. Note occurrence of associated bacterial cells (arrowheads) in the pocket formed at the isthmus region and at the interfaces of mucilage prisms (arrow). (From Fisher and Wilcox 1996, by permission of the *Journal of Phycology*)

can produce about 3 μg of mucilage in 30 days, but mucilage production increases three to four times when cells are grown under low-phosphate or low-nitrate conditions (Domozych et al. 1993; Domozych and Domozych 1993). Mucilage extrusion plays several roles.

A number of zygnematalean algae harbor putatively symbiotic bacteria within the confines of their sheaths (see Figures 20.28 and 20.29) (Gerrath 1993; Fisher and Wilcox 1996; Fisher et al. 1998). In other cases, the mucilage may confer a reduced sinking rate to planktonic forms or serve as an attachment mechanism for periphytic species. Many desmids are capable of motility through mucilage extrusion. Movement is typically quite slow, only about 1 μm per second. Gliding movement is accompanied by continuous secretion of a slime trail that can be visualized at the light microscopic level with the addition of india ink to the preparation. *Closterium* and some other desmids secrete mucilage only from opposing cell tips, in alternate fashion, resulting in a somersault-like movement. The placoderm desmid *Cosmarium*, among others, secretes mucilage from

the older cell wall half, while the younger wall half is lifted from the substrate at a slight angle. Subsequent swelling of the mucilage by water absorption then propels the cell forward (Häder and Hoiczyk 1992). Phototaxis has been observed in many desmids (Gerrath 1993).

Mitosis and development of new semi-cells

In *Closterium*, the first sign of impending mitosis is the appearance of indentations in the two chloroplasts at positions about two-fifths of the distance from the centrally located nucleus. These represent the early stages of chloroplast division by constriction, which is mediated by actin microfilaments (Hashimoto 1992). The nuclear envelope breaks down in prophase, and a precocious cleavage furrow begins to appear at the same time. At telophase, as furrowing continues, the spindle abruptly disintegrates, and transverse microtubules develop near the forming cross-wall (Pickett-Heaps and Fowke 1970), but a phragmoplastlike apparatus is absent from *Closterium*. Upon completion of cross-wall development, the chloroplast constrictions become

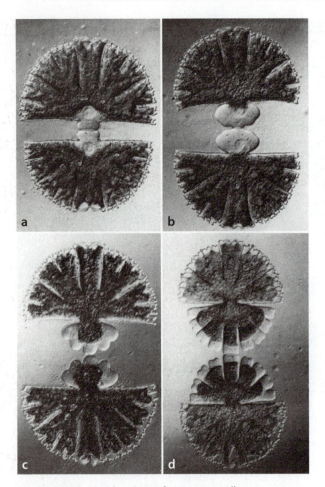

Figure 20.30 Development of new semi-cells in *Micrasterias*. (From Pickett-Heaps 1975)

telophase migration after the formation of a cross-wall; this is necessary to provide a pathway through the constricted isthmus region for expansion of the chloroplast from the old semi-cell into the newly developing semi-cell. An urn-shaped cage of microtubules radiating from a microtubular organizing center often surrounds the moving nucleus, suggesting a mechanism for organellar transport—possibly microtubule-associated motor proteins. In *Euastrum oblongatum*, nuclear movement begins about 80 minutes after cross-wall formation (Url et al. 1993). Usually the nucleus gets out of the way by moving into the expanding semi-cell, but in some cases the nucleus migrates into the older semi-cell. Expansion of the new semi-cell appears to be driven by turgor pressure because plasmolysis stops the expansion process. As the new semi-cell develops, the nucleus migrates back to its usual position in the isthmus, between old and new semi-cells. In *Euastrum oblongatum*, the nucleus begins to move back to the isthmus about 12 hours after the new semi-cell has been completely formed (Url et al., 1993). A new semi-cell is completely formed within about 16 hours in *Micrasterias* (Kiermayer 1981; Meindl 1983).

Lobing of new semi-cells in genera such as *Micrasterias* (see Figure 20.30) seems to occur by greater deposition of primary wall material at the tips of the growing lobes as compared to regions between lobes (Kiermayer and Meindl 1989). The factors influencing this differential distribution of wall material are not well understood. However, it is hypothesized that at the cross-wall development stage, the cell membrane is chemically imprinted with a pattern (Kiermayer 1981) that may later influence localized accumulation of calcium ions and targeted fusion of pectin-bearing Golgi vesicles at the lobes. Cellulose-synthesizing rosettes also appear to be directed to lobe regions, such that cellulose synthesis occurs more rapidly there. Rosettes are thought to arrive at the cell membrane as Golgi-derived vesicles, but the mechanisms by which they are directed to lobe regions are unknown. After primary wall deposition ends, development of the secondary wall of *Micrasterias denticulata* occurs for eight or so hours. Flat Golgi-derived vesicles deliver hexagonal arrays of cellulose-synthesizing rosettes to the cell membrane; these spin out bands of 2–17 adherent cellulose microfibrils (Giddings et al. 1980; Kiermeyer and Meindl 1984). At the beginning of secondary wall deposition, the cell membrane develops concave, circular invaginations that are about 0.2 μm in diameter.

deeper, and a vacuole forms in the resulting groove. The vacuole then increases in size as the chloroplasts continue to constrict. After the daughter cells separate from each other, the chloroplasts complete their division, and the nucleus moves into a central position between them. The cells then elongate on the side closest to the cross-wall so that cell symmetry is reestablished.

In placoderm desmids, often characterized by elaborately lobed semi-cells, new semi-cells develop by expansion after mitosis has been completed (Figure 20.30). The process by which these new semi-cells come to look almost identical to the older ones is still not completely understood. It has primarily been studied in *Closterium*, *Cosmarium*, *Euastrum*, and *Micrasterias* (Waris 1950; Gerrath 1993; Pickett-Heaps 1975). In highly constricted desmids, the nucleus undergoes what is known as post-

These mark the positions where pores will develop. Pore vesicles containing neutral polysaccharides then deposit these materials at the invaginations. These polysaccharides prevent cellulose microfibril deposition across the sites; subsequent disappearance of the plugs occurs when secondary wall development has been completed, leaving open pores in the walls.

Reproduction in Desmidiales

Some evidence indicates that desmidialeans mediate sexual reproduction by chemical communication. A 20 kDa heat-labile, diffusible protein produced by one of the mating types of *Closterium ehrenbergii* induces mitotic divisions that generate gametes in the other mating type (Fukumoto et al. 1997). In this species, vegetative cells were observed to possess two or four times the nuclear DNA level of gametes. There is no increase in DNA level immediately prior to the divisions that give rise to gametes, and thus gametes possess half the DNA of parental vegetative cells. Desmid zygotes often include unfused gamete nuclei that fuse only shortly before or during germination, which involves meiosis. When zygotes undergo meiosis, DNA levels are reduced, as expected, but frequently two or three of the nuclear products do not survive. In the surviving meiotic nuclei, the DNA level is duplicated twice and then the cell divides, partitioning DNA such that each daughter cell nucleus receives at least two copies of the genome. This process restores the typical vegetative DNA level.

In placoderm desmids, the first cells produced from zygotes are known as **gones**. These cells do not generally resemble normal vegetative cells; they are typically less ornamented. When gones divide by mitosis, new semi-cells are produced that exhibit a normal morphology. Early spring collections from nature or laboratory cultures established from recently germinated zygotes may contain some cells having both a typical semi-cell and a gone semi-cell. Asexually produced resting cells have been reported for several genera of placoderm desmids (Gerrath 1993).

Ecology of Desmidiales

Placoderm desmids are particularly common and diverse in oligotrophic (low-nutrient) and dystrophic (highly colored) lakes and ponds (Woelkerling 1976). In nutrient-poor streams, desmids can make up some 2%–10% of the community, and they are persistent residents, rather than forms that have come to be there via incidental drift. More than 200 desmid

Figure 20.31 *Gonatozygon*, a genus that has spines along the length of the cell wall. (Photo: L. W. Wilcox)

species have been observed among stream periphyton (algae attached to substrates), associated with plants such as the moss *Fontinalis*. There, desmids may achieve cell concentrations as high as 10^6 per gram of substrate (Burkholder and Sheath 1984). Some species occur in mesotrophic (higher-nutrient) and eutrophic (high-nutrient) water bodies. *Closterium aciculare* is regarded as an indicator of eutrophic conditions and is sometimes abundant, occasionally growing to bloom proportions. This species is unable to utilize nitrate as a source of combined nitrogen because nitrate reductase is lacking (Coesel 1991), and a requirement for ammonium ion may explain the occurrence of this desmid in highly eutrophic waters. Cells of the very unusual, slow-growing *Oöcardium stratum* live at the tops of branched calcareous tubes in calcareous streams and waterfalls, in association with deposits of tufa and travertine. A number of placoderm desmids occur in soils and other terrestrial habitats, such as on moist rocks and among bryophytes.

Diversity of Desmidiales

Desmidiales appear to form a monophyletic group (McCourt et al. 2000; Turmel et al. 2002; Hall et al. 2008) of about 3000 described species (Gerrath 1993). Genus and species definitions have classically depended upon characters of the cell wall, including wall pores, ridges, knobs, and spines; chloroplast morphology; and zygote structure. A valuable source of classic taxonomic information is the series by Prescott et al. (1975, 1977, 1981, 1982) and Croasdale et al. (1983) on North American desmid taxa. However, several of the classic genera are known to be polyphyletic, and the value of some of the structural features used in taxonomy has been questioned (Hall et al. 2008). A phylogenetic analysis of 39 strains of *Staurastrum* and the related genera *Staurodesmus*, *Cosmarium*, *Xanthidium*, and *Euastrum* was performed by Gontcharov and Melkonian (2005). Representative unicellular placoderm genera

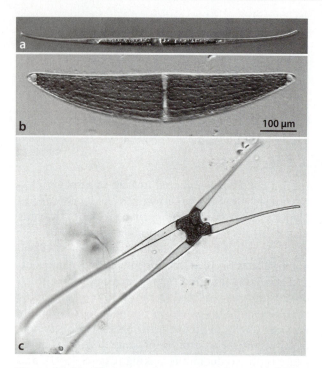

Figure 20.32 *Closterium.* Species vary from needlelike cells (a) to broader cells (b). Note the presence of terminal vacuoles and ridged chloroplast in the latter. (c) Zygote and remains of parental walls of a species of Closterium. (Photos: L. W. Wilcox)

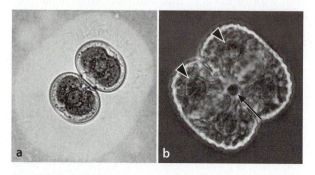

Figure 20.33 *Cosmarium.* Note the extensive mucilage investment (a). The pyrenoids (arrowheads) and nucleus (arrow) are conspicuous in this cell seen under phase-contrast microscopy (b). (Photos: L. W. Wilcox)

are described here first, followed by colonial and filamentous forms.

GONATOZYGON (Gr. *gonatos*, joint or knee + Gr. *zygon*, yoke) (Figure 20.31) occurs as a single cell or a short chain of cells adherent at their poles. Cells are not constricted into semi-cells but bear distinctive

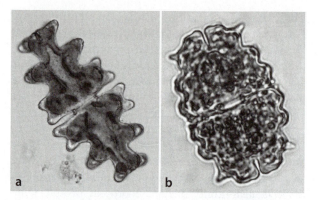

Figure 20.34 *Euastrum,* a lobed, flattened placoderm desmid. (a) *E. pinnatum,* (b) *E. bidentatum.* (Photos: L. W. Wilcox)

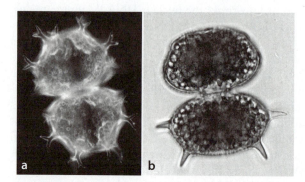

Figure 20.35 *Xanthidium.* This genus is characterized by three-dimensional protuberances on each of the semi-cells (a). In (b), a recently divided cell is shown, where the characteristic spines have yet to develop on the newly formed semi-cell. (Photos: L. W. Wilcox)

spines on wall surfaces. Edges of cells have a squared-off appearance, and the single chloroplast is a flat plate.

CLOSTERIUM (Gr. *klosterion*, small spindle) (Figure 20.32) cells are not constricted into semi-cells, but pores occur in the walls. Vegetative cells are crescent shaped or elongate and somewhat curved. There is a single axial, ridged plastid with several pyrenoids in each semi-cell and a central nucleus. Conspicuous vacuoles occur at the cell tips; these contain barium sulfate crystals (Brook 1980) that move by Brownian motion. The composition and synthesis of pectin and protein components of the cell wall of *C. acerosum* were studied by Baylson et al. (2001). Almost all natural populations of the *Closterium ehrenbergii* species complex have been observed to have distinct mating types segregated in a 1:1 ratio upon zygote

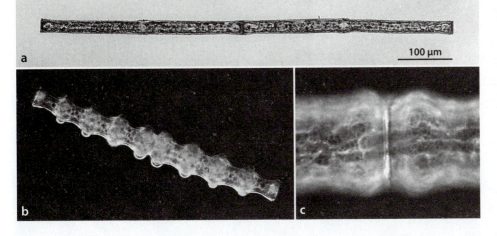

Figure 20.36 *Pleurotaenium.* In (a), two cells have remained connected following cell division. Chains of several cells are sometimes encountered. *Pleurotaenium nodosum* is shown in (b), while (c) is a high-magnification view of the semi-cell junction, showing a ring that is characteristic of some species. (Photos: L. W. Wilcox)

germination. A single gene (*mt*) determines mating type, and the *mt⁻* allele is dominant to *mt⁺* (Kasai and Ichimura 1990; Tsuchikane et al. 2003). MIKC-type MADS-box genes similar to those important in land plant development were identified in the *C. peracerosum–strigosum–littorale* complex; gene expression increased when vegetative cells started to develop into gametangia and decreased following fertilization (Tanabe et al. 2005). About 140 species of *Closterium* have been described.

COSMARIUM (Gr. *cosmarion*, small ornament) (Figure 20.33) occurs as single cells that are deeply divided at the midregion to form a short isthmus and two semi-cells that are rounded in front view but flattened, oval, or elliptical in side view. Walls may be smooth or ornamented; spines are not present on vegetative cells. One or sometimes more axial or parietal chloroplast(s) with pyrenoids occurs in each semi-cell. Sexual reproduction in *Cosmarium turpinii* involves separate mating types, with pairs of cells becoming enclosed by mucilage. Mating cells open at the isthmus, and the emerging protoplasts function as gametes. Their fusion results in the production of a thick-walled, spiny zygote, which can often be found together with the empty parental cell walls, enclosed in the same enveloping mucilage. More than 1000 species have been described, but this genus does not form a monophyletic group (Gontcharov and Melkonian 2005; Hall et al. 2008).

EUASTRUM (Gr. *eu*, good, true, or primitive + Gr. *astron*, star) has flattened cells (Figure 20.34).

There is a characteristic notch in the apices of most species. Semi-cells are typically lobed, and there may be small bumps on the surface that are visible at the light microscopic level in side view. The incision between semi-cells is usually closed, as the semi-cell walls come close together. One or sometimes two chloroplasts with pyrenoids occur in each semi-cell. Chloroplasts are often conspicuously ridged. Zygotes are round and ornamented with spines or other protuberances. There are some 265 described species that do not form a monophyletic group (Hall et al. 2008).

XANTHIDIUM (Gr. *xanthos*, yellow) unicells are mostly characterized by wall protuberances rising from semi-cell surfaces perpendicular to the flat plane of the surface (Figure 20.35). The protuberances may be ornamented with pits or granules or may be pigmented, but these features are difficult to see at the light microscopic level. Another distinctive feature is robust, often paired spines. In general, the wall is less ornamented than that of *Cosmarium*, *Euastrum*, or *Staurastrum*. There are usually four axial or parietal plastids, each with a pyrenoid, per semi-cell. Zygotes have only rarely been observed; they are round, and most bear spines. Some 115 species have been described.

PLEUROTAENIUM (Gr. *pleuron*, rib + Gr. *tainia*, ribbon) (Figure 20.36) is a cylindrical desmid with long, blunt-ended cells. Cells are 4–35 times longer than they are wide. There may be a noticeable ringlike thickening where semi-cells join. The

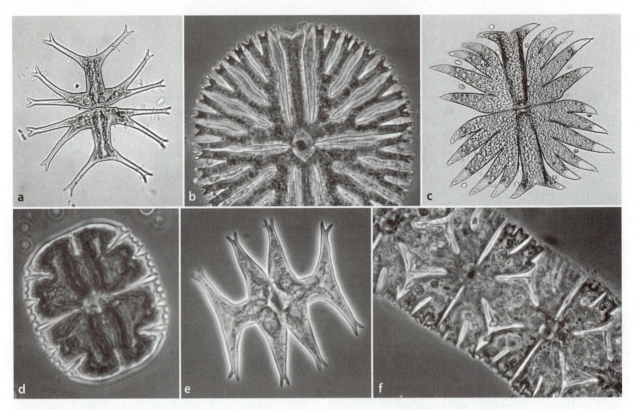

Figure 20.37 Diversity of *Micrasterias*. (a) *M. radiata*, (b) *M. radiosa*, (c) *M. torreyi*, (d) *M. truncata*, (e) *M. pinnatifida*, and (f) *M. foliacea*, which forms filaments by means of interlocking prongs on the ends of semi-cells. (a, c–f: L. W. Wilcox; b: C. Taylor)

chloroplast appears as parietal bands or axial with lamellae, and pyrenoids are present. Sometimes apical vacuoles with crystalline inclusions are present.

MICRASTERIAS (Gr. *mikros*, small + Gr. *aster*, star) is flattened and often highly incised and lobed. Some species look like flattened disks, and others are so highly dissected that they look like stars (Figure 20.37). Each semi-cell has a single large lobed plastid that extends into lobes of the cell. Plastids are studded with numerous pyrenoids. The nucleus lies in the isthmus. Conjugation involves the formation of papillae that allow gamete fusion. *Micrasterias foliacea* is exceptional within the genus for its ability to form filamentlike arrays by overlapping polar lobes and interlocking apical teeth (Lorch and Engels 1979). Zygotes are usually spherical, with spines that are sometimes forked. Gones are very simple in construction.

TRIPLOCERAS (Gr. *triploos*, triple + Gr. *keras*, horn) cells are elongate, 8–20 times longer than they

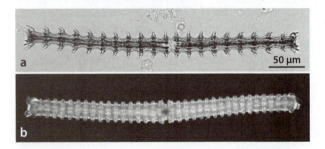

Figure 20.38 Two species of *Triploceras*, one seen with bright-field optics (a) and the other with dark-field optics (b). (Photos: L. W. Wilcox)

are wide, and not noticeably constricted. The cells are ornamented with various kinds of spines and have undulating margins (Figure 20.38).

STAURASTRUM (Gr. *stauron*, cross + Gr. *astron*, star) unicells are highly constricted unicells that are radially symmetrical in end-on (polar) view. The semi-cells are often triradiate or hexaradiate

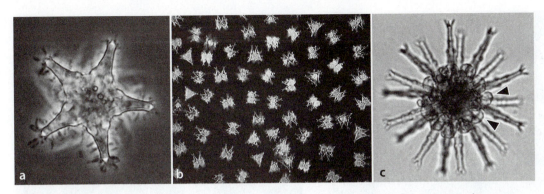

Figure 20.39 *Staurastrum.* (a) An end-on view of one of the semi-cells of a species having five arms. (b) A population of a species having triradiate semi-cells, viewed with dark-field microscopy. (c) A species having nine projections or arms on each semi-cell; note also the infestation of spherical chytrid cells (arrowheads). (Photos: L. W. Wilcox)

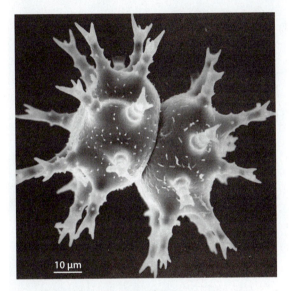

Figure 20.40 A multiple-armed *Staurastrum* species viewed using SEM. (Micrograph: S. Cook, in *Origin of Land Plants.* Graham, L. E. © 1993. Reprinted by permission of John Wiley & Sons, Inc.)

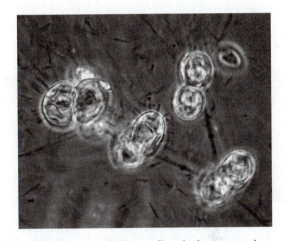

Figure 20.41 *Cosmocladium* cells, which occur at the ends of radiating strands. (Photo: L. W. Wilcox)

and may be highly ornamented with spines and other protuberances (Figures 20.39 and 20.40). The walls are impregnated with polyphenolic compounds that confer decay resistance (Gunnison and Alexander 1975a,b). These materials explain the recovery of fossil remains of *Staurastrum* walls from lake sediment cores that are thousands of years old. At conjugation, the gamete protoplasts escape as the semi-cell walls separate at the isthmus; their fusion generates spiny zygotes. Zygote germination, presumably by meiosis, produces one to four gones. Some 800 species have

been described, primarily on the basis of cell wall characters. Sequencing of the chloroplast genome of *S. punctulatum* revealed substantial differences from those of some other charophyceans and bryophytes in terms of size, gene order, and intron content (Turmel et al. 2005).

COSMOCLADIUM (Gr. *kosmos*, ornament + Gr. *klados*, branch) has relatively small cells that are constricted into semi-cells and compressed. Usually there is one axial chloroplast with pyrenoid in each semi-cell. Individual cells look like small *Cosmarium* cells, but they are interconnected by branching mucilaginous strands to form colonies of variable numbers of cells, the whole enclosed in mucilage (Figure 20.41). This genus diverged just prior to a group of

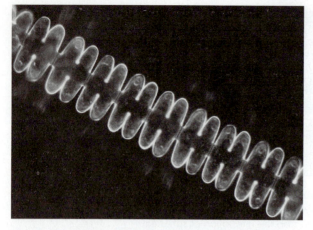

Figure 20.42 *Spondylosium*, a filamentous desmid whose semi-cells are deeply incised. There is a substantial mucilaginous sheath in which bacterial cells often occur. The fine details in the sheath of this specimen represent structure inherent in the sheath. (Photo: L. W. Wilcox)

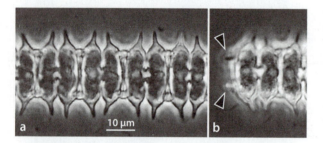

Figure 20.43 *Onychonema*, a filamentous desmid. (a) Arrangement of cells in the filament. (b) Terminal cell with projecting prongs (arrowheads) that interlock with adjacent cells. (Photos: L. W. Wilcox)

filamentous desmids in a phylogenetic study by Hall et al. (2008).

SPONDYLOSIUM (Gr. *spondylos*, vertebra) is a filamentous form consisting of flattened cells that are deeply constricted (Figure 20.42). Cell walls are not highly ornamented. Plastids are axial. Filaments are often twisted and have an extensive sheath. Zygotes are globose and smooth walled or have short spines. The genus is not monophyletic, according to a phylogenetic analysis by Lane et al. (2008).

ONYCHONEMA (Gr. *onyx*, claw + Gr. *nema*, thread) is a twisted filamentous arrangement of small, deeply constricted cells (Figure 20.43). Semi-cells possess distinctive overlapping apical processes

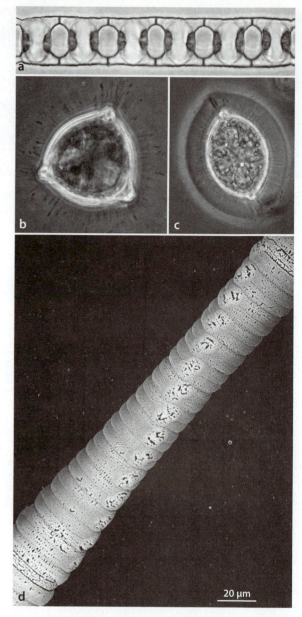

Figure 20.44 *Desmidium*, a filamentous genus. (a) A distinctive species, *Desmidium baileyi*. (b) View from the end of a semi-cell of another species, showing a triradiate structure and numerous attached bacteria within the mucilage sheath. (c) End view of an isolated semi-cell from a biradiately symmetrical species showing sheath. (d) SEM view showing the twisted orientation of adjacent cells, a general feature of *Desmidium*. The mucilaginous sheath was maintained in this preparation. The periodic indentations reflect underlying constrictions between semi-cells and adjacent cells in the filament. (a–c: L. W. Wilcox; d: From Fisher and Wilcox 1996, by permission of the *Journal of Phycology*)

Figure 20.45 *Bambusina*, a filamentous desmid whose semi-cells are not deeply incised. (Photo: L. W. Wilcox)

that can be nearly as long as the semi-cells themselves. There is a wide mucilage sheath. Zygotes are globose with short spines.

DESMIDIUM (Gr. *desmos*, bond) often has a prominent gelatinous sheath (see Figures 20.28 and 20.29). Cells are bi-, tri-, or quadriradiate in end view, depending upon the species. The filaments appear to be spirally twisted because the axes of adjacent cells are offset by a slight angle (Figure 20.44). There are two axial chloroplasts with pyrenoids per cell; there is one plastid per semi-cell. Filaments can fragment. Conjugation tubes are formed between opposing cells of paired filaments, and zygotes form within the tubes or in one of the parental cell walls, depending upon the species. Zygotes are round or ellipsoidal and are smooth walled or have short projections.

BAMBUSINA (named for *Bambusa*, a genus of bamboo [bambu, Indian vernacular name]) cells are cylindrical or barrel shaped and arranged in linear series to form filaments (Figure 20.45). There is only a slight constriction, and the wall region on either side of the isthmus is swollen. Fine striations may be visible at the apices. The axial chloroplasts have

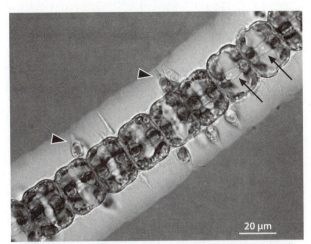

Figure 20.46 *Hyalotheca*. This filamentous desmid often has a wide mucilaginous sheath; it is shown here in an india-ink preparation, which allows the sheath to be better visualized. Sessile chrysophytes (arrowheads) are often found associated with *Hyalotheca* and other filamentous desmids having similar sheaths. Note also the central nuclei in the cells (arrows). (Photo: L. W. Wilcox)

radiating lamellae and a single central pyrenoid. Zygotes are globular or elliptical, and their walls are smooth. According to the phylogenetic analysis of Lane et al. (2008), *Bambusina* is sister to *Hyalotheca*.

HYALOTHECA (Gr. *hyalos*, glass + Gr. *theke*, sheath or box) is a filamentous member of the placoderm desmids (Figure 20.46). Each cell is indented at the midpoint, but the indentation may be very slight. Cells contain two axial, radiately ridged chloroplasts

Figure 20.47 *Chaetosphaeridium.*
(a) The spherical cells of *Chaetosphaeridium* are actually linked to form branched filaments, though this is often difficult to discern. Note the sheathed hairs (arrowhead), a hallmark of the order Coleochaetales.
(b) Delicate colorless strands (arrow) are the remains of the cellular linkages that denote filamentous construction of *Chaetosphaeridium*. Note that each cell bears a hair. (Photos: L. E. Graham)

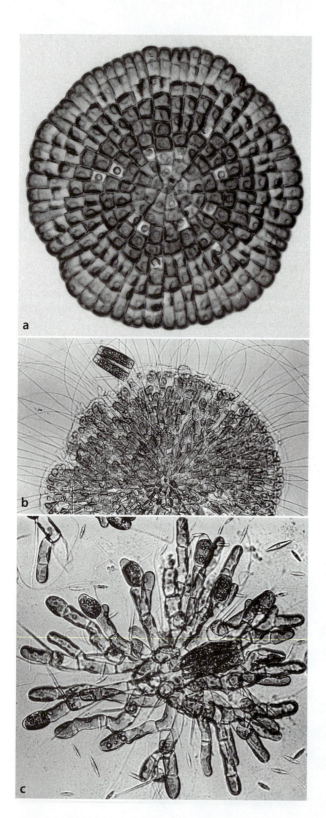

a

b

c

with pyrenoids, one per semi-cell. There is an extensive mucilaginous sheath. Filaments can break apart into fragments, a form of asexual reproduction. Conjugation involves formation of conjugation tubes; zygotes may form in the tubes or within one of the parental cell walls. Zygotes are globose and smooth walled. Meiosis occurs during zygote germination and there are two filaments produced. Cells of filaments can round up and dissociate into thick-walled aplanospores, a form of asexual reproduction.

Coleochaetales

CHAETOSPHAERIDIUM (Gr. *chaite*, long hair + Gr. *sphaira*, ball), with only 4 described species (Thompson 1969) (Figure 20.47), and **COLEOCHAETE** (Gr. *koleos*, sheath + Gr. *chaite*, long hair), including approximately 16 species (Pringsheim 1860; Jost 1895; Printz 1964; Szymánska 1989; Szymánska 2003) (Figure 20.48), are the only widely accepted members of the order Coleochaetales. Molecular data support the concept that Coleochaetales forms a monophyletic group that is part of a monophyletic lineage including Charales and land plants (Lewandowski and Delwiche 2001; Delwiche et al. 2002) (see Figure 20.1). Prominent among the characters that *Chaetosphaeridium* and *Coleochaete* share are sheathed hairs, known as seta cells. These extensions of the cell wall enclose a small amount of cytoplasm and are believed to serve as protection against herbivores (Marchant 1977) (Figure 20.49). The hairs may attain a length greater than 100 times the vegetative cell diameter. A rigid sheath of wall material encloses the bases of hairs, which often break off at the sheath rim. Whereas all *Chaetosphaeridium* cells bear one or more sheathed hairs, only 3%–5% of the cells of

Figure 20.48 Morphological diversity of *Coleochaete*. (a) *Coleochaete orbicularis* has a body that is one cell layer thick. It grows by means of a marginal meristem. (b) A pseudoparenchymatous species of Coleochaete, composed of laterally adherent branched filaments enclosed within a mucilage layer. Note the extensive number of long hairs (setae), which are thought to function in protection from herbivory. (A pair of diatoms lies among the setae.) (c) *Coleochaete pulvinata* is a heterotrichous, branched filament; there is a prostrate system of branches and an erect system of branches, all enclosed by extensive mucilage, which often harbors small diatoms. (Photos: L. E. Graham)

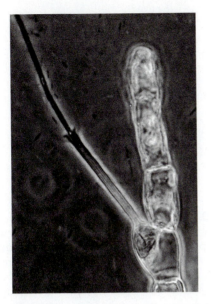

Figure 20.49 Sheathed hair of *Coleochaete pulvinata*. (Photo: L. E. Graham)

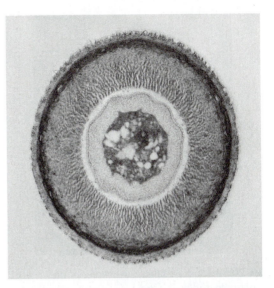

Figure 20.50 Hair cell. Cross-section of one of the hairs (setae) of *Coleochaete pulvinata*, at the level of the sheath (collar), as viewed using TEM. Note the occurrence of multiple cell wall layers and the central column of cytoplasm that runs throughout the length of the hair. Breakage of hairs is thought to release chemicals that are inhibitory to herbivores. (From *Origin of Land Plants*. Graham, L. E. © 1993. Reprinted by permission of John Wiley & Sons, Inc.)

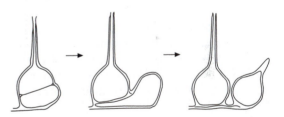

Figure 20.51 Development of *Chaetosphaeridium* filaments. (After Thompson 1969, by permission of the *Journal of Phycology*)

Coleochaete scutata bodies produce hairs, and then only one per cell (McBride 1974). The walls of *Coleochaete* seta cells are composed of several layers (Figure 20.50), whereas those of *Chaetosphaeridium* are of simpler construction. The chloroplasts of *Coleochaete* seta cells typically have a C-shaped appearance and may be observed to rotate. Sometimes the hairs are highly coiled.

Structure and development of Coleochaetales

Although *Chaetosphaeridium* appears, at first glance, to be composed of single cells held together by a gelatinous matrix, it is actually a branched filament (Thompson 1969). Thin colorless branches that are difficult to discern with the light microscope interconnect *Chaetosphaeridium* cells. Division is regularly oblique to the plane of the substrate, giving rise to a new cell located somewhat beneath the parental cell. A bulge develops into which the new protoplast moves, and then the new cell grows into a position next to the parental one. The portion of the cell lying beneath the parental cell collapses, and the new cell develops a hair (Figure 20.51). Lateral branching may take place by the same method. Thompson (1969) did not consider this form of growth to be equivalent to the apical growth exhibited by *Coleochaete*. *Chaetosphaeridium* bodies are covered by abundant mucilage.

Many (if not all) species of *Coleochaete* appear to possess terminal or marginal meristems, meaning that the only cells that undergo vegetative mitotic divisions are those at the tips or edges of bodies. *Coleochaete* species exhibit an unusual degree of body variability. *Coleochaete pulvinata* consists of both radially symmetrical prostrate portions whose branched filaments grow flat against the substrate and radially symmetrical, erect branching systems. Some other species occur only as radially branched prostrate filaments, and yet others are prostrate branched filaments that lack radial symmetry. *Coleochaete orbicularis* and *C. scutata* are radially symmetrical species that grow as

Figure 20.52 Surface cuticle-like material on *Coleochaete*. (SEM: S. Cook)

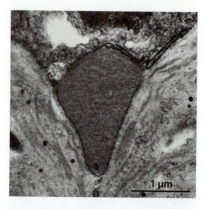

Figure 20.53 *Coleochaete* peroxisome. The peroxisome of *Coleochaete orbicularis* is closely associated with the dividing plastid during mitosis; this is also the case in the monoplastidic hornworts. (Reprinted from Graham, L. E., and Y. Kaneko. 1991. *Critical Reviews in Plant Science* 10:323–342. © CRC Press, Boca Raton, Florida)

a single tissue-like layer of cells. *Coleochaete pulvinata* is covered with an abundant, clear mucilaginous material similar to that of *Chaetosphaeridium*. Surfaces of the tissuelike forms are coated with a ridged material (Marchant and Pickett-Heaps 1973) (Figure 20.52) that exhibits some similarities to cuticles of land plants, especially those of various bryophytes (Cook and Graham 1998).

Chloroplasts occur singly in most cells of *Chaetosphaeridium* and *Coleochaete*, and these cells contain one or more pyrenoids that are similar to those of other charophyceans and hornworts (Graham and Kaneko 1991). The thylakoids of *Coleochaete* are organized into grana.

Cell division in Coleochaetales

Vegetative cell division has been studied at the ultrastructural level in *Coleochaete scutata* (Marchant and Pickett-Heaps 1973) and *C. orbicularis* (Graham 1993; Brown et al. 1994; Cook 2004a). Pairs of centrioles appear at the poles of developing spindles, and the nuclear envelope begins to disintegrate at prometaphase, such that mitosis is open. Despite earlier reports, *Coleochaete* does not display two types of cytokinesis. Rather, cytokinesis consistently involves a phragmoplast similar to that present in land plants. Polarized cytokinesis, in which the phragmoplast and cell plate contact one cell wall and then progress toward the opposite wall, occurs in *Coleochaete* (Cook 2004) and highly vacuolated cells of *Arabidopsis*. As in the cases of land plants and charalean algae, the cross-walls of *Coleochaete* are penetrated by numerous plasmodesmata, whereas these are absent from the cross-walls of earlier- divergent charophyceans. The evolutionary origin of land plant plasmodesmata is significant because these structures are developmentally very important in land plants (Cook et al. 1997). Plasmodesmata may have been a necessary prerequisite to the origin of histogenetic (tissue-producing) meristems (Cook and Graham 1999).

In *Coleochaete*, the plantlike peroxisome becomes closely associated with the plastid and divides (by invagination) at the same time, thus achieving regular partitioning to daughter cells (Figure 20.53). In contrast, as noted earlier, some other charophyceans partition their peroxisomes to daughter cells in different ways.

Asexual reproduction in Coleochaetales

Both *Coleochaete* and *Chaetosphaeridium* produce biflagellate zoospores (lacking eyespots) that serve as asexual reproductive cells. They generate new bodies of the same type from which they were produced and are capable of rapidly generating a clonal population from a single parental body. In *Chaetosphaeridium*, zoospores can be produced from any cell. Zoospore production in *Chaetosphaeridium* involves a precursor cell division, and typically the lower cell differentiates into the zoospore, which escapes by dissolution of the cell wall. Usually only a single zoospore is produced per parental cell of *Chaetosphaeridium*, but sometimes two are generated.

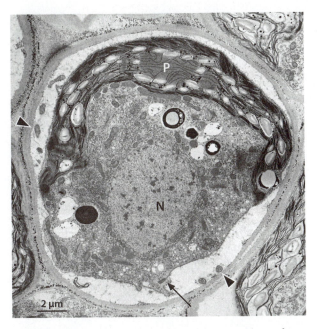

Figure 20.54 *Coleochaete* zoospore prior to release from parental body cell viewed using TEM. Note the single parietal plastid (P) with numerous starch grains, and multilayered structure (arrow). Profiles of the two flagella, which are coiled around the cell, are visible just inside the parental cell wall (arrowheads). (From Graham and McBride 1979)

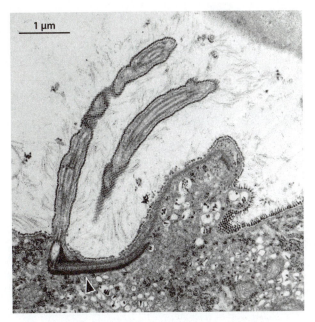

Figure 20.56 *Coleochaete* zoospore. A multilayered structure (arrowhead) is associated with the flagellar bases. Note the presence of numerous scales on the cell and flagellar surface, as well as the numerous long flagellar hairs. These are produced within cytoplasmic vesicles and then transported to the surface. (From Graham and McBride 1979)

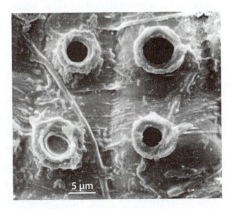

Figure 20.55 *Coleochaete scutata* wall pores through which zoospores were released. (SEM: S. Cook)

Zoospores of *Coleochaete* are always produced singly (Figure 20.54), a precursor division is not involved, and zoospores escape through a specialized discharge pore (Figure 20.55) that is presumed to arise by the localized action of hydrolytic enzymes on the cell wall. Temperature is more influential than irradiance or day length in inducing *Coleochaete*

zoospore production (Graham et al. 1986). Zoospores of *Coleochaete* settle, attach to surfaces, and develop a cell wall beneath the scale layer. A transverse cell division then occurs (reminiscent of the growth pattern in *Chaetosphaeridium*), and the upper cell terminally differentiates into a seta cell. The lower cell continues to divide, and its derivatives serve as the marginal meristem, dividing either radially or circumferentially, depending upon spatial constraints.

In both *Chaetosphaeridium* (Moestrup 1974) and *Coleochaete* (Graham 1993), zoospore surfaces and flagella are coated with scales (Figure 20.56) that resemble the lower layer of body scales of *Mesostigma*. The body and flagella of *Coleochaete* gametes are covered with similar scales. The scale covering gives the surface of motile cells a frosty or granular appearance (Thompson 1969), but its function is not completely understood. Flagellate reproductive cells of *Coleochaete* and *Chaetosphaeridium* also exhibit a single MLS-containing flagellar root (see Figure 20.56). In addition, *Coleochaete* contains another flagellar root composed of just a few microtubules

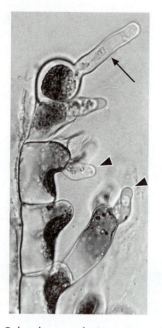

Figure 20.57 *Coleochaete pulvinata* oogonium and antheridia. The oogonium displays an elongate trichogyne (arrow), and small colorless antheridial branches occur nearby (arrowheads). (From *Origin of Land Plants*. Graham, L. E. © 1993. Reprinted by permission of John Wiley & Sons, Inc.)

Figure 20.58 *Coleochaete scutata* antheridia. These structures occur as packets of cells formed in concentric rings between the body periphery and its center. (From *Origin of Land Plants*. Graham, L. E. © 1993. Reprinted by permission of John Wiley & Sons, Inc.)

(Sluiman 1983). A fibrous connective structure links the two flagellar bases, and the flagella emerge laterally, as is generally characteristic of charophycean motile cells.

Sexual reproduction in Coleochaetales

Observations on sexual reproduction have been made on cultured *Chaetosphaeridium globosum* (Thompson 1969). Sexual reproduction involves naked nonmotile egg cells that are larger than vegetative cells and biflagellate sperm cells that are released in pairs from colorless precursor cells. The eggs are expelled before fertilization and held within the body mucilage. Fertilized eggs develop into oval smooth-walled zygotes that are about the same size as egg cells. There is a delay between plasmogamy (cytoplasmic fusion) and karyogamy (nuclear fusion). The zygote wall is several layers thick, and internal layers may be yellow or deep brown in color.

In contrast to *Chaetosphaeridium*, egg cells of *Coleochaete* are not released from the body. Rather, they develop a cell wall protuberance that is relatively short in *Coleochaete orbicularis* but may be quite long in *Coleochaete pulvinata* (Figure 20.57). The

tip of the protuberance (sometimes called a **trichogyne**) disintegrates when the egg is ready for fertilization (Oltmanns 1898) and exudes cytoplasmic contents that appear to attract flagellate sperm. In *C. pulvinata*, the sperm are produced in small colorless branches that occur in groups on precursor cells near egg cells (see Figure 20.57) (Graham and Wedemayer 1984), while in *C. orbicularis* and *C. scutata*, sperm are formed in small cells occurring in packets derived from asymmetric cell divisions (Figure 20.58). Production of sperm is an exception to the general rule that non-peripheral cells do not divide. Sperm release is by localized wall dissolution.

In *Coleochaete*, sexual reproduction occurs in mid- to late summer, with zygote maturation occurring in the fall. When several species co-occur, it appears that sexual reproduction may be temporally separated, perhaps as a species isolation mechanism. Zygotes are not released from parental bodies, similar to retention of zygotes within the female gametangia of early-divergent land plants. Immediately after fertilization, zygote cytoplasm at first shrinks and then begins massive enlargement. During enlargement, surrounding vegetative cells are induced to divide, forming a layer of cortical cells that cover zygotes partially or completely, depending on the species (Figure 20.59). In *C. orbicularis* (Figure 20.60), cortical cells exhibit cellular features similar to those of placental transfer cells at the gametophyte–embryo junction of land plants. In land plants, transfer cells function in the transport of nutrients such as amino

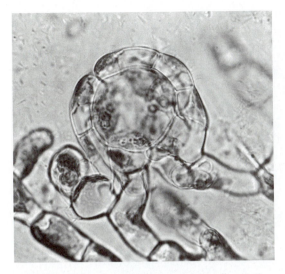

Figure 20.59 *Coleochaete pulvinata* zygote. A putative chemical influence emanating from zygotes induces neighboring cells to grow toward and cover them, forming a protective cortical layer. (From *Origin of Land Plants*. Graham, L. E. © 1993. Reprinted by permission of John Wiley & Sons, Inc.)

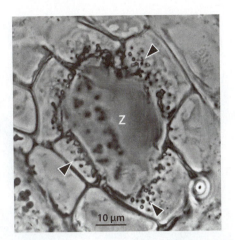

Figure 20.60 *Coleochaete orbicularis* transfer cells. Cells that surround zygotes (Z) have numerous, localized wall projections (arrowheads). Production of such ingrowths appears to be induced by the presence of zygotes and is additional evidence of chemical interaction between parent and progeny. Similar wall ingrowths occur at the interface of parent and progeny tissues in land plants, and the cells in which they occur are known as placental transfer cells. (From Graham and Wilcox 1983)

acids and sugars across the maternal–embryo interface, a process that has been described as **matrotrophy** (Graham 1996; Graham and Wilcox 2000).

In both placental transfer cells of land plants and the cortical cells surrounding *Coleochaete orbicularis* zygotes, elaborate wall ingrowths develop on a localized basis (see Figure 20.60) (Graham and Wilcox 1983). In the case of *Coleochaete*, the wall ingrowths occur only on the vegetative cell walls that are closest to zygotes, suggesting that zygotes have an inductive influence, perhaps by exuding chemical signals. The wall ingrowths vastly increase the surface area of the cell membrane, across which nutrients moving from vegetative cells to zygotes must pass. The ability of vegetative bodies of *Coleochaete* to take up and utilize sugars (Graham et al. 1994) suggests that these algae may possess the requisite cell membrane carbohydrate transporter molecules. Concurrently, zygotes begin to enlarge through accumulation of massive storages of starch and lipid (Figure 20.61). These storage materials are thought to arise in part from maternal contributions, though zygotes maintain green chloroplasts at this stage and therefore are presumably capable of generating photosynthates. The storage materials are needed to fuel production of an unusually large number of meiotic products when zygotes germinate in early spring. At maturity,

Coleochaete zygote walls become lined with a thin layer of material that is similar to sporopollenin in higher plant spores and pollen walls (Delwiche et al. 1989; Graham 1990). Further, the walls of vegetative cells that surround zygotes, and, indeed, any cells within the apparent sphere of influence of zygotes, accumulate highly resistant phenolic compounds (Kroken et al. 1996). Such compounds, together with the sporopollenin layer, are presumed to protect zygotes against microbial attack during the dormant period.

In northern temperate latitudes, *Coleochaete* zygotes are induced to germinate in spring, perhaps by warmer temperatures and longer day lengths. At least some of the zygotes remain in shallow nearshore waters—the same location favored by vegetative bodies—because zygotes are attached to senescent parental bodies, which often remain attached to stable substrates. Though dead, cell walls of parental bodies do not completely decay during this period because they are impregnated with phenolic materials. In at least one species of *Coleochaete*, meiosis occurs during the first divisions of the polyploid zygotes, followed by several mitotic divisions (without further DNA replication) (Hopkins and McBride 1976), yielding 8–32 haploid products. These cells, known as meiospores, develop two flagella, an MLS flagellar

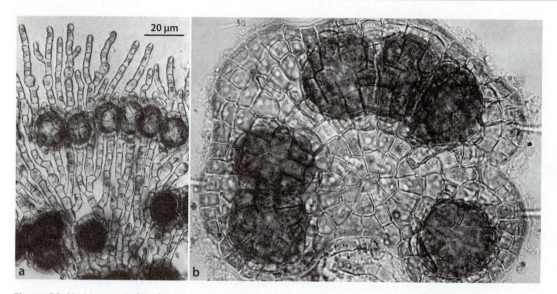

Figure 20.61 Mature Coleochaete zygotes. (a) *Coleochaete pulvinata* with zygotes; the most peripheral ones (top) are the youngest, and the larger, darker zygotes at bottom were formed earlier. (b) *Coleochaete orbicularis* zygotes are the large, dark spheres, which are filled with storage products. Note that in both cases, zygotes are completely covered by a layer of vegetative cells. (a: From Graham 1985; b: L. E. Graham)

root, and a layer of surface scales, and then they escape from zygotes as the wall cracks open (Graham and Taylor 1986a,b). Upon settling in nearby, well-illuminated nearshore habitats, the meiospores seed the development of vegetative populations of *Coleochaete* during the subsequent growing season, completing the life cycle.

Ecology of Coleochaetaleans

Coleochaete and *Chaetosphaeridium* are periphytic, attached to submerged portions of higher plants such as the bulrush *Scirpus*, *Potomogeton*, and the undersides of water lily leaves. They also grow on nonliving substrates such as pebbles near the edges of oligotrophic (low-nutrient) freshwater lakes and ponds, where they are frequently but unpredictably exposed to desiccation as a result of variable wave action and changes in water levels. *Coleochaete nitellarum* can occur on the surfaces of deep-water charalean algae, particularly *Nitella*. Coleochaetaleans are sensitive to the effects of eutrophication; they disappear from water bodies that suffer excessive input of nutrients. They are not known to occur in brackish waters or extreme habitats. In laboratory cultures, one *Coleochaete* species has been demonstrated to utilize exogenous dissolved organic carbon in the form of hexose sugars and sucrose (Graham et al. 1994).

This capability could be useful in low-pH or other environments in which dissolved inorganic carbon levels might be too low to saturate photosynthetic requirements.

Charales

Modern charalean algae are important both ecologically and evolutionarily. They are closely related to the ancestry of land plants (Karol et al. 2001; Qiu et al. 2007) and have the potential to provide information about the features of the ancient charophycean progenitors of embryophytes. For example, a particular RNA polymerase subunit (RNAPIV) that is present in land plants has also been found in several charalean species but not other charophyceans examined using a DNA amplification approach (Luo and Hall 2007).

Charaleans have a long fossil history, based primarily upon calcified reproductive structures that provide useful information about the evolutionary process and patterns of extinction. The ancestors of modern genera are thought to have arisen in the late Triassic (Feist et al. 2005). Charaleans are ecologically important for forming massive growths in both deep and shallow lake and pond waters, sometimes to the point of being regarded as nuisance

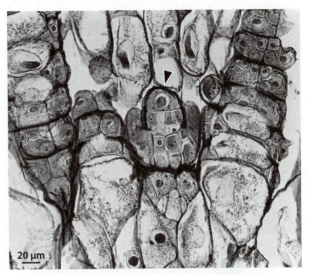

Figure 20.63 Apical meristematic cell of *Chara* (arrowhead). (Photo: L. W. Wilcox)

Figure 20.62 *Chara zeylanica* compared to *Ceratophyllum*, an aquatic flowering plant. These are similar in size and have similar nodal and internodal organization, with whorled branching; thus, they are often confused. *Ceratophyllum* (right) has bifurcating branch tips that distinguish it from *Chara* (left). (From *Origin of Land Plants.* Graham, L. E. © 1993. Reprinted by permission of John Wiley & Sons, Inc.)

weeds. However, charaleans are also an important food for waterfowl and provide a nursery area for fish. A few species, including *Chara evoluta*, occur in brackish waters having salt content of 20–40 ppt, but no modern forms are marine. Some form extensive meadows in fairly deep waters (Stross 1979); *Chara contraria*, for example, has been collected from 150 m in Lake Tahoe. Charaleans are generally considered to be adapted in various ways to low-irradiance, benthic habitats (Andrews et al. 1984). Because many forms accumulate surface layers of calcium carbonate in the form of calcite, the group is known colloquially as stoneworts, muskgrasses, bassweeds, or brittleworts. Calcification gives some forms a white or pale-green appearance (Grant 1990). Charalean algae are now, and for the past few hundred million years have been, major carbonate sediment producers in freshwater lakes because they may be more heavily encrusted by calcium carbonate than aquatic higher plants. Most of the charalean body (except zygotes) readily disintegrates in the benthos, forming marl deposits at rates

that can reach several hundred g m^{-2} $year^{-1}$ (Tucker and Wright 1990).

Charaleans are the largest and most morphologically, developmentally, and reproductively complex group of charophycean green algae. Reaching lengths of 1 m or more, with whorls of branches at nodes, some are regularly confused with similar-appearing aquatic flowering plants such as *Ceratophyllum* (Figure 20.62).

Charalean vegetative structure

Charaleans fundamentally are branched filaments, though the main axis is differentiated at the apex, nodes, and basal region. The erect shoot possesses a single specialized apical meristematic cell; this cell cuts off derivatives from its lower surface only, thereby extending the filament in length (Figure 20.63). (By comparison, the apical meristematic cells of bryophytes have three or four cutting faces and generate tissues.) No centrioles are present at spindle poles in these or other cell divisions in charalean algae, which is consistent with the absence of flagellate zoospores. The immediate derivative of transverse apical cell mitotic division (known as a segment cell) divides again transversely. The uppermost of the resulting two cells will continue to divide to produce a complex node with lateral branches, whereas the lower cell develops into a very long internodal cell without further division. Therefore, the main axis consists of regularly alternating discoidal nodal cells and long cylindrical internodal cells (see Figure 20.63).

Figure 20.64 *Nitella* viewed using fluorescence microscopy. Note the large internodal cells, branches emerging in whorls from much smaller nodal cells, and the numerous autofluorescent, discoid plastids arranged in vertical files. (Photo: L. E. Graham)

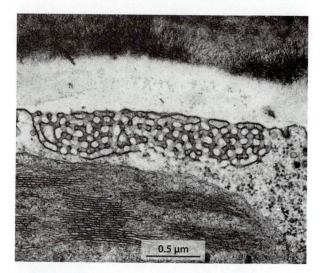

Figure 20.65 *Chara zeylanica* charasome. The charasome is an array of infolded plasma membrane associated with proteinaceous material whose function is unknown. Reprinted from Graham, L. E., and Y. Kaneko. 1991. *Critical Reviews in Plant Science* 10:323–342. © CRC Press, Boca Raton, Florida)

Internodal cells can reach lengths of 15 cm. Internodal cells are so large that microelectrodes can easily be inserted for electrophysiological studies. An excellent review of electrophysiological investigations in charalean algae can be found in Wayne (1994). Internodal cells contain well over 1000 nuclei, which are produced by the replication of a single original nucleus through a process that does not involve the typical mitotic apparatus. The interphase nuclei of vegetative cells in young shoots of *Chara* and *Nitella* can undergo endoreduplication—an increase in the amount of DNA beyond the haploid level (Michaux-Ferrière and Soulié-Märsche 1987). An increased number of nuclei and high levels of nuclear DNA presumably balance the large increase in cell volume, which is mediated by development of a large internal vacuole. The cytoplasm nearest the central vacuole of internodal cells is an ideal site for visualizing cytoplasmic streaming, resulting from actin microfibril activity (Allen 1974; Palevitz and

Hepler 1975; Williamson 1979, 1992). Presumably, such streaming is necessary to achieve mixing and long-distance transport of cell constituents in long cells having large cytoplasmic volumes. It has been hypothesized that such large internodal cells represent adaptation to shady benthic habitats (Raven et al. 1979).

Within the internodal cells, a layer of nonmobile peripheral cytoplasm contains numerous discoid chloroplasts arranged in rows (Figure 20.64) having granalike thylakoid stacks and starch grains but lacking pyrenoids, as well as mitochondria, and peroxisomes containing conspicuous catalase crystals. The plastids are generated by repeated fission. Elaborate invaginations of the cell membrane, known as **charasomes**, may occur along the cell periphery (Figure 20.65); their function is uncertain. The presence of charasomes is apparently not correlated with the ability to use bicarbonate ion in photosynthesis (Lucas et al. 1989) (see Chapter 2). The cell wall of charaleans is composed of distinct layers, including cellulose, and can be relatively thick. Because of the axial, filamentous construction of these algae, much of the rest of the body could be lost should an internodal cell be damaged or destroyed.

The cell plate separating nodal from internodal cells is characterized by very large pores. These

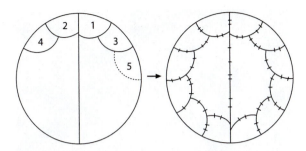

Figure 20.66 *Chara zeylanica.* Structure and development of the nodal complex. A first division gives rise to two equal nodal cells. Each of these then undergoes coordinated, directional, asymmetrical divisions, giving rise to whorls of branch initials. During early stages of development, the branches are of unequal length, reflecting different times since origin. However, branch growth is determinate, and more recently produced branches eventually catch up; in the mature state, all branches of a given whorl are the same length. Short lines denote plasmodesmata. (From Cook et al. 1998 © Springer-Verlag)

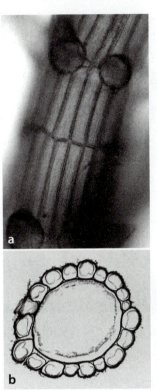

Figure 20.67 *Chara zeylanica* corticating filaments. (a) Corticating filaments grow from both nodes over the surface of internodal cells. (b) Cross-sectional view. (Photos: L. W. Wilcox)

appear to have originated by the coalescence of several plasmodesmata, themselves reported (in at least one species) to have been derived secondarily by the action of wall hydrolytic enzymes (Franceschi et al. 1994; Lucas 1995). These large pores are thought to facilitate passage of materials throughout the vertical axis of the body (Cook et al. 1997).

Nodal initial cells first divide vertically (commonly into two halves), and then each of the resulting cells undergoes a very highly controlled series of asymmetric divisions that produce the branch initials (Figure 20.66). Divisions are synchronized in the two cells making up the node; divisions leading to the formation of branch initials occur in a sequential, radial manner. Charalean branch initials serve as the apical cells for development of branches having the same kind of alternating nodal and internodal cells as the main axis. The branches in turn produce smaller branchlets. Unlike the apical cell of the main axis, which can continue to divide on an indeterminate basis, the branch apices cease division after a determined number of cells have been produced. Branches of two *Chara* species (but not all species examined) have been observed to grow toward light by cell elongation, particularly in high- irradiance conditions. This behavior is suggested to protect gametangia or aid penetration of light into the canopy (Schneider et

al. 2006). Cells of the nodal region may generate other branches that mirror the growth habit of the main axis (i.e., have indeterminate growth). In addition, in most species of *Chara*, basal nodal cells of the branches can generate multiple rows of filaments that grow up or down over the internodal cell surfaces, forming what is known as a corticating layer. Corticating filaments originating from adjacent nodes meet in the middle of the internodal cell (Figure 20.67).

A land-plant-like phragmoplast is present during cytokinesis in charalean algae. In addition to the characteristic longitudinal microtubule array, charalean algae possess actin microfilaments, membrane tubules, coated vesicles, and fenestrated sheets such as are associated with cell plate development in higher plants (Pickett-Heaps 1975; Cook et al. 1997; Braun and Wasteneys 1998). However, charalean algae that have been investigated differ in a number of ways from plants in regulation of the cytokinetic process. These include later dissolution of the phragmoplast, absence of an organellar and ribosome exclusion zone from developing cell plates, cell plate development that is patchy and not regularly centrifugal, co-occurrence of different cell plate

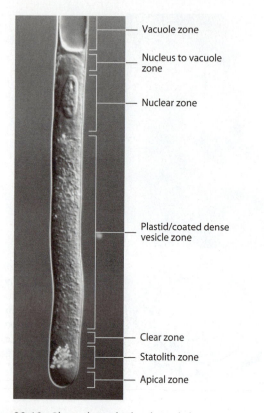

Vacuole zone

Nucleus to vacuole zone

Nuclear zone

Plastid/coated dense vesicle zone

Clear zone

Statolith zone

Apical zone

Figure 20.68 *Chara* rhizoid. The rhizoid shows zonation of the cytoplasm that is related to gravitropism. (From Kiss and Staehelin 1993)

Figure 20.69 *Chara braunii* oogonia. This is an isolated node and its whorl of branches. (Photo: M. Cook)

Figure 20.70 *Chara* antheridia borne on nodes of branchlets. Note the petallike arrays of surface shield cells. (Photo: M. Cook)

developmental stages, and an earlier development of plasmodesmata (Cook et al. 1997). Although some of the literature (Grant 1990) implies that charalean algae produce pre-prophase microtubule bands homologous to those of land plants, to date, such bands have not been demonstrated to occur in any charophycean alga.

Ultrastructural studies of the charalean node suggest that it can be considered to be a tissuelike, or parenchymatous, structure (Pickett-Heaps 1975; Cook et al. 1998). Comparative immunolocalization studies of actin and microtubules in internodal and nodal cells suggest that the latter more closely resemble higher plant meristem cells (Braun and Wasteneys 1998). Moreover, cells in the nodal region of at least one charalean species possess primary plasmodesmata, i.e., those produced during formation of the cell plate at cytokinesis, much like plasmodesmata of land plants (Cook et al. 1997). The basal portion of charalean algae in nature is typically attached to muddy or silty substrates by numerous colorless rhizoids. These are very long cells containing unpigmented plastids and 30–60 barium- and sulfur-containing crystals, the latter having geotropic function. Rhizoids grow at the tip and are not differentiated into nodes and internodes, but they do exhibit a definite polarity, with cells showing at least seven distinct zones (Figure 20.68) (Kiss and Staehelin 1993).

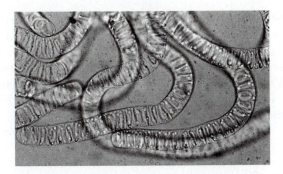

Figure 20.71 *Chara zeylanica* spermatangial filaments. Each of the equal-sized colorless cells will produce a single, elongate, spirally twisted biflagellate sperm. (From *Origin of Land Plants*. Graham, L. E. © 1993. Reprinted by permission of John Wiley & Sons, Inc.)

Reproduction of charaleans

Asexual reproduction in charaleans can occur by means of adventitious (from the shoot) development of new shoots from rhizoids and nodal complexes. Additional asexual structures include bulbils, which are white spherical or star-shaped structures that form on rhizoids of some species and function in dispersal and perennation.

Charaleans have probably the most conspicuous sexual structures of any green algae (Figures 20.69 and 20.70). These specialized structures (gametangia) are of two types—(male) antheridia, where thousands of biflagellate spermatozoids develop, and (female) oogonia, each containing a single egg cell. The antheridia of charalean algae are bright orange at maturity and are visible without the use of a microscope. In monoecious species, antheridia and oogonia are usually borne together at the node of a branchlet. Microscopic examination of antheridia reveals that the orange pigmentation is generated by carotenoid droplets within an outer layer of cells arranged in groups of eight, forming a flowerlike pattern (see Figure 20.70). These are known as the **shield cells**, and their form is used in classification and identification of some species. The shield cells are attached to a columnar-shaped cell known as a **manubrium**, which is also associated, at its other end, with a group of eight cells known as the **primary capitulum**. Cells derived from division of the primary capitulum generate long, unbranched filaments of small cells, each producing a single thin, helically twisted spermatozoid (Figures 20.71 and 20.72). Changes that occur in the configuration of

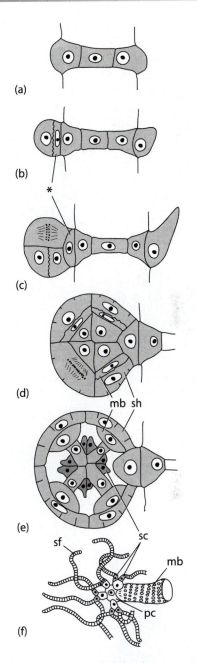

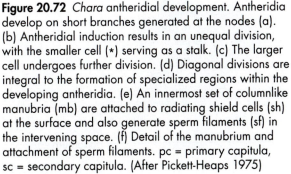

Figure 20.72 *Chara* antheridial development. Antheridia develop on short branches generated at the nodes (a). (b) Antheridial induction results in an unequal division, with the smaller cell (*) serving as a stalk. (c) The larger cell undergoes further division. (d) Diagonal divisions are integral to the formation of specialized regions within the developing antheridia. (e) An innermost set of columnlike manubria (mb) are attached to radiating shield cells (sh) at the surface and also generate sperm filaments (sf) in the intervening space. (f) Detail of the manubrium and attachment of sperm filaments. pc = primary capitula, sc = secondary capitula. (After Pickett-Heaps 1975)

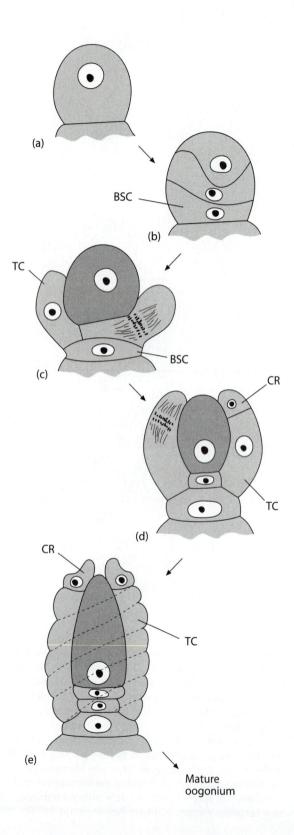

(a)

BSC

(b)

TC

BSC

(c)

CR

TC

(d)

CR

TC

(e)

Mature
oogonium

the flagellar apparatus during sperm development have been studied in *Chara contraria* var. *nitelloides* (Vouilloud et al. 2005). Additional details of sperm structural development can be found in Moestrup (1970) and Pickett-Heaps (1975). When sperm are ready for release, the shield cells separate, allowing the sperm to swim away.

Oogonia arise from branchlet nodal cells as well. A primordial cell divides twice transversely, and the uppermost of the resulting cell stack becomes the egg. The cell just below the egg repeatedly divides, generating a ring of five peripheral cells surrounding a central cell (Figure 20.73). The five peripheral cells elongate to form **tube cells** (also known as sheath cells) (Figure 20.74) that grow upward along the surface of the egg, extending to keep pace with enlargement of the egg. As each tube cell elongates, it takes a counterclockwise helical path, and at its tip, a transverse division(s) gives rise to one or two **coronal (crown) cells** (the number depending on the genus) (Figure 20.75). This process resembles that of zygote cortication in *Coleochaete*, and it has been proposed that both are based on similar cell–cell signal–response events (Graham 1993). As the egg enlarges, it becomes filled with storage products—usually many white starch grains, but lipid droplets are also abundant. Such storage buildup resembles that occurring in *Coleochaete* zygotes, and its role is similar—supporting later zygote germination. At maturity, openings form between the tube cells that allow sperm to reach the egg.

After fertilization a thick, darkly pigmented zygote wall develops that contains a sporopolleninlike layer. Calcification of the concave inner walls of the spiral tube cells of *Chara* also typically occurs after fertilization. In *Tolypella*, calcium carbonate is deposited on the outside of the tube cells; *Nitella* zygotes are not calcified. As bodies are degraded at the

Figure 20.73 Development of oogonia in *Chara*. (a) A cell derived by division of a nodal cell undergoes further division (b) into a basal stalk cell (BSC), a middle cell that generates the five tube cells (TC) and the terminal egg (darker shaded cell). (c–e) Elongation of the tube cells forms a twisted cortical layer that covers the egg cell entirely except for a pore at the top. A final division at the tip of each of the five tube cells gives rise to the five coronal cells (CR) of *Chara*. *Nitella* undergoes two such divisions at the ends of each of the 5 tube cells, giving rise to a total of 10 coronal cells. (After Pickett-Heaps 1975)

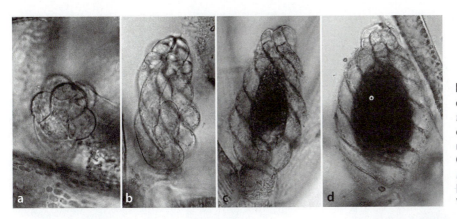

Figure 20.74 Oogonial development in *Nitella tenuissima*. Note the presence of two tiers of coronal cells on mature oogonium. (a–c: L. E. Graham; d: From *Origin of Land Plants*. Graham, L. E. © 1993. Reprinted by permission of John Wiley & Sons, Inc.)

end of the growing season, zygotes (also known as oospores or zygospores), together with their protective tube cells and perhaps a few other vegetative remnants, fall to the sediments. The thick, resistant wall and calcified tube-cell layer (if present) contribute to zygote survival during a period of dormancy. The calcified impressions of the tube cells and enclosed structures may persist in the fossil record; these are known as **gyrogonites**. Fossil gyrogonites reveal that the ancient relatives of modern charaleans often had more than five tube cells; the number has apparently been reduced over time. Another interesting (and unexplained) change has occurred in charalean tube cell orientation over time. Tube cells of older taxa, such as the lower Devonian *Trochiliscus*, were twisted to the right, whereas tube cells of taxa appearing near the end of the Devonian were twisted toward the left (Figure 20.76), as are those of modern forms and intervening ages. A few examples of calcified or silicified remains of antheridia and vegetative parts are also known (Figure 20.77). *Paleonitella* had noncalcified vegetative bodies with nodal organization and was preserved in the geological deposit known as the Rhynie Chert as petrifactions (mineral-impregnated remains) (Kelman et al. 2003).

Zygotes of charaleans are believed to germinate by meiosis, with only one meiotic product surviving. However, this assumption is based entirely upon circumstantial evidence, such as the observation of four nuclei within germinating zygotes (van den Hoek et al. 1995) and the fact that sperm and vegetative nuclei contain the same level of DNA, suggesting that meiosis is not gametic (Shen 1967). Analysis of DNA-level changes or chromosome counts during zygote germination have been difficult to accomplish because the massive amounts of storage photosynthate and thick,

100 µm

Figure 20.75 Mature oogonium of *Chara*. This SEM view shows the five coronal cells. (Micrograph: M. Cook)

dense zygote wall preclude easy observation of nuclear phenomena. Zygote germination occurs when a colorless filamentous "protonema" emerges from a break in the zygote wall. The protonema possesses a colorless primary rhizoid and, under the influence of blue or white light, undergoes transverse divisions to form a short filament with green chloroplasts appearing in the uppermost cells. Germination of charalean zygotes is notoriously difficult to achieve at high levels of efficiency in the laboratory, further adding to difficulties in clarifying the life history of these algae. Cold-temperature and red-light treatments (Takatori and Imahori 1971) are said to increase the rate of zygote germination. Resistance to germination and

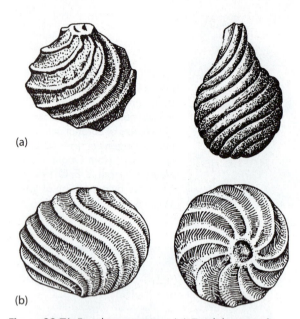

(a)

(b)

Figure 20.76 Fossil gyrogonites. (a) *Trochiliscus* and (b) *Eochara*. Note the difference in direction of surface ornamentation twisting. (From teaching documents of the late Professor L. Grambast, courtesy of M. Feist)

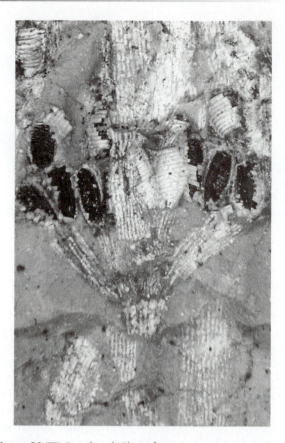

Figure 20.77 Fossilized *Chara* from Argentina. Note the dark zygotes and remnants of cortical and tube cells. (Photo: M. Cook)

germination only after extended storage in wet conditions or treatment with fluctuating water levels, irradiance, or temperature are regarded as adaptations that increase the chances that wetland charaleans will survive one or more unfavorable seasons. Variation among species in zygote germination behavior helps explain seasonal and geographical distribution patterns. For example, *Nitella cristata* var. *ambigua* zygotes germinated well in response to cues for the onset of winter, explaining the occurrence of this winter form (Casanova and Brock 1996).

Charalean diversity

According to Wood and Imahori (1965), there are six genera of living charaleans: *Chara*, *Lamprothamnion*, *Lychnothamnus*, *Nitellopsis*, *Nitella*, and *Tolypella*. Most of the 81–400 species (the number depending upon the expert) belong to *Chara* or *Nitella* (Grant 1990). Species determinations are based on such characters as the arrangement of the gametangia, the presence or absence of cortical layers, and the number and arrangement of cortical cells. Molecular (*rbcL*) data suggest that modern charaleans form a monophyletic group, and there is some support for a monophyletic tribe Chareae (*Chara*, *Lamprothamnion*, *Nitellopsis*,

and *Lychnothamnus*), but monophyly of the tribe Nitelleae (*Nitella* and *Tolypella*) was not supported (McCourt et al. 1996). Molecular sequence data suggest that the genera *Nitella* and *Tolypella* are basal to the more derived Chareae.

CHARA (pre-Linnaean name of unknown origin) is characterized by structures known as **stipules** (or stipulodes), which are single-celled, often sharply tipped structures occurring below the branchlets. The main axes of most species are corticated, but some species, such as *Chara braunii* (Figure 20.78), lack this corticating layer and can be mistaken for *Nitella*. There is a single layer of 5 oogonial coronal cells. (In contrast, *Nitella* possesses two tiers of coronal cells, totalling 10.) *Chara* species are often calcified and thus may have a stony, gray-green appearance. Calcification, together with cortication, gives *Chara* a generally more robust appearance than *Nitella*. Species such as *Chara vulgaris* are regarded

(a)

(b)

Figure 20.78 *Chara braunii*, an uncorticated *Chara* species. (Photo: M. Cook and C. Lipke)

Figure 20.79 *Nitella.* (a) *Nitella flexilis.* (b) *Nitella tenuissima.* Note characteristic dense tufts of branches from nodes that are well separated by internodal cells. (Photos: M. Cook and C. Lipke)

as marl formers because they deposit large amounts of calcium carbonate at the bottoms of water bodies. They primarily occur in relatively high-alkalinity waters. Some forms, including *C. vulgaris*, produce a foul odor that is variously described as "skunky" or "like spoiled garlic."

NITELLA (L. *nitella*, brightness or splendor) is characterized by very regular, symmetrical branching (Figure 20.79). The branches that bear the gametangia are repeatedly forked. Bodies are uncorticated and not typically calcified. The oogonia are either solitary or occur below the antheridia. Oogonia have 10 coronal cells in two tiers of 5 each. The species range greatly in size from meter-long *N. flexilis*, which has very long internodal cells and occupies waters 10–12 m deep (Figure 20.79a), to minute and delicate forms such as *N. tenuissima* (Figure 20.79b), which occurs in shallow, silty waters. In contrast to most *Chara* species, *Nitella* most commonly occurs in soft, slightly acidic waters.

Phytoplankton Ecology

Freshwater diatoms

The word *phytoplankton* comes from two Greek words: *phytos* for "plant" and *planktos* for "wandering." Phytoplankton are photosynthetic microorganisms that are adapted to live and wander in the open surface waters of lakes, rivers, and oceans. The definition includes both prokaryotes (cyanobacteria) and a diverse array of eukaryotes. Cyanobacteria, greens, diatoms, cryptomonads, dinoflagellates, haptophytes, and chrysophyceans are especially rich in planktonic species. Phytoplankton play many important ecological roles in aquatic ecosystems and affect human affairs in many ways. Planktonic algae are the primary producers of aquatic ecosystems and form the base that supports the zooplankton and fish of aquatic food webs. For example, cold, deep nutrient-rich waters rise to the ocean surface along the west coast of South America, where the nutrients support abundant phytoplankton, zooplankton, and a rich fishery. When a climatic change such as El Niño occurs, the nutrient-rich waters are shut off, causing a cascade of events ending in a collapse of the fishery and severe economic repercussions. Phytoplankton also play a major role in the global cycle of carbon dioxide. The phytoplankton in the oceans account for a little less than half (48 Pg C yr^{-1}) of the global net primary production of 105 Pg C yr^{-1} (refer to Glossary of Abbreviations on the following page) (Geider et al. 2001). Photosynthesis of marine phytoplankton

(Photo: L. W. Wilcox)

contributes to the removal of carbon dioxide released by the burning of fossil fuels (about 5.5 ± 0.5 Pg C yr^{-1}) and agricultural activities such as deforestation and land drainage (around 1.6 ± 1.0 Pg C yr^{-1}) (Behrenfeld et al. 2002). These human activities are also responsible for the increasing flow of nutrients into lakes, rivers, and marine coastal waters, where they may trigger harmful algal blooms. These algal blooms may produce toxins, as can cyanobacteria in freshwaters and dinoflagellates in marine coastal waters, or the phytoplankton may die and decompose, depriving the waters of oxygen and producing **dead zones** where marine life is killed by suffocation. The second largest dead zone in the world occurs annually where the Mississippi River enters the Gulf of Mexico. This area is now larger than the state of New Jersey. More than 145 dead zones have been reported globally (43 in coastal U.S. waters), and the number is growing. Phytoplankton ecology is thus relevant to global ecology and human lives.

Phytoplankton ecology has been characterized by the development of a large body of theory and mathematical modeling. Theory and modeling have centered about two topics—competition theory and trophic dynamics. The original source of competition theory lies in the work of Gause (1934) on various

Glossary of Abbreviations	
ANPP	Annual net primary production
CCM	Carbon concentrating mechanism
DIN	Dissolved inorganic nitrogen
DIP	Dissolved inorganic phosphorus
DOC	Dissolved organic carbon
DON	Dissolved organic nitrogen
DOP	Dissolved organic phosphorus
HNF	Heterotrophic nanoflagellates
IDH	Intermediate disturbance hypothesis
MNF	Mixotrophic nanoflagellates
NPP	Net primary productivity
Pg	Petagrams (1 Pg = 10^{15} g)
SA/V	Surface area-to-volume ratio
SRP	Soluble reactive phosphorus
SRSi	Soluble reactive silicon
TP	Total phosphorus

microorganisms in test tubes and flasks. These experiments led to the development of the competitive exclusion principle and later to niche theory. The basic idea is that biological interactions—not physical and chemical external factors—are paramount in community dynamics. Species are assumed to exist close to their maximum density in the environment and to compete for scarce resources. If two species occupy the same niche, one must inevitably drive the other out through competitive displacement. These concepts entered plankton ecology when Hutchinson (1961) wrote his famous paper on the paradox of the plankton. If we assume that species are close to their maximum density in aquatic systems and competitive exclusion is a general rule, how can 50 to 100 species of phytoplankton possibly coexist in only a few milliliters of lake or ocean water? Why are phytoplankton communities so diverse? The controversy concerning this paradox has fueled experimental and theoretical research to the present and has even shaped aquatic management practices.

The second focus of theory and modeling is trophic dynamics. The original concept of trophic dynamics involved bottom-up control. Bottom-up control maintains that phytoplankton populations are fundamentally controlled by nutrients rather than herbivory. Food chains were seen in simple terms. More nutrients means more phytoplankton biomass, and more phytoplankton means more zooplankton to feed fish populations. The concept of top-down control of phytoplankton communities originated with Porter (1977) and arose out of observations of clear-water phases in aquatic systems. Top-down theories assume that phytoplankton populations are controlled by herbivory, which directs species compositions and seasonal patterns of biomass. Research on trophic dynamics led to the discovery and widespread acceptance of the importance of the microbial loop, a complex microbial food web consisting of secondary producers such as bacteria and archaea, small phytoplankton, and heterotrophic flagellates and ciliates that graze upon them. All these concepts have shaped management practices. We discuss both competition theory and trophic dynamics in later sections. We consider phytoplankton population dynamics under two general categories: (1) growth processes, including photosynthesis and nutrient uptake, and (2) loss processes, including perennation (the formation of resting stages), mortality, parasitism, sedimentation, competition, and grazing. But first, we consider briefly the importance of size and scale in phytoplankton ecology and the physical and chemical environment of lakes and oceans, the medium in which phytoplankton ecology occurs.

21.1 Size and Scale in Phytoplankton Ecology

Phytoplankton include a vast range of sizes and forms (Figure 21.1). Individual cells and colonies may possess flagella or lack flagella. *Chlamydomonas* and *Ochromonas* are examples of small flagellate motile cells, while *Chlorella* is a small nonmotile cell. *Scenedesmus* is a small nonmotile colony, and *Gonium* is a small flagellated one. Desmids such as *Staurastrum* lack flagella but can move by jetting mucilage. The dinoflagellate *Ceratium hirundinella* is an exceptionally large (> 100 μm), single-celled, flagellated alga. Large colonial planktonic algae include the buoyant cyanobacterium *Microcystis aeruginosa*, the diatom *Asterionella formosa*, and the flagellate green alga, *Volvox aureus*. The cyanobacterium *Anabaena flos-aquae* is a planktonic filament. According to Sieburth et al. (1978), phytoplankton can be classified into size categories. The picophytoplankton (0.2–2.0 μm) include some cyanobacteria. Many eukaryotic algae and cyanobacteria are in the nanophytoplankton (2.0–20 μm). Planktonic algae that are large enough to be retained by the mesh of standard plankton nets (net phytoplankton) include the microphytoplankton (20–200 μm), mesophytoplankton (0.2–2 mm), and macrophytoplankton (2–20 mm). Why do phytoplankton vary so much in size?

Size in Phytoplankton Ecology

Size is the most important single characteristic affecting the ecology of phytoplankton. As phytoplankton become larger, their volumes increase as the cube of their radius, while their surface areas increase in proportion to only the square of the radius. Consequently, as algal species become larger, their surface area-to-volume ratio (SA/V) becomes smaller, especially if they retain a spherical shape. Table 21.1 presents volumes, surface areas, and SA/V ratios of a number of phytoplankton species (Reynolds 2006). Since phytoplankton growth involves the exchange of materials at the cell surface, single phytoplankton cells and colonies tend to depart from a spherical shape as they become larger. If *Pseudopediastrum* (*Pediastrum*) *boryanum* were a sphere instead of a flat plate, its surface area would be only 3070 μm² instead of 18,200 μm² (Table 21.1). The necessity of maintaining an optimum SA/V ratio has therefore been one factor in determining the varied

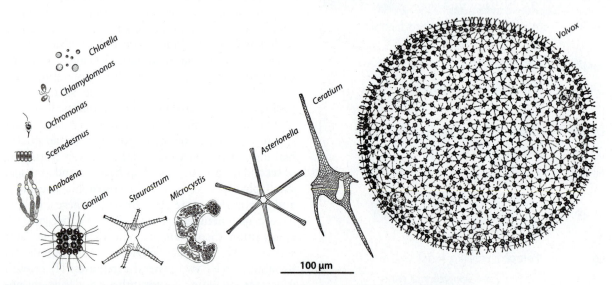

Figure 21.1 Sizes and types of phytoplankton (approximately to scale). (Drawings from Smith, G. M. 1950. *Freshwater Algae of the United States.* McGraw-Hill. Reproduced with permission of the McGraw-Hill Companies, except for *Anabaena*: originally from G. M. Smith, in Prescott, G. W. 1951. *Algae of the Western Great Lakes Area.* Cranbrook Institute of Science, a division of Cranbrook Educational Community)

Table 21.1 Volumes, surface areas, and surface area-to-volume ratios for selected phytoplankton*

Species	Volume (μm^3)	Surface area (μm^2)	SA/V (μm^{-1})
Single cells			
Synechococcus	18	35	1.94
Ankyra judayi	24	60	2.50
Chlorella pyrenoidosa	33	50	1.52
Chrysochromulina parva	85	113	1.33
Stephanodiscus hantzschii	600	404	0.67
Cyclotella meneghiniana	1600	780	0.49
Cryptomonas ovata	2710	1030	0.38
Mallomonas caudata	4200	3490	0.83
Cosmarium depressum	7780	2770	0.36
Synedra ulna	7900	4100	0.52
Ceratium hirundinella	43,740	9600	0.22
Peridinium cinctum	65,500	7070	0.11
Filaments			
Anabaena circinalis	2040	2110	1.03
Aulacoseira subarctica	5930	4350	0.73
Colonies			
Desmodesmus quadricauda	1000	908	0.91
Asterionella formosa	5160	6690	1.30
Fragilaria crotonensis	6230	9190	1.48
Dinobryon divergens	7000	5350	0.76
Pseudopediastrum boryanum	16,000	18,200	1.14
Eudorina unicocca	1.15×10^6	5.31×10^4	0.05
Volvox globator	4.77×10^7	6.36×10^5	0.01

*Data from Reynolds 2006)

shapes of planktonic algae. Nielsen (2006) showed that the growth rates of unicellular green algae and cyanobacteria decreased significantly with increasing size while the growth rates of colonial forms did not change. Small single phytoplankton cells usually have simple geometric shapes and large SA/V ratios. As they become larger, they increasingly become distorted in form from simple geometric shapes to maintain a more favorable SA/V ratio, but form distortion can only go so far, and growth rates decline with size. As cells are added to flat colonies or filaments, the SA/V ratio does not change, and therefore the growth rates do not change.

Size affects more than just phytoplankton growth rates. Small cells can respond to a pulse of nutrients with rapid nutrient uptake and a burst of growth. Large cells, however, can take up and store more nutrients, such as phosphate, than can small cells. These size-related phenomena result in a spectrum of competitive adaptations among phytoplankton species from small, fast-growing single cells to large, slow-growing cells able to store nutrients through periods of scarcity. Between these extremes lies a range of intermediate adaptations. The danger of sinking out of the lighted (euphotic) zone in a body of water is a continual problem. Large cells tend to sink more rapidly through water than small cells. To avoid losses due to sinking, phytoplankton have evolved a wide range of adaptations that increase buoyancy, reduce density, or increase physical-form resistance to sinking; these adaptations are more pronounced in large cells and colonies. In turn, these same adaptations are responsible for much of the morphological diversity within the phytoplankton. Finally, the susceptibility of phytoplankton to losses due to grazing or predation is size dependent. Grazers such as protozoa, rotifers, and crustaceans consume small cells more readily than large cells. Large cells and colonies above about 50 μm in diameter are largely immune to predation by crustaceans (Burns 1968) but are prone to attack by parasites. Intermediate cell and colony sizes may be free from predation by protozoa but susceptible to crustaceans. Many of the adaptations that reduce sinking rates and increase the SA/V ratio also perform anti-grazer roles.

Scale in Phytoplankton Ecology

The importance of cell size is one example of how scale factors into phytoplankton ecology. Different aspects of phytoplankton ecology occur at different scales of time and space. Nutrient regeneration and grazing on phytoplankton can take place at scales of seconds to minutes over distances of millimeters. Lehman and Scavia (1982) showed that zooplankton excrete nutrients in patches that significantly affect nutrient uptake by algae over scales of millimeters. Growth rates operate at scales of hours to days. Phytoplankton patches develop at scales of weeks and kilometers, and successions of species occur at scales of entire seasons across entire lake basins and oceans (Harris 1986). Dynamic processes in lakes and oceans are very similar but operate on different scales. A patch of phytoplankton in a lake may extend for hundreds of meters, whereas a patch in an ocean may be hundreds of kilometers across.

As an example of the importance of scale in phytoplankton ecology, consider the routine problem of determining the abundance of phytoplankton in a body of water. Counts of phytoplankton may be done by flow cytometry or by collecting a sample of water and placing a subsample into a settling chamber. After the algae have settled overnight, they are counted and identified with an inverted microscope. Results are reported as number of cells of each species per unit of volume (ml or l). If an alga were present at a density of one cell per liter, it would likely be missed or considered unimportant. Consider Lawrence Lake in southwestern Michigan, which has a volume of 293,500 m³, equivalent to 293,500,000 liters (Wetzel 1975). In this lake a rare alga existing at an average density of one cell per liter would have a total population of 293,500,000. Many such rare species contribute to the high species diversity of phytoplankton in lakes and oceans.

Even though a phytoplankton species may be rare in a body of water at a particular time, the importance of temporal scale in phytoplankton ecology guarantees that it may not remain so for long. Rates of reproduction in phytoplankton vary from two to three doublings per day to one doubling every week to 10 days. If an alga started at a density of one per liter with a reproduction rate of two doublings per day, it would reach a density of 16,384 cells l^{-1} in 1 week (assuming no losses during this time). It is therefore essential when planning a sampling program to be aware of the time scale at which events occur. In Figure 21.2a, the upper graph shows rapid changes in population densities of chrysophyceans as determined by sampling every 2 to 3 days (Sandgren 1988). Figure 21.2b shows what the same data would have looked like if these populations had been sampled at weekly intervals (a more common practice). The population maxima of several species would have been missed. The phenomenon of missing important phenomena by sampling at an inappropriate scale is called **aliasing**.

For many kinds of species, such as mammals and flowering plants, there is a strong relationship between the number of species present (species richness) and the area sampled. This species–area relationship is expressed as a power law, which is written as:

$$S = cA^z \quad (1)$$

where S is species richness, A is area in km², c is a constant, and z is a scaling exponent. Smith et al. (2005a) found that the same law held for phytoplankton species, and, moreover, it held the same for laboratory microcosms all the way up to oceans across 15 orders of magnitude. This finding means that laboratory studies of phytoplankton can be scaled up to and applied to

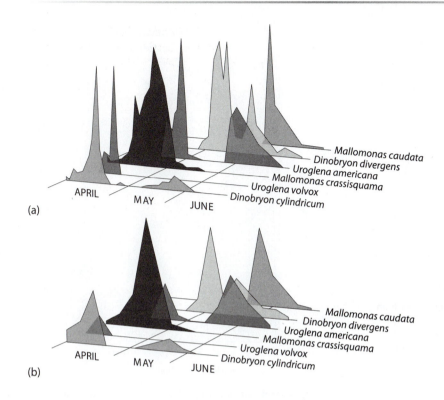

(a)

(b)

Figure 21.2 (a) Patterns in populations of some opportunistic chrysophyceans sampled every two to three days. (Data from Sandgren 1988) (b) A plot of the same data had sampling been conducted only every seven days. Several population maxima are missed.

larger aquatic systems and that phytoplankton diversity patterns are consistent with those of large organisms.

21.2 The Physical Environment

Oceans cover about 71% of the surface of Earth. They are the largest single ecosystem on Earth. While freshwater ecosystems are much less extensive, they can occupy up to 50% of the land area in moist regions, and the system of ponds, lakes, rivers, and wetlands is absolutely essential for terrestrial life. Solar radiation lights and heats the surface waters of aquatic systems, giving rise to a number of distinct zones in a vertical structure. The interaction of winds with solar heating, tides, and Earth's rotation generate many forms of water motion, which in turn affect the distribution and population sizes of phytoplankton.

Water as a Fluid Medium

The physical environment in which phytoplankton reside is determined by the physical properties of water as a molecule and the interaction of water with solar radiation. In a molecule of water, the two hydrogen atoms and single oxygen atom are arranged as if they were at the points of an isosceles triangle, with an obtuse angle of 104.5° at the oxygen atom. Consequently, the hydrogen atoms bear a weak positive charge and the oxygen atom a weak negative charge. Water molecules are capable of forming hydrogen bonds between adjacent molecules. These bonds in turn allow water to act as a liquid crystal—a property that is almost unique among compounds. If water (H_2O) behaved like structurally similar compounds such as H_2S or NH_3, it would be a gas at normal environmental temperatures.

Water is a highly effective solvent for inorganic salts, soluble hydrophilic organic compounds, and a wide array of gases, including O_2, N_2, and CO_2. Therefore, it is an ideal medium for a vast number of chemical interactions. The intermolecular bonding properties of water give it a high specific heat. This means that liquid water can store a large quantity of solar heat and that a large quantity of heat is required to raise the temperature of a body of water. Especially large inputs of heat (540 cal g^{-1}) are required for the phase transition from liquid to gas. Conversely, when a body of water cools, it gives off large amounts of heat to the overlying air and surrounding land. Aquatic environments are very stable thermal environments for phytoplankton, which

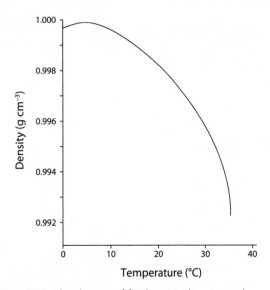

Figure 21.3 The density of freshwater changes with temperature. Maximum density occurs at 4°C. (After Goldman, C. R., and A. J. Horne. 1983. *Limnology.* McGraw-Hill. Reproduced with permission of the McGraw-Hill Companies)

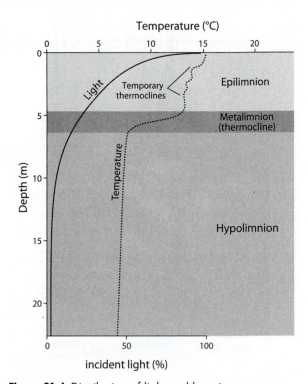

Figure 21.4 Distribution of light and heat in a summer stratified lake. Light declines exponentially with depth. The thermocline divides the lake into upper mixed epilimnion, metalimnion, and lower, colder hypolimnion. Temporary thermoclines are due to warm, calm days. (After Goldman, C. R., and A. J. Horne. 1983. *Limnology.* McGraw-Hill. Reproduced with permission of the McGraw-Hill Companies)

are normally not subjected to sharp temperature changes over short periods of time, as are terrestrial organisms.

Water shows very marked changes in density with temperature (Figure 21.3). Density increases rapidly as temperature falls from 35°C, reaching a maximum density at 4°C. With further cooling, water becomes less dense until it freezes as ice, which floats because it is less dense than liquid water. If, like most other substances, water had its greatest density when it became solid, ice would sink to the bottom of lakes or oceans, which would consequently freeze from the bottom up. Bodies of water would be inhospitable places for life because few large organisms can withstand being frozen solid. Water ice furthermore has a specific heat (0.5) about half that of liquid water, and so ice forms and melts readily. The ease with which it forms means that the underlying warmer and denser water is quickly insulated from the cold air above and therefore remains liquid. Many lakes in Antarctica are covered by permanent ice caps up to 4 m thick; liquid water and phytoplankton survive under the ice. In the oceans, high salinity (on average, 35 g l⁻¹) increases density and depresses the freezing point to −1.91°C, but ice still floats and insulates the underlying waters.

Two final properties of water arise from hydrogen bonding. Surface tension is the tendency of water molecules to bond together at the air–liquid interface. This property makes it possible for a number of species of phytoplankton, including some diatoms and chrysophyceans, to hang from or sit atop the surface film. Collectively they are termed **neuston** (Wetzel 1975). **Viscosity** is the tendency of water to resist flow and impose drag on organisms moving through it. Viscosity increases as temperature decreases. While water is much less viscous than maple syrup, it still has profound effects on the shapes of phytoplankton. If water were significantly less viscous than it is, phytoplankton would have difficulty remaining in suspension.

Light and Heat

Most of the solar radiation entering a lake or an ocean is converted into heat. Light entering a body

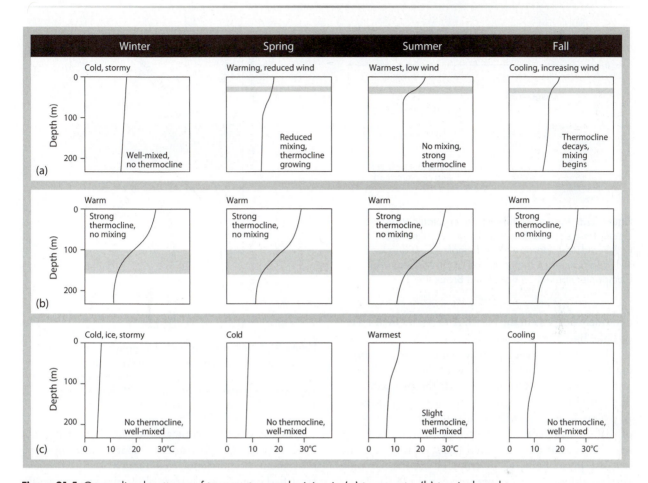

Figure 21.5 Generalized patterns of temperature and mixing in (a) temperate, (b) tropical, and (c) polar oceans as functions of depth and seasonality. Line denotes temperature as a function of depth. Shaded zone indicates thermocline, which prevents mixing in the waters beneath it. (After *Marine Biology* 2nd. Edition, by J. W. Nybakken. © 1987 by Harper Collins Publishers. Reprinted by permission of Addison-Wesley Educational Publishers)

of water declines as a function of depth according to a simple exponential decay function:

$$I_z = I_o e^{-\eta z} \quad (2)$$

Here I_z is the irradiance (µmol quanta m^{-2} s^{-1}) at depth z. I_o is the irradiance at the surface (about 2000 µmol quanta m^{-2} s^{-1}), e the base of natural logarithms, and η the extinction coefficient. The value of the extinction coefficient varies with the wavelength of light, being greater than 2.0 for red and infrared light and less than 0.01 for blue and violet light. Therefore most heat is absorbed near the surface, and only blue light penetrates to any depth. The penetration of light into a lake is shown in Figure 21.4. The depth at which light becomes too dim for photosynthesis (usually set at 1%

of surface irradiance) defines the bottom of the **euphotic zone**. Temperature would follow a similar curve except that wind sets up currents in the water column that mix the heat down into the water column. This mixing process divides a lake or an ocean into distinct layers or strata and is called **stratification**. In lakes the upper warm and less dense layer is termed the **epilimnion**, and the cooler, denser lower layer is known as the **hypolimnion** (see Figure 21.4). Between the upper and lower zones lies a region of sharp change in temperature and density termed the **metalimnion**, which is marked by a decrease in temperature called the **thermocline**. The depth of the epilimnion varies with the area of the lake; it may be as shallow as 2 m or greater than 20 m. Metalimnia are usually several meters in depth. A hypolimnion may be absent entirely in shallow lakes or

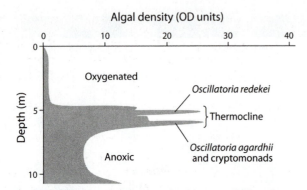

Figure 21.6 Vertical distribution of cryptomonads and *Oscillatoria* species in a stratified kettle lake. (Goldman, C. R., and A. J. Horne. 1983. *Limnology.* McGraw-Hill. Reproduced with permission of the McGraw-Hill Companies, based on Baker and Brook 1971)

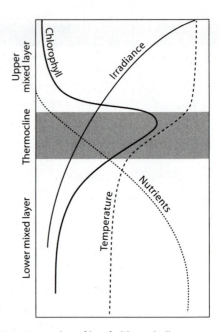

Figure 21.7 Vertical profile of chlorophyll, temperature, irradiance, and nutrients in a stratified water column of the open ocean. (After Falkowski and Raven 1997. Reprinted by permission of Blackwell Science, Inc.)

extremely deep, as in the North American Great Lakes or Lake Baikal in Russia.

In the oceans the upper warm layer is the **epipelagic**. The epipelagic coincides with the euphotic zone and may be 100–200 m deep. All the primary productivity in the oceans occurs in the epipelagic zone. A **pycnocline**, a vertical gradient separating water of different densities due to temperature or salinity, may be present on a seasonal basis, as in temperate oceans; permanent, as in tropical oceans; or only very weakly, as in polar oceans (Figure 21.5). There are four cold lower strata in the oceans. Because they lie in the aphotic zone, all four depend on a constant rain of detritus and living and dead organisms for the support of their organisms.

Stratification has important consequences for phytoplankton. The upper, warmer epilimnion is generally well mixed by various forms of water motion (see below) and well lighted. Phytoplankton may circulate within the epilimnion over 30 minutes to a few hours. Nonmotile species depend on this vertical mixing to maintain position in the lighted water column. Motile species may migrate up to the top of the epilimnion for photosynthesis during the day and back to the darker lower levels to acquire nutrients at night. Freshwater flagellates have been shown to move toward and maintain position in preferred temperatures within a temperature gradient (Clegg et al. 2003). In the metalimnion the mixing time is on the order of weeks (Harris 1986). If the metalimnion is within the euphotic zone, species may be concentrated there and form a chlorophyll maximum (Figure 21.6).

In the oceans, tropical and subtropical regions have a perennial thermocline that retards upward movement of nutrients (see Figure 21.5). In these stratified open ocean waters phytoplankton typically show a deep chlorophyll maximum, which represents a trade-off between decreasing light and increasing nutrients in a more stable environment (Figure 21.7). Phytoplankton biomass in these regions is low year-round, except under specific water motion conditions such as at coastal upwelling zones (see Figure 21.11) where phytoplankton biomass is enhanced. In temperate oceans the epipelagic is recharged with nutrients when the thermocline breaks down. Polar oceans mix year-round and provide a consistent level of nutrients to sustain phytoplankton growth.

Water Motions and Turbulence

Winds, solar heating, tides, and the rotation of Earth interact to generate many different types of water motions. Vertical water motions affect the sinking rates of phytoplankton cells, and horizontal water motions may concentrate or disperse patches of phytoplankton. In addition, there is evidence that species of algae differ in their tolerance of turbulence (Fogg 1991; Willen 1991), and these differences may affect the formation of blooms. Here we focus on a few types of water motions that significantly affect the ecology of phytoplankton.

Consider a column of water that is heated by solar radiation but is not subject to any applied force at the surface. The water molecules at any level in the water column are in random motion that depends on the temperature at that level. If a horizontal force is applied at the surface in the form of a mild wind, the surface layer of water molecules will acquire a velocity in the direction of that wind. Part of the energy of that surface layer motion will be imparted to the layer of water molecules below it, and they will also move in the direction of the wind but at a lower velocity. Each deeper layer of water also moves in the direction of the wind at decreasing velocity until the energy acting at the surface is dissipated. Such a fluid flow is called laminar flow when molecules move in orderly layers one on top of another. If the horizontal force is stronger, however, water molecules are torn from the surface, and the laminar structure breaks up into swirling **eddies**. The resulting turbulent motion moves in the direction of the wind but contains up and down vertical components of velocity in the swirling eddies. The excess energy of turbulent motion is eventually dissipated as smaller and smaller eddies down to about 1 mm in scale. Because laminar flow breaks down at fairly low wind velocities, the surfaces of oceans and lakes are primarily turbulent environments.

The scale of turbulent mixing plays an important role in the growth and sinking of phytoplankton populations. The smallest size of turbulent eddies fall into the range of about 1 mm. For all but the largest colonies and motile phytoplankton, this scale is larger by a factor of 10 to 100 than the size of phytoplankton. As a consequence, most phytoplankton are carried along within the turbulent flow. If the turbulent flow is wholly within the euphotic zone, there may be little effect on phytoplankton growth. With greater mixing depth, light availability for photosynthesis will be reduced, and as a result growth rate and population density will be reduced (Ptacnik et al. 2003; Berger et al. 2006). Conversely, a decrease in mixing depth can act as an enhancement in irradiance and population density. In Upper Lake Constance, the beginning of the first spring phytoplankton bloom is primarily determined by the transition from strong mixing in early spring to weak mixing (Peeters et al. 2007).

Langmuir cells are large-scale turbulent eddies organized as elongate, wind-driven surface rotations in lakes and oceans marked by conspicuous lines of foam called windrows (Figure 21.8a). A minimum wind speed of about 11 km hr^{-1} is necessary for their

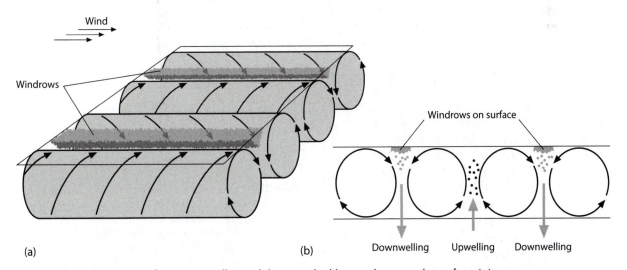

Figure 21.8 The formation of Langmuir cells in a lake is marked by windrows on the surface (a). Water in each cell follows a spiral path in the direction of the wind. Adjacent cells, seen in cross-section in (b), rotate in opposite directions, creating regions of downwelling (where the windrows occur) and upwelling. Foam and buoyant algae (gray circles) accumulate in the downwelling regions where currents converge. Negatively buoyant algae (black circles) accumulate in the upwelling zones beneath the surface.

formation. Within each Langmuir cell, water moves in the direction of the wind but in a spiral pattern. Adjacent cells rotate in opposite directions, creating alternating zones of upwelling and downwelling between them. The foam streaks on the surface occur in the zones of downwelling where bubbles, debris, and buoyant phytoplankton accumulate. Between the rotating cells in the zones of upwelling, negatively buoyant particles and algal cells also concentrate (Figure 21.8b). In both cases phytoplankton are moved through the water in the direction of the wind at a velocity comparable to the downwelling velocity, which may be as high as 10 cm s⁻¹. Langmuir cells often occur on a scale of 5 to 10 m, but may be as large as 200 m (Wüest and Lorke 2003).

A steady wind blowing from the west across a lake for an extended period of time will transmit about 3% of its energy to the water surface (Goldman and Horne 1983) and generate normal surface waves. The same kind of normal waves occur on the surface of the oceans but on a much larger scale. When that flow of water reaches the eastern shore of the lake, waves crash on the beach and also set up a surface current moving north and south along the shore and returning to the west. Internally there will also be an east-to-west return flow along the bottom, where the flow generates mild turbulence in the bottom boundary layer. This return flow acts like a conveyor belt, and it can have a substantial effect on the distribution of phytoplankton. If phytoplankton cells are buoyant near the surface, then the surface flow of the conveyor tends to pile them up on the downwind (eastern) shore. George and Edwards (1976) found that *Microcystis aeruginosa*, a buoyant cyanobacterium, piled up on the downwind shore in Eglwys Nynydd, South Wales (Figure 21.9a). In contrast, the dinoflagellate *Ceratium hirundinella* maintains its position at a lower level in the water column. In Esthwaite Water, English Lake District, Heaney (1976) reported that westerly winds caused this dinoflagellate to accumulate at depth along the western shore due to the subsurface return currents (Figure 21.9b). Clearly, winds can affect the distribution of phytoplankton patches. If sampling were done at a single fixed station, these types of large-scale population movements could result in widely varying population estimates.

In the oceans, water movements occur on much larger scales than in lakes and largely without the edge effects of closed basins. A series of global air cells in the atmosphere interact with the Coriolis effect produced by the rotation of Earth to generate

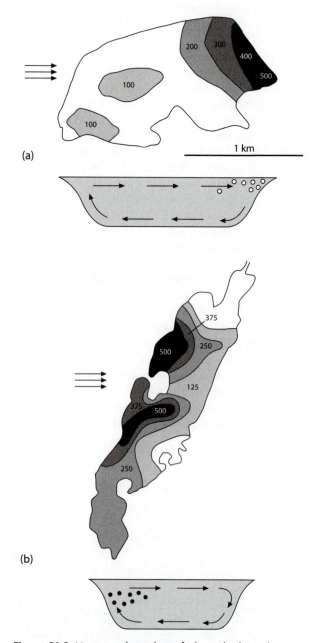

Figure 21.9 Horizontal patches of phytoplankton due to westerly winds. (a) Buoyant *Microcystis aeruginosa* builds up on the downwind (eastern) shore in the Eglwys Nynydd reservoir in southern Wales. Isopleths in μmol chl *a* l⁻¹. (b) *Ceratium hirundinella* favors deeper strata in Esthwaite Water, Lake District, England, and is carried back toward the western shore by subsurface return currents. Isopleths are cells ml⁻¹. Insets show effects of surface currents and subsurface return flows on phytoplankton distribution. (a: After George and Edwards 1976. Reprinted by permission of Blackwell Science, Ltd.; b: After Heaney 1976)

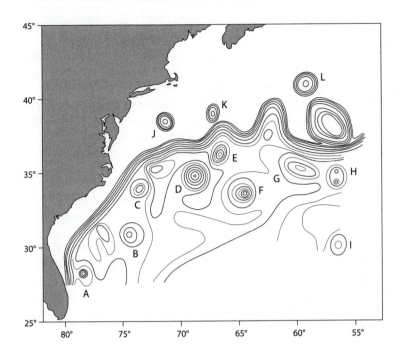

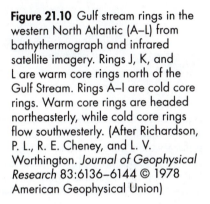

Figure 21.10 Gulf stream rings in the western North Atlantic (A–L) from bathythermograph and infrared satellite imagery. Rings J, K, and L are warm core rings north of the Gulf Stream. Rings A–I are cold core rings. Warm core rings are headed northeasterly, while cold core rings flow southwesterly. (After Richardson, P. L., R. E. Cheney, and L. V. Worthington. *Journal of Geophysical Research* 83:6136–6144 © 1978 American Geophysical Union)

five huge circular ocean currents called **gyres** (see Figure 22.18). The two gyres in the Northern Hemisphere turn clockwise and the three in the Southern Hemisphere rotate counterclockwise. The five gyres cover more than half of the total surface area of the Earth, and the North Pacific Subtropical Gyre is the largest single contiguous biome on Earth. The circulation pattern in the gyres separates them from coastal zones and the adjacent polar oceans. A permanent pycnocline at 200 m separates the surface waters from the nutrient-rich deep waters. As a result surface waters in the gyres are very low in nutrients, phytoplankton biomass, and primary productivity. Growth rates of phytoplankton in the North Atlantic Subtropical Gyre (0.26 d^{-1}) and the South Atlantic Tropical Gyre (0.17 d^{-1}) are very low and nutrient limited (Marañón 2005). Picocyanobacteria such as *Prochlorococcus* and *Synechococcus* contribute from 50%–80% of the phytoplankton biomass.

While phytoplankton patches may extend for hundreds of meters in large lakes, they may extend for hundreds of kilometers in oceans. In the past it was extremely difficult to obtain any picture of phytoplankton distributions across the oceans, but with the advent of remote sensing by orbiting satellites, this has become feasible (see Text Box 2.1). Ueyama and Monger (2005) combined daily standard mapped images from NASA's Sea-viewing Wide Field-of-view

Sensor (SeaWiFS) of surface chlorophyll concentrations with wind data to determine the effects of wind and mixing on the timing of phytoplankton blooms in the North Atlantic. In the subpolar and northern subtropical regions, increased wind-induced mixing delayed bloom development and reduced bloom magnitude. Satellite altimetry also allows tracking of turbulent eddies in oceanic currents. The Gulf Stream meanders north and south as it makes its way across the North Atlantic. As it wanders, it forms large turbulent rotating eddies, varying in diameter from 50 km to over 250 km (Figure 21.10). These eddies move across the ocean, carrying whole ecosystems as discrete parcels that maintain their structure for up to two years. They can be tracked by satellite altimetry because their rotation generates slight elevations or depressions called sea-level anomalies (SLAs). A cyclone is an eddy that rotates in a counterclockwise direction, generating a downwelling current and a negative SLA of as much as 30 cm. A mode water eddy rotates clockwise and generates an upwelling current with a positive SLA of up to 30 cm. Benitez-Nelson et al. (2007) studied a cyclone eddy off Hawaii and found that phytoplankton biomass inside it (dominated by large centric diatoms) formed a deep chlorophyll maximum where primary production was two or three times higher than in the surrounding oligotrophic waters, which were

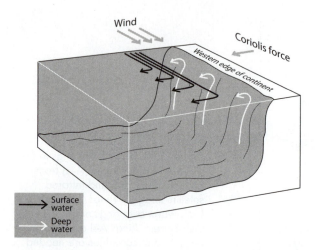

Figure 21.11 Coastal upwelling. The combined action of wind and the Coriolis force of Earth's rotation move surface waters offshore along the western margins of continents (black arrows). This water is replaced by deep water rising to the surface (white arrows), which brings nutrients for phytoplankton growth. (After *Marine Biology* 2nd. Edition, by J. W. Nybakken. © 1987 by Harper Collins Publishers. Reprinted by permission of Addison-Wesley Educational Publishers)

dominated by small phytoplankton. McGillicuddy et al. (2007) found that mode water eddies in the Gulf Stream had persistent diatom blooms whose chlorophyll *a* concentrations were four times those found in the surrounding oligotrophic waters. Eddies appear to be areas of elevated primary production within oceanic gyres and may significantly increase the export of carbon from surface waters (the biological pump).

The second important type of current is that responsible for upwelling zones (Figure 21.11). The most famous of these currents is the Peruvian upwelling. Off the coast of Peru, cold nutrient-rich water rises up from great depth to create an abundant growth of phytoplankton. Zooplankton graze on the phytoplankton and in turn support a rich fishery. When an El Niño event occurs, warm Pacific surface water floods over the cold water and closes it off. The phytoplankton and then the fishery collapse.

Large-scale water movements affect the growth and distribution of phytoplankton populations, but turbulence also affects the cells of individual species. Most turbulence studies have been done on dinoflagellates. It has long been known that certain species of dinoflagellates will not grow if stirred in small culture flasks (White 1976; Pollingher and Zemel 1981).

Cell division is inhibited at low rates of turbulence, and at high rates cells disintegrate. Dinoflagellate red tides are associated with extended periods of relatively calm seas. Among marine red-tide dinoflagellates, *Prorocentrum micans* was more tolerant of turbulence than either *Gonyaulax polyedra* or *Gymnodinium sanguineum* (Thomas and Gibson 1990, 1992). Some dinoflagellates, however, are unaffected by turbulence, and others show higher growth rates in turbulence (Sullivan and Swift 2003). Few other groups of phytoplankton have been examined for turbulence effects. Blooms of the marine diatoms *Chaetoceros armatum* and *Asterionella socialis* occur in the surf zone of beaches (Lewin and Norris 1970). In cultures, some diatoms grew better when subjected to turbulence than when grown in stationary culture (Thomas et al. 1997). These results suggest that turbulence may enhance the growth rates of marine diatoms and may play a role in seasonal succession of species.

21.3 The Chemical Environment

Six elements each contribute more than 1% to the dry mass of phytoplankton cells. In order of decreasing percentage contribution, these six elements are carbon, oxygen, hydrogen, nitrogen, phosphorus, and sulfur. In the case of diatoms, silicon should be added to this list. Carbon, nitrogen, phosphorus, sulfur, and silicon are considered major nutrient elements and may be limiting for phytoplankton growth. Other elements, including sodium, potassium, calcium, magnesium, and chlorine, may be dominant in lakes or oceans but used by phytoplankton in only small amounts. Micronutrients such as zinc, copper, molybdenum, cobalt, manganese, and iron occur in trace amounts in aquatic systems and are used in small amounts by phytoplankton. Molybdenum and iron can be growth-limiting micronutrients. These same micronutrients can be toxic at high levels.

Dynamics of phytoplankton are primarily correlated with spatial and temporal fluxes of major mineral nutrient ions. Redfield (1958) determined that marine phytoplankton growing at their maximum growth rates possessed a characteristic ratio of major nutrients of 106 C : 16 N : 1 P : 0.7 S. Diatoms, and perhaps other algae that require silica, should have a ratio of 106 C : 16 N : 16 Si : 1 P : 0.7 S.

Phytoplankton should show the Redfield ratio when growing close to their maximum growth rate, when they are not nutrient limited. Conversely, phytoplankton will show a ratio of major nutrients different from the Redfield ratio if they are limited by a nutrient and growing at less than their maximum rate. Limitation by a specific nutrient can be shown by a simple test. If the addition of a nutrient causes an increase in the phytoplankton population, that nutrient was limiting population growth.

In this section we first consider the role of dominant ions in lakes and oceans on phytoplankton. The major limiting nutrient elements carbon, nitrogen, phosphorus, and silicon are then discussed. Sulfur as a potential limiting major nutrient was discussed in Chapter 2. In the final section we present the growth-limiting roles of the micronutrients molybdenum and iron.

Salinity and Dominant Ions of Lakes and Oceans

An ocean water sample contains about 35 g of dissolved salts per kg of seawater. The salinity of a body of water is measured in **practical salinity units (psu)**. An ocean water sample with 35 g of salts would have a salinity of 35 psu. Of this amount, sodium accounts for 10.77 g and chloride for 19.35 g. Oceans are dominated by sodium and chloride, followed by magnesium (1.29 g kg^{-1}) and sulfate (2.71 g kg^{-1}). Among the remaining dissolved salts in seawater are four ions of major importance to phytoplankton: bicarbonate, nitrate, phosphate, and silicon dioxide. Major ions in oceans vary by only a few percent from place to place. Lakes are much more diverse in salinity. The Great Salt Lake in Utah has a salinity of 220 psu. An average eutrophic lake has a salinity of 0.18 psu, and oligotrophic lakes have salinities around 0.001–0.006 psu. Lakes are dominated by calcium and bicarbonate, and their concentrations may vary greatly.

Among the relatively abundant ions in aquatic systems, planktonic algae have little use for sodium. In saline lakes, high levels of sodium actually restrict the diversity of algae. Similarly, potassium is not a growth-limiting nutrient (Jaworski et al. 2003), but growth of several chrysophytes has been shown to be inhibited at K^+ levels > 5 mg l^{-1} (Sandgren 1988). Chloride is used in some biochemical transformations such as photolysis of water and ATP production. All phytoplankton cells need magnesium for

energy metabolism in converting ATP to ADP. Magnesium is also the central metal in the chlorophyll molecule. In the oceans coccolith-producing haptophytes require calcium as calcium carbonate in their body scales (Chapter 10). Calcium is also necessary in small amounts for growth and cell wall integrity, and it plays an important role in buffering aquatic systems against pH changes in conjunction with the carbonate buffering system (see below). Calcium-rich lakes are generally well buffered, while calcium-poor lakes, such as are found in Scandinavia, the northeastern United States, and eastern Canada, have little buffering capacity and are susceptible to acidification by nitric and sulfuric acids in acid rain. By participating in a buffering system, calcium plays a selective role among phytoplankton that are sensitive to pH and carbon sources.

Carbon

Although relatively scarce in the atmosphere, carbon dioxide is highly soluble in water. The form that inorganic carbon takes in water depends on the pH of that water. Most lakes have a pH of 6 to 9, where a pH of 7 is considered neutral. Acid waters may have a pH as low as 2, while some alkaline lakes may have a pH > 10. Seawater is slightly alkaline, ranging from pH 7.5 to 8.4. Below pH 6, the bulk of CO_2 dissolved in water is present as soluble carbon dioxide or carbonic acid (H_2CO_3) (Figure 21.12). Above pH 6, bicarbonate ion (HCO_3^-) becomes increasingly important. The normal range of pH for seawater spans the maximum for bicarbonate. Above pH 10, carbonate becomes increasingly important. The carbonate–bicarbonate–CO_2 equilibrium plays a major role in the buffering capacity of lakes and oceans. If an aquatic system has abundant calcium, it will have abundant carbonate. When acid is added to such a lake in the form of protons (H^+), the protons combine with carbonate (CO_3^{2-}) to form bicarbonate (HCO_3^-), and the pH remains stable. Most lakes and all oceans are predominantly bicarbonate solutions.

Phytoplankton cells acquire their carbon through modifications in the enzyme rubisco, various carbon concentrating mechanisms (CCMs), and uptake of organic carbon (Chapter 2). Small lakes and rivers may actually outgas CO_2 (Cole et al. 1994). While it is widely accepted that growth of most phytoplankton populations is likely to be limited by nutrients such as nitrogen, phosphorus, silicon, or iron, there appear to be some circumstances where inorganic

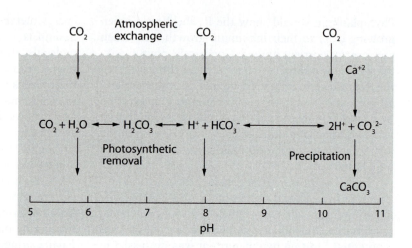

Figure 21.12 Forms of inorganic carbon as a function of pH in lakes and oceans. Most lakes and all oceans are bicarbonate (HCO_3^-) solutions. $(H_2CO_3)^* = H_2O + CO_2$.

carbon may be limiting. In acidic lakes with a pH < 5, inorganic carbon exists as CO_2 only, and it can become limiting to growth if other nutrients are relatively abundant (Hein 1997). Green algae in a series of acidic lakes responded positively to additions of CO_2 or bicarbonate (Fairchild and Sherman 1993). Chrysophytes are often abundant in the phytoplankton of acidic oligotrophic lakes, and several genera of chrysophytes have been shown to be unable to use bicarbonate and to lack any form of CCM (Moroney 2001). Such chrysophytes could become carbon limited. In the oceans bicarbonate concentrations (about 2 mM) are much higher than those of CO_2 (10 µM) (Round 1981). Riebesell et al. (1993) showed that certain marine diatoms were dependent on dissolved CO_2 for growth. Growth rates became carbon limited at levels below 10–20 µM. Similarly, Dason et al. (2004) found that two species of marine dinoflagellate (*Amphidinium carterae* and *Heterocapsa oceanica*) were entirely dependent upon uptake of free CO_2 for inorganic carbon. Despite high levels of bicarbonate ion in ocean waters, these dinoflagellates and diatoms could become carbon limited because of their dependence on dissolved CO_2.

Nitrogen

Nitrogen is present in waters primarily as dissolved dinitrogen gas (N_2), which is largely inert. Dissolved inorganic nitrogen (DIN) includes the ammonium ion (NH_4^+), nitrite ion (NO_2^-), and nitrate ion (NO_3^-). In oceans, 95% of nitrogen occurs as N_2. About two-thirds of the remainder is nitrate ion (0.28–0.56 mg NO_3^- l^{-1}), which turns over very rapidly and is often so scarce in ocean waters, such as the North

Pacific Gyre, the subtropical Atlantic, and the Indian Ocean, as to be undetectable. The amount of DIN may be supplemented by dissolved organic nitrogen (DON) that consists of urea and amino acids. In ocean surface waters, more than 80% of available nitrogen may be DON. In tropical oceans, the filamentous, nitrogen-fixing cyanobacterium *Trichodesmium* is fairly common. For a discussion of nitrogen fixation, refer to Chapters 2 and 6. Cells of the common marine diatom *Rhizosolenia* typically contain two filaments of the nitrogen-fixing cyanobacterium *Richelia*, which effectively acts as a nitrogen-fixing organelle (Chapter 7). Nitrate appears in quantity only in zones of coastal upwelling or pollution, where it is the primary cause of eutrophication of coastal ecosystems (Howarth and Marino 2006). Nitrogen in the form of ammonium or nitrate ions is commonly the limiting mineral nutrient in ocean waters, but in some areas, such as the equatorial Pacific and the Southern Ocean around Antarctica, iron limitation has been demonstrated (see below).

The main source of nitrate in lakes is stream and river discharge. In most temperate oligotrophic and mesotrophic freshwaters, nitrate is present in relative excess and exceeds the supply of phosphorus. Nitrogen fixation plays a relatively minor role in such low-nutrient freshwaters. But in some western U.S. lakes whose basins are predominately volcanic rock, nitrogen is the limiting nutrient (Reuter and Axler 1992). In tropical lakes, nitrogen may be in low supply due to low levels in the surrounding soils of the watershed. In oligotrophic freshwaters nitrate is the major form of DIN, while in eutrophic waters, NH_4^+ and NO_2^- may be present at depth if oxygen is reduced. Nitrite is normally present in insignificant

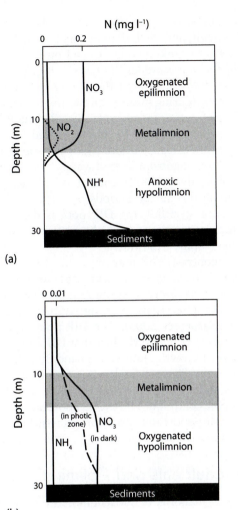

(a)

(b)

Figure 21.13 Generalized distribution of nitrate, nitrite, and ammonia with depth in a eutrophic (a) and an oligotrophic stratified lake (b) in midsummer. (After Goldman, C. R., and A. J. Horne. 1983. *Limnology*. McGraw-Hill. Reproduced with permission of the McGraw-Hill Companies)

amounts, although in eutrophic lakes a narrow layer in the thermocline may contain elevated levels of nitrite (Figure 21.13). If oxygen is present, NO_2^- is converted to NO_3^-, whereas in anoxic waters, it is reduced to ammonia. The supply of phosphorus may exceed that of nitrogen in eutrophic waters. Under these circumstances nitrogen limits phytoplankton growth, and nitrogen-fixing cyanobacteria such as *Anabaena*, *Aphanizomenon*, and *Gloeotrichia* may generate nuisance blooms. Such nitrogen-fixing cyanobacteria are common in freshwaters even when not at bloom levels (Chapter 6).

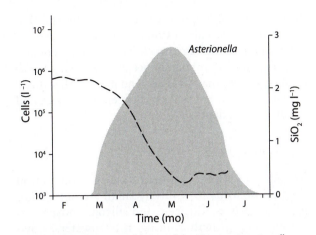

Figure 21.14 The spring bloom of the diatom *Asterionella* in the north basin of Windemere, Lake District, England, and the depletion of the limiting nutrient silica (dashed line). (After Goldman, C. R., and A. J. Horne. 1983. *Limnology*. McGraw-Hill. Reproduced with permission of the McGraw-Hill Companies, based on data in Lund 1964)

Silicon

All planktonic algae require at least small amounts of silicon for protein synthesis. In freshwaters the concentration of soluble reactive silicon (SRSi), which is monosilicic acid, H_4SiO_4, typically ranges from 0.7 to 7 mg Si l⁻¹ (25–250 µM), while in the oceans the maximum concentrations of SRSi (about 3 mg Si l⁻¹) occur at upwelling zones. In both freshwater and marine systems, uptake and growth by diatoms and other algae can substantially reduce levels of SRSi. Diatoms play a major role in the geochemical cycle of silicon. Silicon can be a limiting nutrient for the growth of diatoms in lakes during summer stratification when the pool of silicic acid falls to undetectable levels (< 0.1 µM). The decline of silicon is certainly one major factor in bringing about the decline of the spring diatom bloom in many lakes (Figure 21.14).

Since polymerized silica in the valves of diatoms is dense (2.6 g cm⁻³), it adds considerably to the density of diatoms. Populations of diatoms depend on turbulent water motions for entrainment and maintenance within the euphotic zone. Diatom production may sediment out of the water column, or primary consumers may eat it. Polymerized silica decomposes slowly (on the order of 50 days), and it does not dissolve during gut passage through herbivorous zooplankton. Thus there is essentially no rapid recycling of silicon in the epilimnion of shallow lakes. Diatoms may reach the benthos intact, where their valves will be preserved in the sediments. Dissolved silicon is only restored to surface waters by external inputs and the turnover of the water column at spring and fall mixing. In oceans the mixing depth is much greater than in lakes, and most of the silicon in diatom valves redissolves between the surface and about 1000 m depth.

Phosphorus

As indicated by the Redfield ratio, phytoplankton need phosphorus in relatively small amounts compared to carbon, nitrogen, and silicon. Phosphorus is frequently growth limiting in freshwaters, however, because it is often in short supply in many watersheds and binds to metal cations to form precipitates in lake sediments (Chapter 2). Phosphorus limitation is still generally true of large, deep lakes at higher latitudes. Most of the total phosphorus (TP) in waters is in the form of living or dead particulates. Dissolved phosphorus occurs as either inorganic phosphorous (DIP) or dissolved organic phosphorus (DOP). Most dissolved phosphorus is DOP. DIP is primarily orthophosphate (PO_4^{3-}), with much lower amounts of monophosphate (HPO_4^{2-}) and dihydrogen phosphate ($H_2PO_4^-$). Phytoplankton can only use dissolved inorganic phosphate, which is termed soluble reactive phosphate (SRP). When the supply of SRP is exhausted, phytoplankton can release alkaline phosphatases, which are extracellular enzymes capable of freeing phosphate bound to organic substances. The level of activity of alkaline phosphatases is a widely accepted indicator of phosphate limitation (Rengefors et al. 2003). When brief pulses of SRP do occur, many phytoplankton cells can take up and store excess phosphate as polyphosphate bodies. This so-called luxury consumption is an important mechanism for dealing with phosphate shortages. During nutrient pulses, an algal cell may be able to store enough phosphate to provide for as many as four cell divisions. Finally, many planktonic algae can take up phosphate at extremely low ambient levels, well below the level of detection (see below).

In many lakes the amount of phosphorus present in late winter determines the size of the phytoplankton populations that can develop in the summer. In winter, DIP turnover is relatively slow in all freshwaters—on the order of hours to days—as a result of low populations of algae with low growth rates. DIP represents less than 10% of TP in oligotrophic

freshwaters, where during summer the demand for DIP is high, and the DIP pool may become undetectable. Phytoplankton cells depend on DIP recycling. Using the radioisotopes $^{32}PO_4$ and $^{33}PO_4$, Hudson and Taylor (1996) measured the rate of regeneration of phosphate (during summer) in two oligotrophic lakes. Measured rates ranged from 15 to 205 ng P l^{-1} hr^{-1}. Grazers smaller than 40 µm accounted for 77% of the measured regeneration. Many marine waters are similar to oligotrophic freshwaters with respect to phosphate dynamics. In coastal marine waters, DIP builds up during the periods of vertical mixing. If the ocean stratifies, the DIP pool is depleted, and phytoplankton depend on phosphorus recycling, as in oligotrophic freshwaters.

In contrast, DIP may approach 100% of TP in eutrophic freshwaters. Excess phosphorus enters eutrophic lakes from sources such as sewage, industrial effluents, and runoff from agricultural and urban areas (Carpenter 2005). The DIP pool may exhibit slow turnover due to the fact that DIP input may exceed algal growth requirements and thus build up to high levels. Phosphorus levels build up in the sediments and through internal recycling maintain the eutrophic condition. The excess phosphorus provides a growth opportunity to cyanobacteria that can fix nitrogen and form blooms.

Micronutrients and Colimitation

While the majority of nutrient limitation studies have focused on major mineral nutrient elements such as silicon, phosphorus, and nitrogen, two micronutrient elements—molybdenum and iron—have been shown to be limiting in specific pelagic systems. In 1960 Goldman observed that populations of phytoplankton in Castle Lake, California, were well below the growth that could be supported by the levels of available phosphorus and nitrogen. Addition of a few micrograms of molybdenum caused a significant increase in phytoplankton populations. A later study showed that molybdenum stimulated nitrogen uptake (Axler et al. 1980). Molybdenum acts as a cofactor with the enzymes nitrate reductase (Chapter 2) and nitrogenase in nitrogen fixation by cyanobacteria. Molybdenum has been reported to be limiting in a number of other lakes in California, Alaska, and New Zealand (Goldman 1972).

Iron limitation was first reported in the oceans (Chapter 2). In the subarctic and equatorial Pacific Ocean and the Southern Ocean around Antarctica, sparse phytoplankton populations occur in the presence of relatively abundant levels of nitrogen and phosphorus. In these oceans iron has been shown to limit algal productivity (Martin 1992; Martin et al. 1994). The discovery of iron limitation in the oceans seems to have led to more detailed examination of the concept of nutrient limitation and recent findings that in some ocean areas phytoplankton populations and processes may be limited by more than one mineral nutrient, a circumstance called **colimitation**. Phytoplankton growth in the Ross Sea, one of the most productive areas within the Southern Ocean, is colimited by iron and vitamin B_{12} during the austral summer (Bertrand et al. 2007). Bacteria and archaea in surface waters are the main source of vitamin B_{12} for the phytoplankton. Phytoplankton productivity is limited by nitrogen in the eastern tropical North Atlantic, but nitrogen fixation is colimited by iron and phosphorus, the main source of which is dust from the Sahara Desert (Mills et al. 2004). Colimitation of nitrogen fixation prevents nitrogen-fixing cyanobacteria from fully exploiting a nitrogen-limited environment.

Colimitation has been detected in a few large freshwater lakes. Phytoplankton growth in western Lake Superior is generally phosphorus limited. The addition of phosphorus resulted in only small increases in growth, however, because phytoplankton became limited by iron (Sterner et al. 2004). Nitrogen, phosphorus, and iron may colimit Lake Erie phytoplankton at times. Addition of all three nutrients increased phytoplankton biomass more than did addition of N, P, or Fe alone (North et al. 2007). Nutrient colimitation may prove to be a widespread occurrence in aquatic systems.

21.4 Growth Processes of Phytoplankton Populations

Populations take up space, show various levels of mobility, and are distributed in a variety of patterns in time and space. Populations are characterized by the fact that they grow, that is, they change in size (numbers) and density (number per unit area or volume). The concept of growth rate is essential for describing population dynamics. In field research, growth rate usually means the net rate of change in numbers or biomass and represents the balance between additions due to reproduction and losses due to various sources

of mortality or export. In culture studies, however, phytoplankton cells essentially grow without losses, and so growth rate is equivalent to the rate of reproduction or the "birth rate" in field research. We begin with a description of simple growth processes, called exponential and logistic growth, that make no explicit reference to the cause of the growth they describe. We then show how growth can be described as a function of light level and of nutrient uptake and external supply to form a useful description of phytoplankton growth in natural aquatic systems.

Exponential and Logistic Growth

The simplest way to express the process of population growth is to use the exponential growth equation. Exponential growth is expressed mathematically as:

$$\frac{dN}{dt} = rN \quad (3)$$

In this equation, dN/dt represents the change in numbers or biomass during a unit of time (such as one day). Sometimes dN/dt is referred to as the growth rate, but it is actually the population growth rate or the change in population size in a unit of time. This change in population size is equal to r, the net growth rate (with units of reciprocal time or, in this case, day^{-1}) multiplied by N, the number of individuals in the population, or its biomass. If we isolate r on the right side of the equation by dividing both sides by N, we can see that r is the change in the population size per individual in that population N or the net per capita growth rate. Since the net growth rate is equal to the rate of reproduction minus any losses due to death, the net growth rate is equivalent to:

$$r = \mu - \lambda \quad (4)$$

where μ is the gross growth rate, rate of reproduction, or the birth rate, and λ is the death rate or the loss rate. The net growth rate may be positive or negative. If positive, then the population will increase in numbers or biomass. If negative, the population will decrease in numbers or biomass over time. Note that a population could have a high rate of gross growth (μ) but show no net growth at all. This situation can arise anytime the gross growth rate (μ) is roughly balanced by the loss rate (λ). In that situation (r = 0), a phytoplankton population could be photosynthesizing, taking up nutrients, and dividing at a significant rate, but repeated sampling would show no net change in numbers because losses balance production. Lack of evidence of net growth (r) among phytoplankton does not mean there is no gross growth or that the planktonic algae are not physiologically active.

Equation (3) (above) is an example of a differential equation because of the presence of the term dN/dt. If equation (3) is solved explicitly to remove the differential term dN/dt, the result can be expressed as:

$$N = N_o\, e^{rt} \quad (5)$$

Here the growth rate, r, appears as the exponent of the base, e, of natural logarithms, from which the name exponential growth equation derives. A population of phytoplankton growing according to equation (5) will increase exponentially without an upper limit (Figure 21.15a). If the natural logarithms of both sides of equation (4) are taken, the equation becomes that of a straight line whose slope is the net growth rate (r):

$$\ln N = rt + \ln N_o \quad (6)$$

Thus, the net growth rate can be estimated from the natural logarithms of the population sizes. Equation (6) can also be used to calculate the doubling time of the population (t_d). Equation (6) can be rearranged into the form:

$$\ln N - \ln N_o = \ln \frac{N}{N_o} = rt \quad (7)$$

Now, when N = 2 (N_o), ln (N/N_o) = ln 2 or 0.69, and we get t_d = ln (2)/r. If the net growth rate were 0.69 day^{-1}, then the doubling time would be 0.69/0.69 day^{-1} or 1.0 day. Therefore, an r of 0.69 day^{-1} corresponds to one doubling per day. If the net growth rate were –0.69 day^{-1}, the population would decrease by 50% per day.

No population of phytoplankton can continue to grow indefinitely at an exponential rate as described by equation (5). Even a slow-growing population would soon overflow its habitat. There must be some upper limit to population density due to limits in available resources such as nutrients. This upper limit is called the **carrying capacity** of a population. The first widely used expression that included a carrying capacity was the logistic equation:

$$\frac{dN}{dt} = rN \frac{(K - N)}{K} \quad (8)$$

In the logistic equation, the carrying capacity is represented by the parameter K, which has the same

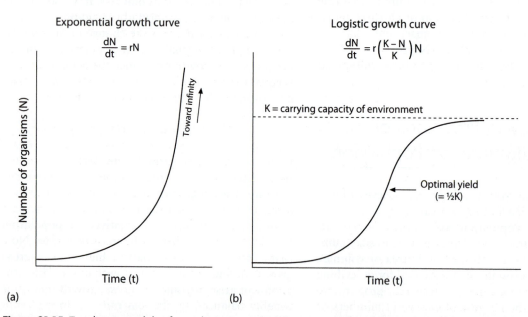

Figure 21.15 Two basic models of population growth: (a) exponential and (b) logistic. (After Wilson and Bossert 1971)

units as N (cells ml^{-1} or mg ml^{-1}). Initially, when N is small, (K–N)/K is close to 1, and the population grows at nearly an exponential rate. As N approaches K, (K–N)/K approaches zero, and the population ceases to grow when at its carrying capacity (Figure 21.15b). The logistic equation accurately describes the growth of an algal population in a batch or stationary culture of definite volume and nutrient content. To determine the value of K, a growth experiment has to be run and the population counted at a number of time intervals. For some practice in working with the exponential and logistic equations, refer to Question Box 21.1. The logistic equation makes no explicit reference to the cause of the carrying capacity, such as the consumption of some resource, and the value of K has little relevance to any natural aquatic system. In a later section we will show how to replace the logistic term (K–N)/K with terms that describe nutrient consumption.

Growth and Light

The most fundamental aspect of phytoplankton ecology is the conversion of light energy into biomass through photosynthesis. Photosynthesis is the biochemical process by which light energy is used to transform inorganic molecules into organic matter. Details of the biochemistry are beyond the scope of this chapter. A recent treatment is available in Falkowski and Raven (2007). The process of photosynthesis can be summarized by the familiar equation:

$$6CO_2 + 6H_2O \rightarrow C_6H_{12}O_6 + 6O_2 \quad (9)$$

Light energy is used to strip protons and electrons from water molecules with the resulting production of oxygen. Those protons and electrons are used to reduce CO_2 to an organic carbon molecule such as glucose, as shown in equation (9). A significant fraction of carbon metabolism may be coupled to nitrate assimilation with the final production of amino acids and proteins (Turpin 1991).

Plant physiologists quantify photosynthesis in two ways. Gross photosynthesis (P_g) is the light-dependent rate of electron flow from water to CO_2 in the absence of respiratory losses (Lawlor 1993). Respiration (R_l) is the flow of electrons from organic carbon to O_2 with the production of CO_2. Photosynthesis occurs only in the light, and respiratory losses in the light reduce the level of gross photosynthesis.

The difference between gross photosynthesis and respiration is called net photosynthesis (P_n). Therefore:

$$P_n = P_g - R_l \quad (10)$$

All three terms are rates and therefore time dependent.

Equation (9) indicates two ways to measure photosynthesis: by carbon uptake and by oxygen evolution. In the past, the most commonly used method was the uptake rate of ^{14}C (usually as $H^{14}CO_3^-$) into organic matter. The resulting rates may be expressed as mg C g hr^{-1}, mg C cell^{-1} hr^{-1}, or mg C (mg chl a)$^{-1}$ hr^{-1}. Their interpretation is complicated by the fact that it is not clear whether gross or net photosynthesis is being measured. If short incubation times are used, the ^{14}C uptake method should approximate gross photosynthesis. If incubation runs to equilibrium, the rates may approach those of net photosynthesis (Falkowski and Raven 2007). Radiocarbon techniques provide no information about respiration. Interpretation is more straightforward if oxygen evolution techniques are used. If photosynthesis is measured with an oxygen electrode, the data represent net photosynthesis since respiration removes oxygen. If a chamber containing algae is covered with a light-tight bag immediately after net photosynthesis measurements, the consumption of oxygen gives a measure of R_l, the respiration rate in the light (Graham et al. 1985). Furthermore, if the chamber remains in the dark for several hours, the dark or basal rate of respiration can also be obtained. Both methods share the problem of requiring confined spaces. If natural phytoplankton assemblages are used, bacteria and microzooplankton may alter the observed rates through their own metabolic activities.

If net photosynthesis or net primary productivity is measured in aquatic systems over entire seasons and at a range of depths, it is possible to obtain a measure of annual net primary production (ANPP) for those systems (Table 21.2). Aquatic systems contribute just under half of global ANPP of 105 Pg C yr^{-1} (1.05 × 10^{17} g C yr^{-1}). Pelagic ocean systems contribute the bulk (95.6%) of aquatic ANPP. Coral reefs, macroalgal beds, salt marshes, and estuaries add only about 4% to the marine total. Coastal ocean systems represent only 11% of the total area of the seas but contribute almost 25% of ocean ANPP. Freshwater systems are productive, but on a global basis they add just 2.8% of global ANPP.

Working with the Exponential and Logistic Equations

Sample calculation

The table and figure to the right present the results from a batch culture experiment in which the cyanobacterium *Anabaena flos-aquae* grew for 120 hours. Use these data to estimate the parameters r and K in the logistic growth equation (8).

Time (hr)	Population (cells ml⁻¹)
0	2,360,742
6	3,335,056
12	5,025,320
15	5,498,578
24	9,478,620
35	13,660,240
48	17,979,420
71	23,399,540
96	23,431,136
120	22,698,608

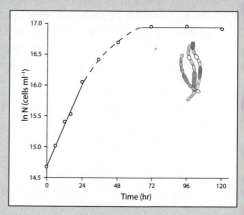

Solution

As shown in equation (6), we can determine the net growth rate (r) of *Anabaena* by taking the natural logarithms of the population sizes. We plot the natural logs against time, as shown in the graph. The first five points make a straight line. Using a calculator, you can run the linear regression of ln (N) on t for these five points. The resulting equation is Y = 0.0578 (X) + 14.68, and the correlation coefficient is 0.99. The slope of this line is the net growth rate, $r = 0.0578$ hr⁻¹. Multiplying this by 24 hr day⁻¹, r becomes 1.38 day⁻¹. The doubling time is given by $t_d = \ln (2)/r$. For *Anabaena*, $t_d = 0.69/1.38$ day⁻¹, which equals 0.5 days. Determining the value of K is much simpler. The values of ln (N) for 71, 96, and 120 hours are roughly the same, and K can be taken as their average. Therefore, K = 16.96, or 23,207,820 cells ml⁻¹.

Practice problems

Work the following problems with a pocket scientific calculator. You will find it essential to plot the data in order to decide which are to be used in making the estimates. Answers are given at the end of the chapter.

1. The data to the right (as cells ml⁻¹) give the results of three replicated batch cultures of *Chlamydomonas reinhardtii*. Calculate the net growth rate, r, for each batch culture. Take the average of the three as the net growth rate of *Chlamydomonas* in this study.

Time (hr)	Culture #1	Culture #2	Culture #3
0	9280	13,890	7860
24	38,550	42,490	24,300
48	100,680	143,480	129,620
71	249,250	355,780	321,440
95	946,230	1,283,390	1,131,070

2. The euglenoid flagellate *Euglena gracilis* grows in a culture medium containing acetate. While *Euglena* can photosynthesize, it can also grow in the dark by using acetate as an energy source. The table contains data on growth of *E. gracilis* in the light and dark. Calculate r for each growth experiment. How much did photosynthesis increase the growth rate of *Euglena* over the rate in the dark? By how much did photosynthesis reduce the doubling time?

Dark-grown		Light-grown	
Time (hr)	Cells ml⁻¹	Time (hr)	Cells ml⁻¹
0	915	0	2830
24	1605	24	4695
48	2590	46	14,910
147	56,454	72	31,120
219	360,464		

3. Growth of *Chlorella pyrenoidosa* was followed in batch culture for 24 days. Use the accompanying table of population counts (cells ml⁻¹) to estimate the logistic parameters r and K.

Day	Cells ml⁻¹	Day	Cells ml⁻¹
1	45,740	10	3,710,040
2	71,904	13	5,266,840
3	85,740	14	8,326,400
4	229,640	15	8,753,200
5	517,580	17	7,576,200
6	570,640	20	10,357,000
7	1,189,320	24	9,341,700
9	2,112,780		

Planktonic algae show a characteristic response curve to increasing light intensity (Figure 21.16). At some low light level, the rate of net photosynthesis (P_n) will just balance the rate of respiration (R). This light level is called the **compensation point**. For any light level greater than the compensation point, the alga will make a net gain in photosynthesis over losses due to respiration. If algal cells are exposed to low light levels for several hours, they will adapt physiologically by increasing their chlorophyll content. Initially, when light is limiting, there is a linear increase in photosynthesis with increasing light. The slope of this linear increase is termed α, a parameter used in modeling. As light levels continue to increase, the rate of increase in photosynthesis declines and the photosynthesis-versus-light curve bends over and levels off at a maximum value (P_m). Phytoplankton species differ in their optimal light levels. Diatoms are especially efficient at photosynthesis under the low light levels and photoperiods that prevail in spring when the water column is mixing and the elevation of the sun is low. This light efficiency is one reason diatoms are numerically abundant in temperate lakes and oceans in spring and fall.

At high light levels approaching the level of full sunlight, many phytoplankton species show **photoinhibition**, a decline in photosynthesis with increasing light. In this case, respiration increases with light while photosynthesis remains constant such that the photosynthesis-versus-light curve declines with increasing light (see Figure 21.16). Phytoplankton will exhibit decreased levels of chlorophyll in their cells at these high levels. Prolonged exposure to such light levels can result in damage to the photosynthetic apparatus in the cell. Surface waters are a turbulent environment, however, and vertical mixing can carry phytoplankton cells from the surface down to 10 m depth and back again in as little as 30 minutes to a few hours. Thus, within as little as 30 minutes an individual algal cell could experience light levels ranging from photoinhibition to below the compensation point and back again. Light, like mineral nutrient elements, can act as a

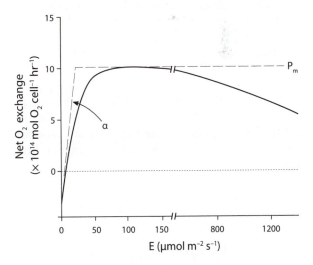

Figure 21.16 Relationship between irradiance, E, and photosynthesis in *Chlamydomonas reinhardtii*. P_m is the maximum rate of light-saturated photosynthesis, and α is the initial slope of the curve. The compensation point lies at the intersection of the P-curve and the zero net–O_2 exchange line (dashed). (Based on Neale 1987, in Geider and Osborne 1992, with kind permission of Kluwer Academic Publishers)

Table 21.2 Global areas, net primary productivities (NPP), and annual net primary production (ANPP) of ocean and freshwater systems (1 Pg = 10^{15} g)

Ecosystem	Area (x 10^6 km^2)	NPP (g C m^{-2} yr^{-1})	ANPP (Pg C yr^{-1})
Oceans			
Tropical & subtropical trades	139.9	93	13.0[a]
Temperate westerlies	129.9	126	16.3[a]
Polar oceans	20.8	310	6.4[a]
Coastal domain	37.4	286	10.7[a]
Coral reefs	0.6	1165	0.7[b]
Macroalgal beds	—	—	0.01[c]
Salt marshes	0.38	1275	0.5[d]
Estuaries	1.4	675	0.9[d]
			Total oceans: 48.5[f]
Freshwater			
Swamps & marshes	2.0	1125	2.2[e]
Lakes & streams	2.5	232	0.6[e]
			Total terrestrial: 56.4[f]

Notes: [a]Longhurst et al. 1995, [b]Crossland et al. 1991, [c]De Vooys 1979, [d]Woodwell et al. 1973, [e]Whittaker and Likens 1973, [f]Field et al. 1998.

limiting factor in phytoplankton growth (Huisman and Weissing 1994).

A large number of models have been proposed for photosynthesis as a function of light (Geider and Osborne 1992; Henley 1993). Two of the most widely used are the exponential function, equation (11), and the hyperbolic tangent function, equation (12):

$$P = P_m \left[1 - \exp\left(\frac{\alpha E}{P_m} \right) \right] \quad (11)$$

$$P = P_m \tanh\left(\frac{\alpha E}{P_m} \right) \quad (12)$$

P_m is the maximum rate of photosynthesis (gross photosynthesis if ^{14}C is used; net photosynthesis if O_2 is measured), and E is the light intensity expressed as μmol quanta m^{-2} s^{-1}. The parameter α is the slope of the initial increase in photosynthesis with light. If net photosynthesis, P_n (in mg O_2/mg dry weight day), is measured, the net specific growth rate, μ_n (day^{-1}), can be calculated from the simple formula:

$$\mu_n = (P_n/C)\,\theta \quad (13)$$

in which C is the mean measured initial carbon content of the phytoplankton alga in mg C/100 mg dry weight, and θ is the photosynthetic quotient (12 mg C/32 mg O_2) (Auer and Canale 1982). These models have been used in the study of coral reefs, macroalgae, and phytoplankton. For example, Fu et al. (2007) used a modified version of equation (11) to study the effects of predicted elevated atmospheric CO_2 levels and ocean temperatures in the year 2100 on growth and photosynthesis in picocyanobacteria. *Synechococcus* showed a fourfold increase in net photosynthesis under predicted global warming conditions of a doubling of atmospheric CO_2 (from the current 380 ppm to 750 ppm) and a 4°C increase in ocean temperatures.

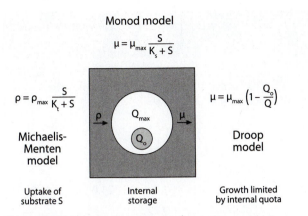

Figure 21.17 Relationships among the various nutrient-based models of phytoplankton growth. The Michaelis-Menten model describes the process of nutrient uptake from the environment. The Droop model describes growth as a function of internal nutrient stores, and the Monod model defines growth in terms of external nutrient supply.

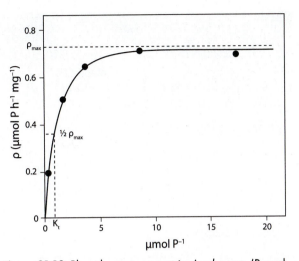

Figure 21.18 Phosphorus transport in *Anabaena*. (Based on Lampert and Sommer 1997. © Georg Thieme Verlag)

Growth and Nutrient Uptake

As phytoplankton grow, they consume mineral resources. Carbon, nitrogen, phosphorus, sulfur, and silicon in the case of diatoms are considered macronutrients required in relatively large amounts. Molybdenum and iron are micronutrients present and needed in only small amounts. The process of nutrient uptake and the use of nutrients in growth can be represented by three different mathematical expressions: one for nutrient uptake, one for internal stores of nutrients, and one for growth as a function of external nutrient levels (Figure 21.17).

Nutrient Uptake

The Michaelis-Menten model, which is based on the kinetics of enzyme function, describes the process by which nutrients are taken up from the environment and transported into the algal cell. Authors differ somewhat in their choice of variables, but the following equation is widely used to describe nutrient transport into algal cells:

$$\rho = \rho_{max}\left(\frac{S}{K_t + S}\right) \quad (14)$$

In this formulation, S is the concentration of mineral nutrient (the substrate) in the water in µM, and ρ is the velocity of the nutrient transporter or the nutrient transport rate in units of µmol of nutrient per cell per minute. If the cell is not the chosen unit for the algal population but some measure of biomass, then the units could be µmol mg⁻¹ min⁻¹. The term ρ_{max} is the maximum velocity of the nutrient transporter; ρ approaches ρ_{max} when the level of the external nutrient S is high and the internal store of that same nutrient (Q) is low. K_t is the half-saturation constant, which equals the value of S where $\rho = \frac{1}{2}\rho_{max}$; that is, where the enzyme-based nutrient transporter is half-saturated with the substrate. S and K_t have the same units of concentration (µmol l⁻¹). Figure 21.18 illustrates the relationships between ρ, ρ_{max}, K_t, and S for *Anabaena* taking up phosphorus. As the concentration of phosphorus (S) increases, the measured level of nutrient uptake, ρ, first increases rapidly and essentially linearly. The curve of ρ then bends over to a plateau as ρ approaches ρ_{max}. The point where ρ reaches $\frac{1}{2}\rho_{max}$ defines the phosphorus concentration that is K_t. For some measured values of K_t and ρ_{max} for phosphorus, refer to Table 21.3 (Sandgren 1988). In this table, ρ_{max} is referred to as V_{max}. In some papers, V_{max} has a different meaning and has units of time⁻¹ (min⁻¹ or hr⁻¹). In this context, V_{max} is the maximum specific uptake rate and:

$$V_{max} = \frac{\rho_{max}}{Q} \quad (15)$$

Table 21.3 shows that phytoplankton species may differ by as much as two orders of magnitude in their uptake rates and half-saturation constants. Nutrient uptake kinetics could therefore play a major role in phytoplankton species growth and interactions.

Table 21.3 Values of parameters in models of growth and phosphorus uptake for P-limited chrysophyte flagellates*

Species	Clone	Mean Cell Volume (μm^3)	μ_{max} (day⁻¹)	K_s (μM)	V_{max} (10^{-9} μmol cell⁻¹ min⁻¹)	K_t (μM)	Q_{max} (10^{-9} μmol cell⁻¹)	Q_o (10^{-9} μmol cell⁻¹)
Dinobryon cylindricum	1	—	0.90	—	—	—	—	2.40
	5	272	0.51	0.014	—	—	18.5	1.77
	7	—	0.58	—	—	—	—	2.15
	13	290	0.75	0.021	—	—	21.0	1.87
					0.34	0.72	—	—
Dinobryon bavaricum		80	—	—	0.10	0.11	—	—
					0.22	0.01	—	—
Dinobryon sociale		—	—	—	2.32	0.39	—	—
Dinobryon divergens		—	—	—	—	0.10–0.27	—	—
Synura petersenii	2b	374	0.51	0.003	5.1	1.19	90.0	3.04
	7c	431	0.76	0.001	21.8	1.35	55.2	1.96
Mallomonas cratis	UW-126	1516	0.55	0.001	14.2	0.36	152.0	7.90
Mallomonas caudata	2j	10625	0.30	—	—	—	—	—

*from Sandgren 1988

Internal Nutrient Stores

As the nutrient transporter pumps nutrients such as P or N into the cells, they go into an internal pool, Q, with units of µmol cell⁻¹ (see Figure 21.17). The alga draws off the internal storage pool as it grows. Droop (1983) advanced the following equation for the relationship between growth rate and internal nutrient quota:

$$\mu = \mu_{max}\left(1 - \frac{Q_o}{Q}\right) \quad (16)$$

Q_o is the minimum cell quota or minimum stored nutrient supply, and Q is the variable internal nutrient store. An alga with internal store at Q_o is at its minimum subsistence quota and cannot divide. The letter µ is the growth rate or rate of reproduction with units of time⁻¹ (day⁻¹ or hr⁻¹). The maximum rate of reproduction is μ_{max}. When Q approaches Q_o, then µ is zero (Figure 21.19). As Q increases above Q_o, µ approaches μ_{max}. Although not part of the original Droop model, an upper limit on the value of Q must exist. Other authors have incorporated

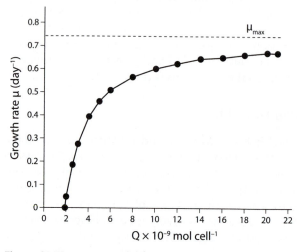

Figure 21.19 Droop model for *Dinobryon cylindricum* using the parameters in Table 21.3.

this term Q_{max} into models of growth, and it has been measured in a number of algae. Table 21.3 gives actual values of Q_o and Q_{max} for phosphorus in chrysophycean flagellates. Large cells have higher

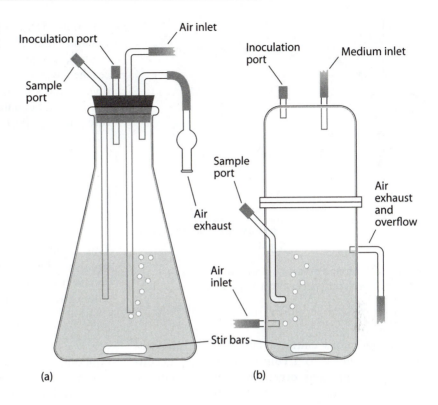

Figure 21.20 Culture vessels for algal growth studies. (a) Batch culture vessel. (b) Continuous culture (chemostat) vessel.

(a)

(b)

values of Q_{max} than small cells because they have more space for storage of polyphosphate. In the absence of any uptake, the large cells listed in Table 21.3 could support up to four cell divisions before Q_{max} would approach Q_o. High values of Q_{max} are the result of luxury uptake.

Growth and External Nutrient Supply

Our final consideration is the process of population growth as a function of nutrient concentration in the surrounding environment. Here we must consider two types of culture: the batch culture and the continuous culture, or chemostat. In the batch culture there is a fixed initial amount of nutrient or substrate S in a fixed volume (Figure 21.20a). The organism X is introduced and grows at the expense of S. The process of depletion of S and growth of X can be described by a Monod (1950) equation, which is similar in form to the Michaelis-Menten equation:

$$\mu = \mu_{max} \frac{S}{K_S + S} \quad (17)$$

Here μ, μ_{max}, and S have been defined previously. K_s is the half-saturation constant for growth as a function of substrate S and is that value of S at which

$\mu = \frac{1}{2}\,\mu_{max}$ (Figure 21.21). When S is very large relative to K_s, then $S/(K_s + S)$ is close to 1, and μ equals μ_{max}. To describe the batch growth process, we write equations for the use of S and the growth of X as follows:

$$\frac{dS}{dt} = -\mu_{max}\left(\frac{S}{K_S + S}\right)\frac{X}{Y} \quad (18)$$

$$\frac{dX}{dt} = \mu_{max}\left(\frac{S}{K_S + S}\right)X \quad (19)$$

In words, equation (18) says the change in S in the batch culture is equal to the consumption of S by the growth of alga X. The parameter Y is the yield coefficient for conversion of S into X. If X were expressed as cells ml^{-1} and S in $\mu mol\ l^{-1}$, then Y would have units of cells $ml^{-1}/\mu mol\ l^{-1}$. Equation (19) then says alga X grows as a Monod function of S. Equation 19 consists of an exponential function $dX/dt = \mu_{max}\ X$ and a Monod term that describes the limitation of algal growth by the supply of a nutrient and the ability of the alga to take up that nutrient. The lower the value of K_s, the more effective the alga is at taking up external nutrient at low levels.

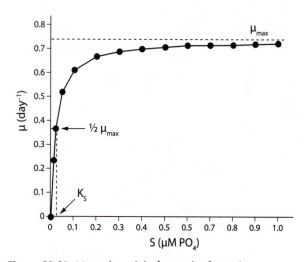

Figure 21.21 Monod model of growth of *Dinobryon cylindricum* with parameters from Table 21.3.

In continuous cultures, nutrients are continuously being added to a culture vessel, and unconsumed nutrients and organisms are continuously being washed out (see Figure 21.20b). The rate at which nutrients are turned over and organisms and used medium are washed out is called the dilution rate, D, with units of time^{-1}. D is calculated as the ratio of the volume flowing into the vessel in a day divided by the volume of the vessel. A value of D = 0.25 day^{-1} therefore means that one-quarter of the volume of the continuous culture vessel is flushed out and turned over every day. In continuous culture, D represents a loss rate such as λ, which we discussed previously with exponential growth. The equations for the same substrate and organism in continuous culture as in equations (18) and (19) are:

$$\frac{dS}{dt} = D\,(S_O - S) - \mu_{max}\left(\frac{S}{K_S + S}\right)\frac{X}{Y} \quad (20)$$

$$\frac{dX}{dt} = \mu_{max}\left(\frac{S}{K_S + S}\right)X - DX \quad (21)$$

In words, equation (20) says that the change in substrate concentration in the continuous culture vessel is equal to the amount delivered (DS_o), minus any residual that is carried out in the overflow ($-DS$), minus the amount consumed by growth of alga X. Equation (21) then says the change in population of alga X is equal to growth (μX) minus loss due to

washout ($-DX$). While the equations may look formidable, they can be solved because in continuous culture, a steady state, or an equilibrium, can be reached where growth equals losses and thus both dS/dt and dX/dt equal zero.

If we set equation (21) equal to zero and substitute μ for $\mu_{max}\,[S/(K_s + S)]$, we get:

$$0 = \mu X - DX \quad (22)$$

And consequently:

$$\mu = D \quad (23)$$

This means that at steady state, the growth of X equals the loss rate or dilution rate, D. If we now set equation (20) equal to zero and substitute D for μ from equation (23), we get:

$$X = Y\,(S_o - S) \quad (24)$$

Equation (24) says that the steady state concentration of alga X equals the amount of substrate consumed times the yield coefficient, which converts the amount of S used into units of alga X. If D is greater than μ_{max}, the population of algae cannot maintain itself against this loss rate and will wash out of the continuous culture vessel.

Continuous cultures have been a major research tool for the study of phytoplankton populations. The mathematical models for nutrient uptake and growth contain parameters that have value in understanding species differences and population dynamics. Researchers who work in continuous culture systems find that the Droop model gives a better fit to results than does the Monod model, but researchers working in natural aquatic systems prefer the Monod model because values of substrate S and phytoplankton population X are a routine part of sampling programs. For some practice working with the Michaelis-Menten, Droop, and Monod models, refer to Question Box 21.2.

21.5 Loss Processes

Recall our earlier discussion of net growth rate being equal to the gross growth rate or rate of reproduction minus the death rate or the sum of all loss processes, equation (4). Loss processes are all those factors that remove or displace phytoplankton in the aquatic environment. Following the formulation of Lampert and Sommer (1997), loss processes (in units of days⁻¹) can be described by the expression:

$$\lambda = \gamma + \sigma + \chi + \delta + \pi + \rho + \omega \quad (25)$$

The Greek letter λ stands for losses, and the remaining Greek letters stand for specific types of loss processes. Greek gamma (γ) stands for grazing—the loss of algal cells to herbivores such as ciliates, rotifers, and crustaceans (see Chapter 3). The letter sigma (σ) stands for sedimentation—the sinking of nonmotile algal cells out of the euphotic zone. The letter chi (χ) represents losses due to competition. Delta (δ) stands for death or mortality, pi (π) is parasitism, and rho (ρ) is perennation. The final symbol, omega (ω), represents loss due to washout. We now discuss each of these processes.

Perennation

Perennation (ρ) is the formation of a resting stage that allows algal species to avoid a period of adverse environmental conditions. Many cyanobacteria form akinetes, a type of asexual spore that develops from vegetative cells (Chapter 6). Akinetes may be resuspended into the water column, where they will germinate (Reynolds 1972). One study found that akinetes could survive and germinate after more than 60 years buried in sediments (Livingstone and Jaworski 1980). *Microcystis*, by contrast, can survive for several years as vegetative cells on sediment surfaces, with neither light nor oxygen (Reynolds et al. 1981). Some centric diatoms such as *Melosira* and *Stephanodiscus* form a resting vegetative cell in which the protoplast contracts into a ball within the frustule; these resting cells can survive on the sediments for several months to a year (see also Chapter 12).

Green algae often produce resting zygotes (Chapters 19 and 20), and dinoflagellates form cysts (Chapter 11). The dinoflagellate *Ceratium hirundinella* forms cysts in response to declining nutrients. Most chrysophyceans are present in the plankton for only a few weeks per year and produce their resting cysts (stomatocysts) when the population is most abundant (Figure 21.22). Resting stages play important roles in both the freshwater and marine environments, but in oceans, only those resting stages that sink over shallow continental shelves have a chance for resuspension.

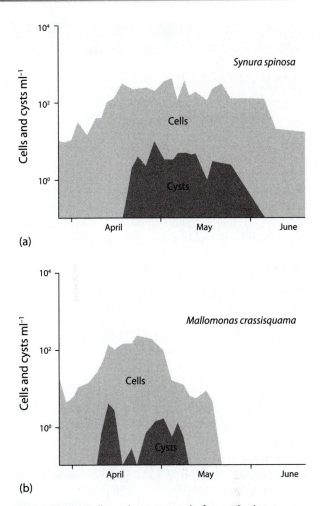

(a)

(b)

Figure 21.22 Cells and cysts per ml of two planktonic chrysophyceans from Egg Lake, Washington (United States) in spring 1976. (After Sandgren 1988)

Mortality and Washout

Most algal cells are either eaten or sink out of the mixing layer, but mortality and cell lysis (δ) are important loss processes. Two causes of cell lysis are viruses and algicidal bacteria, both of which have been studied in the oceans in association with algal blooms. Virus-algal cultures clearly indicate that virus particles are capable of lysing an entire algal population in one to two days (Brussaard 2004). Studies of viral lysis rates in the oceans are limited. Measured lysis rates vary seasonally and spatially. Summer lysis rates averaged 0.41 day⁻¹ in Mediterranean waters along the coast of Spain, while winter rates in the same waters were only 0.061 day⁻¹ (Agustí and Duarte 2000). Culture studies have clearly demonstrated the existence of

Working with the Michaelis-Menten, Droop, and Monod Models of Nutrient Uptake and Growth

Question Box 21.2

Sample calculation 1

Tilman (1976) calculated these growth parameters for the diatoms *Asterionella formosa* and *Cyclotella meneghiniana* under both PO_4 limitation and SiO_2 limitation.

Diatom species	SiO_2 limitation		PO_4 limitation	
	μ_{max} (day^{-1})	K_s ($\mu M\ SiO_2$)	μ_{max} (day^{-1})	K_s ($\mu M\ PO_4$)
Asterionella formosa	1.59	3.9	0.76	0.04
Cyclotella meneghiniana	1.74	1.4	0.76	0.25

Use these data in the Monod equation, equation (17), to plot the growth rates of each alga as a function of the concentration of SiO_2 and separately as a function of the concentration of PO_4. Which alga will dominate under PO_4 limitation? Which will dominate under SiO_2 limitation? These results will be useful in a later section on nutrient competition.

Solution

We choose a convenient range of values of the nutrient concentration of 0 to 10 μM. The Monod equations are:

Diatom species	SiO_2	PO_4
Asterionella formosa	$\mu = 1.59\ [S/(3.9 + S)]$	$\mu = 0.76\ [S/(0.04 + S)]$
Cyclotella meneghiniana	$\mu = 1.74\ [S/(1.4 + S)]$	$\mu = 0.76\ [S/(0.25 + S)]$

From the graphs below, it is clear that if SiO_2 is limiting, *C. meneghiniana* will dominate over *A. formosa*. At every concentration of SiO_2, its growth rate, μ, is higher, and because its K_s value is lower, it can reduce SiO_2 to levels where *A. formosa* cannot grow. Conversely, if PO_4 is limiting, *A. formosa* dominates *C. meneghiniana* at all levels of PO_4. *Asterionella formosa* can reduce the levels of PO_4 to exclude *C. meneghiniana* because its K_s value is so much smaller. Note that as K_s values become smaller, the growth curves rise more rapidly toward μ_{max} and appear more angular. These results give us an important insight into nutrient competition among phytoplankton: If two species are competing for a single nutrient such as silicate, the one with the lower K_s value will always win because it can reduce that nutrient's level to the point where the other species cannot grow.

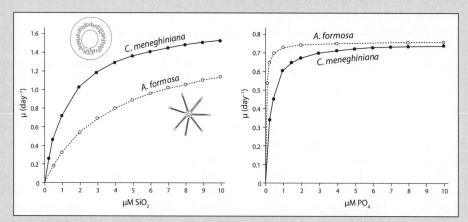

Redrawn with permission from Tilman, D. Ecological competition between algae: Experimental confirmation of resource-based competition theory. *Science* 192:463–465. © 1976. American Association for the Advancement of Science.

Sample calculation 2

The accompanying data (right) show the process of phosphorus uptake by *Cyclotella meneghiniana* as a function of phosphate concentration (adapted from Tilman and Kilham 1976).

Determine the parameters K_t and ρ_{max} in the Michaelis-Menten equation, equation (14), for this set of data. Values of S represent the initial phosphate concentrations. Uptake rates were measured by following the removal of PO_4 and the increase in cell numbers over several hours.

S (μM PO_4)	ρ (μM PO_4 cell^{-1} hr^{-1} $\times$ 10^{-9})
1.00	2.87
1.38	3.40
2.08	5.21
2.86	3.30
4.00	4.65
4.76	4.48
6.66	5.05

Solution

In research applications, the final parameters are determined by nonlinear regression techniques with the use of a computer. You can, however, obtain good estimates of the parameters by using a linear transformation of the Michaelis-Menten equation, where ρ and S are the variables and ρ_{max} and K_t are constants. We use the algebraic axiom that if x = y then it is also true that 1/x = 1/y. The first step is to invert equation (14) to yield:

$$\frac{1}{\rho} = \frac{K_t + S}{\rho_{max} S}$$

We can now separate the two terms on the right-hand side to give:

$$\frac{1}{\rho} = \frac{K_t}{\rho_{max}} \left(\frac{1}{S}\right) + \frac{S}{\rho_{max} S}$$

Canceling out the S in the far-right term we finally derive:

$$\frac{1}{\rho} = \frac{K_t}{\rho_{max}} \left(\frac{1}{S}\right) + \frac{1}{\rho_{max}}$$

This is the equation of a straight line of the form y = mx + b, where y = $1/\rho$, x = 1/S, the slope m = K_t/ρ_{max}, and the intercept b = $1/\rho_{max}$. To derive this equation for our example, we have only to take the inverses of the values of S and ρ, plot them on graph paper, and take the linear regression of $1/\rho$ on 1/S. The intercept will be the inverse of ρ_{max}, and the slope will give us K_t. The inverses are given in the table (right).

x = 1/S	y = 1/ρ
1.00	0.348
0.72	0.294
0.48	0.192
0.35	0.303
0.25	0.215
0.21	0.223
0.15	0.198

If we use all seven points, the resulting linear equation is Y = 0.1529 (X) + 0.1842, with a correlation coefficient of 0.78 (refer to the plot at right). The maximum uptake rate, ρ_{max}, is the inverse of 0.1842, or 5.4×10^{-9} μM PO_4 cell^{-1} hr^{-1}. K_t = 0.1529 ρ_{max}, so K_t = 0.83 μM PO_4. These values are quite close to those reported by Tilman and Kilham (1976) from nonlinear regression (ρ_{max} = 5.1×10^{-9} and K_t = 0.8). In most cases real data are much noisier than this set, and linear regression can at most provide a starting point for a nonlinear regression routine.

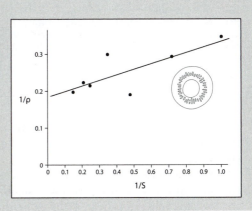

Practice problems

1. Tilman and Kilham (1976) measured the uptake of SiO_2 by the diatom *Cyclotella meneghiniana*. Use the data at right adapted from their paper to estimate the uptake parameters ρ_{max} and K_t. How do the dynamics of SiO_2 uptake compare to those of PO_4 calculated in sample calculation 2?

SiO_2 (μM)	ρ (μM cell^{-1} hr^{-1} × 10^{-9})
0.893	1.66
2.270	4.29
3.860	5.77
7.270	6.72
15.000	8.93
27.300	12.56

2. Ducobu et al. (1998) studied competition for phosphorus between the prochlorophyte *Prochlorothrix hollandica* and the cyanobacterium *Planktothrix agardhii* for phosphorus in continuous culture. The parameters to the right were obtained for the two species to compare the predictions of the Droop model to that of the Monod model.

	Q_o mgP/mgdw	Q_{max} mgP/mgdw	μ_{max} h^{-1}	K_s μg P l^{-1}
Prochlorothrix	0.0016	0.110	0.025	0.015
Planktothrix	0.0027	0.042	0.036	0.170

On one graph plot the curves of the Droop model of growth as a function of internal phosphorus $\mu = \mu'_{max} (Q - Q_o)/Q$, for each species. On a second graph plot the Monod model of growth as a function of external phosphorus, $\mu = \mu_{max}S/(K_s + S)$, for each species. How do these graphs differ from the one plotted in sample calculation 1? Based on these two plots, what phosphorus conditions would favor growth of each species? (*Hint*: Plot Q from Q_o to 0.05 and S from 0.01 to 1 μg P l^{-1}.)

3. Tilman and Kilham (1976) also measured the growth of *Asterionella formosa* as a function of SiO_2 concentration. Use the accompanying data adapted from their paper to estimate the growth parameters μ_{max} and K_s in the Monod equation (17). (*Hint*: The same procedure can be used as with the Michaelis-Menten model of uptake dynamics.)

Note: In the actual data set, the values of μ at low SiO_2 levels are prone to large variation. Therefore, the values of $1/\mu$ are also subject to high variation and were not used in the regression. Answers are found at the end of the chapter.

SiO_2	μ (doublings day^{-1})
3.5	0.508
3.7	0.515
6.5	0.690
7.0	0.700
10.2	0.783
13.2	0.813
18.6	0.833
22.0	0.933

bacteria that can kill marine phytoplankton, and bacteria increase in numbers following algal blooms in the oceans (Mayali and Azam 2004).

Although washout (ω) is a major factor in continuous culture systems, it is not a significant factor for phytoplankton in most lakes or the oceans.

In small lakes or tidal pools, the volume of water may be sufficiently small compared to the potential volume of flushing during storm events that washout can be significant. In such cases the loss rate can be described by a simple exponential decay function (Reynolds 2006).

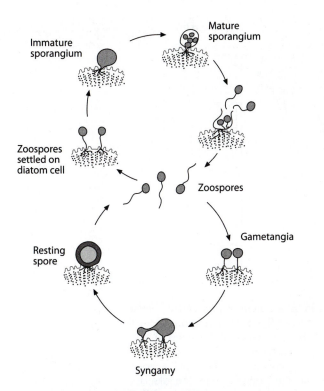

Figure 21.23 Life cycle of the chytrid *Zygorhizidium planktonicum*. (Adapted from van Donk and Ringelberg 1983, by permission of Blackwell Science, Ltd.)

Parasitism

In **parasitism** (π), individuals of one population (the parasite) live on or in members of another population (the host) and feed upon them to their harm. The parasites of macroorganisms seldom kill their hosts in the short term, but the parasites of phytoplankton destroy their hosts. Here we restrict our definition of parasites to eukaryotic organisms. Two groups of such phytoplankton parasites have been well documented: fungal parasites of freshwater phytoplankton and protistan parasites of marine dinoflagellates.

The fungal parasites of freshwater algae are primarily chytrids, which produce free-swimming uniflagellate zoospores that seek out and attach to host cells (Figure 21.23). A mycelial thread penetrates the host cell and supplies nutrients to the enlarging zoospore, which becomes a sporangium. Zoospores are released from the sporangium to complete the cycle. The host cell is killed. Van Donk (1989) followed populations of *Asterionella formosa* and their infection with the chytrid *Zygorhizidium planktonicum* in Lake Maarsseveen

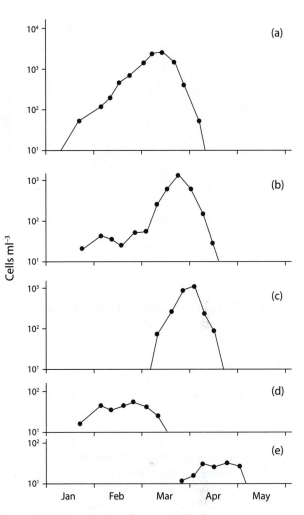

Figure 21.24 Population of *Asterionella formosa* in spring 1982 in the surface waters of Lake Maarsseveen, the Netherlands, divided into five categories. (a) uninfected *A. formosa* cells, (b) cells with zoospores of *Zygorhizidium planktonicum*, (c) *A. formosa* infected with sporangia, (d) cells infected with thick-walled spores, and (e) cells with resting spores of *Z. planktonicum*. Most of the *A. formosa* population was infected by late March (After van Donk 1989. © Springer-Verlag)

(the Netherlands) during the spring from 1978 to 1982 (Figure 21.24). Heavy infection of *A. formosa* allowed the diatoms *Fragilaria crotonensis*, *Stephanodiscus astraea*, and *S. hantzschii* to become abundant. Chytrid infections can therefore affect phytoplankton species succession (Ibelings et al. 2004). A chytrid infection also wiped out more than 75% of a bloom of the desmid *Staurastrum dorsidentifera* in Lake Biwa, Japan (Kagami et al. 2006). Kagami et al. (2004) found that

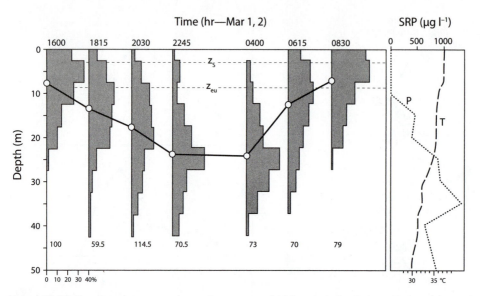

Figure 21.25 Evening descent and morning ascent of *Volvox* in Lake Cahora Bassa, Mozambique, on March 1–2, 1983. The upper horizontal axis gives 24-hour time. The column diagrams show the depth distribution of *Volvox* in percentage of total population. The numbers at the bottoms of the column diagrams give the areal population density in units of 10^3 colonies m^{-2}. The solid lines track the median depth. Z_s marks Secchi-disk transparency in m and Z_{eu} the depth of the euphotic zone. Note that soluble reactive phosphorus (P) is unmeasurable within the euphotic zone. The lake is quite hot by comparison to northern temperate lakes (> 30°C). (From Sommer and Gliwicz 1986)

Daphnia can protect diatoms from chytrid infections by grazing on the fungal zoospores.

In marine systems, dinoflagellates are hosts to protistan parasites, including the flagellate *Parvilucifera infectans* and parasitic dinoflagellates belonging to the genus *Amoebophrya* (Park et al. 2004). *Amoebophrya* species have a life cycle that consists of a motile infective stage, the dinospore, an intracellular growth stage in the host called the trophont, and an extracellular reproductive stage, the vermiform (Coats and Park 2002). Epidemics of *Amoebophrya* coincide with declining blooms of host species. Grazing by ciliates on the dinospore stage may reduce infection of host dinoflagellates (Johansson and Coats 2002).

Sedimentation

If an algal cell sinks below the euphotic zone, it may be lost unless it functions as a perennation stage that can be resuspended in the water column by turbulence. Many planktonic algae can minimize sedimentation losses (σ) by controlling their position in the water column or by reducing their sinking rates.

Swimming and buoyancy

Flagellates can swim at speeds sufficient to maintain their position in the water column. The average swimming speeds of flagellate phytoplankton are about 10 times greater than the average sinking rate, which is on the order of 0.5 m day^{-1} (Sournia 1982). Flagellates may perform extensive daily vertical migrations, often moving from upper waters by day to deeper waters at night. *Peridinium cinctum* migrates 8–10 m daily in Lake Kinneret, Israel (Berman and Rodhe 1971). In the oceans, dinoflagellates may migrate to depths of 10–20 m daily (Eppley et al. 1968). In Lake Cahora Bassa, Mozambique, *Volvox* migrates as much as 20 m into the deeper strata and returns to the surface every day. *Volvox* colonies require high levels of phosphorus for growth. The half-saturation constant (K_s) for P-limited growth is on the order of 19 to 59 µg P l^{-1} (Senft et al. 1981). In the euphotic zone of Lake Cahora Bassa, P-levels are undetectable (Figure 21.25). Thus migration may be an adaptation that allows *Volvox* to take up phosphorus at night before returning to upper waters for photosynthesis in the day (Sommer and Gliwicz 1986).

Many cyanobacteria also control their position in the water column by adjusting the formation of intracellular gas vacuoles and regulating cellular ballast (Chapter 6). Genera that generate nuisance blooms, such as *Microcystis*, *Anabaena*, and *Aphanizomenon*, use such buoyancy mechanisms to adjust their vertical position. During windless periods, daily rhythms of vertical migration have been observed (Reynolds et al. 1987). Vertical migration rates can be high. In Lake George, Uganda, *Microcystis aeruginosa* migrated at speeds greater than 3 m hr^{-1} (Ganf 1975). *Aphanizomenon flos-aquae* reached speeds of 40 cm hr^{-1} to 2.75 m hr^{-1} in the Chowan River of North Carolina (Paerl and Ustach 1982). Buoyancy regulation works best under stable water conditions. Turbulent mixing causes cyanobacteria to circulate along with other phytoplankton, and prolonged mixing can shift dominance away from cyanobacteria and toward other algae (Harris et al. 1980).

Sinking

Phytoplankton cells that can neither swim by using flagella nor regulate their buoyancy are subject to sinking through the water column. Most planktonic algae are only slightly denser than water, at 1.02 to 1.05 g ml^{-1}, but diatoms have densities around 1.3 g ml^{-1}. The silicate of their frustules is very dense, at about 2.6 g ml^{-1}. Only a few substances within algal cells are less dense than water; lipids have densities around 0.86 g ml^{-1}, and gas vacuoles of cyanobacteria have values around 0.12 g ml^{-1}. The sinking dynamics of phytoplankton are described using Stokes' law.

Stokes' law was originally designed to describe the sinking velocity of a metal sphere in oil. Since planktonic algae are often shaped other than as spheres, Ostwald's modification of Stokes' law was formulated to adjust sinking rates for nonspherical phytoplankton cells. The modified Stokes' law has the form:

$$v_s = \left(\frac{2}{9}\right) g\, r_s^2\, (q' - q)\, \upsilon^{-1}\, \phi^{-1} \quad (26)$$

Here, v_s is the sinking velocity in m s^{-1}, g is the gravitational acceleration of Earth (9.8 m s^{-2}), and r_s is the radius (in meters) of a sphere of volume equivalent to that of the algal cell. The term q' is the density of the algal cell expressed in kg m^{-3}, and q is the density of the fluid medium. For water, q has the value

1000 kg m^{-3}. Thus, (q' – q) is the difference between the density of the algal cell and the surrounding water. This difference is positive for algae that are denser than water, but it may be negative for cyanobacteria with gas vacuoles or an alga with a high intracellular concentration of lipids. The term υ (Greek upsilon) is the viscosity of water, which is expressed in units of kg m^{-1} s^{-1}. Water is more viscous at lower temperatures, and thus $\upsilon = 1 \times 10^{-3}$ kg m^{-1} s^{-1} at 20°C but 1.8×10^{-3} kg m^{-1} s^{-1} at 0°C (Lampert and Sommer 1997). The final term, ϕ (Greek phi), stands for form resistance and is dimensionless. Form resistance accounts for the fact that most algae have nonspherical shapes and may possess horns and spines.

In words, equation (26) says that the sinking velocity of an algal cell is proportional to the gravitational force of Earth times a measure of the size of the cell (the radius squared) times the difference between the densities of the cell and its fluid medium. According to the equation, larger cells or colonies or more dense cells such as diatoms should sink more rapidly. Conversely, the more viscous the water and the greater the form resistance of the algal cell, the slower the sinking velocity. Some representative data and a sample calculation are presented in Question Box 21.3, together with a few practice problems.

Planktonic algae have a number of adaptations to reduce sinking rates. Small cell size is the most obvious one. Observed sinking rates of living algal cells and colonies frequently do not correlate well with the predictions of the Ostwald modification of Stokes' law. Living cells do not sink as rapidly as dead or senescent cells of the same species (Figure 21.26). Reynolds (1984) found that the observed sinking velocity of killed cells of *Stephanodiscus astraea* was in accord with the modified Stokes' equation but that the sinking velocity of living cells was not. Living cells seem to have enhanced buoyancy. Some large marine diatoms such as species of *Rhizosolenia* and *Ethmodiscus* are actually positively buoyant. The mechanism of this buoyancy in living cells is still unclear (Moore and Villareal 1996).

Many large phytoplankton cells and colonies, including the desmid *Staurastrum* and cyanobacteria such as *Coelosphaerium* and *Microcystis*, possess a mucilaginous sheath or matrix in which the cells are embedded. Since the density of mucilage is close to that of water, a mucilaginous sheath can reduce the overall cell or colony density to levels closer to that of water, but it cannot make a cell or colony buoyant. At the same time, adding mucilage can increase

Working with the Ostwald Modification of Stokes' Law

The table to the right shows some representative algal species, their cell or colony volumes, the radius of a sphere with a volume equal to that of the algal cell or colony, and the algal density.

Algal species	Volume (µm³)	Radius (µm)	Density (kg m⁻³)
Stephanodiscus astraea	5930[a]	11.23	1091[a]
Asterionella formosa	5160[a]	10.72	1130[a]
Chlorella vulgaris	30[b]	1.9	1095[c]
Microcystis aeruginosa	4.2×10^{6} [a]	100.1	999.4[d]

[a] Reynolds 1984, [b] Bellinger 1974, [c] Oliver et al. 1981, [d] Reynolds et al. 1981

Sample calculation for *Stephanodiscus astraea* at 20°C

From the text, $g = 9.8$ m s^{-2}, $q = 1000$ kg m^{-3}, and $v^{-1} = 10^3$ m s kg^{-1}, the inverse of v. The value of q' is given in the above table. Convert the table values of radius in µm to meters by multiplying by 10^{-6} m µm^{-1}. For the moment we do not concern ourselves with ϕ. Then:

$$v_s = \left(\frac{2}{9}\right) 9.8 \text{ m s}^{-2} (11.23 \times 10^{-6} \text{m})^2 (1091 - 1000 \text{ kg m}^{-3}) 10^3 \text{ m s kg}^{-1} \phi^{-1}$$

Which becomes:

$$v_s = 2.177 \text{ m s}^{-2} (126.1 \times 10^{-12} \text{ m}^2) (91 \text{ kg m}^{-3}) 10^3 \text{ m s kg}^{-1} \phi^{-1}$$

And finally:

$$v_s = 24.98 \times 10^{-6} \text{ m s}^{-1} \phi^{-1}$$

If we multiply this result by 3600 s hr^{-1}, we find that *S. astraea* is predicted to sink at the rate of 0.09 m hr^{-1}. If we multiply the result above by 10^6 µm m^{-1} instead, we get $v_s = 24.98$ µm s^{-1}. Reynolds (1984) reported that the actual sinking rate of killed *S. astraea* cells was 27.62 µm s^{-1}. Thus, ϕ is 24.98/27.62, or 0.904.

Practice problems

Work the following problems with a pocket scientific calculator. Answers are given at the end of the chapter.

1. Calculate the predicted sinking velocity of an 8-celled colony of *Asterionella formosa* according to the Ostwald modification of Stokes' law. Reynolds (1984) reports that the actual sinking velocity of this algal colony is 7.33 µm s^{-1}. What therefore must be the value of ϕ?

2. Calculate the sinking velocity of *Chlorella vulgaris* using equation (26). Assume for this spherical alga a ϕ of 1.0. What does this tell you about the effect of size on sinking rate?

3. Calculate the ascending velocity of a large buoyant colony of *Microcystis aeruginosa* from the data in the table. Assume the colonies are essentially spherical, so $\phi = 1.0$. How does this rate compare to some of the upward migration rates reported in the text for cyanobacteria?

4. In an earlier section on algae in space research (Chapter 4), it was suggested that algae may one day be introduced to Mars as part of a terraforming effort. If *Stephanodiscus astraea* were introduced to a reformed Martian lake, how fast might it sink through a Martian lake's water column? *Hint*: The gravitation force of Mars is 0.38 that of Earth.

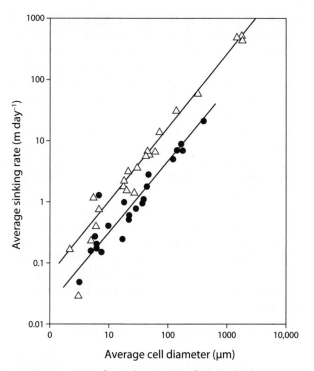

Figure 21.26 Data for sinking rates of phytoplankton. Triangles represent senescent cells; closed circles represent actively growing cells. (Based on Smayda 1970, in Harris 1986, with kind permission of Kluwer Academic Publishers)

cell or colony diameter, which should (according to Stokes' law) increase sinking rate. Reynolds (1984) has shown that there is a trade-off of effects such that adding mucilage decreases sinking up to a point, after which further additions cause sinking rate to rise again.

A further adaptation of phytoplankton to sinking is contained in the Ostwald modification of Stokes' law as the parameter φ for form resistance. According to the original Stokes' equation, larger cells should sink more rapidly. But if larger cells or colonies depart from a spherical shape, they acquire a form resistance that slows their sinking rates. Conway and Trainor (1972) showed that spine-bearing *Scenedesmus* species sink less rapidly than species without spines. Walsby and Xypolyta (1977) removed chitinous fibers from the marine diatom *Thalassiosira weissflogii* with the enzyme chitinase and observed that fiberless cells sank twice as fast as cells with fibers, although the fibers were denser than the cells. Reynolds (1984) showed that many diatom colonies increase their form resistance by adding cells to such an extent that their

actual sinking rates approach a constant rate. In *Asterionella formosa*, the sinking velocity increased to a constant at 6–8 μm s⁻¹ as the number of cells reached eight to nine cells per colony. Beyond this number, additional cells filled in the colony, decreased the form resistance, and increased the sinking rate.

Form-resistant shapes have also been interpreted as a response to grazing pressure. In nearshore marine waters, long-spined *Thalassiosira* are not consumed by microzooplankton such as ciliates, while non-spiny *Thalassiosira* are readily eaten. The long-spined diatoms can be taken by crustacean zooplankton (Gifford et al. 1981). Grazers may reject phytoplankton with mucilaginous sheaths because they are too big to be eaten. Porter (1977) has shown that some algae with mucilaginous sheaths may pass through the guts of grazers unharmed. Horns, arms, spines, or mucilaginous sheaths may serve more than one useful function.

Competition

No subject in phytoplankton ecology generates as much controversy as competition. The idea that competition plays a central role in phytoplankton ecology derives from early ecological theory. Darwin's theory of natural selection proposed that natural populations deplete their resources and press against their carrying capacity. Research by Gause in the 1930s showed that if two species were grown together in the laboratory, the outcome would be depletion of resources and competition in which one of the two species would eventually be eliminated. Gause's work led to the competitive exclusion principle and niche theory, and their application as a major paradigm. But the application of this paradigm to phytoplankton communities led to a paradox, as recognized by Hutchinson (1961). If phytoplankton species were all at or near their carrying capacity and competition were a major force in community structure, how could one explain the coexistence of 50 to 100 species in a milliliter of water under the same apparently homogeneous conditions? That paradox led to a great deal of research in an effort to circumvent the competitive exclusion principle and allow coexistence of many species in a homogeneous environment.

Lotka-Volterra model

The first attempt to formulate mathematically the process of competition between two species was made by Volterra in 1926 and by Lotka in 1932.

Their efforts have come down to us as the Lotka-Volterra competition equations. They are structurally very similar to the logistic growth equation presented earlier:

$$\frac{dN_1}{dt} = \frac{r_1 N_1 (K_1 - N_1 - \alpha N_2)}{K_1} \quad (27)$$

$$\frac{dN_2}{dt} = \frac{r_2 N_2 (K_2 - N_2 - \beta N_1)}{K_2} \quad (28)$$

As in the logistic equation, N_1 and N_2 are the population densities of species 1 and species 2 in units such as cells ml^{-1} or mg ml^{-1}. K_1 and K_2 are the carrying capacities of the two species expressed in the same units, and r_1 and r_2 represent the net per capita growth rates of the two species with units of day^{-1} or hr^{-1}. The new elements are the terms αN_2 and βN_1, which represent the competitive interactions between species 1 and species 2. Species 2 competes against species 1 by reducing the growth of species 1 by acting as if it were αN_2 individuals of species 1. Conversely, species 1 reduces growth of species 2 by the amount βN_1. The parameters α and β are the competition coefficients. Microbiologist G. F. Gause (1934) used the Lotka-Volterra equations to investigate competition between the ciliate protozoans *Paramecium caudatum* and *Paramecium aurelia*. Gause grew these ciliates in 5 ml volumes of physiological salt solution with bacteria as food. Every day, he withdrew 0.5 ml for counting and replaced it with fresh solution. *P. aurelia* always drove down the population of *P. caudatum* (Figure 21.27). But note that *P. aurelia* had not eliminated *P. caudatum* after 16 days. These observations became the key data in the establishment of the competitive exclusion principle. The problem with these equations as a model is that they cannot be used to predict anything. The competition coefficients can only be determined by performing a competition experiment (Grover 1997). The Lotka-Volterra equations say nothing about the mechanism of competition—that is, for what are the two species competing?

A mechanistic model of competition

Tilman (1976) addressed the shortcomings of the Lotka-Volterra model by adapting the Monod equation for nutrient-limited growth to the two species

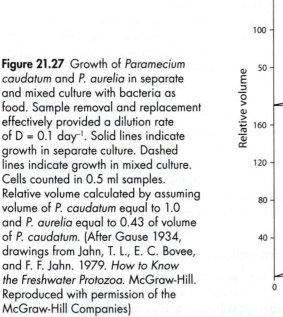

Figure 21.27 Growth of *Paramecium caudatum* and *P. aurelia* in separate and mixed culture with bacteria as food. Sample removal and replacement effectively provided a dilution rate of D = 0.1 day^{-1}. Solid lines indicate growth in separate culture. Dashed lines indicate growth in mixed culture. Cells counted in 0.5 ml samples. Relative volume calculated by assuming volume of *P. caudatum* equal to 1.0 and *P. aurelia* equal to 0.43 of volume of *P. caudatum*. (After Gause 1934, drawings from Jahn, T. L., E. C. Bovee, and F. F. Jahn. 1979. *How to Know the Freshwater Protozoa*. McGraw-Hill. Reproduced with permission of the McGraw-Hill Companies)

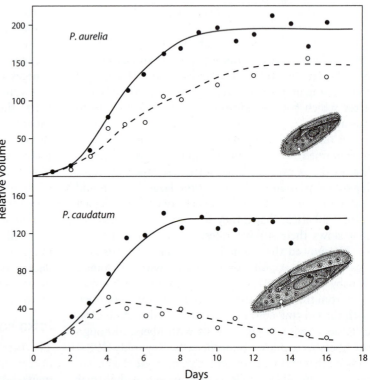

competition problem for algae. In the simplest case, assume that two species are competing for a single nutrient in a continuous culture. The equations for the growth of these two species and the consumption of the limiting nutrient (SiO_2) in continuous (chemostat) culture are given in equations (29) through (31) below.

$$\frac{dS}{dt} = D(S_O - S) - \mu_{m1}\left(\frac{S}{K_{S1} + S}\right)\left(\frac{N_1}{Y_1}\right)$$

$$- \mu_{m2}\left(\frac{S}{K_{S2} + S}\right)\left(\frac{N_2}{Y_2}\right) \quad (29)$$

$$\frac{dN_1}{dt} = \mu_{m1}\left(\frac{S}{K_{S1} + S}\right)N_1 - DN_1 \quad (30)$$

$$\frac{dN_2}{dt} = \mu_{m2}\left(\frac{S}{K_{S2} + S}\right)N_2 - DN_2 \quad (31)$$

These equations are an extension of equations (20) and (21) for growth of a single species in continuous culture. In words, they state that the concentration of nutrient in the chemostat equals the amount flowing in (DS_O) minus the unused nutrient flowing out ($-DS$) minus the consumption of limiting nutrient by the growth of each species. The growth of each species occurs according to the Monod model of nutrient-limited growth (μN) minus the loss of cells due to washout ($-DN$). Tilman (1976) used this model to study growth and competition between the diatoms *Cyclotella meneghiniana* and *Asterionella formosa* in continuous culture under silicate-limited conditions. The predicted outcome of competition between these two diatoms can be seen by plotting their Monod growth curves on the same silicate axis (Figure 21.28). The two growth curves do not cross. The flow rate, D, in continuous culture appears as a straight line parallel to the silicate axis. The straight line for D = 1.1 day^{-1} crosses each Monod growth curve at an equilibrium point where D = μ. Each equilibrium point has a corresponding critical nutrient level R*. For *A. formosa* at D = 1.1 day^{-1}, the corresponding R*$_{1a}$ is about 9 μM, and *A. formosa* washes out of continuous culture at any level of silicate less than this. For *C. meneghiniana* at D = 1.1 day^{-1}, however, R*$_{1c}$ is only 2.5 μM. *C. meneghiniana* wins in competition because it can reduce the nutrient level below that at which *A. formosa* can remain in continuous culture. The same argument applies at D = 0.4

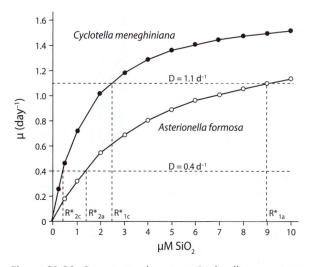

Figure 21.28 Competition between *Cyclotella meneghiniana* and *Asterionella formosa* for silicate in continuous culture. *Cyclotella meneghiniana* always displaces *A. formosa* because it can reduce the level of silicate below that at which *Asterionella* can grow, at any dilution rate (R*$_c$ always < R*$_a$). (Parameters from Tilman 1976)

day^{-1}. *C. meneghiniana* always wins in competition for silicate with *A. formosa* because it can always reduce the critical nutrient level R*$_c$ below that at which its competitor can grow R*$_a$.

Sommer (1986a) tested this idea by inoculating natural phytoplankton from Lake Constance, on the Germany/Switzerland/Austria border, into a chemostat with a culture medium lacking silicate. After several weeks only one species of green alga remained. At the lowest dilution rate only *Mougeotia thylespora* was present. At intermediate dilutions, *Scenedesmus acutus* dominated, and at the highest dilution rate, *Chlorella minutissima* took over the culture vessel (Figure 21.29). Despite uncertainties in the estimation of the half-saturation constants (K_s), these results were in accord with the predictions of Monod kinetics. Thus, unlike the Lotka-Volterra equations, the Monod-based model of competition is predictive.

When many species compete for a single limiting nutrient under steady-state conditions, the mechanistic model predicts that only one species—the one with the lowest value of K_s for that nutrient—can persist. But what happens when there is more than one nutrient limiting growth? Tilman (1976, 1977) developed a Monod-based model to predict the outcome of competition between two species of diatom,

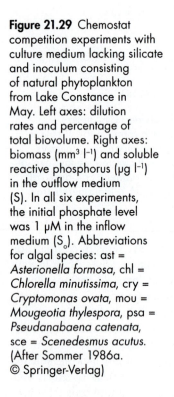

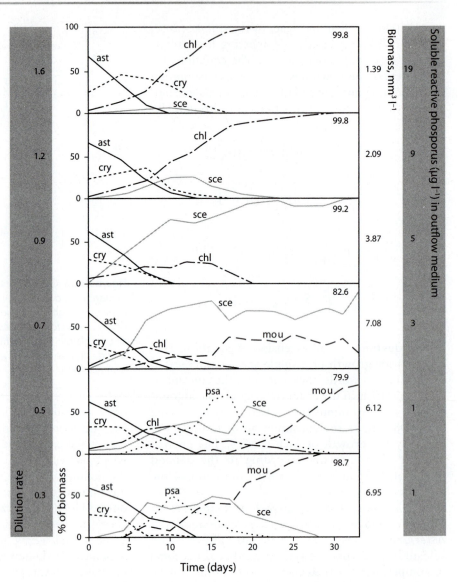

Figure 21.29 Chemostat competition experiments with culture medium lacking silicate and inoculum consisting of natural phytoplankton from Lake Constance in May. Left axes: dilution rates and percentage of total biovolume. Right axes: biomass (mm³ l⁻¹) and soluble reactive phosphorus (µg l⁻¹) in the outflow medium (S). In all six experiments, the initial phosphate level was 1 µM in the inflow medium (S₀). Abbreviations for algal species: ast = *Asterionella formosa*, chl = *Chlorella minutissima*, cry = *Cryptomonas ovata*, mou = *Mougeotia thylespora*, psa = *Pseudanabaena catenata*, sce = *Scenedesmus acutus*. (After Sommer 1986a. © Springer-Verlag)

Asterionella formosa and *Cyclotella meneghiniana*, for two limiting nutrients, silicate and phosphate. Under phosphate limitation, *A. formosa* should win because its K_s for PO_4 (0.04 µM) is lower than that of *C. meneghiniana* (0.25 µM). Under silicate limitation, however, *C. meneghiniana* should dominate because its K_s for silicate (1.4 µM) is lower than that of *A. formosa* (3.9 µM). The Monod growth curves for each diatom species growing separately in continuous culture limited by silicate and phosphate are shown in Figure 21.30. Assume that each continuous culture is operating at a dilution rate of 0.25 day⁻¹. At steady state, D = µ. The dilution rate is represented in each graph by a straight line parallel to the

x-axis (y = 0.25). The intersection of this line with the Monod growth curve gives the minimum amount of PO_4 or SiO_2 (R*) that each diatom needs to grow at the dilution rate D = 0.25 day⁻¹. For a mathematical formula to calculate R*, refer to Question Box 21.4. The values of R* are given in Table 21.4. Consider a graph in which the concentrations of PO_4 are plotted on the y-axis and those of SiO_2 on the x-axis. Such a graph is called a resource plane (Tilman 1982). If $[PO_4]$ < 0.01 µM or $[SiO_2]$ < 1.9 µM, *Asterionella formosa* cannot grow in continuous culture at D = 0.25 day⁻¹ and washes out. These conditions define two lines in the resource plane (x = 1.9 µM SiO_2 and y = 0.01 µM PO_4). In the rectangular area above

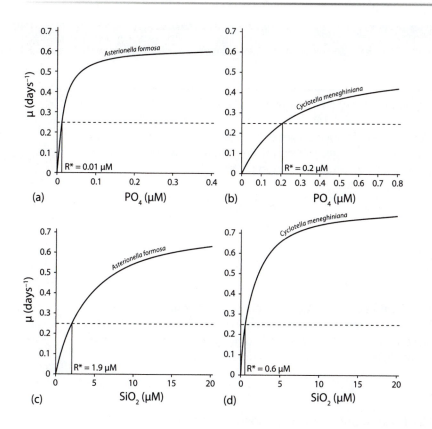

Figure 21.30 Tilman's (1977) mechanistic model of competition. Monod growth curves for *Asterionella formosa* and *Cyclotella meneghiniana* as functions of [PO$_4$] (a and b) and [SiO$_2$] (c and d). (After Lampert and Sommer 1997. © Georg Thieme Verlag)

these lines, *A. formosa* can grow, but below them it washes out (Figure 21.31a). The lines are called zero net growth isoclines, and on these lines, growth of *A. formosa* just balances losses due to washout and dN/dt = 0. Similarly, if [PO$_4$] < 0.2 µM or [SiO$_2$] < 0.6 µM, *C. meneghiniana* cannot grow at D = 0.25 day^{-1}. These values define the zero net growth isoclines for *C. meneghiniana* as x = 0.6 µM SiO$_2$ and y = 0.2 µM PO$_4$ (Figure 21.31b). *Cyclotella meneghiniana* shows positive net growth in the region above these lines and washes out below them.

If we now plot both sets of zero-growth isoclines on the same resource plane, the rectangular areas intersect and define four regions (Figure 21.31c). Both diatoms will wash out in region 1. In region 2, defined by x = 0.6 and x = 1.9 µM SiO$_2$ and above y = 0.2 µM PO$_4$, only *C. meneghiniana* can grow. *Asterionella formosa* would wash out because [SiO$_2$] < 1.9 µM. Region 3 is defined by x > 1.9 µM SiO$_2$ and y between y = 0.01 µM and 0.2 µM PO$_4$; *A. formosa* can grow there, but *C. meneghiniana* cannot because the level of PO$_4$ is too low. Finally, in region 4, where x > 1.9 µM SiO$_2$ and y > 0.2 µM PO$_4$, both diatoms will be able to grow separately.

What happens in region 4 if both diatoms are in continuous culture together? Three outcomes are possible: *C. meneghiniana* displaces *A. formosa*, both diatoms coexist in a stable equilibrium, or *A. formosa* displaces *C. meneghiniana*. What determines the areas within region 4 where these three outcomes may occur? At the intersection of the two rectangles in Figure 21.31c, there is a point of stable coexistence where *A. formosa* is limited by SiO$_2$ and *C. meneghiniana* is limited by PO$_4$. What happens in region 4 above this point depends on the relative amounts

Table 21.4	Values of R* for *A. formosa* and *C. meneghiniana*	
	Nutrient required at D = 0.25 day^{-1} (R*)	
	PO$_4$ (µM)	SiO$_2$ (µM)
A. formosa	0.01	1.9
C. meneghiniana	0.20	0.6

Working with the Mechanistic Model of Competition

Practice problems

1. Tilman (1981) tested his resource competition theory on four species of Lake Michigan (United States) diatoms. The Monod kinetic parameters μ_{max} (day^{-1}), K_s (μM), and cell quotient Q (μmol cell^{-1}), the amount of nutrient necessary to produce one cell of a species, were determined. Q is the inverse of the yield coefficient. The kinetic parameters under phosphate and silicate limitation for two of the diatoms, *Fragilaria crotonensis* and *Synedra filiformis*, are shown in the following table:

Diatom	Parameters under phosphate limitation			Parameters under silicate limitation		
	μ_{max} (day^{-1})	K_s (μM)	Q (μmol cell^{-1})	μ_{max} (day^{-1})	K_s (μM)	Q (μmol cell^{-1})
Fragilaria	0.80	0.011	4.7×10^{-8}	0.62	1.5	9.7×10^{-7}
Synedra	0.65	0.003	1.1×10^{-7}	1.11	19.7	5.8×10^{-5}

The diatoms were all grown in continuous culture at D = 0.25 day^{-1}, as were *A. formosa* and *C. meneghiniana* in the main text. To calculate the minimum concentration of each nutrient that just allows each diatom species to persist in continuous culture at D = 0.25 day^{-1}, we rearrange the terms in the Monod growth equation and substitute in the parameter values from the table above. Thus:

$$\mu = \mu_{max} \left(\frac{S}{K_s + S} \right)$$

At equilibrium, D = μ = 0.25 day^{-1} and S = R*, the nutrient level that allows the diatom to remain in culture with dN/dt = 0. Then substituting in the Monod equation, we get:

$$D = \frac{\mu_{max} R^*}{K_s + R^*}$$

We now solve for R* by multiplying both sides by $(K_s + R^*)$:

$$D (K_s + R^*) = \mu_{max} R^*$$

Then a few algebraic manipulations to isolate R* gives us:

$$DK_s + DR^* = \mu_{max} R^*$$

$$\mu_{max} R^* - DR^* = DK_s$$

$$(\mu_{max} - D) R^* = DK_s$$

$$R^* = \frac{DK_s}{\mu_{max} - D}$$

As a practice problem, graph the four Monod curves for *Fragilaria* and *Synedra*, each limited by phosphate and silicate. Use the above formula for R* to calculate the minimum levels of silicate and phosphate at which each diatom can just maintain a population in continuous culture at D = 0.25 day^{-1}. Then plot the zero-growth isoclines on a resource plane where y = [PO$_4$] and x = [SiO$_2$]. (*Hint*: You will have to make the axis for PO$_4$ extend over a very small range, on the order of 0.05 μM.) Tilman (1981) used the ratios of the nutrient quotients (Q$_p$/Q$_{si}$) as his slopes of the consumption vectors. If you use the ratios of the half-saturation constants, the final figure is not greatly changed. Finally, plot the consumption vectors in your resource plane.

2. Sommer (1986b) studied nitrate and silicate competition among marine diatoms from the frigid waters around Antarctica. He obtained the following values for Monod kinetic constants at 0°C for four species of marine diatoms:

Diatom	μ_{max} (day^{-1})	K_n (µM)	K_{si} (µM)
Corethron criophilum	0.39	0.3	60.1
Nitzschia kerguelensis	0.56	0.8	88.7
Thalassiosira subtilis	0.40	0.9	5.7
Nitzschia cylindrus	0.59	4.2	8.4

Sommer grew these diatoms in continuous culture at $D = \mu = 0.25$ day^{-1}. Use the formula in problem 1 to calculate the values of R* for each diatom and nutrient. Then use this information to plot the zero-growth isoclines in a resource plane where x = µM NO$_3^-$ and y = µM SiO$_2$. You can then use the values of the half-saturation constants to calculate the slopes of the consumption vectors that extend from the intersection points of the zero growth isoclines. (The consumption vectors should be plotted on a separate graph where the axes cover a wider range than that used for the zero-growth isoclines.) Answers are provided at the end of the chapter.

of PO$_4$ and SiO$_2$ delivered to the continuous culture compared to the half-saturation constants of the two diatoms for each nutrient. To make this clearer, refer to Table 21.5, where we list the values for the half-saturation constants for each nutrient and diatom and the ratio of these constants. From the information in Table 21.5 we can then derive the information given in Table 21.6, which defines the nutrient supply ratios that determine whether PO$_4$ or SiO$_2$ is limiting growth of each diatom. If the ratio of supply of phosphate to silicate is less than 0.0103, both diatoms are limited by PO$_4$. *Asterionella formosa* wins in competition by virtue of having the lower value of K_s for PO$_4$. If the ratio of supply is greater than 0.1786, both diatoms are limited by SiO$_2$, and *C. meneghiniana* will

displace *A. formosa* because it has the lower K_s value for silicate and can therefore reduce silicate to a level where its competitor cannot grow. If the supply ratio is greater than 0.0103 but less than 0.1786, *A. formosa* is limited by silicate, and *C. meneghiniana* is limited by phosphate. Note that each diatom is limited by the nutrient that it is least effective at acquiring. Under these conditions each diatom species has more effect on itself than do members of the other species. This is the zone of stable coexistence under steady-state continuous culture conditions.

How do we put this information into our nutrient resource plane to define the zones of competitive displacement and coexistence? We treat the above ratios of nutrient supply as the slopes of consumption

Table 21.5 Values of K_S and ratios of K_S values for *A. formosa* and *C. meneghiniana*

	K_S (µM PO$_4$)	K_S (µM SiO$_2$)	K_S (PO$_4$)/K_S (SiO$_2$)
A. formosa	0.04	3.9	0.01026
C. meneghiniana	0.25	1.4	0.17857

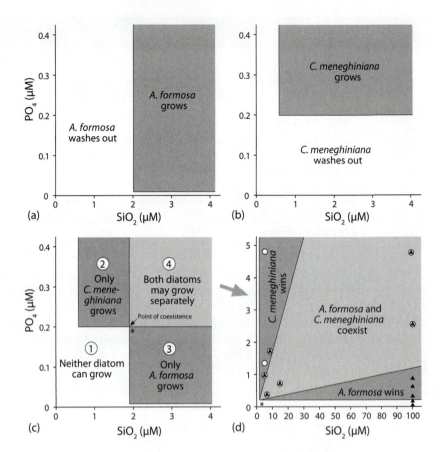

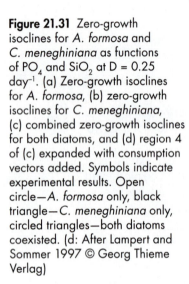

Figure 21.31 Zero-growth isoclines for *A. formosa* and *C. meneghiniana* as functions of PO_4 and SiO_2 at D = 0.25 day^{-1}. (a) Zero-growth isoclines for *A. formosa*, (b) zero-growth isoclines for *C. meneghiniana*, (c) combined zero-growth isoclines for both diatoms, and (d) region 4 of (c) expanded with consumption vectors added. Symbols indicate experimental results. Open circle—*A. formosa* only, black triangle—*C. meneghiniana* only, circled triangles—both diatoms coexisted. (d: After Lampert and Sommer 1997 © Georg Thieme Verlag)

lines or vectors originating at the point where the two rectangles intersect, namely the point x = 1.9 µM SiO_2, y = 0.2 µM PO_4, which is a point of stable coexistence. We wish to construct two straight lines starting at x = 1.9, y = 0.2 with slopes m = 0.0103 for *A. formosa* and m = 0.1786 for *C. meneghiniana*. The following algebraic formula will generate the equations of the straight lines:

$$m = \frac{y - y_1}{x - x_1} \quad (32)$$

Here m is the required slope (the ratio of half-saturation constants) and (x_1, y_1) is the origin (1.9, 0.2). Thus, 0.0103 = (y − 0.2)/(x − 1.9) yields y = 0.0103x + 0.1805 for *A. formosa*, and 0.1786 = (y − 0.2)/(x − 1.9) gives y = 0.1786x − 0.1393 for *C. meneghiniana*. These equations are the consumption vectors that divide region 4 into areas of competitive displacement or coexistence. We have added these lines in Figure 21.31d. Above the upper consumption vector, *C. meneghiniana* displaces *A. formosa*. Below

the lower consumption vector, *A. formosa* outcompetes *C. meneghiniana*. The region between the two consumption vectors is the zone of coexistence where the two diatoms are limited by different nutrients. The results of Tilman's competition experiments have been placed on the figure to show the excellent agreement between prediction and observation. Refer to

Table 21.6 Regions of nutrient limitation defined by the ratios of K_S values

Diatom	Limited by PO_4	Limited by SiO_2
A. formosa	$\dfrac{[PO_4]}{[SiO_2]} < 0.01026$	$0.01026 < \dfrac{[PO_4]}{[SiO_2]}$
C. meneghiniana	$\dfrac{[PO_4]}{[SiO_2]} < 0.17857$	$0.17857 < \dfrac{[PO_4]}{[SiO_2]}$

Question Box 21.4 to try a similar analysis on data for *Fragilaria crotonensis* and *Synedra filiformis*.

Tilman's research on competition between *A. formosa* and *C. meneghiniana* is presented in Tilman (1976, 1977) and Tilman and Kilham (1976), and his mechanistic theory of competition is developed fully in Tilman (1982) and Tilman et al. (1982). Tilman then extended his studies to four species of Lake Michigan diatoms, *A. formosa*, *Fragilaria crotonensis*, *Synedra filiformis*, and *Tabellaria flocculosa*. He found that *A. formosa* and *F. crotonensis* were competitively equal; they coexisted at all ratios of silicate to phosphate because they appear to have similar resource requirements. This result suggests a novel way in which species diversity may increase. Even if two phytoplankton species are morphologically distinct, they can coexist if they are physiologically very similar. It is not known how common this phenomenon might be. Temperature was later found to alter Monod kinetic constants and consequently change the outcome of competitive interactions (Tilman et al. 1981; Mechling and Kilham 1982). In one set of experiments, *A. formosa* displaced *Synedra ulna* below 20°C, but *S. ulna* displaced *A. formosa* above 20°C.

Whereas Tilman, the Kilhams, and their students have primarily focused on freshwater diatoms and silicate and phosphate as limiting nutrients, other researchers have examined different algal groups, including cyanobacteria (Ahlgren 1978; Holm and Armstrong 1981; De Nobel et al. 1997), green algae (Senft et al. 1981; Sommer 1986a; Grover 1989), and chrysophyceans (Lehman 1976; Sandgren 1988). In one marine study, Sommer (1986b) examined Tilman's mechanistic model of competition among four species of marine diatoms from Antarctic waters for limiting supplies of nitrate and silicate (see Question Box 21.4, problem 2). In general, the predictions of the mechanistic model and experimental observations of competition have been in excellent agreement.

Spatial heterogeneity, disturbance, and coexistence

The competitive exclusion principle states that in a homogeneous, well-mixed environment, species competing for the same resources cannot coexist. The mechanistic model predicts that the number of species that can coexist under these conditions at equilibrium cannot be greater than the number of limiting resources unless other mechanisms are involved (Scheffer et al. 2003). The problem with this conclusion is that there are a very limited number of potentially limiting nutrients. The list includes only silicon, nitrogen, phosphorus, carbon, iron, molybdenum, vitamins, and light (Huisman and Weissing 1994). The mechanistic model of competition cannot therefore account for more than about eight species, assuming that all potentially limiting nutrients were limiting at once. Two approaches offer a way around this limitation. One approach recognizes that aquatic systems are not always the homogeneous environments assumed in the competitive exclusion principle. The second approach recognizes that natural aquatic systems are subject to disturbances of various sizes and that these disturbances may prevent phytoplankton populations from coming to equilibrium.

Rather than being homogeneous, well-mixed environments, aquatic systems exhibit various kinds of spatial heterogeneity. Stratification is one type of spatial heterogeneity. Phytoplankton may be adapted to circulate in surface waters above the thermocline, while others may concentrate in the stable waters at the thermocline (see Figure 21.6). Phytoplankton may form blooms in patches of varying horizontal extent and depth (see Figure 21.9). In the oceans also many phytoplankton circulate in the surface waters, while others form a deep chlorophyll maximum (see Figure 21.7). In the oligotrophic waters of the subtropical gyres, water motions called cyclones and mode water eddies form spatially distinct areas of nutrient enrichment and increased phytoplankton diversity and abundance. Coastal ocean regions are more productive and more diverse in species than the oligotrophic open oceans (see Table 21.2). In the littoral zones of lakes, many phytoplankton grow on the sediment surface or attached to rocks and aquatic macrophytes. When these algal species become detached from these surfaces, they join the plankton in the surrounding water and increase the abundance and diversity of phytoplankton. Hansson (1996) found that 32% of the algal taxa in four lakes in northern Wisconsin (United States) were recruited from littoral sediments, and Schweizer (1997) found about 50% of the algal taxa in the epilimnion of Lake Constance were derived from sediments or macrophytes. These detached algae increased the species diversity of the sample stations. Spatial heterogeneity can have a significant effect on species diversity in the phytoplankton.

Disturbances can also have a major effect on species diversity and coexistence. Connell (1978) sought to explain the high diversity of tropical rainforests and coral reefs as due to external disturbances that prevent the community from approaching equilibrium. The highest diversity was associated with intermediate levels of disturbance. According to this intermediate disturbance hypothesis (IDH), low levels of disturbance lead to competitive exclusion and high levels to extinctions. Phytoplankton ecologists examined the role that disturbances have on competition and the number of coexisting species. In these experiments, continuous culture systems often served as controls to compare against the results of various disturbances. Sommer (1985) used natural phytoplankton communities from Lake Constance to compare competition in continuous cultures to competition in cultures in which phosphate or silicate and phosphate over a range of Si:P ratios were added in pulses at 1-week intervals. Pulsed additions of nutrients led to a marked increase in number of persisting species from 1–2 to 9–10 (Table 21.7). Continuous nutrient additions favored diatoms, while pulsed additions promoted green algae (Figure 21.32). In a later study, Gaedeke and Sommer (1986) varied the frequency of pulses from 1 to 14 days. At 1-day intervals, only one species dominated cultures—a result consistent with the competitive exclusion principle. Continuous or frequent pulses favor species that can reduce nutrients to the lowest levels. Species diversity rose when the interval between nutrient additions reached 3 days and reached a maximum at 7 days. Intermediate pulses provide abundant nutrients, allowing species that are not as fast at uptake to store nutrients in excess and persist until the next pulse. At intervals greater than 7 days, diversity declined, a result supporting the IDH. Perturbations had to occur at an interval greater than the average generation time of the algae before they affected species diversity. A

similar study with natural marine phytoplankton yielded maximum diversity at a disturbance interval of 3.5 days (Sommer 1995).

Mixing of the water column increases nutrient levels but also alters competition for light by circulating species into deeper waters, where light levels are reduced. Turbulent mixing is another type of disturbance that favors the species with the lowest critical light intensities for growth. In a field study Flöder and Sommer (1999) tested the IDH on phytoplankton diversity by mixing lake water contained in tubular enclosures 15 m deep within eutrophic Lake Plußee, Germany. Intervals between mixing events ranged from 2 to 12 days, and diversity peaked at a disturbance interval of 6 days. Model simulation studies of mixing and light competition gave a similar result, with peak diversity at a disturbance interval of 7 days (Elliot et al. 2001). Turbulent mixing by air bubblers was used in Lake Nieuwe Meer, a recreational lake in the city of Amsterdam, the Netherlands, to change the dominant phytoplankton populations (Huisman et al. 2004). Lake Nieuwe Meer is a hypereutrophic lake dominated in summer months by a dense surface bloom of the buoyant cyanobacterium *Microcystis aeruginosa*. Given the urban nature of the lake, nutrient reduction was not practical, but the blooms of *Microcystis* posed a threat to human health. Model studies showed that *Microcystis* should not develop a bloom if the mean lake depth exceeds a critical depth and turbulence exceeds a critical level (Figure 21.33). Diatoms and green algae, however, need a certain minimum turbulence to form blooms but are not adversely affected by the level of turbulence that prevents *Microcystis* blooms. When air bubblers mixed the 18 m deep lake, the diatoms and green algae outcompeted *Microcystis* for light and

Table 21.7 Species number in chemostat and pulsed experiments as a function of the Si:P molar ratio*

		Si:P (molar)							
		4:1	10:1	20:1	30:1	40:1	80:1	140:1	≥ 200:1
Chemostat	# of species	1	2	2	3	2	1	–	–
Pulsed P		8	–	6	–	6	–	5	5
Pulsed P + Si		8	8	9	–	10	8	7	6

*Adapted from Sommer 1985.

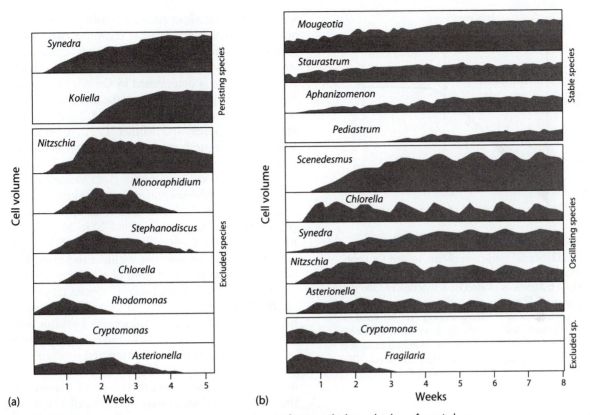

Figure 21.32 (a) Chemostat competition experiment with natural phytoplankton from Lake Constance at Si:P = 20:1. (b) Pulsed P and Si (one week intervals) competition experiment with the same phytoplankton inoculum at Si:P = 20:1. Biomass—$\log_{10}(\mu m^3\ ml^{-1})$—of all species that represent more than 5% of total biomass. Under continuous culture conditions, only two species persisted. Under weekly pulsed P and Si conditions, nine species persisted for over eight weeks. (After Sommer 1985)

dominated the lake. Such experiments demonstrate that turbulence can increase phytoplankton diversity in deep eutrophic and hypereutrophic lakes.

Chaos and coexistence

Perturbation studies have established that intermediate levels of external disturbance can circumvent competitive exclusion by preventing the system of interacting phytoplankton from reaching equilibrium. Tilman's mechanistic model showed that two species can coexist in continuous culture at equilibrium if each is limited by a different nutrient. These same resource competition models generate unexpected results if extended to three species competing for three nutrients. Three-species competition results in sustained oscillations (Huisman and Weissing 1999, 2002). The oscillations arise from the process of competition, not from any external disturbance. Moreover, the oscillations permit additional species

to coexist on three resources. Up to six species can coexist on three resources with oscillations of small amplitude. Nine species may be sustained with large amplitude oscillations. Such sustained oscillations are an example of chaotic behavior. Huisman and Weissing propose that once a phytoplankton community is sufficiently complex to generate non-equilibrium dynamics (chaotic oscillations), the number of coexisting species can exceed the number of limiting resources, even in well-mixed environments.

At present there are a number of potential solutions to the paradox of the plankton. Species coexistence and high species diversity may result from spatial heterogeneity and non-equilibrium dynamics, either through external disturbances such as mixing or through chaotic oscillations arising from the competition process itself. How these different forces apply may depend on the specific aquatic system. Siegel (1998) proposed a model of resource competition by

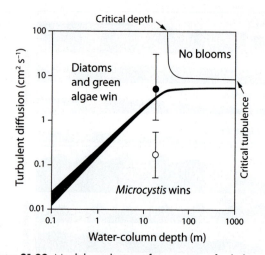

Figure 21.33 Model prediction of competition for light between *Microcystis* and diatoms and green algae. The competitive outcome is plotted as a function of lake depth and turbulence. The open circle shows the condition of Lake Nieuwe Meer with natural mixing and the solid circle the condition with artificial mixing. Error bars indicate the standard deviation of turbulence. Hatched area is a narrow zone of coexistence. (After Huisman et al. 2004, by permission of the Ecological Society of America)

which the intensity of competition would be greatly reduced in oligotrophic systems, where cells are small and populations of phytoplankton are low. External disturbances and grazing would be more significant in determining species diversity. Conversely, in more nutrient-rich systems, competition and chaotic oscillations may be more likely mechanisms of species coexistence.

Grazing

The most dramatic phytoplankton loss process is grazing. Grazing refers to herbivory interactions involving various herbivores that consume archaea, bacteria, and algae. Some planktonic algae are phagotrophic and can ingest archaea and bacteria while others consume eukaryotic algae (refer to Chapter 3). The types of herbivores present depend on the aquatic system. In freshwater systems herbivores tend to be protozoa, rotifers, and crustaceans. Protozoa cover a wide range of sizes, from 2 µm long nanoflagellates to 200 µm long microzooplankton. Protozoa may consume archaea, bacteria, algae, and other protozoa and may be consumed by rotifers and crustaceans. Rotifers are mainly filter feeders that feed on smaller organisms (< 12 µm), but a few specialists feed by raptorial

capture of large algae such as dinoflagellates. Freshwater planktonic crustaceans comprise two distinct groups: the cladocera and the copepods. Cladocerans are nonselective filter feeders. The lower limit for filterable cell size is determined by the size of the mesh formed by the plumose setae on the thoracic legs. This net-like mesh ranges from 0.16 to 4.2 µm in different species (Geller and Müller 1981). Cladocerans with the finest mesh sizes (*Chydorus sphaericus* and *Daphnia magna*) can remove large (1 µm) bacteria; those with the coarsest filters (*Sida cristallina* and *Holopedium gibberum*) miss phytoplankton cells less than 4 µm. Most cladocerans have an upper limit for phytoplankton cell size of around 20 to 30 µm (Sommer et al. 2001). Copepods generally feed by raptorial capture and are selective within a size range from about 20 µm to an upper limit of 50 (Burns 1968) to > 100 µm, depending on copepod species (Sommer et al. 2001). In freshwaters, cladocerans and copepods have contrasting effects; cladocerans consume small phytoplankton cells, flagellates and ciliates, and copepods feed on large phytoplankton and medium-sized (20–40 µm) ciliates (Yoshida et al. 2001; Zollner et al. 2003). Phytoplankton cell, colony, or filament size therefore determines in part the susceptibility to particular grazers.

In the oceans, the herbivores are different from those in freshwaters. Few rotifers occur in the oceans, and only three genera of cladocerans, which are restricted to coastal waters, are present. Copepods are the dominant grazers on marine phytoplankton, but like their freshwater counterparts, they prefer large phytoplankton. Protozoa and an exclusively marine group of organisms called tunicates feed on the smaller size classes of phytoplankton. Tunicates are metazoan zooplankton that form large gelatinous filter-feeding structures capable of gathering particles from picoplankton to the largest phytoplankton chains and colonies. They are present in all marine pelagic systems and appear to be adapted to survive in the most extreme low-resource environments (Sommer and Stibor 2002). Jellyfish (cnidaria) consume tunicates, forming what has been termed a jelly food chain (Sommer et al. 2002).

Freshwater and marine organisms can be grouped into two types of food webs, based on organism size. Organisms less than 20 µm in greatest linear dimension belong in the **microbial food web**. Organisms larger than 20 µm form the **classic food web**, which is often depicted as a simple linear food chain. The microbial food web consists of a variety of microorganisms. The picoplankton (0.2–2.0 μm) include archaea, bacteria, and cyanobacteria. At the boundary between picoplankton and nanoplankton lie some of the smallest free-living eukaryotic cells in existence: 2 μm heterotrophic nanoflagellates (HNFs) and autotrophic algae < 4 μm in diameter. The nanoplankton (2–20 μm) consists of nanophytoplankton, HNFs, mixotrophic nanoflagellates, and nanociliates. In addition to being autotrophic, mixotrophic flagellates can acquire carbon by ingesting small particles or by taking up dissolved organic carbon (DOC). Microzooplankton (20–200 μm) feed on the microbial food web and are in turn eaten by the larger mesozooplankton (0.2–2 mm).

Food webs in the oceans

Up to the 1970s, the classic food chain, consisting of net phytoplankton (> 20 μm), zooplankton, and planktivorous fish, was the dominant paradigm for the oceans. For the past three decades, however, the significance of the microbial loop (now referred to as the microbial food web) has been recognized. The food webs of the oceans can be grouped into three types on the basis of phytoplankton assemblages associated with particular nutrient conditions

(Sommer et al. 2002). The first type has high levels of silicon and nitrogen and occurs seasonally in temperate and boreal seas and coastal upwelling zones with cold waters (Figure 21.34a). These conditions favor a phytoplankton assemblage consisting of large diatoms and nanoflagellates that provide abundant food to copepods and fish. Picophytoplankton represent less than 10% of autotrophic biomass in these cold, nutrient-rich waters (Agawin et al. 2000), but HNFs are the main consumers of bacteria (Vargas et al. 2007). This first type is most similar to the old paradigm of the classic food chain.

The second type of food web is linked to low silicon and nitrogen levels that prevail over vast areas of the subtropical ocean gyres and summer stratified temperate seas (Figure 21.34b). In these oligotrophic waters, algae < 5 μm in diameter dominate the phytoplankton. Low nutrient levels favor small cells with high uptake capabilities and exclude large cells (Sommer 2000). More than 90% of the phytoplankton biomass consists of picoplankton such as the cyanobacteria *Synechococcus* (1–1.5 μm) and *Prochlorococcus* (0.8 μm) and small eukaryotic algae (Sommer and Stibor 2002). Microzooplankton consume about 67% of daily growth of such phytoplankton, which are too

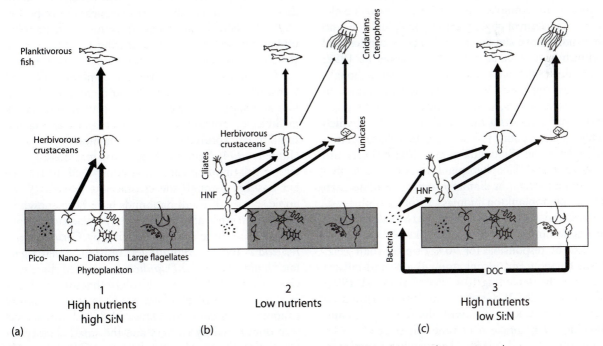

Figure 21.34 Marine food webs. (a) High Si and N conditions favor large diatoms. (b) Low Si and N conditions favor small (< 5 μm) algae. (c) High N and P in coastal areas favor large algae such as dinoflagellates and *Phaeocystis*, which may form toxic blooms. (After Sommer et al. 2002, by permission of Kluwer Academic Publishers)

small for copepods to eat directly (Calbet and Landry 2004). The nanoflagellates may be consumed by ciliates, and the nanoflagellates and ciliates together provide food for copepods. The resulting food chain from phytoplankton to copepods is longer than in the classic food chain and less productive (Berglund et al. 2007).

The third type of marine food web is associated with high nitrogen and phosphorus levels with low silicon, conditions that occur in coastal waters made eutrophic by runoff from terrestrial sources (Figure 21.34c). Small phytoplankton cells (< 5 µm) are subject to heavy grazing pressure, while large algae compete for nutrients at high nutrient levels and are little affected by grazing. Large unicellular dinoflagellates, colonial flagellates such as *Phaeocystis*, or cyanobacteria may form toxic blooms. These large algae may go largely ungrazed and release DOC after death. The DOC fuels bacterial growth that supports HNFs. As in the second type of food web, the copepods obtain food from protists, but here the protists feed mainly as heterotrophs.

Food webs in freshwaters

Although the microbial food web has been considered an essential part of marine pelagic food webs for three decades, many limnologists continue to focus on the classic food chain of phytoplankton, zooplankton, and fish. Studies have shown the presence of a microbial food web in freshwater systems (Sanders et al. 1992; Pace et al. 1998), but the microbial food web is thought to be most important in oligotrophic freshwaters. In unproductive waters, most organic carbon is dissolved (DOC) and supports a microbial food web in which bacteria may account for as much as 90% of planktonic respiration (Biddanda et al. 2001; Roberts and Howarth 2006). Picophytoplankton, defined as phytoplankton < 3 µm in diameter, represent more than 50% of phytoplankton biomass in oligotrophic waters (Bell and Kalff 2001; Callieri and Stockner 2002). Protists dominate the microzooplankton. HNFs were found to be responsible for 90% of protozoan grazing on bacteria and picophytoplankton, and ciliates consumed the remaining 10% (Pernthaler et al. 1996; Callieri et al. 2002). Grazing rates are measured by treating picoplankton cells with fluorescent dyes and following their uptake over time (Sherr et al. 1987; Epstein and Rossel 1995). Mixotrophic nanoflagellates (MNFs) appear to become most abundant in oligotrophic waters when autotrophic algae are nutrient limited. MNFs outcompete HNFs because they are

photosynthetic, and they can acquire nutrients to support photosynthesis by ingesting bacteria (Bergström et al. 2003). As freshwaters increase in nutrients and trophic state, the abundance and biomass of picophytoplankton increases, but their importance relative to other phytoplankton declines as large algae become more abundant. The relative importance of the microbial food web declines with increasing trophic state (Hambright et al. 2007).

In relatively more productive freshwaters, large crustacean zooplankton, particularly *Daphnia*, dominate the planktonic food web. The domination of *Daphnia* arises from two main characteristics: a relatively high level of grazing efficiency across a wide range of particle sizes and the ability to feed on or interfere with microzooplankton that might potentially compete with it. *Daphnia* consume many protists and interfere with rotifers, which consequently decrease in numbers (Gilbert 1989). The potential of *Daphnia* to dominate the food web is dramatically indicated by the existence of a clear-water phase in many mesotrophic and eutrophic lakes in late spring or early summer. In Lake Schöhsee in Germany (Figure 21.35), as the crustaceans *Daphnia* and *Eudiaptomus* increase in numbers, the algae decline, and water clarity increases, as measured by the depth at which a standard disk called a Secchi disk just disappears (Lampert et al. 1986). For a clear-water phase to occur, the phytoplankton loss rate due to grazing (γ) must exceed the phytoplankton reproduction rate (r). *Daphnia* has a strong impact on the structure of planktonic food webs. In Figure 21.36a, *Daphnia* dominates the system and suppresses the protozoa, leaving only large, inedible, or grazing-resistant algae and bacteria too small to be captured by the filtering mesh of *Daphnia* (Jürgens 1994). Some phytoplankton may possess protective gelatinous sheaths that allow them to survive gut passage through the crustaceans (Porter 1977). Some species of small daphniids have been shown to be less effective at depressing phytoplankton populations than large species. Grazing by *Ceriodaphnia reticulata* and *D. ambiqua* left resistant green algae, but *D. mendotae* and *D. pulicaria* removed these species (Tessier et al. 2001). The algae that are abundant in the presence of *Daphnia* are those that *Daphnia* cannot eat. In contrast, Figure 21.36b shows a food web dominated by rotifers and the small cladoceran *Chydorus*. The planktonic algae are small and mostly edible. The protozoa are diverse and graze on bacteria and picoplanktonic algae, leaving grazing-resistant (filamentous and colonial) forms (Jürgens and Güde

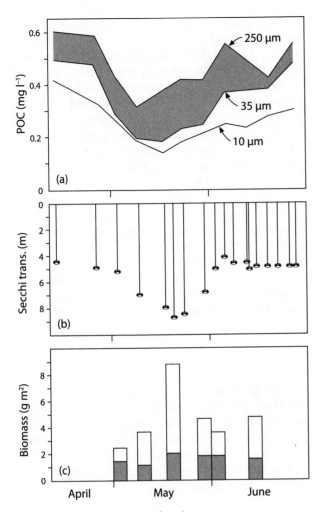

Figure 21.35 Development of a clear-water phase in Lake Schöhsee, Germany, in 1983. (a) Particulate organic carbon (POC) separated by size (< 10 μm, < 35 μm, < 250 μm). (b) Secchi disk transparency (m). (c) Biomass (g m⁻²) of *Daphnia* species (open bar) and *Eudiaptomus* species (shaded bars). (After Lampert et al. 1986)

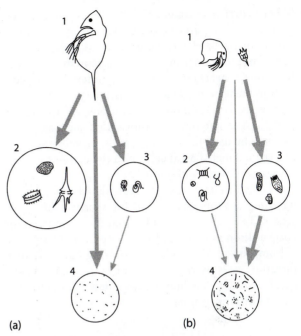

Figure 21.36 Impact of different zooplankton assemblages on the microbial food web. (a) Domination by *Daphnia* leads to grazing-resistant algae, small flagellates, and small bacteria. (b) Domination by *Chydorus* and rotifers produces small edible algae, diverse protozoa, and grazing-resistant bacteria.

1994). Such a food web could arise if planktivorous fish removed most of the large *Daphnia*. In Lake Kinneret, Hambright et al. (2007) found just such a system where intense planktivorous fish predation reduced the crustacean zooplankton to small species such as the cladocerans *Ceriodaphnia*, *Bosmina*, and *Chydorus* and copepods such as *Mesocyclops*. A microzooplankton assemblage that included flagellates, ciliates, and rotifers dominated grazing on bacteria and phytoplankton and nitrogen and phosphorus recycling. Thus, the type of grazing pressure affects the composition of the planktonic food web.

Measuring algal grazing in crustaceans

Grazing in crustacean zooplankton can be measured by two methods: direct cell counts and uptake of radioactive labeled cells. The feeding rate or ingestion rate is defined as the number or mass of cells ingested per individual grazer in a unit of time. Thus:

$$I = C \text{ ind}^{-1} t^{-1} \quad (33)$$

C is expressed in cells or some measure of biomass such as biovolume (μm³) or mg C. The filtering rate, F, of a zooplankton is the volume of water filtered by that grazer to ingest C cells in time t, assuming 100% efficiency in filtration and retention. F has units of ml ind⁻¹ t⁻¹. If cell counts are used, then Gauld's (1951) equation is used:

$$F = \frac{V (\ln C_o - \ln C_t)}{N \times t} \quad (34)$$

The equation assumes an exponential decline in cells or biomass from C_o to C_t as a result of ingestion by

N herbivores over time t. V is the volume of the container in ml. C_o and C_t are measured by direct cell counts or by an electronic particle counter. N is assumed not to change during the measurement period. Because t is necessarily fairly long (a few hours), correction must be made for any increase in number of algal cells. For additional precautions, refer to Peters (1984). Alternatively, a radioisotope such as ^{32}P or ^{14}C may be used to label algal cells. Grazing animals ingest the labeled algal cells, and the activity of each is measured by liquid scintillation.

Filtration rates can be calculated for individual zooplankton species and for separate life cycle stages. Table 21.8 gives individual filtration rates for zooplankton in Blelham Tarn in the United Kingdom (Thompson et al. 1982). Multiplying each of these filtration rates by the number of individuals of each species or stage in a liter, we get a new parameter, the community-grazing rate:

$$G = \Sigma\, F_i\, N_i \quad (35)$$

G represents the losses due to grazing by the entire zooplankton community on the algae used as a tracer and all other phytoplankton with the same relative edibility as the tracer alga. Here it is the same as the γ term in equation (25). Thompson et al. (1982) measured the community-grazing rate on *Chlorella* as tracer alga for all significant herbivores in Blelham Tarn enclosures and found that G varied from 0.2% to 200% day^{-1}, meaning that on some occasions the entire volume of water in an enclosure was filtered twice in one day. Such a rate exceeds the growth rate of many planktonic algae. G may also be measured by using radioactively labeled algal cells as tracers (Haney 1973). Values of G may exceed 100% in lakes for several months (Sterner 1989).

Because of the labor involved, relatively few researchers have followed the procedures of Haney (1973) or Thompson et al. (1982). Furthermore, neither procedure takes into account the relative edibility or selectivity of different phytoplankton species. Assuming that the tracer alga is highly edible and defenseless, its selectivity can be defined as w = 1.0. Relative to it, all other algae would have a selectivity w of 1.0 or less. Figure 21.37 gives the relative selectivity coefficients for *Daphnia magna* feeding on a variety of phytoplankton (Sommer 1988). In this case, the loss rate due to grazing by *D. magna* on each of these algae is:

$$\gamma_i = G\, w_i \quad (36)$$

G is the grazing rate of *D. magna* on a tracer alga, and the w_i are the selectivity coefficients of each of the other algae. To derive a community-grazing rate for all herbivores on all phytoplankton, one would need a similar table for each zooplankton species as well as for many life-cycle stages. For obvious reasons, this has not been done.

Table 21.8 Ranges and means of individual filtration rates for each category of zooplankton in enclosure A in Blelham Tarn, Cumbria, U.K. in 1978*

Species and Stage	Filtration Rates: range (mean) (ml ind^{-1} day^{-1})
Daphnia hyalina V	42.1–62.6 (53.0)
Daphnia hyalina IV	14.0–60.0 (28.4)
Daphnia hyalina III	9.3–29.3 (19.8)
Daphnia hyalina II	5.7–19.3 (11.8)
Daphnia hyalina I	4.0–7.6 (6.8)
Diaptomus gracilis, adult + cop 5	0.5–10.2 (4.4)
Diaptomus gracilis, cops 1–4	0.5–6.7 (2.7)
Chydorus sphaericus	0.6–2.6 (1.0)

*From Thompson et al. 1982.

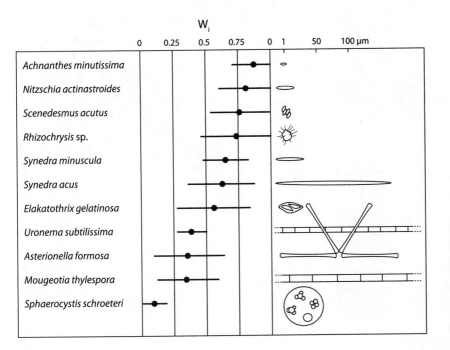

Figure 21.37 Selectivity coefficients (w_i) for various phytoplankton fed on by *Daphnia magna*. Note that large species and those with gelatinous sheaths are not grazed upon well. (After Lampert and Sommer 1997. © Georg Thieme Verlag)

Trophic cascades and biomanipulations

As grazing studies in freshwater systems accumulated, it was recognized that zooplankton could have significant impact on phytoplankton communities (Porter 1977; Bergquist et al. 1985; Bergquist and Carpenter 1986; Jürgens 1994). At the same time, evidence indicated that other trophic levels in freshwater systems could have major impact on the levels below them (Carpenter 1989). Brooks and Dodson (1965) showed that planktivorous fish determine the species composition and size structure of crustacean zooplankton. Tonn and Magnuson (1982) found that piscivorous fish shape the assemblage of planktivorous fish. These studies led to the idea of a **trophic cascade** in which effects cascade from top to bottom down a chain of linked trophic levels (Carpenter et al. 1985). The top predators—piscivorous fish—are seen as keystone species (Paine 1969) with effects extending down many trophic levels below their own position in the food web. The trophic cascade and keystone species concepts led to the hypothesis that, in freshwater systems, control proceeds from the top of the food web downward. Such concepts would generally not apply to the marine pelagic, where trophic cascades are not commonly observed (Micheli 1999). However, Shiomoto et al. (1997) reported a possible trophic cascade from salmon to copepods to phytoplankton in the subarctic North Pacific. The general absence of trophic cascades in the marine pelagic may arise from the absence of *Daphnia* from the food web.

In freshwaters trophic concepts suggested a way to manage cultural eutrophication, which results from human introduction of phosphorus and nitrogen to freshwater systems. Cultural eutrophication is one of the most serious threats to freshwater resources (Smith et al. 2006a; Schindler 2006). Blooms of freshwater algae are a conspicuous indicator of this eutrophication, and they reduce the aesthetic and recreational value of freshwater systems. The trophic cascade and keystone species concepts offered the possibility that a biomanipulation of the upper trophic levels might reduce the phytoplankton community to achieve a clear-water lake with enhanced aesthetic and recreational value.

Potential biomanipulations are diagrammed in Figure 21.38. In a lake from which the fish have been removed, the dominant predators are invertebrates such as *Chaoborus*, which prefers small zooplankton prey. Consequently, large *Daphnia* and copepods dominate and suppress the phytoplankton (Figure 21.38a). Alternatively, large stocks of piscivorous fish could be added to a lake to deplete planktivorous fish. A mixed assemblage of various sizes of zooplankton should result with strong grazing

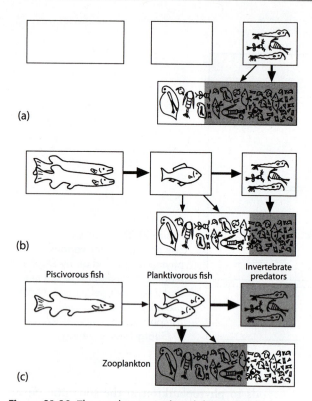

(a)

(b)

Piscivorous fish   Planktivorous fish   Invertebrate predators

Zooplankton

(c)

Figure 21.38 The trophic cascade in lakes. (a) With only invertebrate predators, large zooplankton result. (b) Large populations of piscivorous fish suppress planktivorous fish, leading to large and medium-sized zooplankton. (c) Having few or no piscivorous fish leads to many planktivorous fish and small zooplankton. (After Lampert 1987. © Springer-Verlag)

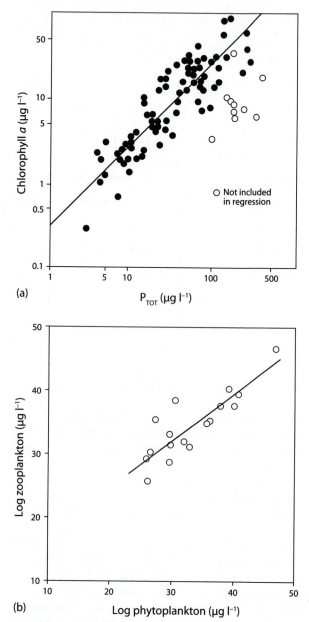

(a)

(b)

Figure 21.39 Bottom-up control data from many lakes of various trophic status. (a) Phytoplankton biomass (as chlorophyll *a*) as a function of phosphorus. (b) Log zooplankton biomass as a function of log phytoplankton biomass. (a: Based on Vollenweider 1982. © Springer-Verlag; b: After McCauley and Kalff 1981)

pressure on the phytoplankton (Figure 21.38b). As Figure 21.38c shows, having few or no piscivorous fish results in numerous planktivorous fish and small zooplankton. Microzooplankton dominate grazing, leading to a food web with many small, edible algae and protozoa.

The keystone species concept is often referred to as the top-down hypothesis of aquatic system control. In its simplest form, the top-down hypothesis predicts that adjacent trophic levels are negatively correlated. In other words, more piscivorous fish means fewer planktivorous fish, which leads to more and larger zooplankton and fewer algae. Conversely, more planktivorous fish would mean fewer zooplankton and more algae. An older, bottom-up hypothesis says that all trophic levels are positively correlated. Thus, more nutrients means more phytoplankton (Figure 21.39a), and more phytoplankton means more zooplankton (Figure 21.39b).

Comparison across many lakes of different trophic status generally supports a bottom-up hypothesis. Biomanipulations within lakes and enclosures, however, often support the top-down hypothesis or a mixed model.

Tests of the trophic cascade hypothesis are difficult to carry out because many trophic levels must be monitored in manipulated and control systems over a significant period of time. Biomanipulations are performed on whole lakes or in enclosures, the latter being amenable to replication. In enclosure experiments, fish may be excluded or added at various densities in combination with nutrient treatments. Fish generally reduce zooplankton (top-down), but nutrients increase phytoplankton (bottom-up) (Vanni 1987). Brett and Goldman (1997) collected data from 11 enclosure studies that employed both nutrient additions and planktivorous fish manipulations. A meta-analysis of these data found that both top-down and bottom-up controls were significant. Zooplankton were under strong planktivore control (top-down) but were only weakly stimulated by nutrients. Planktonic algae were under strong nutrient control (bottom-up) and were moderately controlled by fish through their effects on zooplankton. Specifically, phytoplankton biomass increased, on average, 179% in nutrient treatments but 77% in fish treatments where zooplankton were suppressed. The meta-analysis thus indicated that bottom-up control was stronger than top-down control.

Results from whole-lake studies have been more variable than results from enclosures. Whole-lake studies may involve natural experiments, where a mass mortality of fish or artificial manipulation of fish stocks has occurred. In at least one class of lakes—small, shallow, eutrophic lakes—biomanipulations have been generally successful (Sondergaard et al. 1990; Meijer et al. 1990; van Donk et al. 1990a). Initially, cyanobacterial blooms dominated these lakes in summer. Following removal of planktivorous and benthivorous fish, the zooplankton shifted from rotifers to larger *Daphnia*, and a clear-water phase occurred in spring. Submerged macrophytes reestablished themselves due to improved light conditions at the lake bottoms. The lakes remained clear because macrophytes took up the nutrients. However, when the same manipulations were tried on larger (> 100 ha) shallow eutrophic lakes, results were variable. Lake Breukeleveen, the Netherlands (180 ha), showed no improvement due to wind-induced turbidity that prevented macrophytes from developing (van Donk et al. 1990b), but Lake Christina (1619 ha) in central Minnesota improved dramatically because submerged macrophytes increased (Hanson and Butler 1990). Conditions for successful biomanipulation include removal of at least 75% of fish in one to

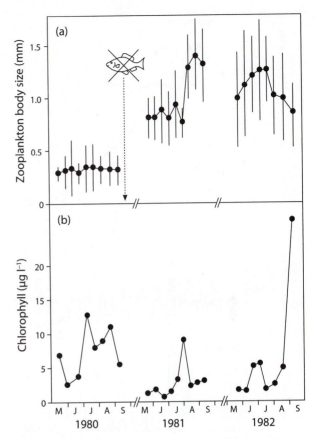

Figure 21.40 Biomanipulation of Round Lake, Minnesota. The entire fish population was eliminated with rotenone in 1980. (a) The average body size of zooplankton increased after the elimination of planktivorous fish. (b) Phytoplankton (as chlorophyll) were reduced in 1981, but a massive bloom of the cyanobacterium *Aphanizomenon flos-aquae* occurred in 1982. (Data from Shapiro and Wright 1984, figure after Lampert and Sommer 1997. © Georg Thieme Verlag)

three years, macrophyte coverage of at least 25% of lake surface area, and reduction of external nutrient inputs prior to the biomanipulation (Hansson et al. 1998; Meijer et al. 1999).

Biomanipulations in other types of lakes have been less predictable. For example, Shapiro and Wright (1984) poisoned the entire fish stock of Round Lake, Minnesota (United States), with rotenone (Figure 21.40). Before this manipulation, planktivorous fish and small zooplankton dominated the lake, and after fish removal, zooplankton size increased and phytoplankton (as chlorophyll *a*) declined. But two years after the manipulation, the grazing-resistant cyanobacterium *Aphanizomenon flos-aquae* bloomed, and chlorophyll levels were higher than previously. In

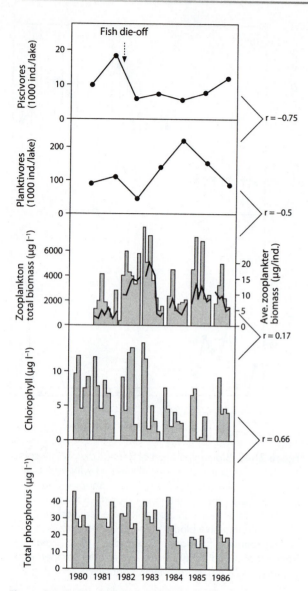

Figure 21.41 Seven years of food web data from Lake St. George, Canada. In the winter of 1981–1982, a winter kill reduced all the fish populations. Zooplankton biomass and body size (center) increased due to reduced planktivore pressure. The planktivores recovered quickly and reduced the zooplankton in 1983. Phytoplankton and phosphorus show little change. Correlation coefficients between adjacent trophic levels are shown on the right. (Data from McQueen et al. 1989, figure after Lampert and Sommer 1997. © Georg Thieme Verlag)

(Figure 21.41). Zooplankton quickly reached maximum body size and biomass, but they had no impact on the phytoplankton. Correlation coefficients showed that piscivorous fish had a strong negative impact on planktivorous fish, which, in turn, had a negative impact on zooplankton. Nutrients had a strong positive impact on phytoplankton. Both top-down and bottom-up control occurred, with the effects diminishing as they passed up or down the food web.

The concept of top-down and bottom-up effects has become increasingly recognized in the literature. In a seven-year study of the interaction of biomanipulations and nutrient enrichment in Peter, Paul, and East and West Long lakes in northern Wisconsin, Carpenter et al. (2001) concluded that a trophic cascade remained in effect at high levels of nutrient enrichment in systems with both three and four trophic levels. Planktivorous fish dominated Peter Lake and reduced crustacean zooplankton mean size. Phytoplankton increased dramatically in response to nutrient enrichment. Piscivorous fish dominated West Long Lake and reduced planktivorous fish stocks. Large-bodied *Daphnia* reduced phytoplankton levels compared to Peter Lake. Grazing, however, was not able to control the cyanobacterial blooms that developed in West Long Lake after nutrient enrichment (Cottingham et al. 1998). Many of the cyanobacteria were colonial and grazing resistant. Carpenter et al. (1996) concluded that potential effects of phosphorus input on phytoplankton were stronger than the potential for controlling phytoplankton through food-web manipulation and recommended that further research define the conditions under which cascades are important and biomanipulations are likely to succeed (Carpenter et al. 2001). If nutrient inputs have a stronger effect on phytoplankton blooms than does food-web manipulation, then effective phytoplankton management will require controlling nutrient inputs through erosion control, wetland preservation or restoration, and effluent treatment.

In summary, both top-down and bottom-up controls are important and interact with each other. Two hypotheses have been proposed to address the interaction between nutrients and food-web effects (Carpenter and Kitchell 1993). The "mesotrophic maximum hypothesis" states that food-web effects are greatest in mesotrophic lakes (Elser and Goldman 1991), although significant food-web effects can occur in oligotrophic lakes (Henrikson et al. 1980).

a seven-year study of Lake St. George (Canada), McQueen et al. (1989) found evidence of both top-down and bottom-up controls. In the winter of 1981–1982, low oxygen under the ice caused a winter fish kill

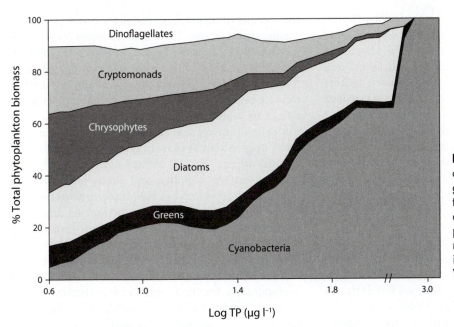

Figure 21.42 Percentage contribution of taxonomic groups of phytoplankton to total summer biomass as a function of average total phosphorus of lakes. As log TP rises, cyanobacteria become increasingly dominant. (After Watson et al. 1997)

The "nutrient attenuation hypothesis" states that as nutrients increase, phytoplankton escape grazer control (McQueen 1990). Watson et al. (1997) examined patterns of phytoplankton taxonomic composition across temperate lakes of varying trophic status (Figure 21.42). In summer, the taxonomic composition of the phytoplankton changes drastically with total phosphorus (TP) concentration. Lakes with low TP (< 5 µg TP l⁻¹) contain very low algal biomass and species from all six major taxonomic groups. Dinoflagellates and chrysophytes may be dominant in acidic lakes (Graham et al. 2004). In mesotrophic lakes with TP ranging from 10 to 30 µg l⁻¹, the taxonomic groups are about evenly represented. As TP increases above about 60 µg TP l⁻¹, fewer taxonomic groups occur until eutrophic and hypereutrophic systems are dominated by cyanobacteria. *Daphnia* cannot survive in dense blooms of cyanobacteria (Gliwicz 1990). Therefore, once lakes are in the upper range of phosphorus enrichment, grazing may have little effect. Recent work suggests, however, that *Daphnia* may be able to reduce the frequency of cyanobacterial blooms by recycling nitrogen (ammonium) and reducing N₂ fixation by cyanobacteria (MacKay and Elser 1998; Chan et al. 2004). Biomanipulations in nutrient-rich lakes may even cause lake quality to deteriorate if they result in the transfer of nutrients from edible algae to inedible algae, as may have happened in the Little Round Lake manipulation and in West Long Lake.

21.6 Phytoplankton and Global Climate Change

In this closing section of the chapter, we discuss changes in phytoplankton populations and interactions that have been observed or are anticipated to arise in response to climate change. We focus on the effects of global warming on the temperature of oceanic surface waters and freshwater lakes. The largest temperature changes have been recorded at high latitudes.

Global warming effects on freshwater systems across North America are summarized in Schindler (1997) and Magnuson et al. (1997). A survey of long-term climatic data from sediments in Arctic lakes across northern Canada and Europe revealed several thousand years of stable conditions, ending in a major change in phytoplankton communities in the last 150 years (Smol et al. 2005). The shift is from benthic diatoms such as *Fragilaria* to planktonic forms such as *Cyclotella*. *Fragilaria* are associated with cold climates and extensive ice cover, with little open water even in summer. The rise of *Cyclotella* is related to reduced ice cover resulting from warmer air temperatures and thermal stratification in open waters. At temperate latitudes global warming is predicted to bring about earlier onset of spring phytoplankton blooms in lakes (Peeters et al. 2007) and earlier occurrence of the clear-water phase resulting from an earlier development of *Daphnia* populations (Scheffer et al. 2001). Meerhoff et al. (2007) suggested that shallow temperate lakes may assume the characteristics of shallow subtropical lakes. Such lakes have abundant submerged and floating aquatic plants with abundant, diverse, but small fish that reduce both the zooplankton and phytoplankton compared to temperate lakes. Small omnivorous fish might replace the trophic cascade of temperate lakes.

Accumulating data suggest that more ominous consequences are arising in the oceans due to global warming. Oceanographers are concerned that global warming, which is predicted to increase upper-ocean stratification, may result in a decrease in phytoplankton abundance and in the transfer of carbon to the deep oceans (the biological pump) (Arrigo et al. 1999). Since 1997 a satellite has carried the Sea-viewing Wide Field-of-view Sensor (SeaWiFS) that is capable of determining global depth-integrated chlorophyll biomass, from which phytoplankton standing stocks and net primary productivity (NPP) can be estimated. The permanently stratified regions of the oceans (the great gyres) have average surface temperatures above 15°C and represent 74% of the ice-free ocean. In these regions SeaWiFS has detected a decline in NPP with rising temperature (Behrenfeld et al. 2006). The oceans are fixing less carbon dioxide from the atmosphere as global temperatures rise. This condition occurs because the phytoplankton depend on vertical mixing to bring up nutrients from deep ocean waters to supply the demands of photosynthesis. Rising surface temperatures reduce mixing, nutrient supplies, and phytoplankton populations and productivity. Fewer phytoplankton cells means fewer cells to sink in the oceans and carry down fixed carbon dioxide into the sediments, where it is sequestered from the atmosphere. As global temperatures rise, the oceans are therefore becoming less capable of removing the excess carbon dioxide that is responsible for the temperature increases. We are living in a rapidly changing world, and rising global temperatures and sea levels and rapidly changing ecosystems—including those dominated by phytoplankton—do not bode well for the future.

Box 21.1 Working with the Exponential and Logistic Equations

1. Culture 1: $Y = 0.04692(X) + 9.2469$ and $r = 0.04692$ hr^{-1}. Culture 2: $Y = 0.04717(X) + 9.5380$ and $r = 0.04717$ hr^{-1}. Culture 3: $Y = 0.05285(X) + 8.9762$ and $r = 0.05285$ hr^{-1}. The average net growth rate of *Chlamydomonas* is 0.04898 hr^{-1} or 1.175 day^{-1}.

2. In the dark, $Y = 0.02797(X) + 6.7091$ so $r = 0.02797$ hr^{-1} or 0.67 day^{-1}. In the light, $Y = 0.03482(X) + 7.8708$ so $r = 0.03482$ hr^{-1} or 0.84 day^{-1}. Photosynthesis increased the net growth rate by 0.17 day^{-1}. The corresponding doubling times are 1.03 days in the dark and 0.83 days in the light. Thus, photosynthesis decreased the doubling time by 0.2 days.

3. The net growth rate can be estimated from a linear regression of the first 7 data points (or the first 9). Refer to the figure (at right) for this decision. The results are:

For n = 7: $Y = 0.5612(X) + 10.0433$ and $r = 0.56$ day^{-1}

For n = 9: $Y = 0.5013(X) + 10.2387$ and $r = 0.50$ day^{-1}

The last 4 or 5 data points can be used to estimate the carrying capacity, K:

For n = 4, K = 8,949,870 and for n = 5, K = 8,821,480.

Which is used is a matter of choice, but a useful rule would be to accept the highest estimates consistent with a reasonable number of data points. An estimate based on two points is not as reliable as one based on three or more.

Box 21.2 Working with the Michaelis-Menten, Droop, and Monod Models of Nutrient Uptake and Growth

1. First take the reciprocals of the values of SiO_2 and ρ (see table to right). Plot these values as shown in the accompanying figure. The regression is $Y = 0.4676(X) + 0.06429$ and the correlation coefficient is 0.993. Now $1/\rho_{max} = 0.06429$ and so $\rho_{max} = 15.5 \times 10^{-9}$ μM $cell^{-1}$

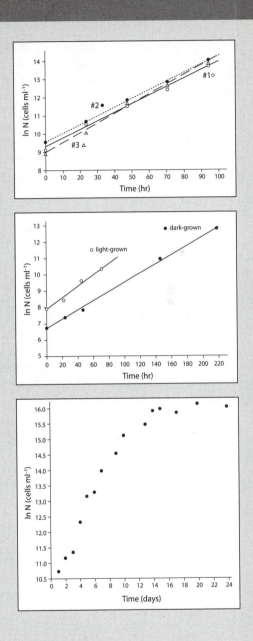

x = 1/S	y = 1/ρ
1.120	0.602
0.440	0.233
0.259	0.173
0.138	0.149
0.066	0.112
0.037	0.080

hr^{-1}. $K_t/\rho_{max} = 0.4676$ and $K_t = 7.3$ µM SiO$_2$. These are quite close to the values reported in Tilman and Kilham (1976), who used nonlinear regression ($\rho_{max} = 15.1 \times 10^{-9}$ and $K_t = 7.5$). The uptake rate for silicate (15.5) is much higher than that for phosphate (5.4), but the half-saturation constant for phosphate uptake is much lower (0.83) than that for silicate (7.3). This means that *C. meneghiniana* can take up phosphate at much lower external concentrations than it can silicate.

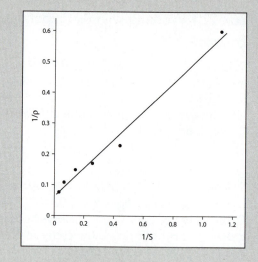

2. The two plots are shown in the accompanying figures. In contrast to the plot in sample problem #1, the graphs in these two plots cross each other. This crossing means that neither species is able to grow at a higher rate than the other over all phosphorus conditions. In the Droop model the two curves cross because *Prochlorothrix* has both a lower value of μ_{max} and a lower value of Q_o than *Planktothrix*. In the Monod model if phosphorus levels are below 0.34 µg P l^{-1} then *Prochlorothrix* should win in competition with *Planktothrix*. The reverse is true at phosphorus levels above 0.34 µg P l^{-1}.

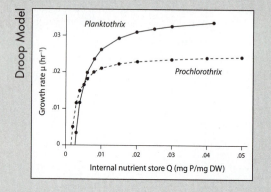

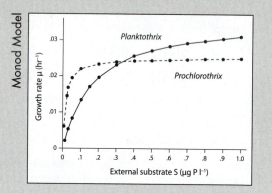

3. Take the reciprocals of the data in the table (right). Then plot the values of $1/S$ and $1/\mu$ on graph paper as shown below. Now take the linear regression of $1/\mu$ on $1/S$ using the equation $1/\mu = (K_s/\mu_{max})(1/S) + 1/\mu_{max}$.

x = 1/S	y = 1/µ
0.2860	1.97
0.2700	1.94
0.1540	1.45
0.1430	1.43
0.0980	1.28
0.0758	1.23
0.0538	1.20
0.0454	1.07

The result is Y = 3.5982(X) + 0.9398, for which the correlation coefficient is 0.994. $1/\mu_{max}$ = 0.9398 so μ_{max} = 1.06 day^{-1}. K_s = 3.5982 (μ_{max}) and therefore K_s = 3.83µM SiO$_2$. The values reported by Tilman and Kilham (1976) were μ_{max} = 1.06 day^{-1} and K_S = 3.94 µM.

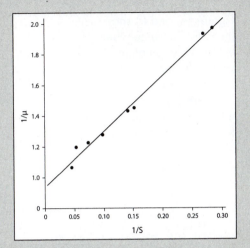

Box 21.3 Working with the Ostwald Modification of Stokes' Law

1. v_s = 32.51 µm s^{-1} φ = 4.43

2. v_s = 0.75 µm s^{-1}. Small cells sink much slower than large ones.

3. v_s = –13.09 µm s^{-1}, or –0.047 m hr^{-1}. This ascent rate is much slower than the rates reported in the text, suggesting that cyanobacteria can achieve much lower densities through gas vacuoles than in this example.

4. Since g on Mars is (0.38) 9.8 m s^{-2}, or 3.72 m s^{-2}, *S. astraea* would sink at a rate of 9.49 µm s^{-1}, or 10.5 µm s^{-1} if you used the observed value of Reynolds (1984).

Box 21.4 Working with the Mechanistic Model of Competition

1. The graphs showing the Monod growth curves for each diatom and limiting nutrient are given below, (a) through (d). The calculated values of R* for *Fragilaria* are 0.005 µM PO$_4$ and 1.0 µM SiO$_2$. R* values for *Synedra* are 0.002 µM PO$_4$ and 5.7 µM SiO$_2$.

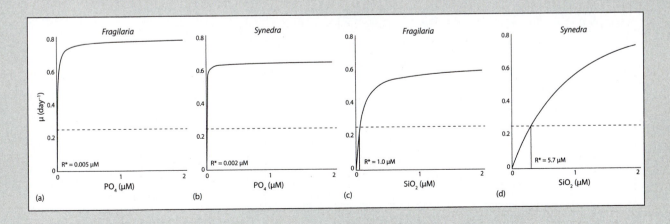

Figures (e) and (f) show the same resource plane with the consumption vectors based on the ratios of the nutrient quotients (e) and on the ratios of the half-saturation constants (f).

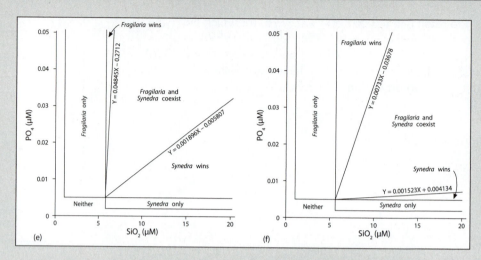

(e)

(f)

2. The values of R* for each species and nutrient are given in the table below. From these values the zero-growth isoclines can be plotted as shown on the right:

Diatom	R* (μM NO^{3-})	R* (μM SiO$_2$)
Corethron criophilum	0.54	107.3
Nitzschia kerguelensis	0.64	71.5
Thalassiosira subtilis	1.50	9.5
Nitzschia cylindrus	3.10	6.2

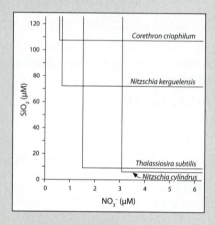

There are three significant intersection points: (0.64, 107.3), (1.5, 71.5), and (3.1, 9.5), where the values are (μM NO$_3^-$, μM SiO$_2$). In addition, there are three other intersection points of the zero-growth isoclines. These intersections have meaning in only two species interactions; in the four-species system, any population at these junctions would fall back to the isoclines further to the left.

There are two consumption vectors radiating from each of the significant intersection points listed above. The vectors have been plotted on the graph at right. The equations of these lines radiating from each point of intersection are:

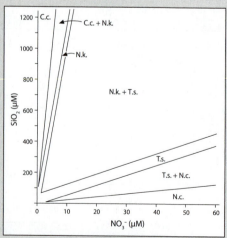

Point of Intersection		
	Corethron criophilum	*Nitzschia kerguelensis*
(0.64, 107.3)	Y = 200.3(X) – 21.9	Y = 110.8(X) + 35.8
	Nitzschia kerguelensis	*Thalassiosira subtilis*
(1.5, 71.5)	Y = 110.8(X) – 94.7	Y = 6.33(X) + 62.0
	Thalassiosira subtilis	*Nitzschia cylindrus*
(3.1, 9.5)	Y = 6.33(X) – 10.12	Y = 2(X) + 3.3

(Consumption vectors based on nutrient quotient)

Chapter 22

Macroalgal and Periphyton Ecology

Postelsia

Macroalgae, also known as seaweeds, and **periphyton** algae (smaller unicellular, colonial, or filamentous algae) typically grow attached to substrata, in contrast to the floating or swimming phytoplankton discussed in the previous chapter. Substrata for macroalgae and periphyton include rocks, coral, carbonate or silica sands, other algae, and aquatic animals and plants. Most seaweeds occur along ocean coastal regions (Figure 22.1), but a few, such as *Cladophora*, *Ulothrix*, and *Bangia*, have colonized freshwaters. Periphytic algae occur in both the oceans and freshwaters. Both macroalgae and the periphyton algae are immensely important, as they provide the community structure and primary productivity that supports a wide array of other organisms. An understanding of the composition and function of periphyton and macroalgal communities is essential to effective management, protection, and restoration of freshwater wetlands, lakes and streams, coastal salt marshes and mangrove forests (mangals), coral reefs (see Text Box 22.1), and rocky ocean coastlines.

On a global basis, coastal ocean waters inhabited by macroalgae are sites of intense primary productivity that support coastal ecosystems, including fisheries. Primary productivity in seaweed communities is equal to or greater than that of the most productive terrestrial plant communities. Productivity of the *Postelsia* community, which inhabits high-energy, wave-exposed sites along

(Photo: L. W. Wilcox)

Algae and Coral Reefs

Text Box 22.1

The modern coral reef originated more than 200 million years ago (Wood 1998). Today, coral reefs occur between the latitudes of 30° N and 30° S and are common throughout the tropics, except the west coasts of Africa and South America, where cold upwelling currents provide increased nutrient levels that favor development of macroalgae. The optimum temperatures for coral reef development are between 23°C and 29°C. Competition between reef-building corals and seaweeds for domination of open, sunlit substrates has been hypothesized to be a direct cause of the restriction of coral reefs to tropical waters (Miller 1998).

Most reef-building (scleractinian) corals contain zooxanthellae and thus have a light requirement of at least 3% of surface irradiance. Although corals may be found at a maximal depth of 145 m, they grow optimally at less than 10 m. Vertical growth rates of coral reefs are 5–10 mm yr^{-1}. Over millions of years, such growth has built structures hundreds of meters thick (in the case of Enewetak Atoll, over 1300 m). There are three major classes of coral reefs: atolls, barrier reefs, and fringing reefs. Atolls are ring-shaped coral islands that nearly or completely surround a central lagoon. Barrier reefs are ridges of coral that parallel the coastline and can be 300–500 m wide and up to 2000 km long (in the case of the Great Barrier Reef). Fringing reefs occur quite close to shore, with little or no water between shore and reef. Living shallow-water coral reefs presently cover about 600,000 km^2.

Coral reefs are important in long-term global carbon cycles; an estimated 700 billion kg of carbon are deposited as calcium carbonate in coral reefs every year. In addition, reefs are major repositories of biodiversity, harboring the greatest vertebrate diversity of any community on Earth. Further, they are among the most productive aquatic communities, offering a wealth of resources, but they are extremely sensitive to overexploitation (Birkeland 1997).

Algae of coral reefs include the essential endosymbiotic zoo-xanthellae of the *Symbiodinium* complex (Chapter 7), turf communities, siphonalean green algae (Chapter 17), and a variety of red algae, notably the reef-consolidating crustose corallines (Chapter 15). Two-thirds of coral reef productivity is attributed to endosymbiotic dinoflagellates and one-third to other algae. The zooxanthellae, which normally occur at densities of 10^6 cells cm^{-2} of coral surface, greatly enhance coral calcification rates (Muller-Parker and D'Elia 1997). Several species of filamentous cyanobacteria function as endolithic microborers because their growth within the coral rock results in its dissolution. The importance of these bioeroders is controversial (Glynn 1997).

Coral reefs are increasingly suffering from anthropogenic effects, including increases in seawater temperature and violent weather, which have been attributed to global climate change, increased UV exposure, overfishing, sedimentation, and nutrient enrichment (Brown 1997). Corals respond to sharp increases (3°C–4°C) for a short time (several days) or more moderate temperature increases over longer periods by

undergoing bleaching—loss of zooxanthellae. If not too severely damaged, coral can "rebrown" by acquiring a new population of zooxanthellae that may be better adapted to the environmental change (Muller-Parker and D'Elia 1997). Stressed corals are vulnerable to pathogenic infestations such as black band disease caused by the cyanobacterium *Phormidium corallyticum* and associated microorganisms (Peters 1997). Nutrient enrichment of coastal waters around coral reefs can shift the community balance from dominance by corals to dominance by macroalgae. Herbivory by fish and various invertebrates such as urchins is essential for maintenance of reef-building corals (Carpenter 1997, Hixon 1997). If overfishing simultaneously occurs on reefs affected by eutrophication, the absence of algal herbivores allows excessive macroalgal growths to smother the reefs (Choat and Clements 1998). Globally, coral reefs are in serious decline, with about 30% already severely damaged and nearly 60% likely to be lost by 2030 (Hughes et al. 2003). For more information on the human threats to coral reefs, see the portions of Section 22.3 dealing with pollution and global environmental change.

the Pacific U.S. coast (Figure 22.2), is estimated to be 14.6 kg m^{-2} yr^{-1}, in contrast to less than 2 kg m^{-2} yr^{-1} for rain forests (Leigh et al. 1987). Kelp-dominated seaweed communities along the eastern United States coast have an estimated primary productivity of 1.75 kg m^{-2} yr^{-1}. An estimated 10% of seaweed biomass is directly consumed by herbivores, whereas 90% enters detrital food webs, thereby contributing indirectly to animal biomass (Branch and Griffiths 1988) (Figure 22.3). Fossil evidence suggests that coastal marine seaweed communities have been in existence for the past 500 million years and have been supporting coastal ecosystems with intense primary productivity throughout this time (Xiao et al. 1998). Freshwater periphyton communities also contribute significantly to primary productivity, particularly in streams or lakes having a relatively large proportion of nearshore substrate to open water. In many bodies of water, periphyton algae contribute more primary productivity than do the phytoplankton (Burkholder and Wetzel 1989).

As a result of human population pressures, coastal ecosystems and inland freshwaters around the world are being seriously affected by inputs of nutrients and contaminants, exposure to increased levels of ultraviolet radiation, overexploitation of resources, loss of species diversity, and colonization by nonnative species. Most studies of nearshore communities have been performed on very small scales, on local levels, and for short time periods. Many ecologists believe that longer-term, regional- and global-scale analyses must be undertaken that include as many of the significant biotic and abiotic ecosystem components as possible. The U.S. National Science Foundation's Long-Term Ecological Research (LTER) program includes several intensively studied aquatic sites, while the Global Ocean Observing System (National Research Council) is an example of an international effort to increase the range of scientific information available to coastal zone planners and policymakers. Remote-sensing techniques such as satellite-based scanners are being used to survey seaweeds and freshwater algae.

Macroalgal and periphyton ecologists are interested in defining the factors that influence algal occurrence and distribution patterns. Coastal marine communities have proven especially valuable as ecological models for determining the relative importance of physical factors, algal morphological and physiological adaptations, and biotic factors such as herbivory and interspecific competition in structuring macroalgal communities. Studies suggest that in both seaweed and periphyton communities, there is a hierarchy of controlling factors, with the limits of algal growth set primarily by the interplay between algal physiological constraints and physical environmental factors. Within these limits, biotic factors—specifically grazing by herbivores and the activities of pathogens—can modify algal distribution patterns. The concept of the keystone species was developed in the marine intertidal zone (Paine 1969), and trophic dynamics are major factors in rocky intertidal, kelp forest, and coral reef communities. The habitats of seaweeds and periphytic algae are highly variable and have a

Figure 22.1 Macroalgae (seaweeds) are commonly seen along coastal areas of the world, including Rialto Beach, Washington (United States). (Photo: L. W. Wilcox)

significant level of disturbance. In such conditions, according to ecological theory (Chapter 21), inter-specific competition cannot go to equilibrium, and species diversity is high. In view of this apparent hierarchy of controlling factors, physical factors and associated algal adaptations are discussed here first, followed by consideration of the roles of herbivory and competition.

Many species of macroalgae occur in marine coastal waters, but in freshwaters the branched, filamentous green alga *Cladophora* is the largest, most conspicuous alga. Attached diatoms, cyanobacteria, and filamentous eukaryotic algae dominate the microscopic periphyton of both freshwaters and marine waters. Although there are some circumstances in which growth of freshwater periphyton resembles that of marine turf-forming seaweeds, the habitats of seaweeds and marine periphyton differ significantly in wave energy, tidal exposure to desiccation and solar radiation, variation in temperature and salinity, and types of herbivores from the habitats of freshwater periphyton. The first part of this chapter focuses on marine macroalgae, the second section is concerned with turf-forming periphytic algae of

Figure 22.2 The annual sea palm, *Postelsia*, is characteristic of high-energy environments where heavy wave action removes sessile animals that would otherwise compete for space. *Postelsia*'s flexible stipe and extensive holdfast are regarded as adaptations to wave drag forces. (Photo: L. W. Wilcox)

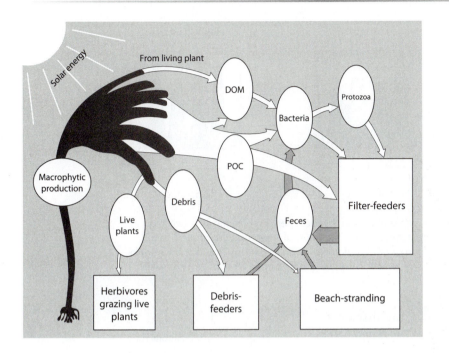

Figure 22.3 The fate of macroalgal primary production. Arrow thickness indicates relative amounts of carbon excreted from living seaweeds in the form of dissolved organic material (DOM), carbon consumed by grazers on living seaweed, and particulate organic carbon (POC) and debris resulting from macroalgal decay. (After Branch and Griffiths 1988)

marine waters, and the third part discusses freshwater periphyton algae.

Part I—Marine Macroalgal Ecology

22.1 Physical Factors and Macroalgal Adaptations

Coastal regions are influenced strongly by tide-related variations in water level and other water motions, including waves and currents. These, combined with weather changes and seasonal variation in solar irradiance levels, subject seaweeds exposed at low tides to frequent variation in light, temperature, nutrient availability, and level of hydration. Seaweeds that are exposed to air (emersed) for long periods lose their source of dissolved nutrients, may become desiccated, and may receive damaging levels of solar radiation. When covered by seawater (immersed) from the incoming tide a few hours later, the same seaweeds may experience greatly reduced irradiance levels and be subjected to stressful mechanical forces from wave action. Water motions also influence the abundance

and activities of herbivores and the extent to which interspecific competition occurs. Hence, our discussion of physical factors will begin with the effects of tides and other water motions on algal distribution patterns, followed by the ways in which seaweeds deal with variation in light environment, salinity changes, desiccation, and nutrient availability.

Tides

Tides are daily changes in sea level that occur along most marine shorelines, where the magnitude of water-level changes can vary geographically from barely detectable to many meters. Rising water levels are called flood tides, and falling levels are ebb tides. Tides are produced by the gravitational pull of the Sun and Moon. The Moon has about twice the pull of the Sun because it is closer to Earth. The Sun and Moon pull on Earth's oceans and raise large tidal bulges. Although tides appear to move in and out, the surface of Earth actually rotates into and out from the tidal bulges, whose magnitude is determined by the relative positions of the Sun and Moon. The strongest gravitational effects occur when Earth, the Sun, and the Moon are in a linear alignment, when there is a full or new moon. Linear alignments result in very high and low "spring" tides, where spring means to "spring forth" rather than a season of the year. When the Sun, Moon, and Earth are oriented

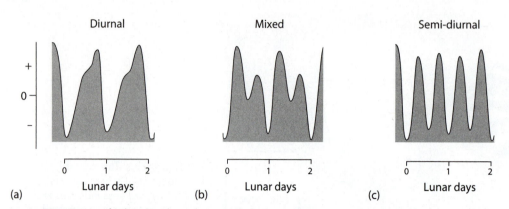

Figure 22.4 Types of tides: (a) diurnal—a single high and low tide per day; (b) mixed—two high and two low tides per day of unequal amplitude; and (c) semidiurnal—two equal high and two equal low tides per day. (Redrawn from *Journal of Experimental Marine Biology* Vol. 62. Swinbanks, D. D. 1982. Intertidal exposure zones: A way to subdivide the shore, pages 69–86. © 1982 with permission from Elsevier Science)

perpendicularly (when the Moon is in the first or third quarter), the gravitational pull of the Sun and Moon counteract each other, and the comparatively weak neap tides occur. Thus there is a lunar monthly cycle of neap and spring tides (Dawes 1998).

The frequency and amplitude of tides are affected by the morphology of the ocean basin, such that different coasts may have distinct types of tidal cycles. Uncommonly, there is a single high and low tide per day; such a tidal cycle can occur in the Gulf of Mexico (Figure 22.4). Along most coastal locations, there are two flood tides and two ebb tides per day. Along the open coast of the Atlantic, these two high and two low tides are equal in amplitude. Along the Pacific and Indian Ocean coasts and in the Gulf of St. Lawrence, the two high and low tides are of unequal magnitude. Weather and large-scale climatic features such as El Niño can also influence tides.

Tidal patterns influence the structure of algal communities by creating an intertidal region—the portion of shoreline lying between a line defined by the highest tides and another defined by the lowest tides. The intertidal region is submerged at high tide and exposed at low tide. Higher areas of the intertidal region are exposed for significantly longer periods of time than lower areas, creating distinct zones within which specific seaweeds may be restricted (Doty 1946). In a worldwide survey, Stephenson and Stephenson (1949, 1972) found that both water levels and the types of organisms present defined these zones, and they developed a system of terminology (Figure 22.5). Their system is still used, with modifications by Lewis (1964),

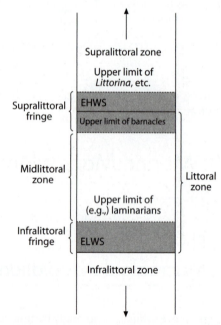

Figure 22.5 Stephenson and Stephenson's (1949, 1972) intertidal zonation system of terminology, based on both water levels—extreme high water of spring tides (EHWS) and extreme low water of spring tides (ELWS)—and organismal distributions, locations of littorinid snails, barnacles, and laminarialean brown seaweeds. (Modified from Stephenson and Stephenson 1949. © British Ecological Society)

who took into account the effects of wave action on exposed versus protected shores (Figure 22.6). Some variation exists among countries in the terms used to denote particular shoreline zones. Here, we use the terms "littoral zone" and "intertidal

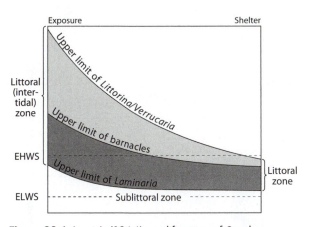

Figure 22.6 Lewis's (1964) modification of Stephenson and Stephenson's scheme, showing variation in the extent and location of intertidal zones between extremely exposed and extremely sheltered sites. (Modified from Lewis 1964)

Figure 22.8 Barnacles (arrowheads), shown here with the brown seaweed *Saccharina*, are characteristic of the intermediate of the three universal intertidal zones. (Photo: L. E. Graham)

Figure 22.7 One of three universal intertidal zones, the black zone (arrowheads) observed in the supralittoral, is the habitat of desiccation-resistant cyanobacteria, lichen, and snails. (Photo: L. E. Graham)

Figure 22.9 *Laminaria* or other brown algae are characteristic of the lowermost of the three universal intertidal zones. The foam shown in the center of this photo is formed by incorporation of air (via wave energy) into colloidal suspensions of polysaccharides (alginates) exuded by the kelp. This is known as kelp foam. (Photo: L. E. Graham)

region" interchangeably to mean the stretch of shoreline between the extreme high water of spring tides (EHWS) and extreme low water of spring tides (ELWS). (The terms *littoral* and *intertidal* are used both as adjectives and as nouns.) The supralittoral zone lies above the littoral and receives seawater spray, whereas the sublittoral zone (also known as the subtidal or infralittoral) is submersed most of the time. The depth of the deepest-growing algae defines the lower limit of the sublittoral. For most geographical regions, this depth is unknown, as survey by submersible is required to make such a determination (Lobban and Harrison 1994).

Stephenson and Stephenson found that there are three nearly universal intertidal zones: (1) an uppermost black strip of highly desiccation-tolerant cyanobacteria, a marine lichen (*Verrucaria*), and littorinid snails (Figure 22.7); (2) an intermediate zone of various seaweeds together with barnacles and limpets (Figure 22.8); and (3) a lowermost zone inhabited by laminarialean brown algae (in temperate or high-latitude environments) or corals (in tropical regions) (Figure 22.9). Aside from these general features, zonation patterns

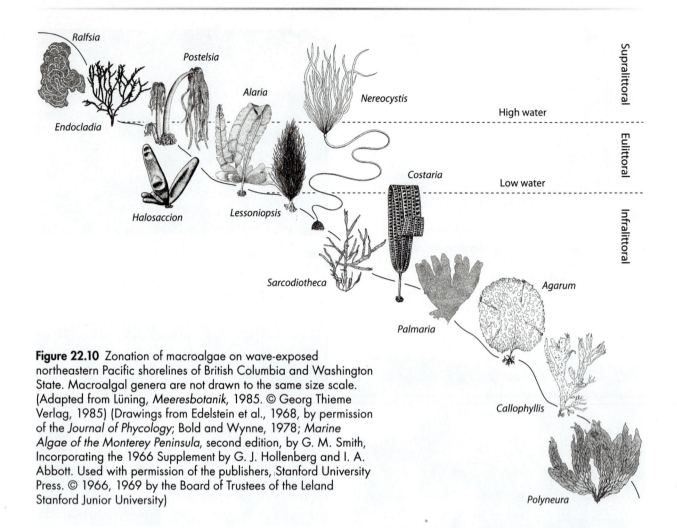

Figure 22.10 Zonation of macroalgae on wave-exposed northeastern Pacific shorelines of British Columbia and Washington State. Macroalgal genera are not drawn to the same size scale. (Adapted from Lüning, *Meeresbotanik*, 1985. © Georg Thieme Verlag, 1985) (Drawings from Edelstein et al., 1968, by permission of the *Journal of Phycology*; Bold and Wynne, 1978; *Marine Algae of the Monterey Peninsula*, second edition, by G. M. Smith, Incorporating the 1966 Supplement by G. J. Hollenberg and I. A. Abbott. Used with permission of the publishers, Stanford University Press. © 1966, 1969 by the Board of Trustees of the Leland Stanford Junior University)

may differ considerably from one biogeographical region to another. Zonation of seaweeds typical of wave-exposed shorelines of British Columbia (Canada) and Washington State (United States) is illustrated in Figure 22.10; seaweeds often found on the wave-exposed southern coast of Newfoundland (Canada) are diagrammed in Figure 22.11a; and seaweed zonation on a Japanese shore (Rikuchu National Park on North Honshu Island) is shown in Figure 22.11b (Lüning 1990). Community analysis is relatively easy to accomplish in the intertidal. At low tide, transect lines can be laid down the shoreline and quadrats used to define small areas for detailed study.

Zonation also occurs in the sublittoral (subtidal) region as a result of light attenuation with increasing depth. Upper, middle, and lower sublittoral zones are characterized by distinct seaweed assemblages. The mid-sublittoral kelp community with the canopy-forming *Macrocystis* and understory brown and red seaweeds is depicted in Figure 22.12. Sublittoral algal communities are less well known and their study considerably more challenging than the algal communities in the intertidal. Although some information can be obtained by dredging up samples from deep water, SCUBA or submersible technology is required for more detailed analysis. Working off the coast of central California with SCUBA and a remotely operated vehicle (ROV), Spalding et al. (2003) reported the discovery of a new and diverse lower sublittoral (30 to 75 m) algal assemblage that formed kelp beds at depths below the well-known giant kelp forests of *Macrocystis* (Figure 22.13).

Waves and Currents

Waves are periodic surge events generated by wind, tides, earthquakes, and landslides. Currents are large

Newfoundland   Japan

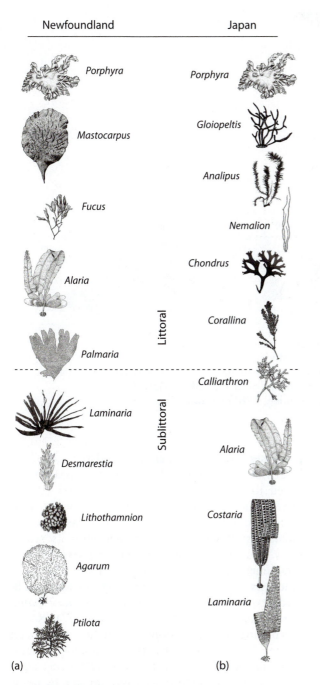

Figure 22.11 Comparison of macroalgal zonation patterns along (a) the northwestern Atlantic (Newfoundland, Canada) and (b) northwestern Pacific (Japan). Supralittoral forms are at the top, and sublittoral forms are at the bottom of the diagram. (Seaweeds are not drawn to the same size scale.) (Adapted from Lüning, *Meeresbotanik*, 1985. © Georg Thieme Verlag, 1985) (Drawings from Smith 1969; Bold and Wynne 1978, 1985; Woelkerling 1988)

or small-scale unidirectional water flows that are strongly influenced by winds and coastal topography. Waves and currents are a tremendous source of energy to nearshore environments, contributing some 10 times more energy than solar radiation. This extraordinary energy input explains the high productivity of coastal macroalgal communities.

Water motions may have positive effects on seaweeds. By constantly moving the fronds of macroalgae, waves and currents reduce self-shading and bring a constant supply of inorganic nutrients to algae. In still waters macroalgae develop a boundary layer of still water around their surfaces, and such boundary layers act as a diffusive barrier to the movement of nutrients from the water column into the algae. Flowing waters reduce the thickness of this boundary layer and increase the rate of nutrient delivery to the macroalgal surfaces. For example, the nitrate uptake rate of *Macrocystis pyrifera* rose by 488% when the current velocity increased from 0 to 3 cm s^{-1} (Wheeler 1982). When phycologists grow seaweeds in culture, aeration or swirling has the same effect—reduction of boundary layers and increase in nutrient uptake and growth rates (Hanisak and Samuel 1987). Waves and currents can also remove sessile animals (such as mussels) that compete with macroalgae for attachment space in the intertidal zone. The annual seaweed *Postelsia* is dependent upon wave removal of barnacles and mussels to open up space in the intertidal for establishment. The tropical sea grass bed inhabitants *Penicillus* and *Codium* reportedly require water motion for normal morphological development; in culture, the normal multiaxial body develops only if the algae are shaken (Ramus 1972). Macroalgae also require waves and currents for dispersal of spores. While most spores settle within a few meters of release, strong currents and waves can disperse spores distances of more than 1 km (Gaylord et al. 2002).

Waves and currents also have negative impacts on macroalgae, including damage, destruction, and removal as a result of mechanical forces. Water-motion velocities can exceed 10 m s^{-1}, and because of the density of water, waves and currents can exert much stronger forces than wind. Existence in such an environment has been compared to that of humans experiencing a hurricane (Lobban and Harrison 1994). Denny (1988), Lobban and Harrison (1994), and Vogel (1994) described mathematical expressions of water motions and the resulting forces operating upon seaweeds. Macroalgal

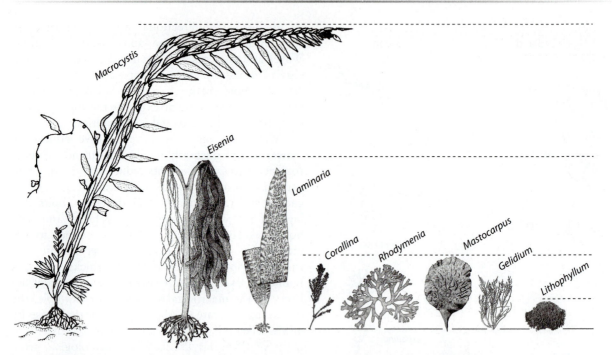

Figure 22.12 Composition of mid-sublittoral macroalgal communities dominated by the giant kelp *Macrocystis*, such as occur on the California (United States) coast. *Macrocystis* is a canopy species that modifies the light environment of smaller understory seaweeds, including the brown algae *Eisenia* and *Laminaria* and the red algae *Corallina*, *Rhodymenia*, *Gigartina*, *Gelidium*, and *Lithophyllum*. (Adapted from Lüning, *Meeresbotanik*, 1985. © Georg Thieme Verlag, 1985) (Drawings from Smith 1969; North 1994; Woelkerling 1988)

surfaces slow down water, setting up fluid or surface shear stress. Shear stress is a force applied parallel to a surface, in contrast to compressive force, which operates perpendicular to surfaces. Drag forces pulling on seaweeds arise from shear stress and are proportional to the square of the velocity of the water. The larger an organism, the more drag force it will experience. Seaweeds are morphologically adapted to cope with these forces. The four main components of macroalgal adaptation are elasticity in tension (stretchiness), elasticity in bending (flexibility), torsion (twisting), and breaking strain (strength). Terrestrial woody plants are rigid, but marine macroalgae are similar to large rubber bands. Many large seaweeds have stretchy bendable bodies; such elasticity and flexibility dissipates the effects of drag forces, allowing seaweeds to grow to greater size than if they were stiff. The flexible stipe of the sea palm *Postelsia* helps it cope with heavy wave action on extremely exposed shorelines (see Figure 22.2). Harder et al. (2006) estimated that four species of wave-exposed seaweeds dissipate 42%–52% of drag force through plastic-viscoelastic

processes (stretching, bending, and twisting). The length of *Nereocystis*'s flexible stipe appears to be tuned to resonate with wave frequencies (Denny 1988). Midribs, such as those in the blades of *Alaria*, may function to reduce tearing across the body. *Fucus* resists mechanical damage with its strong body. Seaweeds of the intertidal possess effective attachment adhesives (Vreeland et al. 1998) and holdfasts that resist separation from substrates. The force required to break the stipes of *Fucus* or *Laminaria* is about 40 kg cm^{-2} (Schwenke 1971). Other morphological adaptations that reduce the effects of drag forces include growth as bushy forms, short turfs, or flat adherent crusts. Flat encrusting algae and microscopic germlings avoid damage by living within the boundary layer of the substrate to which they are attached. Crustose corallines are the only seaweeds that can resist the extreme wave environments of the coral reef crest. Crustose or highly reduced stages in the heteromorphic life histories of various intertidal brown, red, and green seaweeds may help them survive water motions that are especially violent at particular times of the year.

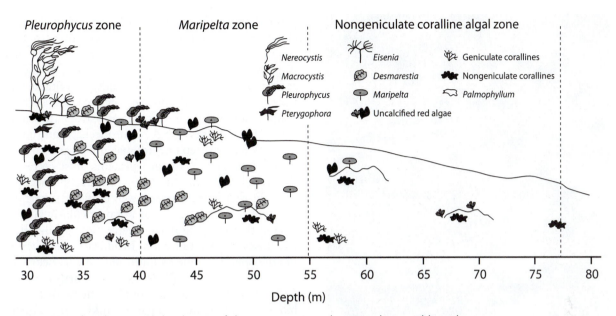

Figure 22.13 Abundance and distribution of deep-water macroalgae in a lower sublittoral community off the central California coast, as determined by diver (30 m) and remotely operated vehicle (ROV) sampling (> 35 m). (From Spalding et al. 2003, by permission of the *Journal of Phycology*)

Phycologists visualize water motions at the surfaces of macroalgae using dyes or flakes of mica. With these methods Hurd and Stevens (1997) studied the responses of single blades and fronds of the giant kelp *Macrocystis*, *Nereocystis*, *Laminaria*, *Costaria*, *Alaria*, *Egregia*, and *Fucus* to currents of various flow rates established in seawater tanks. Their results with three blades of *Macrocystis integrifolia* at four flow velocities are shown in Figure 22.14. At the lowest flow rate, the flow is laminar, and bright patches on the blades indicate areas of boundary layer (Figure 22.14a). A transition from laminar flow to turbulent flow occurs at a current velocity of around 1.5 cm s^{-1} (Figure 22.14b), and full turbulent flow appears at 3.0 cm s^{-1} (Figure 22.14c). At the highest flow rate studied, the three *Macrocystis* blades have become compressed into a streamlined shape, and turbulence appears to be reduced. These results suggest that all seaweeds that are within the general size range of single *Macrocystis* blades will achieve turbulent flow across their surfaces when seawater velocity is greater than 2.5 cm s^{-1}, regardless of the presence of corrugations, undulations, or bulbs (floats or pneumatocysts). Since 2.5 cm s^{-1} is at the lower end of normal ocean current velocities, most large seaweeds will be in turbulent flow conditions most of the time.

Studies of seaweed morphology have revealed correlations between particular structural features and the turbulence of the environment. The sublittoral kelp *Nereocystis* has narrow smooth blades in rapidly moving water and wider ruffled blades in more sheltered habitats. The ruffled blade may be advantageous in creating turbulence that helps reduce surface boundary layers and increase nutrient acquisition, but it tends to tear when flow rates are high. Smooth blades, in contrast, bundle together in high currents to form a streamlined profile that reduces drag forces. The bundled condition, however, contributes to self-shading and reduced nutrient uptake and photosynthesis because individual blades are less exposed to light and water (Koehl and Alberte 1988; Stewart and Carpenter 2003). The cabbagelike intertidal brown alga *Saccharina* likewise has narrow, flat blades in high-energy habitats but puckered (bullate) blades in calmer sites (Armstrong 1987). The conspicuous holes in blades of the brown seaweed *Agarum* and the red *Rhodymenia pertusa* likely represent morphological adaptations that increase turbulence on the blade surface.

There are some habitats in which seaweeds do not experience much mechanical stress and consequently appear relatively delicate or fragile. In the Arctic Ocean, in the stormy winter period, a protective

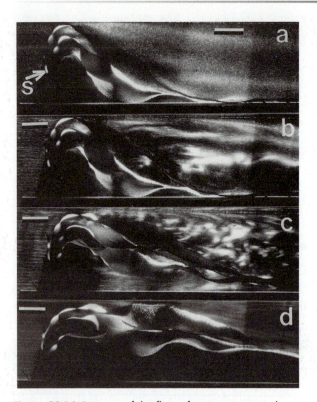

Figure 22.14 Patterns of the flow of seawater around a region of stipe (S, arrow) bearing three blades of *Macrocystis integrifolia* at four different velocities—(a), 0.5, (b), 1.5, (c), 3, and (d), 5 cm s⁻¹—in a recirculating flow tank. Shear is visualized by the use of mica flakes, which minimize drag by aligning with the shear and thus become less reflective (the seawater appears black). Where shear forces are weak, the mica particles are randomly aligned, scatter light, and thus appear white. (Scale bars = 5 cm) (From Hurd and Stevens 1997, by permission of the *Journal of Phycology*)

layer of ice covers the seaweeds. The summer open water period is characterized by relatively calm conditions, and perennial algae may slowly grow to very large sizes, as the bodies do not suffer wave damage (Lüning 1990).

Light

Intertidal seaweeds must cope with both visible light—photosynthetically active radiation (PAR) of wavelengths 400–700 nm—and ultraviolet radiation (UV), particularly UVB (280–320 nm), which may damage DNA and cellular protein, such as the D1 protein essential to photosystem II in chloroplasts. High light conditions cause conformational changes

in or destruction of D1, which can affect the efficiency of electron flow from reaction centers to plastoquinone. Thymine dimers may form in DNA that is exposed to UVB, leading to misreading of coding regions and, if repair does not occur, to heritable mutations. Some algae produce protective UV-absorbing compounds such as β-carotene and aromatic amino acids. Algae possess DNA repair mechanisms, but at high UVB levels, repair rates may not be able to keep up with damage. Damaged proteins must be resynthesized. UVB radiation is a significant component of sunlight that is transmitted well in water. Intertidal and reef surface macroalgae are particularly vulnerable in that they regularly receive biologically effective doses of UVB radiation. In addition, destruction of atmospheric ozone by anthropogenic halocarbons has increased UVB transmission to Earth's surface with the potential for deleterious effects to Earth's biota, including macroalgae (Franklin and Forster 1997).

Several recent studies have revealed that macroalgae such as *Saccharina latissima*, which are attached in the intertidal region, and *Sargassum natans*, a **pelagic** form occupying surface tropical waters, undergo regulated, photoprotective responses to high levels of solar radiation that involve changes in photosynthetic efficiency. Such changes can be monitored with an instrument that measures *in vivo* fluorescence changes in photosystem II (see Text Box 22.2). High light stress causes a phenomenon known as **photoinhibition**, which results in a depression of photosynthetic activity and, at the extreme, photooxidation of chlorophyll. UV radiation exacerbates photoinhibition in brown, red, and green macroalgae, as well as phytoplankton (Sagert et al. 1997). Many algae display a daily pattern of photosynthesis characterized by rising rates in the morning, a decline at noon, and then recovery in the afternoon. The noon decline reflects the effects of photoinhibition, and the subsequent increase represents recovery from photoinhibition. Intertidal algae exposed at low tide and surface pelagic forms experience the highest levels of photoinhibition.

During photoinhibition, energy that has been absorbed, but which cannot be used in photochemical reactions, is dissipated in the form of harmless thermal radiation or fluorescence. In green and brown algae, such energy dissipation can occur through an increase in the zeaxanthin content of photosystem II (PSII) (operation of the protective xanthophyll cycle—see Chapter 1). High light

Fluorescence Methods for Assessing Photosynthetic Competency and Nitrogen Limitation

Text Box 22.2

When illuminated by a beam of light positioned at right angles to a detector, chlorophyll exhibits distinctive red fluorescence having an emission peak at 685 nm. The intensity of this fluorescence varies with that of the incident light and can be measured by means of an instrument known as a fluorometer (Geider and Osborne 1992). The level of fluorescence is an indicator of the functional status of photosystem II (PSII), hence, measuring fluorescence is a common way of assessing the degree of photoinhibition of algal cells or tissues. Nitrogen limitation also results in changes in levels of algal chlorophyll fluorescence. Measurements of algal fluorescence can thus also be used to assess the degree to which natural algal populations are nitrogen limited (Falkowski and Raven 1997).

Immediately upon illumination by light that is photochemically active, such as white light, the fluorescence of chlorophyll *a* increases from essentially zero to

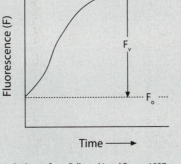

Redrawn from Falkowski and Raven 1997, with permission from Blackwell Science, Inc.

a low level (F_0) (see graph). As the irradiance continues and the sample continues to absorb photons, fluorescence will increase to a maximum level (F_M). The quantity $F_M - F_0$ is known as the variable fluorescence (F_V). The ratio of variable fluorescence to maximum fluorescence (F_V/F_M) is the maximum quantum yield for stable charge separation at PSII—an estimate of the functional status of PSII. (In general, the quantum yield, ϕ, is the ratio of moles of product formed or substrate consumed to the moles of photons absorbed in a photochemical reaction.) The quantum yield for fluorescence (ϕ_f) is the ratio of light emitted as fluorescence to the light absorbed.

Under conditions of maximum quantum yield (F_V/F_M), all PSII reaction centers are considered to be open. Reaction centers are regarded as closed when they are unable to transmit energy to the secondary electron acceptor quinone B (QB). Normally QB is bound to the PSII protein D1, which may be damaged or destroyed under

high irradiance conditions. The greater the number of closed PSII reaction centers, the greater the amount of energy that will be dissipated as fluorescence. Experiments have shown that the quantum yield of variable fluorescence is inversely related to O_2 evolution. In other words, when PSII is maximally functional, maximal O_2 evolution occurs, but minimal variable fluorescence will be detected. If PSII is not optimally functional, O_2 evolution levels will be reduced, and variable fluorescence will increase.

The quantum yield of fluorescence is an estimate of the fraction of photosynthetically competent reaction centers. Changes in the value of F_0 are associated with the onset of photodamage and signal the induction of photoprotective mechanisms (Falkowski and Raven 1997). Fluorescence measurement of intact algal cells and tissues can be made in the laboratory, or fluorometers equipped with light sources (needed for stimulating fluorescence) can be lowered into the ocean. Differences between fluorescent quantum yields in the light versus dark can be used to make instantaneous *in situ* estimates of the rate of photosynthetic electron transport.

induces an increase in the relative proportion of zeaxanthin to violaxanthin. These two carotenoids can be extracted from seaweeds with 100% acetone, and relative amounts of each can be measured by use of high-performance liquid chromatography (HPLC). Correlation between changes in these xanthophylls and the degree of photoinhibition has been demonstrated in the brown seaweeds *Dictyota dichotoma* (Uhrmacher et al. 1995), *Saccharina latissima* (Hanelt et al. 1997), and *Sargassum natans* (Schofield et al. 1998), as well as other algae (Falkowski and Raven 2007).

Another widely employed protective scheme—dynamic photoinhibition—involves an increase in the number of inactive (closed) photosynthetic reaction centers. In this process, photosystem II becomes less efficient in the conversion of light energy into chemical energy because light-harvesting complexes become disconnected from photochemical energy conversion processes during periods of dangerously high irradiance. When irradiance levels decline, recovery of the photosynthetic system occurs. In *Sargassum natans* the efficiency of energy conversion decreased by 50%–60% from predawn levels to noon and then recovered to predawn levels three hours after sunset (Schofield et al. 1998). An analysis of the sensitivity of *Chondrus crispus* from 3.5 to 8.5 m below high tide level revealed that deeper-growing specimens experienced a greater depression of fluorescent yield (see Text Box 22.2) and slower recovery from photoinhibition than did those from shallower water (Sagert et al. 1997). Studies of photoinhibition and recovery after high light stress in the kelp *Saccharina latissima* revealed differences in sensitivity among the various life-history stages. Older sporophytes and gametophytes were relatively less sensitive to high irradiance (55 µmol m^{-2} s^{-1} for a period of two hours) than were young sporophytes, which suffered severe photodamage under these conditions (Hanelt et al. 1997).

Ocean waters are classified into more than a dozen types based upon light transmission characteristics. Turbid coastal waters often contain yellow dissolved organic compounds known as "gelbstoff" (German for "gold substance"). These are humic materials derived from terrestrial plants. Waters containing such compounds strongly absorb blue light, with the result that the irradiance environment is yellow (Figure 22.15b). In clear waters, differential absorption of short and long wavelengths results in an irradiance environment dominated by blue light (Figure 22.15a). Subtidal seaweeds face irradiance environments depleted in both light quantity and quality. In a review of lower depth limits and percent sea surface irradiance (%SSI) of marine macroalgae from the tropics to the high arctic, Markager and Sand-Jensen (1992) concluded that the subtidal zone of mainly leathery algae extended to about 0.5%SSI, an intermediate zone of delicate, foliose algae to about 0.1%SSI, and a lower zone of crustose algae down to about 0.01%SSI.

Macroalgae cope with low irradiance by modifying body structure for optimal light absorbance,

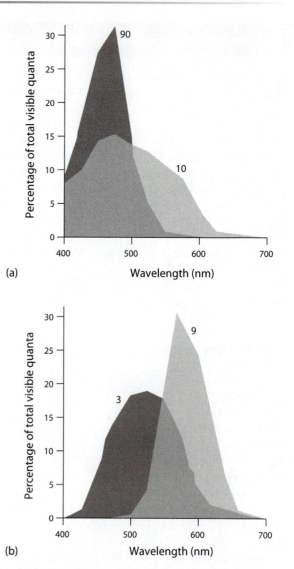

Figure 22.15 (a) The spectral characteristics of clear open-ocean waters at 10 and 90 m depth. The light environment is dominated by blue light. (b) The spectral characteristics of nearshore, coastal waters at 3 and 9 m depth. Blue light is strongly absorbed, and the irradiance environment is dominated by yellow light. (Adapted from Saffo, M. B. *BioScience* 37:654–664 1987. © American Institute of Biological Sciences)

and red, brown, and green macroalgae employ efficient light-harvesting mechanisms (Chapters 14, 15, and 16). The accessory pigments phycoerythrin, fucoxanthin, β-carotene, and siphonoxanthin are able to absorb green light (often more available at depth) and transfer this light energy to chlorophyll-based photosynthetic reaction centers. In all algal groups, β-carotene is sited close to reaction centers of photosystem II, suggesting that efficient energy transfer probably occurs. Red algae that grow in deep waters, such as *Chondrus crispus*, show an increase in phycoerythrin content compared to shallow water forms of the same species, and the efficiency of photosynthesis increases with depth (Sagert et al. 1997). Among brown algae, *Laminaria abyssalis* grows at depths of 40–70 m, where midday light levels may range from only 4–15 µmol photons m^{-2} s^{-1}, while *Laminaria digitata* grows in the upper sublittoral, where it periodically receives 1800–2400 µmol photons m^{-2} s^{-1} and occasional desiccation. If *L. abyssalis* is exposed to high light, it suffers complete and irreversible photoinhibition. In contrast, *L. digitata* shows 60% inhibition of photosynthesis followed by full recovery. As indicated by the quantum yield F_v/F_m (see Text Box 22.2), *L. abyssalis* is nearly twice as efficient in photon capture as *L. digitata*, but *L. abyssalis* has no effective xanthophyll cycle (Rodrigues et al. 2002). To maximize light capture at extremely low light levels, *L. abyssalis* has lost any mechanism for energy dissipation.

An old idea that green algae are adapted to high light levels in shallow waters, brown algae to intermediate depths, and red algae to low light in deep waters has been refuted (Ramus 1983; Saffo 1987). At all depths, algal communities are composed of brown, red, and green algae. Collections made by dredging at 55 m off Hawaii revealed an assemblage of 22 species of red algae, 10 species of green algae, and 5 species of brown seaweeds (Lüning 1990). Similarly, the deep-water macroalgal community off central California (United States) consists of a mix of browns, reds, and greens, although greens are the least abundant (Spalding et al. 2003).

Polar seaweeds, such as the reds *Constantinea subulifera*, *Maripelta rotata*, and *Delesseria sanguinea*, and the brown kelp *Laminaria solidungula*—which grows to depths of 20 m in the Alaskan high Arctic—experience only short annual periods of illumination. At the northernmost locations at which seaweeds can be collected, there may be only 4–10 ice-free weeks during which irradiance levels are sufficient for photosynthesis to form storage products. In fact, 80% of the yearly light supply reaching seaweeds at 10 m depth is received during these few ice-free weeks (Chapman and Lindle 1980). During August in the Arctic, at a depth of 6–7 m, irradiance levels are 100–200 µmol photons m^{-2} s^{-1} in the day and 20–30 µmol photons m^{-2} s^{-1} at night (Dunton and Schell 1986). These light levels allow

production and storage of carbon compounds for use in growth during the long, dark winters. During the nine months that *Laminaria solidungula* is covered by a turbid ice canopy, the alga loses some 30% of its carbon content. But amazingly, at the same time, it produces a new blade and undergoes 90% of its annual growth before the new season of irradiance begins. Reallocation of stored reserve materials allows growth to occur (Dunton and Schell 1986; Dunton 1990). Glycolytic catabolism of stored carbohydrate yields the substrates for so-called dark carbon assimilation, a process involving enzymes such as PEPCK (phosphoenolpyruvate carboxykinase), PEPC (phosphoenol pyruvate carboxylase), and PC (pyruvate carboxylase), none of which have any oxygenase activity (in contrast to rubisco). This process, also known as β-carboxylation (because the β-carbon of 3-carbon substrates is carboxylated) or anaplerotic carbon fixation, occurs in many kinds of organisms. β-carboxylation is a source of compounds needed for growth that are not generated by the Calvin cycle, namely several amino acids, tetrapyrroles, pyrimidines, purines, and lipids. β-carboxylation is measured by following the incorporation of ^{14}C in the dark. Typically, β-carboxylation rates rise when ammonium is added to N-deficient algae. Consequently, β-carboxylation has been used as a test for N limitation in natural populations (Falkowski and Raven 2007).

Light is also an important environmental signal that controls the timing of seaweed growth and reproduction. Sensor pigments, including a phytochromelike red/far-red reversible protein (Lopez-Figueroa et al. 1989) and a cryptochromelike blue-absorbing substance (Lopez-Figueroa and Niell 1989), have been isolated from several seaweeds. These pigments are involved in photoperiodism (dependence of a growth or reproductive response on day length) and photomorphogenesis (dependence upon a particular portion of the spectrum). An example of photomorphogenesis is the blue light induction of oogonia and antheridia in *Laminaria* gametophytes. Short-day (8-hour) (actually long-night) responses of seaweeds include formation by *Laminaria hyperborea* of a new blade in early winter, the spring appearance of the erect body of *Scytosiphon* (Chapter 14), transition from the codiolum phase to the blade phase of *Monostroma* via spore production (Chapter 17), and production of conchospores by the conchocelis phase of *Porphyra* and marine *Bangia* (Chapter 4). Conchosporangial production is suppressed by red light

(660 nm) and induced by a subsequent exposure to far-red light (730 nm). Long-day (16-hour) responses (actually short-night responses) of seaweeds include formation of gametangia (*Sphacelaria*) and production of the gametophyte phase (*Batrachospermum moniforme*) (Lüning 1990).

Salinity and Desiccation

The definition of salinity is the weight of solids that is obtained from drying 1 kg of water. The 10 most abundant ions in seawater are, in decreasing order, chloride, sodium, sulfate, magnesium, calcium, potassium, bicarbonate, bromide, boric acid, and carbonate (Dawes 1998). In marine waters, salinities of 35‰ (parts per thousand, or practical salinity units) are most common, but salinity may range from 10‰–70‰. Tide-pool inhabitants may experience salinity increases as a result of evaporation and salinity decreases resulting from rainfall. Estuarine seaweeds face variation in salinity resulting from freshwater influxes and ocean tides.

An increase in salinity lowers the external water potential, which triggers rapid cell plasmolysis as well as stress responses by the seaweeds. Responses may include uptake of ions (e.g., K^+, Na^+, or Cl^-), water loss, and synthesis of osmotically active carbohydrates such as sucrose in green seaweeds, mannitol in brown algae, and digeneaside in reds. Other osmotically active compounds in seaweeds include amino acids such as proline, quaternary ammonium compounds (betaines), and DMSP (β-dimethylsulfoniopropionate).

Decrease in salinity results in increased external water potential. Responses include uptake of water by the seaweeds with concomitant increase in cell volume and turgor pressure (the internal osmotic pressure against the cell wall or cell membrane) and loss of ions and organic solutes, leading to osmotic adjustment. If plasmolyzed cells are exposed to solutions having a lower ionic content than their cytoplasm, they undergo deplasmolysis; if this is too extensive or rapid, cell damage may result. Whether salinity is increased or decreased, if osmotic adjustments are not made, damage to cell membranes, organelles, and enzymes (and possibly cell death) will result. Extended exposure to higher-than-optimal salinity inhibits cell division and may result in stunted growth and abnormal branching patterns.

Intertidal seaweeds tolerate a wide range of salinities (10‰–100‰), while subtidal forms tolerate a narrower range (18‰–52‰). *Ulva* species are remarkable in that they can be repeatedly plasmolyzed and deplasmolyzed without harm, a feature that explains their success in estuarine environments. Red and brown algae, in contrast, rarely penetrate as far into estuaries as does the green *Ulva*. *Fucus* can recover after loss of as much as 25% of its water content (Lobban and Harrison 1994). The related *Pelvetia canaliculata* apparently requires periodic emersion, since it decays if submerged for periods longer than 6 out of every 12 hours. This seaweed often hosts an endophytic fungus, and it is hypothesized that desiccation keeps the fungus from causing decay of its host (Schonbeck and Norton 1978).

Exposure to air results in dehydration, increased cellular solute concentrations, and loss of the ability to take up dissolved nutrients. The degree of water loss is related to seaweed surface area and volume: Thicker bodies are more resistant to water loss than are thinner forms. The robust body of *Fucus* allows it to remain well hydrated. As long as *Fucus* is well hydrated, photosynthetic rates are as high when emersed as submersed. In general, intertidal algae are regarded as desiccation tolerators (Surif and Raven 1990). Intertidal understory species, however, are able to retain moisture because of protection provided by large intertidal canopy species such as *Saccharina* and *Halosaccion* and are desiccation avoiders. Desiccation is a major structuring force in the rocky intertidal of the Patagonian coast of Argentina (Bertness et al. 2006). Strong dry southern trade winds desiccate the intertidal zone and the only extensive macroalga is the erect coralline red *Corallina officinalis* in the low intertidal of wave-protected bays.

Nutrients

For growth to occur, seaweeds require (in addition to water and light) combined N, inorganic carbon, phosphate, iron, cobalt, manganese, and other elements. Here, our discussion is limited to N, C, and P—the major nutrients known to influence macroalgal growth. In general, the mineral requirements and uptake mechanisms of seaweeds are similar to those of phytoplankton (Chapter 21). Mineral uptake by macroalgae is influenced by irradiance, temperature, water motion, desiccation, and age. Rates of macroalgal mineral uptake are studied by tracking the disappearance of a nutrient from culture medium or by monitoring the uptake and conversion of a radiolabeled substance into algal biomass.

The nutrient that most frequently limits seaweed growth is combined nitrogen. Hanisak (1983) reviewed N uptake by seaweeds and noted that, in general, the rate of uptake of ammonium is greater than that for nitrate. However, at very high concentrations (> 30–50 μM), ammonium can be toxic to macroalgae. Ammonium can be used directly in the synthesis of amino acids. Nitrate may be stored in the cell's vacuole, but before it can be used, it must first be converted to nitrite by means of the cytoplasmic enzyme nitrate reductase. Nitrite must next be reduced to ammonium within the chloroplast, where the chloroplast electron carrier ferredoxin is the source of reductant (Chapter 2). Assays of nitrate reductase activity are sometimes used as an index of the ability of algae to use nitrate. A newer approach is molecular assessment of expression of genes that encode the components of nitrate reductase (by determining mRNA levels).

Marine macroalgae vary in their use of bicarbonate, which is more abundant in seawater than is CO_2. One way to assess this capacity is to measure the natural abundance of the stable carbon isotopes ^{13}C and ^{12}C in seaweed bodies and calculate the ratios of these two substances (Chapter 2). In a large survey of the stable carbon isotopic ratios of seaweeds, Maberly et al. (1992) found that 6 species of green seaweeds, 12 species of brown seaweeds, and 8 species of red macroalgae were able to use bicarbonate but that 6 other rhodophytes were obligate CO_2 users. Obligate CO_2 users may occur among some macroalgae because subtidal red algae may experience light limitation of photosynthesis to such a degree that photosynthesis is not limited by CO_2 availability.

Phosphate can limit the growth of macroalgae under some conditions, particularly during rapid growth, partly because it is required for production of DNA and ATP. Some seaweeds are able to supplement uptake of dissolved inorganic phosphate with phosphate cleaved from organic sources, such as phosphomonoesters. In these seaweeds, phosphorus deficiency induces the production of cell surface alkaline phosphatases that actively release phosphate from extracellular organic compounds. For example, alkaline phosphatase activity was detected in tips of *Fucus spiralis* that had undergone a sharp decrease in the P:N ratio as a result of rapid meristematic growth (Hernandez et al. 1997). An alkaline phosphatase activity assay can be used as an indicator of P limitation in seaweeds.

22.2 Biological Factors and Macroalgal Adaptations

Herbivore Interactions

The herbivores that affect macroalgae are relatively small mesograzers such as amphipods, copepods, and polychaetes, in addition to larger macrograzers including urchins and fish. Interactions between macroalgae and associated invertebrates may be mutually beneficial or harmful to the alga. In high intertidal pools along the California coast, Bracken and Nielsen (2004) found that the excretion of ammonium by algal-associated invertebrates increased the algal species diversity in the tide pools and could provide more than the amount of ammonium required by turfs of *Cladophora columbiana* (Bracken et al. 2007). The invertebrates benefit from the shelter provided by the algae. Similarly, Pfister (2007) showed that the presence of the mussel *Mytilus californianus* increased growth of the red alga *Prionitis lanceolata* in tide pools along the coast of Washington State (United States). Macroalgal growth was four times higher in the presence of the mussels due to increased levels of dissolved inorganic nitrogen (DIN).

Some macroalgae respond to grazing pressure via production of secondary defensive compounds and/or structural defenses (Duffy and Hay 1990). Presence or absence of defenses can influence the direction of seaweed community change. For example, in a mesocosm (tank-based) study, the chemically defended red alga *Hypnea spinella* dominated the attached algal community after grazers (gammarid amphipods) effectively removed non-defended filamentous algae (Brawley and Adey 1981). Macroalgal defensive compounds include terpenes, acetogenins, alkaloids, and phenolics, which (unlike the secondary compounds of terrestrial plants) usually contain halogens such as bromine. Macroalgae often produce more than one defensive compound, particularly when protection against several different types of herbivores is required. Species of the red alga *Laurencia* (some species have been transferred to *Osmundea*) produce more than 500 different types of terpenes (Faulkner 1993). The Hawaiian brown alga *Dictyota mertensii* produces a sterol known as dictyol-H, which is ineffective against amphipods but protects against herbivorous fish (Fleury et al. 1994). Many species of brown seaweed of both tropical and temperate regions produce phlorotannins as herbivore defenses.

Phlorotannins deter feeding and reduce food quality for animals that do consume macroalgae containing them. Targett and Arnold (1998) reviewed the occurrence, diversity, and ecological functions of phlorotannins.

Defensive chemicals may be constitutive (produced continuously) or inducible (produced only when required). An example of an inducible chemical defense is the production by *Fucus* of higher levels of phenolic compounds when herbivores are present than when they are absent (Van Alstyne 1988). Production of defense compounds results in energetic costs to seaweeds. Such compounds are not simple proteins, and therefore multiple biosynthetic steps, enzymes, and genes are required for their synthesis. Seaweeds that produce large amounts of defensive compounds may have lower growth rates than related forms that lack them (DeMott and Moxter 1991). In view of energy costs, producing defense chemicals only when they are needed—in response to induction by herbivores—should be favored (Amsler 2001). However, if herbivory occurs at unpredictable intervals on a nonseasonal basis, the time period required by an alga to synthesize protective compounds may be too long to provide effective protection (Padilla and Adolph 1996). In such cases, constitutive production of defense compounds may be more effective.

Interestingly, marine animals may take advantage of algal secondary compounds to gain protection for themselves, analogous to the way terrestrial monarch butterflies use bad-tasting secondary compounds in their milkweed food plants to acquire protection from predatory birds. For example, the tube-building temperate Atlantic amphipod *Ampithoe* selectively lives and feeds on the chemically defended brown alga *Dictyota menstrualis*. This seaweed exudes diterpene alcohols that discourage fish from consuming both the seaweed and the epizooic amphipod (Duffy and Hay 1994). Given a choice of foods, however, *Ampithoe* prefers the edible *Enteromorpha* (*Ulva*) *intestinalis* to *Dictyota*, although it derives no protection from it (Cruz-Rivera and Hay 2003). Predatory fishes readily consume these amphipods if they are removed from tubular enclosures but not animals that remain enclosed. Hay (1996) summarized these and other examples of marine chemical ecology.

The tough, spiny bodies of the tropical brown alga *Turbinaria* (Figure 22.16) and calcified bodies of the coralline red algae (none of which are chemically defended) represent conspicuous examples of structural defenses (Padilla 1985, 1989).

Figure 22.16 *Turbinaria*, a tropical brown alga related to *Fucus*, is characterized by tough, spiny blades and thus has structural defenses against herbivory. (Photo: R. J. Stephenson)

In some macroalgae, such as *Halimeda*, the timing of growth, secondary defense compounds, and calcification are all involved in herbivory protection. New segments (see Chapter 17) are produced at night, when fish—which find food visually—are less active. The poorly calcified young segments contain more defensive compounds than do older, more extensively calcified regions of the body (Hay 1996). Other aspects of algal structure, including life histories involving heteromorphic alternation of generations, have been interpreted as ecological responses to herbivore interactions. Thornber et al. (2006) showed that herbivores have strong preferences for different life cycle stages of macroalgae. The snail *Tegula funebralis* was three times more likely to graze on gametophyte reproductive tissue of the red alga *Mazzaella flaccida* than on any other tissue type, while the urchin *Strongylocentrotus purpuratus* preferred sporophyte reproductive tissue of the same alga.

Several workers have contributed to the development of form–function theory, a body of thought that seeks to provide ecological explanations of algal morphologies and life histories (Littler and Littler 1980; Littler et al. 1983; Hay 1981; Padilla 1985). The theory allows predictions to be made regarding correlations between morphological type, grazer resistance, food quality, productivity, reproductive success, growth mode, and role in succession. Seaweeds have been classified into functional-form groups such as sheets, filaments, coarsely branched forms, leathery bodies, jointed calcareous forms, and crusts. Often there appears to be a trade-off between the ability to grow rapidly and the ability to resist herbivores (Figure 22.17). For example, fast-growing sheets of *Ulva* and *Porphyra* are vulnerable to grazing, whereas the tough, stony bodies of corallines are grazer resistant but incredibly slow growing. Some corallines are thought to be hundreds of years old. In general, the more highly structured seaweeds are thought to gain persistence in time at the cost of high productivity levels (Littler et al. 1983). However, coralline success depends upon grazer removal of epiphytic diatoms, which can damage coralline surfaces by overgrowth (Steneck 1990).

Competition

Important interactions occur between physical factors, grazing, and levels of interspecific competition in macroalgae. Competition arises when two or more individuals or species use some limiting resource, such as light, space, or nutrients, at the same time. Annual seaweeds—those not persisting from one growing season to the next—are typically not herbivore resistant or good competitors. Rather, they are well adapted, by virtue of fast growth rate, to colonize newly available substrates such as recently formed volcanic islands and shores scoured by waves or ice. Common annuals include *Ulva* and *Porphyra*. Perennials—seaweeds that can live for more than 1 year—can typically outcompete annuals. *Macrocystis pyrifera* lives as long as 7 years, *Cystoseira barbata* and *Constantinea rosa-marina* 18–20 years, and *Ascophyllum nodosum* more than 30 years. Longevity of individuals is estimated by counting annually formed vesicles in *Ascophyllum* and in *Constantinea* by counting scars left by the annual loss of the peltate blade (Lüning 1990). Perennial seaweeds are often well adapted to resist herbivores. Olson and Lubchenco (1990), Carpenter (1990), and Paine (1990)

reviewed data relating to the occurrence of interspecific competition. Harlin (1987) reviewed the role of allelochemical interactions in macroalgae.

Grazing by herbivores may interact with physical factors to alter distribution patterns. Along the south coast of the United Kingdom, the intertidal brown algae *Fucus vesiculosus* and *F. spiralis* form a canopy in wave-protected landward-facing rocky sites. Heavy waves in seaward-facing sites dislodge or prune *Fucus* larger than 10 cm. If the main algal grazer—the limpet *Patella vulgata*—is removed, however, *Fucus* is able to establish in seaward rocky sites (Jonsson et al. 2006).

On the New England coast, the red alga *Chondrus crispus* dominates sheltered subtidal areas, outcompeting the browns *Fucus* and *Ascophyllum*, but, farther north, severe ice scouring in the winter gives the edge to the latter two algae (Lubchenco 1980). In subtidal locations in southern New England, the red alga *Chondrus crispus* benefits from the presence of two species of gastropods. One species removes overgrowth of ascidians from *Chondrus*, and the other removes fouling by bryozoans. Only *Chondrus* with both gastropods present can grow. The snails, in turn, benefit from the association by gaining a refuge from crab predation (Stachowicz and Whitlatch 2005). In some cases both herbivores and potential algal competitors are responsible for structuring the community. In New England mid-littoral tide pools, *Fucus* and its relative *Pelvetia* are noticeably absent unless both herbivores such as the snails *Littorina* and *Achmaea* and the seaweeds *Chondrus crispus*, *Ulva lactuca*, *Rhizoclonium tortuosum*, *Spongomorpha spinescens*, *Polysiphonia* sp., *Dumontia incrassata*, and *Scytosiphon lomentaria* are removed (Lubchenco 1982).

In the rocky intertidal of the Washington State (United States) coast, *Saccharina* is often dominant in areas of moderate wave exposure but is outcompeted by *Laminaria* and *Lessoniopsis* in more exposed areas with heavy wave action. The herbivorous sea urchin *Strongylocentrotus pupuratus* has a major impact on algal communities by overexploiting its prey and clearing areas of all but a few species of coralline algae. The carnivorous starfish *Pycnopodia* and the sea anemone *Anthopleura* consume *Strongylocentrotus*, clearing tidepools of urchins for seaweed recolonization (Dayton 1975). In the same area Paine (2002) found that the less-productive perennials *Saccharina* and *Laminaria* dominated the rocky intertidal when macrograzers were present at normal densities, but more productive annuals *Alaria* and

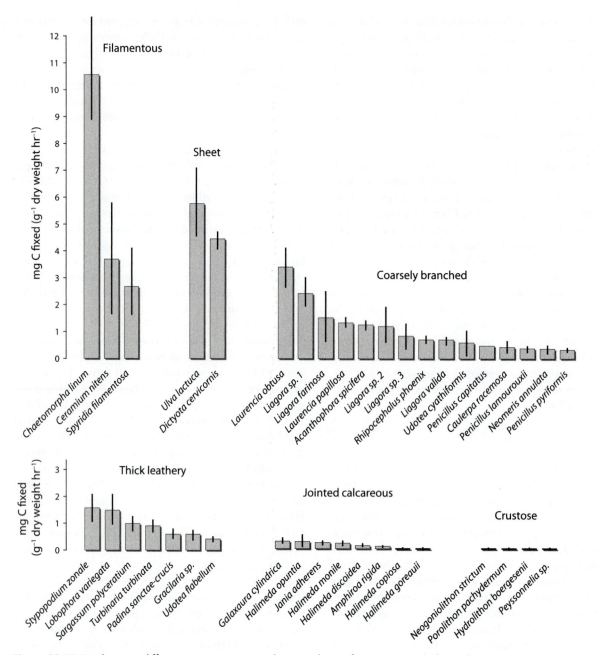

Figure 22.17 Productivity differences among tropical macroalgae of various morphological types. (Adapted from Littler et al. 1983, by permission of the *Journal of Phycology*)

Nereocystis predominated when macrograzers were reduced. These studies emphasize the importance of considering as many direct and indirect physical and biotic factors as possible in attempts to understand the ecological structure of seaweed communities.

Algal epiphytes may compete with their hosts for light and nutrients. Consequently, some seaweeds possess effective methods for ridding themselves of such competitors. Fucalean brown algae and crustose corallines, which often appear free of epiphytes, shed their outer cell layers or the outer portions of cell walls. Other seaweeds undergo continuous erosion of the blade tips or edges, eliminating the epiphyte-ridden older portions. Such strategies also work against animal epiphytes such as bryozoans (Lüning 1990). A study of epiphyte loads on crustose coralline algae in Britain showed that an overlying canopy of larger seaweeds was the most important factor in reducing epiphytes, but herbivory and coralline surface features were also important (Figueiredo et al. 1996).

Intraspecific competition has not been as intensively explored as interspecific competition. Intraspecific competition may result in self-thinning, a process commonly observed in terrestrial plant populations in which biomass increase is coupled to decrease in density over time. In self-thinning, better-adapted individuals survive and utilize available resources more effectively than less-well-adapted individuals of the same species. In an investigation of *Himanthalia elongata*, a brown alga that frequently occurs in dense monospecific stands on rocky north temperate shores, Creed (1995) detected evidence for self-thinning. Young individuals were initially regularly spaced, but over time, spatial patterning became more irregular, with larger individuals surviving better than smaller ones.

The Ecological Roles of Pathogens

Seaweed pathogens include viruses, bacteria, fungi, other algae, and heterotrophic protists (Chapter 3). Determination that a particular pathogen causes a particular disease of seaweeds (or any organism) requires fulfillment of Koch's postulates: (a) The pathogen must always be associated with the disease symptoms, and no symptoms should be present in the absence of the pathogen; (b) the pathogen must be isolated and grown either in laboratory culture or in a susceptible host; (c) when inoculated onto a disease-free host, the isolated pathogen must be able to cause the symptoms associated with the disease; and (d) the pathogen must be re-isolated from the experimentally infected host and shown to be the same organism as the original isolate. Such information is helpful in designing strategies to prevent or reduce effects of the disease. Molecular methods have been successfully applied to the identification of seaweed pathogens (Ashen and Goff 1998). Sussmann et al. (1999) sequenced the internal transcribed spacer regions (ITS1 and ITS2) of nuclear ribosomal DNA from green algal endophytes taken from the red algae *Mazzaella splendens* and *Mastocarpus papillatus* from the coast of British Columbia. The results identified the green algal endophytes as alternate life

history stages of the filamentous green alga *Acrosiphonia*. A molecular investigation of the kelp *Lessoniopsis littoralis* from British Columbia revealed that gametophytes of *Alaria*, *Macrocystis*, and *Nereocystis* colonized older portions of *L. littoralis* as epiphytes and endophytes (Lane and Saunders 2005). The focus here is upon algal pathogens of seaweeds because such pathogenesis may reflect interspecific competition.

Algae (including cyanobacteria) that occur as endophytes ramifying within host tissues may have pathogenic effects and cause noticeable symptoms. Endophytes often cause spots (areas of increased pigmentation) or abnormal twisting of seaweed stipes or blades. Such symptoms may sufficiently affect seaweed appearance to be of concern to those who harvest seaweeds for markets, and they may be ecologically significant if growth is seriously impacted. Algal endophytes of seaweeds include the filamentous green *Acrochaete*, red *Audouinella*, and brown *Streblonema*, which typically grow in host intercellular spaces rather than intracellularly. There are numerous parasitic red algae, however, that infect rhodophycean hosts by the proliferation of parasitic nuclei within host cells (Chapter 15). Movement of photosynthates between the epiphytic red alga *Polysiphonia* and its host, the brown *Ascophyllum*, and from red algal hosts to rhodophycean parasites with reduced pigment content has been documented (Chapter 15). For most hosts and their endophytes, the extent of nutrient transfer, if any, is unknown. By associating with larger hosts, endophytes may obtain protection from desiccation, high irradiance, or herbivory.

A deformative disease of *Mazzaella* (formerly *Iridaea*) *laminarioides* is caused by infection with a species of the cyanobacterium *Pleurocapsa* (Correa and Sánchez 1996). Green patch disease of *M. laminarioides*, which can co-occur with the *Pleurocapsa* infection, is caused by a filamentous green endophyte, the appropriately named *Endophyton*. Infections are present throughout the year, and *Endophyton* can affect 20%–80% of the *Mazzaella* population. *Endophyton*, as well as two endophytic species of the green alga *Acrochaete*, are pathogens of *Chondrus crispus*, causing similar symptoms to those observed in *Mazzaella*. Host specificity of *Acrochaete operculata* appears to be determined by host cell-wall polysaccharides, specifically lambda carrageenans. The sporophytes of *Chondrus crispus* and *Mazzaella cordata* (which possess lambda carrageenans) become infected, but the gametophytic generations (which lack lambda carrageenans) are not susceptible (Correa and Sánchez 1996). This suggests that attachment of propagules of the green algal pathogens may be dependent upon specific chemical attributes of host surfaces.

Other macroalgal diseases caused by endophytes include brown spot disease of cultivated *Undaria pinnatifida* and gall induction in kelps, both of which result from infection with the filamentous brown endophyte *Streblonema* (Correa 1997). *Streblonema* is a fully pigmented form that is found widely within kelps; some *Streblonema* species also occur within red algae such as *Grateloupia*. *Saccharina latissima* often contains the brown endophyte *Laminarionema*, which causes "twisted stipe disease." There is some evidence that brown algal reproductive cells, like those of the green endophytes described above, are able to detect the presence of suitable hosts, probably through chemical means, and to attach to them specifically. However, the biochemical mechanisms underlying specificity are poorly understood. Green and brown algal endophytes/pathogens, as well as numerous cases of red algal parasitism, would seem to offer intriguing opportunities for ecological analysis of interspecific competition, the costs and benefits of the endophytic habit, and pathogen–host evolutionary interactions.

22.3 Macroalgal Biogeography

Biogeography is the study of the geographical distribution patterns of organisms and the mechanisms that produce the patterns. Biogeography is becoming increasingly important as phycologists seek to predict the effects of global climate change on algal distribution and biodiversity. Changes in the global environment, including temperature, ocean pH, UV levels, and nutrient content of coastal waters, are all expected to affect algal biodiversity and associated organisms. Human-aided migrations of seaweeds to new areas—where, in the absence of natural herbivores or pathogens, they sometimes grow to nuisance levels—have also become a major concern. The brief survey of seaweed biodiversity presented here focuses on the geographical sites of highest diversity and biomass of algae as places where global

change could have a large impact and on locations where endemism, and thus extinction rates, could be high.

The present distribution of seaweeds has resulted from both patterns of migration (dispersal from a center of origin) and historical speciation events related to the establishment of geographical barriers to interbreeding. Such barriers include boundaries of dramatic water temperature change and the geological opening or closing of marine basins. Ancient taxon-splitting events (adaptive radiations) can be detected by the use of cladistic (phylogenetic) analyses, which infer the phylogeny of a related group of organisms. In recent years molecular approaches have increasingly been used to deduce the relationships among species (Chapter 5).

Seven major seaweed floristic regions are commonly recognized: Arctic, Northern Hemisphere Cold Temperate, Northern Hemisphere Warm Temperate, Tropical, Southern Hemisphere Warm Temperate, Southern Hemisphere Cold Temperate, and Antarctic. Aside from the Arctic and Antarctic, each of these regions is further subdivided into component areas that are not covered here (see Lüning 1990). These regions are characterized by distinctive climatic features (notably seawater temperature) and by the organisms that inhabit them. The present positions of the continental landmasses and ocean currents affect the regions and their organisms (Figure 22.18). Currents that are particularly important in explaining seaweed occurrence include the California Current on the west coast of North America, the Humboldt Current (or Peru Current) off the western coast of South America, the Benguela Current, which influences the south and west coasts of Africa, and the West Wind Drift, a current that influences all of the Southern Hemisphere continents. The cold-water Humboldt, California, and Benguela currents generate coastal upwelling. Upwelling increases the level of inorganic nutrients and thus increases primary productivity of phytoplankton and seaweeds in these regions. Major ocean currents also set up temperature belts (isotherms) that do not usually run parallel to the equator or Tropics of Cancer and Capricorn. Isotherms are boundaries defined by the same surface water temperature averaged over many years, for a particular month. Along any coast, the northern and southern limits for growth of a particular seaweed species are marked by the February and August isotherms (Lüning 1990).

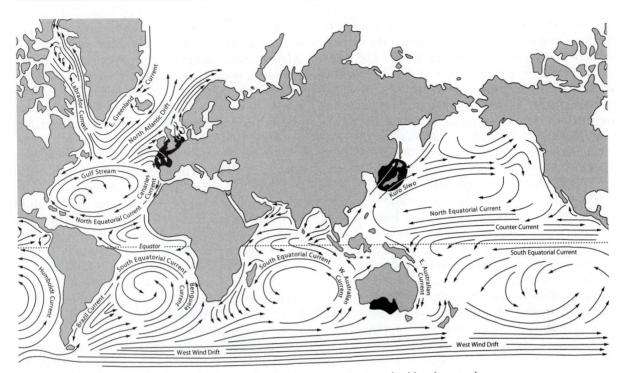

Figure 22.18 Major currents of the world's oceans. The three major macroalgal biodiversity hotspots are indicated by shading of coastal areas around Western Europe, Japan, and Southern Australia. (Current map after Lüning, *Meeresbotanik*, 1985. © Georg Thieme Verlag, 1985)

Every seaweed species has high and low lethal temperature limits set by the tolerance of the least hardy life-history stage. This is often a small cryptic stage capable of persisting for long periods, reproducing only asexually, under conditions that do not permit completion of the sexual life cycle or production of more conspicuous life-history stages. If conditions change in a direction that allows sexual reproduction to occur, the conspicuous life-history stage may appear to have suddenly invaded the area. Some seaweeds have a wide temperature tolerance and are eurythermal, whereas others (termed stenothermal) have narrow temperature tolerances. If a particular geographical region experiences environmental conditions beyond those of stenothermal seaweeds, such seaweeds will not occur there. For example, the brown seaweed *Saccorhiza polyschides*, which is common along shores of the eastern Atlantic Ocean (European coasts), is excluded from western Atlantic shores (North America) because sporophytes die at temperatures above 24°C and below 3°C, and all North American shores experience extremes beyond these temperature limits (Lüning 1990).

Tropical Seaweeds

Many tropical marine genera have a circumglobal distribution. Examples include the green ulvophycean genera *Codium*, *Caulerpa*, and *Acetabularia*; the browns *Dictyota*, *Dictyopteris*, *Padina*, *Sargassum*, and *Turbinaria*; and the reds *Gelidium*, *Pterocladia*, *Galaxaura*, *Liagora*, *Halymenia*, *Grateloupia*, *Jania*, *Amphiroa*, and *Laurencia*. The common occurrence of pantropical seaweeds is explained in terms of the great age of the habitat. Since 180 million years ago, there has been a continuous equatorial belt of warm water that has been interrupted only relatively recently by land bridges (closure of the Mediterranean about 17 million years ago and origin of the Central American Land Bridge some 3–4 million years ago). Thus for many millions of years there was no major barrier to tropical seaweed migrations. The tropical warm water belt is bordered by north and south 20°C winter isotherms, and water temperatures may rise to 30°C. The boundaries of tropical seaweed occurrence coincide with those of reef-forming corals—a habitat of very high animal-species diversity.

For many groups of organisms such as mammals and flowering plants, the tropics contain the highest levels of species richness (Gaston 2000). The tropics might therefore be expected to also harbor the world's largest diversity of seaweeds, but this is not the case. Rather, the three geographical regions that are presently richest in seaweed species are the warm temperate coasts of southern Australia and the cold temperate coasts around Japan and Western Europe (Kerswell 2006). The 5500-km-long coast of southern Australia hosts 1100 seaweed species with a high level of endemism. About 30% of the 800 species of red algae, 20% of the 231 species of browns, and 11% of green seaweed species are restricted to this region (Womersley 1984, 1987). One explanation for this seeming paradox is that in the tropics, corals effectively compete with macroalgae for settling space. Another is that constraints on colonization of remote tropical islands may have contributed to reduced diversity. The number of immigrating species decreases with both island size and increasing distance from continental coasts. The seaweed flora of Easter Island, the most remote island of the Indo-Pacific region, consists of about 170 species, 80% of which are widely distributed in warm waters (Santelices and Abbott 1987), whereas there are more than 300 species in the seaweed flora of the Galapagos Islands (Lüning 1990), which are closer to South America, where some 400 seaweed species occur (Santelices 1980). The species richness of the southern Australian coastline, however, is enhanced by the direct flow into it of the tropical Leeuwin Current that brings tropical flora into the warm temperate zone (Kerswell 2006).

Cold Temperate and Polar Seaweeds

The cold temperate and polar floras of the Northern Hemisphere differ substantially from those of the Southern Hemisphere, presumably because they have been long separated (for at least 65 million years) by the tropical warm belt. Northern and Southern Hemisphere cold temperate and polar seaweeds are thus thought to have arisen in isolation, though there are a few examples of transmigrants. A few seaweeds are thought to have crossed the equator during the Pleistocene, when the mean temperature of the tropical belt was somewhat lower and the mean width somewhat less. *Macrocystis pyrifera* is thought to have crossed the equator along the Pacific Coast of Central America, moving southward against the

Humboldt Current, while *Laminaria ochroleuca* appears to have migrated southward along the West African coast. In the Southern Hemisphere the cold temperate seaweed flora is similar around the world, primarily because the southern continents once had common coastlines as components of Gondwana and because the West Wind Drift facilitates migrations (see Figure 22.18).

In contrast, the cold temperate seaweed floras of the North Atlantic differ greatly from those of the North Pacific, which has much greater diversity and a higher proportion of endemic species. Two-thirds of the 60 or so cold-water North Pacific red algal genera are endemics, compared to only 4 of the 30 cool-water red algae in the Arctic North Atlantic (Lindstrom 1987). Greater Pacific diversity and endemism is explained by the greater age of the North Pacific Basin (Mesozoic) compared to the North Atlantic (no earlier than the Tertiary) and the fact that biotic exchange via the Arctic Ocean has been possible only since the late Tertiary (Lüning 1990) and not at all since the last interglacial (Wilce 1990). The Pacific Basin is thought to have been the center of origin of the Laminariales. Only one of the 10 Pacific *Alaria* species, *A. esculenta*, also occurs in the North Atlantic. Several *Laminaria* species (or closely related sister species) occur in both the North Atlantic and North Pacific, *Saccharina latissima* being a common and conspicuous example (Figure 22.19). The coastal areas around Japan are the second of the three major regions of macroalgal biodiversity. This biodiversity hotspot contains more than 450 genera and as many as 21 endemic genera of macroalgae. As in the southern Australian hotspot, a tropical current, the western North Pacific current of the North Pacific gyre, has significantly enhanced the species richness in this region (Kerswell 2006).

The Arctic Sea has few or no endemic species; its comparatively depauperate flora derives from invasion of cold-adapted North Atlantic species at the end of the Pleistocene. Extremely stressful (ice-scoured) habitats, such as the heads of northern fjords, or sites lacking solid substrata for seaweed attachment, have very sparse algal communities, giving the impression that the Arctic is characterized by few species and low biomass. But more favorable sites with hard substrata, such as the rocky coasts of northwest and east Greenland and much of the eastern coastline of Canada, are more productive. Examples of Arctic green seaweeds include *Ulothrix flacca*, *Rhizoclonium riparium*, *Blidingia minima*, and *Endocladia*

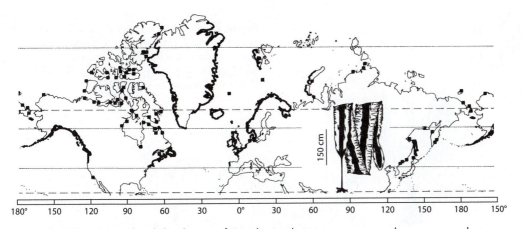

Figure 22.19 Biogeographical distribution of *Saccharina latissima*, a common brown macroalga. This species occurs in both the North Pacific and the North Atlantic. (Adapted from Lüning, *Meeresbotanik*, 1985. © Georg Thieme Verlag, 1985)

viridis. Some Arctic red seaweeds are *Palmaria palmata*, *Clathromorphum compactum*, *Lithothamnion glaciale*, and *Polysiphonia arctica*. Arctic browns include *Sphacelaria arctica*, *Desmarestia aculeata*, *Chorda tomentosa*, and *Saccharina latissima* (Wilce 1990).

The eastern North Atlantic is rich in seaweed species compared to the western North Atlantic because the Gulf Stream and the North Atlantic Drift (see Figure 22.18) bring warm water originating in the tropics and allow migration across the Atlantic of many warm-adapted species that are dispersed along the coasts from Spain to the Netherlands. The Gulf Stream thus supports one-third of the macroalgal biodiversity hotspots (Kerswell 2006). In contrast, the diversity of the northwest Atlantic coast of North America is low because the Labrador Current from the Arctic dominates coastal flows.

The temperate and polar seaweed floras of the Southern Hemisphere are cold-adapted communities that have evolved in parallel to those of temperate and polar floras of the Northern Hemisphere. The distinctness of their algal communities reflects the separation of the coasts of Gondwana and Laurasia 135 million years ago by the tropical warm water belt. Compared with the Northern Hemisphere, where migrations occur mainly along coastlines, ocean currents are more significant in the migration of seaweeds of the Southern Hemisphere. Temperate Southern Hemisphere intertidal waters are regarded as the site of origin of the Fucales because the greatest diversity of fucaleans occurs there (Clayton 1984).

Fucales of tropical and northern waters are thought to have migrated there, probably aided by the air-filled floats common in this group. Approximately 100 species of seaweeds occur in Antarctic waters, and about one-third of these species are endemics. In contrast, the level of endemism in the Arctic is only about 5%. The difference in degree of endemism has been attributed to the long history of isolation of the Antarctic and the absence of coastal connections to cold temperate regions. No laminarialeans occur in the Antarctic, where perennial members of the Desmarestiales, which are thought to have originated in the Southern Hemisphere, dominate the seaweed flora. Dense thickets of the 10-m-long, 1-m-wide *Himantothallus grandifolius* and *Desmarestia* occur at 5–35 m in depth, while the smaller desmarestialean *Phaeurus antarctica* occurs in tide pools and along shores to depths of 10 m (Figure 22.20) (Clayton and Wiencke 1990). As is the case for Arctic seaweeds, Antarctic macroalgae store carbohydrates during the long days of austral summer for use in growth during austral winter.

Seaweed Migration

Seaweeds capable of migrating and colonizing new regions are less vulnerable to extinction than those having very restricted distributions. Seaweed migrations can occur either by short-distance "stepping-stone" transfers from one coast to another close by or by long-distance movements via currents. Migrations are accomplished by production of propagules

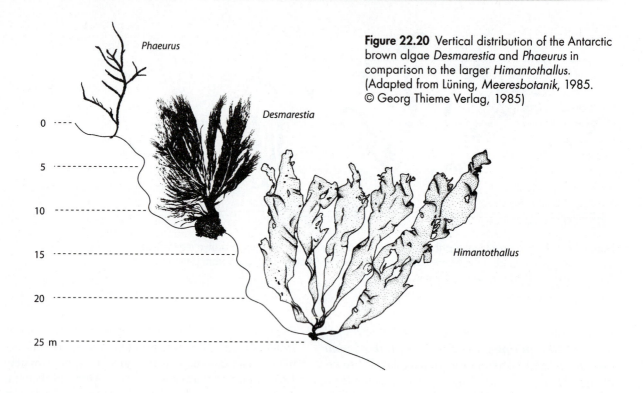

Figure 22.20 Vertical distribution of the Antarctic brown algae *Desmarestia* and *Phaeurus* in comparison to the larger *Himantothallus*. (Adapted from Lüning, *Meeresbotanik*, 1985. © Georg Thieme Verlag, 1985)

such as zoospores or by floatable thalli. Zoospores of the annual colonizers *Ulva*, *Blidingia*, and *Ulothrix* can remain alive in the plankton during migrations over tens of kilometers. These seaweeds were early colonizers of the island of Surtsey, which was formed in 1964 some 35 km from the Icelandic coast, as well as on artificial substrates mounted 35 km away from the North American coast (Amsler and Searles 1980). *Macrocystis pyrifera*, which is equipped with many floats (pneumatocysts), and one floatable species of *Durvillaea* (*D. antarctica*) have been able to migrate throughout the cool temperate Southern Hemisphere via the West Wind Drift. These large seaweeds may have served as rafts for co-migration of epiphytic forms.

During the 20th century an increase in seawater temperature in the Northern Hemisphere was correlated with northward migration of several warm-water macroalgal species (*Dictyota dichotoma*, *Desmarestia ligulata*, and *Gracilaria verrucosa*) as far as Norway, the migration of some Mediterranean brown algae (*Taonia atomaia* and *Dictyopteris membranacea*) to the British Isles, and the spread of *Laminaria ochroleuca* from France to Britain. Such events help phycologists predict the effects of future climate change on seaweed distribution.

More recently, marine ecologists have become greatly concerned about the sudden appearance and spread of nonnative algal species via human activities, such as the dumping of ship ballast waters. Williams and Smith (2007) estimated that the global number of introduced seaweed species stood at 277 as of 2007. The rhodophyte *Asparagopsis armata* has been introduced from Australia to northern Europe, and the ulvophycean *Codium fragile*, a macroalga that fouls ships, has spread from Japan throughout temperate and polar coastal waters of the northern and southern hemispheres (Trowbridge 1998). Although many introduced seaweeds appear to have little effect on native communities, there are a number of nuisance growth-forming migrants. The brown seaweed *Colpomenia* apparently entered European waters from Japan through the exchange of oyster cultures. *Colpomenia*'s gas-inflated bodies have become a nuisance in that their presence on shellfish will often make the shellfish so buoyant that the shellfish float away. Hence the name "oyster thief" is commonly applied to *Colpomenia*. *Sargassum muticum* was probably introduced from Japan to the United States, where it became a nuisance in harbors and along beaches. In Europe, *S. muticum* has spread rapidly and outcompeted native fucaleans and laminarialeans, with serious ecological effects (Ribera and Boudouresque 1995). In Hawaii the introduced red seaweeds *Hypnea musciformis* and *Acanthophora spicifera* occupy the same niche as the native *Hypnea cervicornis*, which has seriously declined. In the 1970s species of the red alga *Kappaphycus* were introduced to Hawaii for the carrageenan industry and *Gracilaria salicornia* for the agar industry. Some 30 years later these species are overgrowing reef habitats and killing corals (Smith et al. 2004a; Conklin and Smith 2005). The introduction and subsequent spread of *Caulerpa taxifolia* in the Mediterranean has had major ecological consequences (Delgado et al. 1996). A second species of *Caulerpa*, *C. filiformis*, has become dominant in Sydney Harbor, Australia, replacing a native species of this genus (Ribera and Boudouresque 1995). An experimental study in tide pools examined the role of different algal functional groups on susceptibility to invasion. A well-developed algal canopy inhibits invasions by other species, and any human activity that reduces native canopy species promotes invasion (Arenas et al. 2006).

Effects of Pollution and Overfishing on Macroalgae

Pollution of and overfishing in marine coastal waters are a direct result of expanding human populations generating increased demand for food, expanding agricultural activities, and increasing urbanization with rising sewage volumes. Pollutants that affect seaweeds include chlorinated hydrocarbons, oil, herbicides, insecticides, heavy metals, radioactive isotopes, chlorine and copper used to reduce ship fouling, and sediments. The most widespread pollution effect, however, is the massive dumping of nutrients from agricultural activities and urban sewage into coastal marine waters. On a worldwide basis, more than 90% of sewage from coastal areas enters the ocean untreated (Lobban and Harrison 1994), where it may produce nuisance growths of phytoplankton and seaweeds such as *Ulva*, *Cladophora*, and the coral-smothering *Dictyosphaeria* (Raffaeli et al. 1998). Oil spills affect seaweeds by reducing photosynthetic rates or interfering with gamete or spore release. Among the heavy metals discharged by industrial activity, mercury is the most toxic; its mechanism of action is inhibition of enzyme activity. In addition, copper can interfere with cell permeability and enzyme function, while cadmium inhibits photosynthesis and protein synthesis. Seaweeds

growing in water released from nuclear power plants can accumulate radioactive metals. Humans should avoid consuming macroalgae and other seafood from the vicinity of nuclear plant outfalls. Sediment loads are increasing in marine coastal waters as a result of agricultural activities that lead to erosion of soils into streams and ultimately into coastal waters. Sediments are especially harmful to coral reefs, where corals may be suffocated and crustose coralline algae buried by sediments. Overfishing removes herbivores and predators that stabilize macroalgal communities. Although pollution and overfishing are widespread, they are considered local problems rather than global because they are amenable to local management such as sewage treatment and fishing regulation. Halpern et al. (2008) reported the development of a global map of human impacts on marine ecosystems. Their map shows that no coral reefs are in a very low impact category, and two of the three macroalgal biodiversity hotspots, the areas around Japan and the west coast of Europe, are among the very highest impacted areas in the oceans.

Increased levels of nutrients in tropical coastal waters have had profound effects on coral reef communities, but for more than a decade the role of nutrients was mired in controversy. As in freshwater systems, where researchers argued for bottom-up (nutrient) control or top-down (grazer) control of phytoplankton (Chapter 21), researchers advocated bottom-up, top-down, or a mixed model of algal control on coral reefs (Hughes et al. 1999; Lapointe 1999). Lapointe showed that nitrogen from land-based sewage sources in southern Florida supported blooms of the invasive green algae *Codium isthmocladum* and *Caulerpa brachypus* on deep coral reefs (Lapointe et al. 2005a,b). Studies of both nutrient and herbivore effects often showed herbivory to be the significant effect with nutrients having only a minor or insignificant effect (Jompa and McCook 2002; Diaz-Pulido and McCook 2003). Smith et al. (2001), however, reported significant results for both nutrients and herbivory, with the greatest total algal biomass of fleshy macroalgae on artificial substrates (stone blocks) subject to increased nutrients and herbivore exclusion. On substrates receiving increased nutrients but exposed to herbivory, calcareous algal biomass was greatest. Herbivory reduces fleshy macroalgae. In the absence of herbivory, the balance of competition between corals and fleshy macroalgae favors the macroalgae, which grow over and kill the corals (Jompa and McCook 2002). The fleshy macroalgae release dissolved organic compounds into the surrounding water, and these compounds enhance the growth of microorganisms. When coral and macroalgae were placed in chambers together but separated by a 0.02 µm filter, corals suffered 100% mortality, apparently due to hypoxia brought about by microbial activity (Smith et al. 2006b). Corals remained healthy in the chambers if macroalgae were absent or if the chambers were treated with the antibiotic ampicillin, which eliminated the microorganisms.

The consensus of current opinion in terms of the roles of top-down and bottom-up control of algae on coral reefs is that both mechanisms are important. Increasing levels of nutrients produce an increase in gross growth of macroalgae, but, in the presence of diverse and abundant herbivores, the increase in gross growth may be entirely consumed without any obvious change in biomass. Overfishing of herbivorous fish and eutrophication have shifted many coral reefs from coral dominance to fleshy macroalgal dominance (Hughes et al. 2003). Caribbean reefs were subject to coastal eutrophication and overfishing for years but showed no major macroalgal blooms until a mass die-off of an abundant herbivore, the sea urchin *Diadema antillarum*, occurred in 1983. Massive blooms and coral loss followed except around the Cayman Islands, where the coastal waters remained oligotrophic (Lapointe 1999). The urchin *Diadema* has not yet recovered in the Caribbean, but in some marine reserves, such as the Exuma Cays Land and Sea Park, where fishing has been banned since 1986, parrotfish thrive and maintain the cover of macroalgae at a level four times less abundant inside the park than outside it (Mumby et al. 2006). Results such as these suggest that the best short-term means to protect coral reefs would be to establish more reserves within which fishing is prohibited.

Overfishing has been the most important single factor in the disturbance of kelp forests (Steneck et al. 2002). Elimination of the top predators in different coastal areas has led to a trophic cascade of events that resulted in the deforestation of large areas of kelp forest, commonly through heavy grazing by sea urchins. The chain of events has been well studied in three North American coastal systems. In coastal Alaskan waters, sea otters were nearly hunted to extinction for the fur trade during the 1700s and 1800s. Sea urchins were released from predation and grazed away the kelp forests (Figure 22.21a). Given legal protection in the 20th century, sea otters slowly recovered, urchin populations declined, and kelp forests returned. In the 1990s, however, killer whales began eating otters in response to a decline in seal populations, urchins increased, and kelp forests began disappearing (Estes et al. 1998). In the western North Atlantic, kelp forests and sea urchins were established, but the top predator was the Atlantic cod. Mechanized commercial fishing eliminated the cod stocks from 1930 to 1940, and sea urchin numbers rose (Figure 22.21b). Kelp forests reached record lows in the 1980s and 1990s, when a fishery based on sea urchin roe developed. Urchins declined rapidly, and kelp returned, but crabs occupied the forests and prevented urchins from returning. In contrast to the first two systems, the kelp forests of southern California contain highly diverse food webs with several herbivorous sea urchins, abalone, snails, and crustaceans plus their predators, lobsters, fish, and otters. Otters had been eliminated by the 1800s, but kelp forests did not begin to disappear until the 1950s and 1960s, when overfishing of sheephead fish, lobsters, and abalone freed sea urchins from predators and competitors (Figure 22.21c). With the help of coastal pollution and El Niño events, urchins grazed down the kelp forests through the 1960s. In the 1970s a fishing industry developed based on export of sea urchin roe to Asia, and the resulting reduction in urchins allowed kelp to recover (Steneck et al. 2002). Unfortunately, California kelp forests are suffering a loss of biodiversity that may make them susceptible to invasion and rapid shifts from forests to barrens. Creating reserves in which fishing is banned may be the best way to protect kelp forest biodiversity for the immediate future.

Global Environmental Change

Global environmental change begins with the rise in atmospheric carbon dioxide concentration that has resulted from human activities such as the burning of fossil fuels and deforestation. The concentration of carbon dioxide has risen from a preindustrial level of around 280 ppm to the present 380 ppm, which exceeds by more than 80 ppm any values determined for the prior 740,000 years (Hoegh-Guldberg et al. 2007). As a result of this rise in atmospheric carbon dioxide during the 20th century, the average temperature of the oceans has risen by 0.74°C, sea level has risen by 17 cm, seawater pH has declined by 0.1 pH unit, and the level of seawater carbonate has fallen by 30 µmol kg^{-1} seawater to 210 µmol kg^{-1} seawater. Atmospheric carbon dioxide is predicted to exceed 480 ppm sometime between 2050 and 2100, at

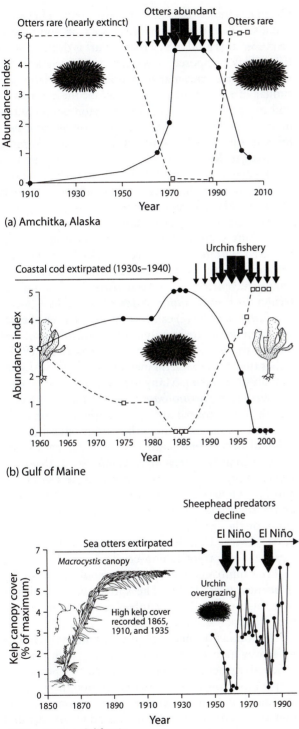

(a) Amchitka, Alaska

(b) Gulf of Maine

(c) Point Loma, California

Figure 22.21 Three cases of temporal trends and trophic cascades in North American kelp forests. Arrows indicate timing and magnitude of major forcing functions. (After Steneck et al. 2002, reprinted with the permission of Cambridge University Press)

which point the seawater concentration of carbonate will fall below 200 µmol kg^{-1} seawater. At that point marine organisms will no longer be able to build carbonate skeletons, and coral reefs will erode even if corals are able to migrate poleward and establish reefs in cooler areas.

A rise in carbon dioxide to 480 ppm is predicted to result in a temperature increase of about 2°C in the oceans. Periodic global climate changes called El Niño–Southern Oscillation (ENSO) events produce higher sea surface temperatures and shut off coastal upwelling zones. The frequency and severity of these ENSO events is increasing with global warming. The most severe ENSO event on record occurred in 1997–1998 and resulted in sea surface temperatures of 3°C–5°C above normal. As a result, ecosystems were disrupted in the Galapagos Islands (Vinueza et al. 2006), kelp forests in California suffered die-offs (Steneck et al. 2002), and coral bleaching occurred worldwide (Loya et al. 2001). In the Galapagos this ENSO event reduced nutrients and increased wave action in the intertidal zone, where the abundance of edible algae declined. As a result, mortality rose among marine iguanas and crabs that depend on edible algae for food. The same ENSO event caused extensive coral bleaching on the reefs around Okinawa, where coral species richness declined 61% and coral cover 85%. Branched corals such as *Acropora* were more heavily affected than massive or encrusting corals (Loya et al. 2001). The same 1997–1998 ENSO event affected Caribbean coral reefs (Mumby et al. 2006) and coral reefs in the western Indian Ocean, where 75%–99% of live coral were lost (Graham et al. 2006). After this event, the reefs of the Seychelles shifted from coral dominated to algal dominated.

Multiple factors are thus interacting to shift coral reefs from coral dominated to algal dominated. Massive beaching events due to rising sea surface temperatures and ENSO events open up large areas of substrate for colonization. In the presence of increased levels of nutrients and removal of herbivores by overfishing, algal colonization of open substrate is increasingly effective. Healthy corals seem able to compete against turf algae (McCook 2001) but not against fleshy macroalgae (Jompa and McCook 2002). In the absence of effective herbivores, turf algae are replaced by fleshy macroalgae (Smith et al. 2001). The macroalgal cover exceeds the capacity of the reduced herbivore populations to graze it down. The resulting macroalgal communities are stable and resistant to a return of corals (Hoegh-Guldberg et al. 2007). The best plan to protect coral reefs may be, once again, to establish marine reserves where fishing is prohibited. Such reserves have been shown to be effective refuges when they are large enough and have existed long enough for abundant populations of herbivores and predators to become reestablished (Mumby et al. 2006). Hughes et al. (2003) have suggested that at least 30% of the world's 600,000 km² of coral reefs should be in reserves where fishing is prohibited to ensure long-term protection of coral reef ecosystem.

In localized coastal areas, especially harbors, bays, or coves with restricted entry, large-scale discharge of heated wastewaters from power plants can have significant effects on the temperature of the receiving waters. This thermal pollution can in turn dramatically alter the coastal communities of macroalgae and invertebrates and give researchers insights into the effects of rising sea surface temperatures due to global warming. Schiel et al. (2004) followed the impact of such thermal pollution on the marine community inside Diablo Cove in central California for nine years before and nine years during operation of a major power plant. Daily discharge of 9.5×10^9 l day^{-1} raised temperatures in the cove by 3.5°C. Whole communities changed in character. In the intertidal, foliose macroalgae declined, leading to an increase in bare rock and red and green turf-forming species (Steinbeck et al. 2005). Grazers such as gastropods and urchins increased in abundance. In the subtidal, southern giant kelp (*Macrocystis pyrifera*) replaced northern bull kelp (*Nereocystis luetkeana*), but this was one of the few cases of a southern species replacing a northern species. The results suggest that rising sea temperatures due to global warming will cause major changes to coastal macroalgal communities but not a simple replacement of northern species with more southern ones.

Part II—Marine Turf-Forming Periphyton

Some marine macroalgae (such as the brown *Dictyota* and the green *Halimeda*) grow as short, tightly compacted patches known as turfs. Turf-forming members of the Dictyotales, for example, often colonize sublittoral regions from which kelp canopies have been removed (Kennelly 1987). Several species of red algae, including *Kappaphycus*, *Hypnea musciformis*, and

Gracilaria salicornia, are invasive in the Hawaiian Islands, where they form dense turfs on coral reefs. However, this kind of macroalgal turf is distinguished from turfs composed of smaller single-celled, colonial, or filamentous marine periphyton. The latter are similar in taxonomic composition (at the class level) and morphological type to periphyton communities in freshwaters. Here, we use the term *marine periphyton turfs* to distinguish them from turf-forming macroalgae. Marine periphyton turfs are common on coral reefs throughout the world, where nutrient-poor ocean waters may have very low or undetectable levels of nitrate (Hackney et al. 1989).

Succession in marine periphyton turf communities begins with a film of attached diatoms and is followed within five to seven weeks by a mixture of diatoms and cyanobacteria. Cyanobacteria may dominate marine periphyton turfs. The cyanobacterium *Hormothamnion enteromorphoides* ("looks like *Enteromorpha*") can form dense blooms of erect tufts that dominate hundreds of square meters of reef flat off the island of Guam (Pennings et al. 1997). Other turf cyanobacteria include *Oscillatoria*, *Lyngbya*, and *Anabaena*. Many turf-forming cyanobacteria, including *Hormothamnion*, are nitrogen fixers, and thus they contribute significantly to community nitrogen resources. Many other kinds of unicellular, colonial, or filamentous algae are found in periphyton turfs, such as the green algae *Cladophora*, *Bryopsis*, *Chaetomorpha*, and *Ulva*; the red algae *Ceramium* and *Polysiphonia*; and the brown algae *Sphacelaria* and *Ectocarpus* (Adey and Goertemiller 1987). In a study of the marine periphyton turfs of coral reefs in St. Croix, U.S. Virgin Islands, the single most abundant eukaryotic, non-crustose member of the turf community was *Ectocarpus rhodochortonoides* (Hackney et al. 1989). The species name indicates that this brown alga resembles the filamentous red alga *Rhodochorton*.

In contrast to macroalgal communities—characterized by high standing crop and very high primary productivity per unit area but low primary productivity per unit biomass—marine periphyton turfs feature low standing crop, moderate primary productivity per unit area, and very high primary productivity per unit biomass. High primary productivity is based on very high surface area-to-volume ratio of the turf algal cells and filaments. This facilitates nutrient uptake in low-nutrient waters. Turf communities are also best developed along shores facing incoming currents, which break up diffusion boundary layers at the turf surface

	← — Intensity of macrograzing pressure — →			
	Low to absent	Moderate to high	High	Extreme
Predominant benthic algal component	Macroalgae	Algal turfs	Crusts	(Bare substrates)
Standing crop	Very high	Low	Very low	
Primary productivity per unit area	Very high	Moderate	Low	
Primary productivity per unit biomass	Low	Very high	Very low	
Predominant types of adaptations	Maintain upright morphology	Growth	Resistance	

Figure 22.22 The relationship between grazing pressure, macroalgal life-form type, and effects on standing crop, primary productivity per unit area, primary productivity per unit biomass, and growth form adaptations. (From Hackney et al. 1989)

and bring in a constant nutrient supply. Adey and Goertemiller (1987) studied productivity of periphyton turfs by suspending screens bearing turf communities at various depths in the ocean. In the Bahamas, the North Equatorial Current (see Figure 22.18) may provide a sufficiently high-energy environment that turf communities escape nutrient limitation, even though the nutrient levels in surrounding waters are exceedingly low.

Moderate to high grazing pressure is characteristic of reef periphyton turf communities, in contrast to relatively low herbivory pressure on macroalgae, which are often well defended (Figure 22.22). Important grazers include herbivorous fish, urchins, crabs, and sea hares (Pennings et al 1997; Choat and Clements 1998). High rates of growth and primary productivity are the main adaptations of marine turf algae to herbivory. Damage to apices of branched, filamentous turf algae, resulting from either herbivory or mechanical damage, increases the extent of branching, yielding a more compact turf. Dense turfs are well adapted to survive desiccation during low tides. After severe damage, the basal portions of turf algae may persist as resting stages until conditions allow regeneration of erect photosynthetic filaments (Hay 1981). Some tropical turf cyanobacteria possess effective

chemical defenses against herbivory, possibly explaining their success despite not being calcified or having tough thalli. For example, *Hormothamnion enteromorphoides* produces a suite of cyclic peptides—principally laxaphycin A—that strongly deter feeding by parrotfish, sea urchins, and crabs but not pufferfish. These compounds also appear to have antifungal activity (Pennings et al. 1997). The cyanobacterium *Lyngbya majuscula* produces secondary metabolites that deter herbivorous fishes. These same secondary metabolites attract the sea hare *Stylocheilus*, a reef grazer that specializes on *L. majuscula* and stores the secondary metabolites to deter its own predators (Nagle and Paul 1999). Chemical defenses may not be limited to cyanobacteria. Extracts from the green algae *Ulva fenestrata* and *Ulvaria obscura* inhibit development of potential competitors such as oysters and *Fucus* (Nelson et al. 2003).

Blooms of marine turf-forming algae are becoming more common on many coral reefs, but the impact of these blooms on reef ecosystems is not yet clear. Cyanobacterial turf blooms on the reefs around Guam have been linked to mass fish kills (Nagle and Paul 1999). The cyanobacterium *Lyngbya majuscula* has been shown to reduce survival of *Acropora* coral larvae and recruitment (establishment and growth) of

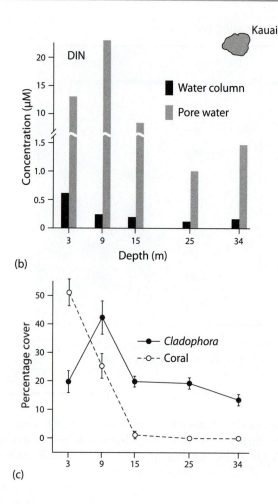

(b)

(c)

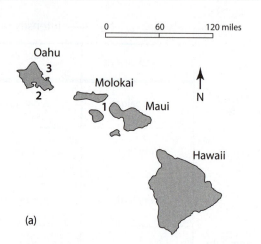

(a)

Figure 22.23 (a) Map of the Hawaiian Islands. Numbers indicate locations of studies of invasive and turf-forming algal species: (1) West Maui, where coastal areas have episodic blooms of the turf-forming green alga *Cladophora sericea*, (2) Waikiki Beach and Diamond Head, Oahu, where the red invasive alga *Gracilaria salicornia* is widespread on coral reefs, and (3) Kaneohe Bay, Oahu, where the invasive red algal species *G. salicornia* and *Kappaphycus* are widespread. Graphs show data from the study site (Kahekili Beach Park) of *Cladophora sericea* blooms. (b) Sediment pore-water samples show high levels of dissolved inorganic nitrogen (DIN) compared to the ocean water column. (c) Distribution with depth of *C. sericea* and corals. (Modified from Smith et al. 2005b. Reprinted with the permission of Inter-Research)

Pocillopora coral larvae in Guam (Kuffner and Paul 2004). Beginning around 2002, the coral reefs of southern Florida developed cyanobacterial turf blooms of *Lyngbya* that remained abundant year-round. These persistent blooms were thought to smother corals and other invertebrates (Paul et al. 2005). Research on the Great Barrier Reef of Australia, however, indicated that the coral *Porites lobata* was able to outcompete a mixed algal turf even on reefs where terrestrial nutrient sources were likely to be strongest. If the corals were damaged, however, the algal turf would occupy the space faster than the coral and competitively exclude coral recruitment (McCook 2001). In Hawaii, episodic blooms of the turf-forming green alga *Cladophora sericea* have been observed for two decades on the coral reefs of West Maui (Figure 22.23a). When these blooms occur, *Cladophora* grows from shore to depths greater than 30 m and may cover beaches in rotting biomass (Smith et al. 2005b). These blooms are likely the result of coastal eutrophication from terrestrial nutrients by groundwater seepage, as indicated by elevated levels of dissolved inorganic nitrogen (DIN) in sediment pore water (Figure 22.23b). At the same time as these *Cladophora* blooms have been occurring, the percentage of coral cover has declined on these West Maui reefs (Figure 22.23c). Research is needed to determine whether the coral decline is directly or indirectly linked to these *Cladophora* turf blooms.

Part III—Freshwater Periphyton

Freshwater periphyton (also known as benthic algae) includes algal unicells, colonies, or filaments that grow attached to substrata in streams and in the littoral zones of lakes. Such communities are very common where light levels are sufficient for their growth. Freshwater periphyton are important because they are the primary source of fixed carbon in shallow lakes and streams and because they are unusually rich in species. They provide an essential source of food for a wide variety of stream and lake micrograzers, mesograzers, and larger animals. The algae of periphyton communities also sequester nutrients such as N and P in wetlands, helping to prevent lake eutrophication. Although macrophytic vegetation (primarily vascular plants) is often credited with this function, macrophytes actually obtain most of their N and P from the sediments, whereas the attached algae are able to harvest N and P from the water column. The algal periphyton on macrophytes traps nutrients and transfers them to the sediments upon their death, thus reducing the rate of nutrient transfer from wetlands to associated lakes (Wetzel 1996). Periphyton communities are also recognized for their ability to stabilize unconsolidated substrata, such as loose sediments and sand, preventing their disaggregation by water movements. For example, sticky mucilages exuded by the diatom *Nitzschia curvilineata* decrease erodibility of sediments in laboratory experiments modeling stream flow (Sutherland et al. 1998).

Periphyton is a complex mixture of microalgae, bacteria, and fungi, often held together in a glue-like mucilaginous matrix produced by the algal or bacterial inhabitants (Figure 22.24). Thus, the term *periphyton* cannot be used as a synonym for the attached algae alone. The matrix also typically includes dead algal cells, fecal pellets generated by micrograzers, calcite particles, which may be produced by some algae, and bound inorganic sediment (Burkholder 1996). Such communities may occur attached to rocks (epilithic), sand (epipsammic), sediments (epipelic), plants or larger algae (epiphytic), or animals (epizooic). In addition, periphyton communities may occur as loose tangles among the branches of larger algae or macrophytes. Attached algae may also break loose from the substrate and occur as a floating mass in the water column. Such floating algae are known as **metaphyton**. Subsurface summer clouds of the filamentous zygnematalean green algae *Spirogyra*, *Mougeotia*, *Temnogametum*, and *Zygnema* are the most common metaphytic forms in lakes—particularly acidic lakes—but metaphytic mats of *Cladophora*, *Chaetophora*, *Oedogonium*, and *Spirogyra* also occur widely in wetlands (Goldsborough and Robinson 1996). Filamentous xanthophyceans (tribophyceans) may also form

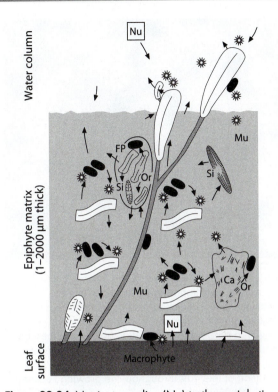

Figure 22.24 Nutrient supplies (Nu) to the periphytic algal community on a macrophyte leaf surface may come from the substrate, the water column, and the epiphytic matrix itself. The epiphytic matrix, defined by the extent of mucosaccharide (Mu) arising from the algae and associated bacteria, is a site of dynamic nutrient regeneration. Fissures in the matrix (not shown) promote exchange of materials between the water column and lower portions of the matrix. Living diatoms are shown in white, whereas diatom frustule remains, including those in fecal pellets (FP), are gray. The latter are a source of silica (Si). Bacterial cells (black ovals) and free phosphatase enzymes (sunbursts) are active in recycling materials from organic matter (Or) and calcium carbonate (Ca). (After Burkholder 1996)

metaphytic clouds in lakes and ponds, particularly during spring and fall.

Freshwater periphytic algae consist largely of cyanobacteria, green algae, and diatoms, but representatives of other groups, such as red algae, chrysophyceans, and xanthophyceans, may also be present. Examples of common periphytic cyanobacteria are *Schizothrix*, *Rivularia*, and *Tolypothrix*; common green members of the periphyton include *Ulothrix*, *Cladophora*, *Rhizoclonium*, *Stigeoclonium*, *Draparnaldia*, and *Coleochaete*; and frequently observed periphytic diatoms include *Epithemia*, *Cocconeis*,

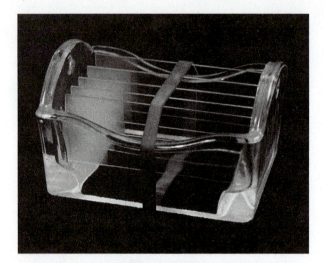

Figure 22.25 A simple and relatively inconspicuous apparatus for growth and analysis of periphyton communities. Glass slides are arranged within a glass holder designed for staining procedures and held in place with a rubber band. The slide holder can then be anchored at various depths along transects within the littoral, and slides can be removed for processing and observation at intervals throughout the growing season. (Photo: L. W. Wilcox)

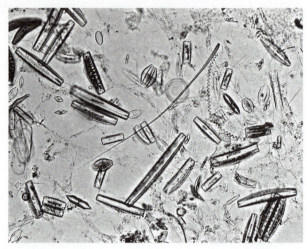

Figure 22.26 A permanent slide preparation derived from application of the method shown in Figure 22.25 to an oligotrophic lake. After removal from the environment, the slide was rinsed to remove sand, sediment, and unattached algae and then dehydrated and stained with Fast Green. Various common members of the freshwater periphyton can be observed. Quadrats can be marked on slides, and periphyton species abundance can be quantified. (Photo: L. W. Wilcox)

Gomphonema, and *Cymbella. Batrachospermum, Lemanea,* and *Audouinella* from the red algae and the xanthophyceans *Tribonema* and *Vaucheria* are also widely encountered. The chrysophycean *Hydrurus* is a common inhabitant of cold mountain streams. Zygnematalean green algae dominate the periphyton communities of acidic *Sphagnum* bogs, with the greatest diversity associated with the higher plant *Utricularia,* which serves as a substrate (Woelkerling 1976). In general, similar genera occur in streams and the shallow nearshore waters of lakes. A dense community of microperiphyton, including cyanobacteria and diatoms, may cover some of the larger filamentous algae, particularly *Cladophora.* In contrast, zygnematalean filaments are commonly algal epiphyte free, as they constantly generate mucilage that prevents adherence of algal epiphytes. Filamentous desmids are occasional exceptions to this generalization in that particular algal epiphytes may be present within their mucilaginous sheaths (see Figure 3.27).

Structure and development of periphyton communities is studied by examination of natural substrates throughout the season or by placing artificial substrates in streams or lakes and retrieving them at intervals (e.g., Sabater et al. 1998). Glass microscope slides or ceramic tiles are often used as artificial substrates that mimic rock surfaces. They may be fastened within a holder that is anchored at the desired depth (Figure 22.25). Algae are typically scraped from tiles for analysis of chlorophyll content and determination of species composition. Slides may be fixed in a 1% solution of glutaraldehyde to preserve cellular structure, followed by dehydration in an alcohol series, staining (Fast Green is recommended), and infiltration with xylenes. Mounting medium is then used to affix cover slips for preparation of permanent slides for microscopic assessment of species composition and spatial relationships (Figure 22.26). For statistical significance, population density estimates require counting and identification of 50–500 cells per sample. Density is often expressed in terms of biovolume rather than cell number because small populations of very large cells may actually prove to be more important than larger populations of very small cells. Lowe and Pan (1996) provided a table of volume estimates for a large number of periphyton species (primarily diatoms) and additional details of periphyton analysis methods.

22.4 The Influence of Physical Factors on Periphyton

The architecture and species complexity of periphyton communities can vary greatly, depending upon the degree of disturbance, current, substrate type, temperature, irradiance, and nutrient levels. Heterogeneous (patchy) distribution of periphyton communities has been attributed to variation in irradiance and substrate characteristics, among other factors such as flow regime, nutrients, and grazing pressure (DeNicola and McIntire 1990a,b). Disturbance frequency, in the form of flooding, wave action on lake shores, currents in streams and rivers, and water motions in general, is probably the strongest influence on periphyton community structure, as is the case for marine intertidal communities, where tides, waves, and currents are major structuring influences. This is in part because water motions strongly affect the balance between biomass accumulation and loss, and they control nutrient supply. In conditions of frequent flooding or strong current, periphyton communities are subjected to shear stresses that can cause disturbance-sensitive species to slough off their substrates and prevent community reestablishment. In contrast, where the current is slower, biomass can accumulate to the point that filamentous forms can generate long, streaming, conspicuous masses, if irradiance and nutrient levels are sufficient. *Cladophora* and *Lemanea* are robust enough to withstand the drag forces of waves and currents, and consequently can attain considerable biomasses. The extensively branched morphology of *Cladophora* provides an enormous surface area for attachment of large numbers of smaller periphytic forms, with positive feedback effects on community productivity. The positive effects of low levels of disturbance and relatively high nitrogen levels on periphyton chlorophyll *a* at 15 sites are shown in Figure 22.27 (Biggs 1996).

As in marine environments, water motions are important in reducing the boundary layer at algal surfaces and in replenishment of nutrients. In currents slower than 10 cm s^{-1}, growth of periphytic algae is limited because the surface boundary layer of unmixed water is not disturbed and continues to restrict nutrient diffusion (Stevenson 1996). Hence, periphytic algal density is often higher in intermediate current velocities than in slow- or fast-moving water. Nevertheless, freshwater attached algae

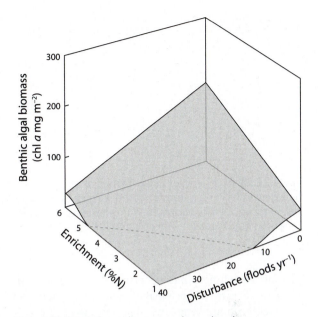

Figure 22.27 A three-dimensional graphical representation (a response surface) showing the effects of N enrichment and disturbance (floods) on the biomass of periphytic algae estimated by measurements of chlorophyll *a*. (After Biggs 1996)

experience lower levels of water turbulence than do phytoplankton. One consequence of this fact is that freshwater periphyton (like marine periphyton) are enriched in the stable carbon isotope ^{13}C (see Chapter 2), as compared to phytoplankton, and a similar difference is apparent in their respective consumers. This has led to the hypothesis that littoral food webs based on attached algal productivity operate independently (are uncoupled) from pelagic food webs based on phytoplankton productivity (France 1995).

Radiotracer evidence suggests that in oligotrophic to mesotrophic waters, nutrients such as phosphorus can be transferred from macrophyte host surfaces to periphytic algae. For example, some periphytic diatoms obtain more than 50% of their phosphorus from *Najas flexilis* tissues (Figure 22.28) (Wetzel 1996). Under higher nutrient conditions, many periphytic algae are thought to obtain most of their nutrients from the water column. An analysis of data from 42 studies on nutrient effects on periphyton revealed that in 23 of these cases, nutrients are most often only secondarily limiting to algal growth, with disturbance, light availability, and grazing playing more significant roles (Borchardt 1996). If light levels are too low to saturate photosynthesis,

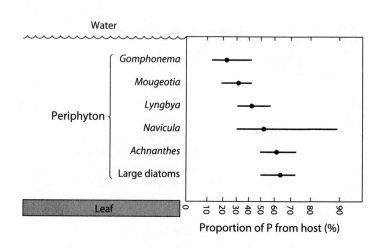

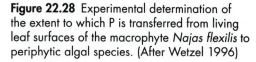

Figure 22.28 Experimental determination of the extent to which P is transferred from living leaf surfaces of the macrophyte *Najas flexilis* to periphytic algal species. (After Wetzel 1996)

nutrient enrichment has no effect on periphytic algal growth rates. This weaker association between water-column nutrients and algal biomass is a major difference between periphytic algae and phytoplankton. As a result of the reduced coupling of nutrient levels in the water to algal growth rates, the biomass of periphytic algae does not always reflect water nutrient levels. Periphytic algae growing in the shade of macrophytes or streamside tree canopies may not exhibit a level of biomass that would be expected on the basis of water nutrient levels, and, conversely, well-illuminated periphyton may be more abundant than predicted by instantaneous measurements of water nutrient levels (Borchardt 1996). Diatoms growing in thin biofilms, however, are generally not light limited in shaded streams with respect to growth rates because diatoms can acclimate to very low light levels (Rier et al. 2006). Light penetration also influences the depth to which periphytic algae can grow in lakes and rivers.

Light limitation at the base of thick periphyton assemblages may stimulate heterotrophic activity by various algae. Tuchman (1996) pointed out that eukaryotic algae are evolutionarily derived from heterotrophic protists and thus are likely equipped with nutrient uptake mechanisms that provide a safety net when conditions are not suitable for photosynthesis. Uptake and utilization of 21 organic compounds, including sugars and amino acids, by many species of periphytic algae—primarily diatoms—has been documented (Tuchman 1996). Increased uptake of the sugar glucose in the dark has been demonstrated in the diatom *Cyclotella* (Hellebust 1971). The periphytic green alga *Coleochaete* is also able to use glucose for growth, though light is required, and DOC

utilization occurs only under conditions of inorganic carbon limitation (Graham et al. 1994).

22.5 The Influence of Biological Factors on Periphyton

Grazing usually results in significant declines in periphyton algal biomass. After a review of the evidence for top-down versus bottom-up control of periphyton communities, Lamberti (1996) concluded that consumers might play a pivotal role in structuring periphyton communities. In addition to grazing as a biological factor, the characteristics of stream bank and lakeshore vegetation can influence periphyton communities.

Grazing

Grazers on periphyton algae include snails, caddisfly larvae, mayfly nymphs, fish, shrimp, chironomid larvae, and tadpoles (see Kupferberg et al. 1994; Peterson et al. 1998). Pleurocerid snails in streams and the widespread crayfish *Orconectes rusticus*, which occurs in the littoral of lakes and sometimes streams, are particularly important. *Orconectes* is one of the few animals that can consume *Cladophora*, thereby making substrate available for other periphyton taxa. Many periphyton grazers are thought to be omnivores that exhibit little specificity in food choice within the size range available to them (Steinman 1996). Grazing modes include raspers and scrapers

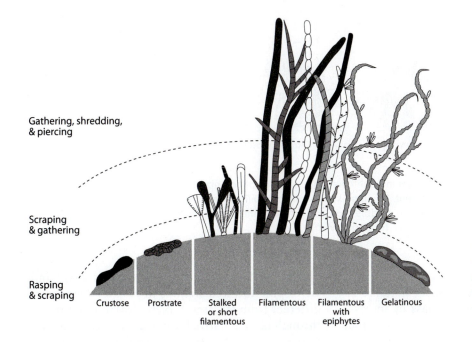

Gathering, shredding,
& piercing

Scraping
& gathering

Rasping
& scraping

Crustose   Prostrate   Stalked   Filamentous   Filamentous   Gelatinous
or short   with
filamentous   epiphytes

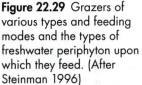

Figure 22.29 Grazers of various types and feeding modes and the types of freshwater periphyton upon which they feed. (After Steinman 1996)

that harvest prostrate or crustose algae and gelatinous colonies; scrapers and gatherers that feed on stalked diatoms and short filaments; and gatherers, shredders, and piercers that obtain filamentous algae and their epiphytes (Figure 22.29). A number of types of periphytic algae can survive passage through the gut of herbivorous snails and other grazers. An analysis of algae from snail feces revealed that more than 60% were able to resume growth (Underwood and Thomas 1990).

Under conditions of intense disturbance or heavy grazing, the periphytic community may include only a few species of closely adherent diatoms, such as the disturbance and grazing-resistant *Cocconeis*. When disturbance and resource levels (light and nutrients) are at medium to low levels, periphyton communities can be dominated by filamentous cyanobacteria, including the nitrogen fixers *Nostoc* and *Tolypothrix*, red algae such as *Audouinella*, and diatoms such as *Epithemia* that have N-fixing endosymbionts. In contrast, where disturbance and grazing are at a medium to low level and resources are moderate to high, a more complex community, including stalked diatoms, filamentous greens, cyanobacteria, and adherent diatoms, may be present (Biggs 1996). In an experiment conducted in streams of the southwestern Ozark Mountains, cyanobacterial mats dominated by the nitrogen fixer *Calothrix* were overgrown by benthic diatoms within 4 to 10 days after exclusion

of herbivorous fish and invertebrates. When exposed to grazers, the diatom growths were removed, and cyanobacterial mats regenerated (Power et al. 1988).

Community development begins with colonization of substrate by bacteria that generate an organic matrix onto which small diatoms can adhere. Erect or stalked diatoms then enter the community, followed by filamentous greens and reds, which together form a layered community. At its maximal development, a peak in biomass is achieved (Figure 22.30). Rier and Stevenson (2006) showed that a Monod model (Chapter 21) best describes the relationship between phosphorus and nitrogen and the development of this peak algal biomass. The time required for development of peak biomass varies from two to many weeks, depending upon the availability of light and nutrients and grazing intensity. Peak biomass may vary from very low levels up to more than 1200 mg chlorophyll *a* m^{-2}. The carrying capacity of the environment is defined as the biomass present when rates of accumulation and loss are balanced. Losses can result from grazing (Cuker 1983), disease (Peterson et al. 1993), parasitism, age, or sloughing. As the community increases in size, it may become vulnerable to removal by water motions. In addition, algae close to the substrate may senesce as a result of light or nutrient limitation, leading to a process known as autogenic (self-generated) sloughing (Biggs 1996). Dense, low-profile communities consisting of

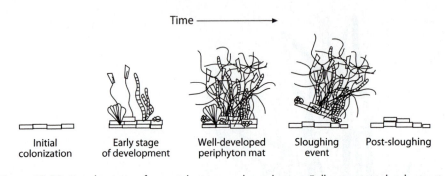

Figure 22.30 Development of a periphyton mat through time. Following initial colonization by diatoms (rectangles), and early development, a complex mat having maximum biomass accumulates. The larger aggregation is more vulnerable than early developmental stages to sloughing as a result of stream turbulence. (After Tuchman 1996)

adnate diatoms and cyanobacterial-dominated mats are resistant to being dislodged by shear stress. High current velocity enhances their biomass by increasing nutrient diffusion to the encrusting periphyton. In contrast, more open growths, such as filamentous green algae, are vulnerable to sloughing in high current velocities (Biggs et al. 1998). Though erect filaments of heterotrichous algae such as *Stigeoclonium* can be broken off in fast currents, the prostrate portions may be retained. *Stigeoclonium* is able to regenerate erect filaments in the post-sloughing phase. Experimental studies suggest that there is a trade-off between productivity and persistence in the case of *Stigeoclonium* (Rosemond and Brawley 1996). Grazing effects on algal succession appear to be highly specific to habitat; currently few generalizations can be made regarding presence or absence of grazer effects on algal succession (Steinman 1996).

Shoreline Vegetation

The nature of shoreline vegetation is another biological factor that can affect a periphyton community. Tree-lined shores of streams, rivers, and lakes produce a different environment in terms of light, temperature, and organic inputs than do shorelines occupied by shrubs or grasses. Thompson and Townsend (2003) compared the stream food webs of native pine-forest streams in Maine and North Carolina to exotic pine forest streams in New Zealand. New Zealand has no native riparian vegetation equivalent to pine trees, and Monterey Pines (*Pinus radiata*) have been introduced to stream margins to establish forest plantations. The algal periphyton community in New Zealand streams

planted with an exotic pine forest was most similar to pine-forested streams in North America and different from streams in New Zealand that passed through land in native bush or grassland. Thus similarity of shoreline vegetation can impose similarity on periphyton assemblages, even in different continental areas.

22.6 Temporal and Spatial Variation

There are three major temporal patterns of periphyton occurrence: (1) relatively constant low biomass under conditions of frequent disturbance, (2) cycles of accumulation and loss linked to less-frequent disturbance, and (3) seasonal cycles resulting from seasonal variance in disturbance, temperature, grazer behavior, and/or irradiance (Biggs 1996). Spatial variation within lakes is primarily attributable to depth-related decreases in water motion and irradiance. The upper littoral is characterized by wave action and high light levels and is dominated by stalked diatoms and filamentous green algae and their epiphytes. *Cladophora*, *Ulothrix*, and *Oedogonium* are common littoral genera, with *Bangia* occurring in the high littoral in sites where there is significant influx of halides from road salting. In contrast, the sublittoral of lakes is characterized by low levels of light and turbulence and is dominated by cyanobacteria together with growths of epipelic and epipsammic diatoms. Though biomass is low, species richness is reportedly higher in the sublittoral than in the upper littoral; reasons for this difference are

Figure 22.31 *Rhizoclonium* generates very long filamentous growths in nutrient-rich streams. (Photo: L. W. Wilcox)

unknown (Lowe 1996). The biomass of periphytic, littoral algae is characteristically low in oligotrophic lakes, whereas it may accumulate to nuisance levels in eutrophic lakes.

Considerable data on the effects of variations in temperature, irradiance, day length, and nutrients on photosynthesis and/or reproduction of freshwater periphytic algae have been obtained by multifactorial studies (Graham 1982; Graham et al. 1982; Hoffmann and Graham 1984; Graham et al. 1984, 1985, 1986; Graham and Graham 1987; Graham et al. 1996a). Genera studied in this fashion include *Cladophora glomerata*, *Ulothrix zonata*, *Spirogyra* sp., *Coleochaete scutata*, and *Bangia atropurpurea*. Irradiance, temperature, and photoperiod explain most of the natural variation in net photosynthesis and reproduction of these periphytic algae. Periphyton occurring in the same habitat (*Cladophora*, *Bangia*, and *Ulothrix*) may partition the environment spatially (by growth at different levels in the littoral) or temporally (dominating at different times of the year). Reproduction of *Ulothrix zonata* is stimulated by long days and results in the loss of most biomass before *Cladophora* becomes well established. Although *Bangia* and *Cladophora* may co-occur, *Bangia* is better able to tolerate high littoral conditions (high irradiance and sporadic desiccation), and thus *Cladophora* occupies the littoral region below *Bangia*.

In streams, spatial variation can occur between pools, runs, or riffle regions and among streams of increasing size (distance from their headwaters).

Riffle areas of streams are characterized by high shear stress and are consequently inhabited by low-growing diatoms such as *Cocconeis*. In contrast, lower velocity runs and pools harbor larger filamentous forms such as *Spirogyra*, *Oedogonium*, and *Cladophora*. *Rhizoclonium* can form conspicuous, long strands in nutrient-rich streams (Figure 22.31). A model of periphytic algal biomass in streams of increasing dimensions—known as the river continuum concept (Vannote et al. 1980)—predicts that biomass should increase as streams coalesce and become wider, such that shading from streamside vegetation is reduced, and then decrease in rivers because light decreases with depth and increased turbidity. Some stream systems appear to behave according to this model, but others do not. Headwater streams are often dominated by cyanobacteria such as *Schizothrix*, red algae including *Batrachospermum*, low-biomass green algae such as *Stigeoclonium*, and diatoms such as *Cymbella*. With increasing downstream enrichment (arising from human land-use practices), nutrient-demanding forms such as *Vaucheria* and *Cladophora* become more abundant. Nuisance growths of *Ulothrix* occur in cold waters, and *Cladophora* and *Rhizoclonium* occur in warmer waters of both streams and lakes. Variations among stream and river systems in biomass and species composition of the periphyton arise from differences in flood frequency and stream enrichment. High biomasses develop only under the combined conditions of low flood frequency and high nutrients. Control of nuisance stream- and lake-edge algae is a major issue in water resources management (Biggs 1996).

22.7 Pollution Effects

Eutrophication of lakes, rivers, and streams with excessive amounts of phosphorus and nitrogen has been responsible for the development and spread of nuisance blooms of periphytic, filamentous algae. The green algae *Cladophora glomerata* and *Ulothrix zonata* became prominent in the 1960s in the North American Great Lakes as a result of increased levels of phosphorus in coastal areas. In addition, the marine alga *Bangia atropurpurea* entered the Great Lakes in the 1960s and began to spread within harbors where elevated levels of salts created a favorable environment (Lin and Blum 1977). By the 1980s nuisance blooms of *C. glomerata* and other periphytic algae had declined in the Great Lakes due to phosphorus abatements, including removal of phosphorus from detergents, improved phosphorus removal at sewage treatment plants, and changes in agricultural practices. After a decade of absence, nuisance blooms of *C. glomerata* have returned to parts of the Great Lakes shores. The return of *Cladophora* seems to be related to the invasion of the Great Lakes by zebra and quagga mussels (*Dreissena polymorpha* and *D. bugensis*), which filter-feed on phytoplankton and may be shuttling phosphorus to inedible *Cladophora*. Marine algae continue to invade the Great Lakes region. Lougheed and Stevenson (2004) reported finding the green macroalga *Enteromorpha flexuosa* (now *Ulva flexuosa*) in several coastal lakes adjacent to Lake Michigan. They suspect that elevated levels of salts from industrial activity affected these lakes and aided the invasion of this marine alga.

Many inorganic compounds (including a variety of heavy metals) inhibit growth of marine and freshwater algae (Genter 1996). In general, these metals are thought to influence algal enzymes, and their effects vary among taxa; levels that inhibit one species may stimulate growth of another. Increases in zinc concentration are associated with a shift from dominance by diatoms to filamentous green algae and then to unicellular green algae. Hoagland et al. (1996) reviewed the inhibitory effects of a wide variety of organic compounds such as polycyclic aromatic hydrocarbons, polyhalogenated biphenyls, herbicides, insecticides, surfactants, detergents, dyes, oils, solvents, and resins on periphytic algae. Substantial variability in sensitivity among attached taxa was the rule. Dramatic shifts in community species composition occur when herbicides reach concentrations in the µg per liter range. It is recommended that both single-species and community bioassays be used to further evaluate the impact of organic compounds on periphytic algae.

Acidification is correlated with a decline in species richness, possibly because of decline in the macrophytes that provide support for periphyton communities. The cyanobacteria are most sensitive to lowered pH. In addition, acidification is correlated with dramatic metaphytic growths of zygnematalean filamentous algae in lakes and slower streams, resulting in increases in algal biomass as compared to unperturbed conditions. For *Mougeotia*, low pH is not optimal for growth. In fact, for one nuisance metaphytic bloom-forming species, the optimum for photosynthesis was pH 8 (Graham et al. 1996b). The reasons for dominance of such algae in acidified waters are not understood (Planas 1996). Tolerance to heavy metals such as aluminum and zinc (which become more available as pH drops) (Graham et al. 1996b), ability to obtain scarce dissolved inorganic carbon under low pH conditions, and release from the constraints of nutrient competition and herbivory all likely contribute to these patterns. Some diatoms, such as *Eunotia* spp., are acid-tolerant and become dominant in fast-flowing streams affected by acidification. Mine drainage is a major cause of stream acidification, and algal periphyton assemblages can be used as bioindicators of the state of acid mine drainage and of progress in bioremediation of affected streams (Verb and Vis 2005).

Terrestrial Algal Ecology

Subaerial algae on lower portion
of wall in Concepción, Chile

(Photo: L. W. Wilcox)

The previous two chapters present the ecology of planktonic and attached algae in freshwater and marine ecosystems. Algae are not, however, restricted to aquatic systems. Cyanobacteria and eukaryotic algae may be found in essentially every type of terrestrial environment on Earth. They occur in every type of soil, from the tropics to the poles, and at every level of precipitation, from deserts through grasslands to forests and agricultural land, including soils polluted by various industrial and mining activities. Cyanobacteria and eukaryotic algae may grow in the surface layers of snowfields at high elevations and in the Arctic and Antarctic in icefields and the upper active layers of permafrost. Cyanobacteria and eukaryotic algae may also grow on and under the surfaces of rocks and beneath translucent rocks, where they form lithic ecosystems. Algae that grow on surfaces that lie above the surface level of soil, water, snow, and ice, where they are exposed to the air, are considered to be subaerial. Thus algae growing on the surfaces of rocks are subaerial and epilithic (on stones), while algae growing beneath the surfaces of rocks or under translucent rocks are endolithic (inside stones) or hypolithic (under stones). Subaerial algae form epilithic communities on the surfaces of natural stone and stone-like substrates, such as brick and concrete, used in human constructions. Subaerial algae also grow as epiphytic communities on plant surfaces, including bark, wood, and leaves. As symbionts with various species of fungi, cyanobacteria and green algae may form lichens, which are found in practically every terrestrial environment, including soils, rocks, the surfaces of other organisms, such as the bark of trees and shrubs, and on the surfaces of human-made structures.

In aquatic ecosystems, most algae are continuously bathed in water and therefore capable of being continuously active metabolically. The only exceptions occur in the intertidal zones of the oceans, where attached macroalgae and periphytic algae undergo periodic exposure, and the shallow littoral zones of large lakes, where persistent winds from a single direction may cause the lake water to pile up on the downwind shore and expose the periphytic algae along the upwind shore for long periods. In contrast, terrestrial algae are not continuously immersed in liquid water. Algae in terrestrial ecosystems can be metabolically active only when there is sufficient water as liquid or vapor to support that activity. The algae in soils are active when liquid water is present in the surface layers as a flow from rain or a film around the soil particles. As the soils dry, some soil algae can move to lower layers and follow the water, while other algae become dehydrated and inactive until the next precipitation falls. In snow and ice fields, algae are active when air temperatures are above freezing and liquid water is present around crystals of snow or ice. If temperatures are continuously below freezing, algae will be very sparse because there will not be sufficient liquid water to permit photosynthesis and growth.

Terrestrial algae must be adapted to survive long periods of metabolic inactivity either in a desiccated state at temperatures above 0°C or frozen at temperatures below 0°C. Moreover, terrestrial algae must be able to survive the transition from a desiccated or frozen state to an active hydrated condition without loss of cellular integrity and viability. These transitions often occur repeatedly over short time spans. Terrestrial algae possess a number of genes that code for specific substances that maintain cellular integrity, structure, and viability through these extreme transitions. These genes are the subject of current research interest because of their potential use in developing agricultural crops that are more resistant to drought or frost. Some subaerial algae

and lichens can remain active without liquid water, provided that the amount of water vapor in the air (the relative humidity) is above 70%. These algae are truly terrestrial in the sense that they are not dependent on liquid water. In Antarctica viable algal cells have been recovered from 18 m-deep layers of permafrost, where they have lain for more than 3 million years. These ancient microbial communities may be the best analog to any potential life on the planet Mars, where permafrost also exists adjacent to the polar zones and may contain remnants of early martian microbial life, if it ever existed.

Terrestrial algae play important roles in every ecosystem. They contribute to the fertility and stability of soils everywhere, through fixation of carbon and nitrogen, release of organic compounds, and binding together of soil particles to reduce soil erosion. In mesic (moist) ecosystems, the contributions of terrestrial algae tend to be overshadowed by the dominant bryophytes in cooler ecosystems and vascular plants in temperate to tropical ecosystems. In more climactically extreme ecosystems, where bryophytes and vascular plants are reduced or excluded entirely, terrestrial algae may be dominant and the only significant source of primary production. In permanent snow and ice fields at high altitudes and latitudes, where temperatures rise above 0°C for some appreciable period of the year, snow algae carry out photosynthesis and grow, generating the only primary productivity. Similarly, in cold deserts, such as the Arctic polar deserts and extremely cold and dry deserts, such as the Antarctic dry valleys, the algae of soils and lithic ecosystems are the only source of primary productivity. Less extreme ecosystems, such as arid and semiarid lands, contain bryophytes and vascular plants, but the vascular plants are sparse, with wide spaces of open land between plants. In these arid systems, soil fertility, water retention, and stability from erosion depend on development of biological soil crusts in the spaces between vascular plants. **Biological soil crusts** are communities made up of bacteria, fungi, algae, and, in the most developed forms, various lichens and bryophytes. Disturbances such as excessive livestock trampling and human commercial and recreational activities degrade biological soil crusts and result in increased erosion and invasion by exotic species. Conservation of these biological soil crusts is an important part of rangeland management.

In the following sections, the terrestrial algae of soil ecosystems are discussed first, followed by the cryophilic algae of terrestrial snow and ice fields. The third section considers subaerial algae, which is a broad category containing lithic algae, algae on living and dead plant surfaces, and algae found on human-made constructions such as buildings. Lichens are discussed in each of the categories in which they occur.

23.1 Soil Algae

Soil algae are the most thoroughly studied of the terrestrial algae. Soil covers most of the terrestrial surface of Earth and is composed of minerals and organic matter. Soils are very diverse. They may be fine or coarse and vary in mineral composition, amount of organic matter, pH, and salinity, and they may be oxygenated or anoxic. In addition, soils are present around the globe and therefore occur across a wide range of climates, from bitterly cold polar deserts to tropical rainforests.

The physical environment of soil imposes constraints on the range of morphological forms of soil algae. Soil algae mainly occur as small spheres or ovals, small packets of spheres or ovals, and filaments, either as singles, in groups, or branched. Green algae and cyanobacteria dominate, along with lesser numbers of diatoms, chrysophyceans, and, rarely, euglenoids (Table 23.1). The few genera of soil diatoms are predominantly pennate, capable of movement, and smaller than freshwater aquatic species of the same genera. Some groups of algae, such as browns and dinoflagellates, are absent from soils, while red algae are represented primarily by the genera *Cyanidium* and *Porphyridium*. *Cyanidium* species occur in hot and very acidic soils (Doemel and Brock 1971). The relatively limited range of morphological forms has led to the conclusion that soil algae consist of relatively few cosmopolitan species. As long as researchers relied on morphological species determinations, studies across different soils and regions seemed to confirm this conclusion, but the application of molecular techniques to soil algal communities has led to a different view. Morphological simplicity is the result of convergent evolution of forms, not a lack of species richness. Among the green algae, different lineages have produced different species of soil algae that are morphologically indistinguishable but molecularly diverse. Because most papers on soil algae published before 2000 used a morphological species approach, the reported species diversity should be regarded at no more

than a generic level at best. Molecular studies have confirmed, however, that many genera and even some species in biological soil crusts are very widespread and occur on almost every continent.

A convenient way to approach the ecology of soil algae is to consider terrestrial ecosystems in two categories: moist ecosystems and more arid ecosystems. Bryophytes and vascular plants dominate the soils of the more moist terrestrial ecosystems, from boreal to tropical, including grasslands, forests, and agricultural lands. In these moist ecosystems, soil algae are present and important to system fertility, but they are clearly not dominant. Cyanobacteria and eukaryotic algae occur as scattered populations in the upper soil layers and do not form soil crusts except in unusual circumstances or following ecosystem disturbance. Moist terrestrial ecosystems with dispersed populations of cyanobacteria and eukaryotic algae form the first category. The second category consists of arid and semiarid lands, ranging in temperature from hot to cold, in which vascular plants are either absent or sparse and widely dispersed. In these ecosystems, soil algae form visible surface crusts across the spaces between vascular plants and contribute significantly to soil fertility, productivity, and stability.

Soil Algae in Moist Terrestrial Ecosystems

In moist terrestrial ecosystems, the pattern of seasonal abundance of soil algae is often related to the seasonal pattern of soil moisture and the dominant bryophytes or vascular plants. The number of algal cells per gram of soil is generally highest in the top few centimeters of soil and drops off rapidly with depth (Petersen 1935). Soil pH is an important factor in determining the relative abundance of cyanobacteria compared to other groups. Cyanobacteria and diatoms prefer neutral to alkaline soils and become scarce at soil pH below 5.0. Continuously wet soils are usually acidic, while arid soils are normally alkaline (Shields and Durrell 1964). Green algae are tolerant of a wider range of pH than are cyanobacteria and diatoms and may therefore dominate acidic soils due to lack of competition. Soil algae are subject to grazing by soil protozoa and nematodes, although the impact of such feeding has not received much attention (Hoffmann 1989).

In a comparative study of soil algae in two abandoned agricultural fields and a nearby temperate deciduous oak forest in New Jersey, Hunt et al. (1979) found 33 genera over a period of 13 months, based

Table 23.1   Frequently occurring soil algae and cyanobacteria from temperate fields and forest in New Jersey (monthly sampling except February)*

Algal group	Genus
Cyanobacteria	*Anabaena*
	Gloeocapsa
	Lyngbya
	Nostoc
	Oscillatoria
	Phormidium
	Plectonema
	Synechococcus
	Synechocystis
Green algae	*Bracteacoccus*
	Characium
	Characiopsis
	Chlamydomonas
	Chlorella
	Chlorococcum
	Chlorosarcina
	Chlorosarcinopsis
	Gleocystis
	Myrmecia
	Nannochloris
	Neochloris
	Oocystis
	Palmellococcus
	Pleurastrum
	Stichococcus
	Tetracystis
	Klebsormidium
	Mesotaenium
Chrysophytes	*Bumilleria*
	Monocilia
Diatoms	*Gomphonema*
	Navicula
Euglenoids	*Euglena*

*From Hunt et al. 1979.

on isolation and identification from cultures (see Table 23.1). The two field sites had been abandoned for different lengths of time—one for only 1 year and the other for 11 years. The deciduous forest was more than 250 years old. Eukaryotic algae were more abundant in the field sites (highest counts of 3.3×10^7 and 2.2×10^7 cells/g soil in the 1- and 11-year-old fields) than in the forest site (1.2×10^5 cells/g soil). The same pattern applied to cyanobacteria, with the highest counts of 8.2×10^5 and 2.3×10^5 cells/g soil at the field sites and 1.6×10^4 cells/g soil in the forest. At all three sites, eukaryotic algae were more abundant than were cyanobacteria (Figure 23.1a and b). Cyanobacteria were particularly sparse in the forest, where the soil pH ranged from 3.9 to 4.6, but they were more abundant in the field sites, where the pH varied from 4.7 to 5.7. All three sites also showed a decline in algal abundance in late spring through early summer, particularly in the forest site. These declines are correlated with the reduction in light levels reaching the soil surface due to the growth of herbaceous plants in the fields and leaf-out of the oak trees and understory dogwoods in the forest. Grondin and Johansen (1995) found similar patterns in a beech–maple forest in northeastern Ohio. Soil algal abundance declined in June and July, when the beech–maple forest canopy was in leaf. Cyanobacteria were very scarce in the beech–maple forest soil, where the pH was 4.5 to 5.0. In regions polluted by acid rain from industrial activity, green algae may be the only soil algae. Lukešová and Hoffmann (1996) found that in heavily acid-impacted soils of the Czech Republic, only green algae occurred in the soils. In limed plots where the acid was neutralized, however, cyanobacteria, diatoms, and xanthophyceans reappeared.

In active agricultural fields in Georgia, Shimmel and Darley (1985) showed that the abundance of algal cells changed dramatically with major farming events. Harvests of crops increased light to the soil surface and produced peaks in soil algal abundance. Herbicide applications caused soil algal populations to decline. Buckley and Schmidt (2003) used ribosomal RNA (rRNA) gene sequences to compare relative abundance of seven common soil bacterial groups and eukaryotes in replicated plots within cultivated fields, abandoned fields, and fields with no prior cultivation. Cultivation significantly changed the microbial community of the plots compared to fields that had never been cultivated. Abandoned fields only assumed the comparable microbial community after five decades of abandonment. Thus

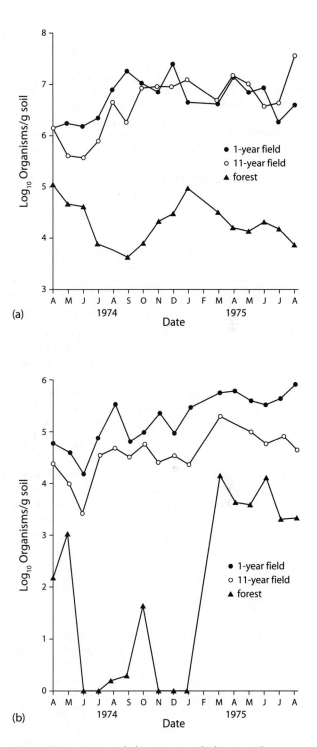

Figure 23.1 Seasonal changes in soil algae numbers. (a) Eukaryotic algae (chlorophytes, charophytes, euglenoids, and chysophyceans). (b) Cyanobacteria. One-year-old field (closed circles), 11-year-old field (open circles), forest (triangles). (From Hunt et al. 1979, by permission of the Ecological Society of America)

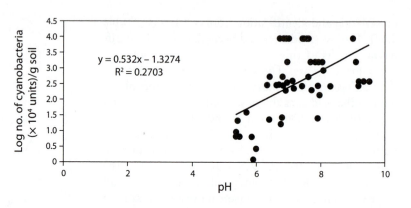

Figure 23.2 Abundance of cyanobacteria as a function of pH in various rice soils from India. (From Nayak and Prasanna 2007. Reprinted with the permission of Szent Istvan Egyetemet, Landscape Architecture and Decision Support System PhD School)

cultivation induces significant changes in soil microbial systems that persist for decades.

In moist temperate regions of Earth, soil algae only form crusts where vascular plants are prevented from forming a closed cover. In cases of disturbance, such as storms, treefall may open areas to colonization, and soil algal crusts may be among the first to enter. Open woodlands, such as the pine barrens of the southeastern United States and discontinuous Mediterranean vegetation around the world, have open spaces for soil algal crusts (Belnap et al. 2001). Smith et al. (2004b) reported that sand dunes in Cape Cod National Seashore supported extensive biological soil crusts composed of filamentous green algae. In dunes, vascular plants are limited by the rapid flow-through of water and unstable substrates. The pH of the pore water in the dunes was acidic (4.99 ± 0.51), and the low pH may have excluded cyanobacteria.

Algal surveys in tropical soils indicate that many of the morphological species present in tropical soils also occur in the soils of other terrestrial ecosystems (Durrell 1964). Tropical soils are often dominated by cyanobacteria because soils there tend to be more alkaline than those in temperate regions. Nayak and Prasanna (2007) showed that the abundance of cyanobacteria in tropical soils in India increased with increasing pH (Figure 23.2). Cyanobacteria have been intensively studied in tropical countries where nitrogen-fixing species are important in agriculture, especially in rice cultivation (Mandal et al. 1999). Free-living nitrogen-fixing cyanobacteria are known to add 20–30 kg N ha^{-1} yr^{-1} to rice-field soils. If the symbiotic association of the water fern *Azolla* and the cyanobacterium *Anabaena* is used in rice fields, the annual addition of nitrogen may be up to 600 kg ha^{-1} yr^{-1} (Nayak et al. 2004). In poor soils in South Africa, inoculation with cyanobacteria improves soil

structure and stability after just a few weeks of incubation (Issa et al. 2007).

Soil Algae in Arid and Semiarid Ecosystems

Arid and semiarid ecosystems represent more than 30% of Earth's land surface. Reduced precipitation restricts the abundance of vascular plants and leaves extensive open areas in which soil algae, together with bacteria, filamentous microfungi, lichens, and bryophytes, may form a visible crust over the surface and extending into the upper soil layers 5 to 10 mm. In the literature, these crust communities have been termed cryptogamic crusts, cryptobiotic crusts, microbiotic crusts, and microphytic crusts, with the different names carrying different implications about the crust communities (Evans and Johansen 1999). The current widely used term is **biological soil crust**, which some researchers abbreviate as **BSC**. The BSC term emphasizes the presence of living organisms without implying any particular taxonomic composition. BSCs have been described from every continent, including Antarctica (Figure 23.3).

The taxonomic composition of BSCs varies with climate. Climate and taxonomic composition interact to produce four types of BSCs that differ in their surface appearance (Belnap et al. 2001). In arid ecosystems that are warm to hot and rarely subject to freezing, the BSC will be smooth and flat. If the soil is acidic and precipitation is somewhat higher, green algae will dominate a smooth crust. If soil pH is alkaline, salt content is relatively high, and precipitation is lower, cyanobacteria will dominate the smooth crust. Under more extreme conditions of heat and aridity, the soil crust of cyanobacteria and algae occurs on the lower surfaces of translucent

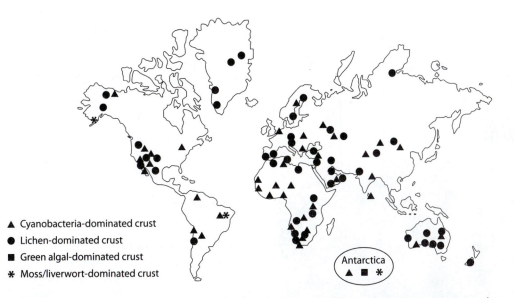

Figure 23.3 Distribution of the sites on Earth where biological soil crusts have been investigated. Each symbol represents a biological soil crust described in a study. The extent of biological soil crusts may be much greater. (From Büdel 2001, by permission of Springer-Verlag)

quartz pebbles among rocks on the desert surface. Such a pebbled surface is called a desert pavement, and the algae in this hypolithic (under rock) system benefit from dew condensation collecting beneath the pebbles (Schlesinger et al. 2003). Under less extreme conditions of higher precipitation, lichens and/or bryophytes join the BSC to form a surface that is termed rugose. Scattered clumps of lichens and mosses produce a low surface (< 2 cm) roughness. Smooth and rugose crusts can be found in the North American Sonoran Desert and the Australian Desert. The other two types of BSC are called rolling and pinnacled. Both occur only in arid lands where soils freeze in winter, as in the Colorado Plateau and the Great Basin of North America. Winter freezing heaves the soil upward. Where the soil surface has an extensive cover (> 40%) of lichens and bryophytes, freezing produces a rolling surface topography up to 5 cm high. Where cyanobacteria dominate and mosses and lichens make up less than 40% of surface cover, freezing produces a pattern of pinnacles that subsequent growth and erosion enhance up to 10 to 15 cm high (Figure 23.4).

Composition of biological soil crusts

Soil crusts from around the world have so far yielded 35 genera of cyanobacteria, 68 eukaryotic genera, 13 genera of lichens with cyanobacterial symbionts

Figure 23.4 A pinnacled biological soil crust from semiarid lands in North America. The black cyanolichen *Collema* is present. (Photo: C. Dott)

(cyanolichens), 69 genera of lichens with green algal symbionts (phycolichens), and 62 genera of mosses and liverworts (Büdel 2001). Taxonomic composition has been extensively studied in North America, Europe, Australia, and Africa, but in other areas there are major gaps in knowledge.

Cyanobacteria are often the dominant autotrophic organisms in BSCs. One of the most common

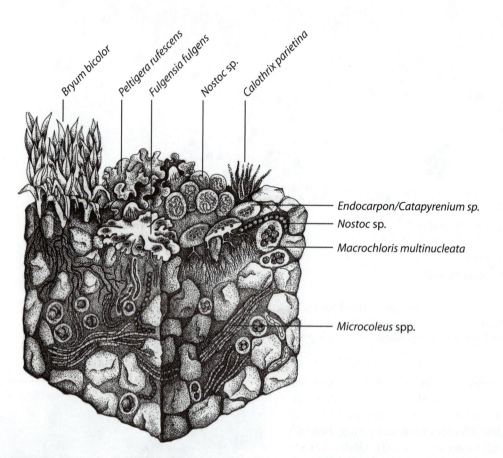

Bryum bicolor   *Peltigera rufescens*   *Fulgensia fulgens*   *Nostoc* sp.   *Calothrix parietina*

Endocarpon/Catapyrenium sp.
Nostoc sp.
Macrochloris multinucleata

Microcoleus spp.

Figure 23.5   Diagram of a biological soil crust with a typical community of species. (From Belnap et al. 2001, by permission of Springer-Verlag)

genera is the nonheterocytous filamentous cyanobacterium *Microcoleus* (see Figure 6.40). Filaments of *Microcoleus* form bundles surrounded by a sheath (Figure 23.5). When the soil is wet, the filaments glide out of their sheaths and move upward through the soil, toward higher light levels. As soil moisture declines, the filaments retreat back to greater depths, where they secrete new sheaths (Belnap and Gardner 1993; Garcia-Pichel and Pringault 2001). The two most frequent species of *Microcoleus* are *M. vaginatus* and *M. chtonoplastes*. *M. chtonoplastes* is common in saline soils, while *M. vaginatus* is widespread in many desert BSCs. Studies on the 16S rDNA sequences of these species revealed that isolates of *M. vaginatus* from Utah, California, Spain, and Israel were essentially identical and formed a cosmopolitan species. *M. chtonoplastes* was not closely related to *M. vaginatus* but instead clustered with some marine

species of cyanobacteria (Garcia-Pichel et al. 2001). In the Sonoran Desert, however, Nagy et al. (2005) found from 16S rDNA sequences that the single most common and abundant microbial group in the BSC closely matched isolates assigned to *Microcoleus steenstrupii*. The *M. steenstrupii* sequences were not related to either *M. vaginatus* or *M. chtonoplastes*. At present, the genus *Microcoleus* appears to be polyphyletic and may need to be split into several genera (Belnap et al. 2001). *Nostoc* species are common nitrogen-fixing cyanobacteria of crusts and are found both within and on top of the soil (see Figure 23.5). Other common filamentous cyanobacteria include the uniseriate genus *Calothrix*; *Scytonema*, which exhibits false branching; and *Stigonema*, a true-branching form. The nonfilamentous colonial genus *Gloeocapsa* is perhaps the most widespread colonial cyanobacterium in BSCs (Büdel 2001).

Eukaryotic algae in BSCs mainly comprise unicellular and filamentous forms. Common green algae include coccoid forms such as *Bracteacoccus*, *Chlorella*, *Myrmecia*, and *Chlorococcum*; colonies such as *Macrochloris* (see Figure 23.5); and filaments such as *Klebsormidium* and *Geminella*. *Klebsormidium* and *Geminella* dominate sand dune crusts in Cape Cod (Smith et al. 2004b). Examination of 18S rDNA sequences from 11 green algal isolates from desert soil crusts in New Mexico, Utah, and Baja, California, revealed unexpected diversity; some simple unicellular greens had sequences linking them to *Scenedesmus* (Lewis and Flechtner 2002). Common pennate diatoms include *Hantzschia*, *Navicula*, and *Nitzschia*. Arid soils containing gypsum (calcium sulfate [$CaSO_4 \cdot 2H_2O$]) have diverse populations of diatoms (Anderson and Rushforth 1976). Representatives of the Chrysophyceae, euglenoids, and Xanthophyceae are only sporadically reported and not in great numbers.

Both lichens and bryophytes have most of their photosynthetic tissue on or above the soil surface. In addition to the separation into phycolichens and cyanolichens, lichens show a range of morphological forms. Crustose lichens that are flat and pressed tightly to the soil surface include *Fulgensia fulgens*, with its well-defined marginal lobes, and the shield-like *Endocarpon*, with a central holdfast composed of strands of hyphae (see Figure 23.5). *Fulgensia* and *Endocarpon* have green algal symbionts. Foliose lichens such as *Peltigera rufescens*, a cyanolichen, appear similar to a leaf. Fruticose lichens are three-dimensional and bushy, with finely branched threads. The fruticose phycolichen *Cladonia rangiferina* is used to simulate trees and shrubs in model train and architectural layouts. Along the Atlantic coast of the central Namib Desert of southwest Africa, fog nurtures BSCs dominated by foliose and fruticose lichens over hundreds of km² (Ullmann and Büdel 2001). Gelatinous lichens such as *Collema* are important soil crust lichens, with symbiotic cyanobacteria that fix nitrogen. Bryophytes indicate more moist habitats for BSCs. *Bryum* and *Ceratodon* are common genera of mosses in BSCs, and *Riccia* is the most important genus of liverwort (Belnap et al. 2001).

Adaptations and distribution of biological crust organisms

BSCs occur in areas where climatic conditions are too extreme for extensive growth of vascular plants. These extreme conditions include extreme high and low temperatures at the soil surface, low levels of precipitation, often with long periods of desiccation, and extreme levels of visible and ultraviolet (UV) irradiance.

Adaptations to hot and dry conditions. Temperatures near 70°C (158°F) have been reported within lichens growing in a steppe in southwest Germany (cited in Belnap et al. 2001). Some of these lichens could survive exposure to temperatures of 90°C to 100°C for up to 30 minutes. Heat resistance is greater in desiccated organisms, and soil crust organisms can endure long periods of desiccation. *Nostoc* species from soil in China recovered after two years of desiccation (Scherer et al. 1984). Viable soil algae of the genera *Nostoc*, *Chlorococcum*, and *Stichococcus* have been obtained from herbarium specimens up to 87 years old (Parker et al. 1969). Because biological soil crust organisms are metabolically active only when wet, soil algae may show a seasonal pattern of abundance related to the availability of water (Johansen et al. 1993). The gelatinous cyanolichens such as *Collema* are adapted to hold on to water when it is present. The filaments of *Microcoleus* species are motile and can migrate up or down in the soil layers to follow water. As the soil dries, the filaments migrate downward and re-form their sheaths to await the next rainfall. Some lichens are able to use water vapor alone to recover from the desiccated state and maintain metabolism.

Adaptation to high irradiances. Organisms in BSCs may be exposed to extreme levels of irradiance, particularly if they lie on the surface. Lichens on the surface are variously pigmented to reflect or absorb incident irradiance and screen their photosynthetic symbionts. Up to 93% of incident irradiance was screened by the fungal cortex of the cyanolichen *Peltula* from South Africa and Mexico (Büdel and Lange 1994). Cyanobacteria on the soil surface of crusts, such as *Nostoc*, *Scytonema*, and *Calothrix*, produce pigments that protect them from excess photosynthetically active radiation (PAR) and UV radiation. These pigments include carotenoids, mycosporin-like amino acids (MAAs), and scytonemins (Garcia-Pichel and Castenholz 1991, 1993). *Nostoc* shows a seasonal pattern in its production of carotenoids and UV protective pigments (Bilger et al. 1997). Bowker et al. (2002) suggested that *Microcoleus* may benefit from the pigments produced by *Nostoc* and *Scytonema* on the surface. *Microcoleus* species do not synthesize UV protective pigments and thus may be damaged by UV radiation in the absence of *Nostoc* or *Scytonema*.

BSC stratification. The organisms in BSCs are often vertically stratified in distinct layers (Garcia-Pichel and Belnap 1996). The lichens and bryophytes have their photosynthetic tissues above or on the soil surface. The algae on the soil surface are nonmotile genera with heavy pigmentation such as *Nostoc*, *Calothrix*, and *Scytonema*. The level of PAR declines rapidly with depth in the soil, such that the level at 1 mm is about 1% of that at the surface (Evans and Johansen 1999). Beneath the soil surface lie the motile algae such as *Microcoleus* and *Oscillatoria*, diatoms, flagellates, and various nonmotile algae, such as *Chlorella*. Hu et al. (2003) looked at the vertical distribution of cyanobacteria and eukaryotic algae in desert soil crusts from China at a microscale level with light and scanning microscopy. *Scytonema* or *Nostoc* dominated the surface layer from 0.02 to 0.05 mm, with a layer from 0.05 to 0.1 mm dominated by the green alga *Desmococcus* (Figure 23.6). *Microcoleus* dominated at the next layer, from 0.1 to 1 mm, and other genera of cyanobacteria such as *Lyngbya* were abundant from 1 to 3.0 mm. Diatoms such as *Navicula* occurred from 3.0 to 4.0 mm. In the phytoplankton of the oceans and lakes, diatoms are known to grow at very low light levels. The presence of diatoms in the lowest layers of these desert soil crusts may well reflect their ability to grow under very low light conditions. Most of the cyanobacteria and algae in soil crusts are located within 4 mm of the surface, although some algae and chlorophyll *a* can be detected down to about 1 cm (Belnap and Gardner 1993).

Ecological functions of biological soil crusts

Biological soil crusts have a number of important ecological functions in arid and semiarid lands. These ecological functions include the stabilization of soil surfaces from erosion, fixation of carbon and nitrogen into the soil, and effects on vascular plants. Of these ecological functions, the stabilization of soil surfaces is probably the most important and most widely recognized.

Soil stabilization. Wind is a major force for erosion in arid and semiarid lands, where sparse vegetation leaves large areas of soil surface exposed (Goudie 1978). BSCs stabilize soil surfaces by binding and cementing together soil particles into aggregates that are more resistant to wind erosion. The

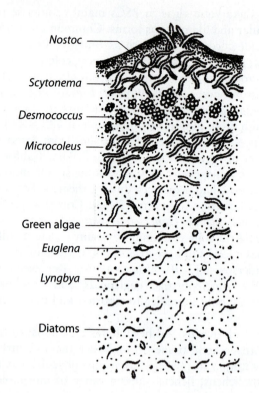

Figure 23.6 Vertical distribution of soil algae and cyanobacteria within a biological soil crust at a desert site in Shapotou, China. (From Hu et al. 2003, by permission of Kluwer Academic Publishers)

nonheterocytous *Microcoleus* species are especially important because they produce a sticky sheath around bundles of their filaments, and this sheath material weaves through the upper layers of the soil and binds soil particles into larger aggregates (Belnap and Gardner 1993). Nonfilamentous algae secrete extracellular organic compounds that cement soil particles together (Bailey et al. 1973; Lynch and Bragg 1985). The combination of sticky filaments and cementing of extracellular organics is more effective than either alone (McKenna-Neuman et al. 1996). Soil fungi and lichens produce hyphae, and bryophytes send rhizoids 4 to 5 mm into the upper soil layers, adding to the aggregation of soil particles. As a result of all this binding of soil particles, pieces of the BSC can be lifted up like pieces of fibrous mat. Because larger aggregates are heavier, they are harder for wind to move and are less subject to erosion.

Stabilization of arid soil surfaces by BSCs is largely based on observations of erosion following disturbances by grazing livestock or range fires (Johansen and St. Clair 1986; Johansen et al. 1984). In Canyonlands National Park, Kleiner and Harper (1972, 1977) reported higher levels of soil erosion in an area subject to light winter grazing than in an adjacent area without grazing. The amount of vascular plant cover was identical at the two sites, but the grazed site had only 5% soil crust cover, whereas the nongrazed site had 38% soil crust cover. Erosion had removed soil from around clumps of perennial bunchgrasses and formed gullies at the grazed site, but erosion was nearly absent at the nongrazed site. Different types of crusts may stabilize the soil surface to different degrees. Moss-dominated crusts are thought to stabilize surfaces to a greater extent than those dominated by lichens, which, in turn, are thought to be more effective than crusts dominated by cyanobacteria or eukaryotic algae (Eldridge and Greene 1994; Belnap and Gillette 1998). Hu et al. (2002), however, conducted wind-tunnel experiments and found that for sandy desert soils in China, a 1-year-old soil crust consisting of only cyanobacteria and algae could withstand a sandstorm of 25 m s^{-1} for more than 8 hours, which was the same level of wind resistance exhibited by much older crusts (up to 42 years of age) with fungi, lichens, and mosses. These results suggest that the cyanobacteria and algae may be the main contributors to soil surface stabilization, while the lichens and mosses affect soil crust structure, thickness, and levels of organic carbon and fixed nitrogen.

The effects of BSCs on erosion by water and the infiltration and retention of water through the crust are more complex than the effects of crusts on wind erosion. Numerous experimental studies using simulated rain on bare and crust-covered soils have shown greater erosion on bare soils than on soils with BSCs. In arid and semiarid lands subject to winter freezing, rolling and pinnacled crusts create a rough topography at the centimeter scale, and the resulting small-scale depressions act as detention ponds to hold back water. In such cool deserts, rough BSCs generally increase water infiltration (Loope and Gifford 1972). Rugose BSCs in warm deserts may function as small-scale detention structures for water as well. Smooth soil crusts appear to decrease water retention and infiltration in warm deserts, as shown for sandy soils in Arizona (Brotherson and Rushforth 1983), Australia (Eldridge 1993), and Israel (Yair 1990). However, studies of sandy soils in Oklahoma and Kansas did not reveal any effect of BSCs on infiltration rates (Booth 1941). Even if the main effect of smooth BSCs in warm and hot deserts is to decrease infiltration of water, the resulting increased surface flow may support sparse clumps of vascular plants (Eldridge et al. 2000).

Soil enrichment. BSCs have been shown to enrich arid soils by increasing the organic carbon content and the amount of fixed nitrogen. The autotrophic organisms within BSCs are capable of high rates of photosynthesis, but they are metabolically active only when they are wet. The sparse vascular plants within arid ecosystems contribute organic matter to the soil beneath them, but they add little to the larger open spaces between them. The carbon contributed by the BSCs maintains the fertility of the plant interspaces and provides energy to heterotrophic populations in the soil. Beymer and Klopatek (1991) showed that BSCs released significant amounts of organic carbon that accumulated in the soil beneath the crust. Jeffries et al. (1993) found that BSCs in the Colorado Plateau of southern Utah had positive net carbon gains only during prolonged wet periods, such as winter and spring, and short-term wet–dry cycles produced a net carbon loss for the crust. They estimated the annual net primary productivity of the semiarid area in southern Utah to be 6.4 to 23 kg C ha^{-1} yr^{-1} for a crust dominated by *Microcoleus* and *Scytonema*. Beymer and Klopatek (1991) reported higher values of annual primary productivity (28 to 350 kg C ha^{-1} yr^{-1}) from soil crusts among a pinyon juniper woodland.

Nitrogen fixation by free-living and lichenized cyanobacteria as well as heterotrophic bacteria has been studied extensively in BSCs (Snyder and Wullstein 1973). Nitrogen concentrations are generally low in desert ecosystems, and there are few vascular plants that host nitrogen-fixing symbionts in their roots. Because nitrogen can limit primary productivity in desert ecosystems (Ettershank et al. 1978; Nobel et al. 1988), nitrogen fixation by cyanobacteria and cyanolichens becomes a major source of fixed nitrogen for desert soils and vascular plants (Evans and Ehleringer 1993). Free-living heterocytous cyanobacteria such as *Nostoc* and *Calothrix* and cyanolichens such as *Collema* are significant sources of fixed nitrogen, but the nonheterocytous *Microcoleus* is also associated with significant nitrogenase activity (Jeffries et al. 1992). *Microcoleus* lacks the dinitrogen reductase gene *nifH* but forms a symbiotic partnership with nitrogen-fixing heterotrophic bacteria. The cyanobacterium provides suitable habitat, in the form of anaerobic microsites within its gelatinous sheath, and a source of organic carbon to the heterotrophic bacteria, which, in turn, supply fixed nitrogen (Steppe et al. 1996).

Estimates of nitrogen fixation by desert BSCs range from less than 1 kg N ha^{-1} yr^{-1} up to 100 kg N ha^{-1} yr^{-1}. Rychert et al. (1978) reported rates of nitrogen fixation from 10 to 100 kg N ha^{-1} yr^{-1} from cold desert ecosystems, while Jeffries et al. (1992) found much lower rates of 0.7 to 3.6 kg N ha^{-1} yr^{-1} on the Colorado Plateau. Estimates from the Sonoran Desert varied from 7 to 18 kg N ha^{-1} yr^{-1}, while a nitrogen fixation rate of 1.3 kg N ha^{-1} yr^{-1} was measured in Australia (Rychert et al. 1978). If BSCs produce an average of 25 kg N ha^{-1} yr^{-1} in most semiarid lands, this amount would be more than sufficient to meet the 10 kg N ha^{-1} yr^{-1} needed by desert vascular plants (West and Skujins 1977). Not all of this fixed nitrogen may be available to support other organisms. Kershaw (1985) found that 19%–28% of ^{15}N label applied to *Nostoc* leaked out into the external ammonia pool. Once in this pool, nitrogen may be lost by direct volatilization of ammonia or by conversion into nitrate, followed by denitrification to gaseous dinitrogen. Johnson et al. (2005, 2007), however, found that these processes did not remove fixed nitrogen, at least in their study sites on the Colorado Plateau. Fixed nitrogen entered the external ammonia pool but was oxidized to nitrates by ammonia-oxidizing bacteria below the crust surface rather than volatilized (Johnson et al. 2005). Rates of denitrification in the same BSCs were inconsequential because of extremely low populations of denitrifying bacteria (Johnson et al. 2007c). The BSCs on the Colorado Plateau therefore function as net exporters of ammonium, nitrate, and organic nitrogen. If these results hold for BSCs in cool deserts or arid lands in general, BSCs are very important to arid land fertility.

Effects on vascular plants. BSCs have a number of effects on vascular plants in arid and semiarid lands. BSCs in the spaces between vascular plants prevent erosion, and, if crusts are removed, erosion can bury vascular plants under eroded soil, isolate them on soil pinnacles held in place by the plant roots, or wash them out entirely. A number of studies have shown that BSCs raise the concentration of most essential minerals, such as K, Ca, Mg, P, Fe, Mn, and S (Harper and Belnap 2001). As discussed previously,

BSCs are the main source of fixed nitrogen in arid ecosystems; their presence increases soil nitrogen by up to 200% (Rogers and Burns 1994; Harper and Belnap 2001). Soil crusts also increase total soil organic carbon by up to 300% (Rogers and Burns 1994). In addition, the presence of soil crusts is correlated with mycorrhizal associations, which increase nutrient availability to plants (Harper and Pendleton 1993). In cool deserts, BSCs increase water retention and infiltration, reduce runoff, and increase the water available for vascular plants. In warm to hot arid lands, soil crusts are generally smooth and increase runoff, but that increased runoff often reaches vascular plants that are positioned in the landscape to intercept surface flows.

Biological soil crusts can directly affect seed germination, seedling establishment, and species richness of vascular plants. These effects vary depending on whether the arid land is warm or cool and whether the vascular plants are perennials, annuals, or invasive species. In lab studies of plants from cool deserts, soil crusts greatly increased germination and establishment of native perennial grasses from southeastern Utah compared to noncrusted soils (St. Clair et al. 1984). Crust removal through livestock trampling decreased survival of perennial grass seedlings (Eckert et al. 1986). BSCs inhibit seed germination and establishment in annuals and invasive species that require soil surface disturbance for seedling establishment (Deines et al. 2007). Vascular plant species richness is higher on crusted soils than on noncrusted soils (Kleiner and Harper 1977; Jeffries and Klopatek 1987) or essentially the same (Anderson et al. 1982a). In cool desert ecosystems, BSCs have either a positive or neutral effect on perennial vascular plants but a negative effect on annuals and invasive species that require disturbance for establishment (Belnap et al. 2001). In warm deserts, however, results are more ambiguous. Eldridge and Greene (1994) found no relation between soil crust cover and shrub seedling establishment in Australian deserts, whereas Graetz and Tongway (1986) reported a positive correlation between shrub cover and crust cover in Australia. Zaady et al. (1997) found that cyanobacterial crusts in Israel inhibited seed germination of some annuals but enhanced others. No firm conclusions can be drawn at this time about the direct relations between warm desert vascular plants and soil crusts. Warm desert soil crusts function in erosion control and mineral retention, however, and these functions have a positive effect on vascular plants.

Disturbances and management

Disturbances to BSC communities arise from mechanical sources, such as livestock grazing, vehicle and human traffic related to recreational uses, and range fires. The more frequent or the more intense the scale of these disturbances, the greater the impact on BSC communities. Lichens and mosses are more susceptible to disturbance damage and require longer time periods to recover from disturbances than do the cyanobacteria and eukaryotic algae. Thus, frequent disturbances simplify soil crust communities and keep them in a state of early succession in which they are dominated by microorganisms. Disturbances that occur during the dry season are more destructive because the organisms cannot begin recovery until water is available. The effects of mechanical damage are most pronounced on soils that are easily eroded and have high relief. Range fires kill BSCs and set off a process of succession from cyanobacteria and eukaryotic algae to mosses and lichens. The general effect of disturbances is to reverse all the ecological functions discussed in the previous section. Disturbances to soil crusts increase erosion and loss of minerals, decrease fixation and input of carbon and nitrogen, and increase germination and establishment of invasive annual species.

Grazing. In pre-settlement times, the arid and semiarid ecosystems west of the Rocky Mountains were subject to low levels of soil disturbance. Populations of large grazing mammals were low, due to limited water supplies and sparse vegetation. Settlement brought in large herds of grazing livestock and increased soil surface disturbance. BSCs persist under grazing disturbance in temperate regions such as the Great Plains of the United States and the Serengeti Plains of east Africa, where precipitation is higher than in arid lands, and animal grazing is intermittent due to migration. Within these plains areas, the cover and diversity of lichens and mosses is lower at sites subject to grazing than at ungrazed sites (Belnap and Lange 2001). In arid and semiarid lands, a number of studies have documented a reduction in BSC cover and species diversity associated with livestock grazing (Anderson et al. 1982a; Beymer and Klopatek 1992). Jeffries and Klopatek (1987) found a dramatic decline in BSC cover as a function of grazing pressure at sites across southern Utah and northern Arizona. The area of soil crust cover was 2129 m² ha⁻¹ on an ungrazed relict site, 1196 m² ha⁻¹ on a lightly grazed

site (winter grazing only for 3 years), 70 m² ha⁻¹ on a heavily grazed site (year-round grazing for more than 100 years), and only 50 m² ha⁻¹ on a recovering site protected for more than 10 years. Grazing is especially damaging to soil crust communities when it occurs from late winter to early spring. At that time, soil water is depleted, and the crusts are easily crushed and eroded (Marble and Harper 1989). Estimated recovery times for areas disturbed by grazing are greater than 20 years (Anderson et al. 1982b; Johansen and St. Clair 1986). Lichens and mosses show longer recovery times than do cyanobacteria and eukaryotic algae.

Mechanical disturbance. In addition to livestock grazing, mechanical disturbance can arise from human foot traffic and vehicular traffic from recreational off-road vehicles and mining activities. Barger et al. (2006) found that human trampling of soil crusts in southeastern Utah increased water runoff losses of organic carbon and nitrogen, compared to intact crusts. Intact late-succession-stage cyanolichen crusts retained nutrients better than early-succession-stage cyanobacterial crusts. Off-road vehicle use in the Namib Desert damages soil crust lichens, and estimates of recovery time, based on the period of time since the most recent disturbance, range as high as 530 years (Lalley and Viles 2008). Vehicles can cause more severe damage to crusts than trampling by livestock or humans because they tear and compact soil surfaces, bury crust organisms, and create channels for increased erosion. Recovery times for BSCs are longer in more arid lands (Johansen 1993).

Although less widespread than trampling by livestock, range fires can be much more damaging to soil crusts in terms of species composition and biomass. Historically, range fires in the semiarid lands west of the Rocky Mountains were infrequent, localized, and of low intensity. Vascular plants were widely spaced, and therefore combustible fuel was sparse. BSCs have too little biomass to carry wildfires across interspaces between plants. The incidence and intensity of range fires across this region have increased due to the invasion of the exotic annual *Bromus tectorum* (cheatgrass). Cheatgrass fills in disturbed soil surfaces between native perennial grasses, creating more homogeneous grasslands and generating a continuous layer of combustible fuel. Cheatgrass now dominates more than 30 million hectares of public land (Knapp 1996), and massive fires have become more common across the Great Basin, Columbia Basin, and Colorado Plateau.

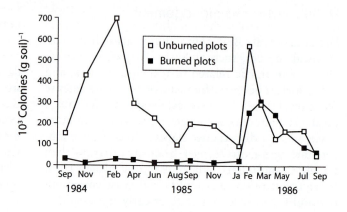

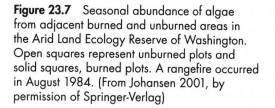

Figure 23.7 Seasonal abundance of algae from adjacent burned and unburned areas in the Arid Land Ecology Reserve of Washington. Open squares represent unburned plots and solid squares, burned plots. A rangefire occurred in August 1984. (From Johansen 2001, by permission of Springer-Verlag)

Following range fires, the BSC community recovers slowly. Based on a number of studies of crust recovery after disturbance, a pattern of succession has been proposed (Anderson et al. 1982b; Johansen et al. 1984; Johansen 2001). In cooler, more moist areas, such as the northern Columbia Basin, Johansen et al. (1993) found that coccoid green algae and diatoms were the first to recover, after just 2 years (Figure 23.7). These algae normally grow below the soil surface and may have escaped the worst of a range fire that occurred in August 1984. Their recovery in February 1986 coincided with the increase in algae on adjacent unburned sites and increased precipitation at both sites. Recovery of cyanobacteria was slower, with *Microcoleus* and *Schizothrix* being the first cyanobacteria to reappear. *Nostoc* species showed a major decrease after fire, perhaps due to their growth on the soil crust surface. At sites in the Great Basin, the differences in algal communities between burned and unburned sites disappeared 3 to 5 years after the range fire (Johansen et al. 1984). Full recovery of the microorganism components of BSCs may take up to 15 years. In an arid area in southern Utah, Callison et al. (1985) found no evidence of crust recovery after 37 years. Lichens and bryophytes recover even more slowly than do cyanobacteria and eukaryotic algae, taking 40 or more years, depending on aridity. The cyanolichen *Collema* is one of the first lichens to reappear after disturbance (Evans and Johansen 1999). In hyperarid deserts, lichen recovery may be so slow as to be indeterminate within the lifetime of a single researcher (Lalley and Viles 2008).

Management. Biological soil crusts are a major factor in soil stabilization and erosion control in arid and semiarid lands. They play a major role in soil mineral retention and are the most significant source of nitrogen input and addition of organic carbon to arid soils. Damage to BSCs can be managed by limiting vehicles to established roads, and visitors to established campsites and hiking trails. For livestock grazing, soil crusts may be protected by managing livestock population levels and dispersing livestock across usable ranges. Soil crusts are least subject to disturbance when soils are frozen or snow covered. Crusts on sandy soils are less subject to damage if wet, while those on clay soils are more resistant if dry. Fire management is difficult. Prescribed fires can be used to reduce fuel buildup, but burning also opens sites to invasion by exotic plants. In the semiarid lands west of the Rocky Mountains, controlled fires must be accompanied by revegetation with native plants to prevent exotic plants from invading. Some research has shown that inoculation of soils with crust slurries or pelletized cyanobacteria can speed up soil crust recovery (St. Clair et al. 1986; Belnap 1993; Buttars et al. 1998). Remote sensing techniques will likely be used to monitor the condition of BSCs across large areas (Karnieli et al. 2001), and measures for crust protection have been included in rangeland management plans (Rosentreter et al. 2001).

Soil Algae in Polar Ecosystems

At the present time, there is more information on the soil algae of Antarctica than for the Arctic (Hansen 2001) and alpine ecosystems (Türk and Gärtner 2001). Many of the soil algae recorded in Antarctica, in the Arctic, and in alpine soils are the same morphological species found in soils of arid and semiarid lands, but the number of species and diversity are reduced in polar and alpine ecosystems. For example, the Russian Arctic has about 302 taxa of soil algae, compared to more than 2000 from soils of the former USSR (Novichkova-Ivanova 1972). The diatom and desmid floras of Antarctica are reduced by comparison to other soil algal floras. In Germany, about 100 species of diatoms in 24 genera have been recorded from soils, but in all terrestrial habitats studied in Antarctica, only about 40 species in 15 genera have been reported. Of these 40 species, only three are seen regularly: *Hantzschia amphioxys*, *Navicula muticopsis*, and *Pinnularia borealis*. In Antarctic terrestrial ecosystems, only about 15 desmid species have been found, while 38 have been reported from soils in the former USSR (Broady 1996). Distinctive soil algal taxa that are absent from Antarctica include the filamentous cyanobacterium *Cylindrospermum* (though present as a human import at one research station), the rhodophyte *Porphyridium purpureum*, xanthophycean *Botrydium*, greens *Protosiphon* and *Trentepohlia*, and *Euglena* (Broady and Smith 1994; Broady 1996). There is some evidence for endemic soil algal species in Antarctica. Two chlorophytes studied in culture, *Fottea pyrenoidosa* from soil on Signy Island (Broady 1976) and a snow alga, *Chloromonas rubroleosa* (Ling and Seppelt 1993), are distinctly different from their closest presumed relatives. *Hemichloris* is an endemic genus found only in endolithic habitats in southern Victoria Land. These algae have been identified by culture studies; molecular studies are needed to verify their status as endemics and determine nearest relatives.

Research in the Antarctic has been encouraged by the treaty status of the continent, which is not the property of any one nation. Antarctica possesses comparatively simple ecosystems and a severe climate that approaches, as far as it is possible on the surface of the Earth, conditions on other solar system bodies where life may exist or have previously existed. The Antarctic biota has been of interest to ecologists studying communities in their simplest form, biochemists interested in adaptation to cold and aridity, and astrobiologists as model systems and testing grounds in the exploration of the solar system and search for life beyond Earth. As a result, the microbiology of the Antarctic cold deserts has been extensively studied.

Antarctica is the coldest, windiest, driest, and highest continent on Earth. It is largely covered in glacial snow and ice to a maximum depth of about 4 km (2.5 mi); only about 280,000 km² or 2% of the continent is ice free (Vishniac 2002). The continent has a wide climactic range and spans 27° of latitude from the northern tip of the Antarctic Peninsula at 63°S to the South Pole. Two climatic zones are widely recognized. The Maritime Antarctic consists of the western half of the Antarctic Peninsula down to about 69°S and adjacent islands together with an arc of islands extending off to the north and east from the tip of the peninsula (Figure 23.8). In this region, the warmest months of austral summer (December to February) have average temperatures above freezing (0–2°C), winter temperatures average around –10°C, and annual precipitation ranges from 350 to 500 mm. Bryophytes and lichens tend to dominate the vegetation in the Maritime Antarctic, where 300–400 lichen species and 400 bryophytes have been recorded (Green and Broady 2001). Cyanobacteria and eukaryotic algae are prominent in the second zone, the Continental Antarctic, which includes the entire main continent of Antarctica. The average temperatures in the warmest months of austral summer are below freezing, winter temperatures are around –20°C, and precipitation ranges from 300 mm at the coast to 50 mm at the South Pole. The only habitats suitable for organisms are ice-free valleys in the shadows of coastal mountain ranges (dry valleys) and the tops of mountains (nunataks) and ridges (moraines) that protrude above the polar ice cap. Such ice-free areas are small in area compared to the polar ice cap and are isolated by large distances from each other. The La Gorce Mountains, for example, located at 86°30′S, are one of the most southern terrestrial areas in the Continental Antarctic. The largest single ice-free area is the McMurdo Dry Valleys or the Ross Desert in the shadows of the Trans-Antarctic Mountains in Victoria Land. Dry valleys are headed by glaciers that provide summer meltwaters to the valleys, along with meltwater from snow banks that accumulate in sheltered areas. In dry valleys, the soils are sandy with abundant rocky fragments, little clay or silt, low levels of organic materials, and abundant salts. At varying depths under the surface material lies a zone of permafrost—frozen soil cemented together with ice. Lichens (125 species) and bryophytes (32 species) are much less diverse in the Continental Antarctic, and in some dry valleys, they are entirely absent.

The majority of work on soil algae in Antarctica has consisted of floristic surveys, with morphological species reported based on microscopic and culture

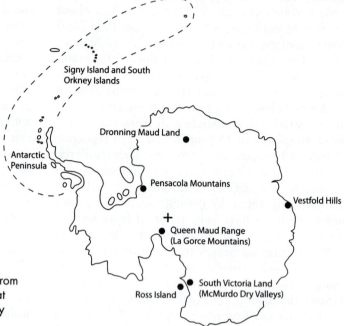

Figure 23.8   Map of Antarctica showing locations from which terrestrial algae have been described and that are discussed in this chapter. (From Broady 1996, by permission of Kluwer Academic Publishers)

studies. Figure 23.8 shows some of the locations where floristic surveys have been conducted. Commonly reported genera of terrestrial algae are listed in Table 23.2 (Broady 1996). Many of the listed genera have been discussed previously in connection with BSCs of arid and semiarid lands in temperate zones. Two exceptions are *Mastigocladus* and *Prasiola*. *Mastigocladus* is a thermophilic cyanobacterium usually found in hot springs (Chapter 6). In Antarctica it occurs only on warm to hot geothermal soils at high altitude on the volcanoes Mt. Erebus and Mt. Melbourne in the Ross Sea area (Melick et al. 1991). In temperate regions, *Prasiola* occurs in cold mountain streams or in the high intertidal of rocky shorelines, but in Antarctica it is a terrestrial alga that grows on moist soils in coastal ice-free areas near penguin rookeries.

The diversity of terrestrial algae appears to be lower in the Continental Antarctic than in the Maritime Antarctic, whether the data are derived from floristic surveys or molecular studies. Broady (1979) reported 85 genera of terrestrial algae from maritime Signy Island, whereas Broady and Weinstein (1998) found only 15 genera of algae in the La Gorce Mountains, which lie 410 km (260 mi) from the South Pole. Lawley et al. (2004) used molecular analysis of 18S rDNA to examine geographic patterns of eukaryote diversity in Antarctic soils from 60°S to 87°S. The

most southerly Continental Antarctic site was three to four times less diverse than the average for the Maritime Antarctic sites. Different study sites gave 18S rDNA sequences with very little overlap between sites and very low relatedness to existing sequences in the databases, suggesting a high degree of site isolation within Antarctica and possible endemic species.

In the Maritime Antarctic, bryophytes and lichens dominate the vegetation in the moist regions, but the most common and widespread terrestrial ecosystems are the fellfields. In fellfields, freezing and thawing shatters rocks, heaves up soil, and sorts out particles into distinct polygons 1 to 2.5 m in diameter that are bordered by large stones and filled by a soil consisting of fine particles. The border stones support small cushions of mosses such as *Ceratodon purpureus*, while the fine soil provides substrate for bacteria, cyanobacteria, and eukaryotic algae. It is generally assumed that the cyanobacteria and eukaryotic algae act to bind together soil particles and stabilize soils, but experimental evidence of this function is lacking for the Antarctic. The microorganisms of soils are metabolically active when liquid water is available. Thus in fellfields on maritime Signy Island, the seasonal cycle of algal growth begins with the first thaw in austral spring (November), when filamentous green algae begin growth under the snow and ice (Figure 23.9). *Zygnema* or *Ulothrix*

Table 23.2 Common genera of Antarctic terrestrial algae*

Algal group	Genus
Cyanobacteria	*Aphanocapsa*
	Chroococcidiopsis
	Gloeocapsa
	Synechococcus
	Lyngbya
	Microcoleus
	Phormidium
	Plectonema
	Pseudanabaena
	Schizothrix
	Anabaena
	Calothrix
	Nodularia
	Nostoc
	Scytonema
	Tolypothrix
	Mastigocladus
	Stigonema
Green algae	*Chlamydomonas*
	Chloromonas
	Bracteacoccus
	Chlorella
	Chlorococcum
	Hemichloris
	Pseudococcomyxa
	Tetracystis
	Coccomyxa
	Stichococcus
	Ulothrix
	Desmococcus
	Prasiococcus
	Pleurococcus
	Prasiola
	Actinotaenium
	Cylindrocystis
	Mesotaenium
Xanthophytes	*Botrydiopsis*
	Monodus
	Heterothrix
	Heterococcus
Diatoms	*Achnanthes*
	Hantzschia
	Navicula
	Pinnularia
Dinoflagellates	*Gloeodinium*

*From Broady 1996.

then decline, and the cyanobacteria *Phormidium* and *Pseudanabaena* replace them after snow and ice disappear from the sites. Subsequent experiments at the same sites on Signy Island showed that late-summer declines in algal populations were greatly reduced when nitrogen was added (as nitrate), which indicates that nitrogen limitation may be responsible for the declines (Davey and Rothery 1992).

In the Continental Antarctic, terrestrial algae occur in the soil of all ice-free areas that have been examined. Cyanobacteria and green algae dominate the soil algae, with xanthophyceans and diatoms as frequent associates. Cavacini (2001) found 63 taxa of cyanobacteria and eukaryotic algae in soil samples from northern Victoria Land, despite there being no visible cover of these organisms. BSCs occur but are widely scattered and require a regular supply of water from snow or glacial melt in the austral summer, in addition to abundant light and shelter from winds. Cyanobacteria may form black, orange, blue-green, or violet BSCs dominated by *Microcoleus*, *Nostoc*, *Phormidium*, or *Leptolyngbya*. At inland sites in the Vestfold Hills, black crusts composed of mucilaginous colonies of *Gloeocapsa* may occur in patches where soil is moist (Green and Broady 2001). In dry valleys, the climate is more arid, bryophytes and lichens may be absent, and the algal community may exist below the soil surface as hypolithic organisms under translucent rocks or as endolithic organisms under the surfaces of porous rocks (see Section 23.3).

In dry valley soils, the concentration of organic carbon is very low (0.01 to 0.5 mg C gm^{-1} soil), and the level of biological activity is so low that physical and biological controls on the flux of CO_2 are of similar scale. As a consequence, it is not possible to measure net primary productivity of dry valley soils by using conventional techniques (Parsons et al. 2004). It is possible, however, to measure photosynthesis in isolated, cultured Antarctic algae. *Prasiola crispa* collected from northern Victoria Land has a net photosynthetic rate of 0.42 mg O_2 gm dry wt^{-1} h^{-1} at 10°C and a lower, but still positive, rate of 0.1 mg O_2 gm dry wt^{-1} h^{-1} at −5°C (Becker 1982). Nitrogen fixation has been studied in Antarctica in more detail than has carbon dynamics. In the ice-free Vestfold Hills, Davey and Marchant (1983) examined the *in situ* nitrogen fixation of *Nostoc* over a 12-month period. Nitrogenase activity was substantial during the austral summer but ceased when soil surface temperatures fell below −7°C. Annual biological nitrogen fixation by *Nostoc* mats and *Nostoc*-moss

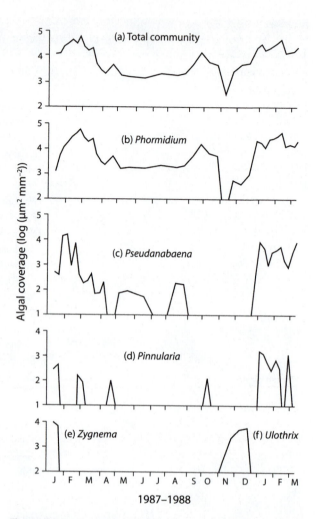

Figure 23.9 Seasonal growth of eukaryotic algae and cyanobacteria at a fellfield site on Signy Island. (a) Total eukaryotic algae plus cyanobacteria, (b) *Phormidium*, (c) *Pseudanabaena*, (d) *Pinnularia*, (e) *Zygnema*, and (f) *Ulothrix*. (From Davey 1991, by permission of Blackwell Publishing)

associations were estimated to add 52 and 119 mg N m^{-2} yr^{-1}, respectively, to terrestrial ecosystems in the Vestfold Hills. These values are equivalent to 0.52 kg and 1.2 kg N ha^{-1} yr^{-1}, values greater than the lower end of the range of estimates reported for Arctic tundra (Fritz-Sheridan 1988) and alpine tundra in Colorado (Wojciechowski and Heimbrook 1984). The few values of nitrogen fixation from Antarctica suggest that soil algal communities are an important source of nitrogen for terrestrial ecosystems. In Continental Antarctica, the algae of soils and crusts are often the only source of primary productivity and represent the main source of food for the few species of soil arthropods that are present (Bokhorst et al. 2007). They are also the primary source of input to another type of soil ecosystem, described next.

Beneath the soil surface in the Arctic and Antarctic, another type of soil algal ecosystem occurs, called the permafrost. Permafrost is perennially frozen soil with an average annual temperature of less than 0°C and with most of its water occurring as ice. Permafrost underlies about 20% of the terrestrial surface area of Earth, and in the northern hemisphere, it may reach depths as great as 1000 m (Gilichinsky 2002). The deeper the layer of permafrost, the older it is. In the Arctic tundra, the seasonal thaw begins in June and proceeds to a maximum depth of 0.3 to 1 m by mid-September. Beneath this seasonal thaw layer is a permafrost table of frozen soil that acts as a barrier against outside physical and biochemical influences. The average annual temperature in the Arctic permafrost ranges from –10°C to –12°C at high elevations to –1°C to –2°C at its southern edge. In Antarctica, the permafrost table is much colder, with average annual temperatures of –24°C to –27°C and an active zone in the McMurdo Dry Valleys of only 10 to 25 cm depth (Gilichinsky et al. 2007). Amazingly, there are 10^7 to 10^8 microbial cells per gram of dry soil (cells/gdw) as determined by epifluorescent counts in Arctic permafrost and up to 10^6 cells/gdw of viable cells, as determined by culture, on various media (Table 23.3). The majority of microorganisms are bacteria, but mycelial fungi, yeasts, cyanobacteria, and eukaryotic algae are also found. The number of viable cells declines with depth and therefore with the age of the permafrost layer. The oldest permafrost layers found to contain viable cells in the Arctic date to the Pliocene, 2 to 3 million years ago. In the Antarctic, the total cell counts and the number of viable cells/gdw of permafrost are lower at comparable ages than in the Arctic samples. The differences in numbers are especially pronounced in the oldest Antarctic layers, which date to the Miocene, at more than 8.1 million years of age, and may be as much as 15 million years old. Viable green algae were recovered from a permafrost layer at 14.1 m in Beacon Valley in the McMurdo Dry Valleys and identified by 18S rDNA sequence analysis as being most closely related to the genus *Nannochloris*. These isolated green algae are presumed to be more than 8 million years old, although more work on dating is needed.

Permafrost represents a uniquely stable environment that permits the long-term survival of ancient microorganisms. Long-term survival seems to be related to the presence of and amount of unfrozen water associated with the soil in the permafrost. At the prevailing stable temperatures in Arctic permafrost, about 3% to 8% of total water is present in liquid form, as thin films around soil particles and as pockets of brine. The microorganisms are associated with these films or brine pockets, where they are protected from ice crystals. The unfrozen water furthermore serves as a nutrient medium and as a means to remove metabolic waste products (Gilichinsky 2002). The pure ice found in polar ice caps lacks these surface films and therefore contains far fewer viable cells (< 10 cells/g) in young ice and no viable cells in ancient ice. In the Antarctic, the average permafrost temperatures are lower, and therefore the amount of unfrozen water as surface films is less. In the McMurdo Dry Valleys, the percentage of unfrozen water is about 2% at –20°C and 1.5% at –30°C, due to the presence of higher salt content than in Arctic permafrost. Studies with radioactively labeled substrates indicate that microbial metabolic activity occurs at temperatures down to –28°C. Microorganisms in permafrost exist at a low level of activity that is sufficient to repair cellular damage from natural levels of radiation in their environment (Gilichinsky 2002).

The permafrost appears to be the only known ecosystem in which microorganisms can remain viable over time periods equivalent to geological epochs. Studies of this system are just beginning. It should be possible to trace changes in species composition of microorganisms back through time to determine the rate of evolution and to develop an accurate biological clock. The system also offers the possibility of examining the basic question of the ultimate duration of life preservation. From an astrobiology perspective, the planet Mars is known to have extensive areas of permafrost that may be as old as 2 to 3 billion years (Baker et al. 1991; Carr 2000; Baker 2004). If microorganisms ever existed on Mars, the depths of its permafrost layers would be the place to search for any traces of them.

23.2 Cryophilic Algae

Cryophilic algae are those found in ice and snow environments. The critical requirement for the existence of snow or ice algae is the presence of liquid water percolating through the snow or ice for a sufficient time period to allow growth, reproduction, and formation

Table 23.3a Total number of microorganisms determined by epifluorescence microscopy in Antarctic and Arctic permafrost sediments at various depths

Antarctic		Arctic	
Depth (m)	Total Count (10^8 cells/g)	Depth (m)	Total Count (10^8 cells/g)
0.5	0.4	12.2	2.4
4.5	0.4	18.8	0.5
6.8	1.4	30.1	3.7
8.2	0.6	32.2	1.5
9.4	1.1	32.7	1.3
10.2	0.1	34.4	1.4
13.0	1.0	37.2	3.2
15.3	2.5	43.5	0.4
16.3	0.8	49.0	0.6
		54.8	0.2
		60.0	2.2
		64.3	0.2

Table 23.3b Number of viable aerobic cells determined by culture on various media in Antarctic and Arctic permafrost sediments of various ages

Antarctic		Arctic	
Age (years)	Viable cells/g	Age (years)	Viable cells/g
present	$3 \times 10^1 - 1 \times 10^6$	$5-10 \times 10^3$	$1 \times 10^5 - 2 \times 10^7$
3×10^4	$8 \times 10^2 - 1 \times 10^4$	$1-4 \times 10^4$	$5 \times 10^4 - 1 \times 10^6$
1.7×10^5	$1 \times 10^2 - 1 \times 10^3$	$1-6 \times 10^5$	$1 \times 10^3 - 3 \times 10^4$
$> 8.1 \times 10^6$	$5 \times 10^1 - 2 \times 10^3$	$0.6-1.8 \times 10^6$	$8 \times 10^2 - 7 \times 10^3$
$2-15 \times 10^6$	$6 \times 10^1 - 5 \times 10^2$	$2-3 \times 10^6$	$2 \times 10^2 - 4 \times 10^3$

From Gilichinsky 2002.

of resting stages (Fritsen 2002). The main habitats for snow algae are snowfields in temperate alpine regions and snowfields associated with glaciers and polar ice caps. The main habitats for ice algae are surface melt-pools on glaciers and ice sheets and cylindrical holes called cryoconite holes, which provide liquid water environments in direct contact with glacial ice. The term *cryoconite* means "cold rock dust" and refers to the dark sediments that accumulate in the holes. Where snow and ice occur under temperatures that are continuously below 0°C, cyanobacteria and algae are essentially absent, and bacteria and fungi are the only microbes present. Accretion ice on lakes and sea ice are also considered cryophilic habitats, but these habitats are basically aquatic rather than terrestrial.

Snow Algae

Snow algae are known from every continent and occupy one of the most extreme environments on our

Figure 23.10 A patch of snow algae (arrowhead) in Colorado. (Photo: L. W. Wilcox)

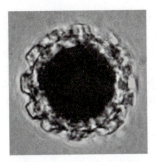

Figure 23.11 Red zygote of *Chlamydomonas nivalis*. (Photo: R. Hoham)

planet. Cells may be exposed to high irradiation levels, high acidity, low temperatures, low nutrients, and extreme desiccation after snowmelt (Hoham 1980). Red-colored snow, which is characteristic of open exposures, has been observed on snow, pack ice, and ice floes in the Arctic and the Antarctic and on snowbanks in high-altitude mountain regions (Figure 23.10). In contrast, green- and orange-colored snows are more commonly found in high-elevation forested regions but may also be associated with high nutrients in open-exposed bird rookeries in Antarctica (Broady 1989).

Though a wide variety of organisms can be found in red-colored snow (Marchant 1998; Hoham and Duval 2001), probably the most commonly encountered red snow alga is the chlorophycean *Chlamydomonas nivalis*. It is found, for example, in Yellowstone Park, Wyoming; the Sierra Nevada, California (Weiss 1983); the Sierra Nevada of Spain; and the Atlas Mountains of Morocco (Duval et al. 1999). *C. nivalis* commonly occurs in snow as 10–50 µm spherical resting cells (Figure 23.11) that are red-pigmented; the green to red biflagellate stage is frequently overlooked. The red pigment is **astaxanthin**, a carotenoid that occurs as an ester in cytoplasmic lipid droplets. The ester group is thought to bind fatty acids, thus allowing accumulation of astaxanthin in the lipid droplets. Astaxanthin has been hypothesized to limit inhibition and photodamage of the photosynthetic apparatus that results from high irradiance levels in high-altitude and high-latitude habitats (Bidigare et al. 1993). This hypothesis is supported by experiments conducted with *Haematococcus*, another chlorophycean that accumulates carotenoids in cytoplasmic droplets.

Chlorophyll fluorescence measurements (see Chapter 22) were used to detect a positive correlation between higher cellular levels of carotenoids and the capacity to withstand high light (Hagen et al. 1994). *C. nivalis* is often found in association with numerous bacteria that have typical Gram-negative walls and occupy a loose fibrous network on the surface of the algal cell wall, suggesting that this microhabitat may be a zone enriched in organic matter (Hoham and Duval 2001). This bacterial–algal association seems to be typical of red-snow populations (Weiss 1983), but little is known about the identity of the bacteria. Similar biofilms are thought to provide freezing protection (Potts 1994). In addition, snow algae are often found in loose association with large populations of bacteria (Thomas and Duval 1995). The interactions between microbes in snow need further study.

Chloromonas is another common snow alga whose vegetative cells or zygotes color green and orange snow (Hoham and Blinn 1979). It is very much like *Chlamydomonas* except that pyrenoids are said to be absent from the plastid. The genus is commonly isolated from soils, snow, or cold peat. In one study of *Chloromonas* from snow, optimal mating occurred in the blue end of the spectrum, under an irradiance level of 95 µmol m^{-2} s^{-1} (Hoham et al. 1998). A molecular phylogenetic analysis of several *Chloromonas* species indicated that this genus probably arose from *Chlamydomonas* but is not monophyletic (Buchheim et al. 1997).

In addition to the widespread species of *Chlamydomonas* and *Chloromonas*, a number of desmids (see Chapter 20), including *Cylindrocystis*, *Mesotaenium*, and *Ancyclonema*, were found in glacial snow and ice in Nepal (Yoshimura et al. 1997), and *Closterium* was reported from a glacier in southern Chile (Takeuchi and Kohshima 2004). Other snow algae occasionally reported include diatoms, cyanobacteria, and heterotrophic euglenoids. Snow communities are now being studied as distinct microbial ecosystems (Hoham and Duval 2001), and snow algal populations can be monitored by using remote-sensing techniques (Painter et al. 2001).

Ice Algae

Ice algae are found in meltpools and cryoconite holes on the surfaces of glaciers and ice sheets in their ablation zones. (*Ablation* refers to any process that produces a net loss of ice mass, such as melting, evaporation, and sublimation.) Dust and sediment blow onto the ice surface in the ablation zone and collect in any local depression. Because dust and sediment are darker than ice, they absorb heat and initiate melting to form a cylindrical hole called a **cryoconite hole**. Once a melthole begins to form on the ice, the hole acts as a sediment trap and accumulates meltwater that is flowing across the ice surface. The meltwater is relatively warm compared to the surrounding ice and also absorbs more heat than does the ice, which increases the rate of melting and the depth of the cryoconite hole. Microbial communities grow in the holes as dark-colored mats, and these mats absorb solar radiation and accelerate melting. Two processes slow the rate of melting and the deepening of the holes. First, even in summer, the sun is at a low solar angle over most glaciers and ice sheets and, as the hole becomes deeper, less solar radiation reaches the bottom and thus less energy is available for melting ice. Second, as the water in the hole grows deeper, the transfer of heat from the surface to the bottom by convection decreases. Cryoconite holes initially develop rapidly and then level off to a constant depth that is determined by the local climate, typically in the range of a few tens of centimeters (Wharton et al. 1985).

Cryoconite holes support a diverse community of microorganisms, including bacteria, cyanobacteria, diatoms, desmids, and protists, as well as metazoans such as tardigrades (Fritsen 2002). Cryoconite holes on the Canada Glacier of the McMurdo Dry Valleys are dominated by cyanobacteria such as *Phormidium*, *Nostoc*, *Microcoleus*, and *Lyngbya*, with lesser numbers of the green algae *Tetracystis* and *Chlamydomonas*, and the diatom *Caloneis*. On several glaciers on Svalbad (79°N) in the High Arctic, cryoconite holes contained a greater microbial diversity than any other habitat associated with the glaciers and adjacent deglaciated terrain (Kastovská

et al. 2005). Cyanobacteria and green algae were the richest in species.

The surfaces of ice sheets, ice shelves, and large glaciers in the Arctic and Antarctic are not smooth but undulate in a pattern of parallel ridges and troughs. Within these troughs, seasonal melting leads to the formation of networks of pools and streams in the ablation zones (Vincent 1988). Open water flows through these networks for one to two months per year, supporting extensive and productive microbial communities. The best-studied such system on an ice shelf is in the McMurdo Sound region of Antarctica, but studies in other areas indicate that the microbial communities are similar. Filamentous mat-forming cyanobacteria such as *Phormidium*, *Lyngbya*, and *Microcoleus* cover the bottoms of the ice-bound pools and lakes. Diatoms such as *Nitzschia*, *Pinnularia*, *Achnanthes*, and *Melosira* also occur in the pools, together with bacteria and protists.

23.3 Subaerial Algae

Subaerial algae occur on the surfaces of substrates, where they are exposed to varying degrees to the open atmosphere. Subaerial algae may be found as epiphytes on the surfaces of plants, both leaves and the bark of trees and shrubs. They also grow on many human-made surfaces, such as buildings and monuments and wooden fences and sheds. The most widespread subaerial algae are those associated with natural rock as **lithic algae**. **Hypolithic algae** occur under translucent rocks in both hot and cold deserts. Lithic algae may also be **epilithic** on the open surfaces of rocks or sheltered to various extents from the open atmosphere by growing in cracks or fissures in rocks as chasmoliths or under the surfaces of rocks in the pore spaces as **endoliths**. Some lithic algae actually bore into rocks and are called **euendoliths**.

Lithic Algae

After soil algae, lithic algae are the most widespread group of terrestrial algae, and they are present in association with rocks in all types of ecosystems where bryophytes and vascular plants do not shade them out. Thus lithic algae may occur even in temperate and tropical areas where exposed rock occurs. Lithic algae become prominent, however, in extremely hot and cold ecosystems, where they may be the only photosynthetic organisms functioning.

Lithic algae in cold deserts

The cold deserts of the Antarctic dry valleys are some of the most severe environments on Earth because of the extreme cold and aridity. Lithic microorganisms occur in epilithic, hypolithic, chasmolithic, and endolithic habitats.

Epilithic algae. Epilithic algae are more common in the Maritime Antarctic, where mosses and epilithic lichens are abundant, than in the colder, drier Continental Antarctic (Seppelt and Broady 1988). Cyanobacteria dominate the epilithic habitat, where they form dark-colored encrustations on rock surfaces that receive films of flowing water for part of the austral summer. The flowing water originates in snowdrifts that accumulate on the wind-sheltered sides of rocks and slowly melt during the brief summer. When meltwater stops flowing, the epilithic algae become desiccated. Epilithic algae are absent from rocks exposed to ocean spray as well as the windward sides of rocks, where abrasion from windborne ice crystals and sand particles scours away any epilithic growth. Epilithic cyanobacteria include single cells such as *Chroococcidiopsis* (see Figure 6.42), mucilaginous colonies such as *Gloeocapsa* and *Myxosarcina*, and filaments such as *Calothrix* (Broady 1981a). Filamentous cyanobacteria usually possess dark yellow-brown sheaths; colonies have violet, red, or yellow-brown mucilage; and *Chroococcidiopsis* has thick, violet-colored cell walls. Dark colors screen the cyanobacteria from high light levels (of PAR) and UV radiation on rock surfaces. The dark colors may also raise the temperatures of the colonies on the rocks to increase metabolic rates. The number of genera of epilithic cyanobacteria declines with increasing distance from the warmer, moister air of coastal desert regions (Wynn-Williams 2000). In the Vestfold Hills near the coast, 18 genera were found (Broady 1986), whereas 10 genera occurred in Dronning Maud Land 200 km (124 mi) inland (Engelskjon 1986) and only 5 genera occurred 1000 km (620 mi) inland at the Pensacola Mountains (Cameron 1972; Parker et al. 1977).

Epilithic lichens. Epilithic lichens are widespread in the Maritime Antarctic, but they are limited to small, scattered bodies in the Continental Antarctic. In the maritime region, fruticose lichens such as *Usnea* grow on the more stable rock surfaces, while on surfaces exposed to ocean spray, crustose lichens such as *Verrucaria* and *Xanthoria* occur. *Lecidea* and *Buellia* appear near bird nesting sites and on inland rock surfaces. In the Continental Antarctic, fruticose lichens are scarce, and most lichens such as *Xantharia*, *Candelaria*, and *Lecanora* occur as small, widely scattered crustose bodies on exposed rocks (Seppelt and Broady 1988). Continental lichens are generally sterile, but a few sexual forms have been found in areas sheltered from winds (Hale 1987). Epilithic lichens belonging to the genera *Lecidea* and *Carbonea* have been recorded as far south as the La Gorce Mountains, located only 420 km (260 mi) from the South Pole (Broady and Weinstein 1998). These lichens contained green algal phycobionts similar to *Trebouxia*.

The physiological hardiness of continental Antarctic lichens cannot be overstated. Desiccated bodies of *Buellia*, *Xanthoria*, and *Lecanora* can become photosynthetically active by uptake of water vapor from the air and have resumed photosynthesis and respiration after cooling to −196°C in liquid nitrogen (Lange and Kappen 1972). Lichens have even recovered after exposure to conditions in outer space (Sancho et al. 2007). The optimum temperature for carbon dioxide uptake may be near 0°C in some species, with photosynthesis occurring at temperatures as low as −15°C (Kappen 2000; Barták et al. 2007). Antarctic lichens are capable of photosynthesis at subfreezing temperatures at low light levels under snow cover. The absence of lichens from extensive areas of the McMurdo Dry Valleys is thought to be due to the permanent lack of water there rather than the extreme cold (Lange and Kappen 1972).

Hypolithic algae. While epilithic cyanobacteria and lichens are totally exposed to the stresses of desiccation, rapid freeze–thaw cycles, and high levels of visible (PAR) and UV radiation, hypolithic (or sublithic) microorganisms benefit from slightly enhanced amounts of soil moisture, reduced light levels, and elevated soil temperatures under translucent quartz rocks. In the Vestfold Hills of Continental Antarctica, extensive areas of dry mineral soil and glacial detritus lack any visible surface growth of photosynthetic organisms, and hypolithic and chasmolithic (see below) cyanobacteria and eukaryotic algae are the only photoautotrophs (Broady 1981b). Dominant taxa include the cyanobacteria *Chroococcidiopsis* and *Plectonema* and the green algae *Desmococcus* and *Prasiococcus*, among some 23 species recorded. Light transmission through white quartz stones ranges from 2.7% of incident light (40 µmol m^{-2} s^{-1} PAR) through a 13 mm thick stone to 0.6% (9 µmol m^{-2} s^{-1} PAR) through an 80 mm stone. While air temperatures vary from 1°C to 4°C just above the ground, the temperatures below the rocks range from 2°C to 12°C. Shade-adapted hypolithic algae therefore benefit from reduced levels of irradiance and a greenhouse effect, which makes the hypolithic habitat the most favorable of the lithic habitats for microbial growth. The hypolithic habitat, however, can occur only where there are large numbers of suitable translucent stones on horizontal surfaces, such as valley floors, ledges, and plateaus that receive drainage from snow melt (Broady 1996; Wynn-Williams 2000).

Chasmolithic algae. Chasmolithic algae grow in cracks, fissures, and exfoliating pieces of translucent rocks. The cyanobacteria *Chroococcidiopsis*, *Plectonema*, and *Gloeocapsa*, as well as the green algae *Desmococcus* and *Prasiococcus*, occur in chasmolithic habitats, as well as in epilithic and hypolithic habitats (Broady 1981c). Chasmolithic habitats support a greater diversity of eukaryotic algae than do endolithic habitats, particularly in coastal locations, where the green algae *Chlorella* and *Stichococcus* and the diatom *Fragilaria* have been recorded (Broady 1996). Chasmolithic algae occur on the protected leeward sides of rocks, and exposure to ocean spray determines the dominant species, *Prasiococcus* dominating in sea-spray areas and *Desmococcus* and *Chroococcidiopsis* dominating away from ocean spray. Chasmolithic habitats represent a transitional stage between epilithic habitats of rock surfaces and the endolithic habitats of rock interiors.

Endolithic algae and lichens. If suitable rocks are present, endolithic cyanobacteria, algae, and lichens can live in places where the external environment is too harsh for epilithic microorganisms and lichens to grow. Endolithic microorganisms require light-colored or translucent rocks such as sandstones, granites, or marbles with sufficient pore spaces between mineral grains to permit their growth under the rock surface. First discovered in the hot Negev Desert of southern Israel, the endolithic communities in the McMurdo Dry Valleys of Antarctica have become the most intensively studied lithic community on Earth (Nienow et al. 2002). In the 1970s, the Antarctic dry valleys were regarded as the closest terrestrial analog to the surface and climate of the planet Mars (Friedmann and

Ocampo 1976). If any microbial life exists on Mars, the endolithic communities of the Antarctic dry valleys would be a good indication of what that life may resemble. Endolithic communities are widespread in the McMurdo Dry Valleys and extend up into the adjacent Trans-Antarctic Mountains. Above 2200 m (7200 ft) elevation, however, the climate becomes too harsh, even for endolithic microbes (Friedmann and Koriem 1989). The disappearance of endolithic communities from the Trans-Antarctic Mountains suggests that life on Mars, if it ever existed, may have become extinct as climatic conditions deteriorated.

Researchers have found five different types of endolithic communities in the porous rocks of Continental Antarctica. The most common is dominated by an endolithic lichen and appears as a layered community consisting of a black upper band extending from just below the rock surface to a depth of 1.5 mm, a white band from 1.5 mm to 4 mm, and a green band below 4 mm (Friedmann 1982). The black and white bands represent the endolithic lichen. The green algal phycobiont (*Trebouxia* or *Pseudotrebouxia*) is present in the black band, where the black pigment screens it from excess radiation and may raise the temperature of the lichen. The green band is a mixture of eukaryotic algae and cyanobacteria. The putative endemic green alga *Hemichloris* (Tschermak-Woess and Friedmann 1984) is often dominant but may be accompanied by the xanthophyte *Heterococcus* and one or more cyanobacteria, such as *Chroococcidiopsis* and *Gloeocapsa*. The other types of endolithic communities are less widespread and contain fewer species. They include a *Hemichloris* community, two types of *Gloeocapsa* communities, and a community in which *Chroococcidiopsis* is the sole photoautotroph (Nienow et al. 2002).

Endolithic microorganisms gain a number of advantages in the endolithic habitat. Overlying rock grains screen out high light levels and UV radiation. The endolithic lichens produce a black layer that reduces light levels even further. Light levels at the bottom of the endolithic community are very low (around 2 µmol m^{-2} s^{-1}), and the entire community appears to be adapted to low light. The subsurface rock layers provide a buffer against the extreme swings in temperature and moisture that can occur on the surface. When air temperatures are below −5°C, solar radiation on north-facing rock surfaces can raise the temperature inside the rock to +15°C. In austral summer, snows can occur, and the resulting meltwater soaks into the rock, where it is retained and can support the metabolic activity of the endolithic community. Nitrogen fixation is rare, and the main source of nitrogen to the endolithic community appears to be atmospheric deposition, which occurs at a rate of about 20 mg N m^{-2} yr^{-1} (Nienow et al. 2002). This source is sufficient because primary production is low, and therefore the demand for nitrogen is low. Friedmann et al. (1993) estimated the net primary production of the endolithic community in the McMurdo Dry Valleys to be between 11 and 17 kg C ha^{-1} yr^{-1}, but they suggested that this production may only apply to about 20% of the surface area in the McMurdo Dry Valleys. Assuming a lower temperature limit for metabolic activity in the endolithic community of –10°C, the community in the McMurdo Dry Valleys can be active for only 700 to 800 hours per year (Kappen and Friedmann 1983). That metabolic activity produces organic compounds that dissolve the rock minerals and cause flaking of rock chips (exfoliation) from the surface. The resulting exposure may disperse endolithic microbes in the wind and transport them to new sites.

Lithic algae in temperate and tropical ecosystems

In temperate and tropical regions, ecologists focus on vascular plant studies, but scattered reports of epilithic and endolithic communities exist from temperate Europe (Allen 1971; Büdel et al. 1991), North America (Matthes-Sears et al. 1997), and South Africa (Wessels and Büdel 1995), as well as from tropical Africa (Büdel et al. 1997) and South America (Büdel et al. 1994). In these warm to hot ecosystems, epilithic and endolithic cyanobacteria and eukaryotic algae are confined to rocky outcrops, cliffs, long continuous cliffs called escarpments, and inselbergs—isolated mountains that rise abruptly out of surrounding plains. Moisture plays a major role in determining the structure and diversity of lithic algal communities in these habitats.

In the temperate United States, diatoms may be a significant part of the epilithic algal community on wet rock outcrops and cliffs. Lowe et al. (2007) reported 41 diatom genera with 223 species on wet rock faces within Great Smoky Mountains National Park. Diatoms often occurred with macroscopic patches of cyanobacteria such as *Nostoc* and *Stigonema*. Diatoms were also the most frequent eukaryotic algae on sandstone cliffs in southeastern Ohio during spring (Casamatta et al. 2002). All sites received sporadic or flowing water. Of 140 algal taxa

found, diatoms accounted for 80, cyanobacteria 43, greens 12, and euglenoids and xanthophytes 5. Wet cliffs had a greater diversity of algal taxa than did dry cliffs. Although diatoms have been reported from a few other moist rock outcrops in North America, they were absent from other sites that were investigated. Matthes-Sears et al. (1997) examined epilithic and endolithic cyanobacteria, eukaryotic algae, and lichens along the cliffs of the Niagara Escarpment in Canada and found that epilithic microorganisms covered an average of 49% of the escarpment face, with 26% of the cover coming from cyanobacteria, 20% from lichens, and only 3% from green algae. Diatoms were not reported. Endolithic microbes occurred under 6% of the surface, and chasmoliths were twice as abundant as endoliths. Fungi were the most common eukaryotic microbes.

In tropical regions, black rock surfaces occur on table mountains and inselbergs in habitats such as rainforests and wet savannas. These black rock surfaces are not barren and are in fact covered with a dense layer of cyanobacteria and cyanolichens. In rainforests, the epilithic cover on inselbergs consists almost entirely of cyanobacteria. The table mountains (tepuis) of the Guayana Uplands of Venezuela are covered in the filamentous cyanobacteria *Stigonema* and *Scytonema* and the colonial cyanobacterium *Gloeocapsa*. The epilithic cyanobacteria are all intensely colored, the filamentous genera with yellow-brown sheaths and the colonial *Gloeocapsa* with layers of red mucilage (Büdel et al. 1994). After rains, the cyanobacterial crusts appear blackish-green. The same genera of cyanobacteria dominate inselbergs in the rainforests of the Ivory Coast in Africa (Büdel et al. 1997). Epilithic cyanolichens such as *Peltula* become more diverse and common on inselbergs in moist savannas, where they replace some of the cyanobacterial crust cover.

Epilithic and endolithic microorganisms are widespread on sandstone cliffs and inselbergs in the tropical dry savannas of South Africa (Büdel 1999). In these relatively dry ecosystems (average annual rainfall about 76 cm), rocky cliffs and inselbergs assume an ochre color due to a covering of cyanolichens of the genus *Peltula*, together with the surface oxidation of the sandstone. Within the sandstones, *Chroococcidiopsis* is the dominant cryptoendolithic cyanobacterium and also the main photobiont in *Peltula*. Lichens generally cover more rock surface area than endolithic communities. The epilithic cyanolichens contain the ultraviolet screening pigment scytonemin and mycosporine-like amino acids, which provide protection from high levels of visible and UV radiation (Büdel et al. 1997). The primary production of these epilithic and endolithic communities can be high, on the order of 27 g C m^{-2} yr^{-1}, a value in the range of the primary productivity of the surrounding savannas. The metabolic activities of *Chroococcidiopsis* raise the pH in the surrounding medium from 8 to 10.5, dissolving away the mineral silicon dioxide in sandstones and causing loosening and exfoliation of rock chips. Cryptoendolithic cyanobacteria are therefore a significant agent of weathering of sandstones and the shaping of sandstone landscapes in these dry savannas (Büdel et al. 2004).

Lithic algae in hot deserts

In hot deserts, epilithic algae are essentially absent due to the extreme temperatures and radiation levels on rock surfaces. The only epilithic organisms are lichens such as *Ramalina* and *Caloplaca*, and even these lichens occur in less-extreme deserts or local habitats within deserts that are cooler and receive higher levels of moisture, such as north-facing slopes (Nash et al. 1977). In parts of the Negev Desert where annual precipitation is greater than 30 cm yr^{-1}, epilithic lichens cover the surfaces of rocks, imparting a gray color to them, and produce a jigsaw-puzzle pattern on rock surfaces (Nienow et al. 2002). Lower amounts of precipitation lead to dominance by endolithic or hypolithic microorganisms.

Hot deserts often contain extensive areas of pebbles embedded in soils called desert pavements. Many of these pebbles are composed of translucent minerals such as quartz that transmit part of the surface irradiance through them and reflect back the rest of the light received. As a result, the undersurface of translucent pebbles has reduced light levels and is cooler than that of dark opaque pebbles that absorb the full incident radiation. Because translucent pebbles are cooler underneath than dark pebbles, they collect moisture from the soil by condensation and retain it for longer periods. In the Mojave Desert of California, the hypolithic environment under quartz pebbles provides a favorable habitat for cyanobacteria such as *Chroococcidiopsis*, green algae (e.g., *Protosiphon*), and in rare occurrences of large quartz rocks, the moss *Tortula*. Schlesinger et al. (2003) found hypolithic communities on all quartz pebbles up to 25 mm thick that were examined along a 50 m transect at a desert pavement field site in the Mojave Desert. The light transmitted through a 25 mm thick pebble is only 2 µmol m^{-2} s^{-1} (PAR), a level close to the compensation point for the community. Consequently, photoautotrophs occurred on the margins of quartz pebbles thicker than 25 mm, where light levels were greater. Mojave desert hypolithic communities have been observed to tolerate temperatures up to 67°C in the field and over 90°C under laboratory conditions.

In hot desert pavements, lichens can occur as euendoliths, boring under the surfaces of limestone pebbles and plates. The euendolithic lichen *Verrucaria rubrocincta* was discovered inside limestone plates in desert pavements of the Sonoran Desert of southwestern Arizona (Breuss 2000). Anatomical studies revealed that this euendolith is distinctly layered, with an upper layer (micrite) of fine-grained calcite and sparse fungal hyphae about 200 µm thick, a green algal photobiont layer 25 to 70 µm thick, and an underlying layer of fungal hyphae (a pseudomedulla) about 1 cm thick (Bungartz et al. 2004). Colonized limestone plates can be recognized by the presence of numerous black circles (ostioles), from which the fruiting bodies (perithecia) of the lichen fungus emerge. The lichen appears to act as a miniature greenhouse, with the micrite forming a roof that screens out high light and UV levels and seals in moisture from winter rains. Another species of euendolithic lichen (*Buellia peregrina*), which has a similar anatomy, has been found on limestone pebbles in the Namib Desert of southwest Africa (Bungartz and Wirth 2007). Frequent fogs and dew supply the moisture that sustains the lichens.

In the temperate hyperarid Atacama Desert of Chile, Warren-Rhodes et al. (2006) were able to track the occurrence of hypolithic cyanobacteria under translucent stones along a gradient of aridity. The hypolithic community consisted of *Chroococcidiopsis* and a few heterotrophic bacteria. The percentage of translucent stones with hypolithic communities fell from 28% (302 of 1093 stones examined) in peripheral areas receiving rainfall of 21 mm yr^{-1} to < 0.1% (3 of 3723 stones examined) in the hyperarid center, which receives < 2 mm yr^{-1}. In this hyperarid core, UV radiation can sterilize the surfaces of rocks because there is no water to allow the metabolic activity necessary to repair UV damage (Cockell et al. 2008). Six years of microclimate data indicate that the limit for hypolithic communities in the Atacama is 5 mm yr^{-1} of rainfall, or about 75 h yr^{-1}, with sufficient water to support photosynthesis.

Under the adverse conditions of hot deserts, microorganisms also occur as endolithic communities under the surfaces of light-colored, porous rocks such as sandstones (Figure 23.12). If the pore spaces between mineral crystals are too small or the rock too dark, endolithic microorganisms are absent. After their initial discovery in the Negev Desert (Friedmann et al. 1967), endolithic communities were found in hot deserts in the southwestern United States, Africa, and Australia (Büdel and Wessels 1991), where the climate on exposed rock surfaces was too severe to allow epilithic algal growth. In all these deserts, the most common endolithic microorganism is *Chroococcidiopsis*, and in the most extreme deserts, it may be the only photoautotroph. Other cyanobacteria such as *Gloeocapsa* and green algae such as *Chlorella* and *Stichococcus* are much less common. Endolithic green algae become more abundant and microbial diversity increases as regional climatic conditions become less severe, as in semi-arid grasslands and woodlands (Bell et al. 1988).

In the hot deserts of the southwestern United States, the biomass of endolithic microorganisms is greatest in the lightest-colored sandstones and least in the darkest colonized sandstones. The endolithic biomass (as chlorophyll *a*) was 87.2 to 106.8 mg m^{-2} in the lightest colonized sandstones in northern Arizona, 26.8 to 56.5 mg m^{-2} in intermediate colored rocks, and only 8.4 to 18.7 mg m^{-2} in the darkest sandstones colonized. The decline in biomass follows the decline in available light at the level of the endolithic community. In white sandstones, the light level at the community layer measures 22 to 66 μmol m^{-2} s^{-1}, but in brown sandstones the light level is < 22 μmol m^{-2} s^{-1} (Bell 1993). Thus, in very dark sandstones, endolithic communities are absent because the level of light at the depth of 1 mm is too low to sustain the microorganisms. Although less than 5% of direct sunlight reaches the level of the endolithic community in the rock, the microorganisms in the upper levels produce pigments that are protective against high light and UV radiation, while those in the lower layers do not. As in other endolithic systems, the metabolic activities of endolithic communities in hot deserts break down the mineral grains of porous rocks, causing exfoliation of rock chips and weathering of surfaces. In extreme desert habitats, the biomass of the endolithic community may equal or exceed that of the surface vascular plants (Bell et al. 1986).

In cold deserts, endolithic microbes gain a prolonged period of metabolic activity from the capacity of large rocks to absorb heat and maintain temperature above that of the surrounding air. In hot deserts,

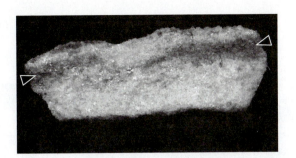

Figure 23.12 A band of endolithic algae (arrowheads) growing within sandstone from northern Arizona. (Photo: L. W. Wilcox, of specimen provided by R. Bell)

this property of rocks works against the endolithic community, except in winter. In the desert Southwest of the United States, for example, the temperature at the endolithic community may be 15°C higher than the air temperature, and internal rock temperatures can be as high as 47°C in Arizona sandstones (Sommerfeld and Bell 1991). Cyanobacteria are particularly tolerant of high temperatures and desiccation, and this tolerance may explain their dominance of endolithic communities in hot deserts. The one advantage microorganisms receive in the endolithic habitat within hot deserts is the acquisition of water. At night, dew condenses on desert rocks and is absorbed into the rock by capillary action. The relative humidity inside the Arizona sandstones was never less than 60% and usually above 80%, a level that should support metabolic activity in some microbes (Bell 1993).

Endolithic and hypolithic habitats represent the last refuge for microbial life. Neither the extreme cold of Antarctic dry valleys nor the extreme heat of hot deserts eliminates these lithic communities. Desiccated Antarctic microorganisms can survive temperatures of –196°C, while hypolithic microorganisms in hot deserts have survived temperatures up to 90°C. The critical factor in the survival and persistence of these lithic communities is the number of hours per year in which water as liquid or vapor is available to support metabolic activity. If the number of hours of available moisture falls below some critical level, the lithic communities disappear, and the habitat becomes virtually sterile.

Lithic algae on human constructions

Humans utilize natural rock in buildings and monuments as well as many forms of manufactured stony materials, such as brick, cement, concrete, plaster, and tile. Exposed to the environment, such building

Figure 23.13 Terrestrial algae on human constructions in Austin, Texas (United States). (a) Dark algal growth on a limestone wall. The capstone on the wall is granite and is uncolonized. (b) Limestone balustrade colonized by algae. (Photos: L. E. Graham)

materials are subject to normal physical and chemical processes of weathering such as rain, wind, and temperature changes, as well as biological weathering due to the activities of microorganisms (Gaylarde and Morton 2002). To lithic cyanobacteria, algae, and lichens, natural rock and manufactured stony materials in human constructions are just lithic substrates for colonization (Figure 23.13). As on natural rock surfaces, temperature, moisture, and exposure to visible light (PAR) and UV radiation are important factors in microbial colonization of building surfaces. The pH of construction materials can affect the composition of lithic communities. Granitic rock and bricks are acidic, but marble, limestone, concrete, and cement are basic. Basic materials favor cyanobacteria. Construction materials differ in porosity (the ratio of open spaces to mass), and endolithic communities may readily colonize porous building materials. Microorganisms growing on buildings and monuments are a mixture of genera found on natural lithic surfaces and in soils. Studies of cyanobacteria, algae, and lichens on human constructions are almost entirely from Europe, Central America, and Asia, where they are concerned with biodeterioration of urban buildings and monuments and particularly cultural heritage sites, such as Greek, Roman, and Mayan ruins (Rindi 2007).

In southern Spain, near the Gibraltar Strait, for example, epilithic and endolithic cyanobacteria,

algae, and lichens have colonized the exposed ruins of the Roman town of Baelo Claudia (Ariño and Saiz-Jimenez 1996). The ruins were built of stone covered in limestone mortar with a pH of 9.0 to 9.2 and a high porosity of 25%. The composition and abundance of lithic communities are related to the orientation of the walls. North-facing walls are not subject to direct sunlight and show stable temperatures and moisture levels. Epilithic cyanobacteria such as *Nostoc* and *Phormidium* and green algae including *Stichococcus* colonize the shaded walls. The filamentous green alga *Trentepohlia* grows on a shaded wall in a corner where it receives runoff water, and it colors the wall orange. Walls facing south and southwest receive direct sunlight and therefore show higher temperatures and less moisture. Endolithic lichens such as *Collema*, *Caloplaca*, and *Verrucaria* dominate these walls along with endolithic cyanobacteria (*Chroococcidiopsis*) and green algae (*Bracteacoccus*). Lithic communities appear to have a strong effect on the deterioration of mortar-covered walls. Growth of lichens leads to intense pitting of walls, and cyanobacteria form small cavities around the cells, indicating dissolution of the limestone. Similar deterioration from cyanobacteria and cyanolichens has been documented from limestone and marble monuments in the Yucatan of Mexico (Ortega-Morales et al. 2000), Rome, and Jerusalem (Danin and Caneva 1990).

A serious form of deterioration occurs with the development of "black crusts" on limestone and marble buildings and monuments in polluted urban areas. In urban areas, the combustion of large amounts of fossil fuels leads to elevated levels of airborne particulates such as fly ash and enhanced levels of sulfur dioxide. The black crust derives its color in part from the fly ash. Deposition of sulfur on limestone structures leads to the conversion of calcium carbonate (limestone) into soluble calcium sulfate and deterioration of the structure (Montana et al. 2008). Various pigments produced by microorganisms can enhance the black color of the crusts. Bacteria and fungi can produce melanins and phenolic polymers, while cyanobacteria can add very dark violet phycobiliproteins and carotenoids mixed with chlorophyll (Saiz-Jimenez 1995). Cyanobacteria and eukaryotic algae also produce extracellular polysaccharides that are sticky and collect airborne particles such as dust, pollen, and fly ash. A thick (> 200 μm), hard crust develops at a rate of about 2 to 5 μm per year, and this black crust is corrosive on limestone and difficult to remove. Black crusts may also develop on limestone surfaces in unpolluted areas, but in these cases the crusts are composed of filamentous cyanobacteria such as *Scytonema* with dark-brown sheaths. These crusts are thin compared to those in polluted areas and are easily removed by gentle scraping (Saiz-Jimenez 1995; Gaylarde et al. 2007).

Not all lithic communities appear to cause corrosion. Green algae tend to be more abundant on buildings in northern Europe than are cyanobacteria, and cooler temperatures and higher humidity are thought to be responsible for this change in dominance. Rindi and Guiry (2002) studied the ecology of the filamentous green alga *Trentepohlia* on limestone, cement, and concrete structures in urban areas of western Ireland. *Trentepohlia* produced extensive streaking on west-facing walls but not on drier south-facing ones. Haubner et al. (2006) found the green algae *Stichococcus* and *Chlorella* from concrete walls in Rostock, Germany, required a relative humidity greater than 90% for photosynthesis and growth. While the change in the color of walls due to green algal growth may be aesthetically unappealing, neither study reported any corrosion resulting from green algal colonization. A few studies have actually reported that colonization by epilithic lichens can provide protection to underlying surfaces. Lichen colonization of rocks usually leads to greater weathering of rock surfaces due to production of organic acids such as oxalic acid (Chen et al. 2000). When physical and chemical weathering processes are faster than biological ones, however, lichen colonization may have a bioprotective function (Carter and Viles 2005). At the Roman city of Baelo Claudia in southern Spain, Ariño et al. (1995) found that lichen colonization actually protected sandstone flagstones paving the forum area from strong physical and chemical weathering.

Although not all lithic communities are corrosive, many lithic communities do pose a threat to culturally significant monuments and buildings. In polluted urban areas, black crusts degrade historically significant buildings and works of art, and their long-term preservation will require major reductions in air pollution. Endolithic microorganisms are particularly destructive to buildings and monuments made from limestone and marble because their metabolic activities dissolve mineral grains and lead to exfoliation of rock flakes. Unless some means of preventing colonization is developed, many historically and culturally important buildings and works of art will be lost to erosion.

Epiphytic Algae

Subaerial cyanobacteria, algae, and lichens that grow on the surfaces of plants or plant parts such as bark or leaves are termed epiphytes. Some cyanobacteria and algae grow inside leaves and are thus termed endophytes. The most common and widespread epiphytic cyanobacteria and algae are those that form symbioses with fungi as lichens (Nienow 2002). Coccoid cyanobacteria with mucilaginous sheaths such as *Chroococcus* and *Gloeocapsa* and sheathed filamentous cyanobacteria such as *Nostoc*, *Phormidium*, and *Scytonema* are the second most widespread group of epiphytic microorganisms, and they tend to dominate on many types of subaerial habitats, particularly in the tropics. Coccoid and filamentous green algae such as *Desmococcus*, *Chlorella*, *Klebsormidium*, and *Trentepohlia* are common as epiphytes on bare wood and tree bark in temperate and tropical ecosystems. The genus *Trentepohlia* actually contains a number of different lineages (López-Bautista et al. 2006); the related genera *Cephaleuros* and *Stomatochroon* are tropical endophytes. Diatoms are a diverse group but a minor component of epiphytic communities, where they are limited to very wet surfaces. Van de Vijver et al. (2004), for example, reported 192 taxa of diatoms on mosses at sub-Antarctic Heard Island. One of the most important aspects of epiphyte ecology is that of cyanobacteria–plant associations in nitrogen fixation. This topic is discussed in the remainder of this chapter.

Epiphytic associations and nitrogen fixation

With the exception of a few studies in temperate and tropical systems, most investigations of cyanobacteria–plant associations focus on polar and subpolar ecosystems. In the most extreme polar environments such as those of Continental Antarctica, cyanobacteria are the dominant photoautotrophs and nitrogen fixers (Vincent 2000). Davey and Marchant (1983) reported that *Nostoc*, in association with a continuous cover of mosses such as *Bryum* and *Grimmia*, fixed nitrogen at a rate of 1.87 to 9.93 µmol N m^{-2} h^{-1} during the summer at the Vestfold Hills. In the Maritime Antarctic, the milder climate permits mosses and liverworts to cover extensive areas, forming carpets, turfs, and cushions; epiphytic cyanobacteria, chlorophytes, diatoms, and xanthophytes typically occur on both the green photosynthetic tissues and the dead material underneath (Broady 1996). *Nostoc* is widespread on mosses in turfs, where it serves as a source of fixed nitrogen. Early in austral summer on Signy Island, *Nostoc* is located on moss shoots near the previous year's apices. As the current apex grows and elongates, the *Nostoc* filaments at the previous apex form motile hormogonia that move up the shoot to the current apex, where they produce filaments and heterocysts. This ascent up the shoot keeps the *Nostoc* filaments in sunlight for photosynthesis and nitrogen fixation (Vincent 1988). On Antarctic Signy and South Orkney Islands, *Nostoc* provides 0.75 to 3.14 µmol N m^{-2} h^{-1} to the moss turfs, with the remainder coming from atmospheric precipitation and bird rookeries (Christie 1987). In the High Arctic of northern Canada, mosses such as *Bryum* and *Grimmia* are also associated with epiphytic cyanobacteria. Karagatzides et al. (1985) reported that *Nostoc* associated with mosses along a lakeshore in the Northwest Territories fixed nitrogen at a rate of 5.91 ± 2.90 µmol N m^{-2} h^{-1} in summer.

Nostoc is also common on mosses in the arctic tundra (Low Arctic), where *Sphagnum* moss is widespread in wetlands and mires. *Nostoc* and other cyanobacteria such as *Anabaena* may occur as epiphytes on *Sphagnum*, but cyanobacteria can also grow as endophytes inside *Sphagnum* hyaline cells, where the cyanobacteria fix nitrogen and methane-producing bacteria use the hydrogen and ammonia from nitrogen fixation (Granhall and Hofsten 1976). Until recently, relatively little was known about cyanobacteria–plant associations in boreal forests and their role in nitrogen fixation. Moss–cyanobacteria associations were reported to be the largest potential source of nitrogen fixation in Alaskan black spruce forests, based on plant cover (Billington and Alexander 1978). In 2002, DeLuca and associates reported that *Nostoc*, in association with the feather moss *Pleurozium schreberi*, fixes between 1.5 and 2.0 kg N ha^{-1} yr^{-1} in the boreal forests of northern Scandinavia. *P. schreberi* is the most common moss on Earth, occurring on every continent except Australia and Antarctica, as well as the most common feather moss in the boreal forest ecosystem, which covers 17% of the terrestrial land area. The rate of nitrogen fixation by the feather moss–*Nostoc* association increases linearly with the time since the last fire disturbance (DeLuca et al. 2007, 2008). This cyanobacteria–moss association represents a newly recognized and highly significant source of nitrogen fixation in the boreal forest ecosystem.

Far less is known about cyanobacteria–plant associations and nitrogen fixation in temperate and tropical areas, where cultural landscapes dominate and extensive areas of moss cover are uncommon (Solheim and Zielke 2003). Nitrogen fixation in cyanobacteria–moss associations has been reported in temperate grasslands (Vlassak et al. 1973; Reddy and Giddens 1981). In temperate and tropical forests, cyanobacteria–moss associations have not been studied closely, but epiphytic lichens formed from cyanobacteria and fungi are significant sources of nitrogen fixation. In the southern Appalachian Mountains of North Carolina, for example, cyanolichens alone add about 0.8 kg N ha^{-1} yr^{-1} to gray beech forests (Becker 1980). The cyanolichen *Lobaria* supplies much of the annual input of fixed nitrogen to old-growth Douglas fir forests in the Pacific Northwest (Dodds et al. 1995). In a montane rain forest in Columbia, cyanolichens are abundant and represent a biomass of 5.7 kg ha^{-1} (Forman 1975). *Nostoc* is the photobiont in 86% of these cyanolichens, and annual fixed nitrogen inputs are estimated to be 1.5 to 8 kg N ha^{-1} yr^{-1}, comparable to the input of nitrogen from precipitation. This source of nitrogen is probably widespread in tropical rain forests and indicates the importance of epiphyte communities to the tropical rain forest ecosystem.

Glossary

accessory pigment A colored molecule that absorbs light energy and transfers it to a reaction center of chlorophyll *a* for use in photosynthesis.

acritarch The general term for unicellular fossils whose relationships are uncertain; many are regarded as resistant cyst stages of planktonic algae, often prasinophyceans or dinoflagellates.

agar A sulfated polygalactan extracted from walls of various red algae that is used as a gelling agent.

akinete A thick-walled spore that functions in asexual reproduction, frequently serving as a resistant stage that undergoes a period of dormancy.

algae Photosynthetic, oxygen-producing aquatic bacteria or protists.

algaenans Decay-resistant polymers of unbranched hydrocarbons present in the cell walls of some algae.

algal bloom (bloom) Visible growth of planktonic algae, often associated with nutrient-enriched waters.

alginic acids Polysaccharides (a mixture of mannuronic and guluronic acids) extracted from the walls of brown algae for industrial applications; may occur in the salt form (alginates or algin).

aliasing Sampling problem resulting from the too-infrequent collection of samples.

allophycocyanin A type of phycobiliprotein produced by cyanobacteria (except chlorophyll *a* and *b*-containing taxa), glaucophytes, red algae, and cryptomonads.

alloxanthin A xanthophyll pigment that is characteristic of photosynthetic cryptomonads.

alternation of generations A life history type in which there are two (or more, in some red algae) multicellular stages that can be distinguished by type of reproductive cell produced and sometimes also by morphological features.

Alveolata A eukaryotic supergroup that includes ciliates, apicomplexans, and dinoflagellates, characterized by membrane sacs known as alveoli at the cell periphery.

alveolates Members of the eukaryotic supergroup Alveolata.

alveoli Membrane sacs located at the periphery of cells of the eukaryotes classified as alveolates.

amoeba A unicellular organism that generates motions via cellular projections.

amoeboid A type of cell organization in which a wall is absent and the protoplasm undergoes rapid shape changes.

amphiesma The covering of dinoflagellate cells, which, in addition to an overlying plasma membrane, consists of membranous alveoli that may contain little or no material or may contain cellulosic thecal plates of varying thickness.

androspores Flagellate cells of Oedogoniales that settle on body cells near oogonia and undergo a few divisions to produce a dwarf male filament, some cells of which produce and release flagellate sperm.

anisogamous A type of sexual reproduction in which gametes are flagellate and structurally distinguishable.

anisogamous reproduction Sexual reproduction involving gametes that are flagellate and structurally distinguishable.

anisogamy A type of sexual reproduction characterized by two types of gametes that differ in size or behavior.

anoxygenic Without production of oxygen; for example, use of hydrogen sulfide rather than water as a source of photosynthetic reductant by some cyanobacteria.

anoxygenic photosynthesis A type of photosynthesis that occurs in some prokaryotes in which water is not used as a reductant and oxygen is not released.

apical growth Growth that occurs by the division of one or more cells located at the tip (apex) of a multicellular body.

apicoplast A non-photosynthetic plastid occurring in the cells of most genera of apicomplexans, a group of parasitic protists that are classified in the Alveolata.

aplanospore A nonflagellate spore that has the genetic potential to produce flagella under appropriate conditions; produced by subdivision of a parental cell.

araphid pennates Diatoms that generally possess bilaterally symmetrical frustules that lack raphes.

archeopyle In dinoflagellates, the exit pore of a germinating cyst.

articulation zone In euglenoids, the point at which ribbonlike pellicular strips join with neighboring strips.

astaxanthin A red carotenoid pigment (3,3'-diketo-4,4'-dihydroxy-β-carotene) produced by some green algae that is commercially useful; also known as haematochrome.

aureochrome A blue-light photoreceptor present in diverse photosynthetic stramenopiles.

autocolony A type of asexual reproductive colony that is a miniature of the adult colony; produced by single cells of the adult.

autospore A type of nonflagellate spore that lacks the genetic potential to produce flagella.

auxiliary cell In the higher red algae, a cell into which the zygote nucleus or one of its mitotic progeny is deposited and that generates the carposporophyte generation by mitotic proliferation.

auxospore A cell produced by diatoms that undergoes enlargement, compensating for the reduction in size that often occurs during population growth; commonly also the zygote of diatoms.

bacterivores Organisms that consume bacteria as prey.

bacterivory The process by which cells ingest and digest bacterial prey.

baeocytes (endospores) Spores of cyanobacteria formed by internal division of vegetative cells. (*compare with* exospores)

bangiophytes An informal term for the Bangiophyceae, a class of red algae.

β-carotene An accessory carotenoid pigment lacking oxygen that occurs in major algal groups; the source of provitamin A.

binary fission The relatively simple process by which prokaryotic cells divide to form two equal-sized progeny cells.

bioassay A procedure that uses organisms and their responses to estimate the effects of physical and chemical agents in the environment.

biological pump The process by which some of the organic materials produced by marine phytoplankton are transported to deep ocean sediments, where they remain for thousands of years.

biological soil crust The community of organisms including bacteria, cyanobacteria, eukaryotic algae, fungi, lichens, and mosses that covers much of the soil surface in arid and semiarid lands.

biological species concept Distinction of species on the basis of breeding incompatibility.

biotic associations Diverse ways in which organisms impact each others' lives.

biphasic alternation of generations A sexual reproductive cycle that involves two body forms that can be distinguished by chromosome level and type of reproductive cell produced.

bipolar A type of centric diatom having elongate shape (i.e., having two poles).

blades A type of algal body in which cells are arranged in flat sheets.

bloom (*see* algal bloom)

blue-green algae (*see* cyanobacteria)

bootstrap value An estimate of the validity of a branch in a phylogenetic tree that is determined by the number of times the branch appears after the data are repeatedly resampled.

branched filaments A type of algal body in which branches emerge from a main filamentous axis.

branches In algae, rows of cells that extend from a main filamentous axis.

canal raphe A tubelike structure extending longitudinally along the valves of some pennate diatoms that opens externally via the raphe slit.

carbon concentrating mechanism (CCM) One of several types of processes that result in increasing the concentration of carbon dioxide in the vicinity of rubisco.

carbon fixation Conversion of carbon dioxide into organic carbon as a result of the light-independent reactions of photosynthesis.

carbon sequestration The burial of calcium carbonate or organic carbon such that these materials are not readily converted back to atmospheric carbon dioxide.

carbonic anhydrase An enzyme that converts carbon dioxide to bicarbonate and vice versa.

carboxysome (polyhedral body) A polygonal structure within the cells of cyanobacteria that contains the enzymes ribulose bisphosphate carboxylase/oxygenase (rubisco) and carbonic anhydrase.

carpogonial branch In red algae, a short branch consisting of a few vegetative cells, terminated by a female gamete, the carpogonium.

carpogonium The nonmotile female gamete of red algae.

carpospore In red algae, the spore released from a carposporangium; usually assumed to be diploid.

carposporophyte (gonimocarp) A multicellular diploid phase in the life cycle of florideophycean red algae that generates carpospores via gonimoblast filaments; in all cases, attached to the female gametophytic body (i.e., not free living).

carrageenans Mucilaginous sulfated polygalactans in the cell walls of red algae that are extracted for use as gelling agents in the food industry.

carrying capacity The number of individual organisms that can be supported with available resources.

CCM (*see* carbon concentrating mechanism)

cell plate A planar array of vesicles containing cell wall material that assembles during early cytokinesis and gives rise by centrifugal extension to new cross-walls; present in certain green and brown algae and in land plants (embryophytes).

central strutted process In certain diatoms, a tubular siliceous structure located at the midpoint of the valve. (*see* fultoportula)

centric diatom A diatom whose frustule is ornamented with radially arranged pores.

Charales (stoneworts) Formal name for the green algal group (order) named for the genus *Chara*.

charasome Latticelike, tubular array of membranes at the cytoplasmic surface of some regions of charalean cells.

charophyceans (charophyte algae, streptophyte algae) An informal term for the paraphyletic group of green algae most closely related to land plants.

charophyte algae (*see* charophyceans)

charophytes An informal term that is sometimes broadly applied to the green algae most closely related to land plants plus land plants, and sometimes applied in a more restricted fashion to the stoneworts (Charales).

chasmolithic Living inside fissures, crevasses, and cavities inside rocks. Chasmoliths are particularly common in cold deserts such as the dry valleys of Antarctica.

chlorarachniophytes A lineage of green algae that is part of the eukaryotic supergroup known as Rhizaria.

chlorophyll *a* The photosynthetic pigment that forms the reaction center of photosystems I and II in most algae and land plants.

chloroxybacteria (*see* cyanobacteria)

chromalveolate hypothesis The hypothesis that haptophytes, cryptomonads, stramenopiles, and alveolates together form a monophyletic group that originated via a single event of secondary endosymbiosis involving a red algal endosymbiont; this hypothesis assumes that modern cryptomonads, stramenopiles, and alveolates that lack plastids arose by plastid loss from ancestors that possessed secondary plastids.

chromatic adaptation   The ability of algae to modify the amounts or proportions of photosynthetic pigments in relation to changes in the light environment.

chromophytes   (*see* photosynthetic stramenopiles)

chrysolaminaran (chrysolaminarin)   (*see* laminaran)

ciliate   The informal name for a type of alveolate protist characterized by numerous cilia, presence of two types of nuclei (a larger macronucleus and a smaller micronucleus), and a cell mouth (cytostome).

cingulum   (a) The transversely oriented girdle region of dinoflagellate cells; (b) the region of overlap of the two halves of diatom frustules.

clade   A group of organisms descended from a common ancestor.

cladocerans   A group (order) of planktonic crustaceans, many of which consume algae as food.

classic food web   An early model of the aquatic food web in which large (> 20 μm) phytoplankton are consumed by zooplankton, and zooplankton are consumed by fish in a simple linear food chain.

clone library   A collection of bacterial colonies that in aggregate contains clones of the DNA present in an original sample.

closed mitosis (intranuclear mitosis)   Mitosis that occurs within an intact nuclear envelope; present in many protists.

coccoid   The morphology of unicellular algae that have cell walls and are often, but not always, spherical in shape.

coccolithophorids   Unicellular members of the haptophyte algae that are characterized by a covering of small, ornate calcium carbonate scales.

coccoliths   Calcium carbonate scales that occur on the surfaces of coccolithophorid algae.

coccosphere   The aggregate of coccoliths forming a coalescent outer covering of coccolithophorid algal cells.

coenobium   A type of protist colony whose shape and cell number is genetically determined, established early in development, and does not change during the life of the organism (though cells commonly enlarge during colony development).

coenocytic   A type of protist body that is multinucleate and without transverse walls (except, in many cases, during reproductive development). (*see also* siphonaceous coenocytes)

colimitation   A state that arises when phytoplankton growth is limited by more than one mineral nutrient at the same time.

colony   A type of protist body consisting of a group of cells held together by mucilage or cell wall material.

compensation point   The depth within the water column at which the photosynthetic rate equals the respiratory rate.

compound root   A flagellar root that consists of more than one type of element (e.g., a group of microtubules plus a banded strand).

conceptacle   A cavity that contains the reproductive cells of some algae (particularly coralline red algae and fucalean brown algae).

conchocelis   A filamentous, sporophytic phase in the life cycle of bangialean red algae that produces conchospores in conchosporangia; commonly occurs in shells or other calcareous materials.

conjugation   Mating of zygnematalean green algae involving nonflagellate gametes.

conjugation tube   A connection between cells of mating filaments of zygnematalean algae, formed by dissolution of the ends of modified branches, which allows gametes to make contact.

connecting cell (connecting filament, ooblast)   In florideophycean red algae, a cell (often long and filament-like) through which a zygote nucleus is transferred from the fertilized carpogonium to an auxiliary cell.

consistency index   An estimate of the degree of homoplasy (parallel or convergent evolution) in a phylogeny.

copepods   Aquatic crustaceans that feed on larger phytoplankton, protozoa, and the juvenile stages of other aquatic crustaceans.

coronal cells   The 5 or 10 cells found at the tips of the tubular cells that form an investment around oogonia of charalean green algae.

cortex   In a fleshy algal body, the layer of cells or tissues lying between the epidermis on the outside and the medulla on the inside.

costa   An elongate, hollow siliceous rib occurring on the frustule of some diatoms, commonly in parallel rows.

crustaceans   Primarily aquatic members of the animal phylum Arthropoda that have chitinous exoskeletons, jointed appendages, and jointed antennae.

cryoconite hole   A cylindrical hole in the surface of an ice sheet or glacier formed by the accumulation of dark sediment in a depression and increased in depth by absorption of heat and melting.

cryophilic algae   Algae found in ice and snow environments.

cryptic speciation   The process by which genetically distinct but morphologically indistinguishable species form.

cryptogamic crusts   One of a number of terms used to describe the biological soil crusts that are communities of organisms consisting of bacteria, cyanobacteria, eukaryotic algae, fungi, lichens, and mosses. Cryptogamic crusts are common in arid and semiarid lands.

cryptomonads   An informal name for members of the phylum Cryptophyta, most of which contain plastids derived from a red alga and which contain the distinctive xanthophyll pigment alloxanthin.

cyanelles (cyanellae)   A term used for the blue-green plastids of glaucophyte algae.

cyanidophytes   An informal name for a group of unicellular, early-diverging red algae.

cyanobacteria (chloroxybacteria, blue-green algae, cyanophytes)   Photosynthetic bacteria that perform oxygenic photosynthesis.

cyanophycean starch Polyglycan granules of glycogen that serve as the carbohydrate storage material of cyanobacterial cells.

cyanophycin Granules consisting of polymers of the amino acids arginine and asparagine that serve as a proteinaceous reserve for cyanobacterial cells.

cyanophytes (*see* cyanobacteria)

cystocarp The carposporophyte (gonimocarp) and enveloping gametophytic tissues (pericarp) of florideophycean red algae.

dead zones Coastal ocean regions that are depleted of oxygen and thus life forms as the result of the decay of large populations of phytoplankton due to ocean eutrophication.

decay values In parsimony, the number of additional steps beyond the shortest tree in which a particular branch collapses.

deep-chlorophyll maxima High levels of chlorophyll that can be measured in subsurface waters, arising from populations of photosynthetic organisms living there.

detritovores Organisms that consume detritus, non-living organic material.

diazotrophy The ability to convert diatomic, gaseous N_2 gas into ammonium ion, which can be used by cells to produce amino acids.

diffuse growth A type of growth that does not involve a localized point of cell division.

dinoflagellate A protist that is usually single celled, characteristically possesses saclike alveoli beneath the cell membrane, swims by means of two distinctive flagella (or, if not motile, produces dinospores having such flagella), and may be photosynthetic or not.

dinokaryon The characteristic nucleus of dinoflagellates.

dinospore The flagellate reproductive cell of non-motile dinoflagellates (analogous to zoospores of various other algae).

dinosporin A decay-resistant compound deposited at the surfaces of the cells of some dinoflagellates and their cyst stages that allows fossilization to occur.

ECM (*see* extracellular matrix)

ecotypes Genetic varieties of a species that may show morphological differences or adaptations to different environments.

eddies The swirling currents of water that make up turbulent motion in aquatic systems.

ejectisome (ejectosome) In cryptomonads, a structure that is explosively discharged from the cell, presumably as a defense mechanism.

endobionts Organisms living within the body of other organisms (e.g., the dinoflagellates that live inside corals).

endoliths Microorganisms such as bacteria, cyanobacteria, eukaryotic algae, and lichens that grow inside the pore spaces between mineral grains within porous translucent rocks.

endoreduplication The process by which the nuclear DNA undergoes repeated rounds of replication without intervening mitotic separation, yielding DNA levels higher than the haploid or diploid state.

endosymbiosis The condition in which one or more organisms live within the cells or body of a host without causing disease or other conspicuous harmful consequences.

environmental genomics A set of procedures that employ phylogenetic information and molecular methods to explore the species diversity of natural microbial communities without first growing the organisms in culture.

epibacteria Bacteria that are attached to surfaces, such as the outsides of microalgal cells or macroalgal bodies.

epibiont An organism that grows on the surfaces of another organism.

epibody A layer of cells that forms the outer surface of a coralline red alga.

epicone In dinoflagellates, the portion of the cell anterior to the cingulum.

epilimnion The upper layer of a stratified water body, whose waters are warmer in summer (and are typically more oxygen rich) than bottom waters (hypolimnion).

epilithic Living on the surfaces of rocks.

epipelagic Upper warm ocean waters.

epipelic Living on the surfaces of mud or sand.

epiphyte An organism that grows on the surfaces of plants or algae.

epitheca (a) In dinoflagellates, the portion of the cell covering (theca) lying anterior to the cingulum; (b) in diatoms, the epivalve plus epicingulum.

epizoic Living on the surfaces of animals.

euendoliths Microorganisms that live under the surfaces of rocks and actively bore their way through the rock. Several desert lichens are known to be euendoliths, as are certain cyanobacteria found on coral reefs.

euglenoid movement A process by which some euglenoids move by means of cellular plasticity.

euglenoids Unicellular protists that primarily occur as flagellates having a distinctive surface composed of proteinaceous pellicular strips and a characteristic storage known as paramylon granules; many but not all conduct photosynthesis.

Euglenozoa A protist supergroup that includes euglenoids and trypanosomes as well as some other protists.

eukaryovory The process of feeding by ingesting whole eukaryotic cells.

euphotic zone The (upper) portion of the water column that receives enough light for photosynthesis to occur.

eutrophic aquatic systems Waters that contain relatively high levels of nutrients such as phosphate and/or combined nitrogen; typically exhibit high levels of primary productivity.

exhaustive search In phylogenetic systematics, the finding of all possible trees that can be constructed from the data. (*compare with* heuristic search)

exospores   In cyanobacteria, spores that are cut off from one end of the parental cell. (*compare with* baeocytes)

extracellular matrix (ECM)   Materials generated by a cell that are secreted from, or produced on, the external surface; includes mucilage, cell walls, and loricas.

extrusome   (*see* trichocysts)

ferritin   Intracellular protein that binds and stores iron.

filament   A type of algal body consisting of a linear array of cells in which neighboring cells share a common wall.

filter feeding   A mode of food collection by herbivores that involves sieving large volumes of water for particles.

flagellar transformation   A maturation process by which the younger flagella of a parental cell become the older flagella of a progeny cell.

flagellates   Unicellular or colonial protists whose cells bear one or more flagella.

floridean starch   In red algae, a branched α-1,4-linked glucose polymer with some α-1,6 linkages that occurs as granules within the cytoplasm.

florideophytes   An informal name for the members of the red algal class Florideophyceae; characterized by a triphasic reproductive cycle that includes a carposporophyte generation.

flow cytometer   An instrument that uses a laser and detectors to measure the optical properties of cells, such as pigment fluorescence.

food quality   The extent to which algae are able to provide essential nutrients when consumed as food.

food vacuoles   Intracellular digestion structures.

food web   A model that illustrates the feeding interactions occurring among diverse types of organisms in a particular habitat.

frustule   In diatoms, the silica enclosure or wall.

fucans   Also known as fucoidins or ascophyllans, polymers of L-fucose and additional sugars that are sulfated.

fucoxanthin   A xanthophyll pigment that confers golden-brown or brown pigmentation to diatoms, chrysophyceans, brown algae, and some other photosynthetic stramenopiles.

fultoportula (pl. fultoportulae) (strutted process)   In diatoms, a complex tubular structure extending from the frustule; often associated with chitin fibril formation.

fusion cell   In red algae, (a) generally, a cell resulting from the coalescence of two or more non-gamete cells; (b) specifically, the cell produced by fusion of an auxiliary cell with one or more neighbors.

gametangium   A container formed of one or more cells, in which gametes are produced.

gametic meiosis   Meiosis occurring during the production of gametes.

gametophore   A branch that bears one or more gametangia.

gametophyte   The multicellular, gamete-producing phase in the life history of organisms having alternation of generations.

gas vacuoles   In cyanobacteria, aggregates of gas vesicles; sometimes used as a synonym for gas vesicles.

gas vesicles   In cyanobacterial cells (and those of some other aquatic bacteria), cylindrical structures whose protein walls are permeable only to gases and increase the buoyancy of the cells.

genicula   The uncalcified, flexible regions occurring as joints between calcified, non-flexible regions of the thalli of jointed coralline red algae and some ulvophycean green algae.

genomics   The application of techniques for study of an organism's entire genome, though research may focus on nuclear, plastid, or mitochondrial DNA sequence information.

girdle   (*see* cingulum)

girdle lamella   A flat sheet composed of three thylakoids that extends just under the plastid envelope in some photosynthetic stramenopile algae.

girdle view   The side view of diatoms.

gland cell   In red algae, a specialized cell that serves in secretion or storage.

glaucophytes   An informal name for a group of eukaryotic algae capable of producing flagella, which have plastids whose envelope is composed of two membranes and which contain phycobilin pigments in phycobilisomes.

gone   In desmid green algae, a cell derived from zygote germination, one or both of whose semi-cells may not possess the typical morphology of the species.

gonidium   In *Volvox* and related colonial green algae, an enlarged, nonmotile cell that can generate new (daughter) colonies.

gonimoblast   In red algae, one or all of the filaments that bear carpospores (the filaments known in aggregate as the carposporophyte).

gonimocarp   (*see* carposporophyte)

green algae   Informal name for the eukaryotic algal group characterized by plastids having a two-membrane envelope and chlorophylls *a* and *b*.

gynandrosporous   Species of oedogonialean algae that produce androspores on the same filament that produces egg cells.

gyres   Huge circular oceanic currents, of which there are five in the oceans—two in the Northern Hemisphere and three in the Southern Hemisphere.

gyrogonites   The calcified zygotes of charalean green algae; includes those occurring in the fossil record and those in modern sediments.

hair cells   In green algae, cells that produce a long, hairlike extension.

haptera   The attachment structures of the kelps.

haptonema   In haptophytes, a flagellumlike structure arising from the cell apex, near the flagella, that contains several microtubules (but not the 9+2 microtubular arrangement of flagella); may function in attachment, feeding, or avoidance responses.

haptophytes   An informal term for a group of mostly unicellular and marine, and sometimes flagellate, species that contain secondary red plastids; includes coccolithophorids.

herbivores   Organisms that consume photosynthetic organisms as food.

heterococcolith   A type of coccolith in which there are crystals of more than one shape or size.

heterocyte (heterocyst)   In some cyanobacteria, a thick-walled, weakly pigmented cell that is the site of nitrogen fixation.

heteromorphic   Morphologically different; in algae, usually applied to distinctive gametophyte and sporophyte phases in sporic life cycles (alternation of generations).

heteromorphic alternation of generations   A type of sexual reproductive cycle involving two or more multicellular phases, at least two of which are morphologically distinct.

heteroplastidic   In certain ulvophycean green algae, having two kinds of plastids—green chloroplasts and colorless, starch-storing amyloplasts (leucoplasts).

heterothallic   Requiring two different clones for sexual reproduction; self-incompatible.

heterotrichous   Having a filamentous body consisting of both an erect and a flat, prostrate (attached) system.

heuristic search   In systematics, a search technique that is not guaranteed to find the optimal solution but that can greatly reduce computational time.(*compare with* exhaustive search)

holdfast   A cell or multicellular structure (*see also* haptera) that functions in attachment to a substrate.

holocarpy   In certain ulvophycean green seaweeds, conversion of all of a body's cytoplasm into reproductive cells, whose release results in the death of the parental alga.

holococcolith   A type of coccolith in which there is a single type of crystal.

homoplasy   Similarity of characters in two or more taxa for reasons other than inheritance.

homothallic   Requiring only one clone for sexual reproduction; self-compatible.

hormogonium   In filamentous cyanobacteria, a few-celled, usually motile filament that functions in asexual reproduction, dispersal, and, in some cases, colonization of a host.

hyalosome   In certain predaceous dinoflagellates, a colorless lenslike region in the eyelike ocellus.

hypnozygote   A thick-walled, resting zygote that may germinate in favorable circumstances following a required period of dormancy.

hypobody   In coralline red algae, a system of lowermost filaments oriented parallel to the substrate.

hypocone   The portion of a dinoflagellate cell posterior to the cingulum.

hypolimnion   The cold, deep layers of stratified lakes.

hypolithic algae   Algae that grow underneath translucent rocks on the bottom surface where moisture collects.

hypotheca   (a) In dinoflagellates, the portion of the wall (theca) lying below the cingulum; (b) in diatoms, the hypovalve plus the hypocingulum.

hystrichosphere   In modern and fossil dinoflagellates, resting cysts commonly bearing distinctive projections and/or markings and an excystment pore (archeopyle).

idioandrospory   The condition of oedogonialean green algal species that produce androspores and eggs on different filaments.

intercalary bands   In dinoflagellates, the regions of the theca that lie between the plates; in diatoms, a portion of the girdle region of the silica frustule.

intercalary meristem   In kelp brown algae, a region of cell division located between stipe and blade tissues.

isogamous   The condition of gametes not being morphologically distinguishable.

isogamous reproduction   A type of sexual reproduction involving gametes that are not morphologically distinguishable.

isogamy   Sexual reproduction involving gametes that are morphologically indistinguishable; structurally similar, but behaviorally distinguishable gametes are described as physiologically anisogamous.

isomorphic alternation of generations   A type of sexual cycle in which there are at least two multicellular stages that are morphologically similar.

K_{cat} The rate of enzyme turnover; the rate of catalysis when the substrate is present in saturating amount.

kerogens   Hydrocarbon-rich sedimentary deposits.

labiate process (lipped process, rimoportula)   In certain diatoms, a siliceous tube extending through the frustule; on the inside it is laterally compressed to form a slit surrounded by liplike (labiate) structures.

laminaran (laminarin)   In brown algae and other photosynthetic stramenopiles, a soluble polysaccharide storage product composed of β–1,3-linked glucose units together with some branch-producing β-1,6-linkages.

lateral conjugation   In some filamentous zygnematalean green algae, the formation of a conjugation tube between adjacent cells of the same filament.

leucoplast   A colorless plastid that, if containing starch, is also known as an amyloplast.

LHC   (*see* light-harvesting complexes)

light-dependent reactions   Photosynthetic processes that involve the use of light energy to split water and produce ATP and NADPH.

light-harvesting complexes (LHCs)   Aggregates of photosynthetic pigments, photoprotective pigments, and proteins that function to harvest light during photosynthesis.

light-independent reactions   Photosynthetic processes that use ATP and NADPH to transform carbon dioxide into organic compounds.

lipopolysaccharides (LPS)   In general, polysaccharides having attached lipids; specifically, toxins produced by certain bacteria and bloom-forming cyanobacteria that can cause fever and inflammation in humans.

lipped process   (*see* labiate process)

lithic algae   Algae that grow in some way in association with rocks or stony substrates such as concrete.

lorica   A protective covering, often mineralized, that surrounds cells of various algae; a form of extracellular matrix (ECM) that is usually more distant from the cell membrane than are cell walls.

LPS   (*see* lipopolysaccharides)

macroalgae   Algae having bodies that are large enough to be seen with the unaided eye.

macroscopic   Large enough to observe with the unaided eye.

manubria   In charalean green algae, columnar cells that connect the pedicel to the shield cells within male gametangia.

marine snow   Particulate aggregates of algal cells or their remains, fecal pellets, bacteria, and heterotrophic protists, held together by mucilage, that are important in the transformation and transport of organic carbon to deep-ocean sediments.

matrotrophy   The provision of nutrients by cells of the parental generation to cells of the next generation that have been retained on the maternal body.

medulla   Cells or tissues occurring in the center of a fleshy multicellular algal body.

meristem   A cell or group of cells that is capable of repeated division and thus adds to the number of cells in a body.

meristoderm   In fleshy brown algae, particularly Laminariales, a surface layer of cells (epidermis) that is capable of dividing (i.e., is meristematic).

mesograzers   Animal herbivores of intermediate size, such as oligochaete worms, freshwater dipteran larvae, and marine amphipods and pteropods.

mesoplankton   A class of plankton consisting of organisms that are between 0.2 mm and 2 mm in diameter.

metaboly (euglenoid movement)   In euglenoids, a form of motility that does not involve flagellar action.

metacentric spindle   In trebouxiophycean green algae, the positioning of centrioles at the midpoint of the mitotic spindle rather than at the poles, as is more common.

metalimnion   The layer of water lying beneath the upper epilimnion that is marked by a steep decline in temperature and increase in density of the water.

metaphyton   Floating algae that have become detached from substrates where they were attached as periphyton. Metaphyton may make up a significant part of the species diversity of the phytoplankton.

microalgae   Algae having bodies so small that a microscope is needed to observe them.

microbial food web   The complex food web consisting of microorganisms less than 20 μm in longest linear dimension and including archaea, bacteria, cyanobacteria, small eukaryotic algae, and small protozoans.

microbial loop   The original term for the microbial food web that recognized the importance of secondary production by archaea and bacteria and their consumption by heterotrophic nanoflagellates and small ciliates to the larger aquatic food web.

microplankton   A class of plankton consisting of organisms that are between 20 μm and 200 μm in diameter.

microsource (scintillon)   In certain marine dinoflagellates, the particulate location of bioluminescence.

microzooplankton   Zooplankton in the size range from 20 μm to 200 μm. Microzooplankton are the main consumers of the microorganisms in the microbial food web.

mixotrophy   A form of nutrition in which both autotrophy and heterotrophy may be utilized, depending on the availability of resources. (*see also* phagotrophy)

MLS   (*see* multilayered structure)

molecular barcodes   DNA sequences accumulated in online databases that can be used to identify organisms in natural samples.

monophyletic   Used to describe a group of organisms that have descended from a single common ancestor. (*compare to* polyphyletic)

morphological species concept   The use of structural differences and similarities to distinguish species and classify them.

mucocyst (muciferous body)   A saclike structure within cells of dinoflagellates and euglenoids from which a thick, rod-shaped mucilage body can be extruded to the cell surface when the organisms are disturbed.

multiaxial   Composed of many similar axial filaments arranged in parallel (also known as fountain-type structure).

multicellularity   Composed of multiple adherent cells that have the capacity to communicate with each other and to specialize.

multilayered structure (MLS)   In the cytoskeleton of flagellate cells of various protists and sperm of nonflowering land plants, a layered structure located near the flagellar basal bodies that includes microtubules; thought to function as a microtubule organizing center.

multipolar   Describing centric diatoms that have angular shapes.

multispecies consortium   An association of several species, often of disparate taxonomic groups, that are involved in metabolic interactions and that occupy the same habitat.

mutualism   A symbiotic association of two or more organisms in which all members benefit.

nanoplankton   A class of plankton consisting of organisms that are between 2 µm and 20 µm in diameter.

NAO   (*see* nuclear associated organelle)

necridia (separation disks)   In filamentous cyanobacteria, dead cells whose production is associated with development of hormogonia and, in some forms, false branching.

nemathecia   In some red algae, a raised wartlike structure on the body surface that contains reproductive structures.

nematocyst   In certain dinoflagellates, a harpoonlike ejectile structure that is morphologically distinct from the more common trichocyst.

neuston   The community of organisms living in surface films.

nitrogen fixation   The process by which many cyanobacteria (and some non-photosynthetic bacteria) transform nitrogen gas into ammonia, fixed nitrogen.

nitrogenase   In cyanobacteria (and some other bacteria), the holoenzyme that performs nitrogen fixation—conversion of diatomic N_2 gas into ammonium ion.

node   A site on an algal body from which branches arise.

nuclear-associated organelle (NAO)   In red algae, a ring-shaped structure that occurs at the spindle poles during cell division.

nucleomorph   In some algae having plastids of secondary origin (cryptomonads, chlorarachniophytes), a plastid-based, double-membrane enclosed structure containing DNA arranged in small chromosomes and other features suggesting origin from a eukaryotic nucleus.

ocellus   (a) In certain marine phagotrophic dinoflagellates, a complex structure consisting of a lens and retinalike cup that is thought to provide a vision system of utility in prey location; (b) in some diatoms, a rimmed perforated plate of elevated silica.

ochrophytes   (*see* photosynthetic stramenopiles)

okadaic acid   A toxin produced by certain dinoflagellates that inhibits serine- and threonine-specific phosphates (which occur widely in eukaryotes); the cause of diarrhetic shellfish poisoning in humans.

oligonucleotide probe   A short piece of DNA that specifically binds the ribosomal RNA of particular species.

oligotrophic aquatic systems   Waters that are low in nutrients such as phosphate and combined nitrogen and consequently low in primary productivity and biomass but typically high in species diversity.

ooblast   (*see* connecting cell)

oogamous   Involving a larger nonmotile egg cell and a smaller motile sperm cell.

oogamous reproduction   Sexual reproduction involving egg and sperm.

oogamy   Sexual reproduction involving syngamy of a small flagellate male gamete and a larger, nonflagellate (or only transiently flagellate) female gamete.

osmotrophy   A form of nutrition in which dissolved organic carbon is imported from the environment into cells.

oxygenic photosynthesis   Photosynthesis that generates oxygen by breaking water.

pallium   A sheetlike extension of cytoplasm, also known as a feeding veil, that is produced by some phagotrophic dinoflagellates for the purpose of capturing and digesting prey.

palmella   A stage produced by algae genetically capable of producing flagella consisting of nonmotile cells embedded in an amorphous mucilaginous matrix.

paramylon   In euglenoids, a β-1,3-glucose polymer that occurs as granules in the cytoplasm.

paraphyletic   In systematics, a group that does not include all the descendants of a common ancestor.

parasite   An organism that lives at the expense of a host.

parasitism   A situation in which a member of one species (the parasite) lives on or in another species (the host) and feeds upon the host to its harm.

parasporangium   In red algae, a sporangium that produces many spores; not equivalent to a tetrasporangium.

paraxonemal rod   A column of protein that stiffens the hair-covered flagella of euglenids and kinetoplastids.

parenchyma   A form of body in which true tissues are produced. (*see also* pseudoparenchyma)

parthenogenesis   Production of a new individual from an unfertilized gamete.

pathogens   Parasites that cause disease symptoms.

peduncle   In dinoflagellates, a microtubule-containing tubular extension of cytoplasm that can be extended to attach to prey cells or tissues and either extract their contents or engulf them.

pelagic   Living in open ocean waters rather than nearshore coastal or inland waters.

pellicle   (a) In euglenoids, an often flexible, protein-containing surface layer consisting of interlocking, helically oriented strips that occurs just inside the cell membrane; (b) in some dinoflagellates, a cellulosic and sporopollenin-containing layer beneath the theca that remains in place when the theca is shed; (c) the surface layers of some other algae.

pennate   One of the major diatom frustule types; exhibits bilateral symmetry of valve ornamentation.

peptidoglycan   A carbohydrate substance cross-linked by peptides that forms much of the cyanobacterial cell wall.

PER   (*see* periplastidal endoplasmic reticulum)

perennation   Living through unfavorable conditions from one favorable period to another.

perialgal vacuoles   Membrane-bound vacuoles that protect enclosed endosymbionts from digestion within host cells.

periaxial cells (pericentral cells)   In uniaxial red algae, cells that are cut off from the main axis that occur in a whorl that surrounds the parental axial cell.

peribody   In coralline red algae, cells produced when the intercalary meristem cuts off derivatives toward the substrate.

pericarp   In florideophycean red algae, a coherent layer of gametophytic filaments that surrounds a carposporophyte; the outermost, presumably protective, layer of a cystocarp.

peridinin   A xanthophyll accessory pigment that colors the plastids of many dinoflagellates golden-brown.

peridinin plastids   Dinoflagellate plastids that contain the pigment peridinin.

periphyton   Organisms that occur on the surfaces of plants, algae, and inorganic substrates in shallow benthic or nearshore littoral habitats.

periplast   In cryptomonads, the cell covering, consisting of intersecting plates lying beneath the cell membrane.

periplastidal compartment   In cryptomonads, haptophytes, and photosynthetic stramenopiles, the space between the outer and inner pairs of plastid envelope membranes (the outer two of which are known as the periplastidal endoplasmic reticulum).

periplastidal endoplasmic reticulum (PER)   An extension of the endoplasmic reticulum that encloses the plastids of some algal groups; sometimes also connected to the nuclear envelope.

perizonium   In pennate diatoms, the silicified outer layer of the auxospore (zygote) wall, consisting of series of bands.

peroxisome   An organelle bound by a single membrane that occurs in the cytoplasm of embryophyte cells and those of some eukaryotic algae and that contains characteristic enzymes such as glycolate oxidase and catalase; this term is also sometimes used more generally as a synonym for microbodies—small single-membrane-bound structures that occur in some algae and contain catalase but not glycolate oxidase.

phagopod   In dinoflagellates, a tubular structure lacking microtubules that can be formed on cells for the purpose of attaching to and extracting the contents of prey cells.

phagotrophs   Organisms that use the process of endocytosis to engulf prokaryotes and/or small algae into cellular digestion structures known as food vacuoles.

phagotrophy (phagocytosis)   A form of nutrition in which particles such as cells are ingested by protists via invagination of the cell surface.

photoautotroph   An organism that obtains its organic nutrients by means of photosynthesis; obligate photoautotrophs are restricted to this form of nutrition.

photoinhibition   The reduction of photosynthetic rates by high irradiance.

photosynthetic heterokonts   (*see* photosynthetic stramenopiles)

photosynthetic stramenopiles (photosynthetic heterokonts, chromophytes, ochrophytes)   An informal aggregate term for the algal lineages that contain photosynthetic representatives and whose flagellate cells display a flagellum coated with multipart tubular hairs and generally also a smooth flagellum; also known as photosynthetic heterokonts, chromophytes, or ochrophytes.

phototaxis   Movement toward (positive phototaxis) or away from (negative phototaxis) light.

phragmoplast   In post-mitotic cells of certain green algae and embryophytes, an array of microtubules arranged perpendicularly to the plane of division that, together with coated vesicles and endoplasmic reticulum, is involved with cytokinesis by centrifugal cell plate formation.

phycobiliproteins   Cyanobacterial phycobilins (open-chain tetrapyrroles) bound to proteins; phycoerythrin, phycocyanin, and allophycocyanin.

phycobilisome   In cyanobacteria, glaucophytes, and red algae, a hemispherical or discoidal structure on the surfaces of thylakoids that contains phycobiliproteins.

phycobiont   The algal component of a lichen.

phycocyanin   A blue-green phycobiliprotein found in cyanobacteria, glaucophytes, cryptomonads, and red algae.

phycocyanobilin   An open-chain tetrapyrrole that, when bound to proteins, constitutes phycocyanin.

phycoerythrin   A red phycobiliprotein found in cyanobacteria, cryptomonads, and red algae.

phycoerythrobilin   An open-chain tetrapyrrole that, when bound to proteins, constitutes phycoerythrin.

phycoplast   In chlorophycean green algae, an array of microtubules oriented parallel to the plane of cytokinesis, through which the developing wall grows.

phylogenetic species concept   A model in which a species is the smallest monophyletic group of organisms that exhibits at least one distinctive, unifying characteristic.

phylogenetic tree   A treelike diagram that represents a hypothesis regarding relationships of a group of organisms.

physodes   In brown algae, cytoplasmic vesicles that contain polyphenolic polymers of phloroglucinol known as phlorotannins that are secreted and likely function to deter herbivores.

phytochrome   A red/far-red light-sensitive pigment that is involved in perception of light and photomorphogenesis in algae and embryophytes.

phytoferritin   An intracellular, semicrystalline array of iron-binding proteins.

phytoplankton   Floating or swimming microscopic algae.

picoplankton   A class of plankton consisting of organisms such as bacteria and certain small eukaryotic algae that are between 0.2 μm and 2 μm in diameter.

pili   Threadlike structures located on the surfaces of at least some cyanobacteria (and many other Gram-negative bacteria) that are associated with twitching motility exhibited by certain cyanobacteria.

pit plug (pit connection)   In red algae, a plug that develops in the wall between adjacent cells and that, in an elongate form, can maintain structural association even when other portions of the cell walls become spatially separated.

plankton   Microscopic organisms that are suspended or swim in the water column.

planozygote   A motile zygote.

plasmodesmata   In certain green and probably all brown algae and embryophytes, protoplasmic connections between adjacent cells; primary plasmodesmata are formed during cytokinesis, whereas secondary plasmodesmata are produced after a new cell wall has been formed.

plurilocular sporangium   In brown algae, a reproductive structure that is subdivided by cell walls into numerous small chambers (locules), each of which produces a single flagellate cell.

pluriseriate   Filaments composed of two or more rows of cells.

pneumatocyst   In some fleshy brown algae, a bulbous, gas-filled structure that functions in buoyancy.

pocket   In euglenoids, a cellular depression from which flagella emerge.

polar rings   In red algae, densely staining rings located at the spindle poles that are thought to play a role in microtubule organization.

polyeder   An angular, polyhedral cell produced by a flagellate meiospore that arises from zygote germination in *Hydrodictyon* and *Pediastrum*; gives rise to new coenobium.

polyphyletic   In systematics, describing a group that contains some members that are actually more closely related to organisms outside the group.

portable plastid hypothesis   The idea that secondary red plastids could have originated more than once because they retain a higher proportion of the original primary endosymbiont's genome than do green plastids.

practical salinity units (psu)   The number of grams of salt contained in a liter of water. A liter of seawater with 35 grams of salt has 35 practical salinity units.

predators   Organisms that attack and feed on other organisms, which are killed and consumed. For example, adult copepods are predators on protozoa and the juvenile stages of other aquatic crustaceans.

primary capitulum   In charalean green algae, cells of antheridia that generate long, unbranched filaments of small cells, each producing a single thin, helically twisted spermatozoid.

primary endosymbiosis   Incorporation of a free-living prokaryote into a host eukaryotic cell, with subsequent transformation into an organelle.

procarp   In certain florideophycean red algae, presence of an auxiliary cell in the carpogonial branch.

prokaryotic   Describes cells lacking a nucleus, membrane-bound organelles (mitochondria, plastids), and endomembrane systems such as endoplasmic reticulum and a Golgi apparatus; prokaryotes also lack sexual reproduction systems involving meiosis and 9+2 flagella (cilia) like those of eukaryotes.

propagule   A multicellular structure that serves in asexual reproduction.

properizonium   In diatoms, a system of delicate siliceous bands on the auxospore that controls the ultimate shape of frustules.

proteomics   The study of types and relative amounts of proteins produced by a cell or other defined structure under defined conditions.

pseudoparenchyma   A form of algal body composed of interwoven, continuous filaments; superficially resembles parenchyma.

pseudoraphe   A siliceous region, uninterrupted by a slit, on one or both valves of some pennate diatoms that occupies the central valve axis and sometimes mimics a true raphe.

psu   (*see* practical salinity units)

pusule   In dinoflagellates, a branched system of tubes, lined by the cell membrane, that opens to the cell exterior.

pycnocline   A vertical gradient that separates water of different densities due to temperature or salinity. (*see also* thermocline)

pyrenoid   A proteinaceous region in the plastids of many types of algae; known in some cases to contain rubisco and commonly associated with formation of storage compounds.

radial centric diatoms   Fossil and modern diatoms having radially symmetrical valves of more or less circular outline.

raphe   An elongate fissure in one or both valves of many pennate diatoms that has both an external opening and an internal opening.

raphid pennates   Pennate diatoms that have a raphe on at least one valve.

raptorial feeding   Obtaining food by seizing prey with specialized appendages.

receptacle   A fertile area of a fleshy body in which reproductive structures develop.

red algae   The informal term for a lineage of algae lacking flagella and centrioles, whose plastids have an envelope consisting of two membranes and contain phycobilin accessory pigments.

refractory carbon   A form of organic carbon that is resistant to microbial, chemical, and physical degradation.

Rhizaria   A eukaryotic supergroup that includes the algae known as chlorarachniophytes as well as many types of non-photosynthetic protists.

rhizoplast   A striated, contractile strand that extends from the flagellar basal bodies into the cell, often connecting with the nuclear surface.

rhizostyle   In cryptomonads, a cluster of posteriorly directed microtubules that extends from the flagellar bases.

rhodolith   Rounded nodule of coralline red algae that develops around the surfaces of stones.

rhodopsin   A type of photoreceptor used by cryptomonads, green flagellates, and dinoflagellates. (A similar photoreceptor occurs in animal eyes.)

rimoportula (*see* labiate process)

rotifers   Aquatic animals characterized by a ring of cilia at their anterior ends. Rotifers mainly employ filter feeding to graze on algae and other microorganisms, but some practice raptorial feeding.

rubisco   Ribulose 1,5-bisphosphate carboxylase/oxygenase, the enzyme that catalyzes incorporation of carbon dioxide into carbohydrate.

saccoderm desmids   Unicellular zygnematalean green algae whose outer walls lack pores and whose cells are not constricted.

sarcinoid   Consisting of a three-dimensional packet of cells.

scalariform conjugation   In filamentous zygnematalean green algae, parallel alignment of two filaments, followed by formation of multiple conjugation tubes between opposed cells, giving a ladderlike appearance.

scintillon   (*see* microsource)

scytonemin   An indole-alkaloid that may color cyanobacterial sheaths yellow or brown and protects cyanobacterial cells from UV radiation.

SDV   (*see* silica deposition vesicle)

secondary endosymbiosis   The incorporation of a photosynthetic eukaryote, whose plastid was derived from a prokaryote, into a eukaryotic host cell.

secondary pit plug (secondary pit connection)   A pit plug formed between a pair of non-sibling cells, by the cutting off of a small cell from one of the two cells (which involves formation of a pit plug) and subsequent fusion of the small cell with the other member of the pair.

segregative cell division   In siphonocladalean green algae, cleavage of the multinucleate protoplast into spherical units that then expand, develop walls, and function as independent regions.

separation disks   (*see* necridia)

seta   A long, thin projection from a cell.

sheath   A layer of polysaccharide mucilage on the body of an alga, particularly filamentous cyanobacteria.

shield cells   In charalean green algae, cells that form the outer walls of the multicellular male gametangia.

siderophores (siderochromes)   Extracellular compounds that bind iron, making it available for import into cells.

signal peptides   Amino acids at the amino ends of polypeptides that target the proteins to specific sites in a eukaryotic cell.

silica deposition vesicle   A membrane-bound intracellular vesicle in which siliceous structures such as diatom frustule components are deposited.

siphonaceous coenocytes   Algal bodies characterized by relatively large, multinucleate cells lacking cross-walls except during the formation of reproductive structures.

somatic cells   In volvocalean green algae, body cells that are not capable of cell division.

sorus   A group of reproductive structures.

specificity   In reference to rubisco, the relative selectivity of CO_2 over O_2

spermatia   The male gametes of red algae; no flagella are present.

spore   In algae, a reproductive spore that may or may not be produced via meiosis; in land plants, cells generated by meiosis that become coated with sporopollenin.

sporic meiosis   Meiosis that occurs during production of spores.

sporophyll   In kelps, a fertile, spore-producing blade.

sporophyte   The spore-producing phase in a life history that involves alternation of spore- and gamete-producing generations.

sporopollenin   In walls of certain green algae and spores of embryophytes, a resistant biopolymer.

stipes   In brown algae, structures that bears blades.

stipules   In charalean green algae, spinelike cells in rings beneath whorls of branches.

stomatocyst (statospore)   A resting spore of chrysophycean and synurophycean algae.

stonewort   A calcified member of the charalean algae (*see also* Charales)

Stramenopila   A eukaryotic supergroup that includes many photosynthetic lineages (algal groups) as well as several early-diverging groups of non-photosynthetic protists; characterized by multipart tubular hairs on one of the flagella borne on motile cells.

stratification   Formation of a surface layer of warm water over deeper, cold water as a result of density differences that develop during warm-season heating.

streptophyte algae   (*see* charophyceans)

striae   In diatoms, rows of pores in the frustule.

stromatolite   A calcareous, layered assemblage of cyanobacteria, occurring as fossils or in modern waters in sheltered areas.

subaerial algae   Algae that grow on the surfaces of substrates, where they are exposed to the atmosphere and sunlight.

sulcus   A longitudinal furrow or depression on the ventral side of a dinoflagellate cell, in which lies the longitudinal flagellum.

sutures   In dinoflagellates, the regions between adjacent plates.

synapomorphies   Shared, derived characters.

tentacle   A slender, contractile feeding structure found in certain dinoflagellates.

tertiary endosymbiosis   The incorporation of a photosynthetic eukaryote, whose plastid was derived from a eukaryotic endosymbiont, into a eukaryotic host cell.

tetrasporangium   In florideophycean red algae, meiosporangium containing four spores produced through meiosis.

tetraspore   In florideophycean red algae, one of the four spores produced within a tetrasporangium.

tetrasporophyte In florideophycean red algae, the life history phase that produces tetraspores.

theca The cover of a dinoflagellate or diatom cell.

thecal plate A portion of the cellulosic theca, or wall, of dinoflagellates.

thermocline The region of greatest rate of vertical temperature change in a stratified water body. (*see also* pycnocline)

thylakoid A flattened, saclike membranous structure in cyanobacterial cells and plastids of eukaryotic algae and plants.

toxins Chemicals from biological sources that kill or disable cells or organisms.

trabecula In some ulvophycean green algae, an extension of the cell wall into the cell lumen, which provides structural support.

transduction The transfer of genes among prokaryotes via viruses.

transit peptides Amino acid sequences at the ends of proteins that foster protein uptake into cell organelles.

transition region (transition zone) In eukaryotic flagella, the zone between the flagellum and its basal body, at the point where the flagellum exits the cell; the specific structure of this region varies among major algal lineages.

trichocysts In dinoflagellates and raphidophyceans, ejectile structures occurring within cytoplasmic membranous vesicles that greatly elongate when discharged, serving a defensive function.

trichogyne In red algae and some charophycean green algae, the elongated apical portion of the female gamete (carpogonium or oogonium) that is receptive to male gametes.

trichome In cyanobacteria, a filament exclusive of the sheath.

trichothallic growth In some brown algae, active cell division occurring in an intercalary position, in a stack of short cells located at the base of filaments.

triphasic alternation of generations The occurrence of three alternating multicellular generations; characteristic of florideophycean red algae.

trophic cascade A chain of effects that proceeds from one level in a food web to another. An increase in phytoplankton, for example, may cause an increase in grazing zooplankton and a further increase in planktivorous fish.

trumpet hyphae In some kelps, colorless, elongate cells in the medulla that are expanded (like the bell of a trumpet) at the cross-walls.

tube cells In charalean green algae, elongate helically twisted tubular cells that surround oogonia.

unialgal culture A culture that contains only one algal species, though other types of organisms, such as bacteria, may be present.

uniaxial construction A body that consists of a single axial filament.

unicell A protist body type consisting of a single cell.

unilocular sporangium In brown algae, a sporangium in which all the spores are produced within a single compartment, usually by meiosis.

uniseriate Having a single row of cells.

utricle In certain ulvophycean green algae, the swollen terminal end of a siphonaceous tube.

valve (a) In diatoms, the flattened, siliceous end wall; (b) in certain dinoflagellates, one of the two larger thecal plates.

valve view Observation of a diatom from the perspective of the valve surface.

vaucheriaxanthin The dominant accessory pigment in the photosynthetic stramenopiles known as raphidophyceans, eustigmatophyceans, and tribophyceans (= xanthophyceans), which tend to have yellow-green or yellow-brown pigmentation.

vegetative cells Non-reproductive cells.

vestibulum In cryptomonads, an anterior depression from which flagella emerge; the anterior end of the cell gullet or furrow.

Viridiplantae A formal term encompassing green algae and land plants.

viscosity A measure of the stickiness or adhesive property of a fluid. Viscosity is measured as the mass in kilograms of a substance that will flow 1 meter in 1 second. Water is more viscous at lower temperatures than at higher temperatures.

xanthophylls Oxygen-containing, yellow-pigmented carotenoids.

zoospore A flagellate spore.

zooxanthellae Unicellular golden-pigmented cells (usually dinoflagellates) that are endosymbiotic in marine animals, including reef-building corals.

zygotic meiosis Meiosis occurring during zygote maturation or germination.

Literature Cited

Abbott, I. A., and G. J. Hollenberg. 1976. *Marine Algae of California*. Stanford University Press, Stanford, CA.

Abraham, W. M., A. J. Bourdelais, J. R. Sabater, A. Ahmed, T. A. Lee, I. Serebriakov, and D. G. Baden. 2005. Airway responses to aerosolized brevetoxins in an animal model of asthma. *American Journal of Respiratory Critical Care Medicine* 171:26–34.

Abrahams, M. V., and L. D. Townsend. 1993. Bioluminescence in dinoflagellates: A test of the burglar alarm hypothesis. *Ecology* 74:258–260.

Abram, N. J., M. E. Gagan, M. T. McCulloch, J. Chappell, and W. S. Hantoro. 2003. Coral reef death during the 1997 Indian Ocean dipole linked to Indonesian wildfires. *Science* 301:952–960.

Ackland, J. C., J. A. West, and J. Pickett-Heaps. 2007. Actin and myosin regulate pseudopodia of *Porphyta pulchella* (Rhodophyta) archeospores. *Journal of Phycology* 43:129–138.

Acquisti, C., J. Kleffe, and S. Collins. 2007. Oxygen content of transmembrane proteins over macroevolutionary time scales. *Nature* 445:47–52.

Adachi, M., T. Kanno, R. Okamoto, A. Shinozaki, K. Fujikawa-Adachi, and T. Nishijima. 2004. *Jannaschia cystaugens* sp. nov., an *Alexandrium* (Dinophyceae) cyst formation-promoting bacterium from Hiroshima Bay, Japan. *International Journal of Systematic and Evolutionary Microbiology* 54:1687–1692.

Adey, W. H., and T. Goertemiller. 1987. Coral reef algal turfs: Master producers in nutrient poor seas. *Phycologia* 26:374–386.

Adey, W. H., C. Luckett, and K. Jensen. 1993. Phosphorus removal from natural waters using controlled algal production. *Restoration Ecology* 1:29–39.

Adl, S. M., et al. 2005. The new higher-level classification of eukaryotes with emphasis on the taxonomy of protists. *Journal of Eukaryotic Microbiology* 52:399–451.

Adolph, S. V., Jung, J. Rattke, and G. Pohnert. 2005. Wound closure in the invasive green alga *Caulerpa taxifolia* by enzymatic activation of a protein cross-linker. *Angewandte Chemie International Edition* 44:2806–2808.

Agawin, N. S. R., C. M. Duarte, and S. Agustí. 2000. Nutrient and temperature control of the contribution of picoplankton to phytoplankton biomass and production. *Limnology and Oceanography* 45:591–600.

Aghajanian, J. G., and M. H. Hommersand. 1980. Growth and differentiation of axial and lateral filaments in *Batrachospermum sirodotii* (Rhodophyta). *Journal of Phycology* 16:15–28.

Agustí, S., and C. M. Duarte. 2000. Strong seasonality in phytoplankton cell lysis in the NW Mediterranean littoral. *Limnology and Oceanography* 45:940–947.

Ahlgren, G. 1978. Growth of *Oscillatoria agardhii* in chemostat culture. II. Dependence of growth constants on temperature. *Internationale Vereinigung für Theoretische und Angewandte Limnologie* 21:88–102.

Ahn, S., D. M. Kulis, D. L. Erdner, D. M. Anderson, and D. R. Walt. 2006. Fiber-optic microarray for simultaneous detection of multiple harmful algal bloom species. *Applied and Environmental Microbiology* 72:5742–5749.

Alberghina, J. S., M. S. Vigna, and V. A. Confalonieri. 2006. Phylogenetic position of the Oedogoniales with the green algae (Chlorophyta) and the evolution of the absolute orientation of the flagellar apparatus. *Plant Systematics and Evolution* 261:151–163.

Alberte, R. S. 1989. Physiological and cellular features of *Prochloron*. In: Lewin, R. A., and L. Cheng (eds.), *Prochloron, a Microbial Enigma*, Chapman & Hall, London, pp. 31–52.

Al-Dhaheri, R. S., and R. L. Willey. 1996. Colonization and reproduction of the epibiotic flagellate *Colacium vesiculosum* (Euglenophyceae) on *Daphnia pulex*. *Journal of Phycology* 32:770–774.

Alexopoulos, C. J., C. W. Mims, and M. Blackwell. 1996. *Introductory Mycology*, 4th edition, John Wiley, New York.

Allard, B., and J. Templier. 2001. High molecular weight lipids from the trilaminar outer wall (TLS)-containing microalgae *Chlorella emersonii*, *Scenedesmus communis* and *Tetraedron minimum*. *Phytochemistry* 57:459–467.

Allemand, D., P. Furia, and S. Benazet-Tambutte. 1998. Mechanisms of carbon acquisition for endosymbiont photosynthesis in Anthozoa. *Canadian Journal of Botany* 76:925–941.

Allen, J. G., and W. Martin. 2007. Out of thin air. *Nature* 445:610–612.

Allen, N. S. 1974. Endoplasmic filaments generate the motive force for rotational streaming in *Nitella*. *Journal of Cell Biology* 63:270–287.

Allen, T. F. H. 1971. Multivariate approaches to the ecology of algae on terrestrial rock surfaces in North Wales. *Journal of Ecology* 59:803–826.

Allwood, A. C., M. C. Walter, B. S. Kamber, C. P. Marshall, and I. W. Burch. 2006. Stromatolite reef from the Early Archaean era of Australia. *Nature* 441:714–718.

Alonso, D. L., C. I. S. del Castillo, E. M. Grima, and Z. Cohen. 1996. First insights into improvement of eicosapentaenoic acid content in *Phaeodactylum tricornutum* (Bacillariophyceae) by induced mutagenesis. *Journal of Phycology* 32:339–345.

Aluwihare, L., D. J. Repeta, and R. F. Chen. 1997. A major biopolymeric component to dissolved organic carbon in surface sea water. *Nature* 387:166–169.

Alverson, A. J., J. J. Cannone, R. R. Gutell, and E. C. Theriot. 2006. The evolution of elongate shape in diatoms. *Journal of Phycology* 42:655–668.

Amado-Filho, G. M., G. Maneveldt, R. C. C. Manso, B. V. Marins-Rosa, M. R. Pascheco, and S. M. P. B. Guimarães. 2007. Structure of thodolith beds from 4 to 55 meters deep along the southern coast of Espírito Santo State, Brazil. *Ciencias Marinas* 33:399–410.

Amano, K., M. Watanabe, K. Kohata, and S. Harada. 1998. Conditions needed for *Chattonella antigua* red tide outbreaks. *Limnology and Oceanography* 43:117–128.

Amsler, C. D. 2001. Induced defenses in macroalgae: The herbivore makes a difference. *Journal of Phycology* 37:353–356.

Amsler, C. D., and R. B. Searles. 1980. Vertical distribution of seaweed spores in a water column offshore of North Carolina. *Journal of Phycology* 16:617–619.

An, S. S., T. Friedl, and E. Hegewald. 1999. Phylogenetic relationships of *Scenedesmus* and *Scenedesmus*-like coccoid green algae as inferred IT-S2 rDNA sequence comparisons. *Plant Biology* 4:418–428.

Anagnostidis, K., and J. Komárek. 1985. Modern approach to the classification system of the cyanophytes I—Introduction. *Algological Studies* 38/39:291–302.

Anagnostidis, K., and J. Komárek. 1988. Modern approach to the classification system of the cyanophytes III—Oscillatoriales. *Algological Studies* 50–53:327–472.

Anagnostidis, K., and J. Komárek. 1990. Modern approach to the classification system of the cyanophytes IV—Stigonematales. *Algological Studies* 59:1–73.

Andersen, R. A. 1987. Synurophyceae classis nov., a new class of algae. *American Journal of Botany* 74:337–353.

Andersen, R. A. 1992. Diversity of eukaryotic algae. *Biodiversity and Conservation* 1:267–292.

Andersen, R. A. 2004. Biology and systematics of heterokont and haptophyte algae. *American Journal of Botany* 91:1508–1522.

Andersen, R. A. 2005. *Algal Culturing Techniques*, The Phycological Society of America, Academic Press, New York.

Andersen, R. A., R. W. Brett, D. Potter, and J. P. Sexton. 1998a. Phylogeny of the Eustigmatophyceae based upon 18S rDNA, with emphasis on *Nannochloropsis*. *Protist* 149:61–74.

Andersen, R. A., D. Potter, and J. C. Bailey. 2002. *Pinguiococcus pyrenoidosus* gen. et sp. nov. (Pinguiophyceae), a new marine coccoid alga. *Phycological Research* 50:57–65.

Andersen, R. A., D. Potter, R. R. Bidigare, M. Latasa, K. Rowan, and C. J. O'Kelly. 1998b. Characterization and phylogenetic position of the enigmatic golden alga *Phaeothamnion confervicola*: Ultrastructure, pigment composition and partial SSU rDNA sequence. *Journal of Phycology* 34:286–298.

Andersen, R. A., G. W. Saunders, M. P. Paskind, and J. P. Sexton. 1993. Ultrastructure and 18S rRNA gene sequence for *Pelagomonas calceolata* gen. et sp. nov. and the description of a new algal class, the Pelagophyceae classis nov. *Journal of Phycology* 29:701–715.

Andersen, R. A., Y. Van de Peer, D. Potter, J. P. Sexton, M. Kawachi, and T. LaJeunesse. 1999. Phylogenetic analysis of the SSU rRNA from members of the Chrysophyceae. *Protist* 150:71–84.

Anderson, D. C., and S. R. Rushforth. 1976. The cryptogamic flora of desert soil crusts in southern Utah. *Nova Hedwigia* 28:691–729.

Anderson, D. C., K. T. Harper, and R. C. Holmgren. 1982a. Factors influencing development of cryptogamic soil crusts in Utah deserts. *Journal of Range Management* 35:180–185.

Anderson, D. C., K. T. Harper, and S. R. Rushforth. 1982b. Recovery of cryptogamic soil crusts from grazing on Utah winter ranges. *Journal of Range Management* 35:355–359.

Anderson, O. R. 1992. Radiolarian algal symbioses. In: Reisser, W. (ed.), *Algae and Symbioses*, BioPress, Bristol, UK, pp. 93–110.

Anderson, R. J. 1994. *Suhria* (Gelidiaceae, Rhodophyta). In: Akatsuka, I. (ed.), *Biology of*

Economic Algae, SPB Academic Publishing, The Hague, The Netherlands, pp. 227–244.

Andreoli, C., I. Moro, N. La Rocca, F. Rigoni, L. Dalla Valle, and L. Bargelloni. 1999. *Pseudopleurochloris antarctica* gen. et sp. nov., a new coccoid xanthophycean from pack-ice of Wood Bay (Ross Sea, Antarctica): Ultrastructure, pigments and 18S rRNA gene sequence. *European Journal of Phycology* 34:149–159.

Andrews, M., R. Box, S. McInroy, and J. A. Raven. 1984. Growth of *Chara hispida*. II. Shade adaptation. *Journal of Ecology* 72:885–895.

Apt, K. E., N. E. Hoffmann, and A. R. Grossman. 1993. The γ-subunit of R-phycoerythrin and its possible mode of transport into the plastid of red algae. *Journal of Biological Chemistry* 268:16208–16215.

Apt, K. E., P. G. Korth-Pancic, and A. R. Grossman. 1996. Stable transformation of the diatom *Phaeodactylum tricornutum*. *Molecular & General Genetics* 252:572–579.

Archibald, J. M., M. B. Rogers, M. Toop, K. Ishida, and P. J. Keeling. 2003. Lateral gene transfer and the evolution of plastid-targeted proteins in the secondary plastid-containing algal *Bigelowiella natans*. *Proceedings of the National Academy of Sciences of the United States of America* 100:7678–7683.

Archibald, P., and H. C. Bold. 1970. Phycological studies XI. The genus *Chlorococcum* Meneghini. *University of Texas Publication* 7015.

Arenas, F., I. Sánchez, S. J. Hawkins, and S. R. Jenkins. 2006. The invisibility of marine algal assemblages: Role of functional diversity and identity. *Ecology* 87: 2851–2861.

Ariño, X., J. J. Ortega-Calvo, A. Gomez-Bolea, and C. Saiz-Jimenez. 1995. Lichen colonization of the Roman pavement at Baelo Claudia (Cadiz, Spain): biodeterioration vs. bioprotection. *The Science of the Total Environment* 167:353–363.

Ariño, X., and C. Saiz-Jimenez. 1996. Colonization and deterioration processes in Roman mortars by cyanobacteria, algae and lichens. *Aerobiologia* 12:9–18.

Armbrust, E. V., et al. 2004. The genome of the diatom *Thalassiosira pseudonana*: Ecology, evolution, and metabolism. *Science* 306:79–86.

Armstrong, S. L. 1987. Mechanical properties of tissues of the brown alga *Hedophyllum sessile* (C. Ag.) Setchell: Variability with habitat. *Journal of Experimental Marine Biology and Ecology* 114:143–151.

Arrigo, K. R., D. H. Robinson, D. L. Worthen, R. B. Dunbar, G. R. DiTullio, M. Van Woert, and M. P. Lizotte. 1999. Phytoplankton community structure and the drawdown of nutrients and CO_2 in the Southern Ocean. *Science* 283:365–367.

Aruga, H., T. Motomura, and T. Ichimura. 1996. Immunofluorescence study of mitosis and cytokinesis in *Acrosiphonia duriuscula* (Acrosiphoniales, Chlorophyta). *Phycological Research* 44:203–213.

Ashen, J. B., and L. J. Goff. 1998. Galls on the marine red alga *Prionitis lanceolata* (Halymeniaceae): Specific induction and subsequent development of an algal bacterial symbiosis. *American Journal of Botany* 85:1710–1721.

Ashida, H., A. Danchin, and A. Yokota. 2005. Was photosynthetic RuBisCo recruited by acquisitive evolution from RuBisCo-like proteins

involved in sulfur metabolism? *Research in Microbiology* 156:611–618.

Auer, M. T., and R. P. Canale. 1982. Ecological studies and mathematical modeling of *Cladophora* in Lake Huron. 3. The dependence of growth rates on internal phosphorus pool size. *Journal of Great Lakes Research* 8:93–99.

Aury, J.-M., et al. 2006. Global trends of whole genome duplications revealed by the ciliate *Paramecium tetraurelia*. *Nature* 444:171–178.

Axler, R. P., R. M. Gersberg, and C. R. Goldman. 1980. Stimulation of nitrate uptake and photosynthesis by molybdenum in Castle Lake, California. *Canadian Journal of Fisheries and Aquatic Science* 37:707–712.

Bach, S. D., and M. N. Josselyn. 1979. Production and biomass of *Cladophora prolifera* (Chlorophyta, Cladophorales) in Bermuda. *Botanica Marina* 22:163–168.

Bachvaroff, T. R., M. V. Sanchez Puerta, and C. F. Delwiche. 2005. Chlorophyll c-containing plastid relationships based on analyses of a multigene data set with all four chromalveolate lineages. *Molecular Biology and Evolution* 22:1722–1782.

Baden, D. G., A. J. Bourdelais, H. Jacocks, S. Michelliza, and J. Naar. 2005. Natural and derivative brevetoxins: Historical background, multiplicity, and effects. *Environmental Health Perspectives* 113:621–625.

Baden, D. G., K. S. Rein, and R. E. Gawley. 1998. Marine toxins: How they are studied and what they can tell us. In: Cooksey, K. E. (ed.), *Molecular Approaches to the Study of the Ocean*, Chapman & Hall, London, pp. 487–514.

Badger, M. R., and G. D. Price. 1994. The role of carbonic anhydrase in photosynthesis. *Annual Review of Plant Physiology and Plant Molecular Biology* 45:369–392.

Bae, Y. M., and J. W. Hastings. 1994. Cloning, sequencing and expression of dinoflagellate luciferase DNA from a marine alga *Gonyaulax polyedra*. *Biochimica Biophysica Acta* 1219:449–456.

Bailey, J. C., R. R. Bidigare, S. J. Christensen, and R. A. Andersen. 1998. Phaeothamniophyceae classis nova: A new lineage of chromophytes based upon photosynthetic pigments, *rbcL* sequence analysis and ultrastructure. *Protist* 149:245–263.

Bailey, J. C., and R. L. Chapman. 1996. Evolutionary relationships among coralline red algae (Corallinaceae, Rhodophyta) inferred from 18S rRNA gene sequence analyses. In: Chaudhary, B. R., and S. B. Agrawal (eds.), *Cytology, Genetics, and Molecular Biology of Algae*, SPB Academic Publishers, Amsterdam, pp. 363–376.

Bailey, J. C., and D. W. Freshwater. 1997. Molecular systematics of the Gelidiales: Inferences from separate and combined analyses of plastid *rbcL* and nuclear SSU gene sequences. *European Journal of Phycology* 32:343–352.

Bailey, D., A. P. Mazurak, and J. R. Rosowski. 1973. Aggregation of soil particles by algae. *Journal of Phycology* 9: 99–101.

Baker, A. C. 2003. Flexibility and specificity in coral–algal symbiosis: Diversity, ecology, and biogeography of *Symbiodinium*. *Annual Review of Ecology, Evolution, and Systematics* 34:661–689.

Baker, A. L., and A. J. Brook. 1971. Optical density profiles as an aid to the study of microstratified phytoplankton populations in lakes. *Archiv für Hydrobiologie* 69:214–233.

Baker, B. J., M. L. Lutz, S. C. Dawson, P. L. Bond, and J. F. Banfield. 2004. Metabolically active eukaryotic communities in extremely acidic mine drainage. *Applied and Environmental Microbiology* 70:6264–6271.

Baker, J. R. J., and L. V. Evans. 1973. The ship-fouling alga *Ectocarpus*. I. Ultrastructure and cytochemistry of plurilocular reproductive stages. *Protoplasma* 77:1–13.

Baker, J. W., J. P. Grover, B. W. Brooks, F. Ureña-Boeck, D. L. Roelke, R. Errera, and R. L. Kiesling. 2007. Growth and toxicity of *Prymnesium parvum* (Haptophyta) as a function of salinity, light, and temperature. *Journal of Phycology* 43:219–227.

Baker, V. R. 2004. A brief geological history of water on Mars. In: Seckbach, J. (ed.), *Origin, Evolution and Diversity of Life*, Kluwer Academic Publishers, Dordrecht, The Netherlands, pp. 619–631.

Baker, V. R., R. G. Storm, V. C. Gulick, J. S. Kargel, G. Komatsu, and V. S. Kale. 1991. Ancient oceans, ice sheets and the hydrological cycle on Mars. *Nature* 352:589–595.

Bakker, F. T., J. L. Olsen, and W. T. Stam. 1995. Evolution of nuclear rDNA ITS sequences in the *Cladophora albida/sericea* clade (Chlorophyta). *Journal of Molecular Evolution* 40:640–651.

Baldauf, S. L., A. J. Roger, I. Wenk-Siefert, and W. F. Doolittle. 2000. A kingdom-level phylogeny of eukaryotes based on combined proteins data. *Science* 290:972–977.

Ballantine, D. L., J. Nelxon Navarro, and D. A. Hensley. 2001. Algal colonization of Caribbean scorpionfishes. *Bulletin of Marine Science* 69:1089–1094.

Ban, S., et al. 1997. The paradox of diatom-copepod interactions. *Marine Ecology Progress Series* 157:287–293.

Bao, Z., et al. 2007. Chemical reduction of three-dimensional silica micro-assemblies into microporous silicon replicas. *Nature* 446:172–175.

Barbier, G., et al. 2005. Comparative genomics of two closely related unicellular thermo-acidophilic red algae, *Galdieria sulphuraria* and *Cyanidioschyzon merolae*, reveals the molecular basis of the metabolic flexibility of *Galdieria sulphuraria* and significant difference in carbohydrate metabolism of both algae. *Plant Physiology* 137:460–474.

Barea-Arco, J., C. Pérez-Martínez, and R. Morales-Baquero. 2001. Evidence of a mutualistic relationship between an algal epibiont and its host, *Daphnia pulicaria*. *Limnology and Oceanography* 46:871–881.

Barger, N. N., J. E. Herrick, J. Van Zee, and J. Belnap. 2006. Impacts of biological soil crust disturbance and composition on C and N loss from water erosion. *Biogeochemistry* 77:247–263.

Barlow, S. B., and R. A. Cattolico. 1981. Mitosis and cytokinesis in the Prasinophyceae. I. *Mantoniella squamata* (Manton and Parke) Desikachary. *American Journal of Botany* 68:606–615.

Barreiro, R., M. Quintela, I. Bárbara, and J. Cremades. 2006. RAPD differentiation of *Grate-

loupia lanceola and the invasive *Grateloupia turuturu* (Gigartinales, Rhodophyta) in the Iberian Peninsula. *Phycologia* 45:213–217.

Barták, M., P. Váczi, J. Hájek, and J. Smykla. 2007. Low-temperature limitation of primary photosynthetic processes in Antarctic lichens *Umbilicaria antarctica* and *Xanthoria elegans*. *Polar Biology* 31:47–51.

Bartlett, R., and R. Willey. 1998. Epibiosis of *Colacium* on *Daphnia*. *Symbiosis* 25:291–299.

Bartley, G. E., and P. A. Scolnick. 1995. Plant carotenoids: Pigments for photoprotection, visual attraction, and human health. *The Plant Cell* 7:1027–1038.

Baschnagel, R. A. 1966. New fossil algae from the Middle Devonian of New York. *Transactions of the American Microbiological Society* 85:297–302.

Bäumer, D., A. Preisfeld, and H. G. Ruppel. 2001. Isolation and characterization of paramylon synthase from *Euglena gracilis* (Euglenophyceae). *Journal of Phycology* 37:38–46.

Baylson, F. A., B. W. Stevens, and D. S. Domozych. 2001. Composition and synthesis of the pectin and protein components of the cell wall of *Closterium acerosum* (Chlorophyta). *Journal of Phycology* 37:796–809.

Beam, C. A., and M. Himes. 1974. Evidence for sexual fusion and recombination in the dinoflagellate *Crypthecodinium* (*Gyrodinium*) *cohnii*. *Nature* 250:435–436.

Beardall, J., and J. A. Raven. 2004. The potential effects of global climate change on microalgal photosynthesis, growth and ecology. *Phycologia* 43:26–40.

Becker, B., D. Becker, J. P. Kamerling, and M. Melkonian. 1991. 2-keto-sugar acids in green flagellates: A chemical marker for prasinophycean scales. *Journal of Phycology* 27:498–504.

Becker, E. W. 1982. Physiological studies on Antarctic *Prasiola crispa* and *Nostoc commune* at low temperatures. *Polar Biology* 1:99–104.

Becker, V. E. 1980. Nitrogen fixing lichens in forests of the southern Appalachian Mountains of North Carolina. *The Bryologist* 83:29–39.

Beech, P. L., and R. Wetherbee. 1990a. Direct observations on flagellar transformation in *Mallomonas splendens* (Synurophyceae). *Journal of Phycology* 26:90–95.

Beech, P. L., and R. Wetherbee. 1990b. The flagellar apparatus of *Mallomonas splendens* (Synruophyceae) at interphase and its development during the cell cycle. *Journal of Phycology* 26:95–111.

Beech, P. L., R. L. Wetherbee, and J. D. Pickett-Heaps. 1990. Secretion and deployment of bristles in *Mallomonas splendens*. *Journal of Phycology* 26:112–122.

Behrenfeld, M. J., W. E. Esaias, and K. R. Turpie. 2002. Assessment of primary production at the global scale. In: Williams, P. J. leB., D. N. Thomas, and C. S. Reynolds (eds.), *Phytoplankton Productivity*, Blackwell Science, Oxford, UK, pp. 156–186.

Behrenfeld, M. J., R. T. O'Malley, D. A. Siegel, C. R. McClain, J. L. Sarmiento, G. C. Feldman, A. J. Milligan, P. G. Falkowski, R. M. Letelier, and E. S. Boss. 2006. Climate-driven trends in contemporary ocean productivity. *Nature* 444:752–755.

Bell, E. M., and J. Laybourn-Parry. 2003.

Mixotrophy in the Antarctic phytoflagellate *Pyramimonas gelidicola* (Chlorophyta: Prasinophyceae). *Journal of Phycology* 39:644–649.

Bell, R. A. 1993. Cryptoendolithic algae of hot semiarid lands and deserts. *Journal of Phycology* 29:133–139.

Bell, R. A., P. V. Athey, and M. R. Sommerfeld. 1986. Cryptoendolithic algal communities of the Colorado Plateau. *Journal of Phycology* 22:429–435.

Bell, R. A., P. V. Athey, and M. R. Sommerfeld. 1988. Distribution of endolithic algae on the Colorado Plateau of Northern Arizona. *The Southwest Naturalist* 33:315–322.

Bell, T., and J. Kalff. 2001. The contribution of picophytoplankton in marine and freshwater systems of different trophic status and depth. *Limnology and Oceanography* 46:1243–1248.

Bellinger, E. 1974. A note on the use of algal sizes in estimates of population standing crop. *British Phycological Journal* 9:157–161.

Belnap, J. 1993. Recovery rates of cryptobiotic crusts: Inoculant use and assessment methods. *Great Basin Naturalist* 53:89–95.

Belnap, J., B. Büdel, and O. L. Lange. 2001. Biological soil crusts: Characteristics and distribution. In: Belnap, J., and O. L. Lange (eds.), *Biological Soil Crusts: Structure, Function, and Management*, Ecological Studies, vol. 150, Springer Verlag, Berlin, pp. 3–30.

Belnap, J., and J. S. Gardner. 1993. Soil microstructure in soils of the Colorado Plateau: The role of the cyanobacterium *Microcoleus vaginatus*. *Great Basin Naturalist* 53:40–47.

Belnap, J., and D. A. Gillette. 1998. Vulnerability of desert soil surfaces to wind erosion: Impacts of soil texture and disturbance. *Journal of Arid Environments* 39:133–142.

Belnap, J., and O. L. Lange. 2001. Structure and functioning of biological soil crusts: a synthesis. In: Belnap, J., and O. L. Lange (eds.), *Biological Soil Crusts: Structure, Function, and Management*, Ecological Studies, vol. 150, Springer Verlag, Berlin, pp. 471–479.

Ben-Amotz, A., A. Katz, and M. Avron. 1982. Accumulation of β-carotene in halotolerant algae: Purification and characterization of β-carotene-rich globules from *Dunaliella bardawil* (Chlorophyceae). *Journal of Phycology* 18:529–537.

Benitez-Nelson, C. R., et al. 2007. Mesoscale eddies drive increased silica export in the subtropical Pacific Ocean. *Science* 316:1017–1021.

Bennamara, A., A. Abourriche, M. Berrada, M. Charrout, N. Chaib, M. Boudouma, and F. X. Garneau. 1999. Methoxybifurcarenone: An antifungal and antibacterial meroditerpenoid from the brown alga Cystoseira tamariscifolia. *Phytochemistry* 52:37–40.

Berger, S. A., S. Diehl, T. J. Kunz, D. Albrecht, A. M. Oucible, and S. Ritzer. 2006. Light supply, plankton biomass, and seston stoichiometry in a gradient of lake mixing depths. *Limnology and Oceanography* 51:1898–1905.

Berger, S., U. Fettweiss, S. Gleissberg, L. B. Liddle, U. Richter, H. Sawitsky, and G. C. Zuccarello. 2003. 18S rDNA phylogeny and evolution of cap development in Polyphysaceae (formerly Acetabulariaceael Dasycladales, Chlorophyta). *Phycologia* 42:506–561.

Berger, S., and M. J. Kaever. 1992. *Dasycladales:*

An Illustrated Monograph of a Fascinating Algal Order, Thieme, Stuttgart, Germany.

Bergey, E. A., C. A. Boettiger, and V. H. Resh. 1995. Effects of water velocity on the architecture and epiphytes of *Cladophora glomerata* (Chlorophyta). *Journal of Phycology* 31:264–271.

Berglund, J., U. Müren, U. Båmstedt, and A. Andersson. 2007. Efficiency of a phytoplankton-based and a bacteria-based food web in a pelagic marine system. *Limnology and Oceanography* 52:121–131.

Bergman, B., C. Johansson, and E. Söderbäck. 1992. The *Nostoc–Gunnera* symbiosis. *New Phytologist* 122:379–400.

Bergquist, A. M., and S. R. Carpenter. 1986. Limnetic herbivory effects on phytoplankton populations and primary production. *Ecology* 67:1351–1360.

Bergquist, A. M., S. R. Carpenter, and J. C. Latino. 1985. Shifts in phytoplankton size structure and community composition during grazing by contrasting zooplankton assemblages. *Limnology and Oceanography* 30:1037–1045.

Bergström, A.-K., M. Jansson, S. Drakare, and P. Blomqvist. 2003. Occurrence of mixotrophic flagellates in relation to bacterioplankton production, light regime and availability of inorganic nutrients in unproductive lakes with differing humic contents. *Freshwater Biology* 48:868–877.

Berkelmans, R., and M. J. H. van Oppen. 2006. The role of zooxanthellae in the thermal intolerance of corals: A "nugget of hope" for coral reefs in an era of climate change. *Proceedings of the Royal Society B* 273:2305–2312.

Berman, T., and W. Rodhe. 1971. Distribution and migration of *Peridinium* in Lake Kinneret. *Internationale Vereiningung für Theoretische und Angewandte Limnologie* 19:266–276.

Berman-Frank, I., A. Kaplan, T. Zohary, and Z. Dubinsky. 1995. Carbonic anhydrase activity in the bloom-forming dinoflagellate *Peridinium gatunense*. *Journal of Phycology* 31:906–913.

Berman-Frank, I., P. Lundgren, Y. Chen, H. Küpper, Z. Kolber, B. Bergman, and P. Falkowski. 2001. Segregation of nitrogen fixation and oxygenic photosynthesis in the marine cyanobacterium *Trichodesmium*. *Science* 294:1534–1537.

Berner, R. A. 1997. The rise of land plants and their effect on weathering and CO_2. *Science* 276:544–546.

Berney, C., and J. Pawlowski. 2006. A molecular time-scale for eukaryote evolution recalibrated with the continuous microfossil record. *Proceedings of the Royal Society B* 273:1867–1872.

Bertness, M. D., C. M. Crain, B. R. Silliman, M. C. Bazterrica, M. V. Reyna, F. Hildago, and J. K. Farina. 2006. The community structure of western Atlantic Patagonian rocky shores. *Ecological Monographs* 76:439–460.

Bertrand, E. M., M. A. Saito, J. M. Rose, C. R. Riesselman, M. C. Lohan, A. E. Noble, P. A. Lee, and G. R. DiTullio. 2007. Vitamin B_{12} and iron colimitation of phytoplankton growth in the Ross Sea. *Limnology and Oceanography* 52:1079–1093.

Betournay, S. A., C. Marsh, N. Donello, and J. W. Stiller. 2007. Selective recovery of microalgae from diverse habitats using "phyto-specific" 16S rDNA primers. *Journal of Phycology* 43:609–613.

Beymer, R. J., and J. M. Klopatek. 1991. Potential contribution of carbon by microphytic crusts in Pinyon-Juniper woodlands. *Arid Soil Research and Rehabilitation* 5:187–198.

Beymer, R. J., and J. M. Klopatek. 1992. Effects of grazing on cryptogamic crusts in Pinyon-Juniper woodlands in Grand Canyon National Park. *American Midland Naturalist* 127:139–148.

Bhattacharya, D., T. Helchen, C. Bebeau, and M. Melkonian. 1995. Comparisons of nuclear-encoded small-subunit ribosomal RNAs reveal the evolutionary position of the Glaucocystophyta. *Molecular Biology and Evolution* 12:415–420.

Biddanda, B., M. Ogdahl, and J. Cotner. 2001. Dominance of bacterial metabolism in oligotrophic relative to eutrophic waters. *Limnology and Oceanography* 46:730–739.

Bidigare, R. R., M. E. Ondrusek, M. C. Kennicutt, R. Iturriaga, H. R. Harvey, R. W. Hoham, and S. A. Macko. 1993. Evidence for a photoprotective function for secondary carotenoids of snow algae. *Journal of Phycology* 29:427–434.

Biebel, P. 1973. Morphology and life cycles of saccoderm desmids in culture. *Nova Hedwigia* 42:39–57.

Biggs, B. J. F. 1996. Patterns in benthic algae of streams. In: Stevenson, R. L., M. L. Bothwell, and R. L. Lowe (eds.), *Algal Ecology: Freshwater Benthic Ecosystems*, Academic Press, New York, pp. 31–56.

Biggs, B. J. F., D. G. Goring, and V. I. Nikora. 1998. Subsidy and stress responses of stream periphyton to gradients in water velocity as a function of community growth form. *Journal of Phycology* 34:598–607.

Bilger, W., M. Bohuschke, and M. Ehling-Schulz. 1997. Annual time course of the contents of carotenoids and UV-protective pigments in the cyanobacterium *Nostoc commune*. *Bibliotheca Lichenologica* 67:223–234.

Billington, M., and V. Alexander. 1978. Nitrogen fixation in a black spruce (*Picea mariana* [Mill] B.S.P.) forest in Alaska. *Ecological Bulletin* 26:209–215.

Bird, D. F., and J. Kalff. 1986. Bacterial grazing by planktonic lake algae. *Science* 231:493–495.

Birkeland, C. 1997. Introduction. In: Birgeland, C. (ed.), *Life and Death of Coral Reefs*, Chapman & Hall, New York, pp. 1–12.

Björk, M., K. Haglund, Z. Ramazanov, and M. Pedersén. 1993. Inducible mechanisms for HCO_3^- utilization and repression of photorespiration in protoplasts and thalli of three species of *Ulva* (Chlorophyta). *Journal of Phycology* 29:166–173.

Black, K., and B. Osborne. 2004. An assessment of photosynthetic downregulation in cyanobacteria from the *Gunnera–Nostoc* symbiosis. *New Phytologist* 162:125–132.

Black, K., R. Parsons, and B. A. Osborne. 2002. Uptake and metabolism of glucose in the *Nostoc–Gunnera* symbiosis. *New Phytologist* 153:297–305.

Bliding, C. 1957. Studies in *Rhizoclonium*. I. Life history of two species. *Botaniska Notiser* 110:271–275.

Blomqvist, S., A. Gunnars, and R. Elmgren. 2004. Why the limiting nutrient differs between temperate coastal seas and freshwater lakes: A matter of salt. *Limnology and Oceanography* 49:2236–2241.

Bodyl, A. 2005. Do plastid-related characters support the chromalveolate hypothesis? *Journal of Phycology* 41:712–719.

Bodyl, A., and K. Moszczynski. 2006. Did the peridinin plastid evolve through tertiary endosymbiosis? A hypothesis. *European Journal of Phycology* 41:435–448.

Boëchat, I. G., and R. Adrian. 2006. Evidence for biochemical limitation of population growth and reproduction of the rotifer *Keratella quadrata* fed with freshwater protists. *Journal of Plankton Research* 28:1027–1038.

Bokhorst, S., C. Ronfort, A. Huiskes, P. Convey, and R. Aerts. 2007. Food choice of Antarctic soil arthropods clarified by stable isotope signatures. *Polar Biology* 30:983–990.

Bold, H. C. 1958. Three new chlorophycean algae. *American Journal of Botany* 45:737–743.

Bold, H. C., and F. J. MacEntee. 1973. Phycological notes. II. *Euglena myxocylindracea* sp. nov. *Journal of Phycology* 9:152–156.

Bold, H. C., and M. J. Wynne. 1978. *Introduction to Phycology*, 1st edition, Prentice Hall, Englewood Cliffs, NJ.

Bold, H. C., and M. J. Wynne. 1985. *Introduction to Phycology*, 2nd edition, Prentice Hall, Englewood Cliffs, NJ.

Bollmann, J., M. Y. Cortés, A. Kleinjne, J. B. Østergaard, and J. R. Young. 2006. *Solisphaera* gen. nov. (Prymnesiophyceae), a new coccolithophore genus from the lower photic zone. *Phycologia* 45:465–477.

Bolton, J. J., and R. J., Anderson. 1994. *Ecklonia*. In: Akatsuka, I. (ed.), *Biology of Economic Algae*, SPB Academic Publishing, The Hague, The Netherlands, pp. 385–406.

Bomber, J. W., M. G. Rubio, and D. R. Norris. 1989. Epiphytism of dinoflagellates associated with the disease ciguatera: Substrate specificity and nutrition. *Phycologia* 28:360–368.

Boo, S. M., and I. K. Lee. 1994. *Ceramium* and *Campylaephora* (Ceramiaceae, Rhodophyta). In: Akatsuka, I. (ed.), *Biology of Economic Algae*, SPB Academic Publishing, The Hague, The Netherlands, pp. 1–33.

Booth, B. C., and H. J. Marchant. 1987. Parmales, a new order of marine chrysophytes, with descriptions of three new genera and seven new species. *Journal of Phycology* 23:245–260.

Booth, W. E. 1941. Algae as pioneers in plant succession and their importance in erosion control. *Ecology* 22:38–46.

Booton, G. C., G. L. Floyd, and P. A. Fuerst. 1998a. Polyphyly of tetrasporalean green algae inferred from nuclear small-subunit ribosomal data. *Journal of Phycology* 34:306–311.

Booton, G. C., G. L. Floyd, and P. A. Fuerst. 1998b. Origins and affinities of the filamentous green algal orders Chaetophorales and Oedogoniales based on 18S rRNA gene sequences. *Journal of Phycology* 34:312–318.

Borchardt, M. A. 1996. Nutrients. In: Stevenson, R. L., M. L. Bothwell, and R. L. Lowe (eds.), *Algal Ecology: Freshwater Benthic Ecosystems*, Academic Press, New York, pp. 184–228.

Borza, T., C. E. Popescu, and R. W. Lee. 2005.

Multiple metabolic roles for the nonphotosynthetic plastid of the green alga *Prototheca wickerhamii*. *Eukaryotic Cell* 4:253–261.

Boucias, D. G., J. J. Becnel, S. E. White, and M. Bott. 2001. In vivo and in vitro development of the protists *Helicosporidium* sp. *Journal of Eukaryotic Microbiology* 48:460–470.

Bouck, G. B., and D. L. Brown. 1973. Microtubule biogenesis and cell shape in *Ochromonas*. *Journal of Cell Biology* 56:340–359.

Bourdelais, A. J., S. Campbell, H. Jacocks, J. Naar, J. L. C. Wright, J. Carsi, and D. G. Baden. 2004. Brevenal is a natural inhibitor of brevetoxin action in sodium channel receptor binding assays. *Cellular and Molecular Neurobiology* 24:553–563.

Bourne, D. G., G. L. Jones, R. L. Blakeley, A. Jones, A. P. Negri, and P. Riddles. 1996. Enzymatic pathway for the bacterial degradation of the cyanobacterial cyclic peptide toxin microcystin LR. *Applied and Environmental Microbiology* 62:4086–4094.

Bourrelly, P. 1968. *Les Algues d'eau douce. Initiation à la Systématique. Tome II: Les Algues jaunes et brunes, Chrysophycées, Phéophycées, Xanthophycées et Diatomées.* Boubée & Cie, Paris.

Bouza, N., J. Caujapé-Castells, M. A. González-Pérez, and P. A. Sosa. 2006. Genetic structure of natural population in the red algae *Gelidium canariense* (Gelidiales, Rhodophyta) investigated by random amplified polymorphic DNA (RAPD) markers. *Journal of Phycology* 42:304–311.

Bowker, M. A., S. C. Reed, J. Belnap, and S. L. Phillips. 2002. Temporal variation in community composition, pigmentation, and F_v/F_m of desert cyanobacterial soil crusts. *Microbial Ecology* 43:13–25.

Bown, P. R., J. A. Lees, and J. R. Young. 2004. Calcareous nannoplankton evolution and diversity. In: Thierstein, H. R., and J. R. Young (eds.), *Coccolithophores—From Molecular Processes to Global Impact*, Springer Verlag, New York, pp. 481–508.

Bown, P., J. Plumb, P. Sánchez-Baracaldo, P. Hayes, and J. Brodie. 2003. Sequence heterogenity of green (Chlorophyta) endophytic algae associated with a population of *Chondrus crispus* (Gigartinaceae, Rhodophyta). *Journal of Phycology* 38:153–163.

Boyer, S. L., J. R. Johansen, V. R. Flechtner, and G. L. Howard. 2002. Phylogeny and genetic variance in terrestrial *Microcoleus* (Cyanophyceae) species based on sequence analysis of the 16S rRNA gene and associated 16S–23S ITS region. *Journal of Phycology* 38:1222–1235.

Boyle, J. A., J. D. Pickett-Heaps, and D. B. Czarnecki. 1984. Valve morphogenesis in the pennate diatom *Achnanthes coarctata*. *Journal of Phycology* 20:563–573.

Bracken, M. E. S., C. A. Gonzalez-Dorantes, and J. J. Stachowicz. 2007. Whole-community mutualism: Associated invertebrates facilitate a dominant habitat-forming seaweed. *Ecology* 88:2211–2219.

Bracken, M. E. S., and K. J. Nielsen. 2004. Diversity of intertidal macroalgae increases with nitrogen loading by invertebrates. *Ecology* 85:2828–2836.

Brahamsha, B. 1996. An abundant cell-surface polypeptide is required for swimming by the nonflagellated marine cyanobacterium *Syn-*

echococcus. *Proceedings of the National Academy of Sciences of the United States of America* 93:6504–6509.

Brakemann, T., W. Schlörmann, J. Marquardt, M. Nolte, and E. Rhiel. 2006. Association of fucoxanthin chlorophyll *a/c*-binding polypeptides with photosystems and phosphorylation in the centric diatom *Cyclotella cryptica*. *Protist* 157:463–475.

Branch, G. M., and Griffiths, C. L. 1988. The Benguela ecosystem. Part V. The coastal zone. *Oceanography and Marine Biology: An Annual Review* 26:395–486.

Brand, J. J., and K. R. Diller. 2004. Application and theory of algal cryopreservation. *Nova Hedwigia* 79:175–189.

Brand, L. E. 1994. Physiological ecology of marine coccolithophores. In: Winter, A., and W. G. Siesser (eds.), *Coccolithophores*, Cambridge University Press, Cambridge, UK, pp. 39–50.

Braun, M., and G. O. Wasteneys. 1998. Reorganization of the actin and microtubule cytoskeleton throughout blue-light-induced differentiation of characean protonemata into multicellular thalli. *Protoplasma* 202:38–53.

Brawley, S. H. 1992. Fertilization in natural populations of the dioecious brown alga *Fucus ceranoides* and the importance of the polyspermy block. *Marine Biology* 113:145–157.

Brawley, S. H., and W. H. Adey. 1981. The effect of micrograzers on algal community structure in a coral reef microcosm. *Marine Biology* 61:167–177.

Brawley, S. H., R. S. Quatrano, and R. Wetherbee. 1977. Fine-structural studies of the gametes and embryo of *Fucus vesiculosus* L. (Phaeophyta). *Journal of Cell Science* 24:275–294.

Brawley, S. H., and R. Wetherbee. 1981. Cytology and ultrastructure. In: Lobban, C. S., and M. J. Wynne (eds.), *The Biology of Seaweeds*, Blackwell Scientific, Oxford, UK, pp. 248–299.

Breglia, S. A., C. H. Slamovits, and B. S. Leander. 2007. Phylogeny of phagotrophic euglenids (Euglenozoa) as inferred from Hsp90 gene sequences. *Journal of Eukaryotic Microbiology* 54:86–92.

Breier, C. E., and E. J. Buskey. 2007. Effects of the red tide dinoflagellate *Karenia brevis* on grazing and fecundity in the copepod *Acartia tonsa*. *Journal of Plankton Research* 29:115–126.

Bremer, K. 1985. Summary of green plant phylogeny and classification. *Cladistics* 1:369–385.

Bremer, K. 1988. The limits of amino acid sequence data in angiosperm phylogeny reconstruction. *Evolution* 42:795–803.

Bremer, K., C. J. Humphries, B. D. Mishler, and S. P. Churchill. 1987. On cladistic relationships in green plants. *Taxon* 36:339–349.

Brett, M. T., and C. R. Goldman. 1997. Consumer versus resource control in freshwater pelagic food webs. *Science* 275:384–386.

Brett, S. J., L. Perasso, and R. Wetherbee. 1994. Structure and development of the cryptomonad periplast: A review. *Protoplasma* 181:106–122.

Breuss, O. 2000. New taxa of pyrenocarpous lichens from the Sonoran region *The Bryologist* 103:705–709.

Bricelj, V. M., and D. L. Lonsdale. 1997. Au-

reococcus anophagefferens: Causes and ecological consequences of brown tides in U.S. mid-Atlantic coastal waters. *Limnology and Oceanography* 42:1023–1038.

Britton, T., C. L., Anderson, D. Jacquet, S. Lundqvist, and K. Bremer. 2007. Estimating divergence times in large phylogenetic trees. *Systematic Biology* 56:741–752.

Broadwater, S. T., and E. A. LaPointe. 1997. Parasitic interactions and vegetative ultrastructure of *Choreonema thuretii* (Corallinales, Rhodophyta). *Journal of Phycology* 33:396–407.

Broadwater, S. T., and J. Scott. 1982. Ultrastructure of early development in the female reproductive system of *Polysiphonia harveyi* Bailey (Ceramiales, Rhodophyta). *Journal of Phycology* 18:427–441.

Broadwater, S. T., and J. L. Scott. 1994. Ultrastructure of unicellular red algae. In: Sechbach, J. (ed.), *Evolutionary Pathways and Enigmatic Algae:* Cyanidium caldarium *(Rhodophyta) and Related Cells*, Kluwer Academic Publishers, Boston, pp. 215–230.

Broady, P. A. 1976. Six new species of terrestrial algae from Signy Island, South Orkney Islands, Antarctica. *British Phycological Journal* 11:387–405.

Broady, P. A. 1979. The terrestrial algae of Signy Island, South Orkney Islands. *British Antarctic Survey Scientific Reports* 98:1–117.

Broady, P. A. 1981a. Ecological and taxonomic observations on subaerial epilithic algae from Princess Elizabeth Land and MacRobertson Land, Antarctica. *British Phycological Journal* 16:257–266.

Broady, P. A. 1981b. The ecology of sublithic terrestrial algae at the Vestfold Hills, Antarctica. *British Phycological Journal* 16:231–240.

Broady, P. A. 1986. Ecology and taxonomy of the terrestrial algae of the Vestfold Hills. In: Pickard, J. (ed.), *Antarctic Oasis: Terrestrial Environments and History of the Vestfold Hills*, Academic Press, Sydney, Australia, pp. 165–202.

Broady, P. A. 1989. Broadscale patterns in the distribution of aquatic and terrestrial vegetation at three ice-free regions on Ross Island, Antarctica. *Hydrobiologia* 172:77–95.

Broady, P. A. 1996. Diversity, distribution and dispersal of Antarctic terrestrial algae. *Biodiversity and Conservation* 5:1307–1335.

Broady, P. A., and R. A. Smith. 1994. A preliminary investigation of the diversity, survivability and dispersal of algae introduced into Antarctica by human activity. *Proceedings of the NIPR Symposium on Polar Biology* 7:185–197.

Broady, P. A., and R. N. Weinstein. 1998. Algae, lichens and fungi in La Gorce Mountains, Antarctica. *Antarctic Science* 10:376–385.

Broady, P. C. 1981c. The ecology of chasmolithic algae at coastal locations of Antarctica. *Phycologia* 20:259–272.

Brocks, J. J., R. Buick, R. E. Summons, and G. A. Logan. 2003. A reconstruction of Archean biological diversity based on molecular fossils from the 2.78 to 2.45 billion-year-old Mount Bruce Supergroup, Hamersley Basin, Western Australia. *Geochimica et Cosmochimica Acta* 67:4321–4335.

Brocks, J. J., G. A. Logan, R. Buick, and R. E. Summons 1999. Archaean molecular fos-

sils and the early rise of eukaryotes. *Science* 285:1033–1036.

Brocks, J. .J., and R. E. Summons. 2003. Sedimentary hydrocarbons, biomarkers for early life. In: Schlesinger, W. H. (ed.), *Treatise on Geochemistry*, vol. 8. Elsevier, Amsterdam, pp. 63–115.

Brodie, J., P. K. Hayes, G. L. Barker, L. M. Irvine, and I. Bartsch. 1998. A reappraisal of *Porphyra* and *Bangia* (Bangiophycidae, Rhodophyta) in the northeast Atlantic based on the *rbcL-rbcS* intergenic spacer. *Journal of Phycology* 34:1069–1074.

Brook, A. J. 1980. Barium accumulation by desmids of the genus *Closterium* (Zygnemaphyceae). *British Phycological Journal* 15:261–264.

Brook, A. J., and D. B. Williamson. 1988. The survival of desmids on the drying mud of a small lake. In: Round, F. E. (ed.), *Algae and the Aquatic Environment*, BioPress, Bristol, UK, pp. 185–196.

Brooks, J. L., and S. I. Dodson. 1965. Predation, body size and composition of plankton. *Science* 150:28–35.

Brotherson, J. D., and S. R. Rushforth. 1983. Influence of cryptogamic crusts on moisture relationships of soils in Navajo National Monument, Arizona. *Great Basin Naturalist* 43:73–78.

Brown, B. E. 1997. Disturbances to reefs in recent times. In: Birkeland, C. (ed.), *Life and Death of Coral Reefs*, Chapman & Hall, New York, pp. 354–379.

Brown, L. M., and J. A. Hellebust. 1980. The contribution of organic solutes to osmotic balance in some green and eustigmatophyte algae. *Journal of Phycology* 16:265–270.

Brown, R. C., B. E. Lemmon, and L. E. Graham. 1994. Morphogenetic plastid migration and microtubule arrays in mitosis and cytokinesis in the green alga *Coleochaete orbicularis*. *American Journal of Botany* 81:127–133.

Brownlee, C., N. Nimer, L. F. Dong, and M. J. Merrett. 1994. Cellular regulation during calcification in *Emiliania huxleyi*. In: Green, J. C., and B. S. C. Leadbeater (eds.), *The Haptophyte Algae*, Clarendon Press, Oxford, UK, pp. 133–148.

Brunberg, A.-K., and P. Blomqvist. 2003. Recruitment of *Microcystis* (Cyanophyceae) from lake sediments: The importance of littoral inocula. *Journal of Phycology* 39:58–63.

Brussaard, C. P. D. 2004. Viral control of phytoplankton populations—A review. *Journal of Eukaryotic Microbiology* 51:125–138.

Buchheim, M., J. A. Buchheim, T. Carlson, and P. Kugrens. 2002. Phylogeny of *Lobocharacium* (Chlorophyceae) and allies: A study of 18S and 26S rDNA data. *Journal of Phycology* 38:376–383.

Buchheim, M. A., J. A. Buchheim, and R. L. Chapman. 1997. Phylogeny of *Chloromonas* (Chlorophyceae): A study of 18S ribosomal RNA gene sequences. *Journal of Phycology* 33:286–293.

Buchheim, M. A., and R. L. Chapman. 1992. Phylogeny of the genus *Carteria* (Chlorophyta) inferred from organismal and molecular evidence. *Journal of Phycology* 28:362–374.

Buchheim, M. A., E. A. Michalopulos, and J. A. Buchheim. 2001. Phylogeny of the Chlorophyceae with special reference to the Sphaeropleales: A study of the 18S and 26S rDNA data. *Journal of Phycology* 37:819–835.

Buchheim, M. A., D. L. Nickrent, and L. R. Hoffman. 1990. Systematic analysis of *Sphaeroplea* (Chlorophyceae). *Journal of Phycology* 26:173–181.

Buchheim, M., et al. 2005. Phylogeny of the Hydrodictyaceae (Chlorophyceae): Inferences from rDNA data. *Journal of Phycology* 41:1039–1054.

Buckley, D. H., and T. M. Schmidt. 2003. Diversity and dynamics of microbial communities in soils from agro-ecosystems. *Environmental Microbiology* 5:441–452.

Büdel, B. 1999. Ecology and diversity of rock-inhabiting cyanobacteria in tropical regions. *European Journal of Phycology* 34:361–370.

Büdel, B. 2001. Synopsis: Comparative biogeography and ecology of soil-crust biota. In: Belnap, J., and O. L. Lange (eds.), *Biological Soil Crusts: Structure, Function, and Management*, Ecological Studies, vol. 150, Springer Verlag, Berlin, pp. 141–154.

Büdel, B., U. Becker, S. Porembski, and W. Barthlott. 1997. Cyanobacteria and cyanobacterial lichens from inselbergs of the Ivory Coast, Africa. *Botanica Acta* 110:458–465.

Büdel, B., and O. L. Lange. 1994. The role of cortical and epinecral layers in the lichen genus *Peltula*. *Cryptogamic Botany* 4:262–269.

Büdel, B., U. Lüttge, R. Stelzer, O. Huber, and E. Medina. 1994. Cyanobacteria of rocks and soils of the Orinoco lowlands and the Guayana uplands, Venezuela. *Botanica Acta* 107:422–431.

Büdel, B., D. Mollenhauer, and R. Mollenhauer. 1991. *Synechococcus elongatus*–cryptoendolithic growth within bleached sandstone from creeks in the midland area Spessart (Germany). *Algological Studies* 64:357–360.

Büdel, B., B. Weber, M. Kühl, H. Pfanz, D. Sültemeyer, and D. Wessels. 2004. Reshaping of sandstone surfaces by cryptoendolithic cyanobacteria: Bioalkalization causes chemical weathering in arid landscapes. *Geobiology* 2:261–268.

Büdel, B., and D. C. J. Wessels. 1991. Rock inhabiting blue-green algae/cyanobacteria from hot arid regions. *Algological Studies* 64:385–398.

Buesseler, K. O., et al. 2008. Ocean iron fertilization—Moving forward in a sea of uncertainty. *Science* 319:162.

Buesseler, K. O., J. E. Andrews, S. M. Pike, and M. A. Charette. 2004. The effects of iron fertilization on carbon sequestration in the southern ocean. *Science* 304:414–417.

Buggeln, R. G., D. S. Fensom, and C. J. Emerson. 1985. Translocation of ^{11}C-photoassimilate in the blade of *Macrocystis pyrifera* (Phaeophyceae). *Journal of Phycology* 21:35–40.

Buick, R. 1992. The antiquity of oxygenic photosynthesis: Evidence from stromatolites in sulphate-deficient Archean lakes. *Science* 255:74–77.

Bungartz, F., L. A. J. Garvie, and T. H. Nash III. 2004. Anatomy of the endolithic Sonoran Desert lichen *Verrucaria rubrocincta* Breuss: Implications for biodeterioration and biomineralization. *The Lichenologist* 36:55–73.

Bungartz, F., and V. Wirth. 2007. *Buellia peregrina* sp. nov., a new, euendolithic calcicolous

lichen species from the Namib Desert. *The Lichenologist* 39:41–45.

Bunt, J. S. 1969. Observations on photoheterotrophy in a marine diatom. *Journal of Phycology* 5:37–42.

Burey, S. C., S. Fathi-Nejad, V. Poroyko, J. M. Steiner, W. Löffelhardt, and H. J. Bohnert. 2005. The central body of the cyanelles of *Cyanophora paradoxa*: A eukaryotic carboxysome? *Canadian Journal of Botany* 83:758–764.

Burger-Wiersma, T., M. Veenhuis, H. J. Korthals, C. C. M. VandeWiel, and L. R. Mar. 1986. A new prokaryote containing chlorophylls *a* and *b*. *Nature* 320:262–264.

Burghardt, I., and H. Wägele. 2004. A new solar powered species of the genus *Phyllodesmium* Ehrenberg, 1831 (Mollusca: Nudibranchia: Aeolidoidea) from Indonesia with analysis of its photosynthetic activity and notes on biology. *Zootaxa* 596:1–18.

Burkholder, J. M. 1996. Interactions of benthic algae with their substrata. In: Stevenson, R. J., M. L. Bothwell, and R. L. Lowe (eds.), *Algal Ecology: Freshwater Benthic Ecosystems*, Academic Press, New York, pp. 253–298.

Burkholder, J. M., et al. 2005. Demonstration of toxicity to fish and to mammalian cells by *Pfiesteria* species: Comparison of assay methods and strains. *Proceedings of the National Academy of Sciences of the United States of America* 102:3471–3476.

Burkholder, J. M., H. B. Glasgow, and A. J. Lewitus. 1998. Physiological ecology of *Pfiesteria piscicida* with general comments on "ambush-predator" dinoflagellates. In: Anderson, D. M., A. D. Cembella, and G. M. Hallegraeff (eds.), *Physiological Ecology of Harmful Algae Blooms*, Springer Verlag, Berlin, pp. 175–191.

Burkholder, J. M., E. J. Noga, C. H. Hobbs, and H. B. Glasgow. 1992. New "phantom" dinoflagellate is the causative agent of major estuarine fish kills. *Nature* 358:407–410.

Burkholder, J. M., and R. G. Sheath. 1984. The seasonal distribution, abundance and diversity of desmids (Chlorophyta) in a softwater, north temperate stream. *Journal of Phycology* 20:159–172.

Burkholder, J. M., and R. G. Wetzel. 1989. Epiphytic microalgae on a natural substratum in a phosphorous-limited hardwater lake: Seasonal dynamics of community structure biomass and ATP content. *Archiv für Hydrobiologie* 83:1–56.

Burki, F., K. Shalchian-Tabrizi, M. Minge, Å. Skjaeveland, S. I. Nikolaev, K. S. Jakobsen, and J. Pawlowski. 2007. Phylogenomics reshuffles the eukaryotic supergroups. *Public Library of Science One* 8:790.

Burns, C. W. 1968. The relationship between body size of filter-feeding cladocera and the maximum size of particles ingested. *Limnology and Oceanography* 13:675–678.

Burzycki, G. M., and J. R. Waaland. 1987. On the position of meiosis in the life history of *Porphyra torta* (Rhodophyta). *Botanica Marina* 30:5–10.

Buskey, E. J., and C. J. Hyaatt. 2006. Use of the FlowCAM for semi-automated recognition and enumeration of red tide cells (*Karenia brevis*) in natural plankton samples. *Harmful Algae* 5:685–692.

Busse, I., D. J. Patterson, and A. Preisfeld. 2003. Phylogeny of phagotrophic euglenids (Euglenozoa): A molecular approach based on culture material and environmental samples. *Journal of Phycology* 39:828–836.

Buttars, S. M., L. L. St. Clair, J. R. Johansen, J. C. Sray, M. C. Payne, B. L. Webb, R. E. Terry, B. K. Pendleton, and S. D. Warren. 1998. Pelletized cyanobacterial soil amendments: Laboratory testing for survival, escapability, and nitrogen fixation. *Arid Soil Research and Rehabilitation* 12:165–178.

Butterfield, N. J. 2000. *Bangiomorpha pubescens* n. gen., n. sp.: Implications for the evolution of sex, multicellularity and the Mesoproterozoic/Neoproterozoic radiation of eukaryotes. *Paleobiology* 26:386–404.

Butterfield, N. J. 2004. A vaucheriacean alga from the middle neoproterozoic of Spitsbergen: Implications for the evolution of Proterozoic eukaryotes and the Cambrian explosion. *Paleobiology* 30:231–252.

Butterfield, N. J., A. H. Knoll, and K. Swett. 1988. Exceptional preservation of fossils in an Upper Proterozoic shale. *Nature* 334:424–427.

Butterfield, N. J., A. H. Knoll, and K. Swett. 1990. A bangiophyte red alga from the Proterozoic of Arctic Canada. *Science* 250:104–107.

Calado, A. J., and Ø. Moestrup. 1997. Feeding in *Peridiniopsis berolinensis* (Dinophyceae): New observations on tube feeding by an omnivorous, heterotrophic dinoflagellate. *Phycologia* 36:47–59.

Calbet, A., and M. R. Landry. 2004. Phytoplankton growth, microzooplankton grazing, and carbon cycling in marine systems. *Limnology and Oceanography* 49:51–57.

Calderon-Saenz, E., and R. Schnetter. 1989. Life cycle and morphology of *Bryopsidella ostreobiformis* (spec. nov.) (Bryopsidaceae, Chlorophyta) from the Mediterranean under culture conditions, with comments on the phylogeny of the *Bryopsis/Derbesia* complex. *Botanica Acta* 102:249–260.

Callieri, C., S. M. Karjalainen, and S. Passoni. 2002. Grazing by ciliates and heterotrophic nanoflagellates on picocyanobacteria in Lago Maggiore, Italy. *Journal of Plankton Research* 24:785–796.

Callieri, C., and J. G. Stockner. 2002. Freshwater autotrophic picoplankton: A review. *Journal of Limnology* 61:1–14.

Callison, J., J. D. Brotherson, and J. E. Bowns. 1985. The effects of fire on the blackbrush (*Coleogyne ramosissima*) community of southwestern Utah. *Journal of Range Management* 38:535–538.

Callow, M. E., J. A. Callow, J. D. Pickett-Heaps, and R. Wetherbee. 1997. Primary adhesion of *Enteromorpha* (Chlorophyta, Ulvales) propagules: Quantitative settlement studies and video microscopy. *Journal of Phycology* 33:938–947.

Callow, M. E., S. J. Coughlan, and L. V. Evans. 1978. The role of Golgi bodies in polysaccharide sulphation in *Fucus* zygotes. *Journal of Cell Science* 32:337–356.

Cameron, R. E. 1972. Farthest south algae and associated bacteria. *Phycologia* 11:133–139.

Campàs, M., and J.-L. Marty. 2007. Highly sensitive amperometric immunosensors for microcystin detection in algae. *Biosensors and Bioelectronics* 22:1034–1040.

Cannell, R. J. P., P. Farmer, and J. M. Walker. 1988. Purification and characterization of pentagalloylglucose, a alpha-glucosidase inhibitor/antibiotic from a freshwater green alga *Spirogyra varians*. *Biochemical Journal* 255:937–941.

Canter-Lund, H., and J. W. G. Lund. 1995. *Freshwater Algae, Their Microscopic World Explored*. BioPress, Bristol, UK.

Capone, D. G., J. P. Zehr, H. W. Paerl, B. Bergman, and E. J. Carpenter. 1997. *Trichodesmium*, a globally significant marine cyanobacterium. *Science* 276:1221–1229.

Cardinale, M., A. M. Puglia, and M. Grube. 2006. Molecular analysis of lichen-associated bacterial communities. *Federation of European Microbiological Sciences Microbial Ecology* 57:484–495.

Cardol, P., et al. 2008. An original adaptation of photosynthesis in the marine green alga *Ostreococcus*. *Proceedings of the National Academy of Sciences of the United States of America* 105:7881–7886.

Carefoot, T. H. 1977. *Pacific Seashores. A Guide to Inter-tidal Ecology*. J. J. Douglas, Vancouver.

Carlton J. T., and J. B. Geller. 1993. Ecological roulette: The global transport of nonindigenous marine organisms. *Science* 261:78–82.

Caron, D. 1990. Carbon utilization by the omnivorous flagellate *Paraphysomonas inperforata*. *Limnology and Ocenography* 35:192–201.

Caron, D. A., K. G. Porter, and R. W. Sanders. 1990. Carbon, nitrogen, and phosphorus budgets for the mixotrophic phytoflagellate *Poterioochromonas malhamensis* (Chrysophyceae) during bacterial ingestion. *Limnology and Oceanography* 35:433–443

Carpenter, E. J., and K. Romans. 1991. Major role of the cyanobacterium *Trichodesmium* in nutrient cycling in the North Atlantic Ocean. *Science* 254:1356–1358.

Carpenter, R. C. 1990. Competition among marine macro-algae: A physiological perspective. *Journal of Phycology* 26:6–12.

Carpenter, R. C. 1997. Invertebrate predators. In: Birkeland, C. (ed.), *Life and Death of Coral Reefs*, Chapman & Hall, New York, pp. 198–229.

Carpenter, S. R. 1989. Temporal variance in lake communities: Blue-green algae and the trophic cascade. *Landscape Ecology* 3:175–184.

Carpenter, S. R. 2005. Eutrophication of aquatic ecosystems: Bistability and soil phosphorus. *Proceedings of the National Academy of Sciences of the United States of America* 102:10002–10005.

Carpenter, S. R., J. J. Cole, J. R. Hodgson, J. F. Kitchell, M. L. Pace, D. Bade, K. L. Cottingham, T. E. Essington, J. N. Houser, and D. E. Schindler. 2001. Trophic cascades, nutrients, and lake productivity: whole-lake experiments. *Ecological Monographs* 71:163–186.

Carpenter, S. R., and J. F. Kitchell. 1993. *The Trophic Cascade in Lakes*. Cambridge University Press, Cambridge, UK.

Carpenter, S. R., J. F. Kitchell, and J. R. Hodgson. 1985. Cascading trophic interactions and lake productivity. *BioScience* 35:634–639.

Carpenter, S. R., J. F. Kitchell, K. L. Cottingham, D. E. Schindler, D. L. Christensen, D. M. Post, and N. Voichick. 1996. Chlorophyll variability, nutrient input, and grazing: evidence from whole-lake experiments. *Ecology* 77:725–735.

Carr, M. H. 2000. Martian oceans, valleys and climate. *Astronomy and Geophysics* 41:320–326.

Carter, N. E. A., and H. A. Viles. 2005. Bioprotection explored: The story of a little known earth surface process. *Geomorphology* 67:273–281.

Carthew, R. W., and J. A. Hellebust. 1982. Transport of amino acids by the soil alga *Stichococcus bacillaris*. *Journal of Phycology* 18:441–446.

Carthew, R. W., and J. A. Hellebust. 1983. Regulation of a glucose transport system in *Stichococcus bacillaris*. *Journal of Phycology* 19:467–473.

Casamatta, D. A., R. G. Verb, J. R. Beaver, and M. L. Vis. 2002. An investigation of the cryptobiotic community from sandstone cliffs in southeast Ohio. *International Journal of Plant Science* 163:837–845.

Casanova, M. T., and M. A. Brock. 1996. Can oospore germination patterns explain charophyte distribution in permanent and temporary wetlands? *Aquatic Botany* 54:297–312.

Casper, E. T., J. H. Paul, M. C. Smith, and M. Gray. 2004. Detection and quantification of the red tide dinoflagellate *Karenia brevis* by real-time nucleic acid sequence-based amplification. *Applied and Environmental Microbiology* 70:4727–4732.

Catenazzi, A., and M. A. Donnelly. 2007. The *Ulva* connection: Marine algae subsidize terrestrial predators in coastal Peru. *Oikos* 116:75–86.

Cavacini, P. 2001. Soil algae from northern Victoria Land (Antarctica). *Polar Bioscience* 14:45–60.

Cavalier-Smith, T. 1981. Eukaryote kingdoms: Seven or nine? *BioSystems* 14:461–481.

Cavalier-Smith, T. 1993. Kingdom protozoa and its 18 phyla. *Microbiology and Molecular Biology Reviews* 57:953–994.

Cavalier-Smith, T. 1994. Origin and relationships of Haptophyta. In: Green, J. C., and B. S. C. Leadbeater (eds.), *The Haptophyte Algae*, Clarendon Press, Oxford, UK, pp. 413–436.

Cavalier-Smith, T. 1999. Zooflagellate phylogeny and the systematics of protozoa. *Biological Bulletin* 196:393–396.

Cavalier-Smith, T. 2003. The tiny enslaved genome of a rhizarian alga. *Proceedings of the National Academy of Sciences of the United States of America* 103:9379–9380.

Cavalier-Smith, T. 2006. The tiny enslaved genome of a rhizarian alga. *Proceedings of the National Academy of Sciences of the United States of America* 103:9379–9380.

Cavalier-Smith, T., and E. E.-Y. Chao. 2003. Phylogeny and classification of phylum Cercozoa (Protozoa). *Protist* 154:341–358.

Cavalier-Smith, T., and E. E. Chao. 1996. 18S rRNA sequence of *Heterosigma carterae* (Raphidophyceae) and the phylogeny of heterokont algae (Ochrophyta). *Phycologia* 35:500–510.

Chan, F., M. L. Pace, R. W. Howarth, and R. M. Marino. 2004. Bloom formation in het-

erocystic nitrogen-fixing cyanobacteria: The dependence on colony size and zooplankton grazing. *Limnology and Oceanography* 49:2171–2178.

Chapin III, F. S., et al. 2000. Consequences of changing biodiversity. *Nature* 405:234–242.

Chapman, A. D., and C. L. Schelske. 1997. Recent appearance of *Cylindrospermopsis* (Cyanobacteria) in five hypereutrophic Florida lakes. *Journal of Phycology* 33:191–195.

Chapman, A. R. O., and J. E. Lindley. 1980. Seasonal growth of *Laminaria solidungula* in the Canadian High Arctic in relation to irradiance and dissolved nutrient concentrations. *Marine Biology* 57:1–5.

Chapman, D. J., and F. T. Haxo. 1966. Chloroplast pigments of Chloromonadophyceae. *Journal of Phycology* 2:89–91.

Chapman, R. L. 1976. Ultrastructure of *Cephaleuros virescens* (Chroolepidaceae: Chlorophyta). I. Scanning electron microscopy of zoosporangia. *American Journal of Botany* 63:1060–1070.

Chapman, R. L. 1980. Ultrastructure of *Cephaleuros virescens* (Chroolepidaceae: Chlorophyta). II. Gametes. *American Journal of Botany* 67:10–17.

Chapman, R. L., and B. H. Good. 1983. Subaerial symbiotic green algae: Interactions with vascular plant hosts. In: Goff, L. J. (ed.), *Algal Symbiosis: A Continuum of Interaction Strategies*, Cambridge University Press, New York, pp. 173–203.

Chapman, R. L., O. L. Borkhsenious, R. C. Brown, M. C. Henk, and D. A. Waters. 2001. Phragmoplast-mediated cytokinesis in *Trentepohlia*: Results of TEM and immunofluorescence cytochemistry. *International Journal of Systematic and Evolutionary Microbiology* 51:759–765.

Chapman, R. L., and M. C. Henk. 1986. Phragmoplasts in cytokinesis of *Cephaleuros parasiticus* (Chlorophyta) vegetative cells. *Journal of Phycology* 22:83–88.

Chapman, R. L., and D. A. Waters. 1992. Epi- and endobiotic chlorophytes. In: Reisser, W. (ed.), *Algae and Symbioses*, BioPress, Bristol, UK, pp. 619–640.

Chasson, W. B., K. G. Johanson, A. R. Sherwood, and M. L. Vis. 2007. Phylogenetic affinities of the form taxon *Chantransia pygmaea* (Rhodophyta) specimens from the Hawaiian Islands. *Phycologia* 46:257–262.

Chen, F., C. A. Suttle, and S. M. Short. 1996. Genetic diversity in marine algal virus communities as revealed by sequence analysis of DNA polymerase genes. *Applied and Environmental Biology* 62:2869–2874.

Chen, J., H.-P. Blume, and L. Beyer. 2000. Weathering of rocks induced by lichen colonization—A review. *Catena* 39:121–146.

Chepurnov, V. A., and D. G. Mann. 2003. Auxosporulation of *Licmophora communis* (Bacillariophyta) and a review of mating systems and sexual reproduction in araphid pennate diatoms. *Phycological Research* 51:1–12.

Cheshire, A. C., J. G. Conran, and N. D. Hallam. 1995. A cladistic analysis of the evolution and biogeography of *Durvillaea* (Phaeophyta). *Journal of Phycology* 31:644–655.

Chesnick, J. M., and E. R. Cox. 1986. Specialization of endoplasmic reticulum architecture in response to a bacterial symbiosis in *Peridin-*

ium balticum (Pyrrhophyta). *Journal of Phycology* 22:291–298.

Chesnick, J. M., and E. R. Cox. 1987. Synchronized sexuality of an algal symbiont and its dinoflagellate host, *Peridinium balticum* (Levander) Lemmermann. *BioSystems* 21:69–78.

Chesnick, J. M., and E. R. Cox. 1989. Fertilization and zygote development in the binucleate dinoflagellate *Peridinium balticum* (Pyrrhophyta). *American Journal of Botany* 76:1060–1072.

Chesnick, J. M., M. Goff, J. Graham, C. Ocampo, B. F. Lang, E. Self, and G. Burger. 2000. The mitochondrial genome of the stramenopile alga *Chrysodidymus synuroideus*. Complete sequence, gene content and genome organization. *Nucleic Acids Research* 28:2512–2518.

Chesnick, J. M., W. H. C. F. Kooistra, U. Wellbrock, and L. K. Medlin. 1997. Ribosomal RNA analysis indicates a benthic pennate diatom ancestry for the endosymbionts of the dinoflagellates *Peridinium foliaceum* and *Peridinium balticum* (Pyrrhophyta). *Journal of Eukaryotic Microbiology* 44:314–320.

Chesnick, J. M., C. W. Morden, and A. M. Schmeig. 1996. Identity of the endosymbiont of *Peridinium foliaceum* (Pyrrophyta): Analysis of the *rbcLS* operon. *Journal of Phycology* 32:850–857.

Chiodini, P. L. 1994. A "new" parasite: Human infection with *Cyclospora cayetanensis*. *Transactions of the Royal Society of Tropical Medicine and Hygiene* 88:369–371.

Chisholm, J. R. M. 2003. Primary productivity of reef-building crustose coralline algae. *Limnology and Oceanography* 48:1376–1387.

Chisholm, S. W., P. G. Falkowski, and J. J. Cullen. 2001. Discrediting ocean fertilization. *Science* 294:309–310.

Chisholm, S. W., S. I. Frankel, R. Goericke, R. J. Olson, J. B. Waterbury, L. West-Johnson, and E. R. Settler. 1992. *Prochlorococcus marinus* nov. gen. et nov. sp.: An oxy-phototrophic marine prokaryote containing divinyl chlorophyll *b*. *Archiv für Microbiologie* 157:297–300.

Cho, G. Y., S. H. Lee, and S. M. Boo. 2004. A new brown algal order, Ishigeales (Phaeophyceaa), established on the basis of plastid protein-coding *rbcL*, *psaA*, and *psbA* region comparisons. *Journal of Phycology* 40:921–936.

Choat, J. H., and K. D. Clements. 1998. Vertebrate herbivores in marine and terrestrial environments: A nutritional ecology perspective. *Annual Review of Ecology and Systematics* 29:375–403.

Chopin, T., C. J. Bird, C. A. Murphy, J. A. Osborne, M. U. Patwary, and J.-Y. Floc'h. 1996. A molecular investigation of polymorphism in the North Atlantic red alga *Chondrus crispus* (Gigartinales). *Phycological Research* 44:69–80.

Chrétiennot-Dinet, M.-J., M.-M. Giraud-Guille, D. Vaulot, J.-L. Putaux, Y. Saito, and H. Chanzy. 1997. The chitinous nature of filaments ejected by *Phaeocystis* (Prymnesiophyceae). *Journal of Phycology* 33:666–672.

Christie, P. 1987. Nitrogen in two contrasting Antarctic bryophyte communities. *Journal of Ecology* 75:73–93.

Ciciotte, S. L., and R. J. Thomas. 1997. Carbon exchange between *Polysiphonia lanosa* (Rhodophyceae) and its brown algal host. *American Journal of Botany* 84:1614–1616.

Cimino, M. T., and C. F. Delwiche. 2002. Molecular and morphological data identify a cryptic species complex in endophytic members of the genus *Coleochaete* Bréb. (Charophyta: Coleochaetaceae). *Journal of Phycology* 38:1213–1221.

Clayton, M. N. 1984. Evolution of the Phaeophyta with particular reference to the Fucales. *Progress in Phycological Research* 3:11–46.

Clayton, M. N. 1987. Isogamy and a fucalean type of life cycle in the Antarctic brown alga *Ascoseira mirabilis* (Ascoseirales, Phaeophyta). *Botanica Marina* 30:447–454.

Clayton, M. N. 1990. Phaeophyta. In: Margulis, L., C. O. Corliss, M. Melkonian, and D. J. Chapman (eds.), *Handbook of the Protoctista*, Jones & Bartlett Publishers, Boston, pp. 698–714.

Clayton, M. N., N. D. Hallam, and C. M. Shankly. 1987. The seasonal pattern of conceptacle development and gamete maturation in *Durvillaea potatorum* (Durvillaeales, Phaeophyta). *Phycologia* 26:35–45.

Clayton, M. N., K. Kevekordes, M. E. A. Schoenwaelder, C. E. Schmid, and C. M. Ashburner. 1998. Parthogenesis in *Hormosira banksii*. *Botanica Marina* 41:23–30.

Clayton, M. N., and C. Wiencke. 1990. The anatomy, life history and development of the Antarctic brown alga *Phaeurus antarcticus* (Desmarestiales, Phaeophyceae). *Phycologia* 29:303–315.

Clegg, M. R., S. C. Maberly, and R. I. Jones. 2003. Behavioural responses of freshwater phytoplanktonic flagellates to a temperature gradient. *European Journal of Phycology* 38:195–203.

Clifton, K. E. 1997. Mass spawning by green algae on coral reefs. *Science* 275:1116–1118.

Cmiech, H. A., G. F. Leedale, and C. S. Reynolds. 1984. Morphological and ultrastructural variability of planktonic cyanophyceae in relation to seasonal periodicity. I. *Gloeotrichia echinulata*: Vegetative cells, polarity, heterocysts, akinetes. *British Phycological Journal* 19:259–275.

Coale, K. H., et al. 2004. Southern Ocean iron enrichment experiment: Carbon cycling in high- and low-Si waters. *Science* 304:408–414.

Coats, D. W., and M. G. Park. 2002. Parasitism of photosynthetic dinoflagellates by three strains of *Amoebophrya* (Dinophyta): Parasite survival, infectivity, generation time, and host specificity. *Journal of Phycology* 38:520–528.

Cockell, C. S., C. P. McKay, K. Warren-Rhodes, and G. Horneck. 2008. Ultraviolet radiation-induced limitation to epilithic microbial growth in arid deserts—Dosimetric experiments in the hyperarid core of the Atacama Desert. *Journal of Photochemistry and Photobiology B: Biology* 90:79–87.

Codd, G. A., J. Lindsay, F. M. Young, L. F. Morrison, and J. S. Metcalf. 2005a. Harmful cyanobacteria. From mass mortalities to management measures. In: Huisman, J., H. C. P. Matthijs, and P. M. Visser (eds.), *Harmful Cyanobacteria*, Springer, New York, pp. 1–19.

Codd, G. A., L. F. Morrison, and J. S. Metcalf. 2005b. Cyanobacterial toxins: Risk management for health protection. *Toxicology and Applied Pharmacology* 203:264–272.

Coesel, P. F. M. 1991. Ammonium dependency in *Closterium aciculare* T. West, a planktonic desmid from alkaline, eutrophic waters. *Journal of Plankton Research* 13:913–922.

Coffroth, M. A., and S. R. Santos. 2005. Genetic diversity of symbiotic dinoflagellates in the genus *Symbiodinium*. *Protist* 156:19–43.

Cogburn, J. N, and J. A. Schiff. 1984. Purification and properties of the mucus of *Euglena gracilis* (Euglenophyceae). *Journal of Phycology* 20:533–544.

Cohen, Y., B. B. Jorgensen, E. Padan, and M. Shilo. 1975. Sulphide-dependent anoxygenic photosynthesis in the cyanobacterium *Oscillatoria limnetica*. *Nature* 257:489–492.

Cohn, S., and R. E. Weitzell. 1996. Ecological characterization of diatom cell motility. 1. Characterization of motility and adhesion in four diatom species. *Journal of Phycology* 32:928–939.

Cole, J. J., N. F. Caraco, G. W. Kling, and T. W. Kratz. 1994. Carbon dioxide supersaturation in the surface waters of lakes. *Science* 265:1568–1570.

Coleman, A. W. 1983. The roles of resting spores and akinetes in chlorophyte survival. In: Fryxell, G. A. (ed.), *Survival Strategies of the Algae*, Cambridge University Press, Cambridge, UK, pp. 1–21.

Coleman, A. W. 1985. Diversity of plastid DNA configuration among classes of eukaryote algae. *Journal of Phycology* 21:1–16.

Coleman, A. W. 2001. Biogeography and speciation in the *Pandorina/Volvulina* (Chlorophyta) superclade. *Journal of Phycology* 37:836–851.

Coleman, M. L., M. B. Sullivan, A. C. Martiny, C. Steglich, K. Barry, E. DeLong, and S. W. Chisholm. 2006. Genomic islands and the ecology and evolution of *Prochlorococcus*. *Science* 311:1768–1770.

Coling, C. E., and J. L. Salisbury. 1992. Characterization of the calcium-binding contractile protein centrin from *Tetraselmis striata* (Pleurastrophyceae). *Journal of Protozoology* 39:385–391.

Collén, J., C. Hervé, I. Guisle-Marsollier, J. J. Léger, and C. Boyen. 2006a. Expression profiling of *Chondrus crispus* (Rhodophyta) after exposure to methyl jasmonate. *Journal of Experimental Botany* 57:3869–3881.

Collén, J., V. Roeder, S. Rousvoal, O. Collin, B. Kloareg, and C. Boyen. 2006b. An expressed sequence tag analysis of thallus and regenerating protoplasts of *Chondrus crispus* (Gigartinales, Rhodophyceae). *Journal of Phycology* 42:104–112.

Comte, K., D. T. Holland, and A. E. Walsby. 2007. Changes in cell turgor pressure related to uptake of solutes by *Microcystis* sp. strain 8401. *Federation of European Microbiological Sciences Microbial Ecology* 61:399–405.

Conklin, E. J., and J. E. Smith. 2005. Abundance and spread of the invasive red algae, *Kappaphycus* spp., in Kane'ohe Bay, Hawai'i and an experimental assessment of management options. *Biological Invasions* 7:1029–1039.

Connell, J. 1978. Diversity in tropical rainforests and coral reefs. *Science* 199:1304–1310.

Conway, K., and F. R. Trainor. 1972. *Scenedesmus* morphology and flotation. *Journal of Phycology* 8:138–143.

Cook, M. E. 2004a. Cytokinesis in *Coleochaete*

orbicularis (Charophyceae): An ancestral mechanism inherited by plants. *American Journal of Botany* 91:313–320.

Cook, M. E. 2004b. Structure and asexual reproduction of the enigmatic charophycean green alga *Entransia fimbriata* (Klebsormidiales, Charophyceae). *Journal of Phycology* 40:424–431.

Cook, M. E., and L. E. Graham. 1998. Structural similarities between surface layers of selected charophycean algae and bryophytes and the cuticles of vascular plants. *International Journal of Plant Science* 159:780–787.

Cook, M. E., and L. E. Graham. 1999. Evolution of plasmodesmata. In: A. van Bel and Kesteren, C. (ed.), *Plasmodesmata: Nanochannels with Megatasks*, Springer Verlag, Berlin, pp. 101–117.

Cook, M. E., L. E. Graham, C. E. J. Botha, and C. A. Lavin. 1997. Comparative ultrastructure of plasmodesmata of *Chara* and selected bryophytes: Toward an elucidation of the evolutionary origin of plant plasmodesmata. *American Journal of Botany* 84:1169–1178.

Cook, M. E., L. E. Graham, and C. A. Lavin. 1998. Cytokinesis and nodal anatomy in the charophycean green alga *Chara zeylanica*. *Protoplasma* 203:65–74.

Coomans, R. J., and M. H. Hommersand. 1990. Vegetative growth and organization. In: Cole, K. M., and R. G. Sheath (eds.), *Biology of the Red Algae*, Cambridge University Press, Cambridge, UK, pp. 275–304.

Correa, J. A. 1997. Infectious diseases of marine algae: Current knowledge and approaches. *Progress in Phycological Research* 12:149–180.

Correa, J. A., and P. Sánchez. 1996. Ecological aspects of algal infectious diseases. *Hydrobiologia* 326/327:89–95.

Cosson, J., E. Deslandes, M. Zinoun, and A. Mouradi-Givernaud. 1995. Carrageenans and agars, red algal polysaccharides. *Progress in Phycological Research* 11:269–324.

Cottingham, K. L., S. R. Carpenter, and A. L. St. Amand. 1998. Responses of epilimnetic phytoplankton to experimental nutrient enrichment in three small seepage lakes. *Journal of Plankton Research* 20:1889–1914.

Cottrell, M. T., and C. A. Suttle. 1995. Dynamics of a lytic virus infecting the photosynthetic marine picoflagellate *Micromonas pusilla*. *Limnology and Oceanography* 40:730–739.

Countway, P. D., R. J. Gast, P. Savai, and D. A. Caron. 2005. Protistan diversity estimates based on 18S rDNA from seawater incubations in the Western North Atlantic. *Journal of Eukaryotic Microbiology* 52:95–106.

Courties, C., A. Vaquer, M. Troussellier, J. Lautier, M. J. Chrétiennot-Dinet, J. Neveux, C. Machado, and H. Claustre. 1994. Smallest eukaryotic organism. *Nature* 370:255.

Cox, E. R., and H. C. Bold. 1966. Phycological studies VII. Taxonomic investigations of *Stigeoclonium*. *University of Texas Publication* 6618.

Cox, P. A., et al. 2005. Diverse taxa of cyanobacteria produce β-N-methylamino-L-alanine, a neurotoxic amino acid. *Proceedings of the National Academy of Sciences of the United States of America* 102:5074–5078.

Coyer, J. A., G. Hoarau, M.-P. Oudot-Le Secq, W. T. Stam, and J. L. Olsen. 2006. A mtDNA-based phylogeny of the brown algal genus *Fucus* (Heterokontophyta: Phaeophyta). *Molecular Phylogenetics and Evolution* 39:209–222.

Coyer, J. A., J. L. Olsen, and W. T. Stam. 1997. Genetic variability and spatial separation in the sea palm kelp *Postelsia palmaeformis* (Phaeophyceae) as assessed with M13 fingerprints and RAPDs. *Journal of Phycology* 33:561–568.

Coyer, J. A., D. L. Robertson, and R. S. Alberte. 1994. Genetic variability within a population and between diploid/haploid tissue of *Macrocystis pyrifera* (Phaeophyceae). *Journal of Phycology* 30:545–552.

Craggs, R. J. 2001. Wastewater treatment by algal turf scrubbing. *Water Science and Technology* 44:427–433.

Craggs, R. J., W. H. Adey, K. R. Jensen, M. S. St. John, F. B. Green, and W. J. Oswald. 1996. Phosphorus removal from waste water using an algal turf scrubber. *Water Science and Technology* 33:191–198.

Craggs, R. J., J. P. Sukias, C. T. Tanner, and R. J. Davies-Colley. 2004. Advanced pond system for dairy-farm effluent treatment. *New Zealand Journal of Agricultural Research* 47:449–460.

Craigie, J. S. 1990. Cell walls. In: Cole, K. M., and R. G. Sheath (eds.), *Biology of the Red Algae*, Cambridge University Press, Cambridge, UK, pp. 221–258.

Crawford, R. M. 1995. The role of sex in the sedimentation of a marine diatom bloom. *Limnology and Oceanography* 40:200–204.

Creed, J. C. 1995. Spatial dynamics of a *Himanthalia elongata* (Fucales, Phaeophyta) population. *Journal of Phycology* 31:851–859.

Croasdale, H., C. E. deM. Bicudo, and G. W. Prescott. 1983. *A Synopsis of North American Desmids. Part II. Desmidiaceae: Placodermae, Section 5*, University of Nebraska Press, Lincoln, NE.

Croft, M. T., A. D. Lawrence, E. Raux-Deery, M. J. Warren, and A. G. Smith. 2005. Algae acquire vitamin B$_{12}$ through a symbiotic relationship with bacteria. *Nature* 438:90–93.

Cronberg, G. 1995. *Mallomonas variabilis*, sp. nov. (Synurophyceae) with stomatocysts found in Lake Konneresi, Finland. In: Sandgren, C. D., J. P. Smol, and J. Kristiansen (eds.), *Chrysophyte Algae. Ecology, Phylogeny and Development*, Cambridge University Press, New York, pp. 333–344.

Cronberg, G. 2005. The life cycle of *Gonyostomum semen* (Raphidophyceae). *Phycologia* 44:285–293.

Crossland, C. J., B. G. Hatcher, and S. V. Smith. 1991. Role of coral reefs in global ocean production. *Coral Reefs* 10:55–64.

Cruz-Rivera, E., and M. E. Hay. 2003. Prey nutritional quality interacts with chemical defenses to affect consumer feeding and fitness. *Ecological Monographs* 73:483–506.

Cuker, B. E. 1983. Grazing and nutrient interactions in controlling the activity and composition of the epilithic algal community of an arctic lake. *Limnology and Oceanography* 28:133–141.

Czygan, F. C., and K. Kalb. 1966. Untersuchungen zur Biogenese der Carotinoide in *Trentepohlia aurea*. *Zeitschrift Pflanzenphysiologie* 55:59–64.

Daft, M. J., S. B. McCord, and W. D. P. Stewart. 1975. Ecological studies on algal-lysing bacteria in fresh waters. *Freshwater Biology* 5:577–596.

Dale, B., C. M. Yentsch, and J. W. Hurst. 1978. Toxicity in resting cysts of the red-tide dinoflagellate *Gonyaulax excavata* from deeper water coastal sediments. *Science* 201:1223–1225.

Danin, A., and G. Caneva. 1990. Deterioration of limestone walls in Jerusalem and marble monuments in Rome caused by cyanobacteria and cyanopilous lichens. *International Biodeterioration* 26:397–417.

Darley, W. M. 1982. *Algal Biology: A Physiological Approach*. Blackwell Scientific, Oxford, UK.

Darley, W. M., and B. E. Volcani. 1969. Role of silicon in diatom metabolism. A silicon requirement for deoxyribonucleic acid synthesis in the diatom *Cylindrotheca fusiformis* Reimann and Lewin. *Experimental Cell Research* 58:334–342.

Dason, J. S., I. E. Huertas, and B. Colman. 2004. Source of inorganic carbon for photosynthesis in two marine dinoflagellates. *Journal of Phycology* 40:285–292.

Daugbjerg, N., Ø. Moestrup, and P. Arctander. 1994. Phylogeny of the genus *Pyramimonas* (Prasinophyceae, Chlorophyta) inferred from the *rbcL* gene. *Journal of Phycology* 30:991–999.

Daugbjerg, N. 1996. *Mesopedinella arctica* gen. et sp. nov. (Pedinellales, Dictyochophyceae) I: Fine structure of a new marine phytoflagellate from Arctic Canada. *Phycologia* 35:435–445.

Daugbjerg, N., and R. A. Andersen. 1997. A molecular phylogeny of the heterokont algae based on analysis of chloroplast-encoded *rbcL* sequence data. *Journal of Phycology* 33:1031–1041.

Daugbjerg, N., and L. Guillou. 2001. Phylogenetic analysis of Bolidophyceae (Heterokontophyta) using *rbcL* gene sequences support their sister group relationship to diatoms. *Phycologia* 40:153–161.

Daugbjerg, N., Ø. Moestrup, and P. Arctander. 1995. Phylogeny of genera of Prasinophyceae and Pedinophyceae (Chlorophyta) deduced from molecular analysis of the *rbcL* gene. *Phycological Research* 43:203–213.

Davey, A., and H. J. Marchant. 1983. Seasonal variation in nitrogen fixation by *Nostoc commune* Vaucher at the Vestfold Hills, Antarctica. *Phycologia* 4:377–385.

Davey, M. C. 1991. The seasonal periodicity of algae on Antarctic fellfield soils. *Holarctic Ecology* 14:112–120.

Davey, M. C., and P. Rothery. 1992. Factors causing the limitation of growth of terrestrial algae in maritime Antarctica during the summer. *Polar Biology* 12:595–601.

Davis, L. S., J. P. Hoffmann, and P. W. Cook. 1990a. Seasonal succession of algal periphyton from a wastewater treatment facility. *Journal of Phycology* 26:611–617.

Davis, L. S., J. P. Hoffmann, and P. W. Cook. 1990b. Production and nutrient accumulation by periphyton in a wastewater treatment facility. *Journal of Phycology* 26:617–623.

Dawes, C. J. 1998. *Marine Botany*, John Wiley, New York.

Dawson, N. S., and P. L. Walne. 1991a. Structural characterization of *Eutreptia pertyi*

(Euglenophyta). I. General description. *Phycologia* 30:287–302.

Dawson, N. S., and P. L. Walne. 1991b. Structural characterization of *Eutreptia* (Euglenophyta). III. Flagellar structure and possible function of the paraxial rods. *Phycologia* 30:415–437.

Dawson, N. S., and P. L. Walne. 1994. Evolutionary trends in euglenoids. *Archiv für Protistenkunde* 144:221–225.

Dayton, P. K. 1975. Experimental evaluation of ecological dominance in a rocky intertidal algal community. *Ecological Monographs* 45:137–159.

de Baar, H. J. W., J. T. M. deJong, D. C. E. Bakker, B. M. Löscher, C. Veth, U. Bathman, and V. Smetacek. 1995. Importance of iron for plankton blooms and carbon dioxide drawdown in the Southern Ocean. *Nature* 373:412–415.

de Brouwer, J. F. C., K. Wolfstein, G. K. Ruddy, T. E. R. Jones and L, J. Stal. 2005. Biogenic stabilization of intertidal sediments: The importance of extracellular polymeric substances produced by benthic diatoms. *Microbial Ecology* 49:501–512.

de Bruin, A., B. W. Ibelings, and E. Van Donk. 2003. Molecular techniques in phytoplankton research: From allozyme electrophoresis to genomics. *Hydrobiologia* 491:47–63.

De Cambiaire, J.-C., C. Otis, C. Lemieux, and M. Turmel. 2006. The complete chloroplast genome sequence of the chlorophycean green alga *Scenedesmus obliquus* reveals a compact gene organization and a biased distribution of genes on the two DNA strands. *BMC Evolutionary Biology* 6:37.

de Koning, A. P., and P. J. Keeling. 2006. The complete plastid genome sequence of the parasitic green alga *Helicosporidium* sp. is highly reduced and structured. *BMC Biology* 4:12.

De Nobel, W. T., J. Huisman, J. L. Snoep, and L. R. Mur. 1997. Competition for phosphorus between the nitrogen-fixing cyanobacteria *Anabaena* and *Aphanizomenon*. *FEMS Microbiology and Ecology* 24:259–267.

de Vargas, C., M.-P. Aubry, I. Probert, and J. Young. 2007. Origin and evolution of coccolithophores: From coastal hunters to oceanic farmers. In: Falkowski, P. C., and A. H. Knoll (eds.), *Evolution of Primary Producers of the Sea*, Academic Press, New York, pp. 251–287.

De Vooys, C. G. N. 1979. Primary production in aquatic environments. In: Bolin, B., E. T. Degens, S. Kempe, and P. Ketner (eds.), *The Global Carbon Cycle*, Scientific Committee on Problems of the Environment (SCOPE) vol. 13, John Wiley & Sons, New York, pp. 259–292.

Deane, J. A., I. M. Strachan, G. W. Saunders, D. R. A. Hill, and G. I. McFadden. 2002. Cryptomonad evolution: Nuclear 18S rDNA phylogeny versus cell morphology and pigmentation. *Journal of Phycology* 38:1236–1244.

Deason, T. R., P. C. Silva, S. Watanabe, and G. L. Floyd. 1991. Taxonomic status of the species of the green algal genus *Neochloris*. *Plant Systematics and Evolution* 177:213–219.

DeBoer, J. A., H. J. Guigli, T. H. Israel, and C. F. D'Elia. 1978. Nutritional studies of two red algae. I. Growth rate as a function of nitrogen source and concentration. *Journal of Phycology* 14:261–266.

Deines, L., R. Rosentreter, D. J. Eldridge, and M. D. Serpe. 2007. Germination and seedling establishment of two annual grasses on lichen-dominated biological soil crusts. *Plant and Soil* 295:23–35.

DeJesus, M. D., F. Tabatabai, and D. J. Chapman. 1989. Taxonomic distribution of copper-zinc superoxide dismutase in green algae and its phylogenetic importance. *Journal of Phycology* 25:767–772.

Delgado, O., C. Rodriguez-Prieto, E. Gacia, and E. Ballesteros. 1996. Lack of severe nutrient limitation in *Caulerpa taxifolia* (Vahl) C. Agardh, an introduced seaweed spreading over the oligotrophic northeastern Mediterranean. *Botanica Marina* 39:61–67.

DeLuca, T. H., O. Zackrisson, F. Gentili, A. Sellstedt, and M.-C. Nilsson. 2007. Ecosystem controls on nitrogen fixation in boreal feather moss communities. *Oecologia* 152:121–130.

DeLuca, T. H., O. Zackrisson, M. J. Gundale, and M.-C. Nilsson. 2008. Ecosystem feedbacks and nitrogen fixation in boreal forests. *Science* 320:1181.

DeLuca, T. H., O. Zackrisson, M.-C. Nilsson, and A. Sellstedt. 2002. Quantifying nitrogen-fixation in feather moss carpets of boreal forests. *Nature* 419:917–920.

Delwiche, C. F. 1999. Tracing the thread of plastid diversity through the tapestry of life. *American Naturalist* 154:S164–S167.

Delwiche, C. F. 2007. The origin and evolution of dinoflagellates. In: Falkowski, P. G., and A. H. Knoll (eds.), *Evolution of Primary Producers of the Sea*, Academic Press, New York, pp. 191–205.

Delwiche, C. F., L. E. Graham, N. Thomson. 1989. Lignin-like compounds and sporopollenin in *Coleochaete*, an algal model for land plant ancestry. *Science* 245:399–401.

Delwiche, C.F., K. G. Karol, M. T. Cimino, and K. J. Sytsma. 2002. Phylogeny of the genus *Coleochaete* (Coleochaetales, Charophyta) and related taxa inferred by analysis of the chloroplast gene *rbcL*. *Journal of Phycology* 38:394–403.

Delwiche, C. F., and J. D. Palmer. 1996. Rampant horizontal transfer and duplication of Rubisco genes in eubacteria and plastids. *Molecular Biology and Evolution* 13:873–882.

DeMott, W., and F. Moxter. 1991. Foraging on cyanobacteria by copepods: Responses to chemical defenses and resource abundance. *Ecology* 72:1820–1834.

DeNicola, D. M., and C. D. McIntire. 1990a. Effects of substrate relief on the distribution of periphyton in laboratory streams. I. Hydrology. *Journal of Phycology* 26:624–633.

DeNicola, D. M., and C. D. McIntire. 1990b. Effects of substrate relief on the distribution of periphyton in laboratory streams. II. Interactions with irradiance. *Journal of Phycology* 26:634–641.

Denny, M. W. 1988. *Biology and the Mechanics of the Wave-Swept Environment*. Princeton University Press, Princeton, NJ.

Derelle, E., et al. 2006. Genome analysis of the smallest free-living eukaryote *Ostreococcus tauri* unveils many unique features. *Proceedings of the National Academy of Sciences of the United States of America* 103:11647–11652.

DeVries, P. J. R., J. Simons, and A. P. VanBeem. 1983. Sporopollenin in the spore wall of *Spir-*

ogyra (Zygnemataceae, Chlorophyceae). *Acta Botanica Neerlandica* 32:25–28.

deVrind-deJong, E. W., P. R. van Emburg, and J. P. M. deVrind. 1994. Mechanisms of calcification: *Emiliania huxleyi* as a model system. In: Green, J. C., and B. S. C. Leadbeater (eds.), *The Haptophyte Algae*, Clarendon Press, Oxford, UK, pp. 149–166.

DeYoe, H. R., R. L. Lowe, and J. C. Marks. 1992. Effects of nitrogen and phosphorus on the endosymbiont load of *Rhopalodia gibba* and *Epithemia turgida* (Bacillariophyceae). *Journal of Phycology* 28:773–777.

DeYoe, H. R., D. A. Stockwell, R. R. Bidigare, M. Latasa, P. W. Johnson, P. E. Hargraves, and C. A. Suttle. 1997. Description and characterization of the algal species *Aureoumbra lagunensis* gen. et sp. nov. and referral of *Aureoumbra* and *Aureococcus* to the Pelagophyceae. *Journal of Phycology* 33:1042–1048.

Diaz-Pulido, G., and L. J. McCook. 2003. Relative roles of herbivory and nutrients in the recruitment of coral-reef seaweeds. *Ecology* 84:2026–2033.

Dietrich, L. E. P., M. M. Tice, and D. K. Newman. 2006. The co-evolution of life and Earth. *Current Biology* 16:R395–R400.

Diouris, M. 1989. Long-distance transport of ^{14}C-labelled assimilates in the Fucales: Nature of translocated substances in *Fucus serratus*. *Phycologia* 28:504–511.

Dodds, W. K., and D. A. Gudder. 1992. The ecology of *Cladophora*. *Journal of Phycology* 28:415–427.

Dodds, W. K., D. A. Gudder, and D. Mollenhauer. 1995. The ecology of *Nostoc*. *Journal of Phycology* 31:2–18.

Dodge, J. D., and R. M. Crawford. 1970. A survey of thecal fine structure in the Dinophyceae. *Botanical Journal of the Linnean Society* 63:53–67.

Doemel, W. N., and T. D. Brock. 1971. The physiological ecology of *Cyanidium caldarium*. *Journal of General Microbiology* 67:17–32.

Doers, M. P., and D. L. Parker. 1988. Properties of *Microcystis aeruginosa* and M. *flos-aquae* (Cyanophyta) in culture: Taxonomic implications. *Journal of Phycology* 24:502–508.

Dolezal, P., V. Likic, J. Tachezy, and T. Lithgow. 2006. Evolution of the molecular machines for protein import into mitochondria. *Science* 313:314–318.

Dominic, B., Zani, S., Y. Chen, M. T. Mellon, and J. P. Zehr. 2000. Organization of the *nif* genes of the nonheterocystous cyanobacterium *Trichodesmium* sp. IMS101. *Journal of Phycology* 36:693–701.

Domozych, C. R., K. Plante, P. Blais, L. Paliulis, and D. S. Domozych. 1993. Mucilage processing and secretion in the green alga *Closterium*. I. Cytology and biochemistry. *Journal of Phycology* 29:650–659.

Domozych, D. S. 1984. The crystalline cell wall of *Tetraselmis convolutae* (Chlorophyta): A freeze fracture analysis. *Journal of Phycology* 20:415–418.

Domozych, D. S., and C. R. Domozych. 1993. Mucilage processing and secretion in the green alga *Closterium*. II. Ultrastructure and immunocytochemistry. *Journal of Phycology* 29:659–667.

Doolittle, W. F., and E. Bapteste. 2007. Pattern pluralism and the Tree of Life hypothesis. *Proceedings of the National Academy of Sciences of the United States of America* 104:2043–2049.

Doran, E., and R. A. Cattolico. 1997. Photoregulation of chloroplast gene transcription in the chromophytic alga *Heterosigma carterae*. *Plant Physiology* 115:773–781.

Doty, M. S. 1946. Critical tide factors that are correlated with the vertical distribution of marine algae and other organisms along the Pacific Coast. *Ecology* 27:315–328.

Douglas, S., et al. 2001. The highly reduced genome of an enslaved algal nucleus. *Nature* 410:1091–1096.

Dow, C. S., and U. K. Swodboda. 2000. Toxin analysis. In: Whitton, B. A., and M. Potts (eds.), *The Ecology of Cyanobacteria*, Kluwer, Dordrecht, The Netherlands, Chapter V.

Dowling, T. E., C. Moritz, J. D. Palmer, and L. H. Rieseberg. 1996. Nucleic acids III: Analysis of fragments and restriction sites. In: Hillis, D. M., C. Moritz, and B. K. Mable (eds.), *Molecular Systematics*, Sinauer, Sunderland, MA, pp. 249–320.

Draisma, S. G. A., J. L. Olsen, W. T. Stam, and W. F. Prud'homme van Reine. 2002. Phylogenetic relationships within the Sphacelariales (Phaeophyceae): *rbcL*, rubisco spacer and morphlogy. *European Journal of Phycology* 37:385–401.

Draisma, S. G. A., W. F. Prud'homme van Reine, W. T. Stam, and J. L. Olsen. 2001. A reassessment of phylogenetic relationships within the Phaeophyceae based on rubisco large subunit and ribosomal DNA sequences. *Journal of Phycology* 37:586–603.

Drenner, R. W., D. J. Day, S. J. Basham, J. D. Smith, and S. I. Jensen. 1997. Ecological water treatment system for removal of phosphorus and nitrogen from polluted water. *Ecological Applications* 7:381–390.

Drew, K. M. 1949. *Conchocelis*-phase in the life history of *Porphyra umbilicalis* (L.) Kuetz. *Nature* 164:748–749.

Dring, M. J., and K. Lüning. 1975. A photoperiodic response mediated by blue light in the brown alga *Scytosiphon lomentaria*. *Planta* 125:25–32.

Dromgoole, F. I. 1990. Gas-filled structure, buoyancy and support in marine macro-algae. *Progress in Phycological Research* 7:169–211.

Droop, M. R. 1983. 25 years of algal growth kinetics. *Botanica Marina* 26:99–112.

Drum, R. W., and H. S. Pankratz. 1964. Pyrenoids, raphes, and other fine structure in diatoms. *American Journal of Botany* 51:405–418.

Drummond, C. S., J. Hall, K. G. Karol, C. F. Delwiche, and R. M. McCourt. 2005. Phylogeny of *Spirogyra* and *Sirogonium* (Zygnematophyceae) based on rbcL sequence data. *Journal of Phycology* 41:1055–1064.

Dube, M. A. 1967. On the life history of *Monostroma fuscum* (Postels et Ruprecht) Wittrock. *Journal of Phycology* 3:64–73.

Ducobu, H., J. Huisman, R. R. Jonker, and L. R. Mur. 1998. Competition between a prochlorophyte and a cyanobacterium under various phosphorus regimes: Comparison with the Droop model. *Journal of Phycology* 34:467–476.

Duff, K. E., B. A. Zeeb, and J. P. Smol. 1995. *Atlas of Chrysophycean Cysts*. Kluwer Academic Publishers, Dordrecht, The Netherlands.

Duffy, J. E., and M. E. Hay. 1990. Seaweed adaptations to herbivory. *BioScience* 40:368–375.

Duffy, J. E., and M. E. Hay. 1994. Herbivore resistance to seaweed chemical defense: The roles of mobility and predation risk. *Ecology* 75:1304–1319.

Duggan, P. S., P. Gottardello, and D. G. Adams. 2007. Molecular analysis of genes in *Nostoc punctiforme* involved in pilus biogenesis and plant infection. *Journal of Bacteriology* 189:4547–4551.

Duggins, D. O, C. A. Simenstad., and J. A. Estes. 1989. Magnification of secondary production by kelp detritus in coastal marine ecosystems. *Science* 245:170–173.

Dunahay, T. G., E. E. Jarvis, and P. E. Roessler. 1995. Genetic transformation of the diatoms *Cyclotella cryptica* and *Navicula saprophila*. *Journal of Phycology* 31:1004–1012.

Dunlap, J. R., P. L. Walne, and J. Bentley. 1983. Microarchitecture and elemental spatial segregation of envelopes of *Trachelomonas leferrei* (Euglenophyceae). *Protoplasma* 117:97–106.

Dunstan, G. A., M. R. Brown, and J. K. Volkman. 2005. Cryptophyceae and Rhodophyceae; chemotaxonomy, phylogeny and application. *Phytochemistry* 21:2557–2570.

Dunton, K. H. 1990. Growth and production in *Laminaria solidungula*: Relation to continuous underwater light levels in the Alaskan High Arctic. *Marine Biology* 106:297–304.

Dunton, K. H., and D. M. Schell. 1986. Seasonal carbon budget and growth of *Laminaria solidungula* in the Alaskan High Arctic. *Marine Ecology Progress Series* 31:57–66.

Durrell, L. W. 1964. Algae in tropical soils. *Transactions of the American Microscopical Society* 83:79–85.

Dutcher, S. K. 1988. Linkage group XIX in *Chlamydomonas reinhardtii* (Chlorophyceae): Genetic analysis of basal body function and assembly. In: A. W. Coleman, L. J. Goff, and J. R. Stein-Taylor (eds.), *Algae as Experimental Systems*, Alan R. Liss, New York, pp. 39–53.

Dutz, J., and M. Koski. 2006. Trophic significance of solitary cells of the prymnesiophyte *Phaeocystis globosa* depends on cell type. *Limnology and Oceanography* 51:1230–1238.

Duval, B., E. Duval, and R. W. Hoham. 1999. Snow algae of the Sierra Nevada, Spain, and High Atlas mountains of Morocco. *International Microbiology* 2:39–42.

Eckert, R. E., F. F. Peterson, M. S. Meurisse, and J. L. Stephens. 1986. Effects of soil-surface morphology on emergence and survival of seedlings in Big Sagebrush communities. *Journal of Range Management* 39:414–420.

Eckhardt, R., R. Schnetter, and G. Siebold. 1986. Nuclear behavior during the life cycle of *Derbesia* (Chlorophyceae). *British Phycological Journal* 21:287–295.

Edelstein, T., L. Chen, and J. McLachlan. 1968. Sporangia of *Ralfsia fungiformis* (Gunn) Setchell and Gardner. *Journal of Phycology* 4:157–160.

Edgar, L. A., and J. D. Pickett-Heaps. 1984a. Diatom locomotion. *Progress in Phycological Research* 3:47–88.

Edgar, L. A., and J. D. Pickett-Heaps. 1984b. Valve morphogenesis in the pennate diatom *Navicula cuspidata*. *Journal of Phycology* 20:47–61.

Edlund, M. B., and E. F. Stoermer. 1997. Ecological, evolutionary, and systematic significance of diatom life histories. *Journal of Phycology* 33:897–918.

Edvardsen, B., W. Eikkrem, J. C. Green, R. A. Andersen, S. Y. M. van der Staay, and L. K. Medlin. 2000. Phylogenetic reconstructions of the Haptophyta inferred from 18S ribosomal DNA sequences and available morphological data. *Phycologia* 39:19–35.

Edvardsen, B., W. Eikrem, K. Shalchian-Tabrizi, I. Riisberg, G. Johnsen, L. Naustvoll, and J. Throndsen. 2007. *Verrucophora farcimen* gen. et sp. nov. (Dictyochophyceae, Heterokonta)— A bloom forming icthyotoxic flagellate from the Skagerrak, Norway. *Journal of Phycology* 43:1054–1070.

Edvardsen, B., and E. Paasche. 1998. Bloom dynamics and physiology of *Prymnesium* and *Chrysochromulina*. In: Anderson, D. M., A. D. Cembella, and G. M. Hallegraeff (eds.), *Physiological Ecology of Harmful Algal Blooms*, Springer Verlag, Berlin, pp. 193–208.

Eggert, A., S. Raimund,, D. Michalik, J. West, and U. Karsten. 2007. Ecophysiological performance of the primitive red alga *Dixoniella grisea* (Rhodellophyceae) to irradiance, temperature and salinity stress: Growth responses and the osmotic role of mannitol. *Phycologia* 46:22–28.

Eigenbrode, J. L., and K. H. Freeman. 2006. Late Archean rise of aerobic microbial ecosystems. *Proceedings of the National Academy of Sciences of the United States of America* 103:15759–15764.

Eisen, J. A., et al. 2006. Macronuclear genome sequence of the ciliate *Tetrahymena thermophila*, a model eukaryote. *Public Library of Science Biology* 4:e286.

Elbrächter, M., and Y.-Z. Qi. 1998. Aspects of *Noctiluca* (Dinophyceae) population dynamics. In: Anderson, D. M., A. D. Cembella, and G. M. Hallegraeff (eds.), *Physiological Ecology of Harmful Algal Blooms*, Springer Verlag, Berlin, pp. 315–335.

Eldridge, D. J. 1993. Cryptogams, vascular plants, and soil hydrological relations: Some preliminary results from the semiarid woodlands of eastern Australia. *Great Basin Naturalist* 53:48–58.

Eldridge, D. J., and R. S. B. Greene. 1994. Assessment of sediment yield by splash erosion on a semi-arid soil with varying cryptogam cover. *Journal of Arid Environments* 26:221–232.

Eldridge, D. J., E. Zaady, and M. Shachak. 2000. Infiltration through three contrasting biological soil crusts in patterned landscapes in the Negev, Israel. *Catena* 40:323–336.

Eller, G., K. Töbe, and L. K. Medlin. 2007. Hierarchical probes at various taxonomic levels in the Haptophyta and a new division level probe for the Heterokonta. *Journal of Plankton Research* 29:629–640.

Elliot, J. A., A. E. Irish, and C. S. Reynolds. 2001. The effects of vertical mixing on a phytoplankton community: a modeling approach to the intermediate disturbance hypothesis. *Freshwater Biology* 46:1291–1297.

Elloranta, P. 1995. Biogeography of chrysophytes in Finnish lakes. In: Sandgren, C. D., J. P. Smol, and J. Kristiansen (eds.), *Chrysophyte Algae. Ecology, Phylogeny and Development*, Cambridge University Press, Cambridge, UK, pp. 214–231.

Elrad, D., and A. R. Grossman. 2004. A genome's eye view of the light-harvesting polypeptides of *Chlamydomonas reinhardtii*. *Current Genetics* 45:61–75.

Elser, J. J., and C. R. Goldman. 1991. Zooplankton effects on phytoplankton in lakes of contrasting trophic status. *Limnology and Oceanography* 36:64–90.

Embley, T. M., and Martin, W. 2006. Eukaryotic evolution, changes and challenges. *Nature* 440:623–630.

Engel, C. R., C. Daguin, and E. A. Serrão. 2005. Genetic entities and mating system in hermaphroditic *Fucus spiralis* and its close dioecious relative *F. vesiculosus* (Fucaceae, Phaeophyceae). *Molecular Ecology* 14:2033–2046.

Engelskjon, T. 1986. Botany of two Antarctic mountain ranges: Gjelsvikfjella and Muhlig-Hofmannfjella, Dronning Maud Land. I. General ecology and development of the Antarctic cold desert cryptogam formation. *Polar Research* 4:205–224.

Eppley, R. W., O. Holm-Hansen, and J. D. Strickland. 1968. Some observations on the vertical migrations of dinoflagellates. *Journal of Phycology* 4:333–340.

Epstein, S. S., and J. Rossel. 1995. Methodology of *in situ* grazing experiments: evaluation of a new vital dye for preparation of fluorescently labeled bacteria. *Marine Ecology Progress Series* 128:143–150.

Ermilova, E. V., Z. M. Zalutskaya, K. Huang, and C. F. Beck. 2004. Phototropin plays a crucial role in controlling changes in chemotaxis during the initial phase of the sexual life cycle in *Chlamydomonas*. *Planta* 219:420–427.

Estes, J. A., M. T. Tinker, T. M. Williams, and D. F. Doak. 1998. Killer whale predation on sea otters linking oceanic and nearshore ecosystems. *Science* 282:473–475.

Ettershank, G., J. Ettershank, M. Bryant, and W. Whitford. 1978. Effects of nitrogen fertilization on primary production in a Chihuahuan desert ecosystem. *Journal of Arid Environments* 1:135–139.

Evans, C., S. V. Kadner, L. J. Darroch, W. H. Wilson, P. S. Liss, and G. Malin. 2007. The relative significance of viral lysis and microzooplankton grazing as pathways of dimethylsulfoniopropionate (DMSP) cleavage: An *Emiliania huxleyi* culture study. *Limnology and Oceanography* 52:1036–1045.

Evans, L. V. 1965. Cytological studies in the Laminariales. *Annals of Botany* 29:541–562.

Evans, R. D., and J. R. Ehleringer. 1993. A break in the nitrogen cycle in aridlands? Evidence from $\delta^{15}N$ of soils. *Oecologia* 94:314-317.

Evans, R. D., and J. R. Johansen. 1999. Microbiotic crusts and ecosystem processes. *Critical Reviews in Plant Sciences* 18:183–225.

Fairchild, G. W., and J. W. Sherman. 1993. Algal periphyton response to acidity and nutrients in softwater lakes: lake comparison vs. nutrient enrichment approaches. *Journal of the North American Benthological Society* 12:157–167.

Falkowski, P. 1997. Evolution of the nitrogen cycle and its influence on the biological sequestration of CO_2 in the ocean. *Nature* 387:272–275.

Falkowski, P. G. 2006. Tracing oxygen's imprint on Earth's metabolic evolution. *Science* 311:1724–1725.

Falkowski, P. G., M. E. Katz, A. H. Knoll, A. Quigg, J. A. Raven, O. Schofield, and F. J. R. Taylor. 2004. The evolution of modern eukaryotic phytoplankton. *Science* 305:354–360.

Falkowski, P. G., and J. A. Raven. 1997. *Aquatic Photosynthesis*. Blackwell Science, Malden, MA.

Falkowski, P. G., and J. A. Raven. 2007. *Aquatic Photosynthesis*, 2nd edition, Princeton University Press, Princeton, N. J.

Farmer, M. A., and R. E. Triemer. 1994. An ultrastructural study of *Lentomonas applanatum* (Preisig) N. G. Euglenida. *The Journal of Eukaryotic Microbiology* 41:112–119.

Fast, N. M., J. C. Kissinger, D. S. Roos, and P. J. Keeling. 2001. Nuclear-encoded, plastid-targeted genes suggest a common origin of apicomplexan and dinoflagellate plastids. *Molecular Biology and Evolution* 18:418–426.

Faugeron, S., E. A. Martinez, J. A. Correa, L. Cardenas, C. Destombe, and M. Valero. 2004. Reduced genetic diversity and increased population differentiation in peripheral and overharvested populations of *Gigartina skottsbergii* (Rhodophyta, Gigartinales) in Chile. *Journal of Phycology* 40:454–462.

Faugeron, S., M. Valero, C. Destombe, E. A. Martinez, and J. A. Correa. 2001. Hierarchical spatial structure and discriminant analysis of genetic diversity in the red alga *Mazzaella laminariodes* (Gigartinales, Rhodophyta). *Journal of Phycology* 37:705–716.

Faulkner, D. J. 1993. Marine natural products. *Natural Products Reports* 10:497–539.

Fawley, K. P., and M. W. Fawley. 2007. Observations on the diversity and ecology of freshwater *Nannochloropsis* (Eustigmatiophyceae), with descriptions of new taxa. *Protist* 158:325–336.

Fawley, M. W., K. P. Fawley, and M. A. Buchheim. 2004. Molecular diversity among communities of freshwater microchlorophytes. *Microbial Ecology* 48:489–499.

Fawley, M. W., and C. M. Lee. 1990. Pigment composition of the scaly green flagellate *Mesostigma viride* (Micromonadophyceae) is similar to that of the siphonous green alga *Bryopsis plumosa* (Ulvophyceae). *Journal of Phycology* 26:666–670.

Fawley, M. W., Y. Yun, and M. Qin. 2000. Phylogenetic analyses of 18S rDNA sequences reveal a new coccoid linage of the Prasinophyceae (Chlorophyta). *Journal of Phycology* 36:387–393.

Feely, R. A., C. L. Sabine, K. Lee, W. Berelson, J. Kleypas, V. J. Fabry, and F. J. Millero. 2004. Impact of anthropogenic CO_2 on the $CaCO_3$ system in the oceans. *Science* 305:362–371.

Feist, M., J. Liu, and P. Tafforeau. 2005. New insights into Paleozoic charophyte morphology and phylogeny. *American Journal of Botany* 92:1152–1160.

Feldmann, J. 1951. Ecology of marine algae. In: Smith, G. M. (ed.), *Manual of Phycology*, Chronica Botanica, Waltham, MA, pp. 313–334.

Felicini, G. P., and C. Perrone. 1994. *Pterocladia*. In: Akatsuka, I. (ed.), *Biology of Economic Algae*, SPB Academic Publishing, The Hague, The Netherlands, pp. 283–344.

Felsenstein, J. 1985. Confidence limits in phylogenies: An approach using the bootstrap. *Evolution* 39:783–791.

Felsenstein, J. 1988. Phylogenies from molecular sequences: Inference and reliability. *Annual Review of Genetics* 22:521–565.

Felsenstein, J. 2004. *Inferring Phylogenies*. Sinauer, Sunderland, PA.

Fenchel, T. 2001. How dinoflagellates swim. *Protist* 152:329–338.

Fensome, R. A., R. A. MacRae, J. M. Moldowan, F. J. R. Taylor and G. L. Williams. 2003. The early Mesozoic radiation of dinoflagellates. *Paleobiology* 22: 329–338.

Fensome, R. A., F. J. R. Taylor, G. Norris, W. A. S. Sargeant, D. I. Wharton, and G. L. Williams. 1993. *A Classification of Living and Fossil Dinoflagellates*. American Museum of Natural History, *Micropaleontology Special Publication #7*, Micropaleontology Press, Hanover, MA.

Field, C. B., M. J. Behrenfeld, J. T. Randerson, and P. Falkowski. 1998. Primary production of the biosphere: integrating terrestrial and oceanic components. *Science* 281:237–240.

Fields, S. D., and R. G. Rhodes. 1991. Ingestion and retention of *Chroomonas* spp. (Cryptophyceae) by *Gymnodinium acidotum* (Dinophyceae). *Journal of Phycology* 27:525–529.

Figueiredo, M. A. deO., J. M. Kain (Jones), and T. A. Norton. 1996. Biotic interactions in the colonization of crustose coralline algae by epiphytes. *Journal of Experimental Marine Biology and Ecology* 199:303–318.

Figueroa, R. I., and K. Rengefors. 2006. Life cycle and sexuality of the freshwater raphidophyte *Gonyostomum semen* (Raphidophyceae). *Journal of Phycology* 42:859–871.

Fisher, M. M., and L. W. Wilcox. 1996. Desmid-bacterial associations in *Sphagnum*-dominated Wisconsin peatlands. *Journal of Phycology* 32:543–549.

Fisher, M. M., L. W. Wilcox, and L. E. Graham. 1998. Molecular characterization of epiphytic bacterial communities on charophycean green algae. *Applied and Environmental Microbiology* 64:4384–4389.

Fleury, B. G., A. Kelecom, R. C. Pereira, and V. L. Teixeira. 1994. Polyphenols, terpenes and sterols in Brazilian Dictyotales and Fucales (Phaeophyta). *Botanica Marina* 37:457–462.

Flöder, S., and U. Sommer. 1999. Diversity in planktonic communities: an experimental test of the intermediate disturbance hypothesis. *Limnology and Oceanography* 44:1114–1119.

Floyd, G. L., H. J. Hoops, and J. A. Swanson. 1980. Fine structure of the zoospore of *Ulothrix belkae* with emphasis on the flagellar apparatus. *Protoplasma* 104:17–31.

Floyd, G. L., and C. J. O'Kelly. 1990. Ulvophyceae. In: Margulis, L., J. O. Corliss, M. Melkonian, and D. J. Chapman (eds.), *Handbook of Protoctista*, Jones & Bartlett Publishers, Boston, pp. 617–635.

Floyd, G. L., C. J. O'Kelly, and D. F. Chappell. 1985. Absolute configuration analysis of the flagellar apparatus in *Cladophora* and *Chaetomorpha* motile cells, with an assessment of the phylogenetic position of the Cladophoraceae (Ulvophyceae, Chlorophyta). *American Journal of Botany* 72:615–625.

Floyd, G. L., K. D. Stewart, and K. R. Mattox. 1971. Cytokinesis and plasmodesmata in *Ulothrix*. *Journal of Phycology* 7:306–309.

Floyd, G. L., K. D. Stewart, and K. R. Mattox.

1972. Cellular organization, mitosis, and cytokinesis in the ulotrichalean alga, *Klebsormidium*. *Journal of Phycology* 8:176–184.

Fogg, G. E. 1991. The phytoplanktonic ways of life. Tansley Review No. 30. *New Phytologist* 118:191–232.

Fogg, M. J. 1995. *Terraforming. Engineering Plant Environments*. Society of Automotive Engineers, Warrendale, PA.

Forman, R. T. T. 1975. Canopy lichens with blue-green algae: a nitrogen source in a Colombian rain forest. *Ecology* 56:1176–-1184.

Foster, K. W., J. Saranak, N. Patel, G. Zarilli, M. Okabe, T. Kline, and K. Nakanishi. 1984. A rhodopsin is the functional photoreceptor for phototaxis in the unicellular eukaryote *Chlamydomonas. Nature* 311:756–759.

Foster, M. S. 2001. Rhodoliths: Between rocks and soft places. *Journal of Phycology* 37:659–667.

Foster, M. S., et al. 2007. Diversity and natural history of a *Lithothamnion muelleri–Sargassum horridum* community in the Gulf of California. *Ciencias Marinas* 33:367–384.

Foster, R. A., E. J. Carpenter, and B. Bergman. 2006a. Unicellular cyanobionts in open ocean dinoflagellates, radiolarians, and tintinnids: Ultrastructural characterization and immunolocalization of phycoerythrin and nitrogenase. *Journal of Phycology* 42:453–463.

Foster, R. A., J. L. Collier, and E. J. Carpenter. 2006b. Reverse transcription PCR amplification of cyanobacterial symbiont 16S rRNA sequences from single non-photosynthetic eukaryotic marine planktonic host cells. *Journal of Phycology* 42:243–250.

Foster, R. A., and J. P. Zehr. 2006. Characterization of diatom-cyanobacteria symbioses on the basis of *nifH, hetR* and 16S rRNA sequences. *Environmental Microbiology* 8:1913–1925.

Fowke, L. C., and J. D. Pickett-Heaps. 1969a. Cell division in *Spirogyra*. I. Mitosis. *Journal of Phycology* 5:240–259.

Fowke, L. C., and J. D. Pickett-Heaps. 1969b. Cell division in *Spirogyra*. II. Cytokinesis. *Journal of Phycology* 5:273–281.

Frada, M., F. Not, I. Probert, and C. de Vargas. 2006. CaCO$_3$ optical detection with fluorescent in situ hybridization: A new method to identify and quantify calcifying microorganisms from the oceans. *Journal of Phycology* 42:1162–1169.

France, R. L. 1995. Differentiation between littoral and pelagic food webs in lakes using stable carbon isotopes. *Limnology and Oceanography* 40:1310–1313.

Franceschi, V. R., B. Ding, and W. J. Lucas. 1994. Mechanism of plasmodesmata formation in characean algae in relation to evolution of intercellular communication in higher plants. *Planta* 192:347–358.

Francis, D. 1967. On the eyespot of the dinoflagellate *Nematodinium. Journal of Experimental Biology* 47:495–502.

Franklin, L. A., and R. M. Forster. 1997. The changing irradiance environment: Consequences for marine macrophyte physiology, productivity and ecology. *European Journal of Phycology* 32:207–232.

Frantz, B. R., M. S. Foster, and R. Riosmena-Rodríguez. 2005. *Clathromorphum nereostratum* (Corallinales, Rhodophyta): The oldest alga. *Journal of Phycology* 41:770–773.

Freshwater, D. W., S. Fredericq, B. S. Butler, and M. H. Hommersand. 1994. A gene phylogeny of the red algae (Rhodophyta) based on plastid *rbcL. Proceedings of the National Academy of Sciences of the United States of America* 91:7281–7285.

Friedl, T. 1995. Inferring taxonomic positions and testing genus level assignments in coccoid green lichen algae: A phylogenetic analysis of 18S ribosomal RNA sequences from *Dictyochloropsis reticulata* and from members of the genus *Myrmecia* (Chlorophyta, Trebouxiophyceae cl. nov.). *Journal of Phycology* 31:632–639.

Friedl, T. 1996. Evolution of the polyphyletic genus *Pleurastrum* (Chlorophyta): inferences from nuclear-encoded ribosomal DNA sequences and motile cell ultrastructure. *Phycologia* 35:456–469.

Friedl, T., and C. J. O'Kelly. 2002. Phylogenetic relationships of green algae assigned to the genus *Planophila* (Chlorophyta): Evidence from 18S rDNA data and ultrastructure. *European Journal of Phycology* 37:373–384.

Friedl, T., and C. Rokitta. 1997. Species relationships in the lichen alga *Trebouxia* (Chlorophyta, Trebouxiophyceae): Molecular phylogenetic analyses of nuclear-encoded large subunit rRNA gene sequences. *Symbiosis* 23:125–148.

Friedmann, E. I. 1982. Endolithic microorganisms in the Antarctic cold desert. *Science* 215:1045–1053.

Friedmann, E. I., L. Kappen, M. A. Meyer, and J. A. Nienow. 1993. Long-term productivity in the cryptoendolithic microbial community of the Ross Desert, Antarctica. *Microbial Ecology* 25:51–69.

Friedmann, E. I., and A. M. Koriem. 1989. Life on Mars: How it disappeared (if it was ever there). *Advances in Space Research* 9:167–172.

Friedmann, E. I., Y. Lipkin, and R. Ocampo-Paus. 1967. Desert algae of the Negev (Israel). *Phycologia* 6:185–200.

Friedmann, E. I., and R. Ocampo. 1976. Endolithic blue-green algae in the dry valleys: Primary producers in the Antarctic desert ecosystem. *Science* 24:1247–1249.

Friedmann, E. I., and W. C. Roth. 1977. Development of the siphonous green alga *Penicillus* and the *Espera* state. *Botanical Journal of the Linnean Society* 74:189–214.

Friedmann, I. 1959. Structure, life-history and sex determination of *Prasiola stipitata* Suhr. *Annals of Botany* 23:571–594.

Friedmann, I., and R. Ocampo-Friedmann. 1995. A primitive cyanobacterium as pioneer microorganism for terraforming Mars. *Advances in Space Research* 15:243–246.

Fritsen, C. H. 2002. Snow and ice environments. In: Bitton, G. (ed.), *Encyclopedia of Environmental Microbiology*, John Wiley, New York, pp. 2872–2880.

Fritz, L., D. Mores, and J. W. Hastings. 1990. The circadian bioluminescence rhythm of *Gonyaulax* is related to daily variation in the number of light emitting organelles. *Journal of Cell Science* 95:321–328.

Fritz-Sheridan, R. P. 1988. Physiological ecology of nitrogen fixing blue-green algal crusts in the upper-subalpine life zone. *Journal of Phycology* 24:302–309.

Frost, T. M., L. E. Graham, J. E. Elias, M. J. Haase, D. W. Kretchmer, and J. A. Kranzfelder. 1997. A yellow-green algal symbiont in the freshwater sponge, *Corvomyenia everetti*: Convergent evolution of symbiotic associations. *Freshwater Biology* 38:395–399.

Fry, B., C. S. Hopkinson, and A. Nolin. 1996. Long-term decomposition of DOC from experimental diatom blooms. *Limnology and Oceanography* 41:1344–1347.

Fryxell, G. A., and G. R. Hasle. 2003. Taxonomy of harmful diatoms. In: Hallegraeff, G. M., D. M. Anderson, and A. D. Cembella (eds.), *Manual on Harmful Marine Microalgae*, UNESCO Publishing, Paris, Chapter 17.

Fu, F.-X., M. E. Warner, Y. Zhang, Y. Feng, and D. A. Hutchins. 2007. Effects of increased temperature and CO$_2$ on photosynthesis, growth, and elemental ratios in marine *Synechococcus* and *Prochlorococcus* (Cyanobacteria). *Journal of Phycology* 43:485–496.

Fujiwara, S., M. Tsuzuki, M. Kawachi, N. Minaka, and I. Inouye. 2001. Molecular phylogeny of the haptophyta based on the rbcL gene and sequence variation in the spacer region of the rubisco operon. *Journal of Phycology* 37:121–129.

Fukumoto, R.-H., T. Fuji, and H. Sekimoto. 1997. Detection and evaluation of a novel sexual pheromone that induces sexual cell division of *Closterium ehrenbergii* (Chlorophyta). *Journal of Phycology* 33:441–445.

Fuller, N. J., et al. 2006. Analysis of photosynthetic picoeukaryote diversity at open ocean sites in the Arabian Sea using a PCR biased towards marine algal plastids. *Aquatic Microbial Ecology* 43:79–93.

Gabrielson, P. W., and D. J. Garbary. 1987. A cladistic analysis of Rhodophyta: Florideophycidean orders. *British Phycological Journal* 22:125–138.

Gabrielson, P. W., and M. H. Hommersand. 1982. The morphology of *Agardhiella subulata* representing the Agardhiellaea, a new tribe in the Solieriaceae (Gigartinales, Rhodophyta). *Journal of Phycology* 18:46–58.

Gabrielson, P. W., T. B. Widdowson, and S. C. Lindstrom. 2006. *Keys to the Seaweeds and Seagrasses of Southeast Alaska, British Columbia, Washington and Oregon*, Phycological Contribution Number 7, PhycoID.

Gaedeke, A., and U. Sommer. 1986. The influence of the frequency of periodic disturbances on the maintenance of phytoplankton diversity. *Oecologia* 71:25–28.

Gaines, G., and M. Elbrächter. 1987. Heterotrophic nutrition. In: Taylor, F. J. R. (ed.), *The Biology of Dinoflagellates*, Blackwell Scientific Publications, Oxford, UK, pp. 224–268.

Galloway, R. E. 1990. Selective conditions and isolation of mutants in salt-tolerant, lipid-producing microalgae. *Journal of Phycology* 26:752–760.

Galway, M. E., and A. R. Hardham. 1991. Immunofluorescent localization of microtubules throughout the cell cycle in the green alga *Mougeotia* (Zygnemataceae). *American Journal of Botany* 78:451–461.

Ganf, G. G. 1975. Photosynthetic production and irradiance-photosynthesis relationships of the phytoplankton from a shallow equatorial lake (L. George, Uganda). *Oecologia* 18:165–183.

Gantt, E. 1975. Phycobilisomes: Light-harvesting pigment complexes. *BioScience* 25:781–788.

Ganzon-Fortes, E. T. 1994. *Gelidiella*. In: Akatsuka, I. (ed.), *Biology of Economic Algae*, SPB Academic Publishing, The Hague, The Netherlands, pp. 149–184.

Garbary, D. J., and P. W. Gabrielson. 1990. Taxonomy and evolution. In: Cole, K. M., and R. G. Sheath (eds.), *Biology of the Red Algae*, Cambridge University Press, Cambridge, UK, pp. 477–498.

Garbary, D., and H. W. Johansen. 1987. Morphogenesis and evolution in the Amphiroideae (Rhodophyta, Corallinaceae). *British Phycological Journal* 22:1–10.

Garcia-Pichel, F., and J. Belnap. 1996. Microenvironments and microscale productivity of cyanobacterial desert crusts. *Journal of Phycology* 32:774–782.

Garcia-Pichel, F., and R. W. Castenholz. 1991. Characterization and biological implications of scytonemin, a cyanobacterial sheath pigment. *Journal of Phycology* 27:395–409.

Garcia-Pichel, F., and R. W. Castenholz. 1993. Occurrence of UV-absorbing, mycosporine-like compounds among cyanobacterial isolates and an estimate for their screening capacity. *Applied and Environmental Microbiology* 59:163–169.

Garcia-Pichel, F., A. López-Cortés, and U. Nübel. 2001. Phylogenetic and morphological diversity of cyanobacteria in soil desert crusts from the Colorado Plateau. *Applied and Environmental Microbiology* 67:1902–1910.

Garcia-Pichel, F., and O. Pringault. 2001. Cyanobacteria track water in desert soils. *Nature* 413:380–381.

Gardner, M. J., et al. 2002. Genome sequence of the human malaria parasite *Plasmodium falciparum*. *Nature* 419:498–511.

Gast, R. J., and D. A. Caron. 1996. Molecular phylogeny of symbiotic dinoflagellates from Foraminifera and Radiolaria. *Molecular Biology and Evolution* 13:1192–1197.

Gast, R. J., T. A. McDonnell, and D. A. Caron. 2000. SrDNA-based taxonomic affinities of algal symbionts from a planktonic foraminifer and a solitary radiolarian. *Journal of Phycology* 36:172–177.

Gast, R. J., D. M. Moran, M. R. Dennett, and D. A. Caron. 2007. Kleptoplasty in an Antarctic dinoflagellate: Caught in evolutionary transition? *Environmental Microbiology* 9:39–45.

Gaston, K. J. 2000. Global patterns in biodiversity. *Nature* 405:220–227.

Gates, R. D., O. Hoegh-Guldberg, M. J. McFall-Ngai, and K. Y. Bil. 1995. Free amino acids exhibit anthozoan "host factor" activity: They induce the release of photosynthate from symbiotic dinoflagellates *in vitro*. *Proceedings of the National Academy of Sciences of the United States of America* 92:7430–7434.

Gauld, D. T. 1951. The grazing rate of planktonic copepods. *Journal of the Marine Biological Association of the UK* 29:695–706.

Gause, G. F. 1934. *The Struggle for Existence*, Williams & Wilkins (Dover ed. 1971), Dover Publications, Minneola, NY.

Gawley, R. E., et al. 2002. Chemosensors for the marine toxin saxitoxin. *Journal of the American Chemical Society* 124:13448–13453.

Gawlik, S. R., and W. F. Millington. 1988. Structure and function of the bristles of *Pediastrum boryanum* (Chlorophyta). *Journal of Phycology* 24:474–482.

Gaylarde, C., and G. Morton. 2002. Biodeterioration of mineral materials. In: Bitton, G. (ed.), *Encyclopedia of Environmental Microbiology*, John Wiley, New York, pp. 516–528.

Gaylarde, C. C., B. O. Ortega-Morales, P. Bartolo-Pérez. 2007. Biogenic black crusts on buildings in unpolluted environments. *Current Microbiology* 54:162–166.

Gaylord, B., D. C. Reed, P. T. Raimondi, L. Washburn, and S. R. McLean. 2002. A physically based model of macroalgal spore dispersal in the wave and current-dominated nearshore. Ecology 83:1239–1251.

Geesink, R. 1973. Experimental investigations on marine and freshwater *Bangia* (Rhodophyta) from The Netherlands. *Journal of Experimental Marine Biology and Ecology* 11:239–247.

Geider, R. J., et al. 2001. Primary productivity of planet earth: Biological determinants and physical constraints in terrestrial and aquatic habitats. *Global Change Biology* 7:849–882.

Geider, R. J., and B. A. Osborne. 1992. *Algal Photosynthesis*, Chapman & Hall, New York.

Geisen, M., C. Billard, A. T. C. Broerse, L. Cros, I. Probert, and J. R. Young. 2002. Life-cycle associations involving pairs of holococco-lithophorid species: Intraspecific variation or cryptic speciation? *European Journal of Phycology* 37:531–550.

Geitler, L. 1955. Über die cytologisch bemerkenswerte Chlorophycee *Chlorokybus atmophyticus*. *Osterreichische Botanische Zeitschrift* 102:20–29.

Gelin, F., I. Boogers, A. A. M. Noordelos, J. S. Sinnighe Damsté, R. Riegman, and J. W. de Leeuw. 1997. Resistant biomacromolecules in marine microalgae of the classes Eustigmatophyceae and Chlorophyceae: Geochemical implications. *Organic Geochemistry* 26:659–675.

Geller, W., and H. Müller. 1981. The filtration apparatus of cladocera: Filter mesh sizes and their implications on food selectivity. *Oecologia* 49:316–321.

Genter, R. B. 1996. Ecotoxicology of inorganic chemical stress to algae. In: Stevenson, R. J., M. L. Bothwell, and R. L. Lowe (eds.), *Algal Ecology: Freshwater Benthic Ecosystems*, Academic Press, New York, pp. 404–468.

George, D. G., and R. W. Edwards. 1976. The effect of wind on the distribution of chlorophyll *a* and crustacean plankton in a shallow eutrophic reservoir. *Journal of Applied Ecology* 13:667–690.

Gerrath, J. F. 1993. The biology of desmids: A decade of progress. *Progress in Phycological Research* 9:79–192.

Gerrath, J. F. 2002. Conjugating green algae and desmids. In: Wehr, J. D., and R. G. Sheath (eds.), *Freshwater Algae of North America, Ecology and Classification*, Academic Press, New York, pp. 353–381.

Gervais, F. 1997. Light-dependent growth, dark survival, and glucose uptake by cryptophytes isolated from a freshwater chemocline. *Journal of Phycology* 33:18–25.

Gerwick, W. H., and N. J. Lang. 1977. Structural, chemical and ecological studies on

iridescence in *Iridaea*. *Journal of Phycology* 13:121–127.

Gerwick, W. H., M. A. Roberts, P. J. Proteau, and J.-L. Chen. 1994. Screening cultured marine algae for anticancer-type activity. *Journal of Applied Phycology* 6:143–149.

Ghadouani, A., B. Pinel-Alloul, K. Plath, G. A. Codd, and W. Lampert. 2004. Effects of *Microcystis aeruginosa* and purified microcystin-LR on the feeding behavior of *Daphnia pulicaria*. Limnology and Oceanography 49:666–679.

Ghirardi, M. L., M. C. Posewitz, P.-C. Maness, A. Dubini, J. Yu, and M. Seibert. 2007. Hydrogenases and hydrogen photoproduction in oxygenic photosynthetic organisms. *Annual Review of Plant Biology* 58:71–91.

Gibb, A. P., R. Aggarwal, and C. P. Swainson. 1991. Successful treatment of *Prototheca* peritonitis complicating continuous ambulatory peritoneal dialysis. *Journal of Infection* 22:183–185.

Gibbs, S. P. 1978. The chloroplasts of *Euglena* may have evolved from symbiotic green algae. *Canadian Journal of Botany* 56:2883–2889.

Gibbs, S. R. 1979. The route of entry of cytoplasmically synthesized proteins into chloroplasts of algae possessing chloroplast ER. *Journal of Cell Science* 35:253–266.

Gibbs, S. R. 1981. The chloroplasts of some algal groups may have evolved from endosymbiotic eukaryotic algae. *Annals of the New York Academy of Sciences* 361:193–208.

Gibson, C. E., J. N. Anderson, and E. Y. Haworth. 2003. *Aulacoseira subarctica*: Taxonomy, physiology, ecology and palaeoecology. *European Journal of Phycology* 38:83–101.

Giddings, T. H., D. L. Brower, and L. A. Staehelin. 1980. Visualization of particle complexes in the plasma membrane of *Micrasterias denticulata* associated with the formation of cellulose fibrils in primary and secondary cell walls. *Journal of Cell Biology* 84:327–339.

Giddings, T. H., and L. A. Staehelin. 1991. Microtubule-mediated control of microfibril deposition: A re-examination of the hypothesis. In C. W. Lloyd (ed.), *The Cytoskeletal Basis of Plant Growth and Form*, Academic Press, London, pp. 85–99.

Giddings, T. H., C. Wasmann, and L. A. Staehelin. 1983. Structure of the thylakoids and envelope membranes of the cyanelles of *Cyanophora paradoxa*. *Plant Physiology* 71:409–419.

Gifford, D. J., R. N. Bohrer, and C. M. Boyd. 1981. Spines on diatoms: Do copepods care? *Limnology and Oceanography* 26:1057–1061.

Gilbert, J. J. 1989. The effect of *Daphnia* interference on a natural rotifer and ciliate community: Short-term bottle experiments. *Limnology and Oceanography* 34:606–617.

Gilbert, S., B. A. Zeeb, and J. P. Smol. 1997. Chrysophyte stomatocyst flora from a forest peat core in the Lena River Region, northeastern Siberia. *Nova Hedwigia* 64:311–352.

Gilichinsky, D. 2002. Permafrost. In: Bitton, G. (ed.), *Encyclopedia of Environmental Microbiology*, John Wiley, New York, pp. 2367–2385.

Gilichinsky, D. A., et al. 2007. Microbial populations in Antarctic permafrost: Biodiversity, state, age, and implication for astrobiology. *Astrobiology* 7:275–311.

Gillott, M. 1990. Cryptophyta (Cryptomonads). In: Margulis, L., J. O. Corliss, M. Melkonian, and D. J. Chapman (eds.), *Handbook of Protoctista*, Jones & Bartlett Publishers, Boston, pp. 139–151.

Gillott, M. A., and S. P. Gibbs. 1980. The cryptomonad nucleomorph: Its ultrastructure and evolutionary significance. *Journal of Phycology* 16:558–568.

Gilson, P. R., V. Su, C. H. Stamovits, M. E. Reith, P. J. Keeling, and G. I. McFadden. 2006. Complete nucleotide sequence of the chlorarachniophyte nucleomorph: Nature's smallest nucleus. *Proceedings of the National Academy of Sciences of the United States of America* 103:9566–9571.

Giordano, M., J. Beardall, and J. A. Raven. 2005. CO_2 concentrating mechanisms in algae: Mechanisms, environmental modulation, and evolution. *Annual Review of Plant Biology* 56:99–131.

Glazer, A. N., and G. J. Wedemayer 1995. Cryptomonad biliproteins—An evolutionary perspective. *Photosynthesis Research* 46:93–105.

Glazer, A. N., C. Chan, R. C. Williams, S. W. Yeh, and J. H. Clark. 1985. Kinetics of energy flow in the phycobilisome core. *Science* 230:1051–1053.

Gliwicz, Z. M. 1990. *Daphnia* growth at different concentrations of blue-green filaments. *Archiv für Hydrobiologie* 120:51–65.

Glynn, P. W. 1997. Bioerosion and coral reef growth: A dynamic balance. In: Birkeland, C. (ed.), *Life and Death of Coral Reefs*, Chapman & Hall, New York, pp. 68–95.

Goddard, R. H., and J. W. La Claire. 1991. Calmodulin and wound healing in the coenocytic green alga *Ernodesmis verticillata* (Kützing) Borgesen. *Planta* 183:281–293.

Goff, L. J. 1979. The biology of *Harveyella mirabilis* (Cryptonemiales, Rhodophyceae). VII. Structure and proposed function of host-penetrating cells. *Journal of Phycology* 15:87–106.

Goff, L. J. 1982. The biology of parasitic red algae. *Progress in Phycological Research* 1:289–370.

Goff, L. J., and A. W. Coleman. 1984. Transfer of nuclei from a parasite to its host. *Proceedings of the National Academy of Sciences of the United States of America* 81:5420–5424.

Goff, L. J., and A. Coleman. 1987. Nuclear transfer from parasite to host: A new regulatory mechanism of parasitism. In: Lee, J. J., and J. K. Frederick (eds.), *Endocytobiology III*. Annals of the New York Academy of Science, New York pp. 402–423.

Goff, L. J., and A. W. Coleman. 1990. DNA: Micro-spectrophotometric studies. In: Cole, K. M., and R. G. Sheath (eds.), *Biology of the Red Algae*, Cambridge University Press, Cambridge, UK, pp. 43–72.

Goff, L. J., D. A. Moon, and A. W. Coleman. 1994. Molecular delineation of species and species relationships in the red algal agarophytes *Gracilariopsis* and *Gracilaria* (Gracilariales). *Journal of Phycology* 30:521–537.

Goff, L. J., D. A. Moon, P. Nyvall, B. Stache, K. Mangin, and G. Zuccarello. 1996. The evolution of parasitism in the red algae; molecular comparisons of adelphoparasites and their hosts. *Journal of Phycology* 32:297–312.

Goff, L. J., and G. Zuccarello. 1994. The evolution of parasitism in red algae: Cellular interactions of adelphoparasites and their hosts. *Journal of Phycology* 30:695–720.

Gojdics, M. 1953. *The Genus Euglena*. University of Wisconsin Press, Madison, WI.

Goldman, C. R. 1960. Molybdenum as a factor limiting primary productivity in Castle Lake, California. *Science* 132:1016–1017.

Goldman, C. R. 1972. The role of minor nutrients in limiting the productivity of aquatic ecosystems. In: Likens, G. E. (ed.), *Nutrients and Eutrophication: The Limiting Nutrient Controversy*, American Society of Limnology and Oceanography Special Symposium, vol. 1, Allen Press, Lawrence, KS, pp. 21–38.

Goldman, C. R., and A. J. Horne. 1983. *Limnology*, McGraw-Hill, New York.

Goldsborough, L. G., and G. G. C. Robinson. 1996. Pattern in wetlands. In: Stevenson, R. J., M. L. Bothwell, and R. L. Lowe (eds.), *Algal Biology. Freshwater Benthic Ecosystems*, Academic Press, New York, pp. 78–120.

Goldstein, S. F. 1992. Flagellar beat patterns in algae. In: Melkonian, M. (ed.), *Algal Cell Motility*, Chapman & Hall, London, pp. 99–154.

Gontcharov, A. A., B. Marin, and M. Melkonian. 2003. Molecular phylogeny of conjugating green algae (Zygnemophyceae, Streptophyta) inferred from SSU rDNA sequence comparisons. *Journal of Molecular Evolution* 56:89–104.

Gontcharov, A. A., and M. Melkonian. 2004. Unusual position of the genus *Spirotaenia* (Zygnematophyceae) among streptophytes revealed by SSU rDNA and *rbcL* sequence comparisons. *Phycologia* 43:105–113.

Gontcharov, A. A., and M. Melkonian. 2005. Molecular phylogeny of *Staurastrum* Meyen ex Ralfs and related genera (Zygnematophyceae, Streptophyta) based on coding and noncoding rDNA sequence comparisons. *Journal of Phycology* 41:887–899.

Gonzalez, M. A., and L. J. Goff. 1989a. The red algal epiphytes *Microcladia coulteri* and *M. californica* (Rhodophyeae, Ceramiaceae). I. Taxonomy, life history and phenology. *Journal of Phycology* 25:545–558.

Gonzalez, M. A., and L. J. Goff. 1989b. The red algal epiphytes *Microcladia coulteri* and *M. californica* (Rhodophyeae, Ceramiaceae). II. Basiphyte specificity. *Journal of Phycology* 15:558–567.

Good, B. H., and R. L. Chapman. 1978. The ultrastructure of *Phycopeltis* (Chroolepidaceae: Chlorophyta). I. Sporopollenin in the cell walls. *American Journal of Botany* 65:27–33.

Goodenough, U., P. A. Detmas, and C. Hwang. 1982. Activation of cell fusion in *Chlamydomonas*: Analysis of wild-type gametes and nonfusing mutants. *Journal of Cell Biology* 92:378–386.

Goodwin, B. C., and C. Briére. 1994. Mechanics of the cytoskeleton and morphogenesis of *Acetabularia*. *International Review of Cytology* 150:225–242.

Gorby, Y. A., et al. 2006. Electrically conductive bacterial nanowiresproduced by *Shewanella oneidensis* strain MR-1 and other microorganisms. *Proceedings of the National Academy of Sciences of the United States of America* 103:11358–11363.

Gordon, A. S., and B. Dyer. 2005. Relative contribution of exotoxin and micropredation to icthyotoxicity of two strains of *Pfiesteria shumwayae* (Dinophyceae). *Harmful Algae* 4:423–431.

Goto, Y., and K. Ueda. 1988. Microfilament bundles of F-actin in *Spirogyra* observed by fluorescence microscopy. *Planta* 173:442–446.

Goudie, A. S. 1978. Dust storms and their geomorphological implications. *Journal of Arid Environments* 1:291–310.

Gould, S. B., M. S. Sommer, K. Hadfi, S. Zauner, P. J. Kroth, and U.-G. Maier. 2006a. Protein targeting into the complex plastid of cryptophytes. *Journal of Molecular Evolution* 62:674–681.

Gould, S. B., M. S. Sommer, P. J. Kroth, G. H. Gile, P. J. Keeling, and U.-G. Maier. 2006b. Nucleus-to-nucleus gene transfer and protein retargeting into a remnant cytoplasm of cryptophytes and diatoms. *Molecular Biology and Evolution* 23:2413–2422.

Goulet, T. L. 2006. Most corals may not change their symbionts. *Marine Ecology Progress Series* 321:1–7.

Graetz, R. D., and D. J. Tongway. 1986. Influence of grazing management on vegetation, soil structure and nutrient distribution and the infiltration of applied rainfall in a semi-arid chenopod shrubland. *Australian Journal of Ecology* 11:347–360.

Graham, J. M. 1991. Symposium introductory remarks: A brief history of aquatic microbial ecology. *Journal of Protozoology* 38:66–69.

Graham, J. M. 2004. The biological terraforming of Mars: Planetary ecosynthesis as ecological succession on a global scale. *Astrobiology* 4:168–195.

Graham, J. M., P. Arancibia-Avila, and L. E. Graham. 1996a. Physiological ecology of a species of the filamentous green alga *Mougeotia* under acidic conditions: Light and temperature effects on photosynthesis and respiration. *Limnology and Oceanography* 41:253–262.

Graham, J. M., P. Arancibia-Avila, and L. E. Graham. 1996b. Effects of pH and selected metals on growth of the filamentous green alga *Mougeotia* under acidic conditions. *Limnology and Oceanography* 41:263–270.

Graham, J. M., M. T. Auer, R. P. Canale, and J. P. Hoffmann. 1982. Ecological studies and mathematical modeling of *Cladophora* in Lake Huron: 4. Photosynthesis and respiration as functions of light and temperature. *Journal of Great Lakes Research* 8:100–111.

Graham, J. M., and L. E. Graham. 1987. Growth and reproduction of *Bangia atropurpurea* (Roth) C. Ag. (Rhodophyta) from the Laurentian Great Lakes. *Aquatic Botany* 28:317–331.

Graham, J. M., L. E. Graham, and J. A. Kranzfelder. 1985. Light, temperature and photoperiod as factors controlling reproduction in *Ulothrix zonata* (Ulvophyceae). *Journal of Phycology* 21:235–239.

Graham, J. M., A. D. Kent, G. H. Lauster, A. C. Yannarell, L. E. Graham, and E. W. Triplett. 2004. Seasonal dynamics of phytoplankton and planktonic protozoan communities in a northern temperate humic lake: Diversity in a dinoflagellate dominated system. *Microbial Ecology* 48:528–540.

Graham, J. M., C. A. Lembi, H. L. Adrian, and D. F. Spencer. 1995. Physiological responses to temperature and irradiance in *Spirogyra*

(Zygnematales, Charophyceae). *Journal of Phycology* 31:531–540.

Graham, L. E. 1982. Cytology, ultrastructure, taxonomy, and phylogenetic relationships of Great Lakes filamentous algae. *Journal of Great Lakes Research* 8:3–9.

Graham, L. E. 1984. An ultrastructure re-examination of putative multilayered structures in *Trentepohlia aurea*. *Protoplasma* 123:1–7.

Graham, L. E. 1985. The origin of the life cycle of land plants. *American Scientist* 73:178–186.

Graham, L. E. 1990. Meiospore formation in charophycean algae. In: Blackmore, S., and R. B. Knox (eds.), *Microspores: Evolution and Ontogeny*, Academic Press, London, pp. 43–54.

Graham, L. E. 1993. *The Origin of Land Plants*, John Wiley, New York.

Graham, L. E. 1996. Green algae to land plants: An evolutionary transition. *Journal of Plant Research* 109:241–251.

Graham, L. E., J. M. Graham, and J. A. Kranzfelder. 1986. Irradiance, daylength and temperature effects on zoosporogenesis in *Coleochaete scutata* (Charophyceae). *Journal of Phycology* 22:35–39.

Graham, L. E., J. M. Graham, W. A. Russin, and J. M. Chesnick. 1994. Occurrence and phylogenetic significance of glucose utilization by charophycean algae: Glucose enhancement of growth in *Coleochaete orbicularis*. *American Journal of Botany* 81:423–432.

Graham, L. E., J. M. Graham, and D. E. Wujek. 1993. Ultrastructure of *Chrysodidymus sunuroideus* (Synurophyceae). *Journal of Phycology* 29:330–341.

Graham, L. E., and Y. Kaneko. 1991. Subcellular structures of relevance to the origin of land plants (Embryophytes) from green algae. *CRC Critical Reviews in Plant Sciences* 10:323–342.

Graham, L. E., F. J. Macentee, and H. C. Bold. 1981. An investigation of some subaerial green algae. *Texas Journal of Science* 23:13–16.

Graham, L. E., and G. E. McBride. 1978. Mitosis and cytokinesis in sessile sporangia of *Tretepohlia aurea* (Chlorophyceae). *Journal of Phycology* 14:132–137.

Graham, L. E., and G. E. McBride. 1979. The occurrence and phylogenetic significance of a multilayered structure in *Coleochaete scutata* spermatozoids. *American Journal of Botany* 66:887–894.

Graham, L. E., and C. Taylor. 1986a. The ultrastructure of meiospores of *Coleochaete pulvinata* (Charophyceae). *Journal of Phycology* 22:299–307.

Graham, L. E., and C. Taylor. 1986b. Occurrence and phylogenetic significance of "special walls" at meiosporogenesis in *Coleochaete*. *American Journal of Botany* 73:597–601.

Graham, L. E., and G. J. Wedemayer. 1984. Spermatogenesis in *Coleochaete pulvinata* (Charophyceae): Sperm maturation. *Journal of Phycology* 20:302–309.

Graham, L. E., and L. W. Wilcox. 1983. The occurrence and phylogenetic significance of putative placental transfer cells in the green alga *Coleochaete*. *American Journal of Botany* 70:113–120.

Graham, L. E., and L. W. Wilcox. 2000. The origin of alternation of generations in land plants: A focus on matrotrophy and hexose

transport. *Philosophical Transactions of the Royal Society of London B* 355:757–767.

Graham, M. H., B. P. Kinlan, L. D. Druehl, L. E. Garske, and S. Banks. 2007. Deep-water kelp refugia as potential hotspots of tropical marine diversity and productivity. *Proceedings of the National Academy of Sciences of the United States of America* 104:16576–16580.

Graham, N. A. J., S. K. Wilson, S. Jennings, N. V. C. Polunin, J. P. Bijoux, and J. Robinson. 2006. Dynamic fragility of oceanic coral reef ecosystems. *Proceedings of the National Academy of Sciences of the United States of America* 103:8425–8429.

Granhall, U., and A. V. Hofsten. 1976. Nitrogenase activity in relation to intracellular organisms in *Sphagnum* mosses. *Physiologica Plantarum* 36:88–94.

Grant, M. C. 1990. Charophyceae (Order Charales). In: Margulis, L., J. O. Corliss, M. Melkonian, and D. J. Chapman (eds.), *Handbook of Proctoctista*, Jones & Bartlett Publishers, Boston, pp. 641–648.

Gratton, L. M., D. Oldach, J. K. Tracy, and D. R. Greenberg. 1998. Neurobehavioral complaints of symptomatic persons exposed to *Pfiesteria piscida* or morphologically related organisms. *Maryland Medical Journal* 47:127–129.

Gray, J., and A. J. Boucot. 1989. Is *Moyeria* a euglenoid? *Lethaia* 22:447–456.

Gray, J., D. Massa, and A. J. Boucot. 1982. Caradocian land plant microfossils from Libya. *Geology* 10:197–201.

Gray, M. W. 1992. The endosymbiont hypothesis revisited. *International Review of Cytology* 141:233–357.

Gray, M. W., G. Burger, and B. F. Lang. 1999. Mitochondrial evolution. *Science* 283:1476–1481.

Green, J. L., B. J. M. Bohannan, and R. J. Whitaker. 2008. Microbial biogeography: From taxonomy to traits. *Science* 320:1039–1042

Green, J. W., A. H. Knoll, S. Golubíc, and K. Swett. 1987. Paleobiology of distinctive benthic microfossils from the Upper Proterozoic limestone-dolomite "series," central East Greenland. *American Journal of Botany* 74:928–940.

Green, T. G. A., and P. A. Broady. 2001. Biological soil crusts of Antarctica. In: Belnap, J., and O. L. Lange (eds.), *Biological Soil Crusts: Structure, Function, and Management*, Ecological Studies, vol. 150, Springer Verlag, Berlin, pp. 133–139.

Greenwood, A. D. 1974. The Cryptophyta in relation to phylogeny and photosynthesis. In: Sanders, J. V., and D. J. Goodchild (eds.), 8th International Conference Electron Microscopy Canberra, Australian Academy of Science, Canberra.

Grell, K. G. 1973. *Protozoology*. Springer Verlag, Berlin.

Gretz, M. R., J. M. Aronson, and M. R. Sommerfeld. 1980. Cellulose in the cell walls of the Bangiophyceae (Rhodophyta). *Science* 207:779–781.

Greuet, C. 1968. Organization ultrastructurale de l'ocelle de deux Péridiniens Warnowiidae, *Erythropsis pavillardi* Kofoid et Swezy et *Warnowia pulchra* Schiller. *Protistologica* 4:202–230.

Greuet, C. 1976. Organisation ultrastructurale du tentacule D' *Erythropsis pavallardi* Kofoid

et Swezy péridinien *Warnowiidae* Lindemann. *Protostologica* 3:335–345.

Griffith, J. K., M. E. Baker, D. A. Rouch, M. G. P. Page, R. A. Skurray, I. T. Paulsen, K. F. Chater, S. A. Baldwin, and P. J. F. Henderson. 1992. Membrane transport proteins: Implications of sequence comparisons. *Current Opinion in Cell Biology* 4:684–695.

Grilli Caiola, M. 1992. Cyanobacterian symbiosis with bryophytes and tracheophytes. In: Reisser, W. (ed.), *Algae and Symbioses*, BioPress, Bristol, UK, pp. 231–254.

Grilli Caiola, M., and S. Pellegrini. 1984. Lysis of *Microcystis aeruginosa* by *Bdellovibrio*-like bacteria. *Journal of Phycology* 20:471–475.

Grolig, F. 1990. Actin-based organelle movements in interphase *Spirogyra*. *Protoplasma* 155:29–42.

Grondin, A. E., and J. R. Johansen. 1995. Seasonal succession in a soil algal community associated with a beech-maple forest in northeastern Ohio, U.S.A. *Nova Hedwigia* 60:1–12.

Gross, H., et al. 2006. Lophocladines, bioactive alkaloids from the red alga *Lophocladia* sp. *Journal of Natural Products* 69:640–644.

Grossman, A. R. 2005. Paths toward algal genomics. *Plant Physiology* 137:410–427.

Grossman, A. R., M. Lohr, and C. S. Im. 2004. *Chlamydomonas reinhardtii* in the landscape of pigments. *Annual Review of Genetics* 38:119–173.

Grover, J. P. 1989. Phosphorus-dependent growth kinetics of 11 species of freshwater algae. *Limnology and Oceanography* 34:341–348.

Grover, J. P. 1997. *Resource Competition*, Chapman & Hall, London.

Grünewald, K., C. Hagen, and W. Braune. 1997. Secondary carotenoid accumulation in flagellates of the green alga *Haematococcus lacustris*. *European Journal of Phycology* 32:387–392.

Grzebyk, D., O. Schofield, C. Vetriani, and P. G. Falkowski. 2003. The Mesozoic radiation of eukaryotic algae: The portable plastid hypothesis. *Journal of Phycology* 39:259–267.

Guerin, M., M. E. Huntley, and M. Olaizola. 2003. *Haematococcus* astaxanthin: Applications for human health and nutrition. *Trends in Biotechnology* 21:210–216.

Gugger, M. F., and L. Hoffmann. 2004. Polyphyly of true branching cyanobacteria (Stigonematales). *International Journal of Systematic and Evolutionary Microbiology* 54:349–357.

Guillou, L., M.-J. Chrétiennot-Dinet, L. K. Medlin, H. Claustre, S. Louseaux-de Goër, and D. Vaulot. 1999. *Bolidomonas*: A new genus with two species belonging to a new algal class, the Bolidophyceae (Heterokonta). *Journal of Phycology* 35:368–381.

Guillou, L., W. Eikrem, M.-J. Chrétiennot-Dinet, F. Le Gall, R. Massana, K. Romari, C. Pedrós-Alió, and D. Vaulot. 2004. Diversity of picoplanktonic prasinophytes assessed by direct nuclear SSU rDNA sequencing of environmental samples and novel isolates retrieved from oceanic and coastal marine ecosystems. *Protist* 155:193–214.

Guiry, M. D. 1978. The importance of sporangia in the classification of the Florideophyceae. In: Irvine, D. E. G., and J. H. Price (eds.), *Modern Approaches to the Taxonomy of Red and Brown Algae*, Systematics Association

Special Volume 10, Academic Press, London, pp. 111–114.

Guiry, M. D., D.-H. Kim, and M. Masuda. 1984. Reinstatement of the genus *Mastocarpus* Kützing (Rhodophyta). *Taxon* 33:53–63.

Gunnison, D., and M. Alexander. 1975a. Basis for the resistance of several algae to microbial decomposition. *Applied Microbiology* 29:729–738.

Gunnison, D., and M. Alexander. 1975b. Resistance and susceptibility of algae to decomposition by natural microbial communities. *Limnology and Oceanography* 20:64–70.

Gustafson, D. E., Jr., D. K. Stoecker, M. D. Johnson, W. F. Van Heukelem, and K. Sneider. 2000. Cryptophyte algae are robbed of their organelles by the marine ciliate *Mesodinium rubrum*. *Nature* 405:1049–1952.

Hackett, J., H., et al. 2004. Migration of the plastid genome to the nucleus in a peridinin dinoflagellate. *Current Biology* 14:213–218.

Hackett, J. D., L. Maranda, H. S. Yoon, and D. Bhattacharya. 2003. Phylogenetic evidence for the cryptophyte origin of the plastid of *Dinophysis* (Dinophysiales, Dinophyceae). *Journal of Phycology* 39:440–448.

Hackett, J. D., H. S. Yoon, S. Li, A. Reyes-Prieto, S. E. Rümmele, and D. Bhattacharya. 2007. Phylogenomic analysis supports the monophyly of cryophytes and haptophytes and the association of Rhizaria with chromalveolates. *Molecular Biology and Evolution* 24:1702–1713.

Hackney, J. M., R. C. Carpenter, and W. H. Adey. 1989. Characteristic adaptations to grazing among algal turfs on a Caribbean coral reef. *Phycologia* 28:109–119

Häder, D.-P., and E. Hoiczyk. 1992. Gliding motility. In: Melkonian, M. (ed.), *Algal Cell Motility*, Chapman & Hall, London, pp. 1–38.

Hadley, R., W. E. Hable, and D. L. Kropf. 2006. Polarization of the endomembrane system is an early event in fucoid zygote development. *BMC Plant Biology* 6:5.

Haferkamp. I., P. Deschamps, M. Ast, W. Jeblick, U. Maier, S. Ball, and H. E. Neuhaus. 2006. Molecular and biochemical analysis of periplastidial starch metabolism in the cryptophyte *Guillardia theta*. *Eukaryotic Cell* 5:964–971.

Hagen, C., W. Braune, and L. O. Björn. 1994. Functional aspects of secondary carotenoids in *Haematococcus lacustris* (Volvocales). III. Action as a "sunshade." *Journal of Phycology* 30:241–248.

Haglund, K. 1997. The use of algae in aquatic toxicity assessment. *Progress in Phycological Research* 12:182–212.

Hale, M. E. 1987. Epilithic lichens in the Beacon sandstone formation, Victoria Land, Antarctica. *The Lichenologist* 19:269–287.

Hall, J. D., K. G. Karol, R. M. McCourt, and C. F. Delwiche. 2008. Phylogeny of the conjugating green algae based on chloroplast and mitochondrial nucleotide sequence data. *Journal of Phycology* 44;467–477.

Hallegraeff, G. M., and C. J. Boalch. 1991. Transport of toxic dinoflagellate cysts via ships' ballast water. *Marine Pollution Bulletin* 32:27–30.

Hallegraeff, G. M., and C. J. Boalch. 1992. Transport of diatom and dinoflagellate resting spores in ships' ballast water: Implications for plankton biogeography and aquaculture. *Journal of Plant Research* 14:1067–1084.

Hallegraeff, G. M., and Y. Hara. 2003. Taxonomy of harmful marine raphidophytes. In: Hallegraeff, G. M., D. M. Anderson, and A. D. Cembella (eds.), *Manual on Harmful Marine Microalgae*, UNESCO Publishing, Paris, pp. 511–584.

Hallmann, A. 2006. Morphogenesis in the family Volvocaceae: Different tactics for turning an embryo right-side out. *Protist* 157:445–461.

Hallmann, A., and S. Wodniok. 2006. Swapped green algal promoters: AphVIII-based gene constructs with *Chlamydomonas* flanking sequences work as dominant selectable markers in *Volvox* and vice versa. *Plant Cell Reports* 25:582–591.

Halpern, B. S., et al. 2008. A global map of human impact on marine ecosystems. *Science* 319:948–952.

Hambright, K. D., T. Zohary, H. Güde. 2007. Microzooplankton dominate the flow and nutrient cycling in a warm subtropical freshwater lake. *Limnology and Oceanography* 52:1018–1025.

Hambrook, J. A., and R. G. Sheath. 1987. Grazing of freshwater Rhodophyta. *Journal of Phycology* 23:656–662.

Han, B., H. Gross, D. E. Goeger, S. L. Mooberry, and W. H. Gerwick. 2006. Aurilides B and C, cancer cell toxins from a Papua New Guinea collection of the marine cyanobacerium *Lyngbya majuscula*. *Journal of Natural Products* 69:572–575.

Han, T.-M., and B. Runnegar. 1992. Megascopic eukaryotic algae from the 2.1-billion-year-old Negaunee iron-formation, Michigan. *Science* 257:232–235.

Hanagata, N., I. Karube, M. Chihara, and P. C. Silva. 1998. Reconsideration of the taxonomy of ellipsoidal species of *Chlorella* (Trebouxiophyceae, Chlorophyta) with establishment of *Watanabea* gen. nov. *Phycological Research* 46:221–229.

Handa, S., M. Nakahara, H. Tsubota, H. Deguchi, and T. Nakano. 2003. A new aerial alga, *Stichococcus ampulliformis* sp. nov. (Trebouxiophyceae, Chlorophyta) from Japan. *Phycological Research* 51:203–210.

Hanelt, D., C. Wiencke, U. Karsten, and W. Nultsch. 1997. Photo-inhibition and recovery after high light stress in different developmental and life-history stages of *Laminaria saccharina* (Phaeophyta). *Journal of Phycology* 33:387–395.

Haney, J. F. 1973. An *in situ* examination of the grazing activities of natural zooplankton communities. *Archiv für Hydrobiologie* 72:87–132.

Hanisak, M. D. 1983. The nitrogen relationships of marine macroalgae. In: Carpenter, E. J., and D. G. Capone (eds.), *Nitrogen in the Marine Environment*, Academic Press, New York, pp. 699–730.

Hanisak, M. D., and M. A. Samuel. 1987. Growth rates in culture of several species of *Sargassum* from Florida, U.S.A. *Hydrobiologia* 151/152:399–404.

Hannaert, V., E. Saavedra, F. Dufflieux, J.-P. Szikora, D. J. Rigden, P. A. M. Michels, and F. R. Opperdoes. 2003. Plant-like traits associated with metabolism of *Trypanosoma* para-

sites. *Proceedings of the National Academy of Sciences of the United States of America* 100:1067–1071.

Hansen, E. S. 2001. Lichen-rich soil crusts of Arctic Greenland. In: Belnap, J., and O. L. Lange (eds.), *Biological Soil Crusts: Structure, Function, and Management*, Ecological Studies, vol. 150, Springer Verlag, Berlin, pp. 57–65.

Hansen, G. I., and S. C. Lindstrom. 1984. A morphological study of *Hommersandia maximicarpa* gen. et sp. nov. (Kallymeniaceae, Rhodophyta) from the North Pacific. *Journal of Phycology* 20:476–488.

Hansen, G., L. Botes, and M. De Salas. 2007. Ultrastructure and large subunit rDNA sequences of *Lepidodinium viride* reveal a close relationship to *Lepidodinium chlorophorum* comb. nov. (= *Gymnodinium chlorophorum*). *Phycological Research* 55:25–41.

Hansen, G., N. Daugbjerg, and P. Henriksen. 2007. *Baldinia anauniensis* gen. et sp. nov.: A "new" dinoflagellate from Lake Tovel, N. Italy. *Phycologia* 46:86–108.

Hansen, P. 1998. Phagotrophic mechanisms and prey selection in mixotrophic phytoflagellates. In: Anderson, D. M., A. D. Cembella, and G. M. Hallegraeff (eds.), *Physiological Ecology of Harmful Algal Blooms*, Springer Verlag, Berlin, pp. 525–537.

Hanson, M. A., and M. G. Butler. 1990. Early responses of plankton and turbidity to biomanipulation in a shallow prairie lake. *Hydrobiologia* 200/201:317–327.

Hansson, L.-A. 1996. Algal recruitment from lake sediments in relation to grazing, sinking, and dominance patterns in the phytoplankton community. *Limnology and Oceanography* 41:1312–1323.

Hansson, L.-A., H. Annadotter, E. Bergman, S. F. Hamrin, E. Jeppesen, T. Kairesalo, E. Luokkanen, P.-Å Nilsson, M. Søndergaard, and J. Strand. 1998. Biomanipulation as an application of food-chain theory: Constraints, synthesis, and recommendations for temperate lakes. *Ecosystems* 1:558–574.

Hanyuda, T., I. Wakana, S. Arai, K. Miyaji, Y. Watano, and K. Ueda. 2002. Phylogenetic relationships within Cladophorales (Ulvophyceae, Chlorophyta) inferred from 18S rRNA gene sequences, with special reference to *Aegagropila linnaei*. *Journal of Phycology* 38:564–571.

Happey-Wood, C. E. 1988. Ecology of freshwater planktonic green algae. In: Sandgren, C. D. (ed.), *Growth and Reproductive Strategies of Freshwater Phytoplankton*, Cambridge University Press, Cambridge, UK, pp. 175–226.

Harder, D. L., C. L. Hurd, and T. Speck. 2006. Comparison of mechanical properties of four large, wave-exposed seaweeds. *American Journal of Botany* 93:1426–1431.

Harlin, M. M. 1987. Allelochemistry in marine macroalgae. *CRC Critical Reviews in Plant Science* 5:237–249.

Harper, J. T., E. Waanders, and P. J. Keeling. 2005. On the monophyly of chromalveolates using a six-protein phylogeny of eukaryotes. *International Journal of Systematic and Evolutionary Microbiology* 55:487–496.

Harper, K. T., and J. Belnap. 2001. The influence of biological soil crusts on mineral uptake by associated vascular plants. *Journal of Arid Environment* 47:347–357.

Harper, K. T., and R. L. Pendleton. 1993. Cyanobacteria and cyanolichens: Can they enhance availability of essential minerals for higher plants? *Great Basin Naturalist* 53:59–72.

Harris, E. 1989. *The Chlamydomonas Handbook*, Academic Press, New York.

Harris, G. P. 1986. *Phytoplankton Ecology*, Chapman & Hall, London.

Harris, G. P., G. D. Haffner, and B. B. Piccinin. 1980. Physical variability and phytoplankton communities II. Primary productivity by phytoplankton in a physically variable environment. *Archiv für Hydrobiologie* 88:393–425.

Harris, M. A., B. F. Cumming, and J. P. Smol. 2006. Assessment of recent environmental changes in new Brunswick (Canada) lakes based on paleolimnological shifts in diatom species assemblages. *Canadian Journal of Botany* 84:151–163.

Harvey, A. S., and W. J. Woelkerling. 2007. A guide to nongeniculate coralline red algal (Corallinales, Rhodophyta) rhodolith identification. *Ciencias Marinas* 33:411–426.

Harwood, D. M., and R. Gersonde. 1990. Lower Cretaceous diatoms from ODP Leg 113 Site 693 (Weddell Sea). Part 2: Resting spores, chrysophycean cysts, an endoskeletal dinoflagellate and notes on the origin of diatoms. In: Barker, P. F., Kennett, J. P., et al. (eds.), Proceedings of the Ocean Drilling Program. Scientific Results, vol. 113, Ocean Drilling Program, College Station, Texas, pp. 329–447.

Haselkorn, R. 1978. Heterocysts. *Annual Review of Plant Physiology* 29:319–344.

Hashimoto, H. 1992. Involvement of actin filaments in chloroplast division of the alga *Closterium ehrenbergii*. *Protoplasma* 167:88–96.

Hasle, G. R. 1994. *Pseudo-nizschia* as a genus distinct from *Nitzschia*. *Journal of Phycology* 30:1036–1039.

Hasle, G. R., and E. E. Syvertsen. 1997. Marine diatoms. In: Tomas, C. R. (ed.), *Identifying Marine Phytoplankton*, Academic Press, San Diego, CA, Chapter 2.

Hatano, K., and K. Maruyama. 1995. Growth pattern of isolated zoospores in *Hydrodictyon reticulatum* (Chlorococcales, Chlorophyceae). *Phycological Research* 43:105–110.

Haubner, N., R. Schumann, and U. Karsten. 2006. Aeroterrestrial microalgae growing in biofilms on facades—Response to temperature and water stress. *Microbial Ecology* 51:285–293.

Haugen, P., D. Bhattacharya, J. D. Palmer, S. Turner, L. A. Lewis, and K. M. Pryor. 2007. Cyanobacterial ribosomal RNA genes with multiple endonuclease-encoding group I introns. *BMC Evolutionary Biology* 7:159.

Hauth, A. M., U. G. Maier, B. R. Lang, and G. Burger. 2005. The *Rhodomonas salina* mitochondrial genome: Bacteria-like operons, compact gene arrangement and complex repeat region. *Nucleic Acids Research* 33:4433–4442.

Hawkes, M. W. 1980. Ultrastructure characteristics of monospore formation in *Porphyra gardneri* (Rhodophyta). *Journal of Phycology* 16:192–196.

Hawkes, M. W. 1990. Reproductive strategies. In: Cole, K. M., and R. G. Sheath (eds.), *Biology of the Red Algae*, Cambridge University Press, Cambridge, UK, pp. 455–476.

Hay, C. H. 1994. *Durvillaea*. In: Akatsuka, I. (ed.), *Biology of Economic Algae*, SPB Academic Publishing, The Hague, The Netherlands, pp. 353–384.

Hay, M. E. 1981. The functional morphology of turf-forming seaweeds: Persistence in stressful marine habitats. *Ecology* 63:739–750.

Hay, M. E. 1996. Marine chemical ecology: What's known and what's next? *Journal of Experimental Marine Biology and Ecology* 200:103–134.

Hay, M. E., Q. E. Kappel, and W. Fenical. 1994. Synergism in plant defenses against herbivores: Interactions of chemistry, calcification and plant quality. *Ecology* 75:1714–1726.

Hayden, H. S., J. Blomster, C. A. Maggs, P. C. Silva, M. J. Stanhope, and R. J. Waaland. 2003. Linnaeus was right all along; *Ulva* and *Enteromorpha* are not distinct genera. *European Journal of Phycology* 38:277–294.

Hayden, H. S., and J. R. Waaland. 2002. Phylogenetic systematics of the Ulvaceae (Ulvales, Ulvophyceae) using chloroplast and nuclear DNA sequences. *Journal of Phycology* 38:1200–1212.

Heaney, S. I. 1976. Temporal and spatial distribution of the dinoflagellate *Ceratium hirundinella* O. F. Müller within a small productive lake. *Freshwater Biology* 6:531–542.

Heckman, D. H., D. M. Geiser, B. R. Eidell, R. L. Stauffer, N. L. Kardos, and B. L. Hedges. 2001. Molecular evidence for the early colonization of land by plants and fungi. *Science* 293:1129–1133.

Hegewald, E., J. Padisák, and T. Friedl. 2007. *Pseudotetraëdriella kamillae*: Taxonomy and ecology of a new member of the algal class Eustigmatophyceae (Stramenopiles). *Hydrobiologia* 586:107–116.

Heil, C. A., P. M. Gilbert, and C. Fan. 2005. *Prorocentrum minimum* (Pavillard) Schiller. A review of a harmful algal bloom species of growing worldwide importance. *Harmful Algae* 4:449–470.

Hein, M. 1997. Inorganic carbon limitation of photosynthesis in lake phytoplankton. *Freshwater Biology* 37:545–552.

Hellebust, J. A. 1971. Glucose uptake by *Cyclotella cryptica*: Dark induction and light inactivation of transport system. *Journal of Phycology* 7:345–349.

Henderson, G. P., L. Gan, and G. J. Jensen. 2007. 3-D ultrastructure of *O. tauri*: Electron cryotomography of an entire eukaryotic cell. *Public Library of Science One* 8:e749.

Henley, W. J. 1993. Measurement and interpretation of photosynthetic light-response curves in algae in the context of photoinhibition and diel changes. *Journal of Phycology* 29:729–739.

Henley, W. J., J. L. Hironaka, L. Guillou, M. A. Buchheim, J. A. Buchheim, M. W. Fawley, and K. P. Fawley. 2004. Phylogenetic analysis of the "Nannochloris-like" algae and diagnoses of *Picochlorum oklahomensis* gen. et sp. nov. (Trebouxiophyceae, Chlorophyta). *Phycologia* 43:641–652.

Hennes, K. P., and C. A. Suttle. 1995. Direct counts of viruses in natural waters and laboratory cultures by epifluorescence microscopy. *Limnology and Oceanography* 40:1050–1055.

Henriksen, K., J. R. Young, P. R. Bown, and S. L. S. Stipp. 2004. Coccolith biomineralization studied with atomic force microscopy. *Palaeontology* 47:725–743.

Henrikson, L., H. G. Nyman, H. G. Oscarson, and J. A. E. Stenson. 1980. Trophic changes without changes in the external nutrient loading. *Hydrobiologia* 68:257–263.

Henry, E. C., and K. M. Cole. 1982a. Ultrastructure of swarmers in the Laminariales (Phaeophyceae). I. Zoospores. *Journal of Phycology* 18:550–569.

Henry, E. C., and K. M. Cole. 1982b. Ultrastructure of swarmers in the Laminariales (Phaeophyceae). II. Sperm. *Journal of Phycology* 18:570–579.

Herfort, L., E. Loste, F. Meldrum, and B. Thake. 2004. Structural and physiological effects of calcium and magnesium in *Emiliania huxleyi* (Lohmann) Hay and Mohler. *Journal of Structural Biology* 148:307–314.

Hernández, I., M. Christmas, J. M. Yelloly, and B. A. Whitton. 1997. Factors affecting surface alkaline phosphatase activity in the brown alga *Fucus spiralis* at a North Sea intertidal site (Tyne Sands, Scotland). *Journal of Phycology* 33:569–575.

Herndon, W. R. 1964. *Boldia*: A new rhodophycean genus. *American Journal of Botany* 51:575–581.

Hessen, D. O., and E. van Donk. 1993. Morphological changes in *Scenedesmus* induced by substances released from *Daphnia*. *Archiv für Hydrobiologie* 127:129–140.

Heywood, P. 1983. The genus *Vacuolaria* (Raphidophyceae). *Progress in Phycological Research* 2:53–86.

Heywood, P. 1990. Raphidophyta. In: Margulis, L., J. O. Corliss, M. Melkonian, and D. J. Chapman (eds.), *Handbook of Protoctista*, Jones & Bartlett Publishers, Boston, pp. 318–333.

Hibberd, D. J. 1973. Observations on the ultrastructure of flagellar scales in the genus *Synura* (Chrysophyceae). *Archiv für Mikrobiologie* 89:291–304.

Hibberd, D. J. 1981. Notes on the taxonomy and nomenclature of the algal classes Eustigmatophyceae and Tribophyceae (synonym Xanthophyceae). *Botanical Journal of the Linnean Society* 82:93–119.

Hibberd, D. J. 1990a. Eustigmatophyta. In: Margulis, L., J. O. Corliss, M. Melkonian, and D. J. Chapman (eds.), *Handbook of Protoctista*, Jones & Bartlett Publishers, Boston, pp. 326–333.

Hibberd, D. J. 1990b. Xanthophyta. In: Margulis, L., C. O. Corliss, M. Melkonian, and D. J. Chapman (eds.), *Handbook of the Protoctista*, Jones & Bartlett Publishers, Boston, pp. 686–697.

Hibberd, D. J., and G. F. Leedale. 1970. Eustigmatophyceae—A new algal class with unique organization of the motile cell. *Nature* 225:758–760.

Hibberd, D. J., and G. F. Leedale. 1971. Cytology and ultrastructure of the Xanthophyceae. II. The zoospore and vegetative cell of coccoid forms, with special reference to *Ophiocytium majus* Naegeli. *British Phycological Journal* 6:1–23.

Hibberd, D. J., and R. E. Norris. 1984. Cytology and ultrastructure of *Chlorarachnion reptans* (Chlorarachniophyta divisio nova, Chlorarachniophyceae, classis nova). *Journal of Phycology* 20:310–330.

Hilenski, L. L., and P. L. Walne. 1983. Ultrastructure of mucocysts in *Peranema trichophorum* (Euglenophyceae). *Journal of Protozoology* 30:491–496.

Hilenski, L. L., and P. L. Walne. 1985a. Ultrastructure of the flagella of the colorless phagotroph *Peranema trichophorum* (Euglenophyceae). I. Flagellar mastigonemes. *Journal of Phycology* 21:114–125.

Hilenski, L. L., and P. L. Walne. 1985b. Ultrastructure of the flagella of the colorless phagotroph *Peranema trichophorum* (Euglenophyceae). II. Flagellar roots. *Journal of Phycology* 21:125–134.

Hill, D. R. A. 1991. A revised circumscription of *Cryptomonas* (Cryptophyceae) based on examination of Australian strains. *Phycologia* 30:170–188.

Hill, D. R. A., and K. S. Rowan. 1989. The biliproteins of the Cryptophyceae. *Phycologia* 28:455–463.

Hill, D. R. A., and R. Wetherbee. 1986. *Proteomonas sulcata* gen. et sp. nov. (Cryptophyceae), a cryptomonad with two morphologically distinct and alternating forms. *Phycologia* 25:521–543.

Hill, D. R. A., and R. Wetherbee. 1989. A reappraisal of the genus *Rhodomonas* (Cryptophyceae). *Phycologia* 28:143–158.

Hill, J., E. Nelson, D. Tilman, S. Polasky, and D. Tiffany. 2006. Environmental, economic and energetic costs and benefits of biodiesel and ethanol biofuels. *Proceedings of the national Academic of Sciences of the USA* 103:11206–11210.

Hillis-Colinvaux. L. 1980. Ecology and taxonomy of *Halimeda*: Primary producer of coral reefs. *Advances in Marine Biology* 17:1–327.

Hillis-Colinvaux, L. 1984. Systematics of the Siphonales. In: Irvine, D. E. G., and D. John (eds.), *Systematics of the Green Algae*, Academic Press, New York, pp. 271–296.

Hippler, M., B. Rimbault, and Y. Takahashi. 2002. Photosynthetic complex assembly in *Chlamydomonas reinhardtii*. *Protist* 153:197–220.

Hiraoka, M., S. Shimada, M. Uenosono, and M. Masuda. 2003. A new green-tide-forming alga, *Ulva ohnoi* Hiraoka et Shimada sp. nov. (Ulvales, Ulvophyceae) from Japan. *Phycological Research* 51:17–29.

Hixon, M. A. 1997. Effects of reef fishes on corals and algae. In: Birkeland, C. (ed.), *Life and Death of Coral Reefs*, Chapman & Hall, New York, pp. 230–248.

Ho, T. S.-S., and M. Alexander. 1974. The feeding of amebae on algae in culture. *Journal of Phycology* 10:95–100.

Hoagland, K. D., J. P. Carder, and R. L. Spawn. 1996. Effects of organic toxic substances. In: Stevenson, R. J., M. L. Bothwell, and R. L. Lowe *(eds.), Algal Ecology: Freshwater Benthic Ecosystems*, Academic Press, New York, pp. 469–496.

Hodell, D. A., and C. L. Schelske. 1998. Production, sedimentation, and isotopic composition of organic matter in Lake Ontario. *Limnology and Oceanography* 43:200–214.

Hodell, D. A., C. L. Schelske, G. L. Fahnenstiel, and L. L. Robbins. 1998. Biologically induced calcite and its isotopic composition in

Lake Ontario. *Limnology and Oceanography* 43:187–199.

Hoef-Emden, K. 2005. Multiple independent losses of photosynthesis and differing evolutionary rates in the genus *Cryptomonas* (Cryptophyceae): Combined phylogenetic analyses of DNA sequences of the nuclear and the nucleomorph ribosomal operons. *Journal of Molecular Evolution* 60:183–195.

Hoef-Emden, K. 2007. Revision of the genus *Cryptomonas* (Cryptophyceae) II: Incongruences between the classical morphospecies concept and molecular phylogeny in smaller pyrenoid-less cells. *Phycologia* 46:402–428.

Hoef-Emden, K., B. Marin, M. Melkonian. 2002. Nuclear and nucleomorph SSU rDNA phylogeny in the Cryptophyta and the evolution of cryptophyte diversity. *Journal of Molecular Evolution* 55:161–179.

Hoef-Emden, K., and M. Melkonian. 2003. Revision of the genus *Cryptomonas* (Cryptophyceae): A combination of molecular phylogeny and morphology provides insights into a long-hidden dimorphism. *Protist* 154:371–409.

Hoeger, S. J., B. C. Hitzfeld, and F. R. Dietrich. 2005. Occurrence and elimination of cyanobacterial toxins in drinking water treatment plants. *Toxicology and Applied Pharmacology* 203:231–242.

Hoegh-Guldberg, O., et al. 2007. Coral reefs under rapid climate change and ocean acidification. *Science* 318:1737–1742.

Hoffman, L. R. 1973. Fertilization in *Oedogonium*. II. Polyspermy. *Journal of Phycology* 9:296–301.

Hoffman, L. R. 1983. *Atractomorpha echinata* gen. et sp. nov., a new anisogamous member of the Sphaeropleaceae (Chlorophyceae). *Journal of Phycology* 19:76–86.

Hoffmann, J. P. 1990. Dependence of photosynthesis and vitamin B_{12} uptake on cellular vitamin B_{12} concentration in the multicellular alga *Cladophora glomerata* (Chlorophyta). *Limnology and Oceanography* 35:100–108.

Hoffmann, J. P. 1998. Wastewater treatment with suspended and nonsuspended algae. *Journal of Phycology* 34:757–763.

Hoffmann, J. P., and L. E. Graham. 1984. Effects of selected physicochemical factors on growth and zoosporogenesis of *Cladophora glomerata*. *Journal of Phycology* 20:1–7.

Hoffmann, L. 1989. Algae of terrestrial habitats. *The Botanical Review* 55:77–105.

Hoffmann, L. J., I. Peeken, K. Locke, P. Assmy, and M. Veldhuis. 2006. Different reactions of southern ocean phytoplankton size classes to iron fertilization. *Limnology and Oceanography* 51:1217–1229.

Hofmann, E., P. M. Wrench, F. P. Sharples, R. G. Hiller, W. Welte, and K. Diederichs. 1996. Structural basis of light harvesting by carotenoids: Peridinin-chlorophyll-protein from *Amphidinium carterae*. *Science* 272:1788–1791.

Hoham, R. W. 1980. Unicellular chlorophytes—Snow algae. In: Cox, E. R. (ed.), *Phytoflagellates*, Elsevier/North Holland, Amsterdam, pp. 61–84.

Hoham, R. W., and D. W. Blinn. 1979. Distribution of cryophilic algae in an arid region, the American Southwest. *Phycologia* 14:133–145.

Hoham, R. W., and B. Duval. 2001. Microbial ecology of snow and freshwater ice with emphasis on snow algae. In: Jones, H. G., J. W.

Pomeroy, D. A. Walker, and R. W. Hoham (eds.), *Snow Ecology, an Interdisciplinary Examination of Snow-covered Ecosystems*, Cambridge University Press, Cambridge, UK, pp. 168–228.

Hoham, R. W., E. M. Schlag, J. Y. Kang, A. J. Hasselwander, A. F. Behrstock, I. R. Blackburn, R. C. Johnson, and S. C. Roemer. 1998. The effects of irradiance levels and spectral composition on mating strategies in the snow alga, *Chloromonas* sp. -D, from the Tughill Plateau, New York State. *Hydrological Processes* 11:1627–1639.

Hoiczyk, E., and A. Hansel. 2000. Cyanobacterial cell walls: News from an unusual prokaryotic envelope. *Journal of Bacteriology* 182:1191–1199.

Holen, D. A., and M. E. Borass. 1995. Mixotrophy in chrysophytes. In: Sandgren, C. D., J. P. Smol, and J. Kristiansen (eds.), *Chrysophyte Algae. Ecology, Phylogeny and Development*, Cambridge University Press, Cambridge, UK, pp. 119–140.

Hollande, A. 1952a. Classe des Eugleniens. *Traité de Zoologie* 1:238–284.

Hollande, A. 1952b. Classe des Cryptomonadines. *Traité de Zoologie* 1:285–308.

Holm, N. P., and D. E. Armstrong. 1981. Role of nutrient limitation and competition in controlling the populations of *Asterionella formosa* and *Microcystis aeruginosa* in semi-continuous culture. *Limnology and Oceanography* 26:622–634.

Holzinger, A., U. Karsten, C. Lütz, and C. Wiencke. 2006. Ultrastructure and photosynthesis in the supralittoral green macroalga *Prasiola crispa* from Spitsbergen (Norway) under UV exposure. *Phycologia* 45:168–177.

Hommersand, M. H., and S. Fredericq. 1990. Sexual reproduction and cystocarp development. In: Cole, K. M., and R. G. Sheath (eds.), *Biology of the Red Algae*, Cambridge University Press, Cambridge, UK, pp. 305–346.

Honda, D., and I. Inouye. 2002. Ultrastructure and taxonomy of a marine photosynthetic stramenopile *Phaeomonas parva* gen. et sp. nov. (Pinguiophyceae) with emphasis on the flagellar apparatus architecture. *Phycological Research* 50:75–89.

Honda, D., A. Yokota, and J. Sugiyama. 1999. Detection of seven major evolutionary lineages in cyanobacteria based on the 16S rRNA gene sequence analysis with new sequences of five marine *Synechococcus* strains. Journal of Molecular Evolution 48:723–739.

Honegger, R. 1992. Lichens: Mycobiont–photobiont relationships. In: Reisse, W. (ed.), *Algae in Symbioses*, BioPress, Bristol, UK, pp. 255–276.

Hong, Y.-K., C. H. Sohn, M. Polne-Fuller, and A. Gibor. 1995. Differential display of tissue-specific messenger RNAs in *Porphyra perforata* (Rhodophyta) thallus. *Journal of Phycology* 31:640–643.

Hooks, C. E., R. R. Bidigare, M. D. Keller, and R. R. L. Guillard. 1988. Coccoid eukaryotic marine ultraplankters with four different HPLC pigment signatures. *Journal of Phycology* 24:571–580.

Hoops, H. J., and G. L. Floyd. 1982. Ultrastructure of the flagellar apparatus of *Pyrobotrys* (Chlorophyceae). *Journal of Phycology* 18:455–462.

Hoops, H. J., G. L., Floyd, and J. A. Swanson. 1982. Ultrastructure of the biflagellate motile cells of *Ulvaria oxysperma* (Kütz.) Bliding and phylogenetic relationships among ulvaphycean algae. *American Journal of Botany* 69:150–159.

Hopkins, A. W., and G. E. McBride. 1976. The life history of *Coleochaete scutata* (Chlorophyceae) studied by a Feulgen microspectrophotometric analysis. *Journal of Phycology* 12:29–35.

Hori, T., and J. C. Green. 1994. Mitosis and cell division. In: Green, J. C., and B. S. C. Leadbeater (eds.), *The Haptophyte Algae*, Clarendon Press, Oxford, UK, pp. 91–110.

Hori, T., I. Inouye, T. Horiguchi, and G. T. Boalch. 1985. Observations on the motile stage of *Halosphaera minor* (Ostenfeld) (Prasinophyceae) with special reference to the cell structure. *Botanica Marina* 28:529–537.

Horiguchi, T. 1996. *Haramonas dimorpha* gen. et sp. nov. (Raphidophyceae), a new marine raphidophyte from Australian mangrove. *Phycological Research* 44:143–150.

Horiguchi, T., and M. Chihara. 1987. *Spiniferodinium galeiforme*, a new genus and species of benthic dinoflagellates (Phytodiniales, Pyrrhophyta) from Japan. *Phycologia* 26:478–487.

Horiguchi, T., and M. Hoppenrath. 2003. *Haramonas viridis* sp. nov. (Raphidophyceae, Heterokontophyta), a new sand-dwelling raphidophyte from cold temperate waters. *Phycological Research* 51:61–67.

Horiguchi, T., and F. Kubo. 1997. *Roscoffia minor* sp. nov. (Peridiniales, Dinophyceae): A new, sand-dwelling, armored dinoflagellate from Hokkaido, Japan. *Phycological Research* 45:65–69.

Horiguchi, T., and Y. Takano. 2006. Serial replacement of a diatom endosymbiont in the marine dinoflagellate *Peridinium quinquecorne* (Peridiniales, Dinophyceae). *Phycological Research* 54:193–200.

Horodyski, R. J., and L. P. Knauth. 1994. Life on land in the Precambrian. *Science* 263:494–498.

Horton, P., and A. Ruban. 2004. Molecular design of the photosystem II light-harvesting antenna: Photosynthesis and photoprotection. *Journal of Experimental Botany* 56:365–373.

Hoshaw, R. W., and R. L. Hilton. 1966. Observations on the sexual cycle of the saccoderm desmid *Spirotaenia condensata*. *Journal of the Arizona Academy of Science* 4:88–92.

Hoshaw, R. W., R. M. McCourt, and J.-C. Wang. 1990. Conjugatophyceae. In: Margulis, L., J. O. Corliss, M. Melkonian, and D. J. Chapman (eds.), *Handbook of Protoctista*, Jones & Bartlett Publishers, Boston, pp. 119–131.

Hotchkiss, A. T., and R. M. Brown. 1987. The association of rosette and globule terminal complexes with cellulose microfibril assembly in *Nitella translucens* var. *axillaris* (Charophyceae). *Journal of Phycology* 23:229–237.

Hotchkiss, A. T., M. R. Gretz, K. C. Hicks, and R. M. Brown. 1989. The composition and phylogenetic significance of the *Mougeotia* (Charophyceae) cell wall. *Journal of Phycology* 25:646–654.

Houdan, A., A. Bonnard, J. Fresnel, S. Fouchard, C. Billard, and I. Probert. 2004. Toxicity of coastal coccolithophores (Prymnesiophyceae, Haptophyta). *Journal of Plankton Research* 26:875–883.

Houle, D., S. B. Gauthier, S. Paquet, D. Planas, and A. Warren. 2006. Identification of two genera of nitrogen-fixing cyanobacteria growing on three feather moss species in boreal forests of Quebec, Canada. *Canadian Journal of Botany* 84:1024–1029.

Howarth, R. W., and R. Marino. 2006. Nitrogen as the limiting nutrient for eutrophication in coastal marine ecosystems: evolving views over three decades. *Limnology and Oceanography* 51:364–376.

Howell, E. T., M. A. Turner, R. L. France, M. B. Jackson, and P. M. Stokes. 1990. Comparison of zygnematacean (Chlorophyta) algae in the metaphyton of two acidic lakes. *Canadian Journal of Fisheries and Aquatic Science* 47:1085–1092.

Hu, C., Y. Liu, L. Song, and D. Zhang. 2002. Effect of desert soil algae on the stabilization of fine sands. *Journal of Applied Phycology* 14:281–292.

Hu, Q., H. Miyashita, I. Iwasaki, N. Kurano, S. Miyachi, M. Iwaki, and S. Itoh. 1998. A photosystem I reaction center driven by chlorophyll d in oxygenic photosynthesis. *Proceedings of the National Academy of Sciences of the United States of America* 95:13319–13323.

Hu, Q., M. Sommerfeld, E. Jarvis, M. Ghirardi, M. Posewitz, M. Selbert, and A. Darzins. 2008. Microalgal triacylglycerols as feedstocks for biofuel production: Perspectives and advances. *The Plant Journal* 54: 621–639.

Hu, C., D. Zhang, Z. Huang, and Y. Liu. 2003. The vertical microdistribution of cyanobacteria and green algae within desert crusts and the development of the algal crusts. *Plant and Soil* 257:97–111.

Huang, K., T. Kunkel, and C. F. Beck. 2004. Localization of the blue-light receptor phototropin to the flagella of the green alga *Chlamydomonas reinhardtii*. *Molecular Biology of the Cell* 15:3605–3614.

Hudson, J. J., and W. D. Taylor. 1996. Measuring regeneration of dissolved phosphorus in planktonic communities. *Limnology and Oceanography* 41:1560–1565.

Huelsenbeck, J. P., B. Larget, R. E. Miller, and F. Ronquist. 2002. Potential applications and pitfalls of Bayesian inference of phylogeny. *Systematic Biology* 51:673–688.

Huelsenbeck, J. P., and B. Rannala. 1997. Phylogenetic methods come of age: Testing hypothesis in an evolutionary context. *Science* 276:227–232.

Hughes, T., A. M. Szmant, R. Steneck, R. Carpenter, and S. Miller. 1999. Algal blooms on coral reefs: What are the causes? *Limnology and Oceanography* 44:1583–1586.

Hughes, T. P., et al. 2003. Climate change, human impacts, and the resilience of coral reefs. *Science* 301:929–933.

Hughey, J. R., P. C. Silva, and M. H. Hommersand. 2001. Solving taxonomic and nomenclatural problems in Pacific Gigartinaceae (Rhodophyta) using data from type material. *Journal of Phycology* 37:1091–1109.

Huisman, J., J. Sharples, J. M. Stroom, P. M. Visser, W. E. A. Kardinaal, J. M. H. Verspagen, and B. Sommeijer. 2004. Changes in turbulent mixing shift competition for light between phytoplankton species. *Ecology* 85:2960–2970.

Huisman, J., and F. J. Weissing. 1994. Light-limited growth and competition for light in well-mixed aquatic environments: an elementary model. *Ecology* 75:507–520.

Huisman, J., and F. J. Weissing. 1999. Biodiversity of plankton by species oscillations and chaos. *Nature* 402:407–410.

Huisman, J., and F. J. Weissing. 2002. Oscillations and chaos generated by competition for interactively essential resources. *Ecological Research* 17:175–181.

Hunt, B. E., and D. F. Mandoli. 1996. A new, artificial seawater that facilitates growth of large numbers of cells of *Acetabularia acetabulum* (Chlorophyta) and reduces the labor inherent in cell culture. *Journal of Phycology* 32:483–495.

Hunt, M. E., G. L. Floyd, and B. B. Stout. 1979. Soil algae in field and forest environments. *Ecology* 60:363–375.

Hurd, C. L., and C. L. Stevens. 1997. Flow visualization around single- and multiple-bladed seaweeds with various morphologies. *Journal of Phycology* 33:360–367.

Huss, V. A. R., and M. L. Sogin. 1990. Phylogenetic position of some *Chlorella* species with the Chlorococcales based upon complete small-subunit ribosomal RNA sequences. *Journal of Molecular Evolution* 31:432–442.

Hutchins, D. A. 1995. Iron and the marine phytoplankton community. *Progress in Phycological Research* 11:1–49.

Hutchins, D. A., and K. W. Bruland. 1998. Iron-limited diatom growth and SI:N uptake ratios in a coastal upwelling regime. *Nature* 393:561–564.

Hutchinson, G. E. 1961. The paradox of the plankton. *American Naturalist* 95:137–145.

Huttenlauch, I., and R. Stick. 2003. Occurrence of articulins and epiplasmins in protists. *The Journal of Eukaryotic Microbiology* 50:15–18.

Ianora, A., et al. 2004. Aldehyde suppression of copepod recruitments in blooms of a ubiquitous planktonic diatom. *Nature* 429:403–407.

Ianora, A., et al. 2006. New trends in marine chemical ecology. *Estuaries and Coasts* 29:531–551.

Ianora, A., S. A. Poulet, A. Miralto, and R. Grottoli. 1996. The diatom *Thalassiosira rotula* affects reproductive success in the copepod *Arcartia clausi*. *Marine Biology* 125:279–286.

Ibelings, B. W., A. de Bruin, M. Kagami, M. Rijkeboer, M. Brehm, and E. van Donk. 2004. Host parasite interactions between freshwater phytoplankton and chytrid fungi (Chytridiomycota). *Journal of Phycology* 40:437–453.

Igarashi, K,. M. Wada, R. Hori, and M. Samejima. 2006. Surface density of cellobiohydrolase on crystalline celluloses: A critical parameter to evaluate enzyme kinetics at a solid–liquid interface. *FEBS Journal* 273:2869–2878.

Igarashi, K., M. Wada, and M. Samejima. 2007. Activation of crystalline cellulose to cellulose III$_I$ results in efficient hydrolysis by cellobiohydrolase. *FEBS Journal* 274:1785–1792.

Iglesias-Rodriguez, M. D., et al. 2008. Phytoplankton calcification in a high-CO_2 world. *Science* 320:336–340.

Iida, K., K. Takishita, K. Ohshima, and Y. Inagaki. 2007. Assessing the monophyly of chlorophyll-*c* containing plastids by multi-gene phylogenies under the unlinked model conditions. *Molecular Phylogenetics and Evolution* 45:227–238.

Imanian, B., and P. J. Keeling. 2007. The dinoflagellates *Durinskia baltica* and *Kryptoperidinium foliaceum* retain functionally overlapping mitochondria from two evolutionarily distinct lineages. *BMC Evolutionary Biology* 7:172.

Inouye, I. 1993. Flagella and flagellar apparatuses of algae. In: Berner, T. (ed.), *Ultrastructure of Microalgae*, CRC Press, Boca Raton, FL, pp. 99–134.

Inouye, I., T. Hori, and M. Chihara. 1990. Absolute configuration analysis of the flagellar apparatus of *Pterosperma cristatum* (Prasinophyceae) and consideration of its phylogenetic position. *Journal of Phycology* 26:329–344.

Inouye, I., and M. Kawachi. 1994. The haptonema. In: Green, J. C., and B. S. C. Leadbeater (eds.), *The Haptophyte Algae*, Clarendon Press, Oxford, UK, pp. 73–90.

Inouye, I., and R. N. Pienaar. 1985. Ultrastructure of the flagellar apparatus in a species of *Pleurochrysis* (class Prymnesiophyceae). *Protoplasma* 125:24–35.

Isailovic, D., I. Sultana, G. J. Phillips, and E. S. Yeung. 2006. Formation of fluorescent proteins by the attachment of phycoerythrobilin to R-phycoerythrin alpha and beta apo-subunits. *Analytical Biochemistry* 358:38–50.

Iseki, M., et al. 2002. A blue-light-activated adenylyl cyclase mediates photoavoidance in *Euglena gracilis*. *Nature* 415:1047–1051.

Ishida, K., B. R. Green, and T. Cavalier-Smith. 1999. Diversification of a chimaeric algal group, the chlorarachniophytes: Phylogeny of nuclear and nucleomorph small-subunit rRNA genes. *Molecular Biology and Evolution* 16:321–331.

Ishida, K.-I., Y. Cao, M. Hasegawa, N. Okuda, and Y. Hara. 1997. The origin of chlorarachniophyte plastids, as inferred from phylogenetic comparisons of amino acid sequences of EF-Tu. *Journal of Molecular Evolution* 45:682–687.

Ishida, K.-I., T. Nakayama, and Y. Hara. 1996. Taxonomic studies on the Chlorarachniophyta. II. Generic delimitation of the chlorarachniophytes and description of *Gymnochlora stellata* gen. et sp. nov. and *Lotharella* gen. nov. *Phycological Research* 44:37–45.

Issa, O. M., C. Defarge, Y. Le Bissonnais, B. Marin, O. Duval, A. Bruand, L. P. D'Acqui, S. Nordenberg, and M. Annerman. 2007. Effects of the inoculation of cyanobacteria on the microstructure and the structural stability of a tropical soil. *Plant and Soil* 290:209–219.

Iwamoto, K., and T. Ikawa. 2000. A novel glycolate oxidase requiring flavin mononucleotide as the cofactor in the prasinophycean alga *Mesostigma viride*. *Plant and Cell Physiology* 41:988–991.

Izaguirre, G. A.-D. Jungblut, and B. A. Neilan. 2007. Benthic cyanobacteria (Oscillatoriaceae) that produce microcystin-LR, isolated from four reservoirs in southern California. *Water Research* 41:492–498.

Jackson, A. E., and R. D. Seppelt. 1997. Physiological adaptations to freezing and UV radiation exposure in *Prasiola crispa*, an Antarctic terrestrial alga. In: Battaglia, B., J. Valencia, and D. W. H. Walton (eds.), *Antarctic Communities: Species, Structure, and Survival*, Cambridge University Press, Cambridge, UK, pp. 226–233.

Jacobs, W. P. 1993. Rhizome gravitropism precedes gravimorphogenesis after inversion of the green algal coenocyte *Caulerpa prolifera* (Caulerpales). *American Journal of Botany* 80:1273–1275.

Jacobs, W. P. 1994. *Caulerpa*. *Scientific American* 271:100–105.

Jacobsen, A., G. Bratbak, and M. Heidal. 1996. Isolation and characterization of a virus infecting *Phaeocystis pouchetii* (Prymnesiophyceae). *Journal of Phycology* 32:923–927.

Jacobshagen, S., and C. Schnarrenberger. 1990. Two class I aldolases in *Klebsormidium flaccidum* (Charophyceae): An evolutionary link from chlorophytes to higher plants. *Journal of Phycology* 26:312–317.

Jacobson, D. M., and D. M. Anderson. 1986. Thecate heterotrophic dinoflagellates: Feeding behavior and mechanisms. *Journal of Phycology* 22:249–258.

Jacobson, D. M., and D. M. Anderson. 1996. Widespread phagocytosis of ciliates and other protists by marine mixotrophs and heterotrophic thecate dinoflagellates. *Journal of Phycology* 32:279–285.

Jacobson, D. M., and R. A. Andersen. 1994. The discovery of mixotrophy in photosynthetic species of *Dinophysis* (Dinophyceae): Light and electron microscopical observations of food vacuoles in *Dinophysis acuminata*, *D. norvegica*, and two heterotrophic dinophysoid dinoflagellates. *Phycologia* 33:97–110.

Jahn, T. L., E. C. Bovee, and F. F. Jahn. 1979. *How to Know the Freshwater Protozoa*. Wm. C. Brown, Dubuque, IA.

Janech, M. G., A. Krell, T. Mock, J.-S. Kang and J. A. Raymond. 2006. Ice-binding proteins from sea ice diatoms (Bacillariophyceae). *Journal of Phycology* 42:410–416.

Janson, S., and E. Granéli. 2003. Genetic analysis of the *psbA* gene from single cells indicates a cryptomonad origin of the plastid in *Dinophysis* (Dinophyceae). *Phycologia* 42:473–477.

Javaux, E. J., A. H. Knoll, and M. R. Walter. 2001. Morphological and ecological complexity in early eukaryotic ecosystems. *Nature* 412:66–69.

Jaworski, G. H. M., J. F. Talling, and S. I. Heaney. 2003. Potassium dependence and phytoplankton ecology: An experimental study. *Freshwater Biology* 48:833–840.

Jeffrey, C. 1982. Kingdoms, codes, and classification. *Kew Bulletin* 37:403–416.

Jeffries, D. L., and J. M. Klopatek. 1987. Effects of grazing on the vegetation of the Blackbrush Association. *Journal of Range Management* 40:390–392.

Jeffries, D. L., J. M. Klopatek, S. O. Link, and H. Bolton. 1992. Acetyline reduction of cryptogamic crusts from a blackbrush community as related to resaturation and dehydration. *Soil Biology and Biochemistry* 24:1101–1105.

Jeffries, D. L., S. O. Link, and J. M. Klopatek. 1993. CO_2 fluxes of cryptogamic crusts I. Response to resaturation. *New Phytologist* 125:163–173.

Jensen, A. 1995. Production of alginates. In: Wiessner, W., E. Schnepf, and R. C. Starr (eds.), *Algae, Environment and Human Affairs*, BioPress, Bristol, UK, pp. 79–92.

Jensen, J. B. 1974. Morphological studies on Cystoseiraceae and Sargassaceae (Phaeophyceae) with special reference to apical organization. *University of California Publications Botany* 68:1–61.

Jeong, H. J., et al. 2007. Feeding by the *Pfiesteria*-like heterotrophic dinoflagellate *Luciella masanensis*. *Journal of Eukaryotic Microbiology* 54:231–241.

Jewson, D. H. 1992. Life cycle of a *Stephanodiscus* sp. (Bacillariophyta). *Journal of Phycology* 28:856–866.

Jezbera, J., K. Hornak, and Karel Simek. 2005. Food selection by bacterivorous protists: Insight from the analysis of the food vacuole content by means of fluorescence in situ hybridization. *Federation of European Microbiological Societies Microbiology Ecology* 52:351–363.

Jickells, T. D., et al. 2005. Global iron connections between desert dust, ocean biogeochemistry, and climate. *Science* 308:67–71.

Johansen, H. W. 1981. *Coralline Algae, A First Synthesis*. CRC Press, Boca Raton, FL.

Johansen, J. R. 1993. Cryptogamic crusts of semiarid and arid lands of North America. *Journal of Phycology* 29:140–147.

Johansen, J. R. 2001. Impacts of fire on biological soil crusts. In: Belnap, J., and O. L. Lange (eds.), *Biological Soil Crusts: Structure, Function, and Management*, Ecological Studies, vol. 150, Springer Verlag, Berlin, pp. 385–397.

Johansen, J. R., J. Ashley, and W. R. Rayburn. 1993. Effects of rangefire on soil algal crusts in semiarid shrub-steppe of the Lower Columbia Basin and their subsequent recovery. *Great Basin Naturalist* 53:73–88.

Johansen, J. R., and L. L. St. Clair. 1986. Cryptogamic soil crusts: recovery from grazing near Camp Floyd State Park, Utah, USA. *Great Basin Naturalist* 46:632–640.

Johansen, J. R., L. L. St. Clair, B. L. Webb, and G. T. Nebeker. 1984. Recovery patterns of cryptogamic soil crusts in desert rangelands following fire disturbance. *The Bryologist* 87: 238–243.

Johansson, M., and D. W. Coats. 2002. Ciliate grazing on the parasite *Amoebophrya* sp. decreases infection of the red-tide dinoflagellate *Akashiwo sanguinea*. *Aquatic Microbial Ecology* 28:69–78.

John, D. M. 1994. Alternation of generations in algae: Its complexity, maintenance and evolution. *Biological Review* 69:275–291.

John, D. M., and C. A. Maggs. 1997. Species problems in eukaryotic algae: A modern perspective. In: Claridge, M. F., H. A. Dawah, and M. R. Wilson (eds.), *Species: The Units of Biodiversity*, Chapman & Hall, London, pp. 83–105.

John, D. M., B. A. Whitton, and A. J. Brook (eds.) 2002. *The Freshwater Algal Flora of the British Isles. An Identification Guide to Freshwater and Terrestrial Algae*. The Natural History Museum, Cambridge University Press, Cambridge, UK.

John, U., R. A. Fensom, and L. K. Medlin. 2003. The application of a molecular clock based on molecular sequences and the fossil record to explain biogeographic distributions within the

Alexandrium tamarense "species complex" (Dinophyceae). *Molecular Biology and Evolution* 20:1015–1027.

Johnson, J. L., M. W. Fawley, and K. P. Fawley. 2007a. The diversity of *Scenedesmus* and *Desmodesmus* (Chlorophyceae) in Itasca State Park, Minnesota, USA. *Phycologia* 46:214–229.

Johnson, M. D., D. Oldach, C. F. Delwiche, and D. K. Stoecker. 2007b. Retention of transcriptionally active cryptophyte nuclei by the ciliate *Myrionecta rubra*. *Nature* 445:426–428.

Johnson, M. D., and D. K. Stoecker. 2005. Role of feeding in growth and photophysiology of *Myrionecta rubra*. *Aquatic Microbial Ecology* 39:303–312.

Johnson, M. D., T. Tengs, D. Oldach, and D. K. Stoecker. 2006. Sequestration, performance, and functional control of cryptophyte plastids in the ciliate *Myrionecta rubra* (Ciliophora). *Journal of Phycology* 42:1235–1246.

Johnson, R. E., and R. M. Sheehan. 1985. Late Ordovician dasyclad algae of the eastern Great Basin. In: Toomey, D. F., and M. H. Nitecki (eds.), *Paleoalgology: Contemporary Research and Applications*, Springer Verlag, Berlin, pp. 79–84.

Johnson, S. L., C. R. Budinoff, J. Belnap, and F. Garcia-Pichel. 2005. Relevance of ammonium oxidation within biological soil crust communities. *Environmental Microbiology* 7:1–12.

Johnson, S. L., S. Neuer, and F. Garcia-Pichel. 2007c. Export of nitrogenous compounds due to incomplete cycling within biological soil crusts of arid lands. *Environmental Microbiology* 9:680–689.

Johnson, Z. I., E. R. Zinser, A. Coe, N. P. McNulty, E. M. S. Woodward, and S. W. Chisholm. 2006. Niche partitioning among *Prochlorococcus* ecotypes along ocean-scale environmental gradients. *Science* 311:1737–1740.

Joint, I., K. Tait, M. E. Callow, J. A. Callow, D. Milton, P. Williams, and M. Cámara. 2002. Cell-to-cell communication across the prokaryote-eukaryote boundary. *Science* 298:1207.

Jompa, J., and L. J. McCook. 2002. The effects of nutrients and herbivory on competition between a hard coral (*Porites cylindrical*) and a brown alga (*Lobophora variegata*). *Limnology and Oceanography* 47:527–534.

Jones, H. L. J., B. S. C. Leadbeater, and J. C. Green. 1994. Mixotrophy in haptophytes. In: Green, J. C., and B. S. C. Leadbeater (eds.), *The Haptophyte Algae*, Clarendon Press, Oxford, UK, pp. 247–264.

Jones, R. H., and K. J. Flynn. 2005. Nutritional status and diet composition affect the value of diatoms as copepod prey. *Science* 307:1457–1459.

Jonsson, P. R., L. Granhag, P. S. Moschella, P. Åberg, S. J. Hawkins, and R. C. Thompson. 2006. Interactions between wave action and grazing control the distribution of intertidal macroalgae. *Ecology* 87:1169–1178.

Jost, L. 1895. Beiträge zur Kenntniss der Coleochaeteen. *Ber d. Deutsch Botanische Gesellschaft* 13:433–452.

Juhl, A. R., V. L. Trainer, and M. I. Latz. 2001. Effect of fluid shear and irradiance on population growth and cellular toxin content of the dinoflagellate *Alexandrium fundyense*. *Limnology and Oceanography* 46:758–764.

Jürgens, K. 1994. Impact of *Daphnia* on planktonic microbial food webs—A review. *Marine Microbial Food Webs* 8:295–324.

Jürgens, K., and H. Güde. 1994. The potential importance of grazing-resistant bacteria in planktonic systems. *Marine Ecology Progress Series* 112:169–188.

Jüttner, F. 2001. Liberation of 5, 8, 11, 14, 17-eicosapentaenoic acid and other polyunsaturated fatty acids from lipids as a grazer defense reaction in epilithic diatom biofilms. *Journal of Phycology* 37:744–755.

Jüttner, F., and S. B. Watson. 2007. Biochemical and ecological control of geosmin and 2-methylisoborneol in source waters. *Applied and Environmental Microbiology* 73:4395–4406.

Jüttner, F., and H. P. Wessel. 2003. Isolation of di(hydroxymethyl)dihydroxypyrrolidine from the cyanobacterial genus *Cylindrospermum* that effectively inhibits digestive glucosidases of aquatic insects and crustacean grazers. *Journal of Phycology* 39:26–32.

Kaczmarska, I., S. S. Bates, J. M. Ehrman, and C. Léger. 2000. Fine structure of the gamete, auxospore and initial cell in the pennate diatom *Pseudo-nitzschia multiseries* (Bacillariophyta). *Nova Hedwigia* 71:337–357.

Kaczmarska, I., J.M. Ehrman, and S.S. Bates. 2001. A review of auxospore structure, ontogeny, and diatom phylogeny. In: Economou-Amilli, A. (ed.) *Proceedings of the 16th International Diatom Symposium* 601:153–168, University of Athens Press, Athens, Greece.

Kaczmarska, I., M. M. LeGresley, J. L. Martin, and J. Ehrman. 2005. Diversity of the diatom genus *Pseudo-nitzschia* Pergallo in the Quoddy region of the Bay of Fundy, Canada. *Harmful Algae* 4:1–19.

Kagami, M., T. B. Gurung, T. Yoshida, and J. Urabe. 2006. To sink or to be lysed? Contrasting fate of two large phytoplankton species in Lake Biwa. *Limnology and Oceanography* 51:2775–2786.

Kagawa, T., and N. Suetsugu. 2007. Photometrical analysis with photosensory domains of photoreceptors in green algae. *Federation of European Biochemical Societies* 581:368–374.

Kagami, M., E. van Donk, A. de Bruin, M. Rijkeboer, and B. W. Ibelings. 2004. *Daphnia* can protect diatoms from fungal parasitism. *Limnology and Oceanography* 49:680–685.

Kain, J. M., and T. A. Norton. 1990. Marine ecology. In: Cole, K. M., and R. G. Sheath (eds.), *Biology of the Red Algae*, Cambridge University Press, Cambridge, UK, pp. 377–422.

Kajikawa, M., et al. 2006. A front-end desaturase from *Chlamydomonas reinhardtii* produces pinolenic and coniferonic acids by ω13 desaturation in methylotrophic yeast and tobacco. *Plant Cell Physiology* 47:64–73.

Kamiya, M., and H. Kawai. 2002. Dependence of the carposporophyte on the maternal gametophyte in three ceramiacean algae (Rhodophyta), with respect to carposporophyte development, spore production and germination success. *Phycologia* 41:107–115.

Kamiya, M., J. A. West, R. J. King, G. C. Zuccarello, J. Tanaka, and Y. Hara. 1998. Evolutionary divergence in the red algae *Caloglossa leprieurii* and *C. apomeiotica*. *Journal of Phycology* 34:361–370.

Kamiya, R. 1992. Molecular mechanisms of flagellar movement. In: Melkonian, M. (ed.), *Algal Cell Motility*, Chapman & Hall, London, pp. 155–198.

Kamjunke, N., T. Henrichs, and U. Gaedke. 2007. Phosphorus gain by bacterivory promotes the mixotrophic flagellate *Dinobryon* spp. During re-oligotrophication. *Journal of Plankton Research* 29:39–46.

Kappen, L. 2000. Some aspects of the great success of lichens in Antarctica. *Antarctic Science* 1:31–34.

Kappen, L., and E. I. Friedmann. 1983. Ecophysiology of lichens in the dry valleys of southern Victoria Land, Antarctica. II. CO_2 gas exchange in cryptoendolithic lichens. *Polar Biology* 1:227–232.

Kappen, L., M. Bölter, and A. Kuhn. 1987. Photosynthetic activity of lichens in natural habitats in the maritime Antarctic. *Bibiotheca. Lichenologica* 25:297–312.

Kapraun, D. F. 1970. Field and cultural studies of *Ulva* and *Enteromorpha* in the vicinity of Port Aransas, Texas. *Contributions in Marine Science* 15:205–283.

Kapraun, D. F. 2007. Nuclear DNA content estimates in green algal lineages: Chlorophyta and Streptophyta. *Annals of Botany* 99:677–701.

Kapraun, D. F., and P. W. Boone. 1987. Karyological studies of three species of Scytosiphonaceae (Phaeophyta) from coastal North Carolina. *Journal of Phycology* 23:318–322.

Kapraun, D. F., Braly, K. S., and D. W. Freshwater. 2007. Nuclear DNA content variation in the freshwater red algal order Batrachospermales and Thorales (Florideophyceae, Nemaliophycidae). *Phycologia* 46:54–62.

Kapraun, D. F., M. G. Gargiulo, and G. Tripodi. 1988. Nuclear DNA and karyotype variation in species of *Codium* (Codiales, Chlorophyta) from the North Atlantic. *Phycologia* 27:273–282.

Kapraun, D. F., and M. J. Shipley. 1990. Karyology and nuclear DNA quantification in *Bryopsis* (Chlorophyta) from North Carolina, USA. *Phycologia* 29:443–453.

Karagatzides, J. D., M. C. Lewis, and H. M. Schulmann. 1985. Nitrogen fixation in the high arctic tundra at Scarpa Lake, Northwest Territories. *Canadian Journal of Botany* 63:974–979.

Karakashian, S. J., and M. A. Rudzinska. 1981. Inhibition of lysosomal fusion with symbiont-containing vacuoles in *Paramecium bursaria*. *Experimental Cell Research* 131:387–393.

Kargul, J., J. Nield, and J. Barber. 2003. Three-dimensional reconstruction of a light-harvesting complex I-photosystem I (LHCI-PSI) supercomplex from the green alga *Chlamydomonas reinhardtii*. *Journal of Biological Chemistry* 278:16135–16141.

Karnieli, A., R. Kokaly, N. E. West, and R. N. Clark. 2001. Remote sensing of biological soil crusts. In: Belnap, J., and O. L. Lange (eds.), *Biological Soil Crusts: Structure, Function, and Management*, Ecological Studies, vol. 150, Springer Verlag, Berlin, pp. 431–456.

Karol, K. C., R. M. McCourt, M. T. Cimino, and C. F. Delwiche. 2001. The closest living relatives of land plants. *Science* 294:2351–2353.

Karpov, S. A., M. L. Sogin, and J. D. Silberman. 2001. Rootlet homology, taxonomy and phylogeny of bicosoecids based on 18S rRNA gene sequences. *Protistology* 2:34–47.

Karsten, G. 1912. Über die Reduktionstellung bei der Auxosporenbildung von *Surirella saxonica*. *Zeitschrift Botanische* 4:417–426.

Karsten, U., C. Bock, and J. A. West. 1995. ^{13}C-NMR spectroscopy as a tool to study organic osmolytes in the mangrove red algal genera *Bostrychia* and *Stictosiphonia* (Ceramiales). *Phycological Research* 43:241–247.

Karsten, U., T. Friedl, R. Schumann, K. Hoyer, and S. Lembcke. 2005. Mycosporine-like amino acids and phylogenies in green algae: *Prasiola* and its relatives from the Trebouxiophyceae (Chlorophyta). *Journal of Phycology* 41:557–566.

Karsten, U., A. S. Mostaert, R. J. King, M. Kamiya, and Y. Hara. 1996. Osmoprotectors in some species of Japanese mangrove macroalgae. *Phycological Research* 44:109–112.

Kasai, F., and T. Ichimura. 1990. A sex determining mechanism in the *Closterium ehrenbergii* (Chlorophyta) species complex. *Journal of Phycology* 26:195–201.

Kashiyama, Y., et al. 2008. Evidence of global chlorophyll d. *Science* 321:658.

Kastovská, K., J. Elster, M. Stibal, and H. Santruckova. 2005. Microbial assemblages in soil microbial succession after glacial retreat in Svalbard (High Arctic). *Microbial Ecology* 50:396–407.

Katsaros, C. I. 1995. Apical cells of brown algae with particular reference to Sphacelariales, Dictyotales and Fucales. *Phycological Research* 43:43–59.

Katsaros, C., and B. Galatis. 1992. Immunofluorescence and electron microscopic studies of microtubule organization during the cell cycle of *Dictyota dichotoma* (Phaeophyta, Dictyotales). *Protoplasma* 169:75–84.

Katsaros, C. I., D. A. Karyophyllis, and B. D. Galatis. 2002. Cortical F-actin underlies cellulose microfibril patterning in brown algal cells. *Phycologia* 41:178–183.

Katsaros, C., D. Karyophyllis, and B. Galatis. 2006. Cytoskeleton and morphogenesis in brown algae. *Annals of Botany* 97:679–693.

Kawachi, M., M. Atsumi, H. Ikemoto, and S. Miyachi. 2002b. *Pinguiochrysis pyriformis* gen. et sp. nov. (Pinguiophyceae), a new picoplanktonic alga isolated from the Pacific Ocean. *Phycological Research* 50:49–56.

Kawachi, M., I. Inouye, D. Honda, C. J. O'Kelly, J. C. Bailey, R. R. Bidigare, and R. A. Andersen. 2002a. The Pinguiophyceae classis nova, a new class of photosynthetic stramenopiles whose members produce large amounts of omega-3 fatty acids. *Phycological Research* 50:31–47.

Kawachi, M., M.-H. Noël, and R. A. Andersen. 2002c. Re-examination of the marine "chrysophyte" *Polypodochrysis teissieri* (Pinguiophyceae). *Phycological Research* 50:91–100.

Kawai, H. 1988. A flavin-like autofluorescent substance in the posterior flagellum of golden and brown algae. *Journal of Phycology* 24:114–117.

Kawai, H., T. Hanyuda, S. G. A. Draisma, and D. G. Müller. 2007. Molecular phylogeny of *Discosporangium mesarthocarpum* (Phaeophyceae) with a reinstatement of the Discosporangiales. *Journal of Phycology* 43:186–194.

Kawai, H., and I. Inouye. 1989. Flagellar autofluorescence in forty-four chlorophyll *c*-containing algae. *Phycologia* 28:222–227.

Kawai, H., S. Maeba, H. Sasaki, K. Okuda, and E. C. Henry. 2003. *Schizocladia ischiensis*: A new filamentous marine chromophyte belonging to a new class, Schizocladiophyceae. *Protist* 154:211–228.

Kawai, H., and M. Tokuyama. 1995. *Laminarionema elsbetiae* gen. et sp. nov., new endophyte of *Laminaria japonica*. *Phycological Research* 43:185–190.

Kebede-Westhead, E., C. Pizarro, and W. W. Mulbry. 2003. Production and nutrient removal by periphyton grown under different loading rates of anaerobically digested flushed dairy manure. *Journal of Phycology* 39:1275–1282.

Keeling, P. J. 2004. Diversity and evolutionary history of plastids and their hosts. *American Journal of Botany* 91:1481–1493.

Keeling, P. J., et al. 2005. The tree of eukaryotes. *Trends in Ecology and Evolution* 20:670–676.

Keller, L. C., E. P. Romijn, I. Zamora, J. R. Yates III, and W. F. Marshall. 2005. Proteomic analysis of isolated *Chlamydomonas* centrioles reveals orthologs of ciliary disease genes. *Current Biology* 15:1090–1098.

Kelley, I., and L. A. Pfiester. 1990. Sexual reproduction in the freshwater dinoflagellate *Gloeodinium montanum*. *Journal of Phycology* 26:167–173.

Kellogg, E. A., and N. D. Juliano. 1997. The structure and function of RuBisCO and their implications for systematic studies. *American Journal of Botany* 84:413–428.

Kelman, R., M. Feist, N. H. Trewin, and H. Hass. 2003. Charophyte algae from the Rhynie chert. *Transactions of the Royal Society of Edinburgh: Earth Sciences* 94:445–455.

Kennelly, S. J. 1987. Inhibition of kelp recruitment of turfing algae and consequences for an Australian kelp community. *Journal of Experimental Marine Biology and Ecology* 112:49–60.

Kenrick, P., and P. R. Crane. 1997. *The Origin and Early Diversification of Land Plants: A Cladistic Study*. Smithsonian Institution, Washington, DC.

Keren, N., R. Aurora, and H. B. Pakrasi. 2004. Critical roles of bacterioferritins in iron storage and proliferation of cyanobacteria. *Plant Physiology* 135:1666–1673.

Kershaw, K. A. 1985. *Physiological Ecology of Lichens*, Cambridge University Press, New York.

Kerswell, A. P. 2006. Global biodiversity patterns of benthic marine algae. *Ecology* 87:2479–2488.

Kessler, E., M. Schäfer, C. Hümmer, A. Kloboucek, and V. A. R. Huss. 1997. Physiological, biochemical, and molecular characters for the taxonomy of the subgenera of *Scenedesmus* (Chlorococcales, Chlorophyta). *Botanica Acta* 110:244–250.

Kiermayer, O. 1981. Cytoplasmic basis of morphogenesis in *Micrasterias*. In: Kiermayer, O. (ed.), *Cytomorphogenesis in Plants*, Springer Verlag, New York, pp. 147–189.

Kiermayer, O., and U. Meindl. 1984. Interactions of the Golgi apparatus and the plasmalemma in the cytomorphogenesis of *Micrasterias*. In: Wiessner, W., D. Robinson, and R. C. Starr (eds.), *Compartments in Algal Cells and Their Interaction*, Springer Verlag, Berlin, pp. 175–182

Kiermayer, O., and U. Meindl. 1989. Cellular morphogenesis: The desmid (Chlorophyceae) system. In: Coleman, A. W., L. J. Goff, and J. R. Stein-Taylor (eds.), *Algae as Experimental Systems*, Liss, New York, pp. 149–167.

Kiermayer, O., and U. B. Sleytr. 1979. Hexagonally ordered "rosettes" of particles in the plasma membrane of *Micrasterias denticulata* Bréb. and their significance for microfibril formation and orientation. *Protoplasma* 101:133–138.

Kies, L. 1976. Untersuchungen zur Feinstructure und taxonomischen Einordnung von *Gloeochaete wittrockiana*, einer apoplastidalen capsalen Alga mit blaugrünen Endosymbionten (Cyanellen). *Protoplasma* 87:419–446.

Kies, L. 1992. Glaucocystophyceae and other protists harboring prokaryotic endocytobionts. In: Reisser, W. (ed.), *Algae and Symbioses*, BioPress, Bristol, UK, pp. 353–379.

Kies, L., and B. P. Kremer. 1990. Glaucocystophyta. In: Margulis, L., J. O. Corliss, M. Melkonian, and D. J. Chapman (eds.), *Handbook of Protoctista*, Jones & Bartlett Publishers, Boston, pp. 152–166.

Kim, D., T. Nakashima, Y. Matsuyama, Y. Niwano, K. Yamaguchi, and T. Oda. 2007. Presence of the distinct systems responsible for superoxide anion and hydrogen peroxide generation in red tide phytoplankton *Chattonella marina* and *Chattonella ovata*. *Journal of Plankton Research* 29:241–247.

Kim, E., and J. M. Archibald. 2008. Diversity and evolution of plastids and their genomes. In: Sandelius, A. S., and H. Aronsson (eds.), *The Chloroplast–Interactions with the Environment*, Springer, New York, available Feb. 2009.

Kim, E., and L. E. Graham. 2008. EEF2 challenges the monophyly of Archaeplastida and Chromalveolata. *Public Library of Science One* 3:e2621.

Kim, E., L. W. Wilcox, M. W. Fawley, and L. E. Graham. 2006. Phylogenetic position of the green flagellate *Mesostigma viride* based on α-tubulin and β-tubulin gene sequences. *International Journal of Plant Science* 167:873–883.

Kim, E., L. Wilcox, L. Graham, and J. Graham. 2004a. Genetically distinct populations of the dinoflagellate *Peridinium limbatum* in neighboring northern Wisconsin lakes. *Microbial Ecology* 48:521–527.

Kim, G. H., and L. Fritz. 1993. Signal glycoprotein with α-D-mannosyl residues is involved in wound-healing processes of *Antithamnion sparsum*. *Journal of Phycology* 29:89–90.

Kim, G. H., I. K. Lee, and L. Fritz. 1995. The wound-healing responses of *Antithamnion nipponicum* and *Griffithsia pacifica* (Ceramiales, Rhodophyta) monitored by lectins. *Phycological Research* 43:161–166.

Kim, S., M. G. Park, W. Yih, and D. W. Coats. 2004b. Infection of the bloom-forming thecate dinoflagellates *Alexandrium affine* and *Gonyaulax spinifera* by two strains of *Amoebophrya* (Dinophyta). *Journal of Phycology* 40:815–822.

Kingston, M. B. 1999. Effect of light on vertical migration and photosynthesis of *Euglena proxima* (Euglenophyta). *Journal of Phycology* 35:245–253.

Kirk, D. L. 1997. *Volvox. Molecular Genetic Origins of Multicellularity and Cellular Differentiation*. Cambridge University Press, Cambridge, UK.

Kirk, M. M., et al. 1999. *regA*, a *Volvox* gene that plays a central role in germ-soma differentiation, encodes a novel regulatory protein. *Development* 126:639–647.

Kirkland, B. L., and R. L. Chapman. 1990. The fossil green alga *Mizzia* (Dasycladaceae): A tool for interpretation of paleoenvironment in the upper Permian Capitan Reef complex, southeastern New Mexico. *Journal of Phycology* 26:569–576.

Kirkwood, A. E., and W. J. Henley. 2006. Algal community dynamics and halotolerance in a terrestrial hypersaline environment. *Journal of Phycology* 42:537–547.

Kiss, J. Z., and L. A. Staehelin. 1993. Structural polarity in the *Chara* rhizoid: A reevaluation. *American Journal of Botany* 80:273–282.

Kitade, Y., et al. 2008. Identification of genes preferentially expressed during asexual sporulation in *Porphyra yezoensis* gametophytes (Bangiales, Rhodophyta). *Journal of Phycology* 44: 113–123.

Klaveness, D. 1981. *Rhodomonas lacustris* (Pascher & Ruttner) Javornicky (Cryptomonadida): Ultrastructure of the vegetative cell. *Journal of Protozoology* 28:83–90.

Klaveness, D. 1982. The *Cryptomonas–Caulobacter* consortium: Facultative ectocommensalism with possible taxonomic consequences? *Nordic Journal of Botany* 2:183–188.

Klaveness, D. 1988. Ecology of the cryptomonadida: A first review. In: Sandgren, C. D. (ed.), *Growth and Reproductive Strategies of Freshwater Phytoplankton*, Cambridge University Press, Cambridge, UK, pp. 105–133.

Klebahn, H. 1896. Beiträge zur Kenntnis der Auxosporenbildung. I. *Rhopalodia gibba* (Ehrenb). O. Müller. *Jahrbuch wissenschaftliche Botanik* 29:595–654.

Kleijne, A. 1992. Extant Rhabdosphaeraceae (coccolithophorids, class Prymnesiophyceae) from the Indian Ocean, Red Sea, Mediterranean Sea, and North Atlantic Ocean. *Scripta Geologica* 100:1–63.

Kleiner, E. F., and K. T. Harper. 1972. Environment and community organization in grasslands of Canyonlands National Park. *Ecology* 53:299–309.

Kleiner, E. F., and K. T. Harper. 1977. Soil properties in relation to cryptogamic ground cover in Canyonlands National Park. *Journal of Range Management* 30:202–205.

Klemer, A. R., J. J. Cullen, M. T. Mageau, K. M. Hanson, and R. A. Sundell. 1996. Cyanobacterial buoyancy regulation: The paradoxical roles of carbon. *Journal of Phycology* 32:47–53.

Knapp, P. A. 1996. Cheatgrass (*Bromus tectorum* L.) dominance in the Great Basin Desert. *Global Environmental Change* 6:37–-52.

Knaust, R., T. Urbig, L. Li, W. Taylor, and J. W. Hastings. 1998. The circadian rhythm of bioluminescence in *Pyrocystis* is not due to differences in the amount of luciferase: A comparative study of three bioluminescent marine dinoflagellates. *Journal of Phycology* 34:167–172.

Kneip, C., P. Lockhart, C. Voß, and U.-G. Maier. 2007. Nitrogen fixation in eukaryotes—New models for symbiosis. *BMC Evolutionary Biology* 7:55.

Knoll, A. 1996. Archean and Proterozoic paleontology. In: Jansonius, J., and D. C. McGregor.

(eds.), *Palynology: Principles and Applications*, American Association of Stratigraphic Palynologists Foundation, pp. 51–80.

Knoll, A. H. 1992. The early evolution of eukaryotes: A geological perspective. *Science* 256:622–627.

Knoll, A. H., R. E. Summons, J. R. Waldbauer, and J. E. Zumberge. 2007. The geological succession of primary producers in the oceans. In: Falkowski, P. G., and A. H. Knoll (eds.), *Evolution of Primary Producers in the Sea*, Academic Press, New York, pp. 133–163.

Knoll, A. J. 2003. *Life on a Young Planet*. Princeton University Press, Princeton, N. J.

Kociolek, J. P., and M. J. Wynne. 1988. Observations on *Navicula thallodes* (Bacillariophyceae), a blade-forming diatom from the Bering Sea. *Journal of Phycology* 24:439–441.

Kodama, Y., M. Nakahara, and M. Fujishima. 2007. Symbiotic *Chlorella vulgaris* of the ciliate *Paramecium bursaria* shows temporary resistance to host lysosomal enzymes during the early infection process. *Protoplasma* 230:61–67.

Koehl, M. A. R., and R. S. Alberte. 1988. Flow, flapping, and photosynthesis of *Nereocystis luetkeana*: A functional comparison of undulate and flat blade morphologies. *Marine Biology* 99:435–444.

Kofoid, C. A., and O. Swezy. 1921. The free-living unarmored Dinoflagellata. *Memoirs of the University of California* 5:1–562.

Kogame, Y., and H. Kawai. 1996. Development of the intercalary meristem in *Chorda filum* (Laminariales, Phaeophyceae) and other primitive Laminariales. *Phycological Research* 44:247–260.

Köhler, K. 1956. Entwicklungsgeschichte, Geschlects-bestimmung und Befruchtung bei *Chaetomorpha*. *Archiv für Protistenkunde* 101:223–268.

Köhler, S., C. F. Delwiche, P. W. Denny, L. G. Tilney, P. Webster, R. J. M. Wilson, J. D. Palmer, and D. S. Roos. 1997. A plastid of probable green algal origin in apicomplexan parasites. *Science* 275:1485–1489.

Kokinos, J. P., T. I. Eglinton, M. A. Goni, J. Boon, P. A. Martoglios, and D. M. Anderson. 1998. Characterization of a highly resistant biomacromolecular material in the cell wall of a marine dinoflagellate resting cyst. *Organic Geochemistry* 28:265–288.

Komárek, J. 1989. Polynuclearity of vegetative cells in coccal green algae from the family Neochloridaceae. *Archiv für Protistenkunde* 137:255–273.

Komárek, J., and K. Anagnostidis. 1986. Modern approaches to the classification of cyanophytes. II: Chroococcales. *Archiv für Hydrobiologie* 73:157–226.

Komárek, J., and K. Anagnostidis. 1989. Modern approaches to the classification of cyanophytes 4—Nostocales. *Archiv für Hydrobiologie* 82:247–345.

Kooistra, W. H. C. F. 2002. Molecular phylogenies of Udoteaceae (Bryopsidales, Chlorophyta) reveal nonmonophyly for *Udotea*, *Penicillus* and *Chlorodesmis*. *Phycologia* 41:453–462.

Kooistra, W. H. C. F., R. Gersonde, L. K. Medlin, and D. G. Mann. 2007. The origin and evolution of the diatoms: Their adaptation to a planktonic existence. In: Falkowski, P. G., and

A. H. Knoll (eds.), *Evolution of Primary Producers in the Sea*, Academic Press, New York, pp. 210–249.

Kooistra, W. H. C. F., and L. K. Medlin. 1996. Evolution of the diatoms (Bacillariophyta). IV: A reconstruction of their age from small subunit rRNA coding regions and the fossil record. *Molecular Phylogenetics and Evolution* 6:391–407.

Koop, H.-U. 1979. The life cycle of *Acetabularia* (Dasycladales, Chlorophyceae): A compilation of evidence for meiosis in the primary nucleus. *Protoplasma* 100:353–366.

Kopp, R. E., J. L. Kirschvink, I. A. Hilburn, and C. Z. Nash. 2005. The Paleoproterozoic snowball Earth: A climate disaster triggered by the evolution of oxygenic photosynthesis. *Proceedings of the National Academy of Sciences of the United States of America* 102:11131–11136.

Kornmann, P. 1938. Zur Entwicklungsgeschichte von *Derbesia* und *Halicystis*. *Planta* 28:464–470.

Kornmann, P. 1984. *Erythrotrichopeltis*, eine neue Gattung der Erythropeltidaceae (Bangiophyceae, Rhodophyta) *Helgolander Meeresunter-suchungen* 38:207–224.

Kornmann, P. 1987. Der Lebenszyklus von *Porphyrostromium obscurum* (Bangiophyceae, Rhodophyta). *Helgolander Meeresunter-suchungen* 41:127–137.

Kosourov, S. E. Patrusheva, M. L. Ghirardi, M. Seibert, and A. Tsygankov. 2007. A comparison of hydrogen photoproduction by sulfur-deprived *Chlamydomonas reinhardtii* under different growth conditions. *Journal of Biochemistry* 128:776–787.

Koziol, A. G., T. Borza, K.-I. Ishida, P. Keeling, R. W. Lee, and D. G. Durnford. 2007. Tracing the evolution of the light-harvesting antennae in chlorophyll *a/b*-containing organisms. *Plant Physiology* 143:1802–1816.

Kraemer, G. P., and D. J. Chapman. 1991 Biomechanics and alginic acid composition during hydrodynamic adaptation by *Egregia menziesii* (Phaeophyta) juveniles. *Journal of Phycology* 27:47–53.

Kraft, G. T., and I. A. Abbott. 2002. The anatomy of *Neotenophycus ichthyosteus* gen. et sp. nov. (Rhodomelaceae, Ceramiales), a bizarre red algal parasite from the central Pacific. *European Journal of Phycology* 37:269–278.

Kraft, G. T., and P. Robins. 1985. Is the order Cryptonemiales defensible? *Phycologia* 24:67–77.

Kraft, G. T., G. W. Saunders, I. A. Abbott, and R. J. Haroun. 2004. A uniquely calcified brown alga from Hawaii: *Newhousia imbricata* gen. et sp. nov. (Dictyotales, Phaeophyceae). *Journal of Phycology* 40:383–394.

Kratz, R. F., P. A. Young, and D. F. Mandoli. 1998. Timing and light regulation of apical morphogenesis during reproductive development in wild-type populations of *Acetabularia acetabulum* (Chlorophyceae). *Journal of Phycology* 34:138–146.

Kremer, B. P. 1983. Carbon economy and nutrition of the alloparasitic red alga *Harveyella mirabilis*. *Marine Biology* 76:231–239.

Krienitz, L., E. H. Hegewald, D. Hepperle, V. A. R. Huss, T. Rohr, and M. Wolf. 2004. Phylogenetic relationship of *Chlorella* and *Parachlorella* gen. nov. (Chlorophyta, Trebouxiophyceae). *Phycologia* 43:529–542.

Krienitz, L., E. Ustinova, T. Friedl, and V. A. R. Huss. 2001. Traditional generic concepts versus 18S rRNA gene phylogeny in the green algal family Selenastraceae (Chlorophyceae, Chlorophyta). *Journal of Phycology* 37:852–865.

Kröger, N., S. Lorenz, E. Brunner, and M. Sumper. 2002. Self-assembly of highly phosphorylated silaffins and their function in biosilica morphogenesis. *Science* 298:584–586.

Kroken, S. B., L. E. Graham, and M. E. Cook. 1996. Occurrence and evolutionary significance of resistant cell walls in charophytes and bryophytes. *American Journal of Botany* 83:1241–1254.

Krupp, J. M., and N. J. Lang. 1985. Cell division and filament formation in the desmid *Bambusina brebissonii* (Chlorophyta). *Journal of Phycology* 21:16–25.

Kubanek, J., P. R. Jensen, P. A. Keifer, M. C. Sullards, D. O. Collins, and W. Fenical. 2003. Seaweed resistance to microbial attack: A targeted chemical defense against marine fungi. *Proceedings of the National Academy of Sciences of the United States of America* 100:6916–6921.

Kübler, J. E., and J. A. Raven. 1995. The interaction between inorganic carbon acquisition and light supply in *Palmaria palmata* (Rhodophyta). *Journal of Phycology* 31:369–375.

Kuffner, I. B., and V. J. Paul. 2004. Effects of the benthic cyanobacterium *Lyngbya majuscula* on larval recruitment of the reef corals *Acropora surculosa* and *Pocillopora damicornis*. *Coral Reefs* 23:455–458.

Kugrens, P. 1980. Electron microscopic observations on the differentiation and release of spermatia in the marine red alga *Polysiphonia hendryi* (Ceramiales, Rhodomelaceae). *American Journal of Botany* 67:519–528.

Kugrens, P., B. L. Clay, C. J. Meyer, and R. E. Lee. 1999. Ultrastructure and description of *Cyanophora biloba*, sp. nov., with additional observations on *C. paradoxa* (Glaucophyta). *Journal of Phycology* 35:844–854.

Kugrens, P., and R. E. Lee. 1988. Ultrastructure of fertilization in a cryptomonad. *Journal of Phycology* 24:385–393.

Kugrens, P., and R. E. Lee. 1990. Ultrastructural evidence for bacterial incorporation and myxotrophy in the photosynthetic cryptomonad *Chroomonas pochmanni* Huber-Pestalozzi (Cryptomonadida). *Journal of Protozoology* 37:263–267.

Kugrens, P., R. E. Lee, and R. A., Andersen. 1986. Cell form and surface patterns in *Chroomonas* and *Cryptomonas* cells (Cryptophyta) as revealed by scanning electron microscopy. *Journal of Phycology* 22:512–522.

Kühlbrandt, W., D. N. Wang, and Y. Fujiyoshi. 1994. Atomic model of plant light-harvesting complex by electron crystallography. *Nature* 367:614–621.

Kühn, S, L. Medlin, and G. Eller. 2004. Phylogenetic position of the parasitoid nanoflagellate *Pirsonia* inferred from nuclear-encoded small subunit ribosomal DNA and a description of *Pseudopirsonia* n. gen. and *Pseudopirsonia mucosa* (Drebes) comb. nov. *Protist* 155:143–156.

Kupferberg, S. J., J. C. Marks, and M. E. Power. 1994. Effects of variation in natural algal and detrital diets on larval anuran (*Hyla regilla*) life-history traits. *Copeia* 1994:446–457.

Küpper, F. J., et al. 2008. Iodide accumulation provides kelp with an inorganic antioxidant impacting atmospheric chemistry. *Proceedings of the National Academy of Sciences of the United States of America* 105:6954–6958.

Kurogi, M. 1972. Systematics of *Porphyra* in Japan. In: Abbott, I. A., and M. Kurogi (eds.), *Contributions to the Systematics of Benthic Marine Algae of the North Pacific*, Japanese Society of Phycology, Kobe, Japan, pp. 167–191.

Kuvardina, O. N., B. S. Leander, V. V. Aleshin, A. P. Myl'nikov, P. J. Keeling, and T. G. Simdyanov. 2002. The phylogeny of colpodellids (Alveolata) using small subunit rRNA gene sequences suggests they are the free-living sister group to apicomplexans. *Journal of Eukaryotic Microbiology* 49:498–504.

La Claire, J. W. 1982a. Cytomorphological aspects of wound healing in selected Siphonocladales (Chlorophyceae). *Journal of Phycology* 18:379–384.

La Claire, J. W. 1982b. Light and electron microscopic studies of growth and reproduction in *Cutleria* (Phaeophyta). III. Nuclear division in the trichothallic meristem of *C. cylindrica*. *Phycologia* 21:273–287.

La Claire, J. W. 1991. Immunolocalization of myosin in intact and wounded cells of the green alga *Ernodesmis verticillata* (Kützing) Borgesen. *Planta* 184:209–217.

La Claire, J. W., R. Chen, and D. L. Herrin. 1995. Identification of a myosin-like protein in *Chlamydomonas reinhardtii* (Chlorophyta). *Journal of Phycology* 31:302–306.

La Claire, J. W., and J. A. West. 1979. Light- and electron-microscopic studies of growth and reproduction in *Cutleria* (Phaeophyta). II. Gametogenesis in the male plant of *C. hancockii*. *Protoplasma* 101:247–267.

Laatsch, T., S. Zauner, B. Stoebe-Maier, K. V. Kowallik, and U. G. Maier. 2004. Plastid-derived single gene minicircles of the dinoflagellate *Ceratium horridum* are localized in the nucleus. *Molecular Biology and Evolution* 21:1318–1322.

Lackey, J. B. 1968. Ecology of *Euglena*. In: Buetow, D. E. (ed.), *The Biology of Euglena*, vol. I, Academic Press, New York, pp. 28–44.

LaJeunesse, T. C. 2001. Investigation of the biodiversity, ecology, and phylogeny of endosymbiotic dinoflagellates in the genus *Symbiodinium* using the ITS region: In search of a "species" level marker. *Journal of Phycology* 37:866–880.

Lake, J. A. 2007. Disappearing act. *Nature* 446:983.

Lalley, J. S., and H. A. Viles. 2008. Recovery of lichen-dominated soil crusts in a hyper-arid desert. *Biodiversity and Conservation* 17:1–20.

Lam, D. W., and F. W. Zechman. 2006. Phylogenetic analysis of the Bryopsidales (Ulvophyceae, Chlorophyta) based on rubisco large subunit gene sequences. *Journal of Phycology* 42:669–678.

Lamberti, G. A. 1996. The role of periphyton in benthic food webs. In: Stevenson, R. J., M. L. Bothwell, and R. L. Lowe (eds.), *Algal Ecology: Freshwater Benthic Ecosystems*, Academic Press, New York, pp. 533–572.

Lampert, W. 1987. Predictability in lake ecosystems: The role of biotic interactions. In: Schulze, E. D., and H. Zwolfer (eds.), *Potentials and Limitations of Ecosystem Analysis*, Ecological Studies 61, Springer Verlag, Berlin, pp. 333–346.

Lampert, W., W. Fleckner, H. Rai, and B. E. Taylor. 1986. Phytoplankton control by grazing zooplankton: A study on the spring clear-water phase. *Limnology and Oceanography* 31:478–490.

Lampert, W., and U. Sommer. 1997. *Limnoecology*, Oxford University Press, New York.

Lancelot, C., M. D. Keller, V. Rosseau, W. O. Smith, and S. Mathot. 1998. Autecology of the marine haptophyte *Phaeocystis* sp. In: Anderson, D. M., A. D. Cembella, and G. M. Hallegraeff (eds.), *Physiological Ecology of Harmful Algal Blooms*, Springer Verlag, Berlin, pp. 209–224.

Lane, C. E., C. Mays, L. D. Druehl, and G. W. Saunders. 2006. A multi-gene molecular investigation of the kelp (Laminariales, Phaeophyceae) supports substantial taxonomic re-organization. *Journal of Phycology* 42:493–512.

Lane, C. E., and G. W. Saunders. 2005. Molecular investigation reveals epi/endophytic extrageneric kelp (Laminariales, Phaeophyceae) gametophytes colonizing *Lessoniopsis littoralis* thalli. *Botanica Marina* 48:426–436.

Lange, O. L., and L. Kappen. 1972. Photosynthesis of lichens from Antarctica. In: Llano, G. A. (ed.), *Antarctic Terrestrial Biology*, Antarctic Research Series 20, American Geophysical Union, Washington DC, pp. 83–95.

Lapointe, B. E. 1999. Simultaneous top-down and bottom-up forces control macroalgal blooms on coral reefs (Reply to the comment by Hughes et al.). *Limnology and Oceanography* 44:1586–1592.

Lapointe, B. E., P. J. Barile, M. M. Littler, and D. S. Littler. 2005b. Macroalgal blooms on southeast Florida coral reefs. II. Cross-shelf discrimination of nitrogen sources indicates widespread assimilation of sewage nitrogen. *Harmful Algae* 4:1106–1122.

Lapointe, B. E., P. J. Barile, M. M. Littler, D. S. Littler, B. J. Bedford, and C. Gasque. 2005a. Macroalgal blooms on southeast Florida coral reefs. I. Nutrient stoichiometry of the invasive green alga *Codium isthmocladum* in the wider Caribbean indicates nutrient enrichment. *Harmful Algae* 4:1092–1105.

Larkum, T., and C. J. Howe. 1997. Molecular aspects of light-harvesting processes in algae. *Advances in Botanical Research* 27:258–330.

LaRoche, J., G. W. M. van der Staay, P. Partensky, A. Ducret, R. Aebersold, R. Li, S. S. Golden, R. G. Hiller, P. M. Wrench, A. W. D. Larkum, and B. R. Green. 1996. Independent evolution of the prochlorophyte and green plant chlorophyll *a/b* light-harvesting proteins. *Proceedings of the National Academy of Sciences of the United States of America* 93:15244–15248.

Larsen, A., and B. Edvardsen. 1998. Relative ploidy levels in *Prymnesium parvum* and *P. patelliferum* (Haptophyta) analyzed by flow cytometry. *Phycologia* 37:412–424.

Larson, A., M. M. Kirk, and D. L. Kirk. 1992. Molecular phylogeny of the volvocine flagellates. *Molecular Biology and Evolution* 9:83–105.

Latasa, M., X. A. G. Morán, R. Scharek, and M. Estrada. 2005. Estimating the carbon flux through main phytoplankton groups in the northwestern Mediterranean. *Limnology and Oceanography* 50:1447–1458.

Latasa, M, R. Scharek, F. Le Gall, and L. Guillou. 2004. Pigment suites and taxonomic groups in Prasinophyceae. *Journal of Phycology* 40:1149–1155.

Laurent, S., H. Chen, S. Bédu, F. Ziarelli, L. Peng, and C.-C. Zhang. 2005. Nonmetabolizable analogue of 2-oxoglutarate elicits heterocyst differentiation under repressive conditions in *Anabaena* sp. PCC 7120. *Proceedings of the National Academy of Sciences of the United States of America* 102:9907–9912.

Lavau, S., G. W. Saunders, and R. Wetherbee. 1997. A phylogenetic analysis of the Synurophyceae using molecular data and scale case morphology. *Journal of Phycology* 33:135–151.

Lavoie, I., S. Campeau, F. Darchambeau, G. Cabana, and P. J. Dillon. 2008. Are diatoms good integrators of temporal variability in stream water quality? *Freshwater Biology* 53:827–841.

Lawley, B., S. Ripley, P. Bridge, and P. Convey. 2004. Molecular analysis of geographic patterns of eukaryotic diversity in Antarctic soils. *Applied and Environmental Microbiology* 70:5963–5972.

Lawlor, D. W. 1993. *Photosynthesis: Molecular, Physiological, and Environmental Processes*, Longman Scientific, London.

Laycock, M. V., J. F. Jellett, D. J. Easy, and M. A. Donovan. 2006. First report of a new rapid assay for diarrhetic shellfish poisoning toxins. *Harmful Algae* 5:74–78.

Le Gall, L., and G. W. Saunders. 2006. A nuclear phylogeny of the Florideophyceae (Rhodophyta) inferred from combined EF2, small subunit and large subunit ribosomal DNA: Establishing the new red algal subclass Corallinophycidae. *Molecular Phylogenetics and Evolution* 43:1118–1130.

Leadbeater, B. S. C., and D. A. N. Barker. 1995. Biomineralization and scale production in the Chrysophyta. In: Sandgren, C. D., J. P. Smol, and J. Kristiansen (eds.), *Chrysophyte Algae: Ecology, Phylogeny and Development*, Cambridge University Press, Cambridge, UK, pp. 141–164.

Leadbeater, B. S. C., and J. Green. 1993. Berner, T. (ed.), *Ultrastructure of Microalgae*, CRC Press, Boca Raton, FL.

League, E. A., and V. A. Greulach. 1955. Effects of daylength and temperature on the reproduction of *Vaucheria sessilis*. *The Botanical Gazette* 117:45–51.

Leander, B. S. 2004. Did trypanosomatid parasites have photosynthetic ancestors? *Trends in Microbiology* 12:251–258.

Leander, B. S., and M. A. Farmer. 2001. Evolution of *Phacus* (Euglenophyceae) as inferred from pellicle morphology and SSU rDNA. *Journal of Phycology* 37:143–159.

Leander, B. S., R. E. Triemer, and M. A. Farmer. 2001. Character evolution in heterotrophic euglenids. *European Journal of Protistology* 37:337–356.

Lebeau, T., and J.-M. Robert. 2003a. Diatom cultivation and biotechnologically relevant products. Part I. Cultivation at various scales. *Applied Microbiology and Biotechnology* 60:612–623.

Lebeau, T., and J.-M. Robert. 2003b. Diatom cultivation and biotechnologically relevant products. Part II. Current and putative products. *Applied Microbiology and Biotechnology* 60:624–632

Lechat, H., M. Amat, J. Mazoyer, A. Buléon, and M. Lahaye. 2000. Structure and distribution of glucomannan and sulfated glucan in the cell walls of the red alga *Kappaphycus alvarezii* (Gigartinales, Rhodophyta). Journal of Phycology 35:891–902.

Lee, D.-H., M. Mittag, S. Sczekan, D. Morse, and J. W. Hastings. 1993. Molecular cloning and genomic organization of a gene for luciferin-binding protein from the dinoflagellate *Gonyaulax polyedra*. Journal of Biological Chemistry 268:8842–8850.

Lee, J. J. 1992a. Symbiosis in Foraminifera. In: W. Reisser, W. (ed.), *Algae and Symbioses*, BioPress, Bristol, UK, pp. 63–78.

Lee, J. J. 1992b. Taxonomy of algae symbiotic in Foraminifera. In: W. Reisser, W. (ed.), *Algae and Symbioses*, BioPress, Bristol, UK, pp. 79–92.

Lee, K. W., and H. C. Bold. 1973. *Pseudocharaciopsis texensis* gen. et sp. nov., a new member of the Eustigmatophyceae. *British Phycological Journal* 8:31–37.

Lee, R. E., P. Kugrens, and P. P. Mylnikov. 1991. Feeding apparatus of the colorless flagellate *Katablepharis* (Cryptophyceae). *Journal of Phycology* 27:725–733.

Lee, S. H., T. Motomura, and T. Ichimura. 1998. Karyogamy follows plasmogamy in the life cycles of *Derbesia tenuissima* (Chlorophyta). *Phycologia* 37:330–333.

Lee, Y.-K., and S.-Y. Ding. 1995. Effect of dissolved oxygen partial pressure on the accumulation of astaxanthin in chemostat cultures of *Haematococcus lacustris* (Chlorophyta). *Journal of Phycology* 31:922–924.

Leedale, G. F. 1967. *Euglenoid Flagellates*, Prentice Hall, Englewood Cliffs, NJ.

Leedale, G. F., B. S. C. Leadbeater, and A. Massalski. 1970. The intracellular origin of flagellar hairs in the Chrysophyceae and Xanthophyceae. *Journal of Cell Science* 6:701–719.

Leflaive, J., G. Lacroix, Y. Micaise, and L. Ten-Hage. 2008. Colony induction and growth inhibition in *Desmodesmus quadrispina* (Chlorococcales) by allelochemical released from the filamentous alga *Uronema confervicolum* (Ulotrichales). *Environmental Microbiology* 10:1536–1546.

Leflaive, J., and L. Ten-Hage. 2007. Algal and cyanobacterial secondary metabolites in freshwaters: A comparison of allelopathic compounds and toxins. *Freshwater Biology* 52:199–214.

Leggat, W. M. R. Badger, and D. Yellowlees. 1999. Evidence for an inorganic carbon-concentrating mechanism in the symbiotic dinoflagellate *Symbiodinium* sp. *Plant Physiology* 121:1247–1255.

Leggat, W., O. Hoegh-Guldberg, S. Dove, and D. Yellowlees. 2007. Analysis of an EST library from the dinoflagellate (*Symbiodinium* sp.) endosymbiont of reef building corals. *Journal of Phycology* 43:1010–1021.

Legrand, C., N. Johansson, G. Johnsen, K. Y. Borsheim, and E. Granéli. 2001. Phagotrophy and toxicity in the mixotrophic *Prymnesium patelliferum* (Haptophyceae). *Limnology and Oceanography* 46:1208–1214.

Lehman, J. T. 1976. Ecological and nutritional studies on *Dinobryon* Ehrenb.: Seasonal periodicity and the phosphate toxicity problem. *Limnology and Oceanography* 21:646–658.

Lehman, J. T., and D. Scavia. 1982. Microscale patchiness of nutrients in plankton communities. *Science* 216:729–730.

Leigh, E. G., R. T. Paine, J. T. Quinn, and T. H. Suchanek. 1987. Wave energy and intertidal productivity. *Proceedings of the National Academy of Sciences of the United States of America* 84:1314–1318.

Leipe, D. D., S. M. Tong, C. L. Goggin, S. B. Siemenda, N. J. Pieniazek, and M. L. Sogin. 1996. 16S-like rRNA sequences from *Developayella elegans*, *Labyrinthuloides haliotis*, and *Proteromonas lacertae* confirm that stramenopiles are a primarily heterotrophic group. *European Journal of Protistology* 32:449–458.

Leitsch, C. E. W., K. W. Kowallik, and S. Douglas. 1999. The atpA gene cluster of *Guillardia theta* (Cryptophyta): A piece in the puzzle of chloroplast genome evolution. *Journal of Phycology* 35:128–135.

Leliaert, F. F. Rousseau, B. De Reviers, and E. Coppejans. 2003. Phylogeny of the Cladophorophyceae (Chlorophyta) inferred from partial LSU rRNA gene sequences: Is the recognition of a separate order Siphonocladales justified? *European Journal of Phycology* 38:233–246.

Lembi, C. A., and W. R. Herndon. 1966. Fine structure of the pseudocilia of *Tetraspora*. *Canadian Journal of Botany* 44:710–712.

Lembi, C. A., N. L. Perlmutter, and D. F. Spencer. 1980. *Life Cycle, Ecology, and Management Considerations of the Green Filamentous Alga, Pithophora*, Technical Report 130, Purdue University Water Resources Center, W. Lafayette, IN.

Lembi, C. A., S. W. O'Neal, and D. F. Spencer. 1988. Algae as weeds: Economic impact, ecology, and management alternatives. In: Lembi, C. A., and J. R. Waaland (eds.), *Algae and Human Affairs*, Cambridge University Press, Cambridge, UK, pp. 455–481.

Lemieux, C., C. Otis, and M. Turmel. 2007. A clade uniting the green algae *Mesostigma viride* and *Chlorokybus atmophyticus* represents the deepest branch of the Streptophyta in chloroplast genome-based phylogenies. *BMC Biology* 5:2.

Leonardi, P. I., E. J. Cáceres, and C. G.Vélez. 1998. Fine structure of dwarf males in *Oedogonium pluviale* (Chlorophyceae). *Journal of Phycology* 34:250–256.

Lesser, M. P., C. H. Mazel, M. Y. Gorbunov, and P. G. Falkowski. 2004. Discovery of symbiotic nitrogen-fixing cyanobacteria in corals. *Science* 305:997–1000.

Lewandowski, J. D., and C. F. Delwiche. 2001. Two chloroplast transfer RNA introns found in *Chaetosphaeridium* support a monophyletic Coleochaetales. *Journal of Phycology* 37:31.

Lewin, J. C. 1953. Heterotrophy in diatoms. *Journal of General Microbiology* 9:305–313.

Lewin, J. C., and R. A. Lewin. 1960. Auxotrophy and heterotrophy in marine littoral diatoms. *Canadian Journal of Microbiology* 6:127–134.

Lewin, J. C., and R. E. Norris. 1970. Surf-zone diatoms of the coasts of Washington and New Zealand (*Chaetoceros armatum* T. West and *Asterionella* spp). *Phycologia* 9:142–149.

Lewin, R. A., and L. Cheng. 1989. Collection and handling of *Prochloron* and its hosts. In: Lewin, R. A., and L. Cheng (eds.), *Prochloron, a Microbial Enigma*, Chapman & Hall, London, pp. 9–18.

Lewis, C. L., and M. A. Coffroth. 2004. The acquisition of exogenous algal symbionts by an octocoral after bleaching. *Science* 304:1490–1492.

Lewis, J. R. 1964. *The Ecology of Rocky Shores*, English Universities Press, London.

Lewis, J., and R. Hallett. 1997. *Lingulodinium polyedrum* (*Gonyaulax polyedra*) a blooming dinoflagellate. *Oceanography and Marine Biology: An Annual Review* 35:97–161.

Lewis, L. A. 1997. Diversity and phylogenetic placement of *Bracteacoccus tereg* (Chlorophyceae, Chlorophyta) based on 18S ribosomal RNA gene sequence data. *Journal of Phycology* 33:279–285.

Lewis, L.A., and V. R. Flechtner. 2002. Green algae (Chlorophyta) of desert microbiotic crusts: Diversity of North American taxa. *Taxon* 51:443–451.

Lewis, L. A., and V. R. Flechtner. 2004. Cryptic species of *Scenedesmus* (Chlorophyta) from desert soil communities of Western North America. *Journal of Phycology* 40:1127–1137.

Lewis, L. A., and R. M. McCourt. 2004. Green algae and the origin of land plants. *American Journal of Botany* 91:1535–1556.

Lewis, M. A. 1990. Are laboratory-derived toxicity data for freshwater algae worth the effort? *Environmental Toxicology and Chemistry* 9:1279–1284.

Lewis, R. J. 1995. Gametogenesis and chromosome number in *Postelsia palmaeformis* (Laminariales, Phaeophyceae). *Phycological Research* 43:61–64.

Lewis, R. J. 1996a. Chromosomes of the brown algae. *Phycologia* 35:19–40.

Lewis, R. J. 1996b. Hybridization of brown algae: Compatibility and speciation. In: Chaudhary, B. R., and S. B. Agrawal (eds.), *Cytology, Genetics and Molecular Biology of Algae*, SPB Academic Publishing, Amsterdam, pp. 275–289.

Lewis, R. J., S. I. Jensen, D. M. DeNicola, V. I. Miller, K. D. Hoagland, and S. G. Ernst. 1997b. Genetic variation in the diatom *Fragilaria capucina* (Fragilariaceae) along a latitudinal gradient across North America. *Plant Systematics and Evolution* 204:99–108.

Lewis, R. J., and M. Neushul. 1995. Intergeneric hybridization among five genera of the family Lessoniaceae (Phaeophyceae) and evidence for polyploidy in a fertile *Pelagophycus × Macrocystis* hybrid. *Journal of Phycology* 31:1012–1017.

Lewitus, A. J., H. B. Glasgow, and J. M. Burkholder. 1999. Kleptoplastidy in the toxic dinoflagellate *Pfiesteria piscicida* (Dinophyceae). *Journal of Phycology* 35:303–312.

Lewitus, A. J., and T. M. Kana. 1994. Responses of estuarine phytoplankton to exogenous glucose: Stimulation versus inhibition of photosynthesis and respiration. *Limnology and Oceanography* 39:182–189.

Lewitus, A. J., and T. M. Kana. 1995. Light respiration in six esturarine phytoplankton species: Contrast under photoautotorophic and mixotrophic growth conditions. *Journal of Phycology* 31:754–761.

Li, R., M. Watanabe, and M. M. Watanabe. 1997. Akinete formation in plankton *Anabaena* spp. (Cyanobacteria) by treatment with low temperature. *Journal of Phycology* 33:576–584.

Li, S., T. Nosenko, J. D. Hackett, and D. Bhattacharya. 2006. Phylogenomic analysis identifies red algal genes of endosymbiotic origin in Chromalveolates. *Molecular Biology and Evolution* 23:663–674.

Liddle, L. B., S. Berger, and H. Schweiger. 1976. Ultrastructure during development of the nucleus of *Batophora oerstedtii* (Chlorophyta, Dasycladaceae). *Journal of Phycology* 12:261–272.

Lin, C. K., and J. L. Blum. 1977. Recent invasion of a red alga (Bangia atropurpurea) in Lake Michigan. *Journal of the Fisheries Research Board of Canada* 34:2413–2416.

Lin, S., H. Zhang, and A. Dubois. 2006. Low abundance distribution of *Pfiesteria piscicida* in Pacific and western Atlantic as detected by mtDNA-18S rDNA real-time polymerase chain reaction. *Journal of Plankton Research* 28:667–681.

Lindell, D., M. B. Sullivan, Z. I. Johnson, A. C. Tolonen, F. Rohwer, and S. W. Chisholm. 2004. Transfer of photosynthesis genes to and from *Prochlorococcus* viruses. *Proceedings of the National Academy of Sciences of the United States of America* 101:11013–11018.

Lindstrom, S. 1987. Possible sister groups and phylogenetic relationships among selected North Pacific and North Atlantic Rhodophyta. *Helgolander Meeresunters* 41:245–260.

Lindstrom, S. C., and L. A. Hanic. 2005. The phylogeny of North American *Urospora* (Ulotrichales, Chlorophyta) based on sequence analysis of nuclear ribosomal genes, introns and spacers. *Phycologia* 44:194–201.

Lindstrom, S. C., L. A. Hanic, and L. Golden. 2006. Studies of the green alga *Percusaria dawsonii* (= *Blidingia dawsonii* comb. nov., Kornmanniaceae, Ulvales) in British Columbia. *Phycological Research* 54:40–56.

Ling, H. U., and R. D. Seppelt. 1993. Snow algae of the Windmill Islands, continental Antarctica. 2. *Chloromonas rubroleosa* sp. Nov. (Volvocles, Chlorophyta), an alga of red snow. *European Journal of Phycology* 28:77–84.

Linton, E. W., and R. E. Triemer. 1999. Reconstruction of the feeding apparatus in *Ploeotia costata* (Euglenophyta) and its relationship to other euglenoid feeding apparatuses. *Journal of Phycology* 35:313–324.

Litaker, R. W., et al. 2003. Identification of *Pfiesteria piscicida* (Dinophyceae) and *Pfiesteria*-like organisms using internal transcribed spacer-specific PCR assays. *Journal of Phycology* 39:754–761.

Litaker, R. W., et al2007. Recognizing dinoflagellate species using ITS rDNA sequences. *Journal of Phycology* 43:344–355.

Little, A. F., M. J. H. van Oppen, and B. L. Willis. 2004. Flexibility in algal endosymbioses shapes growth in reef corals. *Science* 304:1492–1494.

Littler, D. S., and M. M. Littler. 1991. Systemat-

ics of *Anadyomene* species (Anadyomenaceae, Chlorophyta) in the tropical western Atlantic. *Journal of Phycology* 27:101–118.

Littler, M. M., and D. S. Littler. 1980. The evolution of thallus form and survival strategies in benthic marine macroalgae: Field and laboratory tests of a functional form model. *American Naturalist* 116:25–43.

Littler, M. M., and D. S. Littler. 1983. Heteromorphic life-history strategies in the brown alga *Scytosiphon lomentaria* (Lyngb.) Link. *Journal of Phycology* 19:425–431.

Littler, M. M., and D. S. Littler. 1995. Impact of CLOD pathogen on Pacific coral reefs. *Science* 267:1356–1360.

Littler, M. M., D. S. Littler, S. M. Blair, and J. N. Norris. 1985. Deepest known plant life discovered on an uncharted seamount. *Science* 227:57–59.

Littler, M. M., D. S. Littler, S. M. Blair, and J. N. Norris. 1986. Deep-water plant communities from an uncharted seamount off San Salvador Island, Bahamas: Distribution, abundance, and primary productivity. *Deep-Sea Research* 33:881–892.

Littler, M. M., D. S. Littler, and P. R. Taylor. 1983. Evolutionary strategies in a tropical barrier reef system: Functional form groups of marine macroalgae. *Journal of Phycology* 19:229–237.

Litvaitis, M. K. 2002. A molecular test of cyanobacterial phylogeny: Inferences from constraint analyses. *Hydrobiologia* 468:135–145.

Liu, L., and J. W. Hastings. 2007. Two different domains of the luciferase gene in the heterotrophic dinoflagellate *Noctiluca scintillans* occur as two separate genes in photosynthetic species. *Proceedings of the National Academy of Sciences of the United States of America* 104:696–701.

Liu, L., T. Wilson, and J. W. Hastings. 2004. Molecular evolution of dinoflagellate luciferases, enzymes with three catalytic domains in a single polypeptide. *Proceedings of the National Academy of Sciences of the United States of America* 101:16555–16560.

Livingstone, D., and G. H. M. Jaworski. 1980. The viability of akinetes of blue-green algae recovered from the sediments of Rostherne Mere. *British Phycological Journal* 15:357–364.

Lizotte, M. P., D. H. Robinson, and C. W. Sullivan. 1998. Algal pigment signatures in Antarctic sea ice. In: Lizotte, M. P., and K. R. Arrigo (eds.), *Antarctic Sea Ice Biological Processes, Interactions, and Variability*, Antarctic Research Series 73, American Geophysical Union, Washington, DC, pp. 93–105.

Lobban, C. S. 1978a. The growth and death of the *Macrocystis* sporophyte (Phaeophyceae, Laminariales). *Phycologia* 17:196–212.

Lobban, C. S. 1978b. Translocation of ^{14}C in *Macrocystis* (giant kelp). *Plant Physiology* 61:585–589.

Lobban, C. S., and P. J. Harrison. 1994. *Seaweed Ecology and Physiology*. Cambridge University Press, Cambridge, UK.

Lobban, C. S., M. Schefter, A. G. B. Simpson, X. Pochon, J. Pawlowski, and W. Foissner. 2002. *Maristentor dinoferus* n. gen., n. sp., a giant heterotrich ciliate (Spirotrichea: Heterotrichida) with zooxanthellae, from coral reefs on Guam, Mariana Islands. *Marine Biology* 140:411–423.

Lokhorst, G. M. 1996. Comparative taxonomic studies on the genus *Klebsormidium* (Charophyceae) in Europe. *Cryptogamic Studies* 5:1–132.

Lokhorst, G. M. 1999. Taxonomic study of the genus *Microspora* Thuret (Chlorophyceae). An integrated field, culture and herbarium analysis. *Algological Studies* 93:1–38.

Lokhorst, G. M., and W. Star. 1999. The flagellar apparatus structure in *Microspora* (Chlorophyceae) confirms a close evolutionary relationship with unicellular algae. *Plant Systematics and Evolution* 217:11–30.

Lokhorst, G. M., P. J. Segaar, and W. Star. 1989. An ultrastructural reinvestigation of mitosis and cytokinesis in cryofixed sporangia of the coccoid green alga *Friedmannia israelensis* with special reference to septum formation and the replication cycle of basal bodies. *Cryptogamic Botany* 1:275–294.

Lokhorst, G. M., H. J. Sluiman, and W. Star. 1988. The ultrastructure of mitosis and cytokinesis in the sarcinoid *Chlorokybus atmophyticus* (Chlorophyta, Charophyceae) revealed by rapid freeze fixation and freeze substitution. *Journal of Phycology* 24:237–248.

Longhurst, A., S. Sathyendranath, T. Platt, and C. Caverhill. 1995. An estimate of global primary production in the ocean from satellite radiometer data. *Journal of Plankton Research* 17:1245–1271.

Loope, W. L., and G. F. Gifford. 1972. Influence of a soil microfloral crust on select properties of soils under pinyon-juniper in southeastern Utah. *Journal of Soil and Water Conservation* 27:164–167.

López-Bautista, J. M., and R. L. Chapman. 2003. Phylogenetic affinities of the Trentepohliales inferred from small-subunit rDNA. *International Journal of Systematic and Evolutionary Microbiology* 53:2099–2106.

López-Bautista, J. M., F. Rindi, and M. D. Guiry. 2006. Molecular systematics of the subaerial green algal order Trentepohliales: An assessment based on morphological and molecular data. *International Journal of Systematic and Evolutionary Microbiology* 56:1709–1715.

López-Bautista, J. M., D. A. Waters, and R. L. Chapman. 2003. Phragmoplastin, green algae and the evolution of cytokinesis. *International Journal of Systematic and Evolutionary Microbiology* 53:1715–1718.

Lopez-Figueroa, F., P. Lindemann, S. E. Braslavsky, K. Schaffaer, H. A. W. Schneider-Poetsch, and W. Rudiger. 1989. Detection of a phytochrome-like protein in macroalgae. *Botanica Acta* 102:178–180.

Lopez-Figueroa, F., and F. X. Niell. 1989. Red light and blue light photoreceptors controlling chlorophyll *a* synthesis in the red alga *Porphyra umbilicalis* and in the green alga *Ulva rigida*. *Physiologia Plantarum* 76:391–397.

Lorch, D. W., and M. Engels. 1979. Observations on filament formation in *Micrasterias foliacea* (Desmidiaceae, Chlorophyta). *Journal of Phycology* 15:322–325.

Lotka, A. J. 1932. The growth of mixed populations: Two species competing for a common food supply. *Journal of the Washington Academy of Science* 22:461.

Lougheed, V. L., and R. J. Stevenson. 2004. Exotic marine macroalga (*Enteromorpha flexuosa*) reaches bloom proportions in a coastal lake of Lake Michigan. *Journal of Great Lakes Research* 30:538–544.

Love, J., C. Brownlee, and A. J. Trewavas. 1997. Ca^{2+} and calmodulin dynamics during photopolarization in *Fucus serratus* zygotes. *Plant Physiology* 115:249–261.

Lovejoy, C., et al. 2007. Distribution, phylogeny, and growth of cold-adapted picoprasinophytes in Arctic seas. *Journal of Phycology* 43:78–89.

Lowe, R. L. 1996. Periphyton patterns in lakes. In: Stevenson, R. J., M. L. Bothwell, and R. L. Lowe (eds.), *Algal Ecology: Freshwater Benthic Ecosystems*, Academic Press, New York, pp. 57–76.

Lowe, R. L., P. C. Furey, J. A. Ress, and J. R. Johansen. 2007. Diatom biodiversity and distribution on wetwalls in Great Smoky Mountains National Park. *Southeastern Naturalist Special Issue* 1:135–152.

Lowe, R. L., and Y. Pan. 1996. Benthic algal communities as biological monitors. In: Stevenson, R. J., M. L. Bothwell, and R. L. Lowe (eds.), *Algal Ecology: Freshwater Benthic Ecosystems*, Academic Press, New York, pp. 705–739.

Loya, Y., K. Sakai, K. Yamazato, Y. Nakano, H. Sambali, and R. van Woesik. 2001. Coral bleaching: The winners and the losers. *Ecology Letters* 4:122–131.

Lu, M., and G. C. Stephens. 1984. Demonstration of net influx of free amino acids in *Phaeodactylum tricornutum* using high performance liquid chromatography. *Journal of Phycology* 20:584–589.

Lubchenco, J. 1980. Algal zonation in the New England rocky intertidal community: An experimental analysis. *Ecology* 61:333–334.

Lubchenco, J. 1982. Effects of grazers and algal competitors of fucoid colonization in tide pools. *Journal of Phycology* 18:544–550.

Lucas, W. J. 1995. Plasmodesmata: Intercellular channels for macromolecular transport in plants. *Current Opinion in Cell Biology* 7:673–680.

Lucas, W. J., F. Brechnignac, T. Mimura, and J. W. Oross. 1989. Charasomes are not essential for photosynthetic utilization of exogenous HCO_3^- in *Chara corallina*. *Protoplasma* 151:106–114.

Ludwig, M., and S. P. Gibbs. 1985. DNA is present in the nucleomorph of cryptomonads: Further evidence that the chloroplast evolved from a eukaryotic endosymbiont. *Protoplasma* 127:9–20.

Ludwig, M., J. L. Lind, E. A. Miller, and R. Wetherbee. 1996. High molecular mass glycoproteins associated with the siliceous scales and bristles of *Mallomonas splendens* (Synurophyceae) may be involved in cell surface development and maintenance. *Planta* 199:219–228.

Lukešová, A., and L. Hoffmann. 1996. Soil algae from acid rain impacted forest areas of the Krusné Hory Mts. 1. Algal communities. *Vegetatio* 125:123–136.

Lund, J. W. G. 1964. Primary productivity and periodicity of phytoplankton. *Internationale Vereiningung für Theoretische und Angewandte Limnologie* 15:37–56.

Lüning, K. 1985. *Meeresbotanik: Verbreitung, Ökophysiologie und Nützung der Marinen Makroalgen*, Georg Thieme Verlag, Stuttgart, Germany.

Lüning, K. 1990. *Seaweeds: Their Environment, Biogeography and Ecophysiology*, John Wiley, New York.

Luo, J., and B. D. Hall. 2007. A multistep process gave rise to RNA polymerase IV of land plants. *Journal of Molecular Evolution* 64:101–112.

Lürling, M. 2003a. Phenotypic plasticity in the green algae *Desmodesmus* and *Scenedesmus* with special reference to the induction of defensive morphology. *Annals of Limnology–International Journal of Limnology* 39:85–101.

Lürling, M. 2003b. The effect of substances from different zooplankton species and fish on the induction of defensive morphology in the green alga *Scenedesmus obliquus*. *Journal of Plankton Research* 25:979–989.

Lylis, J. C., and F. R. Trainor. 1973. The heterotrophic capabilities of *Cyclotella meneghiniana*. *Journal of Phycology* 9:365–369.

Lynch, J. M., and E. Bragg. 1985. Microorganisms and soil aggregate stability. *Advances in Soil Science* 2:133–171.

Ma, J., and A. Miura. 1984. Observations on the nuclear division in conchospores and their germlings in *Porphyra yezoensis* Ueda. *Japanese Journal of Phycology* 32:373–378.

Maberly, S. C., J. A. Raven, and A. M. Johnston. 1992. Discrimination between ^{12}C and ^{13}C by marine plants. *Oecologia* 91:481–492.

Machlis, L. 1973. The effects of bacteria on the growth and reproduction of *Oedogonium cardiacum*. *Journal of Phycology* 9:342–344.

Machlis, L., G. C. Hill, K. E. Steinback, and W. Reed. 1974. Some characteristics of the sperm attractant in *Oedogonium cardiacum*. *Journal of Phycology* 10:199–204.

MacKay, N. A., and J. J. Elser. 1998. Nutrient recycling by *Daphnia* reduces N_2 fixation by cyanobacteria. *Limnology and Oceanography* 43:347–354.

MacKenzie, L., I. Sims, V. Beuzenberg, and P. Gillespie. 2002. Mass accumulation of mucilage caused by dinoflagellate polysaccharide exudates in Tasman Bay, New Zealand. *Harmful Algae* 1:69–83.

Madsen, T. V., and S. C. Maberly. 1990. A comparison of air and water as environments for photosynthesis by the intertidal alga *Fucus spiralis* (Phaeophyta). *Journal of Phycology* 26:24–30.

Magnuson, J. J., et al. 1997. Potential effects of climate changes on aquatic systems: Laurentian Great Lakes and Precambrian shield region. *Hydrological Processes* 11:825–871.

Maier, I. 1997a. The fine structure of the male gamete of *Ectocarpus siliculosus* (Ectocarpales, Phaeophyceae). I. General structure of the cell. *European Journal of Phycology* 32:241–253.

Maier, I. 1997b. The fine structure of the male gamete of *Ectocarpus siliculosus* (Ectocarpales, Phaeophyceae). II. The flagellar apparatus. *European Journal of Phycology* 32:255–266

Maier, I., and D. G. Müller. 1998. Virus binding to brown algal cells as visualized by DAPI fluorescence microscopy. *Phycologia* 37:60–63.

Maier, I., and C. E. Schmid. 1995. An immunofluorescence study on lectin binding sites in gametes of *Ectocarpus siliculosus* (Ectocarpales, Phaeophyceae), *Phycological Research* 43:33–42.

Maistro, S., P. A. Broady, C. Andreoli, and E. Negrisolo. 2007. Molecular phylogeny and evolution of the order tribonematales (Heterokonta, Xanthophyceae) based on analysis of plastidial genes *rbcL* and *psaA*. *Molecular Phylogenetics and Evolution* 43:407–417.

Malin, G., and M. Steinke. 2004. Dimethyl sulfide production: What is the contribution of the coccolithophores? In: Thierstein, H. R., and J. R. Young (eds.), *Coccolithophores—From Molecular Processes to Global Impact.* Springer Verlag, New York, pp. 127–164.

Malin, G., S. M. Turner, and P. S. Liss. 1992. Sulfur: The plankton/climate connection. *Journal of Phycology* 28:590–597.

Mandal, B., P. L. G. Vlek, and L. N. Mandal. 1999. Beneficial effects of blue-green algae and *Azolla*, excluding supplying nitrogen, on wetland rice fields: A review. *Biology and Fertility of Soils* 28:329–342.

Mandoli, D. F. 1998. Elaboration of body plan and phase change during development of *Acetabularia*: How is the complex architecture of a giant unicell built? *Annual Review of Plant Physiology and Plant Molecular Biology* 49:173–198.

Manhart, J. R., and R. M. McCourt. 1992. Molecular data and species concepts in the algae. *Journal of Phycology* 28:730–737.

Manhart, J. R., and J. D. Palmer. 1990. The gain of two chloroplast tRNA introns marks the green algal ancestors of land plants. *Nature* 345:268–270.

Mann, D. G. 1993. Patterns of sexual reproduction in diatoms. *Hydrobiologia* 269:11–20

Mann, D. G., and H. J. Marchant. 1989. The origins of the diatom and its life cycle. In: Green, J. C., B. S. C. Leadbeater, and W. I. Diver (eds.), *The Chromophyte Algae: Problems and Perspectives, Systematics Association Special Volume 38,* Clarendon Press, Oxford, UK, pp. 307–323.

Manton, I., and H. Ettl. 1965. Observations on the fine structure of *Mesostigma viride* Lauterborn. *Botanical Journal of the Linnean Society* 59:175–184.

Manton, I., and L. S. Peterfi. 1969. Observations on the fine structure of coccoliths, scales and the protoplast of a freshwater coccolithophorid, *Hymenomonas roseola* Stein, with supplementary observations on the protoplast of *Cricosphaera carterae.* Proceeding of the Royal Society of London B 172:1–15.

Maranger, R., D. F. Bird, and N. M. Price. 1998. Iron acquisition by photosynthetic marine phytoplankton from ingested bacteria. *Nature* 396:248–251.

Marañón, E. 2005. Phytoplankton growth rates in the Atlantic subtropical gyres. *Limnology and Oceanography* 50:299–310.

Marble, J. R., and K. T. Harper. 1989. Effect of timing of grazing on soil-surface cryptogamic communities in Great Basin low-shrub desert: A preliminary report. *Great Basin Naturalist* 49:104–107.

Marchant, H. J. 1977. Ultrastructure, development and cytoplasmic rotation of seta-bearing cells of *Coleochaete scutata* (Chlorophyceae). *Journal of Phycology* 13:28–36.

Marchant, H. J. 1998. Life in the snow: Algae and other microorganisms. In: Green, K. (ed.), *Snow (A Natural History; an Uncertain Future),* Australian Alps Liaison Committee, Canberra, Australia, pp. 83–97.

Marchant, H. J., and A. McEldowney. 1986. Nano-planktonic siliceous cysts from Antarctica are algae. *Marine Biology* 92:53–57.

Marchant, H. J., and J. D. Pickett-Heaps. 1972. Ultrastructure and differentiation of *Hydrodictyon reticulatum.* III. Formation of the vegetative daughter net. *Australian Journal of Biological Science* 25:265–278.

Marchant, H. J., and J. D. Pickett-Heaps. 1973. Mitosis and cytokinesis in *Coleochaete scutata. Journal of Phycology* 9:461–471.

Marchant, H. J., and J. D. Pickett-Heaps. 1974. The effect of colchicine on colony formation in the algae *Hydrodictyon, Pediastrum,* and *Sorastrum. Planta* 116:291–300.

Marchant, H. J., J. D. Pickett-Heaps, and K. Jacobs. 1973. An ultrastructural study of zoosporogenesis and the mature zoospore of *Klebsormidium flaccidum. Cytobios* 8:95–107.

Marchant, H. J., and H. A. Thomsen. 1994. Haptophytes in polar waters. In: Green, J. C., and B. S. C. Leadbeater (eds.), *The Haptophyte Algae,* Clarendon Press, Oxford, UK, pp. 209–228.

Margulis, L., M. F. Dolan, and R. Guerrero. 2000. The chimeric eukaryote: Origin of the nucleus from the karyomastigont in amitochondriate protists. *Proceedings of the National Academy of Sciences of the United States of America* 97:6954–6959.

Marin, B., E. C. M. Nowack, G. Glöckner, and M. Melkonian. 2007. The ancestor of the *Paulinella* chromatophore obtained a carboxysomal operon by horizontal gene transfer from a *Nitrococcus*-like γ-proteobacterium. *BMC Evolutionary Biology* 7:85.

Marin, B., E. C. M. Nowack, and M. Melkonian. 2005. A plastid in the making: Evidence for a second primary endosymbiosis. *Protist* 156:425–432.

Marin, B., A. Palm, M. Klingberg, and M. Melkonian. 2003. Phylogeny and taxonomic revision of plastid-containing euglenophytes based on SSU rDNA sequence comparisons and synapomorphic signatures in the SSU rRNA secondary structure. *Protist* 154:99–145.

Markager, S., and K. Sand-Jensen. 1992. Light requirements and depth zonation of marine macroalgae. *Marine Ecology Progress Series* 88:83–92.

Markey, D. R., and R. T. Wilce. 1975. The ultrastructure of reproduction in the brown alga *Pylaiella littoralis.* I. Mitosis and cytokinesis in the plurilocular gametangia. *Protoplasma* 85:219–241.

Marks, J. C., and M. P. Cummings. 1996. DNA sequence variation in the ribosomal internal transcribed spacer region of freshwater *Cladophora* species (Chlorophyta). *Journal of Phycology* 32:1035–1042.

Marshall, J.-A., P. D. Nichols, B. Hamilton, R. J. Lewis, and G. M. Hallegraeff. 2003. Ichthyotoxicity of *Chattonella marina* (Raphidophyceae) to damselfish (*Acanthochromis polycanthus*): The synergistic role of reactive oxygen species and free fatty acids. *Harmful Algae* 4:273–281.

Marshall, W., and J. Laybourn-Parry. 2002. The balance between photosynthesis and grazing in Antarctic mixotrophic cryptophytes during summer. *Freshwater Biology* 47:2060–2070.

Martin, J., and S. E. Fitzwater. 1988. Iron deficiency limits phytoplankton growth in the north-east Pacific subarctic. *Nature* 331:341–343.

Martin, J. H. 1992. Iron as a limiting factor in oceanic productivity. In: Falkowski, P. G., and A. Woodhead (eds.), *Primary Productivity and Biogeochemical Cycles in the Sea,* Plenum Press, New York, pp. 137–155.

Martin, J. H., et al. 1994. Testing the iron hypothesis in ecosystems of the equatorial Pacific Ocean. *Nature* 371:123–129.

Martin, W., et al. 2002. Evolutionary analysis of *Arabidopsis,* cyanobacterial, and chloroplast genomes reveals plastid phylogeny and thousand of cyanobacterial genes in the nucleus. *Proceedings of the National Academy of Sciences of the United States of America* 99:12246–12251.

Martin, W., and E. V. Koonin. 2006. Introns and the origin of nucleus-cytosol compartmentalization. *Nature* 440:41–45.

Martin-Jézéquel, V., M. Hildebrand and M. A. Brzezinski. 2000. Silicon metabolism in diatoms: Implications for growth. *Journal of Phycology* 36:821–840.

Martiny, A. C., M. C. Coleman, and S. W. Chisholm. 2006. Phosphate acquisition genes in *Prochlorococcus* ecotypes: Evidence for genome-wide adaptation. *Proceedings of the National Academy of Sciences of the United States of America* 103:12552–12557.

Martone, P. T. 2006. Size, strength and allometry of joints in the articulated coralline *Calliarthron. The Journal of Experimental Biology* 209:1678–1689.

Massalski, A., and G. F. Leedale. 1969. Cytology and ultrastructure of the Xanthophyceae. I. Comparative morphology of the zoospores of *Bumilleria sicula* Borzi and *Tribonema vulgare* Pascher. *British Phycological Journal* 4:159–180.

Matsunaga, S., S. Watanabe, S. Sakaushi, S. Miyamura, and T. Hori. 2003. Screening effect diverts the swimming directions from diaphototactic to positive phototactic in a disk-shaped green flagella *Mesostigma viride. Photochemistry and Photobiology* 77:324–332.

Matsuo, Y., H. Imagawa, M. Nishizawa, and Y Shizuri. 2005. Isolation of an algal morphogenesis inducer from a marine bacterium. *Science* 307:1598.

Matsuoka, K., and Y. Fukuyo. 2003. Taxonomy of cysts. In: Hallegraeff, G. M., D. M. Anderson, and A. D. Cembella (eds.), *Manual on Harmful Marine Microalgae,* UNESCO Publishing, pp. 563–592.

Matsuzaki, M., et al. 2004.Genome sequence of the ultrasmall unicellular red alga *Cyanidioschyzon merolae* 10D. *Nature* 428:653–657.

Matthes-Sears, U., J. A. Gerrath, and D. W. Larson. 1997. Abundance, biomass, and productivity of endolithic and epilithic lower plants on the temperate-zone cliffs of the Niagara Escarpment, Canada. *International Journal of Plant Science* 158:451–460.

Matthijs, H. C. P., T. Burger-Wiersma, and L. R. Mur. 1989. A status report on *Prochlorothrix hollandica,* a free-living prochlorophyte. In: Lewin, R. A., and L. Cheng (eds.), *Prochlo-*

ron, *a Microbial Enigma*, Chapman & Hall, London, pp. 83–87.

Mattox, K. R., and K. D. Stewart. 1984. A classification of the green algae: A concept based on comparative cytology. In: Irvine, D. E. G., and D. M. John (eds.), *Systematics of the Green Algae*, Academic Press, London, pp. 29–72.

Mayali, X., and F. Azam. 2004. Algicidal bacteria in the sea and their impact on algal blooms. *Journal of Eukaryotic Microbiology* 51:139–144.

Mazumdar, J., E. H. Wilson, K. Masek, C. A. Hunter, and B. Striepen. 2006. Apicoplast fatty acid synthesis is essential for organelle biogenesis and parasite survival in *Toxoplasma gondii*. *Proceedings of the National Academy of Sciences of the United States of America* 103:13192–13197.

McBride, G. E. 1970. Cytokinesis in the green alga *Fritschiella*. *Nature* 216:939.

McBride, G. E. 1974. The seta-bearing cells of *Coleochaete scutata* (Chlorophyceae, Chaetophorales). *Phycologia* 13:271–285.

McCann, K. 2007. Protecting biostructure. *Nature* 446:29.

McCauley, E., and J. Kalff. 1981. Empirical relationships between phytoplankton and zooplankton biomass in lakes. *Canadian Journal of Fisheries and Aquatic Science* 38:458–463.

McCauley, L. A. R., and J. D. Wehr. 2007. Taxonomic reappraisal of the freshwater brown algae *Bodanella*, *Ectocarpus*, *Heribaudiella*, and *Pleurocladia* (Phaeophyceae) on the basis of rbcL sequences and morphological characters. *Phycologia* 46:429–430.

McClintock, J. B., and B. J. Baker. 2001. *Marine Chemical Ecology*. Series in Marine Science, vol. 25, CRC Press, London.

McConnaughey, T. A. 1994. Calcification, photosynthesis, and global carbon cycles. *Bulletin de l'Institut Océanographique, Monaco* 13:137–161.

McConnaughey, T. A. 1998. Acid secretion, calcification, and photosynthetic carbon concentrating mechanisms. *Canadian Journal of Botany* 76:1119–1126.

McConnaughey, T. A., and J. F. Whelan. 1996. Calcification generates protons for nutrient and bicarbonate uptake. *Earth Science Reviews* 42:95–118.

McCook, L. J. 2001. Competition between corals and algal turfs along a gradient of terrestrial influence in the nearshore central Great Barrier Reef. *Coral Reefs* 19:419–425.

McCourt, R. M. 1995. Green algal phylogeny. *Trends in Ecology and Evolution* 10:159–163.

McCourt, R. M., R. W. Hoshaw, and J.-C. Wang. 1986. Distribution, morphological diversity and evidence for polyploidy in North American Zygnemataceae (Chlorophyta). *Journal of Phycology* 22:307–313.

McCourt, R. M., K. G. Karol, J. Bell, K. M. Helm-Bychowski, A. Grajewska, M. F. Wojciechowski, and R. W. Hoshaw. 2000. Phylogeny of the conjugating green algae (Zygnemophyceae) based on rbcL sequences. *Journal of Phycology* 36:747–758.

McCourt, R. M., K. G. Karol, M. Guerlesquin, and M. Feist. 1996. Phylogeny of extant genera in the family Characeae (Charales, Charophyceae) based on rbcL sequences and morphology. *American Journal of Botany* 83:125–131

McDonald, K. 1972. The ultrastructure of mitosis in the marine red alga *Membranoptera platyphylla*. *Journal of Phycology* 8:156–166.

McElhiney, J., and L. A. Lawton. 2005. Detection of the cyanobacterial hepatotoxins microcystins. *Toxicology and Applied Pharmacology* 203:219–230.

McFadden, G. I. 1993. Second-hand chloroplasts: Evolution of cryptomonad algae. *Advances in Botanical Research* 19:190–230.

McFadden, G. I., P. R. Gilson, and D. R. A. Hill. 1994. *Goniomonas* rRNA sequences indicate that this phagotrophic flagellate is a close relative of the host component of cryptomonads. *European Journal of Phycology* 29:29–32.

McFadden, G. I., M. E. Reith, J. Mulholland, and N. Lang-Unnasch. 1996. Plastid in human parasites. *Nature* 381:482.

McGillicuddy Jr., D. J., et al. 2007. Eddy/wind interactions stimulate extraordinary mid-ocean plankton blooms. *Science* 316:1021–1026.

McGinnis, K. M., T. A. Dempster, and M. R. Sommerfeld. 1997. Characterization of the growth and lipid content of the diatom *Chaetoceros muelleri*. *Journal of Applied Phycology* 9:19–24.

McKenna-Neuman, C., C. D. Maxwell, and J. W. Boulton. 1996. Wind transport of sand surfaces crusted with photoautotrophic microorganisms. *Catena* 27:229–247.

McKenzie, C. H., D. Deibel, M. A. Paranjape, and R. J Thompson. 1995. The marine mixotroph *Dinobryon balticum* (Chrysophyceae): Phagotrophy and survival in a cold ocean. *Journal of Phycology* 31:19–24.

McKerracher, L., and S. P. Gibbs. 1982. Cell and nucleomorph division in the alga *Cryptomonas*. *Canadian Journal of Botany* 11:2440–2452.

McLachlan, J. L., M. R. Seguel, and L. Fritz. 1994. *Tetreutreptia pomquetensis* gen. et sp. nov. (Euglenophyceae). A quadriflagellate, phototrophic marine euglenoid. *Journal of Phycology* 30:538–544.

McManus, G. B., H. Zhang, and S. Lin. 2004. Marine planktonic ciliates that prey on macroalgae and enslave their chloroplasts. *Limnology and Oceanography* 49:308–313.

McQueen, D. J. 1990. Manipulating lake community structure: Where do we go from here? *Freshwater Biology* 23:613–620.

McQueen, D. J., M. R. S. Johannes, J. R. Post, T. J. Stewart, and D. R. S. Lean. 1989. Bottom-up and top-down impacts on freshwater pelagic community structure. *Ecological Monographs* 59:289–309.

McQuoid, M. R., and L. A. Hobson. 1996. Diatom resting stages. *Journal of Phycology* 32:889–902.

Mechling, J. A., and S. S. Kilham. 1982. Temperature effects on silicon limited growth of the Lake Michigan diatom *Stephanodiscus minutus* (Bacillariophyceae). *Journal of Phycology* 18:199–205.

Medina-Sánchez, J. M., M. Felip, and E. O. Casamayor. 2005. Catalyzed reported deposition-fluorescence in situ hybridization protocol to evaluate phagotrophy in mixotrophic protists. *Applied and Environmental Microbiology* 71:7321–7326.

Medlin, L., and N. Simon. 1998. Phylogenetic analysis of marine phytoplankton. In: Cooksey, K. E. (ed.), *Molecular Approaches to the*

Study of the Oceans, Chapman & Hall, London, pp. 161–186.

Medlin, L. K., R. M. Crawford, and R. A. Andersen. 1986. Histochemical and ultrastructureal evidence of the labiate process in the movement of centric diatoms. *British Phycological Journal* 21:297–301.

Medlin, L. K., and I. Kaczmarska. 2004. Evolution of the diatoms: V. Morphological and cytological support for the major clades and a taxonomic revision. *Phycologia* 43:245–270.

Meerhoff, M., J. M. Clemente, F. T. de Mello, C. Iglesias, A. R. Pedersen, and E. Jeppesen. 2007. Can warm climate-related structure of littoral predator assemblies weaken the clear water state in shallow lakes? *Global Change Biology* 13:1888–1897.

Meijer, M.-L., I. de Boois, M. Scheffer, R. Portielje, and H. Hosper. 1999. Biomanipulation in shallow lakes in The Netherlands: An evaluation of 18 case studies. *Hydrobiologia* 408/409:13–30.

Meijer, M.-L., M. W. deHaan, A. W. Breukelaar, and H. Buiteveld. 1990. Is reduction of the benthivorous fish an important cause of high transparency following biomanipulation in shallow lakes? *Hydrobiologia* 200/201:303–315.

Meindl, U. 1983. Cytoskeletal control of nuclear migration and anchoring in developing cells of *Micrasterias denticulata* and the change caused by the anti-microtubular herbicide amiprophos-methyl (APM). *Protoplasma* 118:75–90.

Meinesz, A. 1980. Connaisances actuelles et contribution á l'étude de la reproduction et du cycle des Udoteacées (Caulerpales, Chlorophytes). *Phycologia* 19:110–138.

Meireles, L. A., A. C. Guedes, and F. X. Malcata. 2003. Increase of the yields of eicosapentaenoic and docosahexaenoic acids by the microalga *Pavlova lutheri* following random mutagenesis. *Biotechnology and Bioengineering* 81:50–55.

Melick, D. R., P. A. Broady, and K. S. Rowan. 1991. Morphological and physiological characteristics of a non-heterocystous strain of the cyanobacterium *Mastigocladus laminosus* Cohn from fumarolic soil on Mt Erebus, Antarctica. *Polar Biology* 11:81–89.

Melis, A. 2007. Photosynthetic H_2 metabolism in *Chlamydomonas reinhardtii* (unicellular green algae). *Planta* 226:1075–1086.

Melis, A., and T. Happe. 2001. Hydrogen production. Green algae and a source of energy. *Plant Physiology* 127:740–748.

Melkonian, M. 1975. The fine structure of the zoospores of *Fritschiella tuberosa* Iyeng. (Chaetophorineae, Chlorophyceae). *Protoplasma* 86:391–404.

Melkonian, M. 1989. Flagellar apparatus ultrastructure in *Mesostigma viride* (Prasinophyceae). *Plant Systematics and Evolution* 164:93–122.

Melkonian, M., P. L. Beech, C. Katsaros, and D. Schulze. 1992. Centrin-mediated cell motility in algae. In: Melkonian, M. (ed.), *Algal Cell Motility*, Chapman & Hall, London, pp. 179–221.

Melkonian, M., B. Marin, and B. Surek. 1995. Phylogeny and evolution of the algae. In: Arai, R., M. Kato, and Y. Doi (eds.), *Biodiversity and Evolution*, The National Science Museum Foundation, Tokyo, pp. 153–176.

Melkonian, M., and B. Surek. 1995. Phylogeny of the Chlorophyta: Congruence between ultrastructural and molecular evidence. *Bulletin de la Société zoologique de France* 120:191–208.

Menzel, D., H. Jonitz, and C. Elsner-Menzel. 1996. The perinuclear microtubule system in the green alga *Acetabularia*: Anchor or motility device? *Protoplasma* 193:63–76.

Merchant, S. S., M. D. Allen, J. Kropat, J. L. Moseley, J. C. Long, S. Tottey, and A. M. Terauchi. 2006. Between a rock and a hard place: Trace element nutrition in *Chlamydomonas*. *Biochemica et Biophysica Acta* 1763:578–594.

Meskhidze, N., and A. Nenes. 2006. Phytoplankton and cloudiness in the Southern Ocean. *Science* 314:1419–1423.

Meunier, P. C., M. S. Colón-López, and L. A. Sherman. 1998. Photosystem II cyclic heterogeneity and photoactivation in the diazotrophic, unicellular cyanobacterium *Cyanothece* species ATCC 51142. *Plant Physiology* 116:1551–1562.

Meyer-Harms, B., and F. Pollehne. 1998. Alloxanthin in *Dinophysis norvegica* (Dinophysiales, Dinophyceae) from the Baltic Sea. *Journal of Phycology* 34:280–285.

Michaux-Ferrier, N., and I. Soulie-Märsche. 1987. The quantities of DNA in the vegetative nuclei of *Chara vulgaris* and *Tolypella glomerata* (Charophyta). *Phycologia* 26:435–442.

Micheli, F. 1999. Eutrophication, fisheries, and consumer–resource dynamics in marine pelagic ecosystems. *Science* 285:1396–1398.

Michod, R. E. 2007. Evolution of individuality during the transition from unicellular to multicellular life. *Proceedings of the National Academy of Sciences of the United States of America* 104:8613–8618.

Mignot, J.-P. 1967. Structure et ultrastructure de quelques Chloromonadines. *Protistologica* 3:5–23.

Mignot, J.-P., and G. Brugerolle. 1982. Scale formation in chrysomonad flagellates. *Journal of Ultrastructure Research* 81:13–26.

Mignot, J.-P., L. Joyon, and E. G. Pringsheim. 1968. Complements a l'etude cytologique des Cryptomonadines. *Protistologica* 4:493–506.

Milanowski, R., S. Kosmala, B. Zakrys, and J. Kwiatowski. 2006. Phylogeny of photosynthetic euglenophytes based on combined chloroplast and cytoplasmic SSU rDNA sequence analysis. *Journal of Phycology* 42:721–730.

Miller, M. W. 1998. Coral/seaweed competition and the control of reef community structure within and between latitudes. *Oceanography and Marine Biology: An Annual Review* 36:65–96.

Miller, S. R., S. Augustine, T. E. Olson, R. E. Blankenship, J. Selker, and A. M. Wood. 2005. Discovery of a free-living chlorophyll d-producing cyanobacterium with a hybrid proteobacterial/cyanobacterial small-subunit rRNA gene. *Proceedings of the National Academy of Sciences of the United States of America* 102:850–855.

Miller, T. R., K. Hnilicka, A. Dziedzic, P. Desplats, and R. Belas. 2004. Chemotaxis of *Silicibacter* sp. strain TM1040 toward dinoflagellate products. *Applied and Environmental Microbiology* 70:4692–4701.

Milligan, A. J., and F. M. M. Morel. 2002. A proton buffering role for silica in diatoms. *Science* 297:1848–1850.

Millington, W. F. 1981. Form and pattern in *Pediastrum*. *Cell Biology Monographs* 8:94–118.

Mills, M. M., C. Ridame, M. Davey, J. La Roche, and R. J. Geider. 2004. Iron and phosphorus co-limit nitrogen fixation in the eastern tropical North Atlantic. *Nature* 429:292–294.

Mimura, T., R. Müller, W. M. Kaiser, T. Shimmen, and K.-J. Dietz. 1993. ATP-dependent carbon transport in perfused *Chara* cells. *Plant, Cell and Environment* 16:653–661.

Mine, I., N. Takezaki, S. Sekida, and K. Okuda. 2007. Cell wall extensibility during branch formation in the xanthophycean alga *Vaucheria terrestris*. *Planta* 226:971–979.

Mitman, G. G., and J. P. van der Meer. 1994. Meiosis, blade development, and sex determination in *Porphyra* purpurea (Rhodophyta). *Journal of Phycology* 30:147–159.

Miyashita, H., H. Ikemoto, N. Kurano, K. Adachi, M. Chihara, and S. Miyachi. 1996. Chlorophyll *d* as a major pigment. *Nature* 383:402–403.

Mizuno, M., and K. Okuda. 1985. Seasonal change in the distribution of cell size of *Cocconeis scutellum* var. *ornata* (Bacillariophyceae) in relation to growth and sexual reproduction. *Journal of Phycology* 21:547–553.

Mock, T., et al. 2008. Whole-genome expression profiling of the marine diatom *Thalassiosira pseudonana* identified genes involved in silicon bioprocesses. *Proceedings of the National Academy of Sciences of the United States of America* 105:1579–1584.

Moe, R. L., and P. C. Silva. 1981. Morphology and taxonomy of *Himantothallus* (including *Phaeoglossum* and *Phyllogigas*), an Antarctic member of the Desmarestiales (Phaeophyceae). *Journal of Phycology* 17:15–29.

Moeller, P. D. B., K. R. Beauchesne, K. M. Huncik, W. C. Davis, S. J. Christopher, P. Riggs-Gelasco, and A. K. Gelasco. 2007. Metal complexes and free radical toxins produced by *Pfiesteria piscicida*. *Environmental Science and Technology* 41:1166–1172.

Moestrup, Ø. 1970. The fine structure of mature spermatozoids of *Chara corallina*, with special reference to microtubules and scales. *Planta* 93:295–308.

Moestrup, Ø. 1974. Ultrastructure of the scale-covered zoospores of the green alga *Chaetosphaeridium*, a possible ancestor of the higher plants and bryophytes. *Biological Journal of the Linnean Society* 6:111–125.

Moestrup, Ø. 1978. On the phylogenetic validity of the flagellar apparatus in green algae and other chlorophyll *a* and *b* containing plants. *BioSystems* 10:117–144.

Moestrup, Ø. 1991. Further studies of presumedly primitive green algae, including the description of Pedinophyceae class. nov. and *Resultor* gen. nov. *Journal of Phycology* 27:119–133.

Moestrup, Ø. 1995. Current status of chrysophyte "splinter groups": Synurophytes, pedinellids, silicoflagellates. In: Sandgren, C. D., J. P. Smol, and J. Kristiansen (eds.): *Chrysophyte Algae. Ecology, Phylogeny and Development*, Cambridge University Press, Cambridge, UK, pp. 75–91.

Moestrup, Ø., I. Inouye, and T. Hori. 2003. Ultrastructural studies on *Cymbomonas tetramitiformis* (Prasinophyceae). I. General structure, scale microstructure, and ontogeny. *Canadian Journal of Botany* 81:657–671.

Moestrup, Ø., and M. Sengco. 2001. Ultrastructural studies on *Bigelowiella natans*, gen. et sp. nov., a chlorarachniophyte flagellate. *Journal of Phycology* 37:624–646.

Moestrup, Ø., and H. A. Thomsen. 2003. Taxonomy of toxic haptophytes (prymnesiophytes). In: Hallegraeff, G. M., D. M. Anderson, and A. D. Cembella (eds.), *Manual on Harmful Marine Microalgae*, UNESCO Publishing, pp. 433–464.

Mohamed, H. E., and W. Vermaas. 2004. SLR1293 in *Synechocystis* sp. strain PCC 6803 is the C-3',4' Desaturase (CrtD) involved in myxoxanthophyll biosynthesis. *Journal of Bacteriology* 186:5621–5628.

Moldowan, J. M., F. J. Fago, C. Y. Lee, S. R. Jacobson, D. S. Watt, N.-E. Slougui, A. Jeganathan, and D. C. Young. 1990. Sedimentary 24-n-propylcholestanes, molecular fossils diagnostic of marine algae. *Science* 247:309–312.

Moldowan, J. M., and N. M. Talyzina. 1998. Biogeochemical evidence for dinoflagellate ancestors in the Early Cambrian. *Science* 281:1168–1170.

Mollenhauer, D., R. Mollenhauer, and M. Kluge. 1996. Studies on initiation and development of the partner association in *Geosiphon pyriforme* (Kütz.) v Wettstein, a unique endocytobiotic system of a fungus (Glomales) and the cyanobacterium *Nostoc punctiforme* (Kütz.) Hariot. *Protoplasma* 193:3–9.

Molnar, A., F. Schwach, D. J. Studholme, E. C. Thuenemann, and D. C. Baulcombe. 2007. MiRNAs control gene expression in the single-cell alga *Chlamydomonas* reinhardtii. *Nature* 447:1126–1129.

Monod, J. 1950. La technique de culture continue: Theorie et applications. *Annales de l'Institut Pasteur Lille* 79:390–410.

Montana, G., L. Randazzo, I. A. Oddo, and M. Valenza. 2008. The growth of "black crusts" on calcareous building stones in Palermo (Sicily): A first appraisal of anthropogenic and natural sulphur sources. *Environmental Geology*, 10.1007/s00254-007-1175-y.

Montegut-Felkner, A. E., and R. E. Triemer. 1997. Phylogenetic relationships of selected euglenoid genera based on morphological and molecular data. *Journal of Phycology* 33:512–519.

Montsant, A., et al. 2007. Identification and comparative genomic analysis of signaling and regulatory components in the diatom *Thalassiosira pseudonana*. *Journal of Phycology* 43:585–604.

Moon-van der Staay, S., G. W. M. van der Staay, L. Guillou, and D. Vaulot. 2000. Abundance and diversity of prymnesiophytes in the picoplankton community from the equatorial Pacific Ocean inferred from 18S rDNA sequences. *Limnology and Oceanography* 45:98–109.

Moore, D., G. B. McGregor, and G. Shaw. 2004. Morphological changes during akinete germination in *Cylindrospermopsis raciborskii* (Nostocales, Cyanobacteria). *Journal of Phycology* 40:1098–1105.

Moore, J. F., and J. A. Traquair. 1976. Silicon, a required nutrient for *Cladophora glomerata* (L.) Kütz. *Planta* 128:179–182.

Moore, J. K, and T. A. Villareal. 1996. Size-ascent rate relationships in positively buoyant marine diatoms. *Limnology and Oceanography* 41:1514–1520.

Moore, L. R., G. Rocap, and S. W. Chisholm. 1998. Physiology and molecular phylogeny of coexisting *Prochlorococcus* ecotypes. *Nature* 393:464–467.

Moore, R. B., et al. 2008. A photosynthetic alveolate closely related to apicomplexan parasites. *Nature* 451:959–963.

Morden, C. W., C. F. Delwiche, M. Kuhsel, and J. D. Palmer. 1992. Gene phylogenies and the endosymbiotic origin of plastids. *BioSystems* 28:75–90.

Mori, I. C., G. Sato, and M. Okazaki. 1996. Ca^{2+}-dependent ATPase associated with plasma membrane from a calcareous alga, *Serraticardia maxima* (Corallinaceae, Rhodophyta). *Phycological Research* 44:193–202.

Moriya, M., T. Nakayama, and I. Inouye. 2002. A new class of the Stramenopiles, Placididea classis nova: Description of *Placidia cafeteriopsis* gen. et sp. nov. *Protist* 153:143–156.

Mornin, L., and D. Francis. 1967. The fine structure of *Nematodinium armatum*, a naked dinoflagellate. *Journal Microscopie* (Paris) 6:759–772.

Moro, I., N. La Rocca, L. Dalla Valle, E. Moschin, E. Negrisolo, and C. Andreoli. 2002. *Pyramimonas australis* sp. nov. (Prasinophyceae, Chlorophyta) from Antarctica: Fine structure and molecular phylogeny. *European Journal of Phycology* 37:103–114.

Moroney, J. V. 2001. Carbon concentrating mechanisms in aquatic photosynthetic organisms: A report on CCM 2001. *Journal of Phycology* 37:928–931.

Morse, D. M., P. Salois, P. Markovic, and J. W. Hastings. 1995. A nuclear-encoded form II RuBisCO in dinoflagellates. *Science* 268:1622–1624.

Mostaert, A. S., U. Karsten, Y. Hara, and M. M. Watanabe. 1998. Pigments and fatty acids of marine raphidophytes: A chemotaxonomic re-evaluation. *Phycological Research* 46:213–220.

Mostaert, A. S., U. Karsten, and R. J. King. 1995. Physiological responses of *Caloglossa leprieurii* (Ceramiales, Rhodophyta) to salinity stress. *Phycological Research* 43:215–222.

Motomura, T. 1990. Ultrastructure of fertilization in *Laminaria angustata* (Phaeophyceae, Laminariales) with emphasis on the behavior of centrioles, mitochondria and the chloroplasts of the sperm. *Journal of Phycology* 26:80–89.

Motomura, T., T. Ichimura, and M. Melkonian. 1997. Coordinative nuclear and chloroplast division in unilocular sporangia of *Laminaria angustata* (Laminariales, Phaeophyceae). *Journal of Phycology* 33:266–271.

Motomura, T., and Y. Sakai. 1988. The occurrence of flagellated eggs in *Laminaria angustata* (Phaeophyta, Laminariales). *Journal of Phycology* 24:282–285.

Mujer, C. V., D. L. Andrews, J. R. Manhart, S. K. Peirce, and M. E. Rumpho. 1996. Chloroplast genes are expressed during intracellular symbiotic association of *Vaucheria litorea* plastids with the sea slug *Elysia chlorotica*. *Proceedings of the National Academy of Sciences of the United States of America* 93:12333–12338.

Mulkidjanian, A. Y., et al. 2006. The cyanobacterial genome core and the origin of photosynthesis. *Proceedings of the National Academy of Sciences of the United States of America* 103: 13126–13131.

Müller, D. G. 1981. Sexuality and sex attraction. In: Lobban, C. S., and M. J. Wynne (eds.), *Biology of Seaweeds*, Blackwell Scientific, Oxford, UK, pp. 661–674.

Müller, D. G., W. Boland, U. Becker, and T. Wahl. 1988. Caudoxirene, the spermatozoid-releasing and attracting factor in the marine brown alga *Perithalia caudata* (Sporochnales, Phaeophyta). *Phycologia* 24:467–473.

Müller, D. G., G. Gassman, W. Boland, F. Marner, and L. Jaenicke. 1981. *Dictyota dichotoma* (Phaeophyceae): Identification of the sperm attractant. *Science* 212:1040–1041.

Müller, D. G., L. Jaenicke, M. Donike, and A. Akintobi. 1971. Sex attractant in a brown alga: Chemical structure. *Science* 171:815–816.

Müller, D. G., I. Maier, and G. Gassman. 1985. Survey on sexual hormone specificity in Laminariales (Phaeophyceae). *Phycologia* 24:475–477.

Müller, D. G., A. Peters, G. Gassman, W. Boland, F.-J. Marner, and L. Jaenicke. 1982. Identification of a sexual hormone and related substances in the marine brown alga *Desmarestia*. *Naturwissenschaften* 69:290–292.

Müller, D. G., S. Wolf, and E. R. Parodi. 1996. A virus infection in *Myriotrichia clavaeformis* (Dictyosiphonales, Phaeophyceae) from Argentina. *Protoplasma* 193:58–62.

Müller, J., T. Friedl, D. Hepperle, M. Lorenz, and J. G. Day. 2005. Distinction between multiple isolates of *Chlorella vulgaris* (Chlorophyta, Trebouxiophyceae) and testing for conspecificity using amplified fragment length polymorphism and ITS rDNA sequences. *Journal of Phycology* 41:1236–1247.

Müller, K. M., M. C. Oliveira, R. G. Sheath, and D. Bhattacharya. 2001. Ribosomal DNA phylogeny of the Bangiophycidae (Rhodophyta) and the origin of secondary plastids. *American Journal of Botany* 88:1390–1400.

Müller, K. M., R. G. Sheath, M. L. Vis, T. J. Crease, and K. M. Cole. 1998. Biogeography and systematics of *Bangia* (Bangiales, Rhodophyta) based on the Rubisco spacer, *rbcL* gene and 18S rRNA gene sequences and morphometric analysis. I. North America. *Phycologia* 37:195–207.

Muller-Parker, G., and C. F. D'Elia. 1997. Interactions between corals and their symbiotic algae. In: Birkeland, C. (ed.), *Life and Death of Coral Reefs*, Chapman & Hall, New York, pp. 96–113.

Mumby, P. J., et al. 2006. Fishing, trophic cascades, and the process of grazing on coral reefs. *Science* 311:98–101.

Mumford, T. F., and A. Miura. 1988. *Porphyra* as food: Cultivation and economics. In: Lembi, C. A., and J. R. Waaland (eds.), *Algae and Human Affairs*, Cambridge University Press, Cambridge, UK, pp. 87–118.

Murakami, A., H. Miyashita, M. Iseki, K. Adachi, and M. Mimuro. 2004. Chlorophyll d in an epiphytic cyanobacterium of red algae. *Science* 303:1633.

Murata, K., and T. Suzaki. 1998. High-salt solutions prevent reactivation of euglenoid movement in detergent-treated cell models of *Euglena gracilis*. *Protoplasma* 203:125–129.

Murphy, T. P., D. R. S. Lean, and C. Nalewajko. 1976. Blue-green algae: Their excretion of iron-selective chelators enables them to dominate other algae. *Science* 192:394–395.

Murray, S., M. F. Jorgensen, S. Y. W. Ho, D. J. Patterson, and L. S. Jermiin. 2005. Improving the analysis of dinoflagellate phylogeny based on rRNA. *Protist* 156:269–286.

Murray, S., M. Hoppenrath, J. Larsen, and D. J. Patterson. 2006. *Bysmatrum teres* sp. nov., a new sand-dwelling dinoflagellate from Northwestern Australia. *Phycologia* 45:161–167.

Nachname, V. M. O. P. 2001. Identification of a 41 kDa protein embedded in the biosilica of scales and bristles isolated from *Mallomonas splendens* (Synurophyceae, Ochrophyta). *Protist* 152:315–327.

Nagai, S., S. Yamaguchi, C. L. Lian, Y. Matsuyama, and S. Itakura. 2006. Development of microsatellite markers in the noxious red tide-causing algae *Heterosigma akashiwo* (Raphidophyceae). *Molecular Ecology Notes* 6:477–479.

Nagel, G., D. Ollig, M. Fuhrmann, S. Katriya, A. M. Musti, E. Bamberg, and P. Hegemann. 2002. Channelrhodopsin-1: A light-gated proton channel in green algae. *Science* 296:2395–2398.

Nagle, D. G., and V. J. Paul. 1999. Production of secondary metabolites by filamentous tropical marine cyanobacteria: ecological functions of the compounds. *Journal of Phycology* 35:1412–1421.

Nagy, M. L., A. Peréz, and F. Garcia-Pichel. 2005. The prokaryote diversity of biological soil crusts in the Sonoran Desert (Organ Pipe Cactus National Monument, AZ). *FEMS Microbiology and Ecology* 54:233–245.

Nakada, T., S. Suda, and H. Nozaki. 2007. A taxonomic study of *Hafniomonas* (Chlorophyceae) based on a comparative examination of cultured material. *Journal of Phycology* 43:397–411.

Nakahara, H., T. Sakami, M. Chinain, and Y. Ishida. 1996. The role of macroalgae in epiphytism of the toxic dinoflagellate *Gambierdiscus toxicus* (Dinophyceae). *Phycological Research* 44:113–117.

Nakanishi, K., M. Nishijima, M. Nishimura, K. Kuwano, and N. Saga. 1996. Bacteria that induce morphogenesis in *Ulva pertusa* (Chlorophyta) grown under axenic conditions. *Journal of Phycology* 32:479–482.

Nakayama, T, B. Marin, H. D. Krank, B. Surek, and V. A. R. Huss. 1998. The basal position of scaly green flagellates among the green algae (Chlorophyta) is revealed by analyses of nuclear-encoded SSU rRNA sequences. *Protist* 149:367–380.

Nakayama, T., S. Watanabe, and I. Inouye. 1996b. Phylogeny of wall-less green flagellates inferred from 18SrDNA sequence data. *Phycological Research* 44:151–161.

Nakayama, T., S. Watanabe, K. Mitsui, H. Uchida, and I. Inouye. 1996a. The phylogenetic relationship between the Chlamydomonadales and Chlorococcales inferred from 18S rDNA data. *Phycological Research* 44:47–55.

Nakayama, T., M. Yoshida, M-H. Noël, M. Kawachi, and I. Inouye. 2005. Ultrastructure and phylogenetic position of *Chrysoculter rhomboideus* gen. et sp. nov. (Prymnesiophyceae), a new flagellate haptophyte from Japanese coastal waters. *Phycologia* 44:369–383.

Nakayama, Y., K. Fujiu, M. Sokabe, and K. Yoshimura. 2007. Molecular and electrophysiological characterization of a mechanosensitive channel expressed in the chloroplasts of *Chlamydomonas*. *Proceedings of the National Academy of Sciences of the United States of America* 104:5883–5888.

Nanba, N. 1995. Egg release and germling development in *Sargassum horneri* (Fucales, Phaeophyceae). *Phycological Research* 43:121–125.

Nash III, T. H., S. L. White, and J. E. Marsh. 1977. Lichen and moss distribution and biomass in hot desert ecosystems. *The Bryologist* 80:470–479.

Naustvoll, L.-J. 1998. Growth and grazing by the thecate heterotrophic dinoflagellate *Diplopsalis lenticula* (Diplopsalidaceae, Dinophyceae). *Phycologia* 37:1–9.

Navarro, J. N. 1993. Three-dimensional imaging of diatom ultrastructure with high resolution low-voltage SEM. *Phycologia* 32:151–156.

Navaud, O., P. Dabos, E. Carnus, D. Tremousaygue, and C. Hervé. 2007. TCP transcription factors predate the emergence of land plants. *Journal of Molecular Evolution* 65:23–33.

Nayak, S., and R. Prasanna. 2007. Soil pH and its role in cyanobacterial abundance and diversity in rice field soils. *Applied Ecology and Environmental Research* 5:103–113.

Nayak, S., R. Prasanna, A. Pabby, T. K. Dominic, and P. K. Singh. 2004. Effect of urea, bluegreen algae and *Azolla* on nitrogen fixation and chlorophyll accumulation in soil under rice. *Biology and Fertility of Soils* 40:67–72.

Neale, P. J. 1987. Algal photoinhibition and photosynthesis in the aquatic environment. In: Kyle, D. J., C. B. Osmond, and C. J. Aratzen (eds.), *Photoinhibition*, Elsevier, New York, pp. 39–65.

Nedashkovskaya, O. I., et al. 2003. *Mesonia algae* gen. nov., sp. nov., a novel marine bacterium of the family Flavobacteriaceae isolated from the green alga *Acrosiphonia sonderi* (Kütz.) Kornm. 2003. *International Journal of Systematic and Evolutionary Microbiology* 53:1967–1971.

Nedelcu, A. M., T. Borza, and R. W. Lee. 2006. A land plant-specific multigene family in the unicellular *Mesostigma* argues for its close relationship to Streptophyta. *Molecular Biology and Evolution* 23:1011–1015.

Nedelcu, A. M., and R. E. Michod. 2006. The evolutionary origin of an altruistic gene. *Molecular Biology and Evolution* 8:1460–1464.

Negrisolo, E., S. Maistro, N. Incarbone, I. Moro, L. Dalla Valle, P. A. Broady, and C. Andreoli. 2004. Morphological convergence characterizes the evolution of Xanthophyceae (Heterokontophyta): Evidence from nuclear SSU rDNA and plastidial *rbc*L genes. *Molecular Phylogenetics and Evolution* 33:156–170.

Neilson, A. H., and R. A. Lewin. 1974. The uptake and utilization of organic carbon by algae: An essay in comparative biochemistry. *Phycologia* 13:227–264.

Nelsen, M. P., and A. Gargas. 2008. Dissociation and horizontal transmission of codispersing lichen symbionts in the genus *Lepraria* (Lecanorales: Stereocaulaceae). *New Phytologist* 177:264–275.

Nelson, M. J., Y. Dang, E. Filek, Z, Zhang, V. W. C. Yu, K. Ishida, and B. R. Green. 2007. Identification and transcription of transfer RNA genes in dinoflagellate plastid minicircles. *Gene* 392:291–298.

Nelson, T. A., D. J. Lee, and B. C. Smith. 2003. Are "green tides" harmful algal blooms? Toxic properties of water-soluble extracts from two bloom-forming macroalgae, *Ulva fenestrata* and *Ulvaria obscura* (Ulvophyceae). *Journal of Phycology* 39:874–879.

Nelson, W. A., J. Brodie, and M. D. Guiry. 1999. Terminology used to describe reproduction and life history stages in the genus *Porphyra* (Bangiales, Rhodophyta). *Journal of Applied Phycology* 11:407–410.

Nelson, W. A., and G. A. Knight. 1996. Life history in culture of the obligate epiphyte *Porphyra subtumens* (Bangiales, Rhodophyta) endemic to New Zealand. *Phycological Research* 44:19–25.

Neushul, M., and A. L. Dahl. 1972. Zonation in the apical cell of *Zonaria*. *American Journal of Botany* 59:393–400.

Nguyen, B., R. M. Bowers, T. M. Wahlund, and B. A. Read. 2005. Suppressive subtractive hybridization of and differences in gene expression content of calcifying and noncalcifying cultures of *Emiliania huxleyi* strain 1516. *Applied and Environmental Microbiology* 71:2564–2575.

Nicastro, D., C. Schwartz, J. Pierson, R. Gaudette, M. E. Porter, and J. R. McIntosh. 2006. The molecular architecture of axonemes revealed by cryoelectron tomography. *Science* 313:944–948.

Nicholls, K. H. 1995. Chrysophyte blooms in the plankton and neuston of marine and freshwater systems. In: Sandgren, C. D., J. P. Smol, and J. Kristiansen (eds.), *Chrysophyte Algae: Ecology, Phylogeny and Development*, Cambridge University Press, Cambridge, UK, pp. 181–213.

Nicholls, K. H., and D. E. Wujek. 2002. Chrysophycean algae. In: Wehr, J. D., and R. G. Sheath (eds.), *Freshwater Algae of North America, Ecology and Classification*, Academic Press, New York, pp. 471–509.

Nichols, H. W. 1965. Culture and development of Hildenbrandia rivularis from Denmark and North America. *American Journal of Botany* 52:9–15.

Nicholson, N. L. 1970. Field studies on the giant kelp *Nereocystis*. *Journal of Phycology* 6:177–182.

Nield, J., O. Kruse, J. Ruprecht, P. de Fonseca, C. Buchel, and J. Barber. 2000. Three-dimensional structure of *Chlamydomonas reinhardtii* and *Synechococcus elongatus* photosystem II complexes allows for comparison of their oxygen-evolving complex organization. *Journal of Biological Chemistry* 275:27940–27946.

Nielsen, K. J., C. A. Blanchette, B. A. Menge, and J. Lubchenco. 2006. Physiological snapshots reflect ecological performance of the sea palm *Postelsia palmaeformis* (Phaeophyceae) across intertidal elevation and exposure gradients. *Journal of Phycology* 42:548–559.

Nielsen, S. L. 2006. Size-dependent growth rates in eukaryotic and prokaryotic algae exemplified by green algae and cyanobacteria: Comparisons between unicells and colonial growth forms. *Journal of Plankton Research* 28:489–498.

Nienow, J. A. 2002. Subaerial communities. In: Bitton, G. (ed.), *Encyclopedia of Environmental Microbiology*, John Wiley, New York, pp. 3055–3065.

Nienow, J. A., E. I. Friedmann, and R. Ocampo-Friedmann. 2002. Endolithic microorganisms in arid regions. In: Bitton, G. (ed.), *Encyclopedia of Environmental Microbiology*, John Wiley, New York, pp. 1100–1112.

Niiyama, Y. 1989. Morphology and classification of *Cladophora aegagropila* (L.) Rabenhorst (Cladophorales, Chlorophyta) in Japanese lakes. *Phycologia* 28:70–76.

Nimer, N. A., M. D. Iglesias-Rodriguez, and M. J. Merrett. 1997. Bicarbonate utilization by marine phytoplankton species. *Journal of Phycology* 33:625–631.

Nishida, K., F. Yagisawa, H. Kuroiwa, Y. Yoshida, and T. Kuroiwa. 2007. WD40 protein Mda1 is purified with Dnm1 and forms a dividing ring for mitochondria before Dnm1 in *Cyanidioschyzon merolae*. *Proceedings of the National Academy of Sciences of the United States of America* 104:4736–4741.

Nishiguchi, M. K., and L. J. Goff. 1995. Isolation, purification, and characterization of DMSP lyase (dimethylpropiothetin dethiomethylase [4.4.1.3]) from the red alga *Polysiphonia paniculata*. *Journal of Phycology* 31:567–574.

Nishino, T., S. Asakawa, and S. Ogawa. 1996. Fine filaments observed within the cytoplasm surrounding the leading edge of the septum in telophase cells of *Spirogyra* (Zygnematales, Chlorophyta). *Phycological Research* 44:163–166.

Nobel, P. S., E. Quero, and H. Linares. 1988. Differential growth response of agaves to nitrogen, phosphorus, potassium, and boron applications. *Journal of Plant Nutrition* 11:1683–1700.

Norris, R. E., T. Hori, and M. Chihara. 1980. Revision of the genus *Tetraselmis* (Class Prasinophyceae). *The Botanical Magazine, Tokyo* 93:317–339.

North, R. L., S. J. Guildford, R. E. H. Smith, S. M. Havens, and M. R. Twiss. 2007. Evidence for phosphorus, nitrogen, and iron colimitation of phytoplankton communities in Lake Erie. *Limnology and Oceanography* 52:315–328.

North, W. J. 1994. Review of *Macrocystis* biology. In: Akatsuka, I. (ed.), *Biology of Economic Algae*, SPB Academic Publishing, The Hague, The Netherlands, pp. 447–528.

Norton, T. A., M. Melkonian, and R. A. Andersen. 1996. Algal biodiversity. *Phycologia* 35:308–326.

Not, F., K. Valentin, K. Romari, C. Lovejoy, R. Massana, K. Töbe, D. Vaulot, and L. K. Medlin. 2007. Picobiliphytes: A marine picoplanktonic algal group with unknown affinities to other eukaryotes. *Science* 315:253–255.

Novichkova-Ivanova, L. N. 1972. Soil and aerial algae of polar deserts and arctic tundra. In: Wielgolaski, F. E., and T. Rosswall (eds.), *I. B. P. Tundra Biome. Proceedings of the 4th International Meeting on the Biological Productivity of Tundra*, B. P. Tundra Steering Committee, Stockholm, pp. 261–265.

Nowack, E. C. M., M. Melkonian, and G. Glöckner. 2008. Chromatophore genome sequence of *Paulinella* sheds light on acquisition of photosynthesis by eukaryotes. *Current Biology* 18:410–418.

Nozaki, H., M. Iseki, M. Hasegawa, K. Misawa, T. Nakada, N. Sasaki, and M. Watanabe. 2007. Phylogeny of primary photosynthetic eukaryotes as deduced from slowly evolving nuclear genes. *Molecular Biology and Evolution* 24:1592–1595.

Nozaki, H., and M. Ito. 1994. Phylogenetic relationships within the colonial Volvocales (Chlorophyta) inferred from cladistic analysis based on morphological data. *Journal of Phycology* 30:353–365.

Nozaki, H., M. Ito, R. Sano, H. Uchida, M. M. Watanabe, and T. Kuroiwa. 1995. Phylogenetic relationships within the colonial Volvocales (Chlorophyta) inferred from *rbcL* gene sequence data. *Journal of Phycology* 31:970–979.

Nozaki, H., H. Kuroiwa, T. Mita, and T. Kuroiwa. 1989. *Pleodorina japonica* sp. nov. (Volvocales, Chlorophyta) with bacteria-like endosymbionts. *Phycologia* 28:252–267.

Nozaki, H, K. Misawa, T. Kajita, M. Kato, S. Nohara, and M. M. Watanabe. 2000. Origin and evolution of the colonial Volvocales (Chlorophyceae) as inferred from multipole, chloroplast gene sequences. *Molecular Phylogenetics and Evolution* 17:256–268.

Nudelman, M. A., M. S. Rossi, V. Conforti, and R. E. Triemer. 2003. Phylogeny of Euglenophyceae based on small subunit rDNA sequences: Taxonomic implications. *Journal of Phycology* 39:226–235.

Nybakken, J. W. 1987. *Marine Biology: An Ecological Approach*, HarperCollins, New York.

Oakley, B. R., and J. D. Dodge. 1974. Kinetochores associated with the nuclear envelope in the mitosis of a dinoflagellate. *Journal of Cell Biology* 63:322–325.

Oakley, B. R., and J. D. Dodge. 1976. The ultrastructure of mitosis in *Chroomonas salina*. *Protoplasma* 88:241–254.

O'Doherty, D. C., and A. Sherwood. 2007. Genetic population structure of the Hawaiian alien invasive seaweed *Acanthophora spicifera* (Rhodophyta) as revealed by DNA sequencing and ISSR analyses. *Pacific Science* 61:223–233.

Okamoto, N., and I. Inouye. 2005a. A secondary symbiosis in progress? *Science* 310:287.

Okamoto, N., and I. Inouye. 2005b. The katablepharids are a distant sister group of the Cryptophyta: A proposal for Katablepharidophyta divisio nova/Kathablepharida phylum novum based on SSU rDNA and beta-tubulin phylogeny. *Protist* 156:163–179.

Okamoto, N., and I. Inouye. 2006. *Hatena arenicola* gen. et sp. nov., a katablepharid undergoing probable plastid acquisition. *Protist* 157:401–419.

O'Kelly, C. J. 2002. *Glossomastix chrysoplasta* n. gen., n. sp. (Pinguiophyceae), a new coccoidal, colony-forming golden alga from southern Australia. *Phycological Research* 50:67–74.

O'Kelly, C. J., and B. J. Baca. 1984. The time course of carpogonial branch and carposporophyte development in *Callithamnion cordatum* (Rhodophyta, Ceramiales). *Phycologia* 23:407–417.

O'Kelly, C. J., and G. L. Floyd. 1984. Flagellar apparatus absolute orientations and the phylogeny of the green algae. *BioSystems* 16:227–251.

O'Kelly, C. J., S. Watanabe, and G. L. Floyd. 1994. Ultrastructure and phylogenetic relationships of Chaetopeltidales ord. nov. (Chlorophyta, Chlorophyceae). *Journal of Phycology* 30:118–128.

O'Kelly, C. J., and D. E. Wujek. 1995. Status of the Chrysamoebales (Chrysophyceae): Observations on *Chrysamoeba pyrenoidifera*, *Rhizochromulina marina* and *Lagynion delicatulum*. In: C. D. Sandgren, J. P. Smol, and J. Kristiansen (eds.), *Chrysophyte Algae: Ecology, Phylogeny and Development*, Cambridge University Press, Cambridge, UK, pp. 361–372.

O'Kelly, C. J., B. Wysor, and W. K. Bellows. 2004. *Collinsiella* (Ulvophyceae, Chlorophyta) and other ulotrichalean taxa with shell-boring sporophytes form a monophyletic clade. *Phycologia* 43:41–49.

Okuda, K., and R. M. Brown. 1992. A new putative cellulose-synthesizing complex of *Coleochaete scutata*. *Protoplasma* 168:51–63.

Okuda, K., and S. Mizuta. 1993. Diversity and evolution of putative cellulose-synthesizing enzyme complexes in green plants. *Japanese Journal of Phycology* 41:151–173.

Okuda, K., S. Sekida, S. Yoshinaga, and Y. Suetomo. 2004. Cellulose-synthesizing complexes in some chromophyte algae. *Cellulose* 11:365–376.

Oliveira, M. C., J. Kurniawan, C. J. Bird, E. L. Rice, C. A. Murphy, R. K. Singh, R. G. Gutell, and M. A. Ragan. 1995. A preliminary investigation of the order Bangiales (Bangiophycidae, Rhodophyta) based on sequences of nuclear small-subunit ribosomal RNA genes. *Phycological Research* 42:71–79

Oliver, R. L. 1994. Floating and sinking in gas-vacuolate cyanobacteria. *Journal of Phycology* 30:161–173.

Oliver, R. L., and G. G. Ganf. 2000. Freshwater blooms. In: Whitton, B. A., and M. Potts (eds.), *The Ecology of Cyanobacteria: Their Diversity in Time and Space*, Kluwer Academic Publishers, Boston, pp. 149–194.

Oliver, R. L., A. J. Kinnear, and G.G. Ganf. 1981. Measurements of cell density of three freshwater phytoplankters by density gradient centrifugation. *Limnology and Oceanography* 26:285–294.

Olli, K. 1996. Resting cyst formation of *Eutreptiella gymnastica* (Euglenophyceae) in the northern coastal Baltic Sea. *Journal of Phycology* 32:535–542.

Olli, K., A.-S. Heiskanen, and J. Seppälä. 1996. Development and fate of *Eutreptiella gymnastica* bloom in nutrient-enriched enclosures in the coastal Baltic Sea. *Journal of Plankton Research* 18:1587–1604.

Olli, K., and K. Trunov. 2007. Self-toxicity of *Prymnesium parvum* (Prymnesiophyceae). *Phycologia* 46:109–112.

Olsen, G. J., C. R. Woese, and R. Overbeek. 1994. The winds of (evolutionary) change: Breathing new life into microbiology. *Journal of Bacteriology* 176:1–6.

Olson, A. M., and J. Lubchenco. 1990. Competition in seaweeds: Linking plant traits to competitive outcomes. *Journal of Phycology* 26:1–6.

Olson, R. J., S. W. Chisholm, E. R. Zettler, and E. V. Armbrust. 1990. Pigments, size and distribution of *Synechococcus* in the North Atlantic and Pacific Oceans. *Limnology and Oceanography* 35:45–58.

Omata, T., G. D. Price, M. R. Badger, M. Okamura, S. Gohta, and T. Ogawa. 1999. Identification of an ATP-binding cassette transporter involved in bicarbonate uptake in the cyanobacterium *Synechococcus* sp. strain PCC 7942. *Proceedings of the National Academy of Sciences of the United States of America* 96:13571–13576.

O'Neal, S. W., and C. A. Lembi. 1983. Physiological changes during germination of *Pithophora oedogonia* (Chlorophyceae) akinetes. *Journal of Phycology* 19:193–199.

O'Neil, R. M., and J. W. La Claire. 1984. Mechanical wounding induces the formation of extensive coated membranes in giant algal cells. *Science* 225:331–333.

Orr, J. C., et al. 2005. Anthropogenic ocean acidification over the twenty-first century and its impact on calcifying organisms. *Nature* 437:681–686.

Ortega-Morales, O., J. Guezennec, G. Hernández-Duque, C. C. Gaylarde, and P. M. Gaylarde. 2000. Phototrophic biofilms on ancient Mayan buildings in Yucatan, Mexico. *Current Microbiology* 40:81–85.

Ortiz, W., and C. J. Wilson. 1988. Induced changes in chloroplast protein accumulation during heat bleaching in *Euglena gracilis*. *Plant Physiology* 86:554–561.

Oswald, W. J. 1988. Micro-algae and waste-water treatment. In: Borowitzka, M. A., and L. J. Borowitzka (eds.), *Micro-algal Biotechnology*, Cambridge University Press, Cambridge, UK, pp. 305–328.

Ott, D. W., and R. M. Brown. 1972. Light and electron microscopical observations on mitosis in *Vaucheria litorea* Hofman ex C Agardh. *British Phycological Journal* 7:361–374.

Ott, D. W., and R. M. Brown. 1975. Developmental cytology of the genus *Vaucheria*. III. Emergence, settlement and germination of the mature zoospore of *V. fontanalis* (L.) Christianson. *British Phycological Journal* 10:49–50.

Oudot-Le Secq, M.-P., J. Grimwood, H. Shapiro, E. V. Armbrust, C. Bowler, and B. R. Green. 2007. Chloroplast genomes of the diatoms *Phaeodactylum tricornutum* and *Thalassiosira pseudonana*: Comparison with other plastid genomes of the red lineage. *Molecular Genetics and Genomics* 177:427–439.

Ouellette, A. J. A., S. M. Handy, and S. W. Wilhelm. 2006. Toxic *Microcystis* is widespread in Lake Erie: PCR detection of toxin genes and molecular characterization of associated cyanobacterial communities. *Microbial Ecology* 51:154–165.

Pace, M. L., J. J. Cole, and S. R. Carpenter. 1998. Trophic cascades and compensation: differential responses of microzooplankton in whole-lake experiments. *Ecology* 79:138–152.

Padilla, D. K. 1985. Structural resistance of algae to herbivores: A biomechanical approach. *Marine Biology* 90:103–199.

Padilla, D. K. 1989. Algal structural defenses: Form and calcification in resistance to tropical limpets. *Ecology* 70:835–842.

Padilla, D. K., and S. C. Adolph. 1996. Plastic inducible morphologies are not always adaptive: The importance of time delays in a stochastic environment. *Evolutionary Ecology* 10:105–117.

Paerl, H. W. 1984. N₂-fixation (nitrogenase activity) attributable to a specific *Prochloron* (Prochlorophyta)-ascidian association in Palau, Micronesia. *Marine Biology* 81:251–254.

Paerl, H. W. 1992. Epi- and endobiotic interactions of cyanobacteria. In: Reisser, W. (ed.), *Algae and Symbioses*, BioPress, Bristol, UK, pp. 537–565.

Paerl, H. W., P. T. Bland, N. D. Bowles, and M. E. Haibach. 1985. Adaptation to high-intensity, low-wavelength light among surface blooms of the cyanobacterium *Microcystis aeruginosa*. *Applied and Environmental Microbiology* 49:1046–1052.

Paerl, H. W., and J. L. Pinckney. 1996. A mini-review of microbial consortia: Their roles in aquatic production and biogeochemical cycling. *Microbial Ecology* 31:225–247.

Paerl, H. W., R. C. Richards, R. C. Leonord, and C. R. Goldman. 1975. Seasonal nitrate cycling as evidence for complete vertical mixing in Lake Tahoe, California–Nevada. *Limnology and Oceanography* 20:1–8.

Paerl, H. W., and J. F. Ustach. 1982. Blue-green algal scums: An explanation for their occurrence during freshwater blooms. *Limnology and Oceanography* 27:212–217.

Paine, R. T. 1969. A note on trophic complexity and community stability. *American Naturalist* 103:91–93.

Paine, R. T. 1990. Benthic macroalgal competition: Complications and consequences. *Journal of Phycology* 26:12–17.

Paine, R. T. 2002. Trophic control of production in a rocky intertidal community. *Science* 296:736–738.

Paine, R. T., C. J. Slocum, and D. O. Duggins. 1979. Growth and longevity in the crustose red alga *Petrocelis middendorfii*. *Marine Biology* 51:185–192.

Painter, T. H., B. Duval, W. H. Thomas, M. Mendez, S. Heintzelman, and J. Dozier. 2001. Detection and quantification of snow algae with an airborne imaging spectrometer. *Applied and Environmental Microbiology* 67:5267–5272.

Palenik, B., et al. 2007. The tiny eukaryote *Ostreococcus* provides genomic insights into the paradox of plankton speciation. *Proceedings of the National Academy of Sciences of the United States of America* 104:7705–7710.

Palenik, B., and S. E. Henson. 1997. The use of amides and other organic nitrogen sources by the phytoplankton *Emiliania huxleyi*. *Limnology and Oceanography* 42:1544–1551.

Palenik, B., and H. Swift. 1996. Cyanobacterial evolution and prochlorophyte diversity as seen in DNA-dependent RNA polymerase gene sequences. *Journal of Phycology* 32:638–646.

Palevitz, B. A., and P. K. Hepler. 1975. Identification of actin *in situ* at the ectoplasm-endoplasm interface of *Nitella*. *Journal of Cell Biology* 65:29–38.

Palmer, J. D. 2003. The symbiotic birth and spread of plastids: How many times and who-dunit? 2003. *Journal of Phycology* 39:4–11.

Palmer, J. D., and C. F. Delwiche. 1996. Second-hand chloroplasts and the case of the disappearing nucleus. *Proceedings of the National Academy of Sciences of the United States of America* 93:7432–7435.

Papenfuss, G. F. 1935. Alternation of generations in *Ectocarpus siliculosis*. *Botanical Gazette* 96:421–446.

Park, M. G., W. Yih, and D. W. Coats. 2004. Parasites and phytoplankton, with special emphasis on dinoflagellate infections. *Journal of Eukaryotic Microbiology* 51:145–155.

Parke, M., and J. Adams. 1960. The motile (*Crystalolithus hyalinus* Gaarder and Markali) and nonmotile stages in the life history of *Coccolithus pelagicus* (Wallich) Schiller. *Journal of the Marine Biological Association of the United Kingdom* 39:263–274.

Parke, M., I. Manton, and B. Clarke. 1955. Studies on marine flagellates. II. Three new species of *Chrysochromulina*. *Journal of the Marine Biology Association of the UK* 34:579–609.

Parker, B. C. 1965. Translocation in the giant kelp *Macrocystis*. I. Rates, direction, quantity of ¹⁴C-labeled products and fluorescein. *Journal of Phycology* 1:41–46.

Parker, B. C. 1966. Translocation in *Macrocystis*. III. Composition of sieve tube exudate and identification of the major ¹⁴C label and products. *Journal of Phycology* 2:38–41.

Parker, B. C., and E. Y. Dawson. 1965. Non-calcareous marine algae from California Miocene deposits. *Nova Hedwigia* 10:273–295.

Parker, B. C., A. B. Ford, T. Allnutt, B. Bishop, and S. Wendt. 1977. Baseline microbiology for soils of the Dufek Massif. *Antarctic Journal of the United States* 12:24–26.

Parker, B. C., and J. Huber. 1965. Translocation in *Macrocystis*. II. Fine structure of the sieve tubes. *Journal of Phycology* 1:172–179.

Parker, B. C., N. Schamem, and R. Renner. 1969. Viable soil algae from the herbarium of the Missouri Botanical gardens. *Annuals of the Missouri Botanical Gardens* 56:113–119.

Parodi, E. R., and E. J. Cáceres. 1995. Life history of *Cladophora surera* sp. nov. (Cladophorales, Ulvophyceae). *Phycological Research* 43:223–231.

Parrow, M. W., and J. M. Burkholder. 2003a. Estuarine heterotrophic cryptoperidiniopsoids (Dinophyceae): Life cycle and culture studies. *Journal of Phycology* 39:678–696.

Parrow, M. W., and J. M. Burkholder. 2003b. Reproduction and sexuality in *Pfiesteria shumwayae* (Dinophyceae). *Journal of Phycology* 39:697–711.

Parsons, A. N., J. E. Barrett, D. H. Wall, and R. A. Virginia. 2004. Soil carbon dioxide flux in Antarctic dry valley ecosystems. *Ecosystems* 7:286–295.

Parsons, T. J., C. A. Maggs, and S. E. Douglas. 1990. Plastid DNA restriction analysis links the heteromorphic phases of an apomictic red algal life history. *Journal of Phycology* 26:495–500.

Patrick, R., and C. W. Reimer. 1966. *The Diatoms of the U.S. Exclusive of Alaska and Hawaii, Vol I—Fragilariaceae, Eunotiaceae, Achnanthaceae, Naviculaceae*, Monographs 13, Academy of Natural Sciences, Philadelphia.

Patrick, R., and C. W. Reimer. 1975. *The Diatoms of the U.S. Exclusive of Alaska and Hawaii, Vol II—Part 1, Entomoneidaceae, Cymbellaceae, Gomphonemaceae, Epithemiaceae*, Monographs 13, Academy of Natural Sciences, Philadelphia.

Patron, N., Y. Inagaki, and P. Keeling. 2007. Multiple gene phylogenies support the monophyly of cryptomonad and haptophyte host lineages. *Current Biology* 17:887–891.

Patron, N. J., M. B. Rogers, and P. J. Keeling. 2004. Gene replacement of Fructose-1.6-bisphosphate aldolase supports the hypothesis of a single photosynthetic ancestor of chromalveolates. *Eukaryotic Cell* 3:1169–1175.

Patron, N. J., M. B. Rogers, and P. J. Keeling. 2006a. Comparative rates of evolution in endosymbiotic nuclear genomes. *BMC Evolutionary Biology* 6:46.

Patron, N. J., R. F. Waller, and P. J. Keeling. 2006b. A tertiary plastid uses genes from two endosymbionts. *Journal of Molecular Biology* 357:1373–1382.

Patterson, G. M. L., et al. 1993. Antiviral activity of cultured blue-green algae (Cyanophyta). *Journal of Phycology* 29:125–130.

Patterson, G. M. L., and C. M. Bolis. 1997. Fungal cell wall polysaccharides elicit an antifungal secondary metabolite (phytoalexin) in the cyanobacterium *Scytonema ocellatum*. *Journal of Phycology* 33:54–60.

Patterson, G. M. L., and S. Carmeli. 1992. Biological effects of tolytoxin (6-hydro-7-O-methyl scytophycin b), a potent bioreactive metabolite from cyanobacteria. *Archives of Microbiology* 157:406–410.

Patterson, G. M. L., L. K. Larsen, and R. E. Moore. 1994. Bioactive natural products from blue-green algae. *Journal of Applied Phycology* 6:151–157.

Paul, N. A., L. Cole, R. de Nys, and P. D. Steinberg. 2006. Ultrastructure of the gland cells of the red alga *Asparagopsis armata* (Bonnemaisoniaceae). *Journal of Phycology* 42:637–645.

Paul, V. J., R. W. Thacker, K. Banks, and S. Golubic. 2005. Benthic cyanobacterial bloom impacts the reefs of South Florida (Broward County, USA). *Coral Reefs* 24:693–697.

Pazour, G. J., N. Agrin, J. Leszyk, and G. B. Witman. 2005. Proteomic analysis of a eukaryotic cilium. *Journal of Cell Biology* 170:103–113.

Pearson, E. A., and S. N. Murray. 1997. Patterns of reproduction, genetic diversity, and genetic differentiation in California populations of the geniculate coralline alga *Lithothrix aspergillum* (Rhodophyta). *Journal of Phycology* 33:753–763.

Pedersen, P. M. 1981. Phaeophyta, life histories. In: Lobban, C. S., and M. J. Wynne (eds.), *The Biology of Seaweeds*, Blackwell Scientific, Oxford, UK, pp. 194–217.

Peers, G., and N. M. Price. 2006. Copper-containing plastocyanin used for electron transport by an oceanic diatom. *Nature* 441:341–344.

Peeters, F., D. Straile, A. Lorke, and D. Ollinger. 2007. Turbulent mixing and phytoplankton spring bloom development in a deep lake. *Limnology and Oceanography* 52:286–298.

Pel, R., H. Hoogveld, and V. Floris. 2003. Using the hidden isotopic heterogeneity in phyto- and zooplankton to unmask disparity in trophic carbon transfer. *Limnology and Oceanography* 48:2200–2207.

Pennings, S. C., S. R. Pablo, and V. J. Paul. 1997. Chemical defenses of the tropical, benthic marine cyanobacterium *Hormothamnion enteromorphoides*: Diverse consumers and synergisms. *Limnology and Oceanography* 42:911–917.

Peperzak, L., F. Colijn, E. G. Vrieling, W. W. C. Gieskes, and J. C. H. Peeters. 2000. Observations of flagellates in colonies of *Phaeocystis globosa* (Prymnesiophyceae); a hypothesis for their position in the life cycle. *Journal of Plankton Research* 22:2181–2203.

Pérez-Martinez, C., J. Barea-Arco, and P. Sánchez-Castillo. 2001. Dispersal and colonization of the epibiont alga *Korshikoviella gracilipes* (Chlorophyta) on Daphnia pulicaria (Cladocera). *Journal of Phycology* 37:724–730.

Perl, T. M., L. Bedard, J. C. Hockin, E. C. D. Todd, and R. S. Remis. 1990. An outbreak of toxic encephalopathy caused by eating mussels contaminated with domoic acid. *New England Journal of Medicine* 322:1775–1780.

Pernthaler, J., K. Simek, B. Sattler, A. Schwarzenbacher, J. Bobkova, and R. Psenner. 1996. Short-term changes of protozoan control on autotrophic picoplankton in an oligo-mesotrophic lake. *Journal of Plankton Research* 18:443–462.

Peters, A., M. J. H. van Oppen, C. Wiencke, W. T. Stam, and J. L. Olsen. 1997. Phylogeny and historical ecology of the Desmarestiaceae (Phaeophyceae) support a Southern Hemisphere origin. *Journal of Phycology* 33:294–309.

Peters, A. F., and E. Burkhardt. 1998. Systematic position of the kelp endophyte *Laminarionema elsbetiae* (Ectocarpales *sense lato*, Phaeophyceae) inferred from nuclear ribosomal DNA sequences. *Phycologia* 37:114–120.

Peters, E. C. 1997. Diseases of coral-reef organisms. In: Birkeland, C. (ed.), *Life and Death of Coral Reefs*, Chapman & Hall, New York, pp. 114–139.

Peters, R. 1984. Methods for the study of feeding, grazing and assimilation by zooplankton. In: Downing, J. A., and F. H. Rigler (eds.), *A Manual on Methods for the Assessment of Secondary Productivity in Fresh Waters*, Blackwell Scientific, Oxford, UK, pp. 336–412.

Petersen, J., H. Brinkmann, and R. Cerff. 2003. Origin, evolution, and metabolic role of a novel glycolytic GAPDH enzyme recruited by land plant plastids. *Journal of Molecular Evolution* 57:16–26.

Petersen, J., R. Teich, H. Brinkmann, and R. Cerff. 2006. A "green" phosphoribulokinase in complex algae with red plastids: Evidence for a single secondary endosymbiosis leading to haptophytes, cryptophytes, heterokonts, and dinoflagellates. *Journal of Molecular Evolution* 62:143–157.

Petersen, J. I., A. Foopatthanakamol, and B. Ratanasthien. 2006. Petroleum potential, thermal maturity and the oil window of oil shales and coals in Cenozoic rift basins, Central and Northern Thailand. *Journal of Petroleum Geology* 29:337–360.

Peterson, C. G., T. L. Dudley, K. D. Hoagland, and L. M. Johnson. 1993. Infection, growth, and community-level consequences of a diatom pathogen in a Sonoran desert stream. *Journal of Phycology* 29:442–452.

Peterson, C. G., K. A. Vormittag, and H. M. Valett. 1998. Ingestion and digestion of epilithic algae by larval insects in a heavily grazed montane stream. *Freshwater Biology* 40:607–623.

Peterson, J. B. 1935. Studies on the biology and taxonomy of soil algae. *Dansk Botanisk Arkiv* 8:1–180.

Pfiester, L. A. 1988. Dinoflagellate sexuality. *International Review of Cytology* 114:249–272.

Pfiester, L. A., and D. M. Anderson. 1987. Dinoflagellate reproduction. In: Taylor, F. J. R. (ed.), *The Biology of Dinoflagellates*, Blackwell, Oxford, UK, pp. 611–648.

Pfiester, L. A., and R. Lynch. 1980. Amoeboid stages and sexual reproduction of *Cystodinium bataviense* and its similarity to *Dinococcus* (Dinophyceae). *Phycologia* 19:178–183.

Pfiester, L. A., and J. Popovsky. 1979. Parasitic, amoeboid dinoflagellates. *Nature* 24:421–424.

Pfister, C. A. 2007. Intertidal invertebrates locally enhance primary production. *Ecology* 88:1647–1653.

Phillips, D., and J. Scott. 1981. Ultrastructure of cell division and reproductive differentiation of male plants in the Florideophyceae (Rhodophyta). Mitosis in *Dasya ballouviana*. *Protoplasma* 106:329–341.

Phillips, J. A. 1997. Genus and species concepts in *Zonaria* and *Homoeostrichus* (Dictyotales, Phaeophyceae), including the description of *Exallosorus*, gen. nov. *European Journal of Phycology* 32:303–311.

Phillips, J. A., and M. N. Clayton. 1997. Comparative studies on gametangial distribution and structure in species of *Zonaria* and *Homoeostrichus* (Dictyotales, Phaeophyceae) from Australia. *European Journal of Phycology* 32:25–34.

Phillips, N., R. Burrowes, R. Rousseau, B. de Reviers, and G. W. Saunders. 2008. Resolving evolutionary relationships among the brown algae using chloroplast and nuclear genes. *Journal of Phycology* 44:394–405.

Pickett-Heaps, J. D. 1975. *Green Algae: Structure, Reproduction and Evolution of Selected Genera*, Sinauer Associates, Sunderland, MA.

Pickett-Heaps, J. D., and H. J. Marchant. 1972. The phylogeny of the green algae: A new proposal. *Cytobios* 6:255–264.

Pickett-Heaps, J. D., and L. C. Fowke. 1970. Mitosis, cytokinesis, and cell elongation in the desmid, *Closterium littorale*. *Journal of Phycology* 6:189–215.

Pickett-Heaps, J. D., D. R. A. Hill, and R. Wetherbee. 1986. Cellular movement in the centric diatom *Odontella sinensis*. *Journal of Phycology* 22:334–339.

Pickett-Heaps, J. D., and S. E. Kowalski. 1981. Valve morphogenesis and the microtubule center of the diatom *Hantzschia amphioxys*. *European Journal of Cell Biology* 25:150–170.

Pickett-Heaps, J., A.-M. M. Schmid, and L. A. Edgar. 1990. The cell biology of diatom valve formation. *Progress in Phycological Research* 7:1–157.

Pickett-Heaps, J. D., T. Spurck, and D. H. Tippit. 1984. Chromosome motion and the spindle matrix. *Journal of Cell Biology* 99:137–143.

Pickett-Heaps, J. D., and J. A. West. 1998. Time-lapse video observations on sexual plasmogamy in the red alga *Bostrychia*. *European Journal of Phycology* 33:43–56.

Pickett-Heaps, J. D., J. A. West, S. M. Wilson, and D. L. McBride. 2001. Time-lapse videomicroscopy of cell (spore) movement in red algae. *European Journal of Phycology* 36:9–22.

Pickett-Heaps, J. D., and R. Wetherbee. 1987. Spindle formation in the green alga *Mougeotia*: Absence of anaphase A correlates with postmitotic nuclear migration. *Cell Motility and Cytoskeleton* 7:68–77.

Pienaar, R. N. 1994. Ultrastructure and calcification of coccolithophores. In: Winter, A., and W. G. Siesser (eds.), *Coccolithophores*, Cambridge University Press, Cambridge, UK, pp. 13–38.

Pienaar, R. N., H. Sakai, and T. Horiguchi. 2007. Description of a new dinoflagellate with a diatom endosymbiont, *Durinskia capensis* sp. nov. (Peridiniales, Dinophyceae) from South Africa. *Journal of Plant Research* 120:247–258.

Pierce, S. K., S. E. Massey, J. J. Hanten, and N. E. Curtis. 2003. Horizontal transfer of functional nuclear genes between multicellular organisms. *Biological Bulletin* 204:237–240.

Piercey-Normore, M. D., and P. T. DePriest. 2001. Algal switching among lichen symbioses. *American Journal of Botany* 88:1490–1498.

Pillmann, A., G. W. Woolcott, J. L. Olsen, W. T. Stam, and R. J. King. 1997. Inter- and intraspecific genetic variation in *Caulerpa* (Chlorophyta) based on nuclear rDNA ITS sequences. *European Journal of Phycology* 32:379–386.

Pinhassi, J., M. M. Sala, H. Havskum, F. Peters, O. Guadayol, A. Malits, and C. Marrasé. 2004. Changes in bacterioplankton composition under different phytoplankton regimens. *Applied and Environmental Microbiology* 70:6753–6766.

Pinnegar, J. K., and N. V. C. Polunin. 2000. Contributions of stable-isotope data to elucidating food webs and Mediterranean rocky littoral fishes. *Oecologia* 122:399–409.

Pipes, L. D., and G. F. Leedale. 1992. Scale formation in *Tessellaria volvocina* (Synurophyceae). *British Phycological Journal* 27:1–19.

Pipes, L. D., G. F. Leedale, and P. A. Tyler. 1991. Ultra-structure of *Tessellaria volvocina* (Sunurophyceae). *British Phycological Journal* 26:259–278.

Pipes, L. D., P. A. Tyler, and G. F. Leedale. 1989. *Chrysonephele palustris* gen. et sp. nov. (Chrysophyceae), a new colonial chrysophyte from Tasmania. *Nova Hedwigia* 95:81–97.

Planas, D. 1996. Acidification effects. In: Stephenson, R. J., M. L. Bothwell, and R. L. Lowe (eds.), *Algal Ecology: Freshwater Benthic Ecosystems*, Academic Press, New York, pp. 497–532.

Plude, J. L., D. L. Parker, O. J. Schommer, R. J. Timmerman, S. A. Hagstrom, J. M. Joers, and R. Hnasko. 1991. Chemical characterization of polysaccharide from the slime layer of the cyanobacterium *Microcystis flos-aquae* C3-40. *Applied and Environmental Microbiology* 57:1696–1700.

Polanshek, A. R., and J. A. West. 1975. Culture and hybridization studies on *Petrocelis* from Alaska and California. *Journal of Phycology* 11:434–439.

Polanshek, A. R., and J. A. West. 1977. Culture and hybridization studies on *Gigartina papillata* (Rhodophyta). *Journal of Phycology* 13:141–149.

Pollingher, U. 1988. Freshwater armored dinoflagellates: Growth, reproduction strategies, and population dynamics. In: Sandgren, C. D. (ed.), *Growth and Reproductive Strategies of Freshwater Phytoplankton*, Cambridge University Press, Cambridge, UK, pp. 134–174.

Pollingher, U., and E. Zemel. 1981. *In situ* and experimental evidence of the influence of turbulence on cell division processes of *Peridinium cinctum* forma Westii (Lemm.) Lefevre. *British Phycological Journal* 16:281–287.

Pollio, A., P. Cennamo, C. Ciniglia, M. De Stefano, G. Pinto, and V. A. R. Huss. 2005. *Chlamydomonas pitschmanii* Ettl, a little known species from thermoacidic environments. *Protist* 156:287–302.

Polne-Fuller, M., and A. Gibor. 1984. Developmental studies in *Porphyra*. I. Blade differentiation in *Porphyra* perforata as expressed by morphology, enzymatic digestion, and protoplast regeneration. *Journal of Phycology* 20:609–616.

Pombert, J. F., C. Otis, C. Lemieux, and M. Turmel. 2004. The complete mitochondrial DNA sequence of the green alga *Pseudendoclonium akinetum* (Ulvophyceae) highlights distinctive evolutionary trends in the Chlorophyta and suggests a sister group relationship between the Ulvophyceae and the Chlorophyceae. *Molecular Biology and Evolution* 21:922-935.

Poole, L. J., and J. A. Raven. 1997. The biology of *Enteromorpha*. *Progress in Phycological Research* 12:1–148.

Popovsky, J., and L. A. Pfiester. 1982. The life histories of *Stylodinium sphaera* Pascher and *Cystodinedria inermis* (Geitler) Pasher (Dinophyceae), two freshwater facultative predator-autotrophs. *Archiv für Protistenkunde* 125:115–127.

Porcella, R. A., and P. A. Walne. 1980. Microarchitecture and envelope development on *Dysmorphococcus globosus* (Phacotaceae, Chlorophyceae). *Journal of Phycology* 16:280–290.

Pore, R. S. 1986. The association of *Prototheca* with slime flux in *Ulmus americana* and other trees. *Mycopathologia* 94:67–73.

Pore, R. S., E. A. Barnett, W. C. Barnes, Jr., and J. D. Walker. 2003. *Prototheca* ecology. *Mycopathologia* 81:49–62.

Porta, D., and M. Hernandez-Marine. 1998. The soil alga *Jaagiella alpicola* Vischer (Chlorophyta), a new member for the order Pleurastrales. Cryptogamie Algologie 19:291–301.

Porter, K. G. 1977. The plant–animal interface in freshwater ecosystems. *American Scientist* 65:159–170.

Potter, D., G. W. Saunders, and R. A. Andersen. 1997. Phylogenetic relationships of the Raphidophyceae and Xanthophyceae as inferred from nucleotide sequences of the 18S ribosomal RNA gene. *American Journal of Botany* 84:966–972.

Potts, M. 1994. Desiccation tolerance of prokaryotes. *Microbiological Reviews* 58:755–805.

Power, M. E., A. J. Stewart, and W. J. Matthews. 1988. Grazer control of algae in an Ozark mountain stream: Effects of short-term exclusion. *Ecology* 69:1894–1898.

Prechtl, J., C. Kneip, P. Lockhart, K. Wenderoth, and U.-G. Maier. 2004. Intracellular spheroid bodies of *Rhopalodia gibba* have nitrogen-fixing apparatus of cyanobacterial origin. *Molecular Biology and Evolution* 21:1477–1481.

Preisfeld, A., I. Busse, M. Klingberg, S. Talke, and H. G. Ruppel. 2001. Phylogenetic position and inter-relationships of the osmotrophic euglenids based on SSU rDNA data, with emphasis on the Rhabdomonadales (Euglenozoa). *International Journal of Systematic and Evolutionary Microbiology* 51:751–758.

Premkumar, L., H. M. Greenblatt, U. K. Bageshwar, T. Savchenko, I. Gokhman, J. L. Sussman, and A. Zamir. 2005. Three-dimensional structure of a halotolerant algal carbonic anhydrase predicts halotolerance of a mammalian homolog. *Proceedings of the National Academy of Sciences of the United States of America* 102:7493–7498.

Prescott, G. W. 1951. *Algae of the Western Great Lakes Area*, William C. Brown, Dubuque, IA.

Prescott, G. W., C. E. deM. Bicudo, and W. C. Vinyard. 1982. *A Synopsis of North American Desmids, Part II, Desmidiaceae: Placodermae, Section 4,* University of Nebraska Press, Lincoln, NE.

Prescott, G. W., H. T. Croasdale, and W. C. Vinyard. 1975. *A Synopsis of North American Desmids, Part II. Desmidiaceae: Placodermae, Section 1,* University of Nebraska Press, Lincoln, NE.

Prescott, G. W., H. T. Croasdale, and W. C. Vinyard. 1977. *A Synopsis of North American Desmids, Part II. Desmidiaceae: Placodermae, Section 2,* University of Nebraska Press, Lincoln, NE.

Prescott, G. W., H. T. Croasdale, W. C. Vinyard, and C. E. deM. Bicudo. 1981. *A Synopsis of North American Desmids, Part II. Desmidiaceae: Placodermae Section 3,* University of Nebraska Press, Lincoln, NE.

Pringsheim, N. 1860. Beiträge zur Morphologie und Systematik der Algen III. Die Coleochaeteen. *Jahrbuch fuer wissenschaftliche Botanik* 2:1–38.

Printz, H. 1939. Vorarbeiten zu einer Monographie der Trentepohliaceen. *Nytt Magasin f Naturv* 80:137–210.

Printz, H. 1964. Die Chaetophoralean der Binnenge-wässer. Eine systematische Ubersicht. *Hydrobiologia* 24:1–376.

Probert, I., J. Fresnel, C. Billard, M. Geisen, and J. R. Young. 2007. Light and electron microscope observations of *Algirosphaera robusta* (Prymnesiophyceae). *Journal of Phycology* 43:319–332.

Probert, I., and A. Houdan. 2004. The laboratory culture of coccolithophores. In: Thierstein, H. R., and J. R. Young (eds.), *Coccolithophores—From Molecular Processes to Global Impact,* Springer Verlag, New York, pp. 217–250.

Pröschold, T., B. Marin, U. G. Schlösser, and M. Melkonian. 2001. Molecular phylogeny and taxonomic revision of *Chlamydomonas* Ehrenberg and *Chloromonas* Gobi, and description of *Oogamochlamys* gen. nov. and *Lobochlamys* gen. nov. *Protist* 152265–300.

Proskauer, J. 1950. On *Prasinocladus*. *American Journal of Botany* 37:59–66.

Prosperi, C. H. 1994. A cyanophyte capable of fixing nitrogen under high levels of oxygen. *Journal of Phycology* 30:222–224.

Provan, J., S. Murphy, and C. A. Maggs. 2005. Tracking the invasive history of the green algal *Codium fragile* ssp. *tomentosoides*. *Molecular Ecology* 14:189–194.

Provasoli, L. 1958. Nutrition and ecology of protozoa and algae. *Annual Review of Microbiology* 12:279–308.

Provasoli, L., and I. J. Pintner. 1980. Bacteria induced polymorphism in an axenic laboratory strain of *Ulva lactuca* (Chlorophyceae). *Journal of Phycology* 16:196–201.

Ptacnik, R., S. Diehl, and S. Berger. 2003. Performance of sinking and nonsinking phytoplankton taxa in a gradient of mixing depths. *Limnology and Oceanography* 48:1903–1912.

Puel, F., C. Largeau, and G. Giraud. 1987. Occurrence of a resistant biopolymer in the outer walls of the parasitic alga *Prototheca wickerhamii* (Chlorococcales): Ultrastructural and chemical studies. *Journal of Phycology* 23:649–656.

Pueschel, C. M. 1977. A freeze-etch study of the ultrastructure of red algal pit plugs. *Protoplasma* 91:15–30.

Pueschel, C. M. 1980. Pit connections and translocation in red algae. *Science* 209:422–423.

Pueschel, C. M. 1987. Absence of cap membranes as a characteristic of pit plugs of some red algal orders. *Journal of Phycology* 23:150–153.

Pueschel, C. M. 1988. Cell sloughing and chloroplast inclusions in *Hildenbrandia rubra* (Rhodophyta, Hildenbrandiales). *British Phycological Journal* 23:17–23.

Pueschel, C. M. 1989. An expanded survey of the ultrastructure of red algal pit plugs. *Journal of Phycology* 25:625–636.

Pueschel, C. M. 1990. Cell structure. In: Cole, K. M., and R. G. Sheath (eds.), *Biology of the Red Algae*, Cambridge University Press, Cambridge, UK, pp. 7–42.

Pueschel, C. M., and K. M. Cole. 1982. Rhodophycean pit plugs: An ultrastructural survey with taxonomic implications. *American Journal of Botany* 69:703–720.

Qin, S., P. Jiang, and C. Tseng. 2005. Transforming kelp in a marine bioreactor. *Trends in Biotechnology* 23:264–268.

Qin, H., J. L. Rosenbaum, and M. M. Barr. 2001. An autosomal recessive polycystic kidney disease gene homology is involved in intraflagellar transport in *C. elegans* ciliated sensory neurons. *Current Biology* 11:457–461.

Qiu, Y.-L., et al. 2007. A nonflowering land plant phylogeny inferred from nucleotide sequences of seven chloroplast, mitochondrial, and nuclear genes. *International Journal of Plant Sciences* 168:691–708.

Quatrano, R. S. 1973. Separation of processes associated with differentiation of two-celled *Fucus* embryos. *Developmental Biology* 30:209–213.

Quatrano, R. S., and S. L. Shaw. 1997. Role of the cell wall in the determination of cell polarity and the plane of cell division in *Fucus* embryos. *Trends in Plant Science* 2:15–21.

Quinlan, R., M. S. V. Douglas, and J. P. Smol. 2005. Food web changes in arctic ecosystems related to climate warming. *Global Climate Change* 11:1381–1386.

Quinn, P., R. M. Bowers, Z. Zhang, T. M. Wahlund, M. S. Fanelli, D. Olszova, and B. A. Read. 2006. CDNA microarrays as a tool for identification of biomineralization proteins in the coccolithophorid *Emiliania huxleyi* (Haptophyta). *Applied and Environmental Microbiology* 72:5512–5526.

Radmer, R. J. 1996. Algal diversity and commercial algal products. *BioScience* 46:263–270.

Radmer, R. J., and B. C. Parker. 1994. Commercial applications of algae: Opportunities and constraints. *Journal of Applied Phycology* 6:93–98.

Raffaeli, D. G., J. A. Raven, and L. G. Poole. 1998. Ecological impact of green macroalgal blooms. *Oceanography and Marine Biology: An Annual Review* 36:97–126.

Ragan, M. A. 1981. Chemical constituents of seaweeds. In: Lobban, C. S., and M. J. Wynne (eds.), *The Biology of Seaweeds*, Blackwell Scientific, Oxford, UK, pp. 598–626.

Ragan, M. A., C. J. Bird, E. L. Rice, R. S. Gutell, C. A. Murphy, and R. K. Singh. 1994. A molecular phylogeny of the marine red algae (Rhodophyta) based on the nuclear small-subunit rRNA gene. *Proceedings of the National Academy of Sciences of the United States of America* 91:7276–7280.

Rai, A. N., E. Söderbäck, and B. Bergman. 2000. Cyanobacterium–plant symbioses. *New Phytologist* 147:449–481.

Raimondi, P. T., D. C. Reed, B. Gaylord, and L. Washburn. 2004. Effects of self-fertilization in the giant kelp, *Macrocystis pyrifera*. *Ecology* 85:3267–3276.

Rajaniemi, P., et al. 2005. Phylogenetic and morphological evaluation of the genera *Anabaena*, *Aphanizomenon*, *Trichormus* and *Nostoc* (Nostocales, Cyanobacteria). *International Journal of Systematic and Evolutionary Microbiology* 55:11–26.

Rambold, G., T. Friedl, and A. Beck. 1998. Photobionts in lichens: Possible indicators of phylogenetic relationships. *The Bryologist* 101:392–397.

Ramus, J. 1972. Differentiation of the green alga *Codium fragile*. *American Journal of Botany* 59:478–482.

Ramus, J. 1983. A physiological test of the theory of complementary chromatic adaptation. II. Brown, green and red seaweeds. *Journal of Phycology* 19:173–178.

Raven, J. A. 1995. Comparative aspects of chrysophyte nutrition with emphasis on carbon, phosphorus and nitrogen. In: Sandgren, C. D., J. P. Smol, and J. Kristiansen (eds.), *Chrysophyte Algae: Ecology, Phylogeny and Development*, Cambridge University Press, Cambridge, UK, pp. 95–118.

Raven, J. A. 1997a. Putting the C in phycology. *European Journal of Phycology* 32:319–333.

Raven, J. A. 1997b. Inorganic carbon acquisition by marine autotrophs. *Advances in Botanical Research* 27:86–209.

Raven, J. A., A. M. Johnston, and J. J. MacFarlane. 1990. Carbon metabolism. In: Cole, K. M., and R. G. Sheath (eds.), *Biology of the Red Algae*, Cambridge University Press, Cambridge, UK, pp. 171–202.

Raven, J. A., F. A. Smith, and S. M. Glidewell. 1979. Photosynthetic capacities and biological strategies of giant-celled and small-celled macroalgae. *The New Phytologist* 83:249–309.

Raven, J. A., and A. M. Waite. 2004. The evolution of silicification in diatoms: Inescapable sinking and sinking as escape? *New Phytologist* 162:45–61.

Raven, P. H., R. F. Evert, and S. E. Eichhorn. 1999. *Biology of Plants*, 6th edition. W. H. Freeman, Worth Publishers, New York.

Rawitscher-Kunkel, E., and L. Machlis. 1962. The hormonal integration of sexual reproduction in *Oedogonium*. *American Journal of Botany* 49:177–183.

Raymond, J., and R. E. Blankenship. 2003. Horizontal gene transfer in eukaryotic algal evolution. *Proceedings of the National Academy of Sciences of the United States of America* 100:7419–7420.

Reddy, G. B., and J. Giddens. 1981. Nitrogen fixation by moss–algal associations in grassland. *Soil Biology and Biochemistry* 13:537–538.

Redfield, A. C. 1958. The biological control of chemical factors in the environment. *American Scientist* 46:205–221.

Reed, R. H. 1990. Solute accumulation and osmotic adjustment. In: Cole, K. M., and R. G. Sheath (eds.), *Biology of the Red Algae*, Cambridge University Press, Cambridge, UK, pp. 147–170.

Reimers, C. E. 1998. Carbon cycle feedbacks from the sea floor. *Nature* 391:536–537.

Reiss, H. D., C. Katsaros, and B. Galatis. 1996. Freeze-fracture studies in the brown alga *Asteronema rhodochortonoides*. *Protoplasma* 193:46–57.

Reisser, W. (ed.). 1992. *Algae and Symbioses*. BioPress, Bristol, UK.

Rengefors, K., K. C. Ruttenberg, C. L. Haupert, C. Taylor, B. L. Howes, and D. M. Anderson. 2003. Experimental investigation of taxon-specific response of alkaline phosphatase activity in natural freshwater phytoplankton. *Limnology and Oceanography* 48:1167–1175.

Renner, T., and E. R. Waters. 2007. Comparative genomic analysis of the Hsp70s from five diverse photosynthetic eukaryotes. *Cell Stress and Chaperones* 12:172–185.

Rensing, S. A., P. Obrdlik, N. Rober-Kleber, S. B. Müller, C. J. B. Hofmann, Y. Van de Peer, and U.-G. Maier. 1997. Molecular phylogeny of the stress-70 protein family with reference to algal relationships. *European Journal of Phycology* 32:279–285.

Reuter, J. E., and R. P. Axler. 1992. Physiological characteristics of inorganic nitrogen uptake by spatially separate algal communities in a nitrogen-deficient lake. *Freshwater Biology* 27:227–236.

Reynolds, C. 2006. *Ecology of Phytoplankton*, Cambridge University Press, Cambridge, UK.

Reynolds, C. S. 1972. Growth, gas vacuolation and buoyancy in a natural population of a planktonic blue-green alga. *Freshwater Biology* 2:87–106.

Reynolds, C. S. 1984. *The Ecology of Freshwater Phytoplankton*, Cambridge University Press, Cambridge, UK.

Reynolds, C. S. 2007. Variability in the provision and function of mucilage in phytoplankton: Facultative responses to the environment. *Hydrobiologia* 578:37–45.

Reynolds, C. S., G. H. M. Jaworski, H. A. Cmiech, and G. F. Leedale. 1981. On the annual cycle of the blue-green alga *Microcystis aeruginosa* Kütz emend Elenkin. *Philosophical Transactions of the Royal Society of London* B 293:419–477.

Reynolds, C. S., A. E. Walsby, and R. L. Oliver. 1987. The role of buoyancy in the distribution of *Anabaena* s. in Lake Rotaongaio. *New Zealand Journal of Marine and Freshwater Research* 21:525–26.

Rezacova, M., and J. Neustupa. 2007. Distribution of the genus *Mallomonas* (Synurophyceae)—Ubiquitous dispersal in microorganisms evaluated. *Protist* 158:29–37.

Rheinheimer, G. 1991. *Aquatic Microbiology*, Wiley, New York.

Rhodes, L., J. Smith, R. Tervit, R. Roberts, J. Adamson, S. Adams, and M. Decker. 2006. Cryopreservation of economically valuable marine micro-algae in the classes Bacillariophyceae, Chlorophyceae, Cyanophyceae, Dinophyceae, Haptophyceae, Prasinophyceae and Rhodophyceae. *Cryobiology* 52:152–156.

Ribera, M. A., and C. F. Boudouresque. 1995. Introduced marine plants, with special reference to macroalgae: Mechanisms and impact. *Progress in Phycological Research* 11:188–217.

Rice, D. W., and J. D. Palmer. 2006. An exceptional horizontal gene transfer in plastids: Gene replacement by a distant bacterial paralog and evidence that haptophyte and cryptophyte plastids are sisters. *BMC Biology* 4:31.

Richards, T. A., and M. van der Giezen. 2006. Evolution of the Isd11–IscS complex reveals a single α-proteobacterial endosymbiosis for all eukaryotes. *Molecular Biology and Evolution* 23:1341–1344.

Richards, T. A., A. A. Vepritskiy, D. Gouliamova, and S. Nierzwicki-Bauer. 2005. The molecular diversity of freshwater picoeukaryotes from an oligotrophic lake reveals diverse, distinctive and globally dispersed lineages. *Environmental Microbiology* 7:1413–1425.

Richardson, A. O., and J. D. Palmer. 2007. Horizontal gene transfer in plants. *Journal of Experimental Botany* 58:1–9.

Richardson, F. L., and R. E. Brown. 1970. *Glaucosphaera vacuolata*, its ultrastructure and physiology. *Journal of Phycology* 6:165–171.

Richardson, L. 1997. Occurrence of the black band disease cyanobacterium on healthy corals of the Florida Keys. *Bulletin of Marine Science* 61:485–490.

Richardson, P. L., R. E. Cheney, and L. V. Worthington. 1978. A census of Gulf Stream Rings, Spring 1975. *Journal of Geophysical Research* 83:6136–6144.

Richardson, T. L., and G. A. Jackson. 2007. Small phytoplankton and carbon export from the surface ocean. *Science* 315:838–840.

Richerd, S., C. Destombe, J. Cuguen, and M. Valero. 1993. Variation of reproductive success in a haplo-diploid red alga, *Gracilaria verrucosa*: Effects of parental identities and crossing distance. *American Journal of Botany* 80:1379–1391.

Richmond, A. 1988. *Spirulina*. In: Borowitzka, M. A., and L. J. Borowitzka (eds.), *Micro-Algal Biotechnology*, Cambridge University Press, Cambridge, UK, pp. 85–121.

Richmond, A. 1990. Large scale microalgal culture and applications. *Progress in Phycological Research* 7:269–330.

Riding, R. 2004. *Solenopora* is a chaetidid sponge, not an alga. *Palaeontology* 47:117–122.

Riding, R., J. C. W. Cope, and P. D. Taylor. 1998. A coralline-like red alga from the Lower Ordovician of Wales. *Palaeontology* 41:1069–1076.

Riebesell, U., D. A. Wolf-Gladrow, and V. Smetacek. 1993. Carbon dioxide limitation of marine phytoplankton growth rates. *Nature* 361:249–251.

Riebesell, U., I. Zondervan, B. Rost, P. D. Tortell, R. C. Zeebe, and F. M. M. Morel. 2000. Reduced calcification in marine plankton in response to increased atmospheric CO_2. *Nature* 407:634–637.

Rier, S. T., and R. J. Stevenson. 2006. Response of periphytic algae to gradients in nitrogen and phosphorus in streamside mesocosms. *Hydrobiologia* 561:131–147.

Rier, S. T., R. J. Stevenson, and G. D. LaLiberte. 2006. Photo-acclimation response of benthic stream algae across experimentally manipulated light gradient: A comparison of growth rates and net primary productivity. *Journal of Phycology* 42:560–567.

Rindi, F. 2007. Diversity, distribution and ecology of green algae and cyanobacteria in urban habitats. In Seckbach, J. (ed.), *Algae and Cyanobacteria in Extreme Environments*, Springer Verlag, Dordrecht, The Netherlands, pp. 619–638.

Rindi, F., and M. D. Guiry. 2002. Diversity, life history, and ecology of *Trentepohlia* and *Printzina* (Trentepohliales, Chlorophyta) in urban habitats in western Ireland. *Journal of Phycology* 38:39–54.

Rindi, F., J. M. López-Bautista, A. R. Sherwood, and M. D. Guiry. 2006. Morphology and phylogenetic position of *Spongiochrysis hawaiiensis* gen. et sp. nov., the first known terrestrial member of the order Cladophorales (Ulvophyceae, Chlorophyta). *International Journal of Systematic and Evolutionary Microbiology* 56:913–922.

Rindi, F., L. McIvor, A. R. Sherwood, T. Friedl, M. D. Guiry, and R. G. Sheath. 2007. Molecular phylogeny of the green algal order Prasiolales (Trebouxiophyceae, Chlorophyceae). *Journal of Phycology* 43:811–822.

Rines, J. E. B., and P. E. Hargraves. 1988. The *Chaetoceros* Ehrenberg (Bacillariophyceae) flora of Narragansett Bay, Rhode Island, U.S.A. *Bibliotheca Phycologica* 79:1–196.

Robba, L., S. J. Russell, G. L. Barker, and J. Brodie. 2006. Assessing the use of the mitochondrial cox1 marker for use in DNA barcoding of red algae (Rhodophyta). *American Journal of Botany* 93:1101–1108.

Roberts, B. J., and R. W. Howarth. 2006. Nutrient and light availability regulate the relative contribution of autotrophs and heterotrophs to respiration in freshwater pelagic ecosystems. *Limnology and Oceanography* 51:288–298.

Roberts, K. R. 1984. Structure and significance of the cryptomonad flagellar apparatus. I. *Cryptomonas ovata* (Cryptophyta). *Journal of Phycology* 20:590–599.

Roberts, K. R., G. Hansen, and F. J. R. Taylor. 1995. General ultrastructure and flagellar apparatus architecture of *Woloszynskia limnetica* (Dinophyceae). *Journal of Phycology* 31:948–957.

Robinson, N., G. Eglinton, S. C. Brassell, and P. A. Cranwell. 1984. Dinoflagellate origin for sedimentary 4α-methylsteroids and 5α(H)-stanols. *Nature* 308:439–442.

Rodrigues, M. A., C. P. dos Santos, A. J. Young, D. Strbac, and D. O. Hall. 2002. A smaller and impaired xanthophyll cycle makes the deep sea macroalga *Laminaria abyssalis* (Phaeophyceae) highly sensitive to daylight

when compared with shallow water *Laminaria digitata*. *Journal of Phycology* 38:939–947.

Rodríguez, F., E. Derelle, L. Guillou, F. Le Gall, D. Vaulot, and H. Moreau. 2005. Ecotype diversity in the marine picoeukaryote *Ostreococcus* (Chlorophyta, Prasinophyceae). *Environmental Microbiology* 7:853–859.

Rodríguez-Ezpeleta, N., et al. 2005. Monophyly of primary photosynthetic eukaryotes: Green plants, red algae, and glaucophytes. *Current Biology* 15:1325–1330.

Rodríguez-Ezpeleta, N., H. Philippe, H. Brinkmann, B. Becker, and M. Melkonian. 2007. Phylogenetic analyses of nuclear, mitochondrial, and plastid multigene data sets support the placement of *Mesostigma* in the Streptophyta. *Molecular Biology and Evolution* 24:723–731.

Roessler, P. G. 1990. Environmental control of glycerolipid metabolism in microalgae: Commercial implications and future research directions. *Journal of Phycology* 26:393–399.

Rogers, C. E., D. S. Domozych, K. D. Stewart, and K. R. Mattox. 1981. The flagellar apparatus of *Mesostigma viride* (Prasinophyceae): Multilayered structures in a scaly green flagellate. *Plant Systematics and Evolution* 138:247–258.

Rogers, C. E., K. R. Mattox, and K. D. Stewart. 1980. The zoospore of *Chlorokybus atmophyticus*, a charophyte with sarcinoid growth habit. *American Journal of Botany* 67:774–783.

Rogers, M. B., J. M. Archibald, M. A. Field, C. Li, B. Striepen, and P. J. Keeling. 2004. Plastid-targeting peptides from the chlorarachniophyte *Bigelowiella natans*. *Journal of Eukaryotic Microbiology* 51:529–535.

Rogers, M. B., P. R. Gilson, V. Su, G. I. McFadden, and P. J. Keeling. 2007. The complete chloroplast genome of the chlorarachniophyte *Bigelowiella natans*: Evidence for independent origins of chlorarachniophyte and euglenid secondary endosymbionts. *Molecular Biology and Evolution* 24:54–62.

Rogers, S. L., and R. G. Burns. 1994. Changes in aggregate stability, nutrient status, indigenous microbial populations, and seedling emergence, following inoculation of soil with *Nostoc muscorum*. *Biology and Fertility of Soils* 18:209–215.

Rogerson, A., A. S. W. DeFreitas, and A. G. McInnes. 1986. Growth rates and ultrastructure of siliceous setae of *Chaetoceros gracilis* (Bacillariophyceae). *Journal of Phycology* 22:56–62.

Romo, S., and C. Perez-Martinez. 1997. The use of immobilization in alginate beads for long-term storage of *Pseudanabaena galeata* (cyanobacteria) in the laboratory. *Journal of Phycology* 33:1074–1076.

Rose, J. M., and D. A. Caron. 2007. Does low temperature constrain the growth rates of heterotrophic protists? Evidence and implications for algal blooms in cold waters. *Limnology and Oceanography* 52:886–895.

Rosemond, A. D., and S. H. Brawley. 1996. Species-specific characteristics explain the persistence of *Stigeo-clonium tenue* (Chlorophyta) in a woodland stream. *Journal of Phycology* 32:54–63.

Rosenberg, E., O. Koren, L. Reshef, R. Efony, and I. Zilber-Rosenberg. 2007. The role of

microorganisms in coral health, disease and evolution. *Nature Reviews Microbiology* 5:355–362.

Rosentreter, R., D. D. Eldridge, and J. H. Kaltenecker. 2001. Monitoring and management of biological soil crusts. In: Belnap, J., and O. L. Lange (eds.), *Biological Soil Crusts: Structure, Function, and Management*, Ecological Studies, vol. 150, Springer Verlag, Berlin, pp. 457–470.

Rosowski, J. R., and W. G. Langenberg. 1994. The near-spineless *Trachelomonas grandis* (Euglenphyceae) superficially appears spiny by attracting bacteria to its surface. *Journal of Phycology* 30:1012–1022.

Rost, B., and U. Riebesell. 2004. Coccolithophore calcification and the biological pump: Response to environmental changes. In: Thierstein, H. R., and J. R. Young (eds.). *Coccolithophores—From Molecular Processes to Global Impact*. Springer Verlag, New York, pp. 99–126.

Rothhaupt, K. O. 1995. Algal nutrient limitation affects rotifer growth rate but not ingestion rate. *Oceanography and Limnology* 40: 1201–1208.

Rothschild, L. J., L. J. Giver, M. R. White, and R. L. Mancinelli. 1994. Metabolic activity of microorganisms in evaporites. *Journal of Phycology* 30:431–438.

Round, F. E. 1981. *The Ecology of Algae*. Cambridge University Press, Cambridge, UK.

Round, F. E. 1992. Epibiotic and endobiotic associations between chromophyte algae and their hosts. In: Reisser, W. (ed.), *Algae and Symbioses*, BioPress, Bristol, UK, pp. 593–618.

Round, F. E., R. M. Crawford, and D. G. Mann. 1990. *The Diatoms—Biology and Morphology of the Genera*, Cambridge University Press, Cambridge, UK.

Rousseau, F., and B. de Reviers. 1999. Circumscription of the order Ectocarpales (Phaeophyceae): Bibliographical synthesis and molecular evidence. *Cryptogamie Algologie* 20:5–18.

Rublee, P. A., D. L. Remington, E. R. Schaefer, and M. M. Marshall. 2005. Detection of the dinozoans *Pfiesteria piscidida* and *P. shumwayae*: A review of detection methods and geographic distribution. *Journal of Eukaryotic Microbiology* 52:83–89.

Rühland, K. M., J. P. Smol, and R. Pienitz. 2003. Ecology and spatial distributions of surface-sediment diatoms from 77 lakes in the subarctic Canadian treeline region. *Canadian Journal of Botany* 81:57–73.

Russell, C. A., M. D. Guiry, A. R. McDonald, and D. J. Garbary. 1996. Actin-mediated chloroplast movement in *Griffithsia pacifica* (Ceramiales, Rhodophyta). *Phycological Research* 44:57–61.

Rychert, R., J. Skujins, D. Sorensen, and D. Porcella. 1978. Nitrogen fixation by lichens and free-living microorganisms in deserts. In: West, N. E., and J. J. Skujins (eds.), *Nitrogen in Desert Ecosystems*, Dowden, Hutchinson and Ross, Stroudsburg, PA, pp. 20–30.

Rynearson, T. A., and E. V. Armbrust. 2005. Maintenance of clonal diversity during a spring bloom of the centric diatom *Ditylum brightwellii*. *Molecular Ecology* 14:1631–1640.

Rynearson, T. A., J. A. Newton, and V. A. Armbrust. 2006. Spring bloom development, genetic variation, and population succession in the planktonic diatom *Ditylum brightwellii*. *Limnology and Oceanography* 51:1249–1261.

Sabater, S., S. V. Gregory, and J. R. Sedell. 1998. Community dynamics and metabolism of benthic algae colonizing wood and rock substrata in a forest stream. *Journal of Phycology* 34:561–567.

Sáez, A. G., I. Probert, M. Geisen, P. Quinn, J. R. Young, and L. K. Medlin. 2003. Pseudo-cryptic speciation in coccolithophores. *Proceedings of the National Academy of Sciences of the United States of America* 100:7163–7168.

Saffo, M. B. 1987. New light on seaweeds. *BioScience* 37:654–664.

Sagert, S., R. M. Forster, P. Feuerpfiel, and H. Schubert. 1997. Daily course of photosynthesis and photoinhibition in *Chondrus crispus* (Rhodophyta) from different shore levels. *European Journal of Phycology* 32:363–371.

Saito, A., et al. 2003. Gliding movement in *Peranema trichophorum* is powered by flagellar surface motility. *Cell Motility and the Cytoskeleton* 55:244–253.

Saito, K., T. Drgon, J. A. F. Robledo, D. N. Krupatkina, and G. R. Vasta. 2002. Characterization of the rRNA locus of *Pfiesteria piscicida* and development of standard and quantitative PCR-based detection assays targeted to the nontranscribed spacer. *Applied and Environmental Microbiology* 68:5394–5407.

Saiz-Jimenez, C. 1995. Microbial melanins in stone monuments. *The Science of the Total Environment* 167:273–286.

Saker, M. L., M. Vale, D. Kramer, and V. M. Vasconcelos. 2007. Molecular techniques for the early warming of toxic cyanobacteria blooms in freshwater lakes and rivers. *Applied Microbiology and Biotechnology* 75:441–449.

Saldarriaga, J. F., B. S. Leander, F. J. R. "Max" Taylor, and P. J. Keeling. 2003. *Lessardia elongata* gen. et sp. nov. (Dinoflagellata, Peridiniales, Podolampaceae) and the taxonomic position of the genus *Roscoffia*. *Journal of Phycology* 39:368–378.

Salisbury, J. L., and G. L. Floyd. 1978. Calcium-induced contraction of the rhizoplast of a quadriflagellate green alga. *Science* 202:975–977.

Salisbury, J. L., J. A. Swanson, G. L. Floyd, R. Hall, and N. J. Maihle. 1981. Ultrastructure of the flagellar apparatus of the green alga *Tetraselmis subcordiformis* with special consideration given to the function of the rhizoplast and rhizanchora. *Protoplasma* 107:1–11.

Sanchez, E., et al. 2004. Descripción ultraestructural de *Euglena pailasensis* (Euglenozoa) del Volcán Rincón de la Vieja, Guanacaste, Costa Rica. *Revista de Biología Tropical* 52:31–40.

Sanchez-Puerta, M. V., T. R. Bachvaroff, and C. F. Delwiche. 2005. The complete plastid genome sequence of the haptophyte *Emiliania huxleyi*: A comparison to other plastid genomes. *DNA Research* 12:151–156.

Sanchez-Puerta, M. V., J. C. Lippmeier, K. E. Apt, and C. F. Delwiche. 2007. Plastid genes in a non-photosynthetic dinoflagellate. *Protist* 158:105–117.

Sancho, L. G., R. De La Torre, G. Horneck, C. Ascaso, A. De Los Rios, A. Pintado, J. Wierzchos, and M. Schuster. 2007. Lichens survive in space: Results from the 2005 LICHENS experiment. *Astrobiology* 7:443–454.

Sanders, R. W., D. A. Caron, and U.-G. Berninger. 1992. Relationships between bacteria and heterotrophic nanoplankton in marine and fresh waters: An inter-ecosystem comparison. *Marine Ecology Progress Series* 86:1–14.

Sanders, R. W., C. E. Williamson, P. L. Stutzman, R. E. Moeller, C. E. Goulden, and T. Aoki-Goldsmith. 1996. Reproductive success of "herbivorous" zooplankton fed algae and nonalgal food resources. *Limnology and Oceanography* 41:1295–1305.

Sanders, W. B., R. L. Moe, and C. Ascaso. 2005. Ultrastructural study of the brown alga *Petroderma maculiforme* (Phaeophyceae) in the free-living state and in lichen symbiosis with the intertidal marine fungus *Verrucaria tavaresiae*. *European Journal of Phycology* 40:353–361.

Sandgren, C. D. (ed.). 1988. *Growth and Reproductive Strategies of Freshwater Phytoplankton*, Cambridge University Press, Cambridge, UK.

Sandgren, C. D. 1981. Characteristics of sexual and asexual resting cyst (statospore) formation in *Dinobryon cylindricum* Imhof (Chrysophyta). *Journal of Phycology* 17:199–210.

Sandgren, C. D., and J. Flanagin. 1986. Heterothallic sexuality and density dependent encystment in the chrysophycean alga *Synura petersenii* Korsh. *Journal of Phycology* 22:206–216.

Sandgren, C. D., S. A. Hall, and S. B. Barlow. 1996. Siliceous scale production in chrysophyte and synurophyte algae. I. Effects of silica-limited growth on cell silica content, scale morphology, and the construction of the scale layer of *Synura petersenii*. *Journal of Phycology* 32:675–692.

Sandgren, C. D., and W. E. Walton. 1995. The influence of zooplankton herbivory on the biogeography of chrysophyte algae. In: Sandgren, C. D., J. P. Smol, and J. Kristiansen (eds.), *Chrysophyte Algae: Ecology, Phylogeny and Development*, Cambridge University Press, Cambridge, UK, pp. 269–302.

Sands, M., D. Poppel, and R. Brown. 1991. Peritonitis due to *Prototheca wickerhamii* in a patient undergoing chronic ambulatory peritoneal dialysis. *Reviews of Infectious Diseases* 13:376–378.

Santelices, B. 1980. Phytogeographic characterization of the temperate coast of Pacific South America. *Phycologia* 19:1–12.

Santelices, B., and I. A. Abbott. 1987. Geographic and marine isolation: An assessment of the marine algae of Easter Island. *Pacific Science* 41:1–20.

Santelices, B., P. Camus, and A. J. Hoffmann. 1989. Ecological studies for harvesting and culturing *Gymnogongrus furcellatus* (Rhodophyta, Gigartinales) in Central Chile. *Journal of Applied Phycology* 1:171–181.

Santore, U. J. 1985. A cytological survey of the genus *Cryptomonas* (Cryptophyceae) with comments on its taxonomy. *Archiv für Protistenkunde* 130:1–52.

Santos, L. M. A. 1996. The Eustigmatophyceae: Actual knowledge and research perspectives. *Nova Hedwigia* 112:391–405.

Santos, L. M. A., and G. F. Leedale. 1995. Some notes on the ultrastructure of small azoosporic members of the algal class Eustigmatophyceae. *Nova Hedwigia* 60:219–225.

Santos, S. R., T. L. Shearer, A. R. Hannes, and M. A. Coffroth. 2004. Fine-scale diversity and specificity in the most prevalent lineage of symbiotic dinoflagellates (*Symbiodinium*, Dinophyceae) of the Caribbean. *Molecular Ecology* 13:459–469.

Saranak, J., and K. W. Foster. 2005. Photoreceptor for curling behavior in *Peranema trichophorum* and evolution of eukaryotic rhodopsins. *Eukaryotic Cell* 4:1605–1612.

Sarcina, M., and P. J. Casselton. 1995. Degradation of adenine by *Prototheca zopfii* (Chlorophyta). *Journal of Phycology* 31:575–576.

Sasaki, H., H. Kataoka, M. Kamiya, and H. Kawai. 1999. Accumulation of sulfuric acid in Dictyotales (Phaeophyceae): Taxonomic distribution and ion chromatography of cell extracts. *Journal of Phycology* 35:732–739.

Sasaki, M., T. Shida, and K. Tachibana. 2001. Synthesis and stereochemical confirmation of the HI/JK ring system of prymnesins, potent hemolytic and ichthyotoxic glycoside toxins isolated from the red tide alga. *Tetrahedron Letters* 42:5725–5728.

Sato, M., T. Nishikawa, T. Yamazaki, and S. Kawano. 2005. Isolation of the plastid *FtsZ* gene from *Cyanophora paradoxa* (Glaucocystophyceae, Glaucocystophyta). *Phycological Research* 53:93–96.

Sauer, N., and W. Tanner. 1993. Molecular biology of sugar transporters in plants. *Botanica Acta* 106:277–286.

Saunders, G. W. 2005. Applying DNA barcoding to red macroalgae: A preliminary appraisal holds promise for future applications. *Philosophical Transactions of the Royal Society of London B* 360:1879–1888.

Saunders, G. W., and J. C. Bailey. 1997. Phylogenesis of pit-plug-associated features in the Rhodophyta: Inference from molecular systematic data. *Canadian Journal of Botany* 75:1436–1447.

Saunders, G. W., C. J. Bird, M. A. Ragan, and E. L. Rice. 1995. Phylogenetic relationships of species of uncertain taxonomic position within the Acrochaetiales–Palmariales complex (Rhodophyta): Inferences from phenotypic and 18S rDNA sequence data. *Journal of Phycology* 31:601–611.

Saunders, G. W., and M. H. Hommersand. 2004. Assessing red algal supraordinal diversity and taxonomy in the context of contemporary systematic data. *American Journal of Botany* 91:1494–1507.

Savin, M. C., J. L. Martin, M. LeGresley, M. Giewat, and J. Rooney-Varga. 2004. Plankton diversity in the Bay of Fundy as measured by morphological and molecular methods. *Microbial Ecology* 48:51–65.

Sawitzky, H., and F. Grolig. 1995. Phragmoplast of the green alga *Spirogyra* is functionally distinct from the higher plant phragmoplast. *Journal of Cell Biology* 130:1359–1371.

Scheffer, M., S. Rinaldi, J. Huisman, and F. J. Weissing. 2003. Why plankton communities have no equilibrium: Solutions to the paradox. *Hydrobiologia* 491:9–18.

Scheffer, M. S., D. Straile, E. H. van Nes, and H. Hosper. 2001. Climatic warming causes regime shifts in lake food webs. *Limnology and Oceanography* 46:1780–1783.

Scherer, S., A. Ernst, T.-W. Chen, and P. Böger. 1984. Rewetting of drought-resistant blue-green algae: Time course of water uptake and reappearance of respiration, photosynthesis and nitrogen fixation. *Oecologia* 62:418–423.

Schiel, D. R., and M. S. Foster 2006. The population biology of large brown seaweeds: Ecological consequences of multiphase life histories in dynamic coastal environments. *Annual Review of Ecology, Evolution, and Systematics* 37:343–372.

Schiel, D. R., J. R. Steinbeck, and M. S. Foster. 2004. Ten years of induced ocean warming causes comprehensive changes in marine benthic communities. *Ecology* 85:1833–1839.

Schindler, D. W. 1997. Widespread effects of climatic warming on freshwater ecosystems in North America. *Hydrological Processes* 11:1043–1067.

Schindler, D. W. 2006. Recent advances in the understanding and management of eutrophication. *Limnology and Oceanography* 51:356–363.

Schindler, D. W., K. H. Mills, D. F. Malley, D. L. Findlay, J. A., Shearer, I. J. Davies, M. A. Turner, G. A. Linsey, and D. R. Cruikshank. 1995. Long-term ecosystem stress: The effects of years of experimental acidification on a small lake. *Science* 228:1395–1401.

Schlesinger, W. H., J. S. Pippen, M. D. Wallenstein, K. S. Hofmockel, D. M. Klepeis, and B. E. Mahall. 2003. Community composition and photosynthesis by photoautotrophs under quartz pebbles, southern Mojave Desert. *Ecology* 84:3222–3231.

Schmechel, D. E., and D. C. Koltai. 2001. Potential human health effects associated with laboratory exposures to Pfiesteria piscida. *Environmental Health Perspectives* 109:775–779.

Schmid, A.-M. M. 1997. Putative main function of actin bundles in raphid diatoms: Necessity for a new locomotion model. *Phycologia* 36:99 (abstract).

Schmid, A.-M. M. 2003. Endobacteria in the diatom Pinnularia (Bacillariophyceae.) I. "Scattered ct-nucleoids" explained: DAPI–DNA complexes stem from exoplastidial bacteria boring into the chloroplasts. *Journal of Phycology* 39:122–138.

Schmidt, A. R., E. Ragazzi, O. Coppellotti, and G. Roghi. 2006a. A microworld in Triassic amber. *Nature* 444:835.

Schmidt, M., et al. 2006b. Proteomic analysis of the eyespot of Chlamydomonas reinhardtii provides novel insight into its components and tactic movements. *Plant Cell* 18:1908–1930.

Schmitt, R., S. Fabry, and D. L. Kirk. 1992. In search of molecular origins of cellular differentiation in Volvox and its relatives. *International Review of Cytology* 139:189–265.

Schmitz, K. 1981. Translocation. In: Lobban, C. S., and M. J. Wynne (eds.), *The Biology of Seaweeds*, Blackwell Scientific, Oxford, UK, pp. 534–558.

Schneegurt, M. A., D. M. Sherman, and L. A. Sherman. 1997. Growth, physiology, and ultrastructure of a diazotrophic cyanobacterium Cyanothece sp strain ATCC 51142, in mixotrophic and chemoheterotrophic cultures. *Journal of Phycology* 33:632–642.

Schneider, S., C. Ziegler, and A. Melzer. 2006. Growth towards light as an adaptation to high light conditions in Chara branches. *New Phytologist* 172:83–91.

Schnepf, E., and G. Drebes. 1993. Anisogamy in the dinoflagellate Noctiluca? *Helgoland Marine Research* 47:265–273.

Schnepf, E., and M. Elbrächter. 1992. Nutritional strategies in dinoflagellates. A review with emphasis on cell biological aspects. *European Journal of Protistology* 28:3–24.

Schnepf, E., and M. Elbrächter. 1999. Dinophyte chloroplasts and phylogeny. A review. *Grana* 38:81–97.

Schnepf, E., I. Schlegel, and D. Hepperle. 2002. *Petalomonas sphagnophila* (Euglenophyta) and its endocytobiotic cyanobacteria: A unique form of symbiosis. *Phycologia* 41:153–157.

Schoenwaelder, M. E. A. 2002. The occurrence and cellular significance of physodes in brown algae. *Phycologia* 41:125–139.

Schofield, O., T. J. Evans, and D. F. Mille. 1998. Photosystem II quantum yields and xanthophyll-cycle pigments of the macroalga Sargassum natans (Phaeophyceae) responses under natural sunlight. *Journal of Phycology* 34:104–112.

Schonbeck, M, and T. A. Norton. 1978. Factors controlling the upper limits of fucoid algae on the shore. *Journal of Experimental Marine Biology and Ecology* 31:303–313.

Schopf, J. W. 1993. Microfossils of the Early Archean Apex chert: New evidence of the antiquity of life. *Science* 260:640–646.

Schopf, J. W. 1996. Cyanobacteria: Pioneers of the early earth. *Nova Hedwigia* 112:13–32.

Schopf, J. W. 2006. Fossil evidence of Archaean life. *Philosophical Transactions of the Royal Society B* 361:869–885.

Schopf, J. W., A. B. Kudryavtsev, A. D. Czaja, and A. B. Tripathi. 2007. Evidence of Archean life: Stromatolites and microfossils. *Precambrian Research* 158:141–155.

Schornstein, K. L., and J. Scott. 1982. Ultrastructure of cell division in the unicellular red alga Porphyridium purpureum. *Canadian Journal of Botany* 60:85–97.

Schubert, N., Garcia-Mendoza, and I. Pacheco-Ruiz. 2006. Carotenoid composition of marine red algae. *Journal of Phycology* 42:1208–1216.

Schweizer, A. 1997. From littoral to pelagial: comparing the distribution of phytoplankton and ciliated protozoa along a transect. *Journal of Plankton Research* 19:829–848.

Schwender, J., C. Gemünden, and H. K. Lichtenthaler. 2001. Chlorophyta exclusively use the 1-deoxyxylulose 5-phosphate/2-C-methylerythritol 4-phosphate pathway for the biosynthesis of isoprenoids. *Planta* 212:416–423.

Schwenke, H. 1971. Water movements. In: Kinne, O. (ed.), Marine Ecology, Wiley-Interscience, London, pp. 1091–1121.

Scott, J., C. Bosco, K. Schornstein, and J. Thomas. 1980. Ultrastructure of cell division and reproductive differentiation of male plants in the Florideophyceae (Rhodophyta): Cell division in Polysiphonia. *Journal of Phycology* 16:507–524.

Scott, J., and S. Broadwater. 1990. Cell division. In: Cole, K. M., and R. G. Sheath (eds.), *Biology of the Red Algae*, Cambridge University Press, Cambridge, UK, pp. 123–146.

Scott, J. L., and P. S. Dixon. 1973. Ultrastructure of spermatium liberation in the marine red alga Ptilota densa. *Journal of Phycology* 9:85–91.

Scrosati, R., and R. E. DeWreede. 1998. The impact of frond crowding on frond bleaching in the clonal intertidal alga Mazzaella cornucopiae (Rhodophyta, Gigartinaceae) from British Columbia, Canada. *Journal of Phycology* 34:228–232.

Searles, R. B. 1980. The strategy of the red algal life history. *American Naturalist* 115:113–120.

Seckbach, J. 1994. The natural history of Cyanidium (Geitler 1933: past and present perspectives). In: Seckbach, J. (ed.), *Evolutionary Pathways and Enigmatic Algae: Cyanidium Caldarium (Rhodophyta) and Related Cells*, Kluwer Academic Publishers, Boston, pp. 99–112.

Seckbach, J., E. González, and I. M. Wainright. 1992. Peroxisomal function in the Cyanidiophyceae (Rhodophyta): A discussion of phylogenetic relationships and the evolution of microbodies (peroxisomes). *Nova Hedwigia* 55:99–109

Sekiguchi, H., M. Moriya, T. Nakayama, and I. Inouye. 2002. Vestigial chloroplasts in heterotrophic stramenopiles Pteridomonas danica and Ciliophrys infusionum (Dictyochophyceae). *Protist* 153:157–167.

Semple, S., and M. Steel. 2003. *Phylogenetics*. Oxford lecture series in mathematics and its applications, 24. Oxford University Press, Oxford, UK.

Senft, W. H., R. A. Hunchberger, and K. E. Roberts. 1981. Temperature dependence of growth and phosphorus uptake in two species of Volvox (Volvocales, Chlorophyta). *Journal of Phycology* 17:323–329.

Senousy, H. H., G. W. Beakes, and E. Hack. 2004. Phylogenetic placement of Botryococcus braunii (Trebouxiophyceae) and Botryococcus sudeticus isolate UTEX 2629 (Chlorophyceae). *Journal of Phycology* 40:412–423.

Seppelt, R. D., and P. A. Broady. 1988. Antarctic terrestrial ecosystems: The Vestfold Hills in context. *Hydrobiologia* 165:177–184.

Shalchian-Tabrizi, K., et al. 2006a. Telonemia, a new protist phylum with affinity to chromist lineages. *Proceedings of the Royal Society B.* 273:1833–1842.

Shalchian-Tabrizi, K., M. A. Minge, T. Cavalier-Smith, J. M. Nedreklepp, D. Klaveness, and K. S. Jakobsen. 2006b. Combined heat shock protein 90 and ribosomal RNA sequence phylogeny supports multiple replacements of dinoflagellate plastids. *Journal of Eukaryotic Microbiology* 53:217–224.

Shapiro, J., and D. I. Wright. 1984. Lake restoration by biomanipulation: Round Lake, Minnesota, the first two years. *Freshwater Biology* 14:371–83.

Sheath, R. G. 1984. The biology of freshwater red algae. *Progress in Phycological Research* 3:89–157.

Sheath, R. G., and J. A. Hambrook. 1988. Mechanical adaptations to flow in freshwater red algae. *Journal of Phycology* 24:107–111.

Sheath, R. G., and J. A. Hambrook. 1990. Freshwater ecology. In: Cole, K. M., and R. G. Sheath (eds.), *Biology of the Red Algae*, Cambridge University Press, Cambridge, UK, pp. 423–454.

Sheath, R. G., K. M. Müller, M. L. Vis, and T. J. Entwhistle. 1996a. A re-examination of the morphology, ultrastructure and classification of genera in the Lemaneaceae (Batrachospermales, Rhodophyta). *Phycological Research* 44:233–246.

Sheath, R. G., M. K. Müller, A. Whittick, and T. J. Entwhistle. 1996b. A re-examination of the morphology and reproduction of *Nothocladus lindauerii* (Batrachospermales, Rhodophyta). *Phycological Research* 44:1–10.

Shen, E. Y. F. 1967. Microspectrophotometric analysis of nuclear DNA in *Chara zeylanica*. *Journal of Cell Biology* 35:377–384.

Sheppard, C. R. 2003. Predicted recurrences of mass coral mortality in the Indian Ocean. *Nature* 425:294–297.

Sherr, B. F., E. B. Sherr, and R. D. Fallon. 1987. Use of monodispersed, fluorescently labeled bacteria to estimate in situ protozoan bacterivory. *Applied and Environmental Microbiology* 53:958–965.

Sherwood, A. R., and G. G. Presting. 2007. Universal primers amplify a 23S rDNA plastid marker in eukaryotic algae and cyanobacteria. *Journal of Phycology* 43:605–608.

Sherwood, A. R., and R. G. Sheath. 2003. Systematics of the Hildenbrandiales (Rhodophyta): Gene sequence and morphometric analysis of global collections. *Journal of Phycology* 39:409–422.

Shi, T., and P. G. Falkowski. 2008. Genome evolution in cyanobacteria: The stable core and the variable shell. *Proceedings of the National Academy of Sciences of the United States of America* 105:2510–2015.

Shi, Y., W. Zhao, W. Zhang, Z. Ye, and J. Zhao. 2006. Regulation of intracellular free calcium concentration during heterocyst differentiation by HetR and NtcA in *Anabaena* sp. PCC 7120. *Proceedings of the National Academy of Sciences of the United States of America* 103:11334–11339.

Shields, L. M., and L. W. Durrell. 1964. Algae in relation to soil fertility. *The Botanical Review* 30:92–128.

Shimmel, S. M., and W. Marshall Darley. 1985. Productivity and density of soil algae in and agricultural system. *Ecology* 66:1439–1447.

Shin, W., S. Brosnan, and R. E. Triemer. 2002. Are cytoplasmic pockets (MTR/pocket) present in all photosynthetic euglenoid genera? *Journal of Phycology* 38:790–799.

Shiomoto, A., K. Tadokoro, K. Nagasawa, and Y. Ishida. 1997. Trophic relations in the subarctic North Pacific ecosystem: possible feeding effect from pink salmon. *Marine Ecology Progress Series* 150:75–85.

Shiraiwa, Y., and Y. Umino. 1991. Effect of glucose on the induction of the carbonic anhydrase and the change in $K_{1/2}(CO_2)$ of photosynthesis in *Chlorella vulgaris* 11h. *Plant Cell Physiology* 32:311–314.

Shoup, S., and L. A. Lewis. 2003. Polyphyletic origin of parallel basal bodies in swimming cells of chlorophycean green algae (Chlorophyta). *Journal of Phycology* 39:789–796.

Sicko-Goad, L., E. F. Stoermer, and G. Fahnenstiel. 1986. Rejuvenation of *Melosira granulata* (Bacillariophyceae) resting cells from the anoxic sediments of Douglas Lake, Michigan. I. Light microscopy and ^{14}C uptake. *Journal of Phycology* 22:22–28.

Sieburth, J. McN., V. Smetacek, and J. Lenz. 1978. Pelagic ecosystem structure: Heterotrophic compartments of the plankton and their relationship to plankton size fractions. *Limnology and Oceanography* 23:1256–1263.

Siegel, D. A. 1998. Resource competition in a discrete environment: Why are plankton distributions paradoxical? *Limnology and Oceanography* 43:1133–1146.

Siesser, W. G., and A. Winter. 1994. Composition and morphology of coccolithophore skeletons. In: Winter, A., and W. G. Siesser (eds.), *Coccolithophores*, Cambridge University Press, Cambridge, UK, pp. 51–62.

Silva, P. C., J. Throndsen, and W. Eikrem. 2007. Revisiting the nomenclature of haptophytes. *Phycologia* 46:471–475.

Simon, A. G. Glöckner, M. Felder, M. Melkonian, and B. Becker. 2006. EST analysis of the scaly green flagellate *Mesostigma viride* (Streptophyta): Implications for the evolution of green plants (Viridiplantae). *BMC Biology* 6:2.

Simons, J. 1974. *Vaucheria compacta*: A euryhaline estuarine algal species. *Acta Botanica Neerlandica* 23:613–626.

Simpson, A. G. B., and D. J. Patterson. 2006. Current perspectives on high-level groupings of protists. In: Katz, L., and D. Bhattacharya (eds.), *Genomics and Evolution of Microbial Eukaryotes*, Cambridge University Press, Cambridge, UK, Chapter 1.

Sims, P. A., D. G. Mann, and L. K. Medlin. 2006. Evolution of the diatoms: Insights from fossil, biological and molecular data. *Phycologia* 45:361–402.

Sineshchekov, D. A., and E. G. Govorunova. 1999. Rhodopsin-mediated photosensing in green flagellate algae. *Trends in Plant Science* 4:58–63.

Sineshchekov, O. A., E. G. Govorunova, K.-H. Jung, S. Zauner, U.-G. Maier, and J. L. Spudich. 2005. Rhodopsin-mediated photoreception in cryptophyte flagellates. *Biophysical Journal* 89:4310–4319.

Sinninghe Damsté, J. S., et al. 2004. The rise of the rhizosolenid diatoms. *Science* 304:584–587.

Sitte, P. 1993. Symbiogenetic evolution of complex cells and complex plastids. *European Journal of Protistology* 29:131–143.

Siver, P. A. 1995. The distribution of chrysophytes along environmental gradients: Their use as biological indicators. In: Sandgren, C. D., J. P. Smol, and J. Kristiansen (eds.), *Chrysophyte Algae: Ecology, Phylogeny and Development*, Cambridge University Press, Cambridge, UK, pp. 232–268.

Siver, P. A., and A. P. Wolfe. 2005. Eocene scaled chrysophytes with pronounced modern affinities. *International Journal of Plant Sciences* 166:533–536.

Six, C., A. Z. Worden, R. Rodríguez, H. Moreau, and F. Partensky. 2005. New insights into the nature and phylogeny of prasinophyte antenna proteins: *Ostreococcus tauri*, a case study. *Molecular Biology and Evolution* 22:2217–2230.

Skovgaard, A. 1996. Engulfment of *Ceratium* spp. (Dinophyceae) by the thecate photosynthetic dinoflagellate *Fragilidium subglobosum*. *Phycologia* 35:490–499.

Skovgaard, A., and P. J. Hansen. 2003. Food uptake in the harmful alga *Prymnesium parvum* mediated by excreted toxins. *Limnology and Oceanography* 48:1161–1166.

Skulberg, O. M. 1995. Use of algae for testing water quality. In: Wiessner, W., E. Schnepf, and R. C. Starr, *Algae, Environment and Human Affairs*, BioPress, Bristol, UK, pp. 181–199.

Sluiman, H. 1993. Nucleus, nuclear division, and cell division. In: Berner, T. (ed.), *Ultrastructure of Microalgae*, CRC Press, Boca Raton, FL, pp. 221–268.

Sluiman, H. J. 1983. The flagellar apparatus of the zoospore of the filamentous green alga *Coleochaete pulvinata*: Absolute configuration and phylogenetic significance. *Protoplasma* 115:160–175.

Sluiman, H. J. 1989. The green algal class Ulvophyceae. An ultrastructural survey and classification. *Cryptogamic Botany* 1:83–94.

Sluiman, H. J., C. Guihal, and O. Mudimu. 2008. Assessing phylogenetic affinities and species delimitations in Klebsormidiales (Streptophyta): Nuclear encoded rDNA phylogenies and ITS secondary structure models in *Klebsormidium*, *Hormidella*, and *Entransia*. *Journal of Phycology* 44:193–195.

Smayda, T. J. 1970. The suspension and sinking of phytoplankton in the sea. *Oceanography and Marine Biology: An Annual Review* 8:353–414.

Smith, E. C., and H. Griffiths. 1996. The occurrence of the chloroplast pyrenoid is correlated with the activity of a CO_2-concentrating mechanism and carbon isotope discrimination in lichens and bryophytes. *Planta* 198:6–16.

Smith, G. M. 1917. The vertical distribution of *Volvox* in the plankton of Lake Mendota. *American Journal of Botany* 5:78–185.

Smith, G. M. 1950. *The Fresh-Water Algae of the United States*, McGraw-Hill, New York.

Smith, G. M. 1969. *Marine Algae of the Monterey Peninsula, CA* (incorporating 1966 Supplement by G. J. Hollenberg and I. A. Abbott), Stanford University Press, Stanford, CA.

Smith, J. E., C. L. Hunter, E. J. Conklin, R. Most, T. Sauvage, C. Squair, and C. M. Smith. 2004a. Ecology of the invasive red alga *Gracilaria salicornia* (Rhodophyta) on O'ahu, Hawai'i. *Pacific Science* 58:325–343.

Smith, J. E., C. L. Hunter, and C. M. Smith. 2002. Distribution and reproductive characteristics of nonindigenous and invasive marine algae in the Hawaiian Islands. *Pacific Science* 56:299–315.

Smith, J. E., J. W. Runcie, and C. M. Smith. 2005b. Characterization of a large-scale ephemeral bloom of the green alga *Cladophora sericea* on the coral reefs of West Maui, Hawai'i. *Marine Ecology Progress Series* 302:77–91.

Smith, J. E., M. Shaw, R. A. Edwards, D. Obura, O. Pantos, E. Sala, S. A. Sandin, S. Smriga, M. Hatay, and F. L. Rohwer. 2006b. Indirect effects of algae on coral: algae-mediated, microbe-induced coral mortality. *Ecology Letters* 9:835–845.

Smith, J. E., C. M. Smith, and C. L. Hunter. 2001. An experimental analysis of the effects of herbivory and nutrient enrichment on benthic community dynamics on a Hawaiian reef. *Coral Reefs* 19:332–342.

Smith, R. M., and S. B. Williams. 2006. Circadian rhythms in gene transcription imparted by chromosome compaction in the cyanobacterium *Synechococcus elongatus*. *Proceedings* of the National Academy of Sciences of the United States of America 103:8564–8569.

Smith, S. M., R. M. M. Abed, and F. Garcia-Pichel 2004b. Biological soil crusts of sand dunes in Cape Cod National Seashore, Massachusetts, USA. *Microbial Ecology* 48:200–208.

Smith, V. H., B. L. Foster, J. P. Grover, R. D. Holt, M. A. Leibold, and F. deNoyelles, Jr. 2005a. Phytoplankton species richness scales consistently from laboratory microcosms to the world's oceans. *Proceedings of the National Academy of Sciences of the United States of America* 102:4393–4396.

Smith, V. H., S. B. Joye, and R. W. Howarth. 2006a. Eutrophication of freshwater and marine ecosystems. *Limnology and Oceanography* 51:351–355.

Smol, J. P. 1995. Application of chrysophytes to problems in paleoecology. In: Sandgren, C. D., J. P. Smol, and J. Kristiansen (eds.), *Chrysophyte Algae: Ecology, Phylogeny and Development*, Cambridge University Press, Cambridge, UK, pp. 303–329.

Smol, J. P. 2007. Marine sediments tell it like it was. *Proceedings of the National Academy of Sciences of the United States of America* 104:17563–17564.

Smol, J. P., and B. F. Cumming. 2000. Tracking long-term changes in climate using algal indicators in lake sediments. *Journal of Phycology* 36:986–1011.

Smol, J. P., et al. 2005. Climate-driven regime shifts in the biological communities of arctic lakes. *Proceedings of the National Academy of Sciences of the United States of America* 102:4397–4402.

Snyder, J. M., and L. H. Wullstein. 1973. The role of desert cryptogams in nitrogen fixation. *American Midland Naturalist* 90:257–265.

Sœmundsdóttir, S., and P. A. Matrai. 1998. Biological production of methyl bromide by cultures of marine phytoplankton. *Limnology and Oceanography* 43:81–87.

Solheim, B., and M. Zielke. 2003. Associations between cyanobacteria and mosses. In: Rai, A. N., B. Bergman, and U. Rasmussen (eds.), *Cyanobacteria in Symbiosis*, Kluwer Academic Publishers, Boston, pp. 137–152.

Soll, J., and E. Schleiff. 2004. Protein import into chloroplasts. *Nature Reviews Molecular Cell Biology* 5:198–208.

Sommer, M. S., S. B. Gould, P. Lehmann, A. Gruber, J. M. Przyborski, and U.-G. Maier. 2007. Der1-mediated preprotein import into the periplastid compartment of chromalveolates? *Molecular Biology and Evolution* 24:918–928.

Sommer, U. 1985. Comparison between steady state and non-steady state competition: Experiments with natural phytoplankton. *Limnology and Oceanography* 30:335–346.

Sommer, U. 1986a. Phytoplankton competition along a gradient of dilution rates. *Oecologia* 68:503–506.

Sommer, U. 1986b. Nitrate- and silicate-competition among Antarctic phytoplankton. *Marine Biology* 91:345–351.

Sommer, U. 1988. Growth and survival strategies of planktonic diatoms. In: Sandgren, C. D. (ed.), *Growth and Reproductive Strategies of Freshwater Phytoplankton*, Cambridge University Press, Cambridge, UK, pp. 227–260.

Sommer, U. 1995. An experimental test of the intermediate disturbance hypothesis using cultures of marine phytoplankton. *Limnology and Oceanography* 40:1271–1277.

Sommer, U. 2000. Scarcity of medium-sized phytoplankton in the northern Red Sea explained by strong bottom-up and weak top-down control. *Marine Ecology Progress Series* 197:19–25.

Sommer, U., and H. Stibor. 2002. Copepoda-Cladocera-Tunicata: The role of three major mesozooplankton groups in pelagic food webs. *Ecological Research* 17:161–174.

Sommer, U., and Z. M. Gliwicz. 1986. Long range vertical migration of *Volvox* in tropical Lake Cahora Bassa (Mozambique). *Limnology and Oceanography* 31:650–653.

Sommer, U., F. Sommer, B. Santer, C. Jamieson, M. Boersma, C. Becker, and T. Hansen. 2001. Complementary impact of copepods and Cladocerans on phytoplankton. *Ecology Letters* 4:545–550.

Sommer, U., H. Stibor, A. Katechakis, F. Sommer, and T. Hansen. 2002. Pelagic food web configurations at different levels of nutrient richness and their implications for the ratio fish production:primary production. *Hydrobiologia* 484:11–20.

Sommerfeld, M. R., and R. A. Bell. 1991. Cryptoendolithic algae of the semi-desert southwestern United States. *Journal of Phycology* 27:69.

Sondergaard, M., E. Jeppesen, E. Mortensen, E. Dall, P. Kristensen, and O. Sortkjaer. 1990. Phytoplankton biomass reduction after planktivorous fish reduction in a shallow, eutrophic lake: a combined effect of reduced internal P-loading and increased zooplankton grazing. *Hydrobiologia* 200/201:229–240.

Soule, T., V. Stout, W. D. Swingley, J. C. Meeks, and F. Garcia-Pichel. 2007. Molecular genetics and genomic analysis of scytonemin biosynthesis in *Nostoc punctiforme* ATCC 29133. *Journal of Bacteriology* 189:4465–4472.

Sournia, A. 1982. Form and function in marine phytoplankton. *Biological Review* 57:347–394.

Spalding, H., M. S. Foster, and J. N. Heine. 2003. Composition, distribution, and abundance of deep-water (>30 m) macroalgae in central California. *Journal of Phycology* 39:273–284.

Spaulding, S. A., D. M. McKnight, R. L. Smith, and R. Dufford. 1994. Phytoplankton population dynamics in perennially ice-covered Lake Fryxell, Antarctica. *Journal of Plankton Research* 16:527–541.

Spear-Bernstein, L., and K. R. Miller. 1989. Unique location of the phycobiliprotein light-harvesting pigment in the cryptophyceae. *Journal of Phycology* 25:412–419.

Spero, H. J., and D. L. Angel. 1991. Planktonic sarcodines: Microhabitat for oceanic dinoflagellates. *Journal of Phycology* 27:187–195.

St. Clair, L. L., J. R. Johansen, and B. L. Webb. 1986. Rapid stabilization of fire-disturbed sites using a soil crust slurry: Inoculation studies. *Reclamation and Revegetation Research* 4:261–269.

St. Clair, L. L., B. L. Webb, J. R. Johansen, and G. T. Nebeker. 1984. Cryptogamic soil crusts: Enhancement of seedling establishment in disturbed and undisturbed areas. *Reclamation and Revegetation Research* 3:129–136.

Stabenau, H., and U. Winkler. 2005. Glycolate metabolism in green algae. *Physiologia Plantarum* 123:235–245.

Stache-Crain, B., D. G. Müller, and L. J. Goff. 1997. Molecular systematics of *Ectocarpus* and *Kuckuckia* (Ectocarpales, Phaeophyceae) inferred from phylogenetic analysis of nuclear- and plastid-encoded DNA sequences. *Journal of Phycology* 33:152–168.

Stachowicz, J. J., and R. B. Whitlatch. 2005. Multiple mutualists provide complementary benefits to their seaweed host. *Ecology* 86:2418–2427.

Stadler, R., K. Wolf, C. Hilgarth, W. Tanner, and N. Sauer. 1995. Subcellular localization of the inducible *Chlorella* HUP1 monosaccharide-H$^+$ symporter and cloning of a co-induced galactose-H$^+$ symporter. *Plant Physiology* 107:33–41.

Stal, L. J. 1995. Physiological ecology of cyanobacteria in microbial mats and other communities. *New Phytologist* 131:1–32.

Stam, W. T., J. L. Olsen, S. F. Zaleski, S. N. Murray, K. R. Brown, and L. J. Walters. 2006. A forensic and phylogenetic survey of *Caulerpa* species (Caulerpales, Chlorophyta) from the Florida coast, local aquarium shops, and e-commerce: Establishing a proactive baseline for early detection. *Journal of Phycology* 42:1113–1124.

Stanley, S. M., J. B. Ries, and L. A. Hardie. 2005. Seawater chemistry, coccolithophore population growth, and the origin of Cretaceous chalk. *Geology* 33:593–596.

Starr, R. C. 1963. Homothallism in *Golenkinia minutissima*. In: *Studies in Microalgae and Bacteria*. Japanese Society of Plant Physiology, University of Tokyo Press, Tokyo, pp. 3–6.

Starr, R. C., and W. R. Rayburn. 1964. Sexual reproduction in *Mesotaenium kramstai*. *Phycologia* 4:22–26.

Steidinger, K. A. 1993. Some taxonomic and biologic aspects of toxic dinoflagellates. In: Falconer, I. R. (ed.), *Algal Toxins in Seafood and Drinking Water*, Academic Press, London, pp. 1–28.

Steidinger, K. A., and K. Tangen. 1997. Dinoflagellates. In: Tomas, C. R. (ed.), *Identifying Marine Phytoplankton*, Academic Press, San Diego, CA, pp. 387–584.

Stein, J. R. (ed.). 1973. *Handbook of Phycological Methods. Culture Methods and Growth Measurements*. Cambridge University Press, Cambridge, UK.

Steinbeck, J. R., D. R. Schiel, and M. S. Foster. 2005. Detecting long-term change in complex communities: A case study from the rocky intertidal zone. *Ecological Applications* 15:1813–1832.

Steinbrenner J., and G. Sandmann. 2006. Transformation of the green alga *Haematococcus pluvialis* with a phytoene desaturase for accelerated astaxanthin biosynthesis. *Applied and Environmental Microbiology* 72:7477–7484.

Steindler, L., D. Huchon, A. Avni, and M. Ilan. 2005. 16S rRNA phylogeny of sponge-associated cyanobacteria. *Applied and Environmental Microbiology* 71:4127–4131.

Steinkötter, J., D. Bhattacharya, I. Semmelroth, C. Bibeau, and M. Melkonian. 1994. Prasinophytes form independent lineages within the Chlorophyta: Evidence from ribosomal RNA sequence comparisons. *Journal of Phycology* 30:340–345.

Steinman, A. D. 1996. Effects of grazers on freshwater benthic algae. In: Stevenson, R. J., M. L. Bothwell, and R. L. Lowe (eds.), *Algal Ecology: Freshwater Benthic Ecosystems*, Academic Press, New York, pp. 341–374.

Stelter, K., N. M. El-Sayed, and F. Seeber. 2007. The expression of a plant-type ferredoxin redox system provides molecular evidence for a plastid in the early dinoflagellate *Perkinsus marinus*. *Protist* 158:119–130.

Stemler, A. J. 1997. The case for chloroplast thylakoid carbonic anhydrase. *Physiologica Plantarum* 99:348–353.

Steneck, R. S. 1990. Herbivory and the evolution of non-geniculate coralline algae (Rhodophyta, Corallinales) in the North Atlantic and North Pacific. In: Garbary, D. J., and G. R. South (eds.), *Evolutionary Biogeography of the Marine Algae of the North Atlantic*, NATO Advanced Research Workshop Series G, vol. 22, Springer Verlag, Berlin, pp. 107–129.

Steneck, R. S., M. H. Graham, B. J. Bourque, D. Corbett, J. M. Erlandson, J. A. Estes, and M. J. Tegner. 2002. Kelp forest ecosystems: biodiversity, stability, resilience and future trends. *Environmental Conservation* 29:436–459.

Stephenson, T. A., and A. Stephenson. 1949. The universal features of zonation between tide-marks on rocky coasts. *Journal of Ecology* 38:289–305.

Stephenson, T. A., and A. Stephenson. 1972. *Life Between Tidemarks on Rocky Shores*. W. H. Freeman, San Francisco.

Steppe, T. F., J. B. Olson, H. W. Paerl, R. W. Litaker, and J. Belnap. 1996. Consortial N_2 fixation: A strategy for meeting nitrogen requirements of marine and terrestrial cyanobacterial mats. *FEMS Microbiology and Ecology* 21:149–156.

Steppe, T. F., J. L. Pinckney, J. Dyble, and H. W. Paerl. 2001. Diazotrophy in modern marine Bahamian stromatolites. *Microbial Ecology* 41:36–44.

Sterner, R. W. 1989. The role of grazers in phytoplankton succession. In: Sommer, U. (ed.), *Plankton Ecology*, Springer Verlag, Berlin, pp. 107–170.

Sterner, R. W., T. M. Smutka, R. M. L. McKay, Q. Xiaoming, E. T. Brown, and R. M. Sherrell. 2004. Phosphorus and trace metal limitation of algae and bacteria in Lake Superior. *Limnology and Oceanography* 49:495–507.

Stevens, D. R., and S. Purton. 1997. Genetic engineering of eukaryotic algae: Progress and prospects. *Journal of Phycology* 33:713–722.

Stevenson, R. J. 1996. An introduction to algal ecology in freshwater benthic habitats. In: Stevenson, R. J., M. L. Bothwell, and R. L. Lowe (eds.), *Algal Ecology: Freshwater Benthic Ecosystems*, Academic Press, New York, pp. 3–30.

Stewart, H. L., and R. C. Carpenter. 2003. The effects of morphology and water flow on photosynthesis of marine macroalgae. *Ecology* 84:2999–3012.

Stewart, S. A., and M. L. Vis. 2007. Investigation of two species complexes in *Batrachospermum* section *Batrachospermum* (Batrachospermales, Rhodophyta). *Phycologia* 46:380–385.

Stewart, W. D. P. (ed.). 1974. *Algal Physiology and Biochemistry*, University of California Press, Berkeley, CA.

Stewart, W. D. P., G. P. Fitzgerald, and R. H.

Burris. 1967. *In situ* studies on N_2 fixation using the acetylene reduction technique. *Proceedings of the National Academy of Sciences of the United States of America* 58:2071–2078.

Stewart, W. D. P., A. Haystead, and H. W. Pearson. 1969. Nitrogenase activity in heterocysts of blue-green algae. *Nature* 224:226–228.

Stiger, V., and C. Payri. 1997. Strategies of reef invasion by two brown algae in Tahiti (French Polynesia): Reproduction, dispersion, competition. *Phycologia* 36:108.

Stiger, V., T. Horiguchi, Y. Yoshida, A. W. Coleman, and M. Masuda. 2003. Phylogenetic relationships within the genus *Sargassum* (Fucales, Phaeophyceae), inferred from ITS-2 nrDNA, with an emphasis on the taxonomic subdivision of the genus. *Phycological Research* 51:1–10.

Stiller, J. W. 2003. Weighing the evidence for a single origin of plastids. *Journal of Phycology* 39:1283–1285.

Stiller, J. W. 2007. Plastid endosymbiosis, genome evolution and the origin of green plants. *Trends in Plant Science* 12:391–396.

Stiller, J. W., and L. Harrell. 2005. The largest subunit of RNA polymerase II from the Glaucocystophyta: Functional constraint and short-branch exclusion in deep eukaryotic phylogeny. *BMC Evolutionary Biology* 5:71.

Stiller, J. W., D. C. Reel, and J. C. Johnson. 2003. A single origin of plastids revisited: Convergent evolution in organellar genome content. *Journal of Phycology* 39:95–105.

Stimson, J., S. Larned, and K. McDermid. 1996. Seasonal growth of the coral reef macroalga *Dictyosphaeria cavernosa* (Forskål) Borgensen and the effects of nutrient availability, temperature and herbivory on growth rate. *Journal of Experimental Marine Biology and Ecology* 196:53–77.

Stoecker, D. K., A. E. Michaels, and L. H. Davis. 1987. Large proportion of marine planktonic ciliates found to contain functional chloroplasts. *Nature* 326:790–792.

Stoermer, E. F. 1978. Phytoplankton as indicators of water quality in the Laurentian Great Lakes. *Transactions of the American Microscopical Society*. 97:2–16.

Stokes, P. M. 1983. Responses of freshwater algae to metals. *Progress in Phycological Research* 2:87–112.

Stratmann, J., G. Paputsogle, and W. Oertel. 1996. Differentiation of *Ulva mutabilis* (Chlorophyta) gametangia and gamete release are controlled by extracellular inhibitors. *Journal of Phycology* 32:1009–1021.

Stross, R. G. 1979. Density and boundary regulation of the *Nitella* meadow in Lake George, NY. *Aquatic Botany* 6:285–300.

Strzepek, R. F., and P. J. Harrison. 2004. Photosynthetic architecture differs in coastal and oceanic diatoms. *Nature* 431:689–692.

Suda, S. 2004. Taxonomic characterization of *Pyramimonas aurea* sp. nov. (Prasinophyceae, chlorophyta). *Phycologia* 43:682–692.

Suda, S., M. M. Watanabe, and I. Inouye. 1989. Evidence for sexual reproduction in the primitive green alga *Nephoselmis olivacea* (Prasinophyceae). *Journal of Phycology* 25:596–600.

Suda, S., M. M. Watanabe, and I. Inouye. 2004. Electron microscopy of sexual reproduction in *Nephroselmis olivacea* (Prasinophyceae, Chlorophyta). *Phycological Research* 52:273–283.

Suetsugu, N., and M. Wada. 2007. Chloroplast photorelocation movement mediated by phototropin family proteins in green plants. *Biological Chemistry* 388:927–935.

Sulli, C., and S. D. Schwartzbach. 1995. A soluble protein is imported into *Euglena* chloroplasts as a membrane-bound precursor. *Plant Cell* 8:43–53.

Sullivan, J. M., and E. Swift. 2003. Effects of small-scale turbulence on net growth rate and size of ten species of marine dinoflagellates. *Journal of Phycology* 39:83–94.

Sumper, M. 2002. A phase separation model for the nanopatterning of diatom biosilica. *Science* 295:2430–2433.

Sumper, M., E. Brunner, and G. Lehmann. 2005. Biomineralization in diatoms: Characterization of novel polyamines associated with silica. *Federation of European Biochemical Societies* 579:3765–3769.

Surif, M. B., and J. A. Raven. 1990. Photosynthetic gas exchange under emersed conditions in eulittoral and normally submersed members of the Fucales and the Laminariales: Interpretation in relation to C isotope ratio and N and water use efficiency. *Oecologia* 82:68–80.

Surzycki, R., L. Cournac, G. Peltier, and J.-D. Rochaix. 2007. Potential for hydrogen production with inducible chloroplast gene expression in *Chlamydomonas*. *Proceedings of the National Academy of Sciences of the United States of America* 104:17548–17553.

Sussman, A. V., and R. E. DeWreede. 2002. Host specificity of the endophytic sporophyte phase of *Acrosiphonia* (Codiolales, Chlorophyta) in southern British Columbia, Canada. *Phycologia* 41:169–177.

Sussmann, A. V., B. K. Mable, R. E. DeWreede, and M. L. Berbee. 1999. Identification of green algal endophytes as the alternate phase of *Acrosiphonia* (Codiolales, Chlorophyta) using ITS1 and ITS2 ribosomal DNA sequence data. *Journal of Phycology* 35:607–614.

Sutherland, T. F., C. L. Amos, and J. Grant. 1998. The effect of buoyant biofilms on the erodibility of sublittoral sediments of a temperate microtidal estuary. *Limnology and Oceanography* 43:225–235.

Suttle, C. A. 1994. The significance of viruses to mortality in aquatic microbial communities. *Microbial Ecology* 28:237–243.

Suzaki, T., and R. E. Williamson. 1986. Ultrastructure and sliding of pellicular structures during eugenoid movement in *Astasia longa* Pringsheim (Sarcomastigophora, Euglenida). *Journal of Protozoology* 33:179–184.

Svenning, M. M., R. Eriksson, and U. Rasmussen. 2005. Phylogeny of symbiotic cyanobacteria within the genus *Nostoc* based on 16S rDNA sequence analyses. *Archiv Microbiologia* 183:19–26.

Sweeney, B. M. 1982. Microsources of bioluminescence in *Pyrocystis fusiformis* (Pyrrhophyta). *Journal of Phycology* 18:412–416.

Sweeney, B. M. 1987. *Rhythmic Phenomena in Plants*, 2nd edition, Academic Press, San Diego, CA.

Swift, E., W. H. Biggley, and H. H. Seliger. 1973. Species of oceanic dinoflagellates in the genera *Dissodinium* and *Pyrocystis*: Interclonal and interspecific comparisons of the color and photon yield of bioluminescence. *Journal of Phycology* 9:420–426.

Swift, H., and B. Palenik. 1993. Prochlorophyte evolution and the origin of chloroplasts: Morphological and molecular evidence. In: R. Lewin, A. (ed.), *Origins of Plastids*, Chapman & Hall, New York, pp. 123–139.

Swinbanks, D. D. 1982. Intertidal exposure zones: A way to subdivide the shore. *Journal of Experimental Marine Biology* 62:69–86.

Sym, S. D., and R. N. Pienaar. 1991. Light and electron microscopy of a punctate species of *Pyramimonas, P. mucifera* sp. nov. (Prasinophyceae). *Journal of Phycology* 27:277–290.

Sym, S. D, and R. N. Pienaar. 1993. The class Prasinophyceae. *Progress in Phycological Research* 9:281.

Szymánska, H. 1989. Three new *Coleochaete* species (Chlorophyta) from Poland. *Nova Hedwigia* 49:435–446.

Szymánska, H. 2003. *Coleochaete spalikii* Szymanska sp. nov. (Charophyceae, Chlorophyta)—A new member of the *Coleochaete sieminskiana* group. *Nova Hedwigia* 76:129–135.

Tai, V., J. E. Lawrence, A. S. Lang, A. M. Chan, A. I. Culley, and C. A. Suttle. 2003. Characterization of HaRNAV, a single-stranded RNA virus causing lysis of *Heterosigma akashiwo* (Raphidophyceae). *Journal of Phycology* 39:343–352.

Takahashi, F., et al. 2007. AUREOCHROME, a photoreceptor required for photomorphogenesis in stramenopiles. *Proceedings of the National Academy of Sciences of the United States of America* 104:19625–19630.

Takahashi, H., H. Takano, H. Kuroiwa, R. Itoh, K. Toda, S. Kawano, and T. Kuroiwa. 1998. A possible role for actin dots in the formation of the contractile ring in the ultramicroalga *Cyanidium caldarium* RK-1. *Protoplasma* 202:91–104.

Takatori, S., and K. Imahori. 1971. Light reactions in the control of oospore germination of *Chara delicatula*. *Phycologia* 10:221–228.

Takeda, H. 1991. Sugar composition of the cell wall and the taxonomy of *Chlorella* (Chlorophyceae). *Journal of Phycology* 27:224–232.

Takeuchi, N., and S. Kohshima. 2004. A snow algal community on Tyndall Glacier in the southern Patagonia icefield, Chile. *Arctic, Antarctic, and Alpine Research* 36:92–99.

Takishita, K., K. Koike, T. Maruyama, and T. Ogata. 2002. Molecular evidence for plastid robbery (kleptoplastidy) in *Dinophysis*, a dinoflagellate causing diarrhetic shellfish poisoning. *Protist* 153:293–302.

Tam, L.-W., N. F. Wilson, and P. A. Lefebvre. 2007. A CDK-related kinase regulates the length and assembly of flagella in *Chlamydomonas*. *Journal of Cell Biology* 176:819–829.

Tamura, H., I. Mine, and K. Okuda. 1996. Cellulose-synthesizing terminal complexes and microfibril structure in the brown alga *Sphacelaria rigidula* (Sphacelariales, Phaeophyceae). *Phycological Research* 44:63–68.

Tanabe, Y., et al. 2005. Characterization of MADS-box genes in charophycean green algae and its implication for the evolution of MADS-box genes. *Proceedings of the National Academy of Sciences of the United States of America* 102:2436–2441.

Tappan, H. 1980. *The Paleobiology of Plant Protists*, W. H. Freeman, New York.

Targett, N. M., and T. M. Arnold. 1998. Predicting the effects of brown algal phlorotannins on marine herbivores in tropical and temperate oceans. *Journal of Phycology* 34:195–205.

Tartar, A., and D. G. Boucias. 2004. The nonphotosynthetic, pathogenic green alga *Helicosporidium* sp. has retained a modified, functional plastid genome. *Federation of European Microbiological Societies* 233:153–157.

Tartar, A., D. G. Boucias, B. J. Adams, and J. J. Becnel. 2002. Phylogenetic analysis identifies the invertebrate pathogen *Helicosporidium* sp. as a green alga (Chlorophyta). *International Journal of Systematic and Evolutionary Microbiology* 52:273–279.

Tatarenkov, A., L. Bergström, R. B. Jönsson, E. A. Serrão, L. Kautsky, and K. Johannesson. 2005. Intriguing asexual life in marginal populations of the brown seaweed *Fucus vesiculosus*. *Molecular Ecology* 14:647–651.

Tatarenkov, A., R. B. Jönsson, L. Kautsky, and K. Johannesson. 2007. Genetic structure in populations of *Fucus vesiculosus* (Phaeophyceae) over spatial scales from 10m to 88 km. *Journal of Phycology* 43:675–685.

Tatewaki, M. 1972. Life history and systematics in *Monostroma*. In: Abbott, I. A., and M. Kurogi (eds.), *Contributions to the Systematics of Benthic Marine Algae of the North Pacific*, Japanese Society of Phycology, Kobe, Japan, pp. 1–15.

Tatewaki, M., L. Provasoli, and I. J. Pintner. 1983. Morphogenesis of *Monostroma oxyspermum* (Kütz.) Doty (Chlorophyceae) in axenic culture, especially in bialgal culture. *Journal of Phycology* 19:409–416.

Taylor, A. R., N. F. H. Manison, C. Fernandez, J. Wood, and C. Brownlee. 1996. Spatial organization of calcium signaling involved in cell volume control in the *Fucus* rhizoid. *The Plant Cell* 8:2015–2031.

Taylor, A. R, A, and L. C.-M. Chen. 1994. *Chondrus*. In: Akatsuka, I. (ed.), *Biology of Economic Algae*, SPB Academic Publishing, The Hague, The Netherlands, pp. 35–76.

Taylor, F. J. R. 1971. Scanning electron microscopy of thecae of the dinoflagellate genus *Ornithocercus*. *Journal of Phycology* 7:249–258.

Taylor, F. J. R. 1980. On dinoflagellate evolution. *BioSystems* 13:65–108.

Taylor, F. J. R. 1990. Dinoflagellata (Dinomastigota). In: Margulis, L., J. O. Corliss, M. Melkonian, and D. J. Chapman (eds.), *Handbook of Protoctista*, Jones & Bartlett Publishers, Boston, pp. 419–437.

Taylor, F. J. R. 2004. Illumination or confusion? Dinoflagellate molecular phylogenetic data viewed from a primarily morphological standpoint. *Phycological Research* 52:308–324.

Taylor, T. N., and E. L. Taylor. 1993. *The Biology and Evolution of Fossil Plants*, Prentice Hall, Englewood Cliffs, NJ.

Taylor, W. R. 1972. *Marine Algae of the Eastern Tropical and Subtropical Coasts of the Americas*. University of Michigan Press, Ann Arbor, MI.

Tcherkez, G. G. B., G. D. Farquhar, and T. J. Andrews. 2006. Despite slow catalysis and confused substrate specificity, all ribulose bisphosphate carboxylases may be nearly perfectly optimized. *Proceedings of the National Academy of Sciences of the United States of America* 103:7246–7251.

Tchernov, D., M. Y. Gorbunov, C. de Vargas, S. N. Yadav, A. J. Milligan, M. Häggblom, and P. G. Falkowski. 2004. Membrane lipids of symbiotic algae are diagnostic of sensitivity to thermal bleaching in corals. *Proceedings of the National Academy of Sciences of the United States of America* 101:13531–13535.

Teles-Grilo, M. L., J. Tato-Costa, S. M. Duarte, A. Maia, G. Casal, and C. Azevedo. 2007. Is there a plastid in *Perkinsus atlanticus* (Perkinsozoa)? *European Journal of Protistology* 43:163–167.

Tenaud, M., M. Ohmori, and S. Miyachi. 1989. Inorganic carbon and acetate assimilation in *Botryococcus braunii* (Chlorophyta). *Journal of Phycology* 25:662–667.

Tessier, A. J., E. V. Bizina, and C. K. Geedey. 2001. Grazer–resource interactions in the plankton: Are all daphnids alike? *Limnology and Oceanography* 46:1585–1595.

Thierstein, H. R., K. R. Geitzenauer, B. Molfino, and N. J. Shackleton. 1977. Global synchronicity of late Quarternary coccolith datum levels: Validation by oxygen isotopes. *Geology* 5:400–404.

Thomas, W. H., and B. Duval. 1995. Sierra Nevada, California, U.S.A., snow algae: Snow albedo changes, algal-bacterial interrelationships, and ultraviolet radiation effects. *Arctic and Alpine Research* 27:389–399.

Thomas, W. H., and C. H. Gibson. 1990. Quantified small-scale turbulence inhibits a red tide dinoflagellate, *Gonyaulax polyedra* Stein. *Deep-Sea Research* 37:1583–1593.

Thomas, W. H., and C. H. Gibson. 1992. Effects of quantified small-scale turbulence on the dinoflagellate, *Gymnodinium sanguineum (splendens)*: Contrast with *Gonyaulax (Lingulodinium) polyedra* and fishery implication. *Deep-Sea Research* 39:1429–1437.

Thomas, W. H., C. T. Tynan, and C. H. Gibson. 1997. Turbulence-phytoplankton interrelationships. *Progress in Phycological Research* 12:284–324.

Thompson, J. B., S. Schultze-Lam, T. J. Beveridge, and D. J. DesMarais. 1997. Whiting events: Biogenic origin due to the photosynthetic activity of cyanobacterial pico-plankton. *Limnology and Oceanography* 42:133–141.

Thompson, J. M., A. J. D. Ferguson, and C. S. Reynolds. 1982. Natural filtration rates of zooplankton in a closed system: The derivation of a community grazing index. *Journal of Plankton Research* 4:545–560.

Thompson, R. H. 1958. The life cycles of *Cephaleuros* and *Stomatochroon*. Proceedings of the 9th International Botanical Congress (Montreal) 2:397.

Thompson, R. H. 1969. Sexual reproduction in *Chaetosphaeridium globosum* (Nordst.) Klebahn (Chlorophyceae) and description of a species new to science. *Journal of Phycology* 5:285–290.

Thompson, R. H., and D. E. Wujek. 1997. *Trentepohliales. Cephaleuros, Phycopeltis, and Stomatochroon: Morphology, Taxonomy and Ecology*, Science Publishers, Enfield, NH.

Thompson, R. M., and C. R. Townsend. 2003. Impacts on stream food webs of native and exotic forest: An intercontinental comparison. *Ecology* 84:145–161.

Thornber, C., J. J. Stachowicz, and S. Gaines. 2006. Tissue type matters: Selective herbivory on different life history stages of an isomorphic alga. *Ecology* 87:2255–2263.

Throndsen, J. 1997. The planktonic marine flagellates. In: Tomas, C. R. (ed.), *Identifying Marine Phytoplankton*, Academic Press, New York, pp. 591–730.

Tillmann, U. 1998. Phagotrophy by a plastidic haptophyte, *Prymnesium patelliferum*. *Aquatic Microbial Ecology* 14:155–160.

Tillmann, U. 2004. Interactions between planktonic microalgae and protozoan grazers. *Journal of Eukaryotic Microbiology* 51:156–168.

Tilman, D. 1976. Ecological competition between algae: Experimental confirmation of resource-based competition theory. *Science* 192:463–465.

Tilman, D. 1977. Resource competition between planktonic algae: An experimental and theoretical approach. *Ecology* 58:338–348.

Tilman, D. 1981. Tests of resource competition theory using four species of Lake Michigan algae. *Ecology* 62:802–815.

Tilman, D. 1982. *Resource Competition and Community Structure*. Princeton University Press, Princeton, NJ.

Tilman, D., and S. S. Kilham. 1976. Phosphate and silicate growth and uptake kinetics of the diatoms *Asterionella formosa* and *Cyclotella meneghiniana* in batch and semicontinuous culture. *Journal of Phycology* 12:375–383.

Tilman, D., S. S. Kilham, and P. Kilham. 1982. Phyto-plankton community ecology: The role of limiting nutrients. *Annual Review of Ecology and Systematics* 13:349–372.

Tilman, D., M. Mattson, and S. Langer. 1981. Competition and nutrient kinetics along a temperature gradient: An experimental test of a mechanistic approach to niche theory. *Limnology and Oceanography* 26:1020–1033.

Ting, C. S., C. Hsieh, S. Sundararaman, C. Mannella, and M. M. Marko. 2007. Cryo-electron tomography reveals the comparative three-dimensional architecture of *Prochlorococcus*, a globally important marine cyanobacterium. *Journal of Bacteriology* 189:4485–4493.

Tittel, J., Bissinger, V., U. Gaedke, and N. Kamjunke. 2005. Inorganic carbon limitation and mixotrophic growth in *Chlamydomonas* from an acidic mining lake. *Protist* 156:63–75.

Töbe, K., G. Eller, and L. K. Medlin. 2006. Automated detection and enumeration for toxic algae by solid-phase cytometry and the introduction of a new probe for *Prymnesium parvum* (Haptophyta: Prymnesiophyceae). *Journal of Plankton Research* 28:643–657.

Tomitani, A., A. H. Knoll, C. M. Cavanaugh, and T. Ohno. 2006. The evolutionary diversification of cyanobacteria: Molecular-phylogenetic and paleontological perspectives. *Proceedings of the National Academy of Sciences of the United States of America* 103:5442–5447.

Tomo, T., et al. 2007. Identification of the special pair of photosystem II in a chlorophyll d-dominated cyanobacterium. *Proceedings of the National Academy of Sciences of the United States of America* 104:7283–7288.

Tonkin, C. J., B. J. Foth, S. A. Ralph, N. Struck, A. F. Cowman, and G. I. McFadden. 2008. Evolution of malaria parasite plastid targeting sequences. *Proceedings of the National Academy of Sciences of the United States of America* 105:4781–4785.

Tonn, W. M., and J. J. Magnuson. 1982. Patterns in the species composition and richness of fish assemblages in northern Wisconsin lakes. *Ecology* 63:1149–1166.

Trainor, F. R., and C. A. Burg. 1965. *Scenedesmus obliquus* sexuality. *Science* 148:1094–1095.

Trainor, F. R., J. R. Cain, and L. E. Shubert. 1976. Morphology and nutrition of the colonial green alga *Scenedesmus*: 80 years later. *The Botanical Review* 42:5–25.

Tranvik, L. J., K. G. Porter, and J. McN. Sieburth. 1989. Occurrence of bacterivory in *Cryptomonas*, a common freshwater phytoplankter. *Oecologia* 78:473–476.

Triemer, R. E. 1980. Role of Golgi apparatus in mucilage production and cyst formation in *Euglena gracilis* (Euglenophyceae). *Journal of Phycology* 16:46–52.

Triemer, R. E. 1997. Feeding in *Peranema trichophorum* revisited (Euglenophyta). *Journal of Phycology* 33:649–654.

Triemer, R. E., and M. A. Farmer. 1991. An ultrastructural comparison of the mitotic apparatus, feeding apparatus, flagellar apparatus, and cyskeleton in euglenoids and kinetoplastids. *Protoplasma* 164:91–104.

Triemer, R. E., E. Linton, W. Shin, A. Nudelman, A. Monfils, M. Bennett, and S. Brosnan. 2006. Phylogeny of the Euglenales based upon combined SSU and LSU rDNA sequence comparisons and description of *Discoplastis* gen. nov. (Euglenophyta). *Journal of Phycology* 42:731–741.

Triemer, R. E., P. V. Zimba, and M. Rowan. 2003. Identification of euglenoids that produce ichthyotoxin(s) (Euglenophyta). *Journal of Phycology* 39:56–57.

Trowbridge, C. D. 1998. Ecology of the green macroalga *Codium fragile* (Suringar) Hariot 1889: Invasive and non-invasive subspecies. *Oceanography and Marine Biology: An Annual Review* 36:1–64.

Troxler, R. F. 1994. The molecular aspects of pigments and photosynthesis in *Cyanidium caldarium*. In: Sechback, J. (ed.), *Evolutionary Pathways and Enigmatic Algae: Cyanidium caldarium (Rhodophyta) and Related Cells*, Kluwer Academic Publishers, Boston, pp. 263–282.

Tschermak-Woess, E. 1980. *Asterochloris phycobiontica* gen. et spec. nov., der phycobiont der Flechte *Varicellaria carneonivea*. *Plant Systematics and Evolution* 135:279–294.

Tschermak-Woess, E., and E. I. Friedmann. 1984. *Hemichloris antarctica*, gen. et sp. nov. (Chlorococcales, Chlorophyta), a cryptoendolithic alga from Antarctica. *Phycologia* 23:443–454.

Tsekos, I. 1996. The supramolecular organization of red algal cell membranes and their participation in the biosynthesis and secretion of extracellular polysaccharides: A review. *Protoplasma* 193:10–32.

Tsekos, I., K. Okuda, and R. M. Brown. 1996. The formation and development of cellulose-synthesizing linear terminal complexes (TCs) in the plasma membrane of the marine red alga *Erythrocladia subintegra* Rosenv. *Protoplasma* 193:33–45.

Tsuchikane, Y, R.-H, Fukumoto, S. Akatsuka, T. Fujii, and H. Sekimoto. 2003. Sex pheromones that induce sexual cell division in the *Closterium peracerosum–strigosum–littorale*

complex (Charophyta). *Journal of Phycology* 39:303–309.

Tsunakawa-Yokoyama, A., J. Yokoyama, H. Ohashi, and Y. Hara. 1997. Phylogenetic relationship of Porphyridiales (Rhodophyta) based on DNA sequences of 18SrRNA and *psbA* genes. *Phycologia* 36:13.

Tuchman, N. C. 1996. The role of heterotrophy in algae. In: Stevenson, R. J., M. L. Bothwell, and R. L. Lowe (eds.), *Algal Ecology: Freshwater Benthic Ecosystems*, Academic Press, New York, pp. 299–319.

Tucker, M. E., and V. P. Wright. 1990. *Carbonate Sedimentology*, Blackwell Scientific, Oxford, UK.

Türk, R., and G. Gärtner. 2001. Biological soil crusts of the subalpine, alpine, and nival areas in the alps. In: Belnap, J., and O. L. Lange (eds.), *Biological Soil Crusts: Structure, Function, and Management*, Ecological Studies, vol. 150, Springer Verlag, Berlin, pp. 67–73.

Turmel, M., M. Ehara, C. Otis, and C. Lemieux. 2002. Phylogenetic relationships among streptophytes as inferred from chloroplast small and large subunit rRNA gene sequences. *Journal of Phycology* 38:364–375.

Turmel, M., C. Otis, and C. Lemieux. 2003. The mitochondrial genome of *Chara vulgaris*: Insights into the mitochondrial DNA architecture of the last common ancestor of green algae and land plants. *The Plant Cell* 15:1888–1903.

Turmel, M., C. Otis, and C. Lemieux. 2005. The complete chloroplast DNA sequences of the charophycean green algae *Staurastrum* and *Zygnema* reveal that the chloroplast genome underwent extensive changes during the evolution of the Zygnematales. *BMC Biology* 3:22.

Turmel, M., C. Otis, and C. Lemieux. 2006. The chloroplast genome sequence of *Chara vulgaris* sheds new light on the closest green algal relatives of land plants. *Molecular Biology and Evolution* 23:1324–1338.

Turmel, M., J.-F. Pombert, P. Charlebois, C. Otis, and C. Lemieux. 2007. The green algal ancestry of land plants as revealed by the chloroplast genome. *International Journal of Plant Sciences* 168:679–689.

Turner, C. H. C., and L. V. Evans. 1986. Translocation of photoassimilated ^{14}C in the red alga *Polysiphonia lanosa*. *British Phycological Journal* 13:51–55.

Turner, M. A., E. T. Howell, M. Summerby, R. H. Hesslein, D. L. Findlay, and M. B. Jackson. 1991. Changes in epilithon and epiphyton associated with experimental acidification of a lake to pH 5. *Limnology and Oceanography* 36:1390–1405.

Turner, S., T. Berger-Wiersma, S. J. Giovannoni, L. R. Mur, and N. R. Pace. 1989. The relationship of a prochlorophyte *Prochlorothrix hollandica* to green chloroplasts. *Nature* 337:380–385.

Turpin, D. H. 1991. Effects of inorganic N availability on algal photosynthsis and carbon metabolism. *Journal of Phycology* 27:14–20.

Tyler, B. M., et al. 2006. *Phytophthora* genome sequences uncover evolutionary origins and mechanisms of pathogenesis. *Science* 313:1261–1266.

Ucko, M., M. Elbrächter, and E. Schnepf. 1997. A *Crypthecodinium cohnii*-like dinoflagellate feeding myzocytotically on the unicellular red alga *Porphyridium* sp. *European Journal of Phycology* 32:133–140.

Uemori, C., C. Nagasato, A. Kato, and T. Motomura. 2006. Ultrastructural and immunocytological studies on the rhizoplast in the chrysophycan alga *Ochromonas danica*. *Phycological Research* 54:133–139.

Ueno, R., N. Urano, and N. Suzuki. 2003. Phylogeny of the non-photosynthetic green microalgal genus *Prototheca* (Trebouxiophyceae, Chlorophyta) and related taxa inferred from SSU and LSU ribosomal DNA partial sequence data. *Federation of European Microbiological Societies* 223:275–280.

Ueyama, R., and B. C. Monger. 2005. Wind-induced modulation of seasonal phytoplankton blooms in the North Atlantic derived from satellite observations. *Limnology and Oceanography* 50:1820–1829.

Uhrmacher, S., D. Hanelt, and W. Nultsch. 1995. Zeaxanthin content and the photoinhibitory degree of photosynthesis are linearly correlated in the brown alga *Dictyota dichotoma*. *Marine Biology* 123:159–165.

Ullmann, I., and B. Büdel. 2001. Biological soil crusts of Africa. In: Belnap, J., and O. L. Lange (eds.), *Biological Soil Crusts: Structure, Function, and Management*, Ecological Studies, vol. 150, Springer Verlag, Berlin, pp. 107–118.

Underwood, G. J. C., and J. D. Thomas. 1990. Grazing interactions between pulmonate snails and epiphytic algae and bacteria. *Freshwater Biology* 23:505–522.

Urabe, J., T. B. Gurung, T. Yoshida, T. Sekino, M. Nakanishi, M. Maruo, and E. Nakayama. 2000. Diel changes in phagotrophy by *Cryptomonas* in Lake Biwa. *Limnology and Oceanography* 45:1558–1563.

Urbach, E., D. L. Robertson, and S. W. Chisholm. 1992. Multiple evolutionary origins of prochlorophytes within the cyanobacterial radiation. *Nature* 355:267–270.

Url, T., M. Höftburger, and U. Meindl. 1993. Cytochalasin B influences dictyosomal vesicle production and morophogenesis in the desmid *Euastrum*. *Journal of Phycology* 29:667–674.

Usher, K. M., J. Fromont, D. C. Sutton, and S. Toze. 2004a. The biogeography and phylogeny of unicellular cyanobacterial symbionts in sponges from Australia and the Mediterranean. *Microbial Ecology* 48:167–177.

Usher, K. M., J. Kuo, J. Fromont, S. Toze, and D. C. Sutton. 2006. Comparative morphology of five species of symbiotic and non-symbiotic coccoid cyanobacteria. *European Journal of Phycology* 41:179–188.

Usher K. M., S. Toze, J. Fromont, J. Kuo, and D. C. Sutton. 2004b. A new species of cyanobacterial symbiont from the marine sponge *Chondrilla nucula*. *Symbiosis*. 36:183–192.

Ustinova, I., L. Krienitz, and V. A. R. Huss. 2001. *Closteriopsis acicularis* (G. M. Smith) Belcher et Swale is a fusiform alga closely related to *Chlorella kessleri* Fott et Novakova (Chlorophyta, Trebouxiophyceae). *European Journal of Phycology* 36:341–351.

van Alstyne, K. L. 1988. Herbivore grazing increases polyphenolic defenses in the intertidal brown alga *Fucus*. *Ecology* 69:655–663

van de Meene, A., and J. D. Pickett-Heaps.

2004. Valve morphogenesis in the centric diatom *Rhizosolenia setigera* (Bacillariophyceae, Centrales) and its taxonomic implications. *European Journal of Phycology* 39:93–104.

van den Hoek, C., D. G. Mann, and H. M. Jahns. 1995. *Algae: An Introduction to Phycology*, Cambridge University Press, Cambridge, UK.

van de Peer, Y., S. A. Rensing, U.-G. Maier, and R. De Wachter. 1996. Substitution rate calibration of small subunit ribosomal RNA identified chlorarachniophyte endosymbionts as remnants of green algae. *Proceedings of the National Academy of Sciences of the United States of America* 93:7732–7736.

van de Peer, Y., and R. De Wachter. 1997. Evolutionary relationships among the eukaryotic crown taxa taking into account site-to-site variation in 18S rRNA. *Journal of Molecular Evolution* 45:619–630.

Vanderauwera, G., and Dewachter. 1997. Complete large subunit ribosomal RNA sequence from the heterokont algae *Ochromonas danica*, *Nannochloropsis salina*, and *Tribonema aequale* and phylogenetic analysis. *Journal of Molecular Evolution* 45:84–90.

van der Meer, J. P., and M. U. Patwary. 1995. Genetic studies on marine macroalgae: A status report. In: Wiessner, W., E. Schnepf, and R. C. Starr (eds.), *Algae, Environment and Human Affairs*, BioPress, Bristol, UK, pp. 235–258.

van der Meer, J. P., and E. R. Todd. 1980. The life history of *Palmaria palmata* in culture. A new type for the Rhodophyta. *Canadian Journal of Botany* 58:1250–1256.

Van der Stap, I., M. Voß, and W. M. Mooij. 2006. Linking herbivore-induced defences to population dynamics. *Freshwater Biology* 51:424–434.

Van der Stap, I., M. Voß, and W. M. Mooij. 2007. Inducible defences and rotifer food chain dynamics. *Hydrobiologia* 593:103–110.

Van de Vijver, B., L. Beyens, S. Vincke, and N. J. M. Gremmen. 2004. Moss-inhabiting diatom communities from Heard Island, sub-Antarctic. *Polar Biology* 27:532–543.

Van Dok, W., and B. T. Hart. 1997. Akinete germination in *Anabaena circinalis* (Cyanophyta). *Journal of Phycology* 33:12–17.

van Donk, E. 1989. The role of fungal parasites in phytoplankton succession. In: Sommer, U. (ed.), *Plankton Ecology*, Springer Verlag, Berlin, pp. 171–194.

van Donk, E., and K. Bruning. 1995. Effects of fungal parasites on planktonic algae and the role of environmental factors in the fungus-alga relationship. In: Weissner, W., E. Schnepf, and R. Starr *Algae, Environment and Human Affairs*, BioPress, Bristol, UK, pp. 223–234.

van Donk, E., M. P. Grimm, R. D. Gulati, P. G. M. Heuts, W. A. de Kloet, and L. van Liere. 1990b. First attempt to apply whole-lake food-web manipulation on a large scale in The Netherlands. *Hydrobiologia* 200/201:291–301.

van Donk, E., M. P. Grimm, R. D. Gulati, and J. P. G. Klein Breteler. 1990a. Whole-lake food-web manipulation as a means to study community interactions in a small ecosystem. *Hydrobiologia* 200/201:275–289.

van Donk, E., M. Lürling, D. O. Hessen, and G. M. Lokhorst. 1997. Altered cell wall morphology in nutrient-deficient phytoplankton and its impact on grazers. *Limnology and Oceanography* 42:357–364.

van Donk, E., M. Lürling, and W. Lampert. 1999. Consumer induced changes in phytoplankton: Inducibility, costs, benefits, and the impact on grazers. In: Tollrian, R., and C. D. Harvell (eds.), *The Ecology and Evolution of Inducible Defenses*, Princeton University Press, Princeton, NJ, pp. 89–103.

van Donk, E., and J. Ringelberg. 1983. The effect of fungal parasitism on the succession of diatoms in Lake Maarsseveen (The Netherlands). *Freshwater Biology* 13:241–251.

Van Etten, J. L., L. C. Lane, and R. H. Meints. 1991. Viruses and viruslike particles of eukaryotic algae. *Microbiological Reviews* 55:586–620.

Van Etten, J. L., and R. H. Meints. 1999. Giant viruses infecting algae. *Annual Review of Microbiology* 53:447–494.

van Gool, E., and J. Ringelberg. 1996. Daphnids respond to algae-associated odours. *Journal of Plankton Research* 18:197–202.

Van Hannen, E. J., P. Fink, and M. Lürling. 2002. A revised secondary structure model for the internal transcribed spacer 2 of the green algae *Scenedesmus* and *Desmodesmus* and its implication for the phylogeny of these algae. *European Journal of Phycology* 37:203–208.

Van Hannen, E. J., M. Lürling, and E. van Donk. 2000. Sequence analysis of the ITS-2 region to identify strains of *Scenedesmus* (Chlorophyceae). *Journal of Phycology* 36:605–607.

Van Lenning, L., et al. 2003. Pigment signatures and phylogenetic relationships of the Pavlovophyceae (Haptophyta). *Journal of Phycology* 39:379–389.

Van Mooy, B. A. S., G. Rocap, H. F. Fredricks, C. T. Evans, and A. H. Devol. 2006. Sulfolipids dramatically decrease phosphorus demand by picocyanobacteria in oligotrophic marine environments. *Proceedings of the National Academy of Sciences of the United States of America* 103:867–8612.

Vanni, M. J. 1987. Effects of nutrients and zooplankton size on the structure of a phytoplankton community. *Ecology* 68:624–635.

Vannote, R. L., G. W. Minshall, K. W. Cummins, J. R. Sedell, and C. E. Cushing. 1980. The river continuum concept. *Canadian Journal of Fisheries and Aquatic Science* 37:130–137.

van Rijssel, M., C. E. Hamm, and W. W. C. Gieskes. 1997. *Phaeocystis globosa* (Prymnesiophyceae) colonies: Hollow structures built with small amounts of polysaccharides. *European Journal of Phycology* 32:185–192.

Van Tussenbroek, B. I., and J. K. van Dijk. 2007. Spatial and temporal variability in biomass and production of psammophytic *Halimeda incrassaata* (Bryopsidales, Chlorophyta) in a Caribbean reef lagoon. *Journal of Phycology* 43:69–77.

Van Valkenburg, S. D. 1971. Observations on the fine structure of *Dictyocha fibula* Ehrenberg. II. The protoplast. *Journal of Phycology* 7:118–132.

VanWinkle-Swift, K. P., and W. L. Rickoll. 1997. The zygospore wall of *Chlamydomonas monoica* (Chlorophyceae): Morphogenesis and evidence for the presence of sporopollenin. *Journal of Phycology* 33:655–665.

Vargas, C. A., R. Escribano, and S. Poulet. 2006. Phytoplankton food quality determines time windows for successful zooplankton reproductive pulses. *Ecology* 87:2992–2999.

Vargas, C. A., R. A. Martínez, L. A. Cuevas, M. A. Pavez, C. Cartes, H. E. González, R. Escribano, and G. Daneri. 2007. The relative importance of microbial and classical food webs in a highly productive coastal upwelling area. *Limnology and Oceanography* 52:1495–1510.

Vargas, D. R., and L. Collado-Vides. 1996. Architectural models for apical patterns in *Gelidium* (Gelidiales, Rhodophyta): Hypothesis of growth. *Phycological Research* 44:95–100.

Varvarigos, V., B. Galatis, and C. Katsaros. 2005. A unique pattern of F-actin organization supports cytokinesis in vacuolated cells of *Macrocystis pyrifera* (Phaeophyceae) gametophytes. *Protoplasma* 226:241–245.

Veldhuis, M. J. W., C. P. D. Brussaard, and A. A. M. Noordeloos. 2005. Living in a *Phaeocystis* colony: A way to be a successful algal species. *Harmful Algae* 4:841–858.

Venn, A. A., M. A. Wilson, H. G. Trapido-Rosenthal, B. J. Keely, and A. E. Douglas. 2006. The impact of coral bleaching on the pigment profile of the symbiotic alga, *Symbiodinium. Plant, Cell and Environment* 29:2133–2142.

Verb, R. G., and M. L. Vis. 2005. Periphyton assemblages as bioindicators of mine-drainage in unglaciated western Allegheny Plateau lotic systems. *Water, Air, and Soil Pollution* 161:227–265.

Verity, P. G. 1985. Grazing, respiration, excretion, and growth rates of tintinnids. *Limnology and Oceanography* 30:1268–1282.

Verschoor, A. M., I. van der Stap, N. R. Helmsing, M. Lürling, and E. van Donk. 2004. Inducible colony formation within the Scenedesmaceae: Adaptive responses to infochemicals from two different herbivore taxa. *Journal of Phycology* 40:808–814.

Versteegh, G. J. M., and P. Blokker. 2004. Resistant macromolecules of extant and fossil microalgae. *Phycological Research* 52:325–339.

Vesk, M., T. P. Dibbayawan, and P. A. Vesk. 1996. Immunogold localization of phycoerythrin in chloroplasts of *Dinophysis acuminata* and *D. fortii* (Dinophysiales, Dinophyta). *Phycologia* 35:234–238.

Vickerman, K. 1990. Kinetoplastida. In: Margulis, L., J. O. Corliss, M. Melkonian, and D. J. Chapman (eds.), *Handbook of Protoctista*, Jones & Bartlett Publishers, Boston, pp. 215–238.

Vickerman, K., and T. M. Preston. 1976. Comparative cell biology of the kinetoplastid flagellates. *In:* Lumsden, W. H. R., and D. A. Evans (eds.), *Biology of the Kinetoplastida,* vol. I. Academic Press, New York, Chapter 2.

Vila-Costa, M., R. Simó, R. Harada, J. M. Gasol, D. Slezak, and R. P. Kiene. 2006. Dimethylsulfoniopropionate uptake by marine phytoplankton. *Science* 314:652–649.

Villareal, T. A., and E. J. Carpenter. 2003. Buoyancy regulation and the potential for vertical migration in the oceanic cyanobacterium *Trichodesmium. Microbial Ecology* 45:1–10.

Villareal, T. A., S. Woods, J. K. Moore, and K. Culver-Rymsza. 1996. Vertical migration of *Rhizosolenia* mats and their significance to NO_3^- fluxes in the central North Pacific gyre. *Journal of Plankton Research* 18:1103–1121.

Villarejo, A., M. I. Orús, and F. Martinez. 1997.

Regulation of the CO_2-concentrating mechanism in *Chlorella vulgaris* UAM 101 by glucose. *Physiologia Plantarum* 99:293–301.

Villatte, F., H. Schulze, R. D. Schmid, and T. T. Bachmann. 2002. A disposable acetylcholinesterase-based electrode biosensor to detect anatoxin-a(s) in water. *Analytical Bioanalytical Chemistry* 372:322–326.

Vincent, W. F. 1988. *Microbial Ecosystems in Antarctica*, Press Syndicate of the University of Cambridge, Cambridge, UK.

Vincent, W. F. 2000. Cyanobacterial dominance in the polar regions. In: Whitton, B. A., and M. Potts (eds.), *The Ecology of Cyanobacteria: Their Diversity in Time and Space*, Kluwer Academic Publishers, Dordrecht, The Netherlands, pp. 321–340.

Vinueza, L. R., G. M. Branch, M. L. Branch, and R. H. Bustamante. 2006. Top-down herbivory and bottom-up El Niño effects on Galápagos rocky-shore communities. *Ecological Monographs* 76: 111–131.

Vis, M. L., T. J. Entwhisle, J. A. West, and F. D. Ott. 2006. *Ptilothamnion richardsii* (Rhodophyta) is a chantransia stage of *Batrachospermum. European Journal of Phycology* 41:125–130.

Vis, M. L., G. W. Saunders, R. G. Sheath, K. Dunse, and T. J. Entwhistle. 1998. Phylogeny of the Batrachospermales (Rhodophyta) inferred from *rbcL* and 18S ribosomal DNA gene sequences. *Journal of Phycology* 34:341–350.

Vishniac, H. S. 2002. Desert environments–soil microbial communities in cold deserts. In: Bitton, G. (ed.), *Encyclopedia of Environmental Microbiology*, John Wiley, New York, pp. 1023–1029.

Visser, P. M., B. W. Ibelings, L. R. Mur, and A. E. Walsby. 2005. The ecophysiology of the harmful cyanobacterium *Microcystis*. Features explaining its success and measures for its control. In: Huisman, J., H. C. P. Matthijs, and P. M. Visser (eds.), *Harmful Cyanobacteria*, Springer, New York, pp. 109–134.

Vitousek, P. M., H. A. Mooney, J. Lubchenco, and J. M. Melilo. 1997a. Human domination of earth's ecosystems. *Science* 277:494–499.

Vitousek, P. M., J. D. Aber, R. W. Howarth, G. E. Likens, P. A. Matson, D. W. Schindler, W. H. Schlesinger, and D. G. Tilman. 1997b. Human alteration of the global nitrogen cycle: Sources and consequences. *Ecological Applications* 71:737–750.

Vlassak, K., E. A. Paul, and K. Jones. 1973. Assessment of biological nitrogen fixation in grassland and associated sites. *Plant and Soil* 38:637–649.

Vogel, S. 1994. *Life in Moving Fluids*, 2nd edition, Princeton University Press, Princeton, NJ.

Vogeley, L., O. A. Sineshchekov, V. D. Trivedi, J. Sasaki, J. L. Spudlich, and H. Luecke. 2004. *Anabaena* sensory rhodopsin: A photochromic color sensor at 2.0 Å. *Science* 306:1390–1393.

Vollenweider, R. 1982. Eutrophication of waters, monitoring, assessment and control. *OECD*, Paris, France.

Volterra, V. 1926. Variazioni e fluttuazioni del numero d'individui in specie animali conviventi. *Memorie. Reale Accademia Nazionale dei Lincei. Classe di Scienze Fisiche, Matematiche, e Naturali.* Ser. VI, Vol. 2.

von der Heyden, S., E. E. Chao, and T. Cavalier-Smith. 2004. Genetic diversity of goniomonads: An ancient divergence between marine and freshwater species. *European Journal of Phycology* 39:343–350.

Von Stackelberg, M., S. A. Rensing, and R. Reski, 2006. Identification of genic moss SSR markers and a comparative analysis of twenty-four algal and plant gene indices reveal species-specific rather than group-specific characteristics of microsatellites. *BMC Plant Biology* 6:9.

von Stosch, H. A. 1964. Zum problem der sexuellen Fortplanzung in der Peridineengattung *Ceratium. Helgolander wissenschaftliche Meeresuntersuchungen* 10:140–153.

Vonshak, A., S. M. Cheung, and F. Chen. 2000. Mixotrophic growth modifies the response of *Spirulina* (*Arthrospira*) *platensis* (Cyanobacteria) cells to light. *Journal of Phycology* 36:675–679.

Vouilloud, A. A., E. J. Cáceres, and P. I. Leonardi. 2005. Changes in the absolute configuration of the basal/flagellar apparatus and evidence of centrin during male gametogenesis in *Chara contraria* var *nitelloides* (Charales, Charophyta). *Plant Systematics and Evolution* 251:89–105.

Vreeland, V., J. H. Waite, and L. Epstein. 1998. Polyphenols and oxidases in substratum adhesion by marine algae and mussels. *Journal of Phycology* 34:1–8.

Vrieling, E. G., et al. 2007. Salinity-dependent diatom biosilicification implies an important role of external ionic strength. *Proceedings of the National Academy of Sciences of the United States of America* 104:10441–10446.

Vroom, P. S., C. M. Smith, and S. C. Keeley. 1998. Cladistics of the Bryopsidales: A preliminary analysis. *Journal of Phycology* 34:351–360.

Waaland, J. R., and D. Branton. 1969. Gas vacuole development in a blue-green alga. *Science* 163:1339–1341.

Waaland, S. D. 1990. Development. In: Cole, K. M., and R. G. Sheath (eds.), *Biology of the Red Algae*, Cambridge University Press, Cambridge, UK, pp. 259–274.

Waaland, S. D., and R. E. Cleland. 1974. Cell repair through cell fusion in the red alga *Griffithsia pacifica. Protoplasma* 79:185–196.

Waaland, S. D., and J. R. Waaland. 1975. Analysis of cell elongation in red algae by fluorescent labeling. *Planta* 126:127–138.

Wagner, G., and F. Grolig. 1992. Algal chloroplast movements. In: Melkonian, M. (ed.), *Algal Cell Motility*, Chapman & Hall, London, pp. 39–72.

Wagner, R., and T. Pfannschmidt. 2006. Eukaryotic transcription factors in plastids—Bioinformatic assessment and implications for the evolution of gene expression machineries in plants. *Gene* 381:62–70.

Walker, T. L., C. Collet, and S. Purton. 2005. Algal transgenics in the genomic era. *Journal of Phycology* 41:1077–1093.

Wall, D., and B. Dale. 1969. The "hystrichosphaerid" resting spore of the dinoflagellate *Pyrodinium bahamense*, Plate, 1906. *Journal of Phycology* 5:140–149.

Waller, R. F., N. J. Patron, and P. J. Keeling. 2006. Phylogenetic history of plastid-targeted proteins in the peridinin-containing dinoflagellate *Heterocapsa triquetra*. *International Journal of Systematic and Evolutionary Microbiology* 56:1439–1447.

Walne, P. L., and H. J. Arnott. 1967. The comparative ultrastructure and possible function of eyespots: *Euglena granulata* and *Chlamydomonas eugametos*. *Planta* 77:325–353.

Walne, P. L., and N. S. Dawson. 1993. A comparison of paraxial rods in the flagella of euglenoids and kinetoplastids. *Archiv für Protistenkunde* 143:177–194.

Walne, P. L., and P. A. Kivic. 1990. Euglenida. In: Margulis, L., J. O. Corliss, M. Melkonian, and D. J. Chapman (eds.), *Handbook of Protoctista*, Jones & Bartlett Publishers, Boston, pp. 270–287.

Walsby, A. E., and A. Xypolyta. 1977. The form resistance of chitan fibres attached to the cells of *Thalassiosira fluviatilis* Hustedt. *British Phycological Journal* 12:215–223.

Wang, J. T., and A. E. Douglas. 1997. Nutrients, signals, and photosynthetic release by symbiotic algae. The impact of taurine on the dinoflagellate alga *Symbiodinium* from the sea anemone *Aiptasia pulchella*. *Plant Physiology* 114:631–636.

Wang, S.-H., Y.-Z. Qi, and Y.-F. Yang. 2007. Cyst formation: An important mechanism for the termination of *Scrippsiella trochoidea* (Dinophyceae) bloom. *Journal of Plankton Research* 29:209–218.

Wang, Y., J. Lu, J.-C. Mollet, M. R. Gretz, and K. D. Hoagland. 1997. Extracellular matrix assembly in diatoms (Bacillariophyceae). II. 2,6-dichlorobenzonitrile inhibition of motility and stalk production in the marine diatom *Achnanthes longipes*. *Plant Physiology* 113:1071–1080.

Ward, D. M., and R. W. Castenholz. 2000. Cyanobacteria in geothermal habitats. In: Whitton, B. A., and M. Potts (eds.), *The Ecology of Cyanobacteria: Their Diversity in Time and Space*, Kluwer Academic Publishers, Boston, pp. 37–59.

Waris, H. 1950. Cytophysiological studies on *Micrasterias* I. Nuclear and cell division. *Physiologia Plantarum* 3:1–16.

Warren-Rhodes, K. A., K. L. Rhodes, S. B. Pointing, S. A. Ewing, D. C. Lacap, B. Gomez-Silva, R. Amundson, E. I. Friedmann, and C. P. McKay. 2006. Hypolithic cyanobacteria, dry limit of photosynthesis, and microbial ecology in the hyperarid Atacama Desert. *Microbial Ecology* 52:389–398.

Watanabe, M. M., S. Suda, I. Inouye, T. Sawaguchi, and M. Chihara. 1990. *Lepidodinium viride* gen. et sp. nov. (Gymnodiniales, Dinophyta), a green dinoflagellate with a chlorophyll *a*- and *b*-containing endosymbiont. *Journal of Phycology* 26:741–751.

Watanabe, S., and G. L. Floyd. 1989. Ultrastructure of the zoospore of the coenocytic algae *Ascochloris* and *Urnella* (Chlorophyceae) with emphasis on the flagellar apparatus. *British Phycological Journal* 24:143–152.

Watanabe, S., A. Himizu, L. A. Lewis, G. L. Floyd, and P. A. Fuerst. 2000a. *Pseudoneochloris marina* (Chlorophyta), a new coccoid ulvophycean alga, and its phylogenetic position inferred from morphological and molecular data. *Journal of Phycology* 36:596–604.

Watanabe, S., and T. Nakayama. 2007. Ultra-

structure and phylogenetic relationships of the unicellular green algae *Ignatium tetrasporus* and *Pseudocharacium americanum* (Chlorophyta). *Phycological Research* 55:1–16.

Watanabe, S., S. Hirabayashi, S. Boussiba, Z. Cohen, A. Vonshak, and A. Richmond. 1996. *Parietochloris incisa* comb. nov. (Trebouxiophyceae, Chlorophyta). *Phycological Research* 44:107–108.

Watanabe, S., K. Mitsui, T. Nakayama, and I. Inouye. 2006a. Phylogenetic relationships and taxonomy of sarcinoid green algae: *Chlorosarcinopsis, Desmotetra, Sarcinochlamys* gen. nov., *Neochlorosarcina*, and *Chlorosphaeropsis* (Chlorophyceae, Chlorophyta). *Journal of Phycology* 42:679–675.

Watanabe, S., S. Tsujimura, T. Misono, S. Nakamura, and H. Inoue. 2006b. *Hemiflagellochloris kazakhstanica* gen. et sp. nov.: A new coccoid green alga with flagella of considerably unequal lengths from a saline irrigation land in Kazakhstan (Chlorophyceae, Chlorophyta). *Journal of Phycology* 42:696–706.

Watanabe, Y., J. E. J. Martini, and H. Ohmoto. 2000b. Geochemical evidence for terrestrial ecosystems 2.6 billion years ago. *Nature* 408:574–577.

Waterbury, J. B., J. M. Willey, D. G. Franks, F. W. Valois, and S. W. Watson. 1985. A cyanobacterium capable of swimming motility. *Science* 230:74–76.

Watras, C. J., and T. M. Frost. 1989. Little Rock Lake (Wisconsin): Perspectives on an experimental ecosystem approach to seepage lake acidification. *Archives of Environmental Contamination* 18:157–165.

Watson, D. C., and T. A. Norton. 1985. The physical characteristics of seaweed thalli as deterrents to littorine grazers. *Botanica Marina* 28:383–387.

Watson, M. W. 1975. Flagellar apparatus, eyespot and behavior of *Microthamnion kuetzingianum* (Chlorophyceae) zoospores. *Journal of Phycology* 11:439–448.

Watson, M. W., and H. J. Arnott. 1973. Ultrastructural morphology of *Microthamnion* zoospores. *Journal of Phycology* 9:15–29.

Watson, S. B., E. McCauley, and J. A. Downing. 1997. Patterns in phytoplankton taxonomic composition across temperate lakes of different nutrient status. *Limnology and Oceanography* 42:487–495.

Wayne, R. 1994. The excitability of plant cells: With a special emphasis on characean internodal cells. *Botanical Review* 60:265–367.

Webster, K. E., T. M. Frost, C. J. Watras, W. A. Swenson, M. Gonzalez, and P. J. Garrison. 1992. Complex biological responses to the experimental acidification of Little Rock Lake, Wisconsin, USA. *Environmental Pollution* 78:73–78.

Wedemayer, G. J., and L. W. Wilcox. 1984. The ultrastructure of the freshwater, colorless dinoflagellate *Peridiniopsis berolinense* (Lemm.) Bourrelly (Mastigophora, Dinoflagellida). *Journal of Protozoology* 31:444–453.

Wedemayer, G. J., L. W. Wilcox, and L. E. Graham. 1982. *Amphidinium cryophilum* sp. nov. (Dinophyceae), a new freshwater dinoflagellate. I. Species description using light and scanning electron microscopy. *Journal of Phycology* 18:13–16.

Wee, J. L. 1997. Scale biogenesis in synuro-

phycean protists: Phylogenetic implications. *CRC Critical Reviews in Plant Science* 16:497–534.

Weers, P. M. M., and R. D. Gulati. 1997. Growth and reproduction of *Daphnia galeata* in response to changes in fatty acids, phosphorus, and nitrogen in *Chlamydomonas reinhardtii*. *Limnology and Oceanography* 42:1584–1589.

Wehr, J. D. 2002. Brown algae. In: Wehr, J. D., and R. G. Sheath (eds.), *Freshwater Algae of North America: Ecology and Classification*, Academic Press, New York, pp. 757–773.

Weibel, D. B., P. Garstecki, D. Ryan, W. R. Diluzio, M. Mayer, J. E. Seto, and G. M. Whitesides. 2005. Microoxen: Microorganisms to move microscale loads. *Proceedings of the National Academy of Sciences of the United States of America* 102:11963–11967.

Weiss, R. L. 1983. Fine structure of the snow alga (*Chlamydomonas nivalis*) and associated bacteria. *Journal of Phycology* 19:200–204.

Weissflog, J., S. Adolph, T. Wiesemeier, and G. Pohnert. 2008. Reduction of herbivory through wound-activated protein cross-linking by the invasive macroalga *Caulerpa taxifolia*. *Chembiochem* 9:29–32.

Wellman, C. H., P. L. Osterloff, and U. Mohiuddin. 2003. Fragments of the earliest land plants. *Nature* 425:283–285.

Wells, M. L. 1998. Marine colloids, a neglected dimension. *Nature* 391:530–531.

Wessels, D. C. J., and B. Büdel. 1995. Epilithic and cryptoendolithic cyanobacteria of Clarens sandstone cliffs in the Golden Gate Highlands National Park, South Africa. *Botanica Acta* 108:220–226.

West, J. A. 1970. The life history of *Rhodochorton concrescens* in culture. *British Phycological Journal* 5:179–186.

West, J. A. 1972. The life history of *Petrocelis francisciana*. *British Phycological Journal* 7:299–308.

West, J. A., T. A. Klochkova, G. H. Kim, and S. Loiseaux-de Goer. 2006a. *Olpidopsis* sp., an oomycete from Madagascar that infects *Bostrychia* and other red algae: Host species susceptibility. *Phycological Research* 54:72–85.

West, N. E., and J. J. Skujins. 1977. The nitrogen cycle in North American cold-winter semi-desert ecosystems. *Ecologia Plantarum* 12:45–53.

West, N. J., R. Bacchieri, G. Hansen, C. Thomas, P. Lebaron, and H. Moreau. 2006b. Rapid quantification of the toxic alga *Prymnesium parvum* in natural samples by use of a specific monoclonal antibody and solid-phase cytometry. *Applied and Environmental Microbiology* 72:860–868.

Westbroek, P., J. E. Van Hinte, G. J. Brummer, M. Veldhuis, C. Brownlee, J. C. Green, R. Harris, and B. R. Heimdal. 1994. *Emiliania huxleyi* as a key to biosphere–geosphere interactions. In: Green, J. C., and B. S. C. Leadbeater (eds.), *The Haptophyte Algae, Systematics Association Special Volume 51*, Oxford University Press, Oxford, UK, pp. 321–334.

Wetherbee, R. 1979. "Transfer connections": Specialized pathways for nutrient translocation in a red alga? *Science* 204:858–859.

Wetherbee, R., J. L. Lind, J. Burke, and R. S. Quatrano. 1998. The first kiss: Establishment and control of initial adhesion by raphid diatoms. *Journal of Phycology* 34:9–15.

Wetherbee, R., M. Ludwig, and A. Koutoulis. 1995. Immunological and ultrastructural studies of scale development and deployment in *Mallomonas* and *Apedinella*. In: Sandgren, C. D., J. P. Smol, and J. Kristiansen (eds.), *Chrysophyte Algae: Ecology, Phylogeny and Development*, Cambridge University Press, Cambridge, UK, pp. 165–178.

Wettstein, A. 1901. *Handbuch der systematischen Botanik*, Leipzig, Vienna, Austria.

Wetzel, R. G. 1975. *Limnology*, W. B. Saunders, Philadelphia.

Wetzel, R. G. 1996. Benthic algae and nutrient cycling in lentic freshwater ecosystems. In: Stevenson, R. J., M. L. Bothwell, and R. L. Lowe (eds.), *Algal Ecology: Freshwater Benthic Ecosystems*, Academic Press, New York, pp. 641–667.

Wetzel, R. G., and G. E. Likens. 1991. *Limnological Analyses*, 2nd edition, Springer Verlag, New York.

Wharton, R. A., Jr., C. P. McKay, G. M. Simmons, Jr., and B. C. Parker. 1985. Cryoconite holes on glaciers. *Bioscience* 35:499–503.

Wharton, R. A., D. T. Smernoff, and M. A. Averner. 1988. Algae in space. In: Lembi, C. A., and J. R. Waaland (eds.), *Algae and Human Affairs*, Cambridge University Press, Cambridge, UK, pp. 485–510.

Whatley, J. 1993. Chloroplast ultrastructure. In: Berner, T. (ed.), *Ultrastructure of Microalgae*, CRC Press, Boca Raton, FL, pp. 135–204.

Wheeler, W. N. 1982. Nitrogen nutrition of *Macrocystis*. In Srivastava, L. M. (ed.), *Synthetic and Degradative Processes in Marine Macrophytes*, Walter de Gruyter & Co., New York, pp. 121–135.

White, A. W. 1976. Growth inhibition caused by turbulence in the toxic marine dinoflagellate *Gonyaulax excavata*. *Journal of the Fisheries Research Board Canada* 33:2598–2602.

Whittaker, R. H., and G. E. Likens. 1973. Primary production: The biosphere and man. *Human Ecology* 1:357–369.

Whittle, S. J., and P. J. Casselton. 1975. The chloroplast pigments of the algal classes Eustigmatophyceae and Xanthophyceae. I. Eustigmatophyceae. *British Phycological Journal* 10:179–181.

Whitton, B. A., and M. Potts. 2000. Introduction to the cyanobacteria. In: Whitton, B. A., and M. Potts (eds.), *The Ecology of Cyanobacteria: Their Diversity in Time and Space*, Kluwer Academic Publishers, Boston, pp. 1–11.

Wiedner, C., J. Rücker, R. Brüggemann, and B. Nixdorf. 2007. Climate change affects timing and size of populations of an invasive cyanobacterium in temperate regions. *Oecologia* 152:473–484.

Wiencke, C., and M. N. Clayton. 1990. Sexual reproduction, life history, and early development in culture of the Antarctic brown alga *Himantothallus grandifolius* (Desmarestiales, Phaeophyceae). *Phycologia* 29:9–18.

Wilce, R. T. 1990. Role of the Arctic Ocean as a bridge between the Atlantic and the Pacific Ocean: Fact and hypothesis. In: Garbary, D. J., and G. R. Sough (eds.), *Evolutionary Biogeography of the Marine Algae of the North Atlantic*, NATO ASI Series G, Ecological Science, vol. 22, Springer Verlag, Berlin, pp. 323–347.

Wilce, R. T., and J. R. Searles. 1991. *Schmitzia sanctae-crucis* sp. nov. (Calosiphoniaceae, Rhodophyta) and a novel nutritive development to aid in zygote nucleus amplification. *Phycologia* 30:151–169.

Wilcox, L. W. 1989. Multilayered structures (MLSs) in two dinoflagellates, *Katodinium campylops* and *Woloszynskia pascheri*. *Journal of Phycology* 4:785–789.

Wilcox, L. W., and G. L. Floyd. 1988. Ultrastructure of the gamete of *Pediastrum duplex* (Chlorophyceae). *Journal of Phycology* 24:140–146.

Wilcox, L. W., and G. E. Wedemayer. 1984. *Gymnodinium acidotum* Nygaard (Pyrrophyta), a dinoflagellate with an endosymbiotic cryptomonad. *Journal of Phycology* 20:236–242.

Wilcox, L. W., and G. J. Wedemayer. 1985. Dinoflagellate with blue-green chloroplasts derived from an endosymbiotic eukaryote. *Science* 227:192–194.

Wilcox, L. W., and G. J. Wedemayer. 1991. Phagotrophy in the freshwater, photosynthetic dinoflagellate *Amphidinium cryophilum*. *Journal of Phycology* 27:600–609.

Wilcox, L. W., G. J. Wedemayer, and L. E. Graham. 1982. *Amphidinium cryophilum* sp. nov. (Dinophyceae), a new freshwater dinoflagellate. II. Ultrastructure. *Journal of Phycology* 18:18–30.

Wildman, R. B., J. H. Loescher, and C. L. Winger. 1975. Development and germination of akinetes of *Aphanizomenon flos-aquae*. *Journal of Phycology* 11:96–104.

Wilkinson, C. R. 1992. Symbiotic interactions between marine sponges and algae. In: Reisser, W. (ed.), *Algae and Symbioses*, BioPress, Bristol, UK, pp. 111–152.

Will, A., and W. Tanner. 1996. Importance of the first external loop for substrate recognition as revealed by chimeric *Chlorella* monosaccharide/H+ symporters. *Federation of European Biological Societies Letters* 381:127–130.

Willame, R., C. Boutte, S. Grubisic, A. Wilmotte, J. Komárek, and L. Hoffmann. 2006. Morphological and molecular characterization of planktonic cyanobacteria from Belgium and Luxembourg. *Journal of Phycology* 42:1312–1332.

Willen, E. 1991. Planktonic diatoms—An ecological review. *Archiv für Hydrobiologie Supplement* 89:69–106.

Willey, R. L. 1984. Fine structure of the mucocysts of *Colacium calvum* (Euglenophyceae). *Journal of Phycology* 20:426–430.

Willey, R. L., and R. G. Wibel. 1985. A cytostome (cytopharynx) in green euglenoid flagellates (Euglenales) and its phylogenetic implications. *BioSystems* 18:369–376.

Willey, R. L., P. L. Walne, and P. A. Kivic. 1988. Origin of euglenoid flagellates from phagotrophic ancestors. *CRC Critical Reviews in Plant Sciences* 7:313–340.

Williams, S. L., and J. E. Smith. 2007. A global review of the distribution, taxonomy, and impacts of introduced seaweeds. *Annual Review of Ecology, Evolution and Systematics* 38:327–359.

Williamson, R. E. 1979. Filaments associated with the endoplasmic reticulum in the streaming cytoplasm of *Chara corallina*. *European Journal of Cell Biology* 20:177–183.

Williamson, R. E. 1992. Cytoplasmic streaming in characean algae: Mechanisms, regulation by Ca2+, and organization. In: Melkonian, M. (ed.), *Algal Cell Motility*, Chapman & Hall, London, UK, pp. 73–98.

Wilmotte, A. 1994. Molecular evolution and taxonomy of the cyanobacteria. In: Bryant, D. A. (ed.), *The Molecular Biology of Cyanobacteria*, Kluwer Academic Publishers, Amsterdam, pp. 1–25.

Wilmotte, A., and M. Herdman. 2001. Phylogenetic relationships among the Cyanobacteria based on 16S rRNA sequences. In: Garrity, G., D. R. Boone, and R. W. Castenholz (eds.), *Bergey's Manual of Systematic Bacteriology. Volume 1: The Archaea and the Deeply Branching and Phototropic Bacteria*, 2nd edition, Springer Verlag, New York, pp. 487–493.

Wilson, A., G. Ajlani, J.-M. Verbavatz, I. Vass, C. A Kerfeld, and D. Kirilovsky. 2006a. A soluble carotenoid protein involved in phycobilisome-related energy dissipation in cyanobacteria. *The Plant Cell* 18:992–1007.

Wilson, E. O., and W. H. Bossert. 1971. *A Primer of Population Biology*, Sinauer Associates, Stamford, CT.

Wilson, K. M., M. A. Schembri, P. D. Baker, and C. P. Saint. 2000. Molecular characterization of the toxic cyanobacterium *Cylindrospermopsis raciborskii* and design of a species-specific PCR. *Applied and Environmental Microbiology* 66:332–338.

Wilson, S. M., J. D. Pickett-Heaps, and J. A. West. 2006b. Vesicle transport and the cytoskeleton in the unicellular red alga *Glaucosphaera vacuolata*. *Phycological Research* 54:15–20.

Wilson, W. H., et al. 2005. Complete genome sequence and lytic phase transcription profile of a *Coccolithovirus*. *Science* 309:1090–1092.

Windust, A. J., J. L. C. Wright, and H. L. McLachlan. 1996. The effects of the diarrhetic shellfish poisoning toxins, okadaic acid and dinophysistoxin-1, on the growth of microalgae. *Marine Biology* 126:19–25

Winter, A., and W. G. Siesser. 1994. Atlas of living coccolithophores. In: Winter, A., and W. G. Siesser (eds.), *Coccolithophores*, Cambridge University Press, Cambridge, UK, pp. 107–160.

Wodniok, S., A. Simon, G. Glöckner, and B. Becker. 2007. Gain and loss of polyadenylation signals during evolution of green algae. *BMC Evolutionary Biology* 7:65.

Woelkerling, W. J. 1976. Wisconsin desmids. I. Aufwuchs and plankton communities of selected acid bogs, alkaline bogs, and closed bogs. *Hydrobiologia* 48:209–232.

Woelkerling, W. J. 1988. *The Coralline Red Algae: An Analysis of the Genera and Subfamilies of Non-geniculate Corallinaceae*, British Museum (Natural History), Oxford University Press, London, UK.

Wojciechowski, M. F., and M. E. Heimbrook. 1984. Dinitrogen fixation in alpine tundra, Niwot Ridge, Front Range, Colorado. *Arctic and Alpine Research* 16:1–10.

Wolf, F. R., and E. R. Cox. 1981. Ultrastructure of active and resting colonies of *Botryococcus braunii* (Chlorophyceae). *Journal of Phycology* 17:395–405.

Wolf, M., M. Buchheim, E. Hegewald, L.

Krienitz, and D. Hepperle. 2002. Phylogenetic position of the Sphaeropleaceae (Chlorophyta). *Plant Systematics and Evolution* 230:161–171.

Wolf, M., E. Hegewald, D. Hepperle, and L. Krienitz. 2003. Phylogenetic position of the Golenkiniaceae (Chlorophyta) as inferred from 18S rDNA sequence data. *Biologia* 58:433–436.

Womersley, H. B. S. 1984, 1987. *The Marine Benthic Flora of Southern Australia* (2 parts), Government Printer, Adelaide, Australia.

Wong, A., and T. Beebee. 1994. Identification of a unicellular, non-pigmented alga that mediates growth inhibition in anuran tadpoles: A new species of the genus *Prototheca* (Chlorophyceae, Chloroccales). *Hydrobiologia* 277:85–96.

Wong, P.-F., L.-J. Tan, H. Nawi, and S. AbuBakar. 2006. Proteomics of the red alga, *Gracilaria changii* (Gracilariales, Rhodophyta). *Journal of Phycology* 42:113–120.

Wood, R. 1998. The ecological evolution of reefs. *Annual Review of Ecology and Systematics* 29:179–206.

Wood, R. D., and K. Imahori. 1965. *A Revision of the Characeae*, vol. 1, Cramer, Weinheim, Germany.

Wood-Charlson, E. M., L. L. Hollingsworth, D. A. Krupp, and V. M. Weis. 2006. Lectin/glycan interactions play a role in recognition in a coral/dinoflagellate symbiosis. *Cellular Microbiology* 8:1985–1993.

Woods, J. K., and R. E. Triemer. 1981. Mitosis in the octaflagellate prasinophyte, *Pyramimonas amylifera* (Chlorophyta). *Journal of Phycology* 17:81–90.

Woodwell, G. M., P. H. Rich, and C. A. S. Hall. 1973. Carbon in estuaries. In: Woodwell, G. M., and E. V. Pecan (eds.), *Carbon and the Biosphere*. *Proceedings of the 24th Brookhaven Symposium in Biology*, U. S. Atomic Energy Commission Symposium Series 30, NTIS U. S. Department of Commerce, Springfield, VA, pp. 221–240.

Worden, A. Z., J. K. Nolan, and B. Palenik. 2004. Assessing the dynamics and ecology of marine picophytoplankton: The importance of the eukaryotic component. *Limnology and Oceanography* 49:168–179.

Wright, P. J., J. A. Callow, and J. R. Green. 1995b. The *Fucus* (Phaeophyceae) sperm receptor for eggs. II. Isolation of a binding protein which partially activates eggs. *Journal of Phycology* 31:592–600.

Wright, P. J., J. R. Green, and J. A. Callow. 1995a. The *Fucus* (Phaeophyceae) sperm receptor for eggs. I. Development and characteristics of a binding assay. *Journal of Phycology* 31:584–591.

Wüest, A., and A. Lorke. 2003. Small-scale hydrodynamics in lakes. *Annual Review of Fluid Mechanics* 35:373–412.

Wujek, D. E., and R. H. Thompson. 2002. The genera *Uroglena*, *Uroglenopsis*, and *Eusphaerella* (Chrysophyceae). *Phycologia* 41:293–305.

Wujek, D. J., and J. E. Chambers. 1966. Microstructure of pseudocilia of *Tetraspora gelatinosa* (Vauch.) Desv. *Transactions of the Kansas Academy of Science* 68:563–565.

Wujek, D. J., and L. C. Saha. 1995 The genus *Paraphysomonas* from Indian rivers, lakes,

ponds and tanks. In: Sandgren, C. D., J. P. Smol, and J. Kristiansen (eds.), *Chrysophyte Algae: Ecology, Phylogeny and Development*, Cambridge University Press, Cambridge, UK, pp. 373–384.

Wustman, B. A., M. R. Gretz, and K. D. Hoagland. 1997. Extracellular matrix assembly in diatoms (Bacillariophyceae). I. A model of adhesives based on chemical characterization and localization of polysaccharides from the marine diatom *Achnanthes longipes* and other diatoms. *Plant Physiology* 113:1059–1069.

Wustman, B. A., J. Lind, R. Wetherbee, and M. R. Gretz. 1998. Extracellular matrix assembly in diatoms (Bacillariophyceae). III. Organization of fucoglucuronoglactans within the adhesive stalks of *Achnanthes longipes*. *Plant Physiology* 116:1431–1441.

Wynne, M. J. 1981. Phaeophyta: Morphology and classification. In: Lobban, C. S., and M. J. Wynne (eds.), *The Biology of Seaweeds*, Blackwell Scientific, Oxford, UK.

Wynne, M. J. 1988. A reassessment of the *Hypoglossum* group (Delesseriaceae, Rhodophyta), with a critique of its genera. *Helgoländer Meeresuntersuchungen* 42:511–534.

Wynne, M. J. 1996. A revised key to genera of the red algal family Delesseriaceae. *Nova Hedwigia* 112:171–190.

Wynn-Williams, D. D. 2000. Cyanobacteria in deserts—Life at the limit. In: Whitton, B. A., and M. Potts (eds.), *The Ecology of Cyanobacteria: Their Diversity in Time and Space*, Kluwer Academic Publishers, Boston, pp. 341–366.

Xia, B., and I. A. Abbott. 1987. Edible seaweeds of China and their place in the Chinese diet. *Economic Botany* 41:341–353.

Xiao, S., A. H. Knoll, X. Yuan, and C. M. Pueschel. 2004. Phosphatized multicellular algae in the Neoproterozoic Doushantuo Formation, China, and the early evolution of florideophyte red algae. *American Journal of Botany* 91:214–227.

Xiao, S., X. Yuan, M. Steiner, and A. H. Knoll. 2002. Macroscopic carbonaceous compressions in a terminal Proterozoic shale: A systematic reassessment of the Miaohe Biota, South China. *Journal of Paleontology* 76:347–376.

Xiao, S., Y. Zhang, and A. H. Knoll. 1998. Three-dimensional preservation of algae and animal embryos in a Neoproterozoic phosphorite. *Nature* 391:553–558.

Xiaoping, G., J. D. Dodge, and J. Lewis. 1989. Gamete mating and fusion in the marine dinoflagellate *Scrippsiella* sp. *Phycologia* 28:342–351.

Xinyao, L., et al. 2006. Feeding characteristics of an amoeba (Lobosea: *Naegleria*) grazing upon cyanobacteria: Food selection, ingestion and digestion progress. *Microbial Ecology* 51:315–325.

Xiong, J., W. M. Fischer, K. Inoue, M. Nakahara, and C. E. Bauer. 2000. Molecular evidence for the early evolution of photosynthesis. *Science* 289:1724–1730.

Yair, A. 1990. Runoff generation in a sandy area—The Nizzana sands, western Negev, Israel. *Earth Surface Processes and Landforms* 15:597–609.

Yamagishi, T. T. Motomura, C. Nagasato, A. Kato and H. Kawaii. 2007. A tubular mastigoneme-related protein, Ocm1, isolated from the flagellum of a chromophyte alga, *Ochromonas danica*. *Journal of Phycology* 43:519–527.

Yamagishi, Y., and K. Kogame. 1998. Female dominant population of *Colpomenia peregrina* (Scytosiphonales, Phaeophyceae). *Botanica Marina* 41:217–222.

Yamaguchi, K., S. Prieto, M. V. Haynes, W. H. McDonald, Y. R. Yates III, and S. P. Mayfield. 2002. Proteomic characterization of the small subunit of *Chlamydomonas reinhardtii* chloroplast ribosome: Identification of a novel S1 domain containing protein and unusually large orthologs of bacterial S2, S3, and S5. *Plant Cell* 14:2957–2974.

Yildiz-Fitnat, H., P. Davies-John, and A. R. Grossman. 1994. Characterization of sulfate transport in *Chlamydomonas reinhardtii* during sulfur-limited and sulfur-sufficient growth. *Plant Physiology* 104:981–987.

Yoon, H. S., et al. 2008. Broadly sampled multigene trees of eukaryotes. *BMC Evolutionary Biology* 8:14.

Yoon, H. S., and J. W. Golden. 1998. Heterocyst pattern formation controlled by a diffusible peptide. *Science* 282:935–938.

Yoon, H. S., J. D. Hackett, and D. Bhattacharya. 2002a. A single origin of the peridinin- and fucoxanthin-containing plastids of dinoflagellates through tertiary endosymbiosis. *Proceedings of the National Academy of Sciences of the United States of America* 99:11724–11729.

Yoon, H. S., J. D. Hackett, C. Ciniglia, G. Pinto, and D. Bhattacharya. 2004. A molecular timeline for the origin of photosynthetic eukaryotes. *Molecular Biology and Evolution* 21:809–819.

Yoon, H. S., J. D. Hackett, G. Pinto, and D. Bhattacharya. 2002b. The single, ancient origin of chromist plastids. *Journal of Phycology* 38:40.

Yoon, H. S., Müller, R. G. Sheath, F. D. Ott, and D. Bhattacharya. 2006. Defining the major lineages of red algae (Rhodophyta). *Journal of Phycology* 42:482–492.

Yopp, J. H., D. R.Tindall, D. M. Mille, and W. E. Schmid. 1978. Isolation, purification and evidence for a halophilic nature of the blue-green alga *Aphanothece halophytica* Fremy (Chroococcales). *Phycologia* 17:172–178.

Yoshida, M., M.-H. Noël, T. Nakayama, T. Naganuma, and I. Inouye. 2006a. A haptophyte bearing siliceous scales: Ultrastructure and phylogenetic position of *Hyalolithus neolepis* gen. et sp. nov. (Prymnesiophyceae, Haptophyta). *Protist* 157:213–234.

Yoshida, T., T. B. Gurung, M. Kagami, and J. Urabe. 2001. Contrasting effects of a cladoceran (*Daphnia galeata*) and a calanoid copepod (*Eodiaptomus japonicus*) on algal and microbial plankton in a Japanese lake, Lake Biwa. *Oecologia* 129:602–610.

Yoshida, T., and H. Mikami. 1996. *Sorellocolax stellaris* gen et sp. nov., a hemiparasitic alga (Delesseriaceae, Rhodophyta) from the East Coast of Honshu Japan. *Phycological Research* 44:125–128.

Yoshida, T., Y. Takashima, Y. Tomaru, Y. Shirai, Y. Takao, S. Hiroishi, and K. Nagasaki. 2006b. Isolation and characterization of a cyanophage infecting the toxic cyanobacterium *Microcystis aeruginosa*. *Applied and Environmental Microbiology* 72:1239–1247.

Yoshii, Y., T. Hanyuda, I. Wakana, K. Miyaji, S. Arai, K. Ueda, and I. Inouye. 2004. Carotenoid composition of Cladophora balls (*Aegagropila linnaei*) and some members of the Cladophorales (Ulvophyceae, Chlorophya): Their taxonomic and evolutionary implication. *Journal of Phycology* 40:1170–1177.

Yoshii, Y, S. Takaichi, T. Maoka, and I. Inouye. 2003. Photosynthetic pigment composition in the primitive green alga *Mesostigma viride* (Prasinophyceae): Phylogenetic and evolutionary implications. *Journal of Phycology* 39:570–576.

Yoshimura, Y., S. Kohshima, and S. Ohtani. 1997. A community of snow algae on a Himalayan glacier: Change of algal biomass and community structure with altitude. *Arctic and Alpine Research* 29:126–137.

Young, J. R., K. Henriksen, and I. Probert. 2004. Structure and morophogenesis of the coccoliths of the CODENET species. In: Thierstein, H. R., and J. R. Young (eds.), *Coccolithophores—From Molecular Processes to Global Impact*, Springer Verlag, New York, pp. 191–216.

Young, J. R., P. R. Brown, and J. A. Burnett. 1994. Paleontological perspectives. In: Green, J. C., and B. S. C. Leadbeater (eds.), *The Haptophyte Algae*, Clarendon Press, Oxford, UK, pp. 379–392.

Zaady, E., Y. Gutterman, and B. Boeken. 1997. The germination of mucilaginous seeds of *Plantago coronopus*, *Reboudia pinnata*, and *Carrichtera annua* on cyanobacterial soil crust from the Negev Desert. *Plant and Soil* 190:247–252.

Zapata, M., S. W. Jeffrey, S. W. Wright, F. Rodriguez, J. L. Garrido, and L. Clementson. 2004. Photosynthetic pigments in 37 species (65 strains) of Haptophyta: Implications for oceanography and chemotaxonomy. *Marine Ecology Progress Series* 270:83–102.

Zaslavskaia, L. A., J. C. Lippmeier, P. G. Kroth, A. R. Grossman, and K. E. Apt. 2000. Transformation of the diatom *Phaeodactylum tricornutum* (Bacillariophyceae) with a variety of selectable marker and reporter genes. *Journal of Phycology* 36:379–386.

Zaslavskaia, L. A, J. C. Lippmeier, C. Shih, D. Ehrhardt, A. R. Grossmand, and K. E. Apt. 2001. Trophic conversion of an obligate photoautotrophic organism through metabolic engineering. *Science* 292:2073–2075.

Zechman, F. W. 2003. Phylogeny of the Dasycladales (Chlorophyta, Ulvophyceae) based on analyses of rubisco large subunit (*rbcL*) gene sequences. 2003. *Journal of Phycology* 39:819–827.

Zeeb, B. A., K. A. Duff, and J. P. Smol. 1996. Recent advances in the use of chrysophyte stomatocysts in paleoecological studies. *Nova Hedwigia* 114:247–252.

Zhang, X., and M. M. Watanabe. 2001. Grazing and growth of the mixotrophic chrysomonad *Poterioochromonas malhamensis* (Chrysophyceae) feeding on algae. *Journal of Phycology* 37:738–743.

Zhang, Z., B. R. Green, and T. Cavalier-Smith. 1999. Single gene circles in dinoflagellate chloroplast genomes. *Nature* 400:155–159.

Zhao, Y., Y. Shi, W. Zhao, X. Huang, D. Wang, N. Brown, J. Brand, and J. Zhao. 2005. CcbP, a calcium-binding protein from *Anabaena* sp. PCC 7120, provides evidence that calcium ions regulate heterocyst differentiation. *Proceedings of the National Academy of Sciences of the United States of America* 102:5744–5748.

Zingmark, R. G. 1970. Sexual reproduction in the dinoflagellate *Noctiluca miliaris*. *Journal of Phycology* 6:122–126.

Zöllner, E., B. Santer, M. Boersma, H.-G. Hoppe, and K. Jürgens. 2003. Cascading predation effects of *Daphnia* and copepods on microbial food web components. *Freshwater Biology* 48:2174–2193.

Zuccarello, G. C., and G. M. Lokhorst. 2005. Molecular phylogeny of the genus *Tribonema* (Xanthophyceae) using *rbcL* gene sequence data: Monophyly of morphologically simple algal species. *Phycologia* 44:384–392.

Zuccarello, G. C., D. Moon, and L. J. Goff. 2004. A phylogenetic study of parasitic genera placed in the family Choreocolacaceae (Rhodophyta). *Journal of Phycology* 40:937–945.

Zuccarello, G. C., and J. A. West. 2003. Multiple cryptic species: Molecular diversity and reproductive isolation in the *Bostrychia radicans/B. moritziana* complex (Rhodomelaceae, Rhodophyta) with focus on North American isolates. *Journal of Phycology* 39:948–959.

Taxonomic Index

Page numbers in boldface indicate figures and those with "T" indicate tables.
Numbers with "D" indicate generic descriptions.

Subject Index

Page numbers in boldface indicate figures.